The Hindu Kush Himalaya Assessment

Philippus Wester • Arabinda Mishra
Aditi Mukherji • Arun Bhakta Shrestha
Editors

The Hindu Kush Himalaya Assessment

Mountains, Climate Change, Sustainability and People

Editors
Philippus Wester
International Centre for Integrated Mountain Development (ICIMOD)
Kathmandu, Nepal

Aditi Mukherji
International Centre for Integrated Mountain Development (ICIMOD)
Kathmandu, Nepal

Arabinda Mishra
International Centre for Integrated Mountain Development (ICIMOD)
Kathmandu, Nepal

Arun Bhakta Shrestha
International Centre for Integrated Mountain Development (ICIMOD)
Kathmandu, Nepal

https://doi.org/10.1007/978-3-319-92288-1

Published for the Hindu Kush Himalayan Monitoring and Assessment Programme (HIMAP)

Foreword

This assessment report establishes the value of the Hindu Kush Himalaya (HKH) for the 240 million hill and mountain people across the eight countries sharing the region, for the 1.65 billion people in the river basins downstream, and ultimately for the world. Yet, the region and its people face a range of old and new challenges moving forward, with climate change, globalization, movement of people, conflict and environmental degradation. At the same time, we also see incredible potential to meet these challenges in a sustainable manner.

In spite of its importance, relatively less is known about the HKH, its ecosystems and its people, especially in the context of rapid change. Over the last few decades, there has been more research on the region, but the knowledge gathered is often scattered, reaches a limited audience, is sectoral or based on single disciplines and, most importantly, does not reach decision-makers, whether they be in government, in local communities, or in the private sector.

The rationale for this assessment is manifold. The first is about extending the accessible knowledge base. There has been incredible value in bringing together people engaged in generating knowledge about the HKH to collate existing knowledge. Plus, by working across disciplines and countries, the assessment blends insights from different perspectives about the mountains. Global assessments and programmes like the Intergovernmental Panel on Climate Change (IPCC) can now benefit from an important knowledge source about this region, and the book has great value in informing global debates and discourses. Then, there is a value beyond the assessment report itself, in bringing together a network of people who can work across disciplinary and geographical boundaries in the future.

But the main reason for the assessment goes beyond the collation of knowledge. It is to answer a range of policy-oriented questions we all grapple with. Some of these are quite scientific, such as what will happen with climate change, or what the impact of air pollution is. Others are more targeted to actions that people should take, like pathways to sustainable access to energy, or building resilience. The main objective of the assessment thus is to inform decision-makers with the best science and knowledge we have. This assessment has made important strides in this direction. A very important finding of the assessment is that while we have significant knowledge gaps, we know enough to take action.

The publication of the Hindu Kush Himalayan Monitoring and Assessment Programme's (HIMAP) flagship piece—*The Hindu Kush Himalaya Assessment: Mountains, Climate Change, Sustainability and People*—is an important milestone in a larger process that aims to bring together researchers, policy makers and the public to better manage the HKH so that women, men and children can enjoy improved well-being in a healthy mountain environment.

The HIMAP process will continue to engage in science-policy discussions at country and regional levels to enhance cooperation between communities, states and countries in managing the HKH. It will also develop more targeted assessments about specific areas of concern that emerge as we develop more knowledge about the region. Importantly, the authors of the assessment have laid out mountain-specific priorities consistent with the Sustainable

Development Goals (SDGs), and this book provides an important baseline in reaching these goals for the mountains and people of the HKH region. Perhaps the greatest good is that we have an expanding community of practice working together to match and rise above the challenges facing the HKH today.

Kathmandu, Nepal

David Molden
Director General, International Centre
for Integrated Mountain Development

How to cite: P. Wester, A. Mishra, A. Mukherji, A. B. Shrestha (eds) (2019) *The Hindu Kush Himalaya Assessment—Mountains, Climate Change, Sustainability and People* Springer Nature Switzerland AG, Cham.

Acknowledgements

The Hindu Kush Himalayan Monitoring and Assessment Programme (HIMAP) is a long-term, integrated science-policy initiative coordinated by the International Centre for Integrated Mountain Development (ICIMOD) that aims to promote enabling policies, sustainable solutions and more robust regional cooperation in the Hindu Kush Himalaya (HKH) region to sustain mountain environments and livelihoods, by:

- Assessing the current state of knowledge of the HKH region, through comprehensive, thematic and subregional assessments and topical outlooks;
- Increasing the understanding of various drivers of change and their impacts;
- Developing evidence-based policy solutions and recommendations; and
- Engaging decision-makers across sectors and institutions through a series of HKH Science-Policy Forums.

HIMAP is a platform for long-term collaboration and coordination among a broad and diverse group of more than 350 leading researchers, practitioners and policy specialists working in the HKH. Under HIMAP, experts from the region have come together to develop the first HKH Assessment Report, as the first in a series of monitoring and assessment reports. It deals with major issues such as climate change, biodiversity, energy, cryosphere (frozen water), water, food security, air pollution, disaster and resilience, poverty, adaptation, gender and migration. The assessment addresses the social, economic and environmental pillars of sustainable mountain development and will serve as a basis for evidence-based decision-making to safeguard the environment and advance people's well-being in the HKH region.

We are deeply grateful and wish to acknowledge the large network that made this first HKH Assessment Report possible. The commitment, rigour and expertise shown by the Coordinating Lead Authors and Lead Authors, with important support by Contributing Authors, all freely contributing their time, are highly appreciated. The Review Editors ensured the integrity of the review process and played a critical role in supporting the chapter teams and improving the overall quality of the chapters. We deeply appreciate the time, rigour and professionalism of the external expert and government reviewers, which has contributed importantly to the credibility of this report. While not being fully exhaustive, we wish to acknowledge the following individuals, institutions and governments for making this HKH Assessment Report possible:

Overall Coordinator: Philippus Wester.

Chapter Coordinating Lead Authors: Arivudai Nambi Appadurai, Ruchi Badola, Soumyadeep Banerjee, Ram B. Bhagat, Tobias Bolch, Nakul Chettri, Dhrupad Choudhury, Shobhakar Dhakal, Lam Dorji, Rucha Ghate, Giovanna Gioli, Chanda Gurung Goodrich, Shichang Kang, Fawad Khan, Raghavan Krishnan, Shiyin Liu, Arabinda Mishra, Eddy Moors, Aditi Mukherji, M. S. R. Murthy, Nusrat Nasab, Hemant R. Ojha, Arnico Panday, Golam Rasul, Guoyu Ren, Bernadette P. Resurreccion, Joyashree Roy, Abdul Saboor,

Eri Saikawa, Christopher Scott, Bikash Sharma, Eklabya Sharma, Joseph Shea, Arun Bhakta Shrestha, Mandira Singh Shrestha, Tasneem Siddiqui, Yiching Song, Leena Srivastava, Ganesh Thapa, Ramesh Ananda Vaidya, Yanfen Wang, Ning Wu, Jianchu Xu, Fan Zhang.

Chapter Lead Authors: Bhupesh Adhikari, Lipy Adhikari, Ahsan Uddin Ahmad, Mozaharul Alam, Ghulam Muhammad Arif, Elisabetta Aurino, Mohd. Farooq Azam, Aditya Bastola, Luna Bharati, Pashupati Chaudhary, R. P. Chaudhary, Ganesh Chettri, Netra Chhetri, Zhiyuan Cong, Purnamita Dasgupta, Chao Fu, Yang Gao, Ritesh Gautam, Nilabja Ghosh, Anandajit Goswami, Stephan Gruber, Deo Raj Gurung, Christian Huggel, Abid Hussain, S. A. Hussain, Walter Immerzeel, Michiko Ito, Sanjay Jayanarayanan, Deepa Joshi, Ulka Kelkar, Yuba Raj Khatiwada, Bahadar Nawab Khattak, Rajan Kotru, Nagami Kozo, Anil Kulkarni, Clemens Kunze, Huilin Li, Janwillem Liebrand, Chengfan Liu, Hina Lotia, Brijesh Mainali, Ghulam Mohamad Malikyar, Laura Mapstone Scott, David Molden, Rashid Memon, Sanjay Kumar Mohanty, Seinn Seinn Mu, Daanish Mustafa, Thinley Namgyel, Dev Nathan, Orzala Ashraf Nemat, Rabindra Nepal, Andrea Nightingale, Debajit Palit, Dinesh Paudel, Shi Peili, Manfred Perlik, Bharat Pokhrel, Neera Shrestha Pradhan, Rebecca Pradhan, Sunita Pradhan, S. V. R. K. Prabhakar, Anjal Prakash, Pallav Purohit, Atiq Rahman, Rupak Rajbhandari, Bimal Raj Regmi, Yuyu Ren, Long Ruijun, Sajjad Saeed, Shaheen Ashraf Shah, Upasna Sharma, Ankita Shrestha, Krishna K. Shrestha, Bandita Sijapati, Surendra Pratap Singh, E. Somanathan, Md. Abu Syed, Adnan Ahmad Tahir, Pema Thinley, Prakash Tiwari, Ramesh Vellore, Kul Bahadur Wakhley, Robert James Wasson, Muhammad Arif Wattoo, Philippus Wester, Ying Xu, Tandong Yao, Qinglong You, Guoqing Zhang, Linxiu Zhang, Yinsheng Zhang, Robert Zomer, Eric Zusman.

Chapter Contributing Authors: Acknowledged in the individual chapters.

Review Editors: Sara Ahmed, Jayanta Bandopadhyay, Richard Black, Purnamita Dasgupta, Marc Foggin, Koji Fujita, Enamul Haque, Sarala Khaling, Asuncion Lera St. Clair, Valerio Lucarini, Bikash Pandey, Krishna Prasad Pant, Atiq Rahman, Surendra Pratap Singh, Cecilia Tortajada, Nguyen Thi Kim Oanh, Laurie Vasily, Philippus Wester, Linxiu Zhang, Yan Zhaoli.

External Reviewers: Acknowledged in Annex 3.

Editors: Christopher Butler and Nick Moschovakis.

Copy Editors: Diwas K. C, Beth Duncan, Merrill Feitell, Shradha Ghale, Elaine Monaghan, Beatrice Murray, Amy Sellmyer, Bill Wolfe.

Graphics Design: Amy Sellmyer.

Steering Committee: David Molden, Chair (International Centre for Integrated Mountain Development), Yuba Raj Khatiwada (Ministry of Finance, Government of Nepal), Atiq Rahman (Bangladesh Centre for Advanced Studies), Eklabya Sharma (International Centre for Integrated Mountain Development), SP Singh (Central Himalayan Environment Association), Tandong Yao (Institute of Tibetan Plateau Research, Chinese Academy of Sciences), Linxiu Zhang (United Nations Environment Programme and Center for Chinese Agricultural Policy, Chinese Academy of Sciences), Philippus Wester, Member Secretary (International Centre for Integrated Mountain Development).

HIMAP Secretariat: Philippus Wester (Coordinator), Rekha Khatri Thapa, Ritu Meher Shrestha, Bhawana Syangden, Avash Pandey, Nisha Wagle.

Core support for the assessment process leading to the production of this book was provided by the governments of Afghanistan, Australia, Austria, Bangladesh, Bhutan, China, India, Myanmar, Nepal, Norway, Pakistan, Sweden, Switzerland and the UK, through core funding to ICIMOD.

Authors and reviewers participating in the Hindu Kush Himalaya Assessment are affiliated with numerous institutions inside and outside the HKH region. Their institutions have contributed to the assessment through the in-kind support of their staff. We wish to acknowledge the contributions of the following institutions, without implying an endorsement of the outcomes of the assessment:

Asian Development Bank; Afghanistan Research and Evaluation Unit; Aga Khan Agency for Habitat; Aga Khan Development Network; Agriculture and Forestry University, Nepal; Agriculture University Peshawar; Appalachian State University; Arizona State University; Ashoka Trust for Research in Ecology and Environment; Asian Institute of Technology; Bangladesh Centre for Advanced Studies; Bath Spa University; Bhutan Power Corporation; Bjerknes Centre for Climate Research, Norway; Carleton University; Center for Chinese Agricultural Policy, Chinese Academy of Sciences; Center for Earth System Science, Tsinghua University; Central Himalayan Environment Association; Central University of Rajasthan; Central Water Commission, India; Centre for Development and Environment, University of Bern; Centre for Development Studies, India; Centre for Environment and Development, Bhutan; Centre for Global Change, Bangladesh; Centre for the Study of Labour and Mobility, Social Science Baha, Nepal; Chengdu Institute of Biology, Chinese Academy of Sciences; China Center for Agricultural Policy, School of Advanced Agricultural Sciences, Peking University; China Meteorological Administration; China University of Geosciences; Chinese Academy of Meteorological Sciences; Chinese Academy of Sciences; COMSATS Institute of Information Technology; Copenhagen Business School; Department of Foreign Affairs and Trade, Government of Australia; DNV GL; East West University; Emory College; FAO Regional Office of Asia and Pacific; Forest Research Institute, Indian Council of Forestry Research and Education; FutureWater; GB Pant National Institute of Himalayan Environment and Sustainable Development; George Mason University; Global Change Impact Studies Centre; Goethe University; Grand Canyon Trust; Grantham Institute, Imperial College London; Green Park Consultants GPC Ltd; GRID-Arendal; Heidelberg University; Helvetas Swiss Inter-cooperation Nepal; Hiroshima University; HNB Garhwal University; Imperial College London; Indian Institute of Science; Indian Institute of Science Education and Research; Indian Institute of Technology Bombay; Indian Institute of Technology Delhi; Indian Institute of Technology Guwahati; Indian Institute of Technology Tirupati; Indian Institute of Tropical Meteorology; Indian Statistical Institute, Delhi; Institut de Recherche pour le Développement, France; Institut des Géosciences de l'Environnement; Institute for Advanced Sustainability Studies; Institute for Global Environmental Strategies; Institute for Human Development, India; Institute for Social and Environmental Transition; Institute of Economic Growth; Institute of Geographic Science and Natural Resources Science, Chinese Academy of Sciences; Institute of International Rivers and Eco-security, Yunnan University; Institute of Loess Plateau, Shanxi University; Institute of Mountain Hazards and Environment, Chinese Academy of Sciences; Institute of Tibetan Plateau Research, Chinese Academy of Sciences; International Centre for Climate Change and Development, Bangladesh; International Centre for Integrated Mountain Development; International Institute for Applied Systems Analysis; International Institute of Population Studies, India; International Organization for Migration; International Water Management Institute; Jadavpur University; Japan International Cooperation Agency; Jawaharlal Nehru University; Kathmandu University; Katholieke Universiteit Leuven; Keio University; King's College London; Kumaun University; Kunming Institute of Botany, Chinese Academy of Sciences; Lahore University; Lahore University of Management Sciences; Leadership for Environment and Development (LEAD), Pakistan; Li-Bird; Linnaeus University; Management Development Institute; Massachusetts Institute of Technology; Massey University; Ministry of Agriculture and Forests, Royal Government of Bhutan; Ministry of Agriculture, Livestock and Irrigation, Government of Myanmar; Ministry of Economic Affairs, Royal Government of Bhutan; Ministry of Environment and Forest, Government of Bangladesh; Ministry of Environment, Forest and Climate Change,

Government of India; Ministry of Finance, Government of Nepal; Ministry of Forests and Soil Conservation, Government of Nepal; Mountain Research Initiative, University of Bern; Murray-Darling Basin Authority, Government of Australia; Nagoya University; Nanjing University of Information Science and Technology; Nanyang Technological University; NASA Langley Research Center; National Center for Atmospheric Research, USA; National Council of Bhutan; National Environment Commission (NEC), Royal Government of Bhutan; National Environmental Protection Agency of the Islamic Republic of Afghanistan; National Institute of Hydrology, India; National Oceanic and Atmospheric Administration, USA; National Technical University of Athens; National University of Singapore; Norwegian Polar Institute; Norwegian University of Life Sciences; Observer Research Foundation; Oslo University; Oxford Policy Management Limited; Pakistan Agricultural Research Council; Pakistan Institute of Development Economics; NITI Aayog, India; PMAS Arid Agriculture University; Population Council, India; Potsdam Institute for Climate Impact Research (PIK), Germany; Practical Action; Reading University; RECOFTC—The Center for People and Forests, Bangkok, Thailand; REDD Implementation Centre, Government of Nepal; Research Center for Applied Sciences and Technology (RECAST), Tribhuvan University; Room to Read; Royal Society for Protection of Nature, Bhutan; SaciWATERs; Scott Polar Research Institute, University of Cambridge; SOAS University of London; Southasia Institute of Advanced Studies, Nepal; State Key Laboratory of Cryospheric Science, Northwest Institute of Eco-Environment and Resources, Chinese Academy of Science; Stockholm Environment Institute; Sunrise Nepal Food and Beverage Pvt. Ltd.; Swedish University of Agricultural Sciences; Technical University of Denmark; TERI School of Advanced Studies; The Energy and Resources Institute; Tribhuvan University; UN Environment International Ecosystem Management Partnership (UNEP-IEMP); United Nations Development Programme; United Nations Environment Programme; United Nations University; Universität Innsbruck; Universite Grenoble Alpes; University of Agriculture, Faisalabad; University of Arizona; University of Arkansas; University of Birmingham; University of Bremen; University of Central Asia; University of Dhaka; University of Edinburgh; University of Leuven; University of Manitoba; University of Michigan; University of Montana; University of Moratuwa; University of New South Wales; University of Northern British Colombia; University of Oregon; University of Portsmouth; University of Potsdam; University of Queensland; University of the Chinese Academy of Sciences; University of Toulouse; University of Zurich; Paris Diderot University , Sorbonne Paris Cité; Utrecht University; Wageningen University and Research; Water Research Institute, National Research Council, Italy; Wildlife Institute of India; Winrock International; World Agroforestry Centre (ICRAF); World Resources Institute.

Contents

Introduction to the Hindu Kush Himalaya Assessment

Coordinating Lead Author
Eklabya Sharma, International Centre for Integrated Mountain Development (ICIMOD), Kathmandu, Nepal, e-mail: eklabya.sharma@icimod.org (corresponding author)

Lead Authors
David Molden, International Centre for Integrated Mountain Development (ICIMOD), Kathmandu, Nepal, e-mail: david.molden@icimod.org
Atiq Rahman, Bangladesh Centre for Advanced Studies, Dhaka, Bangladesh, e-mail: atiq.rahman@bcas.net
Yuba Raj Khatiwada, Ministry of Finance, Govt. of Nepal, Kathmandu, Nepal, e-mail: dryubaraj@gmail.com
Linxiu Zhang, International Ecosystem Management Partnership, United Nations Environment Programme and Center for Chinese Agricultural Policy, Chinese Academy of Sciences, Beijing, China, e-mail: linxiu.zhang@un.org
Surendra Pratap Singh, Central Himalayan Environment Association, Nainital, India, e-mail: surps@yahoo.com
Tandong Yao, Institute of Tibetan Plateau Research, Chinese Academy of Sciences, Beijing, China, e-mail: tdyao@itpcas.ac.cn
Philippus Wester, International Centre for Integrated Mountain Development (ICIMOD), Kathmandu, Nepal, e-mail: philippus.wester@icimod.org

Note: All authors are HIMAP Steering Committee Members, except Philippus Wester, who is HIMAP Coordinator

Corresponding Author
Eklabya Sharma, International Centre for Integrated Mountain Development (ICIMOD), Kathmandu, Nepal, e-mail: eklabya.sharma@icimod.org

P. Wester et al. (eds.), *The Hindu Kush Himalaya Assessment*,
https://doi.org/10.1007/978-3-319-92288-1_1

Contents

1.1 Global Mountain Perspective

Mountains are large landforms raised above the surface of the earth emerging into peaks and ranges. Mountains occupy 22% of the world's land surface area and are home to about 13% of the world's population (FAO 2015). While about 915 million people live in mountainous region, less than 150 million people live above 2,500 m above sea level (masl), and only 20–30 million people live above 3,000 masl.

About half of all humankind directly depends on mountain resources, primarily water. Mountains support 25% of world's terrestrial biodiversity and include nearly half of the world's biodiversity 'hotspots'. Of the 20 plant species that supply 80% of the world's food, six of those (apples, barley, maize, potatoes, sorghum and tomatoes) originated in mountains (Fleury 1999). In humid parts of the world, mountains provide 30–60% of the fresh water downstream; and in semi-arid and arid environments, they provide 70–95% (Kapos et al. 2000; WCMC-UNEP 2002). Mountains provide goods and services of global significance in the form of water, hydroelectricity, timber, biodiversity and niche products, mineral resources, recreation, and flood management (Schild and Sharma 2011; Molden and Sharma 2013). Mountains are more diverse region rich in ethnicity and languages. In general, poverty is higher in mountain regions and people are often at higher risk than people elsewhere. According to a recent FAO analysis, 39% of mountain populations (urban and rural combined) in developing countries were considered vulnerable to food insecurity in 2012, an increase of 30% compared to 12 years prior (FAO 2015).

Mountain geological formations are fragile and ecosystems are degrading fast because of both natural and anthropogenic drivers of change. Mountains are also places of cultural meaning and refuge. Many mountain inhabitants have settled there to escape religious or political persecution or wars in lowlands. Mountains are also often focal areas of armed conflict. Mountain areas have ecological, aesthetic, and socioeconomic significance, not only for people living there, but for those living beyond—especially those in the lowlands who benefit from the ecological services mountains provide. Thus, mountains, in one perspective, stand as some of the planet's last natural 'islands' in a sea of increasingly anthropogenic influenced lowlands, providing a number of significant ecological functions extending beyond mountain regions (Hamilton 2002).

Mountains also represent unique areas for detecting climate change and assessing climate change impacts (Nogues-Bravo et al. 2008; Dyuergerov and Meier 2005). As climate changes rapidly through elevation over relatively short horizontal distances, so do hydrology, vegetation, ecological conditions, and socio-economic settings (Whiteman 2000; Xu and Melick 2006). This rapid change over distance, in turn, also influences cultural values and societies. In this way, it is important to recognise the complexities of environment-society interactions —culture and environment are mutually reciprocating systems.

The increasing awareness of climate change impacts on mountains, mountain ecosystems, and mountain communities have started drawing attention to mountains during international debates such as the United Nations Conference on Environment and Development in Rio de Janeiro in 1992,

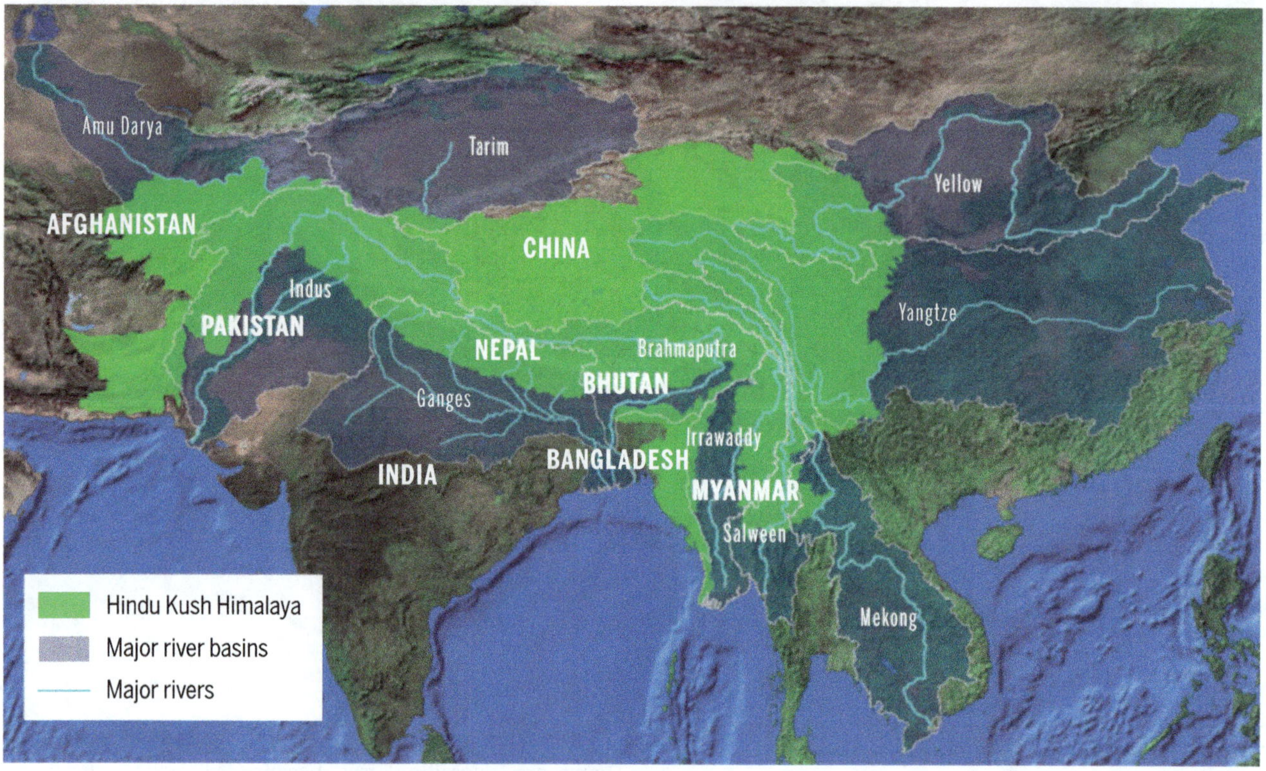

Fig. 1.1 The Hindu Kush Himalayan region and 10 major river basins

the Sendai Framework for Disaster Risk Reduction 2015, the Paris Agreement 2015 of the United Nations Framework Convention on Climate Change, and the Sustainable Development Goals (SDGs) and Targets 2030. There should be more pursuits for mountain perspectives to form an integral part in any discussions about future plans for sustainable development in the context of climate change. That means not just highlighting the vulnerabilities and fragilities inherent to mountain locations, but also emphasizing the resilience and strength that mountain people and communities bring when seeking to deal with these challenges.

1.2 The Hindu Kush Himalaya—A Global Asset

A critically important geo-ecological asset, the Hindu Kush Himalaya (HKH) is the origin of 10 major river basins and encompasses over 4.2 million km^2 area (Bajracharya and Shrestha 2011; Bajracharya et al. 2015) (Fig. 1.1). This HKH area and Tien Shan mountains together form the largest area of permanent ice cover outside of the North and South Poles (hence, the occasional reference to the HKH as the "Third Pole") and is home to four global biodiversity hotspots, 330 important bird areas (Chettri et al. 2008), and hundreds of mountain peaks over 6,000 m. The region provides ecosystem services (e.g., water, food, energy) that directly sustain the livelihoods of 240 million people in the mountain and hills of the HKH. Nearly 1.9 billion people living in the 10 river basins also benefit directly and indirectly from its resources (see Box 1.1), while more than 3 billion people enjoy the food produced in its river basins. The region is also home to some of the most diverse cultures, languages, religions, and traditional knowledge systems in the world.

Box 1.1 Population of the Hindu Kush Himalaya and the ten major rivers basins originating in the Hindu Kush Himalaya

Box Authors: Golam Rasul, Abid Hussain and Sudip Pradhan, ICIMOD

Based on the latest available government data sources and projections, in 2017 the population of the mountain and hills of the Hindu Kush Himalaya is around 240 million people (see Table 1.1). The total population in the ten major river basins with their headwaters in the HKH is around 1.9 billion, including the 240 million in the mountain and hills of the HKH (see Table 1.2).

The ethnic diversity and cultural wealth of the HKH extend from the Hindu Kush valleys in Afghanistan to the diverse hill

Table 1.1 Population in the mountain and hills of the HKH

Countries	Areas included in the HKH region	Population of HKH in million (Year of data sources)	[a]Population in 2017 (million)	[a]Population in 2030 (million)
Afghanistan	All provinces except the provinces of Kandahar, Helmand, Nimroz, Farah, and Herat	**22.85** (2016–17)	22.85[c]	29.91
Bangladesh	Chittagong hills	**1.60** (2011)	1.78	2.27
Bhutan	Entire territory	**0.78**[b] (2017)	0.78[c]	0.96
China	Parts of the provinces of Yunnan (Diqing, Nujiang and Dali prefectures), Sichuan (Ganzi, Aba and Liangshan prefectures), and Gansu (Gannan, Wuwei and Zhangye prefectures); Xinjiang autonomous region (Kashigar, Kezilesu, Hetian and Altai prefectures); Tibet (entire territory), and Qinghai province (entire territory)	**32.51** (2015)	33.29	38.86
India	Entire territory of 11 mountain states (Assam, Uttarakhand, Himachal Pradesh, Manipur, Jammu and Kashmir (Indian administered area), Meghalaya, Mizoram, Nagaland, Sikkim, Tripura, Arunachal Pradesh), and Darjeeling and Kalimpong districts of West Bengal state	**76.98** (2011)	86.27	110.44
Myanmar	Chin, Shan, Rakhine and Kachin states	**11.18** (2014)	11.70	14.24
Nepal	Entire territory	**26.49** (2011)	28.75	34.31
Pakistan	Khyber Pakhtunkhwa province, 24 districts (out of 32) of Balochistan province (excluded districts are Kachhi, Gwadar, Jafarabad, Jhal Magsi, Lasbela and Sohbatpur), Azad Jammu and Kashmir (AJK), Gilgit-Baltistan and Federally administered Tribal Areas (FATA)	**51.47**[d] (2017)	51.47 [c]	72.64
Total		223.86	236.90	303.63

Notes
[a]Exponential projections of populations. The periods of growth rate estimation for HKH areas were 2009–2017 (Afghanistan), 2001–2011 (Bangladesh), 2010–2017 (Bhutan), 2011–2015 (China), 2001–2011 (India), 1983–2014 (Myanmar), 2001–2011 (Nepal), and 1998–2017 (Pakistan). These growth rates are used to project populations for 2017 and 2030
[b]Projected population for 2017 procured from Statistical Year Book of Bhutan, 2016
[c]Population statistics for 2017 are directly collected from official sources of respective countries (Statistical Year Book of Bhutan 2016; Population Census of Pakistan 2017; Statistical Year Book of Afghanistan 2016–17)
[d]Projected population of Azad Jammu and Kashmir and Gilgit-Baltistan for 2017 was added to the population of HKH areas procured from Population Census of Pakistan 2017
Sources (1) Data for latest population statistics are collected from Statistical Year Book of Afghanistan 2016–17; Population and Housing Census (Bangladesh) 2011; Statistical Year Book of Bhutan 2016; Year Books of China, Yunnan Province, Sichuan Province, Gansu Province, Xinjiang Autonomous Region (China) 2016; Population Census (India) 2011; Population and Housing Census of Myanmar 2014 (The Union Report, Volume 2); National Population and Housing Census of Nepal (national report) 2011; Population Census 2017; Govt. of AJK 2014; Govt. of Gilgit-Baltistan 2013
(2) Data for Base Year population (used for growth rate estimation) collected from Statistical Year Book of Afghanistan 2008-09; Population and Housing Census (Bangladesh) 2001; Statistical Year Book of Bhutan, 2016; Year Books of China, Yunnan Province, Sichuan Province, Gansu Province, Xinjiang Autonomous Region (China) 2012; Population Census (India) 2001; Population and Housing Census of Myanmar 2014 (The Union Report, Volume 2); National Population and Housing Census of Nepal (national report) 2011; Population Census (Pakistan) 1998
Assumptions for Projections In- and out-migration rates, fertility rate and mortality rate will remain stable

and mountain systems of Myanmar. Between these areas, we find the arid and semi-arid regions of the Pamir and Karakoram mountains; the high Himalaya of India, Nepal, and Pakistan; the un-spoilt beauty of Bhutan; the Tibetan Plateau of China; and the three parallel rivers of the Salween, Mekong and Yangtze in the far eastern region located in Yunnan Province of China. The HKH features great heterogeneity from north to south and east to west in relation to precipitation, vegetation, and human livelihoods. This variability defy making easy generalizations about the region.

Table 1.2 Area and population of the ten major river basins originating in the HKH

River basins	[a]Area (km^2)	[b]Population in 2010 (million)	[b]Population in 2015 (million)
Amu Darya	645,870	27.19	30.18
Brahmaputra	528,083	64.63	68.07
Ganges	1,001,090	539.43	580.09
Indus	1,116,350	244.31	268.42
Irrawaddy	426,393	40.18	42.87
Mekong	841,337	74.58	77.31
Salween	363,898	18.19	17.88
Tarim	929,254	10.65	11.37
Yangtze	2,066,050	600.92	604.94
Yellow River	1,073,440	192.86	198.02
Total	8,991,765	1,812.95	1,899.14

Notes
[a]The area of individual basins have been calculated from basin boundary shapefile in Albers equal-area conic map projection developed by ICIMOD
[b]The basin-wise population has been calculated from the "Gridded Population of the World adjusted to UN country level population estimates for 2010 and 2015" dataset produced by the Center for International Earth Science Information Network (CIESIN), Columbia University (http://sedac.ciesin.columbia.edu/data/collection/gpw-v4/united-nations-adjusted)

1.3 Key Issues of the HKH

This assessment considers the key issues in the HKH region in the context of related questions that draws regional attention, cooperation and policy solutions. The HKH region is geologically fragile, with young and rising mountains, usually vulnerable to erosion and landslides, even without human interference. The region is undergoing rapid change driven by stressors such as climate change and human conflicts, and factors like globalization, infrastructure development, migration, tourism and urbanization. The outcome of interplay of these complex drivers of change is challenging to predict but will have major consequences, not just in the region but globally. There is a critical need to assess these drivers' potential cost to the HKH environment and human wellbeing as well as the opportunities they may present. Per capita fossil fuel CO_2 emission from the HKH countries is one-sixth of the global average, however the region immensely suffers from the impact of climate change. Climate change is further enhanced by short-lived climate pollutants such as black carbon, which is emitted in large quantities in regions upwind of the HKH where dirty energy sources also have a large impact on health. From a policy standpoint, achieving food, water, energy, and livelihood security in the region will require exploring scenarios based on different assumptions so that the scientific community, policy-makers, the private sector, and community stakeholders can come together and make optimal governance decisions to sustain this global asset. It will also require country-specific recommendations to guide national-level policy-making.

1.4 Overall Objective, Rationale and Key Questions

1.4.1 Overall Objective

This assessment aims to (1) establish the global significance of the HKH, (2) reduce scientific uncertainty on various mountain issues, (3) lay out practical and up-to-date solutions and offer new insights for development of this region, (4) value and conserve existing ecosystems, cultures, societies, knowledge, and distinctive HKH solutions that are important to the rest of the world, (5) addresses contemporary policy questions, and (6) influence policy processes with robust evidence for sustainable mountain development.

1.4.2 Rationale for the Assessment

An assessment is distinct from a review. Whereas research speaks to other scientists in a particular field, assessments critically evaluate current states of knowledge about a topic with an aim to develop policy-oriented solutions, and inform relevant decision-makers across sectors. Assessments are structured to address specific social problems by translating science into forms that are salient, legitimate, and credible to wider audiences (Clark et al. 2006). Nevertheless, assessments also give due importance to reducing scientific uncertainty.

An integral part of assessments is indicating the level of confidence that chapter teams have concerning key findings, based on the available data, evidence and peer-reviewed publications. For the HIMAP Assessment it was decided to follow the four-box model adopted by the Intergovernmental Science-Policy Platform on Biodiversity and Ecosystem Services (IPBES) for the qualitative communication of confidence, which juxtaposes the level of agreement with the quantity and quality of the evidence. In the Chapter Overviews the following four confidence terms are used, in brackets and italics: *(well established)*, *(established but incomplete)*, *(unresolved)* and *(inconclusive)*. For a key finding to be well established, the quantity and quality of the evidence is high as is the level of agreement. For inconclusive both are low, while for unresolved multiple independent studies exist but the conclusions do not agree (low level of agreement). Lastly, established but incomplete pertains to findings where the general level of agreement is high among experts, although only a limited number of studies exist and the level of evidence is low to moderate.

The target audiences for this assessment are those who make decisions on investments and management regarding mountain development, that is, policy-makers, government agencies, foundations, academics, natural resource managers, private-sector investors, and civil-society members. In addition, our assessment aims to inform the general public about important mountain issues so that everyone can help to make better decisions through political processes in HKH countries.

In 2007, the Intergovernmental Panel on Climate Change (IPCC's) fourth assessment report (Pachauri and Reisinger 2007) predicted that climate change will be the most prominent driver of global change in the 21st century and pointed to the lack of consistent long-term monitoring in the HKH. The report called for national, regional, and global efforts to fill this data gap. Little progress was made in the HKH by the time of the IPCC's fifth assessment report (Pachauri and Meyer 2014). While universities, nongovernmental organizations, and scientific organizations have made strides in assembling and consolidating data, information on environment, natural resources and social systems of the HKH collectively remains too fragmented and incomplete to derive any meaningful conclusions about trends and scenarios.

The Hindu Kush Himalayan Monitoring and Assessment Programme (HIMAP), which brings together hundreds of scientists and experts from the region and around the world, aims to address these knowledge gaps and propose a way forward. This comprehensive assessment goes beyond climate change and is expected to greatly assist efforts to address threats and act on opportunities, and gives importance to upscale cutting-edge approaches. HIMAP derives inspiration from the Arctic Monitoring and Assessment Programme, which has systematically generated meaningful data and analysis about key trends and future scenarios on environment and natural resources in the Arctic region.

1.4.3 Key Questions

A set of questions relevant for the assessment was developed first by nalysing the key issues of the HKH region. The assessment was designed in a way that each chapter team considered relevant questions in writing their chapters. HIMAP has considered the following critical questions:

- What are the most important drivers of change in the HKH, what is the role of climate change, and what are their potential impacts on biodiversity, ecosystem services, livelihoods, and water resources?
- What are the most important strategies, policies, and governance arrangements for enhancing community adaptation to drivers of change, including climate change; how can they be out-scaled; and what are their impacts?
- How do gender-equitable and inclusive approaches support sustainable mountain development, and how can these be realized?
- What migration trends exist in the region, what are their present impacts on livelihoods and the environment, how climate change is inducing migration and should migration be taken as adaptive strategy, and what are the options for addressing migration and the likely consequences of those options?
- What is the existing status of the cryosphere, what changes are likely, and what are the possible impacts of those changes?
- What is the current and likely future quantity, variability, and quality of the water in the 10 major river basins of the HKH; what are the potential impacts of change on water availability; and how can negative impacts be mitigated?
- What are the best means of reducing the risk of floods and droughts, and how can they be introduced at various scales, including on Transboundary Rivers?
- Why is air pollution increasing in the HKH, how is the HKH affected by air pollution from within and beyond the HKH, and how can the problem be reduced?
- What are the energy needs and possibilities for the people of the HKH, what are the positive and negative impacts of hydropower development, how effective and safe is hydropower generation as an economic enterprise, and how can hydropower best be sustainably developed in the region?
- How can ecosystems be managed to support soil and biodiversity conservation, and improved livelihoods in the various contexts found in the HKH?
- What ecosystem services do mountains provide, and how can management and supply of these services be compensated?
- What watershed-, landscape-, and forestry-based approaches will best support ecosystem services, food and water security, and community resilience?
- How can the HKH develop a green economy? What technologies (modern, traditional, and indigenous) and approaches are best suited for sustainable mountain development in the region, and how can they be out-scaled?

1.5 Vision

The assessment foresaw the key issues of the region, drew 13 critical questions for addressing the identified issues, and both the issues and questions were used in formulating the vision of the assessment: *To enable a prosperous, peaceful, and poverty-free people; food, energy, environment and water secure people; and climate- and disaster-resilient mountain communities for the region and the world.*

(1) Prosperous—wellbeing in terms of productive and dignified, social (quality of life, social capital, healthy), cultural (identity plus integration), and environment (clean air, water, pollution management, and healthy natural resource base)
(2) Equality of access to opportunities and benefits of resources for everyone regardless gender and social class
(3) Food security—healthy people with access to adequate, affordable, good quality and nutritious food
(4) Energy security—access to adequate amount of energy that is affordable, non-polluting and sustainable, without unduly affecting the present low carbon status.
(5) Water security—access to quality, affordable water; and protection from extreme events such as floods and droughts
(6) Vibrant and bio-diverse ecosystem services for people to support culture and economies; protection and wise use of ecosystem services
(7) Climate and disaster resilient communities and countries; contributing to mitigation and adaptation; means—finance, capacity building, knowledge and technology
(8) Cooperation at all levels (people to people, business to business, government to government) between countries for sustainable and mutual benefits to achieve vision
(9) Unrestricted people movement across the HKH countries
(10) Recognition of the HKH region as a global asset.

1.6 HKH Priorities Contributing to the Sustainable Development Goals

Considering the issues, questions and vision as part of this assessment, we drafted our chapters and key messages in line with the United Nations' Sustainable Development Goals (SDGs). In this way, our "Priorities for Mountains and People of the HKH" reflect the ideals and inspiration of the SDGs. We created this complementarity through a three-step exercise:

(a) Define HKH priorities, align them with SDGs and refer to relevant HIMAP assessment chapters;
(b) Define HKH specific targets for 2030;
(c) Identify suitable indicators derived from the list of proposed SDG indicators; and indicate data availability.

Table 1.3 lays out the relationship between HKH priorities and the SDG framework.

1.7 Conceptual Framing of the Assessment

The concept of the assessment was logically developed step by step by framing of key issues, identifying critical questions to address the issues and visioning of the exercise and identifying nine priorities that could contribute to 2030 SDGs.

Our assessment addresses the environmental, economic and social pillars of sustainable mountain development and will serve as a basis for evidence-based decision-making to safeguard the environment and advance people's wellbeing. This report will not be a one-time publication. It is planned as the first of a series of monitoring and assessment reports about the HKH.

In spite of the vast expanse of mountains and their importance in the world, as a unique and exclusive land form, they have been largely ignored within better known environmental assessments such as the IPCC and Millennium Ecosystem Assessment. In those assessments, mountains are not examined in detail: scientific knowledge is scattered and traditional indigenous knowledge systems are mostly absent. This assessment intends to fill these gaps and provide information for improved decision making in and for the HKH. HIMAP intends to provide a connection of this region in global assessments such as IPCC AR 6 and subsequent ones and IPBES, and intends to contribute to global targets like 2030 SDG goals, 1.5° World after Paris UNFCCC 2015 agreement and Sendai Framework for Disaster Risk Reduction 2015.

The assessment chapters consider status, trends and scenarios on environmental, economic and social systems of the HKH region, and come up with recommendations that build into key policy messages. This assessment focuses on various drivers of change all of which are influenced by impacts of climate change. Mountain people and ecosystems tend to experience change more rapidly and with greater intensity. Mountain regions are no longer isolated from globalization. The HKH's biodiverse resources, rich indigenous knowledge systems, and enormous reservoirs of water provide vibrancy to the region and beyond. Understanding how these features may change over time is extremely important. In response, we devote many pages of this assessment to considering alternative development pathways and discussing ideas for enhancing regional cooperation in the HKH for sustainable mountain development.

Table 1.3 Framework for SDG consistent priorities for mountains and peoples of the HKH

SDG consistent priorities for the HKH region	Targets (related SDG targets in parentheses)	HKH Indicators (SDG indicators listed in parentheses)	HIMAP assessment corresponding chapter	Link with most relevant SDG
End poverty in all its form everywhere in the mountains and ensure that women, men and children of the HKH region lead healthy lives in an inclusive and equitable environment	• Reduce income poverty to zero in mountain areas by 2030 (1.1) • Reduce non-income poverty including health, education, and other basic needs to zero in mountain areas by 2030 (1.2) • Achieve universal health coverage, access to quality healthcare services and access to safe, effective, quality, and affordable essential medicines and vaccines for all people in the mountains (3.8) • All girls and boys in the mountains complete free, equitable and quality primary and secondary education (4.1) • Facilitate orderly, safe, and responsible migration and mobility of people within and between mountain and non-mountain areas (10.7) • By 2030, reduce to less than 3% the transaction costs of migrant remittances and eliminate remittance corridors with costs higher than 5% (10.c)	• Proportion of *mountain* population below the international poverty line (= $1.25 a day), by sex, age, employment status and geographical location (urban/rural) (1.1.1) • Proportion of *mountain* men, women and children of all ages living in poverty in all its dimensions according to national definitions (1.2.2) • Proportion of *mountain* population covered by social protection floors/systems, by sex, distinguishing children, unemployed persons, older persons, migrants, persons with disabilities, pregnant women, new-borns, work-injury victims and the poor and the vulnerable (1.3.1) • Proportion of *mountain* population living in households with access to basic services (sanitation, health, education) (1.4.1) • Proportion of total adult *mountain* population with secure tenure rights to land, with legally recognized documentation and who perceive their rights to land as secure, by sex and by type of tenure (1.4.2) • Mortality rate attributed to household and ambient air pollution (3.9.1) • Proportion of *mountain* youth (aged 15–24 years) not in education, employment or training (8.6.1) • Proportion of the rural *mountain* population who live	• HKH drivers of change (2) • Air pollution (10) • Disaster risk reduction and increasing resilience (11) • Mountain poverty vulnerability and livelihoods (12) • Gender and inclusive development (14) • Migration (15)	**Goal 1**. End poverty in all its forms everywhere **Goal 3**. Ensure healthy lives and promote well-being for all at all ages **Goal 4**. Ensure inclusive and equitable quality education and promote lifelong learning opportunities for all

(continued)

Table 1.3 (continued)

SDG consistent priorities for the HKH region	Targets (related SDG targets in parentheses)	HKH Indicators (SDG indicators listed in parentheses)	HIMAP assessment corresponding chapter	Link with most relevant SDG
		within 2 km of an all-season road (9.1.1) • Proportion of *mountain* population that has convenient access to public transport, by sex, age and persons with disabilities (11.2.1)		
Promote sustainable production systems to assure food security, nutrition security, and income for mountain people, with particular attention to women's changing roles in agriculture	• End all forms of malnutrition in the mountains and improve food and nutrition security, particularly for women and girl children (2.2) • Enable higher incomes for small-scale farmers, including women farmers (2.3) • By 2030, ensure sustainable food production systems and implement resilient agricultural practices *in the mountains* that increase productivity and production, that help maintain ecosystems, that strengthen capacity for adaptation to climate change, extreme weather, drought, flooding and other disasters and that progressively improve land and soil quality (2.4) • Increase investment in rural infrastructure, agricultural research, technology development, and plant and livestock gene banks in the mountains to improve agricultural productive capacity (2.a) • Achieve sustainable management and efficient use of natural resources (12.2)	• Prevalence of undernourishment by sex and age (2.1.1) • Prevalence of malnutrition by sex and age (2.2.2) • Average income of small-scale food producers, by sex and indigenous status (2.3.2)	• Food and nutrition security (9) • Disaster risk reduction and increasing resilience (11) • Mountain poverty vulnerability and livelihoods (12) • Adaptation strategies (13) • Gender and inclusive development (14)	**Goal 2**. End hunger, achieve food security and improved nutrition and promote sustainable agriculture **Goal 12**. Ensure sustainable consumption and production patterns
Achieve gender and social equity through inclusive and transformative change in the mountains	• Eliminate all forms of violence against all women and girls (5.2)	• Proportion of government recurrent and capital spending to sectors that disproportionately benefit	• Sustaining HKH Biodiversity and Ecosystem Services (5) • Meeting Future Energy Needs in the HKH (6)	**Goal 5**. Achieve gender equality and empower all women and girls **Goal 10**. Reduce inequality within and among countries

(continued)

Table 1.3 (continued)

SDG consistent priorities for the HKH region	Targets (related SDG targets in parentheses)	HKH Indicators (SDG indicators listed in parentheses)	HIMAP assessment corresponding chapter	Link with most relevant SDG
	• Ensure women's and marginalized groups' effective participation and equal opportunities for leadership at all levels of decision-making in political, economic and public life (5.5) • Increase number of women in institutions by at least 100%, particularly at the decision-making levels (16.7) • Adopt and strengthen policies and legislation for the promotion of gender and social equality and the empowerment of all women and girls at all levels, with a focus on mountains (5.c) • Eliminate gender disparities in education in the mountains (4.3) • Empower and promote the social, economic and political inclusion of all irrespective of age, sex, race, ethnicity, origin, religion or economic or other status (10.2)	women, the poor and vulnerable groups (1.b.1) • Participation rate of youth and adults in formal and non-formal education and training in the previous 12 months, by sex (4.3.1) • Whether or not legal frameworks are in place to promote, enforce and monitor equality and non—discrimination on the basis of sex (5.1.1) • Proportion of total agricultural population with ownership or secure rights over agricultural land, by sex; and share of women among owners or rights-bearers of agricultural land, by type of tenure (5.a.1) • Growth rates of household expenditure or income per capita among the bottom 40% of the population and the total population (10.1.1) • Proportions of positions (by sex, age, persons with disabilities and population groups) in public institutions (national and local legislatures, public service, and judiciary) compared to national distributions (16.7.1) • Proportion of population who believe decision-making is inclusive and responsive, by sex, age, disability and population group (16.7.2)	• Water availability and use (8) • Food and nutrition security (9) • Air pollution (10) • Disaster risk reduction and increasing resilience (11) • Mountain poverty vulnerability and livelihoods (12) • Adaptation strategies (13) • Gender and inclusive development (14) • Migration (15) • Governance and Institutions (16)	**Goal 16**. Promote peaceful and inclusive societies for sustainable development, provide access to justice for all and build effective, accountable and inclusive institutions at all levels
Ensure a year-round secure water supply in the mountains with universal and affordable access to safe drinking water, sanitation,	• Create secure water supply for key development sectors (agriculture, energy) that are viable year-round (6.4)	• Proportion of *mountain* population (disaggregated by sex, age and social categories) using safely managed drinking water services (6.1.1)	• Climate change in the HKH (3) • Status and change in the cryosphere (4) • Water availability and use (5) • Food and nutrition security (9)	**Goal 6**. Ensure availability and sustainable management of water and sanitation for all

(continued)

Table 1.3 (continued)

SDG consistent priorities for the HKH region	Targets (related SDG targets in parentheses)	HKH Indicators (SDG indicators listed in parentheses)	HIMAP assessment corresponding chapter	Link with most relevant SDG
and water for productive purposes	• Build effective and efficient mechanisms to implement and monitor transboundary cooperation agreements (6.5) • Achieve universal and equitable access to safe and affordable drinking water to all mountain people by 2030 (6.1) • Achieve access to adequate and equitable sanitation services and hygiene education for all in mountain regions (6.2) • Reduce women and children's water collecting time and work load by 2030 • Support and strengthen the participation of mountain communities in water management (6.b)	• Proportion of *mountain* population (disaggregated by sex, age and social categories) using safely managed sanitation services, including a hand-washing facility with soap and water (6.2.1) • Level of water stress: freshwater withdrawal as a proportion of available freshwater resources (6.4.2) • Proportion of transboundary basin area with an operational arrangement for water cooperation (6.5.2) • Proportion of men and women in the decision-making levels in water and climate related institutions	• Disaster risk reduction and increasing resilience (11) • Mountain poverty vulnerability and livelihoods (12) • Adaptation strategies (13) • Gender and inclusive development (14) • Migration (15) • Governance and Institutions (16)	**Goal 13**. Take urgent action to combat climate change and its impacts
Universal access to clean energy in the mountains from sources that are affordable, reliable, and sustainable	• Universal access to clean and affordable energy by the people in the mountains (7.1) • Increase electrification in rural areas (7.1) • Increase use of renewable energy (7.2) • Decrease air pollution (3.9 and 11.6) • Increase access of energy for women decreasing their workload, time and drudgery (7.1)	• Mortality rate attributed to household and ambient air pollution (3.9.1) • Proportion of mountain population (disaggregated by sex and social categories) with access to electricity (7.1.1) • Proportion of mountain population (disaggregated by sex and social categories) with primary reliance on clean fuels and technology (7.1.2) • CO_2 emission per unit of value added (9.4.1) • Annual mean levels of fine particulate matter (e.g. PM2.5 and PM10) in cities (population weighted) (11.6.2)	• Climate change in the HKH (3) • Status and change in the cryosphere (7) • Air pollution (10)	**Goal 7**. Ensure access to affordable, reliable, sustainable and modern energy for all
Halt biodiversity loss, land degradation and sustainably manage forests and other ecosystems in the mountains to enhance ecosystem resilience for sustained flow of services	• Ensure the conservation of mountain ecosystems, including their biodiversity and habitats (15.4) • Take urgent action to minimise human-wildlife conflict and end	• Change in the extent of ecosystems over time (6.6.1) • Proportion of important sites for terrestrial and freshwater biodiversity that are covered by	Sustaining HKH Biodiversity and Ecosystem Services (5)	**Goal 15**. Protect, restore and promote sustainable use of terrestrial ecosystems, sustainably manage forests, combat desertification, and halt and

(continued)

Table 1.3 (continued)

SDG consistent priorities for the HKH region	Targets (related SDG targets in parentheses)	HKH Indicators (SDG indicators listed in parentheses)	HIMAP assessment corresponding chapter	Link with most relevant SDG
	poaching and trafficking of protected species of flora and fauna in the mountains (15.7) • Reduce ecosystem degradation by development projects by 50% and restore degraded ecosystems (15.5) • Include ecosystem values in national accounting practices (15.9) • Increase investment in biodiversity conservation, and ecosystem based adaptation and sustaining services by 50% by 2030 (15.a) • Ensure 100% community participation in biodiversity programmes at the local level • Increase women's participation in decision making processes by 50% in natural resource access and benefit sharing programmes • Establish a mountain specific database for species and ecosystem services (17.18)	protected areas, by ecosystem type (15.1.2) • Proportion of land that is degraded over total land area (15.3.1) • Coverage by protected areas of important sites for mountain biodiversity (15.4.1) • Mountain Green Cover Index (indicator to measure changes of green vegetation in mountain areas, informed by satellite imagery data) (15.4.2) • Red List Index (endangered species) (15.5.1) • Official development assistance and public expenditure on conservation and sustainable use of biodiversity and ecosystems (15.b.1) • Number of countries that have adopted legislative, administrative and policy frameworks to ensure fair and equitable sharing of benefits (of use of genetic or other natural resources) (15.6.1)		reverse land degradation and halt biodiversity loss
Ensure integration between adaptation to climate change, disaster risk reduction and sustainable development for the mountains through evidence-based decision making	• Concerted action to keep global level climate change to 1.5° by 2100 (17.14) • Strengthen resilience and adaptive capacity to climate related hazards and natural disasters in the mountains (13.1) • Reduce mortality rates, especially for women and children due to extreme climate events (1.5) • Reduce economic loss due to extreme climate events (11.5) • Integrate mountain specific climate change measures into	• Number of deaths, missing persons and persons affected by disaster per 100,000 people (disaggregated by sex) (1.5.1) • Direct disaster economic loss in relation to global GDP, including disaster damage to critical infrastructure and disruption of basic services (11.5.2) • Proportion of local governments that adopt and implement local disaster risk reduction strategies in line with the Sendai Framework for	• Status and change in the cryosphere (7) • Disaster risk reduction and increasing resilience (11) • Adaptation strategies (13)	**Goal 13**. Take urgent action to combat climate change and its impacts **Goal 11**. Make cities and human settlements inclusive, safe, resilient and sustainable **Goal 1**. End poverty in all its forms everywhere

(continued)

Table 1.3 (continued)

SDG consistent priorities for the HKH region	Targets (related SDG targets in parentheses)	HKH Indicators (SDG indicators listed in parentheses)	HIMAP assessment corresponding chapter	Link with most relevant SDG
	national policies, strategies and planning (13.2)	Disaster Risk Reduction 2015–2030 (11.b.1)		
Build resilient, equitable and inclusive mountain communities empowered by economic opportunity and investment in mountain infrastructure and connectivity	• Develop sustainable and resilient infrastructure in the mountains to support economic development and human well-being (9.1) • Sustain per capita economic growth in the mountains and at least 7% annual GDP growth (8.1) • Devise and implement mountain specific policies to promote sustainable mountain tourism, which creates local jobs, promotes local culture and products (8.9) • Achieve access to full and productive employment and decent work for all women and men in the mountains, and equal pay for work of equal value (8.5) • Protect labour rights and promote safe and secure working environments for all workers, including migrant workers from mountain areas, in particular women and those in precarious employment (8.8) • Eradicate forced child labour and human trafficking (8.7)	• Annual growth rate of real GDP per capita, *disaggregated for mountain areas* (8.1.1) • *Mountain tourism* direct GDP as a proportion of total GDP and in growth rate (8.9.1) • Proportion of jobs in sustainable *mountain* tourism industries out of total *mountain* tourism jobs (8.9.2) • Average hourly earnings of female and male employees *in mountain areas*, by occupation, age and persons with disabilities (8.5.1) • Unemployment rate, by sex, age and persons with disabilities *in mountain areas* (8.5.2) • Level of national compliance with labour rights (freedom of association and collective bargaining) based on International Labour Organization (ILO) textual sources and national legislation, by sex and migrant status (8.8.2) • Volume of remittances *to mountain areas* (in United States dollars) as a proportion of total GDP (17.3.2)	• Mountain poverty vulnerability and livelihoods (12) • DRR and increasing resilience (11) • Adaptation strategies (13) • Gender and Inclusive Development (14) • Migration (15)	**Goal 8**. Promote sustained, inclusive and sustainable economic growth, full and productive employment and decent work for all **Goal 9**. Build resilient infrastructure, promote inclusive and sustainable industrialization and foster innovation **Goal 11**. Make cities and human settlements inclusive, safe, resilient, and sustainable
Promote a mountain-specific agenda for achieving the SDGs through increased regional cooperation among and between mountain regions and nations	• Cooperate at all levels across the HKH region for sustainable and mutual benefits (17.17) • Enhance regional and international cooperation and access to science, technology and innovation to achieve the SDGs in mountain areas (17.6) • In national, regional, and global decision making institutions and processes, recognize and	• Number of science and/or technology cooperation agreements and programmes between countries, by type of cooperation, *focusing on mountains* (17.6.1) • Extent of use of country-owned results frameworks and planning tools by providers of development cooperation *and*	Governance and Institutions (16)	**Goal 17**. Revitalize the global partnership for sustainable development

(continued)

Table 1.3 (continued)

SDG consistent priorities for the HKH region	Targets (related SDG targets in parentheses)	HKH Indicators (SDG indicators listed in parentheses)	HIMAP assessment corresponding chapter	Link with most relevant SDG
	prioritize the uniqueness of the HKH mountains and its people. Ensure representation in decision-making (17.15) • Allocate significantly greater resources and identify incentives for conservation of benefits from mountain ecosystems (15a and 17.2) • Enhance capacity-building support to HKH countries to increase significantly the availability of high quality, timely, reliable data that is specific to mountain regions, disaggregated by income, gender, age, race, ethnicity, migratory status and disability (17.18) • Equal protection of migrants under effective rule of law and good governance (16.3)	*recognition of the HKH* (17.15.1) • Net official development assistance *to mountain areas in the HKH*, total and to least developed countries, as a proportion of the Organization for Economic Cooperation and Development (OECD) Development Assistance Committee donors' gross national income (GNI) (17.2.1) • Total amount of approved funding *for mountain areas* in developing countries to promote the development, transfer, dissemination and diffusion of environmentally sound technologies (17.7.1) • Proportion of sustainable development indicators produced at the national level *specific to the HKH mountain areas*, with full disaggregation when relevant to the target, in accordance with the Fundamental Principles of Official Statistics (17.18.1) • Level of national compliance with labour rights (freedom of association and collective bargaining) based on International Labour Organization (ILO) textual sources and national legislation, by sex and migrant status (8.8.2)		

1.8 Assessment Process

The International Centre for Integrated Mountain Development (ICIMOD) coordinated HIMAP, constituted the chapter author teams and the process was steered by policy decisions of the Steering Committee. The assessment process involved several rounds of Steering Committee meetings, workshops of Coordinating Lead Authors and Lead Authors including write-shops and peer inter-chapter reviews, subject expert reviews, and open reviews for anyone interested. Science-policy dialogues were organized to develop key policy messages. For this assessment, HIMAP has engaged more than 300 researchers, practitioners, experts, and policy-makers. The publication of the first Comprehensive Assessment of the HKH in 2018 is planned as a wide-ranging, innovative evaluation of the current state of knowledge in the region and of various drivers of change and their impacts, and a set of practically oriented policy recommendations. The process is following these steps:

- Framing of the assessment: A framing workshop and consultations with various experts to define the structure and process of the assessment.
- Drafting of chapters: Based on the experience of other assessments, a network of people with in-depth knowledge of the region to draft the chapters.
- Peer review: Rigorously review the chapter drafts, both by peers and via open review.
- Dissemination: Using multiple channels, to communicate to a wide range of audiences during the process to draw attention while the assessment is still in preparation.
- Engagement with policy-makers: share with policy-makers in the region through various processes.
- Development of a summary document: A summary for decision-makers based on the results of the process.
- Publication and launch: Publication of the first edition of the assessment in 2017.

1.9 Outline of the Assessment

Each chapter of the assessment address three broad themes within its particular confines:

(1) Defining the vision and state of knowledge;
(2) Drivers of change and integrated future scenarios; and
(3) Noting ideas and praxis for sustainable development.

The critical questions were used by each of the chapters to address the key issues of the region. The sixteen chapters include: Introduction—setting the scene: Drivers—local, regional, and global; Climate change in the HKH; Future scenarios; Sustaining HKH biodiversity and ecosystem services; Meeting future energy needs; The cryosphere; Water security—availability, use, and governance; Food and nutrition security; Air pollution; Disaster risk reduction and increasing resilience; Mountain poverty, vulnerability and livelihoods; Adaptation strategies; Gender and inclusive development; Migration; and Governance and institutions.

References

Bajracharya, S. R., & Shrestha, B. (Eds.). (2011). *The status of glaciers in the Hindu Kush-Himalayan region*. Kathmandu: ICIMOD.

Bajracharya, S. R., Maharjan, S. B., Shrestha, F., Guo, W., Liu, S., Immerzeel, W., et al. (2015). The glaciers of the Hindu Kush Himalayas: Current status and observed changes from the 1980s to 2010. *International Journal of Water Resources Development, 31* (2), 161–173.

Bangladesh Bureau of Statistics. (2001). *Population and housing census 2001*. Dhaka, Bangladesh: Statistics and Informatics Division, Ministry of Planning, Government of Bangladesh.

Bangladesh Bureau of Statistics. (2011). *Population and housing census 2011*. Dhaka, Bangladesh: Statistics and Informatics Division, Ministry of Planning, Government of Bangladesh.

Central Bureau of Statistics (2011). *National population and housing census of Nepal (national report) 2011*. Kathmandu, Nepal: National Planning Commission Secretariat, Government of Nepal. Retrieved from http://cbs.gov.np/image/data/Population/National%20Report/National%20Report.pdf.

Central Statistics Organization. (2009). *Statistical year book of Afghanistan 2008–09*. Kabul, Afghanistan: Government of Islamic Republic of Afghanistan. Retrieved from http://cso.gov.af/Content/files/Population%20Full%20chapter.pdf.

Central Statistics Organization (2017). *Statistical year book of Afghanistan 2016–17*. Kabul, Afghanistan: Government of Islamic Republic of Afghanistan. Retrieved from http://cso.gov.af/Content/files/%D8%B3%D8%A7%D9%84%D9%86%D8%A7%D9%85%D9%87%20%D8%A7%D8%AD%D8%B5%D8%A7%D8%A6%DB%8C%D9%88%DB%8C%20%D8%B3%D8%A7%D9%84%201395/Population.pdf.

Chettri, N., Shakya, B., Thapa, R., & Sharma, E. (2008). Status of a protected area system in the Hindu Kush-Himalayas: An analysis of PA coverage. *The International Journal of Biodiversity Science and Management, 4*(3), 164–178.

Clark, W. C., Mitchell, R. B., & Cash, D. W. (2006). Evaluating the influence of global environmental assessments. In R. B. Mitchell, W. C. Clark, D. W. Cash, & N. M. Dickson (Eds.), *Global environmental assessments: Information and influence* (pp. 1–28). Cambridge, USA: MIT Press.

Dyurgerov, M. D., & Meier, M. F. (2005). *Glaciers and changing earth system: A 2004 snapshot*. Boulder: Institute of Artic and Alpine Research, University of Colorado.

FAO. (2015). *Mapping the vulnerability of mountain peoples to food insecurity*. Rome, Italy: Food and Agriculture Organization of the United Nations.

Fleury, J. M. (1999). *Mountain biodiversity at risk* (Vol. 2, pp. 1–6). IDRC Briefing.

Gansu Provincial Bureau of Statistics. (2012). *Statistical yearbook of Gansu Province 2012*. Lanzhou, People's Republic of China: Government of Gansu Province.

Gansu Provincial Bureau of Statistics. (2016). *Statistical yearbook of Gansu Province 2016*. Lanzhou, People's Republic of China: Government of Gansu Province.

Hamilton, L. S. (2002). *Why mountain matters? World Conservation: The IUCN Bulletin 1/2002*.

Kapos, V., Rhind, J., Edwards, M., Price, M. F., & Ravilious, C. (2000). Developing a map of the world's mountain forests. In M. F. Price & N. Butt (Eds.), *Forests in Sustainable Mountain Development: A State-of-Knowledge Report for 2000* (pp. 4–9). Wallingford: CAB International.

Ministry of Home Affairs. (2001). *Population census 2001*. Delhi, India: Office of the Registrar General and Census Commissioner, Ministry of Home Affairs, Government of India. Retrieved from http://www.censusindia.gov.in/2011-common/census_data_2001.html.

Ministry of Home Affairs. (2011). *Population census 2011*. Delhi, India: Office of the Registrar General and Census Commissioner, Ministry of Home Affairs, Government of India. Retrieved from http://www.censusindia.gov.in/2011-common/census_data_2001.html.

Molden, D., & Sharma, E. (2013). ICIMOD's strategy for delivering high-quality research and achieving impact for sustainable mountain development. *Mountain Research and Development, 33*(2), 179–183.

National Bureau of Statistics of China. (2012). *Statistical yearbook of China 2012*. Beijing, China: Government of the People's Republic of China. Retrieved from http://www.stats.gov.cn/tjsj/ndsj/2012/indexeh.htm.

National Bureau of Statistics of China. (2016). *Statistical yearbook of China 2016*. Beijing, China: Government of the People's Republic of China. Retrieved from http://www.stats.gov.cn/tjsj/ndsj/2016/indexeh.htm.

National Statistics Bureau. (2016). *Statistical year book of Bhutan 2016*. Thimphu, Bhutan: Government of Bhutan. Retrieved from http://www.nsb.gov.bt/publication/files/SYB_2016.pdf.

Nogues-Bravo, D., Araujo, M. B., Romdal, T., & Rahbek, C. (2008). Scale effects and human impact on the elevational species richness gradients. *Nature, 453*(8), 216–220.

Pachauri, R. K., & Reisinger, A. (Eds.). (2007). In *Climate change 2007: Synthesis report. Contribution of working groups I, II and III to the fourth assessment report of the intergovernmental panel on climate change*. Geneva, Switzerland: Intergovernmental Panel on Climate Change. Retrieved March 12, 2016, from www.ipcc.ch/report/ar4/.

Pachauri, R. K., & Meyer, L. A. (Eds.). (2014). *Climate change 2014: Synthesis Report. Contribution of working groups I, II and III to the fifth assessment report of the intergovernmental panel on climate change*. Geneva, Switzerland: Intergovernmental Panel on Climate Change. Retrieved March 12, 2016, from www.ipcc.ch/report/ar5/.

Pakistan Bureau of Statistics. (1998). *Population census 1998*. Islamabad: Government of Pakistan. Retrieved from http://www.pbs.gov.pk/content/population-census.

Pakistan Bureau of Statistics. (2017). *Population census 2017*. Islamabad: Government of Pakistan. Retrieved from http://www.pbs.gov.pk/content/population-census.

Planning and Development Department. (2014). *Azad Jammu and Kashmir at a Glance 2014 (report)*. The government of Pakistan. Retrieved from https://pndajk.gov.pk/uploadfiles/downloads/AJK%20at%20a%20glance%202014.pdf.

Planning and Development Department. (2013). *Gilgit-Baltistan Azad Jammu and Kashmir at a Glance 2013*. Muzaffarabad, Pakistan: Statistical Cell, Planning and Development Department, Government of Gilgit-Baltistan. Retrieved from http://www.gilgitbaltistan.gov.pk/DownloadFiles/GBFinancilCurve.pdf.

Schild, A., & Sharma, E. (2011). Sustainable mountain development revisited. *Mountain Research and Development, 31*(3), 237–241.

Sichuan Provincial Bureau of Statistics. (2012). *Statistical yearbook of Sichuan Province 2012*. Chengdu, People's Republic of China: Government of Sichuan Province.

Sichuan Provincial Bureau of Statistics. (2016). *Statistical yearbook of Sichuan Province 2016*. Chengdu, People's Republic of China: Government of Sichuan Province.

The Ministry of Immigration and Population. (2014). *Population and housing census of Myanmar 2014 (State wise reports)*. Nay Pyi Taw, Myanmar: Department of Population, Ministry of Immigration and Population, Government of Myanmar.

UNEP-WCMC. (2002). *Mountain watch: Environmental change and sustainable development in mountains*. Cambridge, UK: UNEP—World Conservation Monitoring Centre.

Whiteman, D. (2000). *Mountain meteorology*. London: Oxford University Press.

Xinjiang Autonomous Region Bureau of Statistics. (2012). *Statistical yearbook of Xinjiang autonomous region 2012*. Ürümqi, People's Republic of China: Government of Xinjiang Autonomous Region.

Xinjiang Autonomous Region Bureau of Statistics. (2016). *Statistical yearbook of Xinjiang autonomous region 2016*. Ürümqi, People's Republic of China: Government of Xinjiang Autonomous Region.

Xu, J., & Melick, D. (2006). Emancipating indigenous knowledge: Can traditional cultures assist Himalayan sustainable development. *Sustainable Mountain Development, 50*, 31–35.

Yunnan Provincial Bureau of Statistics. (2012). *Statistical yearbook of Yunnan Province 2012*. Kunming, People's Republic of China: Government of Yunnan Province.

Yunnan Provincial Bureau of Statistics. (2016). *Statistical yearbook of Yunnan Province 2016*. Kunming, People's Republic of China: Government of Yunnan Province.

Drivers of Change to Mountain Sustainability in the Hindu Kush Himalaya

Coordinating Lead Authors
Yanfen Wang, University of Chinese Academy of Sciences, Beijing, China, e-mail: yfwang@ucas.ac.cn
Ning Wu, International Centre for Integrated Mountain Development, Kathmandu, Nepal; and Chengdu Institute of Biology, Chinese Academy of Sciences, Chengdu, China,
e-mail: wuning@cib.ac.cn (corresponding author)

Lead Authors
Clemens Kunze, International Centre for Integrated Mountain Development, Kathmandu, Nepal,
e-mail: clemens.kunze@icimod.org
Ruijun Long, International Centre for Integrated Mountain Development, Kathmandu, Nepal,
e-mail: ruijun.long@icimod.org
Manfred Perlik, Centre for Development and Environment, University of Bern, Switzerland; and Laboratoire Pacte/LabEx ITEM, University of Grenoble, France, e-mail: manfred.perlik@cde.unibe.ch

Contributing Authors
Neha Bisht, International Centre for Integrated Mountain Development, Kathmandu, Nepal,
e-mail: neha.bisht16@gmail.com
Huai Chen, Chengdu Institute of Biology, Chinese Academy of Sciences, Chengdu, China,
e-mail: chenhuai@cib.ac.cn
Xiaoyong Cui, University of Chinese Academy of Sciences, Beijing, China,
e-mail: cuixy@ucas.ac.cn
Rucha Ghate, International Centre for Integrated Mountain Development, Kathmandu, Nepal,
e-mail: ruchaghate@gmail.com
Srijana Joshi, International Centre for Integrated Mountain Development, Kathmandu, Nepal,
e-mail: srijana.joshi@icimod.org
Abhimanyu Pandey, International Centre for Integrated Mountain Development, Kathmandu, Nepal,
e-mail: abhimanyu.pandey@icimod.org
Jianqiang Zhang, Institute of Mountain Hazard and Environment, Chinese Academy of Sciences, Chengdu, China, e-mail: zhangjq@imde.ac.cn

Review Editor
Marc Foggin, Mountain Societies Research Institute, University of Central Asia, Kyrgyzstan | Tajikistan | Kazakhstan; and Institute of Asian Research, University of British Columbia, Canada,
e-mail: marc.foggin@ucentralasia.org

Corresponding Author
Ning Wu, International Centre for Integrated Mountain Development, Kathmandu, Nepal; and Chengdu Institute of Biology, Chinese Academy of Sciences, Chengdu, China,
e-mail: wuning@cib.ac.cn

P. Wester et al. (eds.), *The Hindu Kush Himalaya Assessment*,
https://doi.org/10.1007/978-3-319-92288-1_2

Contents

Chapter Overview

Key Findings

1. **Looming challenges characterize the HKH as environmental, sociocultural, and economic changes are dynamically impacting livelihoods, environmental conditions, and ultimately sustainability.** Many challenges for sustainability are related to weak governance, natural resource overexploitation, environmental degradation, certain aspects of unregulated or rapid urbanization, and loss of traditional culture. Addressing these problems will require policy and action at local, national, and international levels, including common action among HKH states.
2. **However, for mountain societies of the HKH, some changes may also bring novel opportunities for sustainable development.** A range of opportunities lie in improved connectivity including transportation and communication, which increases access to information, partnerships, and markets. Enhanced access to social services may be enabled and strengthened by economic growth and the advancement of science and technology. Additionally, a growing network of local urban centres may support the transmission of new prosperity to rural populations, as the development of mountain towns and cities often can help—besides their mere economic power—to enhance the political influence of these regions within the national states.
3. **The drivers of change to environmental, sociocultural, and economic sustainability in the HKH are interactive, inextricably linked, and increasingly influenced by regional and global developments.** Among the most important drivers in this intricate network of causes and effects are demographic changes and current governance systems, as well as land use and land cover change, over-exploitation of natural resources, economic growth and differentiation, and climate change.

Policy Messages

1. **To meet the challenges arising from environmental, sociocultural, and economic changes in the HKH, policy approaches must become more holistic and multidimensional.** While the drivers of change continue to demand individual identification and analysis, the complexity of their interactions also calls for more comprehensive and integrated strategies to be adopted—incorporating globally recognized mountain priorities, promoting transboundary cooperation, and encouraging development of mountain-specific responses by government policy makers.
2. **Governments should take strong and timely action to strengthen the sociocultural and environmental dimensions of sustainability, while also fostering responsible economic growth in the mountain regions.** Such interventions should be developed with more inclusive, participatory approaches in natural resource management. Both proximate and ultimate drivers of change must be recognized and incorporated into national development planning. This is the Mountain Agenda.
3. **Regional governments should combine and accelerate efforts to advance sustainable mountain development, especially with a view to benefiting from the global conservation and development agenda, such as the Sustainable Development Goals for 2030.** New financing mechanisms for climate change mitigation and adaptation and for infrastructure development offer valuable opportunities for increased investment in mountain regions. Governments must seize these, with the aim of creating enabling environments and institutions that empower mountain people to share in the regional and global achievements and benefits of inclusive growth and sustainable development.

What are the main ***drivers of change*** affecting mountain sustainability in the HKH? This chapter seeks answers to this critical question. Building on the definition of sustainability given by the United Nations' Agenda 21, we describe environmental, sociocultural, and economic dimensions of change and their drivers in the HKH. Specifically, we aim to outline and describe trends and existing and potential impacts of a varied and sometimes complex set of drivers of mountain sustainability.

A comprehensive analysis of the major drivers leading to changes in the HKH reveals that individual and cumulative impacts are reflected at multiple spatial and temporal scales (***well established***). For example, globalization, climate change, and the rapid spread of invasive alien species exist or occur in many localities and across many scales. As a result, in any given area or community, the interactions and impacts of these and other drivers of change are complex, and often are local and specific in the context of the dominant ecosystem or landscape and the sociocultural and economic environments.

The HKH is among the most discrete yet diverse regions of the world in terms of environmental, sociocultural and economic settings. It represents globally significant biodiversity hotspots and provides numerous ecosystem services to millions of people living within as well as outside the region (***well established***). The HKH region also faces enormous pressures from both global and regional change, and from a combination of both natural and anthropogenic forces (***well established***).

In most areas of the HKH, rapid demographic and economic growth have increased the demand for natural resources leading in many instances to their overexploitation, significant land use and land cover change (LULCC), habitat fragmentation, and unsustainable socioeconomic activities (***well established***). Rapid economic growth has changed levels and patterns of consumption as well as infrastructure investment. For example, dam construction for irrigation may make food production more efficient and create opportunities to export food, while those dams built for hydropower can improve livelihoods locally with the provision of electricity as well as energy for export, and may also transform agricultural communities through diversification of livelihood opportunities such as tourism. However, all large scale investments also have a multitude of intended or unintended effects, some of which are negative, including social and environmental consequences (***established but incomplete***). Similarly, demographic shifts, with people increasingly concentrated in town and cities, are expected to create future challenges through environmental impacts caused especially by a growing demand for food and energy, yet at the same time may also lead to improved quality of and access to social services such as education, health care and waste management (***established but incomplete***).

Technological innovations have markedly affected people's ways of life in the HKH, especially local and indigenous sociocultural practices, enabled through the development of a range of opportunities in remote mountain areas (***established but incomplete***). With a gradual integration into regional and global markets, many rural societies in the region—though not all—are now shifting from subsistence farming to more market-based agricultural production, including cash crops. Such changes have contributed to the conversion of croplands to non-agricultural use, to a decline in traditional ways of life, and also to a more intense use of natural resources—along with rising incomes and enhanced livelihoods, albeit not uniformly amongst all people or groups (***established but incomplete***). Advances in agricultural technology and biotechnology may improve the yield of crops and people's food security, yet arguable with a bundle of uncertainties (***established but incomplete***), while the increased access to global ICT services including smartphones, open access software and cloud computing can make mountain regions more accessible (***established but incomplete***).

Several major drivers, in particular demographic changes, governance systems and institutions, and climate change, are likely to have the most harmful or challenging impacts on sustainable development in the HKH (***established but incomplete***). As HKH countries work to adapt their practices and mitigate these impacts, all their efforts must involve close collaboration among countries. Weak governance and uncertain or unsecure land tenure, in particular, along with political unrest, local conflicts and migration, are also exacerbating environmental degradation through various activities such as poaching, unsustainable timber harvesting, and other forms of over-exploitation of natural resources (***established but incomplete***).

All the above challenges create an urgent need for improved coordination amongst development stakeholders including governments for formulating evidence-based policies and legislation, for developing more effective institutional arrangements, for more transparent decision making, and for greater transboundary cooperation in regional aspects of conservation and development across the HKH.

2.1 Pillars and Drivers of Sustainability in HKH Mountains

Mountain systems have long been admired and protected on the grounds of their serenity, wilderness, and landscape beauty (Antonelli 2015; Foggin 2016; Messerli and Ives 1997; Price 2015). Although direct human influence on the world's mountains is relatively low, when compared to more densely populated lowlands, nonetheless many mountain systems are strongly affected by multiple local and global drivers of change (Price 2015).

Decisions and actions taken by a diverse range of stakeholders operating at many scales in the HKH impact the overall sustainability in mountains. Drivers of change affect sustainability through environmental, sociocultural, or economic dimensions, either individually or in combination. In addition, local processes may have regional and global impacts, and vice versa. The origin and impact of key drivers of change in the HKH—which will be introduced in this chapter—are manifest at three main levels or scales: (1) *direct local impacts* of drivers on land and natural resources and their management, within a framework of coupled social–ecological systems; (2) *regional effects* of local drivers, mediated largely through provision of ecosystem services in the context of highland–lowland linkages; and (3) *regional and global influences* on the HKH through tele-coupled systems, whereby decisions or actions made outside the region have significant impacts within the HKH. In the widely interconnected spheres of influence of each driver, multiple pathways of impact are common, including both direct and indirect associations with mountain sustainability.

There are also three primary dimensions or pillars of sustainability. Ensuring that the needs of the present generation are met without compromising the ability of future generations to meet their own needs requires that the environment be safeguarded from degradation or loss, that development occurs in socially and culturally appropriate and equitable ways, and that development plans and interventions be economically viable. In mountain regions, additional complexities are introduced because of upstream–downstream linkages (e.g., flow of ecosystem services) and the multi-scalar nature of many of the key drivers and/or their impacts (cf. tele-coupled systems) (see Box 2.1).

Box 2.1 What is mountain sustainability?

Sustainability has environmental, sociocultural, and economic dimensions. Healthy ecosystems and environments are necessary for the survival of humans and other organisms, and thus constitute the basis of sustainable development. Yet without sociocultural equity or viable economic plans and initiatives, few development interventions will succeed. The three dimensions or pillars of sustainability are interdependent and mutually reinforcing, and—in the long run—none can exist without the others (Morelli 2011). Moving towards sustainability is a societal challenge that involves national and international legislation, urban and regional development, transport and other sectors, and that equally involves engagement with local and individual ways of life, and, especially in an increasingly urban world, positive choices to promote more ethical consumerism.

In mountain regions such as the HKH, linkages are critical between upstream and downstream ecosystems, people, and production systems, as well as between centres of decision making and remoter mountain communities. Achieving sustainability in the HKH will require not only that drivers of change and their pathways of impact be clearly identified and addressed, but also that regional and transboundary coordination and collaboration are endorsed, promoted, and strengthened across all relevant sectors.

Because human-related drivers affect mountain sustainability at different spatial and temporal scales, both the assessment and management of these drivers are complex matters. Changes in one driver generally result in interactive

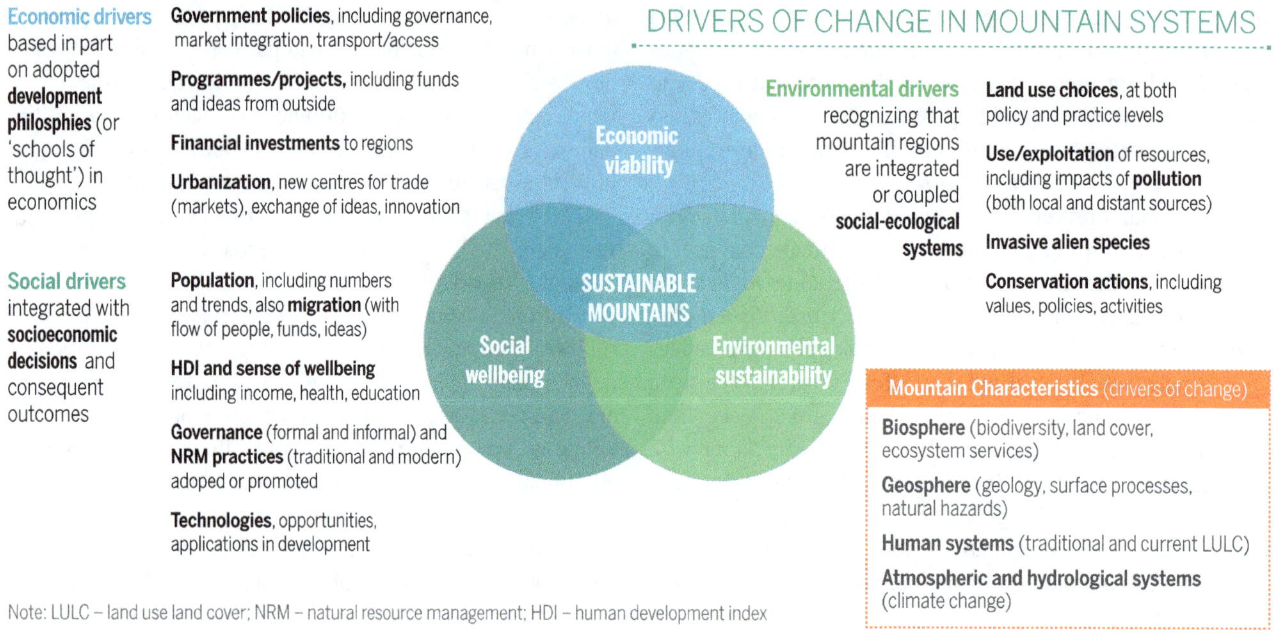

Fig. 2.1 Model of three pillars of mountain sustainability and network of interactive drivers of change (*Source* developed by Marc Foggin for this chapter). *Note* LULC: Land use and land cover; NRM: Natural resource management; HDI: Human development index

or feedback effects on other drivers. That is, for any given change, the effects are always multiple and often interactive; there is no simple one-to-one causal link between one driver of change and a single aspect of mountain social–ecological systems.

Further, certain drivers of change are themselves the result of complex combinations of multiple other factors. Causality is almost always mediated by a range of interacting factors, complicating attempts to establish the proportionality of various contributors to change. Therefore, although the following sections address a suite of drivers of change separately, each in turn, their combined effects should be kept in mind along with their respective effects on the three different dimensions of sustainability—environmental, sociocultural, and economic.

Figure 2.1 provides a conceptual illustration (model) of mountain sustainability and the key drivers of change, broadly organized around the concept of the three pillars of sustainability. Throughout this chapter, however, it will become evident that the impacts of each individual driver almost always range across all of these dimensions.

In summary, this chapter presents a high-level review and assessment of available information about the main *drivers of change* to mountain sustainability in the HKH, with a review of their multidimensional impacts across a range of scales and introducing their major pathways of impact. Notably, not every element in the conceptual model shown in Fig. 2.1 is substantiated with full studies or information directly from the HKH—a number of gaps still exist—but the drivers detailed in this chapter are largely overlapping and consistent with the model.

2.2 Environmental Drivers of Change to Mountain Sustainability

This section focuses on the multiple, interrelated effects of land use and land cover change, levels of resource exploitation, forms of pollution, presence of invasive alien species, mountain hazards, and climate change, as the main (direct) environment-related drivers of change to mountain sustainability.

2.2.1 Land Use and Land Cover Change

Land use and land cover change (LULCC) directly impacts the environment and leads to socioeconomic changes for human communities. LULCC is the primary cause of soil degradation, altering ecosystem functions and services, thus affecting the abilities of ecosystems to support human needs—both locally and downstream. In this way LULCC largely determines (or mediates) both the vulnerability and the resilience of ecosystems and human society to external perturbations such as climate change, national and regional policies, and other aspects of globalization.

Integrated research on LULCC across the HKH is generally lacking. Most published case studies have been carried

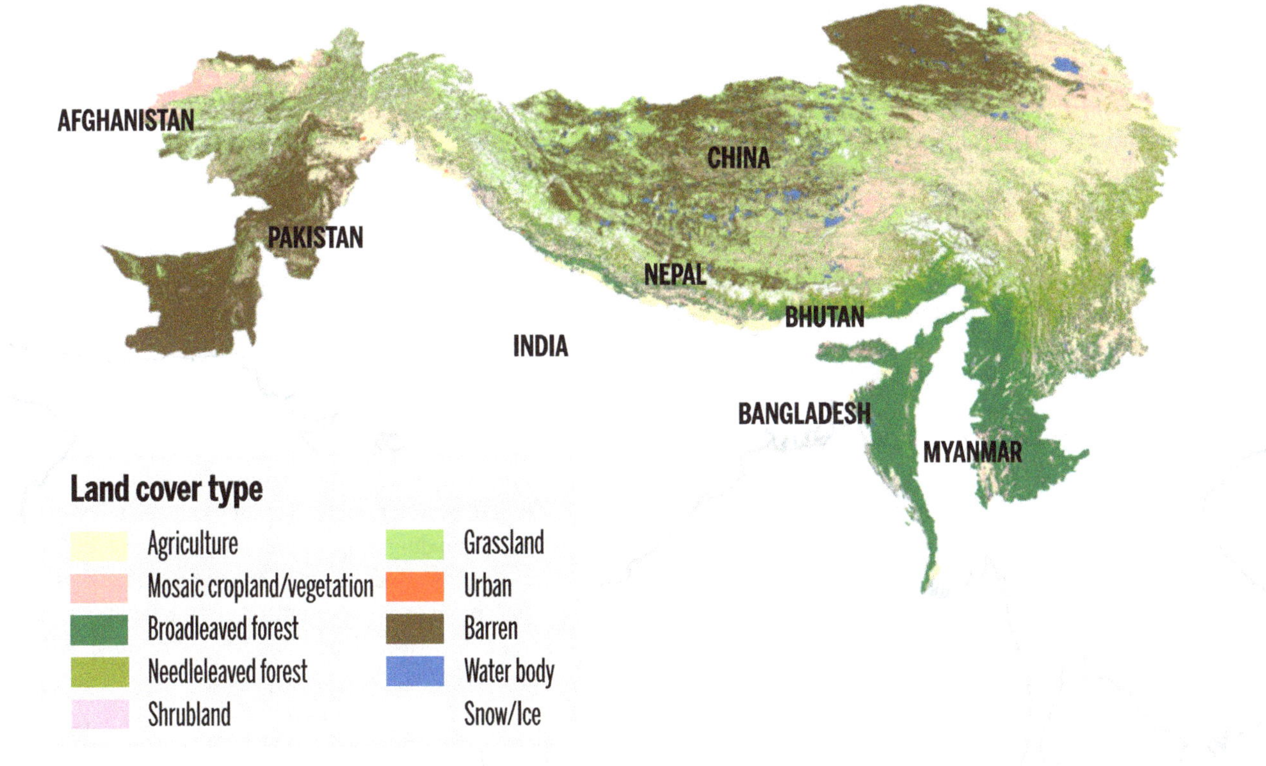

Fig. 2.2 Main land cover types in the Hindu Kush Himalaya

out in the Tibetan Plateau region in China, focusing especially on the vast grasslands. Because most available case studies analyse satellite data that are temporally and spatially variable, regional patterns of LULCC are difficult to distinguish clearly (Harris 2010; Box 2.2).

Box 2.2 General trends of LULCC in the HKH

The HKH encompasses a total land area of 3.4 million km^2, more than half of which is rangeland, specifically grasslands and shrubland (Fig. 2.2). Grasslands are the dominant form and include alpine pastures and meadows. Most of the shrubland used for grazing is distributed above the tree line. In addition, forests occupy approximately 14% of the region, comprising broad-leaved forests, needle-leaved forests and plantations. Around 25% of the area is classified as agricultural land. Wetlands cover under 10% of the HKH.

The most significant LULCC in recent decades has been caused by degradation of grasslands and deforestation at lower altitudes. Other significant transitions have been the transformation of forest and grassland to farmland, shrinkage of wetland, and LULCC related to urban and infrastructure developments (Cui and Graf 2009; Jin et al. 2010). In some cases, however, the recent implementation of large-scale ecological restoration programmes in HKH countries has slowed and sometimes even reversed the trend of degradation (Cai et al. 2015).

HKH rangelands support high floristic richness and large numbers of wild and domesticated animals. Forests provide important altitudinal connectivity for species between lowland and mountain habitats. The croplands have a diversity of mostly mixed farming systems, containing a rich genetic diversity of cultivated plants and livestock breeds (ICIMOD 2009). Wetlands also provide habitats for several globally significant migratory birds (Gujja et al. 2003; Sharma et al. 2010). However, when LULCC occurs, the quality and quantity of these habitats are impacted, affecting a wide range of organisms.

Extensive research has shown that degradation of grassland has spread over the past few decades across the whole high-altitude areas of the HKH (Wu et al. 2013). In the Tibetan Plateau of China, high-altitude grassland is sensitive to disturbance, especially in its drier part, where the removal of turf is irreversible. In the north-east portion of the Tibetan Plateau, many alpine meadows have been seriously degraded since the 1960s (Wang et al. 2008). By some measures, 'degraded grassland' constituted approximately 92% of the total area of usable grassland in 2004, of which severely and moderately degraded grassland accounted for 38% and 54%, respectively (Wang et al. 2008). Additionally, the areas of alpine meadow and swamp meadow decreased by around 4% and 25%, respectively, in the 1970s and 1980s; they decreased by almost another 25% and 35%, respectively, in the 1990s (Zhou et al. 2008).

Wetlands and lakes constitute around 5% of the total area of the Tibetan Plateau, but this area decreased by about 5% between 1990 and 2000, due primarily (84%) to wetland shrinkage (Li and Xue 2010). In contrast, lake expansion was also found in some high-altitude belts (Liu et al. 2009, 2010) such as the source areas, or headwaters, of the Yangtze River, Yellow River, and Mekong River in the central Tibetan Plateau; this is attributed to glacier melting caused by global warming (unpublished data).

In addition, warm–temperate coniferous forest has nearly disappeared from the south-east Tibetan Plateau, mostly due to commercial logging before the end of the 1990s (Cui and Graf 2009). In the high mountainous areas of Pakistan, the forested area decreased by 30% from 1968 to 2007, one-third being caused by agricultural expansion (Qasim et al. 2013). In the same region, agricultural lands in mid-elevation zones expanded by 70% and forests decreased by 50%. At lower altitudes agricultural lands more than doubled—expanding by 130% and causing a decrease of over 30% in the forested area between 1968 and 2007, when annual deforestation rates were 0.80%, 1.28%, and 1.86% at high, mid- and low altitudes, respectively (Qasim et al. 2013). An annual loss of 0.2% in forested area has been reported for the Indian mountains (Reddy et al. 2013) and of 0.3% for Myanmar (Leimgruber et al. 2005). However, forest cover in Bhutan is increasing at an annual rate of 0.22% (Gilani et al. 2015).

Changes in LULCC in HKH, as elsewhere, are caused by a combination of natural drivers such as climate change and human activity such as land conversion. Recent warming trends on the Tibetan Plateau, for example, directly influence the extent of permafrost cover and the degree of snow melting. Human activities also have intensified rapidly on the plateau over the last half century, with significant impact on land use. Desertification on the northern Tibetan Plateau is attributed almost equally to climatic changes and human activities, with climatic factors taking a slight lead (Yang et al. 2004). However, the magnitude and the extent of land cover changes under future global warming scenarios are difficult to assess because of deficits in current global climate models in this topographically complex area (Cui and Graf 2009).

Studies using Landsat images show that climatic change, human activities, animal and insect damage, and government policies all influence LULCC. During one 16-year study period (1994–2010) LULCC was shown mainly to occur slowly, and was linked to both natural and anthropogenic pressures (Song et al. 2009). Although political tools such as key national projects to improve ecological conditions could help with revegetation and slow desertification, the harsh natural conditions and fragility of this high-altitude region make it extremely difficult for degraded land to be rehabilitated (Song et al. 2009).

Policy decisions also influence LULCC in the HKH, as elsewhere. The de facto privatized use of wetlands by individuals or groups of households on the basis of long-term leases or selling, for example, has negatively impacted many aspects of hydrologic function, landscape conditions, and biodiversity in wetlands in the eastern HKH (Narain and Singh 2017; Yan and Wu 2005). The uneven spatial distribution of water resources on private lands, in particular, has led to the practice of actively extracting groundwater, which has lowered the water table. Fencing rangelands to limit grazing on expanding sand dunes also has affected some landscapes negatively, as consequent year-round use of other pastures that previously were grazed only in summer has in effect shrunk the overall activity space of some wildlife species (Yan and Wu 2005).

Market-driven changes in crop and livestock production strategies equally affect land use in HKH. Over the past 10–15 years, five HKH countries (Bhutan, China, India, Nepal, and Pakistan) have seen increasing trends in crop diversification, generally moving towards horticultural and cash crops. Forest transition due to plantations, agricultural intensification, and infrastructure development have led to many large-scale LULCC in mountains (Xu et al. 2009; Sharma et al. 2010). These trends have positive implications for the future development of mountain agriculture in terms of harnessing mountain niche products with their comparative advantages (Tulachan 2001).

In the livestock sector, there is a general decline in the cattle population across the HKH which could, for example, permit greater development of smallholder dairies with improved buffaloes in the Himalayan subtropics, and consequently contribute to enhanced food security and nutrition in mountain households (Tulachan 2001). In Afghanistan, the nature of the topography and arid climate means that vast areas are subject to soil erosion (Saba 2001). Loss of vegetation and soil humus creates ever drier conditions. Land abandonment, poor reclamation schemes, overgrazing, and the removal or destruction of vegetation as fuel have all contributed to widespread desertification (Saba 2001), which further impacts negatively on the pastoral production and livelihoods of local people in arid and semi-arid areas.

2.2.2 Over-Exploitation of Natural Resources

Exploitation of natural resources is one of the major environmental problems and drivers of change in mountains.

Over-exploitation includes the destruction of forest and shrublands for fuelwood and for commercial timber collection, overharvesting of non-timber forest products (NTFPs) such as medicinal plants, overgrazing, overfishing, and unsustainable and/or illegal mining (Dhanai et al. 2015; Shrestha and Bawa 2015; Shrestha and Dhillion 2003; Uniyal et al. 2002).

Generally, demands for natural resources increase in tandem with local and regional socioeconomic development. Demand pressures are further exacerbated by population growth. Together, these two factors are the main (ultimate) driving forces leading to over-exploitation in HKH (Poudel and Shaw 2015; Tsering et al. 2010). In the medium to long term, resource over-exploitation always results in the degradation of ecosystems, often followed by loss of genetic diversity and the extinction of species (Kaur et al. 2012). A decline in the provision of ecosystem services, including the loss of biodiversity, ultimately has disastrous effects on human wellbeing.

2.2.2.1 Extraction of Non-timber Forest Products (NTFPs)

NTFPs contribute significantly to people's livelihoods in the HKH, especially in forest-dependent communities, by strengthening or increasing local food security, incomes, health, and sustainable human development. There is ample evidence of the importance of NTFPs for rural livelihood improvement from across the region (Johnson et al. 2013; Mukul et al. 2010; Negi et al. 2011; Uprety et al. 2010; Yadav and Dugaya 2013). The collection of NTFPs has increased in HKH to meet growing demands in national and international markets, where they constitute increasingly important ingredients in herbal cosmetics, herbal tea, food, and medicines (Banjade and Paudel 2008). For example, unsustainable and illegal harvesting of NTFPs has been reported as one of the major issues in the Kanchenjunga landscape, with challenges arising in sustainable conservation and management of NTFPs (Uprety et al. 2016). In the central and eastern Himalaya, unsustainable collection and trade in *yarsagumba* (or caterpillar fungus, *Cordyceps sinensis*)—the world's most expensive biological resource—is regarded as a major driver of its population decline (Gruschke 2011; Shrestha and Bawa 2013).

2.2.2.2 Unsustainable Grazing

Overgrazing by domestic livestock is one of the main causes of degradation of plant and soil resources. By most estimates, the majority of rangelands in the HKH are regularly grazed beyond their carrying capacity (Dong et al. 2007; 2009; Harris 2010; Ho 2001). Overgrazing not only alters the vegetation composition, but also has other impacts such as soil erosion and degradation, declines in soil nutrient content and carbon storage, and altered stream flow regimes (Aryal et al. 2015; Dong et al. 2010; Sharma et al. 2014; Wen et al. 2013). Livestock grazing in HKH rangelands may be unsustainable (sometimes with irreversible damage to land and plant resources) not only because of total numbers of livestock, but also due to seasonality of grazing patterns, short- and long-term grazing intensities, and the composition of livestock herds (Ho 2001; Li et al. 2014).

2.2.2.3 Illegal Hunting

Illegal hunting (poaching) and trade in wildlife or wildlife parts also are major direct threats to biodiversity, and are particularly acute in the region's protected areas (Katuwal et al. 2015; Khan et al. 2016). The threat of illegal hunting is most severe for vertebrates such as rhinoceros, tiger, musk deer, pangolin, and red panda—driven especially by the demand for wildlife parts and products on the international market (Kumar et al. 2016). Traditional medicine plays an important role in health care in countries such as China, Nepal, and India, and while numerous measures have been adopted to reduce the use of wild species in traditional medicine there is still a growing market demand (Liu et al. 2016). Wildlife poaching, combined with retaliatory killing by local community members following livestock depredation, have the potential to drive species towards local extinction (Simms 2011). In the case of snow leopard alone, Li and Lu (2014) identified a total of 43 instances of traded snow leopard parts across China's borders between 2000 and 2013, primarily pelts and some bones, involving 98 snow leopard individuals.

2.2.2.4 Tourism

The HKH has tremendous potential for mountain tourism, which can provide alternative, environmentally friendly employment opportunities for local communities and contribute positively to their socioeconomic wellbeing. In most of the region, however, tourism development is poorly planned, often even unplanned, and the development of infrastructure such as recreational facilities, guest houses, camping sites, and restaurants often have significant negative impacts on the mountain environment (Dorji 2001; Nepal 2005, 2011). Tourism also contributes to ecosystem degradation through unregulated disposal of solid waste, trampling of soil and vegetation, and locally intensified resource extraction. The infrastructure deemed necessary to sustain tourism also can negatively affect local aesthetic and cultural assets, reducing their value and future tourism income potential (Reinfeld 2003; Zomer and Oli 2011). One survey about the impacts of adventure tourism in Hinku Valley in Makalu-Barun National Park, Nepal, showed visitor numbers, lodges, and other tourist-related infrastructure growing rapidly (e.g., from 1 lodge in 1995 to 129 structures in 2007), leading to accelerated harvest of subalpine timber for lodge construction and shrubs for fuel used in alpine tourist lodges (Byers 2014).

2.2.2.5 Mining

Uncontrolled mining operations also can have detrimental effects on fragile mountain environments, sometimes resulting in large-scale degradation of landscapes and economic crises of sustainability. The impacts of mining on the natural environment and socioeconomic conditions have been studied and documented by several researchers in the region (Huang et al. 2009; IUCN Pakistan 2009; Riaz et al. 2015). Uncontrolled mining practices such as blasting with dynamite have led to vegetation loss, soil erosion, and disturbance to wildlife in Pakistan (Donnelly 2004; Wu et al. 2014a). Coal mining has adversely affected the composition and structure of vegetation in the Nokrek Biosphere Reserve in Meghalaya, India (Sarma and Barik 2011). Large-scale mining, such as in Gyama Valley near Lhasa, China, can impact water quality downstream (Huang et al. 2010). Excessive riverbed mining for gravel and sand in an unplanned, or unregulated manner throughout the Himalaya is also a main reason for, or contributor to, irreparable damage to HKH river ecosystems (Singh et al. 2016).

2.2.3 Pollution

In the popular imagination, the HKH is one of the areas in the world least disturbed by human activity, especially by pollution. Nevertheless, studies from recent decades provide a rather different, thought-provoking perspective and the facts are now challenging our traditional thinking (Wu et al. 2016a). As air pollution will be elaborated in Chap. 11, in this section we only summarize the current status and trends, and the potential effects, of water and soil pollution in the HKH.

2.2.3.1 Water Pollution: Organic and Inorganic

Generally, water pollutants—especially inorganic pollutants—have already reached unacceptable levels in some HKH areas (Wu et al. 2016a). Lead (Pb) pollution in several study sites in major river basins including the Salween, Mekong, Yangtze, and Yarlung Tsangpo rivers has been found to be unacceptably high, and magnesium (Mg), aluminium (Al), and iron (Fe) are found in unsafe concentrations (Huang et al. 2008). In Manipur and Diphu, India, concentrations of arsenic (As) in groundwater have already exceeded the permissible limit (Das and Kumar 2015; Devi et al. 2010). Similarly, arsenic contamination is severe in Nawalparasi, Nepal (Gurung et al. 2005) and downstream areas of two major Himalayan rivers in India, the Ganges and Yamuna, are severely contaminated by sewage water containing high concentrations of several heavy metals (Chakarvorty et al. 2015).

Significant variation in water quality is observed in freshwater bodies including ponds, lakes, and rivers, especially on south slopes of the Himalaya. This may be attributed to the multiple sources of pollution including domestic sewage, industrial effluents, and runoff from activities such as agriculture and mining (Najar and Basheer 2012; Rashid et al. 2017). Accelerated eutrophication has been observed in many lakes, including Dal Lake, Khajjiar Lake, Nainital Lake, Anchar Lake, and Khushalsar Lake in India (Amin et al. 2014; Najar and Basheer 2012; Najar and Khan 2012; Rashid et al. 2017; Saini et al. 2008). In Dal Lake, for example, a recent study revealed that it was seriously polluted by chemical fertilizers and pesticides, thus unintentionally nutrient-enriched with concentrations of orthophosphate phosphorous and nitrate-nitrogen reaching 46 and 557 μg/L, respectively (Rashid et al. 2017). Pesticide use is generally increasing across the region. For example, pesticide use in India has grown by 750% since the mid-1900s (Evans et al. 2012). Microbial and organic matter pollution also threaten the health of water systems in the HKH (Merz et al. 2004; Sood et al. 2008; Tong et al. 2016). All of these forms of water pollution threaten aquatic biodiversity, water security, and ultimately human health, on a very large scale through upstream–downstream linkages in regional water systems.

2.2.3.2 Solid Waste and Soil Pollution

Solid waste has increased dramatically in both rural and urban areas of the HKH (Alam et al. 2008). The rapid speed of urbanization also means that treatment of solid waste has become a challenge to the sustainable development of cities. Dangi et al. (2011) reported in Nepal, for example, around 497 kg per person per day of solid waste is generated from households in Kathmandu Metropolitan City. Although most waste is organic (Dangi et al. 2011; Pokhrel and Viraraghavan 2005), centralized treatment is still very limited because of inadequate capacity or efficiency of facilities. Increasing waste volumes and growing complexity of waste streams, which contain large volumes of hazardous substances, further impact the soils in the HKH and downstream.

On the Tibetan Plateau, the most common heavy metals in soils are manganese (Mn) and chromium (Cr). The natural concentration of arsenic (As) is also very high at around 20 mg/kg (Sheng et al. 2012). Along the Qinghai–Tibet railway, Zhang et al. (2012b) found that top-soils in many areas were contaminated by heavy metals in relatively high concentrations, including Cd and Zn. On the south slope of the Himalaya, very high concentrations of Zn, Cd, and Pb were found in the soil of farmlands near Kathmandu (Zhang et al. 2012a). Organochlorine pesticide pollution also has been observed in the soils of Nepal at levels ranging from 20 to 250 ng/g (Yadav et al. 2016). Above a critical level, all of these substances are toxic to biodiversity including plants and animals, including humans (Osman 2014).

2.2.3.3 Sources of Pollution

Almost all relevant studies suggest that pollution in the HKH has become increasingly serious (Box 2.3). The sources of pollutants can be both anthropogenic and natural, but their relative weightings differ across regions (Wu et al. 2016a, b). In remote areas with few or no anthropogenic activities, the pollutants arrive mainly by long-distance atmospheric transport from industrial regions, as well as by the weathering of parent materials (Cong et al. 2010; Sheng et al. 2012; Wu et al. 2016b). Along railways or highways, the fuel from transport vehicles are usually the most significant contributors of pollutants (Zhang et al. 2012a, b, 2013). Overall, in both rural and urban areas of the HKH, most pollution is produced by human activities including the use of pesticides and fertilizers, vehicle traffic, and increased industrial activities (Atreya et al. 2011; Babel et al. 2011; Chakarvorty et al. 2015; Kannel et al. 2011).

Box 2.3 Multiple potential effects of pollution

Pollution can lead to a series of adverse effects on biodiversity, environment sustainability, and human health. On one hand, some pollutants are shown or predicted to have harmful effects on global biogeochemical cycles, as well as on plant and microbial communities. For example, mercury (Hg) on the Tibetan Plateau may play a critical role in biogeochemical cycles (Loewen et al. 2007). In addition, pollutants will affect different vegetation to varying degrees (Bing et al. 2016), including fragile timberline (Luo et al. 2013; Tang et al. 2015), potentially introducing severe threat to ecological resilience. Furthermore, presence of heavy metals such as copper (Cu), zinc (Zn), lead (Pb) and cadmium (Cd) may lead to the decline of culturable bacteria, actinomyces, ammonifying bacteria, nitrobacteria and cellulolytic bacteria in nature, and thus significantly affect microbial activities (Zhou et al. 2013). On the other hand, pollutants can lead directly to serious human health problems. Heavy metals are usually characterized according to their levels of toxicity, persistence, and bioaccumulation, and are widely recognized as posing severe threats for human health and social sustainability (Cheng 2007; Luo et al. 2015; Nabulo et al. 2010; Wu et al. 2016b). For example, arsenic (As) is a toxic semi-metallic element that can be fatal to humans. The ingestion of inorganic As can result in both cancerous and non-cancerous disorders and can harm the human nervous, dermal, cardiovascular, gastrointestinal, and respiratory systems. Similarly, microbiological contaminations, especially by pathogenic microorganisms, can pose risk to human health (Karkey et al. 2016). Moreover, water pollution may increase the load of waterborne diseases as well as contribute to pollution-induced water shortages, which worsen the water situations in the HKH (Karkey et al. 2016; Schwarzenbach et al. 2010; Zhang et al. 2015a). Restricted access to safe water and improved sanitation, for example, may cause between 15 and 30% of gastrointestinal diseases, and lead to approximately 1.7 million deaths a year globally [3.1% of all deaths and 3.7% of all Disability-Adjusted Life Years (DALYs)] (Ashbolt 2004; World Health Organization 2009). Additionally, a health risk assessment of organochlorine pesticides in the soils of Nepal suggested that exposure through ingestion of soil constitutes one of the most likely pathways leading to increased cancer risk across the country (Amin et al. 2014). Thus, pollution threatens both ecosystem stability and human health.

2.2.4 Invasive Alien Species (IAS)

Biological invasion may result not only in the loss of native species but also in an alteration of the extent or quality of ecosystem services and processes (Hulme et al. 2013; Sakai et al. 2001), which in turn may have drastic negative economic consequences (Pimentel et al. 2001). Direct costs of such invasions include losses due to reduced productivity in agroecosystems, whereas indirect costs are accrued in combating invasive species. Many IAS found in the HKH are included within a list of the world's 100 most harmful or damaging IAS (Lowe et al. 2000), such as *Eichhornia crassipes*, *Lantana camara*, *Chomolaena odorata*, and *Mikania micrantha*. Many other IAS also pose serious environmental problems and challenges to regional efforts to conserve the environment and to meet sustainable development targets.

2.2.4.1 Status of IAS Studies in HKH

There are few studies of IAS in the HKH. Most studies focus on basic inventories and ecological studies of invasive alien plant species (Akter and Zuberi 2009; Khuroo et al. 2007; Kosaka et al. 2010; Qureshi et al. 2014; Tiwari et al. 2005; Weber et al. 2008). Only several studies have focused on invasive alien fauna species (Budha 2015; Sujoy et al. 2010; Wan and Yang 2016).

In Nepal, inventory and assessment work carried out by IUCN Nepal has identified 166 naturalized plant species, of which 25 are recognized as invasive (Shrestha 2016; Tiwari et al. 2005). Recent inventory work carried out in the 'Kailash Sacred Landscape' in western Nepal has specifically noted 15 plant IAS (Bisht et al. 2016; Shrestha et al. 2018). In terms of fauna IAS, 64 species have been reported in Nepal, including 7 mammals, 6 birds, 19 fish, 22 arthropods, 9 molluscs, and one platyhelminth flatworm (Budha 2015).

In Pakistan, Qureshi et al. (2014) have listed 73 species of IAS including many that also are widespread and invasive in other parts of the world, such as *Broussonetia papyrifera*, *Prosopis juliflora*, *Parthenium hysterophorus*, and *L. camara*. All were reported as high-impact invasive land plants. Bhattarai et al. (2014) also reported that the diversity (species richness) of naturalized plant species increased with elevation up to 1100 m asl, and then decreased with further gains in altitude.

Elsewhere, Sekar (2012) identified 190 IAS in 47 families and 112 genera distributed across the Indian Himalaya. A recent survey by Xu et al. (2012) identified 488 IAS in China, with the highest number reported from Yunnan Province, but the exact number of IAS distributed in the Chinese Himalaya could not be confirmed. There is limited information about IAS in Afghanistan, Bhutan, and Myanmar.

2.2.4.2 Pathways of Introduction

A basic understanding of the major dimensions of introduction pathways of alien plants is important for regulating their invasions (Pyšek et al. 2011). Some studies have provided evidence that recent infrastructure developments, particularly the construction of roads, has greatly facilitated invasions and provided both habitat and potential dispersal corridors for exotic plant species in mountainous regions of China, India, and Nepal (Bhattarai et al. 2014; Chen et al. 2012; Kosaka et al. 2010; Xu et al. 2012). Similarly, natural and anthropogenic disturbances (such as recreational hiking) might act together to facilitate both the introduction and spread of exotic species, putting the ecosystems of the region at higher risk of invasion and consequent damage or loss (Dar et al. 2015; Dobhal et al. 2011).

Another major dispersal pathway for many IAS in the HKH is international trade and human movement (Tiwari et al. 2005; Xu et al. 2012). Agricultural inputs, particularly seed stocks originating outside the community, are another source of invasion (Kunwar 2003). Lack of adequate mechanisms to search for IAS and a poor quarantine system render the region vulnerable to the threat of IAS.

2.2.4.3 Major Effects at Ecosystem Level

Among few studies of IAS in the HKH, most focus primarily on IAS status (Barua et al. 2001; Khuroo et al. 2012; Reshi and Khuroo 2012) and distribution patterns (Bhattarai et al. 2014; Khuroo et al. 2010, 2011; Sang et al. 2010; Shabbir and Bajwa 2006). The overall impact of IAS in the HKH has not been properly studied; however, some adverse environmental impacts have been noted, such as the alteration in habitat and species composition that have been experienced through the aggressively invasive character of some species within some of the region's ecosystems (Dobhal et al. 2011; Dogra et al. 2009; Kunwar and Acharya 2013; Reshi et al. 2008). Moreover, the abundance of some IAS species has led to ecological displacement of native species and caused negative impacts on local livelihoods (Rai and Rai 2013; Rai and Scarborough 2015). *P. hysterophorus*, sometimes known as carrot grass, is perhaps the most troublesome and noxious weed in forest, pasture, wasteland and cultivated areas resulting in health-related issues in both humans and livestock (Kohli et al. 2006; Rashid et al. 2014).

Invasive shrubs and vines also pose a major threat to forest ecosystems, having a negative impact on their regeneration, structure, ecosystem functions, recreation opportunities, and wildlife habitat. Increased abundance of the unpalatable *L. camara*, for example, has suppressed regeneration of a number of native species, with detrimental demographic consequences for several important NTFP species in India and Pakistan (Dogra et al. 2009; Dobhal et al. 2010, 2011; Kannan et al. 2013; Kohli et al. 2006; Rashid et al. 2014). In Chitwan National Park, Nepal, *M. micrantha* has invaded vital habitat for the world's largest population of the great one-horned rhinoceros, destroying preferred wildlife habitat as well as jungle hiking trails.

In farming systems, *M. micrantha* also has suppressed the growth and yield of a variety of food and cash crops (Shen et al. 2013, 2015). The invasive plant *Ageratum conyzoides* equally invades agricultural fields (where it interferes with and causes yield reductions in major staple crops) and rangelands (where it outcompetes native grasses, causing scarcity of fodder) (Kohli et al. 2006).

In aquatic ecosystems, the rapid increase and spread of invasive plants also is a common phenomenon, creating great ecological and economic problems in the region (Akter and Zuberi 2009; Masoodi and Khan 2012; Wang et al. 2016; Xu et al. 2006). For example, *E. crassipes* blocks waterways, threatening biodiversity in Ramsar sites and damaging people's livelihoods in many other lake systems by hindering boat traffic and restricting fishing opportunities, with consequent economic losses (Burlakoti and Karmacharya 2004; Ding et al. 2008; Wang et al. 2016). Economically valuable native freshwater fish in India, Nepal, and Pakistan also are under threat from invasive alien fish species such as tilapia (Husen 2014; Khan et al. 2011).

2.2.5 Mountain Hazards

2.2.5.1 Types of Mountain Hazards: Earthquakes, Landslides, and Erosion

Mountain regions are high-risk areas, where hazards can cause damage, destruction, injury, and death at any time.

Compared with many other mountain areas of the world, in the HKH hazards are likely to be more severe as the region consists of young mountains that are still growing, making it inherently more vulnerable to earthquakes, landslides, and erosion. Every year there are on average 76 hazard events in the HKH, with the highest number noted in China (25) and India (18) (Nibanupudi and Rawat 2012).

As the HKH is located in tectonically active zones, its susceptibility to earthquakes is higher than in lowland areas. The 1934 Nepal–Bihar earthquake in Nepal ($M_w = 8.0$), the 2005 Kashmir earthquake in Pakistan ($M_w = 7.6$), the 2008 Wenchuan earthquake in China ($M_w = 7.9$), and the 2015 Gorkha earthquake in Nepal ($M_w = 7.8$) all led to huge losses of life and property, and also triggered vegetation degradation, landslides, rockfalls, and soil erosion (Lu et al. 2012; Ministry of Science, Technology and Environment, Government of Nepal 2015). There is growing scientific consensus globally that a disproportionally high number of natural disasters occur in mountain areas, and that these regions have become increasingly prone to disasters in recent decades (Guha-Sapir et al. 2016; Pathak et al. 2010). In addition, following earthquakes, a variety of mountain hazards including landslide, debris flow, and dam burst floods are more likely to occur over the ensuing one or two decades (Cui et al. 2008).

2.2.5.2 Effects of Mountain Hazards: Casualties, Financial Loss, Farmland Loss, Damaged Roads

Mountain hazards can cause many human casualties and incur huge economic losses for a region, often constraining economic development for many years following events (Zhang et al. 2016). In Nepal, it is estimated that landslides, floods, and avalanches destroy important infrastructure at an annual cost of around USD 9 million, and causing about 300 deaths per year (DWIDP 2005). In Afghanistan, 362 people were killed or reported missing, 192 injured, and 100,000 displaced as a consequence of flash floods in 2005 (Xue et al. 2009). The annual economic losses caused by landslide hazards in the Himalayan area is estimated at more than USD 1 billion, which is about 30% of the total economic losses caused by landslides worldwide (Li 1990).

The constraints of mountain topography mean that most residential regions are located in valleys where flatter and more accessible lands are found. These areas are at high risk of landslides, debris flows, and flooding. Similarly, arable land is very limited in mountainous regions because of constraints of terrain; most farmland is reclaimed land on terraced slopes or along rivers and on alluvial fans. Although necessary for most farming, irrigation also increases soil moisture and changes the stability of slopes, thus leading to increased risk of sliding or collapse of terraced slopes, or of being buried at the foot of slopes.

Road construction in general is a large and complicated endeavour, which must always consider and respond to different geomorphologic and geologic conditions. Construction projects are often at risk from different kinds of geohazards. Road construction introduces high fills, high-cutting slopes, and waste, which disturb the geologic environment along the road and compromise the stability of hill slopes. Even when landslides occur only at one or a few points along the road, they can destroy or block entire sections of road and affect much larger transportation networks and connections. For instance, the Sun Koshi landslide, which occurred in Nepal on 2 August 2014, destroyed the Araniko Highway for only a short distance (2 km). As a result, however, the entire highway—a major trade link between China and Nepal with trade exchange of nearly NPR 38 million (nearly USD 370,000) per day—was blocked for almost 2 months (Zhang et al. 2017).

Various other forms of damage from natural hazards occur as well. For example, tourism is affected by mountain hazards in several ways. Both the natural and cultural landscapes may be damaged or destroyed, including ancient architectural monuments, which may be difficult to recover or rebuild. Mountain hazards can also damage the infrastructure in scenic areas (including transport routes, communication lines, and accommodation); thus the numbers of tourists that can visit an area may be reduced dramatically as a result. The 2015 Gorkha Earthquake in Nepal caused damages and losses to tourism in the amount of NPR 81,242 million (nearly USD 790 million). It is estimated that the number of tourists decreased by 40% in the first year following the earthquake, though is likely to have recovered back to 80% of pre-earthquake numbers in the second year after the earthquake. The overall loss of direct tourism revenue was estimated to be over NPR 47 billion (nearly USD 456 million) from the time of the earthquake to March 2016. A broader accounting of losses in tourism income, including air transport, trekking, tour operations, restaurants, and the costs of debris removal, has estimated tourism-related losses to have amounted to NPR 62.4 billion (nearly USD 606 million) (SafeNEPAL 2016). Similarly, it is estimated that the Jiuzhaigou earthquake in August 2017 in Sichuan, China, which resulted in the temporary closure of Jiuzhaigou National Park, led to the loss of direct tourism revenue in the millions of USD and damage of natural scenic landscapes including forests, lakes and waterfalls (China News Service 2017).

2.2.5.3 Impact of Mountain Hazards Amplified by Climate Change

The frequency and magnitude of mountain hazards will increase under the impact of climate change; and in combination with growth in the population and economy, the risk of mountain hazards will increase dramatically (Cui et al. 2014). For the HKH, climate change is very likely to strongly impact the hydrological cycle, which is predicted to alter rainfall patterns and intensity, as well as the frequency of extreme precipitation events. Extreme rainfall is generally believed to trigger mountain hazards. For example, changing monsoon patterns in South Asia, including increased severity and frequency of storms as projected by climate models, may not only directly threaten agricultural production and the livelihoods of millions of people, but could also destroy critical infrastructure by way of mountain hazards induced by storms (Moors and Stoffel 2013). Increased water flows accelerate river erosion by destabilizing valley slopes, with dramatic effects where slopes are saturated with water after prolonged intense rains, and sediment is then deposited elsewhere. In addition to causing sediment transfer and deposition elsewhere, riverbank erosion changes stream channel morphology and can easily lead to cascading hazards in mountain areas and beyond (Cui et al. 2014).

For its part, global warming reduces snow cover, melts glaciers, and degrades permafrost. Related biophysical systems are thus being affected through enlargement and increased number of glacial lakes, increasing ground instability in permafrost regions, and rock avalanches (Pathak et al. 2010). Glacial lake outburst floods (GLOFs) result in numerous casualties and property damage in the HKH. An investigation in 1999 and 2005 by ICIMOD and its partners showed that there were 8790 glacial lakes covering a total of 801.83 km^2 in the HKH, of which 203 lakes were potentially dangerous and could pose a GLOF threat in the future (Ives et al. 2010). As vegetation development is a slow process at such high altitudes, these sites may remain unprotected against erosion for decades or even centuries (Fusun et al. 2013). Landslides and debris flows thus pose long-term threats to settlement and infrastructure in many places.

2.2.6 Climate Change and Variability

As a key driver of changes in mountain sustainability today, climate change is interacting in complicated ways with many other important drivers—globalization, population growth, urban expansion, and local land use change—all of which can have significant ramifications (see Chap. 3 for full details). Climate change has environmental and social impacts that are likely to increase uncertainty in water supplies and agricultural production for human populations across the HKH. Changes in temperature and precipitation could have serious implications for biodiversity and the goods and services derived from ecosystems (Chettri and Sharma 2016). Climate warming is leading to visible effects in the HKH, with indications of changes in phenology, and with reduced agriculture production in some of the major crops in some parts of the HKH (Hart et al. 2014; Webb and Stokes 2012).

2.2.6.1 Environmental Effects

Our discussion of primary impacts and hazards associated with climate change in the HKH—changes in hydrology, permafrost, and mountain environments—will summarize studies embodying a wider range of climate scenarios and projection. Observed climate changes in the HKH (especially higher regional temperatures) have already affected biological and ecological systems directly and indirectly, coupling with other anthropogenic drivers. Hydrological processes are also altered by climate change, which as a result lead to change in carbon sequestration and nitrogen deposition, as well as in human livelihoods and consumption (Singh et al. 2011).

Hydrology

Changes in hydrology can impose great challenges to human society and natural ecosystems (see Chaps. 3, 7 and 8 for full details). Glaciers in the HKH have been retreating continuously since the 1970s, with an accelerating rate of retreat in the past decades (Bajracharya et al. 2011; Ding et al. 2006; Fujita et al. 2008; Pu et al. 2004; Takeuchi et al. 2009; Yao et al. 2007). Upstream snow and ice reserves in the HKH have been affected substantially by climate change (ICIMOD 2011; Shrestha and Aryal 2011). River discharge is likely to increase for some time because of accelerated melting, which indirectly affects the water availability and food security of large human populations (Immerzeel et al. 2010; Nepal and Shrestha 2015). On the Tibetan Plateau, for example, glacial retreat has caused hydrological changes, including river runoff increasing over 5.5% (Yao et al. 2007), and the water level of most lakes rising by up to 0.2 m yr^{-1} (Zhang et al. 2011), accompanied by surface expansion of many lakes (Liu et al. 2009, 2010). Thorough analysis of the impacts of climate change on future water availability in these basins is thus needed, and could immediately affect climate change policies where transition towards coping with intra-annual shifts in water availability is desirable (Lutz et al. 2014).

Permafrost

The potential impacts of thawing permafrost (see Chap. 7 for full details) remain largely unknown for most of the HKH (Valério et al. 2008). The extensive permafrost is highly sensitive to temperature changes, resulting in significant warming, thawing, thinning, and retreat throughout the HKH in recent decades, especially on the Tibetan Plateau (Yang et al. 2010). Long-term temperature measurements have indicated that the lower altitudinal limit of the permafrost has moved upwards by 25 m at Xidatan in the interior of the plateau during the last 30 years and upwards by 50–80 m along the Qinghai-Kang Highway, located on the eastern edge of the plateau, over the last 20 years (Cheng and Wu 2007). In the Khumbu (Everest) Himalaya, apparently the permafrost lower limit has risen 100–300 m between 1973 and 1991, then stabilized at least until 2004 (Fukui et al. 2007).

Mountain-specific environments

Mountain environments, in particular the HKH, are potentially vulnerable to the impacts of global warming because the combination of high sensitivity to climate change and limited possibilities for species migration to favourable locations renders mountains as "islands" in a "sea" of surrounding lower-lying ecosystems (Busby 1988; Frey et al. 2011; Shrestha et al. 2012). In the high-altitude regions of the HKH, glacial melt is affecting hundreds of millions of rural dwellers who depend on the seasonal flow of water, with more water available in the short term but less in the long run as glaciers and snow cover shrink and disappear. In Nepal, data from 1980 to 2015 show that floods, landslides, and epidemics were the main causes of disaster-related human loss (Carpenter and Grünewald 2016). Many recorded GLOFs in this region have caused severe socioeconomic damage (Hasnain 2007; Richardson and Reynolds 2000). Many risk assessment studies recently carried out in the Himalaya have identified ice avalanches from advanced glacier tongues and ablation of dead ice beneath moraine ridges as potential GLOF triggers (Bolch et al. 2008; Fujita et al. 2008; ICIMOD 2011; Watanabe et al. 2009).

Climate change has synergistic effects with many of the other primary threats or constraints to biodiversity. With enhanced temperature and reduced precipitation, for example, alpine meadows and shrubs may migrate to places higher up the mountains. However, this process will be constrained by environments that do not have soils of sufficient depth for anchorage and nutrient storage. Grabherr et al. (1994) estimated that a 0.5 °C rise in temperature per 100 m elevation could lead to a theoretical shift in altitudinal vegetation belts of 8–10 m per decade (Grabherr et al. 1994). In the eastern Himalaya, this altitudinal shift is expected to be around 20–80 m per decade (based on current estimates of temperature increases of around 0.01–0.04 °C per year) with greater shifts at higher altitudes, as the rate of warming is expected to increase with altitude (ICIMOD 2009). Wetlands will shrink in response to high evaporation, which will be further exacerbated by the expansion of settlements and other human activities. The rising temperature of water bodies renders them more suitable as habitats for invasive species that outcompete native species and synergistically interact with climate change to threaten native organisms (see Sect. 2.2.4).

2.2.6.2 Socioeconomic Effects

Coupled with social, economic, and political stresses, climate change could have serious cascading effects with potentially catastrophic consequences, including adverse impacts on ecosystem services supply, as well as on the agricultural productivity, human health, and livelihoods of the millions of people living in the region (Ariza et al. 2013)—particularly where they are dependent upon natural resources (Rautela and Karki 2015). Declining natural resource availability and uncertainty introduced by climatic variability pose a threat to mountain sustainability in the face of an already declining natural resource base.

Agriculture

Agriculture is the direct or indirect source of livelihood for over 70% of the population of the HKH, and a substantial contributor to national incomes (Tiwari 2000). Changing precipitation patterns, reduced runoff in the major river basins, and increasing temperatures will simultaneously put additional pressure on available water resources and increase agricultural water demand for rainfed and irrigated crops, thus increasing competition for water for agriculture, industry, and human consumption (Hanjra and Qureshi 2010). Agriculture in this region is mostly rainfed (about 60%; World Bank 2016) and therefore vulnerable to changes in timing and frequency of precipitation. Besides water availability, crop yield depends on a number of biophysical processes and variables (such as thermal stress, humidity, solar radiation, nitrogen stress, ozone, and the fertilization effect of CO_2) and their complex, nonlinear interactions (Challinor et al. 2009). Extremes in floods and droughts through much of the upcoming century may destroy the food production base of the region (Bruinsma 2003). Fischer et al. (2002) expected a temperature increase of 1.5–2.5 °C to lead to a decline in the agricultural productivity of crops such as rice, maize, and wheat. Climate changes are predicted to reduce the livelihood assets of rural people, alter the path and rate of national economic growth, and undermine regional food security due to changes in natural systems and impacts on infrastructure.

Human wellbeing

Climate change affects the environment as well as social and economic developments in the region. It exacerbates the difficulties already faced by vulnerable indigenous communities, including political and economic marginalization, loss of land and resources, human rights violations, discrimination, and unemployment (Chavez et al. 2014). The consequences of biodiversity loss from climate change are likely to be worst for the poor and marginalized people who depend almost exclusively on natural resources. Poverty, poor infrastructure (roads, electricity, water supply, education and health care services, communication, and irrigation), reliance on subsistence farming and forest products for livelihoods, substandard health (high infant mortality rate and low life expectancy), and other aspects of development, render the HKH even more vulnerable to climate change as the capacity to adapt is inadequate among the inhabitants (Manandhar et al. 2018, Negi et al. 2012).

Furthermore, climate change, land use transition, and environmental migration in the HKH have impacts on social relations between nations, social classes, ethnic groups, and individual families as they increasingly struggle for access to essential resources. This combination is a significant source of potential as well as actual violent conflicts (Agnew 2011; Aryad et al. 2013; Bhusal and Subedi 2014; Hsiang et al. 2011). The impacts on human health are similar (Sharma 2012), as most emerging human diseases are driven by human activities that modify ecosystems or otherwise spread pathogens into new ecological niches (Taylor et al. 2001). Such modifications or alterations in ecosystems generally lead to large-scale land degradation, changing the ecology of the diseases that influence human health and making people more vulnerable to infections (Collins 2001). Decisions about change of land use, whether in response to climate change or other factors, are thus also human health decisions (Xu et al. 2008).

Migration

Yet another threat to human wellbeing is "forced environmental migration" of populations that can occur as a result of environmental phenomena caused by changes in the Earth's climate (Ramos et al. 2016). In some instances, rural farmers migrate to the city and there are consequent labour shortages in rural areas, leading to socioeconomic difficulties in the rural agroeconomy (Sann Oo 2016). However, livelihoods in the Himalaya have traditionally depended on a multitude of strategies, including available ecosystem goods and services, but typically also encompassing external activities such as trade and labour migration. Yet changes in ecosystem services following environmental or socioeconomic change will clearly affect mountain livelihoods, and therefore are likely to impact migration numbers and patterns (Banerjee et al. 2014). If the incidence and magnitude of extreme events such as droughts and floods increase, there could be large-scale demographic movements and ultimately transformation. The annual rate of increase in migration in the countries of the HKH has been disturbingly high, and the number of internally displaced people also is expected to rise significantly.

2.3 Sociocultural Drivers of Change to Mountain Sustainability

2.3.1 Changing Demographic Situation

Demographic processes such as population growth and decline, changes in age distribution and education, social and spatial mobility (migration), and concentration of activities at market-based cities and local centres (urbanization) are the result of environmental and economic drivers and exert impacts on both. In environmental terms, demographic oscillations increase the demand and consumption of natural resources, and change land use, with the risk of degradation by overuse or underuse. In socioeconomic terms, demographic changes contribute to the transformation of social relations and structural changes in mountain economies. In either case, the changing demographic situation has both quantitative and qualitative impacts.

Sociodemographic changes are in turn affected by a range of factors such as technological innovation, institutional and financial conditions, and climate change. The availability and quality of natural resources, as well as trends in their exploitation, can equally influence demographic processes. The biophysical environment has two functions for human beings: it guarantees their physical existence and livelihoods and is at the same time a resource for economic activities and exchange in a given (and transforming) society. The combination of these two functions generates a continuous tension between use and preservation and defines the human impact on the environment. This is important for the question whether mountains should mainly serve for its inhabiting population or for the downstream consumers and global visitors. Thus, the number of humans, population growth rates, and especially the population distribution are good proxies both of the pressures that human communities face and that they are placing on the Earth. The sociodemographic and environmental interdependencies are complex and remain poorly understood (DeSherbinin et al. 2007). For example, rapid population growth and poverty are often blamed as being the main twin causes of deforestation, yet recent large-scale deforestation in South Asia is largely driven by agricultural enterprises and accompanying road construction along with local and regional migration (DeFries et al. 2010; Rudel et al. 2009).

The total population in the mountains and hills of the HKH region was approximately 225 million in 2015 (United Nations 2015b). Of the eight member countries, India has the highest population distributed within the HKH (50.31 million), followed by Bangladesh (45.55 million), Afghanistan (33.33 million), Pakistan (32.16 million), Nepal (28.83 million), China (20.48 million), Myanmar (13.32

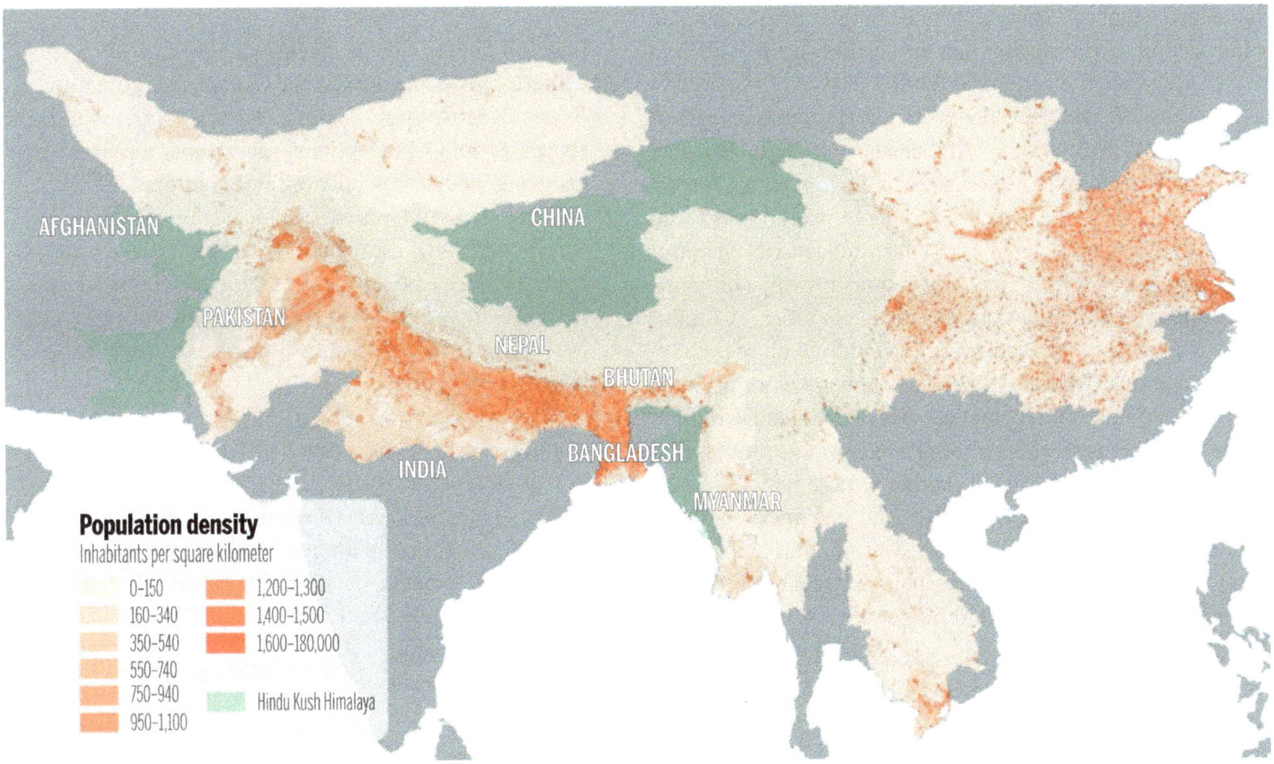

Fig. 2.3 Population density in the HKH

million), and Bhutan (0.78 million). Although China is the most populated country in the world, its HKH population (mainly on the Tibetan Plateau) is much less because of the high altitude and harsh environment with limited agricultural opportunities. Population is most dense along the southern fringe of the HKH including Nepal and India, the Chittagong hill tracts in Bangladesh, and the upper Indus Basin in Pakistan (see Fig. 2.3).

Almost all countries/territories in the region have experienced a steep decline in fertility in recent decades, which has resulted in a lower rate of population growth (World Bank 2015). The mean annual growth rate of most regional countries in 2015 ranged between 1% and 2%, but was higher in Afghanistan (2.8%) and Pakistan (2.1%) and lower in China (0.5%) and Myanmar (0.9%) (United Nations 2015a). During the period from 1960 to 2014, the total fertility rate (births per woman) declined, with the lowest in China (1.6) and the highest in Afghanistan (4.8) and Pakistan (3.6), and the other countries with a total fertility rate of between two and three children (World Bank 2015). Conscious efforts to limit fertility through increased knowledge about contraception, together with decreasing infant mortality, are important factors in controlling high fertility rates (United Nations 2003). Conversely, life expectancy at birth within the eight regional countries rose remarkably in recent decades, exceeding 70 years in China (76), Bangladesh (72), and Nepal (70) and reaching nearly 70 years in Bhutan (69), India (68), and Pakistan (66)[1]; Afghanistan has the lowest life expectancy (60 years). However, it should be noted that, in 1960, life expectancy in none of the eight countries exceeded 50 years. Compared to the global average, people in developing countries normally have shorter lives (United Nations 2012). To a certain extent the observed changes in life expectancy at birth are partial indicators of improving socioeconomic development in this region.

Although the population of the HKH is mostly rural, the urban population increased rapidly in the last three decades due to processes of urbanization, industrialization, and rural–urban migration. The urban population in China rose from 16 to 56% during the period from 1960 to 2014 (United Nations 2014). In 2014 all other countries in the region had a similar percentage—between 30% and 40%—except Nepal (19%) and Afghanistan (27%) (United Nations 2014). While the regional urban population grew rapidly during the 20th century, the next few decades will see an unprecedented scale of urban growth, putting more pressure on land cover, natural resources, ecosystem services, and social services. In most established or formal urban areas, access to water supply and sanitation services is believed to be better than in rural areas. In some peri-urban areas, however, residents have little access to safe drinking water or adequate

[1]See: https://data.worldbank.org/indicator/SP.DYN.LE00.IN.

sanitation services, increasing the danger of water- and sanitation-related diseases (United Nations 2014; World Water Assessment Programme 2009).

In terms of age dependency ratios (i.e., percentage of working-age population), Afghanistan (87%), Pakistan (65%), and Nepal (62%) are ranked at the top, with China at the bottom (37%). This indicates that larger workforces could be provided in these countries if better educational opportunities were available. Rapid population growth, combined with economic difficulties, pushes people to cities. A declining and ageing population induces countries or regions to accept migrants, who are typically willing to work at much lower wages than native workers (Hoermann and Kollmair 2009). The net implication of these demographic processes is clear: the HKH will have substantially more people in vulnerable urban areas in the next 20 years. The migration from rural to urban areas also results in consumption and dietary changes, which may further contribute to increased pressures on many ecosystems (de Sherbinin et al. 2007; Dietz et al. 2007; Romanelli et al. 2015; Tilman et al. 2005; see Sect. 2.4.3 'Urban and peri-urban expansion' in this chapter).

In the HKH, migratory populations include traditional groups of subsistence-level pastoralists and agriculturalists, as well as family groups and individuals seeking greater opportunities and refugees fleeing the consequences of armed conflict or natural disasters (Hoermann and Kollmair 2009; Tiwari and Joshi 2015). People who have become new refugees because of wars, in particular, often pass through camps or informal settlements, leading to the proliferation of informal communities on the fringes of cities (UNHCR 2015). In these situations the arrival of additional people may exacerbate resource scarcity (e.g., water and agricultural land), strain the capacity of the urban infrastructure, and even result in conflicts with local communities. It is believed that climate change, which is predicted to lead to greater frequency and intensity of extreme weather events, is likely to result in increased migration and possibly to an overall increase in the displacement of people in the future (Tiwari and Joshi 2015).

2.3.2 Changing Sociocultural Situation

Cultural drivers of change have existed in the HKH for as long as cultural communities have inhabited the region. Some scholars (Cordaux et al. 2004; Gayden et al. 2007) have understood the HKH to be a great, fixed geographical barrier impeding or restricting the flow of human gene pools (Box 2.4). However, viewing the HKH from a historical/cultural lens, it could be found that the strategic location of this formidable mountain chain is at the intersection of South Asia, East Asia, and Central Asia as a facilitator for the historic emergence of distinct cultural and politico-economic groups.

Box 2.4 Culture as a driver

Culture can be broadly defined as the way a community lives at a certain point in time (Kalman 2009). Culture shapes people's perceptions, judgments, ideas of self and other, and human relationships (LeBaron 2003). To apprehend culture, one has to look at the way living is "performed"—by articulations of shared as well as contested ideas, beliefs, and values through the ways in which relationships between people as well as between people and "things" manifest themselves via practices and representations—at levels both "mundane" and "out of the ordinary" (Kien 2009). In discussing the cultural drivers of change in the HKH, a critical element to highlight is that while culture itself is always in the making, in the process "culture" also affects or changes its own members or participants. New forms of articulations and practices indicate emerging aspirations, desires, capabilities, and forms of resistance among the culture's participants. At the same time, such changes can engender or encourage physical movements of its participants to and away from a particular location. This can be seen, for example, in the phenomenon of rural-to-urban and hills-to-plains migration, wherein changing value systems and agents of modernity effect dislocation. The effects of this process of cultural change can happen in multiple places simultaneously, or more gradually and unevenly. These changes also can affect the ways in which people relate to the surrounding natural landscapes.

Such people and groupings, traditionally reliant largely on locally adapted forms of agropastoralism and trade, generally have deep interdependence with distant populations and societies located on either side of the mountain chain (Bauer 2004; Fürer-Haimendorf 1975; Olschak et al. 1987). For example, among the many Tibeto-Burman cultural and ethnolinguistic communities spread across the HKH, the local material cultures and political economies grew for several centuries in the midst of a complex web of interdependence, with much larger cultural and political spheres extending both to the north and the south of the Himalaya (Gerwin and Bergmann 2012; Shakya 1999). This growth of cultural communities in the HKH was essentially premised upon interactions and movements of people; items of trade, religion, art and folk cultures; and emerging political formations (Cacopardo and Cacopardo 2001; Handa 2001). Such interactions, with cultures, technologies, and polities outside the HKH, have always been the critical cultural drivers of change within the HKH.

However, the pace of changes in the materiality, values and aspirations, and social relations (i.e., the key effects of cultural drivers) among the communities of the HKH has accelerated exponentially since the 20th century. Throughout the region, the 20th century was a period of great political, economic, social, and cultural transformation. It saw the emergence of nation states all over Asia and, with that, the bid to integrate peripheral regions such as those of the HKH into national mainstreams (Bose and Jalal 2004; McCauley 2016). In the last three decades, globalization has also been making deep inroads into the HKH, facilitated by liberalizing economies, better connectivity, and transnational flows of media and tourism (Bauer 2004; Hodge 1991; Hodges et al. 2014; Jodha 2005). At the same time, fluid movements and long-standing relations between borderland communities for trade, pilgrimage, pastoralism, and often also kinship became more and more constricted because of the tightening of borders, and in many places they eventually became completely cut off from one another (Bauer 2004; Harris 2013; Hoon 1996).

Overall, increased connectivity, increasing penetration by the state and markets, geopolitical developments, globalization, and the spread of new technologies have engendered many new kinds of interactions, imaginations, aspirations, and practices in the communities of the HKH. Just as in the past, these processes of interaction between the local community and the world beyond form the cultural drivers of change. Numerous studies from the HKH show that these processes occur in deeply intertwined ways and are heavily interdependent, and thus need to be studied together. Hodge (1991), Kreutzmann (1993, 2004), Banskota et al. (2000), Aggarwal (2004), Bauer (2004, 2005), Bergmann et al. (2008), Ueda and Samdup (2010), Demenge (2012), Harris (2013), Howard (2013), Toffin and Pfaff-Czarnecka (2014), and Grocke (2016) are some good, in-depth studies laying in sharp relief the complex, material ways in which members of various HKH communities experience, contest, and negotiate interactions with the aforementioned agents of change.

The establishment and growth of state-supported institutions of local governance and the implementation of state policies—on sectors ranging from education to health care, natural resource management to primary infrastructure development—are fundamentally contingent upon the improvement of connectivity of these rugged regions, both in terms of physical access and communication technology. Additionally, connectivity has also been to shown to link these mountain communities to national, regional, and global flows of commerce, media, consumptive choices, labour, and tourism (IUCN 2017). As connectivity, the reach of state apparatus, and globalization increase, they cause significant changes in the realms of (1) social relations between different strata of communities, (2) traditional livelihoods and their relation to the local and broader socioecological systems, and (3) power dynamics in terms of gender, social background, and age among community members.

It may be futile, though, to attempt to label the effects of cultural drivers of change in the HKH as completely positive or negative. The interactions enabled by enhanced connectivity occur in the "sticky materiality" of pre-existing socioecological landscapes and thereby have "awkward, unequal, unstable, and creative" effects, often in unintended ways (Tsing 2005). There have been some unquestionably positive developments in many cases, such as improving access to education and general health care, new livelihood opportunities, breaking down of oppressive traditional hierarchies, and improved awareness and practices around maternal health and sanitation. However, the sociocultural drivers of change have also brought in their wake developments such as the loss of traditional knowledge and folk art forms, haphazard road and building construction, breakdown of traditional sociopsychological support systems in the face of difficult times, and a growing sense of alienation (Foggin 2016). With the increasing spread of connectivity, governmentality, and globalization, many parts of the HKH have also faced increasingly volatile and sometimes violent geopolitical conditions over recent decades (Agrawal et al. 2016; ICIMOD 2012). There is therefore a need to create locally sensitive modes of development that help to empower local communities to make decisions regarding their lifeworld, while respecting the opportunities and limitations imposed by the geoclimatically diverse landscapes of the HKH (Hodges et al. 2014).

2.3.3 Governance Systems and Institutions

Mountain areas of the HKH have not remained isolated from development interventions brought in by government and non-government agencies. At the same time, the region has experienced environmental degradation with resulting socioeconomic problems such as water scarcity, crop depredation, and scarcity of fuelwood and fodder. Moreover, this has happened in the last few years despite growing awareness of the importance of maintaining environmental health, especially in mountains. The question that needs to be asked is: what led to this situation? Is it wrong policy adopted by central and provincial governments? Is it changing developmental aspirations of people? Is it the failure of traditional institutions that regulated the use of environmental goods and services until recently, or the failure of modern institutions that replaced them to maintain sustainability? Some scholars feel that it is likely due to a failure of development agencies to consider interventions from a "mountain perspective" (Ives 2013; Price 2015) especially in regard to inaccessibility, marginality, fragility, diversity or heterogeneity, niche or comparative advantage, and human adaptation mechanisms (Jodha 2000, 2005). The autonomy and flexibility of decentralized traditional institutions at the community/local level in the HKH, as well as natural resource endowments, have helped communities adapt to changes in the past, and these attributes should therefore be preserved and strengthened into the future (Foggin 2018).

In the rural areas of HKH natural resources are an important component of sustenance for local communities, and the various institutional arrangements that have evolved over the years represent communities' ways of adapting to the vagaries of nature. However, over time the infrastructural and physical changes have been accompanied by socioeconomic changes, followed by political as well as administrative changes, with a clear impact not only on how natural resources are managed, but also on how local communities look at and treat them. The shift from decentralized ownership, as well as access to centralization, has made this region vulnerable to large demands on goods and services made available by mountain ecosystems (Jodha 2005).

A majority of the inhabited areas in the HKH are facing challenges of sustainable development. This situation has been caused partly by the precarious state of environmental governance, wherein for many reasons regional governance systems are largely unable to address environmental issues. These include fragmented governance, where informal and formal governing institutions are not synchronized; lack of involvement from financial institutions in considering a "bigger picture" before they support external investments; and the proliferation of environmental agreements, often in conflict with trade agreements or practices. All these issues hinder proper functioning of regional governance. For example, high demand for mountain niche products can result in over-exploitation and thereby lead to degradation. With no policy in place and no institutional mechanism to implement, monetary incentives result in the breakdown of local institutions. Similarly, although most areas in the HKH are experiencing the phenomenon of depopulation due to migration, the population of the plains is growing continuously (along with aspirations), which itself has also had growing direct implications for the exploitation of natural resources in the mountains.

Lack of integration of sector policies, which is common to all HKH member countries, has been exacerbated by inadequate institutional capacities in mountain regions, including local governments and local communities (Agrawal et al. 2016; Foggin 2008; ICIMOD 2012). The absence of a gender perspective, and of values of equity and participation in governance, has led to an inability to persuade and influence the public and win the support of public opinion (UNEP 2002, 2008). Generally, environmental problems are embedded in very complex systems, of which our understanding remains quite weak (Underdal 2010). Because of the time lag between human action and environmental effect (hysteresis), the shift from awareness to action sometimes can take as long as a generation. Thus, governance—especially regional governance—has been overlooked by most member countries, although verbal assent can often be heard (Agrawal et al. 2016).

2.3.4 External Knowledge and Technological Innovations

In addition to local tacit knowledge about the environment and its management, which has developed over millennia, the accessibility to and penetration of modern science and technology are amongst the key driving forces behind the three pillars of mountain sustainability. Both the generation and dissemination of traditional and scientific knowledge and associated technologies have significant implications for both conservation and development. Technological implementation can have both positive and negative effects on the environment and human wellbeing (Nelson et al. 2006). This can be seen, for example, in increasing food production, improved accessibility of mountain communities to the external world, and also changes in human consumption patterns. Generally, in the remote mountains of the HKH, inadequate access to appropriate technologies and scientific knowledge is one of the causes (through their absence) of poverty and natural resource degradation (Maikhuri et al. 2011), along with weak systems for communication and dissemination of knowledge. In the following sections, a few innovative areas of technology with major influence on mountain societies are highlighted.

2.3.4.1 Information and Communications Technology (ICT)

ICT is instrumental in reshaping and transforming numerous aspects of the world's economies, governments, and societies, and is becoming one of the fastest-growing industries in the global economy in the 21st century (Baller et al. 2016). Increased availability of information and knowledge-sharing platforms is key to fostering socioeconomic development and also increases awareness about environmental issues, related government legislations, and subsidies (Thapa and Sein 2010; QBE Asia Pacific 2017).

The mountain communities in the HKH suffer from lack of access to basic resources, services, and relevant information (Akhtar and Gregson 2001). However, various agencies working in the region, including governments, non-governmental organizations (NGOs), and public and private entities have shown a growing interest in promoting and using information and technologies to improve people's quality of life and their livelihoods, to link the HKH with markets outside the region, and to reduce the marginalization of mountain communities (Thapa and Sein 2010).

Within the HKH, there are significant differences in terms of the number of mobile subscribers who have used mobile internet services. Highly advanced markets such as China already show above 40% of the population using mobile internet services, followed by India (22%). However, less than 20% of the populations of Myanmar, Bangladesh, Pakistan, Nepal, and Bhutan have used mobile internet services. This naturally leads to a great potential for the emerging markets within the region, in terms of mobile internet subscribers, when identifying growth opportunities (GSMA Intelligence 2015).

Increased availability of information and knowledge-sharing platforms are key to fostering socioeconomic development (Thapa and Sein 2010) as well as enhancing environmental governance, associated with governmental legislation and the development of cleaner technology (IUCN 2017). There is increasing evidence that ICT tools such as radios, internet, TV, and e-magazines are being used for communicating and information sharing (Adhikari et al. 2006). ICT tools also are being used to simplify technical information and to share more accurate and up-to-date information with different communities (Baral et al. 2006). Use of ICT for agro-advisory services and early warnings in disaster situations has also emerged as an effective adaptation measure to deal with the impacts of climate change, weather irregularity, and geo- or biodisaster planning (Baller et al. 2016; QBE Asia Pacific 2017). Another important use for these technologies concerns mitigation of remoteness, whether social or geographic, particularly in relation to increasing people's accessibility to education, healthcare, and natural risk warnings and rescue.

2.3.4.2 Geospatial Technology and Improved Regional Perspectives

Geospatial technology also is recognized as an effective tool for planning, management, and decision making locally, regionally, and globally. The rapid development of geospatial applications has allowed it to be used widely for mapping LULCC; to predict agricultural productivity and natural disasters such as landslides, avalanches, and glacier change (Manfré et al. 2012); and to plan transportation networks and environmental protection (Ingole et al. 2015).

A large temporal and spatial variability in ecological and socioeconomic aspects exists across the HKH with respect to geographical, physical, socioeconomic, and cultural parameters. To understand such large variability in the region and to successfully use that information to predict changes and apply effective natural resource management practices, it becomes essential to use technology that allows access to data over large scales and which provides a platform for integrating geographic data with sociocultural and economic factors.

In the last two decades, ICIMOD and its regional and international partners have successfully applied geospatial and allied technologies for understanding glacier dynamics in the context of climate change, forest fire monitoring with SMS alerts, land cover change assessment for natural resources, deforestation and forest degradation monitoring, vegetation shift and corridor identification in transboundary landscapes, disaster information management and flood early warning systems, and agriculture monitoring for food security analysis (Bajracharya et al. 2007; Li et al. 2017; Qamer et al. 2016; Uddin et al. 2016). GIS platforms and analytic tools also are being used to monitor changes in natural ecosystems (Chaudhary et al. 2014; Gilani et al. 2017).

2.3.4.3 Biotechnological Applications in Agricultural Production Systems

Biotechnology has evolved rapidly to address the existing problems of food, health, and environmental security (Padmanaban 2003) and conserving biodiversity (Joshi et al. 2009). It is being used in medicinal and aromatic plants (Adhikari 2011; Joshi et al. 2009; Juyal et al. 2014), forestry, and agriculture with the development of pest-resistant or drought-resistant seeds, high-yield varieties, biofertilizers, biopesticides, and biofuels. There is no common opinion on this topic, neither between countries nor at the regional level; and not even amongst all the authors of this chapter. The topic is highly contested (see, e.g. for Burkina Faso, Le Monde 2015). In Switzerland, for example, environmental organisations and small scale farmers (who are located especially in mountains) have enforced a moratorium of genetically modified (GM) organisms until 2021. Conversely, several HKH countries see plant and animal GM organisms as powerful tools that hold the promise of improving the productivity, profitability, and sustainability of farm systems, including those in poor districts (Adhikari et al. 2006; Cohen 2005; Delmer 2005).

Biotechnology has potential benefits as well as risks when applied to agricultural crops. Its proponents say that the introduction of genetically modified crops has reduced the spread of diseases, yet there remain concerns regarding the reduction of native crop genetic diversity due to GM plant breeding programmes (Li et al. 2016; Rauf et al. 2010). GM approaches also have generated questions about their possible negative consequences on health, society, and the environment (Akumo et al. 2013; Fisher et al. 2015; Shahzadi et al. 2015). Negative health implications for farmers due to increased exposure to toxins, and to allergic reactions, have also been reported in some studies (Debyani and Neeta 2012; Prakash et al. 2011). Critics also argue that the interactions of GM organisms (GMOs) with existing biodiversity are not yet clarified and therefore unsafe, or at least ill-advised according the precautionary principle. Additionally, GMOs are sterile and farmers therefore must continuously buy seeds from the copyright holding companies, rendering them both financially and technologically dependant. External knowledge also begins to replace regional mountain knowledge (Forster et al. 2013; Public Eye 2017). Furthermore, dependency on external private enterprises endangers the sovereignty of national states in issues of food (De Schutter 2010; UNCTAD 2009). It will be important to look at the biosafety aspects of using new varieties, which may interact with the environment to produce sterile seeds and affect local terrestrial and aquatic populations. A balanced view also demands that even with the new CRISPR (Clustered Regularly Interspaced Short Palindromic Repeats) technologies, coexistence with organic farming must not be inhibited—that is, the option to return to earlier non-GM ways in the future must always be kept open (Niggli 2016).

Extension of agricultural land use coupled with replacement of traditional staple food crops by cash crops, modern and high-yielding varieties, and improved cultivars within existing cropping systems (Agnihotri and Palni 2007; Bardsley and Thomas 2005; Rana et al. 2009; Saxena et al. 2005) is driving the loss of genetic diversity (Xu and Grumbine 2014) and associated traditional farming knowledge. The replacement rate of local seed (landraces) has increased with distribution of modern seed varieties (Rana et al. 2009). In the Humla district of Nepal, rice subsidies over the past 30 years have increased dependence on this staple crop, resulting in the decline in cultivation of indigenous cereal crops such as barley foxtail, panicum and finger millets, buckwheat, amaranth, and beans (Roy et al. 2009).

In the western Himalaya a shift in land use (from subsistence farming to commercial farming), particularly the cultivation of vegetables and fruits as cash crops and the continued use of high-yield crop varieties, has become the principal reason for erosion of genetic diversity (Agnihotri and Palni 2007; Rana et al. 2007; Sharma and Chauhan 2013) and degradation of natural resources (Saxena et al. 2005). Although cash crops have brought substantial economic benefits to local farmers (Saxena et al. 2005), the shift from highly diverse systems to monoculture agriculture also has promoted the excessive and often indiscriminate use of chemicals and pesticides (Kala 2014). Furthermore, the loss of biodiversity could compromise the potential to obtain improved varieties which could help people adapt to changing conditions in the future.

2.4 Economic Drivers of Change to Mountain Sustainability

2.4.1 Economic Growth and Differentiation

Rapid economic developments have occurred in most HKH countries, creating national and local economic opportunities. Increased trade, tourism, resource extraction, and labour migration have opened avenues for income generation and local development; yet large challenges come with these changes (Box 2.5).

Box 2.5 Challenges of economic development in the HKH

Poverty in the HKH is still widespread (Gerlitz et al. 2012) and inequalities between highland and lowland —due to unequal access to markets, inadequate infrastructure, and failed or insufficient policies that do not sufficiently incorporate mountain specificities—are increasing.

Indigenous people and ethnic or religious minorities—a significant part of the population of the HKH —generally benefit less from economic development than the larger population group in urban centres. Instead, indigenous people are often the most underprivileged and vulnerable groups in the mountains (Gerlitz et al. 2012; Lama et al. 2010).

Despite efforts to promote regional economic integration and transboundary exchange, the HKH remains one of the least integrated regions in the world. The borderlands, while creating opportunities for some, continue to constitute significant barriers for exchange. Subsistence and non-monetized economies still play an important role, especially for pastoralists in the high altitudes around these national borders.

Mountainous areas are also often zones of instability and conflict (such as in Kashmir or north Myanmar) (Starr 2006) with significant shadow or conflict economies (for example, human trafficking and illicit trade in timber, precious stones, and drugs).

Informal economic activities, especially in rural parts of the HKH, play a considerable role in local livelihoods but their relative contributions are hard to estimate.

The countries of the HKH have made remarkable economic achievements in the last decades. Since the initiation of market reforms in 1978 and the shift to a market economy, China has experienced rapid economic development with an average annual gross domestic product (GDP) growth of nearly 10%, a rise in GDP from USD 148.4 billion in 1978 to USD 10.9 trillion in 2015, and a rise in GDP per capita from USD 980 in 1990 to USD 14,200 in 2015 (World Bank 2016). Between 1990 and 2005 alone, China lifted 470 million people out of extreme poverty (Wang et al. 2015). South Asia, mainly driven by the economic development of its largest economy, India, has also experienced robust economic growth; and, with a regional average GDP growth of 7.1% in 2016, is now the fastest-growing developing region worldwide.[2]

[2]See: http://www.worldbank.org/en/news/press-release/2016/04/09/south-asia-fastest-growing-region-world-vigilant-fading-tailwinds.

Despite these impressive national growth figures, economic development in the region varies both among and within countries. China's mountainous HKH provinces (Xizang/Tibet Autonomous Region (TAR) and Yunnan, Gansu, Qinghai and Sichuan provinces) lag far behind the coastal provinces both in terms of per capita GDP and rate of poverty reduction (Fang and Ying 2016). In India there is no consistent trend, with per capita income in some mountain states above the national average (e.g., Himachal Pradesh) and others severely below (e.g., Manipur, Assam, Jammu, and Kashmir).

Social instability and prolonged conflicts have notably hampered the performance of some countries (Afghanistan, Pakistan, and Myanmar), and states that are largely mountainous (Nepal and Bhutan) are also lagging behind. Mountain people in the HKH remain economically vulnerable; sustainable development efforts have not produced the desired outcomes; and mountain poverty in the HKH is still widespread (Gerlitz et al. 2015).

The vast majority of mountain people in the HKH are dependent on agriculture and animal husbandry; but growing population numbers have put additional stress on already limited mountain resources and have increased the workforce to the extent that it cannot be absorbed by traditional farming systems. In Nepal, for example, nearly three-quarters of the labour force is dependent on agriculture, which contributes to only 37.4% of GDP (Nepal and Henning 2013), while employment opportunities outside the farm sector are stagnating and the economy is characterized by a low salary structure (Shrestha 2008). In this respect, one livelihood strategy for mountain people is migration to national urban centres, more industrialized Asian countries, and the Gulf States. The remittances migrants send home are significant sources of income for households in mountain areas throughout South Asia. In Nepal, they amount to more than one-third of national GDP and are a major factor contributing to a balance of payment crisis (Shrestha 2008). Additionally, much of the potential labour force is then missing to sustain the traditional agricultural systems, which are degraded by underuse.

In addition to remittances, other external financial contributions are essential for the HKH. For example, in India, several hill states receive special subsidies through the Hill Area Development Programme and other schemes, while the economies of China's mountainous provinces such as Xizang/TAR rely heavily on subsidies and assistance from Beijing (Jin 2015). Official international development cooperation is also an important source of revenue for some HKH countries such as Afghanistan, Bhutan, Bangladesh, and Nepal, with foreign development grants accounting for significant shares of the annual national budget. While large contributions come from outside the region, regional cooperation also is growing. India and China are the largest contributors; in 2015–16 India allocated more than 80% of its USD 1.6 billion foreign aid budget to South Asia, with 63% going to Bhutan alone (Piccio 2015).

Hydropower development (e.g. in Bhutan) and resource extraction generate significant national or local revenues, but they can also lead to environmental degradation (e.g., timber extraction in Pakistan; Ali and Benjaminsen 2004), the withdrawal of the economic basis of pastoralists (see, e.g., Rousselot 2015 for Lesotho in Africa), and local conflicts (Agnew 2011; Global Witness 2015).

Leisure and religious tourism also contribute to the economic development of the HKH and have become significant drivers of socioeconomic and environmental change. Since Bhutan introduced tourism in 1974—with the primary objective of generating foreign exchange revenues (Dorji 2001)—the number of tourists has climbed sharply from 287 to over 7000 in 1999 and more than 155,000 in 2015; tourism now generates annual revenues of over USD 70 million (Tourism Council of Bhutan 2015). In Nepal, with its spectacular mountain scenery and its rich Hindu and Buddhist heritage, tourism has become one of the country's strongest industries. It generates employment opportunities as well as foreign exchange earnings and GDP growth, while also having positive effects on the demand for goods and services, transportation, and communication infrastructure (Gautam 2011; Paudyal 2012). Tourism is an important contributor to the state economy of Himachal Pradesh, India, and to Xizang/TAR, China, where tourist numbers—facilitated by improved rail and air transport to Lhasa in recent years—reached 9.25 million in 2014 and generated revenues of CNY 11.2 billion yuan (USD 1.79 billion) (Xinhua 2015). Herein, we can also clearly see huge risk for potential over-exploitation. Tourism related to summit expeditions (e.g., Mount Everest), alpine trekking, and visiting notable landscape features has become a global export product, but also can create localized conflicts with local populations because of use conflicts (Naitthani and Kanthola 2015; Rao et al. 2000) or problems of over-exploitation (with violent attacks between expedition members at Mount Everest in April 2013). Second homes owned by rich multilocal people from Mumbai, Delhi, and the Gulf States also are sprawling into the unstable hills around Nainital/Uttarakhand, India (Tiwari and Joshi 2016).

Regional and transboundary trade are important for both mountain countries and mountain people along the borders. For example, driven by Chinese demand, *Yarsagumba* or caterpillar fungus (*Cordyceps sinensis*) has become one of the most important sources of cash incomes for mountain people in large parts of the Tibetan Plateau and the India, Nepal, and Bhutan Himalaya (Winkler 2009). Complicated border procedures, insufficient border infrastructure, and a complex socio-political environment, however, mean that regional trade performs well below its potential, and South

Asia remains one of the most un-integrated regions in the world (Ahmed and Ghani 2015).

The South Asian Association for Regional Cooperation (SAARC) and the South Asia Subregional Economic Cooperation (SASEC) Program, aimed at improving regional economic opportunities and fostering connectivity with China and Southeast Asia through Myanmar, still have a long way to go before the mountain people of the HKH can genuinely benefit. Other large regional and international initiatives also attempt to increase regional connectivity and economic integration:

- India's 'Look East' policy has gained new momentum with the opening of Myanmar since 2012;
- Pakistan and China are creating an economic corridor with considerable investments in infrastructure and energy generation in Pakistan's mountainous provinces;
- China announced a modern version of the ancient Silk Road with its 'One Belt, One Road' initiative that aims to connect large parts of Asia with Europe and East Africa through maritime (belt) and overland (road) connections; and
- Bangladesh, China, India, and Myanmar are establishing an economic corridor that will eventually connect Kunming in China's Yunnan Province with Kolkata (India) via Mandalay (Myanmar) and Dhaka (Bangladesh).

It is too early to estimate the full impact of these large initiatives, but they are likely to shape socioeconomic dynamics in the HKH for decades to come. In light of this increasingly rapid globalization and because of the constraining mountain specificities of fragility, marginality, and inaccessibility, further economic integration requires strong and mountain-specific policies to avoid increased dependencies and inequities in highland–lowland exchanges (Jodha 2000, 2005).

As in similar cases in history and in other parts of the world, these economic development processes go hand in hand with a differentiation and polarization in society whereby more active and dynamic families can profit from open and enlarged markets and new demands, while lower-performing individuals or households become increasingly dependent on low-wage jobs or are forced to out-migrate, at least temporarily (e.g., Sherpa society in the Khumbu area) (Jacquemet 2017). We see similar processes in other mountain resort towns, such as several waves of resort development in the European Alps in the late 19th century and from the 1960s to the present, and in the Rocky Mountains in Canada and USA since the 1980s (Glorioso and Moss 2006).

2.4.2 Rapid Infrastructure Development

In 1994, the World Development Report argued that good infrastructure can raise the overall economic performance of a country, contribute to poverty reduction, and help to improve environmental conditions when the provision of service infrastructure responds efficiently to demand (World Bank 1994). Further studies have demonstrated the positive correlation between infrastructure, income levels, and (rural) development (Calderón and Servén 2004; Cook 2005).

In recent decades, increased access to and within mountainous areas in the HKH because of the creation of hard infrastructure has indeed been a powerful driver of change. The expansion of roads and road networks has played the most important role and helped to reshape socioeconomic and political relations within mountain areas, and also between mountains and the downstream centres of population. While it is generally assumed that access (especially through roads) represents opportunities for development, it also can contribute to increases in resource extraction from mountainous areas, growing dependencies on markets in the lowlands, pollution and environmental degradation, out-migration, and (potentially) sociocultural conflict. Also, as the level of infrastructure development varies, the impacts derived and pace of development are distributed unequally between and within regions (Kohler et al. 2001; Kreutzmann 2004).

Comprehensive data on infrastructure in the HKH are still not available. While the South Asian region faces an infrastructure gap of USD 1.7–2.5 trillion until 2020 (Biller et al. 2014), no consistent estimates are available that demonstrate how much of this amount needs to be invested in mountain infrastructure. Also, access to infrastructure within and among South Asian countries is extremely variable, but in general accessibility is lower in countries with large areas of mountainous geography, low rates of development, and long internal conflicts, such as Afghanistan and Nepal (Biller et al. 2014).

Difficult geography, high construction costs, the proximity to sensitive national borders, and a lack of economic incentives for central planners mean that physical infrastructure in the mountains has long been underdeveloped. Early road expansion in the eastern and western Himalaya was guided more by strategic and military interests than by an aim of providing better transport facilities for mountain people. Since the 1990s, however, many of these roads have become accessible to civilians, and the network of roads accessible to motor vehicles has expanded, providing access to hitherto remote and inaccessible mountain valleys, such as Pakistan's Gilgit-Baltistan or north-eastern India (Das 2008; Kreutzmann 2004).

While Nepal had only 2700 km of road in the early 1970s, the official road inventory in 2011 listed approximately 11,000 km of strategic roads and 60,000 km of rural roads. However, only half of the strategic road network and 5% of the rural network are paved (World Bank 2014). Bhutan, which began road development in the early 1960s, now has a road network of 10,578 km built mainly with outside support and comprising 2436 km of national highways and expressways; 1190 km of district roads; 5257 km of rural farm roads; 350 km of urban roads; 667 km of forest roads; and 678 km of other access roads (Asian Development Bank and Royal Government of Bhutan 2014).

The high investment costs in engineering and materials and the geographical conditions in the HKH, with its high mountain passes and frequent river crossings, have hindered road development. In addition, heavy monsoon rainfalls in parts of the HKH, human impact through deforestation or building construction (Haigh and Rawat 2011), and weakened slope stability due to increased road development itself (as in China's western Yunnan Province; Sidle et al. 2014), cause numerous landslides along mountain roads (see Sect. 2.2.5, Mountain hazards). The resulting high cost for road maintenance means that many mountain roads are in poor condition. The summer rainy season and heavy winter snowfalls also make many roads impassable for much of the year. Access and connectivity in the HKH therefore cannot depend only on paved and unpaved mountain roads; it also depends on a network of trails, suspension bridges, and ropeways. Porters and animal transport by donkeys, yaks, and yak crossbreeds also still play an important role in higher elevations and remote areas (Starkey 2001; Wu et al. 2016b). Additionally, air transport is indispensable—especially in remote and otherwise inaccessible areas—often with small airports and untarmacked runways.

The only country in the HKH with established major railways to and in the mountains is China. A spectacular railway with large parts of the tracks on permafrost on the Tibetan Plateau connects Golmud (Qinghai) with Lhasa (Xizang/Tibet) was opened in 2005 and was recently expanded to Xigazê (Shigatse) in 2014.

The potential for hydropower in the HKH exceeds 500 GW (Vaidya 2012). Harnessed properly, this energy could trigger a socioeconomic transformation and have a tremendous impact on the lives of mountain people. However, despite the high potential for hydropower, the energy economy in the region is still weak and characterized by limited access to clean and modern energy services, low per capita consumption, a relatively high consumption of non-commercial biomass, and a high and growing dependency on imported fossil fuels (Asian Development Bank 2013; Shrestha 2013). Energy poverty is high (Rasul and Sharma 2015) and electricity (apart from in China) is available to only half of the population in the HKH (Vaidya 2012). A strategy for strong tourist development increases the needs for energy and may serve as a driver the extension of hydropower installations in Nepal and India (Fort 2017; Naitthani and Kanthola 2015).

In the future, energy poverty could be minimized and a strong driver for change in the mountains could be promoted by:

- Modern technologies that increase the efficiency of fuelwood (still the most important source of energy for many mountain people, both for cooking and heating) and decrease related carbon emissions;
- Decentralized small-scale hydropower development; and
- Better local, national, and regional grid connectivity that could also facilitate hydropower export (but this should be used carefully, and mainly as a means of supra-national cooperation under the aspect of supply security[3]; if it is developed under a merely commercial logic, it risks severe environmental damage and social challenges). The further extension of new sources of energy such as active and passive solar energy also is very important.

2.4.3 Urban and Peri-urban Expansion

Urbanization in the HKH is an independent driver with cross-cutting impacts on the three pillars of mountain sustainability. Urbanization is understood here in a sociological sense, which includes demography, migration, and the differentiation to market economies; it considers especially the socioeconomic processes at play in a given territory (Brenner and Schmid 2014). More succinctly, urbanization encompasses the overall processes of association of people, and concomitantly the movement from generalist towards specialisation—with societies in urban or peri-urban settings largely dependent on broad division of labour and exchange of services.

Urbanization should not be confounded with urban expansion, urban sprawl, or the creation of cities and agglomerations—phenomena which are only part of the comprehensive social process of urbanization. This umbrella term is more comprehensive and includes aspects which are not directly visible, such as temporary outmigration (multi-locality), second homes for leisure, or internet-based

[3]In Bhutan, hydropower export to India is already a major source of revenue. Nepal, on the other hand, is currently a net importer of electricity from India despite an abundance of hydropower resources (Asian Development Bank 2013). For longer discussion on the problem of the impacts of the use of hydropower on local populations, see Rousselot (2015) and Agnew (2011).

activities in areas which at first sight look purely "rural". In this section it is not possible to consider all aspects of urbanization, and we focus mainly on processes related to an ongoing concentration of jobs and people in local centres, cities, or urban agglomerations which often show a spatial sprawl, concomitantly with a relative depopulation in regional peripheries. Over the last three to four decades, human settlements in the HKH have grown both in numbers and in size, with small villages transforming into larger towns and former towns into major metropolitan areas (Ghosh 2007; Pandit 2009).

Urbanization is, on one hand, the consequence of both global and regional socioeconomic driving forces. A global division of labour, production, and trade increases the shift of labour from manual work to more complex organization, logistics, and commercial services, as can be observed in societies of developed countries in North America and Europe (Fourastie 1989 [1949]). There is a widespread, global tendency towards the concentration of populations in dense agglomerations (UNFPA 2007). However, the trajectory of urbanization in the HKH is unique: unlike in the European Alps, for example, the HKH is not comprised of a single or uniform economic space, as multiple national borders and the regional geopolitical constellation of States in HKH play an important role.

Temporary and permanent outmigration from the region with dependencies on foreign remittances is said to be the highest in the world, due largely to enormous disparities in wealth (Karki et al. 2012). In contrast, global outdoor tourism induces the growth of urban agglomerations, resorts towns, and urban infrastructure such as airports. Small and medium-sized local centres in the hills play a specific and particularly important role for the peripheries, as they are privileged places for the introduction of public infrastructure, health services, and technologies, and also for consumer goods. Tourism can thus introduce a market society and simultaneously weakens local agrarian society that has long been based at least partly on subsistence. Both the internal and external frameworks of HKH societies constitute the drivers for an urbanization that follows a combination of both global tendencies and HKH specificities.

On the other hand, urbanization is itself also a driver of change. Existing towns attract people by promising better livelihoods and income, more individual freedom, and larger social networks. The functional division of labour allows cities to create economies of scale (agglomeration economies) which enables cities to concentrate more economic wealth. Agglomeration economies allow them to create jobs, attract reinvestments and remittances, and encourage people to migrate from the rural hinterlands in search for a better life. In cities, space is scarce, and this results in higher buildings made of concrete instead of the locally manufactured brick. Cities change the nutrition regime of the local population who use refrigerated, imported food instead of supplies from local markets. However, the converse also is possible, and agglomeration diseconomies may occur such as increasing pollution, congestion, or insecurity. In such instances, the growth poles may then shift to other agglomerations (cf. Krugman 1991; Marshall 1890; Moriconi-Ebrard 1993; Pumain 1999; Richardson 1995; Storper 1995, among many others).

In order to evaluate and interpret urban development and urbanization it is important that analysis is not restricted to just the high altitudes. Quantitative and qualitative analysis and mapping of the HKH requires the large agglomerations in the foothills to be included, as these lowland areas are primary drivers for regional and international exchange, guarantee accessibility to the mountains, and also attract people from the mountains. These mountain–lowland linkages have become crucial in the mountain ranges of the developed countries with mountain ranges such as the Alps (Perlik 2018); they are even more important in mountain regions where national borders play such critical roles as in HKH. One very significant observation from the European Alps is that the lowland–mountain relationship has continuously to be renegotiated. In particular this relationship defines the demands of the urban population in the lowlands for specific mountain resources, and the need of the mountain population to be protected against natural hazards as well as territorial insecurity. These mutual dependencies and benefits need to be balanced fairly.

In the HKH we see similar spatial processes as in the mountains of the developed countries of the Global North. Similar phenomena affect the environment and society, such as the decline of agricultural subsistence economies as a consequence of the rapid increase in tourism (Jacquemet 2017) or second home residences (Tiwari and Joshi 2016). The latter generally are premised on, or require, integration of mountains in the national and international economy, increased division of labour, and investments in construction and infrastructure. Dynamic urban development also is often observed along international transport routes (for instance along the Karakoram Highway) (Karrar 2013).

There are, however, two important specificities in the HKH. The first is the different development trajectory that has occurred due to the historic path of colonization, post-colonial constellations, and Cold War geo-strategies in the Himalaya. The emergence of the states of China, India, and Pakistan gave rise to the territorialized border regions and the development of a decentralized infrastructure, as well as new trade relations. Today, the Himalaya has a new position as a large hinterland in relation to the hotspots of globalization with their fast-growing economies. The second difference is the time lag with which urbanization is introduced or occurs (hysteresis) in the region. As urbanization takes place in a period of global exchange, where the main manufacturing sites of the world have become centralized outside the mountains, the modern development of mountain areas is narrowed on leisure, tourism, raw materials, and water (Perlik 2018). In some places this delay may to a certain degree prevent some forms or the extent of toxic pollution in mountains, but it also can enlarge the impacts from global tourism, hydropower dams, and abandoned agriculture. Such hysteresis also means that decision centres even for mountain urban agglomerations are almost entirely located in the lowlands, where mountains are often seen as inferior or backward and thus mountain challenges are aggravated.

While urbanization and urban development in the HKH can be explained largely on the basis of similar mechanisms and drivers as seen globally, in the framework of global change and also considering the consequences of the regional HKH trajectory along with certain standard urban patterns (e.g. nation building with a national capital, the effects of global tourism, abandonment and concentration of local markets), certain other processes are specific to the HKH. These include, for example, early forms of territorialization with garrisoned armed forces, creation of centres of international logistics, and the presence and actions of global development and environmental NGOs. Five functional urban patterns can be observed in the HKH.

2.4.3.1 National and Regional Capitals

National and some regional capitals follow their own interests and often compete for status as global cities. They also are perceived as high-end "Central Places" which hold the highest functions and services, in the interest and for the prestige of the whole country. They define national standards of economic activities and the range of consumer goods available. They also are the only diversified places in the frame of urbanization, and they attract all kind of workforces (from unskilled workers to highly qualified specialists) from other parts of the country as well as expat populations. As such, national resources are disproportionally concentrated in these places. As mountain regions rarely hold such capital functions, at least on a global scale, they are often at a relative disadvantage. In the HKH, however, the majority of national capitals lie within the mountain perimeter. Only China, India, and Bangladesh have their national capital outside the mountain ranges, yet China still invests disproportionally in Lhasa and TAR, and India has given Uttarakhand the status of a federal state. In Nepal the mountain part of the country, including Kathmandu, is clearly also considered more important than the lower-lying plains.

2.4.3.2 New Hotspots of Territorialization

Even small and medium-sized towns may be involved in a dynamic growth process whereby—whether by chance or due to the development of new infrastructure—they become connected to national and international decision making. This can be the case when they suddenly find themselves in the centre of a regional conflict, or as they become centres for logistics, administration, or a new border crossing. With the opening of the Karakoram Highway in Pakistan in 1978, for example, a number of adjacent settlements rapidly gained significance. The population of the city of Gilgit grew from 18,000 in 1972 to 57,000 in 1998, and together with its rural hinterlands it has seen a population growth from 145,000 in 1998 to an estimated 222,000 in 2013 (Kreutzmann 1991; UN-HABITAT 2010). Such numbers are a clear indicator of a large in-migration that has taken place because of new infrastructure, and in some instances also due to insecurity and/or resource degradation elsewhere in the country. A similar development takes place in cities and settlements which have become focal points of military deployment or hubs for logistics and foreign aid. Kunduz, in the foothills of the Hindu Kush, had about 74,000 inhabitants in the 1960s, but this grew to 268,893 in 2012 (GoIRA 2015). The city now hosts troops and services related to troops, ranging from independent translators to large companies, and it also is a place for shelter-seeking refugees and NGOs caring for them. Development strategies to reinforce global tourism also result in the enlargement or new construction of international airports (e.g., at Pokhara) (Fort and Adhikari 2018; Rimal et al. 2015).

2.4.3.3 Old Urbanization with Own Production Systems

Shimla in Himachal Pradesh was chosen as the capital of British India during the colonial era because of its milder climate. Following independence Shimla lost this privileged status in favour of New Delhi but remained a centre of wool production. Today it is an important tourist destination. At present Shimla remains an important city with large hinterlands, although with a significantly lower dynamic (143,000 inhabitants in 2001 and 172,000 in 2011) than many other mountain centres in the region. In the last years Shimla has become renowned for its touristic qualities for both domestic and international tourists. The city has remained the critical mass and a sufficient hinterland to be able for restructuration during the process of development in India. Such cities are an example for the long standing impact of defined centres of power, may be due to economic decision making or due to political will. They reflect the successful cases while other towns—today unknown—declined or became integrated in larger agglomerations.

2.4.3.4 Resort Towns for Tourism and Second Homes

With the increase in global tourism and the emergence of urban middle classes in HKH states, resort towns that valorise landscape vistas and amenities and the reputation of prestigious holidays have developed in mountains. They offer tourist services including accommodation, guides for trekking tours, and ropeway infrastructure, and they deliver ambiance and prestige. Some places serve as locations for second homes, as in Nainital, Uttarakhand, frequented by clients from Mumbai and New Delhi as well as from Arab countries (Tiwari and Joshi 2016). The other type of resort towns are small cities near the peaks of renowned mountains, used by trekking tourists and national park visitors. In Joshimath, for instance, the snow resort in nearby Auli hosted the South Asian Winter Games in 2011.

2.4.3.5 Local Centres

The principal towns of a district or subdistrict mainly serve a supply function. They are the places of territorialization at lowest level (for local administration and services) and provide opportunities for local market exchange. Most local centres lie in the Himalayan hills but their distribution is rather uneven (in India they are represented by more than 500 towns, mostly small and medium-sized). Because of their hilly topography, scope for their expansion is limited and the accessibility by roads is difficult (Khawas 2007). With improved accessibility to the plain in the future, the socio-economic structure of these small towns may change. Its inhabitants may search for better jobs in the plain and become out-commuters (and may move definitely) while peasant farmers might abandon their villages in higher altitudes and move to these small towns for the new jobs and better services offered. We must also recognize that temporary outmigration from the hilly zones to the lowlands is already common practice (Benz 2014; Dame 2015).

The recent urbanization processes not only introduce a new pattern of spatial use to the mountains; they also introduce the predominance of a market economy and powerful economic and social actors from outside. These socioeconomic and cultural transformations are the most important factors related to urbanization. The new spatial pattern of land use in the HKH also has led to adopting the current urbanization model based on a large-scale functional division of labour, with a polarization of metropolitan hubs as centres of decision making and surrounding mountains as spaces of leisure. This functional cleavage can be seen well in Uttarakhand, India, where the majority of the territory is state property. The state is following a business model to introduce a professionalized tourism sector, with the aim to attract wealthy foreign tourists to the scenic landscapes of the Nanda Devi region. The concept is based largely on a traditional protectionist National Parks model that generally sees farmers or pastoralists as adversaries to environmental goals, with the administration attempting to prevent traditional activities of local communities (Naitthani and Kainthola 2015). At the same time, India's growing needs as an emerging country have spurred a construction boom including river dams to increase the production of hydro energy, which also is needed for the tourism sector. A concentration of activities on developing and operating tourism resorts and 'keeping out' peasant farmers and pastoralists are both constituent parts of the urbanization process in many places in the HKH, yet from the perspective of sustainable development they are both highly problematic. In this regard, the HKH follows global tendencies, as in Europe and the USA, only considerably delayed (Moss and Glorioso 2014; Perlik 2011). Rural mountain economies are squeezed out in favour of leisure and residences.

2.5 Conclusions

Changes driven by various forces in the HKH are creating constraints as well as new opportunities for sustainable development. We found that environmental and socioeconomic changes in the HKH are driven by a network of interactions among a range of factors or forces. Although some of the elements of these networks are global (e.g., climate change, IAS, globalization, advancement of science and technology) or regional (e.g., pollution, migration), the actual set of interactions that brings about change is more or less specific to each place and related to particular mountain ecosystems or trans-boundary landscapes. There is thus a great spatial diversity amongst drivers of change and their impacts. Considering the entire HKH, rapid demographic change and weak governance systems associated with economic growth and climate change, followed by increasing LULCC, and over-exploitation of natural resources, have had the greatest overall impact on the sustainability of mountain areas (see Table 2.1). At the same time, most of the drivers discussed above are currently increasing in intensity and this trend will continue in the near future.

In the HKH many challenges for sustainability are related to increasing demands on natural resources, which lead to further environmental degradation, unregulated urbanization and infrastructural development, and loss of traditional culture. Both endogenous economic growth and global demands on tangible and intangible mountain products are leading to increased and selective consumption of mountain resources. The transformation from subsistence economies to market-driven economies leads to temporary or permanent emigration to emerging economies abroad which absorb the mainly male workforce while women are overburdened in mountain villages. Spatially this results in a degradation of the cultivated landscape. Extraction of natural resources and establishment of intensive agriculture and plantations are altering the extent and species composition of natural vegetation and reshaping the rural landscape. Growing remittance flows to family members at home, raising their purchasing power, and the general search for a better livelihood are

Table 2.1 Level of influence of key drivers of change in the Hindu Kush Himalaya on three pillars of sustainability

Drivers within three categories		Three pillars of sustainability			Value
		Environmental protection	Sociocultural equity	Economic viability	
Environmental	Land use and land cover change	↗	→	↗	8
	Over-exploitation of natural resources	↗	↗	↗	8
	Pollution	↗	→	↗	6
	Invasive alien species	↗	→	↗	6
	Mountain hazards	→	↗	↗	6
	Climate change and variability	↗	↗	↗	9
Sociocultural	Demographic changes	↗	↗	↗	9
	Sociocultural changes	→	↗	↗	6
	Governance systems and institutions	↗	↗	↗	9
	Technological implementation	↗	↗	↗	6
Economic	Economic growth and differentiation	↗	↗	↗	8
	Infrastructure development	↗	↗	↗	7
	Urban expansion	↗	↗	↗	7

Note: Level of influence: Red: high, orange: medium, yellow: low;
Value of influence: high: 3, medium: 2, low: 1
Trend change: ↗ increase, → stable, ↘ decrease

placing greater demands on local resources as well as imported goods. Increasing levels of international or inter-regional trade increase transport flows and introduce global products which generate huge external diseconomies (for instance, refrigeration, waste treatment, and sewage). At the same time they destroy traditional value chains of local agricultural products and crafts. Addressing these problems will require policy and action at local, national, and international levels, including common action among HKH states.

However, a range of opportunities also lies in the improving connectivity, including transportation and communication (which increase access to information), partnerships, and markets. Road network development is currently expanding accessibility, and international investment in agriculture and forestry in remote areas is driving new trends in rural economic activity. Enhanced access to health and education may be enabled and strengthened by economic growth and the advancement of science and technology. Additionally, a growing network of local urban centres may support the transmission of new prosperity to rural populations, as the development of mountain towns and cities may help—besides their mere economic power—to enhance the political influence of these regions within the national states.

While the drivers of change continue to demand individual identification and analysis, the complexity of their interactions also calls for a more comprehensive and integrated strategy to be adopted, such as promoting trans-boundary cooperation, and encouraging the development of mountain-specific responses by government policy makers. Horizontal analysis of policies and their anticipated impacts, whereby each sector is assessed in light not only of its own priority goals but also those of other sectors, is highly recommended (Foggin and Phillips 2013). Furthermore, advances in knowledge and rapid access to new technology are growing with the influence of globalization. As access to global services improves, the mountain specification of 'inaccessibility' will change to a certain extent. This will, in turn change the demand for specific ecosystem services, and enhance the capacities of local communities to adapt to global change. New financing mechanisms for climate change mitigation and infrastructure development equally offer valuable opportunities for increased investment in mountain regions. Thus, both proximate and ultimate drivers of change must be recognized and incorporated into more comprehensive concepts of development and approaches to development planning.

It has been recognized that an overall transition of mountain systems is taking place in the HKH. Growing global awareness about environment and society has signalled further modification in the stimulus for improved resource governance and institutional reform in the region. Sociocultural changes, infrastructure improvement, and technological implementation are also helping to pave the way for changes in mountain development, especially in remote high-altitude areas. People in the HKH are trying to mitigate the negative impacts of drivers or manage some drivers through more targeted, integrated, and collaborative actions. What is needed at present is a better understanding of changes and interactions, and of resulting trade-offs among different drivers, and enhanced regional interaction and coordination between upstream and downstream regions, populations, and decision-makers. Greater participation of local actors as well as high level transnational cooperation are both necessary to cope with current and potential conflicts and to collaboratively seek and develop viable opportunities for sociocultural, economic, and environmental improvements in the HKH in the future.

References

Adhikari, R. (2011). Biotechnology growth partnership: A potential flagship program in S&T. *Nepal Journal of Biotechnology, 1*(1), 55–58.

Adhikari, A., Upadhyay, M. P., Joshi, B. K., Rijal, D., Chaudhary, P., Paudel, et al. (2006). Multiple approach to community sensitization. In B. R. Sthapit, P. Shrestha, & M. P. Upadhyay (Eds.), *On-farm management of agricultural biodiversity in Nepal: Good practices*. NARC/LI-BIRD/Bioversity International, Nepal.

Aggarwal, R. (2004). *Beyond lines of control: Performance and politics on the disputed borders of Ladakh*. India: Duke University Press.

Agnew, J. (2011). Waterpower: Politics and the geography of water provision. Annals of the Association of American Geographers, https://doi.org/10.1080/00045608.2011.560053. http://dx.doi.org/10.1080/00045608.2011.560053.

Agnihotri, R. K., & Palni, L. M. S. (2007). On-farm conservation of Landraces of Rice (*Oryza sativa* L.) through cultivation in the Kumaun region of Indian Central Himalaya. *Journal of Mountain Science, 4*(4), 354–360.

Agrawal, N. K., Krishnan, A., Leikanger, I. C. P., Sherpa, J. D., Choudhury, D., & Pantha, R. H. (2016). *Action for mountain adaptation: Solutions beyond the boundaries of science, policy, and practice*. Kathmandu: ICIMOD.

Ahmed, S., & Ghani, E. (2015). South Asia's growth and regional integration—An overview. In S. Ahmed, & E. Ghani (Eds.), SOUTH ASIA growth and regional integration (pp. 3–42). Washington, DC: The World Bank.

Akhtar, S., & Gregson, J. (2001). Internet technologies in the Himalayas: Lessons learned during the 1990s. *Journal of Information Science, 21*(1), 9–17.

Akter, A., & Zuberi, M. I. (2009). Invasive alien species in Northern Bangladesh: Identification, inventory and impacts. *International Journal of Biodiversity and Conservation, 1,* 129–134.

Akumo, N, Riedel, H., & Semtanska, I. (2013). Social and economic issues—Genetically modified food. *Agricultural and Biological Sciences, Food Industry*. ISBN 978-953-51-0911-2.

Alam, R., Chowdhury, M. A. I., Hasan, G. M. J., Karanjit, B., & Shrestha, L. R. (2008). Generation, storage, collection and transportation of municipal solid waste—A case study in the city of Kathmandu, capital of Nepal. *Waste Management, 28,* 1088–1097. https://doi.org/10.1016/j.wasman.2006.12.024.

Ali, J., & Benjaminsen, T. A. (2004). Fuelwood, timber and deforestation in the Himalayas. *Mountain Research and Development, 24*(4), 312–318.

Amin, A., Fazal, S., Mujtaba, A., & Singh, S. K. (2014). Effects of land transformation on water quality of Dal Lake, Srinagar, India. *Journal of the Indian Society of Remote Sensing, 42,* 119–128.

Antonelli, A. (2015). Biodiversity: Multiple origins of mountain life. *Nature, 524*(7565), 300–301.

Ariza, C., Maselli, D., & Kohler, T. (2013). *Mountains: Our life, our future. In Progress and perspectives on sustainable mountain development from Rio 1992 to Rio 2012 and beyond. A global synthesis based on 10 regional reports*. Geographica Bernensia.

Arya, D., Bhatt, D., Kumar, R., Tewari, L. M., Kishor, K., & Joshi, G. C. (2013). Studies on natural resources, trade and conservation of kutki (picrorhiza kurroa royle ex benth. scrophulariaceae) from kumaun himalaya. *Scientific Research & Essays, 8*(14), 575–580.

Aryal, S., Cockfield, G., & Maraseni, T.N. (2015). Effect of summer livestock grazing on plant species richness and composition in the Himalayan rangelands. *The Rangeland Journal*, 37, 309–321. https://doi.org/10.1071/RJ14088.

Ashbolt, N. J. (2004). Microbial contamination of drinking water and disease outcomes in developing regions. *Toxicology, 198,* 229–238.

Asian Development Bank. (2013). *Energy outlook for Asia and the Pacific 2013*. Mandaluyong City, Metro Manila, Philippines: Asian Development Bank.

Asian Development Bank & Royal Government of Bhutan. (2014). *Three decades of development partnership: Royal Government of Bhutan and Asian Development Bank*. Mandaluyong, Philippines: Asian Development Bank.

Atreya, K., Sitaula, B. K., Johnsen, F. H., & Bajracharya, R. M. (2011). Continuing issues in the limitations of pesticide use in developing countries. *Journal of Agricultural and Environmental Ethics, 24,* 49–62.

Babel, M. S., Pandey, V. P., Rivas, A. A., & Wahid, S. M. (2011). Indicator-based approach for assessing the vulnerability of freshwater resources in the Bagmati River Basin, Nepal. *Environmental Management, 48,* 1044–1059.

Bajracharya, B., Shrestha, A. B., & Rajbhandari, L. (2007). Glacial lake outburst floods in the Sagarmatha region. *Mountain Research and Development, 27*(4), 336–344.

Bajracharya, S. R., Maharjan, S. B., & Shrestha, F. (2011). Glaciers Shrinking in Nepal Himalaya.

Baller, S., Dutta, S., & Lanvin, B. (Eds.). (2016). *The global information technology report 2016—Innovating in the digital economy*. Geneva: World Economic Forum.

Banerjee, S., Black, R., Kniveton, D., & Kollmair, M. (2014). The changing Hindu Kush Himalayas: Environmental change and migration. The Netherlands: Springer.

Banjade, M. R., & Paudel, N. S. (2008). Economic potential of non-timber forest products in Nepal: Myth or reality? *Journal of Forest and Livelihood, 7,* 36–48.

Banskota, M., Papola, T. S., & Richter, J. (2000). Development in mountain areas of South Asia—Issues and options. Growth, poverty alleviation and sustainable resource management in the mountain areas of South Asia. In *Proceedings of the International Conference Held from 31 January–4 February 2000 in Kathmandu*, Nepal (pp. 1–31). Kathmandu, ICIMOD and ZEL.

Baral, K., Sapkota, T., Bhandari, B., Adhikari, A., Dewan, S., Pageni, P., et al. (2006). Rural radio programme: An effective tool for reaching the unreachable on biodiversity conservation issues. In B. R. Sthapit, P. Shrestha, & M. P. Upadhyay (Eds.), *On-farm management of agricultural Biodiversity in Nepal: Good practices*. NARC/LI-BIRD/Bioversity International, Nepal.

Bardsley, D., & Thomas, I. (2005). In situ agrobiodiversity conservation: Examples from Nepal, Turkey and Switzerland in the first decade of the convention on biological diversity. *Journal of Environmental Planning and Management, 49*(5), 653–674.

Barua, S. P., Khan M. M. H., Reza, A. H. M. A. (2001). The status of alien invasive species in Bangladesh and their impact on the ecosystems. IUCN.

Bauer, K. M. (2004). High frontiers: Dolpo and the changing world of Himalayan pastoralists. Columbia University Press.

Bauer, K. (2005). Development and the enclosure movement in pastoral Tibet since the 1980s. *Nomadic Peoples, 9*(1–2), 53–81.

Baur, I. (2014). *Analyzing and modeling the use of common property pastures in Grindelwald, Switzerland.* Ph.D. Thesis. Munich: LMU. https://edoc.ub.uni-muenchen.de/18295/1/Baur_Ivo.pdf.

Benz, A. (2014). Multilocality as an asset. Translocal development and change among the Wakhi of Gojal. In H. Alff & A. Benz (Eds.), *Tracing connections: Explorations of spaces and places in Asian contexts* (pp. 111–138). Berlin, Germany: Wissenschaftlicher Verlag Berlin.

Bergmann, C., Gerwin, M., Nüsser, M., & Sax, W. S. (2008). Living in a high mountain border region: The case of the 'Bhotiyas' of the Indo-Chinese border region. *Journal of Mountain Science, 5*(3), 209–217.

Bhattarai, K. R., Måren, I. E., & Subedi, S. C. (2014). Biodiversity and invasibility: Distribution patterns of invasive plant species in the Himalayas, Nepal. *Journal of Mountain Science, 11,* 688–696.

Bhusal, J., & Subedi, B. (2014). Climate change induced water conflict in the Himalayas: A case study from Mustang, Nepal. *Ecopersia, 2* (2), 585–595.

Biller, D., Andrés, L., & Dappe, M. H. (2014). Infrastructure gap in South Asia: Inequality of access to infrastructure services. The World Bank.

Bing, H. J., Wu, Y. H., Zhou, J., & Sun, H. Y. (2016). Biomonitoring trace metal contamination by seven sympatric alpine species in Eastern Tibetan Plateau. *Chemosphere, 165,* 388–398.

Bisht, N., Joshi, S., Shrestha, B. B., Yi, S., Chaudhary, R. P., Kotru, R., et al. (2016). *Manual on invasive alien plant species in Kailash Sacred Landscape-Nepal.* Kathmandu: ICIMOD.

Bolch, T., Buchroithner, M. F., Peters, J., Baessler, M., & Bajracharya, S. (2008). Identification of glacier motion and potentially dangerous glacial lakes in the Mt. Everest region/Nepal using spaceborne imagery. *Natural Hazards & Earth System Sciences, 8,* 1329–1340.

Bose, S., & Jalal, A. (2004). Modern South Asia: History, culture, political economy (pp. 1–282). Taylor & Francis e-Library.

Brenner, N., & Schmid, C. (2014). The "urban age" in question. *International Journal for Urban and Regional Research, 38*(3), 731–755.

Bruinsma, J. (2003). World agriculture: Towards 2015/2030. An FAO perspective.

Budha, P. M. (2015). Current state of knowledge on invasive and Alien Fauna of Nepal. *Journal of Institute of Science and Technology, 20,* 68–81.

Bull, A. T. (1996). Biotechnology for environmental quality: Closing the circles. *Biodiversity and Conservation, 5,* 1–25.

Burlakoti, C., & Karmacharya, S. B. (2004). Quantitative analysis of macrophytes of Beeshazar Tal, Chitwan, Nepal. *Himalayan Journal of Sciences, 2,* 37–41.

Busby, J. R. (1988). Potential implications of climate change on Australia's flora and fauna. *Greenhouse planning for climate change* (pp. 387–398). Melbourne, Australia: CSIRO.

Byers, A. (2014). Contemporary human impacts on subalpine and alpine ecosystems of the Hinku Valley, Makalu Barun National Park and Buffer Zone, Nepal. *Himalaya, 33,* 25–42.

Cacopardo, A. M., & Cacopardo, A. S. (2001). *Gates of Peristan. History, religion and society in the Hindu Kush.* Rome: IsIAO.

Cai, H., Yang, X., & Xu, X. (2015). Human-induced grassland degradation/restoration in the central Tibetan Plateau: The effects of ecological protection and restoration projects. *Ecological Engineering, 83,* 112–119.

Calderón, C., & Servén, L. (2004). The effects of infrastructure development on growth and income distribution. The World Bank.

Carpenter, S., & Grünewald, F. (2016). Disaster preparedness in a complex urban system: The case of Kathmandu Valley, Nepal. *Disasters, 40*(3), 411–431. https://doi.org/10.1111/disa.12164.

Chakarvorty, M., Dwivedi, A. K., Shukla, A. D., Kumar, S., Niyogi, A., Usmani, M., et al. (2015). Geochemistry and magnetic measurements of suspended sediment in urban sewage water vis-a-vis quantification of heavy metal pollution in Ganga and Yamuna Rivers, India. *Environmental Monitoring and Assessment, 187,* 17.

Challinor, A. J., et al. (2009). Crops and climate change: Progress, trends, and challenges in simulating impacts and informing adaptation. *Journal of Experimental Botany, 60*(10), 2775–2789.

Chaudhary, S., Chettri, N., Uddin, K., Khatri, T. B., Dhakal, M., Bajracharya, B., et al. (2014). Implications of land cover change on ecosystem, ecosystem services and livelihoods of people: A case study from the Koshi Tappu Wildlife Reserve, Nepal. *Journal of Institute of Forestry, Nepal, 14,* 91–97.

Chavez, R. D., Taulicorpuz, V., Baldosoriano, E., & Al., E. (2014). Climate change and indigenous peoples. *Nordic Journal of Human Rights, 4,* 235–237.

Chen, H., Liu, J., Xue, T., & Wang, R. Q. (2012). Roads accelerate the invasion process of alien species. *Advanced Materials Research, 347–353,* 1483–1487.

Cheng, G., & Tonghua W. (2007). Responses of permafrost to climate change and their environmental significance, Qinghai-Tibet Plateau. *Journal of Geophysical Research, 112*(F2), F02S03. http://doi.wiley.com/10.1029/2006JF000631. 17 Aug. 2016.

Cheng, S. (2007). Heavy metal pollution in China: Origin, pattern and control—A state-of-the-art report with special reference to literature published in Chinese journals. *Environmental Science and Pollution Research, 14,* 489–489.

Chettri, N., & Sharma, E. (2016). Reconciling the mountain biodiversity conservation and human wellbeing: Drivers of biodiversity loss and new approaches in the Hindu-Kush Himalayas. *Proceedings of the Indian National Science Academy, 82*(1), 150–158.

China News Service. (2017). How to recover the Jiuzhai Velley National Park after earthquake remains controversial. http://news.china.com/news100/11038989/20170817/31114873_all.html.

Cohen, J. I. (2005). Poorer nations turn to publicly developed GM crops. *Nature Biotechnology, 33,* 27–33.

Collins, A. E. (2001). Health ecology, land degradation and development. *Land Degradation and Development, 12*(3), 237–250.

Cong, Z. Y., Kang, S. C., Zhang, Y. L., & Li, X. D. (2010). Atmospheric wet deposition of trace elements to central Tibetan Plateau. *Applied Geochemistry, 25,* 1415–1421.

Cook, C. (2005). *Joining the Mainstream: Impact of Transport Investment on Poverty Reduction—RETA 5947.' Presented at the ADBI Workshop on Transport Infrastructure and Poverty Reduction*, ADB Manila, 18–22 July 2005.

Cordaux, R., Weiss, G., Saha, N., & Stoneking, M. (2004). The northeast Indian passageway: A barrier or corridor for human migrations? *Molecular Biology and Evolution, 21*(8), 1525–1533.

Cui, X., & Graf, H. F. (2009). Recent land cover changes on the Tibetan Plateau: A review. *Climate Change, 94,* 47–61.

Cui, P., Wei, F., Chen, X., et al. (2008). Geo-hazards in Wenchuan earthquake area and countermeasures for disaster reduction. *Bulletin of the Chinese Academy of Sciences, 23*(4), 317–323.

Cui, P., Chen, R., Xiang, L., & Su, F. (2014). Risk Analysis of mountain hazards in Tibetan Plateau under global warming. *Advances in Climate Change Research, 10*(2), 103–109.

Dame, J. (2015). Multilokalität im Himalaya: Diversifizierung der Lebenssicherung und neue Mobilität in Ladakh, Nordindien. In J. Poerting, & M. Keck (Eds.), *Aktuelle Forschungsbeiträge zu Südasien*. Geographien Südasiens, Vol. 3, 37–40.

Dangi, M. B., Pretz, C. R., Urynowicz, M. A., Gerow, K. G., & Reddy, J. M. (2011). Municipal solid waste generation in Kathmandu, Nepal. *Journal of Environmental Management, 92,* 240–249. https://doi.org/10.1016/j.jenvman.2010.09.005.

Dar, P. A., Reshi, Z. A., & Shah, M. A. (2015). Roads act as corridors for the spread of alien plant species in the mountainous regions: A case study of Kashmir Valley, India. *Tropical Ecology, 56*(2), 49-56.

Das, P. (2008). Evolution of the road network in Northeast India: Drivers and brakes. *Strategic Analysis, 33*(1), 101–116.

Das, A., & Kumar, M. (2015a). Arsenic enrichment in the groundwater of Diphu, Northeast India: Coupled application of major ion chemistry, speciation modeling, and multivariate statistical techniques. *Clean-Soil Air Water, 43,* 1501–1513.

Das, A., & Kumar, M. (2015b). Arsenic enrichment in the groundwater of Diphu, Northeast India: Coupled application of major ion chemistry, speciation modeling, and multivariate statistical techniques. *Clean-Soil Air Water, 43,* 1501–1513.

De Schutter, O. (2010): Addressing concentration in food supply chains. In *The role of competition law in tackling the abuse of buyer power. Special Rapport on the Right to Food.* Briefing Note 03, December 2010. http://www.ohchr.org/Documents/Issues/Food/BN3_SRRTF_Competition_ENGLISH.pdf.

Debyani, M. R., & Neeta, S. (2012). GM crops in India with reference to Bt cotton: Opportunities and challenges. *Journal of Environmental Research and Development, 7,* 188–193.

DeFries, R. S., et al. (2010). Deforestation driven by urban population growth and agricultural trade in the twenty-first century. *Nature Geoscience, 3*(3), 178–181. http://www.nature.com/doifinder/10.1038/ngeo756.

Delmer, D. P. (2005). Agriculture in the developing world: Connecting innovation in plant research to downstream application. *Proceedings of the National Academy of Sciences, 102,* 15739–15746.

Demenge, J. (2012). *The political ecology of road construction in Ladakh.* University of Sussex.

DeSherbinin, A., et al. (2007). Population and Environment. *Annual Review of Environment and Resources, 32*(1), 345–373. http://www.ncbi.nlm.nih.gov/pmc/articles/PMC2792934/.

Devi, H. T., Ghosh, C. K., Datta, B. K., Dasgupta, R., Mukhopadhayay, S. K., Mandal, T. K., et al. (2010). Arsenic exposure on bovine health and environmental pollution with special emphasis on groundwater system in Manipur. *Indian Journal of Animal Sciences, 80,* 642–646.

Dhanai, R., Negi, R. S., Singh, S., & Parmar, M. K. (2015). Fuelwood consumption by villagers in different altitudinal gradient: A case of Takoligad watershed of Garhwal Himalaya, India. *International Journal of Current Engineering and Technology, 5,* 72–80.

Dietz, T., Rosa, E. A., & York, R. (2007). Driving the human ecological footprint. *Frontiers in Ecology and the Environment, 5* (1), 13–18. http://www.esajournals.org/doi/abs/10.1890/1540-9295(2007)5[13:DTHEF]2.0.CO;2.

Ding, Y., Liu, S., Li, J., & Shangguan, D. (2006). The retreat of glaciers in response to recent climate warming in Western China. *Annals of Glaciology, 43,* 97–105.

Ding, J., Mack, R. N., Lu, P., Ren, M., & Huang, H. (2008). China's booming economy is sparking and accelerating biological invasions. *BioScience, 58,* 317–324.

Dobhal, P. K., Kohli, R. K., & Batish, D. R. (2010). Evaluation of the impact of *Lantana camara* L. invasion, on four major woody shrubs, along Nayar river of Pauri Garhwal, in Uttarakhand Himalaya. *International Journal of Biodiversity and Conservation, 2,* 155–161.

Dobhal, P. K., Kohli, R. K., & Batish, D. R. (2011). Impact of *Lantana camara* L. invasion on riparian vegetation of Nayar region in Garhwal Himalayas (Uttarakhand, India). *Journal of Ecology and The Natural Environment, 3,* 11–22.

Dogra, K. S., Kohli, R. K., & Sood, S. K. (2009). An assessment and impact of three invasive species in the Shivalik hills of Himachal Pradesh, India. *International Journal of Biodiversity and Conservation, 1,* 4–10.

Dong, S. K., Lassoie, J. P., Yan, Z. L., Sharma, E., Shrestha, K. K., & Pariya, D. (2007). Indigenous rangeland resource management in the mountainous areas of northern Nepal: A case study from the Rasuwa District. *The Rangeland Journal, 29,* 149–160.

Dong, S. K., Lassoie, J., Shrestha, K. K., Zhaoli, Y., Sharma, E., & Pariyar, D. (2009). Institutional development for sustainable rangeland resource and ecosystem management in mountainous areas of northern Nepal. *Journal of Environmental Management, 90,* 994–1003.

Dong, S. K., Wen, L., Zhu, L., & Li, X. Y. (2010). Implication of coupled natural and human systems in sustainable rangeland ecosystem management in HKH. *Frontier in Earth Science of China, 4*(1), 42–50.

Donnelly, L. J. (2004). Geological investigations at a high altitude, remote coal mine on the Northwest Pakistan and Afghanistan frontier, Karakoram Himalaya. *International Journal of Coal Geology, 60,* 117–150.

Dorji, T. (2001). Sustainability of tourism in Bhutan. *Journal of Bhutan Studies, 3,* 84–104.

Duan, A. M., Wu, G. X., Zhang, Q., & Liu, Y. M. (2006). New proofs of the recent climate warming over the Tibetan Plateau as a result of the increasing greenhouse gases emissions. *Chinese Science Bulletin, 51*(11), 1396–1400.

DWIDP. (2005). Disaster review 2004, Series XII. Kathmandu: Department of Water Induced Disaster Prevention, Nepal.

Evans A., Hanjra M. A., Jiang Y., Qadir M., & Drechsel P. (2012). Water pollution in Asia: The urgent need for prevention and monitoring. Water Quality. Global Water Forum. UNESCO.

Fang, Y., & Ying, B. (2016). Spatial distribution of mountainous regions and classifications of economic development in China. *Journal of Mountain Science, 13*(6), 1120–1138.

Fischer, G., Shah, M., & van Velthuizen, H. (2002). Climate change and agricultural vulnerability. Vienna, IIASA. http://adapts.nl/perch/resources/climateagri.pdf.

Fischer, K., Ekener-Petersen, E., Rydhmer, L., & Björnberg, K. E. (2015). Social impacts of GM crops in agriculture: A systematic literature review. *Sustainability, 7*(7), 8598–8620.

Foggin, J. M. (2008). Depopulating the Tibetan grasslands: National policies and perspectives for the future of Tibetan herders in Qinghai Province, China. *Mountain Research and Development, 28* (1), 26–31.

Foggin, J. M. (2016). Conservation issues: Mountain ecosystems. *Earth Systems & Environmental Sciences* (Reference Module online). Science Direct, Elsevier Publishing Company. https://doi.org/10.1016/b978-0-12-409548-9.09199-5.

Foggin, J. M. (2018). Environmental conservation in the Tibetan Plateau region: Lessons for China's Belt & Road Initiative in the mountains of Central Asia. *Land, 7,* 52. http://www.mdpi.com/2073-445X/7/2/52.

Foggin, J. M., & Phillips, J. (2013). Horizontal policy analysis: A tool to promote sustainable livelihoods development with implications for ecological resettlement and other major development programs in the Tibetan Plateau region. In Å. Kolås, & Zhaluo (Eds.), Pastoralism in contemporary China: Policy and practice (pp. 3–30). Beijing, China: Social Science Academic Press. [In Chinese].

Forster, D., Andres, C., Verma, R., Zundel, C., Messmer, M. M., et al. (2013). Yield and economic performance of organic and conventional cotton-based farming systems—Results from a field trial in India. *PLoS One, 8*(12), e81039. https://doi.org/10.1371/journal.pone.0081039.

Fort, M. (2017): The mountain environments facing climate change: Disorders or/and opportunities? In *Lecture at the International Conference of LabEx ITEM: The Mountain Environments Facing Climate Change: Disorders or/ and Opportunities?*, 11–13 Jan. 2017, Grenoble.

Fort, M., & Adhikari, B. R. (2018). Pokhara (central Nepal): A dramatic, yet geomorphologically active environment vs. a dynamic, rapidly developing city. In M. J. Thornbush, & C. D. Allen (Eds.), *Urban geomorphology: landforms and processes in cities*. Chap. 9. Elsevier. (forthcoming).

Fourastie, J. (1989 [1949]). *Le Grand Espoir du XXe siècle. Progrès technique, progrès économique, progrès social*. Paris: PUF.

Frey, K. E., Ghimire, B., & Panday, P. K. (2011). Detection of the timing and duration of snowmelt in the Hindu Kush-Himalaya using QuikSCAT, 2000–2008. *Environmental Research Letters, 6*(2), 24007–24013.

Fujita, K., Suzuki, R., Nuimura, T., & Sakai, A. (2008). Performance of ASTER and SRTM DEMs, and their potential for assessing glacial lakes in the Lunana region, Bhutan Himalaya. *Journal of Glaciology, 54,* 220–228.

Fukui, K., Fujii, Y., Ageta, Y., & Asahi, K. (2007). Changes in the lower limit of mountain permafrost between 1973 and 2004 in the Khumbu Himal, the Nepal Himalayas. *Global and Planetary Change, 55*(4), 251–256.

Fürer-Haimendorf, C. V. (1975). *Himalayan traders: Life in highland Nepal*. London: Cox & Wyman Ltd.

Fusun, S., Jinniu, W., Tao, L., Yan, W., Haixia, G., & Ning, W. (2013). Effects of different types of vegetation recovery on runoff and soil erosion on a Wenchuan earthquake-triggered landslide, China. *Journal of Soil and Water Conservation, 68*(2), 138–145.

Gautam, B. P. (2011). Tourism and economic growth in Nepal. *NRB Economic Review, 23*(2), 18–29.

Gayden, T., Cadenas, A. M., Regueiro, M., Singh, N. B., Zhivotovsky, L. A., Underhill, P. A., et al. (2007). The Himalayas as a directional barrier to gene flow. *The American Journal of Human Genetics, 80* (5), 884–894.

Gerlitz, J. Y., Hunzai, K., Hoermann, B. (2012). Mountain Poverty in the Hindu-Kush Himalayas. *Canadian Journal of Development Studies, 33,* 250–265.

Gerlitz, J.-Y., et al. (2015). A multidimensional poverty measure for the Hindu Kush-Himalayas, applied to selected districts in Nepal. *Mountain Research and Development, 35*(3), 278–288.

Gerwin, M., & Bergmann, C. (2012). Geopolitical relations and regional restructuring: The case of the Kumaon Himalaya, India. *Erdkunde*, 91–107.

Ghosh, P. (2007). Urbanization—A potential threat to the fragile Himalayan environment. *Current Science, 93*(2), 126.

Gilani, H., Shrestha, H. L., Murthy, M. S. R., Phuntso, P., Pradhan, S., Bajracharya, B., et al. (2015). Decadal land cover change dynamics in Bhutan. *Journal of Environmental Management, 148,* 91–100.

Gilani H., Qamer F. M., Sohail M., Uddin K., Jain A., & Wu, N. (2017). Review of ecosystem monitoring in Nepal and evolving earth observation technologies. In A. N. Li, W. Deng, & W. Zhao (Eds.), *Land cover change and its eco-environmental responses in Nepal* (pp. 165–184). Singapore: Springer Nature.

Glorioso, R. S., & Moss, L. A. G. (2006). Santa Fe, a fading dream: 1986 profile and 2005 postscript. In L. A. G. Moss (Ed.), *The amenity migrants: Seeking and sustaining mountains and their cultures*. Chap. 5: 73–93. Wallingford/UK: CABI.

GoIRA (Government of the Islamic Republic of Afghanistan). (2015). The state of Afghan cities 2015, Vol. 1. Kabul: GoIRA. http://unhabitat.org/books/soac2015/ (accessed 4.9.2016).

Goodall, S. K. (2004). Rural-to-urban migration and urbanization in Leh, Ladakh. *Mountain Research and Development, 24*(3), 220–227.

Grabherr, G., Gottfried, M., & Pauli, H. (1994). Climate effects on mountain plants. *Nature, 369*(6480), 448.

Grocke, M. U. (2016). *On the road to better health? Impacts of new market access on food security, nutrition, and well-being in Nepal, Himalaya*. Doctoral dissertation, University of Montana.

Gruschke, A. (2011). From Yak Herders to Yartsa Traders. Tibetan Nomads and new market options in Qinghai's Yushu region. *China Tibetology, 3,* 95–118.

Global Witness. (2015). *Jade: Myanmar's big state secret*. London: Global Witness.

GSMA Intelligence. (2015). Mobile internet usage challenges in Asia—Awareness, literacy and local content.

Guha-Sapir, D., Hoyois, P., & Below, R. (2016). *Annual disaster statistical review 2015: The numbers and trends*. Brussels, Belgium: CRED.

Gujja, B., Chatterjee, A., Gautam, P., & Chandan, P. (2003). Wetlands and lakes at the top of the world. *Mountain Research and Development, 23*(3), 219–221.

Gurung, J., Ishiga, H., & Khadka, M. S. (2005). Geological and geochemical examination of arsenic contamination in groundwater in the Holocene Terai Basin, Nepal. *Environmental Geology, 49,* 98–113.

Gyawali, S., Sthapit, B., Joshi, B. K., Mudwari, A., & Bajracharya, J. (2006). Participatory plant breeding: A strategy of on-farm conservation and improvement of landraces. In B. R. Sthapit, P. Shrestha, & M. P. Upadhyay (Eds.), *On-farm management of agricultural biodiversity in Nepal: Good practices*. NARC/LI-BIRD/ Bioversity International, Nepal.

Haigh, M., & Rawat, J. S. (2011). Landslide causes: Human impacts on a Himalayan landslide swarm. *Belgian Journal of Geography, 3–4,* 201–220.

Handa, O. C. (2001). Buddhist Western Himalaya: A politico-religious history (pp. 1–400). Indus Publishing Company.

Hanjra, M. A., & Qureshi, M. E. (2010). Global water crisis and future food security in an era of climate change. *Food Policy, 35*(5), 365–377.

Harris, R. B. (2010). Rangeland degradation on the Qinghai-Tibetan Plateau: A review of the evidence of its magnitude and causes. *Journal of Arid Environments, 74,* 1–12.

Harris, T. (2013). Trading places: New economic geographies across Himalayan borderlands. *Political Geography, 35,* 60–68.

Hart, R., Salick, J., Ranjitkar, S., & Xu, J. (2014). Herbarium specimens show contrasting phenological responses to Himalayan climate. *Proceedings of the National Academy of Sciences of the United States of America, 111*(29), 10615.

Harvey, D. (1982, 2006). *The limits to capital*. London, New York: Verso.

Hasan, A. B., Kabir, S., Reza, A., Zaman, M. N., Ahsan, M. A., Akbor, M. A., et al. (2013). Trace metals pollution in seawater and groundwater in the ship breaking area of Sitakund Upazilla, Chittagong, Bangladesh. *Marine Pollution Bulletin, 71,* 317–324.

Hasnain, S. I. (2007). Impact of climate change on Himalayan glaciers and Glacier lakes. In *Proceedings of Taal, World Lake Conference*.

Ho, P. (2001). Rangeland degradation in North China revisited? A Preliminary statistical analysis to validate non-equilibrium range ecology. *Journal of Development Studies, 37,* 99–133. https://papers.ssrn.com/sol3/papers.cfm?abstract_id=2168755.

Hodge, H. N. (1991). *Ancient futures: Learning from Ladakh*. Oxford: Oxford University Press.

Hodges, J., Foggin, J. M., Long, R. J., & Zhaxi, G. (2014). Globalisation and the sustainability of farmers, livestock-keepers, pastoralists and fragile habitats. *Biodiversity 15*(2–3), 109–118. https://doi.org/10.1080/14888386.2014.931247.

Hoermann, B., & Kollmair, M. (2009). *Labour migration and remittances in the Hindu Kush-Himalayan region*. Kathmandu: ICIMOD.

Hoon, V. (1996). *Living on the move. Bhotiya tribe of Himalaya* (pp. 191–205). New Delhi: Sage Publishers.

Howard, C. A. (2013). *Himalayan journeys: A mobile ethnography and philosophical anthropology: a thesis submitted in partial fulfilment of the requirements for the degree of doctor of philosophy in social anthropology*. Doctoral dissertation. Massey University, Auckland.

Hsiang, S. M., Meng, K. C., & Cane, M. A. (2011). Civil conflicts are associated with the global climate. *Nature, 476*(7361), 438.

Huang, X., Sillanpaa, M., Duo, B., & Gjessing, E. T. (2008). Water quality in the Tibetan Plateau: Metal contents of four selected rivers. *Environmental Pollution, 156,* 270–277.

Huang, X., Sillanpaa, M., Giessing, E. T., & Vogt, R. D. (2009). Water quality in the Tibetan Plateau: Major ions and trace elements in the headwaters of four major Asian rivers. *Science of the Total Environment, 407,* 6242–6254.

Huang, X., Sillanpaa, M., Giessing, E. T., Peranniemi, S., & Vogt, R. D. (2010). Environmental impact of mining activities on the surface water quality in Tibet: Gyama valley. *Science of the Total Environment, 408,* 4177–4184.

Hulme, P. E., Pysek, P., Jaros, V., Pergl, J., & Schaffner, V. M. (2013). Bias and error in understanding plant invasion impacts. *Trends in Ecology & Evolution, 28,* 212–218.

Husen M. A (2014). Impact of invasive alien fish, Nile Tilapia (*Oreochromis niloticus*) on native fish catches of sub-tropical lakes (Phewa, Begnas and Rupa) of Pokhara Valley, Nepal. In G. J. Thapa, N. Subedi, M. R. Pandey, S. K. Thapa, N. R. Chapagain, A. Rana (Eds.), *Proceedings of the international conference on invasive alien species management*. National Trust for Nature Conservation, Nepal.

ICIMOD. (2009). *Mountain biodiversity and climate change*. Kathmandu: ICIMOD.

ICIMOD. (2011). *Glacial lakes and glacial lake outburst floods in Nepal*. Kathmandu: ICIMOD.

ICIMOD. (2012). *A strategy and results framework for ICIMOD*. Kathmandu: ICIMOD.

Immerzeel, W. W., van Beek, L. P., & Bierkens, M. F. (2010). Climate change will affect the Asian water towers. *Science, 328*(5984), 1382–1385.

Ingole, N. A., Ram, R. N., Ranjan, R., & Shankhwar, A. K. (2015). Advance application of geospatial technology for fisheries perspective in Tarai region of Himalayan state of Uttarakhand. *Sustainable Water Resources Management, 1*(2), 181–187. International Disaster Database: www.cred.be.

IUCN (2017). Programmatic strategy for strengthening the presence of IUCN in Asia. Version: May 2017.

IUCN Pakistan. (2009). Environmental fiscal reform in Abbottabad: Mining and quarrying (VI + 26 pp.). IUCN Pakistan, Islamabad, Pakistan.

Ives, J. (2013). Sustainable mountain development—Getting the facts right Jack D. Ives (xv + 293 pp.). Lalitpur, Nepal: Himalayan Association for the Advancement of Science, 2013.

Ives, J. D., Shrestha, R. B., & Mool, Pradeep K. (2010). *Formation of Glacial lakes in the Hindu Kush-Himalayas and GLOF risk assessment*. Kathmandu: ICIMOD.

Jacquemet, E. (2017). Réinventer le Khumbu: la société sherpa à l'ère du « Yak Donald's ». Conférence internationale: La montagne, territoire d'innovation, Jan. 2017, Grenoble, France. https://halshs.archives-ouvertes.fr/halshs-01446577/document 14 Feb. 2017.

Jacquemet, E. (2018). Réinventer le Khumbu: la société sherpa à l'ère du « Yak Donald's ». In M.-C. Fourny (Ed.), La montagne, territoire d'innovation. LabEx ITEM, Collection 'Montagne et Innovation'. Grenoble: Presses Universitaires de Grenoble (forthcoming).

Jeelani, G., & Shah, A. Q. (2006). Geochemical characteristics of water and sediment from the Dal Lake, Kashmir Himalaya: Constraints on weathering and anthropogenic activity. *Environmental Geology, 50,* 12–23.

Jin, W. (2015). *Tibet as recipient of assistance and its sustainable development*. Nottingham.

Jin, J., Lu, S., Li, S., & Miller, L. N. (2010). *Impact of land use change on the local climate over the Tibetan Plateau*. Adv: Meteorol. https://doi.org/10.1155/2010/837480.

Jodha, N. S. (2000). Poverty alleviation and sustainable development in mountain areas: Role of highland–lowland links in the context of rapid globalisation. In M. Banskota, T. S. Papola, & J. Richter (Eds.), *Growth, poverty alleviation, and sustainable resource management in the mountain areas of South Asia* (pp. 541–570). Deutsche Stiftung fur internationale Entwicklung: Feldafing.

Jodha, N. S. (2005). Economic globalisation and its repercussions for Fragile mountains and communities in the Himalayas. In U. M. Huber, H. K. M. Bugmann, & M. A. Reasoner (Eds.), *Global change and mountain regions: An overview of current knowledge* (pp. 583–591). Dordrecht: Springer Netherlands.

Johnson, S. T., Agarwal, R. K., & Agarwal, A. (2013). Non-timber forest products as a source of livelihood option for forest dwellers: Role of society, herbal industries and government agencies. *Current Science, 104,* 440–443.

Joshi, R., Nailwal, T. K., Tewari, L. M., & Shukla, A. (2009). Exploring biotechnology for conserving Himalayan biodiversity. *Life Science Journal, 7*(3), 20–28.

Juyal, D., Thawani, V., Thaledi, S., & Joshi, M. (2014). Ethnomedical properties of Taxus wallichiana Zucc. (Himalayan Yew). *Journal of Traditional and Complementary Medicine*, *4*(3), 159–161.

Kala, C. P. (2014). Changes in traditional agriculture ecosystem in Rawain Valley of Uttarakhand state in India. *Applied Ecology and Environmental Sciences, 2*(4), 90–93.

Kalman, B. (2009). *What is culture?* Crabtree Publishing Company.

Kannan, R., Shackleton, M. C., & Shaanker, U. R. (2013). Reconstructing the history of introduction and spread of the invasive species, Lantana, at three spatial scales in India. *Biological Invasions, 15,* 1287–1302.

Kannel, P. R., Kanel, S. R., Lee, S., & Lee, Y. S. (2011). Chemometrics in assessment of seasonal variation of water quality in fresh water systems. *Environmental Monitoring and Assessment, 174,* 529–545.

Karkey, A., Jombart, T., Walker, A. W., Thompson, C. N., Torres, A., Dongol, S., et al. (2016). The ecological dynamics of fecal contamination and Salmonella Typhi and Salmonella Paratyphi A in municipal Kathmandu drinking water. *PLoS Neglected Tropical Diseases, 10,* 18.

Karki, M., et al. (2012) *Sustainable mountain development 1992, 2012, and beyond Rio + 20. Assessment report for the Hindu Kush Himalayan region*. Kathmandu: ICIMOD.

Karrar, H. H. (2013). Merchants, Markets, and the State. *Critical Asian Studies, 45*(3), 459–480. https://doi.org/10.1080/14672715.2013.829315.

Katuwal, H. B., Neupane, K. R., Adhikari, D., Sharma, M., & Thapa, S. (2015). Pangolins in eastern Nepal: Trade and ethno-medicinal importance. *Journal of Threatened Taxa, 7,* 7563–7567.

Kaur, R., Joshi, S. P., & Srivastava, M. M. (2012). Natural resource degradation in three sub-watersheds of river Tons, Uttarakhand, India. *Tropical Ecology, 53*(3), 333–343.

Khan, H. A., & Weiss, J. (2006). Infrastructure for regional Co-operation. In *Paper Presented in the Joint Workshop of ADBI and IDB in Seoul*, Nov. 15–17.

Khan, A. M., Ali, Z., Shelly, S. Y., Ahmad, Z., & Mirza, M. R. (2011). Aliens; A catastrophe for native fresh water fish diversity in Pakistan. *The Journal of Animal and Plant Sciences, 21,* 435–440.

Khan, B., Ablimit, A., Khan, G., Jasra, A. W., Ali, H., Ali, R., et al. (2016). Abundance, distribution and conservation status of Siberian ibex, Marco Polo and Blue sheep in Karakoram-Pamir Mountain area. *Journal of King Saud University-Science, 28,* 216–225.

Khawas, V. (2007). Environmental challenges and human security in the Himalaya. In *Paper at the National Seminar "Himalayas in the New Millennium: Environment, Society, Economy and Polity"*, January 2007. Centre for Himalayan Studies, North Bengal University, Darjeeling. http://www.mtnforum.org/sites/default/files/publication/files/526.pdf (accessed 4.9.2016).

Khuroo, A. A., Rashid, I., Reshi, Z., Dar, G. H., & Wafai, B. A. (2007). The alien flora of Kashmir Himalaya. *Biological Invasions, 9,* 269–292.

Khuroo, A. A., Weber, E., Malik, A. H., Dar, G. H., & Reshi, Z. A. (2010). Taxonomic and biogeographic patterns in the native and alien flora of Kashmir Himalaya, India. *Nordic Journal of Botany, 28,* 685–696.

Khuroo, A. A., Weber, E., Malik, A. H., Reshi, Z. A., & Dar, G. H. (2011). Altitudinal distribution patterns of the native and alien woody flora in Kashmir Himalaya, India. *Environmental Research, 111,* 967–977.

Khuroo, A. A., Reshi, Z. A., Malik, A. H., Weber, E., Rashid, I., & Dar, G. H. (2012). Alien flora of India: Taxonomic composition, invasion status and biogeographic affiliations. *Biological Invasions, 14,* 99–113.

Kien, G. (2009). Actor-network theory: Translation as material culture. In P. Vannini (Ed.), *Material culture and technology in everyday life: Ethnographic approaches* (pp. 27–44). New York: Peter Lang Publishing.

Kohler, T., et al. (Eds.). (2001). *Mountains of the world: Mountains, energy, and transport*. Bern, Switzerland: Mountain Agenda.

Kohli, R. K., Batish, D. R., Singh, H. P., & Dogra, K. S. (2006). Status, invasiveness and environmental threats of three tropical American invasive weeds (*Parthenium hysterophorus* L., *Ageratum conyzoides* L., *Lantana camara* L.) in India. *Biological Invasions, 8,* 1501–1510.

Kosaka, Y., Saikia, B., Mingki, T., Tag, H., Riba, T., & Ando, K. (2010). Roadside distribution patterns of invasive alien plants along an altitudinal gradient in Arunachal Himalaya, India. *Mountain Research and Development, 30,* 252–258.

Kreutzmann, H. (1991). The Karakoram highway: Impact of road construction on mountain societies. *Modern Asian Studies, 25*(4), 711–736.

Kreutzmann, H. (1993). Challenge and response in the Karakoram: Socioeconomic transformation in Hunza, Northern Areas, Pakistan. *Mountain Research and Development, 13*(1), 19–39.

Kreutzmann, H. (2004). Accessibility for High Asia: Comparative perspectives on northern Pakistan's traffic infrastructure and linkages with its neighbours in the Hindu-Kush-Karakoram-Himalaya. *Journal of Mountain Science, 1*(3), 193–210.

Krugman, P. R. (1991). Increasing returns and economic geography. *Journal of Political Economy, 99*(3), 483–499.

Kumar, K. R., Sahai, A. K., Kumar, K. K., Patwardhan, S. K., Mishra, P. K., Revadekar, J. V., et al. (2006). High-resolution climate change scenarios for India for the 21st century. *Current Science, 9* (3), 334–345.

Kumar, V. P., Rajpoot, A., Mukesh, S. M., Kumar, D., & Goyal, S. P. (2016). Illegal trade of Indian Pangolin (Manis crassicaudata): Genetic study from scales based on mitochondrial genes. *Egyptian Journal of Forensic Sciences*. http://dx.doi.org/10.1016/j.

Kunwar, R. M. (2003). Invasive alien plants and Eupatorium: Biodiversity and livelihood. *Himalayan Journal of Sciences, 1,* 129–133.

Kunwar, R. M., & Acharya, R. P. (2013). *Impact assessment of invasive plant species in selected ecosystems of Bhadaure Tamagi VDC*. IUCN Nepal: Kaski.

Lama, G., Ozawa, M., & Lama, P. (2010). *Geographic isolation and poverty among Indigenous peoples in Nepal*. The Himalayan Research Papers Archive. https://ejournals.unm.edu/index.php/nsc/article/view/624. Accessed 11 Sept. 11, 2016.

LeBaron, Michelle. (2003). *Bridging Cultural Conflicts: A New Approach for a Changing World*. Jossey-Bass, San Francisco, CA.

Le Monde (23 May 2015) Manifestation au Burkina Faso contre les OGM de Monsanto. La société civile burkinabè organise une marche à Ouagadougou pour défendre sa souveraineté alimentaire, dans le cadre de la journée mondiale de résistance aux OGM. http://www.lemonde.fr/afrique/article/2015/05/23/manifestation-au-burkina-faso-contre-les-ogm-de-monsanto_4639357_3212.html. Accessed 12 Dec. 2017.

Leimgruber, P., Kelly, D. S., Steininger, M., Brunner, J., Müller, T., & Songer, M. (2005). Forest cover change patterns in Myanmar (Burma) 1990–2000. *Environmental Conservation, 32*(4), 356–364.

Li, T. (1990). *Landslide management in the mountain area of China* (p. 50), ICIMOD Kathmandu, Occasion Paper no. 15.

Li, J. J., & Fang, X. M. (1999). Uplift of the Tibetan Plateau and environmental changes. *Chinese Science Bulletin, 44*(23), 2117–2124.

Li, J., & Lu, Z. (2014). Snow leopard poaching and trade in China 2000–2013. *Biological Conservation, 176,* 207–211.

Li Q and Xue Y. (2010). Simulated impacts of land cover change on summer climate in the Tibetan Plateau. *Environ Res Lett, 5,* 15102. http://doi.org/10.1088/1748-9326/5/1/015102.

Li, J., Wang, W., Hu, G., & Wei, Z. (2010). Changes in ecosystem service values in Zoige Plateau, China. *Agric. Ecosyst. Environ., 139,* 766–770.

Li, Y., Gongbuzeren, B., & Li, W. J. (2014). *A review of China's rangeland management policies*. IIED Country Report. IIED, London.

Li, C. D., Zhang, Q. G., Kang, S. C., Liu, Y. Q., Huang, J., Liu, X. B., et al. (2015). Distribution and enrichment of mercury in Tibetan lake waters and their relations with the natural environment. *Environmental Science and Pollution Research, 22,* 12490–12500.

Li, D. G., Li, Z. X., Hu, J. S., Lin, Z. X., & Li, X. F. (2016). Polymorphism analysis of multi-parent advanced generation inter-cross (MAGIC) populations of upland cotton developed in China. *Genetics and Molecular Research, 15*(4), gmr15048759.

Li, A. N., Deng, W., & Zhao, W. (Eds.). (2017). *Land cover change and its eco-environmental responses in Nepal*. Singapore: Springer Nature.

Liu, J., et al. (2009). Climate warming and growth of high-elevation inland lakes on the Tibetan Plateau. *Global and Planetary Change, 67*(3–4), 209–217.

Liu, J., Kang, S., Gong, T., & Lu, A. (2010). Growth of a high-elevation large inland lake, associated with climate change and permafrost degradation in Tibet. *Hydrology and Earth System Sciences, 14*(3), 481–489.

Liu, L., Zhang, X. Y., & Zhong, T. Y. (2016a). Pollution and health risk assessment of heavy metals in urban soil in China. *Human and Ecological Risk Assessment, 22,* 424–434.

Liu, Z., Jiang, Z., Fang, H., Li, C., Mi, A., Chen, J., et al. (2016b). Perception, price and preference: Consumption and protection of wild animals used in traditional medicine. *PLoS One, 11*(3), e0145901. https://doi.org/10.1371/journal.pone.0145901.

Loewen, M., Kang, S., Armstrong, D., Zhang, Q., Tomy, G., & Wang, F. (2007). Atmospheric transport of mercury to the Tibetan Plateau. *Environmental Science and Technology, 41,* 7632–7638.

Lowe, S., Browne, M., Boudjelas, S., De Poorter, M. (2000). *100 of the World's worst invasive alien species: A selection from the global invasive species database* (12 pp). Published by The Invasive Species Specialist Group (ISSG) a specialist group of the Species Survival Commission (SSC) of the World Conservation Union (IUCN).

Lu, T., Zeng, H. C., Luo, Y., Wang, Q., Shi, F. S., Sun, G., et al. (2012). Monitoring vegetation recovery after China's May 2008 Wenchuan earthquake using Landsat TM time-series data: A case study in Mao County. *Ecological Research, 27,* 955–966.

Luo, J., Tang, R. G., She, J., Chen, Y. C., Gong, Y. W., Zhou, J., et al. (2013). The chromium in timberline forests in the eastern Tibetan Plateau. *Environmental Science: Processes and Impacts, 15,* 1930–1937.

Luo, J., Tang, R. G., Sun, S. Q., Yang, D. D., She, J., & Yang, P. J. (2015). Lead distribution and possible sources along vertical zone spectrum of typical ecosystems in the Gongga Mountain, eastern Tibetan Plateau. *Atmospheric Environment, 115,* 132–140.

Lutz, A. F., Immerzeel, W. W., Shrestha, A. B., & Bierkens, M. F. P. (2014). Consistent increase in High Asia's runoff due to increasing glacier melt and precipitation. *Nature Climate Change, 4* (7), 587–592.

Maikhuri, R. K., Rawat, L. S., Negi, V. S., Purohit, V. K., Rao, K. S., & Saxena, K. G. (2011). Managing natural resources through simple and appropriate technological interventions for sustainable mountain development. *Current Science, 100*(7), 992–997.

Manandhar, S., Xenarios, S., Schmidt-Vogt, D., Hergarten C., & Foggin, M. (2018). *Climate vulnerability and adaptive capacity of mountain societies in Central Asia.* Research Report 1, Mountain Societies Research Institute, University of Central Asia, Bishkek, Kyrgyzstan. http://ucentralasia.org/Research/Item/1578.

Manfré, L. A., Hirata, E., Silva, J. B., Shinohara, E. J., Giannotti, M. A., Larocca, A. P. C., et al. (2012). An analysis of geospatial technologies for risk and natural disaster management. *ISPRS International Journal of Geo-Information, 1,* 166–185. https://doi.org/10.3390/ijgi1020166.

Marshall, A. (1890). *Principles of economics*. London: Macmillan.

Masoodi, A., & Khan, F. A. (2012). Invasion of alligator weed (*Alternanthera philoxeroides*) in Wular Lake, Kashmir, India. *Aquatic Invasions, 7,* 143–146.

McCauley, M. (2016). *Afghanistan and central Asia: A modern history* (pp. 1–51). London: Routledge.

Merz, J., Nakarmi, G., Shrestha, S., Dahal, B. M., Dongol, B. S., Schaffner, M., et al. (2004). Public water sources in rural watersheds of Nepal's Middle Mountains: Issues and constraints. *Environmental Management, 34,* 26–37.

Messerli, B., & Ives, J. D. (1997). *Mountains of the world: A global priority*. New York: Parthenon.

Ministry of Science, Technology and Environment, Government of Nepal. (2015). *Nepal earthquake 2015: Rapid environmental assessment.* Kathmandu.

Moors, E. J., & Stoffel, M. (2013). Changing monsoon patterns, snow and glacial melt, its impacts and adaptation options in northern India: Synthesis. *Science of the Total Environment, 468–469,* 162–167.

Morelli, J. (2011). Environmental sustainability: A definition for environmental professionals. *Journal of Environmental Sustainability, 1*. Rochester Institute of Technology.

Moriconi-Ebrard, F. (1993). *L'urbanisation du monde depuis 1950.* Paris: Anthropos.

Moss, L. A. G., & Glorioso, R. (Eds.). (2014). *Global amenity migration: Transforming rural culture economy and landscape.* Kaslo, British Columbia: The New Ecology Press.

Mukul, S. A., Uddin, M. B., Rashid, A. Z. M. M., & Fox, J. (2010). Integrating livelihoods and conservation in protected areas: Understanding the role and stakeholder views on prospects for non-timber forest products, a Bangladesh case study. *International Journal of Sustainable Development and World Ecology, 17,* 180–188.

Nabulo, G., Young, S. D., & Black, C. R. (2010). Assessing risk to human health from tropical leafy vegetables grown on contaminated urban soils. *Science of the Total Environment, 408,* 5338–5351.

Naitthani, P., & Kainthola, S. (2015). Impact of conservation and development on the Vicinity of Nanda Devi National Park in the North India. *JAR/RGA (Journal of Alpine Research/Revue de Géographie Alpine), 103*(3). http://rga.revues.org/3100.

Najar, I. A., & Basheer, A. (2012). Assessment of seasonal variation in water quality of Dal Lake (Kashmir, India) using multivariate statistical techniques. In C. A. Brebbia (Ed.), *Water pollution Xi.* WIT transactions on ecology and the environment (pp. 123–134). Southampton: Wit Press.

Najar, I. A., & Khan, A. B. (2012). Assessment of water quality and identification of pollution sources of three lakes in Kashmir, India, using multivariate analysis. *Environmental Earth Sciences, 66,* 2367–2378.

Narain, V., & Singh, A. K. (2017): Lost in transition: Perspectives, processes and transformations in (Peri) urbanizing India. In *Paper Presented at Asian Regional Conference on Peri-Urbanization: Emerging Issues and Practices*, Tongji University, Shanghai, China. 8–10 May 2017.

Negi, V. S., Maikhuri, R. K., & Rawat, L. S. (2011). Non-timber forest products (NTFPs): A viable option for biodiversity conservation and livelihood enhancement in central Himalaya. *Biodiversity and Conservation, 20,* 545–559.

Negi, G. C. S., Samal, P. K., Kuniyal, J. C., Kothyar, B. P., Sharma, R. K., & Dhyani, P. P. (2012). Impacts of climate change on western Himalayan mountain ecosystems: An overview. *Tropical Ecology, 53,* 345–356.

Nelson, G. C., Bennett, E., Berhe, A. A., Cassman, K., DeFries, R., Dietz, T., et al. (2006). Anthropogenic drivers of ecosystem change: An overview. *Ecology and Society, 11*(2), 29.

Nepal, S. (2005). Tourism and remote mountain settlements: Spatial and temporal development of tourist infrastructure in the Mt Everest region, Nepal. *Tourism Geographies: An International Journal of Tourism Space, Place and Environment, 7,* 205–227.

Nepal, S. (2011). Mountain tourism and climate change: Implications for the Nepal Himalaya. *Nepal Tourism & Development Review, 1,* 1–14.

Nepal, R., & Henning, T. (2013). *Remittances and livelihood strategies: A case study in Eastern Nepal.* Kassel: Kassel University Press.

Nepal, S., & Shrestha, A. B. (2015). Impact of climate change on the hydrological regime of the Indus, Ganges and Brahmaputra river basins: A review of the literature. *International Journal of Water Resources Development, 31*(2), 201–218.

Nibanupudi, H. K., & Rawat, P. K. (2012). Environmental concerns for DRR in the HKH region. In A. K. Gupta & S. S. Nair (Eds.), *Ecosystem approach to disaster risk reduction.* New Delhi: National Institute of Disaster Management.

Niggli, U. (2016). Interview with the research director of FibL, Urs Niggli, in the German daily journal TAZ and comments on it. http://www.taz.de/!5290509/; https://www.organicconsumers.org/news/crispr-gmo-technology-needs-no-regulation-says-usda; https://translate.google.com/translate?hl=en&sl=de&u=; http://bio-markt.info/berichte/ein-interview-mit-folgen-fibl-direktor-urs-niggli-bringt-mit-seiner-sicht-auf.html&prev=search. Accessed 12 Dec. 2017.

Nüsser, M. (Ed.). (2014). *Large dams in Asia. Contested environments between technological hydroscapes and social resistance*. Springer.

Olschak, B. C., Bührer, E. M., Gansser, A., & Gruschke, A. (1987). *Himalayas: Growing mountains, living myths, migrating peoples* (pp. 1–20). New York: Facts on File Publishers.

Osman, K. T. (2014). *Soil degradation, conservation and remediation*. Dordrecht: Springer Netherlands. http://link.springer.com/10.1007/978-94-007-7590-9.

Padmanaban, G. (2003). Growth of biotechnology in India. *Current Science, 85*(6), 712–719.

Pandit, M. K. (2009). Other factors at work in the melting Himalaya: follow-up to Xu et al. *Conservation Biology: The Journal of the Society for Conservation Biology, 23*(6), 1346–1347.

Pathak, D., Gajurel, A. P., & Mool, P. K. (2010). *Climate change impacts on hazards in the eastern Himalayas*. Kathmandu: ICIMOD.

Paudyal, S. (2012). Does tourism really matter for economic growth? Evidence from Nepal. *Nepal Rastra Bank Economic Review, 24*(1), 48–66.

Perlik, M. (2011). Alpine gentrification: The mountain village as a metropolitan neighborhood. *JAR/RGA, 99*(1). http://rga.revues.org/1370.

Perlik, M. (2018). *The spatial and economic transformation of mountain regions. Mountains as indicators for a new quality of territorial disparities*. Abington: Routledge (in preparation).

Piccio, L. (2015). *India's 2015–16 foreign aid budget: Where the Money is going?* DEVEX (blog). https://www.devex.com/news/india-s-2015-16-foreign-aid-budget-where-the-money-is-going-85666. Accessed 11 Sept. 2016.

Pimentel, D., McNair, S., Janecka, J., Wightman, J., Simmonds, C., O'Connell, C., et al. (2001). Economic and environmental threats of alien plant, animal, and microbe invasions. *Agriculture, Ecosystems and Environment, 84,* 1–20.

Pokhrel, D., & Viraraghavan, T. (2005). Municipal solid waste management in Nepal: practices and challenges. *Waste Management, 25,* 555–562. https://doi.org/10.1016/j.wasman.2005.01.020.

Poudel, S., & Shaw, R. (2015). Demographic changes, economic changes and livelihood changes in the HKH. In *Mountain Hazards and Disaster Risk Reduction Part of the series Disaster Risk Reduction* (pp 105–123).

Prakash, D., Verma, S., Bhatia, R. & Tiwary, B. N. (2011). Risks and precautions of genetically modified organisms. *ISRN Ecology*. https://doi.org/10.5402/2011/369573.

Price, M. F. (2015). *Mountains: A very short introduction* (p. 152). Oxford, UK: Oxford University Press.

Pu, J. C., et al. (2004). Fluctuations of the Glaciers on the Qinghai-Tibetan Plateau during the past century. *Journal of Glaciology and Geocryology, 26*(5), 517–522.

Public Eye. (2017). *Europe missed the Boat*. https://www.publiceye.ch/en/media/press-release/patents_on_seeds_europe_misses_the_boat/. Accessed 12 Dec. 2017.

Pumain, D. (1999). Quel role pour les viles petites et moyennes des regions peripheriques? *Revue de Géographie Alpine, 2,* 167–184.

Pyšek, P., Jarosik, V., & Pergl, J. (2011). Alien plants introduced by different pathways differ in invasion success: Unintentional introductions as a threat to natural areas. *PLoS One, 6*(9), e24890.

Qamer, F. M., Shehzad, K., Abbas, S., Murthy, M. S. R., Xi, C., Gilani, H., & Bajracharya, B. (2016). Mapping deforestation and forest degradation patterns in Western Himalaya, Pakistan. *Remote Sensing, 8*(5).

Qasim, M., Hubacek, K., Termansen, M., & Fleskens, L. (2013). Modelling land use change across elevation gradients in district Swat, Pakistan. *Regional Environmental Change, 13,* 567–581.

QBE Asia Pacific. (2017). *Asia-Pacific ICT trends 2017—The future of the information communication and technology market in Asia-Pacific*. Insurance Post & QBE Insurance Group.

Qureshi, H., Arshad, M., & Bibi, Y. (2014). Invasive flora of Pakistan: A critical analysis. *International Journal of Biosciences, 4,* 407–424.

Rai, R. K., & Rai, R. (2013). Assessing the temporal variation in the perceived effects of invasive plant species on rural livelihoods: A case of *Mikania micrantha* invasion in Nepal. *Conservation Science, 1,* 13–18.

Rai, R. K., Scarborough, H. (2015). Understanding the effects of the invasive plants on rural forest-dependent communities. *Small-Scale Forestry, 14,* 59–72.

Ramos, M. D. C. P., Ramos, N., & Moreira, A. I. D. R. (2016). *Climate change and forced environmental migration vulnerability of the Portuguese Coastline*. Springer.

Rana, R. B., Garforth, C. J., Sthapit, B. R., Subedi, A., Chaudhary, P., & Jarvis, D. I. (2007). On-farm management of rice genetic diversity: Understanding farmers' knowledge on rice ecosystems and varietal deployment. *Plant Genetic Resources Newsletter, 152,* 58–64.

Rana, J. C., Negi, K. S., Wani, S. A., Saxena, S., Pradheep, K., Kak, A., et al. (2009). Genetic resources of rice in the Western Himalayan region of India: Current status. *Genetic Resources and Crop Evolution, 56*(7), 963–973.

Rao K. S., Nautiyal N., Maikhuri R. K., & Saxena K. G. (2000). Management conflicts in the Nanda Devi Biosphere Reserve, India. *Mountain Research and Development, 20*(4), 320–323. https://doi.org/10.1659/0276-4741(2000)020%5b0320:mcitnd%5d2.0.co;2.

Rashid, M., Abbas, S. H., & Rehman, A. (2014). The status of highly alien invasive plants in Pakistan and their impact on the ecosystem: A review. *Innovare Journal of Agriculture Science, 2,* 1–4.

Rashid, I., Romshoo, S. A., Amin, M., Khanday, S. A., & Chauhan, P. (2017). Linking human-biophysical interactions with the trophic status of Dal Lake, Kashmir Himalaya, India. *Limnologica, 62,* 84–96.

Rasul, G., & Sharma, B. (2015). The nexus approach to water–energy–food security: An option for adaptation to climate change. *Climate Policy, 16*(6), 682–702.

Rauf, S., da Silva, J. T., Khan, A. A., & Naveed, A. (2010). Consequences of plant breeding on genetic diversity. *International Journal of Plant Breeding, 41,* 1–21.

Rautela, P., & Karki, B. (2015). Impact of climate change on life and livelihood of Indigenous people of Higher Himalaya in Uttarakhand, India. *American Journal of Environmental Protection, 3*(4), 112–124.

Reddy, C. S., Sreelekshmi, S., Jha, C. S., & Dadhwal, V. K. (2013). National assessment of forest fragmentation in India: Landscape indices as measures of the effects of fragmentation and forest cover change. *Ecological Engineering*, 60, 453–464.

Reinfeld, M. A. (2003). Tourism and the politics of cultural preservation: A case study of Bhutan. *Journal of Public and International Affairs, 14,* 1–26.

Reshi, Z. A., & Khuroo, A. A. (2012). Alien plant invasions in India: Current status and management challenges. *Proceedings of the National Academy of Sciences, India Section B: Biological Sciences, 82,* 305–312.

Reshi, Z., Rashid, I., Ahmad, K., & Wafai, B. A. (2008). Effect of invasion by *Centaurea iberica* on community assembly of a mountain grassland of Kashmir Himalaya, India. *Tropical Ecology, 49,* 147–156.

Riaz, A., Khan, S., Shah, M. T., Li, G., Gul, N., & Shamshad, I. (2015). *Mercury contamination in the blood, urine, hair and nails of the gold washers and its human health risk during extraction of placer gold along Gilgit*. Pakistan: Hunza and Indus rivers in Gilgit-Baltistan.

Richardson, H. W. (1995). Economies and diseconomies of agglomeration. In H. Giersch (Ed.). *Urban Agglomeration and Economic Growth* (pp. 123–155). Springer.

Richardson, S. D., & Reynolds, J. M. (2000). An overview of glacial hazards in the Himalayas. *Quaternary International, 65–66,* 31–47.

Rimal, B., Baral, H., Stork, N., Paudyal, K., & Rijal, S. (2015). Growing city and rapid land use transition: Assessing multiple hazards and risks in the Pokhara Valley, Nepal. *Land, 4*(4), 957–978.

Romanelli, C., et al. (2015). *Connecting global priorities: Biodiversity and human health—A state of knowledge review*. UNEP, CBD & WHO.

Rousselot, Y. (2015). Upstream flows of water: From the Lesotho highlands to Metropolitan South Africa. *Journal of Alpine Research | Revue de géographie alpine*, 103-3. http://journals.openedition.org/rga/3023. Accessed 12 Dec. 2017.

Roy, R., Schmidt-Vogt, D., & Myrholt, O. (2009). Humla development initiatives for better livelihoods in the face of isolation and conflict. *Mountain Research and Development, 29*(3), 211–219.

Rudel, T. K., et al. (2009). Changing drivers of deforestation and new opportunities for conservation. *Conservation Biology, 23*(6), 1396–1405. http://doi.wiley.com/10.1111/j.1523-1739.2009.01332.x.

Saba, D. S. (2001). Afghanistan: Environmental degradation in a fragile ecological setting. *International Journal of Sustainable Development and World Ecology, 8,* 279–289.

SafeNEPAL. (2016). The Gorkha Earthquake 2015: Economic impact on Nepal tourism. https://ohsnepal.wordpress.com/2016/03/09/part-i-the-gorkha-earthquake-2015-economic-impact-on-nepal-tourism/.

Saini, R. K., Swain, S., Patra, A., Khanday, G. J., Gupta, H., Purushothaman, P., et al. (2008). Water chemistry of three Himalayan Lakes: Dal (Jammu & Kashmir), Khajjiar (Himachal Pradesh) and Nainital (Uttarakhand). *Himalayan Geology, 29,* 63–72.

Sakai, A. K., Allendorf, F., & Holt, J. S. (2001). The population biology of invasive species. *Annual Review of Ecology and Systematics, 32,* 305–332.

Sang, W., Zhu, L., & Axmacher, J. C. (2010). Invasion pattern of Eupatorium adenophorum Spreng in southern China. *Biological Invasions, 12,* 1721–1730.

Sann Oo, K. (2016). Relationship between climate change impact, migration and socioeconomic development. *International Archives of the Photogrammetry Remote Sensing & Spatial Information Sciences, XLI-B4,* 743–746.

Sarma, K., & Barik, S. K. (2011). Coal mining impact on vegetation of the Nokrek Biosphere Reserve, Meghalaya, India. *Biodiversity, 12* (3), 154–164.

Saxena, K. G., Maikhuri, R. K., & Rao, K. S. (2005). Changes in agricultural biodiversity: Implications for sustainable livelihood in the Himalaya. *Journal of Mountain Science, 2*(1), 23–31.

Schwarzenbach, R. P., Egli, T., Hofstetter, T. B., von Gunten, U., & Wehrli, B. (2010). Global water pollution and human health. *Annual Review of Environment and Resources, 35,* 109–136.

Sekar, K. C. (2012). Invasive alien plants of Indian Himalayan region —Diversity and implication. *American Journal of Plant Sciences, 3,* 177–184.

Shabbir, A., & Bajwa, R. (2006). Distribution of parthenium weed (*Parthenium hysterophorus* L.), an alien invasive weed species threatening the biodiversity of Islamabad. *Weed Biology and Management, 6,* 89–95.

Shahzadi, F., Malik, M. F., & Raza, A. (2015). Genetically modified food controversies: A review. *International Journal of Scientific & Engineering Research, 6,* 2072–2089.

Shakya, T. (1999). *The dragon in the land of snows: A history of modern Tibet since 1947*. Columbia University Press.

Sharma, R. (2012). Impacts on human health of climate and land use change in the Hindu Kush-Himalayan region. *Mountain Research and Development, 32*(4), 480–486.

Sharma, H. R., & Chauhan, S. K. (2013). Agricultural transformation in trans Himalayan region of Himachal Pradesh: Cropping pattern, technology adoption and emerging challenges. *Agricultural Economics Research Review, 26,* 173–179.

Sharma, E., Chettri, N., & Oli, K. P. (2010). Mountain biodiversity conservation and management: A paradigm shift in policies and practices in the Hindu Kush-Himalayas. *Ecological Research, 25,* 905–923.

Sharma, L. N., Vetaas, O. R., Chaudhary, R. P., & Maren, I. E. (2014). Ecological consequences of land use change: Forest structure and regeneration across the forest-grassland Ecotone in Mountain Pastures in Nepal. *Journal of Mountain Science, 11,* 838–849.

Shen, S. C., Xu, G. F., Zhang, F. D., Jin, G. M., Liu, S. F., Liu, M. Y., et al. (2013). Harmful effects and chemical control study of *Mikania micrantha* H.B.K in Yunnan, Southwest China. *African Journal of Agricultural Research, 8,* 5554–5561.

Shen, S., Xu, G., Clements, D. R., Jin, G., Liu, S., Zhang, F., et al. (2015). Effects of invasive plant *Mikania micrantha* on plant community and diversity in farming systems. *Asian Journal of Plant Sciences, 14*(1), 27–33.

Sheng, J. J., Wang, X. P., Gong, P., Tian, L. D., & Yao, T. D. (2012). Heavy metals of the Tibetan top soils: Level, source, spatial distribution, temporal variation and risk assessment. *Environmental Science and Pollution Research, 19,* 3362–3370.

Shrestha, B. (2008). contribution of foreign employment and remittance to Nepalese economy. *Nepal Rastra Bank Economic Review, 20,* 1–15.

Shrestha, R. M. (2013). Energy use and greenhouse gas emissions in selected Hindu Kush-Himalayan countries. *Mountain Research and Development, 33*(3), 343–354.

Shrestha, B. B. (2016) Invasive alien plant species in Nepal. In P. K. Jha, M. Siwakoti, & S. R. Rajbhandary (Eds.), *Frontiers of botany* (pp. 269–284). Central Department of Botany, Tribhuvan University, Kathmandu.

Shrestha, A. B., & Aryal, R. (2011). Climate change in Nepal and its impact on Himalayan glaciers. *Regional Environmental Change, 11* (1), 65–77.

Shrestha, U. B., & Bawa, K. S. (2013). Trade, harvest, and conservation of caterpillar fungus (*Ophiocordyceps sinensis*) in the Himalayas. *Biological Conservation, 159,* 514–520.

Shrestha, U. B., & Bawa, K. S. (2015). Harvesters' perceptions of population status and conservation of Chinese caterpillar fungus in the Dolpa region of Nepal. *Regional Environmental Change, 15*(8), 1731–1741.

Shrestha, P. M., & Dhillion, S. S. (2003). Medicinal plant diversity and use in the highlands of Dolakha district, Nepal. *Journal of Ethnopharmacology, 86,* 81–96.

Shrestha, U. B., Gautam, S., & Bawa, K. S. (2012). Widespread climate change in the Himalayas and associated changes in local ecosystems. *PLoS One, 7*(5), e36741.

Shrestha, B. B., Joshi, S., Bisht, N., Yi, S., Kotru, R., Chaudhary, R. P. & Wu, N. (2018). *Inventory and impact assessment of invasive alien plant species in Kailash Sacred Landscape*. ICIMOD Working Paper 2018/2. Kathmandu: ICIMOD.

Sidle, R. C., Ghestem, M., & Stokes, A. (2014). Epic landslide erosion from mountain roads in Yunnan, China—Challenges for sustainable development. *Natural Hazards and Earth System Sciences, 14*(11), 3093–3104.

Simms, A., Moheb, Z., Ali, S. H., Ali, I., & Wood, T. (2011). Saving threatened species in Afghanistan: Snow leopards in the Wakhan

Corridor. *International Journal of Environmental Studies, 68*, 299–312.

Singh, S. P., Bassignana-Khadka, I., Karky, B. S., & Sharma, E. (2011). *Climate change in the Hindu Kush-Himalayas: The state of current knowledge*. Kathmandu: ICIMOD.

Singh, R., Rishi, M. S., & Sidhu, N. (2016). An overview of environmental impacts of riverbed mining in Himalayan terrain of Himachal Pradesh. *Journal of Applied Geochemistry, 18*(4), 473–479.

Song, X., Yang, G., Yan, C., Duan, H., Liu, G., & Zhu, Y. (2009). Driving forces behind land use and cover change in the Qinghai-Tibetan Plateau: A case study of the source region of the Yellow River, Qinghai Province, China. *Environmental Earth Sciences, 59*, 793–801.

Sood, A., Singh, K. D., Pandey, P., & Sharma, S. (2008). Assessment of bacterial indicators and physicochemical parameters to investigate pollution status of Gangetic river system of Uttarakhand (India). *Ecological Indicators, 8*, 709–717.

Starkey, P. (2001). Animal power: Appropriate transport in mountain areas. In T. Kohler, et al. (Eds.), *Mountains of the world: Mountains, energy, and transport*. Bern, Switzerland: Mountain Agenda.

Starr, S. F. (2006). Conflict and peace in mountain societies. In M. Price, L. Jansky, & A. A. Iatsenia (Eds.), *Key issues for mountain areas* (pp. 169–180). Tokyo: United Nations University Press.

Storper, M. (1995). The resurgence of regional economies, ten years later: The Region as a Nexus of untraded Interdependencies. *European Urban and Regional Studies*, *2*(3), 191–221. http://journals.sagepub.com/doi/abs/.

Su, B., Xiao, C., Deka, R., Seielstad, M. T., Kangwanpong, D., Xiao, J., et al. (2000). Y chromosome haplotypes reveal prehistorical migrations to the Himalayas. *Human Genetics, 107*(6), 582–590.

Sujoy, Y. H., Sattagi, H. N., & Patil, R. K. (2010). Invasive alien insects and their impact on agroecosystem. *Karnataka Journal of Agricultural Sciences, 23*, 26–34.

Takeuchi, N., et al. (2009). A shallow ice core re-drilled on the dunde ice cap, Western China: Recent changes in the Asian high mountains. *Environmental Research Letters, 4*(4), 1–6.

Tang, R. G., Luo, J., She, J., Chen, Y. C., Yang, D. D., & Zhou, J. (2015). The cadmium and lead of soil in timberline coniferous forests, Eastern Tibetan Plateau. *Environmental Earth Sciences, 73*, 303–310.

Taylor, L. H., Latham, S. M., & Woolhouse, M. E. (2001). Risk factors for human disease emergence. *Philosophical Transactions of the Royal Society B Biological Sciences, 356*(1411), 983–989.

Thapa, D., & Sein, M. K. (2010). ICT, Social capital and development: the case of a mountain region in Nepal. In *Proceedings of Third Annual SIG GlobDev Workshop*, Saint Louis, USA, December 12, 2010.

Tibet travel arrivals exceed 20 million in 2015 (translated from Chinese). (2016). http://www.tibetcn.com/news/xzly/201601283081.html. Accessed 11 Sept. 2016.

Tilman, D., Polasky, S., & Lehman, C. (2005). Diversity, productivity and temporal stability in the economies of humans and nature. *Journal of Environmental Economics and Management, 49*(3), 405–426. http://linkinghub.elsevier.com/retrieve/pii/S0095069604001160. Accessed 21 July 2014.

Tiwari, A. K. (2000a). *Infrastructure and economic development in Himachal Pradesh*. New Delhi: Indus Publishing.

Tiwari, P. C. (2000b). Land-use changes in Himalaya and their impact on the plains ecosystem: Need for sustainable land use. *Land Use Policy, 17*(2), 101–111.

Tiwari, P. C., & Joshi, B. (2015). Climate change and rural out-migration in Himalaya. *Change and Adaptation in Socio-Ecological Systems, 2*, 8–25.

Tiwari, P. C., & Joshi, B. (2016). *Rapid urban growth in mountainous regions: The case of Nainital, India*. Urbanization and Global Environmental Change Project (UGEC). UGEC Viewpoints. https://ugecviewpoints.wordpress.com/2016/03/29/rapid-urban-growth-in-mountainous-regions-the-case-of-nainital-india/. Accessed 4 Sept. 2016.

Tiwari, S., Adhikari, B., Siwakoti, M., & Subedi, K. (2005). *An inventory and assessment of invasive alien plant species of Nepal*. Nepal: IUCN-The World Conservation Union.

Toffin, G., & Pfaff-Czarnecka, J. (Eds.). (2014). *Facing globalization in the Himalayas: Belonging and the politics of the self* (Vol. 5). SAGE Publications India.

Tong, Y. D., Chen, L., Chi, J., Zhen, G. C., Zhang, Q. Q., Wang, R. N., et al. (2016). Riverine nitrogen loss in the Tibetan Plateau and potential impacts of climate change. *Science of the Total Environment, 553*, 276–284.

Tourism Council of Bhutan. (2015). Bhutan Tourism Monitor 2015. http://tcb.cms.ebizity.net/attachments/tcb_081415_btm-2015—booklet-(web).pdf. Accessed 11 Sept. 2016.

Tse-ring, K., Sharma, E., Chettri, N., & Shrestha, A. (Eds.). (2010). *Climate change vulnerability of mountain ecosystems in the Eastern Himalayas; Climate change impact a vulnerability in the Eastern Himalayas—Synthesis report*. Kathmandu: ICIMOD.

Tsing, A. (2005). *Friction: Ethnography of global encounters* (pp. 1–20). New Jersey: Princeton University Press.

Tulachan, P. M. (2001). Mountain agriculture in the Hindu Kush-Himalaya: A regional comparative analysis. *Mountain Research and Development, 21*(3), 260–267.

Uddin, K., Murthy, M. S. R., Wahid, S. M., & Matin, M. A. (2016). Estimation of soil erosion dynamics in the Koshi basin using GIS and remote sensing to assess priority areas for conservation. *PLoS One, 11*(3), 1–19.

Ueda, A., & Samdup, T. (2010). Chili transactions in Bhutan: An Economic, social and cultural perspective. *Bulletin of Tibetology*, (Special Issue) *45*(2), *46*(1).

UNCTAD World Investment Report. (2009). *Transnational corporations, agricultural production and development*. Geneva.

Underdal, A. (2010). Complexity and challenges of long term environmental governance. *Global Environmental Change, 20*(3), 386–393. https://doi.org/10.1016/j.gloenvcha.2010.02.005.

UNEP. (2002). Civil society statement on international environmental governance. In *Seventh special session of the UNEP Governing Council/GMEF*, Cartagena, Colombia, February 2002.

UNEP. (2008). International environmental governance and the reform of the United Nations. In *XVI Meeting of the Forum of Environment Ministers of Latin America and the Caribbean*.

UNFPA (United Nations Population Fund). (2007). *State of world population 2007: Unleashing the potential of urban growth*. New York: United Nations Population Fund.

UNHCR. (2015). *Global trends forced displacement in 2015*. The UN Refugee Agency.

United Nations. (2003). *Fertility, contraception and population polices*. Department of Economic and Social Affairs, Population Division. New York, United Nations. ESA/A/WP.182.

United Nations. (2012). *Changing levels and trends in mortality: The role of patterns of death by cause*. United Nations Publication.

United Nations. (2014). *World urbanization prospects: The 2014 Revision, Highlights*. Department of Economic and Social Affairs, Population Division (ST/ESA/SER.A/352).

United Nations. (2015a). *World population prospects: The 2015 Revision, custom data acquired via website*. http://esa.un.org/unpd/wpp/DataQuery/.

United Nations. (2015b). *World population prospects: The 2015 revision, key findings and advance table*.

UN-HABITAT (United Nations Human Settlements Programme). (2010). *The state of Asian cities 2010/11. Regional Office for Asia and the Pacific*. Fukuoka. http://unhabitat.org/books/the-state-of-asian-cities-201011/. Accessed 4 Sept. 2016.

Uniyal, S. K., Awasthi, A., & Rawat, G. S. (2002). Current status and distribution of commercially exploited medicinal and aromatic plants in upper Gori valley, Kumaon Himalaya, Uttaranchal. *Current Science, 82,* 1246–1252.

Upreti, B. R., Subedi, B., KC, S., Ghale, Y., & Shivakoti, S. (2016). Understanding dynamics of rural agriculture and employment in Nepal: Evidences from Ilam district of Eastern Nepal. *Nepalese Journal of Agricultural Sciences 1*(4), 132–141. http://www.hicast.edu.np/uploads/colleges/16/582eab6d9e293.pdf. Accessed 12 Dec. 2017.

Uprety, Y., Boon, E. K., Poudel, R. C., Shrestha, K. K., Rajbhandary, S., Ahenkan, A., et al. (2010). Non-timber forest products in Bardiya district of Nepal: Indigenous use, trade and conservation. *Journal of Human Ecology, 30*(3), 143–158.

Uprety, Y., Poudel, R. C., Gurung, J., Chettri, N., & Chaudhary, R. P. (2016). Traditional use and management of NTFPs in Kangchenjunga Landscape: implications for conservation and livelihoods. *Journal of Ethnobiology and Ethnomedicine, 3,* 19. https://doi.org/10.1186/s13002-016-0089-8.

Vaidya, R. (2012). Water and hydropower in the green economy and sustainable development of the Hindu Kush-Himalayan Region. *Hydro Nepal: Journal of Water, Energy and Environment, 10,* 11–19.

Valério, E., Gadanho, M., & Sampaio, J. P. (2008). Reappraisal of the *Sporobolomyces Roseus* species complex and description of sporidiobolus metaroseus Sp. Nov. *International Journal of Systematic & Evolutionary Microbiology, 58*(3), 736–741.

Wan, F.-H., & Yang, N.-W. (2016). Invasion and management of agricultural alien insects in China. *Annual Review of Entomology, 61,* 77–98.

Wang, F., et al. (2015). *China, the millennium development goals, and the post-2015 development agenda*. Beijing.

Wang, H., Zhou, X., Wan, C., Fu, H., Zhang, F., & Ren, J. (2008). Eco-environmental degradation in the northeastern margin of the Qinghai-Tibetan Plateau and comprehensive ecological protection planning. *Environmental Geology, 55*(5), 1135–1147.

Wang, H., Wang, Q., Bowler, P. A., & Xiong, W. (2016). Invasive aquatic plants in China. *Aquatic Invasions, 11,* 1–9.

Watanabe, T., Lamsal, D., & Ives, J. D. (2009). Evaluating the growth characteristics of a glacial lake and its degree of danger of outburst flooding: Imja Glacier, Khumbu Himal, Nepal. *Norsk Geografisk Tidsskrift, 63,* 255–267.

Webb, N. P., & Stokes, C. J. (2012). Climate change scenarios to facilitate stakeholder engagement in agricultural adaptation. *Mitigation and Adaptation Strategies for Global Change, 17*(8), 957–973.

Weber, E., Sun, S. G., & Li, B. (2008). Invasive alien plants in China: Diversity and ecological insights. *Biological Invasions, 10,* 1411–1429.

Wen, L., Dong, S., Li, Y., Li, X., Shi, J., Wang, Y., et al. (2013). Effect of degradation intensity on grassland ecosystem services in the alpine region of Qinghai-Tibetan Plateau, China. *PLoS One, 8*(3), e58432.

Winkler, D. (2009). Caterpillar fungus (*Ophiocordyceps sinensis*) production and sustainability on the Tibetan Plateau and in the Himalayas. *Asian Medicine, 5*(2), 291–316.

World Bank. (1994). *World development report (1994). Infrastructure for development*. Oxford and New York: Oxford University Press for the World Bank.

World Bank. (2014). *Nepal road sector assessment study*.

World Bank. (2015). *Population data in the world*. http://data.worldbank.org/indicator/SP.POP.TOTL.

World Bank. (2016). *World bank data*. http://data.worldbank.org/indicator/. Accessed 10 Sept. 2016.

World Health Organization. (2009). *Global health risks: mortality and burden of disease attributable to selected major risks*. World Health Organization.

World Water Assessment Programme. (2009). *Water in a changing world*. The United Nations World Water Development Report 3. UNESCO.

Wu, Q., Zhang, T. (2008). Recent permafrost warming on the Qinghai-Tibetan Plateau. *Journal of Geophysical Research, 113* (D13), D13108. http://doi.wiley.com/10.1029/2007JD009539.

Wu, Q., Zhang, T., & Liu, Y. (2010). Permafrost temperatures and thickness on the Qinghai-Tibet Plateau. *Global and Planetary Change, 72*(1–2), 32–38.

Wu, N., Rawat, G. S., Joshi, S., Ismail, M., & Sharma, E. (2013). *High-altitude rangelands and their interfaces in the Hindu-Kush Himalayas* (pp. 1–189). Kathmandu: ICIMOD.

Wu, N., Ismail, M., Joshi, S., Qamar, F. M., Phuntsho, K., Weikang, Y., et al. (2014a). *Understanding the transboundary Karakoram-Pamir landscape*. Feasibility and baseline studies #1. Kathmandu: ICIMOD.

Wu, N., Ismail, M., Joshi, S., Yi, S., Shrestha, R. M., & Jasra, A. W. (2014b). Livelihood diversification as an adaptation approach to change in the pastoral Hindu-Kush Himalayan region. *Journal of Mountain Science, 11,* 1342–1355.

Wu, J., Duan, D. P., Lu, J., Luo, Y. M., Wen, X. H., Guo, X. Y., et al. (2016a). Inorganic pollution around the Qinghai-Tibet Plateau: An overview of the current observations. *Science of the Total Environment, 550,* 628–636.

Wu, N., Yi, S. L., Joshi, S., & Bisht, N. (Eds.). (2016b). *Yak on the move—Transboundary challenges and opportunities for yak raising in a changing Hindu Kush Himalayan Region*. Kathmandu, Nepal: International Centre for Integrated Mountain Development.

Xinhua. (2015). *Tibet sees record high tourist arrivals in 2014*. English news, 2015-01-11. www.xinhuanet.com.

Xu, J., & Grumbine, R. E. (2014). Building ecosystem resilience for climate change adaptation in the Asian Highlands. *Climate Change, 5,* 709–718.

Xu, H., Ding, H., Li, M., Qiang, S., Guo, J., Han, Z., et al. (2006). The distribution and economic losses of alien species invasion to China. *Biological Invasions, 8,* 1495–1500.

Xu, J. C., Sharma, R., Fang, J., & Xu, Y. F. (2008a). Critical linkages between land use transition and human health in the Himalayan region. *Environment International, 34*(2), 239–247.

Xu, J. C., Yang, Y., Li, Z. Q., Tashi, N., Sharma, R., & Fang, J. (2008b). Understanding land use, livelihood and health transition of Tibetan nomads: A case from Gangga Township, Dingri County, TAR of China. *EcoHealth, 5*(2), 104–114.

Xu, J., Sharma, R., Fang, J., & Xu, Y. (2008c). Critical linkages between land-use transition and human health in the Himalayan Region. *Environment International, 34*(2), 239–247.

Xu, J. C., Yang, Y., Li, Z., Tashi, N., Sharma, R., & Fang, J. (2008d). Understanding land use, livelihoods, and health transitions among Tibetan Nomads: A case from Gongga Township, Dingri County. *Tibetan Autonomous Region of China. EcoHealth, 5*(2), 104–114.

Xu, J., Grumbine, R. E., Shrestha, A., Eriksson, M., Yang, X., Wang, Y. U. N., et al. (2009). The melting Himalayas: cascading effects of climate change on water, biodiversity, and livelihoods. *Conservation Biology, 23*(3), 520–530.

Xu, H., Qiang, S., Genovesi, P., Ding, H., Wu, J., Meng, L., et al. (2012). An inventory of invasive alien species in China. *NeoBiota, 15,* 1–26.

Xue, X., Guo, J., Han, B. S., Sun, Q. W., & Liu, L. C. (2009). The effect of climate warming and permafrost thaw on desertification in the Qinghai-Tibetan Plateau. *Geomorphology, 108,* 182–190.

Yadav, M., & Dugaya, D. (2013). Non-timber forest products certification in India: Opportunities and challenges. *Environment, Development and Sustainability, 15,* 567–586.

Yadav, I. C., Devi, N. L., Li, J., Zhang, G., & Shakya, P. R. (2016). Occurrence, profile and spatial distribution of organochlorines pesticides in soil of Nepal: Implication for source apportionment and health risk assessment. *Science of the Total Environment, 573,* 1598–1606.

Yan, Z., & Wu, N. (2005). Rangeland privatization and its impacts on the Zoige wetlands on the Eastern Tibetan Plateau. *Journal of Mountain Science, 2,* 105–115.

Yang, M., Wang, S., Yao, T., Gou, X., Lu, A., & Guo, X. (2004). Desertification and its relationship with permafrost degradation in Qinghai-Xizang (Tibet) plateau. *Cold Regions Science and Technology, 39,* 47–53.

Yang, M., et al. (2010a). Permafrost degradation and its environmental effects on the Tibetan Plateau: A review of recent research. *Earth-Science Reviews, 103*(1–2), 31–44.

Yang, M. X., Nelson, F. E., Shiklomanov, N. I., Guo, D. L., & Wan, G. N. (2010b). Permafrost degradation and its environmental effects on the Tibetan Plateau: A review of recent research. *Earth-Science Reviews, 103,* 31–44.

Yao, T., et al. (2007). Recent glacial retreat and its impact on hydrological processes on the Tibetan Plateau, China, and surrounding regions. *Arctic, Antarctic, and Alpine Research, 39*(4), 642–650.

Yasuoka, J., & Levins, R. (2007). Impact of deforestation and agricultural development on anopheline ecology and malaria epidemiology. *American Journal of Tropical Medicine and Hygeine, 76*(3), 450–460.

Yuan, G. L., Wu, M. Z., Sun, Y., Li, J., Li, J. C., & Wang, G. H. (2016). One century of air deposition of hydrocarbons recorded in travertine in North Tibetan Plateau, China: Sources and evolution. *Science of the Total Environment, 560,* 212–217.

Zhang, G., et al. (2011). Monitoring lake level changes on the Tibetan Plateau using ICESat altimetry data (2003–2009). *Remote Sensing of Environment, 115*(7), 1733–1742.

Zhang, F., Yan, X. D., Zeng, C., Zhang, M., Shrestha, S., Devkota, L. P., et al. (2012a). Influence of traffic activity on heavy metal concentrations of roadside farmland soil in mountainous areas. *International Journal of Environmental Research, 9,* 1715–1731.

Zhang, H., Wang, Z. F., Zhang, Y. L., & Hu, Z. J. (2012b). The effects of the Qinghai-Tibet railway on heavy metals enrichment in soils. *Science of the Total Environment, 439,* 240–248.

Zhang, Y. L., Kang, S. C., Li, C. L., Cong, Z. Y., & Zhang, Q. G. (2012c). Wet deposition of precipitation chemistry during 2005–2009 at a remote site (Nam Co Station) in central Tibetan Plateau. *Journal of Atmospheric Chemistry, 69,* 187–200.

Zhang, H., Zhang, Y. L., Wang, Z. F., & Ding, M. J. (2013). Heavy metal enrichment in the soil along the Delhi-Ulan section of the Qinghai-Tibet railway in China. *Environmental Monitoring and Assessment, 185,* 5435–5447.

Zhang, S. P., Hou, L. N., Wei, C. J., Zhou, X. N., & Wei, N. (2015a). Study on water quantity and quality-integrated evaluation based on the natural-social dualistic water cycle. *Polish Journal of Environmental Studies, 24,* 829–840.

Zhang, Y., Wang, L., Hu, Y., Xi, X. F., Tang, Y. S., Chen, J. H., et al. (2015b). Water organic pollution and eutrophication influence soil microbial processes, increasing soil respiration of estuarine wetlands: Site study in Jiuduansha Wetland. *PLoS One, 10,* e0126951.

Zhang, J., Liu, R., Deng, W., et al. (2016). Characteristics of landslide in Koshi River Basin, Central Himalaya. *Journal of Mountain Science, 13,* 1711–1722. https://doi.org/10.1007/s11629-016-4017-0.

Zhang, J., Regmi, A. D., Liu, R., et al. (2017). Landslides inventory and trans-boundary risk management in Koshi river basin, Himalaya. In A. Li, W. Deng, W. Zhao (Eds.), *Land cover change and its eco-environmental responses in Nepal*. Singapore: Springer.

Zhao, W. Y., Li, J. L., & Qi, J. G. (2007). Changes in vegetation diversity and structure in response to heavy grazing pressure in the northern Tianshan Mountains, China. *Journal of Arid Environments, 68*(3), 465–479.

Zhao, N., Lu, X. W., & Chao, S. G. (2014). Level and contamination assessment of environmentally sensitive elements in smaller than 100 mu m street dust particles from Xining, China. *International Journal of Environmental Research and Public Health, 11,* 2536–2549.

Zhao, N., Lu, X. W., Chao, S. G., & Xu, X. (2015). Multivariate statistical analysis of heavy metals in less than 100 μm particles of street dust from Xining,China. *Environmental Earth Sciences, 73,* 2319–2327.

Zhou, H. K., Zhao, X. Q., Zhao, L., Li, Y. N., Wang, S. P., Xu, S. X., et al. (2008). Restoration capability of alpine meadow ecosystem on Qinghai-Tibetan Plateau. *Chinese Journal of Ecology, 27,* 697–704.

Zhou, D. N., Zhang, F. P., Duan, Z. Y., Liu, Z. W., Yang, K. L., Guo, R., et al. (2013). Effects of heavy metal pollution on microbial communities and activities of mining soils in Central Tibet, China. *Journal of Food Agriculture and Environment, 11,* 676–681.

Zomer, R., & Oli, K. P. (Eds.). (2011). *Kailash sacred landscape conservation initiative—Feasibility assessment report*. Kathmandu: ICIMOD.

Unravelling Climate Change in the Hindu Kush Himalaya: Rapid Warming in the Mountains and Increasing Extremes

Coordinating Lead Authors
Raghavan Krishnan, Indian Institute of Tropical Meteorology, India, e-mail: krish@tropmet.res.in
Arun Bhakta Shrestha, International Centre for Integrated Mountain Development, Nepal,
e-mail: arun.shrestha@icimod.org (corresponding author)
Guoyu Ren, China Meteorological Administration and China University of Geosciences, China.
e-mail: guoyoo@cma.gov.cn

Lead Authors
Rupak Rajbhandari, Tribhuvan University, Nepal, e-mail: rupak.rajbhandari@gmail.com
Sajjad Saeed, Katholieke Universiteit Leuven, Belgium; Center of Excellence for Climate Change Research (CECCR), King AbdulAziz University, Jeddah, Saudi Arabia, e-mail: sajjad.saeed@kuleuven.be
Jayanarayanan Sanjay, Indian Institute of Tropical Meteorology, India, e-mail: sanjay@tropmet.res.in
Md. Abu Syed, Bangladesh Centre for Advanced Studies, Bangladesh, e-mail: abu.syed@bcas.net
Ramesh Vellore, Indian Institute of Tropical Meteorology, India, e-mail: rameshv@tropmet.res.in
Ying Xu, China Meteorological Administration, China, e-mail: xuying@cma.gov.cn
Qinglong You, Nanjing University of Information Science & Technology, China,
e-mail: qinglong.you@nuist.edu.cn
Yuyu Ren, National Climate Centre, China Meteorological Administration, China,
e-mail: renyuyu@126.com

Contributing Authors
Ashok Priyadarshan Dimri, Jawaharlal Nehru University, India, e-mail: apdimri@hotmail.com
Arthur Lutz, Future Water, the Netherlands, e-mail: a.lutz@futurewater.nl
Prasamsa Singh, Nagoya University, Japan, e-mail: prasamsasingh@gmail.com
Xiubao Sun, China University of Geosciences, China, e-mail: sun_2005009@126.com
Yunjian Zhan, China Meteorological Administration, China, e-mail: zhanyunjian@foxmail.com

Review Editor
Valerio Lucarini, Reading University, United Kingdom, e-mail: v.lucarini@reading.ac.uk

Corresponding Author
Arun Bhakta Shrestha, International Centre for Integrated Mountain Development, Nepal,
e-mail: arun.shrestha@icimod.org

P. Wester et al. (eds.), *The Hindu Kush Himalaya Assessment*,
https://doi.org/10.1007/978-3-319-92288-1_3

Contents

Chapter Overview

Key Findings

1. **In the future, even if global warming is kept to 1.5 °C, warming in the Hindu Kush Himalaya (HKH) region will likely be at least 0.3 °C higher, and in the northwest Himalaya and Karakoram at least 0.7 °C higher.** Such large warming could trigger a multitude of biophysical and socio-economic impacts, such as biodiversity loss, increased glacial melting, and less predictable water availability—all of which will impact livelihoods and well-being in the HKH. The HKH has seen significant warming in the past decades nearly equal to the global average. Elevation Dependent Warming is widely observed in the region in past as well as future projections.
2. **For the past five to six decades, the HKH have shown a rising trend of extreme warm events; a falling trend of extreme cold events; and a rising trend in extreme values and frequencies of temperature-based indices (both minimum and maximum).** The number of cold nights reduced by 1 night per decade and the number of cold days reduced by 0.5 days per decade, while the number of warm nights increased by 1.7 nights per decade and number of warm days increased by 1.2 days per decade. These changes in extremes will continue and pose even more acute challenges to adaptation.
3. **The HKH is experiencing increasing variability in western disturbances and a higher probability of snowfall in the Karakoram and western**

Himalaya, changes that will likely contribute to increases in glacier mass in those areas. This finding runs counter to many expectations in the scientific community, and more research is needed to understand the reasons for this and its potential future implications.

4. **Consensus among models for the HKH region is weak—a result of the region's complex topography and the coarse resolution of global climate models.** To improve evidence-based adaptation, improved climate models and downscaling strategies capable of capturing changes in extreme events are essential.

Policy Messages

1. **More robust climate change analysis and adaptation planning will not be possible without improved long-term hydrometeorological monitoring in the HKH.** High-altitude areas of the HKH lack long-term observational data, and the available data suffer from large inconsistencies and from high inhomogeneity. Systematic bias is also present through the urbanization effect on meteorological observations, and through the wind effect on precipitation observations.
2. **For accurate cryospheric projections, more reliable projections of elevation-dependent warming are crucial.** Although the evidence for elevation-dependent warming in the HKH is strong, the precise mechanisms underlying this phenomenon involve multiple feedbacks, such as snow-albedo interactions, water vapor—cloud—radiation interactions, aerosol forcing, and warrant further research.
3. **Policies and planning should focus on improved disaster warning systems, management and mitigation measures to address hydrometeorological extremes.** This should include better understanding of hazard and risk, end-to-end monitoring and early warning and response systems.

Historically, the climate of the HKH has experienced significant changes that are closely related to the rise and fall of regional cultures and civilizations. Studies show ***well-established*** evidence that climate drivers of tropical and extra-tropical origin—such as the El Niño-Southern Oscillation (ENSO), the North Atlantic Oscillation (NAO), Indian Ocean Dipole (IOD), the Madden-Julian Oscillation (MJO), and the Arctic Oscillation (AO)—influence the region's weather and climate on multiple spatio-temporal scales.[1]

Although the climate of the HKH has changed significantly in the past, it is projected to change more dramatically in the near future. It is ***well-established*** that the warming in the Tibetan Plateau (TP) has been comparable in magnitude to the averages for the Northern Hemisphere and the same latitudinal zone. This regional warming continued even during the global warming hiatus—the period between 1998 and 2014 when global warming appeared to have slowed down.

Generally, from the last century through the beginning of the current one, the HKH has experienced warming from 1901 to 1940; cooling from 1940 to 1970; and warming from 1970 to the present. From 1901 to 2014, annual mean surface air temperature significantly increased in the HKH, at a rate of about 0.10 °C per decade—while the warming rate over the last 50 years has been 0.2 °C per decade ($p > 0.05$). ***Well-established*** evidence suggests that extreme indices in the region have also changed over this period: occurrences of extreme cold days and nights have declined (days by 0.85 days per decade, nights by 2.40 days per decade), while occurrences of extreme warm days and nights have increased (days by 1.26 days per decade, nights by 2.54 days per decade). Warm nights have increased throughout the region, and extreme absolute temperature indices have changed significantly. Frost days show a significant declining trend in most parts of northern India and the TP. The length of the growing season has increased by 4.25 days per decade—a positive change for agriculture ($p > 0.05$). Observed precipitation trends over the HKH during last five decades are ***inconclusive***.

Evidence exists in the HKH for elevation-dependent warming (EDW), especially in the TP and its surrounding regions (***well-established***). The EDW phenomenon has been reported previously by several studies. However, the driving mechanisms of EDW are ***inconclusive***. These mechanisms call for further investigation—in part because EDW can illuminate cryosphere dynamics, and in part also because EDW makes current efforts to contain global warming all the more important for the HKH. Conference of the Parties (COP21) in Paris in December 2015 agreed to take steps towards limiting the global mean annual surface air temperature increase to well below 2.0 °C above pre-industrial levels, and to pursue efforts towards a target of 1.5 °C. By the end of the century, if average global warming is limited to 1.5 °C above the pre-industrial period, the HKH will warm by 1.80 ± 0.40 °C.

[1] http://www.climate.rocksea.org/research/indian-ocean-warming/.

While the precipitation trends for the HKH are ***established but inconclusive*** over the past century, with some analyses showing that total and extreme precipitation has increased overall over the last five decades, intense precipitation has changed markedly since 1961: rising trends appear in the intensity of annual intense precipitation and also in the frequency of annual intense precipitation day.

Ensemble outputs from the Coordinated Regional Downscaling Experiment (CORDEX) models project significant warming over the HKH region in the future (***well-established***). In the near term (2036–2065), the region is projected to warm by 1.7–2.4 °C for representative concentration pathway 4.5 (RCP4.5) and 2.3–3.2 °C for RCP8.5. In the long term (2066–2095), regional warming is projected to be 2.2–3.3 °C for RCP4.5 and 4.2–6.5 °C for RCP8.5. Increased warming during the winters is also projected. Warming during the winters is projected to warm relatively more. While the Coupled Model Intercomparison Project 5 (CMIP5) general circulation models (GCM) projections differ from (CORDEX) projections in magnitude, they do agree on trends. The TP, the central Himalayan Range, and the Karakoram will see a rise in temperature higher than average of the HKH. It is expected that EDW is projected to continue (***well-established***).

Most scenarios predict that an increase in precipitation is likely (***established but inconclusive***). Monsoon precipitation is projected to increase by 4–12% in the near future and by 4–25% in the long term. Winter precipitation is projected to increase by 7–15% in the Karakoram, but to decline slightly in the Central Himalaya.

Results on future precipitation extremes are ***inconclusive***, across studies and across the region. Increasing precipitation extremes have been projected for the Indus basin, the TP, and the Eastern Himalaya. Increases are also expected in extreme temperatures, in tropical nights, and in the length of the growing season.

Despite the evidence outlined above and throughout this chapter, analyses of past trends are subject to uncertainty because of limitations in what has been observed. One of the sources of uncertainty comes from urbanization, which has caused a systematic bias in historical temperature data (observation sites may be located in urban areas or be engulfed by them as settlements expand). Besides global warming, other regional forcing elements like anthropogenic aerosols and land-use changes appear to have influenced the South Asian monsoon precipitation variations during the post-1950s. Many multi-model projections agree on the direction of future climactic and weather changes; nevertheless, differences appear in the magnitude of the changes and in their spatial distribution. There are also inherent challenges in accurately capturing monsoon precipitation variations in climate models over the HKH region, because of the strong internal dynamics associated with precipitation processes (e.g., aerosol-cloud-radiation interactions, convective and cloud microphysical processes, representation of atmosphere-land-ocean coupling, among others) and the complexities in representing multi-scale interactions.

Climate Change and Sustainable Development Goals

The aim of SDG 13 is to "take urgent action to combat climate change and its impact." Strengthening resilience and adaptive capacity to climate-related hazards and natural disasters in all countries and integrating climate change measures into national policies, strategies and planning are two major targets of SDG 13. By understanding the past and potential future of climate change in the HKH—on the basis of robust scientific analysis—we can better comprehend the region's present and future risks. Our finding that the warming in the HKH region in the future will be greater than the global average warming supports the need to take urgent action. We can more effectively support HKH countries, helping them strengthen resilience and adaptive capacity in the face of climate-related hazards and natural disasters. And we can integrate climate change adaptation measures into national policies, strategies, and plans. It will help in achieving the HKH relevant target on integrating mountain-specific climate change measures into national policies, strategies, and planning.

3.1 Our Understanding of the HKH Climate Needs to Be Improved

The Himalayan climate is mostly alpine but varies significantly with elevation from snow-capped higher elevations to tropical/subtropical climates at lower elevations, and varied vegetation exists over the HKH. Climatically, the HKH play an important role in global weather patterns. They serve as a heat source in summer and heat sink in winter (Wu and Zhang 1998; Yanai et al. 1992; Yanai and Li 1994; Ye 1981). The HKH, together with the elevated TP, exert significant influence on the Asian summer monsoon system (Nan et al. 2009; Wu et al. 2004; Zhou et al. 2009).

The Himalaya are sensitive to climate change and variability (Shrestha and Aryal 2011; Xu et al. 2008). Most of the warming observed during the last few decades of the 20th century is attributed to the increase in anthropogenic greenhouse gas (GHG) concentrations (IPCC 2007, 2013; You et al. 2017). With increased emission of anthropogenic GHG, the cryosphere processes—coupled with the hydrological regimes of this region—are under stress from a warming climate (see Chaps. 7 and 8). Increased temperature causes more evaporation, leading to increased atmospheric moisture content, thus bringing changes in future spatial and temporal precipitation patterns. This can adversely affect the supply of water to humans and agriculture, especially in the dry season. Localized weather events over the complex topography of the HKH pose a greater risk because of cloudbursts, flash floods, snowstorms, high winds, and landslides in the region (See Chap. 11). More frequent flood-inducing rain occurrences at higher altitudes can accelerate glacier melting and flood discharge, thus posing major risks of disasters in the region. Furthermore, changes in water flow regimes have implications for hydropower generation, biodiversity systems/forestry, and the agriculture- and natural resources-based livelihoods of the HKH (see Chaps. 5, 8, and 9). Climate change can have profound consequences for mountain agriculture, agrobiodiversity, and resilience of crop diseases, since many crop species in the HKH—namely wheat, rice, and soybeans—are sensitive to increasing levels of heat-trapping GHG (e.g., Chap. 9; Hoffmann 2013; Porter et al. 2014).

Intense monsoon rainfall in northern India and western Nepal in 2013, which led to landslides and one of the worst floods in history, has been linked to increased loading of GHG and aerosols (Cho et al. 2016). Winter precipitation in the HKH is brought about by synoptic weather disturbances moving from west to east. The passage of the disturbances is directed by a westerly jet stream, which is blocked by the HKH–TP; therefore, winter snowfall is concentrated in the western side of the region (Hasson et al. 2014). Further to the east the jet stream is located south of the Himalaya in winter. Several years of drought conditions in western Nepal since 2000, which culminated in severe drought during 2008–09, have been related to natural variability and anthropogenic influences (Wang et al. 2013).

Robust estimates of the observed variability and long-term changes in climate over the HKH are inadequate owing to sparse and discontinuous observations (Ren and Shrestha 2017). Further, reliable projections of climate over the HKH are crucial for assessment of the impacts of climate change. While people in the region are adapting autonomously to current stresses, each country in the HKH must design and implement effective strategies to adapt to climate change impact to achieve economic and social progress. Adapting to long- and short-term climate-related problems requires a thorough understanding of climate changes in the past and possible changes in the future (You et al. 2017). This chapter presents a broad overview of weather and climate elements pertaining to the HKH, focusing more specifically on the linkage of large-scale drivers to climate variability in the HKH, past and present regional climate variations, and likely projections of future regional climate using high-resolution regional climate models that are capable of resolving the Himalayan topography. While this chapter takes stock of previous studies, the major part of the chapter is based on original analysis, as studies in the HKH domain (Fig. 3.1), as identified by the Hindu Kush Himalayan Monitoring and Assessment Programme (HIMAP) (Sharma et al. 2016), have not been conducted in the past. A concluding section synthesizes major gap areas and future directions for diverse and multidisciplinary solutions for climate change impact in the HKH. During the development of this chapter, the authors published seven articles based on

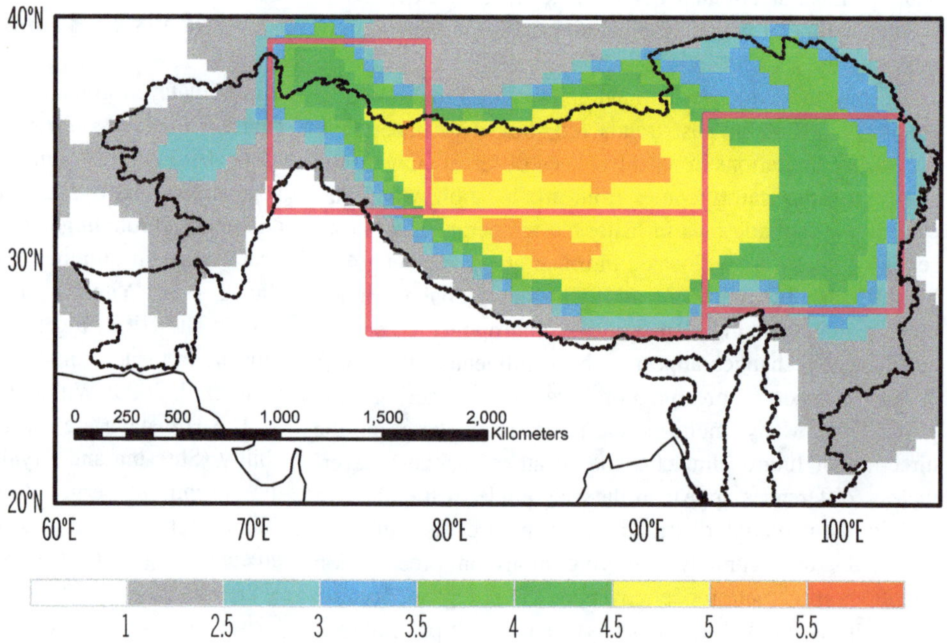

Fig. 3.1 RegCM4 elevation (in km) with three regions of interest defined by grid cells in each box above 2,500 m a.s.l. (non-greyscale): northwestern Himalaya and Karakoram (HKH1); central Himalaya (HKH2); southeastern Himalaya and Tibetan Plateau (HKH3). The Hindu Kush Himalaya (HKH) boundary is shown with a dashed line (*Source* Sanjay et al. 2017b)

the original analysis (Rajbhandari et al. 2017; Ren et al. 2017; Sanjay et al. 2017b; Sun et al. 2017b; Wu et al. 2017; You et al. 2017; Zhan et al. 2017).

3.2 Weather and Climate Mean Conditions in the HKH

3.2.1 Major Features of Climatology: Topographic Control, Seasons, and Liquid and Solid Precipitation

Topographic variations, the annual cycle of seasons, and variability of weather patterns have strong controls on the spatial pattern of temperatures across different geographic regions of the HKH. While the average summer and winter temperatures are about 30 and 18 °C in the southern foothills, the middle Himalayan valleys experience mean summer temperatures between 15 and 25 °C and very cold winters. Regions having elevations above 4800 m experience winter temperatures below freezing point and receive precipitation largely in the form of snow. The mean end-of-summer regional snow line altitude (SLA) zones in the upper Indus Basin of Pakistan range from 3000 to 5000 m a.s.l. (Hasson et al. 2014). Records of observed surface temperature from Pakistani stations in the elevated Karakoram indicate that the average maximum temperature is about 20 °C during July and the average minimum temperature is about −3 °C in February (Kapnick et al. 2014). The regions of Ladakh and Zanskar, situated on the northern flank of the Himalayan range, experience dry conditions, with surface temperatures between 3 and 35 °C in summer and between −20 and −35 °C in winter with accompanying heavy winter snowfall, thus amounting to average annual rainfall of a few centimetres. Hill stations of the western Himalaya like Shimla, Kullu Valley, Kangra, and Chamba; some regions in Uttaranchal such as Kumaon and Garhwal; and areas like Darjeeling and Sikkim in the eastern HKH, largely experience the Indian summer monsoon precipitation. Thus, the Himalayan climate exhibits diverse geographical variability, which is closely linked to the varying topographic distribution of the region (Bookhagen and Burbank 2006).

The bulk of precipitation from the southwestern Indian summer monsoon, falling as frozen precipitation at higher elevations and liquid precipitation at lower elevations and adjacent plains of the Himalaya, constitutes an important ingredient of the major Himalayan river basins and associated hydrological cycle. From the sparse rain gauge network, the southern slopes of the Himalaya typically experienced large annual precipitation totals as high as 400 cm yr^{-1} during the period of 1998–2007 (Bookhagen and Burbank 2010). The contributions of summer and winter monsoon circulation is not evenly distributed over the Himalaya, the summer (winter) rainfall is typically the greatest contributor over the southeastern (northwestern) part of the HKH. Monsoon precipitation is found to be the highest over the Siwalik and Pir Panjal ranges of the lower Himalaya, while it reduces northwards into the high Himalaya, Zanskar, Ladakh, and Karakoram ranges. Rainfall estimates from the Tropical Rainfall Measurement Mission (TRMM) satellite reveal two distinct bands of high rainfall that stretch parallel over the length of the HKH: (1) along the southern margin of the Lesser Himalaya; and (2) along the southern flank of the HKH (Bookhagen and Burbank 2006). Heavy rainfall in the HKH during summer is often associated with west–northwestward-passing synoptic-scale monsoon low-pressure systems and depressions and their interactions with the subtropical westerly winds (e.g., Vellore et al. 2015), as well as during break-monsoon situations (e.g., Dhar et al. 1984; Rao 1976; Vellore et al. 2014). Flooding in the mountainous Himalaya and adjacent low-relief areas is generally attributed to heavy or extreme rainfall events associated with synoptic climate patterns (e.g., Shrestha 2008a).

The western side of the Karakoram Himalaya is prone to large amounts of snowfall in winter from frequent passage of extra-tropical synoptic-scale disturbances known as the western disturbances, emanating from the upper tropospheric westerlies (e.g., Dimri et al. 2015; Madhura et al. 2015; see Box 3.1). Studies have found that teleconnection related to flood events—for example, the Pakistan flood of 2010 and the Russian drought at the same time—is caused by a standing Rossby wave (e.g., Lau and Kim 2012). More than 80% of annual precipitation in the central–eastern part of the Himalaya is from the summer monsoon. Thus, precipitation over the Himalayan river basins is highly variable, and the annual mean precipitation over the Indus, the Ganges, and the Brahmaputra river basins is estimated as 435, 1094, and 2143 mm, respectively (Nepal and Shrestha 2015).

3.2.2 Climate Dominated by Monsoon

The monsoon climate over much of South Asia is dominated by seasonal reversing winds that carry moist air over the Indian Ocean into South Asia during the northern hemisphere summer; whereas cool and dry winds blow southward from the Asian continent towards the Indian Ocean during the winter months. The presence of the HKH topographic barrier restricts upper-level subtropical westerly winds to latitudes poleward of 30° N during the boreal summer months, thus allowing warm and moist summer monsoon circulation to extend northward into the Indian subcontinent. The onset of southwest summer monsoon rains over northeastern India typically occurs around 6 June and later advances towards Bhutan and Nepal. The summer monsoon rains last for more than 4 months over the HKH region and typically withdraw from northeastern India round 15 October (Singh and Ranade 2010). On the other hand, the northern and northwestern regions of India experience the onset and withdrawal of summer monsoon rains typically around 8 July and 19 September, respectively (Singh and Ranade 2010). Thus, the length of the summer monsoon rainy season is significantly longer in the eastern Himalaya as compared to the western Himalaya (Shrestha et al. 2008a).

Summer monsoon precipitation over the Gangetic Valley and along the southern slopes of the Himalaya is strongly influenced by the mean position of the monsoon trough. When the monsoon depressions formed over the Bay of Bengal travel in a west–northwestward direction, precipitation is distributed across the Indian landmass (Rao 1976). During "breaks" in the Indian summer monsoon (i.e., shifting of the monsoon trough towards the southern slope of the Himalaya), the central and eastern Himalaya often experience heavy precipitation together with flooding in the Brahmaputra and its tributaries (Dhar et al. 1984; Vellore et al. 2014), while the plains of the Indian subcontinent suffer deficit rainfall (e.g., Krishnan et al. 2000, 2006, 2009; Rajeevan et al. 2010; Ramamurthy 1969). The monsoon heat low that forms over the northwestern parts of India and Pakistan during the summer plays an important role in the modification of monsoon rainfall over the northwestern parts of the HKH. The eastward-propagating upper-level mid-latitude systems associated with global teleconnection influence the monsoon heat low and associated rainfall over the northwestern HKH and Karakoram region during the summer (Saeed et al. 2010, 2011). While local factors as well as global teleconnections influence the strength of the summer monsoon circulation, rainfall predictions over the HKH beyond a few days are challenging. Intense or extreme rainfall in the HKH is influenced by orographic effects, synoptic scale systems, monsoon convection, and also by a combination of monsoon and extratropical circulation interactions (e.g., Bookhagen and Burbank 2010; Houze et al. 2007, 2011; Medina et al. 2010; Vellore et al. 2014, 2015).

Box 3.1 Westerly disturbance and its linkage with agriculture and glacier dynamics

Winter precipitation in the Himalaya is predominantly from synoptic weather systems known as western disturbances, which propagate eastward from the Mediterranean region (Dimri et al. 2015; Dimri and Chevuturi 2016; Hasson et al. 2014; Madhura et al. 2015). The western disturbances are large-amplitude wave disturbances that travel along the subtropical westerly jet stream and influence HKH precipitation during winter and early spring. Annual winter snowfall amounts are seen to range from a few hundred to several hundreds of centimetres at different elevations, with the maximum over the Karakoram. The western disturbances also significantly impact temperature patterns of the Himalaya besides precipitation. The winter climate conditions over the HKH are favourable for cultivation of wheat and other winter crops. The changes in the westerly disturbance are also believed to increase the mass of some glaciers in the Karakoram and western Himalaya, popularly known as the "Karakoram Anomaly" (e.g., Forsythe et al. 2017; Hewitt 2005), which is likely to continue into the future (e.g., Ridley et al. 2013; Krishnan et al. 2018; see Chap. 7).

3.2.3 Climate Influenced by Large-Scale Climate Elements

Recent studies highlight the role of climate drivers of both tropical and extra-tropical origin in influencing the weather and climate of the HKH on multiple spatio-temporal scales. Examples of such climate drivers are El Niño-Southern Oscillation[2] (ENSO), North Atlantic Oscillation[3] (NAO), Madden–Julian Oscillation[4] (MJO), Arctic Oscillation,[5] Indian Ocean dipole,[6] and Indian Ocean warm pool[7] (e.g., Barlow et al. 2005; Bhutiyani et al. 2009; Cannon et al. 2015, 2017; Yadav et al. 2009, 2010). Significant positive correlations have been identified between NAO and precipitation variability over the Karakoram Himalaya during winter (Archer and Fowler 2004), which are often mediated through the activity of western disturbances (Cannon et al. 2015; Dimri and Chevuturi 2016; Hasson et al. 2014). Likewise, anomalous westward shifts of the summer Pacific anticyclone during La Niña[8] episodes can be critical to severe precipitation activity over the Indo-Pakistani mountainous areas (Mujumdar et al. 2012), besides the classical pathway of ENSO–monsoon teleconnections through the equatorial Walker circulation (Krishna Kumar et al. 2006).

Despite the studies mentioned above, there is considerable ambiguity about how the HKH would respond to background

[2]https://iridl.ldeo.columbia.edu/maproom/ENSO/ENSO_Info.html.
[3]http://www.ldeo.columbia.edu/res/pi/NAO/.
[4]http://www.cpc.ncep.noaa.gov/products/precip/CWlink/MJO/MJO_1page_factsheet.pdf.
[5]https://www.ncdc.noaa.gov/teleconnections/ao/.
[6]http://www.bom.gov.au/climate/enso/history/ln-2010-12/IOD-what.shtml.
[7]http://www.climate.rocksea.org/research/indian-ocean-warming/.
[8]http://www.pmel.noaa.gov/elnino/what-is-la-nina.

changes in large-scale circulation in a warming environment, especially given the spatially heterogeneous temperature increase during recent decades. Several studies have documented the elevation dependency of the climate warming signal which gives rise to pronounced warming at higher elevations rather than at lower elevations (see Diaz and Bradley 1997; Duan et al. 2006; Liu and Chen 2000; Liu et al. 2009a; Shrestha et al. 1999; Thompson et al. 2003). Observations indicate that the annual-mean and winter-mean temperatures at high elevation sites (>2,000 m) of the eastern TP have increased at a rate of about 0.42 °C per decade and 0.61 °C per decade, respectively, during 1961–2006. On the other hand, the low-elevation sites (<500 m) have warmed at a rate of about 0.2 °C per decade during the same period (Liu et al. 2009a). Those trends are significant at 0.001 level.

Madhura et al. (2015) recently reported that the observed pattern of a mid-tropospheric warming trend over west–central Asia in recent decades has consequences for the increasing variability of western disturbances and a higher propensity for heavy winter precipitation events over the western Himalaya. The implications of the climate warming signal on the hydrological cycle of the HKH are not yet adequately clear. In contrast to many places across the globe which have experienced decreases in snowfall amounts and glacial extent during recent decades, the Karakoram Himalaya appear to have slightly gained glacial mass in the early 21st century (see Gardelle et al. 2012; Hewitt 2005). Climate model simulations indicate that changes in winter frozen precipitation over the Karakoram Himalaya appear to shield this region from glacier thickness losses under a warming climate (Kääb et al. 2015; Kapnick et al. 2014). The changes in the westerly disturbance is also attributed to an increase in the mass of some glaciers in the Karakoram and western Himalaya (Forsythe et al. 2017).

3.3 Past HKH Climate Changes Were at Decadal to Greater Than Multi-millennial Time Scales

Box 3.2 Paleoclimate of the HKH: climate was always changing

Knowledge of the paleoclimate is important to finding a climate analogy in the past and understanding causal relationships, which can help in understanding and planning for the future (USGS 2010). The climate of HKH has always been changing. Paleoclimate in the HKH has been inferred from limited proxy sources such as ice cores, tree rings, lake sediments, peats, and cave sediments within the region, as well as from land- and marine-based proxy sources outside the region. There is general consensus that the Asian monsoon was established around 23–7 Ma before present (BP) and is related to the tectonics of the region (Singhvi and Krishnan 2014). Since then, both the precipitation and thermal regimes of the region have gone through significant fluctuation, mainly driven by orbital changes and associated changes in solar radiation (Kutzbach and Otto-Bliesner 1982; Prell and Kutzbach 1987). Around 150–100 ka BP, the monsoon was similar to the present; from 100 to 70 ka BP the monsoon fluctuated considerably, and it was much drier and colder from 70 to 60 ka BP. The monsoon circulation strengthened around 13 ka BP, attributed to orbital factors and increased solar radiation (An et al. 2005; Kutzbach 1981). During the period 20 to 19 ka BP, known as the last glacial maximum (LGM), the region experienced drier monsoons and an overall colder climate (Ashahi and Watanabe 2004).

Many reconstructions have been done of the last 2,000 years of the TP using tree ring data and ice-core records. Yadav et al. (2004) looked at 15–16 tree ring sides and found several cold and warm episodes between 1573 and 1846, including the little ice age (LIA) period between 1560 and 1750. Cook et al. (2003) found linkage between temperature in the eastern Himalaya and volcanic eruptions. Shao et al. (2005) found that annual precipitation underwent a large multi-decadal variability in the northeastern part of the Qaidam Basin, TP, over the past 1,000 years, with the 20th century precipitation obviously higher than that in any other century. Liu et al. (2005) showed that the annual mean temperature experienced remarkable fluctuation in the Qilan Mountains of the TP, with the signals of the LIA and the continuous warming during the 20th century. However, other studies did not find particularly abnormal 20th-century warming in tree ring-based reconstructions in the TP region, despite the fact that the 20th century was indeed warmer than most centuries of the last several hundred years to 1,000 years (e.g., Chu et al. 2005; Yang et al. 2002; Yao 1997). Therefore, whether the 20th century warming was unprecedented in the last 1,000 years in the TP is an unresolved issue.

Fluctuations in the climate in the HKH and beyond have been found to be closely related to the culture and civilization in the region (e.g., An et al. 2005; Dixit et al. 2015; Staubwasser et al. 2003; Weiss and Bradley 2001). Dixit et al. (2015) suggested that the weakening in the monsoon around 4.1 ka BP was related to the decline in the Indus urban culture.

3.3.1 Significant Warming Characterized HKH Surface Air Temperature Trends in Past Decades

Analyses show significant warming in recent decades and the last century, despite the warming rates estimated from various research groups being somewhat different and not uniform in all parts of the HKH. Analyses of surface air temperature change based on varied data sets, including observations and reanalyses, have been conducted for a few of areas of the HKH, including the TP (Du et al. 2001; Duan and Xiao 2015; Fan et al. 2015; Kang et al. 2010; Kuang and Jiao 2016; Liu and Chen 2000; Liu et al. 2006, 2009; Ren et al. 2005, 2017; Wang et al. 2008, 2014, 2016; Yao et al. 2012b; You et al. 2013a, 2016, 2017).

Previous studies have reported large and significant warming in the TP region during the last five to six decades, with night-time warming being especially remarkable. For example, the linear rates of increase over the entire TP during 1955–96 were about 0.16 °C/decade for the annual mean temperature and 0.32 °C/decade for the winter mean temperature, which marginally exceeded the averages for the northern hemisphere and the same latitudinal zone (Liu and Chen 2000). A recent study (Yan and Liu 2014) reported a warming trend of 0.316 °C/decade in annual mean temperature in the TP for the period 1961–2012, which is almost twice the previous estimate by Liu and Chen (2000). Recent studies using updated observations and historical CMIP5 outputs (Kang et al. 2010; You et al. 2013a, 2016) show that the annual mean surface temperatures in the TP have doubled the previous warming rate. This rapid warming in the TP is primarily owing to the warmer conditions for the last decade and during the global warming hiatus period (Kosaka and Xie 2013; You et al. 2016). The warming in the TP from the observation and CMIP5 models is more sensitive and accelerated during the hiatus period (You et al. 2016). Meanwhile, the asymmetric pattern of greater warming trends in minimum temperature than in maximum temperature is found in the TP (Duan and Wu 2006; Liu et al. 2006, 2009a). The accelerated climate warming on the TP has caused significant glacial retreat, snow melt, and permafrost degradation (Kang et al. 2010; Yao et al. 2012a, b), and will also lead to significant changes in the form of precipitation (solid to liquid) and changes in hydrology and water resources on the TP (Immerzeel et al. 2010; Immerzeel and Bierkens 2012; Kuang and Jiao 2016; Yang et al. 2014).

Significant warming of the winter and annual temperature was also observed over the western Himalaya in the last century (Bhutiyani et al. 2007; Kothawale and Rupa Kumar 2005). The temperature increase over the western Himalaya is supported by more rapid growth of tree rings in high-altitude tree-ring chronologies of the region (Borgaonkar et al. 2009). This warming trend over the region in the last few decades is consistent with the increase in northern hemispheric temperatures (Mann et al. 1999) and follows the pattern of global warming in the 20th century. There is greater concern about observed unusually large temperature rises in the high-elevation Himalayan regions—for example, the warming is estimated to be nearly two to three times the global average (Liu and Chen 2000; Shrestha et al. 1999)—and the issue of rising temperatures in a warming world is important in more fragile and delicate cryospheric environments. The warming rate is reported to be rather more substantial in winter compared to other seasons in most parts of the HKH (Bhutiyani et al. 2007; Shrestha and Devkota 2010).

During 1901–2014, annual mean surface air temperature increased significantly in the HKH at a rate of about 0.104 °C/decade. A region-averaged surface air temperature anomaly series of the past century or decades for the HKH as a whole has not been produced. The authors of this chapter made a special analysis of observed climate change in the region by using global land surface air temperature (GLSAT) data sets developed recently by the China Meteorological Administration (CMA). Details of the methodology are given in Annex 1. Based on the CMA GLSAT data set (Ren et al. 2014; Sun et al. 2017a; Xu et al. 2014), annual and seasonal mean surface air temperature in the last 114 years exhibit a significant increase in the entire HKH (Ren et al. 2017). The annual mean surface air temperature series in the HKH since 1901 is shown in Fig. 3.2a, b. For the full period of record (1901–2014), annual temperature trends show significant upward trends ($p < 0.05$), and the increase rates of Tmean, Tmax, and Tmin are 0.104 °C/decade, 0.077 °C/decade, and 0.176 °C/decade, respectively (Table 3.1). For details of the number of stations used, refer to Annex 1. The diurnal temperature range (DTR) shows a significant negative trend of −0.101 °C/decade, due to the much larger rise in minimum temperature than in maximum temperature in the region (Ren et al. 2017). Locally, deviations from the general pattern described above have been found in the Karakoram region, where decreasing temperatures (most notably in summer) have been measured. Possible mechanisms for such an anomaly have been discussed recently in Forsythe et al. (2017).

The trends of annual mean temperature in the HKH show general agreement with the global land surface temperature trends for different periods, with small differences (Table 3.1). During the period 1901–2014, the HKH

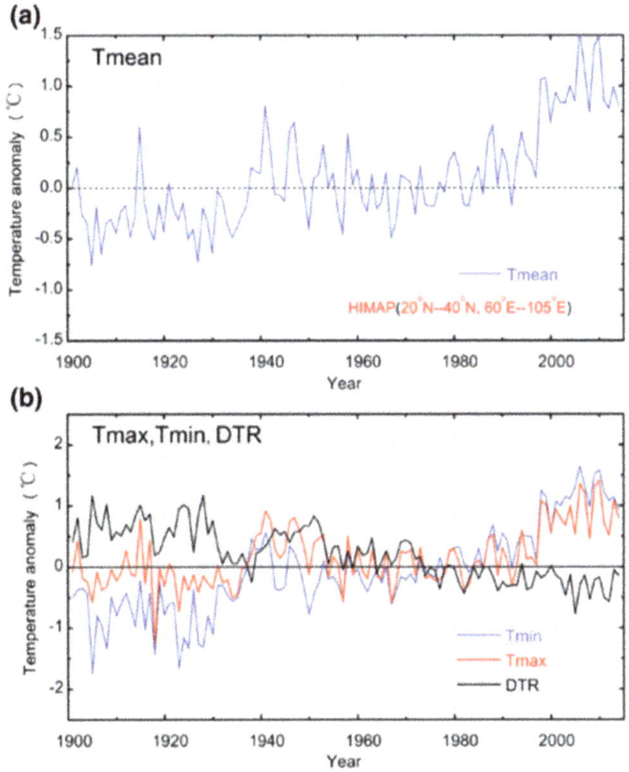

Fig. 3.2 Annual mean temperature anomaly series (°C) relative to 1961–90 mean values for (**a**) Tmean and (**b**) Tmax, Tmin, and DTR for the Hindu Kush Himalaya between 1901 and 2014 (*Data source* CMA GLSAT; Ren et al. 2017)

exhibited trends similar to the global land surface. The annual mean warming rates during the period 1901–2020 was 0.19 °C/decade, while during the period 1951–2014 it was 0.20 °C/decade (Ren et al. 2017). For the period 1951–2014 the trend of annual mean Tmax in the HKH was lower, while the trend of annual mean Tmin was higher than that of the global land surface (Sun et al. 2017b).

The regional average annual mean temperature series of the HKH shows a large decadal to multi-decadal variability during the last more than 100 years. In terms of Tmean change, there are obviously three different stages in the HKH (as shown in Fig. 3.2a). From 1901 to the early 1940s, most of the years showed negative anomalies, and Tmean increased only slightly. From the 1940s to the late 1970s, a relatively cold period for the northern hemisphere, the HKH temperature series showed a significant decreasing trend. After the 1970s, however, a rapid warming appeared in the HKH. The decadal to multi-decadal variations in the warming and cooling episodes are generally consistent with previous studies of the TP and other areas of the HKH (Tang and Ren 2005), as well as in Nepal (Shrestha et al. 1999). In addition, the period from 1998 to 2014 witnessed the warmest years in the past 100 years, in spite of the fact that the annual mean warming trend in this period had slowed down, which is consistent with observations over the globe and northern hemisphere (Trenberth et al. 2014). The warmest 2 years in the period 1901–2014 in the HKH were 2007 and 2010 (Ren et al. 2017).

The annual mean time series of Tmax and Tmin in the HKH exhibited similar decadal to multi-decadal variations with the Tmean series. The annual mean Tmin anomalies were always lower than Tmax before the 1960s, and were generally higher than Tmax afterwards (Table 3.1). The annual mean DTR showed relatively stable change before the 1940s, but a significant decline in the post-1940s period. After the 1960s, although the Tmax anomaly was significantly lower than Tmin, it underwent a large increase compared to the Tmin (Table 3.1). Because of the poor station coverage before the 1940s, the temperature anomaly time series showed strong inter-annual fluctuations, indicating a relatively large sampling uncertainty during the period.

Table 3.1 Annual mean surface temperature trends during 1901–2014 and 1951–2014 in the Hindu Kush Himalaya (HKH) and globally (°C/decade) (Ren et al. 2017; Sun et al. 2017a)

Region	Data source	Period	Trend			
			Tmax	Tmin	DTR	Tmean
HKH	CMA	1901–2014	0.077*	0.176*	–0.101*	0.104*
		1951–2014	0.156*	0.278*	–0.123*	0.195*
Globe (Land + Oceans)	GHCN	1901–2014				0.084*
		1951–2014				0.129*
Globe (Land)	CMA	1901–2014	0.100*	0.142*	–0.036*	0.104*
		1951–2014	0.186*	0.238*	–0.054*	0.202*

*Statistically significant at the 0.05 confidence level
CMA China Meteorological Administration; *DTR* diurnal temperature range; *GHCN* Global Historical Climatology Network

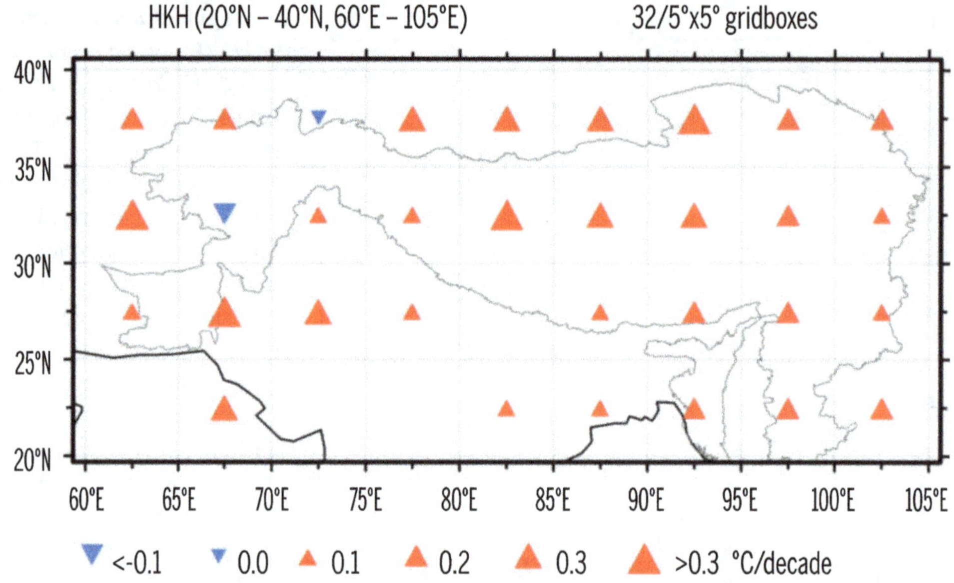

Fig. 3.3 The grid-averaged trends of annual mean temperature in the Hindu Kush Himalaya (HKH) since 1901. The values on the top right corner are the number of grid boxes; white grid boxes indicate the missing data (*Data source* CMA GLSAT; Ren et al. 2017)

The climate warming was more than 0.20 °C/decade in the TP and southern Pakistan. This can be seen in Fig. 3.3, which shows the spatial patterns of trends of annual mean temperature during 1901–2014 in the HKH. Overall, a majority of the grids consistently showed annual warming trends. However, the warming rates exhibit large differences. The larger change occurred in the region of the TP and southern Pakistan, with the warming rates higher than 0.20 °C/decade (Ren et al. 2017). The annual warming rates seemed to increase with the rise in altitude, which is similar to what was found in some analyses (e.g., Liu et al. 2009a), supporting the claim that an altitude-dependent warming trend exists in the HKH (Ren et al. 2017). Northern India and the Sichuan Basin of China showed the weakest warming trend, with annual warming rates below 0.10 °C/decade. Notably, there were fewer available data for the Indian and Nepali regions for the centennial scale analysis, and this may have caused a larger uncertainty in estimating the long-term trend of temperature. A recent study suggests cooling summer temperature over the Karakoram and attributes it to the Karakoram Anomaly (Forsythe et al. 2017).

3.3.2 Precipitation Did Not Show Clear Trends in the Past Decades

Annual and cold-season precipitation in the TP has increased over past decades. The increase has been found to be a part of a broader climatic moistening trend observed in western China, including the TP and northwestern China (Qin et al. 2005; Ren et al. 2000, 2005, 2015; You et al. 2015). The increase in annual precipitation in the northeastern TP seems abnormal in terms of the tree ring-based paleo-reconstruction of precipitation in the last 1,000 years (Shao et al. 2010). Over the last four to five decades, the precipitation increase mainly occurred in winter and spring. Increasing trends in winter precipitation have also been reported over a few stations in the Indus basin since the post-1960s, although there was no spatially coherent pattern of long-term precipitation change over the region (Archer and Fowler 2004). Palazzi et al. (2013) summarized the trends in precipitation in the HKH/Karakoram region, and reported a generally decreasing trend in the Himalaya in summer for the last six decades, but no statistically significant trend was found for winter.

The longer-term (1901–2013) trend of annual precipitation in the entire HKH did not show a positive trend. The authors of this chapter made a special analysis of observed change in precipitation in the region by using Global Land Monthly Precipitation (GLMP) and Global Land Daily Precipitation (GLDP) data sets developed recently by the CMA. Details of the methodology are given in Annex 2. Figure 3.4a displays the regional average annual precipitation standardized anomalies (PSA) and annual precipitation percent anomaly (PPA) from 1901 to 2013 (Zhan et al. 2017). The regional average PSA are fluctuating from one year to another, but the fluctuation became relatively larger from 1930 to 1960, and the overall trend was negative for the HKH. Figure 3.4b shows the spatial distribution of the trends of annual PSA during the period 1901–2013. The trend in the TP was not calculated owing to the lack of precipitation records before 1951. The PSA reduced slightly in southwestern China and most parts of northern India, but increased in the northeastern part of West Asia. All the trends were small and not significant at the 0.05 confidence level. The reduction in annual precipitation in northern India seems consistent with the reported weakening of the Indian summer monsoon over the past century (Ren et al. 2017).

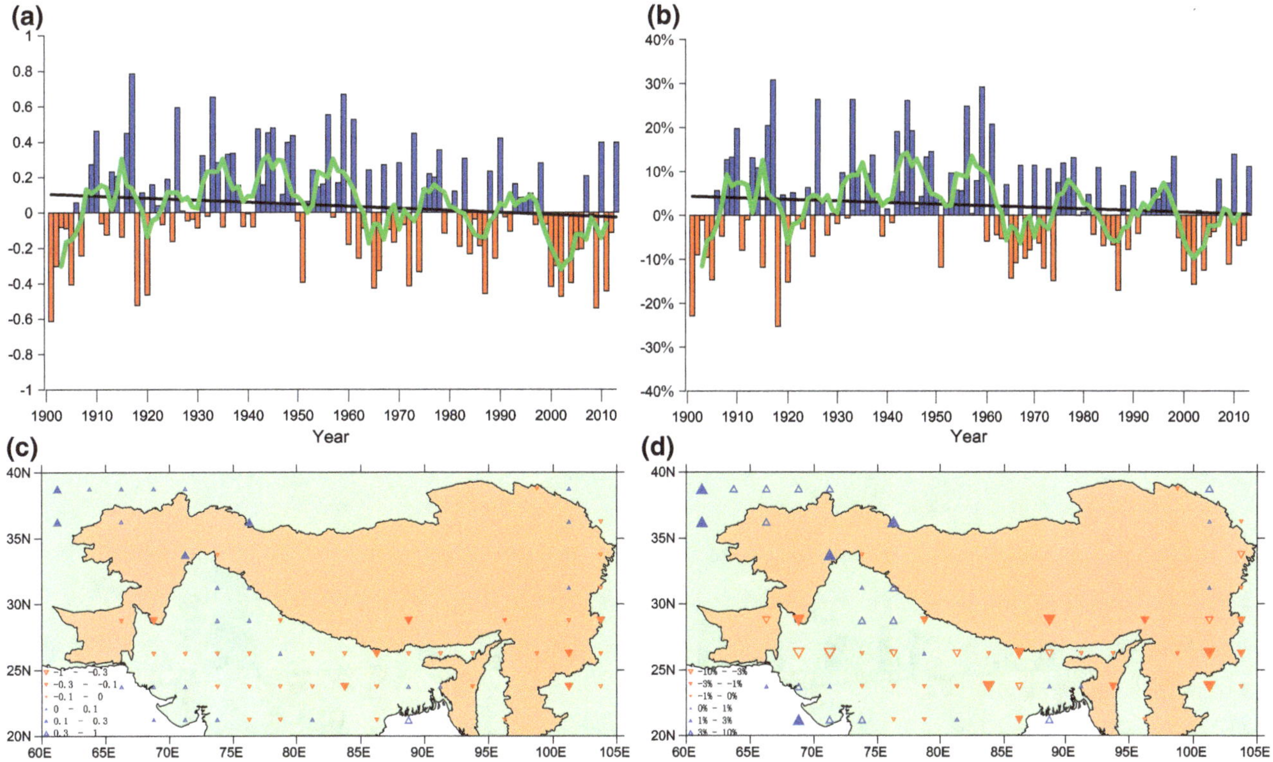

Fig. 3.4 Regional average annual (**a**) precipitation standardized anomalies (PSA), and (**b**) precipitation percentage anomaly (PPA) during 1901–2014 in the HKH (with green line denoting five-year moving average and black line the linear trend); spatial distribution of linear trends in (**c**) PSA, and (**d**) PPA from 1901 to 2014. *Source* Ren et al. (2017) (*Data source* CMA GLMP)

There are some limitations to be noted related to the use of standardized indices in a region where the spatial variation in the statistics is enormous. In particular, trends might be weakly related to actual trends in precipitated water.

In the HKH the annual precipitation and the annual mean daily precipitation intensity of roughly the last 60 years have increased. Figure 3.5 shows the regional average annual PSA, PPA, wet-day anomalies (WDA), and the daily precipitation intensity anomalies (DPIA) from 1951 to 2013 for the HKH (Zhan et al. 2017). During this period, the coverage of stations in the TP improved significantly, which is why it exhibited a different trend from the longer time period shown in Fig. 3.5. The regional average PSA and PPA were mostly positive in the 1950s and fluctuant from the 1960s to the1980s, but increased notably after the 1990s, especially for the PPA. The annual PPA had a rapid upward trend for the time period, with the highest value in 2007. Overall, the trend in regional average PPA increased at a rate of about 5% per decade, which is significant at the 0.01 significance level (Zhan et al. 2017). However, the annual PSA exhibited an insignificant upward trend, despite the fact that it did increase more rapidly from the mid-1980s. As mentioned above, the analysis result of the precipitation change for the region is generally supported by the previous studies, especially for the TP region and the Indus basin (Archer and Fowler 2004; Ren et al. 2015; You et al. 2015; Zhan et al. 2017).

Wet day anomaly (WDA) or precipitation day anomaly experienced a slight and non-significant decline over the last 53 years. Before 1990, the change in regional average WDA was similar to the PSA. However, the WDA did not change notably after 1990, and even reduced slightly after 2005, reaching its lowest level in 2013 in the HKH as a whole. The change in the whole period showed a slight reduction, but the linear trend was only −0.63 days per decade, which does not pass the 0.05 confidence test (Zhan et al. 2017). This result is different from most of the studies conducted for mainland China, which reported a significant decrease in wet days (Ren et al. 2015).

The region-averaged annual mean daily precipitation intensity anomaly (as indicated by DPIA) decreased slightly from the 1950s to the 1980s, with small values of anomalies. However, the DPIA had an abnormally high value period in the early 1990s, although this dropped after 1994. The change during the whole assessment period exhibited a relatively strong downward trend, however, and the change rate was −0.075 mm/d per decade, which is not significant at the 0.05 level (Zhan et al. 2017).

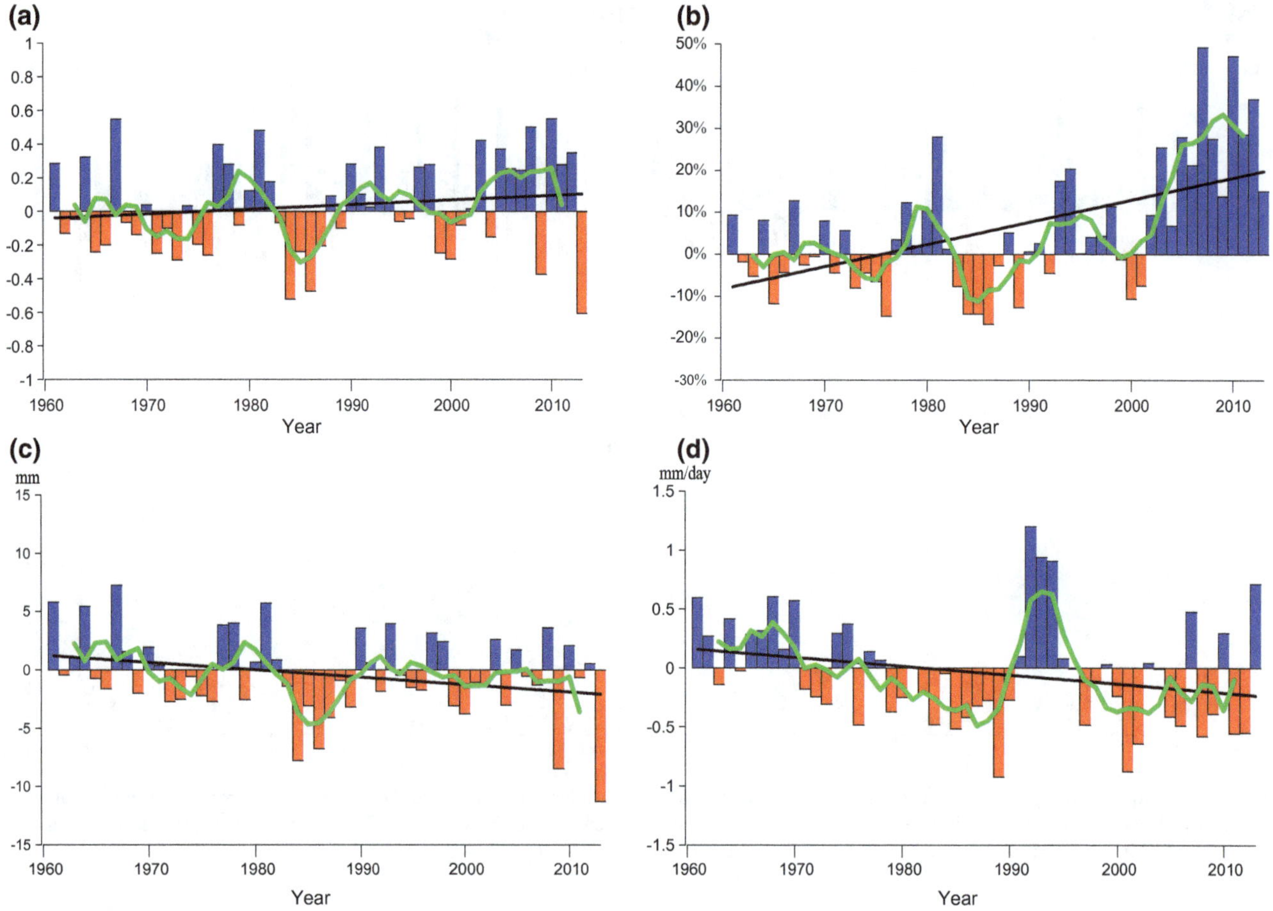

Fig. 3.5 The regional average annual precipitation standardized anomaly (PSA) (**a**) precipitation percentage anomaly (PPA); (**b**) wet day anomaly (WDA) (unit: mm); (**c**) and daily precipitation intensity anomaly (DPIA) (unit: mm/day); (**d**) series over 53 years (1961–2013) in the Hindu Kush Himalaya. Bar indicates values of PSA, PPA, WDA, and DPIA; green lines are 5-year moving averages, and black lines are linear trends (*Data source* CMA GLMP; Zhan et al. 2017)

The annual precipitation undergoes a clear shift in the HKH after 1990. Although the WDA had a fluctuating feature and a weak downward trend, the PPA showed a significant increase during slightly more than the last two decades (Zhan et al. 2017). The temporal characteristics of precipitation variation appear to have entered a mode of greater inter-annual variability and more frequent intense rain and less frequent light rain.

The TP region experienced an increase in all three precipitation indicators, especially for PPA and DPIA, but northern India exhibited an increase in PPA and DPIA but a decrease in WDA over the past 53 years (Zhan et al. 2017). Figure 3.6 shows the spatial distribution of the trends of annual PPA, WDA, and DPIA during the period 1961–2013. The positive PPA trends were statistically significant for some of the grids in the TP region, but the annual PPA generally decreased in southwestern China, the northeastern part of India, and the most northeastern part of the HKH (Ren et al. 2017). The PPA showed a significant increase in a few grids in India, but a significant decrease in some parts of southwestern China and Myanmar.

The WDA had a strong rise in most parts of India and northern TP, but a strong decline in southwestern China and Myanmar. WDA and DPIA exhibited approximately an opposite pattern of long-term trends during the assessment period, especially in southern and southeastern areas of the HKH. In most parts of India and northern TP, for example, the WDA increased notably, but the DPIA decreased remarkably, and southwestern China and Myanmar witnessed a significant decline in WDA and a significant rise in DPIA (Ren et al. 2017).

A more significant increase in annual mean daily precipitation intensity over past decades appeared in the higher-altitude areas, including the TP. Compared to WDA trends, the spatial distribution pattern of the DPIA has a better relationship with the elevation. The annual DPIA increased

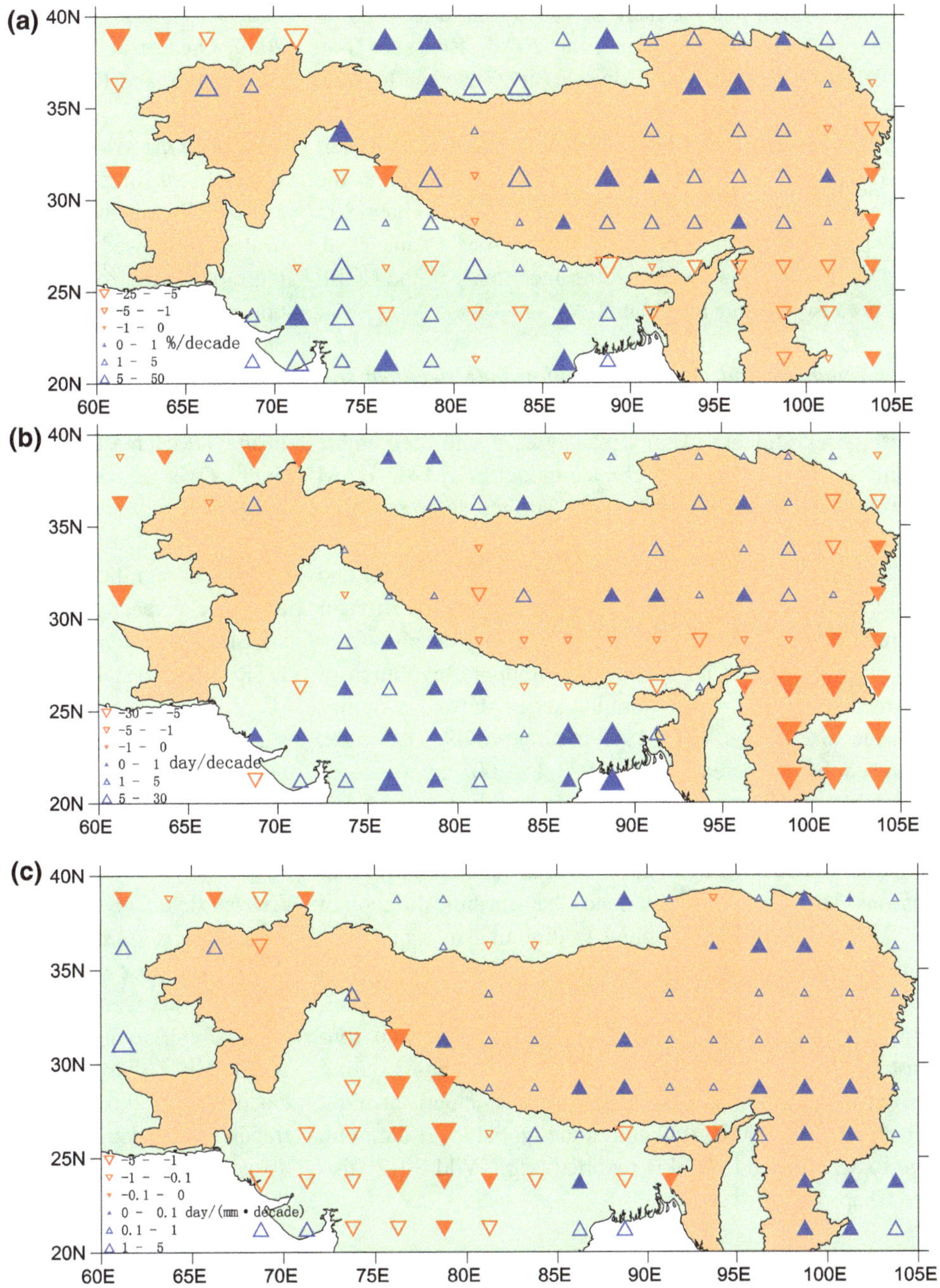

Fig. 3.6 The change trends in annual precipitation percentage anomaly (PPA, unit: % decade^{-1}) (**a**) wet day anomaly (WDA, unit: mm decade^{-1}); (**b**) and daily precipitation intensity anomaly (DPIA, unit: mm/day decade^{-1}); (**c**) in the Hindu Kush Himalaya over 53 years (1961–2013). Filled symbols represent statistically significant data at 0.05 confidence level (*Data source* CMA GLMP; Zhan et al. 2017)

significantly in the highlands region, such as the TP and the Yunnan–Guizhou Plateau, but reduced significantly and consistently in India and probably in other areas south of the Himalaya. It is unclear what caused such a spatial pattern in annual DPIA change in the HKH (Zhan et al. 2017).

3.3.3 Decreasing Near-Surface Wind Speed, Solar Radiation, and Sunshine Duration Indicated by Scanty Data

The near-surface wind speed decreased over most areas of the TP region. No wind and solar radiation observation data were available for the whole region, and only the studies for the TP region were assessed in this chapter. During 1980–2005, both the surface stations and National Centers for Environmental Prediction (NCEP) reanalysis in the TP showed significant decreasing trends, at rates of −0.24/decade and −0.13 m/decade, respectively, especially evident in spring and summer. One of the likely causes of the diminishing wind speed is the asymmetrically decreasing latitudinal surface temperature and pressure gradients over the TP and the surrounding regions, including the Indian Ocean, which may be part of a large-scale atmospheric circulation shift (Ding and Ren 2008; You et al. 2010a). Another major factor affecting the near-surface wind speed was urbanization and the increased sheltering effect around the observational grounds (Liu et al. 2009b; Ren and Zhou 2014; Zhang et al. 2009). However, it is unclear to what extent urbanization had affected the near-surface wind speed decline in the HKH as a whole during the past two and half decades. Furthermore, the increase in wind speed after the 2000s to the present may indicate an overall change in atmospheric circulation mode over East Asia (Yang et al. 2014). It is noted that there is some uncertainty in the NCEP data because of their low spatial.

Solar radiation and sunshine duration both declined significantly in the TP region over the past six decades. Surface observational data, reanalyses, and ensemble simulations with the global climate model ECHAM5-HAM show that the mean annual all-sky surface solar radiation (SSR) series in the TP decreased at a rate of −1.00 Wm^{-2} $decade^{-1}$, primarily in autumn and secondly in summer and winter. Annual clear-sky SSR series exhibit an even stronger decrease of −2.80 Wm^{-2} $decade^{-1}$, especially during winter and autumn (You et al. 2013b). The temporal evolution of the mean annual sunshine duration series showed a significant increase from 1961 to 1982 at a rate of 49.8 h/decade, followed by a decrease from 1983 to 2005 at a rate of −65.1 h/decade, with an overall significant decrease at a rate of −20.6 h/decade during the whole of the 1961–2005 period; this was due mainly to the decline in summer and spring seasons. This confirms the evidence that sunshine duration in the TP ranges from brightening to dimming in accordance with sunshine duration trends in the rest of mainland China (Ding and Ren 2008; Ren et al. 2005; You et al. 2010c). Total and low-level cloud amounts in the TP showed contrasting trends during day and night, with decreases during the day but increases (especially low-level cloud) at night (Duan and Wu 2006), although at the global level dimming due to aerosols and cloud is reported (e.g., Wild 2012; also see Chap. 10).

3.3.4 Significant Changes in the Temperature and Precipitation Extremes in Past Decades

Most parts of the HKH underwent significant long-term changes in frequencies of extreme temperature events over the last decades. In this chapter, assessment of changes in the extreme temperature indices of the TP and the HKH for the period 1961–2014 were mainly based on the CMA global land daily surface air temperature data set. Studies conducted for different parts of the HKH showed a generally significant change in the various extreme temperature indices, with the minimum-temperature-related indices witnessing a significant rise and the maximum-temperature-related indices a significant decline over the last decades (Choi et al. 2009; Qian et al. 2007; Ren et al. 2012; Zhai et al. 1999, 2005; Zhou and Ren 2011).

Almost all the extreme temperature indices in the TP region showed statistically significant trends over the past half century. During the period 1961–2014, temperature extremes in the TP showed patterns consistent with warming, with a large proportion of stations showing statistically significant upward or downward trends for all of the temperature indices analyzed (Sun et al. 2017b). Stations in the northwestern, southwestern, and southeastern part of the TP showed greater trends. Overall, the incidence of extreme cold days and nights in the TP decreased by −0.85 and −2.38 d/decade, respectively. Over the same period, the incidence of extreme warm days and nights increased by 1.26 and 2.54 d/decade, respectively. The number of frost days (FD) and ice days decreased significantly at the rates of −4.32 and −2.46 d/decade, respectively. The length of the growing season has statistically increased by 4.25 d/decade. The annual mean DTR showed a statistically decreasing trend at a rate of −0.20 °C/decade. The extreme temperature indices also exhibited statistically significant increasing trends. In general, minimum temperature indices showed greater warming trends in comparison to maximum temperature indices (Sun et al. 2017b; You et al. 2008a; Zhou and Ren 2011).

Extreme cold events significantly decreased and extreme warm events significantly increased over the whole HKH during the past six decades. Annual mean anomaly time series of the percentile-based (Fig. 3.7a–d) and absolute (Fig. 3.7e–h) temperature extreme indices since 1961 for the whole HKH, based on the CMA GLSAT data set, were calculated for this chapter (Fig. 3.4; Sun et al. 2017b). The trends were calculated only for the grid boxes with at least 40 years of data during the study period, with the last year of the data series no earlier than 2000 (Alexander et al. 2006). Table 3.2 also shows the decadal trend values for extreme indices of the region for the period 1961–2014. Although the percentile-based temperature indices were calculated in percentages, the units were converted into days for ease of understanding, as suggested by Alexander et al. (2006). As in the TP region, the extreme cold events significantly decreased for the whole HKH during the past six decades (Fig. 3.7a, b), while the extreme warm events significantly increased (Fig. 3.7c, d). However, the trends in warm events were larger in magnitude than cold events (Table 3.2), and there was a dramatic increase since the mid-1980s, especially for warm nights (Tn90p). Moreover, the trends of the extreme events related to minimum temperature were greater in magnitude than those related to maximum temperature. Similar results were found in the Koshi Basin by Rajbhandari et al. (2017).

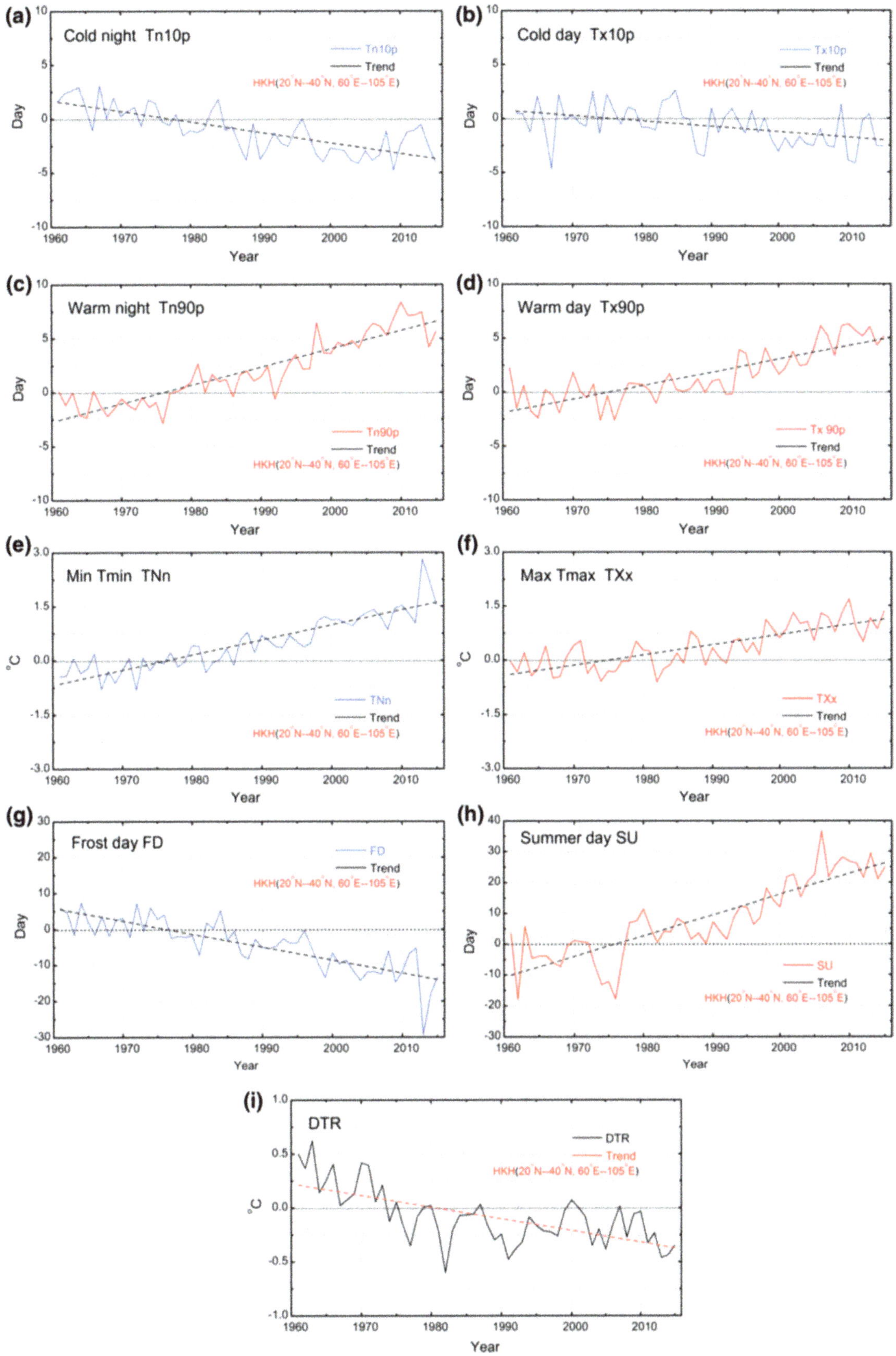

Fig. 3.7 Annual mean anomaly series of extreme temperature indices of the Hindu Kush Himalaya (HKH) for 1961–2014 for (**a**) cold nights (TN10p); (**b**) cold days (TX10p); (**c**) warm nights (TN90p); (**d**) warm days (TX90p); (**e**) monthly maximum value of daily maximum temperature (TXx); (**f**) monthly minimum value of daily minimum temperature (TNn); (**g**) frost days (FD); (**h**) summer days (SU); and (**i**) diurnal temperature range (DTR) (relative to 1961–90 mean values). The trends are calculated only for the grid boxes with sufficient data, as explained in the text (*Data source* CMA GLASAT; Sun et al. 2017b)

Table 3.2 Trends of percentile-based and absolute extreme temperature indices of the Hindu Kush Himalaya between 1961 and 2014

Index	ID	Unit	Trend
Cold night	TN10p	day/decade	−0.977*
Cold day	TX10p	day/decade	−0.511*
Warm night	TN90p	day/decade	1.695*
Warm day	TX90p	day/decade	1.239*
Max Tmax	TXx	°C/decade	0.282*
Min Tmin	TNn	°C/decade	0.419*
Frost day	FD	day/decade	−3.636*
Summer day	SU	day/decade	6.741*
Diurnal temperature range	DTR	°C/decade	−0.108*

(*Data source* CMA GLSAT; Sun et al. 2017a)
*Statistically significant at the 0.05 confidence level

Extreme values and frequencies of the absolute temperature-based indices[9] also experienced a generally significant change in the HKH during the past six decades. For the temperature indices (Fig. 3.7e–h), both extreme values of the highest Tmax (TXx) and the lowest Tmin (TNn) showed increasing trends in the HKH, and the rising rate of TNn was more than twice that of TXx (Sun et al. 2017b; Table 3.2). However, the rising rate of TNn was dominated by the very large positive anomalies after the 1980s. The summer day (SU) frequency also increased, and the rising rate reached 6.74 day/decade, which is related to the dramatic positive anomalies after 1990. During 1989 and 1998, SU increased by about 20 days, which is far more than the average rising rate. Annual FD frequency decreased, with a change rate of −3.63 day/decade. In the HKH as a whole, the annual mean DTR anomaly series showed a significantly decreasing trend before the 1980s and a slight increase after the mid-1980s. The overall decline was mainly a result of the much larger rate in annual mean Tmin increase than of annual mean Tmax (Sun et al. 2017b).

Extreme cold events significantly decreased in most parts of the eastern HKH, especially in southwestern China and TP, while extreme warm events increased over the whole HKH. Figure 3.8 shows the spatial distribution of linear trends of extreme temperature indicators for every grid in the HKH. The cold nights (TN10p) and cold days (TX10p) decreased significantly in most parts of the eastern HKH and showed an upward trend in a few of the gridboxes of the western HKH. The warm nights and warm days (TN90p and TX90p) increased significantly in most parts of the HKH, especially in its western part. The Karakoram, in northwestern Pakistan, is anomalous for its cold nights (TN10p) and cold days (TX10p) (Sun et al. 2017b).

Extreme absolute temperature indices also experienced a generally significant change in the HKH. FD showed significant decreasing trends in most parts of northern India and the TP, and SU showed a significant increase between 60° E and 80° E. The highest Tmax/Min Tmin (TXx/TNn) showed increases over the whole HKH, but the lowest Tmin showed a larger trend of increase in the TP region than did the highest Tmax. Along the Himalaya the DTR had an increasing trend, while most of the other regions showed a significant decrease, with the downward trend especially large in the TP region (Sun et al. 2017b).

Previous studies identified a clear change in extreme precipitation events in the HKH in recent periods. Western China, including the TP region, experienced a major change in extreme precipitation events over the past five decades, as reported in many studies (e.g., Ren et al. 2012; You et al. 2015), consistent with the increase in annual total precipitation during the same period. Some stations over the western Himalaya and the Karakoram region have also shown a significant increase in the number of wet days and extreme rain events during the past few decades (Choi et al. 2009; Klein Tank et al. 2006). In the eastern Himalaya, though, the total amount of precipitation did not change much, and the number of rainy days decreased, which meant a higher amount of rainfall in a short period of time. This torrential rain may cause flash floods and landslides in the eastern Himalaya and hilly regions (Syed and Al Amin 2016).

There has been remarkable change in both the light and intense precipitation since 1961, with annual intense precipitation days (frequency) and annual intense precipitation intensity experiencing increasing trends. Figure 3.9 displays the regional average annual amount, and day and annual intensity anomalies, for the percentile-based light (below the 50th percentile), moderate (between the 50th and 90th percentiles), and intense (beyond the 90th percentile) precipitation in the HKH over the period 1961–2013 (Zhan et al. 2017). Table 3.3 shows the linear trends of the annual amount, days, and intensity of the light, moderate, and intense precipitation, and the test results of the significance. The significant increase in annual intense precipitation amount, days, and intensity (5.28 mm per decade, 0.14 day per decade, and 0.39 mm/day per decade, respectively), and the significant decline in annual light precipitation days and intensity (−0.80 day per decade and −0.02 mm/day per decade, respectively), are notable. No detectable change appears for the moderate precipitation amount, frequency, and intensity over the assessment period (Fig. 3.9). The

[9]The extreme temperature indices used in this chapter are based on Expert Team on Climate Change Detection and Indices (ETCCDI). For more information on the indices refer to http://www.climdex.org/indices.html.

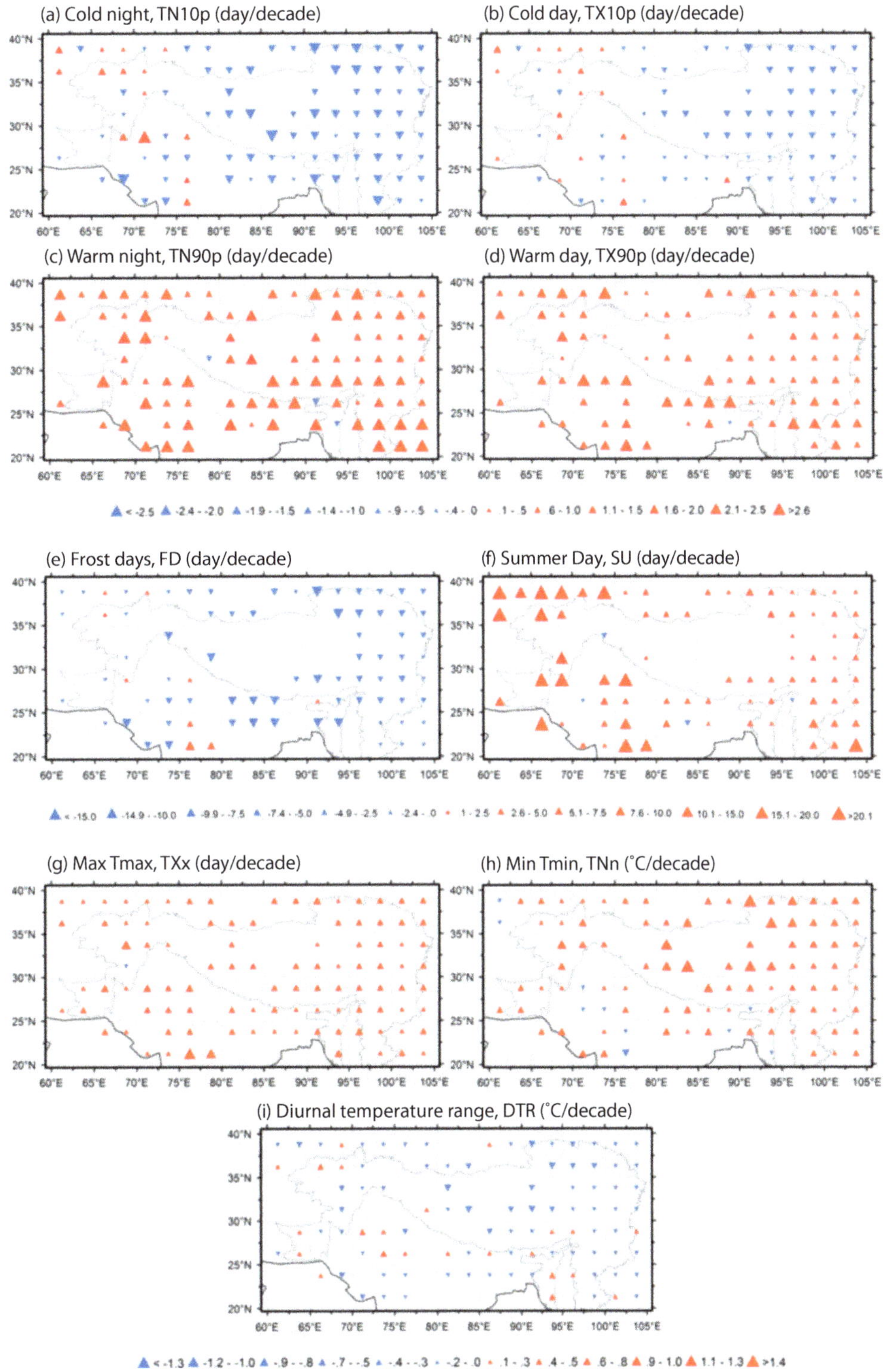

Fig. 3.8 The change trends of extreme temperature indices of the Hindu Kush Himalaya for 1961–2014 for (**a**) cold nights, (**b**) cold days, (**c**) warm nights, (**d**) warm days, (**e**) frost day, (**f**) summer day, (**g**) monthly maximum value of daily maximum temperature, (**h**) monthly minimum value of daily minimum temperature, and (**i**) diurnal temperature range (relative to 1961–90 mean values) (*Data source* CMA GLSAT; Sun et al. 2017b)

definition of light, moderate, and intense rainfall is given in Annex 2, Table 3.11.

Light precipitation frequency and intensity significantly decreased in the HKH over the past 53 years, but its amount had an insignificant upward trend (Table 3.3). The regional average annual light precipitation amount increased over the period 1961–2013, but the change rate was only 0.482 mm per decade, which did not pass the significance test at the 0.05 confidence level. The annual light precipitation days and intensity reduced significantly over the same period, however, with the change rate being

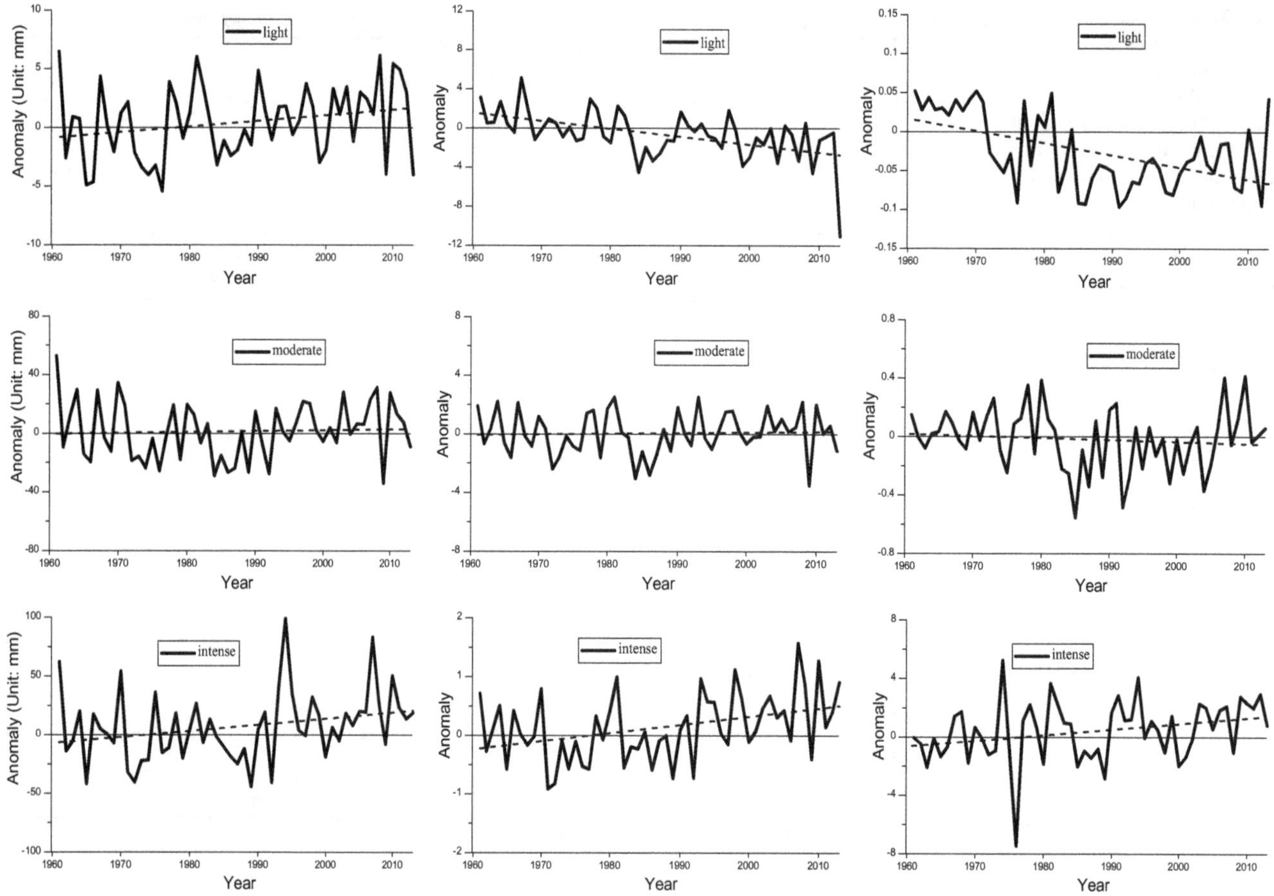

Fig. 3.9 The regional average anomaly time series of extreme precipitation indices for annual amount (unit: mm) (left); days (unit: day) (central); and intensity (unit: mm/day) (right) of light (above), moderate (middle), and intense (below) precipitation over the last 53 years (1961–2013) in the Hindu Kush Himalaya. Dashed lines represent linear trends (*Data source* CMA GLDP; Zhan et al. 2017)

Table 3.3 Linear trends of different categories of precipitation for the period 1961–2013 in the Hindu Kush Himalaya

	Light	Moderate	Intense	Total
PAA (mm decade − 1)	0.482	0.623	5.284*	6.389
WDA (day decade − 1)	−0.800**	0.040	0.141**	−0.629
DPIA (mm/day decade − 1)	−0.016**	−0.013	0.391*	−0.075

(*Data source* CMA GLDP; Zhan et al. 2017)
*Statistically significant at the 0.05 confidence level. **Statistically significant at the 0.01 confidence level. *PAA* precipitation amount anomaly; *WDA* wet day anomaly; *DPIA* daily precipitation intensity anomaly

−0.81 day/decade and −0.02 mm/d per decade, respectively; both were statistically significant at the 0.01 confidence level (Table 3.3). The change may be attributed to the decreasing amplitude of fluctuation in the frequency and intensity after 1985 (Fig. 3.9) with the linear trend of the amount opposite to those of the frequency and intensity. The regional average annual moderate precipitation amount and days increased slightly, while the intensity decreased slightly, but all the changes were non-significant (Zhan et al. 2017).

The regional average annual amount, frequency, and intensity of intense precipitation increased significantly over the period 1961–2013 (Table 3.3). The linear trends for the regional average annual intense precipitation amount, intense precipitation days, and intense precipitation intensity were 5.48 mm/decade, 0.14 day/decade, and 0.39 mm/day/decade, respectively. All trends passed the 0.05 significance test, with the increase in intense precipitation days passing the 0.01 significance test. The increasing trends of annual intense precipitation frequency and intensity in the HKH were approximately consistent with those found in other mid- to high-latitude regions. The annual total precipitation amount in the HKH increased over the 53 years, which was not significant at the 0.05 level, but the annual total precipitation days and intensity exhibited a very weak declining trend (Table 3.3; Zhan et al. 2017).

The annual amount and frequency of light precipitation increased significantly in most parts of India and the northwestern TP. Figure 3.10 shows the spatial distribution of linear trends of standardized precipitation indicators for

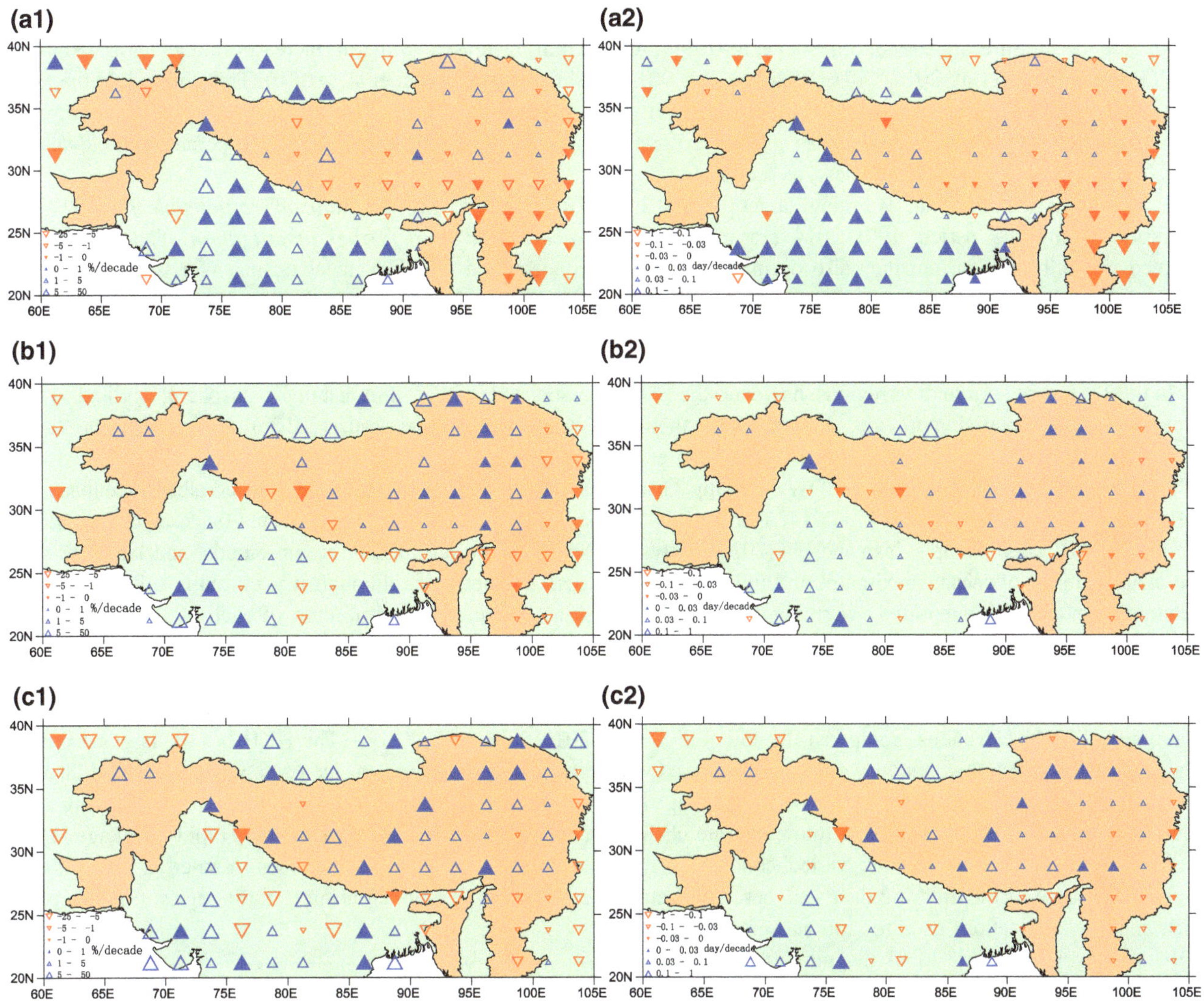

Fig. 3.10 The change trends of annual precipitation amount percent anomaly (PPA) (1/left) and annual precipitation day anomaly (WDA) (2/right) for light (**a**) moderate; (**b**) and intense; (**c**) precipitation over the last 53 years (1961–2013) in the Hindu Kush Himalaya. Filled symbols represent statistically significant trends at the 0.05 confidence level (*Data source* CMA GLDP; Zhan et al. 2017)

every grid in the HKH. The annual PSA of light precipitation increased significantly in most parts of India, and also in the northwestern TP. Southwestern China had a significantly downward trend, but it did not change much in northeastern West Asia. The annual WDA of light precipitation also increased significantly in most parts of India and in the northwestern TP, and it significantly reduced in the southeastern part of the TP and southwestern China (Zhan et al. 2017).

The amount and frequency of intense precipitation and moderate precipitation mostly increased significantly in the TP, but showed heterogeneous change in the other areas of the HKH (Fig. 3.10). The spatial patterns in linear trends of the annual moderate and intense precipitation indicators were similar, and they mostly increased significantly in the TP, especially in areas north of 30° N, and decreased in southwestern China, northern South-east Asia, and northeastern West Asia. In particular, both intense precipitation amount and days increased in the TP, with those trends in most grids being significant. The annual intense precipitation amount and days in the TP increased (Fig. 3.10; Zhan et al. 2017).

3.3.5 Progressively Greater Warming with Elevation

The elevation-dependent warming (EDW) phenomenon in the HKH, particularly in the TP and its surrounding regions, has been reported by many research groups (Duan and Xiao 2015; Guo et al. 2016; Liu and Chen 2000;

Mountain Research Initiative EDW Working Group 2015; Pepin et al. 2015; Shrestha 2008b; Shrestha et al. 1999; Sun et al. 2017b; Wang et al. 2014, 2016; Yan and Liu 2014; Yan et al. 2016; You et al. 2016) but the association of warming trends with elevations is still an unresolved question. The EDW in the HKH seems clear, as seen in Fig. 3.3. The grids with annual mean warming larger than 0.2 °C/decade during the period 1901–2014 are mostly distributed in the TP region, but they also occur in low-lying southern Pakistan and western India. The lowest warming or even cooling is seen in eastern Afghanistan and the Karakoram Mountains.

In the TP and its surrounding regions, however, the EDW seems clearer, as reported in previous studies, though there is still some debate on this (Duan and Xiao 2015; Guo et al. 2016; Mountain Research Initiative EDW Working Group 2015; Pepin et al. 2015; Yan and Liu 2014; Yan et al. 2016; You et al. 2016). For example, Yan and Liu (2014) analysed the change trends of mean, maximum, and minimum temperatures over the TP and reported a warming trend of 0.316 °C/decade in annual mean temperature over the TP for the period 1961–2012, based on data from 73 stations above the elevation of 2,000 m. Liu and Chen (2000) found higher warming at high elevations compared to low elevation regions in the TP. This result was later confirmed by numerical experiments (Chen et al. 2003), suggesting that such a phenomenon may continue in future climate change scenarios (Liu et al. 2009a). The varied conclusions are probably because of the differing data sets, periods of analysis, and lowland stations used for comparison (Kang et al. 2010). Using temperature trend magnitudes at 71 surface stations with elevations above 2,000 m a.s.l. in the eastern and central TP, for example, You et al. (2008b) failed to find an elevation dependency in the trends of temperature extremes in the eastern and central TP; nor did Ding and Zhang (2008) find the EDW phenomenon by analysing the change of temperature for mainland China as a whole. Wang et al. (2014, 2016) and You et al. (2017) indicate that analysis based on simple linear trend may not a good tool for the analysis of EDW on either the global scale or regional scale, probably because the warming in a high-elevation region is related not only to altitude but also to latitude.

Recent study shows that change in extreme cold events in the HKH in the last decades appear to be more sensitive to elevation, with cold nights and cold days decreasing with elevation, and the change in warm extremes showed no detectable relationship with elevation (Sun et al. 2017b). During 1961–2015, FD and minimum Tmin also had a good relationship with elevation, and the trend in FD decreased with an increase in elevation while the trend in minimum Tmin increased with an increase in elevation (Ren and Shrestha 2017; Sun et al. 2017b). This shows that the EDW in the HKH over the last decades may have been represented by the dependence of minimum temperature on elevation.

The exact driving mechanisms responsible for EDW are unclear and need further investigation. Yan et al. (2016) concentrated on EDW mechanisms over the TP, and concluded that the increase in surface net radiation resulted in EDW over the TP. One possible reason for the continued (and accelerated) warming in the TP is the positive feedback associated with a diminishing cryosphere, particularly the snow cover. You et al. (2010b) summarized the factors determining the recent climate warming in the TP: anthropogenic GHG emissions, the snow/ice-albedo feedback, and changes in environmental elements (such as cloud amount, specific humidity, Asian brown clouds, and land use changes). Furthermore, although it is currently difficult to determine the relative contribution of each of these factors, the anthropogenic GHG emission is regarded as the main cause of the climate warming in the HKH, and impacts there are probably larger than in the rest of the world (Kang et al. 2010; You et al. 2010b). The HKH has the largest extent of cryosphere (glaciers and ice caps, snow, river and lake ice, and frozen ground) outside the polar regions (Kang et al. 2010). The glaciers have exhibited a rapid shrinkage in both length and area in recent decades (Yao et al. 2012b), coinciding with the rapid warming in the region (see Chap. 7).

3.4 Climate Models Project Increases in HKH Temperature and Precipitation in the 21st Century

The potential impact of climate change on the various resources of the HKH is of great concern (Ericsson et al. 2009) as a large population across the region relies on monsoon rainfall. Hence, it is necessary to understand the nature of future climate change. Many climate modelling groups have performed future simulations under IPCC's emission scenarios (IPCC 2000; Meinhausen et al. 2011; Panday et al. 2014; Van Vuuren et al. 2011). With global warming, future rainfall and temperature are likely to change in different ways in various ecological regions (Goswami et al. 2006). The impact of climate change on the cryosphere and hydrological regime of the HKH will be marked, because of sharp altitudinal differences, particularly owing to the EDW phenomenon. Eastern HKH river basins are heavily influenced by monsoon rainfall, whereas in western rivers the flow is derived from the melting of snow and glaciers (Immerzeel et al. 2009; Lutz et al. 2014). The changes in temperature and precipitation in the HKH will have serious and far-reaching consequences for climate-dependent sectors such as agriculture, water resources, and health (Shrestha et al. 2015).

3.4.1 Significant Warming Projected, Greater Than Global Average

Based on past studies throughout the HKH, temperature is expected to rise (Kulkarni et al. 2013; Kumar et al. 2011; Rajbhandari et al. 2015, 2016). A study carried out at river-basin scale over the eastern HKH showed the greatest warming over high-altitude regions (Rajbhandari et al. 2016). A similar basin-scale study carried out over the western HKH by Rajbhandari et al. (2015) reported temperature progressively warming over the western Himalaya with minimum temperature trend slightly greater than maximum temperature trend. Kulkarni et al. (2013) reported that significant warming is expected throughout the HKH towards the end of the present century. Overall temperature across the mountainous HKH will increase by about 1–2 °C (in places by up to 4–5 °C) by 2050 (Shrestha et al. 2015).

The HKH region is projected to warm over the 21st century. This is based on the analyses we performed for this chapter on future projections of annual mean surface temperature change relative to 1976–2005, based on a CMIP5 multi-model ensemble mean (a subset of 25 models listed in Table 3.4). These projected changes in surface temperature are higher than the likely ranges reported for global and South Asian regions by the recent IPCC assessment. They indicate continuous warming over the entire HKH in the 21st century (Fig. 3.11a). The projected temperature change is 2.5 ± 1.5 °C for the moderate scenario corresponding to RCP4.5 (Van Vuuren et al. 2011). For the more extreme scenario of RCP8.5, the projected temperature increase is 5.5 ± 1.5 by the end of the 21st century. The simulation data of the CMIP5 models suggest that the projected changes in the surface mean temperature over the HKH are larger compared to the global mean change by the end of the 21st century (Fig. 3.11).

Table 3.4 Coupled Model Intercomparison Project 5 models used for temperature and precipitation projection over the Hindu Kush Himalaya

IPCC ID of the model	RCP4.5	RCP8.5	Country
BCC-CSM1.1	*	*	China
BCC-CSM1.1 M	*	*	China
BNU-ESM	*	*	China
CanESM2	*	*	Canada
CMCC-CM	*	*	Italy
CMCC-CMS	*	*	Italy
CSIRO-ACCESS-1	*	*	Australia
CSIRO-ACCESS-3	*	*	Australia
CSIRO-MK36	*	*	Australia
EC-EARTH	*	*	European Union
FIO-ESM	*	*	China
GFDL-ESM2G	*	*	USA
GISS-E2-R	*	*	USA
HadGEM2-AO	*	*	South Korea
HadGEM2-CC	*	*	UK
HadGEM2-ES	*	*	UK
INMCM4	*	*	Russia
IPSL-CM5A-LR	*	*	France
IPSL-CM5A-MR	*	*	France
IPSL-CM5B-LR	*	*	France
MIROC5	*	*	Japan
MPI-ESM-LR	*	*	Germany
MPI-ESM-MR	*	*	Germany
MRI-CGCM3	*	*	Japan
NCAR-CESM1-BGC	*	*	USA
NCAR-CESM1-CAM5	*	*	USA
NorESM1-M	*	*	Norway
NorESM1-ME	*	*	Norway

For details see Annex 3, Table 3.12

For the first time we used projected changes in near-surface air temperature based on high resolution (0.5° longitude–latitude resolution) dynamically downscaling of CMIP5[10] general circulation models (GCM) using regional climate models (RCM). Regional climate projections over South Asia were generated using the models listed in Annex 3, Table 3.12, based on the protocols of the CORDEX[11] initiative. The CORDEX regional climate models RCM reliably capture the overall increasing trend of surface temperature variations over the South Asian region (Sanjay et al. 2017a). Multi-model ensembles of climate projections over the HKH for the near future (2036–65) and far future (2066–95), based on the CORDEX RCM relative to 1976–2005 (see Annex 3), are shown in Figs. 3.12 and 3.13, respectively. The magnitude

[10]http://cmip-pcmdi.llnl.gov/cmip5/.

[11]http://www.meteo.unican.es/en/projects/CORDEX.

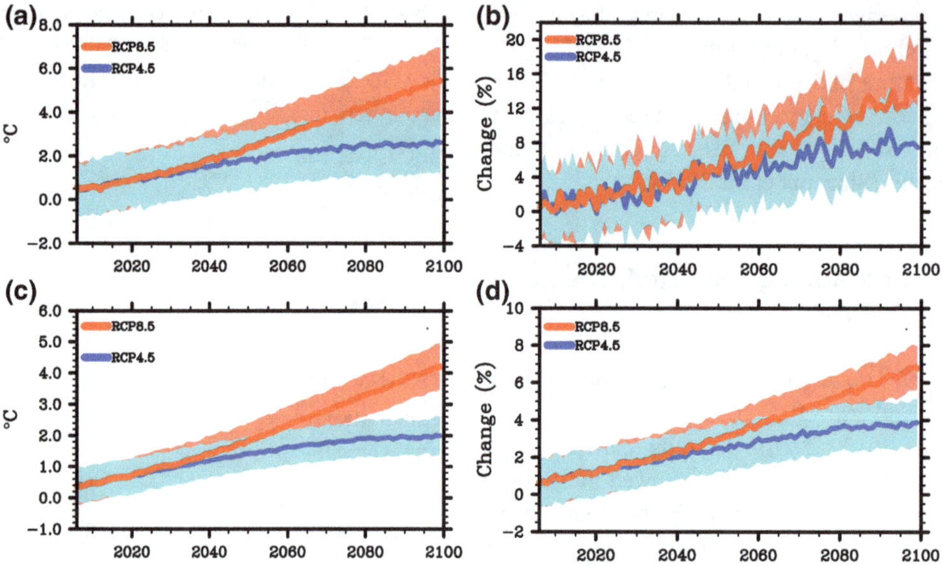

Fig. 3.11 (**a**) Annual mean surface air temperature change (in °C) from 2006 to 2100 relative to 1976–2005 averaged over the Hindu Kush Himalaya for RCP4.5 (blue) and RCP8.5 (red) scenarios. The multi-model ensemble mean is shown by a solid line and the colour shading indicates the multi-model ensemble mean ± one inter-model standard deviation; (**b**) same as (**a**) except for precipitation. The precipitation change is expressed as percent/month; (**c**), (**d**) same as (**a**), (**b**), except showing the global mean change for (**c**) surface air temperature change (in °C) and (**d**) precipitation from 2006 to 2100 relative to 1976–2005

of the projected seasonal warming is found to vary with region, season, averaging period, and scenario. The projected warming differs by more than 1 °C between the eastern and western HKH, with relatively higher values in winter (Sanjay et al. 2017b). The highest warming is projected to be over the central Himalaya for the far-future period with the RCP8.5 scenario (Fig. 3.13f).

The projected temperature changes with RCP4.5 and RCP8.5 scenarios for 2036–65 and 2066–95 relative to 1976–2005 in three HKH subregions (defined by grid cells within each subregion above 2,500 m a.s.l., see Fig. 3.1) suggest that during summer (winter) relatively higher (lower) warming will occur over the hilly regions of the northwestern Himalaya and Karakoram (HKH1) than in the central Himalaya (HKH2) and southeastern Himalaya and TP (HKH3). Table 3.5 presents the seasonal ensemble mean projected changes in near-surface air temperature based on the 13 CORDEX RCM and 10 CMIP5 global climate models (GCM) used to provide initial and lateral boundary conditions for the CORDEX RCM. For the far-future period, the summer season ranges among the three HKH subregions for the higher resolution CORDEX RCM of 2.2–2.6 °C (RCP4.5) and 4.2–4.9 °C (RCP8.5) are lower than the corresponding ranges for the CMIP5 GCM of 2.5–3.3 °C (RCP4.5) and 4.4–5.7 °C (RCP8.5). While the winter season temperature change range among the three HKH subregions in the CORDEX RCM for the RCP4.5 scenario of 3.1–3.3 °C is also lower than that for the corresponding CMIP5 GCM of 3.0–3.6 °C for this period, the CORDEX winter range of 5.4–6.0 °C in the RCP8.5 scenario is higher than the corresponding range for the CMIP5 GCM of 5.1–5.8 °C for the far-future period. Overall, future projections of surface temperature from the CMIP5 models and CORDEX RCM seem to agree on the overall warming trends over the HKH region, although there are differences in magnitude.

We should note that these seasonal ensemble mean projected changes in near-surface temperature over the three HKH subregions for the near-future and far-future periods are relatively higher than the seasonal global means based on the same subset of CMIP5 GCM (Table 3.6). The summer season global mean projected change for the far-future period is 1.9 °C (RCP4.5) and 3.3 °C (RCP8.5), while for the winter season the global mean projected changes are 2.0 °C (RCP4.5) and 3.5 °C (RCP8.5).

A more detailed analysis of future projections over the Tibetan region, based on a larger subset of 24 CMIP5 models (Annex 3), also indicates an increase in temperature over the HKH (Fig. 3.14), with higher magnitudes in the RCP8.5 scenario by the end of the 21st century. The maximum values are projected in the western part of the Tibetan region.

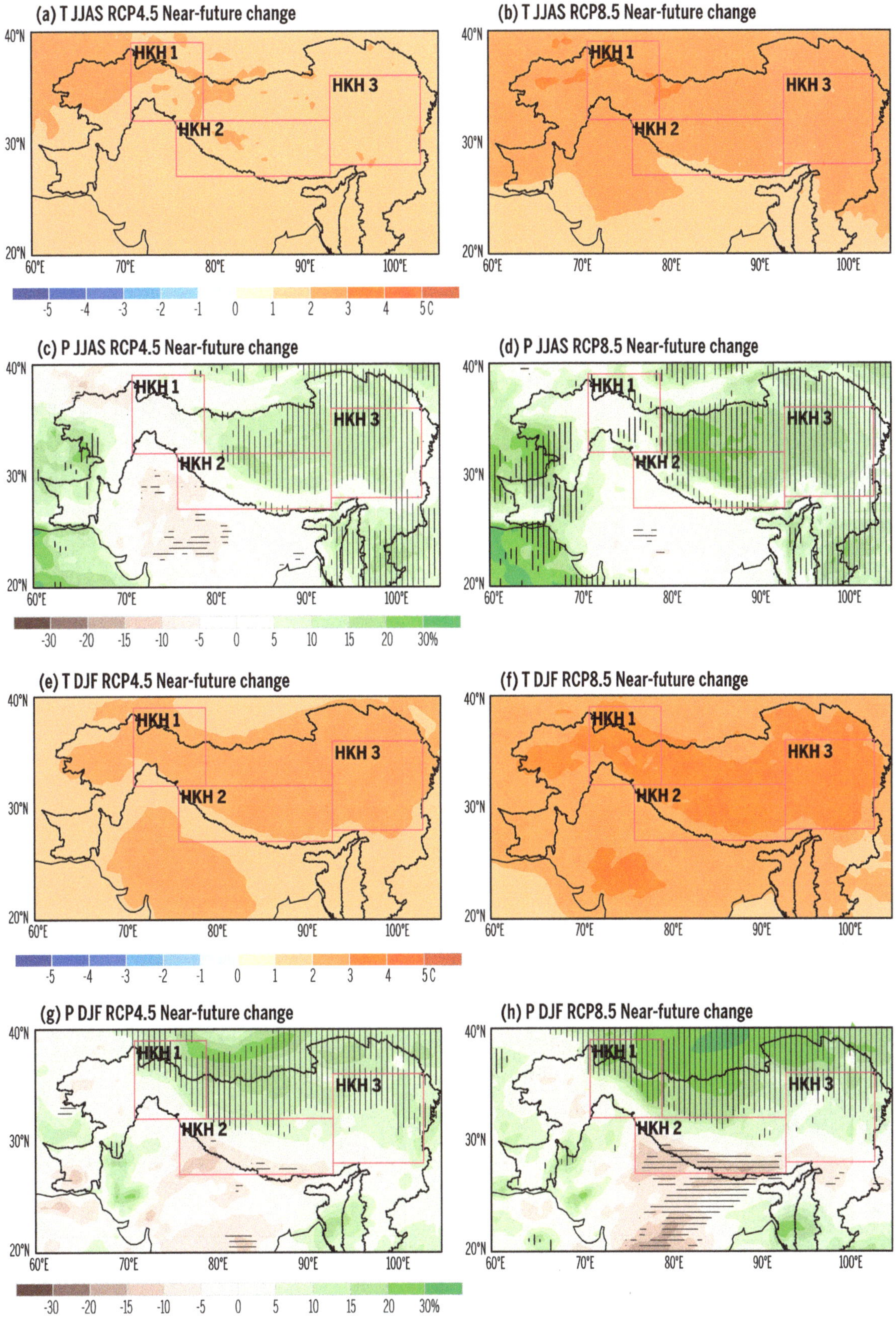

Fig. 3.12 Seasonal ensemble mean climate change in the Hindu Kush Himalaya (HKH) in the near future ([2036–65] with respect to [1976–2005]) for (top panels) surface air temperature (°C) and (bottom panels) total precipitation (%), with scenarios (first and third column) RCP4.5 and (second and fourth column) RCP8.5, during (**a–d**) summer monsoon (JJAS) and (**e–h**) winter (DJF) seasons. Ensemble mean of downscaling CMIP5 GCM with CORDEX South Asia RCM (listed in Table 3.5). Striping in bottom panels indicates where at least 10 of the 13 realizations concur on an increase (vertical) or decrease (horizontal) in RCPs. The HKH boundary is shown with dashed line. The boxes represent the three HKH sub-regions used for detailed analysis (see text) (*Source* Sanjay et al. 2017b)

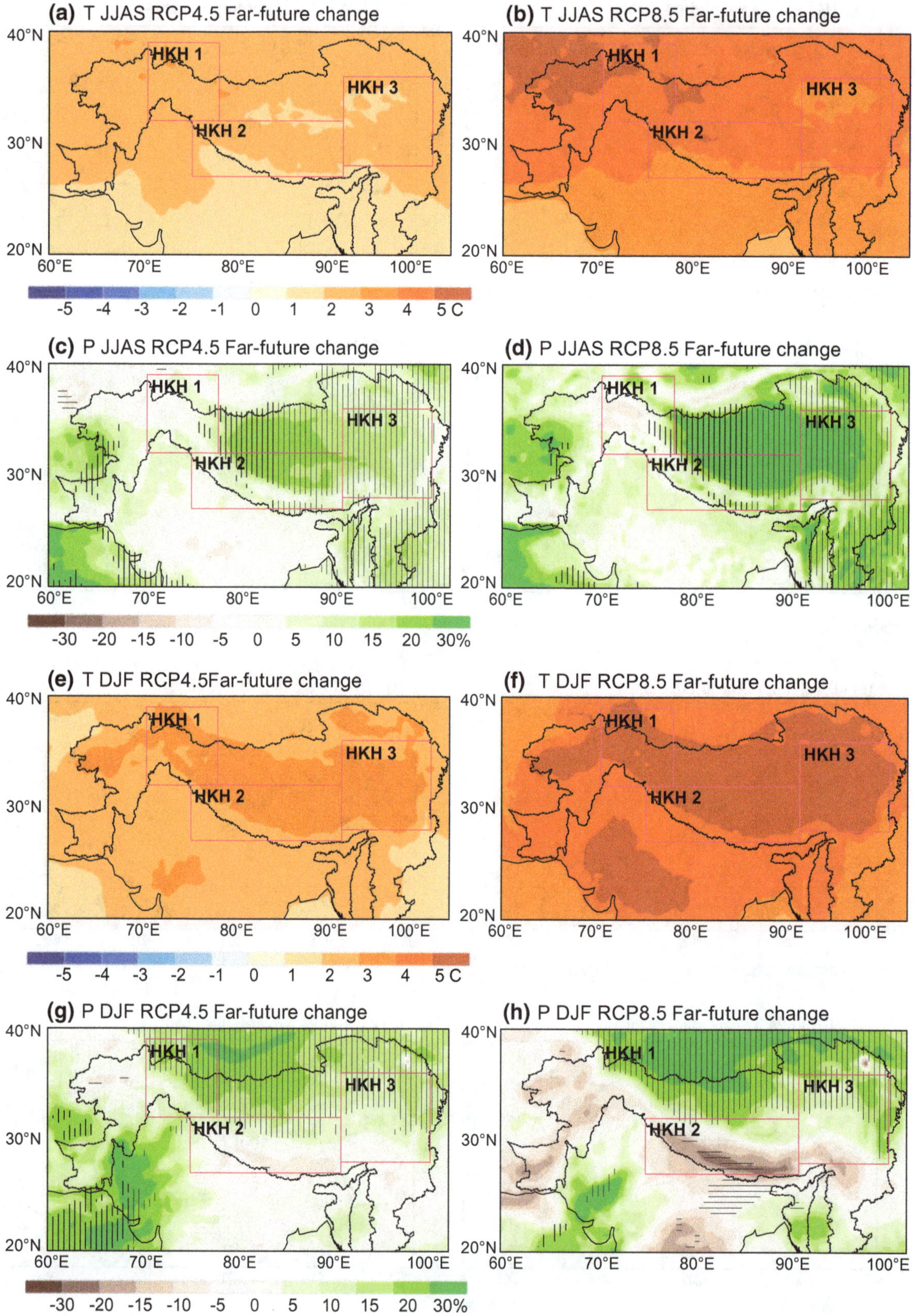

Fig. 3.13 Seasonal ensemble mean climate change in the Hindu Kush Himalaya (HKH) in the far future ([2066–2095]–[1976–2005]) for (top panels) surface air temperature (°C) and (bottom panels) total precipitation (%), with scenarios (first and third column) RCP4.5 and (second and fourth column) RCP8.5, during (**a–d**) summer monsoon (JJAS) and (**e–h**) winter (DJF) seasons. Ensemble mean of downscaling CMIP5 GCM with CORDEX South Asia RCM (listed in Annex 4 and Table 3.5). Striping in bottom panels indicates where at least 10 of the 13 realizations concur on an increase (vertical) or decrease (horizontal) in RCPs. The HKH boundary is shown by a dashed line. The boxes represent three HKH sub-regions used for detailed analysis (see text) (*Source* Sanjay et al. 2017b)

Table 3.5 Seasonal ensemble mean projected changes in near-surface air temperature (°C) relative to 1976–2005 in three Hindu Kush Himalaya (HKH) subregions defined by grid cells within each subregion above 2,500 m a.s.l. (see Fig. 3.1): northwestern Himalaya and Karakoram (HKH1); central Himalaya (HKH2); southeastern Himalaya and Tibetan Plateau (HKH3). The ranges for the 10 general circulation models and 13 regional climate models (listed in Annex 3, Table 3.12) analysed are given in brackets

Scenario	Period	Multi-model ensemble mean	Summer monsoon season (June–September) (°C)			Winter season (December–February) (°C)		
			HKH1	HKH2	HKH3	HKH1	HKH2	HKH3
RCP4.5	2036–65	Downscaled CORDEX RCM	2.0 (1.2, 3.3)	1.7 (1.1, 2.4)	1.7 (1.2, 2.2)	2.3 (1.4, 3.2)	2.4 (1.4, 3.4)	2.4 (1.4, 3.1)
		Driving CMIP5 GCM	2.6 (1.7, 3.3)	2.1 (1.6, 2.7)	2.0 (1.6, 2.4)	2.1 (1.2, 3.2)	2.7 (1.6, 3.9)	2.5 (1.4, 3.3)
	2066–95	Downscaled CORDEX RCM	2.6 (1.4, 3.7)	2.2 (1.4, 3.2)	2.2 (1.7, 2.9)	3.1 (2.2, 4.1)	3.3 (2.3, 4.5)	3.1 (2.0, 4.8)
		Driving CMIP5 GCM	3.3 (2.5, 4.1)	2.7 (2.1, 3.2)	2.5 (1.9, 2.9)	3.0 (2.1, 3.4)	3.6 (2.4, 4.6)	3.3 (2.1, 4.1)
RCP8.5	2036–65	Downscaled CORDEX RCM	2.7 (1.7, 4.3)	2.3 (1.4, 3.2)	2.3 (1.5, 2.9)	3.2 (1.8, 4.4)	3.3 (2.1, 4.6)	3.2 (2.0, 4.6)
		Driving CMIP5 GCM	3.3 (2.5, 4.3)	2.7 (2.0, 3.4)	2.5 (1.9, 3.0)	3.0 (2.2, 3.9)	3.4 (2.3, 4.7)	3.2 (2.1, 4.2)
	2066–95	Downscaled CORDEX RCM	4.9 (3.0, 7.7)	4.3 (3.1, 6.1)	4.2 (3.1, 5.4)	5.4 (3.9, 8.2)	6.0 (4.4, 9.0)	5.6 (4.2, 8.4)
		Driving CMIP5 GCM	5.7 (4.0, 7.1)	4.7 (3.9, 5.6)	4.4 (3.5, 5.3)	5.1 (3.8, 6.3)	5.8 (4.2, 7.8)	5.4 (3.8, 6.9)

Table 3.6 CMIP5 seasonal ensemble global mean projected changes in near-surface surface air temperature (°C) relative to 1976–2005. The range for the 10 GCM analysed (listed in Annex 4, Table 3.13) is given in brackets

CMIP5 global mean projected change	Summer monsoon season (June–September) (°C)		Winter season (December–February) (°C)	
	RCP4.5	RCP8.5	RCP4.5	RCP8.5
2036–2065	1.4 (0.9, 1.9)	1.9 (1.3, 2.4)	1.5 (1.1, 1.9)	2.0 (1.5, 2.5)
2066–2095	1.9 (1.2, 2.4)	3.3 (2.3, 4.1)	2.0 (1.4, 2.4)	3.5 (2.5, 4.2)

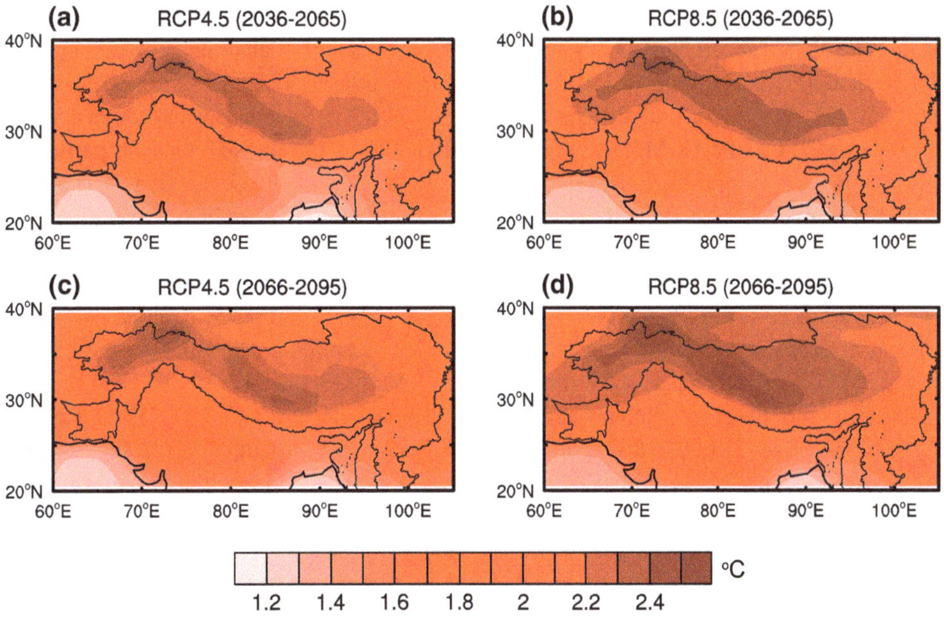

Fig. 3.14 The spatial distributions of annual mean temperature change (°C) over the Hindu Kush Himalaya (top panels) for the period 2036–65 for (**a**) representative concentration pathway 4.5 (RCP4.5) and (**b**) RCP8.5 and (bottom panels) for the period 2066–96 for (**c**) RCP4.5 and (**d**) RCP8.5

Box 3.3 What does a 1.5 °C rise in global average temperature mean for the HKH? Analysis of an ensemble of five GCM runs projecting a global temperature increase of 1.5 °C by the end of the 21st century, compared to the pre-industrial global mean temperature, reveals what this global 1.5 °C increase implies for the High Mountain Asia region. According to this ensemble, for the complete HIMAP domain a 1.5 °C global temperature increase would mean a temperature increase of 1.8 ± 0.4 °C averaged over the region (Table 3.7). Looking solely at the mountain ranges, this enhanced warming is even more pronounced. For the Karakoram, central Himalaya, and southeastern Himalaya, a 1.5 °C global temperature increase would imply regional temperature increases of 2.2 °C ± 0.4 °C, 2.0 °C ± 0.5 °C, and 2.0 °C ± 0.5 °C, respectively.

Table 3.7 Comparison results of models projecting a 1.5 °C increase in near-surface air temperature (°C) globally and for the Hindu Kush Himalaya (HKH) and the three sub-domains. The temperature changes are for the end-of-century from the pre-industrial period (2071–2100 vs. 1851–80)

RCP	Model	Global	HKH	HKH1	HKH2	HKH3
RCP2.6	GISS-E2-R_r1i1p3	1.48	1.82	1.87	1.73	2.35
RCP2.6	MIROC5_r1i1p1	1.48	1.95	2.54	2.46	2.28
RCP2.6	NorESM1-ME_r1i1p1	1.44	1.68	2.05	1.85	1.63
RCP2.6	HadGEM2-AO_r1i1p1	1.57	1.47	2.04	1.49	1.50
RCP2.6	MPI-ESM-MR_r1i1p1	1.58	2.16	2.58	2.42	2.11
MEAN		1.51	1.82	2.22	1.99	1.97
RANGE		0.14	0.69	0.71	0.97	0.85
SD		0.06	0.26	0.32	0.43	0.39

3.4.2 Precipitation Projected to Increase, but with Regional Diversity

There is a divergence among models in projecting future changes in precipitation. No ideal model captures South Asian monsoon behaviour. Unlike temperature, the precipitation response to climate change over the HKH region is subject to larger uncertainties both in the CMIP and CORDEX models (e.g., Choudhary and Dimri 2017; Hasson et al. 2013, 2015; Mishra, 2015; Sanjay et al. 2017b). Evaluating 20 CMIP5 GCM, Hasson et al. (2015) analysed the seasonal cycle of precipitation in the main river basins of the HKH, and emphasized that the inter-model spread on monsoonal precipitation is more in the western parts of the region com-pared to the eastern parts of the region. CMIP5 models are more skillful in terms of simulating patterns, but the intra-seasonal variability remains problematic (Sperber et al. 2013). Climate models have considerable difficulty in simulating the observed pattern because of extremes in topography. Even with multi-model evaluation of global changes in mean precipitation per degree of warming (22 atmosphere–ocean general circulation models (AOGCM) projection), there is an increase in precipitation over the south and southeastern Asian region during the summer season, but over the Himalaya fewer models are in agreement (IPCC 2007). Hasson et al. (2013) reported large uncertainties in the way CMIP3 models describe the future water budget in large HKH river basins, with a clearer tendency for wetter conditions in the eastern basins.

Regional climate model (PRECIS) analysis carried out on a river basin scale (Indus Basin) by Rajbhandari et al. (2015) could not establish any trends in precipitation changes in the future in the western HKH. However, in the eastern Himalaya (Koshi Basin), based on the eight GCM downscaled using the delta method, the authors reported for both RCP4.5 and RCP8.5 a 10–20% increase in rainfall during the summer monsoon and about a 5% increase in the winter season over the trans-Himalayan part of the basin (Rajbhandari et al. 2016). Similarly, Bharati et al. (2016) reported an increase in future precipitation over trans-mountain sub-watersheds in the 2030s and 2050s. A study carried out by Kulkarni et al. (2013) over the HKH using PRECIS reported a 20–40% increase in monsoon rainfall towards the end of the 21st century. The precipitation is not simulated as well at regional scales by global climate models used in CMIP5 experiments for the IPCC Assessment Report 5 (AR5), and the assessment is hampered by observational uncertainties (IPCC 2013). The AR5 highlights that the projected changes in the water cycle at regional scale will be strongly influenced by natural internal variability and may be affected by anthropogenic aerosol emissions. This assessment further notes that while the South Asian summer monsoon winds are likely to weaken, monsoon precipitation is likely to intensify due to the increase in atmospheric moisture (e.g., Kitoh 2017; Krishnan et al. 2013).

Table 3.8 Seasonal ensemble mean projected changes in total precipitation (%) relative to 1976–2005 in three Hindu Kush Himalaya (HKH) subregions defined by grid cells within each subregion above 2,500 m a.s.l. (see Fig. 3.1): northwestern Himalaya and Karakoram (HKH1); central Himalaya (HKH2); southeastern Himalaya and TP (HKH3). The range for the 10 GCM and 13 RCM (listed in Table 3.13 in Annex 4) analysed is given in brackets

Scenario	Period	Multi-model ensemble mean	Summer monsoon season (June–September) (%)			Winter season (December–February) (%)		
			HKH1	HKH2	HKH3	HKH1	HKH2	HKH3
RCP4.5	2036–65	Downscaled CORDEX RCM	−0.1 (−11.6, 19.7)	4.4 (−4.5, 15.1)	6.8 (−5.5, 11.9)	7.0 (−13.9, 21.9)	−0.7 (−18.6, 15.0)	3.1 (−16.4, 19.5)
		Driving CMIP5 GCM	0.8 (−17.1, 35.1)	6.7 (−0.8, 22.1)	4.6 (−1.4, 10.2)	1.0 (−10.2, 18.0)	−7.7 (−21.0, 2.9)	2.1 (−7.8, 14.8)
	2066–95	Downscaled CORDEX RCM	3.5 (−9.8, 29.3)	10.5 (−4.9, 29.9)	10.4 (−3.4, 20.4)	14.1 (−4.9, 34.4)	1.5 (−19.3, 36.7)	3.7 (−19.6, 32.1)
		Driving CMIP5 GCM	−0.3 (−23.2, 34.8)	11.8 (−1.4, 33.0)	7.3 (0.6, 13.0)	6.2 (−6.8, 43.3)	−0.7 (−17.3, 22.0)	5.5 (−14.6, 23.9)
RCP8.5	2036–65	Downscaled CORDEX RCM	3.7 (−13.8, 22.3)	9.1 (−3.2, 23.6)	10.2 (−5.1, 22.2)	12.8 (−12.3, 28.8)	−1.3 (−19.7, 7.0)	0.9 (−13.2, 24.4)
		Driving CMIP5 GCM	3.6 (−16.6, 48.8)	10.7 (−4.1, 28.3)	5.7 (−1.4, 13.8)	5.1 (−10.9, 36.0)	−8.5 (−25.5, 2.1)	0.7 (−18.8, 25.4)
	2066–95	Downscaled CORDEX RCM	3.9 (−14.9, 60.0)	19.1 (−6.0, 40.5)	22.6 (−6.1, 49.0)	12.9 (−30.3, 35.4)	−8.8 (−33.2, 25.9)	0.6 (−16.6, 35.7)
		Driving CMIP5 GCM	5.0 (−17.7, 79.9)	19.1 (−1.1, 34.0)	9.7 (−1.8, 23.8)	6.9 (−20.9, 54.7)	−8.1 (−35.5, 14.2)	6.0 (−26.0, 42.8)

Future projections of summer monsoon precipitation over parts of the HKH, based on the CORDEX RCM, indicate a likely increase by 4–12% in the near future and by 4–25% in the long term (see Table 3.8). On the other hand, winter precipitation is projected to increase by 7%–15% in the Karakoram, but to decline slightly in the central Himalaya. It is important to mention that the CORDEX RCM have inherent limitations in reproducing the observed characteristics of the summer monsoon rainfall variability (Singh et al. 2017), and that future projections of precipitation extremes are unresolved, both across studies and across regions (e.g., Palazzi et al. 2013). Kapnick et al. (2014) emphasize the need to employ higher-resolution climate model simulations, as compared to the CMIP5 models, to better describe the distinct seasonal cycles and resulting climate change signatures of Asia's high-mountain ranges in the HKH.

Future projections of annual mean precipitation change, based on CMIP5 multi-model ensemble mean, have indicated an increase in precipitation over the HKH for both RCP4.5 and RCP8.5 scenarios (Fig. 3.11b). Similar to the surface temperature projections, simulation data from the CMIP5 models reveal enhanced precipitation change over the HKH compared to the global mean change by the end of the 21st century (Fig. 3.11d). The projected change in precipitation is similar for both scenarios until the 2050s, after which the RCP8.5 shows more increase until the end of the 21st century. Apart from a few projections for the near future, most of the projections show an increase in precipitation.

The magnitude and sign of the projected seasonal percent change was found to vary with region, season, averaging period, and scenario used. The spatial pattern of the seasonal ensemble mean projected percent changes in total precipitation based on high-resolution CORDEX South Asia RCM is shown over the HKH for the periods 2036–65 (near future; Fig. 3.12) and 2066–95 (far future; Fig. 3.13) relative to 1976–2005. The striping indicates where at least 10 of the 13 realizations concur on an increase (vertical) or decrease (horizontal) in total precipitation for the RCP. The individual RCM agree on a projected summer season increase in precipitation over the hilly regions in the central Himalaya (HKH2) and southeastern Himalaya and TP (HKH3) for both RCP4.5 and RCP8.5 scenarios in the near-future and far-future periods. There is less model agreement on the southern slopes of the Himalaya, and in the northwestern Himalaya and Karakoram (HKH1) region, during the summer season for both scenarios until the end of the 21st century. The models agree well on the winter season increase in total precipitation over HKH1 for near-future and far-future periods and for both scenarios. The confidence among the models for projected reduction in winter precipitation over HKH2 and HKH3 is low for both scenarios in the two periods.

Table 3.8 shows the projected total precipitation changes in the hilly region (see Fig. 3.1) with RCP4.5 and RCP 8.5 scenarios for near-future and far-future periods. This detailed analysis further confirms that during summer (winter), relatively higher (lower) precipitation increase will occur over

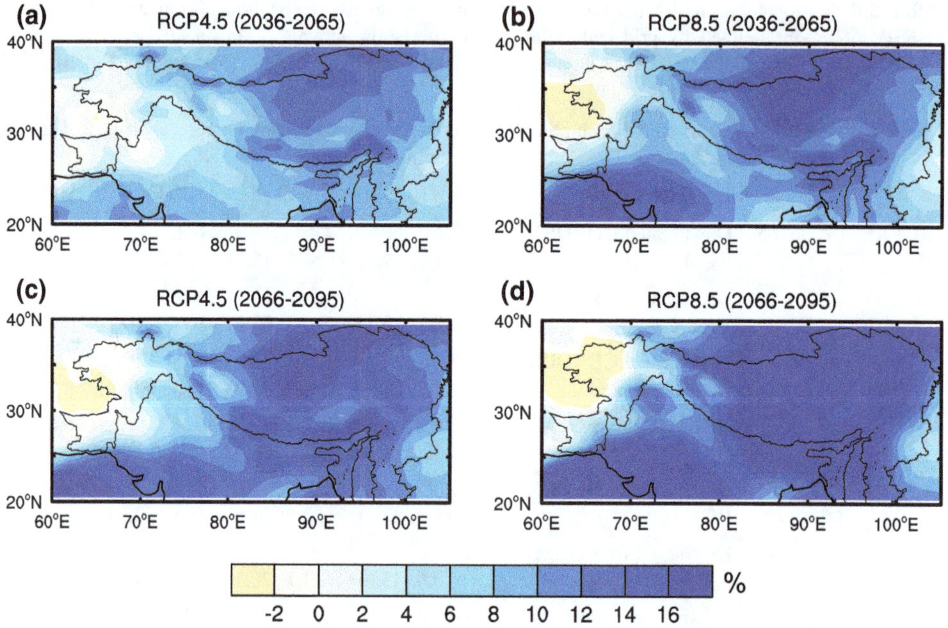

Fig. 3.15 The spatial distributions of annual mean precipitation change (%) over the Hindu Kush Himalayan region for (**a**, **c**) RCP4.5 and (**b**, **d**) RCP8.5 during (**a**, **b**) 2036–65 and (**c**, **d**) 2066–95

HKH2 and HKH3 than in HKH1 for both scenarios by the end of the 21st century. The largest projected seasonal ensemble mean total precipitation increase for the high-resolution CORDEX RCM during the summer monsoon season is about 10% over HKH2 with the RCP4.5 scenario and about 22% over HKH3 with the RCP8.5 scenario. During winter the largest projected increase in precipitation is over HKH1 of about 14% and 13% with RCP4.5 and RCP8.5 scenarios, respectively. However, it can be seen that the spread among the models is large for the high-resolution CORDEX RCM, as well as for their driving CMIP5 GCM (Sanjay et al. 2017b).

Detailed analysis of the CMIP5 multi-model annual precipitation projections over HKH (Annex 3) suggests that the western part of this region tends to be drier and the Tibetan region tends to be wetter in the future (see Fig. 3.15). The large magnitudes of precipitation appear by the end of the 21st century under RCP4.5 and RCP8.5.

3.5 Limitations and Gaps in the Analysis

There is no previous study specifically for the HKH region as defined by HIMAP; therefore, a major part of this chapter is based on original analyses for both the historical period and the future. Large gaps in observations exist in the HKH, especially in Afghanistan, Pakistan, Myanmar, Bhutan, and Nepal, and for the northwestern part of the TP. Monthly and daily pre-1950 data are lacking for most parts of the HKH, and the gap is particularly large for areas outside of India. The daily data post-1950 are also insufficient in some areas, including Afghanistan, Pakistan, Myanmar, Bhutan, and Nepal, and the northwestern part of the TP. The sparseness of observational data is the major source of uncertainties in the estimates of long-term trends of mean and extreme climatic indices in the HKH. Improvement in data coverage is under way, but time is needed to obtain records that are sufficiently long term to allow analysis of past climate change. An effective measure to increase the coverage of data is to recover and digitize the original records from earlier years still kept in paper documents in some countries, in coordination with the activities of Atmospheric Circulation Reconstructions over the Earth (ACRE) China, ACRE SE Asia, and ACRE Indian Ocean.

Uncertainties with the observational studies also result from the gaps in studies, including the data processing and analysing methods. Data inhomogeneities related to station relocations and instrumentation sometimes cause large biases in the estimates of trends of key climatic elements if proper adjustments have not been made. Major uncertainty with the estimates of extreme temperature trends comes from the systematic bias of the historical temperature data series caused by urbanization. Similar to the mean surface air temperature, urbanization also exerts a large effect on the long-term trends of the extreme temperature indices in a subcontinental region like mainland China (Ren et al. 2014; Sun et al. 2016; Zhou and Ren 2009, 2011). The effect of urbanization in annual and seasonal extreme temperature indices series in the TP region has not yet corrected. However, even if we take this effect into account, the baseline warming over the TP and the HKH, probably caused by global and regional drivers including the anthropogenic increase in atmospheric CO_2 concentration, could be still significant. The sparseness of early-year observations in the TP, the Yunnan–Guizhou Plateau, and Pakistan and Afghanistan is also an important source of the uncertainties.

Scenarios or the pathways are the trajectories undertaken by climate models to derive future climate. Models use various pathways to create future climate scenarios; therefore, uncertainties are inherent in the model. In the present study, RCP4.5 and RCP8.5 have been used to analyse the future climate. Although future pathways are uncertain, a comparison of recent global trends in carbon dioxide emissions shows that the current emission is close to RCP8.5 (Friedlingstein et al. 2014; Peters et al. 2012).

The future temperature and precipitation presented in Sect. 3.4 are analyses of outputs from GCM and CORDEX South Asia models. Although many of the multi-model projections agree on the direction of changes, there are differences in magnitude and spatial distribution. CORDEX South Asia model outputs have not been validated over the HKH, and the results provided only for the assessment of the trend and not of the magnitude of the event.

Annex 1: Data and Methods for Analysis of Past Temperature

For this assessment, the source of monthly mean temperature (Tmean), maximum temperature (Tmax), and minimum temperature (Tmin) data is the CMA GLSAT-V1.0 (Xu et al. 2014). The data in the CMA GLSAT-V1.0 have been quality-controlled and homogenized. The record length is 114 years (1901–2014). Only those stations with at least 15 years of records in the base period 1961–90 were selected for analysis and, at the same time, the station's records were required to be greater than or equal to 5 years during the first half and the last half of the analysis period (1901–60 and 1990–2014). The total numbers of stations used for the analysis was therefore 122 for Tmean, 94 for Tmax and 94 for Tmin in the HKH. In addition, if there were 6 months in 1 year without data, that year was not included for calculation. Distribution of the stations used is shown in Fig. 3.16a.

To reduce the biases caused by uneven station density or temporal variations in data coverage, the Climate Anomaly Method (CAM) (Jones and Moberg 2003) was employed. First, each station was assigned to a regular 2.5° × 2.5° or 5° × 5° latitude–longitude grid box. The grid boxes had at least one station. Then, grid box temperature anomalies were calculated by averaging anomalies for the stations within grid boxes. Finally, the HKH average temperature, maximum temperature, and minimum temperature anomalies were calculated by the area-weighted average (using the cosines of the mid-grid latitude as weights) of all grid box anomalies.

The reference period 1961–90 was chosen, mainly because of the better record of spatial coverage. Annual values were calculated by averaging all 12 monthly values. The linear trends of the temperature anomaly series were the linear regression coefficients between temperature and ordinal numbers of time obtained by using the least squares method (e.g., I = 1, 2, 3 … 64 for 1901–2014 or 1951–2014 for extreme analysis). The significance of the linear trends of temperature series was judged by using the two-tailed simple-test method. In this assessment, a trend is considered statistically significant if it is significant at the 5% ($p < 0.05$) level.

The source of daily Tmax and Tmin measurements for current analysis was the global land surface daily air temperature data set V1.0, developed by the CMA. The temperature data set has been quality controlled and includes daily Tmean, Tmax, and Tmin. The record length varied for stations, but a 55-year period (1961–2015) was used for the assessment in the HKH. All stations included in the report were required to have at least 30 years of records for the whole period and at least 15 years of records during the 1961–90 reference period. Stations were also required to have at least 10 months of records for every year with data. The total number of stations used was therefore 478. The spatial distribution and record length of the temperature stations in the HKH are shown in Fig. 3.16, and the data sources and the numbers of stations used in this report are shown in Table 3.9.

Nine temperature-related indices of the 27 extreme climate events indices recommended by the ETCCDI are used in this report. They include cold nights (TN10p), cold days (TX10p), warm nights (TN90p), warm days (TX90p), monthly maximum value of daily maximum temperature (TXx), monthly minimum value of daily minimum temperature (TNn), frost days (FD), summer days (SU), and daily temperature range (DTR). The definitions of the indices can be found in Table 3.10 (You et al. 2015; Zhou and Ren 2011) (Tables 3.11 and 3.12).

Annex 2: Data and Methods for Analysis of Past Precipitation

The sources of monthly and daily precipitation measurement for current analysis are from the CMA GLMP-V1.0 and Global Land Daily Precipitation data set V1.0 (CMA GLDP-V1.0). The data in the GLMP-V1.0 have been quality controlled and the record length is 113 years (1901–2013), while the GLDP-V1.0 data have also been quality controlled but have not been homogenized, and they only have records of 53 years (1961–2013).

In the HKH, stations were selected that have at least 10 years of precipitation records in the base period 1961–90, at least 5 years in the period 1901–50, and 5 years in the

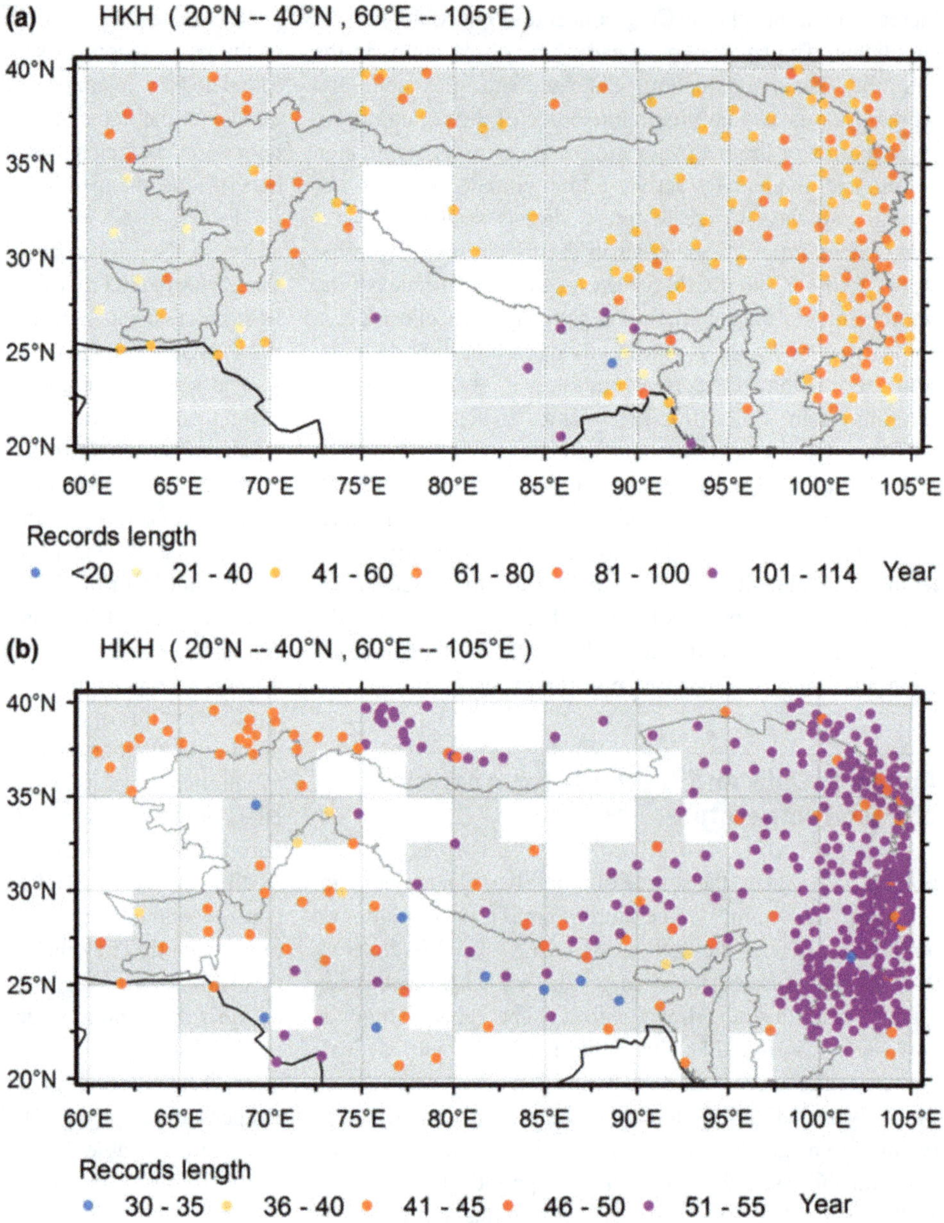

Fig. 3.16 Spatial distribution, record length, and gridboxes (grey area) of stations with daily temperature data in the HKH (20–40° N, 60–105° E). (**a**) 1901–2014 for monthly data; (**b**) 1961–2015 for daily data

Table 3.9 Daily data sources and station numbers between 1951 and 2014

Data set	CMA	GHCND	GSOD	ECA	Russia	Vietnam
Number	387	55	27	6	1	2
Percentage	80.1	11.5	5.6	0.2	0.4	1.2

CMA China Meteorological Administration; *GHCND* Global Historical Climatology Network-Daily; *GSOD* Global Surface Summary of the Day; *ECA* European Climate Assessment

period 1991–2013 for studying the precipitation change in the period 1901–2013. A total of 245 stations was selected, as shown in Fig. 3.17a. The stations are not dense enough, except for India and southwestern China.

For the analysis of extreme precipitation for the period 1961–2013, daily precipitation data were selected with at least 10 years of records in the base period 1961–90 and another 5 years in the period 1991–2013. In total, 1,024 stations were selected for use, and the spatial distribution was plotted in Fig. 3.17b. The stations are uniformly and densely distributed throughout this region, except for Afghanistan, Bangladesh, Bhutan, the Qinghai–Tibet Plateau, Myanmar, Nepal, and Pakistan, where the data coverage is relatively poor.

As the yearly precipitation records in the study region have huge discrepancies, PSA were used to calculate the regional averages of precipitation amount and to compare the changing trends in different parts of the HKH. PPA was

Table 3.10 List of ETCCDI core extreme temperature indices used in this report

	Indicator name	Definitions	Units
DTR	Diurnal temperature range	Monthly mean difference between TX and TN	°C
FD	Frost days	Annual count when TN (daily minimum) <0 °C	Days
SU	Summer days	Annual count when TX (daily maximum) >25 °C	Days
TN10p	Cool nights	Percentage of days when TN <10th percentile	Days
TN90p	Warm nights	Percentage of days when TN >90th percentile	Days
TNn	Min Tmin	Monthly minimum value of daily minimum temp	°C
TX10p	Cool days	Percentage of days when TX <10th percentile	Days
TX90p	Warm days	Percentage of days when TX >90th percentile	Days
TXx	Max Tmax	Monthly maximum value of daily maximum temp	°C

Table 3.11 ETCCDI core extreme precipitation indices used in this chapter

ID	Indicator name	Definitions	Units
ATP	Annual total precipitation	Annual total precipitation when daily rainfall >0.1 mm	mm
PA	Precipitation anomaly	The anomaly (based on 1961–90) of ATP	mm
PSA	Precipitation standardized anomaly	PA divided by ATP standard deviation (on 1961–90)	None
PPA	Precipitation percentage anomaly	PA divided by ATP (on 1961–90)	%
WD	Wet day	Annual total days when daily rainfall >0.1 mm	Day
WDA	Wet day anomaly	The anomaly (based on 1961–90) of WD	Day
SDII	Simple daily intensity index	ATP divided by WD, or average daily precipitation	mm/day
DPIA	Daily precipitation intensity anomaly	The anomaly (based on 1961–90) of SDII	mm/day
LR	Light rain	Precipitation when daily rainfall <50th percentile	mm or d
MR	Moderate rain	Precipitation when daily rainfall is between 50th and 90th percentiles	mm or d
IR	Intense rain	Precipitation when daily rainfall >90th percentile	mm or d

Table 3.12 Basic information on the 24 CMIP5 models

Model	Institution and country	Resolution (long. × lat.)
ACCESS1.0	CSIRO–BOM/Australia	192 × 145
BCC-CSM1.1	Beijing Climate Center, CMA/China	128 × 64
BNU-ESM	GCESS/China	128 × 64
CanESM2	Canadian Centre for Climate Modeling and Analysis/Canada	128 × 64
CCSM4	NCAR/USA	288 × 192
CESM1-BGC	National Science Foundation (NSF)–Department of Energy (DOE)–NCAR/USA	288 × 192
CNRM-CM5	CNRM and CERFACS/France	256 × 128
CSIRO-Mk3-6-0	CSIRO and QCCCE/Australia	192 × 96
GFDL-ESM2G	NOAA/GFDL/USA	144 × 90
GFDL-ESM2 M	NOAA/GFDL/USA	144 × 90
HadGEM2-ES	MOHC/UK	192 × 145
HadGEM2-CC	MOHC/UK	192 × 145
HadCM3	MOHC/UK	96 × 73
INMCM4	Institute of Numerical Mathematics (INM)/Russia	180 × 120
IPSL-CM5A-LR	IPSL/France	96 × 96
IPSL-CM5A-MR	IPSL/France	144 × 143
MIROC5	MIROC/Japan	256 × 128
MIROC-ESM	MIROC/Japan	128 × 64
MIROC-ESM-CHEM	MIROC/Japan	128 × 64
MIROC4 h	MIROC/Japan	640 × 320
MPI-ESM-LR	MPI-M/Germany	192 × 96L56
MPI-ESM-MR	MPI-M/Germany	192 × 96L47
MRI-CGCM3	Meteorological Research Institute/Japan	320 × 160
NorESM1-M	Norwegian Climate Centre (NCC)/Norway	144 × 96

BOM Bureau of Meteorology; *CMA* China Meteorological Administration; *CNRM* Centre National de Recherches Météorologiques; *CSIRO* Commonwealth Scientific and Industrial Research Organisation; *GCESS* Global Change and Earth System Science; *GFDL* Geophysical Fluid Dynamics Laboratory; *IPSL* Institut Pierre-Simon Laplace; *MIROC* Model for Interdisciplinary Research on Climate; *MOHC* Met Office Hadley Centre; *MPI-M* Max Planck Institute for Meteorology; *NCAR* National Center for Atmospheric Research; *NOAA* National Oceanic and Atmospheric Administration

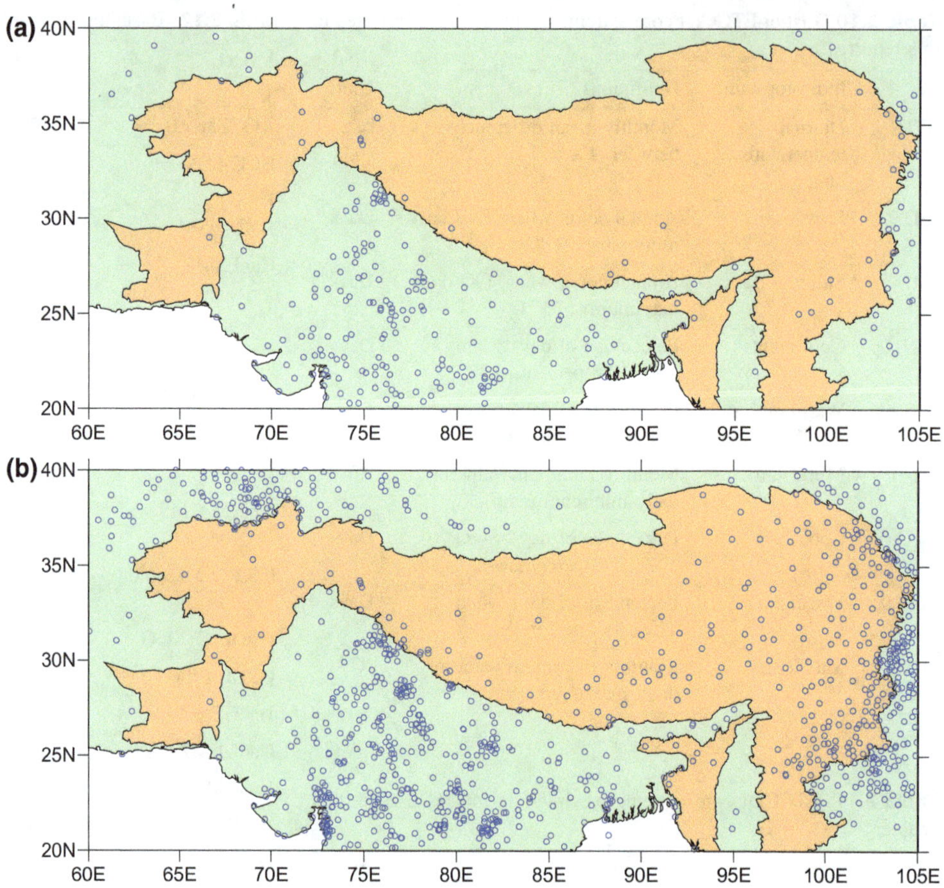

Fig. 3.17 Spatial distribution of precipitation stations in the Hindu Kush Himalaya. (**a**) 1901–2013 for monthly data; (**b**) 1961–2013 for daily data

also used because it is sometimes a more useful indicator for comparing the regional and temporal differences of precipitation amount. For wet days or precipitation days, and simple daily intensity index (SDII) or daily precipitation intensity, however, the WDA and DPIA anomalies were applied for the analysis.

The relative threshold values were used to define the extreme precipitation events in the assessment report. A three-grade system was applied, with the light precipitation day defined as a day with the 24 h precipitation below the 50th percentile value of the reference period (1961–90). The moderate precipitation day was defined as a day with the 24 h precipitation between the 50th and 90th percentile values of the reference period. The intense precipitation day was defined as a day with the 24 h precipitation above the 90th percentile value of the reference period. The precipitation amount of the categories was the accumulated precipitation within the different grades of precipitation days, and the precipitation intensity was the precipitation amount divided by the precipitation days. The definitions of the indices used in this study are given in Table 3.11 (You et al. 2015; Zhang et al. 2011).

Annex 3: Analysis of CMIP5 Data

The projection of temperature and precipitation can help policy makers in their efforts to make decisions. The performance of climate models (for instance, in terms of their resolution and the complexity of the physical processes they can consider) has been improved greatly in recent years. As such, the outputs of these models have provided a solid basis for the projection of climate changes. Many researchers have, through historical analysis, been able to assess the capabilities of the climate models involved in CMIP3 and CMIP5. Based on these assessments, projections have been made with respect to the mean state and climate extremes over different regions in different periods of the 21st century under various emissions scenarios.

In this study, we employed 24 model simulations from CMIP5 RCP4.5 and RCP8.5 (see Table 3.12 for model information). For comparison, the model outputs were interpolated onto a common 1 × 1. The equal weight averages of 24 GCM were taken to be the projected changes, and the changes were relative to the baseline period (1976–2005).

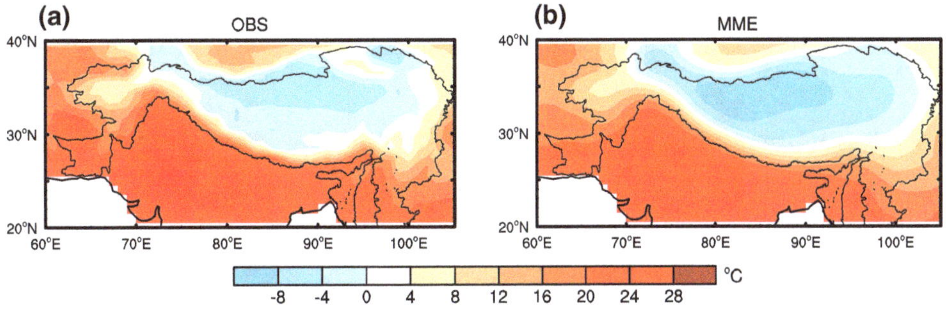

Fig. 3.18 Spatial distributions of mean temperature over the Hindu Kush Himalaya during 1976–2005 (units: °C). (**a**) Observation; (**b**) the multi-model ensemble

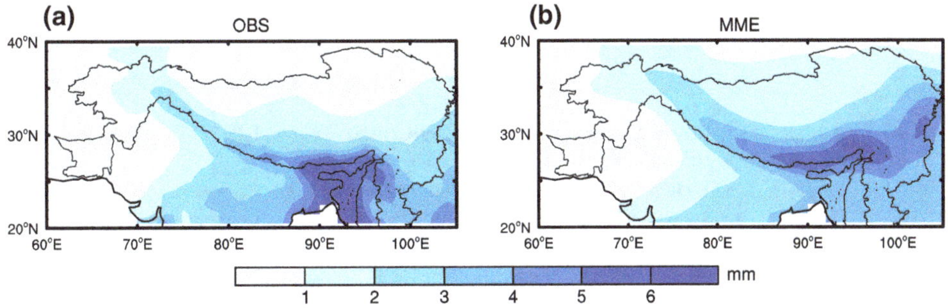

Fig. 3.19 Spatial distributions of mean precipitation over the Hindu Kush Himalaya during 1976–2005 (units: mm); (**a**) observation; (**b**) the multi-model ensemble

The assessments of temperature (Fig. 3.1) and precipitation (Fig. 3.2) over HKH during 1976–2005 were conducted using CRU data and CMIP5 model simulations. Overall, the multi-model ensemble can reproduce the spatial distribution of temperature and precipitation over HKH, but it underestimates for temperature and overestimates for precipitation owing to lower resolution from CMIP5 global models (Figs. 3.18 and 3.19).

Annex 4: Analysis of CORDEX Data

A thorough understanding of the regional climate variability in the present and future time scales would enrich the information available for policy makers and stakeholders and allow them to make informed decisions. Since there is no pre-existing climate assessment of the HKH region, global climate projections available from CMIP using existing state-of-the-art coupled AOGCM need to be downscaled at high spatial resolution to resolve the complexities of the monsoon and other hydrological processes over regional scales. With this objective in mind, the Centre for Climate Change Research (CCCR[12]) at the Indian Institute of Tropical Meteorology (IITM) and various international modelling centres are actively involved in the generation of regional climate change projections for South Asia by participating in CORDEX, the World Climate Research Programme (WCRP) initiative.[13] These ensembles of dynamically downscaled projections of regional climate over the large domain covering South Asia (19.25oE-116.25oE; 15.75oS-45.75oN) using high-resolution (50 km) RCM are archived at CCCR-IITM, the nodal agency for CORDEX South Asia (Sanjay et al. 2017b). A high-end climate data server has been set up at CCCR-IITM for archiving and disseminating these high-resolution regional climate data sets through the CCCR web portal[14] and by publishing on the CCCR-IITM Earth System Grid Federation (ESGF) data node.[15] The list of models used in this study is provided in Table 3.13.

The WCRP's Working Group on Regional Climate and the Working Group on Coupled Modelling, the former coordinating body of CORDEX, and the responsible panel for CMIP5, are gratefully acknowledged. The climate modelling groups (listed in Table 3.12) are thanked sincerely for producing and making available their model output. The authors thank the ESGF infrastructure and the climate data portal hosted at the CCCR, IITM, for providing CORDEX South Asia data.

[12]http://cccr.tropmet.res.in/home/index.jsp.

[13]http://www.cordex.org/.

[14]http://cccr.tropmet.res.in/home/data_portals.jsp.

[15]http://cccr.tropmet.res.in/home/esgf_data.jsp.

Table 3.13 The 13 CORDEX South Asia downscaled regional climate model simulations driven by 10 CMIP5 general circulation models

CORDEX South Asia RCM	RCM description	Contributing CORDEX modelling centre	Driving CMIP5 GCM[a]	Contributing CMIP5 modelling centre
IITM-RegCM4 (six ensemble members)	Abdus Salam ICTP Regional Climatic Model version 4 (RegCM4; Giorgi et al. 2012)	CCCR, IITM, India	CCCma-CanESM2	Canadian Centre for Climate Modelling and Analysis (CCCma), Canada
			NOAA-GFDL-GFDL-ESM2 M	NOAA, GFDL, USA
			CNRM-CM5	CNRM, France
			MPI-ESM-MR	MPI-M, Germany
			IPSL-CM5A-LR	IPSL, France
			CSIRO-Mk3.6	Commonwealth Scientific and Industrial Research Organisation (CSIRO), Australia
SMHI-RCA4 (six ensemble members)	Rossby Centre regional atmospheric model version 4 (RCA4; Samuelsson et al. 2011)	Rossby Centre, Swedish Meteorological and Hydrological Institute (SMHI), Sweden	ICHEC-EC-EARTH	Irish Centre for High-End Computing (ICHEC), European Consortium (EC)
			MIROC-MIROC5	Model for Interdisciplinary Research On Climate (MIROC), Japan Agency for Marine-Earth Sci. and Tech., Japan
			NOAA-GFDL-GFDL-ESM2 M	NOAA, GFDL, USA
			CNRM-CM5	CNRM, France
			MPI-ESM-LR	MPI-M, Germany
			IPSL-CM5A-MR	IPSL, France
MPI-CSC-REMO2009 (one ensemble member)	MPI Regional model 2009 (REMO2009; Teichmann et al. 2013)	Climate Service Center (CSC), Germany	MPI-ESM-LR	MPI-M, Germany

[a]https://verc.enes.org/data/enes-model-data/cmip5/resolution). *CCCR* Centre for Climate Change Research; *CNRM* Centre National de Recherches Météorologiques; *GFDL* Geophysical Fluid Dynamics Laboratory; *ICTP* International Centre for Theoretical Physics; *IITM* Indian Institute of Tropical Meteorology; *IPSL* Institut Pierre-Simon Laplace; *MPIM* Max Planck Institute for Meteorology; *NOAA* National Oceanic and Atmospheric Administration; *RCM* regional climate model

References

Alexander, L. V., Zhang, X., Peterson, T. C., Caesar, J., Gleason, B., Klein Tank, A. M. G., et al. (2006). Global observed changes in daily climate extremes of temperature and precipitation. *Journal of Geophysical Research: Atmospheres, 111*(D5).

An, C. B., Tang, L., Barton, L., & Chen, F. H. (2005). Climate change and cultural response around 4000 cal yr B.P. in the western part of Chinese Loess Plateau. *Quaternary Research, 63*(3), 347–352.

Archer, D. R., & Fowler, H. (2004). Spatial and temporal variations in precipitation in the Upper Indus Basin, global teleconnections and hydrological implications. *Hydrology and Earth System Sciences, 8*(1), 47–61.

Asahi, K., & Watanabe, T. (2004). Paleoclimate of the Nepal Himalayas during the last glacial: Reconstructing from glacial equilibrium-line altitude. *Himalayan Journal of Science, 2*(Special Issue), 100–101.

Barlow, M., Matthew, W., Bradfield, L., & Heidi, C. (2005). Modulation of daily precipitation over southwest Asia by the Madden-Julian oscillation. *Monthly Weather Review, 133*(12), 3579–3594.

Bharati, L., Gurung, P., Maharjan, L., & Bhattarai, U. (2016). Past and future variability in the hydrological regime of the Koshi Basin, Nepal. *Hydrological Sciences Journal, 61*(1), 79–93.

Bhutiyani, M. R., Vishwas, S. K., & Pawar, N. J. (2007). Long-term trends in maximum, minimum and mean annual air temperatures across the Northwestern Himalaya during the twentieth century. *Climatic Change, 85*(1–2), 159–177.

Bhutiyani, M. R., Kale, V. S., & Pawar, N. J. (2009). Climate change and the precipitation variations in the northwestern Himalaya: 1866–2006. *International Journal of Climatology, 30*(4), 535–548.

Bookhagen, B., & Burbank, D. W. (2006). Topography, relief, and TRMM-derived rainfall variations along the Himalaya. *Geophysical Research Letters, 33*(8).

Bookhagen, B., & Burbank, D. W. (2010). Toward a complete Himalayan hydrological budget: Spatiotemporal distribution of snowmelt and rainfall and their impact on river discharge. *Journal of Geophysical Research: Earth Surface, 115*(3), F03019. https://doi.org/10.1029/2009/2009JF001426.

Borgaonkar, H. P., Ram, S., & Sikder, A. B. (2009). Assessment of tree-ring analysis of high-elevation Cedrus deodara D. Don from Western Himalaya (India) in relation to climate and glacier fluctuations. *Dendrochronologia, 27*(1), 59–69.

Cannon, F., Carvalho, L. M. V., Jones, C., & Bookhagen, B. (2015). Multi-annual variations in winter westerly disturbance activity affecting the Himalaya. *Climate Dynamics, 44*, 441–455. https://doi.org/10.1007/s00382-014-2248-8.

Cannon, F., Carvalho, L. M. V., Jones, C., Hoell, A., Norris, J., Kiladis, G. N., et al. (2017). The influence of tropical forcing on extreme winter precipitation in the western Himalaya. *Climate Dynamics, 48* (3), 1213–1232.

Chen, B., Chao, W. C., & Liu, X. (2003). Enhanced climatic warming in the Tibetan Plateau due to doubling CO2: A model study. *Climate Dynamics, 20*, 401–413.

Cho, C., Li, R., Wang, S. Y., Ho, J., & Robert, Y. (2016). Anthropogenic footprint of climate change in the June 2013 Northern India flood. *Climate Dynamics, 46*, 797–805.

Choi, G., Collins, D., Ren, G., Trewin, B., Baldi, M., Fukuda, Y., et al. (2009). Changes in means and extreme events of temperature and precipitation in the Asia-Pacific Network region, 1955–2007. *International Journal of Climatology, 29*, 1906–1925. https://doi.org/10.1002/joc.1979.

Choudhary, A., & Dimri, A. P. (2017). Assessment of CORDEX-South Asia experiments for monsoonal precipitation over Himalayan region for future climate. *Climate Dynamics*. https://doi.org/10.1007/s00382-017-3789-4.

Chu, Z. Y., Ren, G. Y., Shao, X. M., & Liu, H. B. (2005). A preliminary reconstruction of mean surface air temperature over the past 1000 years in China. *Climatic and Environmental Research, 10*(4), 826–836.

Cook, E. R., Krusic, P. J., & Jones, P. D. (2003). Dendroclimatic signals in long tree-ring chronologies from the Himalayas of Nepal. *International Journal of Climatology, 23*, 707–732.

Dhar, O. N., Soman, M. K., & Mulye, S. S. (1984). Rainfall over the southern slopes of the Himalayas and the adjoining plains during breaks in the monsoon. *Journal of Climatolgy, 4*(6), 671–676.

Diaz, H. F., & Bradley, R. S. (1997). Temperature variations during the last century at high elevation sites. *Climatic Change, 36*, 253–279.

Dimri, A. P., & Chevuturi, A. (2016). *Western disturbances—An Indian meteorological perspective* (p. 127). Switzerland: Springer International Publishing.

Dimri, A. P., Niyogi, D., Barros, A. P., Ridley, J., Mohanty, U. C., Yasunari, T., et al. (2015). Western disturbances: A review. *Reviews of Geophysics, 53*(2), 225–246.

Ding, Y. H., & Ren, G. Y. (2008). *An introduction to China climate change science*. Beijing: China Meteorological Press.

Ding, Y. H., & Zhang, L. (2008). Intercomparison of the time for climate abrupt change between the Tibetan Plateau and other regions in China. *Chinese Journal of Atmospheric Sciences, 32*(4), 794–805.

Dixit, Y., Hodell, D. A., & Petrie, C. A. (2015). Abrupt weakening of the summer monsoon in northwest India ~4100 yr ago. *Geology, 42*(4), 339–342.

Du, J. (2001). Change of temperature in Tibetan Plateau from 1961 to 2000. *Acta Geographica Sinica, 56*, 690–698. (in Chinese).

Duan, A. M., & Wu, G. X. (2006). Change of cloud amount and the climate warming on the Tibetan Plateau. *Geophysical Research Letters, 33*(22), L22704.

Duan, A. M., & Xiao, Z. X. (2015). Does the climate warming hiatus exist over the Tibetan Plateau? *Scientific Reports, 5*, 13711.

Duan, A. M., Wu, G. X., Zhang, Q., & Liu, Y. M. (2006). New proofs of the recent climate warming over the Tibetan Plateau as a result of the increasing greenhouse gases emissions. *Chinese Science Bulletin, 51*(11), 1396–1400.

Eriksson, M., Xu, J., Shrestha, A. B., Vaidya, R. A., Nepal, S., & Sandström, K. (2009). *The changing Himalayas: Impact of climate change on water resources and livelihoods in the greater Himalayas*. Kathmandu: ICIMOD.

Fan, X., Wang, Q., Wang, M., & Jimenez, C. V. (2015). Warming Amplification of minimum and maximum temperatures over high-elevation regions across the globe. *PLoS ONE, 10*, e0140213.

Forsythe, N., Fowler, H. J., Li, X.-F., Blenkinsop S., & Pritchard, D. (2017). Karakoram temperature and glacial melt driven by regional atmospheric circulation variability. *Nature Climate Change 7*(9). https://doi.org/10.1038/nclimate3361.

Friedlingstein, P., Andrew, R. M., Rogelj, J., Peters, G. P., Canadell, J. G., Knutti, R., et al. (2014). Persistent growth of CO_2 emissions and implications for reaching climate targets. *Nature Geoscience, 7* (10), 709.

Gardelle, J., Berthier, E., & Arnaud, Y. (2012). Slight mass gain of Karakoram glaciers in the early twenty-first century. *Nature Geoscience, 5*(5), 322–325. https://doi.org/10.1038/NGEO1450.

Giorgi, F., Coppola, E., Solmon, F., Mariotti, L., Sylla, M. B., Bi, X., et al. (2012). RegCM4: Model description and preliminary tests over multiple CORDEX domains. *Climate Research, 52*(1), 7–29.

Goswami, B. N., Venugopal, V., Sengupta, D., Madhusoodanan, M. S., & Xavier, P. K. (2006). Increasing trend of extreme rain events over India in a warming environment. *Science, 314*(5804), 1442–1444.

Guo, D., Yu, E., & Wang, H. (2016). Will the Tibetan Plateau warming depend on elevation in the future? *Journal of Geophysical Research: Atmospheres, 121*(8), 3969–3978.

Hasson, S., Lucarini, V., & Pascale, S. (2013). Hydrological cycle over South and Southeast Asian river basins as simulated by PCMDI/CMIP3 experiments. *Earth System Dynamics, 4*(2), 199–217. https://doi.org/10.5194/esd-4-199-2013.

Hasson, S., Lucarini, V., Khan, M. R., Petitta, M., Bolch, T., & Gioli, G. (2014). Early 21st century snow cover state over the western river basins of the Indus River system. *Hydrology Earth System Sciences, 18*(10), 4077–4100.

Hasson, S., Pascale, S., Lucarini, V., & Böhner, J. (2015). Seasonal cycle of precipitation over major river basins in South and Southeast Asia: A review of the CMIP5 climate models data for present climate and future climate projections. *Atmospheric Research, 180*, 42–63.

Hewitt, K. (2005). The Karakoram anomaly? Glacier expansion and the "elevation effect" Karakoram Himalaya. *Mountain Research and Development, 25*(4), 332–340.

Hoffmann, U. (2013). Section B: Agriculture—A key driver and a major victim of global warming. *Lead Article*, 3, 5. Chapter 1. Hoffmann.

Houze, R. A., Jr., Wilton, D. C., & Smull, B. F. (2007). Monsoon convection in the Himalayan region as seen by the TRMM precipitation radar. *Quarterly Journal of Royal Meteorology Society, 133*(627), 1389–1411.

Houze, R. A., Jr., Rassmussen, K. L., Medina, S., Brodzik, S. R., & Romastshke, U. (2011). Anomolous atmospheric events leading to the summer 2010 floods in Pakistan. *Bulletin of American Meteorological Society, 92I*(2), 291–298. https://doi.org/10.1175/2010BAMS3173.1.

Immerzeel, W. W., & Bierkens, M. F. P. (2012). Asia's water balance. *Nature Geoscience, 5*(12), 841–842.

Immerzeel, W. W., Droogers, P., Jong, S. M. D., & Bierkens, M. F. P. (2009). Large-scale monitoring of snow cover and runoff simulation in Himalayan river basins using remote sensing. *Remote Sensing of Environment, 113*(1), 40–49.

Immerzeel, W. W., Van Beek, L. P. H., & Bierkens, M. F. P. (2010). Climate change will affect the Asian water towers. *Science, 328* (5984), 1382–1385.

IPCC. (2000). *Special report on emmission scenarios* (p. 570). Cambridge: Intergovernmental Panel on Climate Change.

IPCC. (2007). In: S. Solomon, D. Qin, M. Manning, Z. Chen, M. Marquis, K. B. Averyt, et al. (Eds.), *Climate Change 2007: The Physical Science Basis. Contribution of Working Group I to the*

Fourth Assessment Report of the Intergovernmental Panel on Climate Change (p. 996) Cambridge, United Kingdom and New York, USA: Cambridge University Press.

IPCC. (2013). *Climate change. The physical science basis*. Work Group Contribution to the IPCC Fifth Assessment Report (AR5). Stockholm, Sweden: Intergovernmental Panel on Climate Change.

Jones, P. D., & Moberg, A. (2003). Hemispheric and large-scale surface air temperature variations: An extensive revision and an update to 2001. *Journal of Climate, 16*(2), 206–223.

Kääb, A., Treichler, D., Nuth, C., & Berthier, E. (2015). Brief Communication: Contending estimates of 2003–2008 glacier mass balance over the Pamir–Karakoram–Himalaya. *The Cryosphere, 9* (2), 557–564. https://doi.org/10.5194/tc-9-557-2015.

Kang, S. C., Xu, Y. W., You, Q. L., Flugel, W. A., Pepin, N., & Yao, T. D. (2010). Review of climate and cryospheric change in the Tibetan Plateau. *Environmental Research Letters, 5*(1).

Kapnick, S. B., Delworth, T. L., Ashfaq, M., Malyshev, S., & Milly, P. C. D. (2014). Snowfall less sensitive to warming in Karakoram than in Himalayas due to a unique seasonal cycle. *Nature Geoscience, 7*(11), 834–840.

Kitoh, A. (2017). The Asian monsoon and its future change in climate models: A review. *Journal of Meteorological Society of Japan*. https://doi.org/10.2151/jmsj.2017-002.

Klein Tank, A. M. G., Peterson, T. C., Quadir, D. A., Dorji, S., Zou, X., Tang, H., et al. (2006). Changes in daily temperature and precipitation extremes in Central and South Asia. *Journal of Geophysical Research, 111*(D16). https://doi.org/10.1029/2005jd006316.

Kosaka, Y., & Xie, S. (2013). Recent globalwarming hiatus tied to equatorial Pacific surface cooling. *Nature, 501*(7467), 403–407.

Kothawale, D. R., & Rupa Kumar, K. (2005). On the recent changes in surface temperature trends over India. *Geophysical Research Letters, 32*(18). https://doi.org/10.1029/2005gl023528.

Krishna Kumar, K., Rajagapalan, B., Hoerling, M., Bates, G., & Cane, M. (2006). Unraveling the mystery of Indian monsoon failure during El Nino. *Science, 314*(5796), 115–119.

Krishnan, R., Zhang, C., & Sugi, M. (2000). Dynamics of breaks in the Indian summer monsoon. *Journal of Atmospheric Sciences, 57*(9), 1354–1372.

Krishnan, R., Ramesh, K. V., Samala, B. K., Meyers, G., Slingo, J. M., & Fennessy, M. J. (2006). Indian Ocean-Monsoon coupled interactions and impending monsoon droughts. *Geophysical Research Letters, 33*(8), L08711. https://doi.org/10.1029/2006GL025811.

Krishnan, R., Kumar, V., Sugi, M., & Yoshimura, J. (2009). Internal feedbacks from monsoon–midlatitude interactions during droughts in the Indian summer monsoon. *Journal of Atmospheric Sciences, 66*(3), 553–578.

Krishnan, R., Sabin, T. P., Ayantika, D. C., Kitoh, A., Sugi, M., Murakami, H., et al. (2013). Will the South Asian monsoon overturning circulation stabilize any further*? Climate Dynamics, 40* (1–2), 187–211. https://doi.org/10.1007/s00382-012-1317-0.

Krishnan, R., Sabin, T. P., Vellore, R., Mujumdar, M., Sanjay, J., Goswami, B. N., et al. (2016). Deciphering the desiccation trend of the South Asian monsoon hydroclimate in a wamming world. *Climate Dynamics, 47*(3–4), 1007–1027. https://doi.org/10.1007/s00382-015-2886-5.

Krishnan, R., Sabin, T. P., Ranade, M., Vellore, R., Mujumdar, M., Sanjay, J., et al. (2018). Non-monsoonal precipitation response over the Western Himalayas to climate change. *Climate Dynamics*, https://doi.org/10.1007/s00382-018-4357-2.

Kuang, X., & Jiao, J. J. (2016). Review on climate change on the Tibetan Plateau during the last half century. *Journal of Geophysical Research: Atmospheres, 121*(8), 3979–4007.

Kulkarni, A., Patwardhan, S., Kumar, K. K., Ashok, K., & Krishnan, R. (2013). Projected climate change in the Hindu Kush-Himalayan region by using the high-resolution regional climate model PRECIS. *Mountain Research and Development, 33*(2), 142–151.

Kumar, K. K., Patwardhan, S. K., Kulkarni, A., Kamala, K., Rao, K. K., & Jones, R. (2011). Simulated projections for summer monsoon climate over India by a high-resolution regional climate model (PRECIS). *Current Science, 101*(3), 312–326.

Kutzbach, J. E. (1981). Monsoon climate of the early Holocene: Climate experiment with the earth's orbital parameters for 9000 years ago. *Science, 214*(4516), 59–61.

Kutzbach, J. E., & Otto-Bliesner, B. L. (1982). The sensitivity of the African-Asian monsoonal climate to orbital parameter changes for 9000 years B.P. in a low-resolution general circulation model. *Journal of the Atmospheric Sciences, 39*(6), 1177–1188.

Lau, W. K., & Kim, K.-M. (2012). The 2010 Pakistan flood and the Russia heat wave: Teleconnection of extremes. *Journal of Hydrometeorology, 13*(1), 392–403. https://doi.org/10.1175/JHM-D-11-016.1.

Liu, X., & Chen, B. (2000). Climatic warming in the Tibetan Plateau during recent decades. *International Journal of Climatology, 20* (14), 1729–1742.

Liu, X. H., Qin, D. H., Shao, X. M., Chen, T., & Jiawen, R. (2005). Temperature variations recovered from tree-rings in the middle Qilian Mountain over the last millennium. *Science in China Series D: Earth Sciences, 48*(4), 521–529.

Liu, X., Yin, Z. Y., Shao, X., & Qin, N. (2006). Temporal trends and variability of daily maximum and minimum, extreme temperature events, and growing season length over the eastern and central Tibetan Plateau during 1961–2003. *Journal of Geophysical Research:Atmospheres, 111,* D19109.

Liu, X., Cheng, Z., Yan, L., & Yin, Z. Y. (2009a). Elevation dependency of recent and future minimum surface air temperature trends in the Tibetan Plateau and its surroundings. *Global and Planetary Change, 68*(3), 164–174.

Liu, X. F., Jiang, Y., Ren, G. Y., Liang, X. H., & Zhang, C. W. (2009b). Influence of urbanization and observational setting change on data series of near-surface wind speed in Hebei Province, China. *Plateau Meteorology, 28*(2), 433–439.

Lutz, A. F., Immerzeel, W. W., Shrestha, A. B., & Bierkens, M. F. P. (2014). Consistent increase in High Asia's runoff due to increasing glacier melt and precipitation. *Climate Dynamics, 4*(7), 587–592. https://doi.org/10.1038/nclimate2237.

Madhura, R. K., Krishnan, R., Revadekar, J. V., Mujumdar, M., & Goswami, B. N. (2015). Changes in western disturbances over the Western Himalayas in a warming environment. *Climate Dynamics, 44*(3–4), 1157–1168. https://doi.org/10.1007/s00382-014-2166-9.

Mann, M. E., Raymond, S. B., & Malcolm, K. H. (1999). Northern hemisphere temperatures during the past millennium: Inferences, uncertainties, and limitations. *Geophysical Research Letters, 26*(6), 759–762.

Medina, S., Houze, R. A., Kumar, A., & Niyogi, D. (2010). Summer monsoon convection in the Himalayan region: Terrain and land cover effects. *Quarterly Journal of the Royal Meteorological Society, 136*(648), 593–616. https://doi.org/10.1002/qj.601.

Meinshausen, M., Smith, S. J., Calvin, K., Daniel, J. S., Kainuma, M. L. T., Lamarque, J. F., et al. (2011). The RCP greenhouse gas concentrations and their extensions from 1765 to 2300. *Climatic Change, 109*(1), 213.

Mishra, V. (2015). Climatic uncertainty in Himalayan water towers. *Journal of Geophysical Research: Atmospheres, 120*(7), 2680–2705. https://doi.org/10.1002/2014JD022650.

Mountain Research Initiative EDW Working Group. (2015). Elevation-dependent warming in mountain regions of the world. *Nature Climate Change, 5*(5), 424–430.

Mujumdar, M., Preethi, B., Sabin, T. P., Ashok, K., Saeed, S., Pai, D. S., et al. (2012). The Asian summer monsoon response to the La

Nina event of 2010. *Meteorological Applications, 19*(2), 216–225. https://doi.org/10.1002/met.1301.

Nan, S., Zhao, P., Yang, S., & Chen, J. (2009). Springtime tropospheric temperature over the Tibetan Plateau and evolution of the tropical Pacific SST. *Journal of Geophysical Research: Atmospheres, 114* (10). https://doi.org/10.1029/2008jd011559.

Nepal, S., & Shrestha, A. B. (2015). Impact of climate change on the hydrological regime of the Indus, Ganges and Brahmaputra river basins: A review of the literature. *International Journal of Water Resources Development, 31*(2), 201–218. https://doi.org/10.1080/07900627.2015.1030494.

Palazzi, E., Hardenberg, J., & Provenzale, A. (2013). Precipitation in the Hindu-Kush Karakoram Himalaya: Observations and future scenarios. *Journal of Geophysical Research: Atmospheres, 118,* 85–100.

Panday, P. K., Thibeaultb, J., & Freyc, K. E. (2014). Changing temperature and precipitation extremes in the Hindu Kush-Himalayan region: An analysis of CMIP3 and CMIP5 simulations and projections. *International Journal of Climatology, 35*(10), 3058–3077. https://doi.org/10.1002/joc.4192.

Pepin, N. C., Bradley, S., Diaz, H. F., Baraer, M., Caceres, E. B., Forsythe, N., et al. (2015). Elevation-dependent warming in mountain regions of the world. *Nature Climate Change, 5*(5), 424–430.

Peters, G. P., Andrew, R. M., Boden, A. T., Canadell, J. G., Quéré, C. L., Marland, G., et al. (2012). The challenges to keep global warming below 2 °C. *Nature Climate Change, 3*, 4–6.

Porter, J. R., Xie, L., Challinor, A. J., Cochrane, K., Howden, S. M., Iqbal, M. M., et al. (2014). Food security and food production systems. In: C. B. Field, V. R. Barros, D. J. Dokken, K. J. Mach, M. D. Mastrandrea, T. E. Bilir, M. Chatterjee, K. L. Ebi, Y. O. Estrada, R. C. Genova, B. Girma, E. S. Kissel, A. N. Levy, S. MacCracken, P. R. Mastrandrea, L. L. White (Eds.), *Climate Change 2014: Impacts, Adaptation, and Vulnerability. Part A: Global and Sectoral Aspects. Contribution of Working Group II to the Fifth Assessment Report of the Intergovernmental Panel on Climate Change.* (pp. 485–533). Cambridge University Press, Cambridge, United Kingdom and New York, NY, USA.

Prell, W. L., & Kutzbach, J. E. (1987). Monsoon variability over the past 150,000 years. *Journal of Geophysical Research, 92,* 8411–8425.

Qian, W.-H., Fu, J., Zhang, W., & Lin, X. (2007). Changes in mean climate and extreme climate in China during the last 40 years. *Advances in Earth Science, 23*(7), 674–683.

Qin, D. H., Ding, Y. H., & Su, J. L. (2005). Assessment of climate and environment changes in China (I): Climate and environment changes in China and their Projection. *Advances in Climate Change Research, 1,* 4–9.

Rajbhandari, R., Shrestha, A. B., Kulkarni, A., Patwardhan, S. K., & Bajracharya, S. R. (2015). Projected changes in climate over the Indus river basin using a high resolution regional climate model (PRECIS). *Climate Dynamics, 44*(1), 339–357. https://doi.org/10.1007/s00382-014-2183-8.

Rajbhandari, R., Shrestha, A. B., Nepal, S., & Wahid, S. (2016). Projection of future climate over the Koshi River Basin based on CMIP5 GCMs. *Atmospheric and Climate Sciences, 6*(2), 190–204. https://doi.org/10.4236/acs.2016.62017.

Rajbhandari, R., Shrestha, A. B., Nepal, S., Wahid, S., & Ren, G.-Y. (2017). Extreme climate projections over the transboundary Koshi River Basin using a high resolution regional climate model. *Advances in Climate Change Research, 8*(3), 199–211.

Rajeevan, M., Gadgil, S., & Bhate, J. (2010). Active and break spells of the Indian summer monsoon. *Journal of Earth System Science, 119* (3), 229–248.

Ramamurthy, K. (1969). *Monsoon of India: Some aspects of the 'break' in the Indian southwest monsoon during July and August.* Forecasting Manual 1–57 No. IV 18.3, India Meteorological Department, Poona, India.

Rao, Y. P. (1976). *Southwest monsoon (meteorological monograph).* India Meteorological Department, New Delhi, p. 366.

Ren, G., & Shrestha, A. B. (2017). Climate change in the Hindu Kush Himalaya. *Advance in Climate Change Research 8*(3), 130–140.

Ren, G. Y., & Zhou, Y. Q. (2014). Urbanization effect on trends of extreme temperature indices of national stations over mainland China. 1961–2008. *Journal of Climate, 27*(6), 2340–2360.

Ren, G. Y., Wu, H., & Chen, Z. H. (2000). Spatial patterns of change trend in rainfall of China. *Quarterly Journal of Applied Meteorology, 11*(3), 322–330. (in Chinese).

Ren, G.-Y., Guo, J., Xu, M. Z., Chu, Z. Y., Zhang, L., Zou, X. K., et al. (2005). Climate changes of China mainland over the past half century. *Acta Meteorological Sinica, 53*(6), 942–956. (in Chinese).

Ren, G., Ding, Y., Zhao, Z., Zheng, J., Wu, T., Tang, G., et al. (2012). Recent progress in studies of climate change in China. *Advance in Atmospheric Sciences, 29*(5), 958–977.

Ren, G., Ren, Y., Li, Q., & Xu, W. (2014). An overview on global land surface air temperature change. *Advances in Earth Science, 29*(8), 934–946. (in Chinese).

Ren, Y.-Y., Parker, D., Ren, G.-Y., & Dunn, R. (2015). Tempo-spatial characteristics of sub-daily temperature trends in mainland China. *Climatic Dynamics, 46*(9–10), 2737–2748. https://doi.org/10.1007/s00382-015-2726-7.

Ren, Y.-Y., Ren, G.-Y., Sun, X.-B., Shrestha, A. B., You, Q.-L., Zhan, Y.-J., et al. (2017). Observed changes in surface air temperature and precipitation in the Hindu Kush Himalayan region during 1901–2014. *Advance in Climate Change Research, 8*(3). https://dx.doi.org/10.1016/j.accre.2017.08.001.

Ridley, J., Wiltshire, A., & Mathison, C. (2013). More frequent occurrence of westerly disturbances in Karakoram up to 2100. *Science of the Total Environment, 468–469,* S31–S35.

Saeed, S., Müller, W. A., Hagamenn, S., & Jacob, D. (2010). Circumglobal wave train and summer monsoon over northwestern India and Pakistan: The explicit role of the surface heat low. *Climate Dynamics, 37*(5), 1045–1060. https://doi.org/10.1007/s00382-010-0888-x.

Saeed, S., Müller, W. A., Hagemann, S., Jacob, D., Mujumdar, M., & Krishnan, R. (2011). Precipitation variability over the South Asian monsoon heat low and associated teleconnections. *Geophysical Research Letters, 38*(8). https://doi.org/10.1029/2011gl046984.

Sanjay, J., Ramarao, M. V. S., Mujumdar, M. & Krishnan, R. (2017a). In: M. Rajeevan & S. Nayak (Eds.), *Regional climate change scenarios. Chapter 16 in the book: Observed climate variability and change over the Indian region* (pp. 285–304). Springer Geology. https://doi.org/10.1007/978-981-10-2531-0.

Sanjay, J., Krishnan, R., Shrestha, A. B., Rajbhandari, R., & Ren, G. Y. (2017b). Downscaled climate change projections for the Hindu Kush Himalayan region using CORDEX South Asia regional climate models. *Advances in Climate Change Research, 8*(3), 185–198. https://doi.org/10.1016/j.accre.2017.08.003.

Samuelsson, P., Jones, C., Willen, U., Ullerstig, A., and co-authors. (2011). The Rossby Centre Regional Climate model RCAS3: model description and performance, *Tellus, 63A*, 4–23.

Shao, X., Huang, L., Liu, H., Liang, E., Fang, X., & Wang, L. (2005). Reconstruction of precipitation variation from tree rings in recent 1000 years in Delingha, Qinghai. *Science in China, Series D: Earth Sciences, 48,* 939–949.

Shao, X. M., Xu, Y., Yin, Z. Y., Liang, E., Zhu, H., & Wang, S. (2010). Climatic implications of a 3585-year tree-ring width chronology from the northeastern Qinghai-Tibetan Plateau. *Quaternary Science Reviews, 29*(17), 2111–2122.

Sharma, E., Molden, D., Wester, P., & Shrestha, R. M. (2016). The Hindu Kush Himalayan monitoring and assessment programme:

Action to sustain a global asset. *Mountain Research and Development, 36*(2), 236–239.

Shrestha, A. B. (2008a). *Resouce Manual on Flash Flood Risk Management. Module 2: Non-structural Measures*. (includes CD insert), ICIMOD, p. 91.

Shrestha, A. B. (2008b). Climate Change in the Hindu Kush-Himalayas and its impacts on water and hazards. *APMN Bulletin (Newsletter of the Asia Pacific Mountain Network), 9*, 1–5.

Shrestha, A. B., & Aryal, R. (2011). Climate change in Nepal and its impact on Himalayan glaciers. *Regional Environmental Change, 11*, S65–S77.

Shrestha, A. B., & Devkota, L. P. (2010). *Climate change in the Eastern Himalayas: Observed trends and model projections* (p. 20).

Shrestha, A. B., Wake, C. P., Mayewski, P. A., & Dibb, J. E. (1999). Maximum temperature trends in the Himalaya and its vicinity: An analysis based on temperature records from Nepal for the period 1971–94. *Journal of Climate, 12*, 2775–2786.

Shrestha, A. B., Agrawal, N. K., Alfthan, B., Bajracharya, S. R., Maréchal, J., & Van Oort, B. (2015). *The Himalayan Climate and Water Atlas: Impact of climate change on water resources in five of Asia's major river basins*. GRID-Arendal and CICERO: ICIMOD.

Singh, N., & Ranade, A. (2010). *Determination of onset and withdrawal dates of summer monsoon across India using NCEP/NCAR reanalysis* (pp. 1–78). IITM Research Report No. RR-124, Indian Institute of Tropical Meteorology, Pune. ISSN 0252-1075.

Singh, S., Ghosh, S., Sahana, A. S., Vittal, H., & Karmakar, S. (2017). Do dynamic regional models add value to the global model projections of Indian monsoon? *Climate Dynamics, 48*(3–4), 1375–1397. https://doi.org/10.1007/s00382-016-3147-y.

Singhvi, A. K., & Krishnan, R. (2014). Past and the present climate of India. In: V. Kale (Ed.), *Landscapes and landforms of India. World geomorphological landscapes* (pp. 15–23). Dordrecht: Springer

Sperber, K. R., Annamalai, H., Kang, I. S., Kitoh, A., Moise, A., Turner, A., et al. (2013). The Asian summer monsoon: An intercomparison of CMIP5 vs. CMIP3 simulations of the late 20th century. *Climate Dynamics, 41*(9–10), 2711–2744.

Staubwasser, M., Sirocko, F., Grootes, P. M., & Segl, M. (2003). Climate change at the 4.2 ka BP termination of the Indus valley civilization and Holocene south Asian monsoon variability. *Geophysical Research Letter, 30*(8), 7.1–7.4.

Sun, Y., Zhang, X. B., Ren, G. Y., Zwiers, F. W., & Hu, T. (2016). Contribution of urbanization to warming in China. *Nature Climate Change, 6*(7), 706–709. https://doi.org/10.1038/nclimate2956.

Sun, X. B., Ren, G. Y., Xu, W. H., Li, Q. X., & Ren, Y. Y. (2017a). Global land-surface air temperature change based on the new CMA GLSAT dataset. *Science Bulletin, 62*(4), 236–238. https://doi.org/10.1016/j.scib.2017.01.017.

Sun, X.-B., Ren, G.-Y., Shrestha, A. B., Ren, Y.-Y., You, Q.-L., Zhan, Y.-J., et al. (2017b). Changes in extreme temperature events over the Hindu Kush Himalaya during 1961–2015. *Advances in Climate Change Research, 8*(3), 157–165. https://doi.org/10.1016/j.accre.2017.07.001.

Syed, M. A., & Al Amin, M. (2016). Geospatial modeling for investigating spatial pattern and change trend of temperature and rainfall. *Climate, 4*(2), 21.

Tang, G. L., & Ren, G. Y. (2005). Reanalysis of surface air temperature change of the last 100 years over China. *Climatic and Environmental Research, 10*(4), 791–798.

Teichmann, C., Eggert, B., Elizalde, A., Haensler, A., Jacob, D., Kumar, P., et al. (2013). How does a regional climate model modify the projected climate change signal of the driving GCM: A study over different CORDEX regions using REMO. *Atmosphere, 4*(2), 214–236.

Thompson, L. G., Mosley-Thompson, E., Davis, M. E., Lin, P. N., Henderson, K., & Mashiotta, T. A. (2003). Tropical glacier and ice core evidence of climate change on annual to millennial time scales. *Climate Change, 59*(1–2), 137–155.

Trenberth, K. E., Fasullo, J. T., Branstator, G., & Phillips, A. S. (2014). Seasonal aspects of the recent pause in surface warming. *Nature Climate Change, 4*(10), 911–916.

USGS. (2010). *Why study paleoclimate?* (p. 2). Reston, VA: United States Geological Survey.

Van Vuuren, D. P., Edmonds, J., Kainuma, M., Riahi, K., Thomson, A., Hibbard, K., et al. (2011). The representative concentration pathways: An overview. *Climatic Change, 109*(1), 5–31.

Vellore, R., Krishnan, R., Pendharkar, J., Choudhury, A. D., & Sabin, T. P. (2014). On anomalous precipitation enhancement over the Himalayan foothills during monsoon breaks. *Climate Dynamics, 43* (7), 2009–2031. https://doi.org/10.1007/s00382-013-2024-1.

Vellore, R. K., Kaplan, M., Krishnan, R., Lewis, J., Sabade, S., Deshpande, N., et al. (2015). Monsoon-extratropical circulation interactions in Himalayan extreme rainfall. *Climate Dynamics, 46* (11–12), 3517–3546 (p. 30). https://doi.org/10.1007/s00382-015-2784-x.

Wang, B., Bao, Q., Hoskins, B., Wu, G. X., & Liu, Y. M. (2008). Tibetan plateau warming and precipitation changes in East Asia. *Geophysical Research Letters, 35*(14), L14702.

Wang, S.-Y., Yoon, J.-H., Gillies, R. R., & Cho, C. (2013). What caused the winter drought in western Nepal during recent years? *Journal of Climate, 26*(21), 8241–8256. https://doi.org/10.1175/JCLI-D-12-00800.1.

Wang, Q., Fan, X., & Wang, M. (2014). Recent warming amplification over high elevation regions across the globe. *Climate Dynamics, 43* (1–2), 87–101.

Wang, Q., Fan, X., & Wang, M. (2016). Evidence of high-elevation amplification versus Arctic amplification. *Scientific Reports, 6*. https://doi.org/10.1038/srep19219.

Weiss, H., & Bardley, R. S. (2001). What drives societal collapse? *Science, 291*, 609–610. https://doi.org/10.1175/bams-d-11-00074.1.

Wild, M. (2012). Enlightening global dimming and brightening. *Bulletin of American Meteorological Society, 93*(1), 27–37.

Wu, G. X., & Zhang, Y. S. (1998). Thermal and mechanical forcing of the Tibetan Plateau and Asian monsoon onset. Part I: Situating of the onset. *Chinese Journal of Atmospheric Science, 22*(6), 825–838. (in Chinese).

Wu, G. X., Mao, J. Y., Duan, A. M., & Qiong, Z. (2004). Recent progress in the study on the impacts of Tibetan Plateau on Asian summer climate. *Acta Meteorologi Sinica, 62*(5), 528–540. (in Chinese).

Wu, J., Xu, Y., & Gao, X.-J. (2017). Projected changes in mean and extreme climates over Hindu Kush Himalayan region by 21 CMIP5 models. *Advance in Climate Change Research, 8*(3), 176–184. https://doi.org/10.1016/j.accre.2017.03.001.

Xu, J., Shrestha, A. B., Vaidya, R., Eriksson, M., Nepal, S., & Sandstrom, K. (2008). *The changing Himalayas. Impact of climate change on water resources and livlihoods in the Greater Himalayas*. Kathmandu: ICIMOD.

Xu, W. H., Li, Q. X., Yang, S., & Yan, X. (2014). Overview of global monthly surface temperature data in the past century and preliminary integration. *Advances in Climate Change Research, 5*(3), 111–117. https://doi.org/10.1016/j.accre.2014.11.003.

Yadav, R. R., Park, W.-K., Singh, J., & Dubey, B. C. L. (2004). Do the western Himalayas defy global warming? *Geophysical Research Letters, 31*(17).

Yadav, R. K., Rupa Kumar, K., & Rajeevan, M. (2009). Increasing influence of ENSO and decreasing influence of AO/NAO in the recent decades over northwest India winter precipitation. *Journal of*

Geophysical Research Atmospheres, 114(D12112). https://doi.org/10.1029/2008jd011318.

Yadav, R. K., Yoo, J. H., Kucharski, F., & Abid, M. A. (2010). Why is ENSO influencing northwest India winter precipitation in recent decades? *Journal of Climate, 23*(8), 1979–1993.

Yan, L. B., & Liu, X. D. (2014). Has climatic warming over the Tibetan Plateau paused or continued in recent years? *Journal of Earth, Ocean and Atmospheric Sciences, 1,* 13–28.

Yan, L. B., Liu, Z., Chen, G., Kutzbach, J. E., & Liu, X. (2016). Mechanisms of elevation-dependent warming over the Tibetan plateau in quadrupled CO2 experiments. *Climatic Change, 135,* 509–519.

Yanai, M. H., & Li, C. (1994). Mechanism of heating and the boundary layer over the Tibetan Plateau. *Monthly Weather Review, 122,* 305–323.

Yanai, M., Li, C., & Song, Z. (1992). Seasonal Heating of the Tibetan Plateau and its effects on the evolution of the Asian summer monsoon. *Journal of the Meteorological Society of Japan Ser. II, 70,* 319–351.

Yang, B., Braeuning, A., Johnson, K. R., & Yafeng, S. (2002). General characteristics of temperature variation in China during the last two millennia. *Geophysical Research Letters, 29*(9), 38:1–4.

Yang, K., Wu, H., Qin, J., Lin, C., Tang, W., & Chen, Y. (2014). Recent climate changes over the Tibetan Plateau and their impacts on energy and water cycle: A review. *Global and Planetary Change, 112,* 79–91.

Yao, T. D. (1997). A record of climatic environment over the past 2000 years from Guliya ice core, China. *Science of Quaternary, 1,* 52–61.

Yao, T., Thompson, L. G., Mosbrugger, V., Zhang, F., Ma, Y., Luo, T., et al. (2012a). Third Pole Environment (TPE). *Environmental Development, 3*(1), 52–64.

Yao, T., Thompson, L., Yang, W., Yu, W., Gao, Y., Guo, X., et al. (2012b). Different glacier status with atmospheric circulations in Tibetan Plateau and surroundings. *Nature Climate Change, 2,* 663–667.

Ye, D. Z. (1981). Some characteristics of the summer circulation over the Qinghai-Xizang (Tibet) Plateau and its neighborhood. *Bulletin of the American Meteorology Society, 62*(1), 14–19.

You, Q. L., Kang, S. C., Anguilar, E., & Yan, Y. P. (2008a). Changes in daily climate extremes in the eastern and central Tibetan Plateau during 1961–2005. *Journal of Geophysical Research: Atmospheres, 113*(7).

You, Q. L., Kang, S. C., Pepin, N., & Yan, Y. P. (2008b). Relationship between trends in temperature extremes and elevation in the eastern and central Tibetan Plateau, 1961–2005. *Geophysical Research Letters, 35*(4), L04704. https://doi.org/10.1029/2007GL032669.

You, Q. L., Kang, S. C., Flugel, W. A., Pepin, N., Yan, Y. P., & Huang, J. (2010a). Decreasing wind speed and weakening latitudinal surface pressure gradients in the Tibetan Plateau. *Climate Research, 42*(1), 57–64.

You, Q. L., Kang, S. C., Flugel, W. A., Sanchez-Lorenzo, A., Yan, Y. P., Huang, J., et al. (2010b). From brightening to dimming in sunshine duration over the eastern and central Tibetan Plateau (1961–2005). *Theoretical and Applied Climatology, 101*(3), 445–457.

You, Q. L., Kang, S. C., Pepin, N., Flugel, W. A., Yan, Y. P., Behrawan, H., et al. (2010c). Relationship between temperature trend magnitude, elevation and mean temperature in the Tibetan Plateau from homogenized surface stations and reanalysis data. *Global and Planetary Change, 71,* 124–133.

You, Q. L., Fraedrich, K., Ren, G., Pepin, N., & Kang, S. (2013a). Variability of temperature in the Tibetan Plateau based on homogenized surface stations and reanalysis data. *International Journal of Climatology, 33*(6), 1337–1347.

You, Q. L., Sanchez-Lorenzo, A., Wild, M., Folini, D., Fraedrich, K., Ren, G., et al. (2013b). Decadal variation of surface solar radiation in the Tibetan Plateau from observations, reanalysis and model simulations. *Climate Dynamics, 40*(7–8), 2073–2086.

You, Q. L., Min, J., Zhang, W., Pepin, N., & Kang, S. (2015). Comparison of multiple datasets with gridded precipitation observations over the Tibetan Plateau. *Climate Dynamics, 45,* 791–806.

You, Q. L., Min, J., & Kang, S. (2016). Rapid warming in the Tibetan Plateau from observations and CMIP5 models in recent decades. *International Journal of Climatology, 36,* 2660–2670.

You, Q. L., Ren, G. Y., Zhang, Y. Q., Ren, Y. Y., Sun, X. B., Zhan, Y. J., et al. (2017). An overview of studies of observed climate change in the Hindu Kush Himalayan (HKH) region. *Advances in Climate Change Research, 8*(3), 141–147. https://doi.org/10.1016/j.accre.2017.04.001.

Zhai, P. F., Sun, A., Ren, F., Liu, X., Gao, B., & Zhang, Q. (1999). Changes of climate extremes in China. *Climatic Change, 42*(1), 203–218.

Zhai, P. M., Zhang, X. B., Wan, H., & Pan, X. (2005). Trends in total precipitation and frequency of daily precipitation extremes over China. *Journal of Climate, 18*(7), 1096–1108.

Zhan, Y.-J., Ren, G.-Y., Shrestha, A. B., Rajbhandari, R., Ren, Y.-Y., Sanjay, J., et al. (2017). Change in extreme precipitation events over the Hindu Kush Himalayan region during 1961–2012. *Advance in Climate Change Research, 8*(3). https://dx.doi.org/10.1016/j.accre.2017.08.002.

Zhang, A. Y., Ren, G. Y., Guo, J., & Wang, Y. (2009). An analysis of upper-air wind speed in mainland China, 1980–2006. *Plateau Meteorology, 28*(3), 680–687.

Zhang, X., Alexander, L., Hegerl, G. C., Jones, P., Tank, A. L., Peterson, T. C., et al. (2011). Indices for monitoring changes in extremes based on daily temperature and precipitation data. *WIREs Climate Change, 2*(6), 851–870.

Zhou, Y. Q., & Ren, G. Y. (2009). The effect of urbanization on maximum, minimum temperatures and daily temperature range in North China. *Plateau Meteorology, 28*(5), 1158–1166. (in Chinese).

Zhou, Y. Q., & Ren, G. Y. (2011). Change in extreme temperature events frequency over mainland China during 1961–2008. *Climate Research, 50,* 125–139.

Zhou, X., Zhao, P., Chen, J., Chen, L., & Li, W. (2009). Impacts of thermodynamic processes over the Tibetan Plateau on the Northern Hemispheric climate. *Science in China, Series D: Earth Sciences, 52*(11), 1679–1693.

4 Exploring Futures of the Hindu Kush Himalaya: Scenarios and Pathways

Coordinating Lead Authors
Joyashree Roy, Jadavpur University, Kolkata, India, e-mail: joyashreeju@gmail.com
Eddy Moors, VU University, Amsterdam, The Netherlands and IHE Delft, The Netherlands, e-mail: e.moors@un-ihe.org
M. S. R. Murthy, International Centre for Integrated Mountain Development, Kathmandu, Nepal, e-mail: murthy.manchi@gmail.com (corresponding author)

Lead Authors
Prabhakar S. V. R. K., Natural Resources and Ecosystem Services Group, Institute for Global Environmental Strategies, Hayama, Japan, e-mail: prabhakar@iges.or.jp
Bahadar Nawab Khattak, COMSATS Institute of Information Technology, Abbottabad, Pakistan, e-mail: bahadar@ciit.net.pk
Peili Shi, Institute of Geographic Sciences and Natural Resources Research, Chinese Academy of Sciences, China, e-mail: shipl@igsnrr.ac.cn
Christian Huggel, Department of Geography, University of Zurich, Switzerland, e-mail: christian.huggel@geo.uzh.ch
Vishwas Chitale, International Centre for Integrated Mountain Development, Kathmandu, Nepal, e-mail: vishwas.chitale@icimod.org

Contributing Authors
Mohan Poudel, REDD Implementation Centre, Govt. of Nepal, Kathmandu, Nepal, e-mail: mohanprasadpoudel@gmail.com

Review Editors
Laurie Vasily, International Centre for Integrated Mountain Development, Kathmandu, Nepal, e-mail: laurie.vasily@icimod.org
Atiq Rahman, Bangladesh Centre for Advanced Studies, Dhaka, Bangladesh, e-mail: atiq.rahman50@gmail.com

Corresponding Author
M. S. R. Murthy, International Centre for Integrated Mountain Development, Kathmandu, Nepal, e-mail: murthy.manchi@gmail.com

P. Wester et al. (eds.), *The Hindu Kush Himalaya Assessment*,
https://doi.org/10.1007/978-3-319-92288-1_4

Contents

Chapter Overview

Key Findings

This is a precarious moment for the Hindu Kush Himalaya (HKH). Environmentally, socially, and economically, there is no single likely future for the HKH. Between now and 2080, the HKH may run *downhill*, or the region may continue doing *business as usual* and muddling through, or it may advance toward *prosperity*.

Evidence-based actions to reduce disaster risk, to mitigate and adapt to climate change and to adopt good governance, are central to ensuring prosperity in the HKH by 2080, as well as collaboration among state and non-state actors. Two potential pathways—*large-scale investment in sustainable development with regional cooperation,* and *bottom-up investment with local and national cooperation*—both involve substantial collaboration at different levels (regional, national, or sub-national).

Policy Messages

To avert the downhill scenario for the HKH in 2080, institutional mechanisms must confront the main challenges and resolve conflicts at various levels, and among various social groups. Globally, actions towards climate change mitigation and adaptation are urgently needed. Regionally, actions for sustainable livelihoods and economic growth should consider maintaining and improving the diversity and uniqueness of transboundary HKH natural assets, socio-cultural richness, ecosystem services as well as the need for political collaboration and information sharing.

If decision-makers, governments, institutions, and communities in the HKH continue business as usual, the HKH will face significant risk. Strategic action must be taken to change continued inadequate implementation of environmental protections; the ongoing suboptimal use of water and biodiversity resources; the continuance of unplanned urbanization in HKH; and the failure to adequately reduce greenhouse gas emissions.

To achieve prosperity in 2080, it is important to consider two potential pathways for the HKH—large-scale sustainable development investment with regional cooperation, and bottom-up investment with local and national cooperation. Both paths critically presuppose cooperation and coordination. Large-scale investment would rely on high-level decisions, made across national boundaries, to capitalize on emerging and unique economic opportunities in the regional market. Bottom-up investment would mobilize local and national investments and development initiatives, managed across various levels of society and government and with the collaboration of multiple stakeholders.

The two paths toward prosperity in the HKH are not mutually exclusive—they can and need to be integrated. Decision makers may combine elements of either path at various stages, making trade-offs between the benefits and risks associated with different actions at different levels.

To illuminate future uncertainties and inform strategic plans, this chapter presents three qualitative scenarios for the status of the HKH through 2080. The scenarios (Box 4.1) emerged from a participatory visioning exercise for scenario development (also see Sect. 4.1.2) conducted by the chapter team and HIMAP secretariat between January and September 2016. Over six successive workshops, decision makers and scientists representing HKH countries determined what would constitute a *prosperous* HKH scenario for 2080—along with its less desirable alternatives, *business as usual* and *downhill* (Fig. 4.1).

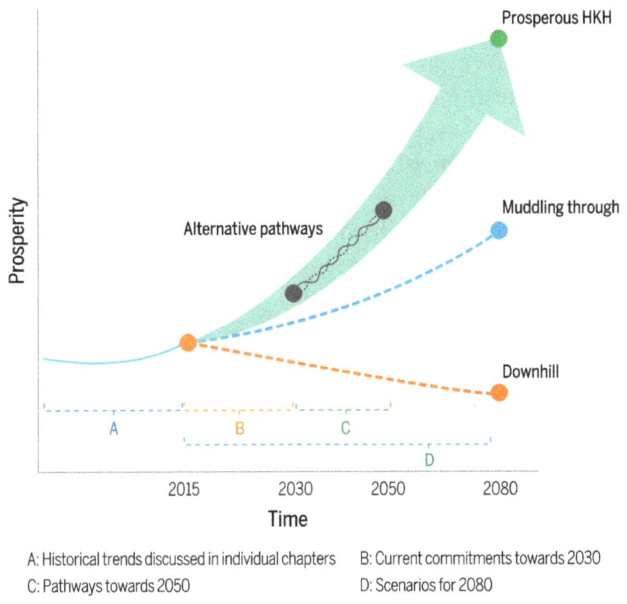

Fig. 4.1 Three scenarios depicting the possible prosperity development for the HKH in 2080: downhill, business as usual, prosperous

Box 4.1 How do scenarios contribute to envisioning the future?

In conventional planning approaches, which are still used in many countries and institutions, one planning period is followed by the next, and each planning period can be relatively independent of the earlier planning periods. This can create gaps and a lack of consistency between consecutive planning phases as the objectives may change drastically from one planning phase to the next. Additionally, the relatively short lifetime of conventional planning periods (5 years), makes them less suitable for exploring uncertainty in future societal and environmental conditions likely to occur in the longer term.

To enable short-term developmental actions to be consistent with longer-term societal and environmental transformation pathways towards a desired future, long-term future scenarios are used. It is important to note that a scenario is not a forecast or prediction; rather it is a plausible story about the future with a logical plot and narrative governing the manner in which events unfold.

Scenario development generally summarizes discourses such as those on poverty, nutrition, food security, and demographic change into a "shared vision" taking into consideration various stakeholders.

No regional quantitative model simulations for future social economic scenarios of the HKH region have been developed previously. As a result, as part of this overall HIMAP assessment, this chapter team made the first step and developed qualitative scenarios. We facilitated a participatory process by interviewing experts and stakeholders to develop such storylines to help in generating qualitative scenarios for HKH. Also, given the unique character of the HKH a regional level scenario development process allowed for a shared vision for the future which individual country-level scenarios cannot. Therefore, the HKH scenarios are presented in the form of storylines based on inputs of all scenario-development workshop participants.

While present trends and policy agreements such as the SDGs are governing country-level commitments for development during 2015–2030, forward-looking alternative pathways can assist in guiding policy for the decades from 2030 to 2050 that will ideally orient the HKH toward prosperity in 2080.

What will the prosperous scenario for the HKH look like, and what pathways lead to it? These questions were at the centre of a participatory scenario-development process that was facilitated by this chapter team. Through this process, participants identified challenges and opportunities, and created the storylines which are described here. Working backward from the three 2080 scenarios—and with reference to currently available national commitments to the Sustainable Development Goals (SDG) and its targets for 2020 and 2030—the participants identified two broad potential pathways to prosperity and the necessary actions that can put the HKH on one of those pathways (see Fig. 4.1).

Participants in this visioning process developed three storylines, one for each possible future scenario in the region. In the worst or **downhill** storyline, regional conflicts over resource sharing persist—and even multiply as scarcity increases. People and institutions do not benefit from emerging opportunities for efficient resource use. Communities remain isolated from the larger market systems. Mountain livelihoods do not include inclusive growth through new innovations, skills and practices. Ecosystems are degraded and biodiversity loss continues, mitigation efforts fail, and fossil fuels remain the dominant energy source. Climate change impacts reflect the Intergovernmental Panel on Climate Change (IPCC)'s worst case scenario—global temperature rising by substantially more than 2.0 °C. In short, the **downhill** scenario encompasses strong climate change, a socially, economically and politically unstable region, and strong ecosystem degradation.

In the **business as usual** storyline, today's economic growth patterns persist. Business and industries strive to keep meeting economic targets, while most meet only the minimum required standards for the environment and sustainability. There is some cooperation among HKH countries, although the cooperation is neither envisioned nor realized in all the ways in which it could succeed, nor in all the sectors where it could do so. The value of ecosystems is recognized by some, but not as broadly or in as many quarters as possible. Although some climate change mitigation efforts are put in place, they do not proceed rapidly or effectively enough to meet the 1.5 °C target set at the UNFCCC Conference of Parties 21 (COP21), held in Paris in December 2015. In short the business as usual scenario envisions strong climate change, medium social, economic and political instability, and medium ecosystem degradation.

In the **prosperous** storyline, regional cooperation across sectors and across governing institutions enables mountain and downstream people to utilize a full range of ecosystem services, to reduce disaster risks and to enjoy sustainable livelihoods and economic growth. The diversity and uniqueness of the region's natural resource assets, political life, and collaborative capacities are embraced. Biodiversity flourishes and the health of ecosystems improves. Climate change mitigation efforts largely succeed, as the regional economy shifts to clean and renewable sources for most of its energy needs. The impact of climate change reflects the IPCC's moderate scenario. So in brief the prosperous scenario represents HKH facing weak climate change, a socially, economically and politically stable region, and low ecosystem degradation.

The approach in Fig. 4.2 shows the three steps for deriving alternative pathways, using the 2080 HKH scenarios (Step 1), the existing commitments towards 2030 (Step 2) and the alternative pathways (Step 3). This approach leads to a description of the actions necessary to achieve progress between 2030 and 2050, while keeping in mind the compatibility with the desired and undesired scenarios for 2080.

Participants in the consultative workshops acknowledged there are many possible and overlapping pathways toward the prosperous scenario but chose to consider two in great depth. These two pathways differ in the scale of actions, in the size of investment needs, in decisions about policy, in choices about technology, and in the inclusion of developmental actors.

Pathway to prosperity 1: *Large-scale sustainable development investments with regional cooperation*. The HKH looks to large or centralised projects in developing its natural resources. Water is harnessed for food and energy in ways that address gender inequality and persistent poverty. Human resources are mobilized on a large scale. National

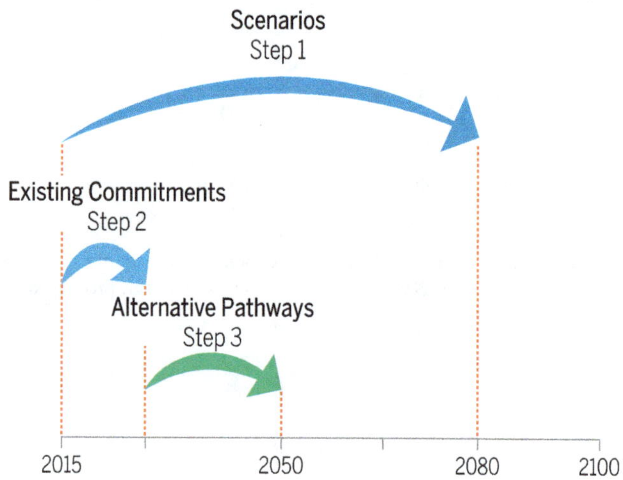

Fig. 4.2 Schematic representation of the steps followed in deriving Alternative Pathways towards 2050

and international funding is provided through collaboration among state, corporate, and non-state actors. Countries increasingly share resources, improve communication, and acquire infrastructure for greater mobility. Institutions gain the strength to govern this cooperation within and outside the HKH. Examples of steps along this pathway today include:

- The South Asian Free Trade Area (SAFTA) (SAARC 2016)
- The Indus Water Treaty (Government of India 2016a)
- The India Bhutan Hydropower Treaty (Government of India 2016b).

Pathway to prosperity 2: *Bottom-up investments with cooperation across multiple levels.* Water and energy, while critical, are developed through smaller-scale and decentralized programmes that promote national self-reliance. Activity is coordinated among many non-state development and social actors, both nationally and sub-nationally. Multi-level governance structures are built to maintain local diversities; to fund projects, and to monitor progress toward scenario goals across actors. Examples of steps along this pathway are:

- Farmer-managed irrigation systems in Nepal;
- Local spring water management projects in Sikkim (Parvez 2017) and Uttarakhand, India, which include micro hydroelectricity projects supported by policy initiatives and entails upstream-downstream cooperation among villages for equitable and robust water allocation.

These two pathways toward a prosperous scenario for the HKH in 2080 are not mutually exclusive. Decision makers may combine actions from each pathway at various stages, as they weigh the benefits and risks and consider associated trade-offs. What is most certain is that for the HKH to seize its unique and emerging opportunities while meeting its challenges, time is of the essence. If actions through 2030 do not pick up speed, but remain at the levels now envisioned, the region will not build the cooperation and multilevel governance structures that are needed to develop its natural resources in ways conducive to prosperity.

4.1 HKH Scenarios for 2080

Making the HKH prosperous in the future is the target. But there are some important questions: What path needs to be followed to make the HKH region prosperous in a time-bound manner? What additional and new actions need to be adopted? What needs to be avoided? What actions decided now can keep the options for a prosperous HKH open in the years to come? To answer these questions, it is necessary to understand the possible pathways that can emerge due to ongoing (in reference period 2015) and likely future decision-making processes locally, nationally, regionally, and globally. This will help us understand how human wellbeing and ecosystem sustenance can continue in the HKH. In the face of uncertainty surrounding various drivers of development, such as technology, demography, socioeconomic decisions, governance systems, and resource availability, there can be no single answer to the above questions. But it is imperative to explore them because a range of plausible futures helps in the design of future agenda setting, course of action through uncertainty reduction, consensus building for strengthening cooperation, and identification of gaps between desired policy setting and current policy settings.

Currently, there are no HKH specific quantitative models for future scenarios (see Box 4.2 and Sect. 4.1.1 for Asia-specific scenarios reflected in global reports that have some relevance for the HKH). In keeping with likely global futures, there are no comprehensive future scenarios in the literature covering environmental change, climate change, and social and economic development for the HKH region. However, there are standalone studies addressing climate change, economic development projections, and future outlooks which can provide limited guidelines to understanding the possible environmental and developmental scenarios for the region. This chapter brings together HKH-specific observed trends and expert knowledge (see Box 4.1 for the concepts and method we followed in this chapter) to develop qualitative scenarios and derive likely pathways.

As scenarios can be arranged in various ways, we restrict ourselves in this chapter to presenting future long-term 2080 scenarios (also see Box 4.1 and Fig. 4.1) for the HKH with in-depth development of three of the innumerable plausible futures. Here, 2080 is used as a key year to represent the 30-year period from 2070 to 2100 to assure independence from present policies and, by doing so, create more flexibility for non-linear changes and more space for creativity in identifying opportunities. Higher or lower degradation of the HKH's ecosystem resources, more or less political instability within the region, and good or poor progress in global negotiation outcomes in the context of climate change will have important implications for opportunities and challenges associated with climate-related vulnerability, human wellbeing, and sustainability within the region. Based on the consultative workshops of this scenario-development exercise and a review of the literature, ecosystem health, climate change, and weak governance leading to political unrest or local conflicts were considered the most important challenges for the HKH. Participants developed the storylines for three HKH 2080 scenarios based on the likely nature of these challenges.

Most current action plans and projected pathways to prosperity emerging from existing policies and commitments end in 2030. To prepare pathways toward prosperity beyond 2030, it is useful to set clear goals for 2050 based on a prosperous future scenario. Based on consultative workshops, the "prosperous HKH" scenario was selected as best representing a desired sustainable future for the HKH (Fig. 4.1). Assessing available knowledge to determine how the goals for 2050 can be achieved will help decision makers to develop near- to medium-term decisions, policies, and actions with effective entry points and timelines. To ascertain long-term sustainability through potential pathways developed for 2050, this chapter uses the three scenarios of 2080 to enable the assessment of desired and undesirable development strategies and actions beyond 2030, while moving toward 2080 through 2050.

4.1.1 Global Scenarios and Regional Outlooks

Global likely futures for the Asian region are mentioned in reports by the OECD, FAO, and The World Bank, among others. Important regional and global outlooks include the United Nations Environment Program's (UNEP) Global Environmental Outlooks (GEO 1 to 5) (UNEP 1997, 1999, 2002, 2007, 2012), the Asian Development Bank's Asian Century and its Energy Outlook (ADB 2013, 2015), IPCC Regional Scenarios (Parry et al. 2007; IPCC 2013), and Global water futures (Gallopín 2012). While these are broad-based outlooks covering the economy and environment, there are sector-specific outlooks for energy, water, land, and forests. Most of these outlooks present a reasonably positive future falling between increased economic and environmental wellbeing, though the outlooks vary in their focus and proximity to economic or environmental wellbeing. For example, the outlook of GEO 3 and 4 are more proximate to economic wellbeing, while FAO Forests 2020 (FAO 2010) is more proximate to environmental wellbeing.

The Global Environmental Outlooks of UNEP are among the more prominent environmental outlooks available at the global scale which provide useful insights into the path the world may take during the period presented. Table 4.1 presents an assessment of how accurately one outlook is able to portray the possible future by comparing the outlook's projections with the actual trend. It is clear from Table 4.1 that GEO 1 was able to accurately project the calorie intake but not the food dependency for the Asia and Pacific region, which is a broader target than the calorie intake per person. In addition, the outlook also deviated from the actual trend in the expansion of agricultural area and life expectancy. Several factors may have contributed to these gaps, which need to be addressed in building possible scenarios, which include shifts in policy and socioeconomic trends due to improved human standards over the years in the areas of education, empowerment, and inclusion. It is worth noting that the extensive stakeholder consultation-based scenario building process used in this HIMAP report is more comprehensive, as a qualitative storyline-building exercise can handle multiple goals and priorities. Use of scale for scoring (Sect. 4.1 and Fig. 4.3), the relative distance from target for any goal based on stakeholder perception and experience also provides scope for defining a range rather than an exact value.

However, some of the major conclusions from the global assessment reports on possible future outlooks remain relevant for the HKH (e.g., desirable directions of change in fuel mix and the technology mix for power generation in the future). For the HKH, the results from the Global Energy

Table 4.1 Checking the reality with scenarios: How well did the GEO 1 scenario (UNEP 1997) represent the future?

Projected scenario (2015–2050)	Observed trend	Proposed factors	Actual factors[a]
Calorie intake to reach up to 2600 kcal per day per capita and animal protein share to 13% of calories	Calorie intake reached 2665 kcal per day per capita in 2009 (Asian Development Bank 2013)	Increase in food supply	Increase in food supply and economic wellbeing
Increasing dependence on food imports globally and in Asia Pacific	The Asia and Pacific region is a net exporter with a balance standing at 4% of the total merchandise by value in 2014 (The World Bank 2014)	Land degradation and insufficient access to technologies	Increased production coupled with trade restrictions
Share of agricultural land to increase to 15% of global land area from a baseline year of 1990	Agriculture land area declined slightly to 12.6% of global land area from 1993 (FAOSTAT 2012)	Increasing food demand, deforestation	Competition for land from other economic sectors
Life expectancy to increase to 55 years (India)	The life expectancy was 66 years in 2013 (The World Bank 2013)	Improvement in individual health, nutrition and reduction in poverty	Better than expected improvements in health, nutrition and poverty reduction

[a]Based on authors' observations

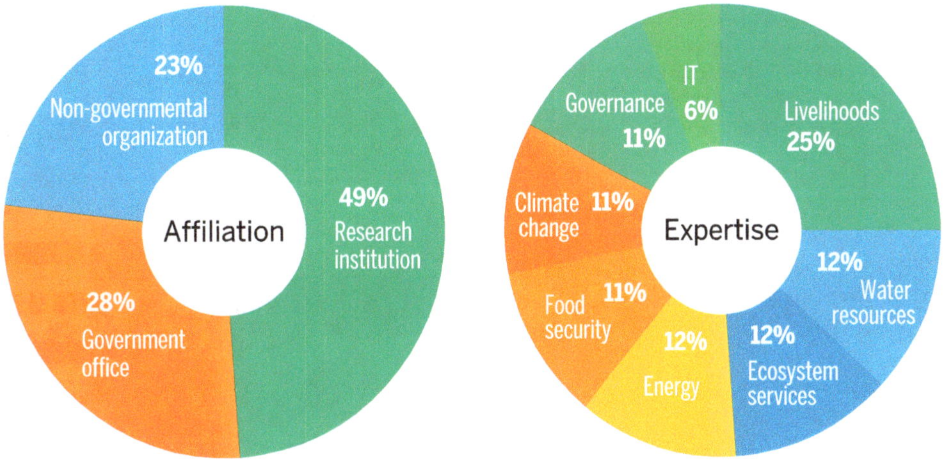

Fig. 4.3 Breakdown of workshop participants' affiliation and subject expertise (*GOV* Government offices, *RS* Research Scientists, *NGO* Non-Government Organization, *FS* Food Security, *WR* Water Resources, *EC* Ecosystems, *LH* Livelihoods, *IT* Information and Technology)

Assessment (Johansson et al. 2012) also remain relevant. In 2015, SDG's major goals, targets, and indicators are specified and implementation mechanisms and pathways are left to be decided by local decision makers. Goal 13 of the SDGs requires urgent action to combat climate change and its impacts. So, IPCC and GEA transformation pathways become relevant.

Equally relevant for the HKH are global SDG indicators: these emphasise the need for disaster risk reduction (DRR); the inclusion of women, youth, and local marginalized groups; the need to protect, restore, and promote a sustainable terrestrial ecosystem; sustainable forest management; avoiding desertification; reversal of land degradation; halting biodiversity loss; and meeting water and sanitation needs by the specific monitoring of green cover indices of mountain ecosystems and changes in the water-related ecosystem.

In keeping with likely global future scenarios, there are no comprehensive future scenarios in the literature covering environmental change, climate change, and social and economic development specifically for the HKH. However, there are studies addressing climate change, economic development projections, and outlooks that can provide, though in a limited sense, guidelines for understanding the possible environmental and developmental scenarios for the region.

Natural Resources: The FAO-OECD indicates a positive outlook for agriculture for countries in the HKH, showing an increase in production of rice and wheat until 2025 (OECD/FAO 2016). These outlooks have limitations since all of them cover the Asia and Pacific region as a whole and do not provide HKH-specific information.

Social: The projections available indicate continued population growth in the HKH countries with rapid projected growth in Afghanistan, followed by Pakistan and India. The region is projected to experience a rapid urbanization in Nepal, followed by Afghanistan and India (The World Bank 2016).

Economic: The sources in the literature, which mostly focusses on the short term, do not agree on the extent of growth in the region. The outlook by ADB indicates positive growth, while that of the International Monetary Fund (IMF) indicates deceleration in the short run (International Monetary Fund 2016; Asian Development Bank 2016). The outlook drawn for Southeast Asia, India, and China projects that the region will experience mild to moderate economic growth during 2016–2020 (OECD 2016). The Global Energy Outlook (OECD/IEA 2016) projects that South Asia will not be able to achieve full energy access targets by 2030 in a no-policy support scenario and will only be able to achieve rural energy access goals with substantial fuel price support policies such as subsidies (Pachauri et al. 2012).

Since these are sector-specific outlooks, it is unclear from the literature how the climate and economic outlooks for the region would interact with other sectors in the future, hence it is difficult to draw an overall picture for the HKH. However, it is possible to build a composite outlook based on a combination of individual economic, environmental, and climate change outlooks available for the region. To overcome the lack of an integrated regional vision on the future of the HKH, a HKH-specific participatory visioning exercise was conducted as part of the current HIMAP assessment (see Box 4.1).

4.1.2 Challenges and Opportunities for the HKH

To build the storyline and pathways for the HKH scenarios, the main drivers in the region were determined on the basis of reviews (see other chapters in this assessment report) and through intensive consultative workshops conducted sequentially from January to August 2016. Participants in these workshops were scientists, decision makers, development practitioners, HIMAP Steering Committee Members, and Chapter Lead Authors from all HIMAP chapters.

Fig. 4.4 (**a**) Challenges and (**b**) opportunities as identified by the workshop participants, expressed as a percentage of the 237 responses

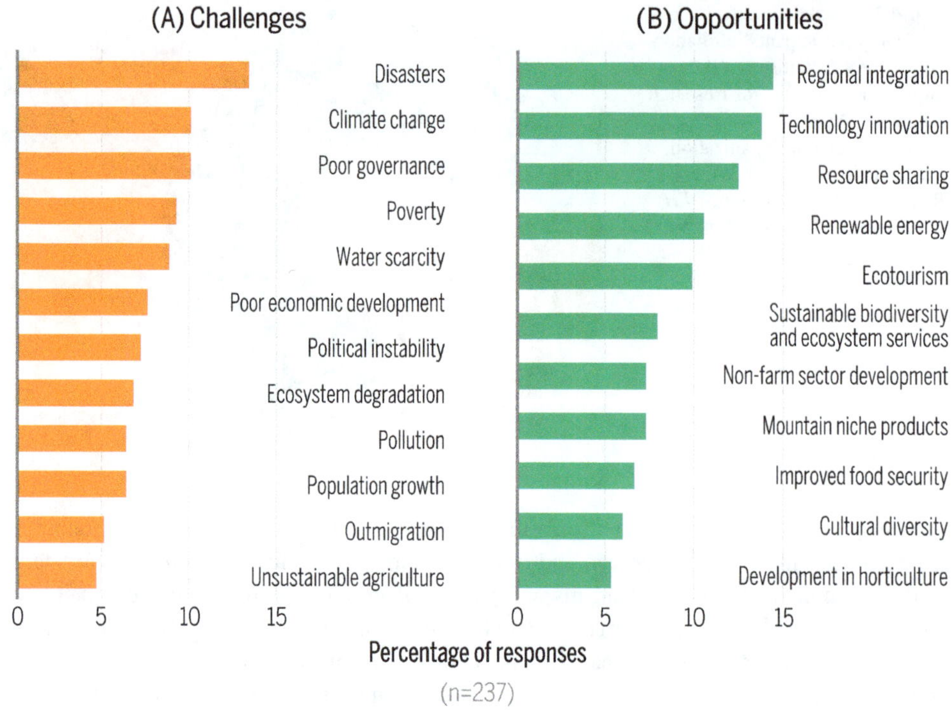

Science experts participated in four workshops and decision makers from government and non-government organizations participated in two different workshops. Participants included representatives of all eight HKH countries and professionals working on the region from the rest of the world. Figure 4.3 provides a breakdown of the affiliations and subject expertise of the participants from all the consultative workshops. A total of 139 people from 74 different institutions participated: 49% from research institutions, 28% from government offices, and 23% from NGOs. The participants represented eight broad fields of expertise: 25% were working in livelihoods; 12% in ecosystems, energy, or water resources; 11% in governance, forest, or climate change; and 6% in information technology.

During these workshops, the starting point was identification of the challenges for the region. Each time the objective was to enrich the results of the previous workshops using the output from those workshops. The science groups, facilitated by the chapter team, categorized the identified challenges and opportunities into the three scenarios. After discussing the challenges and opportunities for the HKH, the scenarios were presented and discussed in all workshops. In the workshop attended by the decision makers, participants were also asked to give their views on actions to attain the desired future and actions to prevent developments that might lead to an undesired future. Figure 4.3 gives an overview of the stakeholders involved in the workshops.

The participants from all workshops provided 237 responses addressing challenges and opportunities for the HKH. Based on a prioritization exercise in post-workshop analysis, this chapter team clustered the challenges and opportunities as presented below in Fig. 4.4a, b. The participants' prioritization of the challenges identifies disasters as the most significant challenge in the region (15%), followed by climate change and poor governance (11%), poverty (9%), and water scarcity (8%) (see Fig. 4.4a). In terms of opportunities, the participants considered the most important to be regional collaboration/integration (14%), technology innovation (13%), resource sharing (12%), and renewable energy (11%) (see Fig. 4.4b). For storyline development, the chapter team clustered these challenges and opportunities under five major themes: climate, natural resources, governance, social, and economy.

4.1.3 Storyline for HKH Scenario: "Downhill"

Compared to the 2015 baseline, the HKH will not be able to prosper by 2080 due to lost opportunities and multiple challenges that will increase over time. Owing to climate change, increasing political unrest, and strong ecosystem degradation, the socioeconomic condition of people in the HKH will deteriorate. Participants' perception of their status and expected change in any component under any theme are expressed by scores on a 1–5 scale (where 1 = poor, 5 = very good) for the "downhill" scenario for 2080. These are presented in Fig. 4.5. Details of the storyline for this scenario are given below.

Climate: Strong climate change will increase extreme events such as flooding, droughts, glacier melting, and river flow, resulting in water scarcity when it is needed and too much water at other times. There will be high vulnerability to climate-induced factors and less adaptation capability.
Natural Resources: Strong ecosystem degradation will threaten biodiversity and the development of improved crop systems for arresting biodiversity loss. Ecological and social resilience will decline and unplanned urbanization will increase, resulting in a loss of biodiversity. Resource use will intensify as a result of migration and tourism.
Governance: More political unrest in the region will result in conflicts over natural resources (e.g., water) and there will be little funding available/allocated for adaptation capacity. Institutional, political, and overall governance systems will be weakened and unable to function well at local, national, and regional levels and will fall short of global expectations. This will undermine the regional integration and cooperation in managing natural resources, leading to further conflict and mismanagement of the resources; the trust in multilateral organizations will be weakened, resulting in a poor response to regional issues with global significance.
Social: More national, regional, and global inequality will lead to serious food and nutrition insecurity, social tensions, and a high rate of migration (rural to urban and highland to lowland). Water scarcity for drinking and irrigation will drive people to migrate, with the poor left behind, leading to high mortality and suicide rates, high population density, and more urbanization. As a result, there will be a high fertility rate, high child mortality, huge refugee camps, and increasing inequalities, as well as a loss of traditional values, norms, and local knowledge.
Economy (including infrastructure and energy): High resource use intensity in production and consumption processes, a lack of innovation and efficient production processes, fossil fuel dependency, rural energy crises, and vulnerability will lead to a decline in economic progress. Competition for meagre resources and poor livelihoods will increase conflict and reduce motivation for innovation and cleaner, renewable energy. Communication will improve and competition will be high, leading to huge gaps in demand and supply.

4.1.4 Storyline for HKH Scenario: "Business as Usual"

Present economic growth patterns persist. Business continues as usual without sufficient change to meet existing and expected challenges to economic growth, and without opening an alternative pathway toward a prosperous HKH. The region has been able to achieve cooperation in some sectors with potential for more. Ecosystems values are slowly being recognized. Mitigation efforts are in place but are not sufficient to meet the 1.5 °C target set at COP21.

The changes that participants perceive will take place in the different thematic domains for each of the components by 2080, as compared to the 2015 baseline, are expressed using scores on a 1–5 scale and are presented in Fig. 4.5. Details of the storyline for this "business as usual" scenario are given below.

Climate: The region will experience moderate to strong climate change with temperatures rising beyond the Paris Agreement 2015, as well as more rainfall and extreme weather events. It will also lead to rapidly diminishing glaciers and moderate to severe ecosystem changes.
Natural Resources: Ecosystem health will experience an unstable equilibrium; there will be larger challenges in containing environmental degradation, promoting biodiversity conservation, and ensuring upstream and downstream linkages.
Governance: Implementing NDCs and achieving SDGs and overall growth remains a challenge because of conflicts of interest and inadequate cooperation in the region, as well as difficulties faced by international/global institutions. With scientific evidence playing a greater role in policy processes and the prospect of key alliances being formed for resource sharing, the possibility of addressing challenges will emerge to a certain extent.
Social: Despite SDGs being in place, livelihoods will still suffer from disparities, including food insecurity and associated malnutrition. Increasing floods and limited success in reducing vulnerability and adaptation to climate change will continue to make life more difficult. More people will start living in cities. Availability of funds following from the Paris Agreement 2015 will offer opportunities for technology transfer and improve food security.
Economy (including infrastructure and energy): The focus on mountain niche product development and adoption of ICT and remote sensing technology will increase. There may be massive infrastructure development that is not in sync with "green" approaches in some parts of the HKH. Even the green development taking place is not sensitive to the gender dimension, at times leading to slower economic progress. More alternative energy development (water + wind + solar) with benefit-sharing mechanisms will evolve, leading to fewer power cuts and enhancing access to power.

4.1.5 Storyline for HKH Scenario: "Prosperous HKH"

Regionally and across sectors, cooperative efforts will allow mountain people to prosper through sustainable livelihoods

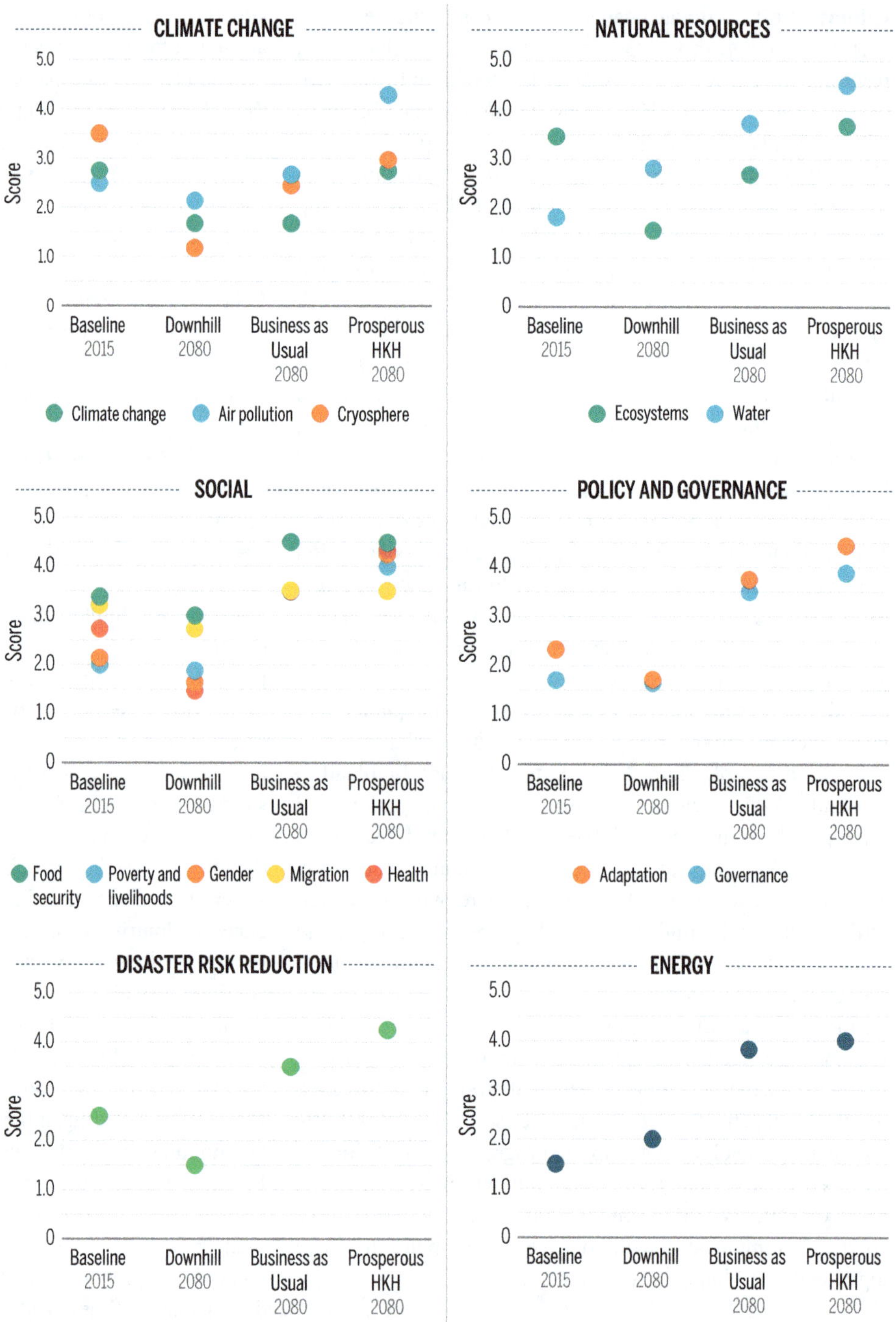

Fig. 4.5 Comparative scores for various components of the thematic areas in different scenarios. *Note* Each numerical value is an outcome of the consultative workshops. In the diagram, each value shows the average of all the scores allocated by participants on a 1–5 scale (where 1 represents "poor" and 5 represents "very good") for a particular component (please see the legends where applicable). While scoring, the initial score was given to the baseline 2015 status of each component and then changes are reflected in 2080 scenario scores, which vary by scenario category

and economic growth. These efforts will embrace the diversity and uniqueness of the region's natural resources, its political life, and its capacities for collaboration. Biodiversity and ecosystems values will improve. Global mitigation efforts will have largely succeeded, and clean and renewable energy sources will dominate the energy supply. Climate change will follow a moderate IPCC scenario. Participants used scores on a 1–5 scale to rank the changes that they expect to take place in the different thematic domains in 2080 compared to the 2015 baseline, which are presented in Fig. 4.5. Details of the storyline for this "prosperous HKH" scenario are given below.

Climate: Since climate change is a global phenomenon, climate in the HKH may not completely stabilize by 2080, as stabilization of climate change and its impacts may follow several years after GHG emissions are reduced, going by IPCC assessments. However, it is expected that the HKH will have mainstreamed climate change mitigation and adaptation to the fullest extent possible and that decisions will be made based on risk assessments. There will be a widespread global understanding of the urgency and means of addressing climate change.

Natural Resources: Reduced poverty will have synergies with the region's biodiversity and ecosystems, with the region enjoying healthy ecosystems. Conservation principles are well integrated into policy and individual decision making, leading to better conservation of natural resources. With reduced pressure on the land, land degradation will be diminished and the improved soil productivity will benefit the region's agricultural production.

Governance: Governance systems in the region will evolve and political unrest will be replaced by stability that will usher in regional cooperation, leading to acceptable and amicable solutions to the issues in the region. As a result, regional institutions will become strong, with necessary resources for providing better governance. Migration will continue but under ideal circumstances of regional cooperation and favourable labour laws. Governments and the private sector will collaborate, leading to better development and implementation of policy solutions, and overall efficiency and effectiveness. Regional cooperation initiatives such as SAARC and intercontinental initiatives like BRICS would have been strengthened, playing a pivotal role in addressing transboundary issues such as air and water pollution and trade of resources such as energy, regionally harmonized food standards, and enhanced regional transportation, communication, and information exchange facilities.

Social: As a result of reduced climatic impacts and better governance for vulnerability reduction, the mountain region will experience a significant reduction in poverty. At the same time, the region will experience sustained cultural diversity. Population growth will stabilize and people will have 100% access to education, while a substantial reduction in inequality, along with enhanced social equity and empowerment of women, will lead to an equitable society. As a result, societies will show high aspirations for continued improvement and for a high standard of living.

Economy (including infrastructure and energy): The region will move toward a low carbon economy with significant decoupling of economic growth from fossil fuel use. Energy use efficiency will become one of the major criteria for production and consumption of products and services in all sectors. The behavioural attitude, to offset the technical efficiency gain through an absolute energy demand increase, will not undermine the full benefits of energy efficiency improvements, due to a better mix of policy interventions (including fiscal instruments), higher levels of economic prosperity, and no unmet demand for energy (Roy 2000; Sorrell et al. 2009). Scientific advancement will lead to rapid technology development in the areas of health, energy, information technology, food and agriculture, natural resource conservation, financial inclusion through tele-banking, etc. With free trade, there will be an impetus for local economies to become vibrant and integrated. Better regional cooperation will lead to the implementation of strategies such as a regional electrical grid and regional transportation corridors.

The comparative evaluation of the scenarios is done based on quantitative scores given by participants. Scores on a 1–5 scale (where 1 is the lowest and 5 is the highest) were given by participants based on their individual assessments of the baseline status of each component under different thematic areas. For example, the climate theme represents three components: climate change, air pollution, and cryosphere. Participants, based on their knowledge, used a 1–5 scale to score the 2015 (baseline) status of the components and, based on their own storylines, scored the same component across various scenarios. This is summarized in Fig. 4.5. The economy thematic area is represented by one component, energy, because participants identified energy as a major driver of the economic wellbeing of the HKH.

4.2 HKH Baseline in 2015

After the first step of envisioning long-term future challenges and opportunities, the second step of "Back casting" (Fig. 4.2) information on the baseline 2015 status of the different indicators was collected from the other chapters to determine the existing historic trends. In addition, the policies presently in place are analyzed to extract possible barriers and bridges for the challenges identified in the HKH 2080 scenarios. The 2080 scenarios defined by storylines will need to be achieved starting from the baseline of 2015. At present, the HKH is undergoing rapid demographic, economic, social, and political changes, hence any quantitative scenario based on past trends will need to be viewed with caution. However, this chapter builds scenarios following the qualitative process and scoring, which gives enough flexibility that visions and imaginations about future possible transformations can be built into the storylines. This chapter focuses on HKH-specific scenarios, which cannot be done in an isolated way. Global and larger regional developments will influence the future of the HKH region. The likely global emissions future is represented mainly by the Representative Concentration Pathways (RCPs) and Shared Socio-Economic Pathways (SSPs) (see Box 4.2). A summary of 2015 baselines for each country in the HKH (in the case of India, Pakistan, Afghanistan, Myanmar, Bangladesh, and China, this is not exclusively HKH-specific, as these countries fall both in the HKH and non-HKH regions) is prepared based on the synthesis of different chapters in this assessment report, vision documents, and SDG documents of the countries. Defining the baseline helps in understanding the challenges of each country in the region to attain HKH-wide prosperity. In the absence of HKH-specific information for many large countries, the countrywide baseline status is summarized in Sect. 4.2.1.

Box 4.2 How are global climate change emissions scenarios and shared socio-economic pathways related?

For more than a decade, the dominant climate change-related scenarios were based on the IPCC Special Report on Emission Scenarios (Nakicenovic and Swart 2000). These scenarios were developed through a process that involved:

- A literature review;
- Generation of qualitative storylines based on the outcome of the literature review and further interactions; and
- Development of quantitative scenarios using the qualitative storylines.

The SRES scenarios have been highly influential in climate change research and policy, but several years ago a process was initiated to develop a new scenario framework that is more comprehensive and builds on a set of global and regional scenarios (Nakicenovic et al. 2014). The logic of the scenarios has been distinctly changed and now no longer uses emission scenarios but defines the future along several lines and pathways.

Representative Concentration Pathways (RCPs) define the emission pathways (Moss et al. 2010). On the other hand, Shared Socio-Economic Pathways (SSPs) describe plausible alternative trends in the evolution of social, economic, political, and natural systems over the 21st century, globally and for major world regions. SSPs comprise two elements, narrative storylines and quantified measures of development (O'Neill et al. 2014). SSPs themselves do not include outcomes in terms of emissions or impacts; rather, they quantify factors considered to drive the outcomes, such as population or economic growth. Quantification of the consequences is done by scenarios produced based on SSPs. As a minimum, SSPs include assumptions on future demographics, economic development, and the degree of global integration.

4.2.1 National Considerations

Afghanistan: Afghanistan tops the list of eight countries of the HKH in poverty measured by head count ratio. Government policies and budget spending on poverty can be seen as major barriers. In 2015 as regards MDG goals on net enrolment ratio significant improvement was documented while health had modest improvements. Carbon dioxide emissions increased. Food security, water, sanitation, and energy-related goals show marginal progress. The National Priority Programs and Aid Management Policy (AMP) had been targeting improvement aid effectiveness, cooperation, accountability development (Islamic Republic of Afghanistan 2012).

Bangladesh: Substantial decline in head count poverty ratio from 60% during 1990 to 29% by 2015 was recorded. With annual population growth at 1.32% and significant progress in water and energy sectors major challenges are identified in malnutrition and food security, disaster vulnerability. Girl child education in rural areas, improvement in rural income, agriculture production, awareness building are identified as means to address the challenges in reducing malnutrition. Scientific assessment, policy interface, institutional and human capacity building for public sector are seen as enabler to accelerate key developmental projects (Peoples' Republic of Bangladesh 2012).

Bhutan: The SAARC Development Goals, Bhutan's country report (Royal Government of Bhutan 2013) mentions progress in poverty, inequality (especially women and children) and hunger reduction, infrastructure development. The percentage of people living below poverty is 11.2% in 2012, Stunting and malnutrition of children under five years reduced by 6.5% and 4.3% respectively during 1999–2010. The rural houses electrified improved from 54 to 87% during the period 2012–2017. The country has achieved 100% telephone connectivity. The rising unemployment and mountainous landlocked economy are the critical challenges.

China: The poverty head count ratio of China reduced from 66.6% to 6.5% during the period 1990–2012 (World Bank 2012). The maternal mortality rate reported at 20.1 per 100,000 births in 2015 compares with 88.8 per 100,000 in 1990. The forest cover and forest reserves increased by 32.78 million hectares and 2.681 billion cubic meters respectively during the 2005–2015. Increase in share of non-fossil fuels by 11.2% per GDP unit and reduction in carbon dioxide emission per GDP unit by 33.8% was reported in 2014. Greater interconnectivity and interdependence has strengthened China as a community of shared destiny. China has the highest per capita CO_2 emissions among HKH countries. Need for further adoption of green technologies is identified as a big challenge (The People's Republic of China 2016).

India: Eradication of extreme poverty in India still continues due to its complex relations with various social, economic and environmental components. Improvements in net enrolment ratio had led to enhanced focus on higher education. The eradication of hunger and control of malnutrition continue be a daunting task due to low level of progress so far. Food security is seen as outcome of landuse change triggered by urbanisation process, infrastructure development and inconsistencies in monsoon pattern. Anaemia is prevalent among 50% of women in the age group 15 to 49 and more than one third of are reported with low body mass index. Rank in Global Gender Gap Index is at 114 out of 142 countries. Gender equality at the workplace is also far off the track (World Economic Forum 2014). Access to safe drinking water continue to be a huge challenge (Government of India 2015a).
Myanmar: Economic stagnation, which is an outcome of more than three decade long economic sanctions, fall in foreign investment, poor macroeconomic management, public-sector debt. Introduction of reforms in 2011 onwards helped the process of recovery (ADB 2014). During 2000–2012, social indicators such as reduction in population below poverty line, infant mortality rate, and population with access to safe water and sanitation improved (Table 4.2). Per capita carbon dioxide emissions increased from 0.1 to 0.2 tons. Structurally agriculture dominates with 31% of GDP and 50% of employment with a very low per capita income. Opportunities are from the strategic location, abundance of natural resources, tourism potential and youthful workforce. Emphasis is on growth-oriented support for agriculture and natural resources followed by actions on manufacturing for domestic and export markets.
Nepal: Extreme poverty is declining, with a large change from 42% in 1995 to 23.8% in 2015 however, with large degree of spatial variability across the country by gender and social groups. The adult female literacy rate tripled from 1990 to 2011. The proportion of population with access to basic water supplies and improved sanitation facility moved up from 46% in 1990 to 83.6% in 2014 and from 6% in 1990 to 70.3% in 2014 respectively (Government of Nepal 2015).
Pakistan: Pakistan will become the fifth most populous nation in the world in 2050 growing at 2012–13 growth rate of 2% per annum. Twenty nine percent of the population lives below the poverty line, half of the population is unemployed, 77 out of 1,000 children die before their first birthday, and one in three does not have enough nutritional food. Slow growth, limited technological innovation, low agriculture inputs, and little government investment in the agriculture sector make the situation worse. Despite significant efforts energy shortages have impacted the quality of livelihoods, as well as business and agriculture. The exogenous developments such as climate change, reduced aid flows, stagnated import demands from developed countries are effecting the growth (Government of Pakistan 2013).

4.3 Tracking Trends and Commitments for 2030: SDG's and NDC's in the HKH

In the third step (Fig. 4.2), the existing trends and policies are extrapolated toward 2030. To achieve this, literature review and two main interactive processes are followed. One interactive process involved scientists engaged with other HIMAP chapters and a second with policy makers from the region. Section 4.2 provided a synthesis of historical trends mentioned in the different chapters of this volume and reflects the status of the 2015 baseline, while this section first summarizes the SDG targets (Sect. 4.3.1, Table 4.2) for various countries and not specific to HKH, as they are available at country level only, and then continues to discuss thematic trends toward 2030 in Sect. 4.3.1. Possible directions toward 2030 are taken from different mandates of national governments reflected in national vision plans, SDGs, INDCs, and national communications. Table 4.2 shows that most of the countries have mentioned different targets and hence need financial requirements and implementation mechanisms, along with scientific strategies of mitigation and adaptation, to achieve the desired goals.

4.3.1 Existing Sectoral Commitments Toward 2030 Sectoral/Thematic Considerations

Based on the literature Table 4.3 explains the existing sectoral commitments of the countries towards policies and plans related to climate change. Compliance and implementation of (I)NDCs stands critical to address multiple SDGs especially to SDGs 13, 14, and 15.

The interaction with scientists engaged with other HIMAP chapters had as objective to elicit information using the following three major questions:

a. What are the main targets of your chapter theme? Mention the specific indicators behind the target.
b. What is the direction (on a quantitative scale of 1–5) of observed change in the indicator based on the past trends and using, for example, INDCs and SDGs relevant for the region and other policy documents or policy analysis documents? What is the likely direction of trends up to 2030?
c. Based on your answers to questions 1 and 2, what are the chances (on a quantitative scale of 1–5) that 2030 targets (e.g., 100% poverty reduction) will be achieved? What are the factors affecting the chances (e.g., weak policy enforcement, lack of resources, global climate change, conflicts with other sectors, reinforcing trends with other sectors, and so on)?

Table 4.2 SDG targets of HKH countries

SDGs/Country	Target 1.1 Poverty head count ratio (% population)	Target 2.2 Stunting (%)	Target 3.2 Infant mortality	Target 4.1 Gross enrolment ratio	Target 6.1% of people with improved water access	Target 6.2% of people with improved sanitation access	Target 7.1% of population with electricity access[2]
Afghanistan	35.8 (20)	40.9[1] (0)	74 (46)	116 (110)	31 (61.5)	8 (66)	43 (100)
Bangladesh	31.5 (0)	40 (25)[3]	39 (NA)	111.8[4] (NA)	86 (100) 2015	89 (100) 2015	59.6 (100)
Bhutan	12 (NA)	33.5 (NA)	47 (NA)	102.1 (NA)	96 (NA)	58.4 (NA)	87 (NA)
China	6.5 (0)[5]	9.4[6] (7)	12.2 (6)	103.9[7] (90%) HSE	Improved Efficiency	Improved Efficiency	100 (NIL)
India	21.9[8] (0)	48[9] (0)	43.2 (12.96)	21.1 (30) GEH	16[10] (90% piped water supply)	38.81 (rural), (100)	78.7 (100)
Myanmar	25 (NA)	28.6[11] (NA)	41.1 (NA)	114.2 (NA)	84.1 (NA)	77.3 (NA)	52.4 (100)
Nepal	23.8 (0)	40.5% (1)	33 (1)	96.5 (99.5) NER	49.5 (99% piped water)	70.3 (100%)	76% (99%)
Pakistan	29.5[8] (14)	44 (0)	74 (47)	93.5[7] (100%)	NA	48% (90%)	67 (90%)
	Legend						
	Unbracketed value			2010–2013			
	Bracketed value		Afghanistan values indicate for 2020	2025–2030	India sanitation value indicate for 2020		
	Infant mortality			No. of deaths of children less than one year of age per 1000 live births			
	Stunting %						
	NER			Net enrolment ratio			
	HSE			High school education			
	GEH			Gross enrolment ratio in higher education			
	NA			Not available			

Sources
Primarily table prepared based on following country reports and other sources if used are shown in superscript numbers
Islamic Republic of Afghanistan: Afghanistan Millennium Development Goals Report, 2012
Royal Government of Bhutan: SAARC Development Goals, Bhutan Country Report, 2013
Peoples' Republic of Bangladesh: Rio+20 Bangladesh National Sustainable Development, 2012
Peoples' Republic of China: China's National Plan on Implementation of the 2030 Agenda for Sustainable Development, 2016
Government of India: Achieving the Sustainable Development Goals in India A Study of Financial Requirements and Gaps, 2015
The Republic of the Union of Myanmar: Myanmar Unlocking the Potential, Country Diagnostic Study, ADB Report, 2014
Government of Nepal: Sustainable Development Goals 2016–2030 National (Preliminary) Report, 2015
Government of Pakistan: Pakistan 2025 One Nation One Vision, 2013
[1]Afghanistan Survey 2013. National nutrition survey of Afghanistan, Survey Report.
[2]World Bank Data 2013. https://data.worldbank.org/indicator/EG.ELC.ACCS.ZS
[3]ICAD 2018. Bangladesh: Country Profile on Nutrition and Child Stunting Trends.
[4]World Bank 2010. Trading Economics data. Gross er 111.8 https://tradingeconomics.com/bangladesh/school-enrollment-primary-percent-gross-wb-data.html
[5]World Bank 2012. http://povertydata.worldbank.org/poverty/country/CHN
[6]World Bank 2010. https://data.worldbank.org/indicator/SH.STA.STNT.ZS?locations=CN
[7]World Data Atlas 2013. https://knoema.com/atlas
[8]Asian Development Bank 2011. Poverty Data. https://www.adb.org/countries/main
[9]UNICEF Report 2013. https://in.reuters.com/article/health-nutrition-unicef/about-48-percent-of-children-in-india-are-stunted-unicef-idINDEE95607N20130607
[10]Govt. of India Report 2017. https://www.hindustantimes.com/india-news/6-3-crore-indians-do-not-have-access-to-clean-drinking-water/story-dWIEyP962FnM8Mturbc52N.html
[11]Frontier Myanamr 2017. https://frontiermyanmar.net/mm/node/4278

Table 4.3 Main nationally determined contributions in the countries in HKH

NDC	Afghanistan	Bangladesh	Bhutan	China	India	Nepal	Myanmar	Pakistan
Fossil fuel subsidy reform	Not mentioned	Not mentioned	Not mentioned	Not mentioned	Mentioned	Not mentioned	Not mentioned	Not mentioned
Cost of mitigation	6,620 million US $	27,000 Million US $	(partial) costs not indicated	(partial) costs not indicated	834,000 million US $	(partial) costs not indicated	(partial) costs not indicated	(partial) costs not indicated
Cost of adaptation	11,000 million US $	40,000 Million US $	(partial) costs not indicated	(partial) costs not indicated	205,000 million US $	(partial) costs not indicated	(partial) costs not indicated	(partial) costs not indicated
Mitigation finance	Fully conditional to contribution	Partly conditional to contribution	Mentioned	Mentioned	Partly conditional to contribution	Mentioned	Fully conditional to contribution	Partly conditional to contribution
Adaptation finance	Partly conditional to contribution	Partly conditional to contribution	Mentioned	Mentioned	Partly conditional to contribution	Mentioned	Partly conditional to contribution	Partly conditional to contribution
Historical responsibilities	Not mentioned	Mentioned own historical responsibility	Mentioned own historical responsibility	Concept mentioned	Quantified own historical responsibility	Mentioned own historical responsibility	Not mentioned	Not mentioned
Section on fairness	Not contextualizing emissions	Contextualizing emissions	No section	No section	Contextualizing emissions	No section	Contextualizing emissions	No section
Adaptation	Adaptations actions and strategies included; NAP Mentioned	Adaptation actions mentioned; NAP Mentioned	Adaptations actions and strategies included; NAP mentioned	Adaptation actions mentioned	Adaptation actions included	Adaptations actions and strategies included; NAP mentioned	Adaptations actions included; NAP mentioned	Not mentioned
Loss and damage	Not mentioned	Mentioned	Not mentioned	Mentioned	Mentioned	Mentioned	Mentioned	Not mentioned
Assessment and review	No references	No references	No references	Section on NDC tracking and review	Reference to informal assessment review	No Reference	Section on NDC tracking and self-assessment of NDC implementation	No Reference
Size of NDC	8 pages	15 pages	9 pages	20 pages	38 pages	12 pages	18 pages	2 pages

Note Synthesized from NDC Country Documents, and prepared by Giovanna Gioli, ICIMOD

The main targets and indicators listed by the chapter scientists were shared with policy makers, who were asked to answer two questions:

a. Are indicators as identified sufficient from a decision making point of view, or is there scope for enrichment?
b. What is the current (i.e., 2015 baseline) status on a quantitative scale (scale of 1–5) of each indicator? Are there variations across the HKH?

The assessment based on these two stage consultations, along with a literature review, are presented below thematically.

Climate: The increasing changes in mean annual temperature, high uncertainty around impact of climate change, regional differences such as higher warming over higher altitudes and faster warming in the eastern part of the HKH region are the increasing challenge of controlling such changes. Very high uncertainty in the total rainfall and monsoon periods with regional differences are also expected (Lutz et al. 2014). Neglect of linking air pollution and climate change issues have led to an absence of improved research in this area. Climate change, in conjunction with the challenge of the sustainable use of natural resources, poses a greater impediment for development in the HKH. Most of the countries have submitted INDCs giving commitments to take steps towards climate change mitigation and adaptation. With increasing extreme weather events, disaster proneness is expected to increase by 2030 with a higher loss of assets. However, with greater recognition of an increasing likelihood of disasters, emergency response and mechanisms for addressing disasters, the magnitude of life losses is going to be reduced. Appreciable attention has been given by all the countries under SDG 13 (Target 13.3) to gradually establish climate forecasting systems, improving coverage of meteorological early warnings, disaster prevention and reduction systems, and strengthen climate resilience. Additionally, the coupling of local and national mechanisms with appropriate communication and coordination systems, vitalizing local institutions, and appropriate policy instruments is necessary to achieve the desired goal of disaster risk reduction.

Natural Resources: Most of the HKH countries have among the highest levels of hunger and malnutrition in the world. It is projected that hunger and malnutrition will be totally eradicated for all by 2030, in particular for the poor, infants, and people in vulnerable situations, as part of SDG2. To achieve this through sustainable food production systems, implementation of resilient agricultural practices, addressing impacts of climate extreme events and disasters, land and soil quality management, and other related parameters of enhancing productivity will play important role. The supply side key challenges to agriculture include optimal land allocation and water use, minimising land degradation, urbanisation trend with increasing affluence and changing tastes of the population. The improvements in financial and credit support systems, agriculture marketing, effective extension services and post-harvest losses are also identified as very important factors.

Food availability and accessibility are also targets of strengthened local food systems and stable food supply from the plains. Improvement in local people's income, planned adaptations to climate change, and higher chances of collaborative food trade in the region also promote food availability and accessibility. The percentage of people who face hunger will also be substantially reduced due to effective food procurement, storage, and distribution systems and ensured equity in inter- and intra-household distribution of food. Reducing the percentage of people with malnutrition is going to be a difficult task unless improved food intake, people's knowledge on nutrition and health, dietary diversity and intake of micronutrients, health services (child- and mother-care and immunizations) and infrastructure (hospitals and health centres), and health environment (hygiene, safe drinking water, and improved sanitation) are all improved. The overall 2030 food security goals can be achieved with the support of effective governance systems and implementation, good infrastructure, and regional cooperation.

Water availability is expected to be on an increasing trend with increased natural flow under higher precipitation and melt conditions. However, the uncertainty over spatial and temporal distribution and increasing trends remains a challenge. Upstream water use for realizing national hydro potential will continue to increase and water quality will deteriorate over time. The higher water use for increased agriculture and urbanization might have implications for water availability outside the HKH. Due to lack of management of infrastructure and land use, the volume of reservoirs and spring flows will reduce. Against this background, universal provision of drinking water (Target 6.1) and sanitation coverage and the quality of water under SDG 12 as planned by 2030 is going to be critical. In countries like Afghanistan, Bhutan, India, and Pakistan, meeting the goal of improved sanitation is going to be a formidable challenge, especially in India, which aims to achieve 100% improvement by 2018. Similarly, most of the countries are far from providing quality water at international standards and ensuring universal access to piped water supply.

Many more large water-sharing agreements that take into account use of regional hydro potential, ecosystem flows, upstream-downstream issues, and sectoral water allocation issues in the region will emerge to facilitate meeting these challenges. Bilateral agreement (e.g., between India and other countries), stronger centralized maintenance of transboundary waters, rising investment in urban infrastructure, and strong, self-reliant local management are going to be greater driving factors of water governance. The policy implications and incentives for the trajectory of hydropower development, regulation and maintenance of ecosystem flows, and effective use of pre-monsoon flows for hydro and agricultural use are going to be critical. Improving water quality, electricity for withdrawing water, wastewater treatment, technical knowledge gaps and systems, infrastructure, O&M cost, enforcement, soil erosion, and dam sedimentation are other intrinsic factors regulating the growth trends toward 2030.

The trends of ecosystem degradation by 2030 are going to be different across the region, with countries like Myanmar, Afghanistan, and Pakistan experiencing higher levels of degradation. Community-based initiatives like REDD+ in Nepal and Bhutan are going to play a significant role in the conservation of ecosystems and ensuring ecosystem flows. With accelerated climate change, there will be shifts in the growth and distribution patterns of high-value products at higher altitudes, affecting ecosystem services. In this context, countries like China, Myanmar, and Bhutan have made commitments to mitigate climate change impacts through forest-based enhanced carbon sink capacity to achieve the goals under SDG 13 and 15.

Social Systems: In order to avoid inequitable transformation, the targets of Goal 1 on Eradication of Poverty need to be supported with intricate and balanced actions among climate, sustainability, and poverty. Accordingly, the Goal (2) on food security, Goal (3) on access to healthcare, Goal (4) on education, Goal (6) on water and Goal (7) energy for all also covers issues related to poverty. Additionally, infrastructure development enabling capital formation and employment generation are addressed under Goals 9 and 11. As part of Goal 8 inclusive economic growth to enable income opportunities are targeted towards poverty reduction. While we discussed possible directions and challenges of these goals in the earlier sections, the HKH faces region-specific social and demographic challenges in bringing overall social change for poverty eradication, eliminating gender discrimination, and enhancing health and education, as explained below.

The current trends of increasing income due to people migrating into higher income jobs in both urban areas and outside the region, upcoming service sectors, tourism, and M/SMEs, and the consequent increase in purchasing power are expected to continue. Migration will also increase even if the population growth rate remains constant. On the other hand, great uncertainty exists about movement beyond 2020 and money flows due to extrinsic factors such as labour policies in the Gulf countries, a slowdown in Chinese growth, reduction in the construction boom, and so forth. Major challenges include increasing vulnerability to climate change, market dependency, difficulties of small farmers in securing finance for purchasing seeds and shifts in land use toward high-value crops. Male outmigration pushes women directly into activities relating to economic development in addition to their traditional roles. This change comes on top of the current top-down policy systems, in which there is lesser engagement of people from various social strata. This situation is going to be a challenge for achieving the 2030 SDGs unless a more inclusive policy-making process is followed. Therefore, there is a rising need for new dimensions of social inclusion in decision making processes by, among others, including this new female workforce in developmental activities. This inclusion will increase the acceptance by the local people and will help reduce poverty to achieve the SDGs by 2030.

The institutional setting in the region, especially in the mountains, is very spread out and weak at the subnational level. It is necessary to have decentralized adaptive governance with institutional settings that enable effective governance mechanisms for adaptation by 2030, especially at the subnational level. Ecological fragility is very strongly distributed in the region. At the lower level, more holistic and science-based policies are needed; for example, REDD + might take away land and lead to more poverty for local inhabitants. Development of more resilient mountain communities, adaptation strategies that ensure equal engagement of women and men, building capacities of institutions involved in planning and implementation, and access to global finance systems should be addressed by 2030 to minimize social disparities and inequalities.

Considering the minimal progress made in combating gender discrimination over the last decade in the HKH, institutional and policy mechanisms are identified by all countries to end discrimination against all women and girls covering all forms (SDG 5). The elements like women's and girls' rights to education, social welfare, labour security and medical care are identified as important for achieving this goal. Raising public awareness and commitment on gender equality and complete eradication of discrimination and prejudice against women and girls is also considered crucial.

Economy: As part of SDG 6, most of the countries envisages to provide reliable, affordable, sustainable, and modern energy services for all by 2030. Continued reforms and restructuring of urban and rural power grids, focusing on upgrading of grids for small towns and rural centres, and ensuring full rural power coverage are advocated as critical steps. Advanced welfare-oriented energy policy that promotes benefit-sharing mechanisms, accelerating poverty reduction programmes through photovoltaic technology, and energy development in poverty-stricken areas are considered important. Increasing the share of non-fossil fuels in primary energy consumption, optimizing the energy mix by improving fossil-fuel efficiency, and increasing the share of clean energy consumption are some of the measures for ensuring that non-fossil fuels and natural gas become the main energy sources. Improved hydropower generation will leapfrog electrification, but political issues still impede the possibility of shifting from biomass to electrification.

These efforts will cost a lot for countries like Bangladesh, requiring an investment of USD 26.6 billion up to 2030 to cut coal production-based carbon emissions. On the other hand, the projected cost for India of meeting the goals of SDG 6 is USD 854 billion with estimated gap of USD 406 billion along with high dependence on foreign imports. The dependence on alternative energy sources beyond fossil fuels is also being necessitated due to falling prices of renewable energy and rise of coal market prices. The improved financial services, affordable credit, integration into value chains and markets are identified as essential elements to promote small and medium scale (SMEs) under SDG 9/Target 9.3. In this context, the SDG report of India and Bhutan National Development Plan clearly identified that inadequate access to financial resources is one of the main challenges for the growth and development of SMEs, especially during the periods of economic crisis. The need for development, transformation, and upgrading of SMEs through mass entrepreneurship and innovation is emphasised as part of China SMEs developmental plans. The provision of intellectual support and building effective platforms of international cooperation are also identified as critical for the sustainable development of SMEs.

Country level challenges: Pakistan plans to progress into Upper Middle-Income countries by 2025 through addressing SDG goals on zero poverty and hunger, universal access to education health services, clean water and sanitation and modern energy services. The people-centred innovative, coordinated, green, open, and shared development is considered as important in China to implement the 2030 agenda and promote sustainable development. The concept of inclusiveness, equitability along with the spirit of a welfare and prosperity are reported as fundamentals for Nepal to become middle-income country by 2030. Myanmar plans to have an effective financial system and services for all people and positive shifts towards from agriculture to services sector and industry to achieve SDG goals of 2030. It is also to be noted that most of the reports have limited focus on HKH mountain systems.

Box 4.3 Summary of (I)NDCs related to climate change mitigation of HKH countries

Afghanistan: Reduce 2030 GHG emissions by 13.6% as against business as usual (BAU) 2030 scenario, with external support as condition (Islamic Republic of Afghanistan 2015).
Bangladesh: Unconditional and conditional GHG emission reduction scenario by 2030 from BAU levels in the power, transport, and industry sectors. The unconditional scenario targets to reduce emissions by 5% based on existing resources. A conditional 15% reduction is targeted subject to appropriate international support (Government of the People's Republic of Bangladesh 2015).
Bhutan: Intends to remain carbon neutral by maintaining emissions below the estimated 6.3 million tons of annual forest carbon sequestration. The export of electricity from clean hydropower projects is also expected to offset up to 22.4 million tons of CO_2 per year by 2025 in the region (INDC, Bhutan 2015).
China: 60–65% reduction in CO_2 emissions per unit of GDP (carbon intensity) against 2005 levels and 4.5 billion cubic metres increase in its forest carbon stock volume from 2005 levels (Republic of China 2015).
India: Reduction of 20–25% in emissions intensity of its GDP by 2020 with reference to 2005 base levels as a voluntary goal having no binding mitigation obligations (Government of India 2015b).
Myanmar: Section 4.2.1—Mitigation actions, envisages plans with a condition of international support to achieve sustainable development needs for fulfilling its contribution to reduce future emissions as part of global action (The Republic of the Union of Myanmar 2015).
Nepal: Nepal intends to pursue and support the efforts to limit the rise in temperature to well below 2 °C and develop Low Carbon Economic Development Strategies to promote future economic development through a low carbon economy (Government of Nepal 2016).
Pakistan: It intends to reduce GHG emissions of 2030 in relation to BAU in 2030 in different sectors such as Energy Supply: 37%, Energy Demand: 22% Transportation: 8%; Agriculture and Forestry: 5.5% (Government of Pakistan 2015).

4.4 Two Paths Toward a Prosperous HKH

Potential pathways leading toward the desired prosperous HKH are myriad and overlapping, but participants in the consultative workshops developed qualitative assessments on how and why the HKH future might unfold in certain directions.

The outcome of these consultations was integrated with the trends, possibilities, and trade-offs coming from related chapters of this assessment. A two-step interactive process with chapter teams of this assessment and decision makers has provided the necessary data to develop the pathways. Interaction with researchers engaged with writing the chapters of this assessment and decision makers revolved around two major questions:

- What direction are the key indicators on challenges (e.g., access to water, energy) and opportunities (e.g., water availability, biodiversity) likely to take in each HKH 2080 scenario as compared to the 2015 baseline, and how strong is this change on a scale of 1–5?
- What additional strategies can help in this transition?

Analysis of the responses clearly showed that there are diverging views among scientists, experts, and decision makers on how the HKH may prosper. These different points of view clearly indicate that there can be more than one pathway toward a prosperous future.

Pathways usually differ in terms of trade-offs, opportunities, and challenges, but still lead to the same outcome. Pathways try to answer questions such as: "What policy actions can create an enabling condition for progress?", "What kind of technological options might be available to make this progress possible?", or "Which economic sectors, natural resources, and developmental actors can play a central role in ensuring that progress is made in the desired direction?" A pathway is described here as a set of actions and combinations of actions that a decision maker (individual, country, business, policy maker) can take.

Here, two broad pathways are presented to facilitate envisioning of the future and to support an effective decision-making process toward a prosperous HKH. The pathways differ from each other in the trade-offs among policy choices, institutional arrangements, and technology choices that are to be made. However, some actions are common to both pathways. Actual effectiveness of the pathways will depend on multiple issues like social acceptance, political agreements, and investment decisions. Based on the consultative workshop participants' feedback, the characteristic actions, trade-offs, benefits, and risks associated with the pathways are described below.

Pathway 1—Large scale investments including regional cooperation: It is possible to get to the prosperous scenario using the HKH's naturally available resources by according higher priority to centralized large-scale power, communication, infrastructure, and development projects. Water and energy are going to be critical for ensuring energy access and food security, and for eliminating poverty. However, this will require large-scale financial and skilled human resource mobilization, collaboration of both state and large corporate, non-state actors for financing the actions, and national and international funding. Widespread communication and mobility that enhance infrastructure, trade and cross-border resource sharing, and institutions for governing the collaboration and cooperation across regions and within the region will be the essential features of this pathway.

Pathway 2—Bottom-up investments with cooperation across multiple levels: It is also possible to get to the prosperous scenario using the HKH's critical resources—energy and water—through higher prioritization of small-scale, decentralized, local or national developmental actions and projects. However, this would require the involvement of a large number of developmental and social actors nationally and sub-nationally. Also, a multilevel governance structure will be needed to address local diversities, to mobilize financial resources for a large number of projects of different scales, and (more than for Pathway 1) to monitor the progress and ensure it is in line with the scenario goals.

Based on the assessment of the feedback received from consultation workshops and the content of this assessment report, it is clear that there cannot be one single best possible set of actions or a single pathway for decision makers to minimize the challenges and seize the opportunities for the HKH in keeping with the long-term goal of prosperity. Multiple actions will need to be followed simultaneously.

It is crucial that visioning and strategic planning start now to enable the implementation of actions beyond 2030 by decision makers. To decide on the choice of pathways, there is need for prior analysis, understanding, and recognition of the HKH's unique and emerging opportunities and challenges. The challenges in the HKH are large, and overcoming them can help the HKH prosper in a sustainable manner. At the same time, there are numerous opportunities for the region. Based on the information on the variety of opportunities and supplementary resources needed to implement these opportunities, detailed actions and the way forward are summarised in Table 4.4. The description of the set of actions for alternative pathways toward 2050 provides a guideline for decision makers in the region. However, each action has varying sources and levels of risks. It is important to be aware of the risks so that risk mitigation strategies can be developed and implemented. Awareness of these risks can also help in guiding the actual implementation decision.

The two alternative pathways are not completely mutually exclusive, so it is possible for the alternative pathways to have some common actions. Actions have thus been categorized under three headings: "Specific actions: how pathways differ", "Common set of actions", and "Actions to be avoided". The two pathways are mainly differentiated by these specific actions.

4.4.1 Major Emerging Opportunities in the HKH

Major opportunities are emerging in the region due to climatic change response strategies, historical developmental actions, and multi-dimensional commitments made by national governments for the period until 2030 to attain SDG targets. The opportunities are summarised below:

Water resource availability is expected to improve in the HKH (Chaps. 3, 7 and 8): Increased flow is expected mainly because of increased precipitation, although glacial melt will decrease. However, this increased flow will happen with greater seasonal variability across countries and within countries in the region and with a changing pattern over time. An exception is the Indus basin because in the Indus river system, the main contribution is from snow and ice melt and to a lesser degree from rainfall.

Multiple new trade potentials from the HKH are going to increase (Chaps. 6, 8, 9 and 12): As the income of mountain people improves, the demand level and pattern will change. However, in the case of a disaster or crisis combined with the rising demand, food security can only be achieved through increased trade and a better distribution system. Electricity trade and water transfer within and between the HKH and non-HKH regions will increase with increasing economic activity, population, and income growth.

Non-farm activity-based livelihood diversification is strongly emerging (Chaps. 5, 6, 9, 12 and 15): People are changing their livelihood by moving toward high-paying cash crop farming and the service sector and by migrating to other regions. Dietary diversity resulting from income increase drives the need for processing and manufacturing industries. Demand for the aesthetic value of the HKH's unique cultural diversity, ecosystem, and services has given rise to high demand for a well-regulated tourism industry and infrastructure managed by skilled human resources.

Energy demand is going to grow significantly, while the potential for renewable and sustainable energy is also increasing (Chaps. 6, 8 and 9): The accelerated development of hydropower can help countries in attaining transformative change in achieving a carbon-free 24 × 7 power supply. This will help to avoid small steps and facilitate leapfrogging in attaining a secure and reliable power supply, clean cooking, and space heating, which will eliminate indoor air pollution and protect the health of women and children. This is not in conflict with a decentralized/micro grid for the high mountains.

Community-based institutions to manage natural resources (Chaps. 8 and 16): These have emerged as a major trend globally, and the HKH already has a long history of working with such institutions.

Need for cooperation across various levels in the region is emerging quickly (Chap. 16).

4.4.2 Two Pathways and Distinguishing Actions

Identifying strategic actions to seize the emerging opportunities is crucial for achieving prosperity in a time-bound manner. Table 4.4a, b list actions identified during the consultative workshops as crucial for a prosperous HKH. Knowledge of varying benefits, investment, and governance needs and the multiple sources and nature of risks associated with actions and their implementation strategies help in making decisions. These are listed in Table 4.4a, b for each of the pathways.

Pathway 1 consists of strategic actions based on globally available technology and knowledge to meet leapfrog growth imperatives in the HKH region. It takes into account mountain-specific resources, cultural diversity, and niche product potentials, but also identifies the need for cross-regional cooperation, institutions, and finance. Demand for access to modern energy and energy security in the region indicates the need for energy system transformation. One way to leapfrog on the supply side with energy security in the HKH region could be to achieve 100% hydropower share (Chap. 6) for the HKH region in the longer run. This will help to promote modern industrial growth and advance mountain mobility and human wellbeing through such changes as cooking fuel leapfrogging from biomass to clean, renewable electric supply, along with self-reliance. Necessary actions have to be taken now to ensure a larger allocation of water for hydropower generation in the HKH region (Chaps. 6 and 8) in the coming years. This can be done by prioritizing large-scale cross-border infrastructure building for large-scale hydropower projects (Chap. 16) with a regional grid and grid integration. For example, the HKH grid can be connected to India's northern regional grid or national grid, or the national grids of other countries. The need for large knowledge capital, technology, and financial capital can be met through more global cooperation between the HKH and non-HKH regions (Chap. 6). Governance of such projects will need cross-border harmonization through power distribution arrangements, power

Table 4.4 a: Specific distinguishing actions for Pathway 1, b: Specific distinguishing actions for Pathway 2

(a)

Pathway 1

Actions	Benefits				Need		Risk
	Economic	Social	Environmental/climate	Cross sectoral	Finance and human resources	Governance	Source
Large hydro power generating capacity	Leapfrog in economic prosperity for the region as a whole, high potential for power trade	New skill development, diversified livelihood options	Air pollution reduction Both adaptation and mitigation	Large water storage to manage seasonal variability and strategic cross-sector allocation	Large corporate, global finance, sustained climate finance	HKH institution, regional tariff, cross-border policy coordination	Lack of transboundary sustainable political cooperation; lack of cross-sector water sharing formal arrangements; lack of ecosystem-based design of reservoirs/power plants; public acceptance, silt accumulation
HKH and non-HKH electric grid	Very high economic prosperity for the region and beyond	New skill, non-farm diversified livelihood options	Unplanned local resource extraction will decrease	Reliable power supply for all sectors	Large corporate, global finance, climate finance	HKH electric distribution corporation	Transboundary sustainable political cooperation; lack of ecosystem-based design
HKH ICT (information and communications technology) network	Boost to regional and local economic growth	New skill, non-farm diversified livelihood options	Connectivity across mountainous terrain without ecological impact	Extent of market cutting across sectors and regions	Large corporations, global finance, climate finance	HKH communications corporation	Transboundary sustainable political cooperation; lack of biodiversity-sensitive design
Cross-border trade corridors e.g., silk route re-development	Income, consumption, production leapfrogs as per comparative advantage, benefit to large-scale tourism industry	Food security, energy security, health service, social interdependence, non-farm livelihood generation	Comparative advantage will lead to biodiversity conservation, enhance payment for ecosystem service	Multiple opportunities across sectors emerge	Regional, global	HKH trade authority	Transboundary sustainable political cooperation; lack of biodiversity-sensitive design in transport corridor development
Large water storage and supply	Income, consumption, production leapfrog	Food security, energy security, non-farm water sector livelihood generation	Less GLOF, less flash floods, pump storage facility	Multiple opportunities across sectors emerge	Regional, global	HKH water council	Transboundary sustainable political cooperation; lack of ecosystem-sensitive development
Large water treatment facilities	Leapfrog in water resource management	Water security, non-farm water sector livelihood generation	Reduction in waste disposal	Multiple opportunities across sectors emerge	Regional, global	HKH water council	Transboundary sustainable political cooperation; lack of ecosystem sensitive development

(continued)

Table 4.4 (continued)

(a)

Pathway 1

Actions	Benefits				Need		Risk
	Economic	Social	Environmental/climate	Cross sectoral	Finance and human resources	Governance	Source
Large-scale urbanization	Leapfrog in economic growth centers	Non-farm water sector livelihood generation	Reserve nature for biodiversity conservation	Multiple opportunities across sectors emerge	Local, national, regional, and global	National urban development authorities	Lack of ecosystem-sensitive development
Large contract farming	Leapfrog in farm-level activity and income	Income, livelihood security	Investment in environmental management	Farming based industrial/trade growth	Local, national, regional, and global	National farming development authorities	Lack of ecosystem-sensitive development; lack of public acceptance, possibility of food crop reduction, crop monoculture

(b)

Pathway 2

Actions	Benefits				Need		Risk
	Economic	Social	Environmental/climate	Cross sectoral	Financial and human resources	Governance	Source
Distributed small hydro power generating capacity	Incremental national, local economic prosperity through self-sufficiency	Traditional skill utilization	Air pollution reduction Both adaptation and mitigation	Water flow uninterrupted	Small to medium national scale finance, programmatic finance by bundling, climate finance	Community level, local, national, multilevel coordination for tariff, etc. to ensure equity	Lack of local capacity for multi-level governance; lack of upstream- downstream water sharing arrangements; lack of ecosystem-based design
Micro grids	Local economic prosperity	Lack of ecosystem-sensitive development	Small infrastructure with less environmental impact	Reliable power supply for target group	Specialized medium-scale global finance, climate finance	Private, local electric distribution companies	Without multilevel governance, inequality may arise across social groups; not a tried and tested technology; maintenance will need local skill building
National ICT (information and communications technology) network	Incremental national growth	Lack of ecosystem-sensitive development	National connectivity in mountainous terrain improves without ecological impact	Extent of market cutting across sectors	National/global investment negotiated competitively	National institutions	Lack of local/national skill, national negotiation capacity
National culture based products, tourism	Incremental progress	Traditional skill, non-farm livelihood	Environmental conservation	Tourism related infrastructure expansion	Local, national	Local and national institutions	Lack of capacity to integrate with the rest of the world

(continued)

Table 4.4 (continued)

(b)							
Pathway 2							
Actions	Benefits				Need		Risk
	Economic	Social	Environmental/climate	Cross sectoral	Financial and human resources	Governance	Source
Decentralized water storage and supply	Incremental progress	Traditional systems to be revived	Environmental conservation	Local infrastructure expansion	Local, national	Local, national	New modern technology to be developed; lack of local/national skill
Decentralized water treatment	Incremental Progress	Traditional systems may be revived	Environmental conservation	Local infrastructure expansion	Local, national	Local, national	New modern technology to be developed; lack of local/national skill
Small settlement planning	Less displacement cost	Less displacement and migration	No change in large-scale land use pattern	Local infrastructure expansion	Local, national	Local, national regulations	Localized environmental impact might go unregulated
Small farming practices	Incremental progress	Continuation of traditional practices	No change in large-scale land use pattern	Local infrastructure expansion	Local, national	Local, national regulations	Localized environmental impact might go unregulated

Source Various chapters of this assessment report and workshop outcomes purchase agreements, power tariffs, incentive designs, and policies. Many mountain tracts will need micro grids with small hydro-electricity (hydel) generation. There will be a need for new HKH financial institutions like BRICs bank and HKH regional bank.

With high priority accorded to hydropower generation, water resource sharing with other competing demand sectors will need special attention and concrete actions. More allocation of water for the power sector will imply a trade-off with water allocation for increased demand for water in urban areas, agriculture, and food (Chap. 8). However, with the penetration of new wastewater treatment technology and water quality management of available surface and groundwater available water supply will increase. Adoption of water-efficient technologies for irrigation and cropping patterns shifting away from paddy cultivation will reduce demand for water in agriculture sector. These sectors can be managed with a relatively smaller quantity of water (Chap. 8). Appliance standard policies need to be in place to achieve water efficiency. There is a need for cross-country learning, best practice sharing and technical knowledge sharing (Chap. 6). Building larger water storage systems to manage seasonal variability in water supply could be a solution. These storage systems can be used for more hydro power generation, which implies scope for wider power trade. The latter will need major new economic cooperation in the region. More cross-country collaboration and human resource sharing between the HKH and non-HKH areas of the countries will also be necessary to ensure access to sustainable energy and water in the longer run.

Pathway 2 is also comprised by various strategic actions, although at a different scale. To meet the rapidly growing demand for access to modern energy and energy security of the region through more renewable energy, there can be decentralized generation and micro grids, one of which will continue to be hydropower generation, mostly with small hydel projects. The current centralized grid will continue, but newer decentralized generation will dominate the action agenda at the national and sub-national scale (Chap. 6). The energy supply source mix will be diversified through solar and advanced biomass use, with less water allocated for hydropower generation. Water allocation may result from bilateral or trilateral upstream–downstream water sharing agreements within and across borders. Such projects will need multi-level governance and more country-specific power tariff rationalization and national incentive designs. Small, decentralized water storage systems managed by the community and a multi-level governance mechanism should be in place to avoid the risk of failure.

Energy need for the expansion of small and medium enterprises (Chap. 6) will need special grid setups for reliable, dedicated supply. To ensure self-reliance in energy supply for each economic activity, special monitoring and maintenance skills will be required and appropriate capacity will have to be built. More multi-level governance within each country will be needed to ensure self-sufficiency and reliability in supply. Power service delivery from multiple sources (e.g., solar, biomass, hydro) will need a multi-skill development effort. Financial capital and institutions will need to deal with multiple investors. However, the need for knowledge capital, technology, and financial capital can be met through wider cooperation between the HKH and non-HKH regions (Chap. 16). Overarching national regulations and norms will have an important role and will need to be developed. However, upstream-downstream resource sharing issues will need to be resolved across and within the countries in the HKH.

4.4.3 Two Pathways but Common Actions

The actions described as part of each pathway can of course be used along either pathway. It is likely that irrespective of the pathway decision makers decide to follow, these actions will need to be implemented to bring prosperity to the HKH. Table 4.5 lists the actions that are likely required for both pathways.

Table 4.5 Actions common to both pathways

Actions	Challenges to overcome
Build better health infrastructure	Malnutrition, poor health, poverty
Create food banks	Increase food security across seasons
Investing high-skill creation at various levels	Creation of new non-farm activities
Develop migration laws across regions to facilitate mobility	Improve livelihood diversification and remittance flow
Create appliance standards for efficient irrigation practices to increase penetration rate	Enhancement of energy as well as water use efficiency and enhancement of water storage capacity
Promote specific products that reflect local diversity	Increase in food and nutrition security
Establish a seed bank	Loss of biodiversity
Improve mobility sector	Improvement of trade and mountain supply system, which implies cross-country collaboration, trade negotiations, policies, pricing to ensure cross-border movement

4.4.4 Actions to Avoid Downhill and Business as Usual (BAU) Scenarios

Clearly conscious evidence-based strategic decisions, mechanisms, and actions are needed no matter which of the potential pathways is adopted to make sure the HKH does not move toward undesirable scenarios. Complex challenges require sustained effort and progressive visioning over time to ensure progress toward prosperity, and it is clear that some actions will certainly undermine progress toward prosperity (Table 4.6).

Table 4.6 Actions to be avoided

Actions to be avoided	Benefits/avoided risks
Water related conflicts	Risk of lack of economic growth and loss of livelihood is avoided. Conflict resolution mechanisms based on scientific evidence should be in place at local, national, and regional levels
Poor maintenance of water storage infrastructure	Risk of reduction of water volume and reservoir capacity because of increased sedimentation
Closed deal of Indus treaty	Lost opportunity to implement efficient and equitable water sharing based on new HKH-specific scientific evidence and changing realities
Energy supply scarcity	Inefficient management of water supply and health infrastructure systems
Use of inefficient technology	Increased waste across all sectors and missed opportunity of low-cost resource availability, e.g., energy supply and demand sectors, water, irrigation
Unhealthy and non-diverse food habits	Health risk. Changing income and food preference towards vegetables and non-paddy crops can be an incentive for choice of "climate proof" crops
Inequitable food distribution	Increased risk of conflicts because of lack of food
Top-down policy	No buy-in by the local people if top-down policy is used without engagement with the public
Gender bias	Risk of inequity and ineffective uptake of projects. Equal engagement of men and women is needed in climate adaptation projects for them to be effective
Low adaptation capacity	Risk of ineffective adaptation projects. If institutions and capacity for adaptation in the mountains remain very minimal, vulnerability to climate change will remain high
Degradation of ecosystems	Loss of biodiversity and quality of life. Unless degradation of ecosystem is avoided, decentralized bottom-up actions and governance will emerge

4.5 Beyond 2050 to 2080: Knowledge Gaps and Ways Forward

The decisions and choice of actions discussed in Sect. 4.4 are going to be important in realizing the 2080 scenario of a prosperous HKH. Unless the suggested actions are implemented, the downhill or business as usual scenarios are likely to prevail and will undermine prosperity beyond 2050. For the two pathways mentioned, there can be many combinations of actions by which prosperity can be achieved.

Decisions are going to be strategic through various levels of cooperation and at clear intervals. Otherwise, the chances of achieving prosperity will be uncertain as spiralling conflicts can have a cumulative effect, leading to a pathway that diverges from long-term prosperity. The most important decision on the level of cooperation will also be guided by the decision of the chosen pathway, technology, and resource sharing. However, regional cooperation on resource sharing will not guarantee prosperity for the region unless global mitigations are guaranteed. The expert and stakeholder consultations clearly brought out the highest pessimism in this context. However, the Paris Agreement, if adequately implemented, can help in achieving the prosperous scenario.

As this chapter has developed future long-term scenarios and considers transformation toward long-term prosperity, the knowledge gaps described here also focus on the long-term context. As this effort is a first of its kind, long-term scenario development for the HKH, it is clear that a more extensive process covering many more stakeholders needs to be undertaken. Such a process will further improve the understanding of the possible strategies that take into consideration technological, socio-political, and economic aspects. In this context, more refined indicator development through a continuous process of participatory engagement in the region might be useful.

The scenario and pathway building reflected in this chapter has been primarily a qualitative process, but one based on bottom-up, participatory, consultative shared narratives and scoring. There is no top-down regional impact assessment model being used to quantify the qualitative scenarios and pathways. Although qualitative storyline building and strategy identification through wider participation in the region remains a major tool for continued use in the future, developing a regional assessment model for quantitative assessment specifically for the HKH would also be advisable to better inform decisions. For example, risks associated with various actions that emerged through the consultative process across pathways are the least understood. An emission assessment of large-scale interventions, an economic assessment for identifying the least costly and most beneficial pathway, and for understanding adaptation and mitigation costs along the pathways, will help to improve future decision making.

References

Afghanistan Survey. (2013). National nutrition survey of Afghanistan, Survey Report. https://reliefweb.int/sites/reliefweb.int/files/resources/Report%20NNS%20Afghanistan%202013%20(July%2026-14).pdf.

Asian Development Bank. (2011). Poverty Data. https://www.adb.org/countries/main

Asian Development Bank. (2013). *Food security in Asia and the Pacific*. Manila: Asian Development Bank.

Asian Development Bank. (2014). *Myanmar unlocking the potential. Country diagnostic study*. Mandaluyong City, Philippines: Economics and Research Department. https://www.adb.org/sites/default/files/publication/42870/myanmar-unlocking-potential.pdf.

Asian Development Bank. (2015). *Energy outlook for Asian and Pacific*. Manila: Asian Development Bank.

Asian Development Bank. (2016). *Asian development outlook 2016 Asia's potential growth*. Manila: Asian Development Bank.

FAO. (2010). *Asia-Pacific forests and forestry to 2020*. Bangkok: Food and Agriculture Organization of the United Nations. http://www.fao.org/docrep/012/i1594e/i1594e00.htm.

FAO. (2012). *Statistical yearbook*. Rome: Food and Agriculture Organization of the United Nations. http://www.fao.org/docrep/015/i2490e/i2490e00.htm.

Frontier Myanmar. (2017). https://frontiermyanmar.net/mm/node/4278.

Gallopín, G. C. (2012). *Five stylized scenarios. Global water futures, 2050*. France: UNESCO. http://unesdoc.unesco.org/images/0021/002153/215380e.pdf.

Government of India. (2015a). *Achieving the sustainable development goals in India. A study of financial requirements and gaps (2015)*. New Delhi: Ministry of Environment, Forest and Climate Change, Government of India.http://www.devalt.org/images/L3_ProjectPdfs/AchievingSDGsinIndia_DA_21Sept.pdf?mid=6&sid=28.

Government of India. (2015b). *Intended nationally determined contributions—India*. Delhi: Ministry of Environment, Forest, and Climate Change. http://www4.unfccc.int/ndcregistry/PublishedDocuments/India%20First/INDIA%20INDC%20TO%20UNFCCC.pdf.

Government of India. (2016a). *The Indus Waters Treaty*. New Delhi: Ministry of External Affairs. http://mea.gov.in/bilateral-documents.htm?dtl/6439/Indus.

Government of India. (2016b). *Energy statistics*. New Delhi: Ministry of Statistics and Programme Implementation. http://www.mospi.nic.in/sites/default/files/publication_reports/Energy_statistics_2016.pdf?download=1.

Government of India Report. (2017). https://www.hindustantimes.com/india-news/6-3-crore-indians-do-not-have-access-to-clean-drinking-water/story-dWIEyP962FnM8Mturbc52N.html.

Government of Nepal. (2015). *Sustainable development goals 2016–2030 national (preliminary) report*. Kathmandu: National Planning Commission. http://www.np.undp.org/content/dam/nepal/docs/reports/SDG%20final%20report-nepal.pdf.

Government of Nepal. (2016). *Intended nationally determined contributions—Nepal*. Kathmandu: Ministry of Population and Environment. http://www4.unfccc.int/Submissions/INDC/Published%20Documents/Nepal/1/Nepal_INDC_08Feb_2016.pdf.

Government of Pakistan. (2013). *Pakistan 2025 one nation one vision*. Islamabad: Ministry of Planning, Development and Reform. http://fics.seecs.edu.pk/Vision/Vision-2025/Pakistan-Vision-2025.pdf.

Government of Pakistan. (2015). Intended nationally determined contributions—Pakistan. Islamabad: Ministry of Climate Change. http://www4.unfccc.int/ndcregistry/PublishedDocuments/Pakistan%20First/Pak-INDC.pdf.

Government of People's Republic of Bangladesh. (2012). *Rio+20 Bangladesh National Sustainable Development*. Peoples' Republic of Bangladesh, Bangladesh Secretariat Dhaka 1000, Bangladesh: Ministry of Environment and Forests. https://policy.asiapacificenergy.org/sites/default/files/Rio%2B20_Bangladesh_reduced.pdf.

Government of the People's Republic of Bangladesh. (2015). *Intended nationally determined contributions—Bangladesh*. Dhaka: Ministry of Environment and Forests (MOEF) Government of the People's Republic of Bangladesh. http://www4.unfccc.int/ndcregistry/PublishedDocuments/Bangladesh%20First/INDC_2015_of_Bangladesh.pdf.

ICAD. (2018). Bangladesh: Country Profile on Nutrition and Child Stunting Trends. https://ec.europa.eu/europeaid/bangladesh-nutrition-country-fiche-and-child-stunting-trends_en.

International Monetary Fund. (2016). *World economic outlook: Subdued demand: Symptoms and remedies*. Washington DC: International Monetary Fund.

IPCC. (2013). Climate change 2013: The physical science basis. In T. F. Stocker, D. Qin, G.-K. Plattner, M. Tignor, S. K. Allen, J. Boschung, A. Nauels, Y. Xia, V. Bex, & P. M. Midgley (Eds.), *Contribution of Working Group I to the Fifth Assessment Report of the Intergovernmental Panel on Climate Change* (1535 pp.). Cambridge, United Kingdom and New York, NY, USA: Cambridge University Press. https://www.ipcc.ch/pdf/assessment-report/ar5/wg1/WG1AR5_Frontmatter_FINAL.pdf.

Islamic Republic of Afghanistan. (2012). *Millennium development goals report*. Kabul, Afghanistan: Ministry of Economy. http://www.af.undp.org/content/afghanistan/en/home/library/mdg/MDGs-report-2012.html.

Islamic Republic of Afghanistan. (2015). *Intended nationally determined contributions—Afghanistan*. Kabul: Islamic Republic of Afghanistan. http://www4.unfccc.int/ndcregistry/PublishedDocuments/Afghanistan%20First/INDC_AFG_20150927_FINAL.pdf.

Johansson, T. B., Patwardhan, A. P., Nakićenović, N., & Gomez-Echeverri, L. (Eds.). (2012). *Global energy assessment: Toward a sustainable future*. Cambridge, UK: Cambridge University Press.

Lutz, A. F., Immerzeel, W. W., Shrestha, A. B., & Bierkens, M. F. P. (2014). Consistent increase in High Asia's runoff due to increasing glacier melt and precipitation. *Nature Climate Change, 4* (7), 587–592.

Moss, R. H., Edmonds, J. A., Hibbard, K. A., Manning, M. R., Rose, S. K., Van Vuuren, D. P., et al. (2010). The next generation of scenarios for climate change research and assessment. *Nature, 463* (7282), 747–756.

Nakićenović, N., & Swart, R. (2000). *Special report on emissions scenarios: A special report of Working Group III of the Intergovernmental Panel on Climate Change*. Cambridge, UK: Cambridge University Press.

Nakicenovic, N., Lempert, R. J., & Janetos, A. C. (2014). A framework for the development of new socio-economic scenarios for climate change research: Introductory essay. *Climatic Change, 122*(3), 351–361.

O'Neill, B. C., Kriegler, E., Riahi, K., Ebi, K. L., Hallegatte, S., Carter, T. R., et al. (2014). A new scenario framework for climate change research: The concept of shared socioeconomic pathways. *Climatic Change, 122*(3), 387–400.

OECD. (2016). *Economic outlook for Southeast Asia, China and India 2016: Enhancing regional ties*. Paris: Organization for Economic Co-operation and Development. http://www.oecd.org/dev/asia-pacific/SAEO2016_Overview%20with%20cover%20light.pdf.

OECD/FAO. (2016). *OECD-FAO agricultural outlook 2016–2025*. Paris: OECD Publishing. http://dx.doi.org/10.1787/agr_outlook-2016-en.

OECD/IEA. (2016). *World energy outlook*. France: International Energy Agency. https://www.iea.org/publications/freepublications/publication/WorldEnergyOutlook2016ExecutiveSummaryEnglish.pdf.

Pachauri, S., Rao, N. D., Nagai, Y., & Riahi, K. (2012). *Access to modern energy: Assessment and outlook for developing and emerging regions*. Laxenburg, Austria: IIASA.

Parry, M., Palutikof, O. F., van der Linden, Paul, J., & Hanson, C. E. (2007). *Climate change 2007: Impacts, adaptation and vulnerability*. Cambridge, UK: Cambridge University Press.

Parvez, A. (2017). How a spring revival scheme in India's Sikkim is defeating droughts. Gangtok, India: Inter Press Service. http://www.ipsnews.net/2017/02/how-a-spring-revival-scheme-in-indias-sikkim-is-defeating-droughts/.

People's Republic of China. (2015). *Intended nationally determined contributions—China*. Beijing: National Development and Reform Commission. http://www4.unfccc.int/ndcregistry/PublishedDocuments/China%20First/China%27s%20First%20NDC%20Submission.pdf.

Roy, J. (2000). The rebound effect: Some empirical evidence from India. *Energy Policy, 28*(6), 433–438.

Royal Government of Bhutan. (2013). *SAARC development goals—Bhutan Country Report*. Bhutan: Gross National Happiness Commission. http://www.indiaenvironmentportal.org.in/files/file/SDG-Country-report-Bhutan-2013.pdf.

Royal Government of Bhutan. (2015). *Intended nationally determined contributions—Bhutan*. Thimphu: National Environment Commission. http://www4.unfccc.int/ndcregistry/PublishedDocuments/Bhutan%20First/Bhutan-INDC-20150930.pdf.

SAARC (2016). The South Asian Free Trade Area (SAFTA), 12th SAARC summit, Islamabad, Pakistan, http://saarc-sec.org/digital_library/detail_menu/agreement-on-south-asian-free-trade-area-safta.

Sorrell, S., Dimitropoulos, J., & Sommerville, M. (2009). Empirical estimates of the direct rebound effect: A review. *Energy Policy, 37* (4), 1356–1371.

The People's Republic of China. (2016). *China's national plan on implementation of the 2030 agenda for sustainable development*. The People's Republic of China, Beijing: Ministry of Foreign Affairs. http://www.fmprc.gov.cn/web/ziliao_674904/zt_674979/dnzt_674981/qtzt/2030kcxfzyc_686343/P020170414689023442403.pdf.

The Republic of the Union of Myanmar. (2015). *Intended nationally determined contributions—Myanmar*. Nay Pyi Taw: Ministry of Environment Conservation and Forestry. http://www4.unfccc.int/ndcregistry/PublishedDocuments/Myanmar%20First/Myanmar%27s%20INDC.pdf.

The World Bank. (2010). *Trading Economics data*. Gross er 111.8 https://tradingeconomics.com/bangladesh/school-enrollment-primary-percent-gross-wb-data.html.

The World Bank. (2010). https://data.worldbank.org/indicator/SH.STA.STNT.ZS?locations=CN.

The World Bank. (2012). http://povertydata.worldbank.org/poverty/country/CHN.

The World Bank. (2013). *Life expectancy at birth, total*, India. https://data.worldbank.org/indicator/SP.DYN.LE00.IN.

The World Bank Data. (2013). https://data.worldbank.org/indicator/EG.ELC.ACCS.ZS.

The World Bank. (2014). *Food exports (% of merchandise exports)*, Asia Pacific Region. https://data.worldbank.org/indicator/TX.VAL.FOOD.ZS.UN.

The World Bank. (2016). *Population, total*. https://data.worldbank.org/indicator/SP.POP.TOTL.

UNEP. (1997). *Global environmental outlook 1*. United Nations Environment Program Global State of the Environment Report 1997. Hertfordshire, United Kingdom: SMI Books.

UNEP. (1999). *Global environmental outlook 2000*. United Nations Environment Program, Nairobi, Kenya.

UNEP. (2002). *Global environmental outlook 2000*. United Nations Environment Program, Nairobi, Kenya.

UNEP. (2007). *Global environmental outlook 4*. United Nations Environment Program, Nairobi, Kenya.

UNEP. (2012). *Global environmental outlook 5*. Environment for the future we want. United Nations Environment Program, Nairobi, Kenya.

UNICEF Report. (2013). https://in.reuters.com/article/health-nutrition-unicef/about-48-percent-of-children-in-india-are-stunted-unicef-idINDEE95607N20130607.

World Data Atlas. (2013). https://knoema.com/atlas.

World Economic Forum. (2014). *Gender gap index of India*. The Global Gender Gap Report, Switzerland. http://reports.weforum.org/global-gender-gap-report-2014/economies/#economy=IND.

Sustaining Biodiversity and Ecosystem Services in the Hindu Kush Himalaya

Coordinating Lead Authors
Jianchu Xu, World Agroforestry Centre, East and Central Asia Regional Programme; Kunming Institute of Botany, Kunming Yunnan Province, P.R. China, e-mail: J.C.Xu@cgiar.org; jxu@mail.kib.ac.cn
Ruchi Badola, Wildlife Institute of India, Chandrabani, Dehra Dun, Uttarakhand, India, e-mail: ruchi@wii.gov.in
Nakul Chettri, International Centre for Integrated Mountain Development, Kathmandu, Nepal, e-mail: Nakul.Chettri@icimod.org (corresponding author)

Lead Authors
Ram P. Chaudhary, Research Center for Applied Sciences & Technology (RECAST), Tribhuvan University, Kirtipur, Kathmandu, Nepal, e-mail: ram@cdbtu.wlink.com.np
Robert Zomer, World Agroforestry Centre, East and Central Asia Regional Programme; Kunming Institute of Botany, Kunming Yunnan Province, P.R. China, e-mail: r.zomer@mac.com
Bharat Pokhrel, HELVETAS, Swiss Inter-cooperation Nepal, Programme Coordination Office, Lalitpur, Kathmandu, Nepal, e-mail: bharat.pokharel@helvetas.org.np
Syed Ainul Hussain, Wildlife Institute of India, Chandrabani, Dehra Dun, Uttarakhand, India, e-mail: hussain@wii.gov.in
Sunita Pradhan, Ashoka Trust for Research in Ecology and the Environment, Regional Office Eastern Himalaya, Gangtok, Sikkim, e-mail: sunita.pradhan@atree.org
Rebecca Pradhan, Royal Society for Protection of Nature, Thimphu, Bhutan, e-mail: rebecca.pradhan@gmail.com

Contributing Authors
Pratikshya Kandel, International Centre for Integrated Mountain Development, Kathmandu, Nepal, e-mail: pratikshya.kandel@icimod.org
Kabir Uddin, International Centre for Integrated Mountain Development, Kathmandu, Nepal, e-mail: kabir.uddin@icimod.org
Pariva Dobriyal, Wildlife Institute of India, Chandrabani, Dehra Dun, Uttarakhand, India, e-mail: pariva@wii.gov.in
Upma Manral, Wildlife Institute of India, Chandrabani, Dehra Dun, Uttarakhand, India, e-mail: upma@wii.gov.in
Amanat K. Gill, Wildlife Institute of India, Chandrabani, Dehra Dun, Uttarakhand, India, e-mail: amanat@wii.gov.in
Uma Partap, International Centre for Integrated Mountain Development, Kathmandu, Nepal, e-mail: uma.partap@icimod.org
Kamal Aryal, International Centre for Integrated Mountain Development, Kathmandu, Nepal, e-mail: kamal.aryal@icimod.org

Review Editors
Sarala Khaling, Ashoka Trust for Research in Ecology and the Environment, Regional Office Eastern Himalaya, Gangtok, Sikkim, e-mail: sarala.khaling@atree.org
Yan Zhaoli, Chengdu Institute of Biology, Chinese Academy of Sciences, Chengdu, China, e-mail: zhaoliy@cib.ac.cn
S.P. Singh, Central Himalayan Environment Association (CHEA) Nainital, Uttarakhand, India, e-mail: singh.js1@gmail.com

Corresponding Author
Nakul Chettri, International Centre for Integrated Mountain Development, Kathmandu, Nepal, e-mail: Nakul.Chettri@icimod.org

P. Wester et al. (eds.), *The Hindu Kush Himalaya Assessment*,
https://doi.org/10.1007/978-3-319-92288-1_5

Contents

Chapter Overview

Key Findings

1. **The mountain ecosystems of the Hindu Kush Himalaya (HKH) are diverse with one of the highest diversity of flora and fauna providing varied ecosystem services to one fourth of humanity.** With four out of 36 global biodiversity hotspots the HKH is a cradle for 35,000+ species of plants and 200+ species of animals. At least 353 new species—242 plants, 16 amphibians, 16 reptiles, 14 fish, two birds, and two mammals, and at least 61 invertebrates—have been discovered in the Eastern Himalaya between 1998 and 2008, equating to an average of 35 new species finds every year.
2. **The HKH has numerous seeds of good practices in conservation and restoration of degraded habitat along with community development which need upscaling and out scaling.** These participatory and community-based approaches have had large ecological, economic, and social positive impacts. Substantial degraded forest areas are regenerating, as decentralized practices reverse deforestation trends. Local communities have gained institutional space to decide for themselves

on issues related to forests, income, inclusion, and social justice.

3. **Global and regional drivers of change on biodiversity and ecosystem loss are prevalent and increasing in the HKH.** These drivers include land use and land cover change, pollution, climate change, invasive species, solid waste, habitat degradation, and overexploitation of resources, among others, impacting biodiversity, ecosystem services, and human wellbeing.

Policy Messages

1. **Regional efforts will enhance the resilience of HKH ecosystems to extreme events while conserving biodiversity and promoting sustainable development.** Climate change and other drivers are altering the structure and population of some HKH ecosystems and species, including their distribution range, with risks to biodiversity and resilience. Because many of these critical HKH ecosystems are transnational, regional cooperation is essential for translating conservation and development challenges into sustainable development opportunities. Attaining the Sustainable Development Goals (SDGs) will depend on such cooperation.
2. **The mountain ecosystems of the HKH need an integrated and transboundary ecosystem approach at the landscape scale for conservation and sustainable development.** It should be managed as a mosaic of integrated socio-ecological systems across political boundaries. Efforts are needed to build on existing traditional practices, promote regional cooperation, and increase national and global investments.
3. **Investments in mountain ecosystems should be made where they are most needed to conserve biodiversity, alleviate poverty, and enhance sustainable livelihoods.** A large population in the HKH region still lives in poverty and is highly dependent on ecosystem services for livelihoods, especially in remote areas and developing nations. Because of varying priorities and resource availability, HKH countries are at different levels of investment in managing the mountain ecosystems. Therefore, more investment should be set aside for enhancing resilience with win-win trade-offs in the remote areas and developing countries.

Mountains make up 24% of the world's land area, are home to 20% of the world's population, provide 60–80% of the world's fresh water, and harbour 50% of the world's biodiversity hotspots (***well-established***). The United Nations recognized the importance of mountain ecosystems, both for conserving biological diversity and for sustaining humanity, in Chap. 13 of Agenda 21. More generally, ecosystem diversity, species diversity, genetic diversity, and functional diversity all play key roles in the ecosystem services that benefit people and communities (***well-established***).

All these types of diversity are fundamental for the mountains of the HKH. With its unique high mountains and numerous micro climates, the HKH contains varied ecological gradients that set the stage for species evolution. The result is the youngest global mountain biome and one of the most ecologically diverse ecosystems in the world (***well-established***). Between 1998 and 2008, an average of 35 new species were discovered each year in the Eastern Himalaya alone (***well-established***).

The ecological diversity of the HKH has long been modified by extraction, trade, culture, and land use (***established but incomplete***). Of the region's population, 70–80% live in rural areas, while 60–85% subsist directly through ecosystem services (***well-established***). Now, however, the region is being subjected to pressures that are more aggressively unfriendly to ecosystems. Climate change is one of these pressures; unprecedented development is another (***established but incomplete***).

Global and regional drivers of biodiversity loss—such as land use change and habitat loss, pollution, climate change, and invasive alien species—are prevalent and increasing in the HKH (***established but incomplete***). For example, by the year 2100 the Indian Himalaya could see nearly a quarter of its endemic species wiped out (***inconclusive***). Countries in the region already place a premium on functional ecosystems and ecosystem services: more than 39% of all land in the HKH lies within a protected area network (***well-established***). Even so, ecosystems are in stress or subject to risks from a changing climate, from varying government policies, and from expanding markets—at all levels (***established but incomplete***).

Broadly, ecosystem services have four kinds of value:

- Social—for public benefit
- Cultural—for aesthetic and communal significance
- Ecological—for environmental conservation and sustainability
- Economic—for livelihoods through goods and services production.

We generally know less about social and cultural value in the HKH than about ecological and economic value (***established but incomplete***). All four kinds of value, however, have received little attention—either qualitative or quantitative—compared with such widely researched topics in the region as carbon, water, and hydropower (***well-established***). For example, recreation is a growing principal livelihood activity in the Himalaya. Some analyses acknowledge the positive economic gains, but also negative impacts on biodiversity and ecosystem services (***established but incomplete***). However, many of these studies focus on a small area and lack the holistic view needed to inform policy decisions (***well-established***).

Better management of HKH ecosystem services entails learning more about the state and trends of coupled socio-ecological systems. The diverse landscapes of the region provide multiple services with complex, dynamic interrelations. Some studies based on integrated systems analysis (most emerging from hydrology and geology) have traced upstream-downstream links at both the catchment and the basin scale. Common drivers, affecting multiple ecosystem processes and interactions among ecosystem services, can create both synergies and trade-offs between ecosystem health and the flow of services (***established but incomplete***). Trade-off analysis is thus critical for integrating ecosystem services into landscape planning, management, and decision making—especially in looking at alternate paths to sustainable land use (***well-established***).

Recent decades have seen considerable development in concepts of biodiversity conservation—from perspectives that focused on species while excluding people, to new approaches centred on people and communities (***well-established***). As a result, biodiversity conservation in the HKH has changed along with natural resource management. Participatory models have emerged and been accepted in various sectors, with the region generally adopting the 'ecosystem approach' advocated by the United Nations Convention on Biological Diversity (1992). Traditional ecological knowledge, beliefs, and culture have contributed substantially towards meeting conservation goals (***established but incomplete***).

These participatory and community-based approaches have had large ecological, economic, and social positive impacts. Substantial degraded forest areas are regenerating, as decentralized practices reverse deforestation trends. Local communities have gained institutional space to decide for themselves on issues related to forests, income, inclusion, and social justice. As people make their own decisions rather than reacting to orders from government officials, rural residents have been able to avail themselves of more local economic opportunities. Progressive policies have driven this paradigm shift (***established but incomplete***).

And yet the challenges facing the region could have cascading effects, especially for communities highly dependent on ecosystem services (***established but incomplete***). The transformative changes to date were driven mainly by a changing climate and land use change. As a result, changes to production systems are required to address potential resource crises arising from a growing population and increasing demand. Special attention must now be paid to governance effectiveness and implementation of evolving policies (***established but incomplete***).

Despite successes in community-based conservation and development, conserving the global assets of the HKH remains a challenge (***established but incomplete***). The HKH ecosystems provide crucial ecosystem services to 1.9 billion people, more than any other mountain system (***well-established***). As they continue to provide these services both within and outside the region, how can their biodiversity be sustained and the continued flow of services assured? The solution will be to manage the HKH as a mosaic of integrated socio-ecological systems across political and sectoral boundaries, linking upstream and downstream conservation action with local climate adaptation strategies (***established but incomplete***).

We still need to improve our understanding of biodiversity and ecosystem functions and services in the HKH (***well-established***). Only with improved technical knowledge, policies, and practices can environmental security be assured. It should also be strengthened through integration of traditional practices with science-based conservation, regional cooperation, and national and global investments (***established but incomplete***).

Biodiversity, Ecosystems and the Sustainable Development Goals

Building the social and ecological resilience of HKH mountain ecosystems will be essential for attaining Sustainable Development Goal (SDG) 15: *Protect, restore and promote sustainable use of terrestrial ecosystems, sustainably manage forests, combat desertification, and halt and reverse land degradation and halt biodiversity loss.* Most specifically relevant is Target 15.4: "By 2030, ensure the conservation of mountain ecosystems, including their biodiversity, in order to enhance their capacity to provide benefits that are essential for sustainable development."

In addition, sustaining the flow of HKH ecosystem services can help attain SDGs 1 (alleviating poverty), SDG 2 (zero hunger), SDG 5 (addressing gender and social equity), SDG 6 (water, sanitation, and water for productive purposes), and SDG 7 (access to clean energy). Here, inclusive and transformative change is needed—recognizing the role of mountain communities in providing ecosystem services, adding new incentives, generating opportunities. Policies should develop and support markets for mountain niche products, and should enable investment in the mountains.

5.1 Mountain Biodiversity and Ecosystem Services: A Major Global Asset Under Threat

Mountains make up 24% of the world's land area, are home to 20% of the world's population, provide 60–80% of the world's freshwater, and harbour 50% of globally recognized biodiversity hotspots (Mittermeier et al. 2004; Rodríguez-Rodríguez et al. 2011; Maselli 2012). Mountain ranges act as barriers to some organisms and bridges to others, and therefore facilitate species isolation, speciation, extinction, and migration (Körner and Ohsawa 2005). The mountain ecosystems provide key livelihood resources such as food, timber, fibre, and medicine and a wide range of services such as fresh air and water, climate regulation, carbon storage, and the maintenance of aesthetic, cultural, and spiritual values (Grêt-Regamey et al. 2008; Schild 2008; Bhat et al. 2013; Sandhu and Sandhu 2014; Ahmad and Nizami, 2015; Hamilton 2015). The natural and semi-natural vegetation cover on mountains helps to stabilize headwaters, prevent flooding, and maintain steady year-round flows of water by facilitating the seepage of rainwater into aquifers, vital for maintaining human life in the densely populated areas downstream. As a result, mountains have often been referred to as 'water towers' (Schild 2008; Mukherji et al. 2015; Molden et al. 2016). Recognizing the importance of mountains for biodiversity and sustaining ecosystem services, Chap. 13 of Agenda 21 (1992) has recognized mountains as a significant habitat for support of all forms of living organisms, animals (including humans), and plants (UN 1992).

Driven by plate tectonics, the mountains of the HKH have unique ecosystems with altitudinal variation giving rise to numerous micro climates and diverse ecological gradients. The HKH is the youngest and one of the most diverse ecosystems among the global mountain biomes, with extreme variations in vegetation, climate, and ecosystems resulting from altitudinal, latitudinal, and soil gradients (Xu et al. 2009a; Sharma et al. 2010). This diverse biophysical habitat sets the stage for a rich biodiversity and species evolution (Miehe et al. 2014; Hudson et al. 2016). The region is the source of 10 major river systems with productive landscapes and strong upstream downstream linkages (Xu et al. 2009a), and includes all or part of four global biodiversity hotspots—Himalaya, Indo-Burma, mountains of Southwest China, and mountains of Central Asia (Mittermeier et al. 2004; Chettri et al. 2010)—which contain a rich variety of gene pools and species with high endemism and novel ecosystem types (Fig. 5.1.) In addition, the region supports more than 60 different ecoregions, many of them Global 200 ecoregions (Olson and Dinerstein 2002). The ecosystem services from the HKH sustain 240 million people in the region and benefit some 1.7 billion people in the downstream river basin areas (see Box 1.1) and have been well recognized by many scholars (Quyang 2009; Xu et al. 2009a; Molden et al. 2014a; Sharma et al. 2015).

The natural and semi-natural landscapes of the HKH have been altered, modified, and influenced by human history, culture, and traditional practices for thousands of years (Deterra 1937; Ives and Messerli 1989; Goldewijk et al. 2010, 2011; Ellis 2015). The HKH has witnessed human intervention since circa 5000 years BP, bringing crop diversity, cattle farming, and cultural congruence from east and west and leading to the creation of dynamic landscapes (Gurung 2004a; Miehe et al. 2009, 2014; Chen et al. 2015a). These dynamic landscapes have brought higher biological diversity through use, diversification, and promotion of plants, animals, agrobiodiversity, and traditional knowledge (Xu et al. 2005; Uprety et al. 2016). The diverse social networks across the region and rapid development of trade facilitated the exchange of cultures, knowledge, and materials (Chaudhary et al. 2015a, b). They evolved into coupled socio-ecological systems which have been significant not only for the people living in the mountain areas, but also for those beyond who benefit from the ecosystem services (Blaikie and Muldavin 2004; Nepal et al. 2014a, b).

The HKH is now being subjected to further change, including climate change (Shrestha et al. 2012) and unprecedented development that is environmentally unfriendly (Pandit et al. 2007; Grumbine and Pandit 2013; Xu and Grumbine 2014a). There are examples of both negative and positive impacts of the various drivers resulting in change in wildlife population, plant phenology, and ecosystem productivity across the region (Bawa and Seidler 2015; Singh and Borthakur 2015; Chaudhary et al. 2016a, b). Moreover, an increase in the number and severity of natural disasters as well as a breakdown of traditional systems of management is an indicator of the decreasing resilience of the HKH system (Elalem and Pal 2015). Global drivers of biodiversity loss, namely land use land cover change, habitat change, overexploitation, pollution, invasive alien species, and climate change, are prevalent in the HKH (Maxwell et al. 2016) and are increasing (Chettri and Sharma 2016). Solid waste management and haphazard development are bringing additional challenges (Kala 2014; Posch et al. 2015). Although there are a number of examples of best practices in community-based

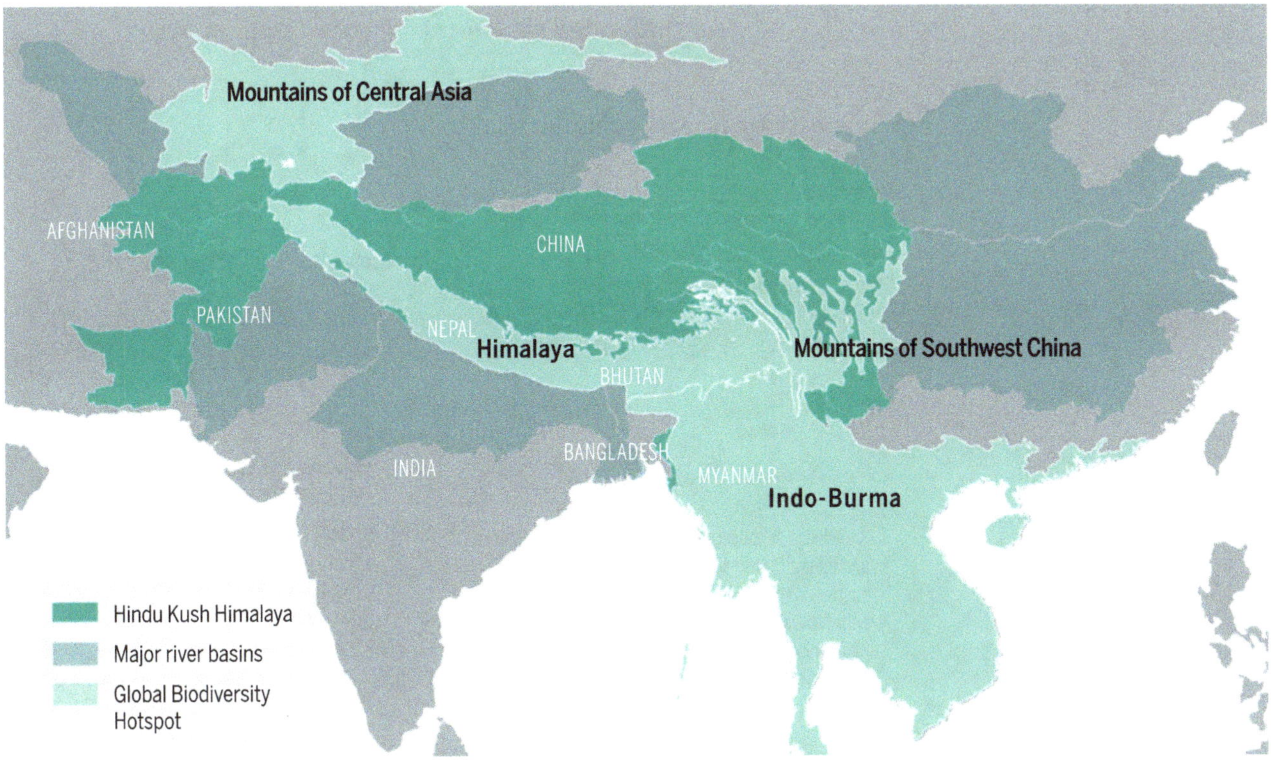

Fig. 5.1 Four global biodiversity hotspots and 10 river basins of the HKH

conservation and development initiatives, the challenges still exist in terms of good governance (Pai and Dutta 2006; Sharma et al. 2010). Sectoral and piecemeal approaches are among the limiting factors as they do not create many incentives for local communities to conserve biodiversity and water resources (Rasul 2014). Despite this, the Himalaya continue to provide ecosystem services that sustain societies both within the region and beyond. The diverse cultures and traditions manifested by over 1,000 ethnic groups (Turin 2005) continue to nurture the ecosystems at various spatial scales. In addtion, innovative practices have also promoted the production, restoration, and conservation of ecosystems and the services that they provide (Banskota et al. 2007; Sharma et al. 2007a, b, c; Aase et al. 2013).

In order to ensure the sustainability of ecosystems and continuity of ES, it is critical that biodiversity be managed as a mosaic of integrated socio-ecological systems. This should encompass systems across political and sectoral boundaries, and link upstream and downstream conservation action with local climate adaptation (Xu and Grumbine 2014a). To gain a better knowledge of the ecosystem services that people depend on for benefits and values, it is necessary to understand the state and dynamics of biodiversity and ecosystem functions. This understanding could be strengthened through deeper understanding of traditional practices and integrated with science-based conservation, regional cooperation, and national and global investments on policy and practices for environmental security (Fig. 5.2). In this chapter we move forward with this integrated approach to biodiversity conservation and ecosystem services and try to visualize their provision spatially and temporally. The content is basically drawn from learning and good practices from the past and lessons from the present, and we try to sow seeds for the sustainability of these services in the future with recommendations for future policy and practice.

Due to the extremely wide scope of the chapter and large volume of literature available, we have attempted to focus only on the key thematic areas of biodiversity and ecosystem services, with some examples to illustrate the trends observed across the HKH. We have structured our chapter to (1) contextualize the state of biodiversity and ecosystem status; (2) highlight the status and trends in biodiversity and ecosystem services; (3) document the current state of the coupled socio-ecological system; (4) highlight conservation and management practices; and (5) identify gaps and suggest strategic directions for mountain sustainability.

5.2 The Rich Biodiversity of the HKH Region

The Convention on Biological Diversity gives a formal definition of biodiversity in Article 2: "biological diversity means the variability among living organisms from all sources including, inter alia, terrestrial, marine, and other

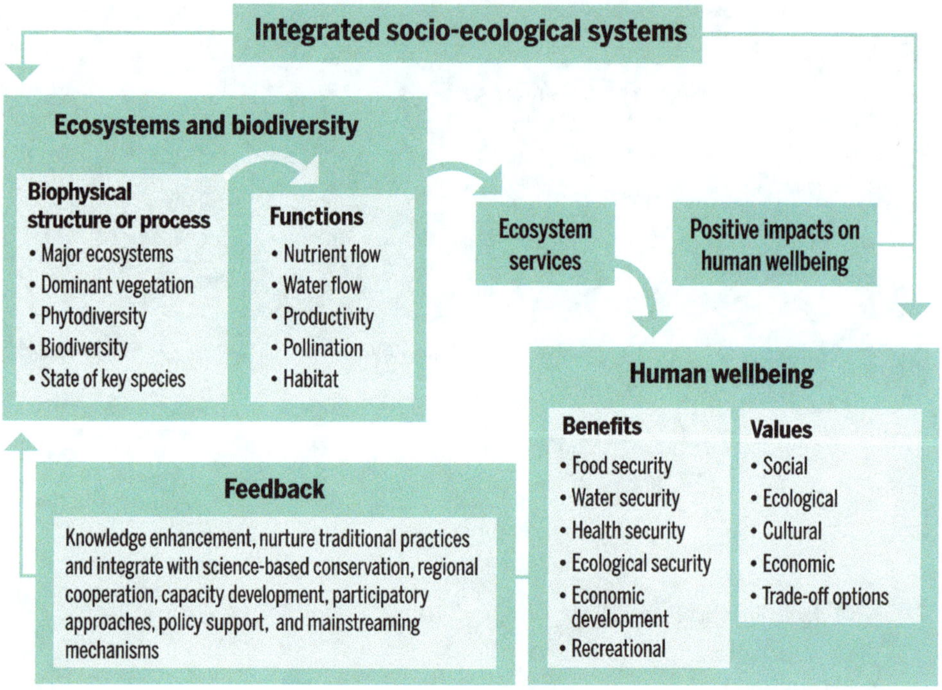

Fig. 5.2 Linkages between ecosystems, biodiversity, and human wellbeing (adapted from de Groot et al. 2010)

aquatic ecosystems and the ecological complexes of which they are part; this includes diversity within species, between species, and of ecosystems". The HKH exhibits high levels of diversity and heterogeneity, partly in response to the high climatic variability and rugged topography. Very high levels of biodiversity and species richness (Myers et al. 2000; Sharma et al. 2010; Zomer and Oli 2011) are the result of a combination of several biophysical and geographical factors (Ives and Messerli 1989; Miehe et al. 2015a), including altitudinal zonation along a long and steep elevation gradient that ranges from near sea level at the base of the foothills up to the highest mountains in the world, with a prominent rain shadow zone on the Tibetan Plateau; a moisture and precipitation gradient that goes from generally wetter in the east to semi-arid and drier zones in the west; and the blocking effect of the high mountain barrier which allows tropical conditions to flourish in deep valleys by blocking cold continental northern winds from protruding further south, even though physically these mountains are outside the tropics. Rainfall in the HKH is primarily fed by the Indian summer monsoon, which weakens as it moves from the eastern reaches of the Himalaya northwest towards the Hindu Kush, and by winter storms which bring moisture in from the Arabian Sea (Barlow et al. 2005). The interaction of these two main precipitation regimes with the altitudinal gradient results in pronounced spatial, but also temporal, gradients in temperature and precipitation throughout the HKH, and consequently high levels of ecosystem diversity.

Additionally, the steep slopes of the HKH have high levels of erosion (Ives and Messerli 1989), which provide nutrient-rich sediments that help to sustain ecosystem flow to the 10 Asian river basins emanating from these highland water towers and impact the agricultural productivity of the floodplains. Equally, the high levels of erosion also result in habitat degradation and biodiversity loss (Xu et al. 2009a).

The mountain building, driven by plate tectonics, have created a diverse landscape, climate variability, ecological gradients, and physical habitats that set the stage for ecosystem differentiation (Hua 2012) and species evolution (Hoorn et al. 2013; Tremblay et al. 2015). The initial uplift of the Himalayan and Hindu Kush ranges from the mid to late Eocene, and the more recent (early Pliocene) uplift of the Hengduan Mountains, have enabled the development of a large number of recognized 'biodiversity hotspots' (Hughes and Atchison 2015; Mittermeier et al. 2015; Rodríguez-Rodríguez et al. 2011). The high levels of species richness found here are derived from both endemic speciation from local ancestors and migration of organisms from distant locales, noting that this region represents a congruence of two different floristic realms (Palearctic and Indomalayan) (Olson and Dinerstein 2002). Likewise, the region is also one of the most productive and intensively cultivated mountain regions in the world, with a high population density and ethnic diversity, which likewise has led to high levels of agrobiodiversity, farming system and agro-ecosystem differentiation, and domestication of many important food plants and animals (Gorenflo et al. 2012; Pandit et al. 2014; Karan 2015).

The bioclimatic zones range from hot tropical moist to lush green and humid valleys in the central and eastern midhills along the ranges, extensive mountain forests,

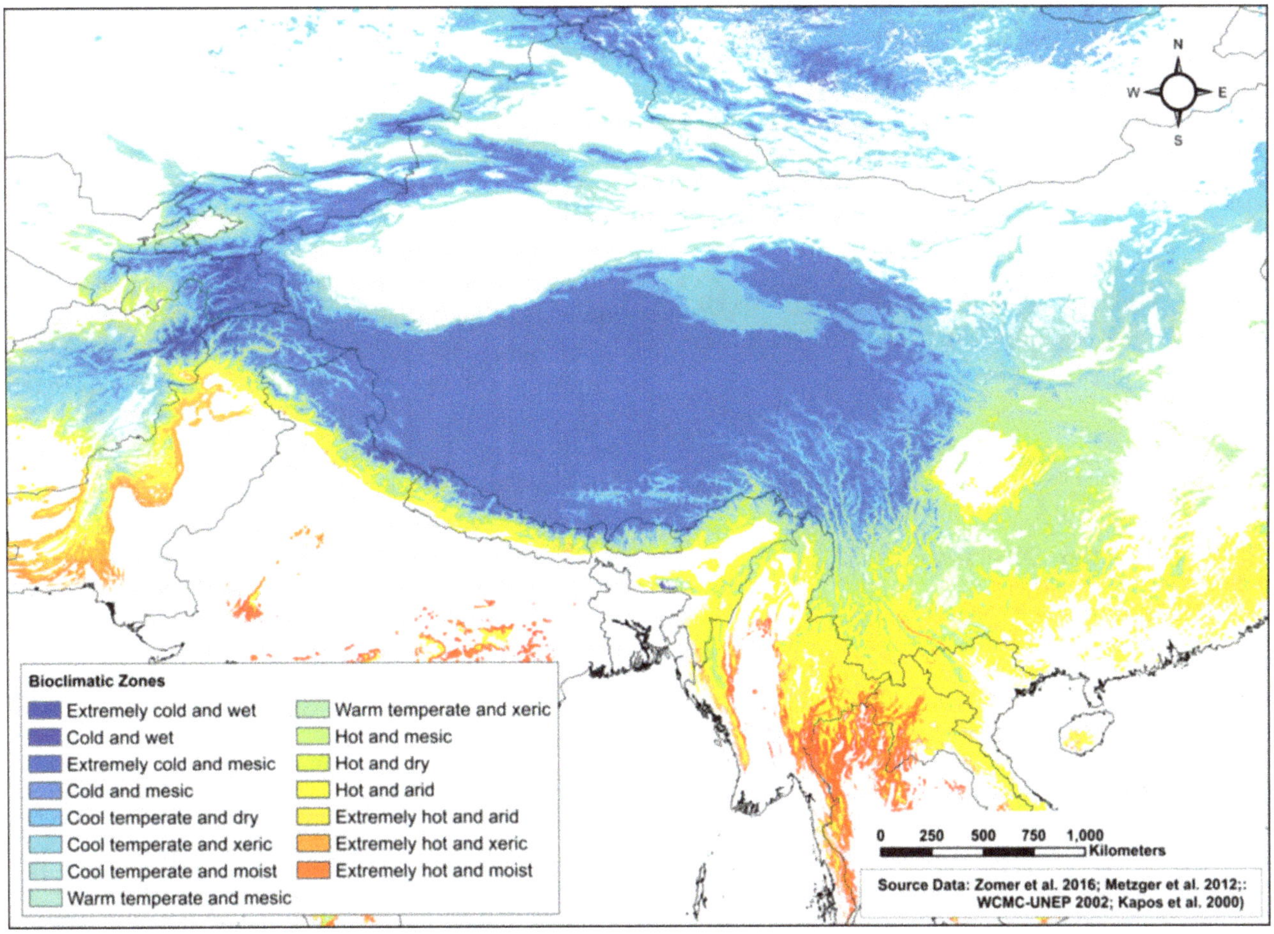

Fig. 5.3 Major bioclimatic zones, based on the Global Environmental Stratification (Metzger et al. 2013), found within the mountainous and highland regions of Asia. Mountains and highland areas are identified based on Kapos et al. (2000) using data from WCMC-UNEP (2002)

moist alpine meadows, remote and arid trans-Himalayan valleys, the cold Tibetan Plateau and vast areas of high altitude grasslands and steppe, and arid and semi-arid regions in the west, as well as extensive areas of permanent snow and ice. Zomer et al. (2016) found that 105 of 125 bioclimatic strata worldwide (Metzger et al. 2013), each representing a broad set of unique but homogenous bioclimatic conditions, were found within the HKH and its associated downstream river basins (Fig. 5.3). This bioclimatic and geographic heterogeneity has given rise to an array of ecosystems, biomes, and forest types (ranging from moist tropical broadleaf to temperate oak forests, alpine conifers, alpine pastures, and high altitude grasslands) providing habitat for a diversity of wildlife (including tiger, Asian elephant, musk deer, blue sheep, snow leopard, Tibetan antelope, and many other rare and endangered species). Many of the highest levels of diversity worldwide, in terms of plants and animals, are found in these mountain ranges, in three major global biodiversity hotspots—Himalaya, Indo-Burma, and the mountains of Southwest China (Allen et al. 2010). This area represents an important habitat for a high proportion of endemic and threatened mammals (Hoffmann et al. 2010); fish, molluscs, and dragonflies (Allen et al. 2010); birds (Dunn et al. 2016); and agrobiodiversity (Chettri et al. 2010), for which the HKH is renowned.

The Indo-Burma hotspot is one of the most significant hotspots, with rich diversity and a high proportion of endemism (Table 5.1). However, high levels of endemism are found throughout these mountains (Myers et al. 2000) amongst a vast array of plants, mammals, birds, reptiles, and other taxa, many of which are threatened or endangered. Recently, the region has been mentioned prominently within various listings of crisis ecoregions, endemic bird areas, megadiverse countries, and Global 200 ecoregions (see Brooks et al. 2006). However, the predictive models indicate that about 70–80% of the original habitat has already been lost and that loss may increase to 80–87% by 2100 (see Jantz et al. 2015 and Table 5.2).

Table 5.1 Distribution of total and endemic (in parentheses) species in the four biodiversity hotspots in the HKH

Biodiversity	Himalaya	Indo-Burma	Mountains of Southwest China	Mountains of Central Asia
Plants	10,000 (3,136)	13,500 (7,000)	12,000 (3,500)	5,500 (1,500)
Mammals	300 (12)	433 (73)	237 (5)	143 (6)
Birds	977 (15)	1,266 (64)	611 (2)	489 (0)
Reptiles	176 (48)	522 (204)	92 (15)	59 (1)
Amphibians	105 (42)	286 (154)	90 (8)	7 (4)
Freshwater fish	269 (33)	1,262 (553)	92 (23)	27 (5)

Source Conservation International (2016)

5.2.1 Ecosystem Diversity

The variation in species richness and diversity are mainly driven by ecosystem diversity (Tews et al. 2004). When there is a mosaic of habitats comprised of different ecosystems such as forest, grassland, water bodies, agriculture and so on, species diversity increases due to interspecific facilitation (Cardinale et al. 2002). This is one of the reasons biodiversity is not equally distributed across the HKH; the western part of the region is comparatively homogenous with arid and semi-arid vegetation (Fig. 5.3). A recent analysis identified the dominant terrestrial ecosystems to be high altitude grassland (54%), followed by forest (20%), shrubland (15%), and agricultural land (5%), with the remaining 6% composed of barren land, rocky outcrops, built-up areas, snow cover, and water bodies (Xu et al. 2009a).

In socio-ecological systems, people are directly or indirectly dependent on their surrounding ecosystems; forest, rangeland (alpine), agriculture, and wetland ecosystems play an important role in the HKH in this context. The most widespread ecosystem in the HKH is rangeland; it is mostly distributed in the western Himalaya and Tibetan Plateau, and provides habitat for many globally significant plants and animals including one of the highest densities of domesticated animals such as yak (*Bos grunniens*) (Schaller 1998; Foggin 2012). The forest ecosystem, with about 20% of the land cover, is one of the most important ecosystems both for local communities and for wildlife living in tropical and temperate conditions. Forest has provided fodder, fuel, medicine, fibre, and many other services for people, and a habitat and corridor for wildlife, for millennia (Uprety et al. 2016; Wang et al. 2016). Although the agriculture ecosystem only covers 5% of the total area, it is key to the direct provision of food and food security and nutrient supply (Rasul and Sharma 2015). The majority of people are subsistence farmers with comparatively small landholdings compared to their counterparts in lowland areas (Hussain et al. 2016). The region has one of the largest number of high altitude wetlands in the world, around 36 of which are designated as Ramsar sites (Upadhaya et al., 2009). These wetland ecosystems are a repository for a wide range of flora and fauna including threatened and endemic species (Jain et al. 2000; Savillo 2009; Sharma et al. 2016), and many of them are vital to culture and tourism (Maharana et al. 2000; Anand et al. 2012).

The ecosystem diversity is further supported by elevation, micro climate, and aspect variations leading to gradients of forest and other ecosystems along the altitudinal variation (see Fig. 5.4). The gradient from tropical (<500 m) to alpine ice-snow (>6000 m), with a principal vertical vegetation regime composed of tropical and subtropical rainforest, temperate broadleaf deciduous or mixed forest, and temperate coniferous forest, including high altitude cold shrub or steppe and cold desert, brings more ecosystem diversity. The variation in ecosystem functions and processes provides different ecosystem services to people and different opportunities to support livelihoods (Miehe et al. 2015b). Some of the ecosystems such as wetlands provide more than 85% of gross household income locally (Sharma et al. 2015).

Table 5.2 Current and future (seven climate-change scenarios) estimates of loss of area in individual biodiversity hotspots relative to the year 1500

	Current estimates of loss (%)		Year 2100 estimates of loss[a] (%)						
Hotspot	Mittermeier et al. (2004)	Modelled 2005	RCP2.6	RCP2.6	RCP4.5	RCP6.0	RCP6.0	RCP8.5	RCP8.5
Himalaya	75	77	93	91	91	95	93	86	92
Indo-Burma	95	38	71	64	63	68	64	83	61
Mountains of Central Asia	80	69	76	77	77	72	77	77	76
Mountains of Southwest China	92	66	97	97	97	83	97	78	96
Total	86	70	84	78	77	82	80	87	80

[a]See Jantz et al. (2015) for descriptions of the land-use change scenarios
RCP = representative concentration pathway

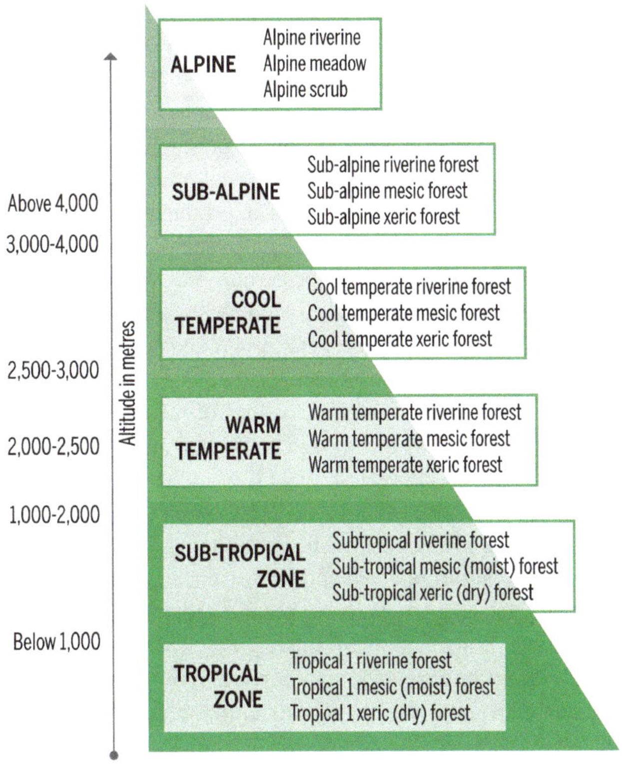

Fig. 5.4 Vegetation zones and dominant forest types found across the HKH (Chettri et al. 2010)

5.2.2 Species Diversity

The variation in ecosystems, land use, and landcover across the HKH is reflected in the species distribution, with high diversity and richness in megadiverse countries like China, India, and Myanmar and comparatively less diversity and richness in arid and semi-arid regions such as Afghanistan (Table 5.3). At least 353 new species were discovered in the Eastern Himalaya between 1998 and 2008, an average of 35 new species finds per year. The discoveries included 242 plants, 16 amphibians, 16 reptiles, 14 fish, 2 birds, 2 mammals, and at least 61 invertebrates (Thompson 2009). In addition, the HKH region—mostly the Eastern Himalaya—is also known for some iconic species such as *Rhododendron*. With seven species of *Rhododendron* have been reported in the western Himalaya, with increasing diversity shown towards the Eastern Himalaya (Nepal, Sikkim and Darjeeling in India, Bhutan, and North East India), and the highest number of species in China (Milleville 2002; Pradhan et al. 2003; Singh 2009; Shu 2005). As many as 46 *Rhododendron* species have been classified as rare or threatened in the India area of the Eastern Himalaya alone (Menon et al. 2012). In addition, orchids, medicinal and aromatic plants, and wild edible plants also play an important role in livelihoods and local economies and are abundant in the region (Kalita et al. 2014; MoEF 2014). Interestingly, the region also contains, wholly or partially, three of Vavilov's eight centres of origin of cultivated plants (Simpson and Ogorzaly 1986).

5.2.3 Genetic Diversity

Genetic diversity, defined as the variety of alleles and genotypes present in a population, is a fundamental source of biodiversity (Hughes et al. 2008). It is critical for the survival and adaptability of a species, helping organisms to cope with current environmental variability, reducing the potentially deleterious effects of close relative breeding, and increasing disease resistance (Frankham 2005). Maintaining diversity at a genetic level also holds significance for species

Table 5.3 Reported species richness in the countries of the HKH

Country	Area (km^2)[9]	Floral diversity		Faunal diversity				
		Angiosperms	Gymnosperms	Mammals	Birds	Reptiles	Amphibians	Fish
Afghanistan[1]	652,230	3,500–4,500	NA	137–150	428–515	92–112	6–8	101–139
Bangladesh[2]	144,000	3,723	7	128	650	154	49	712
Bhutan[3]	38,394	5,603	NA	200	700	124	61	91
China[4]	9,596,960	34,984	NA	556	1,300	1,186	380	279
India[5]	2,387,590	17,926	74	423	1,233	526	342	3,022
Myanmar[6]	676,577	11,800	NA	251	1,000	279	82	350
Nepal[7]	147,181	6,973	26	208	867	123	117	230
Pakistan[8]	882,000	5,757	38	198	696	177	22	>1,000 marine; 198 freshwater

Sources [1]ANBSAP (2013); [2]DoE (2015); [3]MoAF (2014); [4]MoEPC (2014); [5]MoEF (2004, 2008); [6]MoECF (2011); [7]MoFSC (2014); [8]CCD (2014); [9]Chettri and Sharma (2016) (except Nepal)
NA = data not available
Except for Bhutan and Nepal, the numerical data are for the whole country and not segregated for the HKH region

evolution, and through agricultural biodiversity contributes to sustaining and strengthening food, nutrition, health, and livelihood security (Notter 1999; Esquinas-Alcázar 2005). The origin of the chicken and its domestication (*Gallus gallus domesticus*) is thought to be in the region, particularly in India and China (Liu et al. 2006). The huge variety of traditional crops and cultivars used in the subsistence farming system, including swidden agriculture, is very little known outside the region. These species harbour an enormous genetic diversity (landrace/varieties) of both regional and global significance. For example, 2,500 landraces of rice (*Oryza sativa*) have been identified in Nepal (Gupta et al. 1996). The number could increase significantly if reviewed; the western Himalaya alone adds 100 types of basmati rice (Salgotra et al. 2015). Taro (*Colocasia esculenta*) is another widely distributed food crop, which is believed to be from the Eastern Himalaya (Xu et al. 2001). Studies have shown that the indigenous crop varieties traditionally cultivated and maintained by farmers contain high levels of genetic diversity and can serve as potential genetic resources for improving yield, resistance to pests and pathogens, and agronomic performance, thereby helping to maintain future food security in the light of the changing climate (Brush 1995; Hoisington et al. 1999; Mandel et al. 2011).

Despite being a repository of genetic resources of the global significance, the region has received comparatively little attention in terms of genetic research and in situ conservation. For example, in a recent review of biodiversity research, Kandel et al. (2016) noted that only around 2% of the research on the Kangchenjunga Landscape is at genetic level, compared to 20% at the ecosystem level and 78% at the species level. The identification and recording of species are still at an early stage, the necessary baseline data for identifying genetic diversity are not available, and the constant monitoring needed to examine population dynamics as a function of changing climate impacts is sorely lacking. The scant amount of genetic-level research could be due to very limited financial resources, lack of institutional capacity, an inadequate knowledge base, lack of accessible sophisticated technologies, and restrictive government policies towards such research in the region (Grajal 1999; Bubela and Gold 2012).

5.2.4 Functional Diversity

Functional diversity is a component of biodiversity that generally concerns the range of things that organisms do in communities and ecosystems. A variety of definitions exist, such as "the functional multiplicity within a community" (Tesfaye et al. 2003) and "the value and range of those species and organismal traits that influence ecosystem functioning" (Tilman et al. 2001). But the term is frequently used without definition or reference, normally considering phenotypic trait and/or trophic levels (Tilman et al. 2001). The framework of this paper (see Fig. 5.2) indicates that functional diversity among the individual species within an ecosystem is vitally important for the production and flow of ecosystem services. The physiological interaction among biotic and abiotic elements with a production function is instrumental in the derivation of ecosystem services (see de Groot et al. 2010). Among these, nutrient production from decomposition, water from evapotranspiration, and food from pollination services are important ecosystem functions resulting from functional diversity (see Fig. 5.2). In recent years monitoring the functional values of the ecosystems for monitoring has been recognized for better understanding of the ecosystems (Gagic et al. 2015).

Containing some of the youngest mountain ecosystems on Earth, the HKH ecosystem continues to be shaped and reshaped by anthropogenic and geological processes, leading to diversity. This diversity is manifested by thousands of phenotypic traits and their interaction at different trophic levels supporting species evolution and richness. A very high diversity and wide range of climatic zones are found across this highly heterogeneous region, associated with steep elevational gradients and continental, oceanic, and latitudinal influences. Out of 125 bioclimatic strata identified in the Global Environmental Stratification (GEnS) worldwide, more than 105 are found within the HKH and its associated downstream basins (Metzger et al. 2013). This wide range of bioclimatic diversity, combined with the heterogeneous terrain and topographic and orographic effects, and along with the confluence of several major floristic zones across the region, has enabled a rich and highly diverse biodiversity to develop, with a high degree of endemism (Fig. 5.4). This diversity is likewise reflected in the many cultures and languages found across and along these mountains.

Functional diversity plays a pivotal role in the provision of ecosystem services, which can also be considered as the benefit from nature for people's wellbeing (Díaz et al. 2015). Functional diversity at ecosystem, species, and genetic levels is fundamental for the lives of the majority of the rural communities living in the HKH. In many parts, 70–80% of the population live in rural areas, and the majority (60–85%) are still directly or indirectly dependent on this diversity for their livelihoods (Sharma et al. 2015).

Human colonizers of the Himalaya over past millennia devised a wide range of foraging systems, from nomadism to shifting cultivation, from sedentary agriculture to fishery. The earliest human settlement identified on the Tibetan Plateau was established some 5,200 years before the present (Chen et al. 2015a) (see Fig. 5.5). Ecological and societal feedback shape the flow of services and may promote, reduce, or unravel such bundles during the constant

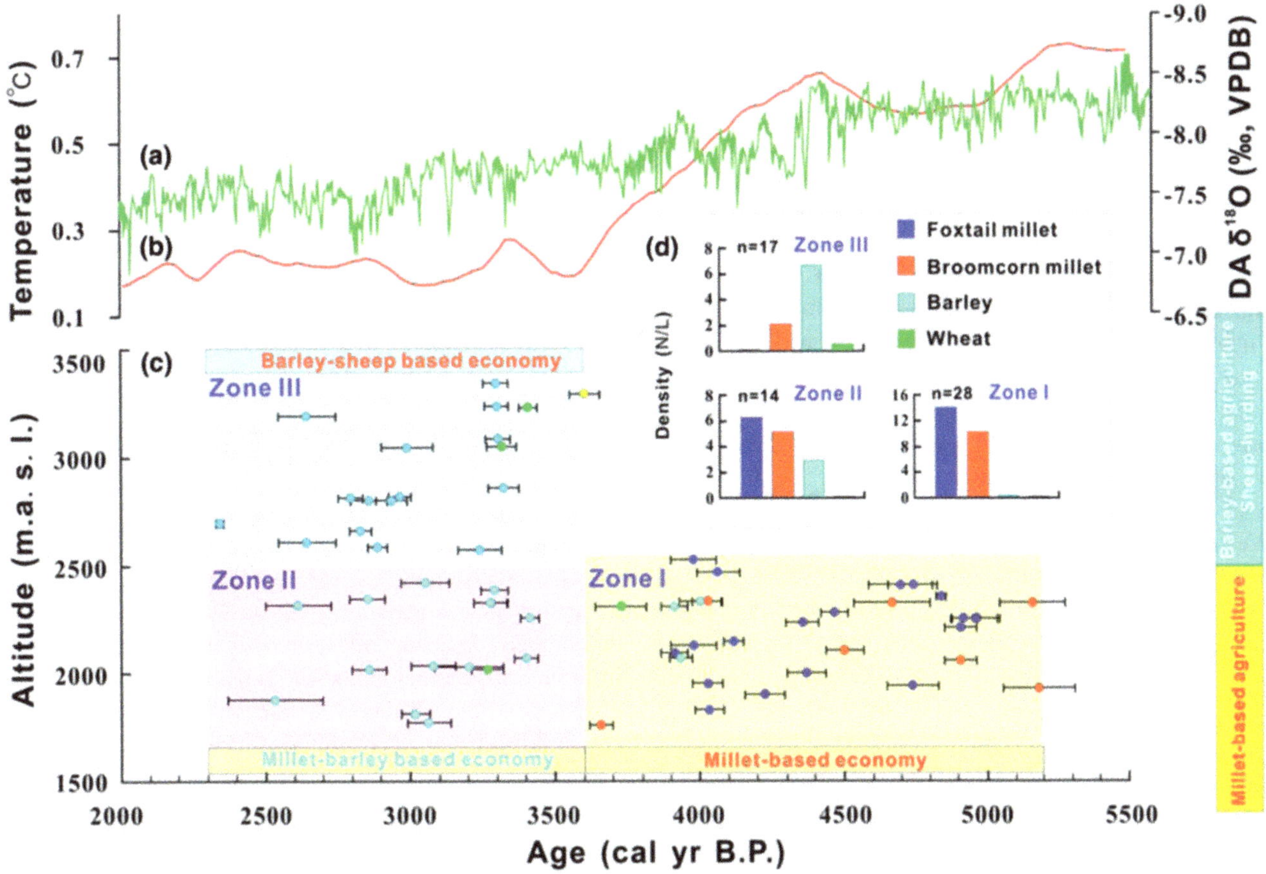

Fig. 5.5 Climatic records, radiocarbon dates, and charred cereal grain records from 53 investigated sites from different archaeological cultures on the northeastern Tibetan Plateau. **a** Asian summer monsoon changes indicated by Dongge Cave speleothem oxygen isotopes. **b** Northern Hemisphere (30° to 90° N) temperature record compared to 1961–90 instrumental mean temperature. **c** Calibrated AMS radiocarbon dates of charred grains (solid symbols with $2s$ error bar) from the 53 sites at different elevation; Zone I includes 25 sites below 2500 masl dated between 5200 and 3600 calendar years BP; zones II and III include 12 sites below and 17 sites above 2500 masl dated between 3600 and 2300 calendar years BP. Circle colours indicate crops as in (**d**), with the addition of capers indicated in yellow. **d** Density variation of crop remains from flotation samples from zones I, II, and III; N = number of charred grains, n = number of flotation samples (Reproduced from Chen et al. 2015a)

negotiation of different trade-offs. In the HKH, cultural diversity is a key contributor to shaping the ecosystems and biodiversity (Turin 2005; Gorenflo et al. 2012). Both ecosystems and cultures have adapted to exist in these relatively remote habitats; for example, the villagers in the Pamir integrated the human body into the seasons and rhythms of their ecological cycle to generate 'calendars of the human body' (Kassam et al. 2011). These coupled socio-ecological systems—facilitating the material, energy, and information flows not only between natural systems and social systems but also among different social systems—may be regarded as cultural landscapes (Taylor and Lennon 2011). In them, a range of cultural beliefs and mores combine taboos, language, technical practices, knowledge transfer, and customary institutions for social consent and governance (Xu et al. 2005). The result may be called traditional ecological knowledge.

5.3 Ecosystem Services—The Source of Human Wellbeing

The diverse ecosystems of the HKH are important natural capital and play a critical role in protecting the life-support systems in the HKH and beyond (Maharana et al. 2000; Kubiszewski et al. 2013; Sharma et al. 2015). A large proportion of the population in the region still lives in poverty (Gerlitz et al. 2012) and is highly dependent on ecosystem services for their livelihood and daily requirements (Paudyal et al. 2015; Sharma et al. 2015; Chaudhary et al. 2016a, b). Biodiversity has a great significance for the societal benefits derived from ecosystems, as manifested in a myriad customs, traditions, and sacred values (Zomer and Oli 2011), and as a result, there is an inextricable link in the HKH between biodiversity, livelihoods, and culture (Aase et al.

2009; Xu et al. 2009b). Degradation of these values affects the availability and accessibility of ecosystem services for people, which ultimately increases the demand for these resources leading to more pressure on the ecosystems and human society (Badola et al. 2014; Paudyal et al. 2015; Sharma et al. 2015; Chaudhary et al. 2016a, b).

While the importance of protecting mountain ecosystems has been widely accepted (UN 1992), conventional conservation approaches have become a matter of debate and the concept of ecosystem services has risen to prominence (Singh 2002; Naidoo et al. 2008; Chaudhary et al. 2015a, b). Over the past two decades research and publications on ecosystem services have grown exponentially and the concept has been discussed and mainstreamed in many decision-making processes (Chaudhary et al. 2015a, b; Díaz et al. 2015). The idea of ecosystem services dates back to Westman (1977), who suggested that the social value of the benefits that ecosystems provide could potentially be quantifiedso that society can make more informed policy and management decisions (Grêt-Regamey et al. 2012). Many of these ecosystems are long-term premium assets, with benefits that reach far beyond their source (see, for example, Fig. 5.6). They can have multiple functions at different times and scales (Creed et al. 2016). However, one of the challenges that policy makers and managers face while addressing the threats to the ecological integrity of the Himalayas, is the fact that there is still not enough information available about ecological status and human impacts in the region to enable prediction of the losses that will occur as a result of the impacts of natural and human-induced disturbance (Chettri et al. 2010; Thompson and Warburton 1985). Globally, ecosystem values can be broadly categorized into four value systems (Körner 2004):

- Social value: both marketed and non-marketed biodiversity used for social benefits and development
- Cultural value: diverse cultures, species, and landscapes as historical treasures of society
- Ecological value: the interdependence, interaction and co-evaluation of species for maintaining ecological processes and functions
- Economic value: directly generating livelihoods from quality and quantity of desired products as well as providing insurance against failure of crops and livestock.

5.3.1 Social Value of Ecosystem Services

Over the past two decades, the ecosystem service concept—the benefits that humans obtain from ecosystems (MA 2005)—has gained importance among scientists, managers, and policymakers as a way to communicate societal dependence on ecological life-support systems integrating perspectives from both the natural and social sciences (Chaudhary et al. 2015a, b). Although the methodology for ecosystem valuation is still debated, the interdependence of human wellbeing and ecosystem health and biodiversity has now been recognized (Díaz et al. 2015). Biodiversity and the ecosystem services derived from diverse ecosystems have been recognized as important sources since time immemorial of societal development in the HKH (Rai et al. 1994; Awasthi et al. 2003; Chen et al. 2015a). The social value of ecosystem services is critical, as people have relied substantially on these diverse ecosystems for food, shelter, medicine, and so on (Pei 1995; Luck et al. 2009; Joshi and Negi 2011). Transformative change has been possible in many mountain

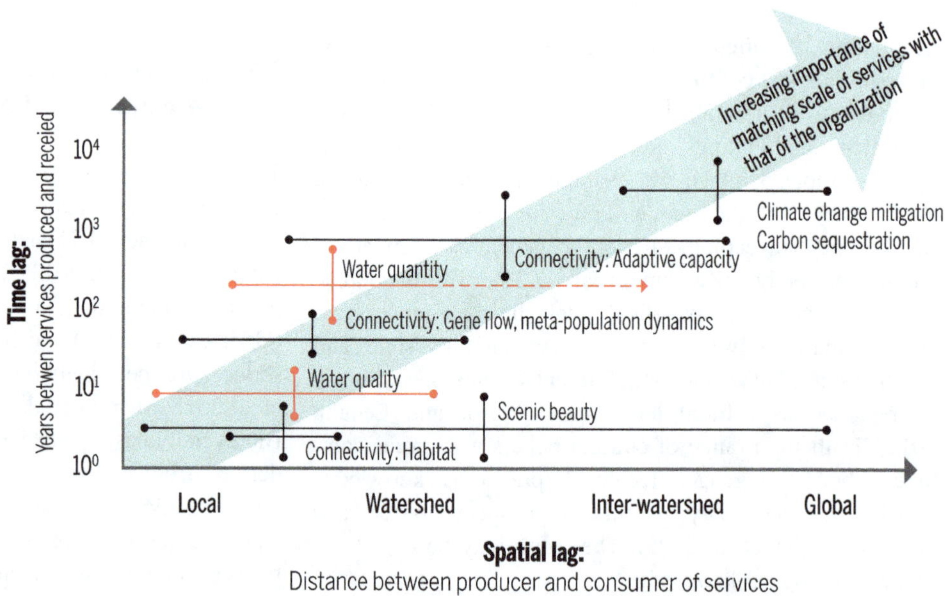

Fig. 5.6 Forest aquatic ecosystem services affect people long after the time and far from where forest management decisions are made. The vertical axis shows the time lag in terms of multi-decadal recovery and the scale of impacts ranging from local to national and global (*Source* Creed et al. 2016)

areas based on the dependence on varied ecosystems for subsistence livelihoods. For example, the community living around Koshi Tappu, a Ramsar site in Nepal, indicated 85% dependency on various ecosystem services (Sharma et al. 2015). Similarly, rangeland and forest ecosystems have provided diverse ecosystem services in most of the rural areas in the HKH (Badola et al. 2010; Dong et al. 2010; Joshi and Negi 2011; Pant et al. 2012). Efforts have been made to understand the complexity of socio-ecological linkages in rangeland (Dong et al. 2010), wetland (Chaudhary et al. 2016a, b), and forest ecosystems (Joshi and Negi 2011). In summary, the ecosystem services derived from the various ecosystems in the HKH have a high value for social development and poverty alleviation.

5.3.2 Cultural Value of Ecosystem Services

Cultural services are defined as the "nonmaterial benefits people obtain from ecosystems" and include the "cultural diversity, spiritual and religious values, knowledge systems, educational values, inspiration, aesthetic values, social relations, sense of place, cultural heritage values, recreation and ecotourism" of an ecosystem (MA 2005). These services are often considered subjective, intangible, and difficult to quantify in monetary terms, and thus, are often neglected or completely excluded from valuation (Chan et al. 2012; Daniel et al. 2012). The failure to recognize and integrate cultural services into ecosystem assessment might lead to biased and misleading trade-off assessments, ecosystem management, and landscape planning (Schaich et al. 2010). The cultural linkage of an ecosystem elicits a positive attitude from the local community towards conservation; systems which have associated cultural beliefs are less disrupted and better maintained (Gao et al. 2013). Cultural services are crucial for sustaining the psychological aspects of human wellbeing and contribute significantly to the overall value of a system to societies.

The HKH is home to varied ethnic communities with vast socioeconomic and cultural diversity (Turin 2005). This cultural diversity is associated with the management of the landscape and natural resources, and has been a part of the co-evolution of society and ecosystems with inextricable links between rural livelihoods, land use, human health, and climate change (Wilkes 2008; Xu et al. 2008). Religious beliefs and rituals, traditions, and customs of local communities often have embedded conservation ethics and have influenced the biophysical conditions of an ecosystem. These landscapes include sacred groves and forest streams, holy mountain peaks, traditional agroforestry systems, and sacred lakes. Many such landscapes in the region have been well studied for their biodiversity and tourism value (for example, Maharana et al. 2000; Anthwal et al. 2010), but very few studies have assessed their cultural services (Sharma et al. 2007a, b, c).

5.3.3 Ecological Value of Ecosystem Services

The ecological value of the HKH is well appreciated and reported (Chettri et al. 2008; Sharma et al. 2010). Various publications rationalize the ecological value of globally significant species, diversity in ecosystems, and ecosystem functions (Myers et al. 2000; Mittermeier et al. 2004; Brooks et al. 2006). Attempts have also been made to understand the distribution of species (Acharya et al. 2011; Bhardwaj et al. 2012) and ecosystems (Chettri et al. 2012; Zomer et al. 2014). The region's ecological value cannot be overstated considering its species richness and diversity, particularly as the habitat for some of the most fascinating and globally significant species in the world. Moreover, the ecosystem of the HKH includes Mount Everest and is the highest biome in the world, unmatched by any other mountain systems. The value of this diversity has long made the HKH one of the most favoured destinations for naturalists, geologists, and explorers (Kandel et al. 2016). In recent years, the contribution of forest, rangeland, and wetland ecosystems to carbon sequestration and soil conservation have been widely acknowledged (Upadhyay et al. 2005; Banskota et al. 2007).

5.3.4 Economic Value of Ecosystem Services

In recent years, efforts have been made to rationalize the significance of ecosystem services in terms of economic value (Costanza et al. 2016). Although research on ecosystem services has progressed significantly, the proportion of research on actual valuation in economic terms is negligible. This is mainly due to limitations in the methodology and the geographical complexities prevailing in the HKH (Rasul et al. 2011a). However, the growing body of literature clearly states that the economic value of both marketed and non-marketed goods is high for rural communities who depend largely on ecosystem services. One of the most comprehensive assessments is from Bhutan, and although preliminary, is revealing. The total estimated value of ecosystem services was approximately USD 15.5 billion per year, significantly greater than the gross domestic product of USD 3.5 billion per year, while 53% of the total benefits accrue to people outside Bhutan, and 47% to those inside (Kubiszewski et al. 2013). The wetland ecosystem is critical for many local communities. For example, some wetlands provide 85% of the total household income (see Sharma et al. 2015). The forested ecosystem is equally important, contributing 80% of household income through provisioning services in some places (Pant et al. 2012).

5.3.5 Changing Ecosystem Services of the HKH Region

Meta-analysis considering the available literature review, trend of research on ecosystem services in the HKH was made using search engine such as google scholar. About 400 peer review articles were collected using search word such as 'ecosystem services' and the name of the countries. The results showed, in general, knowledge base and understanding of ecosystem services is increasing (Fig. 5.7). However, despite a wide range of studies on thematic subjects such as hydropower, water, and carbon storage, there are very few qualitative or quantitative assessments of ecosystem services. Although the number of publications has increased as shown by the increasing trend, the overall ratio between numbers of publications reporting decline or an increase in the value of the ecosystem services provided by the HKH has also changed over the last decade. There is also a regional bias in the studies among the countries in the region; most are from India, followed by China, Nepal, Pakistan, Bangladesh, and Bhutan (Fig. 5.8). We did not find any mountain-specific studies from Myanmar, while studies from Afghanistan were limited to reviews. Most of the studies from India have concluded that the ecosystem services provided by the Indian Himalaya have degraded due to rapid developmental activities and population growth. However, studies from China indicate that the flow of ecosystem services has increased after implementation of the Natural Forest Conservation Programme, Sloping Land Conversion Programme, and Grain for Green Programme (Song et al. 2014). This may also be an outcome of the increased number of studies from China over the last decade and innovative programmes.

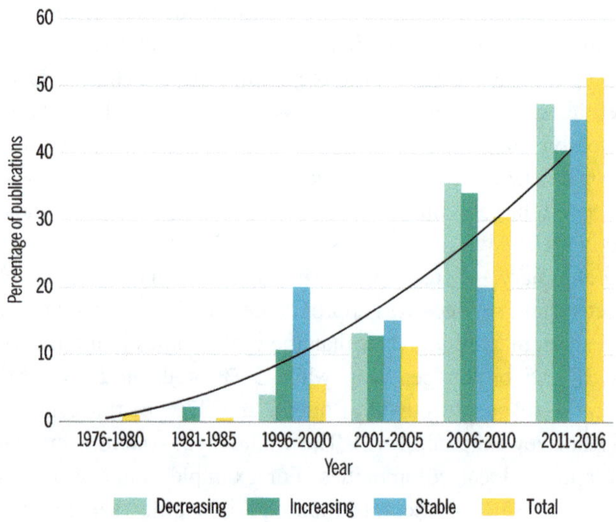

Fig. 5.7 Trend in number of publications on ecosystem services in the HKH, and number of these that predict/suggest/report a decline, increase, or stable state of ES

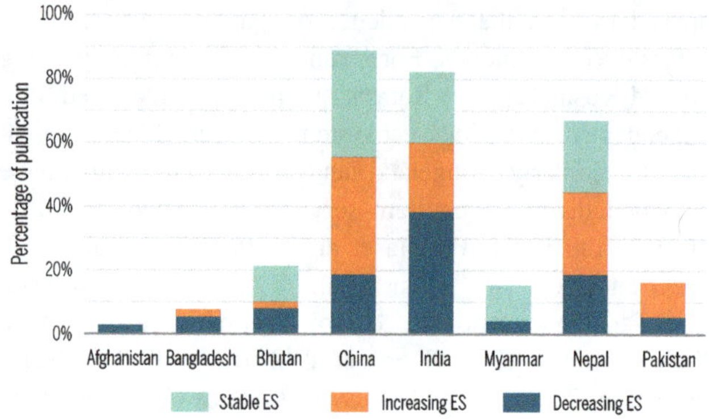

Fig. 5.8 Percentage of publications (1977–2016) and trends predicted in ecosystem services value in the HKH by country

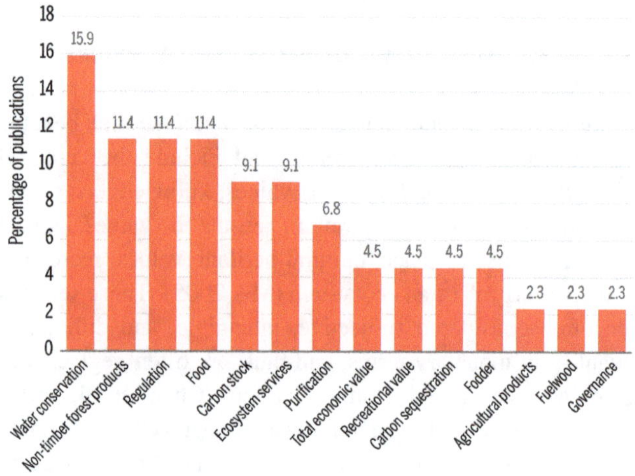

Fig. 5.9 Percentage of publications on different categories of ecosystem services in the HKH

Most of the studies have assessed the benefits people are deriving directly from the natural ecosystems in terms of freshwater, fuelwood, fodder, non-timber forest products (NTFPs), and hydropower. The Himalaya are known as a water tower, contributing immensely to the freshwater needs of the region (Schild 2008; Molden et al. 2014a). Water conservation is the most studied ecosystem service; very few publications have focused on the governance and management of the services in the region (Fig. 5.9).

The overall trend in the provision of ecosystem services increased in recreational value and related activities in the region as most of the areas have become accessible to the visitors. The recreational value is one of the major sources of income for livelihood activities in the area (Nepal 1997; Maharana et al. 2000; Badola et al. 2010). Both NTFPs and water conservation show a stable trend, while biodiversity value and overall ecosystem services show a declining trend (Fig. 5.10). One of the relatively understudied aspects is the role of Himalayan ecosystems in health security (Xu et al. 2008; Sarkar 2011). This is important in view of the climate

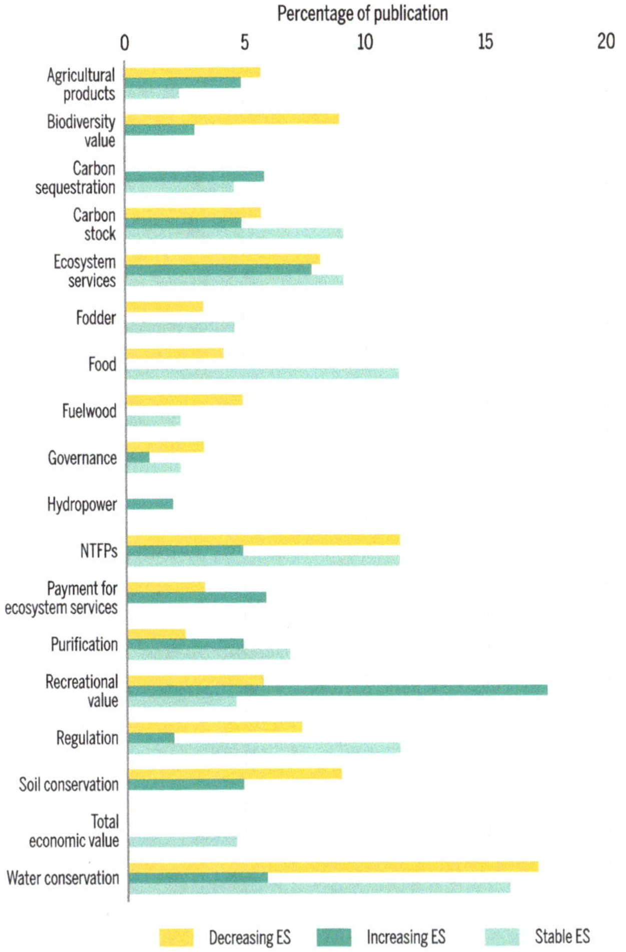

Fig. 5.10 Reported trends within different categories of ecosystem services theme in the HKH

change-induced expansion of vector-borne diseases, as pathogens take advantage of new habitats at elevations that were formerly unsuitable, and diarrheal diseases result from changes in freshwater quality and availability (Ebi et al. 2007; Dangi et al. 2017).

A review of the studies revealed that most focused on a limited topic and lacked a holistic view that would aid better understanding of upstream-downstream linkages; thus they are of limited significance for policy decisions. Large scale studies looking at catchment and basin scales have mainly emerged from the fields of hydrology and geology. An understanding of hydrological dynamics is crucial for sustainable planning and management of water resources in the HKH. However, the lack of hydro-meteorological data in the region, especially for high elevation areas, hinders the process of understanding the system dynamics (Nepal et al. 2014a, b). A huge difference also exists between basins in the extent to which climate change is predicted to affect water availability and food security (Immerzeel et al. 2010). The issue of scale is very important for such assessments. Different issues are relevant at different scales, and the nature of the impacts varies as the scale changes.

The scenarios for the economic values of terrestrial ecosystems services in the HKH countries (Kubiszewski et al. 2016) show a discouraging trend (Table 5.4). The data analysis revealed a 34% decrease in the value under the Fortress World (FW) scenario compared to the 2011 baseline, and a 28% decrease in the Market Forces (MF) scenario; Policy Reform (PR) could reduce the trend to −3%. At the country level, Afghanistan showed the greatest losses in ecosystem services value under both the FW and MF

Table 5.4 Terrestrial values for the ecosystem services in the HKH countries in 2011 (as base) and percentage change estimated under each of four future scenarios till 2050

Country	GDP.PPP (million USD)	ESV (million USD/yr)	SI-MF (million USD/yr)	MF % change from 2011 (%)	S2-FW (million USD/yr)	FW % change from 2011 (%)	S3-PR (million USD/yr)	PR % change from 2011 (%)	S4-GT (million USD/yr)	GT % change from 2011 (%)
Afghanistan	49,338	198,662	56,919	−71	45,434	−77	178,554	−10	271,418	37
Bangladesh	395,684	145,974	107,655	−26	69,847	−52	146,427	0	175,642	20
Bhutan	5,040	14,862	13,255	−11	11,766	−21	14,936	0	17,804	20
China	13,810,256	3,586,924	2,596,138	−28	2,314,370	−35	3,494,582	−3	4,524,762	26
India	5,845,362	1,825,052	1,562,620	−14	1,357,683	−20	1,833,906	0	2,203,965	21
Myanmar	51,920	369,447	305,517	−17	261,775	−29	370,543	0	443,431	20
Nepal	55,504	62,749	54,994	−12	48,631	−22	63,655	1	75,404	20
Pakistan	750,693	294,519	157,302	−47	137,519	−53	264,412	−10	413,554	40
Total	20,963,797	6,498,189	4,854,400	−28	4,247,025	−39	6,367,015	−0.3	8,125,980	26

Source Kubiszewski et al. (2016)
GDP.PPP = gross domestic product, purchasing power parity; ESV = ecosystem services value; Future scenarios: S1-MF = Market Forces; S2-FW = Fortress World; S3-PR = Policy Reform; S4-GT = Great Transition

scenarios (77% and 71%, respectively), followed by Bangladesh and China. Under the PR scenario, Afghanistan and Pakistan showed considerable losses of 10% in ecosystem services values. However, these two countries also showed the greatest gains under the Great Transitions (GT) scenario with a gain of 40% in Pakistan and 37% in Afghanistan. Of the eight countries in the region, Bhutan showed the least loss in ecosystem services value under MF (11%) and FW (21%), a 20% gain under GT, and no change under the PR scenario.

It is undeniable that Himalayan ecosystems provide crucial and valuable ecosystem services to a large part of humanity—more than any other mountain system (Grêt-Regamey et al. 2012). But the lack of data in the Himalaya is hindering understanding of the socio-ecological processes. A concerted effort should be made to fill the knowledge gaps through more focused and coordinated collection of relevant data, especially from the high mountain areas. It is important to improve scientific understanding of ecosystem structures and functioning and drivers of change as a basis for formulating comprehensive ecosystem management approaches and strategies that link to human wellbeing and poverty alleviation. Information on the status of human wellbeing and its linkages with the condition of natural resources is required at the scale where a holistic approach can be taken to address the issues. This knowledge would enable informed decision making, especially where trade-offs among conservation, livelihoods, development, and culture are involved, so that an increase in the supply of one service (such as food or fibre) at the expense of others (such as clean water and self-regulation of pests and diseases) is done with some knowledge of spatio-temporal consequences and who would face them.

5.3.6 Trade-offs and Synergies—Implications for Development

The diverse landscapes of the HKH provide multitudinous services that interrelate in a 'complex dynamic' manner (Birch et al. 2014; Måren et al. 2013; Paudyal et al. 2015). Common drivers affecting multiple ecosystem processes and the interactions among ecosystem services can result in trade-offs and synergies between ecosystem services (Bennett et al. 2009). Trade-offs between ecosystem services arise when one service increases at the cost of another (Ziv et al. 2012), while synergies occur when both services increase or decrease in tandem (Bennett et al. 2009; Haase et al. 2012). Trade-off analysis is a key issue that must be considered when integrating ecosystem services in landscape planning, management, and decision making (de Groot et al. 2010), particularly when analysing alternate pathways leading to sustainable land use in the future (Rounsevell et al. 2012). Any ecosystem management practice that fails to take this into account when attempting to maximise the production of one or more ecosystem services can bring about substantial declines in the provision of other ecosystem services (MA 2005; Bennett et al. 2009). According to the concept of Jointness in Production, there are two different causes for interactions between ES, namely, biological interdependencies and economic interdependencies (Abler 2004; Baumgärtner and Quaas 2010). In landscapes where the provision of ecosystem services is strongly influenced by human activities and vice versa, both these interdependencies are of particular consequence.

The complexity of interactions among ecosystem services is high in managed mountain ecosystem services (Grêt-Regamey et al. 2008), where marginally or periodically productive sites may be relatively more sensitive to climate and socioeconomic shifts (Sharma et al. 2009). The heterogeneity of topography and other landscape characteristics in mountain ecosystems makes spatial dynamics more important than temporal, and climate change may influence the elements of the ecosystem differently (Shrestha et al. 2012). Therefore, when managing trade-offs in the provision of mountain ES, the spatial distribution of ES, the trade-off dynamics over time, and their interaction with structural changes in agriculture and forestry must be accounted for (see Box 5.1 and Table 5.5 for examples).

Box 5.1 Trade-offs: hydropower development and ecosystems in the HKH

Perennial in nature, the Himalayan rivers are regarded as important sources of hydropower generation and have a cumulative hydropower potential that exceeds 500 gigawatts (Pandit et al. 2014). Access to energy for sustenance, agriculture, industries, and other economic activities is critical to the growing population of the HKH region. The large number of rivers that flow out of this region have the potential to fulfil the energy demand and contribute to the wellbeing of the people. There are more than 550 hydropower projects in existence or under construction in the Bhutanese, Indian, Nepalese, and Pakistani Himalayan regions (Pandit et al. 2014). However, it is imperative for hydropower generation in these fragile ecosystems to take into consideration the wide scale impacts that are known to be prevalent while altering these natural resources. For example, most of the dams and hydropower plants located/proposed in the biodiversity rich area of the Indian Himalaya have potential impacts that include forest loss, species extinction, habitat fragmentation, loss of ecosystems, and loss of species diversity (Pandit and Grumbine 2012). Elsewhere, constructing dams on rivers is known to change downstream ecological processes and set in motion complex chain reactions that transform floodplain

Table 5.5 A few examples of synergies and trade-offs between ES

Driver	Service A	Service B	Shared driver	Response type	Interaction type	Synergy or trade-off
Large-scale afforestation and short rotation coppice plantations	Biodiversity	Biomass production	Yes	Opposite	None	Trade-off
Large-scale afforestation and short rotation coppice plantations	Biomass production	Soil organic carbon	Yes	Similar	Unidirectional (Positive)	Synergy
Substitution of extensively used grassland	Food	Biodiversity	Yes	Opposite	None	Trade-off
Substitution of farmland with forest	Carbon sequestration	Protection from gravitational hazard	Yes	Similar	None	Synergy
Natural forest conversion for exotic tree plantations	Biodiversity	Carbon storage	Yes	Similar	None	Synergy
Natural forest regeneration on agricultural and grazing land	Biodiversity	Carbon storage	Yes	Similar	None	Synergy
Agriculture expansion/Fertilizer use	Agricultural production	Water quality	Yes	Opposite	None	Trade-off
Maintaining forest patches close to coffee plantations	Pollination	Agricultural production	No	NA	Unidirectional (Positive)	Synergy
Trail building	Cultural tourism	Agricultural production	No	NA	None	None
Afforestation	Carbon sequestration	Water quantity	Yes	Opposite	Unidirectional (Negative)	Trade-off

Source Bennett et al. (2009)
NA = data not available

vegetation dynamics (Wieringa and Morton 1996). With impacts ranging from changes in river geomorphology and hydrology (Brandt 2000) to impairment of the ecological integrity of rivers through the extirpation of species and loss of ecosystem services (Richter et al. 2003), river regulation is the most substantial and widespread anthropogenic effect on riverine ecosystems (Pandit and Grumbine 2012).

5.4 Conservation and Management Practices

In recent decades, the HKH has witnessed significant conceptual development in biodiversity conservation, from 'people exclusionary' and 'species-focused' to 'people-centred community-based' approaches. The United Nations Conference on Environment and Development (UNCED) in 1992 placed a premium on people's participation and promotion of this conceptual shift in both natural resources management and biodiversity conservation (UN 1992). In response, participatory approaches evolved as the accepted means in various sectors in the HKH (Sharma et al. 2010). The classical approach of biodiversity conservation, which started with an emphasis on the conservation of flagship species (Yonzon 1989; Wikramanayake et al. 1998), evolved to landscape level conservation, with the understanding that 'conservation and management of biodiversity are impossible without people's participation' (Chettri et al. 2010; Zomer and Oli 2011; Bajracharya et al. 2015). Since the 1980s, de-centralization and devolution of authority for biodiversity conservation have been evident in governments' efforts across the HKH (Desai et al. 2011; Sharma et al. 2010; Sunam et al. 2015). During the process, it was realised that biodiversity management by local people is more effective when the utility value and benefit to communities is evident (see Gurung and Seeland 2008). After the late 1990s, conservation approaches in the HKH took on a new dimension with the concept of linking the existing protected areas with biological corridors (Sherpa and Norbu 1999). This approach, while addressing the biophysical advantages of corridors for migration and habitat contiguity, also supports species refugia for restoration, and shifting of species and habitat types in response to environmental pressures such as climate change. Subsequently, the concept of landscape-level conservation approaches evolved in the region, generally adopting the ecosystem approach advocated by the Convention on Biological Diversity (CBD) (see Sharma et al. 2010). These evolutions comprised both ex situ and in situ approaches. In addition, inclusion of traditional ecological knowledge, belief, and culture also contributed substantially in addressing the conservation goal.

5.4.1 Flagship and Keystone Species Conservation

As a result of its ecosystem diversity, the HKH is one of the most biodiverse areas in the world with the highest number of species and endemism (Myers et al. 2000; Mittermeier et al. 2004). Large mammal in situ conservation has a long history. It has been practised since 1950 with species like the greater one-horned rhinoceros (*Rhinoceros unicornis*), Asian elephant (*Elephas maximus*), and Bengal tiger (*Panthera tigris tigris*) in the tropical lowlands, red panda (*Ailurus fulgens*) and Asian black bear (*Ursus tibetanus*) in the temperate region, and snow leopard (*Panthera uncia*) and Himalayan musk deer (*Moschus leucogaster*) in the alpine region. The status of many of these species is facing additional challenges (see Table 5.6). Despite tireless efforts towards conservation, the majority of species are being driven towards extinction. The exceptions are Tibetan antelope or chiru (*Pantholops hodgsonii*) and the giant panda (*Ailuropoda melanoleuca*), which were removed from the endangered species list by the International Union for Conservation of Nature (IUCN) in 2016.

Some of the iconic species in the HKH have been researched more than others (Kandel et al. 2016). Understanding of the ecology of the greater one-horned rhinoceros has considerably improved (Dinerstein and Price 1991; Pradhan et al. 2008); the population is either stable or has increased in its present range countries (Thapa et al. 2013). However, the historical range that extended along the floodplains of the Ganges, Brahmaputra, and Sindh Rivers from Pakistan to the Indo-Burma border (Amin et al. 2006) is now restricted to nine populations in protected areas (PAs) in India and Nepal (Menon 1966). With the exception of the populations in Chitwan National Park in Nepal and Kaziranga National Park and Jaldapara Wildlife Sanctuary in India, the populations each number less than 150 individuals. Notably, Nepal has made significant progress in reversing the decreasing trend of rhinoceros (Fig. 5.11) and has also celebrated three consecutive years of zero poaching (Acharya 2015). Likewise, periodic status reports from the snow leopard home range countries have added to the limited knowledge about this elusive species (Karmacharya et al. 2011; Ale et al. 2014; Alexander et al. 2016). New dimension of species range with sub-species was recently added for now leopard (Janecka et al. 2017). However, analysis of the impact of climate change on the range of snow leopards predicts a contraction in suitable habitat and a fragmentation of distribution—both of which could cause a significant contraction in the range of the species (Forrest et al. 2012; Li et al. 2016). Red panda is another flagship species in the temperate forest, confined to three global biodiversity hotspots—Himalaya, Indo-Burma, and the mountains of southwest China (Kandel et al. 2015). Although it is one of the most researched species in the Eastern Himalaya (Kandel et al. 2016), very little is known about its ecology and distribution (Yonzon 1989; Wei et al. 1999; Choudhary 2001; Pradhan et al. 2001; Groves 2011; Dorji et al. 2012). Red panda has been sighted in five of the eight HKH countries (Bhutan, China, India, Nepal, and Myanmar), and has an estimated habitat area 32,600 km^2 within the region (Kandel et al. 2015). However, unlike snow leopard, tiger, and rhinoceros, the red panda has yet to receive global attention for conservation.

5.4.2 Protected Areas Management

The HKH has made significant progress in the establishment of PAs in recent decades (Chettri et al. 2008). As of 2007, there were 488 PAs (IUCN category I–VI) within the HKH, covering more than 1.6 million km^2 or about 39% of the region's terrestrial area (Table 5.7), with significant growth witnessed over the last three decades (Fig. 5.12). Although PA coverage has been identified as a key indicator for assessing progress in reaching the Aichi targets (Secretariat of CBD 2014), many scholars have pointed out that the percentage of area protected in a given country or biome is not a strong indicator of actual conservation needs or effective action (Oli et al. 2013). In particular, this indicator overlooks the fact that biodiversity is unevenly distributed across the region. For example, the Brahmaputra Basin in the Eastern Himalaya is significant for both aquatic and terrestrial biodiversity with a high level of endemic species and high proportion of rare, endangered, and threatened species compared to the western Himalaya (see Allen et al. 2010). More significant, perhaps, is the fact that actual implementation of conservation measures within PAs varies greatly across the region. In Myanmar, for example, human-induced pressure and lack of financial and skilled human resources are impinging on the effective management of PAs (Rao et al. 2002). Bawa (2006) also points out that local challenges, such as a lack of economic opportunities, interdisciplinarity in conservation actions, institutional development, skilled human resources, and large-scale conservation approaches hinder conservation. In spite of these challenges, half of the HKH countries—Bhutan, China, Nepal, and Pakistan—have either reached, crossed, or are heading towards the global target of 17% area covered by PAs by 2020 (see Table 5.7).

5.4.3 Conservation Through Traditional Knowledge

Management of resources to sustain the flow of ecosystem services was, and still is, widely practised by many communities in the Himalaya as part of their traditional

Table 5.6 Current population trends of key mammal species in the HKH (1996 and 2016)

Common name	Scientific name	Reported distribution	IUCN category (IUCN 1996)	IUCN category (IUCN 2016–17)	Current population trend (IUCN 2016–17)
Alpine musk deer	*Moschus chrysogaster*	China, eastern Nepal, Bhutan, and north-eastern India	NT	EN	↓
Himalayan musk deer	*Moschus leucogaster*	Across the Himalaya of Bhutan, northern India (including Sikkim), Nepal, and China (southwest Xizang)	NT	EN	↓
Black musk deer	*Moschus fuscus*	China (north-western Yunnan and south-eastern Tibet), northern Myanmar, north-eastern India (Arunachal Pradesh), Bhutan and eastern Nepal	NT	EN	↓
Forest musk deer	*Moschus berezovskii*	Most of the alpine regions in eastern Nepal, Bhutan, North-east India and Southern and South-eastern China	NT	EN	↓
Elk (MacNeill's deer)	*Cervus canadensis ssp. macneilli*	Central and SW China (N Qinghai, Gansu, Shaanxi, W Sichuan and E Xizang)	NE	NE	?
Elk (Tibet red deer)	*Cervus canadensiswallichi*	SW China (SE Xizang), Bhutan	NE	NE	?
Kashmir stag	*Cervus hanglu*	North India (Kashmir)	EN	LC	?
White-lipped deer	*Cervus albirostris*	China (Gansu, Qinghai, Yunnan)	VU	VU	?
Argali	*Ovis ammon*	Afghanistan, China India, Nepal and Pakistan (above 3000 m)	VU	NT	↓
Argali (Marco Polo sheep)	*Ovis ammon polii*	North western edge particularly in Pamir mountains(Afghanistan, Pakistan, China)	NE	NE	?
Argali (Tibetan argali)	*Ovis ammon ssp. hodgsoni*	Central and eastern high mountains	NE	NE	?
Mouflon	*Ovis orientalis*	Afghanistan, North-western India (Kashmir) and Pakistan	VU	VU	↓
Blue sheep	*Pseudois nayaur*	Central and eastern high mountains in Bhutan, China, India, Myanmar, Nepal and Pakistan	–	LC	?
Dwarf blue sheep	*Pseudois nayaur ssp. schaeferi*	China (Sichuan, Tibet [or Xizang], Yunnan)	EN	EN	↓
Przewalski's gazelle	*Procapra przewalskii*	Around Qinghai Lake in China	CR	EN	↓
Goitered gazelle	*Gazella subgutturosa*	Afghanistan, northwest China and Pakistan	NT	VU	↓
Tibetan gazelle	*Procapra picticaudata*	Qinghai-Tibet Plateau in China and and Ladakh and Sikkim in India	NT	NT	↓

(continued)

Table 5.6 (continued)

Common name	Scientific name	Reported distribution	IUCN category (IUCN 1996)	IUCN category (IUCN 2016–17)	Current population trend (IUCN 2016–17)
Tibetan antelope (chiru)	*Pantholops hodgsonii*	Qinghai-Tibet Plateau in China and north-eastern Ladakh in India	VU	NT	↑
Brown bear	*Ursus arctos*	Widely distributed across Afghanistan, China, India and Pakistan and small population in Nepal	LC	LC	→
Giant panda	*Ailuropoda melanoleuca*	Sichuan, Shaanxi and Gansu provinces in China	EN	VU	↑
Red panda	*Ailurus fulegns*	Along a narrow elevation Himalayan belt (2500-4000 m) in Bhutan, China, India, Myanmar and Nepal	EN	EN	↓
Dhole	*Cuon alpinus*	Across Himalaya in Bhutan, China, India and Nepal	VU	EN	↓
Gray wolf	*Canis lupus*	All the HKH countries except Bangladesh	LC	LC	
Eurasian lynx	*Lynx lynx*	Higher elevations in Afghanistan, China, India, Nepal and Pakistan	LC	LC	→
Red fox	*Vulpes vulpes*	All the HKH countries	LC	LC	→
Siberian ibex	*Capra sibirica*	North western edge inAfghanistan, China, India and Pakistan,	LC	LC	?
Snow leopard	*Panthera uncia*	Across Afghanistan, Bhutan, China, India, Nepal and Pakistan, primarily in higher elevation	EN	VU	↑
Pallas's cat	*Otocolobus manul*	Across Afghanistan, Bhutan, China, India (Jammu & Kahmir), Nepal and Pakistan, primarily in higher elevation	LC	NT	↓
Takin	*Budorcas taxicolor*	Bhutan, China, northeast India (Arunachal Pradesh and Sikkim) and northern Myanmar	VU	VU	↓
Tibetan fox	*Vulpes ferrilata*	Tibetan Plateau including Nepaland India	LC	LC	?
Kiang	*Equus kiang*	Southern edge and Himalaya in China, India, Nepal and Pakistan	LC	LC	→
Wild yak	*Bos mutus*	China and India (Ladakh), primarily scattered populations in Tibetan plateau and Kunlun mountains in China	VU	VU	↓
Greater one-horned rhino	*Rhinoceros unicornis*	Himalayan foothills of India and Nepal	EN	EN	↓

Sources Chettri et al. (2012) and IUCN (2016, 2017). CR—critically endangered; EN—endangered; VU—vulnerable, NT—near threatened; LC—least concern; NE—not evaluated. ↑ = Increasing; ↓ → = decreasing; = Stable and ? = unknown

Fig. 5.11 Population of the greater one-horned rhinoceros in Nepal from 1950 to 2015 (*Source* Acharya 2015)

Table 5.7 Number and area of PAs in the HKH (as of 2007)

Country	Number of PAs	PA coverage (km²)	% of PA coverage with respect to country	% of PA coverage with respect to total area of HKH
Afghanistan	6	2,461	0.44	0.06
Bangladesh	5	632	1.70	0.01
Bhutan	10	16,396	42.71	0.38
China	221	1,522,172	15.15	35.50
India	135	62,417	8.99	1.46
Myanmar	16	23,967	5.32	0.56
Nepal[a]	20	34,357	23.34	0.80
Pakistan	76	18,721	11.85	0.44
Total	489	1,681,123	NA	39.21

Sources Sharma et al. (2010); [a]CBS (2014)

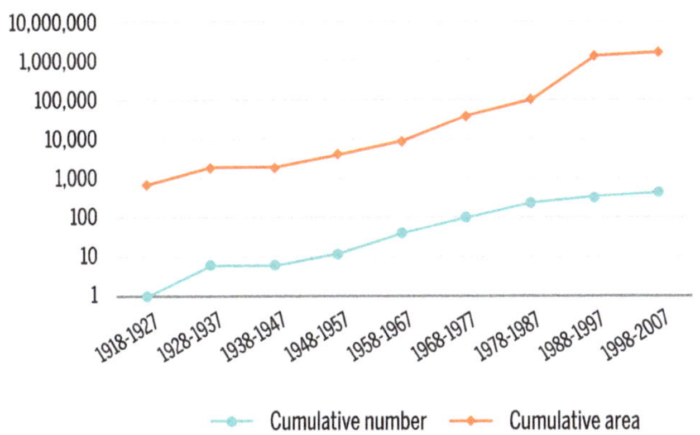

Fig. 5.12 Trend (log value) in number and coverage of protected areas in the HKH from 1918 to 2007 (Chettri et al. 2008)

ecological knowledge system (Dudley et al. 2009; Singh et al. 2011). The blending of cultural, religious, and traditional knowledge systems has contributed substantially to the overall goal of conservation of species, ecosystems, and genetic diversity (Salick et al. 2007; Anthwal et al. 2010). There are numerous plants, animals, and ecosystems (forests, ponds, rivers, mountain peaks) across the region which have been conserved effectively through traditional practices. For example, there are numerous examples in the Himalayas of conservation of sacred groves maintaining significantly higher diversity compared to other areas (Khumbongmayum et al. 2005; Arora 2006). Traditional conservation practices are also regulated by customary laws through participatory decision-making processes. This is evident, for example, in the grazing and other resource management practices regulated through Dzumsa in North Sikkim (Acharya and Sharma 2012), and in the northern part of Humla District, Nepal, in the Kailash Sacred Landscape (Zomer and Oli 2011; Basnet and Chaudhary 2017). Such practices focus on management of ecosystems (Anderson et al. 2005), species (Mehra et al. 2014), or resource use patterns (Acharya and Sharma 2012). Many of these practices are categorised as community-conserved areas (CCA) or key biodiversity areas (KBA) and are being mainstreamed in national conservation practices. These CCAs and KBAs are increasing the focus of conservation interventions outside formal PA systems.

Religious and cultural beliefs related to natural resources have also played an important role in the conservation of the resources in the Himalaya, with the use and exploitation of certain plant and animal species prohibited in many areas (Negi 2005). For example, some of the forests in Garhwal and Kumaon in Uttarakhand, such as the Hariyali Sacred Landscape and Haat Kali Sacred Grove, cannot be used by anyone because they are dedicated to a deity and the forests and streams originating from them are considered sacred (Negi 2005; Sinha and Mishra 2015). In Chhyangru village in Byas Village Development Committee in the Api Nampa Conservation Area, Nepal, people decided to establish a temple in the forest called 'Shyanchho' to prevent further deforestation and degradation and conserve the forest as sacred forest (Chaudhary et al. 2017). Buddhist beliefs have been influential in conserving natural landscapes as ecocultural landscapes, including high elevation lakes and their basins in Ladakh and Sikkim (Maharana et al. 2000; Chandola 2012). In China, there are a few examples where cultural diversity manifested in social and cultural values of natural resources has played an important role in conservation (Anderson et al. 2005; Brandt et al. 2013a). Similarly, the Apatani ecocultural landscape in Arunachal Pradesh is protected by a mix of social and religious institutions which make use of traditional ecological knowledge in sustainable resource management. Hence, there is a growing recognition of a form of environmental governance that acknowledges the role of local communities and their traditional practices to restrain unrelenting forest degeneration while ensuring ecological and economic benefits for the community (Paul and Chakrabarti 2011; MoAF 2014).

Table 5.8 Landscape initiatives in the HKH

Landscape initiative	Geographical coverage	Main themes	Source
Bhutan biological conservation complex	Bhutan	Protected areas and conservation corridors	Sherpa and Norbu (1999), NCD (2004)
Everest complex	China and Nepal	Regional cooperation, information sharing, and developing decision-making tools	Sherpa et al. (2003), Bajracharya et al. (2010)
Terai arc landscape	Nepal and India	Community-based conservation in protected areas and conservation corridors	Gurung (2004b), MoFSC (2015a)
Kangchenjunga landscape	Eastern Nepal, Sikkim and north Bengal in India, and western Bhutan	Conservation and development in protected areas and conservation corridors	Sharma and Chettri (2005)
Kailash sacred landscape	Western Nepal, Uttarakhand in India, and Tibet autonomous region in China	Conservation and development around sacred sites and in protected areas	Zomer and Oli (2011)
Far-Eastern himalaya	Arunachal Pradesh in India, Kachin State in Myanmar, and Yunnan in China	Conservation and development in biodiversity hotspots	Guangwei (2002), Shakya et al. (2011)
Karakoram pamir	Afghanistan, China, Pakistan, and Tajikistan	Conservation and development in arid ecosystems	Ning et al. (2014)
Chitwan-Annapurna landscape	Nepal from Chitwan national park in the south to Manaslu, Langtang, and Annapurna in the north covering 19 districts	Conservation and development in the Gandaki river basin from tropical lowland Terai to alpine high mountain and cold and dry trans-Himalaya	MoFSC (2015b)
Sacred himalayan landscape, Nepal	Koshi basin covering Langtang national park to Makalu Barun national park	Protected area management, river basin management, community forest and cultural conservation	MoFSC (2016)
Western mountain landscape	Fourteen mountain districts in the Mid-western and Far-western development regions in Nepal	Chure, mid-mountains, high-mountains, protected area management, community forest, protection forest, Karnali, Bheri and Seti river basin	MoFSC (2017)

Adapted and modified from Chettri and Sharma (2016)

5.4.4 The Landscape Approach—Recognizing Complexity and Understanding Linkages

Landscape-level biodiversity conservation is an evolving concept in the HKH. The concept has emerged primarily out of recognition that strict protection through a network of PAs (national parks, sanctuaries, wildlife reserves, and others) is an essential but insufficient strategy for biodiversity conservation (Wikramanayake et al. 2004). Now, the focus has shifted from preserving isolated patches of sustained wilderness in the form of PAs to maintaining landscape integrity (Chettri et al. 2008; Wikramanayake et al. 2011) and seeing—and conserving—ecosystems as part of larger agro-ecological and sociocultural landscape (Zomer and Oli 2011). These efforts, including protection through policy and practices, have resulted in an increase in population of many wildlife species (Acharya 2015). The transboundary landscape approach is also gaining prominence in many areas in the search for solutions to reconcile conservation and development trade-offs (see Sharma et al. 2007b; Zomer and Oli 2011). In addition, the approach is addressing the need for regional cooperation, knowledge and information sharing, and opportunities for cross-learning and capacity building from best practices. There are a number of landscape-level initiatives for biodiversity conservation in the HKH at different development levels (Table 5.8). The majority of these initiatives have looked at ways of reconciling conservation with development with a focus on community wellbeing. However, differences in conservation policies and practices have led to differences in reaching conservation goals (see Box 5.2).

Box 5.2 Results of diverse policies and practices

The Kangchenjunga Landscape, a transboundary complex shared by Bhutan, India, and Nepal, has 19 protected areas (see ICIMOD, WCD, GBPNIHESD, RECAST 2017), 17 in India and one each in Bhutan and Nepal. Nine are transboundary in nature, with differences in their protected category on different sides of the border. They include the Kangchenjunga Conservation Area in Nepal which is contiguous with the Khangchendzonga National Park and Biosphere Reserve in India; the Buxa Tiger Reserve in India which is connected to unreserved areas in Bhutan; the Jigme Khesar Strict Nature Reserve in Bhutan which is connected to Neora Valley National Park and Pangolakha Wildlife Sanctuary in India; and the Singhalila National Park in India which is contiguous with unreserved areas in eastern Nepal. These variations in protected area management systems have led to variations in effectiveness in terms of performance and reaching conservation goals (Oli et al. 2013).

5.4.5 Participatory Forest Resources Management Practices

The rapid depletion and degradation of forest resources in the last few decades has resulted in a serious threat to rural livelihoods and environmental sustainability in the HKH. Increasing awareness of the critical situation amongst all actors has led governments to endorse a set of new forest policies and legal frameworks over the last 15–20 years that have enabled forest management practices to flourish. Different types of management system have been introduced—community forestry (CF) and leasehold forestry (LHF) in Nepal, joint forest management (JFM) in India, community-based natural resources management (CBNRM) and an integrated conservation and development programme (ICDP) in Bhutan, and others—but in all the main objective was to address the issue of degrading forest resources through decentralized governance systems. Bangladesh, Bhutan, India, and Nepal, in particular, have shifted their forestry policies from highly centralized approaches to a participatory type of forest management.

Participatory forest management (PFM) began in Bangladesh in the early 1980s under the banner of social forestry (SF). By 2005, more than 40,000 ha of forestland were under SF, which has now become an integral part of the country's forest management (Chowdhury and Koike 2010). In Bhutan, CBNRM, which includes SF and CF, was initiated in the early 1980s. The Ninth Five Year Plan (2002–2007) emphasized PFM as a primary driver of forest management; the number of community forest management groups increased from 31 in 2006 to 677 in 2016, benefiting 28,311 households (Rasul et al. 2011b; MoAF 2016). JFM in India brought an important breakthrough in the relationship between the Forest Department and the community, and is now the primary driver of forest management in India, spreading over 28 states and covering 21.44 million ha of forest (Mukerji 2006). The ecodevelopment projects of the Great Himalayan National Park, Nanda Devi Biosphere Reserve, Hemis National Park, and Rajaji National Park in the Indian Himalaya are among the best examples (Badola et al. 2014). CF was introduced in Nepal in the late 1970s; now, 18,500 community forest user groups are managing more than 1 million ha of forest, accounting for about one-third of total forest area (MoFSC 2015a). In China, the Collective Forest Reform implemented in 2004 has given complete rights to local farmers to manage forestlands (Xu et al. 2010). Pakistan formulated the National Forest Policy (MoCC 2015) to expand, protect, and promote sustainable use of forests, protected areas, natural habitats, and watersheds in order to restore their ecological functions and improve livelihoods and human health, in line with the national priorities and international agreements. The forest-rich areas of North West Frontier Province (now Khyber Pakhtunkhwa) in Pakistan saw an improvement in natural and financial capital for local people with the

Table 5.9 Community-based forest resource management practices and their coverage in the HKH

Country	Terms used to denote participatory forest management	% forest in total area of country	Area of community forestry	
			million ha	% of country's forest area
Afghanistan	Community-based forest management	2.0	NA	0.0
Bangladesh	Social forestry	17.5	NA	0.0
Bhutan	Community based natural resources management	70.4	0.07	2.5
China	Community-owned forests or collective forestlands	NA	NA	58.0
India	Joint forest management	3	21.0	28.5
Myanmar	Community forestry	27.1	NA	NA
Nepal	Community forestry	44.7	1.79	30.3
Pakistan	Community-based forest management of Guzara forests	NA	0.8	20.0

NA = data not available

Forestry Sector Project that institutionalized participatory forest management in the area (Ali et al. 2007).

The outcomes of these participatory and community-based approaches have largely had positive ecological, economic, and social impacts. Ecologically, regeneration of substantial areas of degraded forests has improved and there has been a visible impact in reversing the trend towards deforestation and forest degradation (Gurung et al. 2013). Socially, they have given institutional space for local communities to make their own decisions on a range of issues related to forests, income, inclusion, social justice, and so on. Economically, they have contributed to the economic opportunities available to local rural people (Xu et al. 2010; Birch et al. 2014). Bhutan, China, India, and Nepal have made good progress and Myanmar is trying hard to move, whereas Afghanistan, Bangladesh, and Pakistan have yet to make progress to achieve a visible impact from community forestry (Table 5.9). However, some of the approaches have limitations and shortcomings, such as elite capture, social disparity, inequitable benefit-sharing, and exclusion of poor and marginalized groups (Gurung et al. 2013). Special attention is needed to make participatory forestry inclusive with equitable benefit-sharing and a pro-poor focus.

5.4.6 National and International Policies and Legislations—Support for Biodiversity Conservation

The HKH has witnessed a significant paradigm shift for biodiversity conservation during the last few decades. The alarming forest condition of the Himalaya (Ives and Messerli 1989; Pandit et al. 2007), increasing human population and overexploitation of biodiversity (Sandhu and Sandhu 2014; Chettri and Sharma 2016), and the increasingly changing alpine meadow of the Tibetan Plateau (Klein et al. 2004; Luo et al. 2010) have highlighted numerous conservation and management challenges for biodiversity conservation in the region (see Sharma et al. 2009; Shrestha et al. 2012). However, with adaptive management and appropriate policies, the perception of conservation has changed significantly over the pre- and post-1992 periods (see Fig. 5.9). Since the new thinking in conservation and environmental management began in 1972 at the Stockholm conference, a number of key influential global developments have taken place. The Convention on Biological Diversity of 1992 (UN 1992) was an important milestone in the history of biodiversity conservation. Although a number of conventions were already in place before 1992 (such as the Ramsar Convention, World Heritage Convention, Convention on International Trade in Endangered Species of Wild Fauna and Flora, International Treaty on Plant Genetic Resources, and others), the CBD was instrumental in focusing on conservation and sustainable use of biodiversity (see Fig. 5.13).

The founding of the International Geosphere-Biosphere Programme and Millennium Development Goal agendas during 1992–2000 brought additional support. The perspective of biodiversity conservation totally changed when the Millennium Ecosystem Assessment (MA) processes advanced and the CBD initiated supporting programmes on PAs and mountain biodiversity during 2001–2010. With advancement of the MA, global communities accepted more strongly the concept of ecosystem services as a means of rationalizing the significance of biodiversity to human wellbeing, and the Intergovernmental Science-Policy Platform on Biodiversity and Ecosystem Services (IPBES) was established. Progressive change continued in the global arena with the 2020 Aichi targets in 2010 and Sustainable Development Goals and UNFCCC Paris Agreement in 2015 (Fig. 5.13).

In the HKH, several policies were formulated and practised before 1992, beginning with the Forest Act 1927, which is still instrumental in India and Pakistan. This was followed by the Wildlife Protection Act of 1972 in India and the Islamabad Wildlife (Protection, Preservation, Conservation, and Management) Ordinance 1979 in Pakistan. In Nepal, the National Parks and Wildlife Conservation Act of 1973 was the pioneering legal instrument, while the first formal conservation instrument in China was the law on the Protection of Wildlife 1988. Since 1999, China has been addressing deforestation progressively with the 'Green for Grain' policy and practices, with commendable success (Box 5.3). The first national policy for Bhutan was the forest policy formulated in 1974; which was followed by the Forest and Nature Conservation Act 1995, which dedicates two chapters to biodiversity conservation, with several sections on PAs and conservation of wildlife.

Box 5.3 Grain for green

The Grain-for-Green policy, the largest land reforestation/afforestation program in China, was launched in 1999 to mitigate land degradation by returning steeply sloping cultivated land to forest or grassland. Although the initiative showed variations, it has contributed significantly leading to a doubling of forest cover and increased carbon sequestration (Chen et al. 2015b). The value of nutrient cycling, regulating gases, organic material provision, and soil conservation increased by 64, 39, 40, and 18%, respectively (Song et al. 2015). The initiative is expected to sequester 110.45 Tg C by 2020, and 524.36 Tg C by the end of the century, with economic benefits ranging from USD 8.84–44.20 billion between 2000 and 2100, which may exceed the current total investment (USD 38.99 billion) on the program (Liu et al. 2014). The results indicate that a large-scale initiative with policy support can bring significant change towards addressing climate change.

GLOBAL POLICIES	KEY NATIONAL POLICIES	KEY NATIONAL LAWS	REMARKS
2010-2016			
IPBES established (2010) CBD: Strategic Plan 2010-2020 with Aichi targets (2010) UNFCCC: Paris Agreement (2015) Sustainable Development Goals (2015)	Pakistan: National Climate Change Policy (2012)		ES mainstreamed through IPBES Mountain Ecosystem recognized as important ecosystem by SGD 15
2001-2010			
Millennium Ecosystem Assessment launched (2001) Earth Summit (2002) MEA Framework launched (2003) International Treaty on Plant Genetic Resources for Food and Agriculture (2004) MEA report for policy makers launched (2005) Programme of Work on Protected Area (2004) Programme of Work on Mountain Biodiversity (2004)	India: The Wildlife Action Plan (2002-2016) India: National Action Plan on Climate Change (2008) India: National Biodiversity Strategy and Action Plan (2009) Nepal: Biodiversity Strategy (2002) Pakistan: National Environment Policy (2005) Pakistan: National Forest Policy (2008) Pakistan: National Vision 2030 on forest biodiversity conservation (2008)	India: Wildlife Protection Act (1972) and Amendment Act (2002) India: Biological Diversity Act (2002) Nepal: National Parks and Wildlife Conservation Act, Nepal (1972) and Amendment (2002)	Landscape approach mainstreaming Connectivity and corridor concept embedded in practice Transboundary cooperation in conservation brought into practice Importance of wetland ecosystems highlighted Recognition of community forest practices Biodiversity conservation linked to human wellbeing
1992-2000			
Convention on Biological Diversity (1992) International Geosphere and Biosphere Programme (1997) Millennium Development Goals (2000)	Bhutan: Biodiversity Action Plan (1994) Bhutan: National Forest Policy (1995) China: National Biodiversity Action Plan (1999) Nepal: Environment Policy and Action Plan (1993)	Bhutan: Forest and Nature Conservation Act (1995) India: Panchayati Raj Act (1996) India: Wildlife Protection Act (1972) and Amendment Act (1991) Nepal: Nepal Forest Act (1993) Pakistan: Environmental Protection Action (1997)	Emphasis on government managed to people managed Centralized funding to decentralized funding at the local level Collaborative management of forest, wetlands, and protected areas Participatory and inclusive conservation Emphasis on benefit sharing institutional growth Punitive action to participatory action Concept of landscape and corridor emerged
Pre-1992			
Ramsar Convention (1971) World Heritage Convention (1972) CITES: Washington Convention (1972)	India: Forest Policy (1952) India: National Forest Policy (1988) India: National Conservation Strategy and Policy Statement (1992) Myanmar: Forest Policy (1995) Nepal: New Forest Policy (1978) Nepal: National Conservation Strategy (1988) Pakistan: Forest Policy Statement (1991) Pakistan: National Conservation Strategy (1992) Pakistan: National Biodiversity Action Plan (1998)	China: Forest Law (1984) India: Indian Forest Act (1927) and amendments (1980) India: Environment Protected Act (1986) India: Wildlife Protection Act (1972) Nepal: Legal Code 'mulikhi ain' (1854) Nepal: Forest Nationalization Act (1957) Nepal: Forest Act (1962), amendment (1968) Nepal: National Parks and Wildlife Conservation Act, Nepal (1972) Pakistan Forest Act (1927) Pakistan: Islamabad Wildlife Ordinance (1979)	The word 'ecosystem services' coined (1981) The word 'biodiversity' coined (1986) Disempowerment of people from resources Resources brought under eminent domain Government control over biological resources No people's participation Motive mainly on revenue generation

Fig. 5.13 Development of modern biodiversity conservation at the international level and its impact on conservation policies and laws of HKH countries

Many of the earlier versions of policies and laws prohibit human influences on an ecosystem once the area has been declared as a national park or wetland of international importance. Even areas outside PAs may be under forest or wetland law in which usufruct rights were denied to the local communities. As a result, the degradation of Himalayan ecosystem was strongly predicted during this time (Ives and Messerli 1989). To address Himalayan degradation and other conservation issues, from 1992, the policies and legislation in the Himalayan countries were amended and improved several times and new laws for local resources management were promulgated, mainly driven by the global development and establishment of numerous conventions, including the CBD. The Panchayat (Extension to Scheduled Areas) Act 1996 and Biological Diversity Act 2002 in India, and the Conservation Area Rules 1996, National Parks and Wildlife Conservation Act (1972) and its amendment (1993), Forest Act (1993), Forest Policy (2015), and Buffer Zone Management Regulation 1996 in Nepal are good examples of policy changes towards conservation. Each case represented, in several fundamental ways, a devolution of control from government authorities to local communities—a major shift. As an adaptive social process, these moves strived to create sufficient future forest opportunity to satisfy potentially competitive/conflicting interests that would diminish the forest if left unresolved (Singh et al. 2010). The change has been further boosted by the declaration of the National Mission for Green India targeting the afforestation of 6 million hectares of degraded forest lands and expanding forest cover from 23 to 33% of India's territory through people's involvement, and enlarging the landscape under conservation (GoI 2008). The development of modern biodiversity conservation at the international level and its impact on the conservation policies and laws of HKH countries (Fig. 5.13) has brought about a new paradigm in conservation and sustaining ecosystem services in the HKH.

5.5 Sustaining Ecosystems—Challenges, Opportunities, and Strategies

The local communities in the HKH face uncertainties and challenges as they strive to use, nurture, and sustain the diverse ecosystems at landscape scale where they live and on which they depend. The most powerful contemporary forces that influence both biological and cultural diversity and livelihoods include government policies, expansion of regional and global markets, and climate change, among others. Some of these forces are positive, others are not (Xu and Grumbine 2014a). With increasing awareness of upstream and downstream linkages, biodiversity and ecosystem services have been recognized for their ecological, sociocultural, and economic values (Rasul 2014). While acknowledging that some indigenous communities are more resilient than others, the knowledge, innovations, and practices of local groups can be strengthened with appropriate assistance from partnerships with governments, non-governmental organizations, and the commercial sector to ensure equitable access and sustainable management of natural resources. These partnerships must be based on participatory processes, intercultural dialogues, and co-designing and co-production of hybrid knowledge among different stakeholders (Xu and Grumbine 2014b). They are also an important instrument for ecological restoration and poverty alleviation, especially in the upstream headwater areas (Grêt-Regamey et al. 2012; Sandhu and Sandhu 2014; Bawa and Seidler 2015; Feng et al. 2016). There are examples where increasing financial investment is being made for conservation and restoration of ecosystems. China has launched the Natural Forest Conservation Program (NFCP) and Sloping Land Conversion Program (SLCP), the largest payment for ecosystem service programs in human history (Uchida et al. 2005; Chen et al. 2015b). The cumulative total investment through the NFCP and SLCP exceeds USD 50 billion, and the SLCP alone benefits more than 120 million farmers in 32 million households (Ouyang et al. 2016). Ecosystem sustainability has been increasingly integrated into environmental, biodiversity conservation, and sustainable development agendas (Maselli 2012; Secretariat of CBD 2014; Díaz et al. 2015). The current trends of environmental degradation and the effects of projected climate change in the region necessitate the inclusion of local knowledge and institutions into environment management and conservation practices driven by cutting edge science.

This global natural capital and the Himalayan 'water tower' have been facing multiple drivers of change (see Chettri and Sharma 2016; Chap. 2). Traditional drivers such as deforestation, habitat loss, habitat degradation and fragmentation, overharvesting of biological resources, illegal hunting and poaching, monoculture plantation, agricultural intensification and loss of genetic resources, shifting cultivation with insufficient fallow period, human-wildlife conflict and livestock and crop depredation, invasive alien species, and atmospheric pollution are often recognized as having an adverse impact on conservation and sustainable use of biodiversity and ecosystem services (Aryal and Kerkhoff 2008; Zomer and Oli 2011; Chettri and Sharma 2016; ICIMOD, WCD, GBPNIHESD, RECAST 2017). However, the effects of indirect drivers related to socioeconomic and sociocultural factors, urbanization, poverty, and poor governance and weak institutional frameworks are poorly understood and managed. The HKH has become a centre of rapid and unplanned development through mining operations, urbanization, haphazard construction of dams for hydropower, and poorly planned and constructed roads resulting in changes in this vulnerable and fragile region (Pandit and Grumbine 2012; Pandit et al. 2014). Though at varied scales, forest degradation has been continuous across the HKH except Bhutan (Pandit et al. 2007; Uddin et al. 2015; Chakraborty et al. 2016). The impacts of climate change add to the severity of the problems in the area and have been prominently featured as one of the drivers (Xu et al. 2009a; Shrestha et al. 2012), and alien and invasive species have visibly altered the overall composition of the ecological variables (Kohli et al. 2004; Bhattarai et al. 2014). The direction and trends predicted for climate change and bioclimatic conditions generally indicate accelerated change and major disruption for the region (Shrestha et al. 2012). Rising temperatures, changes in precipitation patterns, and reduction in the volume of glaciers (Xu et al. 2009a; Panday et al. 2014) create a host of cascading effects and will have a major impact on ecosystems, biodiversity, and livelihoods throughout the region (Immerzeel et al. 2013; Zomer et al. 2014, 2015, 2016). The changes have resulted in a decrease in the resilience capacity of the natural systems which is impacting human wellbeing.

There is increasing evidence of the impacts of climate change on biodiversity and ecosystems. Instances of changes in phenology (Ranjitkar et al. 2013; Hart et al. 2014) and shifts of species towards higher elevation (Valley 2003; Joshi et al. 2012; Brandt et al. 2013b; Telwala et al. 2013) have been reported. Such changes are likely to be profound with extraordinary levels of biotic perturbation (Shrestha et al. 2012). Overall, the ability of species to respond to climate change will largely depend on their ability to 'track' the shifting climate by colonizing new territory or to modify their physiology and seasonal behaviour (such as periods of flowering or mating) to adapt to the changed conditions (Thuiller et al. 2008). The efficiency of species' responses under climate change is likely to be highly idiosyncratic and difficult to predict (La Sorte and Jetz 2010). The heterogeneity of the mountain terrain provides both biological refugia and natural dispersal corridors, but can also present a variety of challenges to the migration of species. Niches for shifting mountain species along elevational gradients decrease in size with increasing elevation, or disappear at the mountains' top (Körner 2007). Likewise, rapid changes in seasonal variations, such as the timing and length of the growing season, or warmer winter temperatures, perturb ecosystem functioning, disrupting finely tuned pollinator interactions (as when cycles between the insect and the plant it specializes on become unsynchronized), affecting emergence or migration of either predator or prey species, or allowing for the overwintering and survival of pests and pathogens.

Many of these impacts can be expected to manifest before 2050 (Mora et al. 2015; Zomer et al. 2016). The cascading effects will also impact agricultural and pastoral systems (Maikhuri et al. 2001). Agricultural systems, mountain communities, and mountain livelihoods are susceptible and will be profoundly affected. Local communities, highly dependent upon ES, may be able to adapt through the expansion of cropping systems into new areas, the introduction of new varieties or new technologies, or the modification of existing production practices, and by relying on traditional ecological knowledge for coping with variability and maintaining socio-ecological resilience. The highly diverse and environmentally fine-tuned agrobiodiversity of this region may both provide options and be threatened, including the many genetic lines and landraces of various important food crops and livestock breeds found in the HKH. Although conditions may possibly improve for production in places (e.g., warmer and wetter), erratic and highly variable patterns of rainfall, increases in extreme events, occurrence of drought, or changes in the intensity and duration of the monsoon will create major adaptation challenges (Ramesh and Goswami 2007).

The magnitude and speed of these bioclimatic changes are likely to have an impact on the conservation effectiveness of the many PAs and other conservation efforts within the HKH (La Sorte and Jetz 2010; Zomer et al. 2015). For example, ecological conditions within PAs may change beyond limits conducive for the species currently found there as species ranges shift. The ability to survive, adapt to, or benefit from the changes is species- and site-specific and depends on factors such as population dynamics, seed dispersal mechanisms, habitat availability and/or fragmentation, and physiological adaptability (Corlett and Westcott 2013). Improving our understanding of the responses of the species found in the HKH is imperative if conservation strategies and policies designed to meet these challenges are to be effective. This is equally true for maintaining agricultural production for food security and contributing to sustainable development in the HKH, particularly concerning the traditional mountain agricultural systems found in the region, which are generally highly adapted to specific climatic niches within the highly heterogeneous mountainous terrain.

The challenges could be translated into opportunities through development and implementation of appropriate strategies. The core issue identified from the analysis so far indicates that transformative changes have mainly been driven by either a constantly changing climate or land use change to enhance production systems to address resource crises due to the increasing demand of a growing population. The effectiveness of governance systems and implementation of evolving policies also needs special attention. The following broad strategic direction could be useful for long-term resolution.

5.5.1 An Integrated Approach—The Landscape as a Socio-Ecological System

Humans, unlike any other multicellular species in Earth's history, have emerged as a global force that is transforming the ecology of the entire planet. It is no longer possible to understand, predict, or successfully manage ecological patterns, processes, or change without understanding why and how humans reshape these over the long term. To investigate, understand, and address the ultimate causes of anthropogenic ecological change, not just the consequences, human sociocultural processes must become as much a part of ecological theory and practice as biological and geophysical processes are now. The 240 million people living in the HKH are strongly linked to the ecosystem health of the region that ensures the continuous flow of services for their subsistence livelihood (Xu et al. 2009a). This intricate linkage between ecosystem and human wellbeing has been shaping the ecosystems and helping in supporting diversity (Gorenflo et al. 2012). More important, this diversity, which also addresses poverty, needs socio-ecological understanding (Gerlitz et al. 2012).

The state of biodiversity and ecosystem services in the HKH is rapidly changing due not only to increasing synergistic effects of anthropogenic and natural drivers but also to weak governance and institutions' limited capacity to cope with such changes. An integrated, gender- and socially-inclusive, and national and inter-regional enabling policy approach can help in reforming governance policies and institutional and legal frameworks, and promote conservation and sustainable use of biodiversity and ecosystem services. The diverse ecosystems support tourism, agricultural production, water security, and clean energy development downstream. To sustain these services, and to ensure food, water, and energy security in the HKH, management of ecosystems including forests, wetlands, and rangelands is crucial (Rasul 2014). With increasing demand for and scarcity of land, water, energy, and natural resources for competing uses, the challenge is to minimize trade-offs and maximize synergies. Though the region has long witnessed a sectorial approach for conservation and development, it is essential to address the complexity of the HKH ecosystem and people's dependency through a cross-sector coordinated approach. Mountain tourism is one example where biodiversity conservation, cultural preservation, socioeconomic development, and environmental aspects could be better linked and coordinated (Nepal 2013). The socio-ecological systems are best dealt with by using a landscape approach (Sayer 2009). Strongly grounded in transdisciplinarity, the landscape approach has the potential to maximize synergy and secure integrated actions by multiple stakeholders (MoFSC 2016; ICIMOD, WCD, GBPNIHESD, RECAST 2017). The principles of the landscape approach, including transdisciplinarity, have been widely shared and applied across the countries of the HKH with varying degrees of success (Chettri and Sharma 2016). These principles have also been endorsed by the intergovernmental process dealing with biodiversity conservation and climate change mitigation and adaptation (Sayer et al. 2013). This approach could be used at different scales considering the linkages for raising the adaptive capacity of people and resilience of ecosystems. The river basin, landscape, and ecosystem approaches practised in the region could be further strengthened for better synergy and coordination (Rasul 2014; Chettri and Sharma 2016).

5.5.2 Building Knowledge—Science in Support of Decision-Making

Science-based knowledge development in the HKH is currently undergoing a remarkable transformation driven by (1) new technologies for linking, producing and processing ecological information; (2) increasing economic interest in natural resources from the mountains; (3) the rising prominence of markets, even in remote villages; (4) greater awareness of ecological crises; and (5) efforts to expand the participation of communities in ecological governance (Oli et al. 2013; Molden et al. 2014b). There has been considerable work on the production of ecological knowledge—local knowledge, indigenous knowledge, gendered knowledge, and ethno-ecological knowledge (Khadka and Verma 2012; Uprety et al. 2016). This work has identified issues of competing knowledge, access, and representation of different social groups. There has also been considerable discussion about how to integrate local knowledge, scientific knowledge, and decision sciences, especially state policies. The last decades in the HKH have seen the emergence of new technologies (information technology, geo-spatial tools, and participatory approaches), new actors (youths, academic institution), new values, new institutions (civil society, international development partners), and new territories that have come to play an integral role in ecological decision-making and participation. There is an increasing imperative to explore how diverse ecological knowledge can enter into evolving governance practice. The key question is, how does this knowledge transformation influence the varying capacities of these actors to shape their worlds through these valuations and understanding, and through ecological governance in the HKH and beyond? There is an increasing trend to bring together scientific scholars, graduate students, and community-oriented partners to mobilize ecological knowledge in the HKH. For example, the National Natural Science Foundation of China has launched a regular funding mechanism for scientific research in ecology on the edge of the Tibetan Plateau and Himalaya, and the National Mission for Sustaining the Himalayan Ecosystem (NMSHE) and National Mission for Himalayan Studies (NMHS) in India have been launched to broaden the scientific knowledge base in the Indian Himalayan Region to inform policy and practice (GoI 2016, 2017).

5.5.3 Regional Cooperation for Regional Challenges and Opportunities

The ecosystems of the HKH are diverse and the distribution among countries heterogeneous, but the majority of them are contiguous across borders (Xu et al. 2009a) and many of the globally significant ecoregions are shared by different countries (Olson and Dinerstein 2002). Even the environmental strata are contiguous across countries providing upstream and downstream linkages (Zomer et al. 2016). Many iconic species also have a wider habitat and range across different countries (Dorji et al. 2012; Forrest et al. 2012; Kandel et al. 2015). Many of the HKH countries have also developed a strong interdependence in trade, culture, and tourism (Chettri 2011). Traditional barter systems are still prevalent in remote areas of Nepal and China (Chaudhary et al. 2015a, b). There are also instances of wildlife trade and human-wildlife conflict across borders (Rao et al. 2011; Acharya et al. 2016), while numerous PAs are also transboundary in nature (IUCN 2005; Chettri et al. 2008), and many conservation issues demand regional cooperation. Pereira et al. (2010) have indicated a scenario in which biodiversity will continue to decline over the 21st century; however, the range of projected change is much broader than most studies suggest, partly because there are major opportunities to intervene through better policies, but also because of large uncertainties in projections. This strongly suggests the need to fill the gaps in research-based knowledge, and develop management and communication strategies focusing on multiple sources and approaches that lead to policy intervention. To translate this range of conservation and development challenges into opportunities for sustainable development, regional cooperation among the countries sharing such critical ecosystems will be essential.

5.5.4 National and Global Investment—Securing Future of Biodiversity and Ecosystem Services

Investment in conservation is a wise trade-off for sustaining the continuous flow of ES, enhancing ecosystem resilience, and ensuring a bright future for coming generations. However, the investment is not happening where it is most needed (Wilson et al. 2016). The HKH hosts four of the world's 36 biodiversity hotspots (Mittermeier et al. 2011) and is vulnerable to various drivers of change including climate change. However, the investment in conservation and management has faced disparity and been given low priority even by developing countries (Watson et al. 2016). It was observed that China and Nepal are among the top in investing on gross domestic expenditure on research and development (Katsnelson 2016). It is high time for conservation communities to plan investment based on the priority areas where biodiversity is declining at an accelerated rate (Waldron et al. 2013). Though the conservation investment by HKH countries is at different stages due to varying priorities and resource availability, the countries need to set aside more investment as a trade-off for sustainable development goals.

5.6 Conclusion

The HKH region is the source of ten major river systems and includes all or parts of four global biodiversity hotspots. The rich biodiversity and diverse ecosystems play a critical role in sustaining the wellbeing of the 240 million people of the region, and the goods and services from the mountain ecosystem are estimated to benefit a further 1.7 billion people in the downstream areas. The diverse ecosystems provide services with four values: social—for public benefit, cultural—for aesthetic and communal significance, ecological—for environmental conservation and sustainability, and economic—for livelihoods through production of goods and services. However, these services are poised for major changes in the current scenario of threats, chiefly as a result of climate change; local, regional and global market forces; and the socio-political environment prevalent in individual countries. There are other factors like cross-sectoral policies and strategies that will have a potential impact on ecosystems, while the lack of interdisciplinary understanding and knowledge, governance systems, consumption patterns of a large and growing urban population, and others have further impacted biodiversity and the functions and flow of services from ecosystems. Equally, these threats and their drivers have provide the stakeholders of the HKH with an opportunity for closer regional co-operation at all levels, especially for sharing knowledge, practice, and experience to develop robust strategies for managing the socio-ecological systems that are dependent on the sustained flow of ecosystem services. There has been an unprecedented effort made towards finding innovations and going beyond conventional approaches in managing these life-supporting systems (e.g., from a species to a landscape and ecosystem approach). Yet much remains to be done to reach a stage where we can claim that there is adequate resilience in the ecosystems and the communities to withstand the threats brought about by large local, regional, and global changes.

Regional co-operation needs further strengthening at the government, civil society, private sector, and community levels. Policies and their implementation have yet to reflect the integration that is required across different sectors to address some of the challenges mentioned above. Decision makers at all levels need to be empowered and equipped with information and knowledge that is holistic, interdisciplinary, and exemplified with best practices from elsewhere. Further, there need to be large-scale studies and research to enhance knowledge and information about the four values that biodiversity and ecosystem services provide for informed decision making.

One of the critical factors for achieving some of the aspirations mentioned above is to have more investment in the HKH region from donors, governments, and the private sector to ensure the sustainability of the assets while pursuing the larger goals of poverty alleviation, economic development, and overall human wellbeing.

References

Aase, T., Chaudhary, R. P., & Vetaas, O. R. (2009). Farming flexibility and food security under climatic security: Manang, Nepal Himalaya. *Area*, 1–9. https://doi.org/10.1111/j.1475-4762.2009.00911.x.

Aase, T. H., Chapagain, P. S., & Tiwari, P. C. (2013). Innovation as an expression of adaptive capacity to change in Himalayan farming. *Mountain Research and Development, 33*(1), 4–10.

Abler, D. (2004). Multifunctionality, agricultural policy, and environmental policy. *Agricultural and Resource Economics Review, 33*, 8–17.

Acharya. (2015). *A walk to a zero poaching of Rhino*. Department of National Park and Wildlife Conservation, Ministry of Forest and Soil Conservation, Government of Nepal.

Acharya, B. K., & Sharma, G. (2012). The traditional Dzumsa system and their role in resource management in cultural landscape in North Sikkim. *Cultural Landscape*, 175.

Acharya, B. K., Sanders, N. J., Vijayan, L., & Chettri, B. (2011). Elevational gradients in bird diversity in the Eastern Himalaya: An evaluation of distribution patterns and their underlying mechanisms. *PLoS ONE, 6*(12), e29097. https://doi.org/10.1371/journal.pone.0029097.

Acharya, K. P., Paudel, P. K., Neupane, P. R., & Köhl, M. (2016). Human-wildlife conflicts in Nepal: Patterns of human fatalities and injuries caused by large mammals. *PLoS ONE, 11*(9), e0161717.

Ahmad, A., & Nizami, S. M. (2015). Carbon stocks of different land uses in the Kumrat valley, Hindu Kush region of Pakistan. *Journal of Forestry Research, 26*(1), 57–64.

Ale, S. B., Shrestha, B., & Jackson, R. (2014). On the status of snow leopard *Panthera uncia* (Schreber, 1775) in Annapurna, Nepal. *Journal of Threatened Taxa, 6*(3), 5534–5543. http://dx.doi.org/10.11609/JoTT.o3635.5534-43.

Alexander, J. S., Zhang, C., Shi, K., & Riordan, P. (2016). A granular view of a snow leopard population using camera traps in Central China. *Biological Conservation, 197*, 27–31.

Ali, T., Ahmad, M., Shahbaz, B., & Suleri, A. (2007). Impact of participatory forest management on vulnerability and livelihood assets of forest-dependent communities in Northern Pakistan. *The International Journal of Sustainable Development & World Ecology, 14*(2), 211–223.

Allen, D. J., Molur, S., & Daniel, B. A. (2010). *The status and distribution of freshwater biodiversity in the Eastern Himalaya*. IUCN.

Amin, R., Thomas, K., Emslie, R. H., Foose, T. J., & Van Strien, N. (2006). An overview of the conservation status of and threats to rhinoceros species in the wild. *International Zoo Yearbook, 40*, 96–117.

Anand, A., Chandan, P., & Singh, R. B. (2012). Homestays at Korzok: Supplementing rural livelihoods and supporting green tourism in the Indian Himalayas. *Mountain Research and Development, 32*(2), 126–136.

ANBSAP. (2013). *Afghanistan's national biodiversity strategy and action plan*. Government of the Islamic Republic of Afghanistan.

Anderson, D. M., Salick, J., Moseley, R. K., & Xiaokun, O. (2005). Conserving the sacred medicine mountains: A vegetation analysis of Tibetan sacred sites in Northwest Yunnan. *Biodiversity and Conservation, 14*(13), 3065–3091.

Anthwal, A., Gupta, N., Sharma, A., Anthwal, S., & Kim, K. H. (2010). Conserving biodiversity through traditional beliefs in sacred groves in Uttarakhand Himalaya, India. *Resources, Conservation and Recycling, 54*(11), 962–971.

Arora, V. (2006). The forest of symbols embodied in the Tholung sacred landscape of North Sikkim. *India. Conservation and Society, 4*(1), 55.

Aryal, K. P., & Kerkhoff., E. E. (2008).*The right to practice shifting cultivation as a traditional occupation in Nepal: A case study to apply ILO Conventions Nos. 111 (Employment and Occupation) and 169 (Indigenous and Tribal Peoples)*. Kathmandu: International Labour Office, 2008. ISBN 978-92-2-121452-6.

Awasthi, A., Uniyal, S. K., Rawat, G. S., & Rajvanshi, A. (2003). Forest resource availability and its use by the migratory villages of Uttarkashi, Garhwal Himalaya (India). *Forest Ecology and Management, 174*(1), 13–24.

Badola, R., Hussain, S. A., Mishra, B. K., Konthoujam, B., Thapliyal, S., & Dhakate, P. M. (2010). An assessment of ecosystem services of Corbett Tiger Reserve, India. *The Environmentalist, 30*(4), 320–329.

Badola, R., Hussain, S. A., Dobriyal, P., & Barthwal, S. (2014). *An integrated approach to reduce the vulnerability of local communities to environmental degradation in Western Himalayas*. India: Study Report, Wildlife Institute of India.

Bajracharya, B., Uddin, K., Chettri, N., Shrestha, B., & Siddiqui, S. A. (2010). Understanding land cover change using a harmonized classification system in the Himalaya: A case study from Sagarmatha National Park. *Nepal. Mountain Research and Development, 30*(2), 143–156.

Bajracharya, S. B., Chaudhary, R. P., & Basnet, G. (2015). Biodiversity conservation and protected area management in Nepal. *Nepal—An introduction to the natural history, ecology and human environment in the Himalayas: A companion to the Flora of Nepal* (pp. 473–486). Edinburgh, UK: Royal Botanic Garden.

Banskota, K., Karky, B., & Skutsch, M. (2007). Reducing carbon emissions through community-managed forests in the Himalaya. International Centre for Integrated Mountain Development (ICIMOD).

Barlow, M., Wheeler, M., Lyon, B., & Cullen, H. (2005). Modulation of daily precipitation over Southwest Asia by the Madden-Julian oscillation. *Monthly Weather Review, 133*(12), 3579–3594.

Basnet, G., & Chaudhary, R. P. (2017). Indigenous system of pastureland management: A case of Limi in the Kailash sacred landscape, Nepal. In M. Karki, R. Hill, W. Alangui, K. Ichikawa, & P. Bridgewater (Eds.), *Knowing our lands and resources* (pp. 86–93). Paris: United Nations Educational Scientific and Cultural Organization (UNESCO).

Baumgärtner, S., & Quaas, M. (2010). What is sustainability economics? *Ecological Economics, 69*(3), 445–450.

Bawa, K. S. (2006). Globally dispersed local challenges in conservation biology. *Conservation Biology, 20*(3), 696–699.

Bawa, K. S., & Seidler, R. (2015). Deforestation and sustainable mixed-use landscapes: A view from the Eastern Himalaya. *Annals of the Missouri Botanical Garden, 100*(3), 141–149.

Bennett, E. M., Peterson, G. D., & Gordon, L. J. (2009). Understanding relationships among multiple ecosystem services. *Ecology Letters, 12*(12), 1394–1404.

Bhardwaj, M., Uniyal, V. P., Sanyal, A. K., & Singh, A. P. (2012). Butterfly communities along an elevational gradient in the Tons valley, Western Himalayas: Implications of rapid assessment for insect conservation. *Journal of Asia-Pacific Entomology, 15*(2), 207–217.

Bhat, J. A., Iqbal, K., Kumar, M., Negi, A. K., & Todaria, N. P. (2013). Carbon stock of trees along an elevational gradient in temperate forests of Kedarnath Wildlife Sanctuary. *Forest Science and Practice, 15*(2), 137–143.

Bhattarai, K. R., Måren, I. E., & Subedi, S. C. (2014). Biodiversity and invasibility: Distribution patterns of invasive plant species in the Himalayas. *Nepal. Journal of Mountain Science, 11*(3), 688–696.

Birch, J. C., Thapa, I., Balmford, A., Bradbury, R. B., Brown, C., Butchart, S. H., et al. (2014). What benefits do community forests provide, and to whom? A rapid assessment of ecosystem services from a Himalayan forest, Nepal. *Ecosystem Services, 8*, 118–127.

Blaikie, P. M., & Muldavin, J. S. (2004). Upstream, downstream, China, India: the politics of environment in the Himalayan region. *Annals of the Association of American Geographers, 94*(3), 520–548.

Brandt, S. A. (2000). Classification of geomorphological effects downstream of dams. *CATENA, 40*(4), 375–401.

Brandt, J. S., Wood, E. M., Pidgeon, A. M., Han, L. X., Fang, Z., & Radeloff, V. C. (2013a). Sacred forests are keystone structures for forest bird conservation in southwest China's Himalayan Mountains. *Biological Conservation, 166*, 34–42.

Brandt, J. S., Haynes, M. A., Kuemmerle, T., Waller, D. M., & Radeloff, V. C. (2013b). Regime shift on the roof of the world: Alpine meadows converting to shrublands in the Southern Himalayas. *Biological Conservation, 158*, 116–127.

Brooks, T. M., Mittermeier, R. A., da Fonseca, G. A., Gerlach, J., Hoffmann, M., Lamoreux, J. F., et al. (2006). Global biodiversity conservation priorities. *Science, 313*(5783), 58–61.

Brush, S. B. (1995). In situ conservation of landraces in centers of crop diversity. *Crop Science, 35*(2), 346–354.

Bubela, T., & Gold, E. R. (2012). *Genetic resources and traditional knowledge: Case studies and conflicting interests*. Cheltenham: Edward Elgar Publishing.

Cardinale, B. J., Palmer, M. A., & Collins, S. L. (2002). Species diversity enhances ecosystem functioning through interspecific facilitation. *Nature, 415*(6870), 426–429.

CBS. (2014). *Environmental Statistics of Nepal—2013*. Kathmandu: Central Bureau of Statistics.

CCD. (2014). *Fifth national report—Pakistan*. Government of Pakistan: Climate Change Division.

Chakraborty, A., Sachdeva, K., & Joshi, P. K. (2016). Mapping long-term land use and land cover change in the central Himalayan region using a tree-based ensemble classification approach. *Applied Geography, 74*, 136–150.

Chan, K. M., Guerry, A. D., Balvanera, P., Klain, S., Satterfield, T., Basurto, X., et al. (2012). Where are cultural and social in ecosystem services? A framework for constructive engagement. *BioScience*, *62*(8), 744–756.

Chandola, S. (2012). *An assessment of human wildlife interaction in the Indus Valley, Ladakh, Trans-Himalaya*. Doctoral dissertation, Ph. D. thesis submitted to Saurashtra University, Rajkot (Unpublished).

Chaudhary, R. P., Uprety, Y., Joshi, S. P., Shrestha, K. K., Basnet, K. B., Basnet, G., et al. (2015a). *Kangchenjunga Landscape Nepal: From conservation and development perspectives*. Kathmandu, Nepal: Ministry of Forests and Soil Conservation (MoFSC), Government of Nepal; Research Centre for Applied Science and Technology (RECAST), Tribhuvan University; and International Centre for Integrated Mountain Development (ICIMOD).

Chaudhary, S., McGregor, A., Houston, D., & Chettri, N. (2015b). The evolution of ecosystem services: A time series and discourse-centered analysis. *Environmental Science & Policy, 54*, 25–34.

Chaudhary, R. P., Uprety, Y., and Rimal, S. K. (2016a). Deforestation in Nepal: Causes, consequences and responses. In R. Sivanpillai & J. F. Shroder (Eds.), *Biological and environmental hazards and disasters* (pp. 335–372). Elsevier Inc.

Chaudhary, S, Chettri N, Uddin, K, Khatri TB, Dhakal, M, Bajracharya, B., et al. (2016). Implications of land cover change on ecosystem services and people's dependency. A case study from the Koshi Tappu Wildlife Reserve, Nepal. *Ecological Complexity*. https://doi.org/10.1016/j.ecocom.2016.04.002.

Chaudhary, R. P., Bhattarai, S. H., Basnet, G., Bhatta, K. P., Uprety, Y., Bhatta, L. D., et al. (2017). Traditional practice and knowledge of indigenous and local communities in Kailash Sacred Landscape, Nepal. *ICIMOD Working Paper 2017/1*. Kathmandu: ICIMOD.

Chen, F. H., Dong, G. H., Zhang, D. J., Liu, X. Y., Jia, X., An, C. B., et al. (2015a). Agriculture facilitated permanent human occupation of the Tibetan Plateau after 3600 B.P. *Science 345*(6219), 248.

Chen, Y., Wang, K., Lin, Y., Shi, W., Song, Y., & He, X. (2015b). Balancing green and grain trade. *Nature Geoscience, 8*(10), 739.

Chettri N. (2011). Role of actors and institutions in cross border tourism development. In *Proceedings on integrated tourism concepts to contribute to sustainable development in mountain regions*, 15–22 June 2009 (pp. 154–170). Kathmandu & Jomsom, Nepal: ICIMOD and InWent.

Chettri, N., & Sharma, E. (2016). Reconciling the mountain biodiversity conservation and human wellbeing: Drivers of biodiversity loss and new approaches in the Hindu-Kush Himalayas. *Proceedings of the Indian National Science Academy, 82*, 53–73.

Chettri, N., Shakya, B., Thapa, R., & Sharma, E. (2008). Status of protected area system in the Hindu Kush Himalaya: an analysis of PA coverage. *International Journal of Biodiversity Science and Management., 4*(3), 164–178.

Chettri, N., Sharma, E., Shakya, B., Thapa, R., Bajracharya, B., Uddin, K., et al. (2010) *Biodiversity in the Eastern Himalayas: Status, trends and vulnerability to climate change*. Climate change impact and vulnerability in the Eastern Himalayas—Technical report 2 (p. 23). Kathmandu: ICIMOD.

Chettri, N., Sharma, E., Shrestha, A. B., Zhoali, Y., Hua, Q., & Bajracharya, B. (2012). Real world protection for the third pole and its people. In F. Huettmann (Ed.), *Protection of Polar Regions* (pp. 113–133). Japan: Springer.

Choudhary, A. U. (2001). An overview of the status and conservation of the red panda in India, with reference to its global status. *Oryx, 35*, 250–259.

Chowdhury, M. S. H., & Koike, M. (2010). An overview on the protected area system for forest conservation in Bangladesh. *Journal of Forestry research, 21*(1), 111–118.

Conservation International. (2016). *Biodiversity hotspots*. Retreived 22 September 2016.

Corlett, R. T., & Westcott, D. A. (2013). Will plant movements keep up with climate change? *Trends in Ecology & Evolution, 28*(8), 482–488.

Costanza, R., Howarth, R. B., Kubiszewski, I., Liu, S., Ma, C., Plumecocq, G., et al. (2016). Influential publications in ecological economics revisited. *Ecological Economics, 123,* 68–76.

Creed, I. F., Weber, M., Accatino, F., & Kreutzweiser, D. P. (2016). Managing forests for water in the Anthropocene—The best kept secret services of forest ecosystems. *Forests, 7*(3), 60.

Dangi, M. B., Chaudhary, R. P., Rijal, K., Stahl, P. D., Belbase, S., Gerow, K. G., et al. (2017). *Impacts of environmental change on agroecosystems and livelihoods in Annapurna Conservation Area.* Nepal: Environmental Development (In press).

Daniel, T. C., Muhar, A., Arnberger, A., Aznar, O., Boyd, J. W., Chan, K. M., et al. (2012). Contributions of cultural services to the ecosystem services agenda. *Proceedings of the National Academy of Sciences, 109*(23), 8812–8819.

de Groot, R. S., Alkemade, R., Braat, L., Hein, L., & Willemen, L. (2010). Challenges in integrating the concept of ecosystem services and values in landscape planning, management and decision making. *Ecological Complexity, 7*(3), 260–272.

Desai, B. H., Oli, K. P., Yang, Y., Chettri, N., & Sharma, E. (2011). *Implementation of convention on biological diversity: A retrospective analysis in the Hindu Kush Himalayan Countries* (p. 33). Kathmandu, Nepal: ICIMOD.

Deterra, H. (1937). Cenozoic cycles in Asia and their bearing on human prehistory. *Proceedings of the American Philosophical Society,* 289–308.

Díaz, S., Demissew, S., Carabias, J., Joly, C., Lonsdale, M., Ash, N., et al. (2015). The IPBES conceptual framework—Connecting nature and people. *Current Opinion in Environmental Sustainability, 14,* 1–16.

Dinerstein, E., & Price, L. (1991). Demography and habitat use by greater one-horned rhinoceros in Nepal. *The Journal of Wildlife Management,* 401–411.

DoE. (2015). *Fifth national report of Bangladesh to the convention on biological diversity.* Ministry of Environment and Forests, Government of the Peoples Republic of Bangladesh, Dhaka.

Dong, S., Wen, L., Zhu, L., & Li, X. (2010). Implication of coupled natural and human systems in sustainable rangeland ecosystem management in HKH region. *Frontiers of Earth Science in China, 4* (1), 42–50.

Dorji, S., Rajaratnam, R., & Vernes, K. (2012). The Vulnerable red panda Ailurus fulgens in Bhutan: Distribution, conservation status and management recommendations. *Oryx, 46*(04), 536–543.

Dudley, N., Higgins-Zogib, L. I. Z. A., & Mansourian, S. (2009). The links between protected areas, faiths, and sacred natural sites. *Conservation Biology, 23*(3), 568–577.

Dunn, J. C., Buchanan, G. M., Stein, R. W., Whittingham, M. J., & McGowan, P. J. (2016). Optimising different types of biodiversity coverage of protected areas with a case study using Himalayan Galliformes. *Biological Conservation, 196,* 22–30.

Ebi, K. L., Woodruff, R., von Hildebrand, A., & Corvalan, C. (2007). Climate change-related health impacts in the Hindu Kush-Himalayas. *EcoHealth, 4*(3), 264–270.

Elalem, S., & Pal, I. (2015). Mapping the vulnerability hotspots over Hindu-Kush Himalaya region to flooding disasters. *Weather and Climate Extremes, 8,* 46–58.

Ellis, E. C. (2015). Ecology in an anthropogenic biosphere. *Ecological Monographs, 85*(3), 287–331.

Esquinas-Alcázar, J. (2005). Protecting crop genetic diversity for food security: political, ethical and technical challenges. *Nature Reviews Genetics, 6*(12), 946–953.

Feng X., Fu B., Piao S., Wang S., Ciais P., Zeng Z., et al. (2016). Revevetation in China's Loss Plateau is approaching sustainable water resource limits. *Nature Climate Change.* https://doi.org/10.1038/nclimate3092.

Foggin, M. (2012). Pastoralists and wildlife conservation in western China: Collaborative management within protected areas on the Tibetan Plateau. *Pastoralism: Research, Policy and Practice, 2*(1), 1.

Forrest, J., Wikramanayake, E., Shrestha, R., Areendran, G., Gyeltshen, K., Maheshwari, A., et al. (2012). Conservation and climate change: assessing the vulnerability of snow leopard habitat to tree line shift in the Himalaya. *Biological Conservation* (150), 129–135.

Frankham, R. (2005). Genetics and extinction. *Biological Conservation, 126* (2), 131–140.

Gagic, V., Bartomeus, I., Jonsson, T., Taylor, A., Winqvist, C., Fischer, C., et al. (2015). Functional identity and diversity of animals predict ecosystem functioning better than species-based indices. *Proceedings of the Royal Society of London B: Biological Sciences, 282* (1801), 20142620.

Gao, H., Ouyang, Z., Chen, S., & van Koppen, C. S. A. (2013). Role of culturally protected forests in biodiversity conservation in Southeast China. *Biodiversity and Conservation, 22*(2), 531–544.

Gerlitz, J. Y., Hunzai, K., & Hoermann, B. (2012). Mountain poverty in the Hindu-Kush Himalayas. *Canadian Journal of Development Studies/Revue canadienne d'études du développement, 33*(2), 250–265.

GoI. (2008). *National action plan on climate change.* Government of India, New Delhi.

GoI. (2016). *National mission for sustaining the Himalayan ecosystem.* Government of India, New Delhi.

GoI. (2017). *National mission on Himalayan studies.* Government of India, New Delhi.

Goldewijk, K. K., Beusen, A., & Janssen, P. (2010). Long-term dynamic modeling of global population and built-up area in a spatially explicit way: HYDE 3.1. *The Holocene.* 1–9.

Goldewijk, K. K., Beusen, A., Van Drecht, G., & De Vos, M. (2011). The HYDE 3.1 spatially explicit database of human-induced global land-use change over the past 12,000 years. *Global Ecology and Biogeography, 20*(1), 73–86.

Gorenflo, L. J., Romaine, S., Mittermeier, R. A., & Walker-Painemilla, K. (2012). Co-occurrence of linguistic and biological diversity in biodiversity hotspots and high biodiversity wilderness areas. *Proceedings of the National Academy of Sciences, 109*(21), 8032–8037.

Grajal, A. (1999). Biodiversity and the nation state: Regulating access to genetic resources limits biodiversity research in developing countries. *Conservation Biology, 13*(1), 6–10.

Grêt-Regamey, A., Walz, A., & Bebi, P. (2008). Valuing ecosystem services for sustainable landscape planning in Alpine regions. *Mountain Research and Development, 28*(2), 156–165.

Grêt-Regamey, A., Brunner, S. H., & Kienast, F. (2012). Mountain ecosystem services: Who cares? *Mountain Research and Development, 32*(S1), S23–S34.

Groves, C. (2011). The taxonomy and phylogeny of Ailurus. In A. R. Glatston (Ed.). *Red panda: Biology and conservation of the first panda* (pp. 101–124). William Andrew.

Grumbine, R. E., & Pandit, M. K. (2013). Threats from India's Himalaya dams. *Science, 339*(6115), 36–37.

Guangwei, C. (2002). *Biodiversity in the Eastern Himalayas: Conservation through dialogue.* Summary report of workshops on biodiversity conservation in the Hindu Kush-Himalayan ecoregion. International Centre for Integrated Mountain Development (ICIMOD).

Gupta, S. R., Upadhaya, M. P., & Katsumoto, T. (1996). *Status of plant genetic resources in Nepal.* Paper presented at the 19th summer crops workshop, held at National Rice Research Programme, Parwanipur, Bara, Nepal, 27–29 February 1995. Parwanipur, Nepal: National Rice Research Programme.

Gurung, H. (2004a). *Mountain reflections, patterns and development*. Kathmandu: Mandala Publications.

Gurung, P. C. (2004b). Terai arc landscape: A new paradigm in conservation and sustainable development. In *Managing mountain protected areas: Challenges and responses for the 21st century* (pp. 80–86). Andromeda, Colledara, Italy.

Gurung, D. B., & Seeland, K. (2008). Ecotourism in Bhutan: Extending its benefits to rural communities. *Annals of Tourism Research, 35* (2), 489–508.

Gurung, A., Bista, R., Karki, R., Shrestha, S., Uprety, D., & Oh, S. E. (2013). Community-based forest management and its role in improving forest conditions in Nepal. *Small-scale Forestry, 12*(3), 377–388.

Haase, D., Schwarz, N., Strohbach, M., Kroll, F., & Seppelt, R. (2012). Synergies, trade-offs, and losses of ecosystem services in urban regions: an integrated multiscale framework applied to the Leipzig-Halle Region. *Germany. Ecology and Society, 17*(3), 22.

Hamilton, L. S. (2015). When the sacred encounters economic development in mountains. *George Wright Forum 32*(2), 132–140.

Hart, R., Salick, J., Ranjitkar, S., & Xu, J. (2014). Herbarium specimens show contrasting phenological responses to Himalayan climate. *Proceedings of the National Academy of Sciences, 111*(29), 10615–10619.

Hoffmann, M., Hilton-Taylor, C., Angulo, A., Böhm, M., Brooks, T. M., Butchart, S. H., et al. (2010). The impact of conservation on the status of the world's vertebrates. *Science, 330*(6010), 1503–1509.

Hoisington, D., Khairallah, M., Reeves, T., Ribaut, J. M., Skovmand, B., Taba, S., et al. (1999). Plant genetic resources: What can they contribute toward increased crop productivity?. *Proceedings of the National Academy of Sciences, 96*(11), 5937–5943.

Hoorn, C., Mosbrugger, V., Mulch, A., & Antonelli, A. (2013). Biodiversity from mountain building. *Nature Geoscience, 6*(3), 154–154.

Hua, Z. (2012). Biogeographical divergence of the Flora of Yunnan, Southwestern China initiated by the uplift of Himalaya and extrusion of Indochina block. *PLoS ONE, 7*(9), e45601. https://doi.org/10.1371/journal.pone.0045601.

Hudson, A. M., Olsen, J. W., Quade, J., Lei, G., Huth, T. E., & Zhang, H. (2016). A regional record of expanded Holocene wetlands and prehistoric human occupation from paleowetland deposits of the western Yarlung Tsangpo valley, southern Tibetan Plateau. *Quaternary Research*.

Hughes, C. E., & Atchison, G. W. (2015). The ubiquity of alpine plant radiations: From the Andes to the Hengduan Mountains. *New Phytologist, 207*(2), 275–282.

Hughes, A. R., Inouye, B. D., Johnson, M. T., Underwood, N., & Vellend, M. (2008). Ecological consequences of genetic diversity. *Ecology Letters, 11*(6), 609–623.

Hussain, A., Rasul, G., Mahapatra, B., & Tuladhar, S. (2016). Household food security in the face of climate change in the Hindu-Kush Himalayan region. *Food Security, 8*(5), 921–937.

ICIMOD, WCD, GBPNIHESD, RECAST. (2017). Kangchenjunga landscape conservation and development strategy and regional cooperation framework. *ICIMOD Working Paper 2017/2*. Kathmandu: ICIMOD.

Immerzeel, W. W., Van Beek, L. P., & Bierkens, M. F. (2010). Climate change will affect the Asian water towers. *Science, 328*(5984), 1382–1385.

Immerzeel, W. W., Pellicciotti, F., & Bierkens, M. F. P. (2013). Rising river flows throughout the twenty-first century in two Himalayan glacierized watersheds. *Nature Geoscience, 6*(9), 742–745.

IUCN. (2005). Benefits beyond boundaries. In *Proceeding of the Vth World Park Congress* (Ix+ 306 pp.). Gland, Switzerland: IUCN; UK: Cambridge.

IUCN. (2016). *The IUCN red list of threatened species*. Version 2016-2. www.iucnredlist.org. Retrieved on 15 October 2016.

Ives, J. D., & Messerli, B. (1989). *The Himalayan dilemma: Reconciling development and conservation*. Psychology Press.

Jain, A., Rai, S. C., & Sharma, E. (2000). Hydro-ecological analysis of a sacred lake watershed system in relation to land-use/cover change from Sikkim Himalaya. *CATENA, 40*(3), 263–278.

Janecka, J. E., Zhang, Y., Li, D., Munkhtsog, B., Bayaraa, M., Galsandorj, N., et al. (2017). Range-wide snow leopard phylogeography supports three subspecies. *Journal of Heredity*, esx044.

Jantz, S. M., Barker, B., Brooks, T. M., Chini, L. P., Huang, Q., Moore, R. M., et al. (2015). Future habitat loss and extinctions driven by land-use change in biodiversity hotspots under four scenarios of climate-change mitigation. *Conservation Biology, 29*(4), 1122–1131.

Joshi, G., & Negi, G. C. (2011). Quantification and valuation of forest ecosystem services in the Western Himalayan region of India. *International Journal of Biodiversity Science, Ecosystem Services & Management, 7*(1), 2–11.

Joshi, P. K., Rawat, A., Narula, S., & Sinha, V. (2012). Assessing impact of climate change on forest cover type shifts in Western Himalayan Eco-region. *Journal of Forestry Research, 23*(1), 75–80.

Kala, C. P. (2014). Deluge, disaster and development in Uttarakhand Himalayan region of India: Challenges and lessons for disaster management. *International Journal of Disaster Risk Reduction, 8*, 143–152.

Kalita, P., Tag, H., Sarma, H. N., & Das, A. K. (2014). Evaluation of nutritional potential of five unexplored wild edible food plants from Eastern Himalayan biodiversity hotspot region (India). *International Journal of Biological, Food, Veterinary Agricultural Engineering, 8*, 207–210.

Kandel, K., Huettmann, F., Suwal, M. K., Regmi, G. R., Nijman, V., Nekaris, K. A. I., et al. (2015). Rapid multi-nation distribution assessment of a charismatic conservation species using open access ensemble model GIS predictions: Red panda (*Ailurus fulgens*) in the Hindu-Kush Himalaya region. *Biological Conservation, 181*, 150–161.

Kandel, P., Gurung, J., Chettri, N., Ning, W., & Sharma, E. (2016). Biodiversity research trends and gap analysis from a transboundary landscape. *Eastern Himalayas. Journal of Asia-Pacific Biodiversity, 9*(1), 1–10.

Kapos, V., Rhind, J., Edwards, M., Ravilious, C., Price, M. F., & Butt, N. (2000). *Developing a map of the world's mountain forests, Forests in sustainable mountain development: A state of knowledge report for 2000*. Wallingford: Task Force on Forests in Sustainable Mountain Development; CABI Publishing. https://doi.org/10.1079/9780851994468.0004.

Karan, P. P. (2015). Land, Life, and Environmental Change in the Himalaya. In *Spatial diversity and dynamics in resources and urban development* (pp. 223–233). Netherlands: Springer.

Karmacharya, D. B., Thapa, K., Shrestha, R., Dhakal, M., & Janecka, J. E. (2011). Noninvasive genetic population survey of snow leopards (Panthera uncia) in Kangchenjunga conservation area, Shey Phoksundo National Park and surrounding buffer zones of Nepal. *BMC Research Notes, 4*(1), 1.

Kassam, K. A., Bulbulshoev, U., & Ruelle, M. (2011). Ecology of time: Calendar of the human body in the Pamir Mountains. *Journal of Persianate Studies, 4*(2), 146–170.

Katsnelson, A. (2016). Big science spenders. *Nature, 537*, S2.

Khadka, M., & Verma, R. (2012). *Gender and biodiversity management in the Greater Himalayas: Towards equitable mountain development*. International Centre for Integrated Mountain Development (ICIMOD).

Khumbongmayum, A. D., Khan, M. L., & Tripathi, R. S. (2005). Sacred groves of Manipur, northeast India: Biodiversity value, status and strategies for their conservation. *Biodiversity and Conservation, 14*(7), 1541–1582.

Klein, J. A., Harte, J., & Zhao, X. Q. (2004). Experimental warming causes large and rapid species loss, dampened by simulated grazing, on the Tibetan Plateau. *Ecology Letters, 7*(12), 1170–1179.

Kohli, R. K., Dogra, K. S., Batish, D. R., & Singh, H. P. (2004). Impact of invasive plants on the structure and composition of natural vegetation of northwestern Indian Himalayas. *Weed Technology, 18,* 1296–1300.

Körner, C. (2004). Mountain biodiversity, its causes and function. *Ambio*, 11–17.

Körner, C. (2007). Climatic treelines: Conventions, global patterns, causes. *Erdkunde, 61,* 316–324.

Körner, C., & Ohsawa, M. (2005). Mountain system. Ecosystems and human well being, current states and trends. In *Millennium ecosystem assessment.* Washington, DC: Island.

Kubiszewski, I., Costanza, R., Dorji, L., Thoennes, P., & Tshering, K. (2013). An initial estimate of the value of ecosystem services in Bhutan. *Ecosystem Services, 3,* e11–e21.

Kubiszewski, I., Anderson, S. J., Costanza, R., & Sutton, P. C. (2016). The future of ecosystem services in Asia and the Pacific. *Asia & the Pacific Policy Studies, 3*(3), 389–404.

La Sorte, F. A., & Jetz, W. (2010). Projected range contractions of montane biodiversity under global warming. *Proceedings of the* Royal *Society of London B: Biological Sciences*, rspb20100612.

Li, J., McCarthy, T. M., Wang, H., Weckworth, B. V., Schaller, G. B., Mishra, C., et al. (2016). Climate refugia of snow leopards in High Asia. *Biological Conservation*, *203*, 188–196.

Liu, Y. P., Wu, G. S., Yao, Y. G., Miao, Y. W., Luikart, G., Baig, M., et al. (2006). Multiple maternal origins of chickens: out of the Asian jungles. *Molecular Phylogenetics and Evolution*, *38*(1), 12–19.

Liu, D., Chen, Y., Cai, W., Dong, W., Xiao, J., Chen, J., et al. (2014). The contribution of China's grain to green program to carbon sequestration. *Landscape Ecology*, *29*(10), 1675–1688.

Luck, G. W., Kai, C., & Fay, J. P. (2009). Protecting ecosystem services and biodiversity in the world's watersheds. *Conservation Letters,* 179–188.

Luo, C., Xu, G., Chao, Z., Wang, S., Lin, X., Hu, Y., et al. (2010). Effect of warming and grazing on litter mass loss and temperature sensitivity of litter and dung mass loss on the Tibetan plateau. *Global Change Biology*, *16*(5), 1606–1617.

MA. (2005). *Millennium ecosystem assessment findings*. Millennium Ecosystem Assessment.

Maharana, I., Rai, S. C., & Sharma, E. (2000). Environmental economics of the Khangchendzonga National Park in the Sikkim Himalaya, India. *GeoJournal, 50*(4), 329–337.

Maikhuri, R. K., Rao, K. S., & Semwal, R. L. (2001). Changing scenario of Himalayan agroecosystems: Loss of agrobiodiversity, an indicator of environmental change in Central Himalaya, India. *The Environmentalist, 21*(1), 23–39.

Mandel, J. R., Dechaine, J. M., Marek, L. F., & Burke, J. M. (2011). Genetic diversity and population structure in cultivated sunflower and a comparison to its wild progenitor. *Helianthus Annuus L Theoretical and Applied Genetics, 123*(5), 693–704.

Måren, I., Bhattarai, K. R., & Chaudhary, R. P. (2013). Forest ecosystem services and biodiversity: The resource flux from forests to farm in the Himalayas. *Environmental Conservation, 41*(1), 73–83. https://doi.org/10.1017/s0376892913000258.

Maselli, D. (2012). Promoting sustainable mountain development at the global level: The Swiss development cooperation's involvement. *Mountain Research and Development, 32*(S1), S64–S70.

Maxwell, S. L., Fuller, R. A., Brooks, T. M., & Watson, J. E. (2016). Biodiversity: The ravages of guns, nets and bulldozers. *Nature, 536,* 143–145.

Mehra, A., Bajpai, O., & Joshi, H. (2014). Diversity, utilization and sacred values of ethno-medicinal plants of Kumaun Himalaya. *Tropical Plant Research, 1,* 80–86.

Menon, V. (1996). *Under Siege: Poaching and protection of greater one-horned rhinoceroses in India.* Cambridge, UK: Traffic International.

Menon, S., Latif Khan, M., Paul, A., & Peterson, A. T. (2012). Rhododendron species in the Indian Eastern Himalayas: New approaches to understanding rare plant species distributions.

Metzger, M. J., Bunce, R. G., Jongman, R. H., Sayre, R., Trabucco, A., & Zomer, R. (2013). A high-resolution bioclimate map of the world: a unifying framework for global biodiversity research and monitoring. *Global Ecology and Biogeography, 22*(5), 630–638.

Miehe, G., Miehe, S., & Schlütz, F. (2009). Early human impact in the forest ecotone of southern High Asia (Hindu Kush, Himalaya). *Quaternary Research, 71*(3), 255–265.

Miehe, G., Miehe, S., Böhner, J., Kaiser, K., Hensen, I., Madsen, D., et al. (2014). How old is the human footprint in the world's largest alpine ecosystem? A review of multiproxy records from the Tibetan Plateau from the ecologists' viewpoint. *Quaternary Science Reviews, 86,* 190–209.

Miehe, G., Pendry, C., & Chaudhary, R. P. (2015a). *Nepal—An introduction to the natural history, ecology and human environment in the Himalayas: A companion to the Flora of Nepal*. Edinburgh, UK: Royal Botanic Garden.

Miehe, G., Miehe, S., Bohner, J., Baumler, R., Ghimire, S. K., Bhattarai, K., et al. (2015b). Vegetation ecology. *Nepal—An introduction to the natural history, ecology and human environment in the Himalayas: A companion to the Flora of Nepal* (pp. 385–472). Edinburgh, UK: Royal Botanic Garden.

Milleville, R. D. (2002). *The Rhododendrons of Nepal* (p. 136). Lalitpur: Himal Books.

Mittermeier, R. A., Gil, R. P., Hoffman, M., et al. (2004). *Hotspots revisited. Earth's biologically richest and most endangered terrestrial ecoregions*. Mexico City: CEMEX/Agrupación Sierra Madre.

Mittermeier, R. A., Turner, W. R., Larsen, F. W., Brooks, T. M., & Gascon, C. (2011). Global biodiversity conservation: the critical role of hotspots. In *Biodiversity Hotspots* (pp. 3–22). Berlin, Heidelberg: Springer.

Mittermeier, R. A., van Dijk, P. P., Rhodin, A. G., & Nash, S. D. (2015). Turtle hotspots: an analysis of the occurrence of tortoises and freshwater turtles in biodiversity hotspots, high-biodiversity wilderness areas, and turtle priority areas. *Chelonian Conservation and Biology, 14*(1), 2–10.

MoAF. (2014). *National biodiversity strategies and action plan, Bhutan 2014.* Ministry of Agriculture and Forest, Royal Government of Bhutan, Thimphu, Bhutan.

MoAF. (2016). *Forest facts and figures*. Ministry of Agriculture and Forest, Royal Government of Bhutan, Thimphu, Bhutan.

MoCC. (2015). *National forest policy 2015*. Ministry of Climate Change, Government of Pakistan, Islamabad, Pakistan.

MoECF. (2011). *National biodiversity strategy and action plan—Myanmar.* Ministry of Environmental Conservation and Forestry, Myanmar.

MoEF. (2004). *National biodiversity strategy and action plan for Bangladesh.* Ministry of Environment and Forests, Government of the People's Republic of Bangladesh, Dhaka.

MoEF. (2008). *National biodiversity action plan.* Ministry of Environment and Forests, Government of India.

MoEF. (2014). *India's fifth national report to the convention on biological diversity*. Ministry of Environment and Forests, Government of India.

MoEPC. (2014). *China's fifth national report on the implementation of the convention on biological diversity*. The Ministry of Environmental Protection of Cal Protection of China.

MoFSC. (2014). *Nepal's 5th national report to convention on biological diversity*. Ministry of Forests and Soil Conservation, Government of Nepal.

MoFSC. (2015a). *Strategy and action plan 2015–2025, Terai Arc Landscape, Nepal*. Ministry of Forests and Soil Conservation, Singha Durbar, Kathmandu, Nepal.

MoFSC. (2015b). *Strategy and action plan 2016–2025, Chitwan-Annapurna Landscape, Nepal*. Ministry of Forests and Soil Conservation, Singha Durbar, Kathmandu, Nepal.

MoFSC. (2016). *Conservation landscapes of Nepal*. Ministry of Forest and Soil Conservation. Government of Nepal, Kathmandu, Nepal.

MoFSC. (2017). *Western mountain landscape level programme's feasibility study and strategic plan*. Department of Forests, Ministry of Forests and Soil Conservation, Nepal.

Molden, D. J., Vaidya, R. A., Shrestha, A. B., Rasul, G., & Shrestha, M. S. (2014a). Water infrastructure for the Hindu Kush Himalayas. *International Journal of Water Resources Development, 30*(1), 60–77.

Molden, D., Verma, R., & Sharma, E. (2014b). Gender equality as a key strategy for achieving equitable and sustainable development in mountains: The case of the Hindu Kush-Himalayas. *Mountain Research and Development, 34*(3), 297–300.

Molden, D. J., Shrestha, A. B., Nepal, S., & Immerzeel, W. W. (2016). Downstream implications of climate change in the Himalayas. In *Water security, climate change and sustainable development* (pp. 65–82). Singapore: Springer.

Mora, C., Caldwell, I. R., Caldwell, J. M., Fisher, M. R., Genco, B. M., & Running, S. W. (2015). Suitable days for plant growth disappear under projected climate change: potential human and biotic vulnerability. *PLoS Biology, 13*(6), e1002167.

Mukerji, A. K. (2006). Evolution of good governance through forest policy reform in India. In P. Bhattacharya, A. K. Kandya, & K. N. Krishna Kumar (Eds.), *JFM at crossroads: Future strategy and action programme for institutionalizing community forestry, Workshop PrePrints* (pp. 17–24).

Mukherji, A., Molden, D., Nepal, S., Rasul, G., & Wagnon, P. (2015). Himalayan waters at the crossroads: Issues and challenges. *International Journal of Water Resources Development, 31*(2), 151–160.

Myers, N., Mittermeier, R. A., Mittermeier, C. G., Da Fonseca, G. A., & Kent, J. (2000). Biodiversity hotspots for conservation priorities. *Nature, 403*(6772), 853–858.

Naidoo, R., Balmford, A., Costanza, R., Fisher, B., Green, R. E., Lehner, B., et al. (2008). Global mapping of ecosystem services and conservation priorities. *Proceedings of the National Academy of Sciences, 105*(28), 9495–9500.

NCD. (2004) *Bhutan biological conservation complex: A landscape conservation plan—Way forward*. Bhutan: Nature Conservation Division (NCD) and WWF.

Negi, C. S. (2005). Socio-cultural and ethnobotanical value of a sacred forest, Thal Ke Dhar, central Himalaya. *Indian Journal of Traditional Knowledge, 4*(2), 190–198.

Nepal, S. K. (1997). Sustainable tourism, protected areas and livelihood needs of local communities in developing countries. *The International Journal of Sustainable Development & World Ecology, 4*(2), 123–135.

Nepal, S. K. (2013). Mountain tourism and climate change: Implications for the Nepal Himalaya. *Nepal Tourism and Development Review, 1*(1), 1–14.

Nepal, S., Flügel, W. A., & Shrestha, A. B. (2014a). Upstream-downstream linkages of hydrological processes in the Himalayan region. *Ecological Processes, 3*(1), 1.

Nepal, S., Krause, P., Flügel, W. A., Fink, M., & Fischer, C. (2014b). Understanding the hydrological system dynamics of a glaciated alpine catchment in the Himalayan region using the J2000 hydrological model. *Hydrological Processes, 28*(3), 1329–1344.

Ning, W., Ismail, M., Joshi, S., Qamar, F. M., Phuntsho, K., Weikang, Y., et al. (2014). *Understanding the transboundary Karakoram-Pamir landscape*. Kathmandu, Nepal: ICIMOD.

Notter, D. R. (1999). The importance of genetic diversity in livestock populations of the future. *Journal of animal science, 77*(1), 61–69.

Oli, K. P., Chaudhary, S., & Sharma, U. R. (2013). Are governance and management effective within protected areas of the Kanchenjunga landscape (Bhutan, India and Nepal). *Parks, 19*(1), 25–36.

Olson, D. M., & Dinerstein, E. (2002). The Global 200: Priority ecoregions for global conservation. *Annals of the Missouri Botanical Garden*, 199–224.

Ouyang, Z., Zheng, H., Xiao, Y., Polasky, S., Liu, J., Xu, W., et al. (2016). Improvements in ecosystem services from investments in natural capital. *Science, 352*(6292), 1455–1459.

Pai, R., & Datta, S. (2006). Measuring milestones. In *Proceedings of the National Workshop on Joint Forest Management (JFM)*, 17 October 2006, New Delhi.

Panday, P. K., Williams, C. A., Frey, K. E., & Brown, M. E. (2014). Application and evaluation of a snowmelt runoff model in the Tamor River basin, Eastern Himalaya using a Markov Chain Monte Carlo (MCMC) data assimilation approach. *Hydrological Processes, 28*(21), 5337–5353.

Pandit, M. K., & Grumbine, R. E. (2012). Potential effects of ongoing and proposed hydropower development on terrestrial biological diversity in the Indian Himalaya. *Conservation Biology, 26*(6), 1061–1071.

Pandit, M. K., Sodhi, N. S., Koh, L. P., Bhaskar, A., & Brook, B. W. (2007). Unreported yet massive deforestation driving loss of endemic biodiversity in Indian Himalaya. *Biodiversity and Conservation, 16*(1), 153–163.

Pandit, M. K., Manish, K., & Koh, L. P. (2014). Dancing on the roof of the world: ecological transformation of the Himalayan landscape. *BioScience, 64*(11), 980–992.

Pant, K. P., Rasul, G., Chettri, N., Rai, K. R., & Sharma, E. (2012). Value of forest ecosystem services: a quantitative estimation from the Kangchenjunga landscape in eastern Nepal. *ICIMOD Working Paper*, (2012/5).

Paudyal, K., Baral, H., Burkhard, B., Bhandari, S. P., & Keenan, R. J. (2015). Participatory assessment and mapping of ecosystem services in a data-poor region: case study of community-managed forests in central Nepal. *Ecosystem Services, 13*, 81–92.

Paul, S., & Chakrabarti, S. (2011). Socio-economic issues in forest management in India. *Forest Policy and Economics, 13*(1), 55–60.

Pei, S. (1995). *Banking on biodiversity*. Report on the regional consultations on biodiversity assessment in the Hindu Kush Himalaya. Kathmandu, Nepal: ICIMOD.

Pereira, H. M., Leadley, P. W., Proenca, V., Alkemade, R., et al. (2010). Scenarios for global biodiversity in the 21st century. *Science, 330*, 1496–1501.

Posch, E., Bell, R., Weidinger, J. T., & Glade, T. (2015). Geomorphic processes, rock quality and solid waste management-examples from the Mt. Everest region of Nepal. *Journal of Water Resource and Protection, 7*(16), 1291.

Pradhan, S., Saha, G. K., & Khan, J. A. (2001). Ecology of the red panda *Ailurus fulgens* in the Singhalila National Park, Darjeeling, India. *Biological Conservation, 98*(1), 11–18.

Pradhan, R., Argent, G., & McFarlane, M. (2003). Wild rhododendrons of Bhutan. In *Rhododendrons in horticulture and science*. Papers presented at the International Rhododendron Conference, Edinburgh, UK, 17–19 May 2002. (pp. 37–41). Edinburgh: Royal Botanic Garden.

Pradhan, N. M., Wegge, P., Moe, S. R., & Shrestha, A. K. (2008). Feeding ecology of two endangered sympatric megaherbivores: Asian elephant *Elephas maximus* and greater one-horned rhinoceros

Rhinoceros unicornis in lowland Nepal. *Wildlife Biology, 14*(1), 147–154.

Quyang, H. (2009). The Himalayas—Water storage under threat. *Sustainable Mountain Development, 56,* 3–5.

Rai, S. C., Sharma, E., & Sundriyal, R. C. (1994). Conservation in the Sikkim Himalaya: traditional knowledge and land-use of the Mamlay watershed. *Environmental Conservation, 21*(01), 30–34.

Ramesh, K. V., & Goswami, P. (2007). Reduction in temporal and spatial extent of the Indian summer monsoon. *Geophysical Research Letters, 34*(23).

Ranjitkar, S., Luedeling, E., Shrestha, K. K., Guan, K., & Xu, J. (2013). Flowering phenology of tree rhododendron along an elevation gradient in two sites in the Eastern Himalayas. *International Journal of Biometeorology, 57*(2), 225–240.

Rao, M., Rabinowitz, A., & Khaing, S. T. (2002). Status review of the protected-area system in Myanmar, with recommendations for conservation planning. *Conservation Biology, 16*(2), 360–368.

Rao, M., Zaw, T., Htun, S., & Myint, T. (2011). Hunting for a living: Wildlife trade, rural livelihoods and declining wildlife in the Hkakaborazi National Park, North Myanmar. *Environmental Management, 48*(1), 158–167.

Rasul, G. (2014). Food, water, and energy security in South Asia: A nexus perspective from the Hindu Kush Himalayan region. *Environmental Science & Policy, 39,* 35–48.

Rasul, G., & Sharma, B. (2015). The nexus approach to water–energy–food security: An option for adaptation to climate change. *Climate Policy*, 1–21.

Rasul, G., Chettri, N., & Sharma, E. (2011a). *Framework for valuing ecosystem services in the Himalayas*. ICIMOD.

Rasul, G., Thapa, G. B., & Karki, M. B. (2011b). Comparative analysis of evolution of participatory forest management institutions in South Asia. *Society & Natural Resources, 24*(12), 1322–1334.

Richter, B., Matthews, R., Harrison, D., & Wigington, R. (2003). Ecologically sustainable water management: Managing river flows for ecological integrity. *Ecological Applications, 13,* 206–224.

Rodríguez-Rodríguez, D., Bomhard, B., Butchart, S. H., & Foster, M. N. (2011). Progress towards international targets for protected area coverage in mountains: A multi-scale assessment. *Biological Conservation, 144*(12), 2978–2983.

Rounsevell, M. D., Pedroli, B., Erb, K. H., Gramberger, M., Busck, A. G., Haberl, H., et al. (2012). Challenges for land system science. *Land Use Policy*, *29*(4), 899–910.

Salgotra, R. K., Gupta, B. B., Bhat, J. A., & Sharma, S. (2015). Genetic diversity and population structure of basmati rice (Oryza sativa L.) germplasm collected from North Western Himalayas using trait linked SSR markers. *PloS ONE, 10*(7), e0131858.

Salick, J., Amend, A., Anderson, D., Hoffmeister, K., Gunn, B., & Zhendong, F. (2007). Tibetan sacred sites conserve old growth trees and cover in the eastern Himalayas. *Biodiversity and Conservation, 16*(3), 693–706.

Sandhu, H., & Sandhu, S. (2014). Linking ecosystem services with the constituents of human well-being for poverty alleviation in eastern Himalayas. *Ecological Economics, 107,* 65–75.

Sarkar, S. (2011). Climate change and disease risk in the Himalayas. *Himalayan Journal of Sciences, 6*(8), 7–8.

Savillo, I. T. (2009). Present status of Ramsar sites in Nepal. *International Journal of Biodiversity and Conservation, 1*(5), 146–150.

Sayer, J. (2009). Reconciling conservation and development: Are landscapes the answer? *Biotropica, 41*(6), 649–652.

Sayer, J., Sunderland, T., Ghazoul, J., Pfund, J. L., Sheil, D., Meijaard, E., et al. (2013). Ten principles for a landscape approach to reconciling agriculture, conservation, and other competing land uses. *Proceedings of the National Academy of Sciences, 110*(21), 8349–8356.

Schaich, H., Bieling, C., & Plieninger, T. (2010). Linking ecosystem services with cultural landscape research. *Gaia-Ecological Perspectives for Science and Society, 19*(4), 269–277.

Schaller, G. B. (1998). *Wildlife of the Tibetan steppe*. Chicago: University of Chicago Press.

Schild, A. (2008). The case of the Hindu Kush-Himalayas: ICIMOD's position on climate change and mountain systems. *Mountain Research and Develpment, 28*(3/4), 328–331.

Secretariat of the Convention on Biological Diversity. (2014) *Global Biodiversity Outlook 4* (155 pp.). Montréal.

Shakya, B., Uddin, K., Chettri, N., Bajracharya, B., & Sharma, E. (2011) Use of geo-spatial tools in the management of potential habitats outside the protected areas in the transboundary Brahmaputra Salween landscape. In *Contribution of ecosystem restoration to the objectives of the CBD and a healthy planet for all people*, Abstracts of posters presented at the 15th meeting of the Subsidiary Body on Scientific, Technical and Technological Advice of the Convention on Biological Diversity, 7–11 November 2011, Montreal, Canada, Technical Series No. 62, pp 98–100. Montreal, Canada: Secretariat of the Convention on Biological Diversity.

Sharma, E., & Chettri, N. (2005). ICIMOD's transboundary biodiversity management initiative in the Hindu Kush-Himalayas. *Mountain Research and Development, 25*(3), 280–283.

Sharma, E., Bhuchar, S., Xing, M., & Kothyari, B. P. (2007a). Land use change and its impact on hydro ecological linkages in Himalayan watersheds. *Tropical Ecology, 48*(2), 151–161.

Sharma, R., Xu, J., & Sharma, G. (2007b). Traditional agroforestry in the Eastern Himalayan region: Land management system supporting ecosystem services. *Tropical Ecology, 48*(2), 189.

Sharma, E., Chettri, N., Gurung, J., & Shakya, B. (2007c). *Landscape approach in biodiversity conservation: A regional cooperation framework for implementation of the convention on biological diversity in Kangchenjunga landscape* (p. 29). Kathmandu, Nepal: ICIMOD.

Sharma E., Chettri N., Tse-ring K., Shrestha A. B., Jing F., Mool P., et al. (2009). *Climate change impacts and vulnerability in the Eastern Himalayas* (p. 27). Kathmandu, Nepal: ICIMOD.

Sharma, E., Chettri, N., & Oli, K. P. (2010). Mountain biodiversity conservation and management: A paradigm shift in policies and practices in the Hindu Kush-Himalayas. *Ecological Research, 25* (5), 909–923.

Sharma, B., Rasul, G., & Chettri, N. (2015). The economic value of wetland ecosystem services: Evidence from Koshi Tappu wildlife reserve. *Nepal. Ecosystem Services., 12,* 84–93.

Sharma, B. K., Haokip, T. P., & Sharma, S. (2016). Loktak Lake, Manipur, northeast India: a Ramsar site with rich rotifer (Rotifera: Eurotatoria) diversity and its meta-analysis. *International Journal of Aquatic Biology*, *4*(2).

Sherpa, M., & Norbu, U. P. (1999). Linking protected areas for ecosystem conservation: A case study from Bhutan. *Parks, 9,* 35–45.

Sherpa, L. N., Peniston, B., Lama, W., & Richard, C. (2003). *Hands around Everest: Transboundary cooperation for conservation and sustainable livelihoods*. Kathmandu, Nepal: International Centre for Integrated Mountain Development (ICIMOD).

Shrestha, U. B., Gautam, S., & Bawa, K. S. (2012). Widespread climate change in the Himalayas and associated changes in local ecosystems. *PLoS ONE, 7*(5), e36741.

Shu D. J. (2005). *Flora of China*. Rhododendron. Vol 15-260-445. http://flora.huh.harvard.edu/china/PDF/PDF14/Rhododendron.pdf. Retrieved 21 September 2016.

Simpson, B. B., & Ogorzaly, M. C. (1986). *Economic botany: Plants in our world*. New York: McGraw-Hill.

Singh, S. P. (2002). Balancing the approaches of environmental conservation by considering ecosystem services as well as biodiversity. *Current Science, 82*(11), 1331–1335.

Singh, K. K. (2009). Notes on the Sikkim Himalayan rhododendrons: A taxa of great conservation importance. *Turkish Journal of Botany, 33*(4), 305–310.

Singh, B., & Borthakur, S. K. (2015). Phenology and geographic extension of lycophyta and fern flora in nokrek biosphere reserve of Eastern Himalaya. *Proceedings of the National Academy of Sciences, India Section B: Biological Sciences, 85*(1), 291–301.

Singh, S. P., Singh, V., & Skutsch, M. (2010). Rapid warming in the Himalayas: Ecosystem responses and development options. *Climate and Development, 2*(3), 221–232.

Singh, S. P., Bassignana-Khadka, I., Karky, B. S., & Sharma, E. (2011).*Climate change in the Hindu Kush-Himalayas: the state of current knowledge*. International Centre for Integrated Mountain Development (ICIMOD).

Sinha, B., & Mishra, S. (2015). Ecosystem services valuation for enhancing conservation and livelihoods in a sacred landscape of the Indian Himalayas. *International Journal of Biodiversity Science, Ecosystem Services & Management, 11*(2), 156–167.

Song, X., Peng, C., Zhou, G., Jiang, H., & Wang, W. (2014). Chinese grain for green program led to highly increased soil organic carbon levels: A meta-analysis. *Scientific Reports, 4,* 4460.

Song, W., Deng, X., Liu, B., Li, Z., & Jin, G. (2015). Impacts of grain-for-green and grain-for-blue policies on valued ecosystem services in Shandong Province, China. *Advances in Meteorology, 2015,* 1–10.

Sunam, R. K., Bishwokarma, D., & Darjee, K. B. (2015). Conservation Policy Making in Nepal: Problematising the Politics of Civic Resistance. *Conservation and Society, 13*(2), 179.

Taylor, K., & Lennon, J. (2011). Cultural landscapes: a bridge between culture and nature? *International Journal of Heritage Studies, 17* (6), 537–554.

Telwala, Y., Brook, B. W., Manish, K., & Pandit, M. K. (2f013). Climate-induced elevational range shifts and increase in plant species richness in a Himalayan biodiversity epicentre. *PLoS One, 8* (2), e57103.

Tesfaye, M., Dufault, N. S., Dornbusch, M. R., Allan, D. L., Vance, C. P., & Samac, D. A. (2003). Influence of enhanced malate dehydrogenase expression by alfalfa on diversity of rhizobacteria and soil nutrient availability. *Soil Biology & Biochemistry, 35*(8), 1103–1113.

Tews, J., Brose, U., Grimm, V., Tielbörger, K., Wichmann, M. C., Schwager, M., et al. (2004). Animal species diversity driven by habitat heterogeneity/diversity: the importance of keystone structures. *Journal of Biogeography, 31*(1), 79–92.

Thapa, K., Nepal, S., Thapa, G., Bhatta, S. R., & Wikramanayake, E. (2013). Past, present and future conservation of the greater one-horned rhinoceros Rhinoceros unicornis in Nepal. *Oryx, 47* (03), 345–351.

Thompson, C. (2009). New Species Discoveries–the Eastern Himalayas where Worlds Collide.

Thompson, M., & Warburton, M. (1985). Decision making under contradictory certainties: how to save the Himalayas when you can't find out what's wrong with them. *Journal of Applied Systems Analysis, 12*(1), 3–34.

Thuiller, W., Albert, C., Araújo, M. B., Berry, P. M., Cabeza, M., Guisan, A., et al. (2008). Predicting global change impacts on plant species' distributions: future challenges. *Perspectives in Plant Ecology, Evolution and Systematics, 9*(3), 137-152.

Tilman, D., Fargione, J., Wolff, B., D'Antonio, C., Dobson, A., Howarth, R., et al. (2001). Forecasting agriculturally driven global environmental change. *Science, 292*(5515), 281–284.

Tremblay, M. M., Fox, M., Schmidt, J. L., Tripathy-Lang, A., Wielicki, M. M., Harrison, T. M., et al. (2015). Erosion in southern Tibet shut down at ~ 10 Ma due to enhanced rock uplift within the Himalaya. *Proceedings of the National Academy of Sciences, 112* (39), 12030–12035.

Turin, M. (2005). Language endangerment and linguistic rights in the Himalayas: A case study from Nepal. *Mountain Research and Development, 25*(1), 4–9.

Uchida, E., Xu, J., & Rozelle, S. (2005). Grain for green: cost-effectiveness and sustainability of China's conservation set-aside program. *Land Economics, 81*(2), 247–264.

Uddin, K., Chaudhary, S., Chettri, N., Kotru, R., Murthy, M., Chaudhary, R. P., et al. (2015). The changing land cover and fragmenting forest on the Roof of the World: A case study in Nepal's Kailash Sacred Landscape. *Landscape and Urban Planning, 141,* 1–10.

UN. (1992). Report of the United Nations Conference on Environment and Development. Rio de Janeiro 3–14 June 1992. A/CONF.151/26 (Vol. 1).

Upadhaya, S., Chalise, L., & Pandel, R. P. (2009). High altitude Ramsar sites of Nepal. *The Initiation, 3,* 135–148.

Upadhyay, T. P., Sankhayan, P. L., & Solberg, B. (2005). A review of carbon sequestration dynamics in the Himalayan region as a function of land-use change and forest/soil degradation with special reference to Nepal. *Agriculture, Ecosystems & Environment, 105* (3), 449–465.

Uprety, Y., Poudel, R. C., Gurung, J., Chettri, N., & Chaudhary, R. P. (2016). Traditional use and management of NTFPs in Kangchenjunga Landscape: implications for conservation and livelihoods. *Journal of Ethnobiology and Ethnomedicine, 12*(1), 1.

Valley, H. P. (2003). Upward shift of Himalayan pine in western Himalaya. *India. Current Science, 85*(1), 135.

Waldron, A., Mooers, A. O., Miller, D. C., Nibbelink, N., Redding, D., Kuhn, T. S., et al. (2013). Targeting global conservation funding to limit immediate biodiversity declines. *Proceedings of the National Academy of Sciences, 110*(29), 12144–12148.

Wang, L., Young, S. S., Wang, W., Ren, G., Xiao, W., Long, Y., et al. (2016). Conservation priorities of forest ecosystems with evaluations of connectivity and future threats: Implications in the Eastern Himalaya of China. *Biological Conservation, 195,* 128–135.

Watson, J. E., Jones, K. R., Fuller, R. A., Marco, M. D., Segan, D. B., Butchart, S. H., et al. (2016). Persistent disparities between recent rates of habitat conversion and protection and implications for future global conservation targets. *Conservation Letters*, 9(6), 413–421.

WCMC-UNEP. (2002). Mountains of the World—Dataset. World Conservation Monitoring Center—UNEP. Cambridge, U.K. https://www.arcgis.com/home/item.html?id=a068913914cd4fecb302b9207a532d1a.

Wei, F., Feng, Z., Wang, Z., & Hu, J. (1999). Current distribution, status and conservation of wild red pandas Ailurus fulgens in China. *Biological Conservation, 89*(3), 285–291.

Westman, W. E. (1977). How much are nature's services worth? *Science, 197*(4307), 960–964.

Wieringa, M. J., & Morton, A. G. (1996). Hydropower, adaptive management, and biodiversity. *Environmental Management, 20*(6), 831–840.

Wikramanayake, E. D., Dinerstein, E., Robinson, J. G., Karanth, U., Rabinowitz, A., Olson, D., et al. (1998). An Ecology-Based Method for Defining Priorities for Large Mammal Conservation: The Tiger as Case Study. *Conservation Biology, 12*(4), 865–878.

Wikramanayake, E. D., McKnight, M., Dinerstein, E., Joshi, A., Gurung, B., & Smith, J. L. D. (2004). Designing a conservation landscape for tigers in human-dominated environments. *Conservation Biology, 18,* 839–844.

Wikramanayake, E., Dinerstein, E., Seidensticker, J., Lumpkin, S., Pandav, B., Shrestha, M., et al. (2011). A landscape-based conservation strategy to double the wild tiger population. *Conservation Letters, 4*(3), 219–227.

Wilkes, A. (2008). *Towards mainstreaming climate change in grassland management policies and practices on the Tibetan Plateau* (No. 67). Working paper.

Wilson, K. A., Auerbach, N. A., Sam, K., Magini, A. G., Moss, A. S. L., Langhans, S. D., et al. (2016). Conservation Research Is Not Happening Where It Is Most Needed. *PLoS Biology, 14*(3), e1002413.

Xu, J., & Grumbine, R. E. (2014a). Building ecosystem resilience for climate change adaptation in the Asian highlands. *Wiley Interdisciplinary Reviews: Climate Change, 5*(6), 709–718.

Xu, J., & Grumbine, R. E. (2014b). Integrating local hybrid knowledge and state support for climate change adaptation in the Asian highlands. *Climatic Change, 124,* 93–104.

Xu, J., Yang, Y., Ayad, W. G., & Eyzaguirre, P. B. (2001). Genetic Diversity in Taro (*Colocasia esculenta* Schott, Araceae) in China: An Ethnobotanical and Genetic Approach. *Economic Botany, 55*(1), 14–31.

Xu, J., Ma, E., Tashi, D., Fu, Y., Lu, Z., & Melick, D. (2005). Integrating Sacred Knowledge for Conservation: Cultures and Landscapes in Southwest China. *Ecology and Society, 10*(2), 7.

Xu, J., Sharma, R., Fang, J., & Xu, Y. (2008). Critical linkages between land-use transition and human health in the Himalayan region. *Environment International, 34*(2), 239–247.

Xu, J., Grumbine, R. E., Shrestha, A., Eriksson, M., Yang, X., Wang, Y. U. N., et al. (2009a). The melting Himalayas: cascading effects of climate change on water, biodiversity, and livelihoods. *Conservation Biology, 23*(3), 520–530.

Xu, J. C., Lebel, L., & Sturgeon, J. (2009b). Functional links between biodiversity, livelihoods and culture in a Hani swidden landscape in Southwest China. *Ecology and Society, 14*(2), 20.

Xu, J., White, A., & Lele, U. J. (2010). *China's forest tenure reforms: Impacts and implications for choice, conservation, and climate change*. Rights and Resources Initiative.

Yonzon, P. B. (1989). *Ecology and conservation of the red panda in the Nepal-Himalayas* (Doctoral dissertation, University of Maine).

Ziv, G., Baran, E., Nam, S., Rodríguez-Iturbe, I., & Levin, S. A. (2012). Trading-off fish biodiversity, food security, and hydropower in the Mekong River Basin. *Proceedings of the National Academy of Sciences, 109*(15), 5609–5614.

Zomer, R., & Oli, K. P. (Eds.). (2011). *Kailash Sacred Landscape Conservation Initiative- Feasibility assessment report*. Kathmandu: ICIMOD.

Zomer, R. J., Trabucco, A., Metzger, M. J., Wang, M., Oli, K. P., & Xu, J. (2014). Projected climate change impacts on spatial distribution of bioclimatic zones and ecoregions within the Kailash Sacred Landscape of China, India. *Nepal. Climatic change, 125*(3–4), 445–460.

Zomer, R. J., Xu, J., Wang, M., Trabucco, A., & Li, Z. (2015). Projected impact of climate change on the effectiveness of the existing protected area network for biodiversity conservation within Yunnan Province, China. *Biological Conservation, 184,* 335–345.

Zomer, R. J., Wang, M., Trabucco, A., & Xu, J. C. (2016). Projected climate change impact on hydrology, bioclimatic conditions and terrestrial ecosystems in the Asian highlands. (ICRAF Working Paper 222). World Agroforestry Centre East and Central Asia, Kunming, China. 56 pp.

6 Meeting Future Energy Needs in the Hindu Kush Himalaya

Coordinating Lead Authors
Shobhakar Dhakal, Associate Professor and Head, Department of Energy, Environment and Climate Change, School of Environment, Resources and Development, Asian Institute of Technology, Bangkok, Thailand. e-mail: shobhakar@ait.ac.th; shobhakar.dhakal@gmail.com
Leena Srivastava, Vice Chancellor, TERI School of Advanced Studies, New Delhi, India. e-mail: leena@terisas.ac.in
Bikash Sharma, former Senior Environmental Economist, Livelihoods, ICIMOD, Kathmandu, Nepal. e-mail: bikasharma99@gmail.com (corresponding author)

Lead Authors
Debajit Palit, Senior Fellow & Director, Rural Energy & Livelihoods Division, TERI, New Delhi, India. e-mail: debajitp@teri.res.in
Brijesh Mainali, Researcher/Lecturer, Department of Built Environment and Energy Technology, Linnaeus University, Växjö, Sweden. e-mail: brijesh.mainali@lnu.se
Rabindra Nepal, Senior Lecturer in Economics, Tasmanian School of Economics and Finance, University of Tasmania, Hobart, Australia. e-mail: Rabindra.Nepal@utas.edu.au
Pallav Purohit, Research Scholar, Air Quality & Greenhouse Gases (AIR) Program, International Institute for Applied Systems Analysis (IIASA), Laxenburg, Austria. e-mail: purohit@iiasa.ac.at
Anandajit Goswami, Management Development Programme Coordinator, TERI School of Advanced Studies, New Delhi, India. e-mail: anandajit.goswami@terisas.ac.in
Ghulam Mohd Malikyar, Ecosystem Health Expert, Deputy Director General for Technical Affairs of the National Environmental Protection Agency of the Islamic Republic of Afghanistan, Kabul, Afghanistan. e-mail: malikyargh@gmail.com
Kul Bahadur Wakhley, former Chief Executive Officer, Bhutan Electricity Authority, Ministry of Economic Affairs, Royal Government of Bhutan, Thimphu, Bhutan. e-mail: kbwakhley@hotmail.com; kbwakhley@gmail.com

Contributing Authors
Ranjan Parajuli, Research Scholar, Ralph E. Martin Department of Chemical Engineering, University of Arkansas, Fayetteville, Arkansas, USA. e-mail: rparajul@uark.edu
Muhammad Indra al Irsyad, PhD candidate, School of Earth and Environmental Science, University of Queensland, St Lucia, Queensland, Australia. e-mail: m.alirsyad@uq.edu.au
Yuwan Malakar, Research Scholar, Energy and Poverty Research Group, School of Chemical Engineering, The University of Queensland, St Lucia - Queensland, Australia. e-mail: y.malakar@uq.edu.au
Bhupendra Shakya, PhD candidate, School of Photovoltaic and Renewable Energy Engineering (SPREE), The University of New South Wales (UNSW), Sydney, Australia. e-mail: b.shakya@unsw.edu.au

Review Editor
Bikash Raj Pandey, Director Clean Energy, Winrock International, Arlington, Virginia, USA. e-mail: BPandey@winrock.org

Corresponding Author
Bikash Sharma, former Senior Environmental Economist, Livelihoods, ICIMOD, Kathmandu, Nepal. e-mail: bikasharma99@gmail.com

P. Wester et al. (eds.), *The Hindu Kush Himalaya Assessment*,
https://doi.org/10.1007/978-3-319-92288-1_6

Contents

Chapter Overview

Key Findings

1. **The Hindu Kush Himalaya (HKH), despite having huge hydropower potential of ~500 GW, remains energy poor and vulnerable**. More than 80% of the rural population in HKH countries, a large part of whom live in mountain areas, rely on traditional biomass fuels for cooking and about 400 million people in HKH countries still lack basic access to electricity.
2. **Measures to enhance energy supply in the HKH have had less than satisfactory results** because of low prioritization and a failure to address challenges of remoteness and fragility. However, this is slowly changing, with an emphasis on innovative business models for remote off-grid areas including mountains.
3. **Inadequate data and analyses are a major barrier to designing context-specific interventions.** This weakness deters the addressing of the special challenges faced by mountain communities:

scale economics, access to infrastructure and resources, poverty levels and capability gaps, and also thwarts the large-scale replication of successful innovative demonstration/pilot projects that have been implemented in the region.

Policy Messages

1. **Quantitative targets (accompanied by quality specifications of alternative energy options) based on an explicit recognition of the full costs and benefits should be the basis of designing policies, prioritizing actions and strengthening investments**. To do this, the institutions and mechanisms for systematic data collection and analyses need to be strengthened. Capacity building, and empowering institutions involved in designing and monitoring the progress of outcome-based energy policies and programmes, must be undertaken.
2. **Governments of the HKH countries need to prioritize use of locally available energy resources**. Policies that seek to maximize energy independence through promotion of community-based and off-grid renewable energy solutions, perhaps through private sector participation, are essential. Large hydropower projects with well-defined benefit-sharing mechanisms would be beneficial. Vibrant markets need to be created to meet the strong, latent demand for decentralized sustainable energy in the mountains.
3. **A high-level, empowered, regional mechanism should be established to strengthen regional energy trade and cooperation**. This mechanism should be equipped to recognize, and take, an integrated approach to meet development needs, as exemplified by the sustainable development goals (SDGs) and their interlinkages.

As mentioned in earlier chapters, the HKH regions form the entirety of some countries, a major part of other countries, and a small percentage of yet others. Because of this, when we speak about meeting the energy needs of the HKH region we need to be clear that we are not necessarily talking about the countries that host the HKH, but the clearly delineated mountainous regions that form the HKH within these countries. It then immediately becomes clear that energy provisioning has to be done in a mountain context characterized by low densities of population, low incomes, dispersed populations, grossly underdeveloped markets, low capabilities, and poor economies of scale. In other words, the energy policies and strategies for the HKH region have to be specific to these mountain contexts.

This chapter critically examines the energy outlook of the HKH in its diverse aspects, including demand-and-supply patterns; national policies, programmes, and institutions; emerging challenges and opportunities; and possible transformational pathways for sustainable energy. We set out to answer three broad questions:

1. How does the abundance of renewable energy resources in the HKH inform, and be informed by, national, regional, and global energy and climate policies?
2. How can energy-deprived mountain communities be empowered to meet their growing energy demands in an environmentally benign and sustainable way?
3. How can the enabling environment for regional integration and cooperation be strengthened to seize climate mitigation, adaptation, and development opportunities and make rapid, meaningful progress?

Despite its vast potential of hydropower and other renewables, the HKH region remains energy poor and vulnerable. Climate change is posing a new challenge to energy security on account of receding treelines, the contribution of biomass burning to short-lived carbon pollutants, altered rainfall patterns, and unpredictable water stocks and flows among other factors. The challenge for the HKH is to simultaneously address the issues of energy security, climate change, and poverty while attaining multiple SDGs. Success will require urgent action and massive investment for energy transformation. If technology access remains inadequate and energy infrastructure continues to be of poor quality, the HKH will neither be able to respond to SDG 7 (affordable and clean energy) nor to any of the related SDGs (***well-established***).

The HKH, by virtue of its massive stock of snow and ice, represents the third pole of Planet Earth. The snow melt from the Himalayan mountains, and the rainfall catchment they provide, feed and supplement the water flows of a number of the rivers that originate from this region. Both the felling of trees for biomass production and the burning of biomass for energy provision have an adverse impact on the water balance of the region. How the region handles its water, energy and forest resources would have a significant impact on the health of the Himalayan mountain ecosystems, the health and wellbeing of its populations, on the vulnerability of its people and those living downstream to glacier melting and associated impacts, and on the opportunities to mitigate climate change impacts at a wider scale (***well-established***). The world needs to engage with the HKH to define an ambitious new energy vision: one that involves building an inclusive green society and economy, with mountain communities enjoying modern, affordable, reliable, and sustainable energy to improve their lives and the environment. Specifically, the region requires a radical energy transformation that ensures *universal electricity access, through grid-connected and off-grid power for lighting and productive uses, and the complete replacement of traditional, inefficient sources for cooking and heating energy with clean sustainable energy options that are efficient, reliable, affordable, and demand-driven (****well-established****).*

Translating this new energy vision into action will require a reform agenda much more ambitious than today's efforts. By tapping into the full potential of hydropower and other renewables, the HKH can overcome its energy poverty and attain energy security, while mitigating and adapting to climate change. Success, however, will critically depend on removing policy, institutional, and capacity barriers that now perpetuate energy poverty and vulnerability in mountain communities.

As there are no "one-size-fits-all" solutions for ensuring universal access to modern energy in mountain areas, the HKH regions must create an enabling environment to ensure access to energy services in a technology- and resource-neutral manner, albeit bearing in mind the leverage energy access provides for meeting other development goals. (***well-established***).

A priority action agenda for sustainable energy transformation

This chapter presents a four-point priority action agenda for a sustainable energy transition to ensure energy security for all, in a climate-resilient manner, through hydropower and other renewables. It also highlights the urgent need to customize SDG 7 targets and indicators to the specific needs and priorities of the HKH region.

Underpinning the implementation of our proposed HKH energy vision and action agenda is a crucial guiding principle: that sustainable energy is a shared responsibility. To accelerate progress and make it meaningful, all key stakeholders must partner with one another and work synergistically for a sustainable energy transition.

The action agenda calls for:

- **Making mountain-specific energy policies and programmes an integral part of national energy development strategy**. A coherent mountain-specific policy framework needs to be well integrated in the national development strategy and translated into action. A prerequisite will be to establish a regional mountain-specific energy data generation and management system that supports national institutions in periodic energy data gathering and in assessing various aspects of energy access.
- **Establishing monitorable quantitative and qualitative targets for each energy end use and tracking the progress of different attributes of clean energy access**. This needs a mountain-specific multi-tier assessment framework, with appropriate energy access indicators, to facilitate a demand-driven approach to energy service provisioning and to prioritize interventions with the largest co-benefits for SDGs.
- **Scaling up current investments and ensuring access to finance through capacity building at different levels**. Innovative and affordable international, regional, national, and local funds will need to be mobilized to support energy planning and infrastructure investments. Incentives for commercial lending, de-risking investments, and leveraging private finance will be needed, to accelerate rapid diffusion and scale-up of appropriate, customized business models for off-grid renewable energy solutions, by empowering local communities.
- **Accelerating the pace of regional trade and cooperation in sustainable energy through a high-level, empowered, regional mechanism**. This can be achieved through integrated regional energy planning, trade, and establishment of institutional mechanisms. These will also ensure that environmental externalities are managed and that revenue-sharing mechanisms for the local areas are in place. Climate change-induced human security threats have the potential to drive energy cooperation and facilitate multilateral agreements on the basis of a common framework.

Meeting Future Energy Needs and the SDGs

Energy is an essential input for the achievement of most SDGs. Transition to sustainable energy, therefore, should be the region's major priority and must be central to national development strategies. An a priori recognition of interlinkages between energy and other context-specific SDGs would allow countries to maximize the developmental benefits of energy provisioning. This is, however, a complex exercise requiring capacity building, governance and institutional changes, and effective design of incentive systems.

That said, SDG 7 and its three targets will need to be customized to the specific needs of each HKH country. Such customization could follow the sectorally integrated Energy Plus approach, maximizing synergies while managing trade-offs.

- Target 7.1—"By 2030, ensure universal access to affordable, reliable and modern energy services"
  HKH needs a multi-tier, mountain-specific assessment framework that captures improved quantity, quality, and reliability of electricity supply. This should assure the availability, efficiency, affordability, safety, health (reduced emissions of short-lived climate pollutants (SLCPs)), and convenience (reduced drudgery for women and children) of modern, clean cooking facilities, and allow progress on energy poverty to be more visible and measurable for national, regional, and global public policy makers.
- Target 7.2—"By 2030, increase substantially the share of renewable energy in the global energy mix"
  The objective could be to move towards a fossil fuel-free and efficient energy future. Such renewable energy, in both centralized and decentralized forms, must serve local populations and meet their demands and fuel growth. Given the low current consumption levels, and the opportunity to provide customized energy, the objective of governments should be to meet all the energy demands of the region.
- Target 7.3—"By 2030, double the global rate of improvement in energy efficiency"
  Policies should focus on the traditionally biomass-dependent residential sector, which has the highest energy use but the lowest energy efficiency. Each country should develop mountain-specific energy efficiency indicators (and should not use energy intensity as a proxy, a practice that can generate misleading results). These indicators should use final energy demand at the most disaggregated end-use level, to accurately reflect energy efficiency improvements and to monitor progress towards the target.

6.1 Introduction

6.1.1 Setting the Scene: The Transition Challenge

Recognizing that energy is fundamental to human development, SDG 7 states that access to affordable, reliable and sustainable energy is crucial not only for future energy security, but for also meeting other SDGs. Although energy is essential for all development needs, current energy systems continue to face major pressures to address energy security, climate change, and energy poverty (Van Vuuren et al. 2012). Given the important role energy security plays for dimensions of human security, a robust energy strategy is key to any country's future development. Prevailing notions of energy, however, are largely supply-biased and growth-centred and fail to respond to the energy development nexus, especially in the HKH.

Mountain areas of the HKH are hotspots for energy poverty-related development challenges. The region also faces a host of issues around energy security and climate change impacts, while having the potential to mitigate the emissions of greenhouse gases (GHGs), in particular SLCPs. The mountain areas of the eight HKH countries (Afghanistan, Bangladesh, Bhutan, China, India, Myanmar, Nepal, and Pakistan) host 9% of the collective population. It is these areas (rather than HKH countries) that are referred to as the HKH region in this chapter. More than 80% of the rural population in the HKH countries lack a modern energy source for cooking (IEA and World Bank 2015), although these country-level data gloss over the realities of energy poverty in the HKH region. Heavy reliance on traditional solid fuel used in traditional ways entrenches poverty, and erodes indoor air quality and environmental sustainability. Caught between poverty and environmental degradation, mountain communities in most of the HKH find it increasingly difficult to meet their daily energy service needs in a sustainable manner. For mountain people in general, and women in particular, the foremost priority is to meet household energy needs for cooking and space heating. The endless cycle of gathering and burning biomass in homes causes enormous damage to the environment, triggers widespread harm to human health, and results in serious social deprivation (Sharma et al. 2005).

Excessive biomass usage and the resulting deforestation and biomass loss are widespread in the HKH, and have a major impact on ecosystem services, climate change, and human health. Sustainable energy has now emerged as the centrepiece of mitigation and adaptive responses to the impacts of both SLCPs and GHGs. Black carbon is considered an SLCP and is now recognized as a major concern in the climate-sensitive HKH, linked with enhanced glacier melting alongside other impacts on public health, environment, and livelihood security (Menon et al. 2002; Ramanathan and Carmichael 2008; Ramanathan et al. 2007). For these and other reasons, expanding access to sustainable energy for reducing these emissions promises significant co-benefits for the HKH.

6.1.2 Transformation in Energy Systems: Multidimensional Linkages

The HKH is interconnected biophysically and socioeconomically across sectors, and this affects the energy supply, demand, underlying drivers, and the solutions themselves. The costs and benefits of energy are not always confined to the HKH, and this requires extended thinking about benefit sharing. One such example is hydroelectricity production in the mountains and hills of the HKH region, which feeds largely into the plain areas. We also find significant sectoral interdependencies in energy, climate, water, and food, and these are numerous, growing, multidimensional and dynamic (Beniston 2013; Hoff 2011; Hussey and Pittock 2012; Marsh and Sharma 2007; Rasul and Sharma 2016; Rockström et al. 2009; State of the Planet Declaration 2012). Sustaining Himalayan freshwater ecosystem services is critical for food, water, and energy security in the HKH and downstream river basins (Rasul 2014a, b). These interdependencies make it crucial for policy makers to understand the cross-sectoral policy linkages, and their effects at multiple scales, while taking into account the transboundary nature of HKH ecosystems and rivers. A regional perspective will be needed so appropriate strategies for energy system transformation can be found, allowing a common solution to be reached by determining and resolving trade-offs, and by exploiting potentials for synergy.

6.1.3 Framework and Roadmap

Sustainable energy development is an evolving concept (Sovacool et al. 2011). Adequacy, reliability, quality, and guarantee of energy resources and carriers on the supply side, and accessibility and affordability on the demand-side serve as important boundaries for sustainable energy transitions (Fig. 6.1). Between supply and demand are enablers who can make and influence current developments and the transitions needed through policy, regulation, design, and implementation of financing solutions, market development, and equitable solutions. If a reliable energy system is not in place, households cannot obtain access to modern fuel, even if they can afford and use such fuel (Fig. 6.1, area C). Regardless of the availability and acceptability of energy technologies (Fig. 6.1, area B), the vast majority of rural people will be unable to afford them because of lack of funds or credit facilities. Even when modern forms of energy are affordable and available (Fig. 6.1, area A) households may not use them if they are not culturally acceptable or if they are less convenient, less reliable, and more expensive than

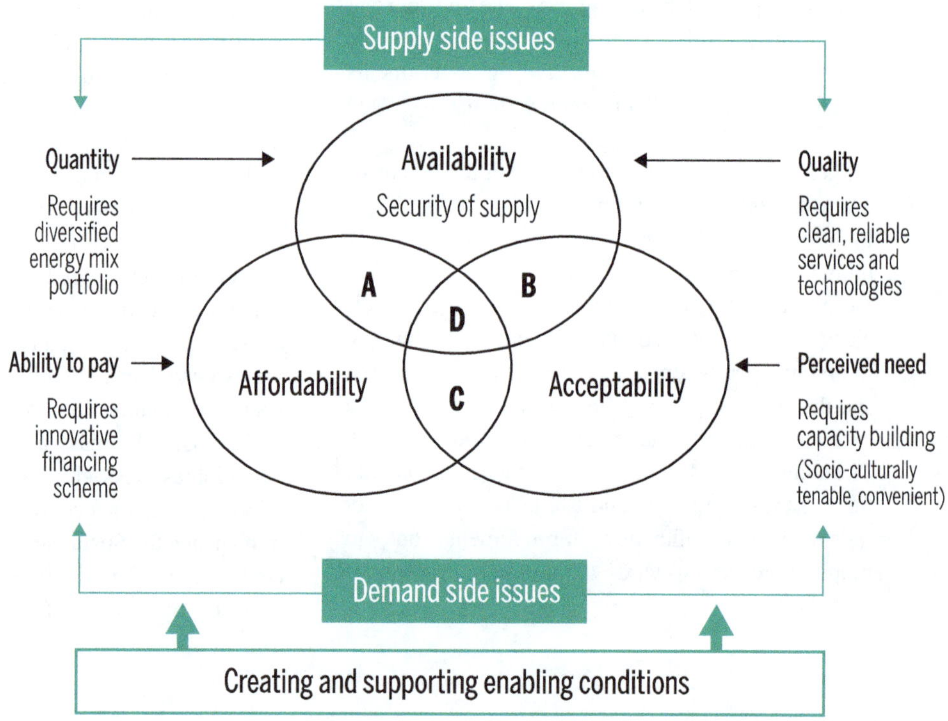

Fig. 6.1 Framework for addressing sustainable energy transition (Sharma 2009)

traditional biomass. Scaling up modern energy services depends on the existence of effective governance and policy frameworks that create an enabling environment for increased adoption and investment in energy access.

We can see there are multiple possible visions of energy transition pathways that could reach the goal of sustainable energy development for the HKH. Each pathway will come with its own barriers and opportunities, and the need for specific transformation in multiple domains. The framework for assessing sustainable energy adopted in this chapter is, therefore, based on the three pillars of sustainable development: social, economic, and environmental. It uses a set of underlying indicators to study the current situation, policies and action; find the gaps and challenges in options and action; provide pathways to sustainable energy futures; and offer insights to the key elements of enabling conditions for decision makers.

Section 6.2 examines the energy supply and demand trends, and the transition pattern, using the available information on the HKH countries. This section also draws on insights into the energy profile of the HKH and discusses the underlying drivers. Section 6.3 examines the key national energy policies, programmes, and institutions in the wider context of national development. Section 6.4 identifies and assesses the challenges and opportunities for sustainable energy facing the HKH countries. Section 6.5 explores the future pathways for sustainable energy solutions through alternative combinations of options and strategies. Section 6.6 offers insights on national, regional, and global linkages of HKH energy resource bases. It also explores avenues for seizing policy opportunities, investing in energy infrastructure, innovative financing mechanisms, capacity development and governance reforms and regional integration and cooperation, among others.

6.2 Energy Demand and Supply—Trends and Patterns

Data on energy demand and supply for mountain areas of HKH countries (except in Bhutan and Nepal, which are fully within the HKH) are largely inadequate and of poor quality. Given the significant differences in energy access, economic development, and institutional structures across countries, it is difficult to accurately capture HKH-specific demand-and-supply trends and patterns. This highlights an urgent need to establish a sound energy information base. Equally, understanding the risks and vulnerability of energy systems to socioenvironmental change is essential to ensure proper design of strategies and interventions for improved energy security.

6.2.1 Biophysical and Socioeconomic Context Shaping Energy Demand

Mountain-specific characteristics such as inaccessibility, fragility, and marginality combined with the "isolated enclave" nature of mountain economies and communities lead to different manifestations of energy demand patterns and trends (Papola 2002). The following paragraphs explain the context of energy demands as depicted in Fig. 6.2.

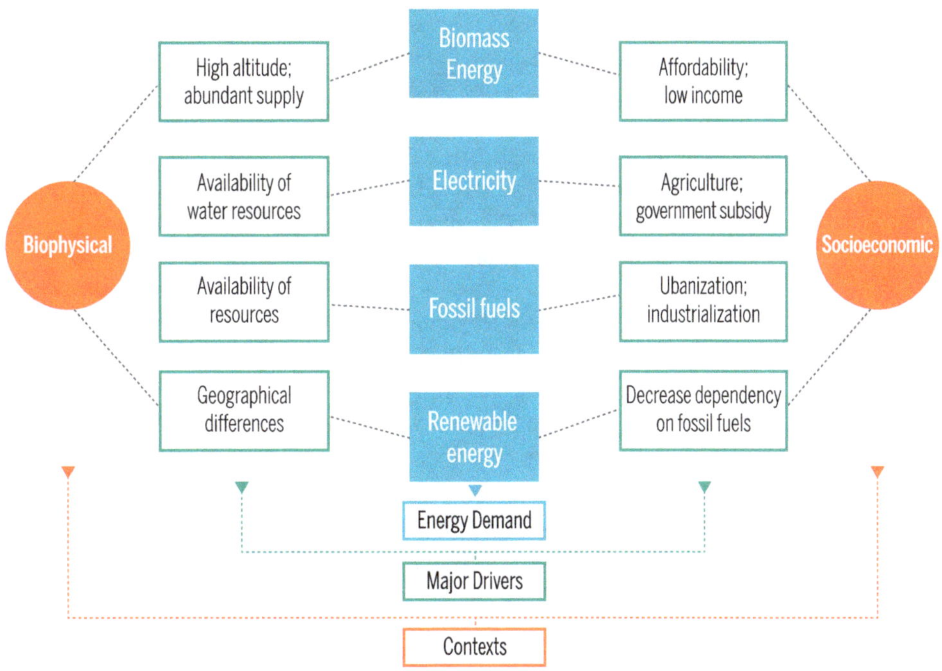

Fig. 6.2 Biophysical and socioeconomic contexts of energy demand in the HKH

Table 6.1 Energy access in HKH countries during 1990–2014

Country	Access to electricity (% of population)				Access to clean fuels and technologies for cooking (% of population)			
	2000	2010	2012	2014	2000	2010	2012	2014
Afghanistan	38	43	69	90	23	19	18	17
Bangladesh	32	55	59	62	11	10	10	10
Bhutan	69	82	92	100	38	60	64	68
China	98	100	100	100	46	54	56	57
India	60	76	80	79	24	32	33	34
Myanmar	47	49	51	52	4	8	8	9
Nepal	27	67	76	85	7	21	23	26
Pakistan	75	91	94	98	24	39	42	45

Source IEA and World Bank (2017)

Energy access: About 400 million people in HKH countries, a large percentage of whom live in mountain areas, still lack basic access to electricity and rely on traditional biomass fuels for cooking (Table 6.1). At the same time, the impacts of climate change on hydropower and biomass resources may bring greater uncertainty to the region (Kaltenborn et al. 2010). However, access to modern energy forms is crucial to support water, food, and livelihood security and to reduce the vulnerability of mountain communities to the impacts of climate change.

Middle- and high-altitude climatic conditions: Due to the generally cool to cold weather throughout the HKH, the need for space and water heating ranks high, as fuel is needed for cooking and lighting. Use of biomass as a cooking fuel serves the additional purpose of heating in the middle- and high-altitude regions of the HKH (Bhatt and Sachan 2004; Palit and Garud 2010; Santner and Jussel 2003). Because of this, smokeless stoves have failed to catch on in Bhutan because they are poor sources of space heating (Palit and Garud 2010), and the demand for traditional fuels for space heating remains high (Rahut et al. 2016). However, experimentation with new technologies is also taking place in the region; one such example is fuel-efficient stoves with attached water warming facilities, in Pakistan.

Abundant primary energy supply: The HKH comprises 20% forests, 15% shrub lands, and 39% grasslands (Schild 2008). The region is also known as the "water tower" of Asia, with an estimated hydroelectricity generation potential of more than 500 GW (Vaidya 2013). The primary energy base of the HKH is, therefore, biomass, hydro, solar, and wind, although solar and wind potentials remain largely unquantified. Yadama et al. (2012) suggest that the possibility of using biomass increases with easy access to forests.

Supply of modern energy: The demand for modern energy is linked with creating improved economic opportunity, rising income, proliferation of new technologies, reducing indoor pollution, modernizing agriculture, entrepreneurship development, and others. Although it is widely recognized that access to modern and clean energy services contributes to tackling poverty (Practical Action 2014), people living in the HKH have limited access to these forms of energy (Shrestha 2013). Despite large-scale availability of fossil fuel reserves and resources beyond the mountainous region of the HKH (Table 6.2), these countries also import fossil fuels to meet their domestic demand, contributing to a growing share of fossil fuels in the total primary energy supply mix of the HKH (Fig. 6.3 and Table 6.3).

Affordability: Low income is a key determinant for continual use of biomass for cooking (Hosier and Dowd 1987). Not only are modern fuels more expensive, but the associated costs of appliances are also high (Bhattacharyya 2006). In urban households, the decreasing share of biomass use is often attributed to rising incomes (Duan et al. 2014). Given

Table 6.2 Proven fossil fuel reserves in HKH countries, 2015

HKH countries	Coal (Million tonnes)	Oil (thousand million tonnes)	Natural gas (Trillion cubic metres)
Bangladesh	N.A.	0.0038[a]	0.2
China	114,500	2.5	3.8
India	60,600	0.8	1.5
Myanmar	N.A.	0.068[a]	0.5
Pakistan	2070	0.051[a]	0.5

Sources BP Statistical Review of World Energy (2016); [a]CIA, The World Factbook (2016)
N.A.: Data not available

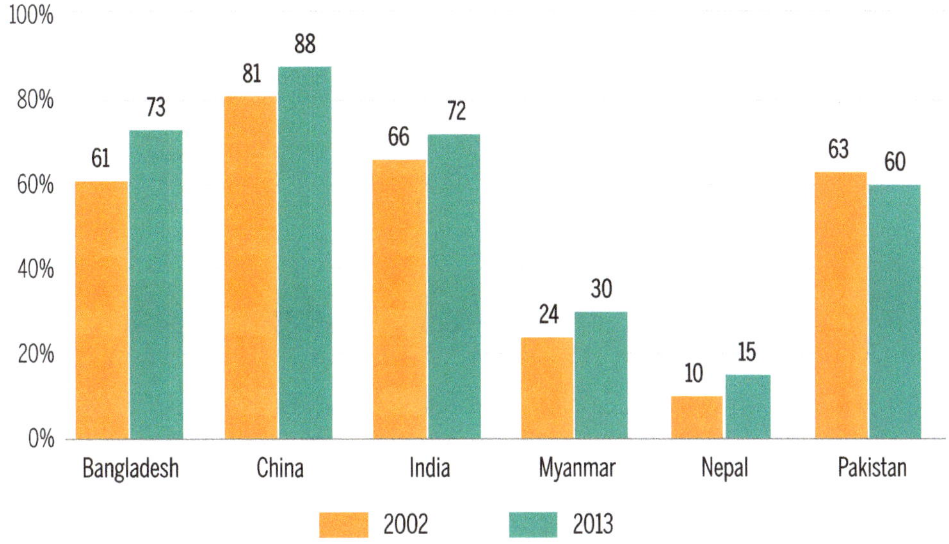

Fig. 6.3 Percentage of fossil fuels in the total primary energy supply in six HKH countries (Based on IEA data from OECD/IEA 2015 World Energy Outlook © OECD/IEA 2015 World Energy Outlook, www.iea.org/statistics, Licence: www.iea.org/t&c; as modified by ICIMOD)

Table 6.3 Electricity generated by fossil fuels in HKH countries (installed capacity)

HKH countries	Year	% of the total	HKH countries	Year	% of the total
Afghanistan	2012	35.4	India	2016	69.3
Bangladesh	2013	97.7	Myanmar	2012	24.8
Bhutan	2015	0.5	Nepal	2016	6.2
China	2016	64	Pakistan	2015	67.7

Source CIA, The World Factbook (2016) (accessed June 2017 at https://www.cia.gov/library/publications/the-world-factbook/index.html)

Table 6.4 Urban population (% of total population) in HKH countries

HKH countries	2007	2015	HKH countries	2007	2015
Afghanistan	23.6	26.7	India	29.9	32.7
Bangladesh	28.2	34.3	Myanmar	29.9	34.1
Bhutan	32.5	38.6	Nepal	15.8	18.6
China	45.2	55.6	Pakistan	35.4	38.8

Source World Bank (2017). https://data.worldbank.org/indicator/SP.URB.TOTL.IN.ZS?end=2015&start=1960

the scale of poverty in HKH countries, the unaffordability of modern fuels and associated appliances may have led to a higher demand for biomass energy. However, in recent years, income in the region has risen (owing in large part to tourism development, cash crops, access to information, remittance income, and entrepreneurship development), and so has the affordability of modern fuels and appliances. In some parts of the region, especially in Nepal, foreign remittance has been a key factor behind the rise in household income.

Urbanization and industrialization: Most studies confirm that urbanization and industrialization increase energy demand but the relationship varies at different stages of development (Li and Lin 2015). In China, Yang et al. (2016) confirm that urbanization rates are a valid predictor of electricity consumption. The share of urban population in HKH countries is rising (Table 6.4), and will eventually lead to greater demand for modern forms of energy (Duan et al. 2014) (see Fig. 6.4). It is also evident that outmigration from the HKH is relatively high; people shift to urban areas in search of better income-generating options (Wu et al. 2014). In urban areas, more people own technologies/appliances (e.g., cars) that demand more fossil fuels. Industries are also growing to meet the demands of growing populations and to achieve economic growth. Urbanization may reduce the use of biomass as urban households are more likely to opt for electrical appliances. But, if electricity is produced from fossil fuel, then it is unlikely to help in moving towards clean energy transition.

Diversification and intensification of agriculture: The region is diversifying and intensifying agriculture to improve food production. This demands more irrigation, and then more electricity to power the irrigation. For example, in Balochistan, Pakistan, the demand for electricity for irrigation has increased to support growing more types of fruit in this arid region (Khair 2013). As many HKH countries provide subsidies for electricity used in groundwater extraction, this leads to overexploitation of groundwater as well as increased electricity demand (Khair 2013; Rasul and Sharma 2016). However, one must note that not all agricultural intensification is electricity-based, and the energy intensity of agriculture in hills and mountains will not be as high as in the plains, because agriculture in the hills and mountains continues to be mostly rainfed (see Chap. 8).

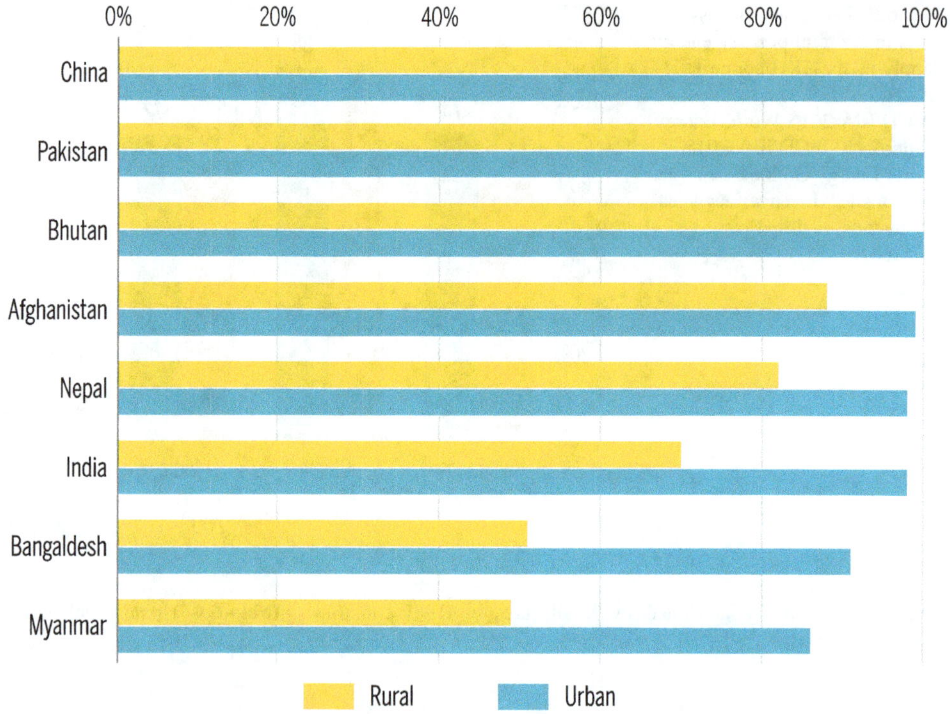

Fig. 6.4 Percentage of urban and rural population with access to electricity in 2014 (*Source* IEA and World Bank 2017)

6.2.2 Energy Demand and Consumption Characteristics

There has been no realistic assessment of the underlying energy demand-and-supply patterns and trends in the HKH region since 1999, when Rijal carried out a comprehensive energy assessment in the HKH regions of China, India, Pakistan and Nepal (Rijal 1999). We must, therefore, rely primarily on national-level data to draw these inferences.

National-Level Final Energy Consumption Pattern in HKH Countries: Table 6.5 provides a summary of the national-level total final energy consumption (TFEC) trends by fuel type and sector for 2013. The total energy consumption of HKH countries as a whole (excluding Afghanistan and Bhutan) grew at an average annual rate of 4.6% over the period 1995–2013. Between 2008 and 2013, Afghanistan recorded the highest annual average growth rate of 16.5% followed by Bhutan (9.7%). China and India dominated the other HKH countries in total energy consumption, consuming around 75% and 20% overall, respectively. Bhutan and China had the highest overall per capita energy usage, and both these countries had made rapid progress in securing access to basic energy for their rural and mountain populations.

In the absence of data, and in terms of fuel composition, the practice in Bhutan and Nepal (which are largely located in the HKH region) can be used to infer patterns for the region as a whole. As such, while traditional biomass at an aggregate level contributes only 16% of TFEC, signalling a transition from traditional fuels towards modern fuels in the HKH, its use is still particularly high in Nepal (80%), Myanmar (70%), and Bhutan (57%). Similarly, the sectoral composition of energy consumption reveals that about 45% of TFEC in HKH countries as a whole is used in the industrial sector. However, the share of residential sector TFEC is highest in Nepal (83%), followed by Myanmar (70%), and Bhutan (58%).

HKH-specific energy consumption pattern: As discussed above, an important characteristic of household energy use in the HKH is heavy dependence on traditional biomass fuels (wood, agricultural residues, and animal dung). For example, approximately 90% of the cooking energy needs in rural areas in India are met through traditional sources of energy (Misra et al. 2005). In Nepal (Table 6.6) more than 85% of total domestic energy needs are met through traditional fuels. Several studies over time (ADB 2012; Rijal 1999; Shrestha 2013; UNDP 2013; Ramji et al. 2012) revealed this trend, whereby extraction of fuelwood exceeded the sustainable supply. Although the availability of Renewable Energy Technologies (RETs) is improving, their high up-front cost—and subsidized commercial fuel—make them unattractive.

The cold climate in much of the HKH leads to cooking, space heating, and water boiling being other major energy end uses of the residential sector in the HKH. Figure 6.5 shows the distribution of energy consumption in the residential sector of Nepal. The figure indicates that cooking (61%) and heating (14%) account for the largest share of

Table 6.5 Structure of total final energy consumption (TFEC) in HKH countries in 2013

	Afghanistan[a]		Bangladesh		Bhutan[a]		China		India		Myanmar		Nepal		Pakistan		Total	
	2013	%	2013	%	2013	%	2013	%	2013	%	2013	%	2013	%	2013	%	2013	%
By sector																		
Industry	1239	35.8	6079	24.6	163	8.1	957,004	49.2	179,091	33.9	1894	12.4	614	6.0	17,144	23.7	1,163,228	44.7
Transport	0	0	2893	11.7	0	0.0	258,301	13.3	74,795	14.2	1369	9.0	738	7.3	13,188	18.3	351,284	13.5
Residential	1293	37.3	12,545	50.9	1172	58.5	398,093	20.5	182,882	34.6	10,730	70.4	8469	83.3	36,157	50.1	651,341	25.1
Commercial and public service	611	17.6	443	1.8	188	9.4	72,260	3.7	20,556	3.9	133	0.9	204	2.0	2158	3.0	96,553	3.7
Agriculture/forestry	300	8.7	1141	4.6	0	0.0	39,230	2.0	23,754	4.5	718	4.7	132	1.3	726	1.0	66,001	2.5
Other (non-specified)	22	0.6	42	0.2	479	23.9	21,074	1.1	10,828	2.0	258	1.7	7	0.1	86	0.1	32,796	1.3
Non-energy use	0	0	1525	6.2	0	0.0	197,528	10.2	36,433	6.9	147	1.0	0	0.0	2782	3.9	238,415	9.2
Total (ktoe)	3465	100	24668	100	2002	100	1943490	100	528339	100	15249	100	10164	100	72241	100	2599618	100
By fuel source																		
Coal	943	27.2	652	2.6	22	1.1	718,501	37.0	103,473	19.6	246	1.6	445	4.4	3697	5.1	827,979	31.9
Oil product	1857	53.6	3632	14.7	186	9.3	446,758	23.0	150,022	28.4	2701	17.7	1135	11.2	13,680	18.9	619,971	23.8
Natural gas	148	4.3	7875	31.9	0	0.0	94,141	4.8	26,612	5.0	775	5.1	0	0.0	18,287	25.3	147,838	5.7
Biofuel and waste	258	7.4	8835	35.8	1135	56.7	196,856	10.1	171,242	32.4	10,778	70.7	8280	81.5	29,792	41.2	427,176	16.4
Electricity	260	7.5	3668	14.9	659	32.9	386,971	19.9	76,544	14.5	749	4.9	304	3.0	6785	9.4	475,940	18.3
Other	0	0	0	0	0	0	100,263	5.2	446	0.1	0	0	0	0	0	0	100,709	3.9
Total (ktoe)	3466	100	24662	100	2002	100	1943490	100	528339	100	15249	100	10164	100	72241	100	2599613	100
Per capita (toe/capita)	0.14		0.15		2.65		1.42		0.42		0.29		0.37		0.39			
AAGR	15.19		3.44		9.69		5.12		3.63		2.07		2.37		2.74		4.67[b]	

Source Based on data from IEA (2016). [a]Based on data from United Nations Statistics Division 2015

[b]Includes data from six countries only (excluding Afghanistan and Bhutan)

AAGR, average annual growth rate 2008–13 for Afghanistan and Bhutan and to 1995–2013 for the six other countries; ktoe, kilotonne of oil equivalent; toe, tonne of oil equivalent

Table 6.6 Structure of total final energy consumption (TFEC) in Nepal 2011–12

		Fuelwood (%)	Dung and agri-residue (%)	Coal (%)	Petroleum product (%)	Electricity (grid) (%)	Electricity (renewables) (%)
	National shares (%)	71.10	8.70	3.90	12.30	2.80	1.20
Residential	80.30	94.20	100.00	0.00	15.30	45.70	100.00
Industrial	7.80	2.60	0.00	92.30	10.50	38.30	0.00
Commercial and transport	10.50	2.70	0.00	7.20	65.10	12.40	0.00
Agricultural	1.20	0.00	0.00	0.00	9.00	2.20	0.00
Others	0.10	0.00	0.00	0.00	0.00	1.40	0.00

Source WECS (2014)

Table 6.7 Trend in energy efficiency in HKH countries

	Primary energy intensity[a] (megajoules per 2011 USDPPP)			Change in energy intensity[b] (%)					Avoided energy consumption[c] (petajoules)
				Primary energy	Final energy				
					Agricultural	Industry	Services	Residential	
	2010	2012	2012–14	2012–14	2012–14	2012–14	2012–14	2012–14	2013–14
Afghanistan	2.94	2.98	2.64	–6.00	4.93	5.69	5.19	0.3	–4.59
Bangladesh	3.44	3.3	3.13	–2.61	–5.82	–4.28	–1.04	1.79	–13.45
Bhutan	12.55	11.56	11.06	–2.19	2.71	8.01	2.92	–1.41	–2.94
China	8.68	8.19	7.43	–4.74	0.5	–4.95	–3.58	2.68	–2906.04
India	5.35	5.2	4.94	–2.49	3.19	0.42	–5.17	1.33	–558.27
Myanmar	3.15	3.1	3.24	2.29	27.03	Na	Na	0.16	15.12
Nepal	7.97	7.27	7.67	2.69	6.06	1.33	–0.25	6.02	–7.31
Pakistan	4.87	4.67	4.43	–2.58	4.5	–3.42	–2.03	–0.77	–84.83

Sources IEA and World Bank (2017)
Na refers to not available
[a]Primary energy intensity—an imperfect proxy indicator to measure energy efficiency—is the ratio of total primary energy supply (TPES) to gross domestic product (GDP), measured at purchasing power parity (PPP) in constant 2011 US dollars (USD). This indicates how much energy is used to produce one unit of economic output. A lower ratio indicates that less energy is used to produce one unit of economic output
[b]Refers to the average annual growth rate of primary energy intensity between two years. Negative values represent improvements in energy intensity (less energy is used to produce one unit of economic output), and vice versa
[c]Refers to energy saved because of energy efficiency improvement during the period. A negative value means reduced energy use (avoided energy use) because of energy intensity reduction

energy consumption in Nepal. This might be the case for the mountain regions of other HKH countries as well, though no data are available to confirm this.

In Bhutan, as in other countries, the choice of fuel type for cooking is affected by a multiplicity of factors including income level, household wealth, age, gender, head of household educational level, access to electricity, and location (Rahut et al. 2016). Based on these factors, people choose and shift the type of fuel used along the energy ladder (Cai and Jiang 2008). Electricity is so versatile in meeting most energy services that it is also an important means of supply. Furthermore, the population in rural areas is almost double that in urban areas of Bhutan. In Nepal, a significant proportion (45%) of electricity supplied in the residential sector is from the grid. In the past almost all electricity was supplied from domestic hydropower plants. Power shortages in recent years have led to back-up diesel generators and electricity imported from India becoming prevalent too.

Some common cottage industries in the HKH include agro processing, saw mills, potteries, blacksmiths, dairies, workshops, and bakeries. Process heat demand in these industries is generally fulfilled by fuelwood and other biomass, petroleum products, coal, and electricity. Motive power is met by petroleum products and electricity (Rijal 1999). The key commercial sectors in the HKH are local businesses, hotels, and restaurants. Tourism is an emerging sector for employment and income generation in the HKH, and will increase the commercial share of energy consumption as well as of liquid fuel consumption and electricity demand.

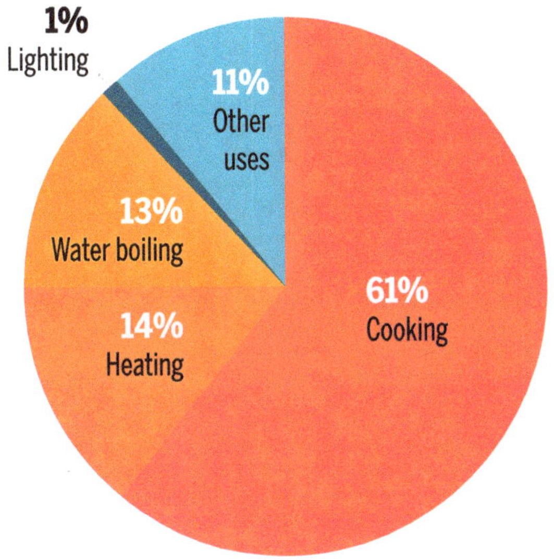

Fig. 6.5 Utilization of Energy in residential sector, Nepal (*Source* WECS 2014)

The agricultural sector also requires significant energy. In Nepal, its energy consumption is 9.9% of the total petroleum product consumption and 2.2% of total electricity consumption (Table 6.6). The end uses to which this energy is applied in Nepal are tillage (52%), irrigation (32%), and threshing (13%) (WECS 2014). The higher demand for electricity and petroleum products reflects the gradual shift from traditional subsistence agriculture to high-value commercial products such as horticulture, medicinal plants, herbs, and vegetables (Rasul 2014a). Some parts of the HKH have also started deploying solar photovoltaic (PV)-based irrigated systems in recent years (Mukherji et al. 2017).

Energy Efficiency: The pace of improvement in energy intensity as an imperfect proxy of energy efficiency varies across HKH countries (Table 6.7). From 2012 to 2014, primary energy intensity declined in all HKH countries except Myanmar and Nepal, with the highest improvement recorded in Afghanistan and China. Over the same period, China avoided TFEC of 2906 petajoules, mainly as a result of efficiency improvement in two energy-intensive sectors (industry and service). India avoided TFEC of 558 petajoule through significant efficiency improvement in the service sector, resulting in the second highest energy savings worldwide after China (IEA and World Bank 2017). Interestingly, in 2012–14 the intensity of final energy use in the residential sector increased in all HKH countries, implying that there is scope for improving energy efficiency by moving away from the highly inefficient combustion of biomass fuel primarily used in this sector. This is especially needed in the rural areas of the HKH region.

6.2.3 Ensuring Sustainable Energy Supply

Most HKH countries rely heavily on commercial energy sources (Fig. 6.1) and over the last few years the share of the supply of commercial energy has increased in all HKH countries (Table 6.8).

The HKH has large potential for hydroelectricity that could help to reduce the dependency on fossil fuel-based electricity in HKH countries (Vaidya 2013). Compared to its potential, installed capacity in the region remains very low. Yupapin et al. (2011) report that Myanmar has the potential for installed capacity of more than 39 GW of hydroelectricity, but has only installed 0.32 GW. Likewise, Nepal has only been able to install less than 2% of its total potential (Ale and Shrestha 2008). In these countries the share of hydropower in total electricity generation is high (Table 6.10). If hydropower were tapped more fully, it would be possible to meet the domestic demand, as well as a significant share of regional demand for electricity through regional cooperation (Rahman et al. 2011; Srivastava and Misra 2007).

The latest national-level data show that the share of renewable energy use in TFEC varies across HKH countries, ranging from 17% in Afghanistan and China to 86% in Bhutan. Traditional solid fuel dominates renewable energy

Table 6.8 Trends in the share of energy sources in total energy consumption in HKH countries

HKH countries	Year	Commercial energy (%)	Trend	Renewable energy (%)	Trend
Afghanistan	2008	75	↑	25	↓
	2011	85		15	
Bangladesh	2002	61	↑	39	↓
	2013	73		27	
Bhutan	2008	10	↑	90	↓
	2011	14		86	
China	2008	82	↑	18	↓
	2013	89		11	
India	2002	67	↑	33	↓
	2013	74		27	
Myanmar	2002	24	↑	76	↓
	2013	30		70	
Nepal	2002	10	↑	90	↓
	2013	15		85	
Pakistan	2008	59	↑	41	↓
	2013	62		38	

↑ arrow shows an increasing and ↓ arrow shows a decreasing trend
Source United Nations Statistics Division 2013 (for Afghanistan and Bhutan); IEA (2015) (for the other countries)

Table 6.9 Share of renewable energy in total final energy consumption (TFEC) in HKH countries

	Share in total final energy consumption (%)										Total final energy use (petajoules)
	Total renewable energy		Solid biofuels		Hydro	Liquid biofuels	Wind	Solar	Geothermal	Other (biogas, renewable waste, marine)	
			Traditional use	Modern use							
	1990	2014	2014	2014	2014	2014	2014	2014	2014	2014	2014
Afghanistan	15.9	16.8	8.82	0	7.93	0	0	0	0	0	146
Bangladesh	71.7	37.5	37.26	0	0.18	0	0	0.04	0	0	1018
Bhutan	95.9	86.7	74.81	0.15	11.7	0	0	0	0	0	61
China	34.1	17.1	10.21	0.19	4.12	0.09	0.61	1.2	0.26	0.42	76,546
India	58.7	36.5	26.6	7.6	1.62	0.07	0.46	0.17	0	0.02	21,550
Myanmar	90.9	68.5	63.21	2.09	3.22	0	0	0	0	0	700
Nepal	95.1	84.4	78.19	0.87	2.9	0	0	0	0	2.41	483
Pakistan	57.5	47.2	39.17	4.89	3.1	0	0.04	0	0	0	2991
All HKH countries		28	20.12	2.97	3.15	0.07	0.49	0.72	0.14	0.28	105,509

Source IEA and World Bank (2017)

Table 6.10 Electricity generation from various sources in HKH countries (2015)

Country	Total electricity production (Gigawatt-hours)	Sources of electricity production (as % of total)			
		Hydropower	Other renewable sources[a]	Fossil fuels[b]	Nuclear power
Afghanistan	1034	86.1	0.0	13.9	0.00
Bhutan	7748	100.0	0.0	0.0	0.00
Bangladesh	59,011	1.0	0.3	98.8	0.00
China	5,814,573	19.4	3.9	73.7	2.94
India	1,354,382	9.0	2.7	85.5	2.76
Myanmar	15,970	59.5	0.0	40.5	0.00
Nepal	3503	99.8	0.2	0.0	0.00
Pakistan	115,073	29.5	1.3	65.2	4.00

Note [a]It includes non-hydro renewables such as solar, wind, geothermal etc.
[b]It includes coal, natural gas, oil etc.
Source Based on data from United Nations Statistics Division for 2015, available online at https://unstats.un.org/unsd/energy/yearbook/2015/t32.pdf

use in all HKH countries, with the share of modern renewable use in the TFEC being marginal (Table 6.9).

In addition to hydropower, there is a large potential for electricity generation from other renewable sources in the HKH. As shown in Table 6.10, large amounts of electricity are being produced from hydro, but renewable sources—such as wind or decentralized PV—and small-scale hydropower projects have been increasing in the region in recent years.

Extending national grid electricity to sparsely populated, remote, and low-density settlements in the HKH is difficult and might not contribute much in improving access to electricity in these areas. Small-scale hydropower projects may be more suitable for the HKH. These systems could help in ensuring sustainable energy and improve rural economies and, in the process, check or reduce outmigration to urban centres (KC et al. 2011). For instance, in Afghanistan, nearly 50% of the beneficiaries of small-scale community-based renewable energy projects in selected areas reported that the projects helped them to improve their living conditions (Shoaib and Ariaratnam 2016).

Loka et al. (2014) found that decentralized micro-grid PV systems could benefit remote villages because they are both viable and cost effective. Proietti et al. (2015) successfully tested the possibility of electricity generation from wind in high-altitude remote regions of Nepal. Chauhan and Saini (2015) found a potential for integrating various sources—such as hydropower, biogas, solar, and wind—into off-grid energy systems in the mountainous regions of India. In addition to electricity generation, solar energy can potentially be used for cooking and heating (Badran et al. 2010). Thus, off-grid energy solutions (such as PVor mini-hydro) often offer better options for providing electricity in remote areas. Small-scale hydro (micro, mini) can be cost effective in some situations but in others it can be quite costly. PV are now often cheaper, although significant untapped economic potential for mini- and micro-hydro does appear to exist in the HKH. The use of solar power for cooking can reduce the dependency on other fuels (Yettou et al. 2014), such as firewood and fossil fuels.

To summarize: universal electricity access (through a rational combination of grid-connected and off-grid power) and elimination of inefficient use of traditional sources for cooking and heating (through a sustainable supply of clean, affordable, and demand-responsive energy and technology options) should be highlighted as a priority in national energy agenda. This would also entail recognizing that in the short-to-medium term, displacing traditional fuels completely is unlikely owing to economic and cost imperatives; but economic potential exists for greater use of small-scale renewable energy options. Some strategic options for sustainable energy development in the HKH may include:

- establishing an HKH-specific energy database for evidence-based policy making;
- optimizing effective use of biomass through advanced cookstoves, upgrading the quality of biomass fuels, and increasing access to cleaner fuels such as liquefied petroleum gas (LPG) and electricity to meet cooking and heating needs while addressing indoor air pollution;
- maximizing use of new and renewable decentralized resources and technologies, not only to sustain and increase economic activities, but also to reduce human drudgery, particularly that of women and children;
- initiating large-scale development of hydropower to generate revenue for alleviating the poverty of mountain communities while ensuring overall development of mountain areas within environmental and social limits; and
- developing an integrated approach for bridging demand-and-supply gaps. This could support the principles of energy self-sufficiency, energy efficiency improvements, diversification of revenue sources for mountain communities, pricing, market and regulatory frameworks, and opportunities for mountains to become quality exporters of clean energy.

6.3 National Energy Policies, Programmes, Institutions, and Markets

Energy policies and programmes follow many diverse models and exhibit differences in terms of institutions and markets within the HKH. This assessment highlights that:

- several examples of success and policy experiences with regard to electricity access, decentralized energy, and energy efficiency could be scaled up. In particular, micro-hydro, biogas, improved cook stove (ICS) implementation, and solar programmes have made the HKH a leader in renewable energy and energy efficiency;
- existing national policy frameworks primarily focus on electrification (power sector) with limited attention paid to clean energy for cooking;
- in most HKH countries there is no separate energy policy framework for off-grid-based rural electrification, although some countries do have renewable energy development agencies, off-grid schemes, and programmes;
- generally, the HKH lacks supportive policy, legal, and institutional frameworks and innovations in mountain-specific technology and financing, or enhanced multi-stakeholder capacity building at all levels, for scaling-up successful energy programmes in off-grid mountain areas; and
- there is no one-size-fits-all solution for ensuring universal access to modern forms of energy given the diverse sources of energy supply, a wide portfolio of technologies, and a variety of institutional and local circumstances. Therefore, policies and programmes must be tailored according to local needs, resources, and existing institutional arrangements and capabilities to deliver co-benefits.

6.3.1 Energy in Relation to National Development Strategies

A review of national policy papers[1] indicates that access to energy, energy security, and regional cooperation are the key stated energy objectives for every HKH country. The countries also share a commitment to electrifying all communities and increasing their use of renewable energy. The target year for complete electrification differs from country to country and has often been pushed back. For example,

[1]National policy papers reviewed include, among others, Energy Sector Strategy (2007/08–2012/13), Rural Renewable Energy Policy (2013) and National Development Strategy (2008–2013) from Afghanistan; Power Sector Master Plan (2016), Renewable Energy Policy (2008), Perspective Plan 2021 and National Energy Policy (2004) from Bangladesh; Integrated Energy Policy (2006), Electricity Act 2003, and National Electricity Policy (2011) from India; Intended Nationally Determined Contributions and Sustainable Development Goals (2016) —National Report (2016–2030) from Nepal; Pakistan Energy Vision 2035; Pakistan National Power Policy 2013; Myanmar National Energy Policy 2013 (draft) and National Sustainable Development Strategy of Myanmar 2009.

India's target is to achieve power for all by 2019, whereas it is 2030 for Myanmar (World Bank 2015). Similarly, Bangladesh envisions achieving universal access to electricity by 2021 (Government of the People's Republic of Bangladesh 2012) and 100% clean cooking solutions by 2030 (Government of the People's Republic of Bangladesh 2013). The Nepalese government's SDG report states that 99% of households will have access to electricity by 2030 and only 10% will be using solid fuels for cooking (Government of Nepal 2016). Afghanistan's target is to achieve near 100% access in the urban areas and 65% in rural areas by 2032 (ADB 2013).

The region has several examples of on-grid and off-grid electrification that have achieved varying degrees of success. Some HKH countries have demonstrated scaling-up off-grid initiatives that are globally recognized as successful. China has been a leader in parallel development of off-grid and on-grid electrification to achieve 100% electrification, as well as having one of the largest cookstove and biogas programmes in the world. Unlike many other countries following the top-down approach to rural electrification, China's success is rooted in a bottom-up approach with strong government commitment, active local participation, technological flexibility and diversity, strong emphasis on rural development, and capacity building and training, all of which provide lessons for other countries in the HKH region (Bhattacharyya and Ohiare 2012). Likewise, Bangladesh's IDCOL is recognized as the world's fastest growing off-grid electrification programme and has already deployed 4 million solar home systems using innovative financial and technology packages. This has demonstrated that electrification programmes targeted at poor communities can be scalable and sustainable (Vinci et al. 2015). India has a number of programmes including the Jawaharlal Nehru National Solar Mission (JNNSM) for mini-grids and off-grid applications. Unlike Bangladesh, India adopts a slightly different approach to promotion and channels finance to banks; however, not as many people are taking loans as had been hoped primarily due to perceived difficulty in rectifying non-functional devices. In Nepal more than 10% of the population has access to off-grid energy through micro-hydro, solar home systems, and domestic biogas. Afghanistan and Pakistan have included off-grid mini-grid programmes under rural infrastructure programmes such as the National Solidarity Programme and Pakistan Poverty Alleviation programme. There is still room for improvement, but there is now sufficient experience from other HKH countries to evaluate these initiatives and to provide policy recommendations about which programmes work best, and under which conditions. Most countries in the HKH now have a policy framework for rural electrification. Experience shows that while a market-driven approach to off-grid renewable energy deployment is key to bringing technology, financing, and product innovation, the governments have an important role in ensuring that the appropriate environment is established for private sector investment. This will accelerate the expansion of modern energy services in a financially sustainable manner (Vinci et al. 2015).

Despite the progress and efforts made so far, existing governance arrangements and policy frameworks for the energy sector are insufficient to address the broad range of technical, capacity, and policy and regulatory barriers which have hindered promotion of decentralized sustainable energy solutions in the HKH. Rahman et al. (2012) observe that many feasible hydropower projects are located in remote and mountainous areas (where other infrastructure, such as accessible roads and high-voltage transmission lines, does not exist), highlighting the gaps in policy coherence. Further, there are no specific initiatives to improve the overall household connection level (Palit and Chaurey 2011). The sectoral division of responsibilities without full coordination for integrated planning has generally been the rule rather than the exception in most energy policy planning processes. This has limited the scope for treating energy, water, and food in a combined way to maximize synergy and manage trade-offs (Sharma and Banskota 2005). In some cases, ignorance of mountain specificities is reported to have led to improper choice of energy technologies and institutions (Rijal 1999; Sharma 2007). Social inclusion and gender sensitivity in energy policy analyses and design is still not fully integrated into mainstream energy development activities in the region (Box 6.1). Finally, existing national development policy strategies pursued by all HKH countries see energy more as an advancement of electrification, and place limited emphasis on energy for clean cooking, heating, and mechanical power solutions in rural areas.

Box 6.1 Engendering energy and empowering women

Throughout the Himalayan region, women face the burden of fetching heavy loads of fuelwood and spend hours in drudgery to meet their daily energy and water needs. However, the importance of bringing a gender perspective to energy policy analyses and design is still not fully appreciated in the region. A recent analysis of selected countries revealed that electricity policies rarely address gender issues explicitly and that policies are mostly gender-blind. For energy to become an instrument for poverty alleviation and sustainable development, a fundamental readjustment of public policies is needed. This should recognize the differentiated needs of women and men and, at the same time, also focus on integrating women's roles and needs in decision making for supply chains and using energy. Gender mainstreaming alone may encounter challenges because women may not be in a position to participate on an equal basis because of their heavy workloads at home, poor access to finance, low literacy, and entrenched gender norms (EFEWEE 2016; Sharma and Banskota 2005).

6.3.2 National Energy Policy Framework, Programmes, and Markets

Energy for lighting and enterprises

A review of the electrification programmes in the HKH indicates that most countries in the region have established organizations dedicated to rural electrification or formulated schemes with supportive legislation to extend electrification primarily through central grid extension, either using a government-owned utility model or rural electricity cooperatives. In most cases, electrification plans for mountain communities are included within the national energy plans of the respective countries, although in many cases their special requirements have not been taken into account when formulating the plans.

All HKH countries are targeting universal access to electricity through a two-pronged approach: (1) rapid extension of the national grid; and (2) off-grid electricity for areas the grid may not reach. Off-grid technologies have been used either through the creation of local mini-grids based on solar and/or micro-hydro, or by disseminating household-level technology such as solar home systems for lighting and powering other low-load devices. For example, Afghanistan has acknowledged its national grid will not reach 20–25% of the population within the next 20 years. These underserved populations will be supplied through decentralized renewable energy sources. In India, the national rural electrification programme also includes a decentralized distributed generation component for 3500 villages, mostly in the eastern Himalayan mountain provinces. However, most countries focus on physical infrastructure rather than on the quality, availability, and affordability of supply: the latter are equally important but vary from place to place.

The electricity grid has been extended mainly by the utility-based delivery model (e.g., India, Bhutan), by rural electricity cooperatives (e.g., Bangladesh), or by community rural electrification (e.g., Nepal). China and India have national programmes that subsidize the extension of grid electricity to rural areas. For areas where grid extension was found to be economically daunting, households are covered by off-grid solutions, implemented primarily by state agencies. Bangladesh and Nepal have been following both community-led and private sector models (with partial subsidy) for expanding their rural electrification programme through grid extension and off-grid interventions. In Nepal, the state-owned Nepal Electricity Authority has achieved considerable success in promoting electrification in close cooperation with community-run local cooperatives (Box 6.2). Nepal used an 80/20 (government/local) financing model for grid-extended rural electrification for many years. According to Nepal's 2016 renewable energy subsidy policy, the subsidy amount differs by technology and region, but generally covers 40% of the total costs, with approximately 30% coming from credit and 30% from private sector investment, communities, households, in-kind, and/or cash (MoPE 2016).

Box 6.2 Community Rural Electrification Programme (CREP) in Nepal

CREP is a grid-based rural electrification programme, was launched in 2003 to expand grid-based rural electrification, primarily using hydro resources and involving the community. Community involvement was sought to achieve operational efficiency in the distribution sector, which had seen high system losses and poor revenue collection for some years. Consumer associations, typically in the form of village cooperatives, take responsibility for managing, maintaining, and expanding the rural distribution of electricity. In addition to local management, communities also provide a portion of the required funding (around 20%) in cash and kind, and the remainder comes from government sources. More than 230 cooperatives in various parts of the country have reportedly entered into agreement for electricity distribution with Nepal Electricity Authority (NEA) (Palit and Chaurey 2011).

Progress in the use of off-grid solutions for rural electricity supply in the HKH has been mixed. The most common technologies are solar PV and micro-hydro systems. While micro-hydro has been used to supply local mini-grids, solar PV technology applications cover solar lanterns, solar home systems, and solar PV-based mini-grids.

In terms of country coverage, Nepal and China have used micro hydropower-based mini-grids extensively. Nepal's Rural Energy Development Programme (REDP) is widely cited as one of the best examples of off-grid electrification programmes, and is centred around the decentralized and participatory decision making and a holistic development approach. India has implemented almost all off-grid electrification technologies, but solar PV has been the preferred modality in the mountains. Small hydro projects in the Indian mountains, on the other hand, feed into the national grid.

Most off-grid electrification programmes in the HKH have been grant-based and donor-driven, and continue to be so in Myanmar, Nepal, and Pakistan. Markets have emerged, however, for stand-alone solar systems in Bangladesh and India (Palit and Bhattacharyya 2014). Palit and Chaurey (2011) indicate that while community-based models are often adopted for mini-grid-based electrification, fee-for-service, leasing, and consumer financing were used to promote individual solar home systems. Bangladesh has covered approximately 4 million off-grid households with solar home systems using an innovative consumer financing model; this consists of a combination of soft financing, institutional development, and product buy-down grants. Local grassroots organizations with experience in micro-financing were engaged to assist this approach. Other countries—such as Afghanistan, Bhutan, Myanmar, and Pakistan—have developed off-grid systems relatively slowly, although they have made progress in recent years. The role of the private sector is paramount in providing cost-effective solutions that do not depend on public subsidies alone; these are often run by local entrepreneur-driven supply chains (for example, the Lighting India programme).

Energy for cooking and heating

While electricity access programmes in HKH countries have received the necessary attention from policy planners and governments, energy access for cooking continues to be an overlooked but critical issue in all countries of the region. Policy makers have also failed to influence a shift from biomass-based cooking in rural areas. From India, Srivastava and Rehman (2006) observe that subsidies on LPG and kerosene have mostly benefited middle- and high-income rural households and the urban poor. Palit et al. (2014) find that India's 2010 National Biomass Cookstove Initiative (NBCI) is neither well-organized, nor has received the resources necessary for it to be successful in comparison to the national rural electrification programme.

Today, most cash-poor rural households, especially the mountain community, continue to use biomass (available at zero cash outlay) instead of commercial cooking fuels, even when those commercial fuels are subsidized. Part of the reason is that the time saved by using commercial cookstoves does not necessarily result in increased income-generating time for rural populations. Also, a lack of awareness about the health benefits of cleaner cooking fuels is not yet appreciated locally because of ingrained gender biases. Electricity, on the other hand, is regarded as an aspiration which can also provide people the opportunity to earn more. Cleaner cooking devices, however, are considered expensive and thus usually avoided (Palit et al. 2014), again as a result of clear gender biases where women's health and safety receive lower attention. A study by Kishore and Spears (2014), using national survey data from India, shows that having a son increases the use of clean cooking fuel in Indian households.

Although there has been limited success in shifting from solid biomass fuels to other cleaner sources, some HKH countries have promoted biogas as an alternative cooking energy, and China and Nepal have successful biogas programmes. However, in Bhutan, the shift from biomass is towards electric devices for cooking (Palit and Garud 2010). Induction cookstoves were introduced into one of the Indian mountain states some years ago, but Banerjee et al. (2016) observe that the intervention has largely replaced LPG and not necessarily firewood from the fuel mix (Box 6.3). While evidence from the HKH does not yet appear in the published literature, there is anecdotal evidence about the increase in use of rice cookers in rural areas of HKH. In Vietnam, 55% of households in rural areas started using rice cookers after they were connected to the grid (World Bank 2011). In the Khumbu Bijuli Company supplied area in Nepal, 54% of the customers of micro-hydro have sufficient power subscription (1.25 kW) to meet the amount of the electricity used for cooking during off-peak hours. In 2016 the Government of India launched a large-scale clean cooking programme, Pradhan Mantri Ujjwala Yojana (PMUY), targeted to provide 80 million LPG connections to underprivileged households. The connection is provided in the name of a woman member of the household, and half of the up-front total capital cost (equivalent to total ~USD50) required for a new connection is waived, while the balance can be an interest-free loan. While PMUY has enabled 32 million poor women to access subsidized LPG within 20 months since its launch in September 2016, initial reports on programme performance indicate on average only four refills per household. Kar et al. (2018) observe that primary LPG users (i.e., 4–5 cylinders per year) account for only 38% of UJJWALA beneficiaries, and the majority (42%) are more occasional users using 2–3 cylinders annually, or rare users (20%) who have not come back for refills.

Box 6.3 Induction stoves as an option for clean cooking

Clean cooking initiatives generally consider improved biomass stoves, biogas, or LPG as potential carriers. Electricity is rarely considered a fuel option as there is a shortage of infrastructure. In induction cooking, a cooking pot is heated by magnetic induction instead of thermal conduction. A survey of 1000 households in Himachal Pradesh, India, indicates that 84% of the beneficiary households using LPG as a secondary cooking fuel shifted to the induction stoves and that only 5% of the beneficiary households displayed a shift in their primary cooking from firewood to induction stoves. The study concludes that underprivileged households are less likely to use commercial fuels for cooking, either because the expenditure on electricity is likely to increase, or because they can only afford the minimal cash outlay required for biomass. However, there may be ideal areas in mountain regions to introduce electricity as a cooking fuel option, particularly where cheap and reliable electricity from hydropower is available (Banerjee et al. 2016). Of households surveyed in this study, 42% used a 1300 W device for an average of 2 h a day (78 KWh per month), a cost of USD4.10 per month.

Despite these advances, solid fuels will remain an important cooking fuel in the near future, especially in the mountains. For this reason, creating more energy-efficient cookstoves will be an important project, particularly as this technology faces several barriers to more widespread adoption (Khandelwal et al. 2016; Palit and Bhattacharyya 2014). National experiences with stove programmes in the HKH indicate that strong technical and administrative capacity, sound programme design, sustained national-level attention, and high-level government support are imperative for successful cookstove programmes (Kandlikar et al. 2009). The Chinese National Improved Stove Program (NISP) deployed 120 million improved cookstoves from 1983 to 1996, and is widely regarded as one of the world's most successful Chinese projects in energy efficiency and rural development (Smith et al. 2007). Conversely, the Indian National Programme for Improved Chulhas (NPIC) was considered a failure (Hanbar and Karve 2002) despite distributing 90 million improved cookstoves between 1983 and 2002. India's failure is generally attributed to its top-down approach and dissemination of non-user-friendly stoves (though with much higher efficiency level than traditional stoves) and stoves with lower life (e.g., mud stoves). China's success was owing to strong administrative, technical, and outreach competence, and resources situated at the local level (Sinton et al. 2004). China has integrated the Clean Stove Initiative into its action plan for energy conservation and emissions reduction contained in its Twelfth Five-Year Plan (World Bank 2013).

Given the mixed results of many programmes in the past, it is now widely accepted that developing sustainable biomass-based cookstoves requires consideration of not just the thermal performance of the stove. The behaviour of users and their participation in the design, marketing, and maintenance of cookstoves; commercial scalability; and monitoring of effectiveness through education and outreach efforts to promote use also need to be taken into account (World Bank 2013). Limited affordability also hinders the promotion of cookstoves, and so suitable financing mechanisms to support adoption need to be explored (Palit and Bhattacharyya 2014). In terms of a policy framework, governments could support the development of performance and quality standards of advanced biomass stoves and remove import duties or value-added tax (VAT) so that attractive technologies can be imported from other countries in the region. A mountain-specific requirement is that international standards such as the ongoing International Organization for Standardization (ISO) process for improved cookstoves (ICS) includes performance of space heating stoves within the multi-tier framework, along with cookstoves benchmarking efficiency, indoor and outdoor emissions, and safety.

Fuel stacking, when households simultaneously use multiple fuels for the same purpose, is common in the region and also needs to be considered. Cheng and Urpelainen (2014) observe that, owing to constraints on the availability of LPG, stacking of LPG and traditional biomass has grown rapidly in India over the past two decades. The same phenomenon has also been observed in a national-level study by Jain et al. (2015) in India and by Palit and Garud (2010) in Bhutan, as well as in a local level study by Banerjee et al. (2016) in Himachal Pradesh.

Institutional arrangements and governance mechanisms

In all HKH countries, the commercial energy sector is regulated by the government and, in some cases, the government is directly responsible for supply, albeit through multiple agencies. For instance, the Ministry of Energy and Water (MEW) in Afghanistan manages and operates the power sector through nine departments and four public sector organizations.[2] In China, the National Development and Reform Commission (NDRC), its main development planning body, manages the energy sector through the National Energy Bureau. In addition, the National Energy Administration of China is responsible for the development of policy instruments, standards, laws, and regulations. In India, both federal and provincial governments have jurisdiction over electricity. While the jurisdiction of provincial governments includes generation, intrastate transmission, and distribution, the federal government's purview includes policy formulation, generation, and interstate transmission. In the rural electrification sector, principal actors have traditionally been state electricity utilities, because they were responsible for distribution in the states. However, the Indian federal government is now directly implementing a rural

[2]Da Afghanistan Breshna Sharkat, or DABS, is the largest of these Afghan organizations and is responsible for the generation, transmission, and distribution of electricity; operation and maintenance of assets; and sales of electricity and revenue collection.

electricity infrastructure development programme to provide access to universal electricity in a time-bound manner.

Government institutions have also played a key role in promoting off-grid electrification. Almost all countries have dedicated agencies for promoting the renewable energy sector. Nepal started to develop the off-grid market after establishing the Alternate Energy Promotion Centre (AEPC) in 1999. Under the AEPC, donor-supported programmes—the Energy Sector Assistance Programme (ESAP) and REDP—substantially helped to promote the supply of off-grid energy. In India, the Ministry of New and Renewable Energy is the nodal ministry for promotion of renewable energy-based off-grid interventions. Bangladesh, China, and Pakistan also have agencies dedicated to the development of the sector. Although Afghanistan does have a specialized agency, its Ministry of Power has implemented several off-grid networks based on micro-hydro, small diesel units, or renewable energy sources.[3] Most of the micro-hydro and solar home systems in Afghanistan have been installed under the National Solidarity Programme, which was a programme under the Ministry of Reconstruction and Rural Development (MRRD). In Myanmar, off-grid electricity is handled by the Department of Rural Development (DRD) under the Ministry of Agriculture.

6.3.3 Cross-Regional Experiences and Recommendations

Challenges to national energy policy can be categorized in four ways: technical, financial, regulatory, and institutional. According to those categories, we make the following recommendations:

The energy sector in each country is governed individually by legal, regulatory, and policy frameworks with little or no coordination between the energy sector institutions and regulators. While institutional structures exist in all the countries in the form of relevant energy ministries and energy subsector institutions, the ability of the institutions to deliver the desired impacts varies from country to country. Regional coordination between countries is also lacking, and this can multiply regulatory risks for future cross-border energy transactions. This lack of coordination also hinders the exchange of ideas and experience from which HKH countries could mutually benefit.

The rate of success for promoting rural electrification depends on the government's commitment to creating and supporting an enabling environment. Bhutan, Pakistan, China, and India provide examples where we can see that an enabling policy development and targeted approach have helped to increase the electrification rate substantially. All rural electrification projects examined have involved a significant subsidy (especially capital) component. However, different approaches have been adopted for grid-based and off-grid electrification.

Off-grid electrification, through mini-grids or otherwise, has been developed mainly through community-centred projects. The top-down approach has been adopted primarily by extending the grid to rural areas, with planning and implementation undertaken by federal- or provincial-level agencies. The approach to off-grid electricity development thus lacks an organized delivery model, which has hindered its scaling-up despite the huge potential in the region. Based on rural electrification experience in India, Palit et al. (2014) observe that an appropriate institutional structure should be a mixture of both participatory and multi-level approaches, while local issues could be better addressed through a participatory mode of governance, policy, regulatory, and financing at appropriate intermediary and/or higher levels.

Merely fixing targets for village electrification is not sufficient unless effective implementation and monitoring are in place to connect rural households and also to ensure sustained electricity supply. While the level of village electrification in Bangladesh and India is high, the actual number of connected households is relatively low as not all households in the villages are electrified. The key issue is thus to improve household-level connection and to provide a sustained electricity supply to rural areas in line with demand, and not just a satisfactory overall rate of extending the electricity infrastructure to the villages.

Sharing information on energy sector policies and plans, and the respective legal, institutional, and regulatory frameworks and lessons are expected to result in strong confidence building among HKH countries. Such sharing platforms are currently limited and are needed. Enhancing institutional capacity for improved coordination and implementation of the joint activities to promote regional energy cooperation will also be important.

6.4 Challenges and Opportunities for Sustainable Energy

6.4.1 Key Sustainable Energy Concerns in the HKH

As a cornerstone of growth and human development, SDG 7 aims to provide access to modern energy to all people by 2030 in an efficient manner, using modern forms of energy. However, despite all the efforts made thus far, providing

[3]Khan Muhammad Alamyar, Renewable Energy for Sustainable Development, Economic Policy Directorate, Ministry of Economy, pers. comm., April 2014.

Table 6.11 Indicators for evaluating energy sustainability in HKH

Dimension	Indicators	Theme	Measurement description (Units)
Economic	Per capita electricity consumption	Per capita use	Ratio of total final electricity consumption to total population (kWh per capita)
	Household income share spent on fuel and electricity	Affordability	Share of household expenditure on fuels and electricity to the total household income (%)
	Cooking/heating energy conversion efficiency	End-use efficiency	Share of useful energy to the consumption of final energy (%)
	T&D losses	Delivery efficiency	Transmission and distribution line losses (%)
	Share of renewable energy in electricity generation	Renewability	Contribution of renewable energy generation in the total HKH electricity supply (%)
	Diversification of energy in HKH (by fuel types/sources)	Diversification	Diversified energy mix and share of import energy in the total primary energy (%)
Social	HKH population without electricity	Accessibility	Percentage of population without electricity (%)
	Share of HKH population still using traditional solid fuels		Percentage of HKH population/households using traditional solid fuels for cooking/heating and other household energy uses
	Disparity in electricity distribution	Disparity	Ratio of electricity use of lower quintile to electricity use of upper quintile
	Disparity in clean energy distribution		Ratio of clean fuel use of lower quintile to clean fuel use of upper quintile
	Unplanned interruptions/year	Reliability	Number of days of power supply interruption (unplanned) each year
Environmental	GHG emissions from energy production and use per capita	Global impact	Annual GHG emissions from energy production and use per capita (kg/capita)
	Impact of HAP from energy systems	Local impact	DALYS per 1000 people
	Annual rate of change in forest area		Extent of forest land (%)

Source Geng and Ji (2014), Gupta (2008), Mainali et al. (2014), Sovacool (2013), Vivoda (2010)
DALYS, disabled adjusted life years; GHG, greenhouse gas; HAP, household air pollution; T&D, transmission and distribution

sustainable energy access in hilly and mountain regions remains a challenge (Mainali and Silveira 2013). Lack of access to technology, infrastructure, finance, and subsidies, and the high up-front costs of new technologies, pose additional major challenges. Limited investment in the renewable sector in most HKH countries has impeded harnessing of the full potential capacity of renewable energy sources (Pode et al. 2016).

The framework we employ to define and measure sustainable energy determines the way we design and implement any strategies for sustainable energy solutions. Adopting a systematic framework to assess and monitor energy systems has received as much attention in energy policy strategies, especially for achieving SDG 7, to ensure that they are equally robust enough to adapt to anticipated climate-related impacts. Such a framework must be holistic and rooted in the three dimensions of sustainability (social, economic, and environmental) while aligning with the three pillars of sustainable energy (i.e., the targets of SDG 7 and sustainable energy for all (SE4ALL)—access to modern energy, energy efficiency, and renewable energy. Table 6.11 outlines such a framework, using a set of underlying indicators to capture accessibility, affordability, disparity, security, renewability, and environmental concerns (Geng and Ji 2014; Gupta 2008; Mainali et al. 2014; Sovacool 2013; Vivoda 2010).

In the social dimension, two key indicators that describe accessibility are populations with access to electricity and using non-solid fuels. A distinct positive correlation exists between the use of electricity and the human development index (HDI), especially for low- and medium-income countries (Gómez and Silveira 2010). The correlations are even stronger in low-income HKH countries (Table 6.12).

In terms of economics, affordability and secure energy supply are key concerns. Without affordability, people in

Table 6.12 Electricity per capita and HDI of HKH countries (1990–2012)

HKH countries	1990		2000		2010		2012	
	KWh/Cap	HDI	KWh/Cap	HDI	KWh/Cap	HDI	KWh/Cap	HDI
Bangladesh	48	0.386	102	0.468	241	0.546	276	0.563
China	511	0.501	993	0.588	2944	0.699	3475	0.718
India	273	0.428	395	0.496	644	0.586	724	0.600
Myanmar	43	0.352	74	0.425	122	0.520	153	0.528
Nepal	35	0.384	59	0.451	103	0.531	119	0.540
Pakistan	278	0.399	373	0.444	467	0.522	452	0.532
Correlation coefficient	0.927		0.915		0.966		0.965	

Source UNDP (2015), World Bank (2016a, b)
HDI, human development index

Table 6.13 CO_2 emissions from fossil energy in 2016

	Afghanistan	Bangladesh	Bhutan	China	India	Myanmar	Nepal	Pakistan	Total
$MtCO_2$	12	82	1.1	10,151	2431	24	9.1	189	12,899
%	0.1	0.6	0	78.7	18.8	0.2	0.1	1.5	100

Source Global Carbon Atlas (2017). http://www.globalcarbonatlas.org/en/CO2-emissions
$MtCO_2$, million ton of carbon dioxide

Table 6.14 Renewable share in total electricity generation in HKH countries

	Afghanistan	Bangladesh	Bhutan	China	India	Myanmar	Nepal	Pakistan
Share of renewable energy in total electricity generation (%)	85.3	1.3	100.0	22.6	15.4	62.4	100.0	30.2

Source Based on World Bank database for the year 2014, available online at https://data.worldbank.org/indicator/EG.ELC.RNEW.ZS?locations=AF&view=chart

these regions cannot benefit from available energy. The poor normally spend a significant share of their income on electricity and cooking fuels (Mainali et al. 2012, 2014; Pachauri and Jiang 2008). Diversifying energy mixes and supply sources, and increasing the share of renewable energy in electricity generation, can be important energy supply solutions.

Environmentally, reducing carbon intensity and air pollution-related health hazards remain central concerns throughout the HKH (see Chap. 10). Currently, about 80% of GHG emissions from the energy sector in HKH countries come from China, followed by India (19%) (Table 6.13).

A study from Nepal also points out contradictory policies (e.g., subsidies for both renewable energy and fossil fuels) that may, in fact, impede the transition towards sustainable energy (KC et al. 2011). Sheikh (2010), in a study in Pakistan, identifies a lack of policies to encourage private companies in the renewable energy sector as one of the impediments for the transition. Despite all the efforts made, providing access to sustainable energy in the hilly and mountain regions still remains a challenge (Mainali and Silveira 2013). Among HKH countries, the share of renewable energy in total electricity generation is high in Bhutan (100%) and Nepal (100%) but lower in other countries such as Bangladesh (1.3%) and India (15.4%) (Table 6.14).

Figure 6.6 shows the performance of HKH countries in terms of regulatory indicators for sustainable energy (RISE) score. This is grounded in 27 indicators to capture the quality of policies and regulations for the three pillars of SDG 7 and SE4ALL—energy access, renewable energy, and energy efficiency (Banerjee et al. 2017). Of the possible maximum of 100, scores range from 81 in China to less than 25 in Afghanistan, where the energy efficiency performance scoring is the lowest in all HKH countries.

6.4.2 Links Between Energy and the SDGs: Synergies and Trade-offs

Energy is crucial for achieving almost all SDGs. The synergies and trade-offs between energy and other SDG goals happen at multiple levels. McCollum et al. (2017) and Nilsson et al. (2016) define the nature and extent of such

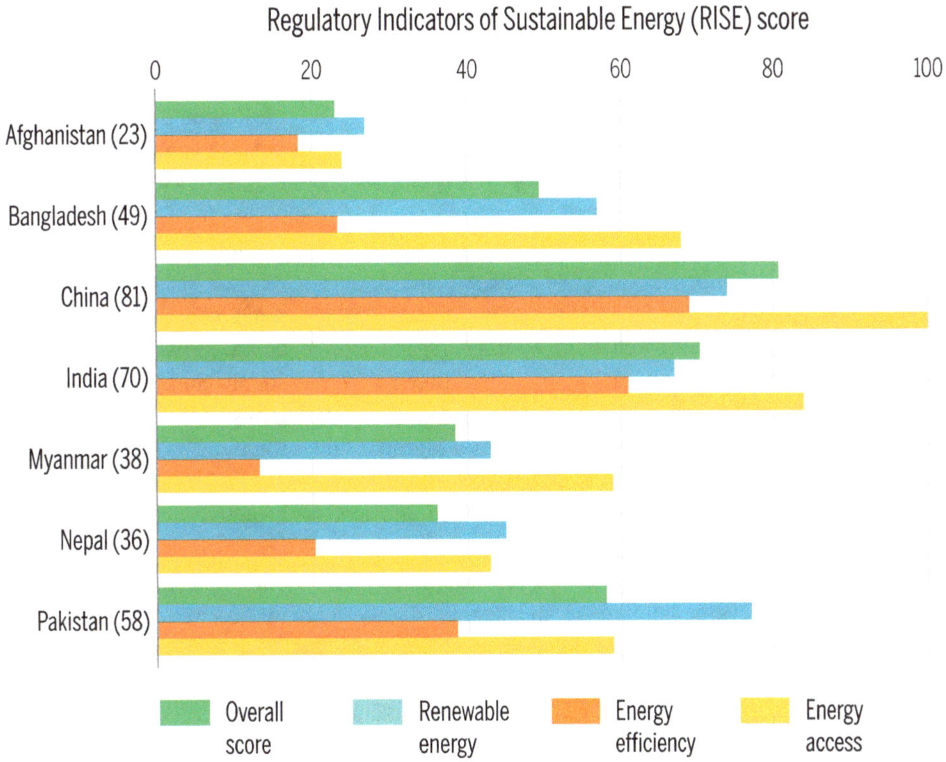

Fig. 6.6 Performance of HKH countries in regulatory indicators of sustainable energy (RISE) (score value in 2016). (*Source* IEA and World Bank 2017)

linkages from synergies (indivisible, reinforcing, and enabling), to no interactions, to trade-offs (constraining, counteracting, and cancelling). Globally, McCollum et al. (2017) show that access to clean energy enables countries to meet all other SDGs. Goal 7 (energy) reinforces all SDGs except Goals 4 (education), 5 (gender), 8 (work and economy), and 10 (inequality). Goal 11 (cities) is inseparable from Goal 7 (energy). Globally, Goal 7 (energy) also presents some level of trade-off with Goals 2 (hunger) and Goal 15 (land) in particular, and constrains the seven other goals. It is important to understand the interlinkages among various SDGs to create synergies, and also to understand that the hierarchy of needs among these goals is important (Mainali et al. 2018). Electrification may not be the first priority if people are facing hunger and need clean water. Failures have been seen in Nepal and Peru where energy programmes wrongly targeted poor villages where hunger needed to be addressed before energy.

In the HKH region, access to clean energy and energy efficiency reduces indoor air pollution (Goal 3), which affects women (Goal 5) and children disproportionately. Women in the HKH are more vulnerable to energy poverty than men. The gender inequality index (GII) published by UNDP in 2012 shows huge gender inequality in HKH countries (ranging from 0.464 in Bhutan to 0.617 in India) (ENERGIA 2015). Access to improved or modern energy sources can reduce the physical burden associated with carrying wood, and frees up valuable time which could be used for other productive purposes, while reducing health risks posed by indoor pollution. However, this link is very poorly addressed in energy policies as well as in the gender policies of the region. Harnessing abundant renewable energy resources reduces fossil fuel use within and outside the region and thus supports climate change mitigation (Goal 13). The HKH has hydroelectricity potential of over 500 GW (Vaidya 2013) of which a large part could be exported with a suitable benefit-sharing mechanism to promote local livelihoods and infrastructure for economic development (Goal 8). Use of biomass as an energy resource, however, must be complemented with relevant technologies, or might lead to the production of black carbon (Goal 13) and unsustainable land use changes (Goal 15). Greater access to energy could provide opportunities for entrepreneurship and stimulate economic development. Access to electricity and clean energy improves livelihoods and the economy, which could help in improving adaptive capacity (Perera et al. 2015).

SDGs can affect the energy SDG Goal. Climate change (Chap. 3) could have several impacts on the HKH region's water availability for hydropower and energy infrastructure, and thus for energy security. Energy intensity could increase owing to the need to pump water (as a result of water table drawdown). Land degradation and potential consequences for agricultural productivity could affect the supply of biomass (ICIMOD 2011).

Understanding how these multiple targets and SDGs cut across different sectors, and are linked, may help in the design and appraisal of common strategies and cross-sectoral policies for integrated development programmes in the region (Mainali et al. 2018).

6.4.3 Assessing the Vulnerability of Vital Energy Services and Systems

Supply-side vulnerability: security of primary energy supply sources: Vulnerabilities in sustainable energy systems in the HKH are often interlinked with hydropower development and losses in agricultural productivity linked to the primary supply of biomass sources. The great dependency of primary energy supplies on biomass indicates a high risk for fuel security in extreme climatic events. Another important mechanism that may discourage the use of clean and modern energy sources is the pricing of energy (IEA 2015). Subsidizing energy is an important instrument for addressing price-induced vulnerabilities in the supply of energy sources to end-users in this region. In spite of this, there are claims that subsidies on energy were generally disproportionately beneficial to richer households and were drawn from other development-related expenditures (Parajuli et al. 2014; TERI and IISD 2012). Globally, studies have shown that energy subsidies generally benefit the rich more than the poor, because of how subsidies are designed and implemented. Capped lifeline tariff subsidies up to a maximum consumption level can be effective in dealing with energy poverty, although the rich also get this subsidy. Overall tariff price subsidies are inefficient policy instruments. Subsidized loans for the poor to access modern energy services (e.g., PV/biogas technologies) or preferential taxation treatment for specific technologies (reduced import taxes and duties), and programmes and policies to support private sector service providers (for example, Lighting Asia) can often be more efficient instruments if designed and applied correctly.

The degree of energy self-sufficiency (indicated in terms of the ratio of alternative energy technologies to the TPES) indicates the capacity of a country to withstand shocks that result from vulnerabilities in the energy supply. Even large countries like India and China are considered poor and vulnerable when it comes to this measure of energy self-sufficiency (Sovacool et al. 2011). Furthermore, the energy import dependence of a country also helps us understand its vulnerability to potential energy shocks. From 1996–2009 in Nepal, the annual growth in commercial energy imports (covering all forms of fossil fuels and excluding electricity) was 2.5%. This is often related to the amount of export earnings spent on imports. For example, in Nepal in the fiscal year 2015–16, the import of petroleum products was equivalent to 97% of the total value of exports (TEPC 2017). However, energy self-sufficiency and reliability or vulnerability are not always positively correlated. For example, a country could be fully self-sufficient through hydropower generation, but be highly vulnerable to prolonged droughts. Energy supply diversity is essential for reducing vulnerability, but so is increased integration and connectivity with wider energy systems.

Demand-side vulnerability: security of energy services in end-use sectors: While supply of energy in HKH countries remains a challenge, demand in these countries has continued to grow. The scale of electricity shortages (gaps between supply and peak demand) in most HKH countries is enormous. For example: 1990 MW for Bangladesh, 10,296 MW for India, 336 MW for Nepal, and 5230 MW for Pakistan; these figures represented 32, 12, 44, and 45% of the total installed capacity of the respective countries, respectively, and shortages are increasing over time (Rahman et al. 2011). Such shortages affect various sectors of the economy. In HKH countries, the contribution of groundwater to irrigation is three-fifths of the region's irrigation water (Shah 2010) and this could increase energy demand as the population grows and the water table falls. Parajuli et al. (2014) stress that the growth in the agricultural sector's contribution to Nepal's GDP was largely influenced by access to energy and the status of energy consumption in that sector. Demand-side vulnerability could be much more complex, emanating from regulatory, perverse incentives from subsidies and very poor energy efficiency (a neglected area in HKH countries and policy settings).

6.4.4 Overcoming Barriers and Seizing Potential Opportunities

Despite all the concerns and vulnerabilities identified above, the HKH region has several opportunities to enhance its energy security in a sustainable manner.

Studies have shown that quicker transition from traditional to modern fuels in cooking can be achieved by coupling a well-targeted subsidy policy along with availability of easy credit (Ekholm et al. 2010; Mainali et al. 2012). The financial burden of such a policy needs to be evaluated alongside the benefits it can bring—namely, reductions in adverse impacts on health and GHG emissions, and general improvements in people's socioeconomic situations. The opportunity that arises here should not be missed.

For energy importing countries, diversification of sources of imported energy can reduce risks (Geng and Ji 2014). Energy resilience is possible through different measures, such as trade and product diversification with respect to petroleum products, increasing storage capacity to meet the regular energy/fuel demand, increasing the stake in decentralized energy, establishing better energy and power connections across borders, and improving energy conservation initiatives. These require structural changes in both demand-and-supply management.

It is critically important to develop strategic plans and policies for meeting the demand for sustainable energy with the aim to address the challenge of increasing electricity deficits, avoiding the dominance of a single fuel type in electricity generation, and decreasing the dependency on imports (ADB 2012). The region has abundant renewable energy resources but these are unevenly distributed geographically (Saroha and Verma 2013), increasing the threat of inter-country conflict (ESCAP 2013). Harnessing available energy resources through mutual cooperation (e.g., technology development, investments, and trade) can strengthen the energy security of the region (ESCAP 2013; Taliotis et al. 2016).

Although biomass is technically renewable, heavy reliance on its traditional use is responsible for serious health effects and environmental problems. Reduced dependency on traditional biomass through more efficient renewable energy technologies is desirable, even though it reduces the share of renewable energy overall. However, improved energy efficiency measures through deployment of efficient cooking and heating technologies, and renovation of buildings, can also increase the renewable energy share by reducing overall final energy consumption (denominator of energy intensity) through avoided energy consumption.

Clearly, integrated thinking around renewable energy and energy efficiency is essential for a transition to a more sustainable energy future in the HKH region. Increasing the share of renewable energy requires not only rapid energy efficiency improvement measures, but also the adoption of major transformative policies. More specifically, a successful energy efficiency strategy requires a number of supply-side and demand-side measures, including:

- careful energy planning through establishing solid data on energy end uses for proper estimation of energy efficiency potential, and setting proper priority energy efficiency objectives and targets;
- development of energy conservation policies such as standards and appliance labelling, as well as mandatory energy audits, while also strengthening the compliance and enforcement mechanism with sufficient dedicated resources and strong institutional capacity to meet these standards;
- a combination of information, awareness, and incentives to encourage consumers to adopt energy-efficient technologies, and producers to invest in technology innovation and meet energy performance standards;
- introducing cost-reflective electricity tariffs through phasing out subsidies on energy prices for all consuming sectors to unlock the potential of energy efficiency; and
- leveraging private investment in energy efficiency by establishing funding mechanisms to jump-start energy efficiency financing to help an energy service company (ESCO) overcome the initial high set-up costs.

There are major challenges in meeting the SDGs (Weitz et al. 2014) and there is a lack of coordination among various line agencies within the energy sector and across the other sectors (water, sanitation, agriculture etc.). However, the opportunities for the HKH region are immense as international technology and financial mechanisms for operationalizing the Paris Agreement on Climate Change are being devised. To benefit from these developments, proper institutional structures and mechanisms must be established for creating a bridge that links multiple global initiatives and targets.

6.5 Future Energy Scenarios and Pathways

Given the current energy scenario, and the identified barriers and vulnerabilities, HKH countries must have a future plan for the region that empowers them in meeting their sustainable energy goal as well as all related SDGs, and to contribute to the climate goals. Future scenarios and transitions must drive sustainable development and emerging concerns such as low-carbon development, which requires a better understanding of possible pathways.

As energy transitions are often slow and lack a one-size-fits-all-solution, reaching such a sustainable future could entail multiple pathways depending on the local context, policy ambitiousness, and political commitment. The lack of data and knowledge gaps on resource availability, economic feasibility studies of sustainable energy technologies, and demand profile could be key limitations in the HKH.

The HKH's future long-term scenario could aim to fully harness hydro and other renewable energies with efficient and clean technologies to allow a sustainable energy transition. For clean cooking energy transition (and especially for transition to 100% electric cooking) the region must aspire to ensure complete replacement of traditional, inefficient sources of energy with clean, sustainable energy options that are efficient, reliable, affordable, and demand-responsive. For electricity access, decentralized (off-grid) electrification is the cornerstone of sustainable solutions in the HKH because of the remoteness of settlements where grid extension is not always feasible.

A broad range of barriers (policies, technical, economic, institutional, regulatory, sociocultural, and environmental) continue to hinder transition to sustainable energy in the HKH in the absence of mountain-specific national energy development strategies. The region must seize the emerging opportunities to confront these barriers in the context of the implementation of SE4ALL, SDGs, and the Paris Climate Agreement.

6.5.1 Future Energy Scenarios of HKH Countries

Of the eight member countries of the HKH, China and India are shifting the centre of gravity of the global energy system towards Asia. However, the situation in the HKH—which accounts for 9% of the population of these countries and less than 7% of GDP—may continue to remain precarious. Using country-level data from China and India to derive sensible directions for individual HKH countries may result in distortions, but nevertheless represents a worthwhile start in beginning the process and identifying knowledge gaps. Five HKH countries have more than 50% of their population residing in mountainous regions. Nepal and Bhutan are fully contained within the HKH and 83% of Afghanistan's population is situated in the Himalaya.

When exploring future energy scenarios of the HKH it is also important to bear in mind the implications of climate pledges on energy choices made by these countries. Bhutan intends to remain carbon neutral with emissions of GHGs not exceeding the estimated 6.3 Mt CO_2eq carbon sequestration by forests. Nepal has committed to achieve 80% electrification through renewable energy sources by 2050, while reducing dependency on fossil fuels by 50% and maintaining 40% of the total area of the country under forest cover. Afghanistan outlined its intended contribution of reducing its GHG emissions by 13.6% below the 2030 business as usual (BAU) scenario (UNFCCC 2015). Myanmar has provided a list of policy actions in the energy and forestry sectors. Pakistan showed commitment to reduce up to 20% of its 2030 projected GHG emissions. China and India committed to 30–45% (from 2007) and 33–35% (from 2005) reductions in GHG emissions per unit GDP by 2030. Many of these targets are conditional: subject to affordability, provision of international climate finance, transfer of technology, and capacity building.

Table 6.15 summarizes the growth projections of the Asian Development Bank (ADB) for the countries of the HKH up to 2035. Unfortunately, no projections exist for the delimited HKH region per se, and these are for the HKH countries *as a whole*. Total primary energy demand under the BAU case is projected to grow by 2035 in all HKH countries, with the highest annual average rate being recorded in Afghanistan (6.5%), followed by Bangladesh (3.7%).

Not only will the total primary energy demand be greater in the BAU scenario, but the share of fossil fuels in the primary energy mix is also likely to increase in all countries by 2035. Table 6.16 presents the future carbon and energy intensity outlook to 2035. While the energy and emissions intensity decreases in BAU and alternative scenarios, Afghanistan's emission intensity increases in the outlook period 2010–35 (ADB 2015). Increasing the share of renewable energy sources and introducing efficient technologies under the alternative scenario will reduce the overall energy demand, improve energy security, and diminish GHG emissions.

Figure 6.7 presents primary energy consumption from 2015 to 2040 under current policies (CP), new policies (NP), and SDG scenarios[4] for the countries of the HKH region; estimates are based on the IEA (2017) and GAINS model.

Total primary energy consumption (TPES) in HKH countries will increase by a factor of 1.7 in the CP scenario from 2015 to 2040. TPES will increase by a factor of 1.5 in the NP scenario, and 1.2 in the SDG scenario from 2015 to 2040, primarily because of energy efficiency measures and large-scale renewable energy penetration in China and India. At the national level, TPES will be reduced by 33% in China and 30% in India because of sustainable energy strategies.

Many studies have projected future energy use, GHG emissions, and air pollutants for individual HKH countries (Amann et al. 2008; Klimont et al. 2009; IEA 2015; Mir et al. 2016; Purohit et al. 2010, 2013). However, none assess future energy use and GHG emissions in the HKH mountain region alone. The global and national scenario of GHG mitigation relies heavily on initiatives such as rapid deployment of renewable energy, greater use of biomass, deployment of best-practice technologies to boost energy efficiency, and employing emerging technologies such as carbon capture and storage. All these have direct relevance for the HKH, which already needs to modernize the biomass sector and harness clean, decentralized energy sources.

6.5.2 Pathways Towards Rural Electrification

Energy consumption is also expected to grow significantly as industrialization, urbanization, and economic growth increase. HKH countries must meet their growing energy demand in an environmentally benign and sustainable

[4]The NP scenario is designed to show where existing policies (as well as announced policy intentions) might lead the energy sector. The CP scenario provides a point of comparison by considering only those policies and measures enacted into legislation by mid-2017. The SDG scenario examines what it would take to achieve the three main energy-related components of the "2030 Agenda for Sustainable Development" adopted in 2015 by Member States of the United Nations. The three energy-related goals are: (1) to achieve universal energy access to modern energy by 2030; (2) to take urgent action to combat climate change; and (3) to dramatically reduce the pollutant emissions that cause poor air quality (IEA 2017).

Table 6.15 HKH country energy outlook through 2035 under BAU and alternative scenarios

	Afghanistan	Bangladesh	Bhutan[a]	China	India	Myanmar	Nepal[a]	Pakistan
GDP (constant 2000 USD billion) 2035	47.7	243.0	3.8	15,871.9	3877.0	135.9	20.0	269.6
GDP growth rate % (2010–35)	6.8	4.4	5.6	6.6	5.7	7.6	3.7	3.4
Population 2035	59.0	187.1	0.9	1381.6	1579.8	55.0	42.0	245.9
BAU case								
Primary energy demand (Mtoe) 2035	4.3 (0.07)	77.6 (0.41)	1.7 (1.84)	4218.1 (3.05)	1441.6 (0.91)	30.3 (0.55)	16.6 (0.4)	145.8 (0.59)
Primary energy demand growth % 2010–35	6.5	3.7	0.8	2.3	3.0	3.1	2.0	2.2
Sectoral energy demand share (%)								
Industry	6	29.1	15.1	33.6	31	11.5	6	33
Transport	30.1	19.3	5.8	16.3	19.7	23.4	9.1	22.6
Other sectors	63.9	45.6	79.1	43.8	41.1	64.6	84.9	40.5
Alternative case								
Primary energy demand (Mtoe) 2035	3.8 (0.06)	68.8 (0.37)	1.5 (1.58)	3418.7 (2.47)	1239.2 (0.78)	29.2 (0.53)	16.3 (0.39)	130.9 (0.53)
Primary energy demand growth % 2010–35	6.0	3.2	0.2	1.4	2.4	3.0	1.9	1.8
Sectoral energy demand share (%)								
Industry	6.2	29.3	22.2	31.6	30.5	11.4	5.8	33.5
Transport	30.9	19.1	9.5	16.3	19.9	21.1	8.6	23
Other sectors	62.9	45	68.4	45	40.7	66.9	85.6	39.2

Figures in parenthesis indicate primary energy demand per capita (toe/person)
Figures may not add up to total because of rounding and other sector
BAU, business as usual; GDP, gross domestic product; Mtoe, million tonnes of oil equivalent
[a]Bhutan and Nepal are fully in the HKH while the rest are partially covered. *Source* ADB (2015)

Table 6.16 Future carbon intensity and energy intensity outlook in HKH countries to 2035

	BAU		Alternative case	
	2035	AAGR (%) 2010–35	2035	AAGR (%) 2010–35
CO_2 intensity (t CO_2/constant 2000 USD million)				
Afghanistan	224	3.1	194	2.5
Bangladesh	793	0.9	547	–0.6
Bhutan	333	–3.4	294	–3.9
China	759	–4.3	526	–5.7
India	957	–2.3	687	–3.6
Myanmar	414	0.2	368	–0.2
Nepal	411	–0.6	385	–0.9
Pakistan	1,081	–0.3	788	–1.6
Primary energy intensity (toe/constant 2000 USD million)				
Afghanistan	89	–0.3	80	–0.8
Bangladesh	319	–0.6	283	–1.1
Bhutan	445	–4.6	383	–5.2
China	266	–4	215	–4.8
India	372	–2.6	320	–3.2
Myanmar	223	–4.2	215	–4.3
Nepal	833	–1.7	815	–1.8
Pakistan	541	–1.2	485	–1.6

AAGR, average annual growth rate; BAU, business as usual; toe, tonne of oil equivalent
Source ADB (2015)

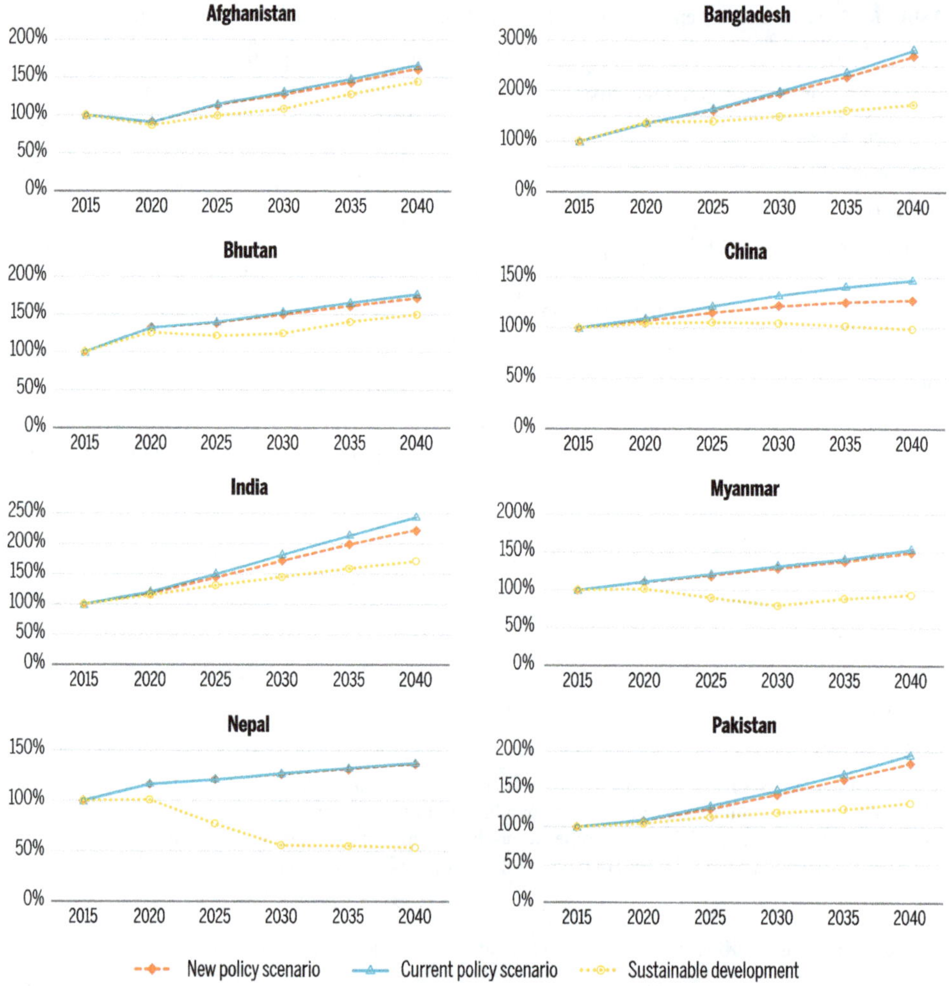

Fig. 6.7 Total primary energy consumption in current and alternative policy scenarios in HKH countries. (*Source* Author estimation based on IEA (2017) and the GAINS model (http://gains.iiasa.ac.at/models/index.html))

manner. To be truly sustainable, an energy system must (Dincer and Acar 2015):

- have minimal or no negative environmental or social impact;
- cause no natural resource depletion;
- be able to supply current and future population energy demand;
- be equitable and efficient;
- protect air, land, and water;
- have little or no net carbon or other GHG emissions; and
- result in safety today without burdening future generations.

In the HKH, discussions on electrification pathways tend to be between the merits of centralized on-grid versus decentralized off-grid electrification, with both routes being promoted in parallel. There has been no comprehensive and systematic analysis to find a balance between the two. Mainali and Silveira (2013) compare three pathways for rural electrification: (1) off-grid renewable energy technologies for individual households, mainly solar; (2) mini-grids (with micro-hydro, diesel generators, and solar, in some cases); and (3) grid extension. Their analysis shows that micro-hydro-based mini-grid technology is the most competitive alternative to electrify isolated and remote rural areas of the HKH. The choice of technology and the centralized on-grid and decentralized off-grid electrification pathway adopted in Nepal seems to be functional. In Afghanistan, the technological pathways for rural electrification are not well defined. A micro-hydro-based mini-grid has been suggested as a more sustainable option in Afghanistan than diesel generators, which were promoted in the transitional phase (Mainali and Silveira 2013).

Because of the challenges of grid extension in mountain areas due to sparse settlement and smaller load demand, the literature favours decentralized renewable electricity in the majority of cases. However, in some scenarios, it is also possible that off-grid solutions are just an intermediate pathway towards grid electricity. Energy conservation and efficiency can also make an important contribution since they reduce supply burdens and help to provide enhanced energy access. For example, Pode et al. (2016) describe a successful rice husk biomass project for sustainable fuel to power rural electrification in Myanmar. Other options such as wind and gasifiers are also discussed as potential options in mountain areas but their applications are not significant. Large hydropower projects are well suited for urban centres of the HKH and provide opportunities for electricity trade with neighbouring regions. Access to renewable energy accelerates growth in the tourism industry, herbal/medicinal products market, cash crops, and other mountain small and medium-sized enterprises (SMEs).

6.5.3 Pathways Towards Sustainable Clean Cooking Energy

The HKH needs to chart its own energy transition pathway for cooking in the future. A meaningful cooking energy transition addressing multiple economic, social, and environmental needs (through a sustainable supply of demand-responsive clean, efficient, reliable, affordable energy options) needs to be started and made operational. The challenge is especially daunting for rural areas, given the isolated nature of rural settlements that require innovative technology interventions for efficiency, fuel shifts, innovations in finance, institutional development, and the design of appropriate policy instruments. To achieve a sustainable cooking and heating energy transition, the HKH must aspire to a 100% renewable future with multiple pathways as a long-term goal. A more detailed quantitative analysis is necessary but, based on the available limited information, we surmise that the HKH could achieve a 100% renewable future through two key pathways. These only differ on whether cooking services are to be fully provided by electricity or also by improved biomass and other sources as well as by technologies.

Pathway 1. Electric future: access to renewable electricity and electric cooking: As discussed above, the HKH has significant renewable energy potential to achieve 100% electrification in the near future.[5] Cross-border hydropower imports from Bhutan, Myanmar, and Nepal to Bangladesh would also be vital to maintain the country's long-term energy security. There is enormous diversity in the types and amounts of fuels used for cooking in households in HKH countries and regions. Leapfrogging from solid fuels (fuelwood, agricultural residues, dung cakes, coal, etc.) to electric or induction stoves[6] using renewable electricity for cooking will skip steps in the energy ladders. Unelectrified households in rural areas are well suited to off-grid renewable electricity (such as solar PV, hydro, and biomass gasifiers), based on resource availability. Switching to electric stoves will have a significant impact on health (particularly of women and young children) through reduced household air pollution.[7] Yangka and Diesendorf (2016) observe that the electric cooking scenario also complements the vision of HKH countries for reducing deforestation and carbon emissions. This also helps to eliminate indoor air pollution, and to mitigate climate change by controlling SLCPs from biomass use. Additionally, switching to energy-efficient light-emitting diode (LED) bulbs and appliances in households using grid electricity will provide significant opportunities for climate change mitigation. The case of Bhutan is interesting here, as it has surpassed even China in providing access to clean cooking (IEA and World Bank 2017).

Pathway 2. Access to renewable electricity and transition to clean fuels/technology for cooking: This pathway differs from the first only in its assumption that full electric cooking could be unrealistic due to high electricity costs, small-scale and limited electricity supply from decentralized sources, expensive grid extension, abundant availability of alternative biomass-based cooking fuels, and other socioeconomic factors. Even in urban areas, electric cooking is not popular, and biogas and improved cookstoves are preferred.

Providing cooking and heating services simultaneously by biomass in rural mountain areas, and the implications for efficiency loss and associated indoor air pollution, needs careful planning. Sharma (2009) observes the trade-off between cooking and space heating efficiency from a single stove in the high mountains, and highlights the need for a housing energy efficiency retrofit policy to overcome this. Solar, micro- and mini-hydro, and biomass gasifiers are promising rural electricity technologies in the HKH. Transition from traditional to advanced biomass cookstoves could yield enormous gains for the health and welfare of the

[5]For example, Nepal has enormous hydropower potential (43 GW) of which 2% is tapped (Dhakal and Raut 2010; KC et al. 2011), while Bhutan's vision of 100% renewable energy supply is based on its existing untapped hydropower (30 GW). The HKH of India has huge hydropower potential (114,398 MW), wind power (6883 MW), and solar power (5–7 kWh/m^2/day) (CEA 2016; NIWE 2016). There is an estimated hydropower potential of 41,722 MW in Pakistan (Mirza et al. 2008), the technical potential of grid-connected solar (PV of 50,174 MW) in Bangladesh (Mondal and Islam 2011), and electricity generation potential of 314,500 MW from solar, wind, and hydro in Afghanistan (Sahel 2014), sufficient for a pathway towards a 100% renewable electricity system.

[6]For example, as part of a programme on "access to clean cooking alternatives in rural India", induction stoves were introduced in nearly 4000 rural households in Himachal Pradesh, one of the few highly electrified states in India (Banerjee et al. 2016).

[7]In 2015, 1.2 million premature deaths in China and 1 million premature deaths in India were attributed to household air pollution (OECD/IEA 2016).

weakest and most vulnerable sections of society. The HKH has a large potential of biogas[8] for cooking. Transition to biogas provides clean fuel for cooking and lighting, and organic manure to rural and semi-urban households.

6.5.4 A Sustainable Energy Future: Barriers and Opportunities

Sustainable energy transition needs to be accelerated in the HKH, and transformative change is required. While there has been progress in certain areas—particularly at national and regional levels—the HKH faces many policy, regulatory, and technical hurdles in adopting sustainable energy technologies because energy pathways for the region are nascent and immature. Increased financial assistance, technological knowledge, and capacity building are critically important to support clean energy infrastructure. Fortunately, energy transition needs and available options are in line with individual countries' national targets for the SDGs (universal access), Energy for All (clean energy access), and the Paris Agreement (climate change and complete decarbonization of energy systems). The HKH has, therefore, a great opportunity to benefit from support measures formulated globally for meeting these goals globally: these include technology transfer, financing, and capacity building.

Furthermore, financial innovation and private sector involvement are two main factors with the potential to increase the penetration of off-grid renewable energy technologies to enhance access (Palit and Chaurey 2011). Benecke (2008) observes that contextual factors specifically related to political stability, good governance in terms of human rights, participatory and regulatory frameworks, and sociohistorical conditions matter in the successful implementation of new renewable energy options.

Confronting barriers: The barriers to developing a sound and sustainable energy future can be categorized in the following ways (Balachandra 2011; Burns 2011; Ershad et al. 2016; Mondal et al. 2010; Yaqoot et al. 2016):

Capital related:

(1) The availability of capital investment, supported by a viable policy framework, to construct the infrastructure needed to exploit the energy source and transmission potential and/or implement decentralized technological solutions.
(2) Allocation of adequate budgets when programmes are supported nationally, and ensuring timely financial flows.

Governance related:

(1) Restructuring ministries involved in the sector, and policy and institutional coherence, especially of regulatory reform and development of viable regulatory bodies.
(2) Designing taxation regimes and other financial incentives to encourage energy investment and renewable energy; withdrawal of subsidies currently being given to fossil fuels.
(3) Extensive training and capacity building at managerial and technical levels throughout the sector; cooperation with international agencies.
(4) Local community and awareness generation for sustainable management of renewable resources.

Market related:

(1) Developing market-oriented policy implementation and technology promotion.
(2) Lack of standardized power purchase agreements (PPAs) for power generation.
(3) Encouraging private sector participation in an environment of public sector dominance.
(4) Unavailability of skilled manpower for maintenance, unavailability of spare parts, high cost, and lack of access to credit.
(5) Unfair energy pricing (involved accounting of full cost recovery); and ensuring energy utilities operate as financially viable entities, and that the rich are not subsidized.
(6) Lack of adequate training on operation and maintenance of decentralized RE systems.

Available evidence and data:

(1) The lack of an integrated energy strategy, including for RE projects.
(2) Lack of data (comprehensive resource assessments and feasibility studies on potential).
(3) Suitable, replicable business models for the range of contexts and applications in the region.
(4) Lack of information or awareness.

Seizing opportunities: Approximately USD35 billion of climate finance flows through official international public financing outlets, of which a large portion is allocated to the sustainable energy sector. The newly created Green Climate Fund has over USD10 billion in pledges. The demand for

[8]The HKH area of India has a potential of 759,500 family-size biogas plants, of which 223,857 had been installed by March 2015 (MNRE 2016).

climate financing has, so far, lagged behind supply, and donors are looking for bankable projects to fund. Considering the large potential for sustainable energy in the HKH, there is a need to help countries understand and navigate this new international carbon financing opportunity. Some HKH countries can continue to use the Clean Development Mechanism (CDM) of the Kyoto Protocol, despite the lower price of current carbon offset projects. HKH countries must be aware of emerging market mechanisms under the Paris Agreement, as well as opportunities in bilateral and internal markets.

Recent studies on energy requirements in HKH countries reveal that China, India, Pakistan, Bangladesh, and Afghanistan have energy demands surpassing their domestic supply. Bhutan, Myanmar, and Nepal have energy resources—hydropower in particular, far in excess of their domestic needs—that can be traded within the region. Bangladesh, India, and Pakistan have potential for significant electricity and gas trade within South Asia. Enabling energy trade between HKH countries is a great opportunity for the region to increase its energy access and security. Regional energy trade and cooperation may be advanced by individual national governments, but first it is necessary to engage in meaningful multilateral energy trade as a cornerstone of regional integration and connectivity in HKH countries.

6.6 Energy Cooperation in HKH: National, Regional, and Global Linkages

Faced with an energy deficit and rising demand, HKH countries need to seek external cooperation to enhance their energy security and related environmental performance. Key gaps in gaining greater energy independence, identified in earlier sections, relate to: access to modern energy forms (Sect. 6.2); viability of conventional solutions (Sect. 6.5); proven, replicable business models (Sect. 6.4); capacities and capabilities; and poorly designed policies and delivery mechanisms (Sect. 6.3). Energy cooperation for addressing these gaps can take various forms, depending on the problems and possible solutions, and can be undertaken at various scales.

At the same time, the impacts of climate change in the region can negatively affect the flows of energy and related services bi-directionally, making all regions and countries of the HKH into stakeholders for climate action. Therefore, large-scale adoption and deployment of clean and renewable energy through harmonized regional policies and programmes will need to play a central role as they act as both mitigating and adaptive responses to climate change-induced security threats.

In the long term, regional energy security depends heavily on how an innovative multi-governance approach for regional cooperation can be fostered and sustained. Mechanisms are needed to encourage economic cooperation despite continuing political differences. The success of the electricity trade between Bhutan and India demonstrates that, when there is enlightened political leadership and mutual benefit, even very large hydroelectric projects in remote regions can be developed quickly, to the advantage of both parties (Biswas 2011; Ebinger 2011). An overnight shift in the dynamics of regional relationships that have been shaped for decades is not possible. However, by engaging in regional energy trade and cooperation through a mutual understanding of interests, long-term benefits can be ensured by means of dialogue at all levels of governance.

6.6.1 Energy Challenges and Regional Cooperation

The rich energy resources of the region are unevenly distributed and largely untapped. None of the HKH countries will be able to meet their energy needs entirely from their own domestic resources; they face a stark choice between rapid development and energy self-sufficiency as they cannot achieve both without energy interdependence and a collaborative approach (McMillan 2008; Mahmud 2012). For example, Nepal and Bhutan have great untapped hydropower potential while India, Pakistan, and Bangladesh have large reserves of gas and coal. Energy resource-surplus countries such as Nepal and Bhutan could benefit from export-led growth by fully exploiting their hydropower potential while simultaneously meeting demands in energy-deficient countries such as India, Afghanistan, Pakistan, and Bangladesh where energy requirements exceed their economically viable hydropower potential (Iftikhar et al. 2015; World Bank 2008). Differences in daily load patterns and seasonal variation in electricity demand, as well as generation, provide ample opportunities for HKH countries to optimize the use of regional resources and system operation by exploiting cross-border opportunities—even in the current circumstances (Box 6.4).

Box 6.4 Seasonal mismatch between electricity demand and supply

Variability in energy supply and demand during peak and off-peak periods across days and seasons provides a real opportunity to engage in electricity trading even in situations where, at an aggregate level, shortages might exist. For example, within Bangladesh, average sizeable generation capacities of 1200 MW remain unused during off-peak hours even though it faces shortages of power during peak hours. The available capacity of off-peak hours can be a ready source for regional cooperation for import–export of electricity within neighbouring countries (Lama 2004). Peak months for hydropower generation are August–September while the lean season is January–June. In Nepal and Bhutan, generation from hydropower plants is low during peak demand time (December–January) and its supply capacity reaches its peak during the lean demand season (August–September). This mismatch in the seasonality of energy supply and demand leads to a need for complementary cross-border power trade (Nanda and Goswami 2008).

The Indian system faces a deficit of energy and capacity during the hot summer months. The peak season begins in May and runs until August and September, with an overlapping lean demand season in Nepal and Bhutan. Indian thermal power plants complement and balance the lean dry months of the hydro plants in winter and the pre-monsoon season. The same Indian thermal power plants can be redesigned and restructured to address fluctuations in Nepal and Bhutan (Nanda and Goswami 2008), and it has been found viable to import electricity and other products from India for countries such as Nepal, Bhutan, and Bangladesh (World Bank 2007). This can be successful if cross-border transmission infrastructure projects are initiated through a memorandum of understanding (MOU). One such MOU was recently signed for transmission lines between India and Nepal.

An analysis by ADB (2015) of partially or fully integrated SAARC power systems indicates that cross-border power trading is both technically feasible and economically attractive, with the estimated economic benefits of six cross-border transmission interconnections[9] ranging from USD105 million to USD1,840 million (Wijayatunga et al. 2015). A study by USAID has clearly demonstrated how a regionally interconnected power system can bring a number of technical, operational, economic, and environmental benefits to a country, as well as to the region (USAID 2013). Cooperation can also facilitate integrated planning for sustainable transboundary water resource management to bring additional benefits such as flood control, irrigation, and navigation through multipurpose river projects (Price and Mittra 2016; Rasul 2014b).

Despite the tremendous benefits to be derived from regional cooperation, examples of cross-border hydropower cooperation are limited except between India and Bhutan (Rahman et al. 2011). India has assisted Bhutan in building almost 96% of the country's current hydropower capacity, and electricity production now accounts for around one-fifth of Bhutan's GDP while its per capita GDP has increased fivefold from 1992 to 2013 (World Bank 2014). Although the specific political relationship between the two countries cannot be replicated elsewhere in the region, the processes and outcome of cooperation are pertinent to other countries, such as Nepal (Price and Mittra 2016).

6.6.2 Models for Energy Cooperation

Electricity: The power sector offers the most obvious and promising area for regional energy cooperation, given the enormous potential for hydropower. While SAARC member states have agreed on the basic idea of sharing electricity through a common regional energy grid to promote regional energy sustainability, the technicalities of realizing this objective still need to be resolved (Dawn 2014). The Bangladesh, Bhutan, India, and Nepal (BBIN) initiative for creating an energy cooperation model has resulted in the emergence of five models of power exchange in the BBIN sub-region (Lama 2016), providing scope for generation and transmission system expansion on a regional basis (Nexant 2001). India also needs to quickly develop policies and implement them so it can act as a transiting country for power transmission in the HKH between Nepal, Bangladesh, Bhutan, and Myanmar.

Currently, there is a constraint in the chicken-neck area (Siliguri corridor that connects West Bengal to Northeast) of India for constructing power transmission lines. To overcome this, India and Bangladesh are discussing the construction of transmission lines through Bangladesh, and a certain percentage of the power will be transferred through Bangladesh. Singh et al. (2015) establish that Bhutan and Nepal have the potential to cost-effectively supply electricity from hydroelectric resources in excess of their own demand, while Bangladesh and India are likely to become more dependent on higher-cost coal and natural gas to generate electricity.

The World Bank (2015) study quantifies the potential economic benefits that the HKH could reap if the countries engage in regional electricity trade and cooperation. According to this study, unrestricted electricity trade provision through optimal expansion of electricity generation capacities and transmission interconnections in the region would save USD226 billion in electricity supply costs over the period 2015–40. To achieve these benefits, the region has to add an estimated 95,000 MW of new cross-border transmission interconnection capacity (Timilsina et al. 2015). The cooperation benefits pertain to direct sector-level gains from generation and transmission assets at the regional level, rather than country by country.

[9]This refers to the Bhutan–India additional grid reinforcement; the India–Nepal 400 kV transmission link; and the Bangladesh–India high voltage direct current transmission link.

Fuels and technologies: Existing bilateral energy cooperation has been limited to some hydropower, and there is scope for possible regional collaboration in other energy resources like natural gas. The HKH, except for Bangladesh, is poorly endowed with reserves of natural gas. However, the real benefit in this sector will accrue from a region-wide integrated gas pipeline network to import gas from outside the region, jointly developed and owned by the participating nations (USAID 2006). India's difficulties and apprehensions for establishing gas pipelines through Bangladesh could be mitigated to some extent if India considers creating a regional network—including Nepal and Bhutan—rather than bilateral arrangements (Nanda and Goswami 2008). This energy trading will help countries diversify their energy usage and enable them to reduce their dependence on traditional biomass. A number of proposed gas pipeline projects under discussion could lay the foundation for a regional gas grid, but investment requirements and security concerns must first be addressed before this concept can be fully explored and implemented (USAID 2006).

Biomass: Given a variety of bioenergy technologies emerging as equally competitive with conventional power generation options, there is scope for regional collaboration in further market development and demonstration to stimulate new and larger investments, while taking advantage of economies of scale. This calls for joint R&D efforts to develop commercially viable and efficient renewable energy technologies such as waste-to-energy, biogasification, and biofuel, which India and China have been actively pursuing (Srivastava and Misra 2007; Srivastava et al. 2013). Biomass-based energy through efficient gasification technology can help deliver energy to rural households in the region, and that energy can be traded through integrated grid networks. However, the technology needs first to be developed, and countries such as India and China need to share their technology in biogas, gasifiers, and cookstoves with other countries of the region. An efficient technology to use the biomass for energy generation can reduce the stress on forest resources and improve ecosystem services, and lead to reductions in GHG emissions, with efficient utilization of biomass resources.

6.6.3 Role of Multi-level Governance in Securing Sustainable Energy in HKH

Several studies offer valuable empirical and theoretical insights into the huge potential benefits of regional energy cooperation, especially in hydropower (ESCAP 2013; Iftikhar et al. 2015; Lama 2016; Price and Mittra 2016; Rahman et al. 2011; Timilsina et al. 2015; USAID 2013; Wijayatunga et al. 2015; World Bank 2008). However, these lack a holistic approach and solution-oriented, multi-pronged strategic options for improved regional energy security cooperation (Mahmud 2012; Srivastava and Misra 2007; Srivastava et al. 2013). As a result, far too little has been done to bring about the much-needed change for sustainable energy security cooperation in the region.

Existing energy policies in the region lack integrated energy planning, trade, and institutional mechanisms to leverage complementary regional energy resources. Even modest efforts to encourage regional energy trade are historically blocked by longstanding disputes, political exigencies, and mistrust between the countries of the region (Ebinger 2011; USAID 2006). However, it is rightly stressed that delays in decision making to ensure stronger and mutually beneficial cooperation efforts are associated with high costs, not only for the energy sector, but also for the development agenda of the region (Srivastava and Misra 2007; Srivastava et al. 2013). It is, therefore, critical for all stakeholders in HKH countries to graduate from the bilateral approach to a multilateral approach to develop a regional energy market.

International experience from a number of regional power sector cooperation initiatives demonstrates that nations with political differences have also come together for regional power sector cooperation and, in the process, realized various technical, economic, and environmental benefits of regional cooperation (USAID 2013). Experience, particularly from the South African Power Poll (SAPP) and the Greater Mekong Sub-Region (GMS), (which have a long history of bilateral arrangements leading to the development of regional power sector cooperation) has special relevance for the HKH (See Box 6.5).

Box 6.5 Regional power sector cooperation: International experience

A number of initiatives have led to regional power sector cooperation, including GMS, SAPP, the South East Asia, Europe, Gulf Coast Countries (GCC), and the Nile Basin Initiative (NBI). Their experiences show that key drivers and motivation for such regional cooperation include reliable system operation, regionally coordinated investment in generation, enhanced regional energy security, lower reserve requirements, and optimized system operations that take into account daily and seasonal variations in demand and generation. The SAPP experience demonstrates the feasibility of trade in power, and reliable and economical operation of the integrated system, even in the presence of historically rooted political differences, provided that complementary power sources, an active regional organization for economic cooperation, and political will to support increased regional energy trade is present (USAID 2013).

New ways and mechanisms need to be explored to create innovative multi-governance institutional models for energy cooperation. This calls for a multi-pronged strategy for regional energy cooperation, ranging from several softer options aimed at confidence building, to larger-scale cooperation on information collection and sharing, knowledge networking, and technology cooperation for rural energization, building greater energy efficiency, and enlisting private sector participation (Srivastava and Misra 2007; Srivastava et al 2013). Any successful regional cooperation effort will also depend on a smart policy for tariff setting, given the central role tariffs will play in determining the comparative advantages in regional trade negotiations (Srivastava and Misra 2007). Critically, several barriers related to both domestic policy reforms and regional political climate will need to be addressed in the pursuit of regional energy cooperation in the HKH. At the domestic level, some policy-related barriers include insufficient installation of generation capacity to match growing demand, weak financial performance of utilities due to technical and non-technical losses, poor operational efficiency of installed capacity, politically distorted tariffs, limited private sector participation owing to low financial incentives and political uncertainties, and partial domestic power sector reforms to harmonize policy and regulatory framework (CUTS International 2016).

As regional cooperation entails transboundary activities and shared control over resources, addressing existing trust gaps among governments and other key players (private sector, development partners, civil society) is a critically important first step (CUTS International 2016). Frequent meetings and discussions at appropriate levels and with relevant agencies need to take place at multilateral, bilateral, and regional levels for trust and confidence building. An integrated regional energy master plan must be developed, focusing on joint development mechanisms for maximizing investments in shared energy infrastructure (USAID 2006), and sharing with neighbouring countries through modes of bilateral/trilateral/multilateral cooperation.

Development partners can play an effective role in providing technical assistance to pursue regional and international energy trade, mobilizing international private investments, and developing the capacity of energy sector organizations. Bangladesh is currently taking a pioneering role in trying to push forward a trilateral MOU with India and Bhutan. The governments of Bangladesh, India and Bhutan have signed a MoU to jointly participate in the construction of hydropower project in Bhutan's Lhuentse district[10] The Bangladeshi government recently approved an equity investment of USD1 billion, for the Dorjilung 1125 MW hydropower project in Lhuentse district in eastern Bhutan, for power that will be exported to Bangladesh.

Relevant institutions with a dedicated regional agency comprising representatives from government, private, and non-governmental sectors with expertize in energy must be initiated to steer inclusive and sustained dialogues at multiple levels for regional cooperation and development. India and Bangladesh have created a Joint Expert Technical Committee (JETC) to study cross-border power transfer by constructing a 700 kV HVDC line through Bangladesh from the eastern part of India (Arunachal Pradesh) with a 10% power sharing arrangement for Bangladesh. Setting up appropriate back-to-back arrangements by engaging commercial entities (public or private) in the business of electricity trading and lagged energy exchange arrangements[11] could be other possible options for cooperation (Srivastava and Misra 2007).

Within the structure of bilateral and multilateral energy cooperation, a regional cooperation model structure can be considered. The structure will need to ensure that various institutions and rules can be designed and implemented to address the social, economic, and environmental domains of sustainability in the region. These institutions can also feed into larger regional and global debates on sustainability. Hence, the institutional model of the HKH for energy cooperation has to constantly evolve by being a contributor/receptor of national, regional, and global sustainability issues. This body would need to grow and change over time to remain in tune with regional and worldwide needs.

6.6.4 Climate Change, Energy Resilience, and Regional Energy Cooperation

Climate change-induced security threats, particularly those emanating from water stress and natural calamities, are becoming major problems for regional security dynamics in the HKH. They will intensify the loss of livelihood due to food, water, and energy scarcity in the HKH and

[10] Asianpower, 12 July 2017, available online at https://asian-power.com/project/news/bhutan-bangladesh-and-india-jointly-develop-hydropower-project.

[11] Under such arrangements Bangladesh, for example, could begin to provide gas to India for a stipulated period, in exchange for which India would be obligated to provide Bangladesh the contracted amount of energy in future.

downstream river basins, and this loss will exacerbate displacement, migration, and loss of biodiversity (Muniruzzaman 2012). The enormous national, regional, and global linkages of HKH energy resource bases—together with the complex interdependencies in water, food, and energy security—highlight the urgent need for intersectorally integrated and holistic solutions from an ecosystem perspective (Dharmadhikary 2008; Rasul and Sharma 2016).

It is important to assess and quantify various externalities associated with regional energy cooperation, given its implications for large-scale environmental degradation, the costs of climate change, and indoor pollution problems (Srivastava and Misra 2007). As a vital component of sustainable hydropower development, appropriate benefit-sharing mechanisms need to be explored to reach a win–win solution where the benefits derived from hydropower projects are shared with mountain communities in a fair and equitable manner, beyond the compensation and mitigation of project impacts (Shrestha et al. 2016). Benefit sharing from hydropower projects is becoming increasingly important for sustainable hydropower development to ensure that the rights and interests of the affected populations are recognized. Unlike compensation, which is usually covered under the project costs, benefit sharing typically takes a small percentage from the revenues generated by the project. Mekong countries have experience with one or more forms of benefit sharing in the hydropower sector. Nepal and China have laws that allocate a portion (1–3%) of gross revenue from hydropower projects to permanent local development and reconstruction funds in reservoir areas; in India, allocation ranges from 10 to 14%. A more recent study has identified a variety of benefit-sharing models and practices evolving in Nepal (Shrestha et al. 2016). Among them, equity share in hydropower projects for local citizens is an innovative market-based strategy of benefit sharing. This mechanism was designed in response to the pronounced demand from local citizens for ownership of shares in hydropower projects, as incentives among stakeholders, to avoid costly conflicts and contestations. Payment for environmental services (PES) schemes[12] is one way to reward mountain communities for the vital services they provide, and should therefore be promoted at a global scale to make mountain social–ecological systems more resilient in the face of climate change. Capacity-building programmes, sustained financial support, and knowledge exchange are needed for technology development, leading to a reduction in dependence on fossil fuels and lowered GHG emissions. This is important in the HKH, as impacts of climate change there will impact the agriculture and livelihoods of the region, affecting the economic, environmental, and social domains of sustainability.

Although the impending crisis is recognized, there is a general lack of policy direction and political will in tackling these multifaceted issues. Mutual distrust hinders effective action, and sovereignty concerns overrule regional interests. Environmental cooperation generally lags far behind economic cooperation in the HKH (Morton 2008). Domestic reforms are under way, but alone are inadequate. Disputes and rivalries must be put aside to achieve regional energy cooperation, as entrenched mistrust has perpetuated parochialism for decades (Ebinger 2011). However, there are good reasons to believe that climate change-induced human security threats have the potential to drive energy cooperation and facilitate multilateral agreements on the basis of a common framework.

6.7 Way Forward

The HKH has a huge diversity of contexts and, despite being rich in energy resources, remains energy poor and vulnerable. The urgent challenge for meeting the energy need of the HKH region is to build on past successes to seize the huge untapped renewable resource opportunities on the one hand. On the other hand, the need is to accelerate the rapid diffusion and scale-up of appropriate, customized business models for off-grid renewable energy solutions by creating an enabling policy environment for clean energy investment in a sustainable manner.

As there is no one-size-fits-all solution for ensuring universal access to modern energy in mountain areas, the HKH region must pursue multi-goal energy perspectives to chart its own energy transition pathway, drawing lessons from demonstrated examples of success across the region on how to jump-start and leapfrog to reach a meaningful energy transition. This section sets out four priority areas for immediate action to make sustainable energy transition a reality.

- **Make mountain-specific energy policies and programmes an integral part of national energy development strategy**. Policies based on anecdotal information or that merely mimic national energy strategies for the local HKH region are unlikely to address mountain-specific energy challenges. Establishing a comprehensive energy database (on supply and demand, resources, technologies in use, and good practices from the region) is a necessary first step towards devising evidence-based strategies for sustainable energy provisioning. To achieve this, a regional mountain-specific data management system supporting national institutions in periodic energy data gathering

[12] They consist of payment for the maintenance or provision of an environmental service by the users to the providers of the service.

with innovative survey methodologies (using information technology based surveys, remote sensing, etc.) and analysis to assess various aspects of energy access is of utmost importance.

- **Establish monitorable quantitative and qualitative targets for each energy end use**. A mountain-specific multi-tier assessment framework with appropriate energy access indicators must be established. This must allow proper measurement and tracking of progress of different attributes of access to clean energy (e.g., availability, quality, affordability, health, convenience, safety, and efficiency) for each end use while at the same time promoting integrated and holistic approaches to other related SDGs. The diversity of contexts in this region may also require nuanced signalling and incentivizing of appropriate technology choices. Such targets must give specific attention to meeting cooking energy targets, given their implications for forest degradation, human health, and the environment. Such an outcome-oriented strategy would facilitate a demand-driven approach to energy service provisioning. It would prioritize interventions with the largest co-benefits on SDGs related to poverty elimination, better health and education, reduced emissions of SLCPs, reduced drudgery for women and children, greater gender equality, decent jobs, etc.
- **Scale up current investments and ensure access to finance through capacity building at different levels**. Mobilizing innovative and affordable international, regional, national, and local funds is crucial for supporting energy planning, infrastructure investments, and creating incentives for commercial lending, generating soft loans, and providing technical assistance. The engagement of the private sector in this effort is critical, for both large-scale energy generation and infrastructure projects and smaller-scale decentralized energy provisioning. Governments can encourage this through the creation of markets and using public funding and financing for de-risking investments and leveraging private finance. Promoting off-grid renewable energy requires customized business and affordable financing models. While engaging and empowering local communities is critical to the sustainability of the off-grid sector, capacity-building efforts must also specifically address themselves to effective design and management of public–private partnerships as well as to private or community-based entrepreneurship.
- **Accelerate the pace of regional trade and cooperation in sustainable energy through a high-level, empowered, regional mechanism**. With the exception of a few bilateral hydropower exchanges, the region has little else to show as successful cooperation efforts. However, a great opportunity exists to leverage the complementary energy resources in the region to secure sustainable energy. This can be achieved through integrated regional energy planning, trade, and establishment of institutional mechanisms which would also ensure that environmental externalities are managed and revenue-sharing mechanisms for local areas are in place. Concerted regional approaches also help in reducing costs, generating economies of scale, attracting investment, boosting financial capacity, transferring technology and knowledge, and accelerating the deployment of renewable energy. Climate change-induced human security threats have the potential to drive energy cooperation and facilitate multilateral agreements on the basis of a common framework.

References

ADB. (2012). *Energy trade in South Asia: Challenges and opportunity*. Mandaluyong City, Philippines: Asian Development Bank.

ADB. (2013). Islamic Republic of Afghanistan—Power Sector Master Plan; Asian Development Bank, Manila. Retrieved October 12, 2016, from https://www.adb.org/sites/default/files/project-document/76570/43497–012-afg-tacr.pdf.

ADB. (2015). *Energy outlook for Asia and the Pacific*. Manila, Philippines: Asian Development Bank.

Afghanistan National Development Strategy Secretariat. (2008). *Energy sector strategy (2007/08–2012/13)*. Kabul: Islamic Republic of Afghanistan. http://mew.gov.af/Content/files/Energy_Sector_Strategy-English.pdf.

Ale, B. B., & Shrestha, S. B. (2008). Hydrogen energy potential of Nepal. *International Journal of Hydrogen Energy, 33*(15), 4030–4039.

Amann, M., Kejun, J., Jiming, H., Wang, S., Xing, Z., Xiang, D. Y., et al. (2008). GAINS Asia. Scenarios for cost-effective control of air pollution and greenhouse gases in China. International Institute of Applied Systems Analysis (IIASA), Luxenberg, Austria.

Badran, A. A., Yousef, I. A., Joudeh, N. K., Al Hamad, R., Halawa, H., & Hassouneh, H. K. (2010). Portable solar cooker and water heater. *Energy Conversion and Management, 51*(8), 1605–1609.

Balachandra, P. (2011). Dynamics of rural energy access in India: An assessment. *Energy, 36*(9), 5556–5567.

Banerjee, M., Prasad, R., Rehman, I. H., & Gill, B. (2016). Induction stoves as an option for clean cooking in rural India. *Energy Policy, 88*, 159–167.

Banerjee, S. G., Moreno, F. A., Sinton, J. E., Primiani, T., & Seong, J. (2017). Regulatory indicators for sustainable energy: A global scorecard for policy makers. Washington, DC, World Bank. http://documents.Worldbank.org/curated/en/538181487106403375/pdf/112828-REVISED-PUBLIC-RISE-2016-Report.pdf.

Benecke, G. E. (2008). Success factors for the effective implementation of renewable energy options for rural electrification in India–Potentials of the clean development mechanism. *International Journal of Energy Research, 32*(12), 1066–1079.

Beniston, M. (2013, February 20). Mountain water resources in a changing climate. Presented at Mountains under Watch Conference, Val d'Aosta, Italy. www.muw2013.it/ session-details.

Bhatt, B. P., & Sachan, M. S. (2004). Firewood consumption pattern of different tribal communities in Northeast India. *Energy Policy, 32* (1), 1–6.

Bhattacharyya, S. C. (2006). Energy access problem of the poor in India: Is rural electrification a remedy? *Energy Policy, 34*(18), 3387–3397.

Bhattacharyya, S. C., & Ohiare, S. (2012). The Chinese electricity access model for rural electrification: Approach, experience and lessons for others. *Energy Policy, 49,* 676–687.

Biswas, A. K. (2011). Cooperation or conflict in transboundary water management: Case study of South Asia. *Hydrological Sciences Journal, 56*(4), 662–670.

BP Statistical Review of World Energy. (2016). BP Statistical Review of World Energy June 2016. Retrieved from https://www.bp.com/content/dam/bp/pdf/energy-economics/statistical-review-2016/bp-statistical-review-of-world-energy-2016-full-report.pdf.

Burns, R. K. (2011). Afghanistan: Solar assets, electricity production, and rural energy factors. *Renewable and Sustainable Energy Reviews, 15*(4), 2144–2148. https://doi.org/10.1016/j.rser.2010.12.002.

Cai, J., & Jiang, Z. (2008). Changing of energy consumption patterns from rural households to urban households in China: An example from Shaanxi Province, China. *Renewable and Sustainable Energy Reviews, 12*(6), 1667–1680.

CEA. (2016). Status of Hydro Electric Potential Development, Central Electricity Authority (CEA), Ministry of Power, Govt. of India, New Delhi. Retrieved October 22, 2016, from http://www.cea.nic.in/reports/monthly/hydro/2016/hydro_potential_region-02.pdf.

CIA. The World Factbook. (2016). Retrieved June, 2017, from https://www.cia.gov/library/publications/the-world-factbook/index.html.

Chauhan, A., & Saini, R. P. (2015). Renewable energy based off-grid rural electrification in Uttarakhand state of India: Technology options, modelling method, barriers and recommendations. *Renewable and Sustainable Energy Reviews, 51,* 662–681.

Cheng, C. Y., & Urpelainen, J. (2014). Fuel stacking in India: Changes in the cooking and lighting mix, 1987–2010. *Energy, 76,* 306–317.

CUTS International. (2016). Promoting Energy Cooperation among South Asian Countries. *SDIP Advocacy Brief* NO. 5 Sustainable Development Investment Portfolio (SDIP).

Dawn. (2014, November 27). South Asian Leaders Reach Last-Minute Energy Deal at SAARC summit. The DAWN. http://www.dawn.com/news/1147270.

Dhakal, S., & Raut, A. K. (2010). Potential and bottlenecks of the carbon market: The case of a developing country, Nepal. *Energy Policy, 38*(7), 3781–3789.

Dharmadhikary, S. (2008, December). Mountains of Concrete, International Rivers Organisation.

Dincer, I., & Acar, C. (2015). A review on clean energy solutions for better sustainability. *International Journal of Energy Research, 39* (5), 585–606.

Duan, X., Jiang, Y., Wang, B., Zhao, X., Shen, G., Cao, S., et al. (2014). Household fuel use for cooking and heating in China: Results from the first Chinese Environmental Exposure-Related Human Activity Patterns Survey (CEERHAPS). *Applied Energy, 136,* 692–703.

Ebinger, C. K. (2011). *Energy and security in South Asia: Cooperation or conflict?* Brookings Institution Press.

EFEWEE. (2016). Exploring Factors that Enhance and restrict Women's Empowerment through Electrification (EFEWEE), Scoping study report. http://www.energia.org/cms/wp-content/uploads/2016/07/RA1-Scoping-Report.pdf.

Ekholm, T., Krey, V., Pachauri, S., & Riahi, K. (2010). Determinants of household energy consumption in India. *Energy Policy, 38*(10), 5696–5707.

ENERGIA. (2015). Regional Gender Assessment of Energy Policies and Programmes in South Asia. The International Network on Gender and Sustainable Energy (ENERGIA), ETC Foundation in the Netherlands.

Ershad, A. M., Brecha, R. J., & Hallinan, K. (2016). Analysis of solar photovoltaic and wind power potential in Afghanistan. *Renewable Energy, 85,* 445–453.

ESCAP. (2013). Regional cooperation for energy access and energy security in South and South-West Asia. Prospects and Challenges. South and South-West Asia Development Papers 1302. Economic and Social Commission for Asia and the Pacific.

Global Carbon Atlas. (2017). http://www.globalcarbonatlas.org/en/CO2-emissions.

Geng, J. B., & Ji, Q. (2014). Multi-perspective analysis of China's energy supply security. *Energy, 64,* 541–550.

Gupta, E. (2008). Oil vulnerability index of oil-importing countries. *Energy Policy, 36,* 1195–1211.

Gómez, M. F., & Silveira, S. (2010). Rural electrification of the Brazilian Amazon–Achievements and lessons. *Energy Policy, 38* (10), 6251–6260.

Government of Nepal. (2016). Intended Nationally Determined Contributions (INDC); Ministry of Population and Environment; Kathmandu. http://www4.unfccc.int/Submissions/INDC/Published%20Documents/Nepal/1/Nepal_INDC_08Feb_2016.pdf.

Government of Nepal. (2015). Sustainable Development Goals 2016–2030; National (Preliminary) Report; National Planning Commission, Kathmandu. http://www.np.undp.org/content/dam/nepal/docs/reports/SDG%20final%20report-nepal.pdf.

Government of the People's Republic of Bangladesh. (2012). Perspective Plan of Bangladesh 2010–2021. General Economics Division, Planning Commission, Government of the People's Republic of Bangladesh, Dhaka. Retrieved October 12, 2016, from, http://bangladesh.gov.bd/sites/default/files/files/bangladesh.gov.bd/page/6dca6a2a_9857_4656_bce6_139584b7f160/Perspective-Plan-of-Bangladesh.pdf.

Government of the People's Republic of Bangladesh. (2013). Country Action Plan for Clean Cookstoves; Ministry of Power, Energy and Mineral Resources, Government of the People's Republic of Bangladesh, Dhaka. Retrieved October 12, 2016, from http://cleancookstoves.org/resources/235.html.

Hanbar, R. D., & Karve, P. (2002). National Programme on Improved Chulha (NPIC) of the Government of India: An overview. *Energy for Sustainable Development, 6*(2), 49–55.

Hoff, H. (2011). Understanding the nexus: Background paper for the Bonn 2011 Nexus Conference.

Hosier, R. H., & Dowd, J. (1987). Household fuel choice in Zimbabwe: An empirical test of the energy ladder hypothesis. *Resources and Energy, 9*(4), 347–361.

Hussey, K., & Pittock, J. (2012). The energy–water nexus: Managing the links between energy and water for a sustainable future. *Ecology and Society, 17*(1).

ICIMOD. (2011). Regional Assessment Report for Rio+20: Hindu Kush Himalaya and SE Asia Pacific Mountains. From Rio 1992 to 2012 and beyond: Sustainable Mountain Development Hindu Kush Himalaya (HKH) Region. ICIMOD, Kathmandu, Nepal (pp. 1–71). Retrieved 15 May, 2016, from http://www.uncsd2012.org/content/documents/HKH_30_9_2011_0%5b1%5d.pdf.

IEA. (2015). World Energy Outlook 2015. International Energy Agency (IEA), Paris.

IEA. (2016). Energy Balances. Retrieved 14 July, 2016, from https://www.iea.org/statistics/statisticssearch.

IEA. (2017). World Energy Outlook 2017. International Energy Agency (IEA), Paris.

Iftikhar, M., Najeeb, N., Mohazzam, F., & Khan, A. K. (2015). Sustainable Energy for All in South Asia: Potential, Challenges, and Solutions. Sustainable Development Policy Institute. Working Paper #151. Sustainable Development Policy Institute (SDPI), Islamabad, Pakistan.

IEA and World Bank. (2015). Sustainable Energy for All 2015—Progress Toward Sustainable Energy. World Bank, Washington, DC. https://doi.org/10.1596/978-1-4648-0690-2.

IEA and World Bank. (2017). Sustainable Energy for All 2017—Progress toward Sustainable Energy (Summary). World Bank, Washington, DC.

Jain, A., Ray, S., Ganesan, K., Aklin, M., Cheng, C. Y., & Urpflainen, J. (2015). *Access to clean cooking energy and electricity—Survey of states*. Report. New Delhi, India: Council on Energy, Environment and Water (CEEW).

Kaltenborn, B. P., Nellemann, C., & Vistnes, I. I. (Eds). (2010). High mountain glaciers and climate change—Challenges to human livelihoods and adaptation. United Nations Environment Programme, GRID-Arendal. www.grida.no.

Kandlikar, M., Reynolds, C. C., & Grieshop, A. P. (2009). *A perspective paper on black carbon mitigation as a response to climate change*. Copenhagen: Copenhagen Consensus Center.

Kar, A., Singh, D., & Zerriffi H. (2018). Has India's Ujjwala Program enabled millions to shift to Liquefied Petroleum Gas as a primary cooking fuel? http://erdelab.forestry.ubc.ca/2018/04/.

Kc, S., Khanal, S. K., Shrestha, P., & Lamsal, B. (2011). Current status of renewable energy in Nepal: Opportunities and challenges. *Renewable and Sustainable Energy Reviews, 15*(8), 4107–4117.

Khair, M. (2013). The efficacy of groundwater markets on agricultural productivity and resource use sustainability: Evidence from the upland Balochistan region of Pakistan. Unpublished Ph.D. thesis, Charles Sturt University, Wagga Wagga.

Khandelwal, M., Hill, M. E., Greenough, P., Anthony, J., Quill, M., Linderman, M., et al. (2016). Why Have Improved Cook-Stove Initiatives in India Failed? World Development.

Kishore, A., & Spears, D. (2014). Having a son promotes clean cooking fuel use in urban India: Women's status and son preference. *Economic Development and Cultural Change, 62*(4), 673–699.

Klimont, Z., Cofala, J., Xing, J., Wei, W., Zhang, C., Wang, S., et al. (2009). Projections of SO2, NOx and carbonaceous aerosols emissions in Asia. *Tellus Series B: Chemical and Physical Meteorology, 61 B*(4), 602–617.

Lama, M. (2016). BBIN Initiatives: Options for Cross-Border Power Exchange.

Lama, M. P. (2004, August 20–21). Energy Cooperation in South Asia, SAFMA Regional Conference, Dhaka.

Loka, P., Moola, S., Polsani, K., Reddy, S., Fulton, S., & Skumanich, A. (2014). A case study for micro-grid PV: Lessons learned from a rural electrification project in India. *Progress in Photovoltaics: Research and Applications, 22*(7), 733–743.

Li, K., & Lin, B. (2015). Impacts of urbanization and industrialization on energy consumption/CO_2 emissions: Does the level of development matter? *Renewable and Sustainable Energy Reviews, 52,* 1107–1122.

Mahmud, K. (2012). Power grid interconnection of south Asian region to retain sustainable energy security and figure out the energy scarcity. *Global Journal of Research in Engineering, 12*(10-F).

Mainali, B., & Silveira, S. (2013). Alternative pathways for providing access to electricity in developing countries. *Renewable energy, 57,* 299–310.

Mainali, B., Pachauri, S., & Nagai, Y. (2012). Analyzing cooking fuel and stove choices in China till 2030. *Journal of Renewable and Sustainable Energy, 4*(3), 031805.

Mainali, B., Pachauri, S., Rao, N. D., & Silveira, S. (2014). Assessing rural energy sustainability in developing countries. *Energy for Sustainable Development, 19,* 15–28.

Mainali, B., Luukkanen, J., Silveira, S., & Kaivo-oja, J. (2018). Evaluating synergies and trade-offs among Sustainable Development Goals (SDGs): Explorative analyses of development paths in South Asia and Sub-Saharan Africa. *Sustainability, 10,* 815.

Marsh, D. M., & Sharma, D. (2007). Energy-water nexus: An integrated modeling approach. *International Energy Journal, 8*(4), 235–242.

McCollum, D., Gomez Echeverri, L., Busch, S., Pachauri, S., Parkinson, S., Rogelj, J., et al. (2017). Connecting the Sustainable Development Goals by their energy inter-linkages. IIASA Working Paper. IIASA, Laxenburg, Austria.

McMillan, J. (2008, September). Energy security in South Asia: Can interdependence breed stability? In *Strategic forum* (Vol. 232). National Defense University. Washington DC: Institute for National Strategic Studies.

Menon, S., Hansen, J., Nazarenko, L., & Luo, Y. (2002). Climate effects of black carbon aerosols in China and India. *Science, 297* (5590), 2250–2253.

Ministry of Energy and Water & Ministry of Rural Rehabilitation and Development. (2013). Afghanistan Rural Renewable Energy Policy, Kabul: Islamic Republic of Afghanistan. http://www.red-mew.gov.af/wp-content/uploads/2014/11/4.3-Rural-Renewable-Energy-Policy.pdf.

Ministry of Foreign Affairs. (2008). Afghanistan National Development Strategy, 2008–2013: Strategy for Security, Governance, Economic Growth & Poverty Reduction; Kabul: Islamic Republic of Afghanistan. Retrieved October 2016, from http://mfa.gov.af/en/page/6547/afghanistan-national-development-strategy/afghanistan-national-development-strategy-ands.

Ministry of Power, Energy and Mineral Resources. (2004). *National energy policy*. Dhaka: Government of the People's Republic of Bangladesh. http://www.asialeds.org/sites/default/files/resource/file/NEP_2004_fulldoc.pdf.

Ministry of Power, Energy and Mineral Resources. (2008). *Renewable energy policy of Bangladesh*. Dhaka: Government of the People's Republic of Bangladesh. https://www.iea.org/media/pams/bangladesh/Bangladesh_RenewableEnergyPolicy_2008.pdf.

Ministry of Power, Energy and Mineral Resources. (2016). *Power system master plan 2016*. Dhaka: Government of the People's Republic of Bangladesh. http://powerdivision.portal.gov.bd/sites/default/files/files/powerdivision.portal.gov.bd/page/4f81bf4d_1180_4c53_b27c_8fa0eb11e2c1/(E)_FR_PSMP2016_Summary_revised.pdf.

Ministry of Power. (2011a). The Electricity Act 2003; Power Compendium—A compilation of Act, Polices, Rules, Guidelines 2011. Government of India, New Delhi. http://www.cea.nic.in/reports/isacpower/power_compendium.pdf.

Ministry of Power. (2011b). National Electricity Policy, Power Compendium—A compilation of Act, Polices, Rules, Guidelines 2011. Government of India, New Delhi. http://www.cea.nic.in/reports/isacpower/power_compendium.pdf.

Ministry of Water and Power. (2013). National Power Policy 2013, Government of Pakistan, Islamabad. https://www.scribd.com/document/187041200/National-Power-Policy-2013-Pakistan.

Mir, K. A., Purohit, P., Goldstein, G. A., & Balasubramanian, R. (2016). Analysis of baseline and alternative air quality scenarios for Pakistan: An integrated approach. *Environmental Science and Pollution Research, 23*(21), 21780–21793.

Mirza, U. K., Ahmad, N., Majeed, T., & Harijan, K. (2008). Hydropower use in Pakistan: Past, present and future. *Renewable and Sustainable Energy Reviews, 12*(6), 1641–1651.

Misra, N., Chawla, R., Srivastava, L., & Pachauri, R. K. (2005). *Petroleum pricing in India: Balancing efficiency and equity*. Delhi: The Energy and Resources Institute.

MNRE (Ministry of New and Renewable Energy). (2016). Annual Report, 2016–2017. Retrieved from http://mnre.gov.in/file-manager/annual-report/2016-2017/EN/pdf/1.pdf.

Mondal, M. A. H., & Islam, A. S. (2011). Potential and viability of grid-connected solar PV system in Bangladesh. *Renewable energy, 36*(6), 1869–1874.

Mondal, M. A. H., Kamp, L. M., & Pachova, N. I. (2010). Drivers, barriers, and strategies for implementation of renewable energy technologies in rural areas in Bangladesh—An innovation system analysis. *Energy Policy, 38*(8), 4626–4634.

MoPE. (2016). Renewable Energy Subsidy Policy 2073 BS, Ministry of Population and Environment, Government of Nepal.

Morton, K. (2008). China and environmental security in the age of consequences. *Asia-Pacific Review, 15*(2), 52–67.

Mukherji, A., Chowdhury, D. R., Fishman, R., Lamichhane, N., Khadgi, V., & Bajracharya, S. (2017). Sustainable financial solutions for the adoption of solar powered irrigation pumps in Nepal's terai.

Muniruzzaman, A. N. M. (2012). Climate change and regional security. In T. Delinic & N. Pandey (Eds.), *Regional environmental issues: Water and disaster management*. Centre for South Asian Studies (CSAS) and Adenauer Stiftung (KAS).

Nanda, N., & Goswami, A. (2008). Energy cooperation in South Asia. *Man and Development, 30*(1), 109–120.

National Commission for Environmental Affairs, Ministry of Forestry. (2009). National Sustainable Development Strategy of Myanmar. http://extwprlegs1.fao.org/docs/pdf/mya152933.pdf.

Nexant. (2001). The four borders project: Reliability improvement and power transfer in South Asia, a pre-feasibility study. Prepared for USAID-SARI/ Energy Program.

Nilsson, M., Griggs, D., & Visbeck, M. (2016). Map the interactions between sustainable development goals. *Nature, 534,* 320–322.

NIWE. (2016). Estimation of Installable Wind Power Potential at 80 m level in India. National Institute of Wind Energy (NIWE), Chennai. Retrieved October 22, 2016, from niwe.res.in/department_wra_est.php.

OECD/IEA. (2016). Energy and Air Pollution: World Energy Outlook Special Report. International Energy Agency (IEA), Paris.

Pachauri, S., & Jiang, L. (2008). The household energy transition in India and China. *Energy Policy, 36*(11), 4022–4035.

Pakistan Energy Vision 2035. https://www.sdpi.org/publications/files/Pakistan%20Energy%202035-FINAL%2020th%20October%202014.pdf.

Planning Commission. (2012). *Perspective plan of Bangladesh 2021—Making vision 2021 a reality*. Dhaka: Government of the People's Republic of Bangladesh. http://bangladesh.gov.bd/sites/default/files/files/bangladesh.gov.bd/page/6dca6a2a_9857_4656_bce6_139584b7f160/Perspective-Plan-of-Bangladesh.pdf.

Planning Commission. (2006). Integrated Energy Policy, Government of India, New Delhi. http://planningcommission.gov.in/reports/genrep/rep_intengy.pdf.

Palit, D., & Bhattacharyya, S. (2014). Adoption of cleaner cookstoves: Barriers and way forward.

Palit, D., & Chaurey, A. (2011). Off-grid rural electrification experiences from South Asia: Status and best practices. *Energy for Sustainable Development, 15*(3), 266–276.

Palit, D., & Garud, S. (2010). Energy consumption in the residential sector in the Himalayan kingdom of Bhutan. *Boiling Point, 58,* 34–36.

Palit, D., Bhattacharyya, S. C., & Chaurey, A. (2014). Indian approaches to energy access. In *Energy poverty: Global challenges and local solutions* (pp. 237–256).

Papola, T. S. (2002). Poverty in mountain areas of the Hindu Kush-Himalayas: Some basic issues in measurement, diagnosis and alleviation. International Centre for Integrated Mountain Development. Kathmandu, Nepal.

Parajuli, R., Østergaard, P. A., Dalgaard, T., & Pokharel, G. R. (2014). Energy consumption projection of Nepal: An econometric approach. *Renewable Energy, 63,* 432–444.

Perera, N., Boyd, E., Wilkins, G., & Phillips, R. (2015). Literature Review on Energy Access and Adaptation to Climate Change. Evidence on demand.

Pode, R., Pode, G., & Diouf, B. (2016). Solution to sustainable rural electrification in Myanmar. *Renewable and Sustainable Energy Reviews, 59,* 107–118.

Practical Action. (2014). Poor people's energy outlook 2014: Key messages on energy for poverty alleviation. Retrieved from Rugby, UK.

Price, G., & Mittra, S. (2016). *Water, Ecosystems and Energy in South Asia Making Cross-Border Collaboration Work*. Asia Programme. Research Paper. The Royal Institute of International Affairs, Chatham House: London.

Proietti, S., Sdringola, P., Castellani, F., Garinei, A., Astolfi, D., Piccioni, E., et al. (2015). On the possible wind energy contribution for feeding a high altitude Smart Mini Grid. *Energy Procedia, 75,* 1072–1079.

Purohit, P., Amann, M., Mathura, R., Gupta, I., Marwan, S., Varma, V., et al. (2010). Gains Asia: Scenarios for cost-effective control of air pollution and greenhouse gases in India. International Institute for Applied Systems Analysis (IIASA), Ladenburg, Austria.

Purohit, P., Munir, T., & Rafaj, P. (2013). Scenario analysis of strategies to control air pollution in Pakistan. *Journal of Integrative Environmental Sciences, 10*(2), 77–91.

Rahman, S. H., Wijayatunga, P. D., Gunatilake, H., & Fernando, P. (2011). Energy trade in South Asia: Opportunities and challenges. Asian Development Bank, Philippines.

Rahman, S. H., Wijayatunga, P. D., Gunatilake, H., & Fernando, P. (2012). Energy trade in South Asia: Opportunities and challenges. Asian Development Bank, Manila. http://www.adb.org/sites/default/files/publication/29703/energy-trade-south-asia.pdf.

Rahut, D. B., Behera, B., & Ali, A. (2016). Household energy choice and consumption intensity: Empirical evidence from Bhutan. *Renewable and Sustainable Energy Reviews, 53*(C), 993–1009.

Ramanathan, V., & Carmichael, G. (2008). Global and regional climate changes due to black carbon. *Nature Geoscience, 1*(4), 221–227.

Ramanathan, V., Li, F., Ramana, M. V., Praveen, P. S., Kim, D., Corrigan, C. E., et al. (2007). Atmospheric brown clouds: Hemispherical and regional variations in long range transport, absorption, and radiative forcing. *Journal of Geophysical Research: Atmospheres, 112*(D22).

Ramji, A., Soni, A., Sehjpal, R., Das, S., & Singh, R. (2012). Rural energy access and inequalities: An analysis of NSS data from 1999-00 to 2009-10.

Rasul, G. (2014a). Food, water, and energy security in South Asia: A nexus perspective from the Hindu Kush Himalayan region. *Environmental Science and Policy, 39,* 35–48.

Rasul, G. (2014b). Why Eastern Himalayan countries should cooperate in transboundary water resource management. *Water Policy, 16*(1), 19–38.

Rasul, G., & Sharma, B. (2016). The nexus approach to water–energy–food security: An option for adaptation to climate change. *Climate Policy, 16*(6), 682–702.

Rijal, K. (1999). Energy use in mountain areas: Trends and patterns in China, India, Nepal and Pakistan. ICIMOD, Kathmandu, Nepal.

Rockström, J., Steffen, W., Noone, K., Persson, Å. Chapin III, F. S., Lambin, E., et al. (2009). Planetary boundaries: Exploring the safe operating space for humanity. *Ecology and society, 14*(2).

Sahel, B. K. (2014). Afghanistan has potential to generate 314,500 MW electricity—Minister. Afghanistan Times, 14th November 2014.

Santner, M., & Jussel, R. (2003). Rural Stoves for Bhutan. Interdisciplinary Research Institute for Development Cooperation, Johannes Kepler University, Linz, Austria.

Saroha, S., & Verma, R. (2013). Cross-border power trading model for South Asian regional power pool. *International Journal of Electrical Power and Energy Systems, 44*(1), 146–152.

Schild, A. (2008). ICIMOD's position on climate change and mountain systems: The case of the Hindu Kush-Himalayas. *Mountain Research and Development, 28*(3), 328–331.

Shah, T. (2010). Taming the anarchy: Groundwater governance in South Asia. Routledge.

Sharma, B. (2009). Sustainable Energy for the Himalayan Rangelands. Information sheet #9/09, ICIMOD.

Sharma, B. (2007, November 21–23). Renewable energy and energy efficiency in the Himalayan Region: Opportunities and challenges. In *7th meeting of the Global Forum for Sustainable Energy (GFSE) energy efficiency for developing countries—Strong policy and new technologies*, Vienna, Austria.

Sharma, B., & Banskota, K. (2005) *Women, energy and water in the Himalayas: Project learning*. Nairobi: UNEP and Kathmandu: ICIMOD.

Sharma, B., Banskota, K., & Luitel, S. (2005). *Women, energy and water in the Himalayas: Policy guidelines*. Nairobi: UNEP and Kathmandu: ICIMOD.

Shoaib, A., & Ariaratnam, S. (2016). A study of socioeconomic impacts of renewable energy projects in Afghanistan. *Procedia Engineering, 145,* 995–1003.

Shrestha, P., Lord, A., Mukherji, A., Shrestha, R. K., Yadav, L., Rai, N. (2016). *Benefit sharing and sustainable hydropower: Lessons from Nepal*. ICIMOD Research Report 2016/2. Kathmandu: Nepal.

Shrestha, R. M. (2013). Energy use and greenhouse gas emissions in selected Hindu Kush-Himalayan countries. *Mountain Research and Development, 33*(3), 343–354.

Singh, A., Jamasb, T., Nepal, R., & Toman, M. (2015). Cross-border electricity cooperation in South Asia.

Sinton, J. E., Smith, K. R., Peabody, J. W., Yaping, L., Xiliang, Z., Edwards, R., et al. (2004). An assessment of programs to promote improved household stoves in China. *Energy for Sustainable Development, 8*(3), 33–52.

Smith, K. R., Dutta, K., Chengappa, C., Gusain, P. P. S., Masera, O., Berrueta, V., et al. (2007). Monitoring and evaluation of improved biomass cookstove programs for indoor air quality and stove performance: Conclusions from the Household Energy and Health Project. *Energy for Sustainable Development, 11*(2), 5–18.

Sovacool, B. K. (2013). An international assessment of energy security performance. *Ecological Economics, 88,* 148–158.

Sovacool, B. K., Mukherjee, I., Drupady, I. M., & D'Agostino, A. L. (2011). Evaluating energy security performance from 1990 to 2010 for eighteen countries. *Energy, 36*(10), 5846–5853.

Srivastava, L., & Misra, N. (2007). Promoting regional energy co-operation in South Asia. *Energy Policy, 35*(6), 3360–3368. https://doi.org/10.1016/j.enpol.2006.11.017.

Srivastava, L., Misra, N., & Hassan, S. (2013, April). Promoting Regional Energy Cooperation in South Asia (Revised Draft), Commonwealth Secretariat.

Srivastava, L., & Rehman, I. H. (2006). Energy for sustainable development in India: Linkages and strategic direction. *Energy Policy, 34,* 643–654.

State of the Planet Declaration. (2012, March 26–29). Planet under pressure: New knowledge towards solutions conference, London. International Council for Science (ICSU), Paris.

Taliotis, C., Shivakumar, A., Ramos, E., Howells, M., Mentis, D., Sridharan, V., et al. (2016). An indicative analysis of investment opportunities in the African electricity supply sector—Using TEMBA (The Electricity Model Base for Africa). *Energy for Sustainable Development, 31,* 50–66.

TEPC. (2017). Foreign Trade Statistics of Nepal 2016 /2017. Kathmandu: Trade and Export Promotion Centre, Government of Nepal, Ministry of Industry, Commerce and Supply.

TERI and IISD. (2012). A citizens' guide to energy subsidies in India (pp. 1–44). The Energy and Resources Institute, and International Institute for Sustainable Development. Retrieved May 20, 2016, from https://www.iisd.org/gsi/sites/default/files/ffs_india_czguide.pdf.

Timilsina, G. R., Toman, M., Karacsonyi, J., & de Tena Diego, L. (2015). How much could South Asia benefit from regional electricity cooperation and trade? Policy Research Working Paper; No.7341. World Bank, Washington, DC.

UNDP. (2013). 'SE4All, Nepal: Rapid assessment and gap analyses, GoN, UNDP.

UNDP. (2015). Human Development Report 2015: Work for Human Development. United Nations Development Programme, 1 UN Plaza, New York, NY 10017, USA.

UNFCCC. (2015). Intended Nationally Determined Contributions (INDC). United Nations Framework Convention on Climate Change (UNFCCC), Bonn. http://unfccc.int/focus/indc_portal/items/8766.php.

United Nations Statistics Division. (2015). Energy Statistics Yearbook, New York. https://unstats.un.org/unsd/energy/yearbook/default.htm.

USAID. (2013). Prospects for regional cooperation on cross-border electricity trade in south Asia. Integrated Research and Action for Development (Tirade).

USAID. (2006). Regional energy security for south Asia. Regional report. http://pdf.usaid.gov/pdf_docs/Pnads866.pdf.

U Thoung Win. (2013). Myanmar National Energy Policy 2013 (draft). http://www.dmr.go.th/ewt_dl_link.php?nid=79472&filename=roya.

Vaidya, R. (2013). Water and hydropower in the green economy and sustainable development of the Hindu Kush-Himalayan region. *Hydro Nepal: Journal of Water, Energy and Environment, 10,* 11–19.

Van Vuuren, D. P., Nakicenovic, N., Riahi, K., Brew-Hammond, A., Kammen, D., Modi, V., et al. (2012). An energy vision: The transformation towards sustainability—Interconnected challenges and solutions. *Current Opinion in Environmental Sustainability, 4* (1), 18–34.

Vinci, S., Nagpal, D. N., Hodges, T., & Ferroukhi, R. (2015). Accelerating off-grid renewable energy IOREC 2014: Key findings and recommendations. In *Second international renewable energy conference*.

Vivoda, V. (2010).Evaluating energy security in the Asia-Pacific region: A novel methodological approach. *Energy Policy 38*, 5258–5263.

WECS. (2014). Energy Data Sheet, Water and Energy Commission Secretariat (WECS), Government of Nepal.

Weitz, N., Nilsson, M., Huber-Lee, A., & Hoff, H. (2014). Cross-sectoral integration in the sustainable development goals: A nexus approach. In *SEI discussion brief.* Stockholm: Stockholm Environment Institute.

Wijayatunga, P., Chattopadhyay, D., & Fernando, P. N. (2015). *Cross-Border Power Trading in South Asia: A Techno Economic Rationale*. ADB South Asia Working Paper Series No. 38. Manila: Asian Development Bank.

World Bank. (2007). *South Asia Energy: Potential and Prospects for Regional Trade*. Washington DC: World Bank.

World Bank. (2008). Potential and Prospects for Regional Energy Trade in the South Asia Region. Energy Sector Management Assistance Programme. World Bank. Washington D.C.

World Bank. (2011). *State and people, central and local, working together: The Vietnam rural electrification experience*. Washington, DC: World Bank.

World Bank. (2013). China: Accelerating household access to clean cooking and heating. In *East Asia and Pacific clean stove initiative series*. Washington, DC: World Bank.

World Bank. (2014). Bhutan: Country Snapshot. http://www.worldbank.org/content/dam/Worldbank/document/SAR/bhutan-countrysnapshot.

World Bank. (2015). Myanmar: Achieving universal access to electricity by 2030. Energy Sector Management Assistance Programme /the World Bank, Washington DC. Retrieved October 12, 2016, from http://pubdocs.worldbank.org/en/384351442415891708/pdf/Myanmar-Electrification-Plan-Sept-2015.pdf.

World Bank. (2016a). Access to Energy. Retrieved June 15, 2016, from http://databank.worldbank.org/data/reports.aspx?source=2andTopic=5.

World Bank. (2016b). World Bank database for Electric power consumption (kWh per capita). http://data.worldbank.org/indicator/EG.USE.ELEC.KH.PC.

World Bank. (2017). World Bank database for urban population. https://data.worldbank.org/indicator/SP.URB.TOTL.IN.ZS?end=2015&start=1960.

Wu, N., Ismail, M., Joshi, S., Yi, S. L., Shrestha, R. M., & Jasra, A. W. (2014). Livelihood diversification as an adaptation approach to change in the pastoral Hindu-Kush Himalayan region. *Journal of Mountain Science, 11*(5), 1342–1355.

Yadama, G. N., Peipert, J., Sahu, M., Biswas, P., & Dyda, V. (2012). Social, economic, and resource predictors of variability in household air pollution from cookstove emissions. *PloS one, 7*(10), e46381.

Yang, B., Li, Y., Wei, H., & Lu, H. (2016). Is urbanisation rate a feasible supplemental parameter in forecasting electricity consumption in China? *Journal of Engineering*.

Yangka, D., & Diesendorf, M. (2016). Modeling the benefits of electric cooking in Bhutan: A long term perspective. *Renewable and Sustainable Energy Reviews, 59,* 494–503.

Yaqoot, M., Diwan, P., & Kandpal, T. C. (2016). Review of barriers to the dissemination of decentralized renewable energy systems. *Renewable and Sustainable Energy Reviews, 58,* 477–490.

Yettou, F., Azoui, B., Malek, A., Gama, A., & Panwar, N. L. (2014). Solar cooker realizations in actual use: An overview. *Renewable and Sustainable Energy Reviews, 37,* 288–306.

Yupapin, D. P., Pivsa-Art, D. S., Ohgaki, D. H., Kyaw, W. W., Sukchai, S., Ketjoy, N., et al. (2011). 9th eco-energy and materials science and engineering symposium energy utilization and the status of sustainable energy in union of Myanmar. *Energy Procedia, 9,* 351–358. https://doi.org/10.1016/j.egypro.2011.09.038.

Status and Change of the Cryosphere in the Extended Hindu Kush Himalaya Region

Coordinating Lead Authors
Tobias Bolch, Institut für Geographie, University of Zurich, Zurich, Switzerland, e-mail: tobias.bolch@geo.uzh.ch
Joseph M. Shea, International Centre for Integrated Mountain Development, Kathmandu, Nepal, Geography Program, University of Northern British Columbia, Prince George, Canada, and Centre for Hydrology, University of Saskatchewan, Saskatoon, Canada, e-mail: joseph.shea@unbc.ca (corresponding author)
Shiyin Liu, Institute of International Rivers and Eco-security, Yunnan University, Kunming, China, e-mail: shiyin.liu@ynu.edu.cn

Lead Authors
Farooq M. Azam, Indian Institute of Technology, Indore, India, National Institute of Hydrology, Roorkee, India, e-mail: farooqaman@yahoo.co.in
Yang Gao, Chinese Academy of Sciences, Beijing, China, e-mail: Yanggao@itpcas.ac.cn
Stephan Gruber, Geography and Environmental Studies, Carleton University, Ottawa, Canada, e-mail: stephan.gruber@carleton.ca
Walter W. Immerzeel, International Centre for Integrated Mountain Development, Kathmandu, Nepal, Department of Physical Geography, Utrecht University, Utrecht, the Netherlands, e-mail: w.w.immerzeel@uu.nl
Anil Kulkarni, Divecha Center for Climate Change, Centre for Atmospheric & Oceanic Sciences, Indian Institute of Science, Bangalore, India, e-mail: anilkul54@gmail.com
Huilin Li, Northwest Institute of Eco-Environment and Resources, Chinese Academy of Sciences, China, e-mail: lihuilin@lzb.ac.cn
Adnan A. Tahir, Department of Environmental Sciences, COMSATS University Islamabad (CUI)–Abbottabad Campus, Abbottabad, Pakistan, e-mail: uaf_adnan@hotmail.fr
Guoqing Zhang, Institute of Tibetan Plateau Research, Chinese Academy of Sciences, Beijing, China, e-mail: guoqing.zhang@itpcas.ac.cn
Yinsheng Zhang, Institute of Tibetan Plateau Research, Chinese Academy of Sciences, Beijing, China, e-mail: yszhang@itpcas.ac.cn

Contributing Authors
Argha Bannerjee, Indian Institute of Science Education Research, Pune, India, e-mail: argha@iiserpune.ac.in
Etienne Berthier, CNRS, LEGOS, University of Toulouse, e-mail: etienne.berthier@legos.obs-mip.fr
Fanny Brun, CNRS, LEGOS, University of Toulouse, Universite Grenoble Alpes, Grenoble, France, e-mail: fanny.brun@univ-grenoble-alpes.fr
Andreas Kääb, Department of Geosciences, Oslo University, Oslo, Norway, e-mail: andreas.kaab@geo.uio.no
Phillip Kraaijenbrink, Department of Physical Geography, Utrecht University, Utrecht, the Netherlands, e-mail: P.D.A.Kraaijenbrink@uu.nl
Geir Moholdt, Norwegian Polar Institute, Tromsø, Norway, e-mail: geir.moholdt@npolar.no
Lindsey Nicholson, Department of Atmospheric and Cryospheric Sciences, Universität Innsbruck, Innsbruck, Austria, e-mail: lindsey.nicholson@uibk.ac.at
Nicholas Pepin, Department of Geography, University of Portsmouth, UK, e-mail: nicholas.pepin@port.ac.uk
Adina Racoviteanu, Department of Geography and Earth Sciences, Aberystwyth University, Aberystwyth, UK, e-mail: racovite@gmail.com

Electronic supplementary material
The online version of this chapter (https://doi.org/10.1007/978-3-319-92288-1_7) contains supplementary material, which is available to authorized users.

P. Wester et al. (eds.), *The Hindu Kush Himalaya Assessment*,
https://doi.org/10.1007/978-3-319-92288-1_7

Review Editor
Koji Fujita, Graduate School of Environmental Studies, Nagoya University, Nagoya, Japan, e-mail: cozy@nagoya-u.jp

Corresponding Author
Joseph M. Shea, International Centre for Integrated Mountain Development, Kathmandu, Nepal, Geography Program, University of Northern British Columbia, Prince George, Canada, and Centre for Hydrology, University of Saskatchewan, Saskatoon, Canada, e-mail: joseph.shea@unbc.ca

Contents

Chapter Overview

Key Findings

1. **The cryosphere—snow, ice, and permafrost—is an important part of the water supply in the extended Hindu Kush Himalaya region. Observed and projected changes in the cryosphere will affect the timing and magnitude of streamflows across the region, with proportionally greater impacts upstream.** Cryospheric change will have modest impacts on total annual streamflows in large river systems but will strongly affect the timing and seasonal distribution of runoff, which is relevant for both ecology and economy. As snow and ice melt provide a reliable source of meltwater during warmer months, the cryosphere helps to buffer against changes in streamflows due to climate change and monsoon variability.
2. **There is high confidence that snow-covered areas and snow volumes will decrease in most regions over the coming decades in response to increased temperatures, and that snowline elevations will rise.** The greatest changes in snow accumulations will be observed in regions with higher mean annual temperatures. Projected changes in snow volumes and snowline elevations (+400 to +900 m) will affect seasonal water storage and mountain streamflows.
3. **Glaciers have thinned, retreated, and lost mass since the 1970s, except for parts of the Karakoram, eastern Pamir, and western Kunlun. Trends of increased mass loss are projected to continue in most regions, with possibly large consequences for the timing and magnitude of glacier melt runoff and glacier lake expansion.** Glacier volumes are projected to decline by up to 90% through the 21st century in response to decreased snowfall, increased snowline elevations, and longer melt seasons. Lower emission pathways should, however, reduce the total volume loss.
4. **There is high confidence that permafrost will continue to thaw and the active layer (seasonally thawed upper soil layer) thickness will increase.** Projected permafrost degradation will destabilize some high mountain slopes and peaks, cause local changes in hydrology, and threaten transportation infrastructure.

Policy Messages

1. **To reduce and slow cryospheric change, international agreements must mitigate climate change through emission reductions.** Lower emission pathways will reduce overall cryospheric change and reduce secondary impacts on water resources from mountain headwaters.
2. **To better monitor and model cryospheric change and to assess spatial patterns and trends, researchers urgently need expanded observation networks and data-sharing agreements across the extended HKH region.** This should include in situ and detailed remote sensing observations on selected glaciers, rapid access to high-resolution satellite imagery, improved and expanded snow depth and snow water equivalent measurements, and ground temperatures and active layer thickness measurements in different regions, aspects and elevations.
3. **Improved understanding of cryospheric change and its drivers will help reduce the risk of high-mountain hazards.** Glacier lake outburst floods (GLOFs), mass movements (rockfalls, avalanches, debris flows), and glacier collapses present significant risks to mountain residents. This risk can be minimized with improved observations and models of cryospheric processes.

The cryosphere is defined by the presence of frozen water in its many forms: glaciers, ice caps, ice sheets, snow, permafrost, and river and lake ice. In the extended Hindu Kush Himalaya (HKH) region, which includes the Pamirs, Tien Shan, and Tibetan Plateau ranges, the cryosphere is a key freshwater resource, playing a vital and significant role in local and regional hydrology and ecology. Industry, agriculture, and hydroelectric power generation rely on timely and sufficient delivery of water in major river systems; changes in the cryospheric system may also pose challenges for disaster risk reduction in the extended HKH region.

Surveying the status and trends of the extended HKH cryosphere requires a detailed and comprehensive analysis of all its parts. The response of these components to climate change varies in both space and time. Recent glacier mass and area changes exhibit important regional variations, and the response times of snow cover and permafrost to climate change are vastly different. This chapter summarizes the current status of cryospheric components in the extended HKH, examines patterns and impacts of change, and synthesizes cryospheric change projections in response to representative concentration pathway (RCP) scenarios. We further identify significant knowledge gaps and consider the policy relevance of cryospheric research.

Snow is an important seasonal water storage component in the extended HKH and, in many areas, a critical source of streamflow for irrigation (Chap. 8). Yet snowfall totals at high elevations are poorly known. Long-term measurements of snow water equivalent (SWE) and solid precipitation are needed to test gridded data sets (i.e., regionalized precipitation data based on physical modelling and/or interpolation based on existing in situ measurements) or remote sensing-derived precipitation data sets. As snow cover in the extended HKH is highly variable and satellite-derived records are short, observed trends in snow cover are generally weak and inconsistent between studies and regions. Future projections point towards reduced snow cover and lower basin-wide SWE, although regional differences exist due to synoptic setting.

Glacier meltwater provides a regular and reliable source of streamflow in glacierized river basins. Such predictability is especially vital in the post-monsoon season and for regions with lower summer precipitation. According to a compilation of glacier mass and area change studies, glaciers in most regions are shrinking and losing mass. However, glaciers in the eastern Pamir, Karakoram, and western Kunlun Shan have lost less mass than others, and since at least 2000 have had balanced mass budgets or even slight mass gains.

Projections of glacier change are consistent across studies: glacier mass loss will accelerate through the 21st century, and higher-emission scenarios will result in even greater mass loss. The rise of regional equilibrium line altitudes (ELAs) will eventually result in the complete disappearance of debris-free lower elevation glaciers, and will increase volume losses from glaciers with high-elevation accumulation areas.

Glacial lakes occur frequently in the extended HKH, and numerous new lakes will form in response to cryospheric change. Since the 1990s, glacial lakes show a clear increase both in number and in area. Both the largest total glacial lake area and the greatest absolute glacial lake growth rate appear in the eastern and central Himalaya. Several glacial lakes in the extended HKH are potentially hazardous—and as glaciers continue to retreat, the risk of dangerous glacial lake outburst floods (GLOFs) may increase further. More assessments, projections, and impact studies are needed to clarify GLOF risks across the extended HKH.

Permafrost exists beneath large parts of the extended HKH —yet its occurrence and importance is not widely known in the region. As permafrost cannot easily be seen, it is easy to ignore. Yet permafrost can shape many climate impacts on cold regions. Although detailed data are scarce, existing measurement sites indicate permafrost warming, with an increase in the depth of the active layer. Thawing permafrost can reduce ground stability and cause a range of problems, from undermining engineered structures, to increased occurrence of rockfalls, and increased outburst potential of glacier lakes. In addition, thawing permafrost affects the hydrological cycle: water stored in ground ice may be released or near-surface soil water availability may decline as the active layer thickens.

Continued neglect of permafrost in large parts of the extended HKH could limit future capacity for adaptation and risk reduction programmes, as relevant environmental trends will be misjudged or missed completely. Progress will depend on learning more about the physical characteristics of permafrost in the extended HKH and about its causal links with human activity. Furthermore, observation networks are required for developing and testing simulation tools specific to extended HKH conditions.

Information about ***lake and river ice*** has been identified as a suitable proxy for mean air temperatures and their variability. However, very few studies address changes in lake ice coverage, freeze dates, and break-up dates. In situ observations are almost non-existent. Remote sensing-based studies show ice coverage declining in larger lakes since 1980, but no clear trend emerges for the period 2000–12. While no future projections exist, it is likely that with a continuous increase in air temperature, ice coverage will decrease further.

Hydrological trends related to cryospheric change are difficult to identify for three reasons: confounding influences on discharge, scarcity of long-term data sets, and high interannual variability in discharge that masks any temporal trends. Nevertheless, large volumes of snow and ice in the extended HKH are important regional water supplies—more so as one looks further upstream. An increase in air temperature will reduce snowpack accumulations and result in earlier and lower snowmelt runoff volumes. Medium- and long-term changes in glaciers and permafrost will reduce summer melt contributions.

Our synthesis of the scientific literature on cryospheric change in the extended HKH suggests that more resources are needed for cryospheric change impact studies—both regional-and sector-specific. High-level international agreements should promote systematic data collection, data sharing, training of local and regional scientists and technicians, and the development of cryosphere-related hazard warning systems. Above all, stronger global commitments are needed to reduce greenhouse gas emissions, as all projections indicate that lower emission futures result in reduced cryospheric change.

Reliable, freely available, and continuous in situ measurements of glaciers, snowpacks, permafrost, meteorology, and hydrology are lacking within the extended HKH. In addition, we identify the following critical research gaps for which urgent action is needed:

1. Better SWE observations and estimates are needed for high mountain areas; long-term snow course monitoring sites and strategies should be developed.
2. Glacier volume change estimates prior to 2000 are currently unavailable for several regions—a gap that researchers can now fill, however, using declassified satellite imagery.
3. We lack detailed studies of high-elevation snow accumulation and snowmelt processes and scenarios concerning future snowpack properties.
4. Well-documented, reliable, and long-term hydrological observations are needed across different climate zones and elevations to assess uncertainty and variability in discharge observations, and to improve and develop models of hydrological change
5. Although many studies quantify glacier area and volume change, few try to diagnose the reasons for these changes. Better models of future glacier change will require a closer focus on regional glacier sensitivity and the causes of regional glacier change.

Cryosphere and the Sustainable Development Goals With its fundamental role in regulating the supply of water in the extended HKH, the cryosphere is important to achieving several Sustainable Development Goals (SDGs). These include the following:

1. SDG 6—Ensure availability and sustainable management of water and sanitation for all. Snow and ice melt contributions to streamflow vary across the extended HKH, and they are more important upstream than downstream. Yet they provide a reliable water source, and they can buffer against years or months of drought or low precipitation.
2. SDG 7—Ensure access to affordable, reliable, sustainable, and modern energy for all. The cryosphere contributes to hydroelectric power generation. Snow and ice are short- and medium-term water reservoirs; their depletion may constrain hydroelectric power generation during low-flow seasons and affect long-term energy production.
3. SDG 13—Take urgent action to combat climate change and its impacts. The future of the cryosphere depends in part on international agreements to mitigate climate change through emission reductions. Such reductions will ultimately reduce the total impact of climate change on the extended HKH cryosphere.

7.1 Situating the Cryosphere in the Hindu Kush Himalaya and Tibetan Plateau-Pamir Region

The Hindu Kush-Himalaya and Tibetan Plateau-Pamir includes the world's highest mountains and probably the greatest range of climatic conditions that can be observed over a distance of a few hundred kilometres anywhere on the planet. Influenced by both the summer monsoon and westerly low pressure systems, the occurrence of snow, ice, and permafrost is governed by the mountain topography and how it interacts with prevailing atmospheric circulation patterns. This section outlines the regional definitions used to examine cryospheric change and provides a short introduction to the climatic characteristics of the region. For detailed information regarding the climate, we refer to Chap. 3 on climate change.

7.1.1 Defining the Extended HKH Region from a Cryosphere Perspective

This assessment investigates the high mountain regions of the Hindu Kush, Karakoram, Himalaya, and Tibetan Plateau. The whole of the Pamir is also included in the assessment since parts of the Pamir are frequently included in studies addressing the Tibetan Plateau, and the remaining parts provide important additional information to understand the variability of the cryospheric changes. For simplicity, we refer to the entire region as the extended HKH region, and define it in this chapter to include the Hindu Kush-Karakoram-Himalaya-Tibetan Plateau-Pamir mountains.

To provide a suitable regional overview and highlight regional variations in cryospheric trends, we divide the extended HKH into 22 subregions (Fig. 7.1). The divisions were based on existing regional delineations (Gurung 1999; Shi et al. 2008; Shroder 2011) updated according to the topographical and climatological characteristics (e.g.,

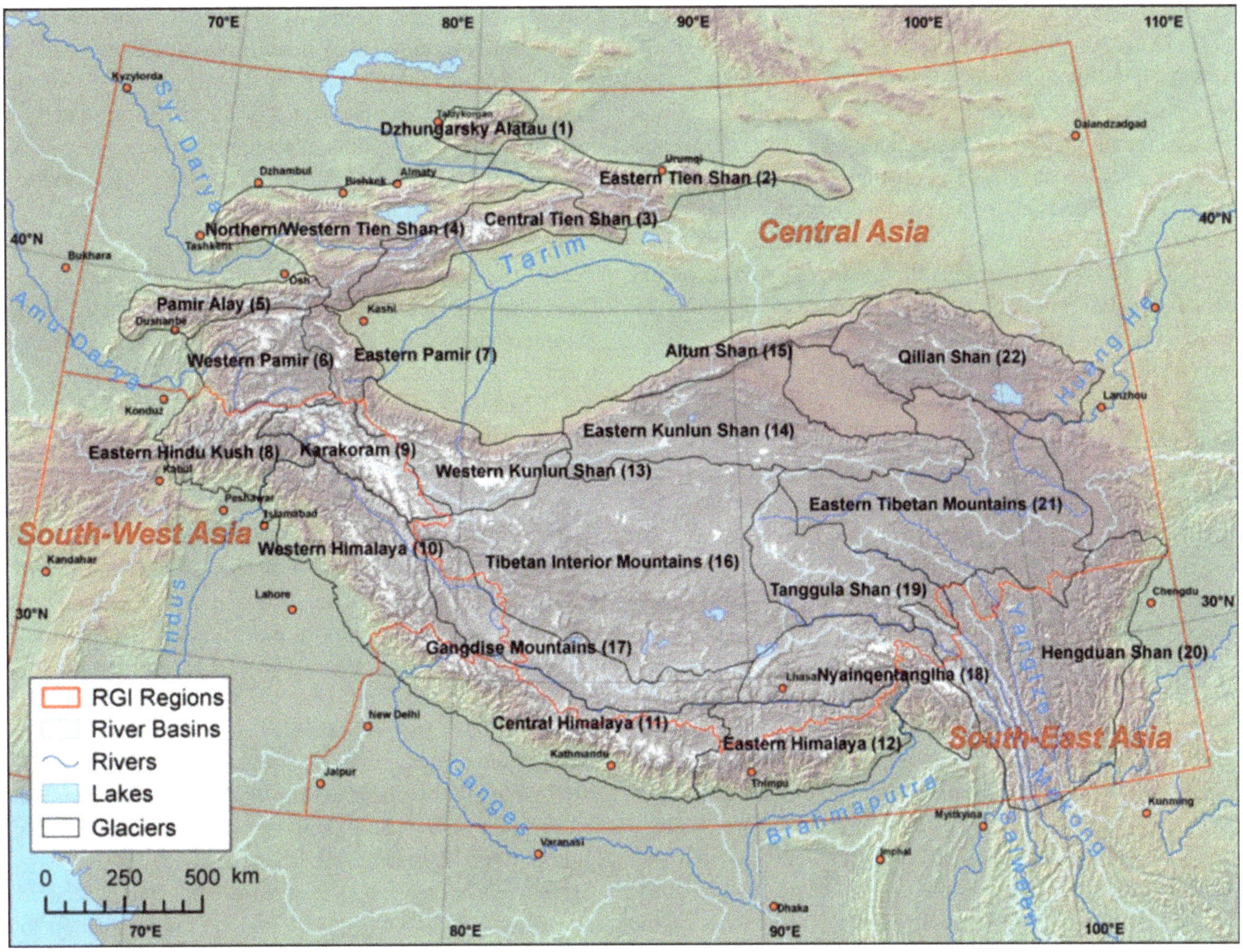

Fig. 7.1 High mountains of Asia showing the extended HKH subregions (06–22) used in this assessment (black), the Randolph Glacier Inventory (RGI) regions (red), and major rivers (blue)

Maussion et al. 2014) of the different mountain ranges. Regional glacier modelling studies typically use the larger region definitions given by the Randolph Glacier Inventory (RGI; Pfeffer et al. 2014), and these are also used here (Fig. 7.1).

7.1.2 Regional Climate

The extended HKH is characterized by extreme topographic and climatic heterogeneity. Interactions between atmospheric circulation and the world's tallest mountains produce strong horizontal and vertical gradients in both temperature and precipitation. In general, the climate is influenced by the East Asian and Indian monsoon systems, which deliver the bulk of precipitation in the central and eastern Himalayas between June and September (Bookhagen and Burbank 2006), with annual precipitation totals of up to 3,000 mm (Maussion et al. 2014; Bookhagen and Burbank 2010). In the western part of the region, westerly disturbances in winter account for high snowfall ratios (over 80%) in annual precipitation rates of 1,000–2,000 mm (Barlow et al. 2005; Bookhagen and Burbank 2010; Hewitt 2011; Maussion et al. 2014), but winters in the monsoon-dominated central and eastern areas are dry (Shrestha et al. 2000). As a result of the topography, mean annual temperatures can vary up to 40 °C over distances of a few hundred kilometres, and strong gradients of precipitation exist across the mountain ranges (Shrestha et al. 2012).

While precipitation typically increases with elevation due to orographic forcing (Barry 1992), the extreme elevations and heterogeneity encountered in the extended HKH results in vertical precipitation gradients that vary both in space (Andermann et al. 2011) and time (Collier and Immerzeel 2015). In monsoon-dominated regions, the greatest precipitation rates are typically observed at elevations around 3,000 masl (Putkonen 2004; Bookhagen and Burbank 2006; Shrestha et al. 2012). However, in the Upper Indus basin, both observations and inverse glacier mass balance modelling suggest that precipitation above 5,000 masl can be five-to-ten times greater than that received in the valleys below (Hewitt et al. 1988; Winiger et al. 2005; Hewitt 2005; Duan et al. 2015b; Immerzeel et al. 2015). As the dynamics and mechanisms of precipitation delivery (orographic versus synoptic versus convective) vary throughout the region, vertical precipitation gradients cannot be generalized spatially or even seasonally. The spatial variability in climatic regimes could account for the spatial heterogeneity of glacier response (Sakai et al. 2015; Sakai and Fujita 2017). However, knowledge of precipitation gradients through remote sensing and/or ground-based measurements is critical for evaluating basin-wide snow water equivalences.

Trends in temperature and precipitation vary spatially across the extended HKH region (Malik et al. 2016) and are examined in detail in Chap. 3. Climatic change will likely be amplified by elevation-dependent warming (see Box 7.1), placing additional pressures on the extended HKH cryosphere.

7.2 Snow

Snow is a critical component of the extended HKH cryosphere. Northern hemisphere snow cover plays an important role in global hydrology and surface energy balance (Groisman et al. 1994), and regulates the global climate (Frei et al. 2012). Snow is also a critical short-term water storage mechanism (Zhang et al. 2012), with a potentially high sensitivity to future climate change (Barnett et al. 2005; Immerzeel et al. 2009). People in the river basins of the extended HKH region rely to a varying degree on snowmelt for drinking water, irrigation, and hydropower generation (Bookhagen and Burbank 2010; A. Lutz et al. 2014), while snow cover on the Tibetan Plateau has a profound influence on atmospheric circulation and the Asian monsoon (Wu and Zhang 1998; Bansod et al. 2003; Qian et al. 2011). Changes in seasonal snow cover will directly impact the availability and seasonal distribution of fresh water in mountain river basins (Beniston 2003; Mukhopadhyay and Khan 2015) and will also affect permafrost via alterations in the thermal coupling between atmosphere and subsurface. This section explores how snow is measured, and summarizes the observed and projected changes in snow cover in the extended HKH region.

7.2.1 Monitoring of Snow

Snow is quantified through in situ measurements of snow depth, snow water equivalent (SWE), or as a precipitation volume. Annual snow accumulation rates can be estimated from glacier mass balance records (Sect. 7.3) and firn or ice cores. Remotely-sensed snow measurements include snow-covered area (SCA), snow-covered fraction (SCF), and derived quantities such as snow cover duration (SCD). Snowline elevations are also sensitive indicators of climate change, and are highly correlated with glacier equilibrium line altitudes (ELAs) and mass balance. Unfortunately, the lack of in situ observations represents a serious limitation to snow-related studies in the extended HKH (Rohrer et al. 2013; Salzmann et al. 2014).

7.2.1.1 Field Measurement

Field measurements are challenging to maintain in high mountain areas due to poor accessibility and challenging weather conditions (e.g., Germann et al. 2006; Palazzi et al. 2013; Shea et al. 2015a). There are a number of specific problems in measuring snowfall in the extended HKH. First, solid precipitation is difficult to measure in windy environments, and gauge undercatch can be a significant problem (Førland et al. 1996), but instruments that continually monitor SWE (e.g., snow pillows) directly are rare in the region (Hasson et al. 2014) and are also susceptible to errors. Second, most weather stations in the extended HKH lie below 5,000 masl (Shrestha et al. 1999; Fowler and Archer 2006; Ma et al. 2008) and are thus not representative of glacier accumulation areas or the elevation of maximum snow accumulation (Harper and Humphrey 2003; Immerzeel et al. 2015). Finally, available in situ data is usually fragmented and held by different stakeholders in the regional countries and not easily accessible (Dahri et al. 2016).

Point glacier mass balance measurements may be used to assess seasonal snow accumulations at high elevation locations, though this is complicated by wind erosion/drift, sublimation, and access difficulties. Ice cores and shallow firn cores provide an important method for evaluating past snow accumulation rates and can provide century-scale series of climatic data at high elevations where observations are sparse or non-existent (Kaspari et al. 2008; Duan et al. 2015a). However, possible locations for ice cores are limited, and the data are subject to the same complications as point mass balance measurements. Climate reconstructions have been made from ice cores taken from the Everest region (Kaspari et al. 2008), the Pamir (Aizen et al. 2009), and in several locations on the Tibetan Plateau (Thompson et al. 2000; Aizen et al. 2006).

7.2.1.2 Remote Sensing Measurement

Remote sensing offers one of the most promising sources of information on snow in the extended HKH, due to the large spatial extent of the region and the inaccessible terrain (Immerzeel et al. 2009). Optical remote sensing techniques have been used previously in the region to measure snow-covered area or fraction, snow cover duration, and albedo (Hall et al. 2006; Pu and Xu 2009; Frei et al. 2012; Brun et al. 2015). Active microwave sensors can provide

Box 7.1 Accelerated warming at high elevations

Arctic amplification, whereby enhanced warming over time (*dT/dt*) is evident at high latitudes (Serreze and Barry 2011; Hinzman et al. 2005), is well accepted amongst the scientific community. There is also strong theoretical reasoning indicating that this enhancement should apply at high elevations (Mountain Research Initiative EDW Working Group 2015). Elevation dependent warming (EDW) is encouraged by many physical mechanisms (Fig. 7.2). Changes in surface characteristics caused by snow/ice melt and treeline-migration will tend to decrease surface albedo and enhance *dT/dt* in a particular elevation band as the snowline/treeline moves upslope (Fig. 7.2a). A warmer atmosphere can hold more moisture, which will encourage increased latent heat release, especially at high elevations (Fig. 7.2b). A moistening of the atmosphere will also increase downwelling of longwave radiation since water vapour is a greenhouse gas, but the effects of a given moistening will be increased at high elevations, which are currently very dry (Fig. 7.2c). Blackbody longwave radiation emissions are proportional to the fourth power of temperature;

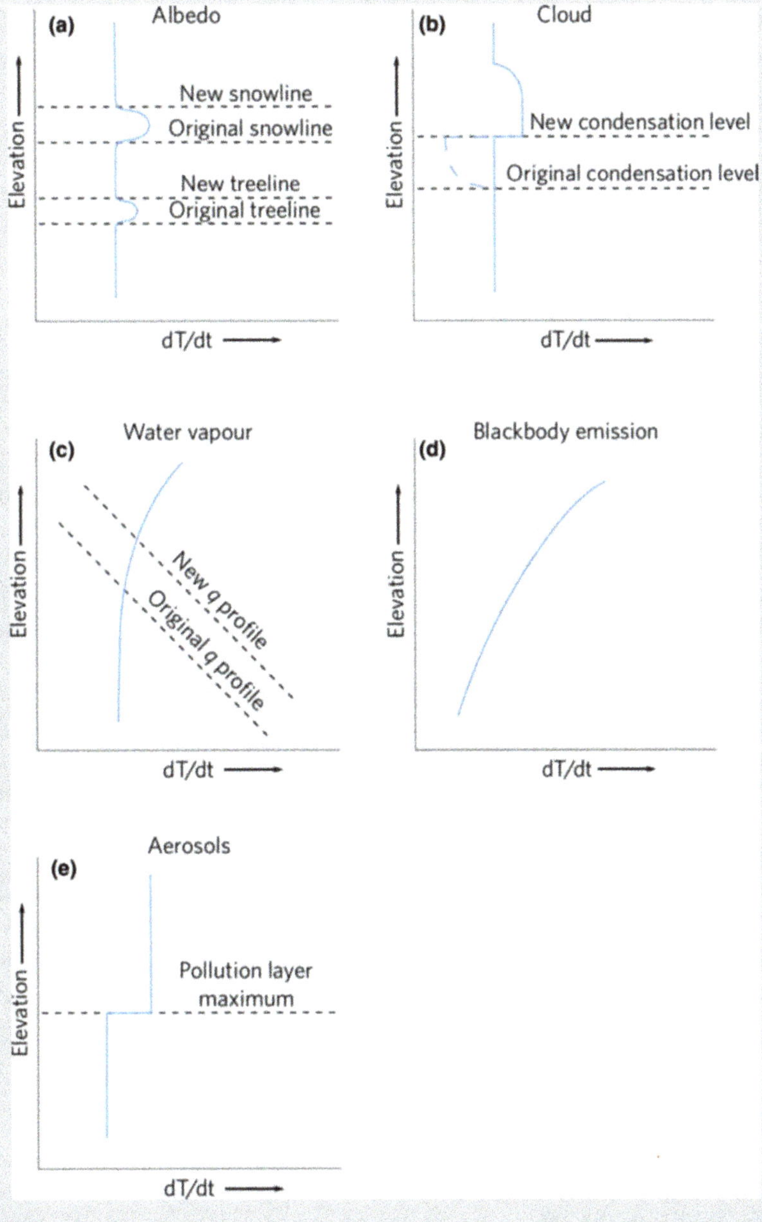

Fig. 7.2 Schematic diagrams of five mechanisms of elevation dependent warming. dT/dt is the change in temperature over time (adapted from Mountain Research Initiative EDW Working Group 2015)

near-surface air temperature sensitivities will therefore be greater at the lower temperatures common in both high latitudes and high mountains (Fig. 7.2d). Finally, aerosols, in particular black carbon and dust, cause both global dimming at lower elevations (meaning less effect in high mountains) and, when deposited on snow and ice, encourage more rapid melt of glaciers (Fig. 7.2e). Combined, these five factors result in amplified warming observed in high-mountain environments in comparison to their adjacent lowlands.

Unfortunately, observational evidence of EDW is limited, primarily because of a lack of high elevation data. Surface stations are biased towards low elevations, satellites measure local surface temperature (which shows extreme local variability), and models, although very powerful tools, are usually not of high enough resolution to capture the complexities of high elevation terrain. Thus, it is possible that future climate changes in mountain regions are currently underestimated.

information about snow characteristics (Koskinen et al. 1997; Foster et al. 2011), snowmelt onset, and snow accumulation (Tedesco and Miller 2007).

Snowline elevations in the extended HKH have been monitored with a combination of optical imagery and digital elevation models (Thakuri et al. 2014; Wu et al. 2014; Kronenberg et al. 2016; Spiess et al. 2016). For temperate glaciers, snowline elevations at the end of the melt season are directly related to glacier mass balance (Dumont et al. 2012) and water resources (Winiger et al. 2005). But snowline elevations vary spatially and temporally, and are prone to rapid fluctuations in response to individual storm events.

From a water resources standpoint, distributed measurements of SWE are of primary importance as SWE relates directly to the water available for melt and streamflow. Passive microwave (PM) remote sensing, despite being subject to considerable uncertainties (Chang et al. 1987; König et al. 2001; Qin et al. 2006; Frei et al. 2012; Tiwari et al. 2016), offers a potentially useful tool to measure SWE (Smith and Bookhagen 2016). The typical resolution of PM sensors (e.g., Special Sensor Microwave/Imager, or SSM/I) is large (8–60 km), but it has been used to monitor snowmelt onset in the western Pamirs (Vuyovich and Jacobs 2011).

Spaceborne radar has been used to measure precipitation rates and totals across the extended HKH (Anders et al. 2006; Bhatt and Nakamura 2006; Bookhagen and Burbank 2006; Yin et al. 2008). These studies indicate that the spatial pattern and magnitude of annual precipitation are consistent from year to year. Unfortunately, snowfall is not well captured by the Tropical Rainfall Measurement Mission (TRMM) platform, and only limited information can be obtained from satellites about high-elevation snowfall rates and accumulated SWE (Anders et al. 2006).

7.2.1.3 Regional Differences in Snow Cover

Snow cover varies considerably both spatially and temporally over the extended HKH, but no study has systematically evaluated snow cover metrics (SCA, SCF, SCD) for the entire region. The average snow-covered area of the extended HKH ranges between 10 and 18%, with mean maximum and mean minimum values of 21–42% in winter and 2–4% in summer (Immerzeel et al. 2009; Gurung et al. 2011). About 59% of the Tibetan Plateau is snow-covered in the winter, with the southeastern portion having the deepest and most persistent snow (Qin et al. 2006; Li et al. 2009). However, snow cover on the Tibetan Plateau is often shallow and patchy, and of short duration (Robinson and Dewey 1990; Qin et al. 2006).

Westerly-affected regions such as the Karakoram and Pamir have extensive winter snow cover (Xiao et al. 2007; Pu and Xu 2009; Zhou et al. 2013; Shen et al. 2015; Tahir et al. 2016), while snow cover in monsoon-dominated areas is limited to high elevations (Immerzeel et al. 2009; Dahri et al. 2011; Savoskul and Smakhtin 2013; Hasson et al. 2014; Singh et al. 2014). The snow-covered area in the upper Indus Basin (UIB) varies between 4 and 57%, with the maximum occurring in spring; westerly-affected catchments within the UIB have a higher mean annual snow-covered area, at 51%, than monsoon-dominated catchments, at 20% (Hasson et al. 2014). The UIB is generally more snow-covered than the entire extended HKH (Immerzeel et al. 2009), with an average annual snow-covered area value of ~34% (12% in summer and ~57% in spring). The snow-covered area in three entire large river basins (Indus, Ganges, and Brahmaputra) ranges from 85% during snow accumulation periods to approximately 10% during ablation periods (Singh et al. 2014; Tahir et al. 2016).

7.2.2 Observed Changes

Information extracted from ice cores on the Tibetan Plateau shows decreased snow accumulation rates since the 1960s at many high-elevation sites (Thompson et al. 2000; Hou et al. 2002; Kehrwald et al. 2008; Kang et al. 2015; An et al. 2016). Coarse-resolution SSM-I products also show

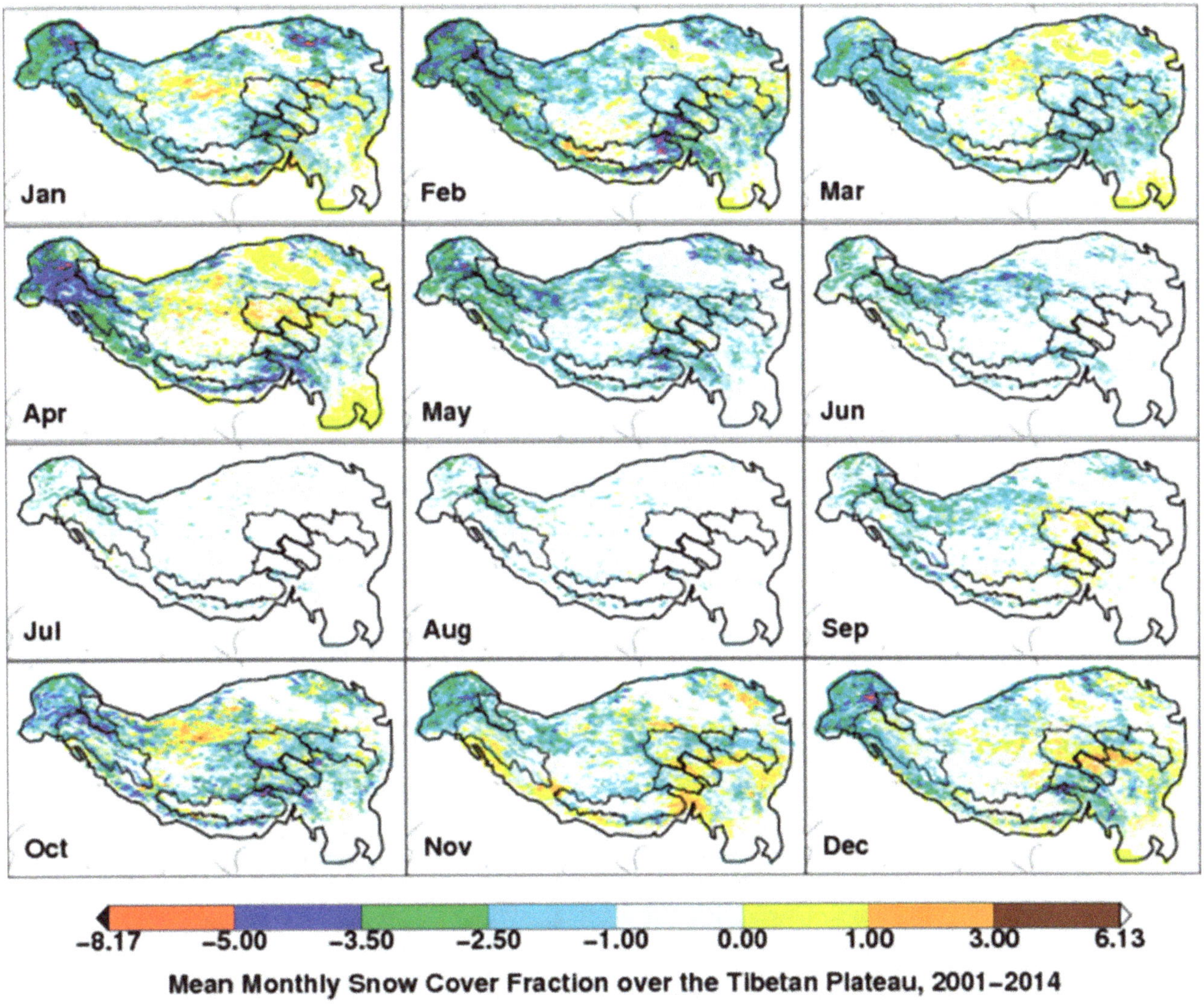

Fig. 7.3 Snow cover fraction trends (%/yr) over the extended HKH from 2000 to 2014 (*Source* Li et al. 2017)

decreases in snow cover duration between 1966 and 2001 at all elevations across the extended HKH (Rikiishi and Nakasato 2006). Increases in winter SCA/SCF fraction have been observed in westerly-dominated basins since 2000 by most of the studies, and are either decreasing or stable elsewhere (Immerzeel et al. 2009; Hasson et al. 2014; Wang et al. 2015c; Tahir et al. 2016; Fig. 7.3). However, a recent study found a slightly decreasing trend of winter snow cover in the west based on similar data but a slightly longer period (2000–2014) (Li et al. 2017). This indicates the need for further investigations to resolve discrepancies.

7.2.2.1 Hindu Kush-Himalaya-Karakoram

The reported changes in snowline elevation, snow cover, and snow accumulation rates in the extended HKH are inconsistent. High interannual variability in snow cover and relatively short (<15 years) datasets also prevent trend detection.

Annual snow-covered area trends over the entire extended HKH since 2000 are negative, but not significant (Gurung et al. 2011); slight increases (but still insignificant) have been observed in autumn in the central and eastern Himalayas, and in spring and summer in the western Himalayas (Gurung et al. 2011; Maskey et al. 2011; Wang et al. 2013a). At the river basin scale, snow-covered area increased in the Indus basin, but decreased in the Ganges and Brahmaputra basins between 2000 and 2011 (Singh et al. 2014). Late winter and spring snow cover, in particular, has decreased in the UIB since 2000 (Immerzeel et al. 2009; Mukhopadhyay 2012; Li et al. 2017).

End-of-summer snowline elevations in Himachal Pradesh rose by approximately 400 m between 1980 and 2007, a rate of 14.8 m/yr (Pandey et al. 2013), while end-of-summer snowline elevations in the UIB lowered by 20 m/yr between 2000 and 2011 (Hasson et al. 2014).

Increased snow accumulation has been reported in the eastern extended HKH (Zhou et al. 2007), but this contrasts with results derived from ice cores from the Everest region which showed decreased snow accumulation totals since the 1970s (Kaspari et al. 2008). Modelled snowfall in the extended HKH region has decreased since 1861 (Kapnick et al. 2014). Glacier albedo, which depends on pollution (see Box 7.2), snow cover, and snowline elevation, has decreased at multiple sites in the extended HKH since 2000 (Ming et al. 2012; Brun et al. 2015; Xu et al. 2016). A positive trend in albedo was noted at Chhota Shigri Glacier (see Sect. 7.3), however, MODIS observations of albedo should be interpreted with caution.

Box 7.2 Black carbon, dust, aerosols
The Hindu Kush-Himalaya lies downwind of large sources of pollution produced by electricity generation, biomass burning, and brick factories. Short-lived climate pollutants (SLCPs) such as black carbon, dust, and other aerosols, affect snow and ice melt rates through direct and indirect effects. Direct effects include the darkening of the surface due to deposition, which will increase melt rates on snow in particular. However, black carbon and dust in the atmosphere also lead to local warming of the air mass. The net effect of SLCPs may be up to an order of magnitude greater than that observed from CO_2 alone (Ramanathan and Carmichael 2008; Jacobi et al. 2015). SLCPs can lead to earlier melt onset, advanced snowline retreat, and an increase in annual glacier melt, but these impacts depend on the timing of deposition and local meteorology (Ménégoz et al. 2014). Knowledge of the temporal and spatial distribution of SLCPs, and in particular their distribution with elevation, will lead to improved estimates of their impact on snow and ice melt. Chapter 10 (Air Pollution) provides detailed information on air pollution sources and impacts.

7.2.2.2 Pamir

Snowmelt is of particular importance to river basins in central Asia (Aizen et al. 1995). Though records of snow cover are short, significant decreasing trends have been observed in snow cover duration in the Amu Darya basin (Zhou et al. 2013). Variations in snow cover duration can be caused by either earlier melt onset dates or delayed snow cover onset (Dietz et al. 2013). A decrease in maximum snow depth and snow cover duration has accompanied a general warming trend in the Muztag Ata region (Unger-Shayesteh et al. 2013). Confirmation of reduced snow accumulation in eastern Pamir since the 1970s was obtained from ice core measurements (Duan et al. 2015a).

7.2.2.3 Tibetan Plateau

Recent decreases in average annual and winter snow-covered fraction and snow depth have been observed on the Tibetan Plateau (Ma 2008; Wang et al. 2013b, 2015a; Xu et al. 2016; Fig. 7.3). As TP snow cover is generally thin and patchy (Qin et al. 2006), it is assumed to be highly sensitive to climate change (Pu et al. 2007; Li et al. 2008; Kang et al. 2010). Longer snow records based on station data suggest that snow cover increased between the 1950s and late 1990s, with increased inter-annual variability since the mid-1980s (Qin et al. 2006), and it has been speculated that this may be due to changes in atmospheric water vapour content. Analyses of station data from 1951 to 2004 indicate that snow cover at the margins of the Tibetan Plateau is more sensitive to climate change than in the interior (Ma et al. 2010). Increases in both temperature and precipitation have resulted in an elevation-dependent snow cover response in eastern Tibet, with decreased snow cover duration at lower elevations and increased snow cover duration at higher elevations (Gao et al. 2012; Wang et al. 2015a).

7.2.3 Projected Changes

There are relatively few projections of future snowpack behaviour in the extended HKH. At the basin scale, high emission climate scenarios (RCP8.5) from the most recent Coupled Model Intercomparison Project (CMIP5) show snowfall reductions of 30–50% in the Indus Basin, 50–60% in the Ganges basin, and 50–70% in the Brahmaputra Basin by 2071–2100 (Viste et al. 2015). A 50% reduction in average basin SWE has also been projected for the Upper Indus by the 2050s (Bocchiola et al. 2011) under a 'business-as-usual' emission scenario (SRES A2). Snow depth reductions of 25–50% and 17–39% have also been projected for the Himalayas and Hindu Kush-Karakoram, respectively (Terzago et al. 2014), though an earlier modelling study of the Spiti River showed only a 1–7% decrease in SWE in response to a +2 °C temperature increase (Singh and Kumar 1997). Projected decreases in end of winter SWE (Diffenbaugh et al. 2013) across the Himalaya may not be relevant to local hydrology, particularly in monsoon-dominated regions, since winter snow only comprises a small fraction of all snow here.

By 2100, snowline elevations are projected to rise between 400 and 900 m (4.4 to 10.0 m/yr) in the Indus, Ganges, and Brahmaputra basins under RCP8.5 emission scenarios (Viste et al. 2015). Coarse resolution general circulation models (GCMs) project a similar rate of rise in the freezing line altitude (zero degree isotherm) for the region (Ghatak et al. 2014). Chevallier et al. (2014) project a snowline rise of 4–7 m/yr for the Pamir region. We note that these values are higher than the average 150 m increase in snowline that corresponds to a 1 °C increase in temperature in mountain regions (Beniston 2003).

In western China, snowfall totals may increase until mid-century (Li et al. 2008), while decreases are expected in eastern China and Inner Mongolia. Future decreases in snow cover and depth on the Qinghai-Tibetan Plateau (QTP) are dependent on emissions scenarios, with accelerated snow losses under high-emissions scenarios (Wei and Dong 2015).

7.2.4 Recommendations

Differences in reported snow cover trends may be due to differences in data sets, methodologies, areas covered, and time periods. Future snow cover research should aim for systematic basin-scale studies and consistent methodologies. Remote sensing missions and products developed for sustained monitoring are critical to reduce the uncertainties associated with the present lack of coherence in observing systems.

Our review identifies a number of key areas that have not been addressed in the published literature with respect to snow. These include:

- More robust estimates of regional snowline elevation change based on time-series of optical satellite imagery (as opposed to end-of-season snapshots, which are not suitable for monsoon climates)
- Regional studies of the rain-snow limit: how does it vary spatially and temporally? What are the implications of a temperature increase on solid/liquid precipitation volumes?
- Detailed studies of pollution transport and the impacts of regional pollution on the surface energy balance of high-elevation snowpacks (Box 7.2) are required to examine the cascading effects on snow hydrology and glacier mass balance.
- Attribution of snow cover extent and duration changes to meteorological conditions should be considered. What drives the spatial variability in snowpack changes?
- Improvement in the resolution and frequency of microwave SWE monitoring from satellites

7.3 Glaciers

Glaciers are important indicators of climate change (IPCC 2013). Glacier meltwater contributes to the discharge of major Asian rivers (Immerzeel et al. 2010) and feeds endorheic lakes on the Tibetan Plateau (Zhang et al. 2013). The importance of the glacier melt contribution varies across the region, but generally increases upstream (Lutz et al. 2014). Glaciers store meteoric water in the form of ice and release water in the warmer months, and are an important source of water in drier regions and seasons (Kaser et al. 2010). Changes in glaciers might also increase risks, especially from glacial lake outburst floods (Richardson and Reynolds 2000; Sect. 7.4).

This section introduces existing methods to measure glacier area and estimate their volume and then provides a synthesis of glacier changes throughout the extended HKH, followed by a summary of projected future change.

7.3.1 Measuring Glacier Change

Our knowledge of glacier changes in the extended HKH has increased substantially in recent years and a number of studies provide an overview of the changes in the region (Bolch et al. 2012; Kääb et al. 2012; Yao et al. 2012; Gardner et al. 2013; Brun et al. 2017). Glacier fluctuations are not only a reaction to climatic forcing (Leysinger Vieli and Gudmundsson 2004; Oerlemans 2005; Leclercq and Oerlemans 2012; Lüthi et al. 2015; Hewitt 2005): glacier responses are also affected by topographic features such as the length, area, glacier surface slope, glacier bedrock slope, and surroundings (e.g., proglacial lake, type of bedrock sediment), supraglacial features (debris-cover, lakes, ice cliffs, and so on), subglacial hydrology, and surge dynamics (in the case of surge-type glaciers). As a result, changes in glacier length and area through advance or retreat are an indirect, delayed, and filtered signal to changes in climate. In contrast, the glacier mass balance (i.e., the annual change in volume or mass) is a relatively direct and undelayed response to the annual atmospheric conditions (Haeberli and Hoelzle 1995).

Comparison of glacier outlines obtained from different periods in time, usually from remote sensing data, allows quantification of glacier length and area changes (Paul et al. 2004; Shangguan et al. 2007; Bolch et al. 2010a; Bhambri et al. 2011). Area changes of debris-covered glaciers need special consideration as glaciers with thick debris-cover typically exhibit lower rates of shrinkage than debris-free glaciers (Scherler et al. 2011; Nuimura et al. 2012; Pieczonka and Bolch 2015).

Glacier mass changes can be measured in the field and by remote sensing. In situ measurements consist of ablation measurements with ablation stakes and accumulation measurements using snow pits (Kaser et al. 2003). The results of these measurements are then extrapolated to the entire glacier (Hock and Jensen 1999), with glacier-wide errors of typically

0.30 m w.e./yr if the measurement locations are well distributed over the entire glacier (Wagnon et al. 2013). In situ measurements provide detailed information but are laborious and can therefore only be obtained for relatively few glaciers. As reported by Vincent et al. (2013) and Gardner et al. (2013) glaciological mass balance data may be negatively biased due to poor spatial sampling. For this reason, we exclude data sets that have known errors from our analysis.

A widely-used approach to estimate glacier volume change and subsequently the mass change is the geodetic approach, or the difference in surface elevation measured at two or more points in time (Finsterwalder 1954; Bamber and Rivera 2007; Berthier et al. 2007; Bolch et al. 2010a). The major advantage of the geodetic approach is that large regions can be investigated simultaneously, but it is constrained by the availability of suitable elevation data or imagery. A global comparison of direct and geodetic mass balance measurements yields root mean square errors of approximately 0.4 m w.e./yr (Cogley 2009).

The Gravity Recovery and Climate Experiment (GRACE) programme has provided additional estimates of glacier mass loss or gain in the extended HKH region through analysis of changes in the earth's gravity field. Though gravimetric changes do not only depend on glacier mass changes (e.g., Reager et al. 2016), GRACE data provide useful insights on the interannual variability of mass balance (e.g., Yi and Sun 2014). However, large interannual variability in the gravity signal for the extended HKH region results in large uncertainties in mass loss estimates: -4 ± 20 Gt over the period 2003–2010 (Jacob et al. 2012).

Modelling provides a final avenue for examining glacier mass change. For a detailed discussion on glacier mass balance models and limitations, see Sect. 7.3.4 (Glacier projections).

7.3.1.1 Glacier Area

Knowledge about glacier cover is crucial, but only since the release in 2012 of the first Randolph Glacier Inventory (RGI, www.glims.org/rgi) (Arendt et al. 2012; Pfeffer et al. 2014) has a complete inventory of glaciers in the entire extended HKH region been freely available. Earlier information about glacier coverage exists, but the quality varies and the data is often not available in digital format.

Prior to the satellite era, information about glacier area was based on manual delineation from aerial photography or topographic maps. Freely available satellite imagery has greatly facilitated glacier mapping. This is especially true for the satellites which also provide short wave infrared acquisitions (such as Landsat TM, launched in 1984), as this wavelength allows automated identification of clean ice and snow (Paul et al. 2002). Automated classifications of debris-covered ice and seasonal snow on glacier-free areas are more difficult, and manual adjustments are needed to improve the quality. Current inventories are based solely on manual delineations (Nuimura et al. 2015) or on automated classification with manual post-processing (Bajracharya and Shrestha 2011; Frey et al. 2012). Satellite images also allow the identification of moraines from the Little Ice Age (LIA) and thus provide information on the LIA extent of glaciers (Baumann et al. 2009; Loibl and Lehmkuhl 2014). Uncertainties in glacier inventories result mainly from the resolution of the selected images, their quality with respect to snow and cloud cover, the existence of cast shadows, and the presence of debris cover (Paul et al. 2013).

Publicly available glacier inventories now exist for the entire High Mountain Asia (HMA) region (GAMDAM inventory, Nuimura et al. 2015), as well as larger subregions such as NW India (Frey et al. 2012), China (Guo et al. 2015), eastern Himalaya (Nagai et al. 2016; Ojha et al. 2017), and the whole of Hindu Kush-Karakoram-Himalaya (Bajracharya and Shrestha 2011). Recent inventories still contain significant differences in area; these can arise from different methodological approaches, difficulties such as snow and cloud cover on available satellite scenes, the treatment of debris-covered glaciers, and, in particular, different definitions of the glaciers. For example, GAMDAM (Nuimura et al. 2015) largely omitted ice on steep headwalls, while these areas are included in Frey et al. (2012). The available recent estimates of glacier area for the entire extended HKH (Table 7.1) vary from ~76,580 km^2 (GAMDAM, Nuimura et al. 2015) to 81,140 km^2 (RGI5.0, Arendt et al. 2015).

Table 7.1 Recent estimates of glacier area and volume for the extended HKH region

Himalaya[a]	Karakoram[b]	Pamir[c]	Tibetan-Plateau[d]	References
Glacierized area (km^2)				
21,973	21,205	NA	NA	Cogley (2011)
22,829	17,946	NA	NA	Bolch et al. (2012)
19,991	18,563	10,403	27,622	Nuimura et al. (2015)
NA	NA	NA	31,573	Guo et al. (2015)
26,688	19,962	13,071	39,822	RGI 3.2
20,070	21,475	10,681	28,912	RGI 5.0
Ice volume (km^3)				
1,212	1,683	NA	NA	Huss and Farinotti (2012)
1,297	1,869	NA	NA	Frey et al. (2014)

[a]Regions 10–12
[b]Region 9
[c]Regions 6–7
[d]Regions 13–22 as defined in Fig. 7.1
NA = not available

Between 10 and 25% of the total glacierized area in the Karakoram and Himalayan ranges is debris-covered (Scherler et al. 2011; Bajracharya and Shrestha 2011). In the upper Indus basin ~18% of the total glacier area is debris-covered (Khan et al. 2015); in the Ganges basin this is ~14%, and in the Brahmaputra basin this is ~12% (Bajracharya and Shrestha 2011). Lower topographic relief, and, on average, lower glacier ablation rates, limit the number of debris-covered glaciers on the TP.

Surging glaciers exhibit recurrent periods of elevated flow velocity (or surges) at rates of one to two orders of magnitude greater than during their periods of inactivity (or quiescence). Surging glaciers are especially common in the Karakoram (Copland et al. 2011; Paul 2015) and Pamir (Kotlyakov et al. 2008), but also occur in western Kunlun (Hewitt 2007; Kotlyakov et al. 2008; Hewitt 2011; Quincey et al. 2011; Sevestre and Benn 2015). The cause of glacial surging in the extended HKH region is largely unknown, although links have been made to climate, internal dynamics, and subglacial hydrology (Yasuda and Furuya 2013, 2015). Analyses of glacier change should distinguish carefully between surging and non-surging glaciers, as surging glaciers will exhibit patterns of length, area, and elevation changes that are inconsistent with those of nearby non-surging glaciers.

7.3.1.2 Glacier Volume

There are essentially two different methods to estimate glacier volume: (1) volume-area scaling approaches (Bahr et al. 1997; Cogley 2011), where ice volume is calculated based on an empirical formula in relation to the glacier area, as larger glaciers tend to be thicker; and (2) distributed estimates of glacier thickness (Huss and Farinotti 2012; Linsbauer et al. 2012; Farinotti et al. 2017). Empirical volume-area scaling approaches are potentially strongly misleading and inappropriate, particularly for the composite glaciers found in the extended HKH region, and we caution against their use in estimating glacier volumes. Distributed approaches are based on the assumption that ice thickness can be related to glacier slope, flow velocity, and basal shear stress. Freely available surface elevation data can thus be used to estimate current and future glacier thicknesses and bed topographies. All methods of ice volume estimation tend to have large uncertainties (Frey et al. 2014; Farinotti et al. 2017) and are calibrated to the limited number of existing thickness measurements (Gärtner-Roer et al. 2014).

Depending on the inventory and method used, total glacier volumes are estimated to range between 3,000 and 4,700 km^3 (Table 7.1) (Frey et al. 2014). According to these available estimates, the volume of glacier ice in the Karakoram is likely to be larger than the volume in the Himalaya.

7.3.1.3 In Situ Mass Balance Measurements and Reference Glaciers

International organizations such as the United Nations (UN) and the International Council for Science (ICSU) recommend the collection of detailed, direct, and long-term standardized measurements of seasonal glacier mass changes. To bridge the gap between detailed process studies and global coverage, field-based glacier measurements should be combined with observations of glacier length change, geodetic mass balance, meteorology, and streamflow variations. And ideally, these measurements should be representative of a particular region (International Hydrological Decade 1970; Fountain et al. 2009). Monitoring and reporting agencies for glacier mass balance include the World Glacier Monitoring Service (WGMS) as part of the Global Terrestrial Network for Glaciers (GTN-G) within the Global Climate Observing System (GCOS), both of which follow guidelines set out by Haeberli et al. (2007).

Although measurements should be representative of a region, glaciers are usually chosen for detailed measurement based on logistical constraints rather than representation. Glaciers with mass balance observations tend to be located in accessible regions at lower elevation ranges. This, in addition to the difficulties involved in accessing accumulation areas (Vincent et al. 2013), may lead to negatively biased mass balance measurements (Fujita and Nuimura 2011). There are only three sets of long-term (>30 years) in situ mass balance measurements in all of High Mountain Asia—in the Pamir-Alay (Abramov Glacier) and Tien Shan (Urumqi Glacier No. 1 and Tuyuksu Glacier). Despite the large area of ice cover in the extended HKH, there are no long-term measurements available from anywhere in the region. In total, only about 30 glaciers, covering an area of less than 120 km^2 out of the total glacierized area of 80–100,000 km^2, have had direct glaciological measurements made for one or more years.

Several mass balance studies were initiated in the extended HKH in the 1970s but were discontinued after some years. Direct mass balance measurements were made on only four glaciers during the 1990s (AX010, Rikha Samba, Kangwure, and Dokriani). These measurements were either short (<5 years) or discontinuous (e.g., Dokriani Glacier) and provide an incomplete picture of glacier change. The longest time series of in situ mass balance measurements in the extended HKH is that for the Xiao Dongkemdi Glacier on the Tibetan Plateau, where measurements started in 1989. The longest mass balance time series in the Himalaya is Chhota Shigri, where measurements started in 2002 (Table 7.2). There are several other glaciers in the extended HKH which have published data for several years of investigations, are still being investigated, and/or have been selected for continuous measurement.

Table 7.2 Glaciers with 5 or more years of in situ mass balance data where measurements are still ongoing

Glacier name	Region	Country	Latitude	Longitude	Measurement period(s)	Sources
Chhota Shigri	Lahaul-Spiti	India	32°28'N	77°52'E	Since 2002	Azam et al. (2016); WGMS
Phuche Glacier	Ladakh	India	34°16'N	77°33'E	Since 2010	Thayyen et al. (2015)
Rikha Samba	Hidden Valley	Nepal	28°50'N	83°30'E	1998–1999 Since 2011	Fujita and Nuimura (2011); WGMS
Yala	Langtang Valley	Nepal	28°14'N	85°36'E	1996, 1998 Since 2011	Fujita and Nuimura (2011), Baral et al. (2014); WGMS
Mera	Dudh Koshi	Nepal	27°43'N	86°52'E	Since 2007	Wagnon et al. (2013), Sherpa et al. (2017); WGMS
Pokalde	Dudh Koshi	Nepal	27°55'N	86°50'E	Since 2009	Wagnon et al. (2013), Sherpa et al. (2017); WGMS
West Changri Nup	Dudh Koshi	Nepal	28.0°N	86.8°E	Since 2010	Sherpa et al. 2017; WGMS
Qiyi	Qilian Shan	China	39°15'N	97°45'E	1975–1977 1984–1988 2002–2003 Since 2006	Yao et al. (2012)
Xiao Dongkemadi	Tanggula Shan	China	33°10'N	92°08''E	Since 1989	Fujita et al. (2000), Yao et al. (2012); WGMS
Parlung Glacier No. 94	SE-Tibet	China	29°23'N	96°59'	Since 2006	Yao et al. (2012); WGMS

Table 7.2 shows all glaciers with measurements for more than five years which have an ongoing measurement programme. We refer to the Supplementary Table S7.1 for all available measurements.

7.3.2 Observed Changes

7.3.2.1 Hindu Kush-Himalaya-Karakoram

Since the mid-18th century, glaciers in the extended HKH have been, on average, in retreat (Iwata 1976; Mayewski and Jeschke 1979; Bräuning 2006). A regional increase in the number of glaciers that were stationary or advancing was observed between 1920 and 1940, with the maximum advance of most Karakoram glaciers observed in 1940 (Mayewski and Jeschke 1979).

Since the 1950s, only reductions in glacier area (or shrinkage) have been observed (Fig. 7.4). Based on a compilation of area change studies, eastern Himalaya glaciers have tended to shrink faster than glaciers in the central or western Himalaya. A clear recession of glaciers has also been observed on the northern slopes of the Himalaya (Nie et al. 2010; Li et al. 2011; Shangguan et al. 2014; Xiang et al. 2014). Rates of area change range between −0.1%/yr for the Chandra-Bhaga basin (northwestern Indian Himalaya) between 1980 and 2010 to more than −1.0%/yr for the Poiqu basin (on the northern slopes of central Himalaya) between 1986 and 2001. Smaller glaciers are shrinking faster on average than larger ones (Bhambri et al. 2011; Bolch et al. 2010a; Thakuri et al. 2014; Ojha et al. 2016), although the smaller glaciers of Ladakh show lower rates of retreat than other Himalayan glaciers (Schmidt and Nüsser 2012).

While glaciers in the Hindu Kush mountains have also experienced significant length reductions since 1973 (Haritashya et al. 2009; Sarikaya et al. 2012), we were unable to identify any studies that specifically examined area change in this region.

Reported shrinkage rates for the heavily debris-covered glaciers in the Khumbu Himalaya vary depending on the time window and the glaciers investigated. Reported rates of area change vary between −0.12 ± 0.05%/yr for the period 1962–2005 (20 glaciers; Bolch et al. 2008) and −0.27 ± 0.06%/yr for the period 1962–2011 (29 glaciers; Thakuri et al. 2014).

In contrast to the Himalayan glaciers, a complicated picture emerges in the Karakoram, where large, often rapid, advances and retreats have occurred, more or less out of phase with one another (Hewitt 2011; Bhambri et al. 2013; Paul 2015). On average, glacier areas in the Karakoram have not changed significantly (Bhambri et al. 2013; Minora et al. 2016; Fig. 7.4). Individual glaciers, however, have shown large area increases or decreases due to surge behaviour, while non-surge-type glaciers were relatively stable over the last decade. Given the context of glacier retreat throughout the rest of the extended HKH region, this behaviour has been designated the 'Karakoram anomaly' (Hewitt 2005). In contrast, Liu et al. (2006) reported a significant area loss at a rate of −0.13%/yr over the period 1968–1999 for glaciers on the north slope of the Karakoram in the Yarkand Basin.

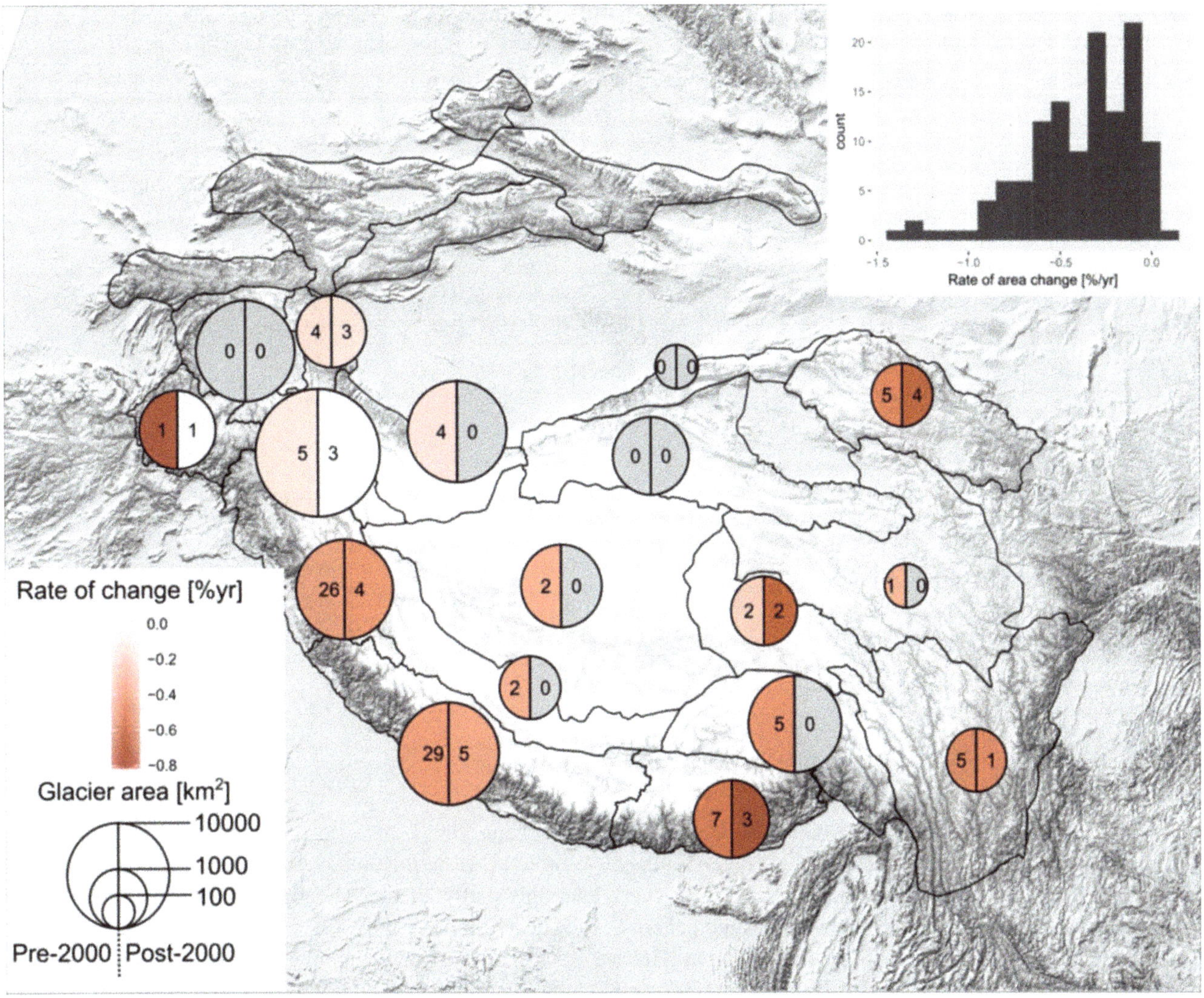

Fig. 7.4 Glacier area change rates (%/yr) for different regions in the extended HKH. Pre-2000 and post-2000 means are given by the colour in the left and right half-circles, respectively, together with the number of studies available for each period. Circle diameter is scaled to the current glacierized area in each region. A histogram of all observed change rates is given in the upper right

However, recent information about mass changes did not show significant differences between the northern and southern slopes (Kääb et al. 2015; Brun et al. 2017); there are no data available at the present date on area changes after 2000.

Trends in glacier area change vary across the extended HKH region. Aggregated trends (Fig. 7.4; Table 7.3) are not area weighted, and may not agree with individual studies due to sampling bias. Over the past six decades, average rates of area change have increased in the western and eastern Himalaya, and remained constant or decreased in the central Himalaya (Fig. 7.4; Table 7.3). Individual studies, in contrast, show increased area losses in the central Himalaya (Bolch et al. 2008; Bhambri et al. 2011; Thakuri et al. 2014) and decreased losses in the upper Indus basin in Ladakh and Ravi basin, both in the western Himalaya (Schmidt and Nüsser 2012; Chand and Sharma 2015). Area loss has led to greater supraglacial debris cover in the Himalaya (Bolch et al. 2008; Schmidt and Nüsser 2009; Kamp et al. 2011; Nuimura et al. 2012; Thakuri et al. 2014), while debris-covered area has remained constant in the Karakoram (Herreid et al. 2015). Area loss also leads to glacier fragmentation; the number of glaciers in the Himalaya is reported to have increased over the past five decades (Bhambri et al. 2011).

Mass Changes

Glaciers in the extended HKH show mass loss since at least the 1970s (Fig. 7.5). Individual rates of mass change and uncertainty estimates for each study and region are given in Supplementary Table S7.2. ICESat (Ice, Cloud, and land Elevation Satellite) laser altimetry and DEM (Digital Elevation Model) differencing revealed an average mass loss of −0.2 m w.e./yr for the first decade of the 21st century

Table 7.3 Average glacier mass and area change in the extended HKH for different regions and time periods, with number of studies in parentheses. Averages are calculated from all study types (geodetic, glaciological, modelled) (See Supplementary Tables S7.1 and S7.2 for details.)

Region ID (see Fig. 7.1)	Region	Glacier mass balance rate (m w.e./yr)		Glacier area change rate (%/yr)	
		~ 1970–2000	~ 2000–2010	~ 1970–2000	~ 2000–2010
6	Western Pamir	–	−0.26 (2)	–	
7	Eastern Pamir	−0.12 (2)	−0.02 (10)	−0.13 (4)	−0.12 (3)
8	Hindu Kush	–	−0.30 (5)	−0.79 (1)	+0.01 (1)
9	Karakoram	−0.10 (3)	−0.06 (12)	−0.11 (5)	+0.02 (3)
10	Western Himalaya	−0.33 (5)	−0.50 (11)	−0.38 (26)	−0.34 (4)
11	Central Himalaya	−0.39 (10)	−0.42 (14)	−0.41 (29)	−0.37 (5)
12	Eastern Himalaya	−0.17 (1)	−0.66 (10)	−0.55 (7)	−0.81 (3)
13	Western Kunlun	−0.07 (2)	+0.08(9)	−0.13 (4)	–
14	Eastern Kunlun	–	−0.40 (4)	–	–
15	Altun Shan	–	+0.10 (1)	–	–
16	Tibetan Interior	–	−0.01 (7)	−0.32 (2)	–
17	Gangdise	–	−0.52 (2)	−0.34 (2)	–
18	Nyainqentanglha	–	−0.58 (9)	−0.44 (5)	–
19	Tanggula Shan	−0.21 (3)	−0.29 (4)	−0.2 (2)	−0.69 (2)
20	Hengduan Shan	−0.30 (3)	−0.85 (6)	−0.46 (5)	−0.47 (1)
21	Eastern Tibetan Mtns.	–	−0.64 (3)	−0.32 (1)	–
22	Qilian Shan	−0.23 (8)	−0.40 (6)	−0.51 (5)	−0.66 (4)
	Regions 6–22	−0.26 (41)	−0.37 (123)	−0.35 (98)	−0.42 (25)

(Kääb et al. 2012; Gardner et al. 2013; Gardelle et al. 2013), which is approximately half of the global average (Gardner et al. 2013; Marzeion et al. 2014) for this period. However, mass changes are not uniform but show contrasting patterns. The greatest rates of mass loss in the extended HKH post-2000 are found in the eastern Himalaya (−0.6 m w.e./yr) and western Himalaya (−0.6 m w.e./yr). Moderate losses are observed in the central Himalaya (−0.4 m w.e./yr), and the Hindu Kush (−0.3 m w.e./yr). In contrast, glaciers in the Karakoram showed neutral mass balances or even slight mass gains after 2000 (Table 7.3).

Mass balance estimates for the period before 2000 are sparse, but the use of declassified satellite imagery for pre-2000 DEM generation is leading to increased understanding (Bolch et al. 2008; Maurer et al. 2016; Bhattacharya et al. 2016). Available studies seem to confirm that pre-2000 mass balance patterns are similar to post-2000 patterns (Table 7.3), although mass loss rates have increased in the Himalaya and the eastern Tibetan Plateau. The inferred stability of Karakoram glaciers, based on stable debris-covered areas since the 1970s (Herreid et al. 2015), has recently been confirmed by geodetic studies in the region (Bolch et al. 2017; Zhou et al. 2017). Recent publications have found that this anomaly can not only be expanded to include the eastern Pamir and the western Kunlun Shan (Holzer et al. 2015; Lin et al. 2017; Neckel et al. 2014; Wu and Zhang 2008), but also that the region of mass gain is centred at western Kunlun, while the Karakoram and Pamir are only partly affected (Kääb et al. 2015; Brun et al. 2017; see also Table 7.3, Fig. 7.5, and Sect. 7.3.3.2).

Glaciers in the Khumbu region (central Himalaya) showed a mass loss of −0.3 m w.e./yr between 1970 and 2007 (Bolch et al. 2011a), while glaciers in the eastern Himalaya lost −0.2 m w.e./yr for a similar period, 1974–2006 (Maurer et al. 2016).

Results from in situ measurements and modelling confirm the predominantly negative mass balance during the last five decades with a higher mass loss since roughly 1995 (Bolch et al. 2012). Debris-covered glaciers (Box 7.3) in the Langtang and Khumbu regions of the central Himalaya also lost mass in recent years, although at a lower rate than debris-free glaciers (Immerzeel et al. 2014; Pellicciotti et al. 2015; Shea et al. 2015b; Thompson et al. 2016; Vincent et al. 2016). However, there is evidence for a period of balanced conditions for Lahaul-Spiti. For example, Vincent et al. (2013) concluded that Chhota Shigri Glacier was on average in balance within the period 1988 to 2002, based on elevation measurements in 1988 and 2010 and in situ mass balance measurements starting 2002. Modelling studies of the Chhota Shigri (western Himalaya) and Mera (central Himalaya) glaciers also suggest a period of balanced conditions between mid-1980 and mid/late-1990 (Azam et al. 2014; Shea et al. 2015a).

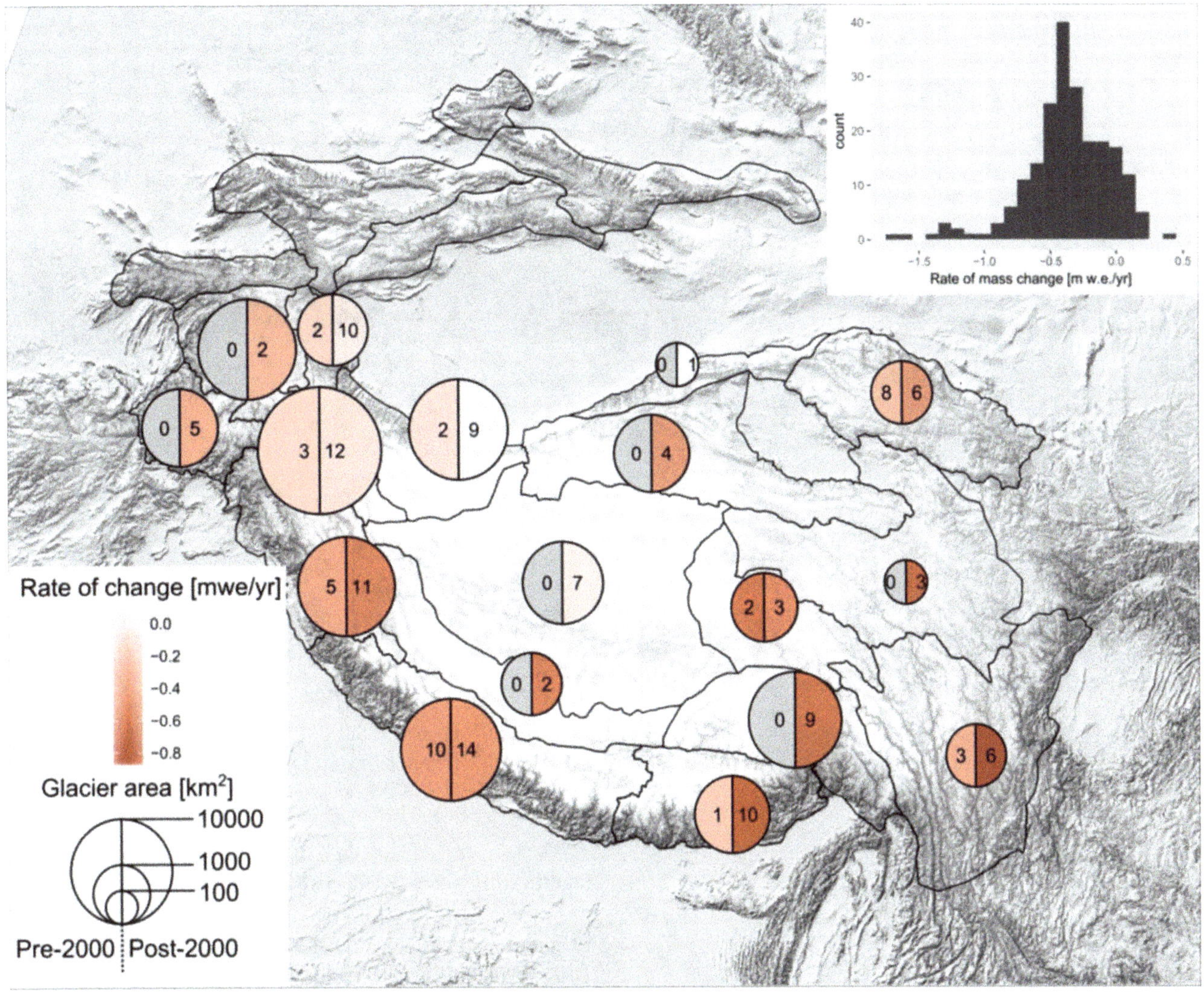

Fig. 7.5 Summary of 164 glaciological, geodetic, and modelling estimates of rates of glacier mass change (m w.e./yr) by glacier region. The arithmetic average of data from all studies pre- and post-2000 is shown by the shading of half circles (pre-2000 left, post-2000 right), together with the number of studies available for each period; periods without data are coloured in grey. A histogram of all reported regional mass change rates is given in the upper right

Box 7.3 Debris-covered glaciers

Rapid weathering of high-relief glacial valleys in the extended HKH supplies a sizeable amount of rock debris onto the glaciers via rock and mixed snow/ice avalanches. This debris travels downstream with the ice flow, and as the surrounding ice melts away, it builds up as a thick layer of surface rock material in the lower part of the glacier. The debris layer acts like an insulating blanket that protects the ice beneath from melting. Yet debris-covered glaciers also contain varying concentrations of bare ice cliffs (Fig. 7.6) and supraglacial ponds. The role of ice cliffs and ponds in the overall melt rate of debris-covered glaciers was first examined by Sakai et al. (2000a, 2002) and is an active area of research. The termini of debris-covered glaciers respond slowly to climate change, the position does not change very much and a long tongue of slow-flowing debris-covered ice is preserved at the lower part of the glacier. Despite significant recent progress through remote-sensing/field measurements (Immerzeel et al. 2014) and model studies of the complex coupled system of debris, ice cliffs, and ponds (Miles et al. 2016; Buri et al. 2016; Banerjee 2017), significant challenges still remain in understanding these glaciers: we cannot yet predict the long-term changes in the debris cover area or thickness, or the effect of such changes on the future of glaciers and their meltwater production at the scale of the whole Himalaya.

Fig. 7.6 Ice cliff on debris-covered Khumbu Glacier, Nepal Himalaya, November 2016 (photo: J. Shea)

7.3.2.2 Pamir

Area and Length Changes

Frequent surges and contrasting length changes are not only common in the Karakoram but also in central and northeast Pamir (Dolgoushin and Osipova 1975; Kotlyakov et al. 2008; Osipova and Khromova 2010; Holzer et al. 2016; Wendt et al. 2017), and a recent surge was also reported in eastern Pamir (Shangguan et al. 2016). In contrast to the Karakoram, however, existing studies on glacier area and length changes suggest a continuous area loss in northeastern and central Pamir. For example, Shchetinnikov (1993) reports an area loss of about 0.5%/yr from 1957–60 to 1978–80. Published results addressing the glacier area changes during recent decades found the least negative rates of glacier retreat (<0.1%/yr) in central Pamir, where the highest peaks and the largest glaciers are located, while shrinkage rates of >0.25%/yr are observed in north, west, and south Pamir (Khromova et al. 2006; Aizen and Aizen 2014; Holzer et al. 2016; Zhang et al. 2016). The area change rates in eastern Pamir appear to be consistent (−0.13%/yr) for the past 30 years. In contrast, the glacier area at Muztag Ata, the highest mountain massif in the Pamir, has not changed significantly since the 1970s (Holzer et al. 2015). However, as yet there has been no detailed peer-reviewed assessment of glacier area changes in the whole of the Pamir based on consistent datasets and methodologies, and the above-mentioned values should be interpreted with care.

Mass Changes

Mass change estimates in the Pamir vary depending on the region under consideration. Gardelle et al. (2013) report slight mass gain (+0.16 ± 0.14 m w.e./yr) for central Pamir (2000 to 2010) based on DEM differencing, but Kääb et al. (2015) found significant mass loss between 2003 and 2008 (−0.48 ± 0.14 m w.e./yr) based on ICESat laser altimetry data. Recalculated mass loss estimates for the regional definitions shown above (Fig. 7.1) based on the data by Kääb et al. (2015) show strong mass loss in western Pamir (−0.44 ± 0.10 m w.e./yr) and balanced conditions in the east (−0.02 ± 0.25 m w.e./yr). Gardner et al. (2013) reported mass loss of −0.11 ± 0.19 m w.e./yr, assuming an average density of 850 kg m^{-3}. A recent geodetic estimate based on optical stereo data for 2000 to 2016 gave a slightly negative mass budget of −0.08 ± 0.06 m w.e./yr in western Pamir and a possible positive budget of +0.07 ± 0.10 m w.e./yr in eastern Pamir (Brun et al. 2017). Similar results were published by Lin et al. (2017) for 2000–2014 based on TerraSAR-X/TanDEM-X and SRTM data (−0.11 ± 0.07 m w.e./yr for western and +0.12 ± 0.07 m w.e./yr for eastern Pamir). As the above studies investigate different time periods, the temporal variability in mass changes will lead to varying estimates of rates of mass change. The most detailed investigations are available for the Fedchenko glacier, by far the largest glacier in Pamir (Lambrecht et al. 2014). The glacier experienced only a slight mass loss of 0.07 km^3 w.e./yr (specific mass loss <0.001 m w.e./yr) between 1928 and 2000, with about 75% of the loss occurring during the second half of the 20th century. Differences in mass loss estimates based on the same dataset can be attributed to the regional aggregation, the assumptions for SRTM radar beam penetration into snow and ice, and the time period analysed.

In situ measurements and modelling results for the Abramov Glacier in Pamir-Alay, north of the Pamir mountains (Fig. 7.1), reveal clear mass loss since the beginning of the measurements in the 1960s. Reanalysis of available in situ data from 1971–94 found an average mass balance of −0.46 ± 0.06 m w.e./yr; modelled mass balances of −0.51 ± 0.17 m w.e./yr were obtained for 2000–2011 (Zemp et al. 2015; Barandun et al. 2015). These results contrast with Gardelle et al. (2013) who found no significant mass changes for the same glacier based on the geodetic method. A geodetic study of Muztag Ata glaciers revealed, on average, balanced conditions since 1973 (Holzer et al. 2015). This agrees with in situ measurements of a small glacier in this mountain massif where in situ measurements from 2005–2014 showed an average value of +0.10 m w.e./yr (Holzer et al. 2015).

From the available studies and data, we conclude that glaciers in western Pamir experienced a moderate mass loss, while glaciers in eastern Pamir were in balance or had only slight mass losses for several decades.

7.3.2.3 Tibetan Plateau

Area Changes

Compilations of area changes for over 3,100 glaciers in western China since the Little Ice Age maximum (Wang 1991; Liu et al. 2000, 2003; Wang and Ding 2002; Su and Shi 2002) indicate that glaciers in western China have lost a total area of ~16,000 km^2, or ~27%, to the middle of the 20th century (1950s through the 1980s) (Liu et al. 2006). Glacial shrinkage exhibits a regional pattern, with smaller area change rates (−21%) in the inner region of the Tibetan Plateau and higher rates (−47.7%) in the upper Indus river basin located in China. Remote sensing analyses for SE Tibet show similar rates of area loss (−27%) since the LIA maximum (Loibl and Lehmkuhl 2014). A complete loss of glaciers less than 0.6 km in length was observed in Bugyai Kangri, in the southeast Tanggula Range (Wang and Ding 2002).

Glacier area changes on the TP have been calculated at a range of scales: river basins (Wei et al. 2014; Wang et al. 2015a), mountain ranges (Bolch et al. 2010b; Sun et al. 2015), and mountain sections/parts (Jiang et al. 2013; Tian et al. 2014). First maps are based on photogrammetry (Liu et al. 2016) and extend through the Landsat data acquired after the mid-1970s (Nie et al. 2010; Wang et al. 2012). A clear recession of glaciers has been observed since the 1950s/1970s/1980s in western Hengduan Shan (Liu et al. 2016), the A'Nyêmaqên Range in the Yellow River source area (Wang et al. 2015b; Liu et al. 2002), and the Qilian mountains (Sun et al. 2015), with annual shrinkage rates of approximately −0.5%/yr.

Glaciers in the western Kunlun (north-eastern Tibetan Plateau) have been on average relatively stable (Fig. 7.4), an eastern continuation of the stable conditions observed in the eastern Pamir and Karakoram. A recent study investigating the changes of the debris-free parts of the glaciers of entire extended HKH region for the period ~1976 until 2013 is in general in line with previous findings with highest area loss in south-eastern and the lowest in the in the north-eastern Tibetan Plateau (Ye et al. 2017). The combination of advancing and retreating glaciers has resulted in a negligible change in glacier area during 1970–2001 and 1990–2011 in the latter region. (Shangguan et al. 2007; Li et al. 2013; Bao et al. 2015). Four glaciers have been recognized as surging during 2003 and 2011 (Yasuda and Furuya 2013). Surging glaciers have also been found in the central Kunlun Mountains (Guo et al. 2013) and the Purogangri Ice Cap in central Tibet (Neckel et al. 2013). The retreat of small glaciers and glaciers with lower median elevations has resulted in the complete loss of some of the small glaciers and the disintegration of larger glaciers into two or more smaller glaciers (Meiping et al. 2015).

Glaciers in different mountain ranges show varying responses. The available data on glacier area changes in western Kunlun (Bao et al. 2015), Mount Geladandong (Wang et al. 2013a, b) and Bugyai Kangri in the Tanggula range (Liu et al. 2016), Mount Tuanjie in the Qilian Mountains (Xu et al. 2013), and Mount Gongga (Pan et al. 2012) show that the decadal trend in glacier area change differs for each mountain range (Fig. 7.2).

Mass Changes

Region-wide mass balance estimates are available for the period after 2000 from remote sensing methods, mainly based on ICESat laser altimetry data for the period 2003 to 2009, but also on DEM differencing (Table 7.3, Fig. 7.5). While the studies deviate in absolute numbers, they agree on general trends. The most negative mass balances are found in southeast TP (East Nyainqentanglha), while balanced budgets or even positive mass balances are reported for the western Kunlun Shan (Li et al. 2011; Zhang et al. 2012; Neckel et al. 2014; Brun et al. 2017). Results for glaciers in inner or central Tibet slightly deviate: while Neckel et al. (2014) report possible slight mass gains for 2003–2009, Brun et al. (2017) report slight mass losses for 2000–2016.

This general pattern is supported by in situ measurements (Yao et al. 2012): a strongly negative mass balance (−0.9 m w.e./yr for 2006–2014) is reported for Parlung glacier No. 94 in Hengduan Shan, and a less negative mass balance (−0.6 m w.e./yr, for 2006–2010) for Xiaodongkemadi glacier in Tanggula Shan. A mass balance of −0.5 m w.e./yr was observed for Qiyi Glacier in Qilian Shan.

Observed mass balance data prior to 2000 is available for only a few glaciers (Qiyi, Meikuang, Xiaodongkemadi, and Kangwure). Observed and simulated mass balance data for individual glaciers and regions/sub-regions also show a distinctive pattern (Fig. 7.5). Mass balances of glaciers in Qilian Shan, eastern/western Kunlun Shan, and Tanggula Shan were near zero (or positive) until the early 1990s, and then became increasingly negative over time. Glacier mass changes in the Hengduan Shan have fluctuated dramatically since 1960 and have been increasingly lower than values in other regions since 2000. Positive mass balances observed from geodetic studies in western Kunlun Shan in 2006–2010 contradict modelled negative mass balances. There is a conflict between the mass balance in western Kunlun Shan during 2006–2010 (positive, obtained by DEMs) and the area change as shown in Fig. 7.4 (negative, simulated). Essentially, the former is more convincing as all geodetic studies (using DEMs and/or ICESat data) reveal positive balances (Supplementary Table S7.2).In addition, it needs to be considered that area changes in the highly continental mountain ranges might show a strongly delayed signal due to long response times to climate forcing.

7.3.3 Glacier Projections

Changes in temperature and precipitation affect seasonal snowpacks, equilibrium line altitudes, glacier mass balances, and volume at annual timescales, and glacier area and length in the long term. Projections of future glacier change typically combine mass balance models, estimates of future temperature and precipitation change, and physically-based or empirical snow and ice melt models. The challenges in modelling glacier change are substantial, in particular for debris-covered glaciers.

Projections of future glacier change in the region have been done at regional scale (Marzeion et al. 2012; Giesen and Oerlemans 2013; Radić et al. 2014; Zhao et al. 2014; Huss and Hock 2015; Kraaijenbrink et al. 2017), at catchment scale (Shea et al. 2015b; Soncini et al. 2015, 2016), and at the scale of individual glaciers (Naito et al. 2000; Adhikari and Huybrechts 2009; Rowan et al. 2015).

Recent projections typically use a number of different climate scenarios (e.g., RCP4.5 and RCP8.5) and thus a range of volume change estimates and a multi-model mean can be used to assess the climate-based uncertainty (Wiltshire 2014). Structural uncertainty, or the uncertainty related to how simplified models can capture the wide range of processes and physics that determine glacier evolution, is more difficult to assess. Most modelling studies parametrize ice flow either through volume length (or area) scaling or through slightly more advanced approaches. However, to accurately understand the interaction between glaciers and climate, fully coupled ice flow models are required that explicitly simulate the ice transport from the accumulation areas to the ablation tongues. In the extended HKH, only a few studies exist that make a first attempt to do this (Miller et al. 2012; Rowan et al. 2015; Shea et al. 2015b). A third level of uncertainty is derived from the initial glacier volume conditions, which are largely unknown as measurements are very scarce and existing estimates deviate significantly (Frey et al. 2014).

Projections of mass balance alone are insufficient for the purpose of examining future glacier change. Glaciers respond to climate change by reducing (or expanding) their area and volume. Glacier-wide mass balances are a function of glacier area, and projections that do not account for glacier dynamics will tend towards increasingly negative (and unrealistic) values. In response to increased equilibrium line altitudes and melt rates and decreased snow accumulation totals, mass balances throughout the region are projected to be increasingly negative for static geometries (Chaturvedi et al. 2014; Kulkarni et al. 2016).

7.3.3.1 Near-Term Glacier Change (2030)

In the near term, glaciers in the extended HKH are projected to see increased volume loss, though some scenarios and ensemble members show limited volume change (Figs. 7.7, 7.8). Regional scale studies show volume losses between −9 and −32% by 2030 under moderate warming scenarios (RCP4.5), and between −8.7 and −26.1% for higher-emission scenarios (RCP8.5) (Giesen and Oerlemans 2013; Marzeion et al. 2012; Radić et al. 2014). While overall rates of loss are similar between the emission scenarios, climate model variability can lead to greater mass loss for the lower emission scenarios in the near term.

Glacier-specific studies that explicitly incorporate debris-cover feedbacks show much more conservative

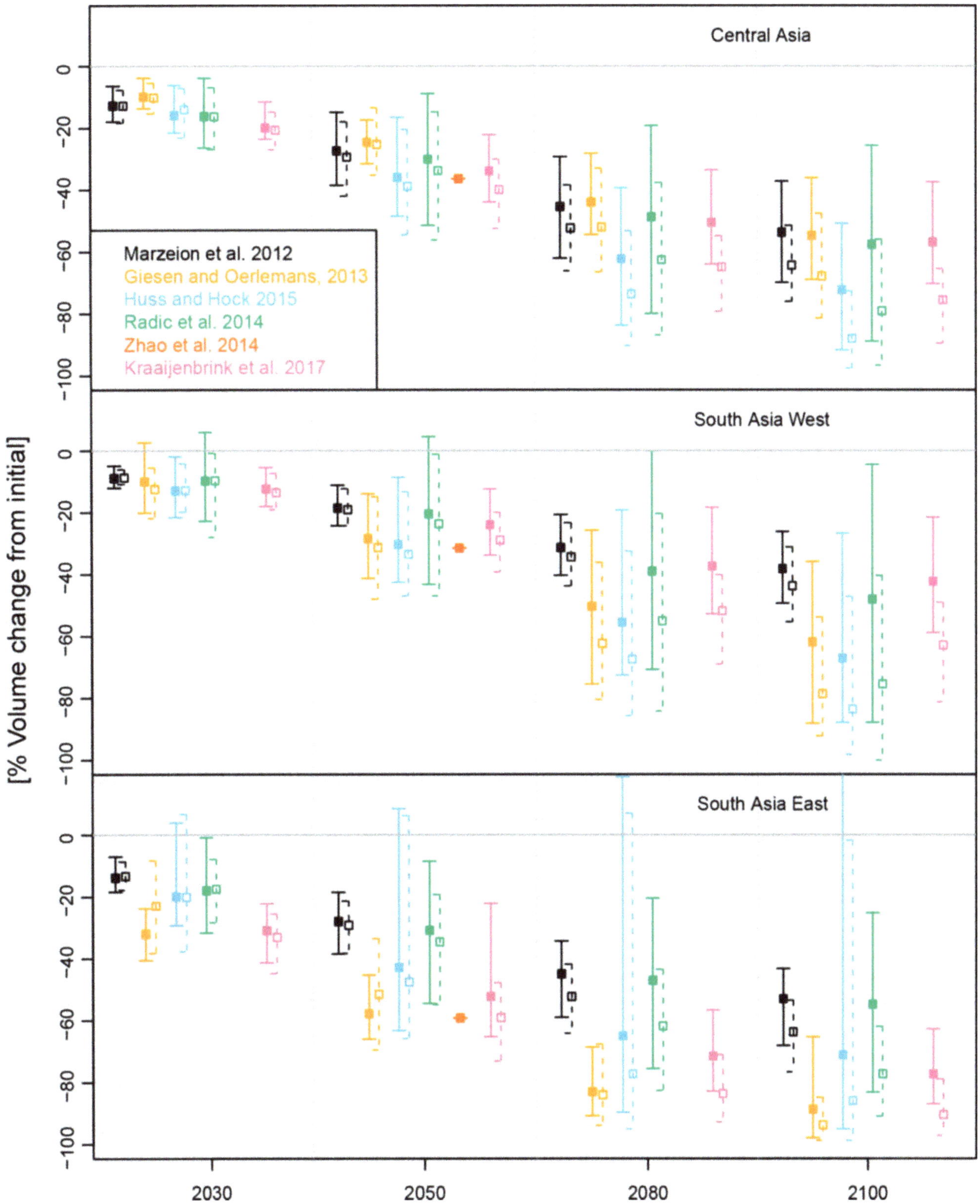

Fig. 7.7 Regional glacier volume change projections for RGI regions 13 (Central Asia), 14 (South Asia West), and 15 (South Asia East). Error bars show mean and range of multiple climate simulations for 2030, 2050, 2080, and 2100. Solid lines show results from RCP4.5 scenarios and dashed lines from RCP8.5 scenarios. *Sources* Marzeion et al. (2012), Giesen and Oerlemans (2013), Radić et al. (2014), Zhao et al. (2014), Huss and Hock (2015), Kraaijenbrink et al. (2017). See Supplementary Table S7.3 for details

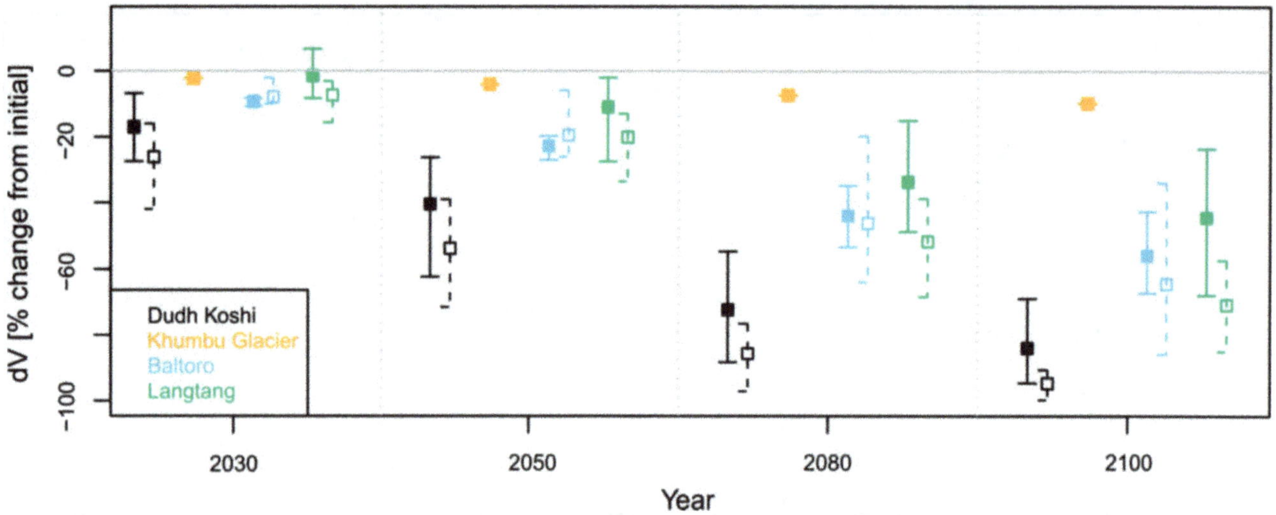

Fig. 7.8 Volume change projections for basins and glaciers in extended HKH. Symbols and error bars show mean and range of multiple climate simulations for 2030, 2050, 2080, and 2100. Solid lines show RCP4.5 and dashed lines RCP8.5 scenarios, where reported. *Sources* Baltoro and Langtang (Immerzeel et al. 2013), Khumbu (Rowan et al. 2015), Dudh Koshi (Shea et al. 2015b)

reductions in glacier volumes; for example, −2.2% for the Khumbu Glacier by 2030 (Rowan et al. 2015). Basin-scale studies in the Khumbu valley (Shea et al. 2015b; Soncini et al. 2016) show greater rates of near-term volume loss that are also in line with regional estimates (Fig. 7.8).

7.3.3.2 Medium-Term Glacier Change (2050)

By mid-century, differences in the climate forcing (e.g., RCP4.5 vs RCP8.5) begin to show in the different glacier volume projections. Rates of mass loss appear to be higher in the eastern Himalaya, although again the spread in projections highlights the strong uncertainty in climate futures. For the RGI region Central Asia which includes the Pamir and the Tibetan Plateau, ensemble mean glacier volume change ranges between −24.6 and −35.9% for RCP4.5 and −25.3 and −38.8% for RCP8.5. RGI region South Asia (West), which includes the Karakoram region, could see volume reductions between −18.6 and −30.3% (RCP4.5) and −19.1 and −35.9% (RCP8.5). Eastern portions of the extended HKH could lose between one quarter and one half of their current glacier volume by 2050.

Mid-century projections of glacier change also begin to diverge regionally. Volume declines projected for the Karakoram region begin to accelerate mid-century (Immerzeel et al. 2013; Li et al. 2016), and similar patterns are observed for west Kunlun (Kraaijenbrink et al. 2017). Other regions show sustained rates of mass loss (Rupper et al. 2012). This contrast is consistent with observations of recent mass loss in most extended HKH regions and anomalous mass gains (or steady states) in west Kunlun, Karakoram, and eastern Pamir (e.g., Sec 7.3.3). Under future temperature rise, the mass gains of these anomalous regions cannot be sustained and volume declines commence as a result (Fig. 7.9).

7.3.3.3 Long-Term Glacier Change (2080 and 2100)

Projected end-of-century changes in ice volume are pronounced in all regions. A modelling study in the Pamir projects a loss of approximately 45% by 2100 (Ohara et al. 2014), while the most negative scenarios in the eastern Himalaya point towards a near-total loss of glaciers (−63.7 to −94.7%). Losses of a similar order of magnitude are expected in regions with predominantly small, sensitive glacier tongues, such as the inner Tibetan Plateau and the Qilian Shan. As several studies have noted, these volume decreases are large in part because of the distribution of glaciers in the region and the lack of large high-elevation accumulation plateaus (Shea et al. 2015b). Relative mass losses in the Karakoram and West Kunlun Shan (~35% under RCP4.5 scenarios) are limited compared to other regions in the extended HKH (Kraaijenbrink et al. 2017); this is a function of the current and projected mass balance rates, the existing ice volumes, and the regional climatic differences. Projected absolute ice losses in these regions are still large and relevant for sea-level rise, as the existing ice volumes in the region comprise a large portion of the total ice volume in extended HKH. Even if warming can be limited to the ambitious target of +1.5 °C, volume losses of more than one-third are projected for extended HKH glaciers, with more than half of glacier ice lost in the eastern Himalaya (Kraaijenbrink et al. 2017).

Equilibrium line altitudes in High Asia are projected to increase by up to 800 m by 2100 (8.8 m/yr, Fig. 7.10), depending on future emission scenarios (Huss and Hock 2015). Similar rates of ELA increase (8–12 m/yr) are projected for the Everest region under high emission scenarios (Shea et al. 2015b). Glacier tongues covered by thick debris

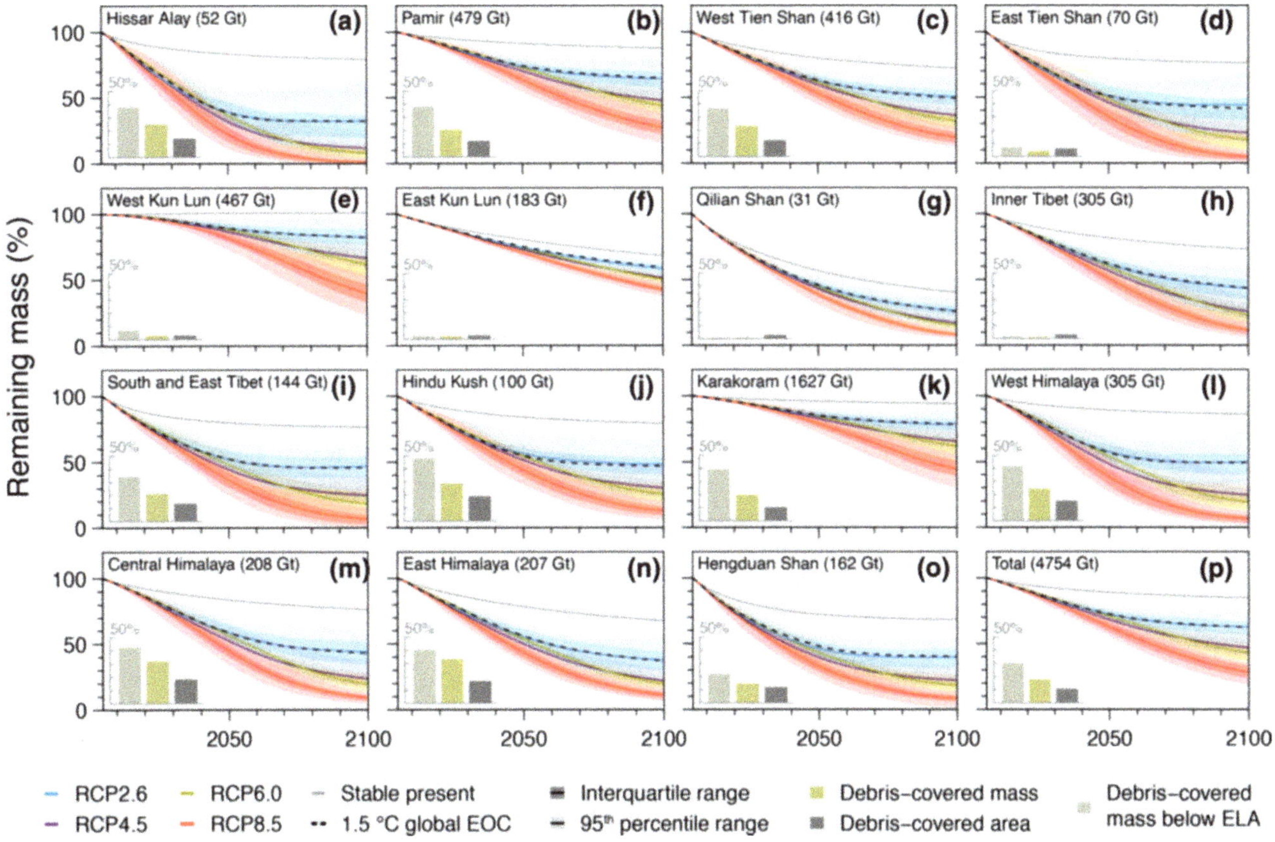

Fig. 7.9 Mass change projections and prevalence of debris cover for all RGI sub-regions (a-o) that are encompassed by the RGI regions 13 (Central Asia), 14 (South Asia West), and 15 (South Asia East). Panel p shows the projections aggregated over the entire region. Mass loss projections are shown for all four RCP scenarios, for a stable present-day climate experiment, as well as for a selection of six climate models that have a temperature rise of 1.5 °C at the end of the century. The error bands indicate both climate model uncertainty and glacier model parameter variability. The bar charts denote the relative abundance of debris cover in each sub-region (*Source* Kraaijenbrink et al. 2017)

which are insulated from melt, are expected to separate from their accumulation areas as the greatest melt rates occur immediately above debris-covered zones (Naito et al. 2000; Rowan et al. 2015; Shea et al. 2015b; Soncini et al. 2016). These changes will result in unsustainable negative mass balances, and will lead to the gradual disappearance of lower elevation glaciers.

7.3.4 Recommendations

The general agreement between our compilations of glacier change studies (cf. Figs. 7.4 and 7.5) and the regional mass change studies (e.g., Gardner et al. 2013; Gardelle et al. 2013; Neckel et al. 2014; Kääb et al. 2015; Brun et al. 2017) supports the utility of glacier and basin scale area and mass change studies. To improve our understanding of glacier

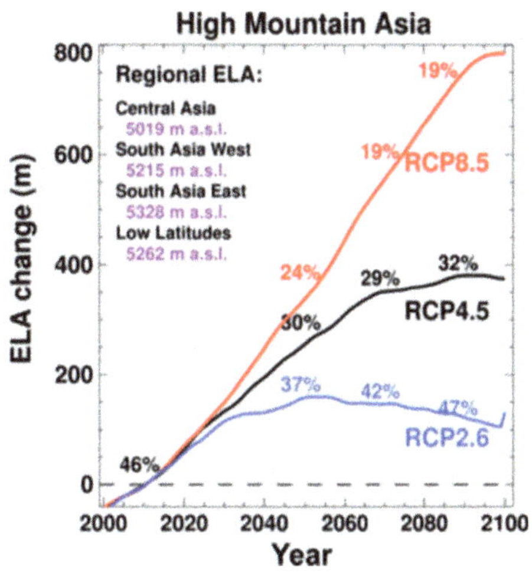

Fig. 7.10 Changes in equilibrium line altitudes (ELAs) for RGI regions in high mountain Asia relative to 2010 for three emission scenarios (*Source* Huss and Hock 2015)

change in the region and reduce uncertainty in future projections, we recommend the following:

- Develop estimates of glacier mass change prior to 2000, which are currently missing in many regions, using declassified stereo imagery
- Collect field-based measurements of glacier thickness and debris thickness (e.g. by ground-penetrating radar (GPR)) in representative glacier regions to evaluate glacier volume estimates and debris cover characteristics
- Develop and test physically based and empirical models of debris-covered glacier melt components that include heat conduction through debris, ice cliff backwasting, and supraglacial ponds, and scale up to larger glacierized regions
- Develop fully coupled ice flow models taking debris sources, transport and deposition into account
- Model intercomparison projects to assess different sources of uncertainty: initial conditions, forcing data, and model structures.

7.4 Glacial Lakes

Glacial lakes are lakes that are fed by glacier meltwater. This can include lakes on top of glaciers (supra-glacial), lakes in front of glaciers (pro-glacial), and glacier-fed lakes that are disconnected from glaciers but close to glacier termini. In contrast, non-glacier-fed lakes have no glacier in their watershed and are fed only from precipitation and snow meltwater.

Supra-glacial lakes are often ephemeral and drain when they connect to the englacial drainage system ('perched lakes' after Benn et al. 2001). However, if there is no developed englacial drainage system or the lakes reach the glacier bed, supra-glacial lakes can grow and coalesce; when these water bodies collect behind the terminal moraine they form pro-glacial lakes ('base-level lakes' after Benn et al. 2001). Such glacial lakes usually form and grow concomitant with glacier recession, but ice-dammed lakes can also form when a surging glacier blocks a river valley (Hewitt and Liu 2010). Pro-glacial lakes pose a serious risk when the moraine (or glacier) dam fails and the water drains catastrophically as a glacial lake outburst flood or GLOF. Such events can have significant and long-lasting human and physical impacts (Richardson and Reynolds 2000; Hewitt and Liu 2010). With degrading high-mountain permafrost (see Sect. 7.5) there is also an increasing probability of rock ice/avalanches from steep slopes reaching glacial lakes and triggering GLOF events (Haeberli et al. 2016). In this section we present the current state of knowledge on the occurrence of glacial lakes and their past and projected changes.

7.4.1 Occurrence

An inventory of glacier-fed lakes that was manually mapped based on Landsat satellite imagery found 4,260 lakes (with an area >0.003 km^2) covering a total area of 557 km^2 in 2010 (Zhang et al. 2015a, b, Fig. 7.11). However, a recent inventory for the same region based on Landsat 8 OLI data from 2014 revealed more than 8200 glacial lakes and an overall lake area of more than 830 km^2 (Chen et al. 2017). Differences may be explained by the different data sets and investigated periods, different methods, but also the difficulty to identify turbid and ice covered lakes and lakes in shadow. However the existing studies agree that the majority of glacial lakes are distributed in the extended HKH. Lakes in the TP are on average larger and are mainly located in the interior TP (Zhang et al. 2013).

7.4.2 Observed Change

Change analysis of glacier-fed lakes in the extended HKH region using Landsat TM/ETM + satellite data for three periods (1990, 2000, and 2010) showed a clear increase both in number (3,350 to 3,641 to 4,260) and area (440.2 to 467.6 to 556.9 km^2 (Zhang et al. 2015a, b). This study showed further that (1) small lakes (<0.2 km^2) are more sensitive to climate change; (2) lakes closer to glaciers and at higher elevations, particularly those connected to glacier termini, have undergone greater changes in area; and (3) glacier-fed lakes are dominant in both number and area (>70%) and exhibit faster expansion trends compared to non-glacier-fed lakes. Similar results were observed by Ashraf et al. (2017), for the western Himalaya/Karakoram/Hindu Kush region. The total increase in glacier-fed lake area for the entire extended HKH was greatest in the central and eastern Himalaya, and only one region (Nyainqentanghla) exhibited a decrease in glacier-fed lake area (Zhang et al. 2015a, b; Table 7.4).

A study conducted by ICIMOD in 2011 identified 1466 'glacial lakes' in Nepal, it included supraglacial and moraine-dammed lakes only (ICIMOD 2011). In contrast, for the same region Ives et al. (2010) found 2323 glacial lakes in 2001, which suggests a decrease in the number of glacial lakes. A closer reading suggests that the 2001 inventory number is based in part on survey data from the 1950s and 1960s. Differences in datasets and methodology for glacial lake identification are problematic when evaluating changes in glacial lake area and number. Several studies with consistent datasets and methods have subsequently found increasing glacial lake areas and numbers in most regions of the extended HKH (Gardelle et al. 2011; Mergili et al. 2013).

More critically, a subset of glacial lakes have been identified as critical or potentially dangerous based on

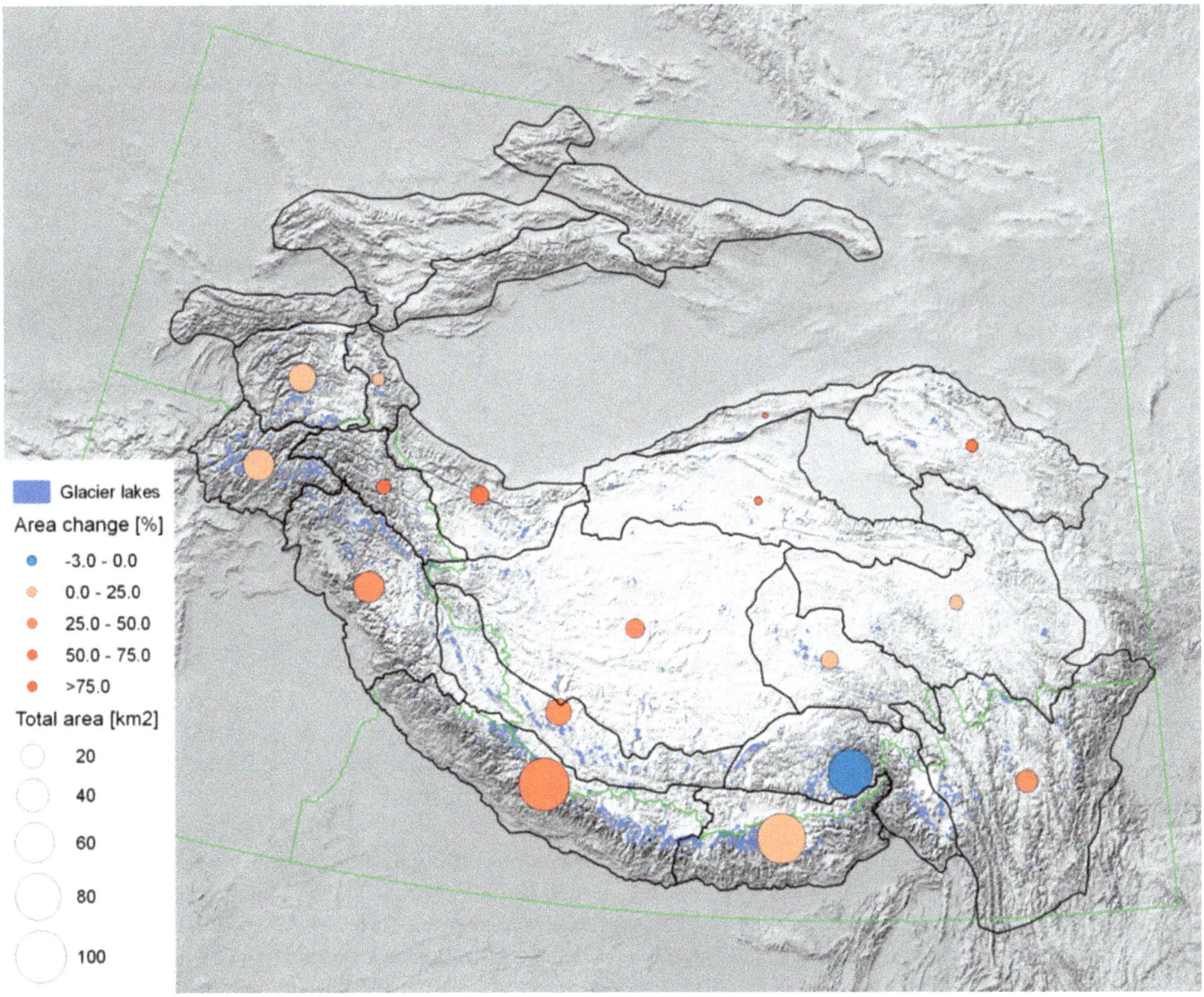

Fig. 7.11 Distribution and total area (centroid size) of ice-contact and glacier-fed lakes in the extended HKH, and percentage change in total lake area between 1990 and 2010 (centroid colour) (data source Zhang et al. 2015a, b). Only one region has experienced a decline in glacier lake area (blue, Nyainqentanghla region)

criteria that include rapid lake expansion or height change, moraine dam condition, glacier condition, and potential for mass movements into the lake (ICIMOD 2011; Rounce et al. 2016; Ives et al. 2010; Fujita et al. 2013). In Nepal, 21 glacial lakes present a potentially dangerous risk (ICIMOD 2011). The upper Indus basin, which includes parts of the Hindu Kush, Karakoram, and western Himalaya sub-regions, contains 2420 glacier lakes, of which 52 are potentially dangerous (Ashraf et al. 2012). These inventories present an important database of glacial lakes and provide a basis for long-term monitoring and evaluation of potential outburst flood disasters caused by glacial lakes (Fujita et al. 2008, 2009).

The number of glacial lakes identified as ‘potentially dangerous’ has increased (Richardson and Reynolds 2000; Bolch et al. 2008; Ives et al. 2010; Liu et al. 2013), but it is unclear if the frequency of GLOF events is on the rise. Historical records indicate that 33 GLOF events occurred in the Himalaya before 2000 (Richardson and Reynolds 2000). The most severe disaster, the 1981 Cirenmaco GLOF, destroyed the China-Nepal Friendship Bridge and killed 200 people (Wang et al. 2015). The inventory of glacial lakes from Zhang et al. (2015a, b) serves as a basis for further analyses of priority lake and hazard assessments.

7.4.3 Projections

To our knowledge, few quantitative studies of glacial lake development exist. However, using simplified physically-based models, the locations of overdeepening and potential future lakes can potentially be predicted (e.g., Linsbauer et al. 2016), and it is generally expected that existing lakes will further grow and new lakes will develop

Table 7.4 Number and area of glacier-fed lakes in 1990, 2000, and 2010 (Zhang et al. 2015a, b)

Region ID (see Fig. 7.1)	Region name	1990		2000		2010	
		Number	Area (km^2)	Number	Area (km^2)	Number	Area (km^2)
6	Western Pamir	220	26.4	244	29.0	265	32.8
7	Eastern Pamir	21	3.7	26	4.1	28	4.4
8	Eastern Hindu Kush	333	32.8	384	38.6	393	39.8
9	Karakoram	40	5.7	55	5.8	77	9.6
10	Western Himalaya	301	27.6	379	35.1	400	38.8
11	Central Himalaya	644	84.8	694	100.6	796	119.4
12	Eastern Himalaya	579	92.4	606	91.9	728	115.9
13	Western Kunlun Shan	53	11.6	47	11.4	71	17.6
14	Eastern Kunlun Shan	15	0.9	22	0.9	38	2.8
15	Altun Shan	2	0.03	7	0.08	8	0.16
16	Tibetan Interior Mountains	72	14.7	89	11.8	107	18.9
17	Gangdise Mountains	308	19.7	394	24.7	428	28.9
18	Nyainqentanglha	511	93.4	424	79.3	544	90.2
19	Tanggula Shan	57	9.5	50	9.4	78	11.7
20	Hengduan Shan	152	17.0	163	16.7	242	24.2
21	Eastern Tibetan Mountains	26	5.4	30	5.8	30	5.8
22	Qilian Shan	21	0.12	31	4.0	32	4.6

with ongoing glacier shrinkage. Conceptual studies show that glacial lakes will likely form where flat, crevasse-free surfaces indicate compressing flow through bed overdeepenings, followed by steeper and crevassed surfaces which indicate extending flow over bedrock riegels (e.g., Quincey et al. 2007; Frey et al. 2010; Maanya et al. 2016). These overdeepenings will be filled with water if not filled with sediment. As demonstrated by Rounce et al. (2016), the formation or growth of glacial lakes must be assessed along with changes in other objective hazards (moraine stability, mass movement exposure) for a full evaluation of glacier lake hazards. It should be noted that numerous new lakes will form with continued deglaciation (Sect. 7.3.4), and hazardous lake risk assessment should evolve continually. In combination with the glacier shrinkage and lake formation or expansion, permafrost degradation on oversteepened rock slopes will increase the probability of impact waves and far-reaching flood waves from large rock/ice avalanches (Haeberli et al. 2016; Bolch et al. 2010b).

7.4.4 Recommendations

Further development of hazardous lake risk assessments should incorporate projected changes in lake volume, glacier extent, extreme climatic events, and stability-related hazards, such as permafrost degradation. Repeat detailed UAV surveys of moraine-dammed glacier lakes will enable identification of small-scale changes to dam integrity.

7.5 Permafrost

Permafrost, like glaciers and snow cover, not only occurs at high latitudes but also in cold, high-elevation environments at lower latitudes (Haeberli et al. 2010). It is commonly defined as soil or rock remaining below 0 °C for more than two consecutive years. Besides this thermal definition, permafrost is often understood as a ground material: soil or rock that is frozen (i.e., contains ice) for a long time. The ice in the subsurface can exist in pores of rock or soil, as well as in discrete layers. When this ice melts, in response to human disturbance or environmental change, ground characteristics such as mechanical strength and hydraulic permeability undergo drastic changes as well. With seasonal changes, a surficial layer of ground undergoes freezing and thawing. In permafrost conditions, the portion of the ground that thaws during summer is termed the active layer. Because the permafrost is located beneath this layer, it is not visible at the terrain surface and its distribution and changes cannot readily be observed based on remote sensing products. For this reason, the understanding of glaciers and snow and their changes is much better developed in many regions than the understanding of permafrost. In this section, we provide background information about the permafrost research and distribution in the extended HKH, followed by a summary of the current knowledge about past changes in ground temperature and active layer thickness in permafrost-dominated areas. The last section outlines the limited knowledge about future trends related to permafrost.

Within the extended HKH, permafrost research has been limited mainly to the Tibetan Plateau (TP). It is mostly focused on an engineering corridor where the construction and maintenance of roads, railways, and other linear infrastructure requires detailed understanding and characterization of the permafrost environment. Some permafrost research was also conducted in the Tajik Pamir during the Soviet time (Gorbunov 1990).

Permafrost research on the TP started in the early 1960s (Wu and Zhang 2008; Zhang et al. 2008a, b). After the mid-1970s, large-scale and systematic permafrost investigations were widely conducted along the Qinghai-Tibetan Highway (Cheng and Wang 1983) and over the past two decades, numerous institutions and individual investigators have established extensive monitoring networks (Wu and Zhang 2008; Zhao et al. 2010; Wu et al. 2012). Monitoring of the active layer and permafrost temperatures provides data and information for climate change studies and for infrastructure operations and maintenance. Studies investigate how changes in the active layer and permafrost impact hydrological processes, the carbon cycle, vegetation and ecosystems, geomorphological processes (such as the expansion of thermokarst, thaw lakes, or thaw slumps), and engineered infrastructures on the TP and in surrounding regions.

Outside China, there has been no systematic collection of permafrost data in the extended HKH region to date, and even within the TP, there has been no systematic collection in many remote mountain areas. Existing research has largely focused on local regions, such as the TP (Fort and van Vliet-Lanoe 2007; Jakob 1992; Barsch and Jakob 1998; Owen and England 1998; Regmi 2008; Hewitt 2014), western Himalaya (Fort 2003), Bhutan (Iwata et al. 2003), Pamir-Alai (Gorbunov and Titkov 1989), or systematic remote mapping of rock glaciers (Schmid et al. 2015). Further, only a few observations have been made of ground ice (Rastogi and Narayan 1999; Gruber et al. 2017) or ground temperature (Shiraiwa 1992; Klimeš and Doležal 2010) exist.

Knowledge about the distribution and characteristics of permafrost, however, is important in the extended HKH. This assessment provides an overview of relevant permafrost research in and beyond the extended HKH and infers relevant processes and impacts.

7.5.1 Occurrence

Due to its subterranean nature, the spatial extent of permafrost cannot be quantified using visual observations. Expert estimation (Gorbunov 1978) or computer models for generating permafrost maps are thus required to visualize where permafrost is expected to exist (Gruber 2012). Remote sensing data and products (e.g., land surface temperature) and cryogenic landforms (e.g., rock glaciers) can provide additional information about permafrost occurrence (Westermann et al. 2014; Schmid et al. 2015).

The occurrence of permafrost is strongly driven by surface climatology. Often, a mean annual air temperature (MAAT) below −1 °C is used as a first-order identification of areas with some proportion of permafrost. Shaded locations may have permafrost at this temperature, whereas sun-exposed slopes will only have permafrost at a much lower MAAT. Naturally, MAAT decreases with increasing elevation. Aridity also influences the occurrence of permafrost. Thick winter snow cover insulates the ground from a cool atmosphere leading to higher ground temperatures than at similar locations with less snow. Thus little snow cover implies favourable conditions for permafrost. Furthermore, arid areas can have debris slopes with permafrost at MAATs where glaciers would exist in areas with more precipitation. As a consequence, the abundance of permafrost beneath debris and soil increases with aridity. Steep rock slopes shed snow and are less affected by snow and glacier cover. Accordingly, permafrost occurs similarly in steep bedrock in both arid and humid regions. As a general rule, the relative proportion of permafrost and glaciers changes with aridity, with dry landscapes usually having more permafrost than glaciers and humid landscapes having more glaciers. In the extended HKH, these general patterns are overprinted with regional differences caused by differing seasonality and trends in temperature and precipitation; detailed maps are so far only available for parts of the TP (Ran et al. 2012).

For the Pamir, permafrost has been described down to 3,800 masl in alpine meadows, as well as lacustrine and fluvial sediments, while many dry slopes remain largely free of permafrost to above 4,300 masl (Gorbunov 1990). A number of maps exist for the TP, (Ran et al. 2012); while a recent map (Gruber 2012) indicates large areas of permafrost for the remainder of the extended HKH. This map is based on gridded climate data and established rules for estimating zones of permafrost occurrence based on MAAT. It has an approximate resolution of 1 km and contains some estimates of uncertainty.

A recent review for the whole extended HKH (Gruber et al. 2017) estimates the permafrost area to exceed that of glaciers in nearly all countries (Fig. 7.12). The total area of permafrost estimated across the region is 1,000,000 (500,000–1,500,000) km^2, compared to roughly 90,000 km^2 covered by glaciers; without the large contribution of the Tibetan and Pamir plateaus located within China, the expected permafrost area is 95,000 (30,000–180,000) km^2, and that of glaciers about 55,000 km^2.

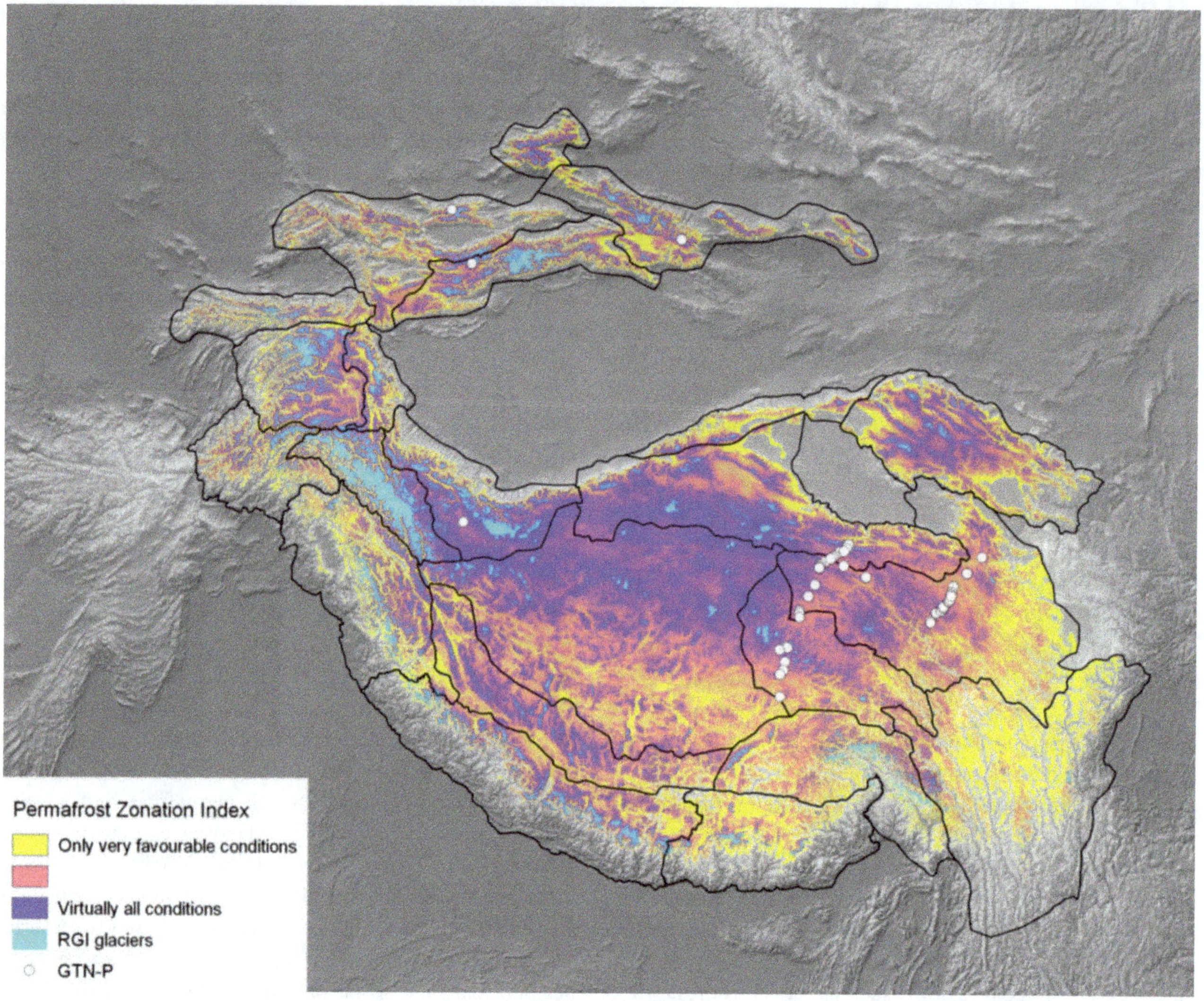

Fig. 7.12 Modelled permafrost zonation and borehole sites in the extended HKH. Boreholes listed in the GTN-P database (GTN-P 2016) are shown, but there are many more in China. The Permafrost Zonation Index (PZI) is based on Gruber (2012), glacier outlines are from the Randolph Glacier Inventory (Pfeffer et al. 2014)

7.5.2 Observed Change

Observed permafrost changes have been reported for a limited number of locations, with systematic observations only available for locations on the TP and in the Tien Shan. Changes reported from the TP are good exemplars, as they mirror many cold landscapes in the extended HKH in terms of latitude, aridity, and elevation. At the same time, many regions in the extended HKH have environmental conditions (especially precipitation amounts and terrain steepness) that differ strongly from the sites for which observational data exists.

As a general trend, most permafrost in the extended HKH is likely to have undergone warming and thaw during the past decades Zhao et al. (2010). With global climate change, this is dictated by physical processes connecting the subsurface and the warming atmosphere. Consistent observations of permafrost warming and thaw in other permafrost regions worldwide corroborate this (Romanovsky et al. 2010). Finally, subsurface temperatures and the occurrence of permafrost are driven by the same energy and mass balance at the ground surface as glaciers. The loss of glacier ice, which is now documented relatively well throughout the extended HKH (Sect. 7.3; Brun et al. 2017; Bolch et al. 2012), can thus be regarded as a rough proxy indicating subsurface warming and thaw. However, it should be noted that glacier mass and permafrost changes cannot be directly related as, for example, timing and amount of snow cover can affect glaciers and ground temperatures via differing processes.

Active-layer thickness has been monitored on the TP since the early 1990s. The primary goal of these measurements is to support engineering infrastructure and operations. Observations show that active layer thickness increased during the period from 1995–2007 (Wu et al. 2010) to 2006–2010 (Wu et al. 2012). Often, the active layer thickness increased more for sites close to roads than for those farther away. The majority of permafrost boreholes on the TP have temperatures between 0 and −2 °C. Borehole temperatures along the Qinqhai-Tibetan Highway and railway also show an increasing trend (Wu and Zhang 2008; Zhao et al. 2010), although with locally differing magnitudes. In the Tien Shan, systematic measurements of permafrost temperature and active-layer thickness started in the 1970s. Marchenko et al. (2007) report an increase of borehole temperature of 0.3–0.6 °C and an increase in active-layer thickness by more than 20% at the observed sites compared to the 1970s.

Overall, permafrost temperatures have increased more for cold permafrost and less or non-significantly when relatively close to the freezing point. This is because the majority of energy input into permafrost from the surface will be consumed for phase change when close to 0 °C. The observed thickening of the active layer and increase of permafrost temperatures on the TP corroborates the notion of widespread permafrost degradation in the extended HKH.

Mountain slopes with thawing permafrost are expected to undergo destabilization with continued warming and thaw (e.g., Huggel et al. 2013; Krautblatter et al. 2013). This will likely cause increased probability of rock falls, landslides, or debris flows, as well as lead to unforeseen events from locations that have previously shown little mass-movement activity. These mechanisms have been shown elsewhere (Deline et al. 2015) and are expected to be equally relevant in the extended HKH region (Fort 2015; Gruber et al. 2017). Slope failures can impact infrastructure and settlements both directly and indirectly, for example when impacting glacial lakes, which are often located below or in proximity to steep slopes (Haeberli et al. 2016). Furthermore, permafrost thaw can lead to ground movement via thaw settlement, and changed geotechnical characteristics via ground-ice melt. This is important because it can lead to problems with infrastructure built in permafrost environments, which requires special attention during planning, construction, and maintenance (Doré et al. 2016; Bommer et al. 2010).

7.5.3 Projections

Projected climate change in the extended HKH differs among regions, along with differing regional climate and circulation regimes (see Chap. 3). Furthermore, the strong vertical climate gradients observed today are likely to cause climate change to vary with elevation. Given how little we know about the distribution and characteristics of mountain permafrost outside the TP, few detailed projections of change are available. On the TP, a reduction in permafrost area by 9–14% in 2050 and by 13–46% in 2100 has been derived from model scenarios (Nan et al. 2005). A more recent projection suggests even greater future reductions in near-surface permafrost −39% by 2050 and 81% by 2100—accompanied by a marked increase in active layer thickness (Guo et al. 2012). While possible issues with model structure (cf. Burn and Nelson 2006; Delisle 2007) and resolution (cf. Fiddes et al. 2015) may affect the numbers presented, the trend to widespread and strong permafrost thaw is virtually certain to be robust. It is thus to be expected that strong atmospheric warming will result in active layer deepening and at least partial permafrost thaw in nearly the entire permafrost area of the extended HKH (cf. Gruber et al. 2017).

7.5.4 Recommendations

Increased permafrost monitoring and research in the mountainous parts of extended HKH is required in order to better understand possible impacts, to provide evidence of permafrost change, to enable the evaluation of numerical models, and to support the development of adaptation measures. Monitoring and research activities can be based on, and should integrate well with, ongoing efforts internationally.

7.6 River and Lake Ice

Information on lake and river ice has been identified as a suitable proxy for mean air temperatures and their variability. Measures of river and lake ice change include ice cover duration, as well as freeze-up and break-up dates. In line with the general temperature increase, several studies show that the average duration of lake ice occurrence has significantly decreased in the northern hemisphere during the last decades (Magnuson et al. 2000; Che et al. 2011; Dibike et al. 2011). Lake surface temperatures have also increased in the region (O'Reilly et al. 2015).

7.6.1 Occurrence

There are many large lakes on the Tibetan and the Pamir Plateaus which freeze in winter. The largest lakes are Selin Co, Nam Co, and Karakul Lake. However, investigations using moderate resolution remote sensing imagery (MODIS; spatial resolution of 500 m) for the period 2000–2010 revealed that several of the lakes in Tibet do not freeze up completely (Kropáček et al. 2013). There are no lakes of similar sizes in the extended HKH, and although some studies mention the existence of lake ice (e.g. Sakai et al. 2000b), to our knowledge there are no studies addressing lake freezing and thawing in the extended HKH regions outside TP.

7.6.2 Observed Change and Projections

The freeze-up date and thawing dates of all larger lakes on the TP were observed over the period 2000–2010 (Fig. 7.13). They were highly heterogeneous and no clear trend could be determined in the ice cover duration during this period, even when lakes were clustered into groups with similar characteristics (both climatically and lake characteristics) (Kropáček et al. 2013). Passive microwave data for the period 1978–2013 showed a decrease in the length of ice coverage of Nam Co lake, one of the largest lakes in Tibet, by ~20 days, with freezing onset ~9 days later and thawing date ~9 days earlier (Ke et al. 2013). Similar trends were also found for Qinghai lake (Che et al. 2009).

Very few studies address changes in river ice such as changes in freeze and break-up dates and maximum ice thickness. This is probably partly due to the fact that it is difficult to make comparable measurements in mountain rivers due to water turbulence and changing river beds. No projections exist, to our knowledge. Studies of the Yellow River indicate later freeze-up dates and earlier break-up dates in the 1990s compared to the average for 1950–1990s (Xiao et al. 2007; Jiang et al. 2008). River-ice duration along the Ningxia-Inner Mongolia reach in the upper Yellow River was reduced by about seven days between 1954 and 2013 (Si et al. 2015).

7.6.3 Recommendations

Baseline inventories of lake freeze-up, ice cover duration, and break-up should be compiled for lakes and the large rivers in the extended HKH region using remote sensing products and techniques. In addition, few in-site measurement sites should be established.

7.7 Cryosphere and Water Resources

The mountains of the extended HKH are a source of water for more than a billion people. Climatic change in the source areas of Asia's main rivers will affect snow, glaciers, and permafrost and may potentially have a large societal impact (Immerzeel et al. 2012; Bolch 2017). The cryosphere is a key component of the hydrological cycle (Mark et al. 2015), yet the hydrology is only marginally resolved due to the intricacy of monsoon dynamics (Turner and Annamalai 2012), the poorly quantified dependence on the cryosphere (Immerzeel et al. 2010), and the physical constraints of doing research in high-elevation and generally inaccessible terrain.

Within the extended HKH, the Indus basin has the largest human dependence on snow and ice melt (Kaser et al. 2010). However, there are very large inter-basin differences in snow

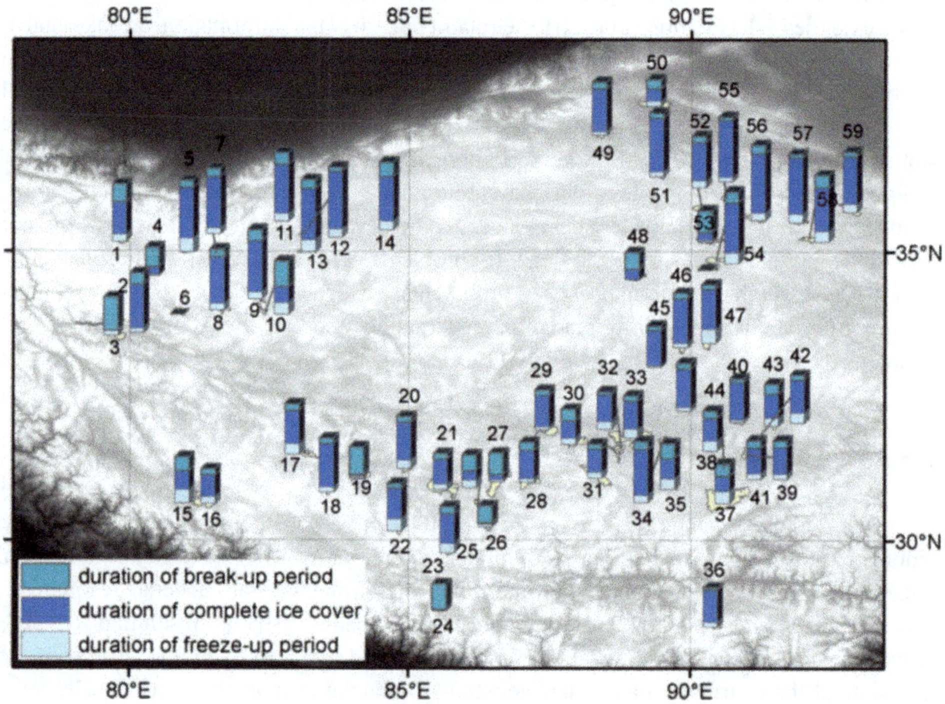

Fig. 7.13 Ice cover duration of the studied lakes on the TP showing duration of ice cover represented by the total height of each bar, and duration of freeze-up period, complete ice cover, and break-up period by the heights of the respective coloured sections *Source* Kropáček et al. (2013)

and ice contributions to streamflow (Immerzeel et al. 2010; Schaner et al. 2012; Farhan et al. 2014; Zhang et al. 2015a, b). A number of methods have been used in the extended HKH to quantify glacier contributions to streamflow (La Frenierre and Mark 2014). These include direct discharge measurements (Thayyen and Gergan 2010), hydrological modelling (Singh and Jain 2002; Nepal et al. 2013; Lutz et al. 2014; Ragettli et al. 2015), glaciological balance (Zhang et al. 2008a, b), and isotopic analyses (Racoviteanu et al. 2013; Wilson et al. 2015; Box 7.4).

However, quantitative comparisons between studies and basins are complicated by the fact that the definition of 'glacier contribution to streamflow' is inconsistent (O'Neel et al. 2014), with a key inconsistency being the way in which rain and snowmelt on top of a glacier surface is dealt with. In most cases glacier runoff is considered as the runoff from the total glacierized area, and includes all runoff leaving at the glacier terminus. This includes rain, and snow and ice melt. However, in the strict sense, only the ice which melts should be counted as glacier run-off, as rain and snowmelt would also runoff without the presence of a glacier (Radić and Hock 2014).

There is also an additional spatial dependence with regards to cryosphere contribution to streamflow: the relative importance of glacier and snowmelt runoff decreases with distance from the glaciers and snow reserves in the basin (Lutz et al. 2014; Kaser et al. 2010).

Box 7.4 Determining streamflow contributions

Despite recent studies on the hydrology of the Himalaya (Immerzeel et al. 2009, 2010; Kaser et al. 2010; Thayyen and Gergan 2010; Lutz et al. 2014), we still lack a consensus on the role of snow and ice in the hydrologic regime of Himalayan catchment basins. Our current understanding of the contribution from different sources is based primarily on modelling. However, hydrological models come in a wide range of complexities. When hydrological models are calibrated against total basin discharge, the wide range of processes and assumptions leads to a fundamental problem in hydrology: equifinality means it is possible to get the right answer for the wrong reasons (Beven 2006).

Direct measurements of hydrochemistry (including stable water isotope tracers such as $\delta^{18}O$, the ration between ^{18}O and ^{16}O, and impurities) have the potential to identify different sources of streamflow (snow or ice melt, rainfall, groundwater). At present, such isotope studies are scarce in the extended HKH, due to the difficulty of regularly acquiring water samples in remote terrain.

Previous extended HKH isotopic studies have identified moisture sources and atmospheric circulation patterns in Ladakh (Mayewski et al. 1983), Central Asia, and the Tibetan Plateau (Aizen et al. 1996; Garzione et al. 2000; Pande et al. 2000; Kang et al. 2002; Niu et al. 2016). In the western Himalaya (upper Indus basin), stable water isotope measurements were used together with historical hydrological records for river discharge to assess water balance in the basin (Karim and Veizer 2002). Only small watershed-scale studies have been conducted.on the southern slopes of the Himalaya, (Zhang et al. 2001; Racoviteanu et al. 2013; Wilson et al. 2015; Williams et al. 2016).

Differences in the hydrochemistry of melt from debris-free and debris-covered glaciers (Garzione et al. 2000; Williams et al. 2016) may also lead to improved estimates of melt contributions from debris-covered glaciers. Field campaigns for water isotope studies that are currently emerging in different regions of the extended HKH (Williams et al. 2013; Jeelani et al. 2016; Rai et al. 2016) will provide new opportunities for a full understanding of water source contributions to river discharge and regional climate in this area (Fig. 7.14).

7.7.1 Observations

As detailed snow information is lacking and difficult to obtain (Miller et al. 2012), the main focus in the extended HKH has historically been glacier change and its long-term impact on catchment hydrology (Singh and Bengtsson 2005; Yao et al. 2007; Akhtar et al. 2008; Kehrwald et al. 2008; Xu et al. 2009). However, recent modelling studies have shifted to include a focus on snow (Lutz et al. 2014; Tahir et al. 2016; Wulf et al. 2016), as it is an important component of the annual water budget (Archer 2003; Stewart 2009; Bookhagen and Burbank 2010) and will be affected by climate change in the near term. The role of permafrost in mountain hydrology, particularly in the extended HKH, is largely unknown.

Sufficiently long time series are required to identify changes and trends in cryospheric and hydrological components, but these are rarely available for river discharge. Interannual variability in both meteorological conditions and total catchment or basin discharge is also possibly greater than any climate change signal (Forsythe et al. 2014). While there is confidence in field-based or remotely-sensed observations of cryospheric change, discharge measurements in remote

Fig. 7.14 Glacier meltwater from Langshisha Glacier in Langtang valley, Nepal (*Photo* J. Shea)

mountain areas are also prone to measurement errors. These uncertainties pertain to sensor type (e.g., reading gauge, radar, pressure transducer), river type, and the availability of an accurate rating curve under representative flow conditions (di Baldassarre and Montanari 2009). To our knowledge, no systematic evaluation of discharge rating curves has yet been conducted, which leads to significant uncertainty in runoff observations and any trend analyses.

Increased temperatures have resulted in earlier snowmelt and a shift towards greater flow fractions in winter on the Tibetan Plateau (Ye et al. 2005). Increases in spring temperatures and spring discharge volumes have been observed at lower elevation stations in the Indus (Khattak et al. 2011; Sharif et al. 2013), although some stations show no significant trends. Decreased flows have been observed in snowmelt dominated basins in Nepal (Shrestha and Aryal 2011).

Increased discharge and a correlation with increased summer temperature has been observed in glacierized basins in the Pamir (Chevallier et al. 2014) and on the Tibetan Plateau (Yao et al. 2007; Zhang et al. 2007; Liu et al. 2010). Discharge has decreased in the heavily glacierized Hunza and Shyok catchments of the Karakoram (Fowler and Archer 2006; Khattak et al. 2011; Sharif et al. 2013), where temperatures have also decreased and glacier mass balances have been neutral or slightly positive since the 1970s. The limit and timing of glacier meltwater increases, known as 'peak water' (Baraer et al. 2012), has not been explored in detail in the extended HKH.

Given the multiple factors influencing streamflow, it is not possible to provide conclusive regional evidence of either declining or increasing streamflow trends. Trends change in space and time within a basin. For example, analysis of summer streamflows in the upper Indus basin shows periods of both decline and increase in runoff for the same stations for different decades, and also contradicting trends for the same decade but for different stations (Reggiani et al. 2017). These observations are explained by the conspiring effects of seasonal temperature and precipitation changes and variable dependence of runoff on glacier and snow resources.

7.7.2 Projections of Cryospheric and Hydrological Change

In general, overall annual flows are not projected to change significantly with climate change (Lutz et al. 2014), especially in major rivers where monsoon rainfall largely determines total runoff. Shifts in the timing and magnitude of streamflows in mountain rivers, however, will become apparent as climatic and cryospheric changes progress. Basin-scale projections of glacier and hydrological change (e.g., Zhang et al. 2015a, b) also illustrate that variability in projected climate fields can mask changes in glacier contributions to streamflow.

Projected changes in overall runoff are the compound effect of different processes, and few studies have attempted to isolate and quantify these effects. Conceptually, an increase in temperature will result in a range of hydrological impacts (Box 7.5):

- More precipitation will fall in the form of rain instead of snow. This will (a) cause the snow line to migrate upwards and result in less seasonal snow storage and a reduction in glacier accumulation, and (b) result in increased rainfall runoff during snowfall seasons. In general, this will reduce the buffering capacity of a catchment and lead to a higher susceptibility to both extreme runoff and prolonged low flows.
- Seasonal snowpack melt rates will increase. Snowmelt will occur earlier in the season and make a larger contribution to total runoff in early spring causing increased probability of floodings,
- Glacier shrinkage will cause glacier contributions to river runoff to increase initially, and then decrease. Glacier melt seasons will be extended and higher temperatures will enhance glacier melt rates. This impact is transient, however, and ice volumes will eventually be sufficiently reduced to limit total meltwater production. More than half the basins in the extended HKH are expected to have reduced glacier melt contributions by 2100 (Shea and Immerzeel 2016).
- The sign and magnitude of changes in evapotranspiration are less clear. The role of evapotranspiration in high-elevation water cycles is unstudied, but evapotranspiration rates are a function of temperature, solar radiation, and wind speed, and are limited primarily by available soil moisture. All things being equal, an increase in temperature will lead to an increased evapotranspiration rate, which may affect river runoff during warmer seasons.

An increase in precipitation will also have various compound hydrological impacts:

- More precipitation implies more rainfall, which will cause more runoff in the rivers, a more capricious runoff pattern, and an increased susceptibility to extreme runoff.
- More rainfall will also cause an increase in soil moisture, which will enhance evapotranspiration rates in relatively dry seasons and thus cause a reduction in streamflow in the shoulder seasons of the monsoon.
- More precipitation implies more snowfall and thus more glacier accumulation. This will contribute to more positive glacier mass balances and a longer period ahead before ice volumes become so small that the contribution to river runoff diminishes.
- More snowfall implies more extensive seasonal snowpacks, which results in a longer snowmelt season with a larger overall amount. This then results in an attenuation of the hydrograph and a reduction in the susceptibility to extremes.

In reality, all these factors may conspire and together they determine the future hydrograph. Given the uncertainties, in particular with regards to future precipitation changes, it is difficult to draw accurate regional conclusions; however, average runoff is likely to be balanced for the coming decades (Immerzeel et al. 2013; Lutz et al. 2014) and dealing with runoff extremes is the largest challenge we face (Lutz et al. 2016).

Although the body of literature regarding future hydrological projections is scarce, the near-, medium-, and long-term projections that are available are presented in the following sections.

7.7.2.1 Near-Term Change

Seasonal snowmelt is the dominant source of streamflow in the pre-monsoon (Apr–Jun) in nearly all catchments across the region, with particular importance in the west (Singh and Bengtsson 2005; Bookhagen and Burbank 2010; Wulf et al. 2016). The most prominent near-term change will likely be observed as a shift in snowmelt timing and magnitude.

The Satluj river basin (western Himalaya), for example, is projected to see increased spring discharge (+6%) and decreased summer discharge (-7%) in response to a +1 °C temperature increase (Singh and Bengtsson 2005). Projected increases in summer runoff in the Indus are linked to increased snowfall and snowmelt totals, although uncertainty in climate model outputs is large (Ragettli et al. 2013). Other modelling studies have shown shifts in the timing and magnitude of snowmelt runoff across the extended HKH (Hagg et al. 2006, 2007; Tahir et al. 2011). Decreased snowmelt contributions in lower elevation western basins may reduce water availability during critical irrigation periods (see Chap. 8), but greater low flows in winter and earlier spring snowmelt runoff could potentially benefit hydroelectric power generation (see Chap. 6).

7.7.2.2 Medium- and Long-Term Change

Glacier melt occurs primarily during the monsoon (Jun–Sep), when seasonal snowpacks have retreated to higher elevations and bare ice is exposed to annual maximums of mean daily air temperature and relatively high solar radiation (Box 7.5; Archer 2003). Smaller volumes of ice melt are also produced beneath debris-covered glaciers when energy is transferred from the surface to the debris/ice interface (Benn et al. 2012). Melt conditions for both clean and debris-covered glaciers can occur between March and October, and the hydrological impacts of glacier change will thus be felt primarily during these months. The magnitude and significance of hydrological change due to glacier change will depend on climatic setting, and will likely be observed on decadal-to-century timescales (Immerzeel et al. 2012; Lutz et al. 2014; Shea and Immerzeel 2016).

The projected increase in temperatures and decline in glacier areas and volumes (Sect. 7.3) will have compensating effects: glacier melt and runoff will increase initially, but glacier melt contributions will be reduced as glacier area and volumes adjust (Rees and Collins 2006; Immerzeel et al. 2012; Bliss et al. 2014; Lutz et al. 2014; Shea and Immerzeel 2016). A notable exception to this pattern is suggested for the western parts of the extended HKH, where large-scale modelling projects no significant decline in glacier melt until 2100 (Bliss et al. 2014). The loss of glaciers in extended HKH catchments will lead to a reduced buffering capacity, reduced summer melt contributions, and increased streamflow variability (Fountain and Tangborn 1985).

Permafrost degradation, which occurs on decadal-to-century timescales, will lead to an increase in the active layer depth, a greater near-surface water reservoir, and possible increases in drainage (Wang et al. 2000; Niu et al. 2011; Torre Jorgenson et al. 2013).

7.7.3 Recommendations

To have confidence in both observations and projections of hydrological change related to cryospheric change, improved and expanded high-elevation monitoring sites are necessary, and should be complemented with remote sensing and modelling efforts. This could include the following:

- Dedicated collaborative research catchments across a range of climatic and elevational zones with hydrological, meteorological, and glaciological observations
- Detailed studies on high elevation snow accumulation, sublimation, and redistribution
- Synoptic and time-series surveys of stable isotopes to estimate streamflow components (Box 7.4), and modelled streamflow contributions.

Box 7.5 Climatological and hydrological change in glacierized watersheds

As an example of the potential impacts of hydrology on glacierized watersheds, we use the mean daily runoff observed in the Langtang river, Nepal. Above the gauging station, the basin is approximately 25% glacierized, and detailed hydrological modelling indicates that snowmelt, rainfall, and glacier melt account for 40, 36, and 24% of annual runoff, respectively (Ragettli et al. 2015). Based on a simple increase in temperature, we impose three separate scenarios on the modelled streamflow contributions (Fig. 7.15):

- More rain, less snow: increased temperatures lead to a 30% reduction in snowfall. This is added to rainfall derived runoff.
- Increased snow and ice melt: increased temperatures result in a 50% increase in melt.
- Longer melt season: snowmelt season starts earlier, but total melt contribution remains the same.

As all three impacts will likely be observed simultaneously, in addition to changes in precipitation, the specific attribution of hydrological change in glacierized watersheds is a difficult task. Further, the hydrological response will vary between basins depending on snow accumulation gradients, topography, glacier extent, state, and debris cover.

7.8 Policy Relevance

We identify three key areas of policy relevance with regards to the status, trends, and projections for the extended HKH cryosphere.

1. *Improved resource allocation for region- and sector-specific studies of the impacts of cryospheric change (e.g., impacts of glacier and snowpack changes on hydroelectricity, and GLOF hazards in the Himalaya)*

In most regions of the extended HKH, glaciers are thinning, losing mass, and retreating, except for western Kunlun and parts of the Karakoram and Pamir where glaciers have been relatively stable over the past decades. In addition, the limited data suggests seasonal snow cover is reducing and permafrost is thawing. Cryospheric change will ultimately influence the seasonal availability of water in the extended HKH river systems. However, our understanding on these issues is limited, and we need to improve our data collection, analyses, and modelling strategies to provide sector- and region-specific details of projected changes. The speed and extent of changes will vary from region to region and may lead to conflicts due to competing demands from different sectors, such as domestic, agriculture, hydropower, and industrial usages. To avoid conflict, and to develop just and equitable water resource distribution policies, targeted studies are needed.

2. *High-level international agreements for systematic data collection, data sharing, training, and hazard warnings related to the cryosphere*

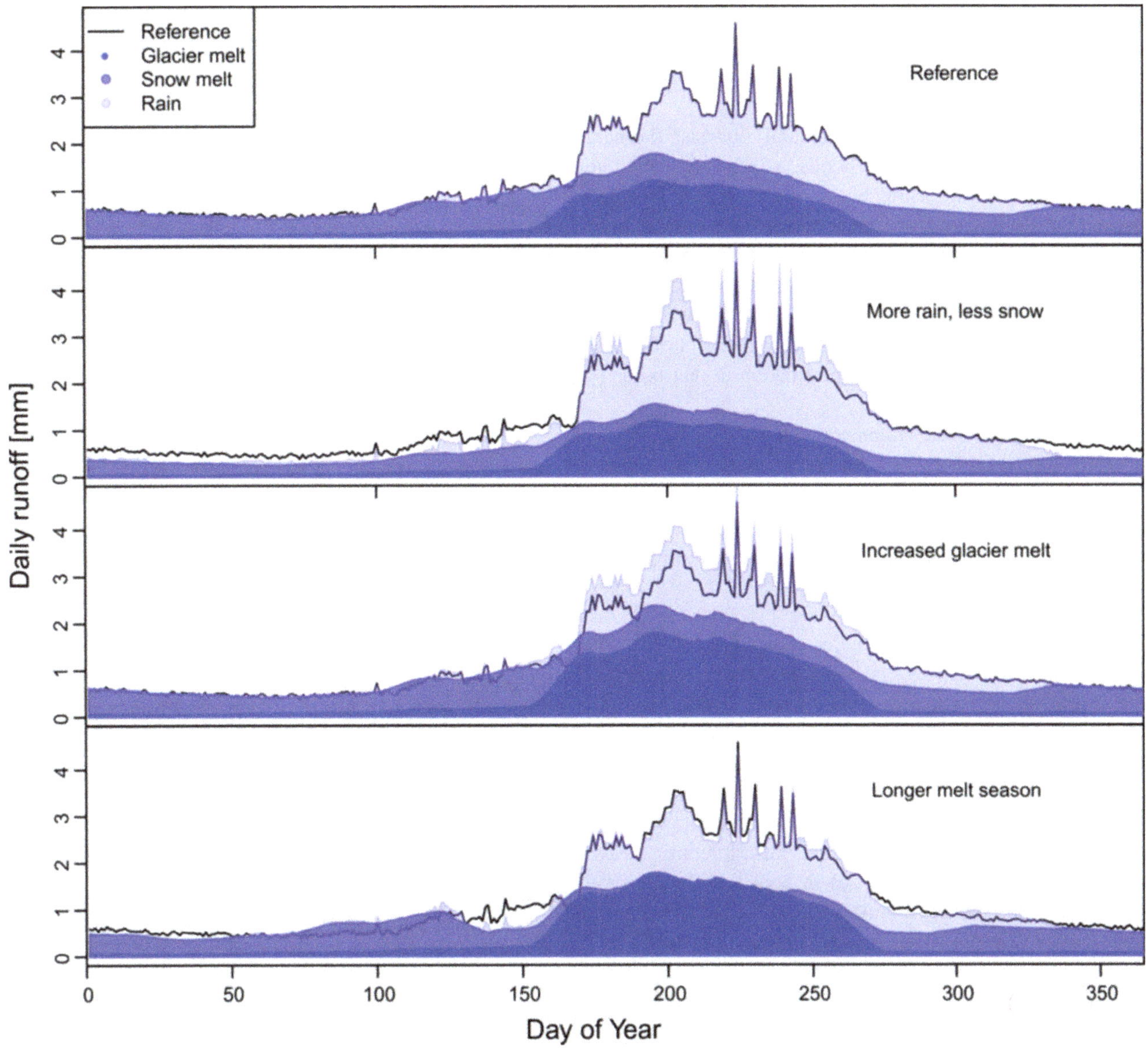

Fig. 7.15 Conceptual changes in streamflow contributions in the Langtang Valley, Nepal

Numerous rivers, such as the Indus, Ganges, and Brahmaputra originate in the high mountains of extended HKH, and the river water is then shared among various countries in the region by agreement, for example the Mahakali Treaty between Nepal and India, the 30-year Ganges (Ganga) water-sharing arrangement between India and Bangladesh, and the Indus water treaty between India and Pakistan. According to the Indus treaty, control over the three 'eastern' rivers (the Beas, Ravi, and Sutlej) was given to India, and the three 'western' rivers (the Indus, Chenab, and Jhelum) to Pakistan. However, in the future, glaciers and snow cover could change differently in the so-called eastern and western basins, leading to substantial change in the availability of water in each basin. This may lead to renegotiation of existing water-sharing treaties. To make informed decisions, we will need a better understanding of the changes in snow- and glacier-stored water and its melting processes. This will require proper international agreements for systematic and standardized data collection and sharing on the cryosphere in extended HKH. These agreements should follow guidelines and standards from existing high-level organizations such as the World Glacier Monitoring Service (WGMS, www.wgms.ch), Glacier and Permafrost Hazards in Mountains (GAPHAZ, www.gaphaz.org), and the Global Terrestrial Observing System (GTOS, www.fao.org/gtos/) as part of the Global Climate Observing System (GCOS, public.wmo.int/en/programmes/global-climate-observing-system).

Obtaining access to existing and newly gathered data will be crucial and will provide longer time series for many regions. It will also call for improvement in the existing pool of trained scientific personnel and the sharing of best practices in data collection and analysis. It should also be noted, a significant amount of data exists in private hands, such as hydropower companies.

Retreating glaciers can also potentially increase the hazard from glacial lake outburst floods, as new lakes may be formed due to overdeepening of the newly-exposed land and permafrost degradation will increase hazard potential from destabilizing mountain slopes. To minimize damage to human life, property, and the environment, a better trans-boundary hazard warning system is needed. This will require new international agreements and the development of proper infrastructure to share data and information in real time.

3. *Stronger global commitments and actions to reduce GHG emissions due to the recognition that climate change and carbon emissions are global phenomena that significantly affect the extended HKH cryosphere, and hence, also livelihoods in one of the most populated regions* (is +1.5C enough?)

All projections point towards increased cryospheric changes in the coming decades and centuries in the extended HKH. Decreased snowpacks, accelerated glacier mass loss, and changes in permafrost, rivers, and lakes will occur in response to increased temperatures. These cryospheric changes will directly affect populations both within and downstream of the mountains through changes in hydrology, natural hazard risks, and potential infrastructure losses. Notwithstanding the uncertainties in both the climate projections and the model formulations, high-emission scenarios will result in greater cryospheric impacts than low-emission scenarios. This potential for climate mitigation represents a key policy recommendation: the international community needs to make stronger and more accountable commitments to restrict GHG emissions and meet emission scenarios as given in RCP2.6 or lower.

Acknowledgements The authors acknowledge the contributions of all researchers working in this field, agencies that provide access to field-based and remote sensing datasets, and organizations that work to standardize cryospheric measurements and reporting. We also thank the researchers and colleagues who have graciously provided us with their data and model outputs (Rianne Giesen, Valentina Radic, Ben Marzeion, Daniel Farinotti) and information (Phuntsho Tshering, Inka Koch, Miriam Jackson). We appreciate the specific and general comments of our reviewers and colleagues, and the support of the review editor Koji Fujita. Any remaining errors or omissions are ours.

References

Adhikari, S., & Huybrechts, P. (2009). Numerical modelling of historical front variations and the 21st-century evolution of glacier AX010, Nepal Himalaya. *Annals of Glaciology, 50*(52), 27–34.

Aizen, V. B., et al. (1996). Isotopic measurements of precipitation on central Asian glaciers (southeastern Tibet, northern Himalayas, central Tien Shan). *Journal of Geophysical Research: Atmospheres, 101*(D4), 9185–9196.

Aizen, V. B., et al. (2006). Climatic and atmospheric circulation pattern variability from ice-core isotope/geochemistry records (Altai, Tien Shan and Tibet). *Annals of Glaciology, 43,* 49–60.

Aizen, V. B., et al. (2009). Stable-isotope and trace element time series from Fedchenko glacier (Pamirs) snow/firn cores. *Journal of Glaciology, 55*(190), 275–291.

Aizen, V. B., & Aizen, E. M. (2014). The Central Asia climate and cryosphere/water resources changes. In *Materials of the International Conference "Remote- and Ground-based Earth Observations in Central Asia"* (pp. 240–245).

Aizen, V. B., Aizen, E. M., & Melack, J. M. (1995). Climate, snow cover, glaciers, and Runoff in the Tien Shan, Central Asia. *Journal of the American Water Resources Association, 31*(6), 1113–1129.

Akhtar, M., Ahmad, N., & Booij, M. J. (2008). The impact of climate change on the water resources of Hindukush-Karakorum-Himalaya region under different glacier coverage scenarios. *Journal of Hydrology, 355*(1–4), 148–163.

An, W., et al. (2016). Possible recent warming hiatus on the northwestern Tibetan Plateau derived from ice core records. *Scientific Reports, 6,* 32813.

Andermann, C., Bonnet, S., & Gloaguen, R. (2011). Evaluation of precipitation data sets along the Himalayan front. *Geochemistry, Geophysics, Geosystems 12*(7).

Anders, A. M. et al. (2006). Spatial patterns of precipitation and topography in the Himalaya. In S. D. Willett, N. Hovius, M. T. Brandon & D. Fisher (Eds.), *Tectonics, climate, and landscape evolution: Geological society of America special paper 398* (pp. 39–53).

Archer, D. (2003). Contrasting hydrological regimes in the upper Indus Basin. *Journal of Hydrology, 274*(1), 198–210.

Arendt, A. A. et al. (2015). *Randolph glacier inventory–a dataset of global glacier outlines: Version 5.0.*

Arendt, A. A. & Others. (2012). *Randolph glacier inventory (RGI), Vers. 1.0: a dataset of global glacier outlines.*

Ashraf, A., Naz, R., & Iqbal, M. B. (2017). Altitudinal dynamics of glacial lakes under changing climate in the Hindu Kush, Karakoram, and Himalaya ranges. *Geomorphology, 283,* 72–79.

Ashraf, A., Naz, R., & Roohi, R. (2012). Glacial lake outburst flood hazards in Hindukush, Karakoram and Himalayan Ranges of Pakistan: Implications and risk analysis. *Geomatics, Natural Hazards and Risk, 3*(2), 113–132.

Azam, M. F., et al. (2016). Meteorological conditions, seasonal and annual mass balances of Chhota Shigri Glacier, western Himalaya, India. *Annals of Glaciology, 57*(71), 328–338.

Azam, M. F., et al. (2014). Reconstruction of the annual mass balance of Chhota Shigri glacier, Western Himalaya, India, since 1969. *Annals of Glaciology, 55,* 69–80.

Bahr, D. B., Meier, M. F., & Peckham, S. D. (1997). The physical basis of glacier volume-area scaling. *Journal of Geophysical Research, 102*(B9), 20355.

Bajracharya, S. R., & Shrestha, B. (2011). In S. R. Bajracharya & B. Shrestha (Eds.), *The status of glaciers in the Hindu Kush-Himalayan region.* ICIMOD: ICIMOD.

Di Baldassarre, G., & Montanari, A. (2009). Uncertainty in river discharge observations: A quantitative analysis. *Hydrology and Earth System Sciences, 13,* 913–921.

Bamber, J. L., & Rivera, A. (2007). A review of remote sensing methods for glacier mass balance determination. *Global and Planetary Change, 59*(1–4), 138–148.

Banerjee, A. (2017). Brief communication: Thinning of debris-covered and debris-free glaciers in a warming climate. *The Cryosphere, 11* (1), 133–138.

Bansod, S. D., et al. (2003). Thermal field over Tibetan Plateau and Indian summer monsoon rainfall. *International Journal of Climatology, 23*(13), 1589–1605.

Bao, W., et al. (2015). Glacier changes during the past 40 years in the West Kunlun Shan. *Journal of Mountain Science, 12*(2), 344–357.

Baraer, M., et al. (2012). Glacier recession and water resources in Peru's Cordillera Blanca. *Journal of Glaciology, 58,* 134–150.

Baral, P., et al. (2014). Preliminary results of mass-balance observations of Yala Glacier and analysis of temperature and precipitation gradients in Langtang Valley. *Nepal. Annals of Glaciology, 55*(66), 9–14.

Barandun, M., et al. (2015). Re-analysis of seasonal mass balance at Abramov glacier 1968–2014. *Journal of Glaciology, 61*(230), 1103–1117.

Barlow, M., et al. (2005). Modulation of Daily Precipitation over Southwest Asia by the Madden–Julian Oscillation. *Monthly Weather Review, 133*(12), 3579–3594.

Barnett, T. P., Adam, J. C., & Lettenmaier, D. P. (2005). Potential impacts of a warming climate on water availability in snow-dominated regions. *Nature, 438*(7066), 303–309.

Barry, R. G. (1992). *Mountain weather and climate* (2nd ed.). New York: Routledge.

Barsch, D., & Jakob, M. (1998). Mass transport by active rockglaciers in the Khumbu Himalaya. *Geomorphology, 26*(1–3), 215–222.

Baumann, S., Winkler, S., & Andreassen, L. M. (2009). Mapping glaciers in Jotunheimen, South-Norway, during the "Little Ice Age" maximum. *The Cryosphere, 3*(2), 231–243.

Beniston, M. (2003). Climatic change in mountain regions: A review of possible impacts. *Climatic Change, 59*(1), 5–31.

Benn, D. I., et al. (2012). Response of debris-covered glaciers in the Mount Everest region to recent warming, and implications for outburst flood hazards. *Earth-Science Reviews, 114*(1–2), 156–174.

Benn, D. I., Wiseman, S., & Hands, K. A. (2001). Growth and drainage of supraglacial lakes on debrismantled Ngozumpa Glacier, Khumbu Himal Nepal. *Journal of Glaciology, 47*(159), 626–638.

Berthier, E., et al. (2007). Remote sensing estimates of glacier mass balances in the Himachal Pradesh (Western Himalaya, India). *Remote Sensing of Environment, 108*(3), 327–338.

Beven, K. (2006). A manifesto for the equifinality thesis. *Journal of Hydrology, 320*(1), 18–36.

Bhambri, R., Bolch, T., Chaujar, R. K., et al. (2011). Glacier changes in the Garhwal Himalaya, India, from 1968 to 2006 based on remote sensing. *Journal of Glaciology, 57,* 543–556.

Bhambri, R., et al. (2013). Heterogeneity in glacier response in the upper Shyok valley, northeast Karakoram. *Cryosphere, 7*(5), 1385–1398.

Bhatt, B. C., & Nakamura, K. (2006). A climatological-dynamical analysis associated with precipitation around the southern part of the Himalayas. *Journal of Geophysical Research: Atmospheres, 111* (D2), D02115.

Bhattacharya, A. et al. (2016). Overall recession and mass budget of Gangotri Glacier, Garhwal Himalayas, from 1965 to 2015 using remote sensing data. *Journal of Glaciology*, 1–19.

Bliss, A., Hock, R., & Radić, V. (2014). Global response of glacier runoff to twenty-first century climate change. *Journal of Geophysical Research (Earth Surface), 119,* 717–730.

Bocchiola, D., et al. (2011). Prediction of future hydrological regimes in poorly gauged high altitude basins: The case study of the upper Indus, Pakistan. *Hydrology and Earth System Sciences, 15*(7), 2059–2075.

Bolch, T. (2017). Hydrology: Asian glaciers are a reliable water source. *Nature, 545*(7653), 161–162.

Bolch, T., et al. (2017). Brief communication: Glaciers in the Hunza catchment (Karakoram) have been nearly in balance since the 1970s. *The Cryosphere, 11*(1), 531–539.

Bolch, T., et al. (2012). The state and fate of Himalayan Glaciers. *Science, 336*(6079), 310–314.

Bolch, T., Pieczonka, T., & Benn, D. I. (2011a). Multi-decadal mass loss of glaciers in the Everest area (Nepal Himalaya) derived from stereo imagery. *Cryosphere, 5*(2), 349–358.

Bolch, T., et al. (2011b). Identification of potentially dangerous glacial lakes in the northern Tien Shan. *Natural Hazards, 59*(3), 1691–1714.

Bolch, T., Menounos, B., & Wheate, R. (2010a). Landsat-based inventory of glaciers in western Canada, 1985–2005. *Remote Sensing of Environment, 114*(1), 127–137.

Bolch, T., et al. (2010b). A glacier inventory for the western Nyainqentanglha range and the Nam Co Basin, Tibet, and glacier changes 1976–2009. *Cryosphere, 4*(3), 419–433.

Bolch, T., et al. (2008). Planimetric and volumetric glacier changes in the Khumbu Himal, Nepal, since 1962 using Corona, Landsat TM and ASTER data. *Journal of Glaciology, 54*(187), 592–600.

Bommer, C., Phillips, M., & Arenson, L. U. (2010). Practical recommendations for planning, constructing and maintaining infrastructure in mountain permafrost. *Permafrost and Periglacial Processes, 21*(1), 97–104.

Bookhagen, B., & Burbank, D. W. (2006). Topography, relief, and TRMM-derived rainfall variations along the Himalaya. *Geophysical Research Letters, 33*(8), L08405.

Bookhagen, B., & Burbank, D. W. (2010). Toward a complete Himalayan hydrological budget: Spatiotemporal distribution of snowmelt and rainfall and their impact on river discharge. *Journal of Geophysical Research: Earth Surface, 115*(3), 1–25.

Bräuning, A. (2006). Tree-ring evidence of "Little Ice Age" glacier advances in southern Tibet. *The Holocene, 16*(3), 369–380.

Brun, F., et al. (2017). A spatially resolved estimate of High Mountain Asia glacier mass balances from 2000 to 2016. *Nature Geoscience, 10*(9), 668–673.

Brun, F., et al. (2015). Seasonal changes in surface albedo of Himalayan glaciers from MODIS data and links with the annual mass balance. *Cryosphere, 9*(1), 341–355.

Buri, P., et al. (2016). A grid-based model of backwasting of supraglacial ice cliffs on debris-covered glaciers. *Annals of Glaciology, 57*(71), 199–211.

Burn, C. R., & Nelson, F. E. (2006). Comment on "A projection of severe near-surface permafrost degradation during the 21st century" by David M. Lawrence and Andrew G. Slater. *Geophysical Research Letters*, 33(21), L21503.

Chand, P., & Sharma, M. C. (2015). Glacier changes in the Ravi basin, North-Western Himalaya (India) during the last four decades (1971–2010/13). *Global and Planetary Change, 135,* 133–147.

Chang, A. T. C., Foster, J. L., & Hall, D. K. (1987). Nimbus-7 SMMR derived global snow cover parameters. *Annals of Glaciology, 9,* 39–44.

Chaturvedi, R. K., et al. (2014). Glacial mass balance changes in the Karakoram and Himalaya based on CMIP5 multi-model climate projections. *Climatic Change, 123*(2), 315–328.

Che, T., Li, X., & Dai, L. (2011). Monitoring freeze-up and break-up dates of Northern Hemisphere big lakes using passive microwave remote sensing data. In *2011 IEEE International Geoscience and Remote Sensing Symposium* (pp. 3191–3193), IEEE.

Che, T., Li, X., & Jin, R. (2009). Monitoring the frozen duration of Qinghai Lake using satellite passive microwave remote sensing low frequency data. *Chinese Science Bulletin, 54*(13), 2294–2299.

Chen, F., Zhang, M., Tian, B., & Li, Z. (2017). Extraction of glacial Lake outlines in Tibet plateau using landsat 8 imagery and google earth engine. *IEEE Journal of Selected Topics in Applied Earth Observations and Remote Sensing, 10*(9), 4002–4009.

Cheng, G., & Wang, S. (1983). Distributive regularities of high ice-content permafrost along Qinghai-Xizang Highway. In

Proceedings of the 4th International Conference on Permafrost (pp. 142–146).

Chevallier, P., et al. (2014). River flow regime and snow cover of the Pamir Alay (Central Asia) in a changing climate. *Hydrological Sciences Journal, 59*(8), 1491–1506.

Cogley, J. G. (2009). Geodetic and direct mass balance measurements: Comparison and joint analysis. *Annals of Glaciology, 50*(50), 96–100.

Cogley, J. G. (2011). Present and future states of Himalaya and Karakoram glaciers. *Annals of Glaciology, 52*(59), 69–73.

Collier, E., & Immerzeel, W. W. (2015). High-resolution modeling of atmospheric dynamics in the Nepalese Himalaya. *Journal of Geophysical Research: Atmospheres, 120*(19), 9882–9896.

Copland, L., et al. (2011). Expanded and recently increased glacier surging in the Karakoram. *Arctic, Antarctic, and Alpine Research, 43*(4), 503–516.

Dahri, Z. H., et al. (2016). An appraisal of precipitation distribution in the high-altitude catchments of the Indus basin. *Science of the Total Environment, 548–549,* 289–306.

Dahri, Z.H. et al., 2011. satellite-based snowcover distribution and associated snowmelt runoff modeling in Swat River Basin of Pakistan. In *Proceedings of Pakistan Academy of Sciences, Islamabad, Pakistan* (pp. 19–32).

Deline, S., et al. (2015). Ice loss and slope stability in high-mountain regions. In W. Haeberli, C. Whiteman, & J. F. Shroder (Eds.), *Snow and Ice-Related hazards, risks, and disasters* (pp. 521–561). Saint Louis: Elsevier Science.

Delisle, G., 2007. Near-surface permafrost degradation: How severe during the 21st century? *Geophysical Research Letters*, *34*(9).

Dibike, Y. et al. (2011). Response of Northern Hemisphere lake-ice cover and lake-water thermal structure patterns to a changing climate. *Hydrological Processes*, *25*(19), n/a-n/a.

Dietz, A. J., Kuenzer, C., & Conrad, C. (2013). Snow-cover variability in central Asia between 2000 and 2011 derived from improved MODIS daily snow-cover products. *International Journal of Remote Sensing, 34*(11), 3879–3902.

Diffenbaugh, N. S., Scherer, M., & Ashfaq, M. (2013). Response of snow-dependent hydrologic extremes to continued global warming. *Nature climate change, 3*(4), 379–384.

Dolgoushin, L. D., & Osipova, G. B. (1975). Glacier surges and the problem of their forecasting. In *Snow and Ice-Symposium-Neiges et Glaces (Proceedings of the Moscow Symposium, August 1971; Actes du Colloque de Moscou, août 1971)*, IAHS-AISH (pp. 292–304).

Doré, G., Niu, F., & Brooks, H. (2016). Adaptation methods for transportation infrastructure built on degrading permafrost. *Permafrost and Periglacial Processes, 27*(4), 352–364.

Duan, K., Xu, B., & Wu, G. (2015a). Snow accumulation variability at altitude of 7010 m a.s.l. in Muztag Ata Mountain in Pamir Plateau during 1958–2002. *Journal of Hydrology, 531*(9), 912–918.

Duan, K., Xu, B., & Wu, G. (2015b). Snow accumulation variability at altitude of 7010 m a.s.l. in Muztag Ata Mountain in Pamir Plateau during 1958-2002. *Journal of Hydrology, 531,* 912–918.

Dumont, M., et al. (2012). Linking glacier annual mass balance and glacier albedo retrieved from MODIS data. *Cryosphere, 6*(6), 1527–1539.

Dyurgerov, M. B., & Meier, M. F. (2005). *Glaciers and the changing Earth system: A 2004 snapshot* (Vol. 58). Boulder, CO: Institute of Arctic and Alpine Research, University of Colorado.

Farhan, S. Bin, et al. (2014). Hydrological regimes under the conjunction of westerly and monsoon climates: A case investigation in the Astore Basin, Northwestern Himalaya. *Climate Dynamics, 44* (11–12), 3015–3032.

Farinotti, D., et al. (2017). How accurate are estimates of glacier ice thickness? Results from ITMIX, the Ice thickness models intercomparison eXperiment. *The Cryosphere, 11*(2), 949–970.

Fiddes, J., Endrizzi, S., & Gruber, S. (2015). Large-area land surface simulations in heterogeneous terrain driven by global data sets: application to mountain permafrost. *The Cryosphere, 9*(1), 411–426.

Finsterwalder, R. (1954). Photogrammetry and glacier research with special reference to glacier retreat in the eastern Alps. *Journal of Glaciology, 2,* 306–315.

Førland, E. J., Allerup, P., Dahlström, B., Elomaa, E., Jónsson, T., Madsen, H., Perälä, J., Rissanen, P., Vedin, H., & Vejen, F. (1996). Manual for operational correction of Nordic precipitation data. *DNMI report, 24*, 96.

Forsythe, N., et al. (2014). Application of a stochastic weather generator to assess climate change impacts in a semi-arid climate: The Upper Indus Basin. *Journal of Hydrology, 517,* 1019–1034.

Fort, M. (2003). Are high altitude, lava stream-like, debris mixtures all rock glaciers? *Zeitschrift für Geomorphologie, 47,* 11–29.

Fort, M. (2015). Natural hazards versus climate change and their potential impacts in the dry, northern Himalayas: Focus on the upper Kali Gandaki (Mustang District, Nepal). *Environmental Earth Sciences, 73*(2), 801–814.

Fort, M., & van Vliet-Lanoe, B. (2007). Permafrost and periglacial environment of Western Tibet. *Landform Analysis, 5,* 25–29.

Foster, J. L., et al. (2011). A blended global snow product using visible, passive microwave and scatterometer satellite data. *International Journal of Remote Sensing, 32*(5), 1371–1395.

Fountain, A. G., et al. (2009). The "benchmark glacier" concept—Does it work? Lessons from the North Cascade Range, USA. *Annals of Glaciology, 50*(50), 163–168.

Fountain, A. G., & Tangborn, W. V. (1985). The effect of glaciers on streamflow. *Water Resources Research, 21*(4), 579–586.

Fowler, H. J., & Archer, D. (2006). Conflicting signals of climatic change in the Upper Indus Basin. *Journal of Climate, 19*(17), 4276–4293.

Frei, A., et al. (2012). A review of global satellite-derived snow products. *Advances in Space Research, 50*(8), 1007–1029.

La Frenierre, J., & Mark, B. G. (2014). A review of methods for estimating the contribution of glacial meltwater to total watershed discharge. *Progress in Physical Geography.*

Frey, H., et al. (2010). A multi-level strategy for anticipating future glacier lake formation and associated hazard potentials. *Natural Hazards and Earth System Science, 10*(2), 339–352.

Frey, H., et al. (2014). Estimating the volume of glaciers in the Himalayan– Karakoram region using different methods. *The Cryosphere, 8*(6), 2313–2333.

Frey, H., Paul, F., & Strozzi, T. (2012). Compilation of a glacier inventory for the western Himalayas from satellite data: Methods, challenges, and results. *Remote Sensing of Environment, 124,* 832–843.

Fujita, K., Ageta, Y., Pu, J., & Yao, T. (2000). Mass balance of Xiao Dongkemadi Glacier on central Tibetan Plateau from 1989 to 1995. *Annals of Glaciology, 31,* 159–163.

Fujita, K., et al. (2008). Performance of ASTER and SRTM DEMs, and their potential for assessing glacial lakes in the Lunana region, Bhutan Himalaya. *Journal of Glaciology, 54*(185), 220–228.

Fujita, K., et al. (2013). Potential flood volume of Himalayan glacial lakes. *Natural Hazards and Earth System Science, 13*(7), 1827–1839.

Fujita, K., et al. (2009). Recent changes in Imja Glacial Lake and its damming moraine in the Nepal Himalaya revealed by *in situ* surveys and multi-temporal ASTER imagery. *Environmental Research Letters, 4*(4), 45205.

Fujita, K., & Nuimura, T. (2011). Spatially heterogeneous wastage of Himalayan glaciers. *Proceedings of the National Academy of Sciences, 108*(34), 14011–14014.

Gao, J., et al. (2012). Spatiotemporal distribution of snow in eastern Tibet and the response to climate change. *Remote Sensing of Environment, 121,* 1–9.

Gardelle, J., et al. (2013). Region-wide glacier mass balances over the Pamir-Karakoram-Himalaya during 1999-2011. *The Cryosphere, 7* (4), 1263–1286.

Gardelle, J., Arnaud, Y., & Berthier, E. (2011). Contrasted evolution of glacial lakes along the Hindu Kush Himalaya mountain range between 1990 and 2009. *Global and Planetary Change, 75*(1–2), 47–55.

Gardner, A. S., et al. (2013). A reconciled estimate of glacier contributions to sea level rise: 2003 to 2009. *Science, 340*(6134), 852–857.

Gärtner-Roer, I., et al. (2014). A database of worldwide glacier thickness observations. *Global and Planetary Change, 122,* 330–344.

Garzione, C. N. et al. (2000). Predicting paleoelevation of Tibet and the Himalaya from δ18O vs. altitude gradients in meteoric water across the Nepal Himalaya. *Earth and Planetary Science Letters*, 183(1), pp. 215–229.

Germann, U., et al. (2006). Radar precipitation measurement in a mountainous region. *Quarterly Journal of the Royal Meteorological Society, 132*(618), 1669–1692.

Ghatak, D., Eric, S., & James, M. (2014). Role of snow-albedo feedback in higher elevation warming over the Himalayas, Tibetan Plateau and Central Asia. *Environmental Research Letters, 9*(11), 114008.

Giesen, R. H., & Oerlemans, J. (2013). Climate-model induced differences in the 21st century global and regional glacier contributions to sea-level rise. *Climate Dynamics, 41*(11–12), 3283–3300.

Gorbunov, A. (1990). Permafrost in the arid mountains of middle asia —the eastern Pamirs, USSR. *Permafrost and Periglacial Processes, 1,* 309–312.

Gorbunov, A. P. (1978). Permafrost investigations in high-mountain regions. *Arctic and Alpine Research, 10*(2), 283–294.

Gorbunov, A. P., & Titkov, S. N. (1989). *Kamennye Gletchery Gor Srejnej Azii (Rockglaciers of Central Asian Mountains)*. Irkutsk: Akademia Nauk SSSR.

Groisman, P. Y., et al. (1994). Changes of snow cover, temperature, and Radiative heat balance over the Northern Hemisphere. *Journal of Climate, 7*(11), 1633–1656.

Gruber, S. (2012). Derivation and analysis of a high-resolution estimate of global permafrost zonation. *The Cryosphere, 6*(1), 221–233.

Gruber, S., Fleiner, R., Guegan, E., Panday, P., Schmid, M.-O., et al. (2017). Review article: Inferring permafrost and permafrost thaw in the mountains of the Hindu Kush Himalaya region. *The Cryosphere, 11*(1), 81–99.

Guo, D., Wang, H., & Li, D. (2012). A projection of permafrost degradation on the Tibetan Plateau during the 21st century. *Journal of Geophysical Research: Atmospheres, 117*(D5), n/a-n/a.

Guo, W., et al. (2013). The 2008/09 surge of central Yulinchuan glacier, northern Tibetan Plateau, as monitored by remote sensing. *Annals of Glaciology, 54*(63), 299–310.

Guo, W., et al. (2015). The second Chinese glacier inventory: Data, methods and results. *Journal of Glaciology, 61*(226), 357–372.

Gurung, D. R., et al. (2011). *Snow-cover mapping and monitoring in the HinduKush-Himalayas.* Kathmandu, Nepal: ICIMOD.

Gurung, H. (1999). *Mountains of Asia.* Kathmandu: ICIMOD.

Haeberli, W., et al. (2007). Integrated monitoring of mountain glaciers as key indicators of global climate change: The European Alps. *Annals of Glaciology, 46,* 150–160.

Haeberli, W., et al. (2010). Mountain permafrost: development and challenges of a young research field. *Journal of Glaciology, 56* (200), 1043–1058.

Haeberli, W., Schaub, Y., & Huggel, C. (2016). Increasing risks related to landslides from degrading permafrost into new lakes in de-glaciating mountain ranges. *Geomorphology.*

Haeberli, W., & Hoelzle, M. (1995). Application of inventory data for estimating characteristics of and regional climate-change effects on mountain glaciers: A pilot study with the European Alps. *Annals of Glaciology*, 206–212.

Hagg, W., et al. (2007). Modelling of hydrological response to climate change in glacierized Central Asian catchments. *Journal of Hydrology, 332*(1–2), 40–53.

Hagg, W. et al. (2006). Runoff modelling in glacierized Central Asian catchments for present-day and future climate. *Hydrology Research, 37*(2).

Hall, D. K. et al. (2006). Estimation of snow extent and snow properties. In *Encyclopedia of hydrological sciences*. Wiley.

Haritashya, U. K., et al. (2009). Space-based assessment of glacier fluctuations in the Wakhan Pamir, Afghanistan. *Climatic Change, 94*(1–2), 5–18.

Harper, J. T., & Humphrey, N. F. (2003). High altitude Himalayan climate inferred from glacial ice flux. *Geophysical Research Letters, 30*(14).

Hasson, S., et al. (2014). Early 21st century snow cover state over the western river basins of the Indus River system. *Hydrology and Earth System Sciences, 18*(10), 4077–4100.

Herreid, S., et al. (2015). Satellite observations show no net change in the percentage of supraglacial debris-covered area in northern Pakistan from 1977 to 2014. *Journal of Glaciology, 61*(227), 524–536.

Hewitt, K. (2011). Glacier change, concentration, and elevation effects in the Karakoram Himalaya, Upper Indus Basin. *Mountain Research and Development, 31*(3), 188–200.

Hewitt, K. (2014). *Glaciers of the Karakoram Himalaya.* Dordrecht, The Netherlands: Springer.

Hewitt, K., et al. (1988). Hydrological investigations at Biafo Glacier, Karakorum Range, Himalaya; an important source of water for the Indus River. *Annals of Glaciology, 13,* 103–108.

Hewitt, K. (2005). The Karakoram anomaly? Glacier expansion and the "elevation effect". *Karakoram Himalaya. Mountain Research and Development, 25*(4), 332–340.

Hewitt, K. (2007). Tributary glacier surges: An exceptional concentration at Panmah Glacier, Karakoram Himalaya. *Journal of Glaciology, 53*(181), 181–188.

Hewitt, K., & Liu, J. (2010). Ice-Dammed lakes and outburst floods, Karakoram Himalaya: Historical perspectives on emerging threats. *Physical Geography, 31*(6), 528–551.

Hinzman, L. D., et al. (2005). Evidence and implications of recent climate change in northern Alaska and other Arctic regions. *Climatic Change, 72*(3), 251–298.

Hock, R., & Jensen, H. (1999). Application of kriging interpolation for glacier mass balance computations. *Geografiska Annaler, 81A,* 611–619.

Holzer, N., et al. (2015). Four decades of glacier variations at Muztagh Ata (eastern Pamir): A multi-sensor study including Hexagon KH-9 and Pléiades data. *Cryosphere, 9*(6), 2071–2088.

Holzer, N., et al. (2016). Glacier variations in the Trans Alai massif and the Lake Karakul catchment (northeastern Pamir) measured from space. *Climate change, glacier response, and vegetation dynamics in the Himalaya* (pp. 139–153). Cham: Springer International Publishing.

Hou, S. G., et al. (2002). Recent change of the ice core accumulation rates on the Qinghai-Tibetan Plateau. *Chinese Science Bulletin, 47* (20), 1746–1749.

Huggel, C., et al. (2013). Detecting potential climate signals in large slope failures in cold mountain regions. In C. Margottini (Ed.), *Landslide science and practice* (Vol. 7, pp. 361–367). Berlin, Heidelberg: Springer.

Huss, M., & Farinotti, D. (2012). Distributed ice thickness and volume of all glaciers around the globe. *Journal of Geophysical Research (Earth Surface), 117,* 4010.

Huss, M., & Hock, R., 2015. A new model for global glacier change and sea-level rise. *Frontiers in Earth Science, 3*(54).

ICIMOD. (2011). *Glacial lakes and glacial lake outburst floods in Nepal.* Kathmandu: ICIMOD.

Immerzeel, W. W., et al. (2015). Reconciling high-altitude precipitation in the upper Indus basin with glacier mass balances and runoff. *Hydrology and Earth System Sciences, 19*(11), 4673–4687.

Immerzeel, W. W., Kraaijenbrink, P. D. A., Shea, J. M., Shrestha, A. B., Pellicciotti, F., Bierkens, M. F. P., et al. (2014). High-resolution monitoring of Himalayan glacier dynamics using unmanned aerial vehicles. *Remote Sensing of Environment, 150,* 93–103.

Immerzeel, W. W., Pellicciotti, F., & Bierkens, M. F. P. (2013). Rising river flows throughout the twenty-first century in two Himalayan glacierized watersheds. *Nature Geoscience, 6*(9), 742–745.

Immerzeel, W. W., et al. (2012). Hydrological response to climate change in a glacierized catchment in the Himalayas. *Climatic Change, 110*(3–4), 721–736.

Immerzeel, W. W., van Beek, L. P. H., & Bierkens, M. F. P. (2010). Climate change will affect the Asian water towers. *Science, 328,* 1382–1385.

Immerzeel, W. W., Droogers, P., & Bierkens, M. F. P. (2009). Large-scale monitoring of snow cover and runoff simulation in Himalayan river basins using remote sensing. *Remote Sensing of Environment, 113,* 40–49.

International Hydrological Decade. (1970). Combined heat, ice and water balances at selected glacier basins. *UNESCO Technical Papers in Hydrology, 5,* 20.

IPCC, 2013. *AR5—Summary for Policymakers.*

Ives, J. D., Shrestha, R. B., & Mool, P. (2010). *Formation of glacial lakes in the Hindu Kush-Himalayas and GLOF risk assessment.* Kathmandu: ICIMOD.

Iwata, S. (1976). Late Pleistocene and holocene region, moraines Khumbu in the Himal * Sagarmatha (Everest). *Journal of the Japanese Society of Snow and Ice, 38*(Special), 109–114.

Iwata, S., et al. (2003). Rock glaciers and the lower limit of mountain permafrost in the Bhutan Himalayas. *Zeitschrift für Geomorphologie N. F. Supplementalband, 130,* 129–143.

Jacob, T., et al. (2012). Recent contributions of glaciers and ice caps to sea level rise. *Nature, 482*(7386), 514–518.

Jacobi, H. W., et al. (2015). Black carbon in snow in the upper Himalayan Khumbu Valley, Nepal: Observations and modeling of the impact on snow albedo, melting, and radiative forcing. *Cryosphere, 9*(4), 1685–1699.

Jakob, M. (1992). Active rock glaciers and the lower limit of discontinuous alpine permafrost, Khumbu Himalaya, Nepal. *Permafrost and Periglacial Processes, 3,* 253–256.

Jeelani, G. et al. (2016). Estimation of snow and glacier melt contribution to Liddar stream in a mountainous catchment, western Himalaya: An isotopic approach. *Isotopes in Environmental and Health Studies*, 1–18.

Jiang, S., et al. (2013). Response of glacier variation to climate change in the Bukatage Ice Cap from 1973 to 2010. *Journal of Arid Land Resources & Environment, 27*(3), 47–52.

Jiang, Y., et al. (2008). Long-Term changes in ice phenology of the yellow river in the past decades. *Journal of Climate, 21*(18), 4879–4886.

Kääb, A., et al. (2015). Brief Communication: Contending estimates of 2003–2008 glacier mass balance over the Pamir-Karakoram-Himalaya. *The Cryosphere, 9,* 557–564.

Kääb, A., et al. (2012). Contrasting patterns of early twenty-first-century glacier mass change in the Himalayas. *Nature, 488* (7412), 495–498.

Kamp, U., Byrne, M., & Bolch, T. (2011). Glacier fluctuations between 1975 and 2008 in the Greater Himalaya Range of Zanskar, southern Ladakh. *Journal of Mountain Science, 8*(3), 374–389.

Kang, S., et al. (2002). Stable-isotopic composition of precipitation over the northern slope of the central Himalaya. *Journal of Glaciology, 48*(163), 519–526.

Kang, S., et al. (2010). Review of climate and cryospheric change in the Tibetan Plateau. *Environmental Research Letters, 5*(1), 15101.

Kang, S., et al. (2015). Dramatic loss of glacier accumulation area on the Tibetan Plateau revealed by ice core tritium and mercury records. *The Cryosphere, 9*(3), 1213–1222.

Kapnick, S. B., et al. (2014). Snowfall less sensitive to warming in Karakoram than in Himalayas due to a unique seasonal cycle. *Nature Geosci, 7*(11), 834–840.

Karim, A., & Veizer, J. (2002). Water balance of the Indus River Basin and moisture source in the Karakoram and western Himalayas: Implications from hydrogen and oxygen isotopes in river water. *Journal of Geophysical Research, 107*(D18), 4362.

Kaser, G., Fountain, A. G., & Jansson, P. (2003). *A manual for monitoring the mass balance of mountain glaciers*, Paris: IHP-VI Technical Documents in Hydrology, No. 59, UNESCO.

Kaser, G., Grosshauser, M., & Marzeion, B. (2010). Contribution potential of glaciers to water availability in different climate regimes. *Proceedings of the National Academy of Science, 107,* 20223–20227.

Kaspari, S., et al. (2008). Snow accumulation rate on Qomolangma (Mount Everest), Himalaya: Synchroneity with sites across the Tibetan Plateau on 50–100 year timescales. *Journal of Glaciology, 54*(185), 343–352.

Ke, C.-Q., Tao, A.-Q., & Jin, X. (2013). Variability in the ice phenology of Nam Co Lake in central Tibet from scanning multichannel microwave radiometer and special sensor microwave/imager: 1978 to 2013. *Journal of Applied Remote Sensing, 7*(1), 73477.

Kehrwald, N. M., et al. (2008). Mass loss on himalayan glacier endangers water resources. *Geophysical Research Letters, 35*(22), 1–6.

Khan, A., Naz, B. S., & Bowling, L. C. (2015). Separating snow, clean and debris covered ice in the Upper Indus Basin, Hindukush-Karakoram-Himalayas, using Landsat images between 1998 and 2002. *Journal of Hydrology, 521,* 46–64.

Khattak, M. S., Babel, M. S., & Sharif, M. (2011). Hydro-meteorological trends in the upper Indus River basin in Pakistan. *Climate Research, 46*(2), 103–119.

Khromova, T. E., et al. (2006). Changes in glacier extent in the eastern Pamir, Central Asia, determined from historical data and ASTER imagery. *Remote Sensing of Environment, 102*(1), 24–32.

Klimeš, L., & Doležal, J. (2010). An experimental assessment of the upper elevational limit of flowering plants in the western Himalayas. *Ecography, 33,* 590–596.

König, M., Winther, J.-G., & Isaksson, E. (2001). Measuring snow and glacier ice properties from satellite. *Reviews of Geophysics, 39*(1), 1–27.

Koskinen, J. T., Pulliainen, J. T., & Hallikainen, M. T. (1997). The use of ERS-1 SAR data in snow melt monitoring. *IEEE Transactions on Geoscience and Remote Sensing, 35*(3), 601–610.

Kotlyakov, V. M., Osipova, G. B., & Tsvetkov, D. G. (2008). Monitoring surging glaciers of the Pamirs, central Asia, from space. In *Annals of Glaciology*, 125–134.

Kraaijenbrink, P. D. A. et al. (2017). Impact of a global temperature rise of 1.5° Celsius on Asia's glaciers. *Nature, 549*(7671), 257–260.

Krautblatter, M., Funk, D., & Günzel, F. K. (2013). Why permafrost rocks become unstable: a rock-ice-mechanical model in time and space. *Earth Surface Processes and Landforms, 38*(8), 876–887.

Kronenberg, M. et al. (2016). Mass-balance reconstruction for Glacier No. 354, Tien Shan, from 2003 to 2014. *Annals of Glaciology, 57* (71), 92–102.

Kropáček, J., et al. (2013). Analysis of ice phenology of lakes on the Tibetan Plateau from MODIS data. *The Cryosphere, 7*(1), 287–301.

Kulkarni, A., et al. (2016). Projected climate change in the Hindu Kush—Himalayan region by using the high-resolution regional climate model PRECIS. *Mountain Research and Development, 33*(2), 142–151.

Lambrecht, A., et al. (2014). The evolution of Fedchenko glacier in the Pamir, Tajikistan, during the past eight decades. *Journal of Glaciology, 60*(220), 233–244.

Leclercq, P. W., & Oerlemans, J. (2012). Global and hemispheric temperature reconstruction from glacier length fluctuations. *Climate Dynamics, 38*(5–6), 1065–1079.

Leysinger Vieli, G. J.-M. C., & Gudmundsson, G. H. (2004). On estimating length fluctuations of glaciers caused by changes in climatic forcing. *Journal of Geophysical Research, 109*(F01007), F01007.

Li, C. et al. (2017). Spatiotemporal variation of snow cover over the Tibetan Plateau based on MODIS snow product, 2001–2014. *International Journal of Climatology*.

Li, H., et al. (2016). Water resources under climate change in Himalayan Basins. *Water Resources Management, 30*(2), 843–859.

Li, M., et al. (2009). Snow distribution over the Namco lake area of the Tibetan Plateau. *Hydrology and Earth System Sciences, 13*(11), 2023–2030.

Li, X., et al. (2008). Cryospheric change in China. *Global and Planetary Change, 62*(3–4), 210–218.

Li, Z., et al. (2011). Climate and glacier change in southwestern China during the past several decades. *Environmental Research Letters, 6* (4), 45404.

Lin, H., et al. (2017). A decreasing glacier mass balance gradient from the edge of the Upper Tarim Basin to the Karakoram during 2000–2014. *Scientific reports, 7*(1), 6712.

Linsbauer, A., et al. (2016). Modelling glacier-bed overdeepenings and possible future lakes for the glaciers in the Himalaya-Karakoram region. *Annals of Glaciology, 57*(71), 119–130.

Linsbauer, A., Paul, F., & Haeberli, W. (2012). Modeling glacier thickness distribution and bed topography over entire mountain ranges with GlabTop: Application of a fast and robust approach. *Journal of Geophysical Research (Earth Surface), 117,* 3007.

Liu, B. et al. (2013). Wet precipitation chemistry at a high-altitude site (3,326 m a.s.l.) in the southeastern Tibetan Plateau. *Environmental Science and Pollution Research, 20*(7), 5013–5027.

Liu, C. H., et al. (2000). Glacier resources and their distributive characteristics in China—A review on Chinese Glacier Inventory. *Journal of Glaciology and Geocryology, 22*(2), 106–112.

Liu, Q., et al. (2016). Recent glacier and glacial lake changes and their interactions in the Bugyai Kangri, southeast Tibet. *Annals of Glaciology, 57*(71), 61–69.

Liu, Q., et al. (2010). Recent shrinkage and hydrological response of Hailuogou glacier, a monsoon temperate glacier on the east slope of Mount Gongga, China. *Journal of Glaciology, 56*(196), 215–224.

Liu, S., et al. (2003). Glacier changes since the Little Ice Age Maximum in the western Qilian Mountains, northwest China. *Journal of Glaciology, 49*(164), 117–124.

Liu, S. et al. (2006). Glacier retreat as a result of climate warming and increased precipitation in the Tarim river basin, northwest China. *Annals of Glaciology*, 91–96).

Liu, S. Y., et al. (2002). Glacier variation from maximum of the little ice age in the western Qilian Mountains, Northwest China. *Journal of Glaciology and Geocryology, 24*(3), 227–233. (in Chinese).

Loibl, D. M., & Lehmkuhl, F. (2014). Glaciers and equilibrium line altitudes of the eastern Nyainqêntanglha Range, SE Tibet. *Journal of Maps*, (June), 1–14.

Lüthi, Z. L., et al. (2015). Atmospheric brown clouds reach the Tibetan Plateau by crossing the Himalayas. *Atmospheric Chemistry and Physics, 15*(11), 6007–6021.

Lutz, A. F., et al. (2014). Consistent increase in High Asia's runoff due to increasing glacier melt and precipitation. *Nature Climate Change, 4*(7), 587–592.

Lutz, A. F. et al. (2016). Climate change impacts on the upper indus hydrology: Sources, shifts and extremes. *PLoS One* (Vol. 11(11)).

Ma, L., et al. (2010). Sensitivity of the number of snow cover days to surface air temperature over the Qinghai-Tibetan Plateau. *Advances in Climate Change Research, 1*(2), 76–83.

Ma, L. (2008). In *The Temporal and Spatial Characteristics of Snow Depth over the Qinghai-Tibetan Plateau in the Recent 50 Years and its Relationship with Factors of Atmospheric Circulation (in Chinese)*. Chinese Academy of Meteorological Sciences/Graduate University of Chinese Academy of Sciences.

Ma, Y., et al. (2008). Roof of the World: Tibetan observation and research platform. *Bulletin of the American Meteorological Society, 89*(10), 1487–1492.

Maanya, U. S., et al. (2016). Identification of potential glacial lake sites and mapping maximum extent of existing glacier lakes in Drang Drung and Samudra Tapu glaciers, Indian Himalaya. *Current Science, 111*(3), 553–560.

Magnuson, J. J. et al. (2000). Historical trends in Lake and River Ice cover in the Northern Hemisphere. *Science, 289*(5485).

Malik, N., Bookhagen, B., & Mucha, P. J. (2016). Spatiotemporal patterns and trends of Indian monsoonal rainfall extremes. *Geophysical Research Letters, 43*(4), 1710–1717.

Marchenko, S. S., Gorbunov, A. P., & Romanovsky, V. E. (2007). Permafrost warming in the Tien Shan Mountains, Central Asia. *Global and Planetary Change, 56*(3), 311–327.

Mark, B. G. et al. (2015). Glaciers as water resources. *The High-Moutain Cryosphere*, 184–203.

Marzeion, B., et al. (2014). Attribution of global glacier mass loss to anthropogenic and natural causes. *Science, 345*(6199), 919–921.

Marzeion, B., Jarosch, A. H., & Hofer, M. (2012). Past and future sea-level change from the surface mass balance of glaciers. *The Cryosphere, 6*(6), 1295–1322.

Maskey, S., Uhlenbrook, S., & Ojha, S. (2011). An analysis of snow cover changes in the Himalayan region using MODIS snow products and in-situ temperature data. *Climatic Change, 108*(1), 391.

Maurer, J. M., Rupper, S. B., & Schaefer, J. M. (2016). Quantifying ice loss in the eastern Himalayas since 1974 using declassified spy satellite imagery. *The Cryosphere Discussions*, (April), 1–20.

Maussion, F., et al. (2014). Precipitation Seasonality and Variability over the Tibetan Plateau as Resolved by the High Asia Reanalysis. *Journal of Climate, 27*(5), 1910–1927.

Mayewski, P. A., & Jeschke, P. A. (1979). Himalayan and Trans-Himalayan Glacier Fluctuations Since AD 1812. *Arctic and Alpine Research, 11*(3), 267.

Mayewski, P. A., Lyons, W. B., & Ahmad, N. (1983). Chemical composition of a high altitude fresh snowfall in the Ladakh Himalayas. *Geophysical Research Letters, 10*(1), 105–108.

Ménégoz, M., et al. (2014). Snow cover sensitivity to black carbon deposition in the Himalayas: From atmospheric and ice core measurements to regional climate simulations. *Atmospheric Chemistry and Physics, 14*(8), 4237–4249.

Mergili, M., Müller, J. P., & Schneider, J. F. (2013). Spatio-temporal development of high-mountain lakes in the headwaters of the Amu Darya River (Central Asia). *Global and Planetary Change, 107,* 13–24.

Miles, E. S., et al. (2016). Refined energy-balance modelling of a supraglacial pond, Langtang Khola, Nepal. *Annals of Glaciology, 57*(71), 29–40.

Miller, J. D., Immerzeel, W. W., & Rees, G. (2012). Climate change impacts on glacier hydrology and River discharge in the Hindu Kush—Himalayas A Synthesis of the Scientific Basis. *Mountain Research and Development, 32*(4), 461–467.

Ming, J., et al. (2012). Darkening of the mid-Himalaya glaciers since 2000 and the potential causes. *Environmental Research Letters, 7* (1), 14021.

Minora, U., et al. (2016). Glacier area stability in the Central Karakoram National Park (Pakistan) in 2001–2010: The "Karakoram Anomaly" in the spotlight. *Progress in Physical Geography, 40* (5), 629–660.

Mountain Research Initiative EDW Working Group. (2015). Elevation-dependent warming in mountain regions of the world. *Nature Climate Change, 5*(5), 424–430.

Mukhopadhyay, B. 2012. Signature and hydrologic consequences of climate change within the upper–middle Brahmaputra Basin. *Hydrological Processes.*

Mukhopadhyay, B., & Khan, A. (2015). A reevaluation of the snowmelt and glacial melt in river flows within Upper Indus Basin and its significance in a changing climate. *Journal of Hydrology, 527,* 119–132.

Nagai, H., Fujita, K., Sakai, A., Nuimura, T., & Tadono, T. (2016). Comparison of multiple glacier inventories with a new inventory derived from high-resolution ALOS imagery in the Bhutan Himalaya. *The Cryosphere, 10*(1), 65–85.

Naito, N. et al. (2000). Numerical simulation of recent shrinkage of Khumbu Glacier, Nepal Himalayas. In *Debris-covered Glaciers: Proceedings of an International Workshop Held at the University of Washington in Seattle, Washington, USA, 13–15 September 2000.* pp. 245–254.

Nan, Z., Li, S., & Cheng, G. (2005). Prediction of permafrost distribution on the Qinghai-Tibet Plateau in the next 50 and 100 years. *Science in China Series D, 48*(6), 797.

Neckel, N., Braun, A., Kropáček, J., & Hochschild, V. (2013). Recent mass balance of the Purogangri Ice Cap, central Tibetan Plateau, by means of differential X-band SAR interferometry. *The Cryosphere, 7*(5), 1623–1633.

Neckel, N., et al. (2014). Glacier mass changes on the Tibetan Plateau 2003–2009 derived from ICESat laser altimetry measurements. *Environmental Research Letters, 9*(1), 14009.

Nepal, S., et al. (2013). Understanding the hydrological system dynamics of a glaciated alpine catchment in the Himalayan region using the J2000 hydrological model. *Hydrological Processes, 28*(3), 1329–1344.

Nie, Y., et al. (2010). Monitoring glacier change based on remote sensing in the Mt. Qomolangma National Nature Preserve, 1976–2006. *Acta Geographica Sinica, 65*(1), 13–28.

Niu, H. et al. (2016). Chemical compositions of snow from Mt. Yulong, southeastern Tibetan Plateau. *Journal of Earth System Science, 125* (2), 403–416.

Niu, L., et al. (2011). Effect of permafrost degradation on hydrological processes in typical basins with various permafrost coverage in Western China. *Science China Earth Sciences, 54*(4), 615–624.

Nuimura, T., et al. (2012). Elevation changes of glaciers revealed by multitemporal digital elevation models calibrated by GPS survey in the Khumbu region, Nepal Himalaya, 1992–2008. *Journal of Glaciology, 58*(210), 648–656.

Nuimura, T., et al. (2015). The GAMDAM glacier inventory: A quality-controlled inventory of Asian glaciers. *Cryosphere, 9*(3), 849–864.

O'Neel, S., et al. (2014). Assessing streamflow sensitivity to variations in glacier mass balance. *Climatic Change, 123*(2), 329–341.

O'Reilly, C. M. et al. (2015). Rapid and highly variable warming of lake surface waters around the globe. *Geophysical Research Letters, 42*(24), 10,773–10,781.

Oerlemans, J. (2005). Extracting a Climate Signal from 169 Glacier Records. *Science, 308,* 675–677.

Ohara, N., et al. (2014). Dynamic Equilibrium Glacier modeling under evolving climate condition. *World Environmental and Water Resources Congress, 2014,* 608–615.

Ojha, S., et al. (2016). Glacier area shrinkage in eastern Nepal Himalaya since 1992 using high-resolution inventories from aerial photographs and ALOS satellite images. *Journal of Glaciology, 62,* 1–13.

Ojha, S., et al. (2017). Topographic controls on the debris-cover extent of glaciers in the Eastern Himalayas: Regional analysis using a novel high-resolution glacier inventory. *Quaternary International, 455,* 82–92.

Owen, L. A., & England, J. (1998). Observations on rock glaciers in the Himalayas and Karakoram mountains of northern Pakistan and India. *Geomorphology, 26*(1–3), 199–213.

Osipova, G., & Khromova, T. (2010). Elektronnaja baza dannykh "Pulsiruyuzhchne ledniki Pamira" (= Electronic database of surging glaciers in the Pamirs). *Лёд и Снег (Ice and Snow), 4,* 15–24.

Palazzi, E., et al. (2013). Precipitation in the hindu-kush karakoram himalaya: Observations and future scenarios. *Journal of Geophysical Research Atmospheres, 118*(1), 85–100.

Pan, B. T., et al. (2012). Glacier changes from 1966-2009 in the Gongga Mountains, on the south-eastern margin of the Qinghai-Tibetan Plateau and their climatic forcing. *Cryosphere, 6* (5), 1087–1101.

Pande, K., et al. (2000). Stable isotope systematics of surface water bodies in the Himalayan and Trans-Himalayan (Kashmir) region. *Journal of Earth System Science, 109*(1), 109–115.

Pandey, P., Kulkarni, A. V., & Venkataraman, G. (2013). Remote sensing study of snowline altitude at the end of melting season, Chandra-Bhaga basin, Himachal Pradesh, 1980–2007. *Geocarto International, 28*(4), 311–322.

Paul, F., et al. (2013). On the accuracy of glacier outlines derived from remote-sensing data. *Annals of Glaciology, 54*(63), 171–182.

Paul, F., et al. (2004). Rapid disintegration of Alpine glaciers observed with satellite data. *Geophysical Research Letters, 31*(21), 12–15.

Paul, F. (2015). Revealing glacier flow and surge dynamics from animated satellite image sequences: Examples from the Karakoram. *Cryosphere, 9*(6), 2201–2214.

Paul, F. et al. (2002). The new remote-sensing derived Swiss glacier inventory. 1. Methods. *Annals of Glaciology, 34*(September 1985), 355–361.

Pellicciotti, F., et al. (2015). Mass-balance changes of the debris-covered glaciers in the Langtang Himal, Nepal, from 1974 to 1999. *Journal of Glaciology, 61*(226), 373–386.

Pfeffer, W. T., et al. (2014). The Randolph Glacier Inventory: A globally complete inventory of glaciers. *Journal of Glaciology, 60,* 537–552.

Pieczonka, T., & Bolch, T. (2015). Region-wide glacier mass budgets and area changes for the Central Tien Shan between ~ 975 and 1999 using Hexagon KH-9 imagery. *Global and Planetary Change, 128,* 1–13.

Pu, Z., & Xu, L. (2009). MODIS/Terra observed snow cover over the Tibet Plateau: Distribution, variation and possible connection with the East Asian Summer Monsoon (EASM). *Theoretical and Applied Climatology, 97*(3–4), 265–278.

Pu, Z., Xu, L., & Salomonson, V. V. (2007). MODIS/Terra observed seasonal variations of snow cover over the Tibetan Plateau. *Geophysical Research Letters, 34*(6), 1–6.

Putkonen, J. K. (2004). Continuous Snow and rain data at 500 to 4400 m Altitude near Annapurna, Nepal, 1999–2001. *Arctic, Antarctic, and Alpine Research, 36*(2), 244–248.

Qian, Y., et al. (2011). Sensitivity studies on the impacts of Tibetan Plateau snowpack pollution on the Asian hydrological cycle and monsoon climate. *Atmospheric Chemistry and Physics, 11*(5), 1929–1948.

Qin, D., Liu, S., & Li, P. (2006). Snow cover distribution, variability, and response to climate change in Western China. *Journal of Climate, 19*(9), 1820–1833.

Quincey, D. J., et al. (2007). Early recognition of glacial lake hazards in the Himalaya using remote sensing datasets. *Global and Planetary Change, 56*(1), 137–152.

Quincey, D. J., et al. (2011). Karakoram glacier surge dynamics. *Geophysical Research Letters, 38*(18), 2–7.

Racoviteanu, A. E., Armstrong, R., & Williams, M. W. (2013). Evaluation of an ice ablation model to estimate the contribution of melting glacier ice to annual discharge in the Nepal Himalaya. *Water Resources Research, 49*(9), 5117–5133.

Radić, V., et al. (2014). Regional and global projections of twenty-first century glacier mass changes in response to climate scenarios from global climate models. *Climate Dynamics, 42*(1–2), 37–58.

Radić, V., & Hock, R. (2014). Glaciers in the Earth's hydrological cycle: Assessments of Glacier Mass and Runoff changes on global and regional scales. *Surveys in Geophysics, 35*(3), 813–837.

Ragettli, S., et al. (2013). Sources of uncertainty in modeling the glaciohydrological response of a Karakoram watershed to climate change. *Water Resources Research, 49*(9), 6048–6066.

Ragettli, S., et al. (2015). Unraveling the hydrology of a Himalayan catchment through integration of high resolution in situ data and remote sensing with an advanced simulation model. *Advances in Water Resources, 78,* 94–111.

Rai, S. P., et al. (2016). Isotopic characteristics of cryospheric waters in parts of Western Himalayas, India. *Environmental Earth Sciences, 75*(7), 600.

Ramanathan, V., & Carmichael, G. (2008). Global and regional climate changes due to black carbon. *Nature Geoscience, 1,* 221–227.

Ran, Y., et al. (2012). Distribution of Permafrost in China: An overview of existing Permafrost maps. *Permafrost and Periglacial Processes, 23*(4), 322–333.

Rastogi, S.P. & Narayan, S., 1999. Permafrost in the Tso Kar Basin, Ladakh. In *Proceedings of the Symposium for Snow, Ice and Glaciers, March 1999, Geological Survey of India Special Publication* (pp. 315–319).

Reager, J. T., et al. (2016). A decade of sea level rise slowed by climate-driven hydrology. *Science, 351*(6274), 699–703.

Rees, H. G., & Collins, D. N. (2006). Regional differences in response of flow in glacier-fed Himalayan rivers to climatic warming. *Hydrological Processes, 20,* 2157–2169.

Reggiani, P., et al. (2017). A joint analysis of river runoff and meteorological forcing in the Karakoram, upper Indus Basin. *Hydrological Processes, 31*(2), 409–430.

Regmi, D. (2008). Rock Glacier distribution and the lower limit of discontinuous mountain permafrost in the Nepal Himalaya. In *Proceedings of the Ninth International Conference on Permafrost (NICOP), June 29–July 3, 2008, Alaska Fairbanks* (pp. 1475–1480).

Richardson, S. D., & Reynolds, J. M. (2000). An overview of glacial hazards in the Himalayas. *Quaternary International, 65,* 31–47.

Rikiishi, K., & Nakasato, H. (2006). Height dependence of the tendency for reduction in seasonal snow cover in the Himalaya and the Tibetan Plateau region, 1966–2001. *Annals of Glaciology, 43* (1), 369–377.

Robinson, D. A., & Dewey, K. F. (1990). Recent secular variations in the extent of Northern Hemisphere snow cover. *Geophysical Research Letters, 17*(10), 1557–1560.

Rohrer, M., et al. (2013). Missing (in-situ) snow cover data hampers climate change and runoff studies in the Greater Himalayas. *Science of the Total Environment, 468–469,* S60–S70.

Romanovsky, V. E., Smith, S. L., & Christiansen, H. H. (2010). Permafrost thermal state in the polar Northern Hemisphere during the international polar year 2007–2009: a synthesis. *Permafrost and Periglacial Processes, 21*(2), 106–116.

Rounce, D. R., et al. (2016). A new remote hazard and risk assessment framework for glacial lakes in the Nepal Himalaya. *Hydrology and Earth System Sciences, 20*(9), 3455–3475.

Rowan, A. V., et al. (2015). Modelling the feedbacks between mass balance, ice flow and debris transport to predict the response to climate change of debris-covered glaciers in the Himalaya. *Earth and Planetary Science Letters, 430,* 427–438.

Rupper, S., et al. (2012). Sensitivity and response of Bhutanese glaciers to atmospheric warming. *Geophysical Research Letters, 39*(19), 1–6.

Sakai, A., et al. (2000a). Role of supraglacial ponds in the ablation process of a debris-covered glacier in the Nepal Himalayas. *International Association of Hydrological Sciences, 264,* 119–130.

Sakai, A., Chikita, K., & Yamada, T. (2000b). Expansion of a moraine-dammed glacial lake, Tsho Rolpa, in Rolwaling Himal, Nepal Himalaya. *Limnology and Oceanography, 45*(6), 1401–1408.

Sakai, A., Nakawo, M., & Fujita, K. (2002). Distribution characteristics and energy balance of ice cliffs on debris-covered glaciers, Nepal Himalaya. *Arctic, Antarctic, and Alpine Research, 34*(1), 12.

Sakai, A., et al. (2015). Climate regime of Asian glaciers revealed by GAMDAM glacier inventory. *The Cryosphere, 9*(3), 865–880.

Sakai, A., & Fujita, K. (2017). Contrasting glacier responses to recent climate change in high-mountain Asia. *Scientific Reports, 7*(1), 13717.

Salzmann, N. et al. (2014). Data and knowledge gaps in glacier, snow and related runoff research—A climate change adaptation perspective. *Journal of Hydrology, 518*(PB), 225–234.

Sarikaya, M. A., et al. (2012). Space-based observations of Eastern Hindu Kush glaciers between 1976 and 2007, Afghanistan and Pakistan. *Remote Sensing Letters, 3*(1), 77–84.

Savoskul, O.S., & Smakhtin, V. (2013). *Glacier systems and seasonal snow cover in six major Asian river basins: Hydrological role under changing climate*, Colombo, Sri Lanka: International Water Management Institute (IWMI).

Schaner, N., et al. (2012). The contribution of glacier melt to streamflow. *Environmental Research Letters, 7*(3), 34029.

Scherler, D., Bookhagen, B., & Strecker, M. R. (2011). Spatially variable response of Himalayan glaciers to climate change affected by debris cover. *Nature Geoscience, 4*(3), 156–159.

Schmid, M. O., et al. (2015). Assessment of permafrost distribution maps in the Hindu Kush Himalayan region using rock glaciers mapped in Google Earth. *Cryosphere, 9*(6), 2089–2099.

Schmidt, S., & Nüsser, M. (2012). Changes of High Altitude Glaciers from 1969 to 2010 in the Trans-Himalayan Kang Yatze Massif, Ladakh, Northwest India. *Arctic, Antarctic, and Alpine Research, 44*(1), 107–121.

Schmidt, S., & Nüsser, M. (2009). Fluctuations of Raikot Glacier during the past 70 years: a case study from the Nanga Parbat massif, northern Pakistan. *Journal of Glaciology, 55,* 949–959.

Serreze, M. C., & Barry, R. G. (2011). Processes and impacts of Arctic amplification: A research synthesis. *Global and Planetary Change, 77*(1–2), 85–96.

Sevestre, H., & Benn, D. I. (2015). Climatic and geometric controls on the global distribution of surge-type glaciers: implications for a unifying model of surging. *Journal of Glaciology, 61*(228), 646–662.

Shangguan, D., et al. (2016). Characterizing the May 2015 Karayaylak Glacier surge in the eastern Pamir Plateau using remote sensing. *Journal of Glaciology, 62*(235), 944–953.

Shangguan, D., et al. (2014). Glacier changes in the Koshi River basin, central Himalaya, from 1976 to 2009, derived from remote-sensing imagery. *Annals of Glaciology, 55*(66), 61–68.

Shangguan, D., et al. (2007). Glacier changes in the west Kunlun Shan from 1970 to 2001 derived from Landsat TM/ETM + and Chinese glacier inventory data. *Annals of Glaciology, 46,* 204–208.

Sharif, M., et al. (2013). Trends in timing and magnitude of flow in the Upper Indus Basin. *Hydrology and Earth System Sciences, 17*(4), 1503–1516.

Shchetinnikov, A. S. (1993). Izmenenie pazmerov olodenia Pamiro-Alaya za 1957–1980 gody (Investigations on glaciation in Pamir-Alay between 1957 and 1980). *Data of Glaciological Studies, 76,* 77–83.

Shea, J. M., et al. (2015a). A comparative high-altitude meteorological analysis from three catchments in the Nepalese Himalaya. *International Journal of Water Resources Development, 31*(2), 174–200.

Shea, J. M., et al. (2015b). Modelling glacier change in the Everest region, Nepal Himalaya. *Cryosphere, 9*(3), 1105–1128.

Shea, J. M., & Immerzeel, W. W. (2016). An assessment of basin-scale glaciological and hydrological sensitivities in the Hindu Kush—Himalaya. *Annals of Glaciology, 57*(71), 308–318.

Shen, S. S. P., et al. (2015). Characteristics of the Tibetan Plateau snow cover variations based on daily data during 1997–2011. *Theoretical and Applied Climatology, 120*(3), 445–453.

Sherpa, S. F., et al. (2017). Contrasted surface mass balances of debris-free glaciers observed between the southern and the inner parts of the Everest region (2007–15). *Journal of Glaciology, 63* (240), 637–651.

Shi, Y., et al. (2008). *Concise Glacier Inventory of China*. Shanghai, China.: Shanghai Popular Science Press.

Shiraiwa, T. (1992). Freeze-thaw activities and rock breakdown in the Langtang Valley, Nepal Himalaya. *Environmental science, Hokkaido University: Journal of the Graduate School of Environmental Science, 15*(1), 1–12.

Shrestha, A. B., et al. (1999). Maximum temperature trends in the Himalaya and its vicinity: An analysis based on temperature records from Nepal for the period 1971–94. *Journal of Climate, 12*(9), 2775–2786.

Shrestha, A. B., et al. (2000). Precipitation fluctuations in the Himalaya and its vicinity: An analysis based on temperature records from Nepal. *International Journal of Climate, 20,* 317–327.

Shrestha, A. B., & Aryal, R. (2011). Climate change in Nepal and its impact on Himalayan glaciers. *Regional Environmental Change, 11* (1), 65–77.

Shrestha, D., Singh, P., & Nakamura, K. (2012). Spatiotemporal variation of rainfall over the central Himalayan region revealed by TRMM precipitation radar. *Journal of Geophysical Research Atmospheres, 117*(22), 1–14.

Shroder, J. (2011). In V. Singh, P. Singh & U. Haritashya (Eds.), *Encyclopedia of snow, ice and glaciers*. Springer Science & Business Media.

Si, Y. et al. (2015). Impacts of climate variability on river ice phenology of Ningxia-Inner Mongolia reach in the upper Yellow River. In *E-proceedings of the 36th IAHR World Congress 28 June–3 July, 2015, The Hague, The Netherlands*.

Singh, P., & Bengtsson, L. (2005). Impact of warmer climate on melt and evaporation for the rainfed, snowfed and glacierfed basins in the Himalayan region. *Journal of Hydrology, 300*(1), 140–154.

Singh, P., & Jain, S. K. (2002). Snow and glacier melt in the Satluj River at Bhakra Dam in the western Himalayan region. *Hydrological Sciences Journal, 47*(1), 93–106.

Singh, P., & Kumar, N. (1997). Impact assessment of climate change on the hydrological response of a snow and glacier melt runoff dominated Himalayan river. *Journal of Hydrology, 193*(1–4), 316–350.

Singh, S. K., et al. (2014). Snow cover variability in the Himalayan-Tibetan region. *International Journal of Climatology, 34*(2), 446–452.

Smith, T., & Bookhagen, B. (2016). Assessing uncertainty and sensor biases in passive microwave data across High Mountain Asia. *Remote Sensing of Environment, 181,* 174–185.

Soncini, A. et al. (2016). Future hydrological regimes and glacier cover in the Everest region: The case study of the upper Dudh Koshi basin. *Science of The Total Environment.*

Soncini, A. et al. (2015). *Future hydrological regimes in the Upper Indus Basin: A case study from a high-altitude Glacierized Catchment.* http://dx.doi.org/10.1175/JHM-D-14-0043.1.

Spiess, M., Schneider, C., & Maussion, F. (2016). MODIS-derived interannual variability of the equilibrium-line altitude across the Tibetan Plateau. *Annals of Glaciology, 57*(71), 140–154.

Stewart, I. T. (2009). Changes in snowpack and snowmelt runoff for key mountain regions. *Hydrological Processes, 23*(1), 78–94.

Su, Z., & Shi, Y. (2002). Response of monsoonal temperate glaciers to global warming since the Little Ice Age. *Quaternary International, 97–98,* 123–131.

Sun, M., Liu, S., & Yao, X. (2015). Glacier changes in the Qilian Mountains in the past half century: Based on the revised first and second Chinese glacier inventory. *Acta Geographica Sinica, 70*(9), 1402–1414.

Tahir, A. A. et al. (2016). Comparative assessment of spatiotemporal snow cover changes and hydrological behavior of the Gilgit, Astore and Hunza River basins (Hindukush–Karakoram–Himalaya region, Pakistan). *Meteorology and Atmospheric Physics*, 1–19.

Tahir, A. A., et al. (2011). Modeling snowmelt-runoff under climate scenarios in the Hunza River basin, Karakoram Range, Northern Pakistan. *Journal of Hydrology, 409*(1–2), 104–117.

Tedesco, M., & Miller, J. (2007). Observations and statistical analysis of combined active–passive microwave space-borne data and snow depth at large spatial scales. *Remote Sensing of Environment, 111* (2–3), 382–397.

Terzago, S. et al. (2014). Snowpack changes in the Hindu-Kush Karakoram Himalaya from CMIP5 Global Climate Models. *Journal of Hydrometeorology.*

Thakuri, S. et al. (2014). Tracing glacier changes since the 1960s on the south slope of Mt. Everest (central Southern Himalaya) using optical satellite imagery. *The Cryosphere*, *8*(4), 1297–1315.

Thayyen, R. J., & Gergan, J. T. (2010). Role of glaciers in watershed hydrology: a preliminary study of a Himalayan catchment. *The Cryosphere, 4*(1), 115–128.

Thayyen, R. J., Rai, S. P., & Goel, M. K. (2015). *Glaciological studies of Phuche glacier, Ladakh Range*. Roorkee: India.

Thompson, L. G., et al. (2000). A high-resolution millennial record of the South asian monsoon from himalayan ice cores. *Science, 289* (5486), 1916–1920.

Thompson, S., et al. (2016). Stagnation and mass loss on a Himalayan debris-covered glacier: Processes, patterns and rates. *Journal of Glaciology, 62*(233), 467–485.

Tian, H., Yang, T., & Liu, Q. (2014). Climate change and glacier area shrinkage in the Qilian mountains, China, from 1956 to 2010. *Annals of Glaciology, 55*(66), 187–197.

Tiwari, S., Kar, S. C., & Bhatla, R. (2016). Interannual variability of snow water equivalent (SWE) over Western Himalayas. *Pure and Applied Geophysics, 173*(4), 1317–1335.

Torre Jorgenson, M., et al. (2013). Reorganization of vegetation, hydrology and soil carbon after permafrost degradation across heterogeneous boreal landscapes. *Environmental Research Letters, 8*(3), 35017.

Turner, A. G., & Annamalai, H. (2012). Climate change and the South Asian summer monsoon. *Nature Climate Change, 2*(8), 587–595.

Unger-Shayesteh, K. et al. (2013). What do we know about past changes in the water cycle of Central Asian headwaters? A review. *Global and Planetary Change, 110*, Part, 4–25.

Vincent, C., et al. (2013). Balanced conditions or slight mass gain of glaciers in the Lahaul and Spiti region (northern India, Himalaya) during the nineties preceded recent mass loss. *The Cryosphere, 7*(2), 569–582.

Vincent, C., et al. (2016). Reduced melt on debris-covered glaciers: Investigations from Changri Nup Glacier, Nepal. *Cryosphere, 10* (4), 1845–1858.

Viste, E., Sorteberg, A., & Renssen, H. (2015). Snowfall in the Himalayas: an uncertain future from a little-known past. *The Cryosphere, 9*(3), 1147–1167.

Vuyovich, C., & Jacobs, J. M. (2011). Snowpack and runoff generation using AMSR-E passive microwave observations in the Upper Helmand Watershed, Afghanistan. *Remote Sensing of Environment, 115*(12), 3313–3321.

Wagnon, P., et al. (2013). Seasonal and annual mass balances of Mera and Pokalde glaciers (Nepal Himalaya) since 2007. *The Cryosphere, 7*(6), 1769–1786.

Wang, S., et al. (2000). Permafrost degradation on the Qinghai–Tibet Plateau and its environmental impacts. *Permafrost and Periglacial Processes, 11*(1), 43–53.

Wang, F. et al. (2015). *China, the millennium development goals, and the post-2015 development agenda*, Beijing.

Wang, K., et al. (2015a). Research for glaciers and climate change of Anyemaqen Mountain nearly 30 years. *Research of Soil & Water Conservation, 22*(3), 300–303308. (in Chinese).

Wang, W. et al. (2015b). Integrated hazard assessment of Cirenmaco glacial lake in Zhangzangbo valley, Central Himalayas. *Geomorphology*.

Wang, W., et al. (2015c). Spatio-temporal change of snow cover and its response to climate over the Tibetan Plateau based on an improved daily cloud-free snow cover product. *Remote Sensing, 7*(1), 169.

Wang, N., & Ding, L. (2002). Study on the glacier variation in Bujiagangri section of the east Tanggula range since the Little Ice Age. *Journal of Glaciology and Geogryology, 24*(3), 234–244. (in Chinese).

Wang, T., et al. (2013a). Declining snow cover may affect spring phenological trend on the Tibetan Plateau. *Proceedings of the National Academy of Sciences, 110*(31), E2854–E2855.

Wang, X., et al. (2013b). Glacier and glacial lake changes and their relationship in the context of climate change, Central Tibetan Plateau 1972–2010. *Global and Planetary Change, 111*, 246–257.

Wang, X. et al. (2012). Glacier temporal-spatial change characteristics in western Nyainqentanglha Range, Tibetan Plateau 1977–2010. *Earth Science/Diqiu Kexue, 37*(5).

Wang, Z. (1991). The glacier and environment in the middle sector of Tianshan and the eastern sector of Qilianshan since the Little Ice Age. *Acta Geographica Sinica, 46*, 160–168.

Wei, J., et al. (2014). Surface-area changes of glaciers in the Tibetan Plateau interior area since the 1970s using recent Landsat images and historical maps. *Annals of Glaciology, 55*(66), 213–222.

Wei, Z., & Dong, W. (2015). Assessment of simulations of snow depth in the Qinghai-Tibetan Plateau using CMIP5 multi-models. *Arctic, Antarctic, and Alpine Research, 47*(4), 525–611.

Wendt, A., Mayer, C., Lambrecht, A., & Floricioiu, D. (2017). A glacier surge of Bivachny Glacier. *Pamir Mountains, observed by a time series of high-resolution digital elevation models and glacier velocities, Remote Sensing, 9*(4), 2072–4292.

Westermann, S. et al. (2014). Remote sensing of permafrost and frozen ground. In *Remote sensing of the cryosphere* (pp. 307–344). Chichester, UK: Wiley.

Williams, M. W., et al. (2016). Using geochemical and isotopic chemistry to evaluate glacier melt contributions to the Chamkar Chhu (river), Bhutan. *Annals of Glaciology, 57*(71), 339–348.

Wilson, A. M., et al. (2015). Use of a hydrologic mixing model to examine the roles of meltwater, precipitation and groundwater in the Langtang River basin, Nepal. *Annals of Glaciology, 57*(71), 155–168.

Wiltshire, A. J. (2014). Climate change implications for the glaciers of the Hindu Kush, Karakoram and Himalayan region. *Cryosphere, 8* (3), 941–958.

Winiger, M., Gumpert, M., & Yamout, H. (2005). Karakorum-Hindukush-western Himalaya: Assessing high-altitude water resources. *Hydrological Processes, 19*(12), 2329–2338.

Wu, G., & Zhang, Y. (1998). Tibetan Plateau Forcing and the timing of the Monsoon onset over South Asia and the South China Sea. *Monthly Weather Review, 126*(4), 913–927.

Wu, H., et al. (2014). Regional glacier mass loss estimated by ICESat-GLAS data and SRTM digital elevation model in the West Kunlun Mountains, Tibetan Plateau, 2003–2009. *Journal of Applied Remote Sensing, 8*(1), 83515.

Wu, Q., & Zhang, T. (2008). Recent permafrost warming on the Qinghai-Tibetan Plateau. *Journal of Geophysical Research, 113* (D13), D13108.

Wu, Q., Zhang, T., & Liu, Y. (2010). Permafrost temperatures and thickness on the Qinghai-Tibet Plateau. *Global and Planetary Change, 72*(1), 32–38.

Wu, Q., Zhang, T., & Liu, Y. (2012). Thermal state of the active layer and permafrost along the Qinghai-Xizang (Tibet) Railway from 2006 to 2010. *The Cryosphere, 6*(3), 607–612.

Wulf, H., Bookhagen, B., & Scherler, D. (2016). Differentiating between rain, snow, and glacier contributions to river discharge in the western Himalaya using remote-sensing data and distributed hydrological modeling. *Advances in Water Resources, 88*, 152–169.

Xiang, Y., Gao, Y., & Yao, T. (2014). Glacier change in the Poiqu River basin inferred from Landsat data from 1975 to 2010. *Quaternary International, 349*, 392–401.

Xiao, C., et al. (2007). Observed changes of cryosphere in China over the second half of the 20th century: an overview. *Annals of Glaciology, 46*, 382–390.

Xu, J. et al., 2013. Recent changes in glacial area and volume on Tuanjiefeng peak region of Qilian Mountains. In Q. China & Sun (Eds.), *PLoS One*, (Vol. 8(8), p. e70574).

Xu, J., et al. (2009). The melting Himalayas: Cascading effects of climate change on water, biodiversity, and livelihoods. *Conservation Biology, 23*(3), 520–530.

Xu, Y., Ramanathan, V., & Washington, W. M. (2016). Observed high-altitude warming and snow cover retreat over Tibet and the Himalayas enhanced by black carbon aerosols. *Atmospheric Chemistry and Physics, 16*(3), 1303–1315.

Yao, T., et al. (2012). Different glacier status with atmospheric circulations in Tibetan Plateau and surroundings. *Nature Climate Change, 2*(9), 663–667.

Yao, T., et al. (2007). Recent Glacial Retreat and its impact on Hydrological processes on the Tibetan Plateau, China, and surrounding regions. *Arctic, Antarctic, and Alpine Research, 39* (4), 642–650.

Yasuda, T., & Furuya, M. (2015). Dynamics of surge-type glaciers in West Kunlun Shan, Northwestern Tibet. *Journal of Geophysical Research: Earth Surface, 120*(11), 2393–2405.

Yasuda, T., & Furuya, M. (2013). Short-term glacier velocity changes at West Kunlun Shan, Northwest Tibet, detected by Synthetic Aperture Radar data. *Remote Sensing of Environment, 128,* 87–106.

Ye, B. et al. (2005). The Urumqi River source Glacier No. 1, Tianshan, China: Changes over the past 45 years. *Geophysical Research Letters, 32,* 21504.

Ye, Q., Zong, J., Tian, L., Cogley, J. G., Song, C., & Guo, W. (2017). Glacier changes on the Tibetan Plateau derived from Landsat imagery: Mid-1970s–2000-13. *Journal of Glaciology, 63*(238), 273–287.

Yi, S., & Sun, W. (2014). Evaluation of glacier changes in high-mountain Asia based on 10 year GRACE RL05 models. *Journal of Geophysical Research: Solid Earth, 119*(3), 2504–2517.

Yin, Z.-Y., et al. (2008). An assessment of the biases of satellite Rainfall Estimates over the Tibetan Plateau and Correction methods based on Topographic analysis. *Journal of Hydrometeorology, 9*(3), 301–326.

Zemp, M. et al. (2015). *Global glacier change bulletin no. 1 (2012–2013),* Zurich, Switzerland.

Zhang, G., Yao, T., et al. (2015a). An inventory of glacial lakes in the Third Pole region and their changes in response to global warming. *Global and Planetary Change, 131,* 148–157.

Zhang, Y., et al. (2015b). Glacier runoff and its impact in a highly glacierized catchment in the southeastern Tibetan Plateau: Past and future trends. *Journal of Glaciology, 61*(228), 713–730.

Zhang, G., et al. (2013). Increased mass over the Tibetan Plateau: From lakes or glaciers? *Geophysical Research Letters, 40*(10), 2125–2130.

Zhang, G. et al. (2012). Snow cover dynamics of four lake basins over Tibetan Plateau using time series MODIS data (2001–2010). *Water Resources Research, 48*(10), p.n/a-n/a.

Zhang, T., et al. (2008a). Statistics and characteristics of permafrost and ground-ice distribution in the Northern Hemisphere. *Polar Geography, 31*(1–2), 47–68.

Zhang, Y., et al. (2008b). Glacier change and glacier runoff variation in the Tuotuo River basin, the source region of Yangtze River in western China. *Environmental Geology, 56*(1), 59–68.

Zhang, X. et al. (2001). Variation of precipitation $\delta^{18}O$ in Langtang Valley Himalayas. *Science in China Series D: Earth Sciences, 44* (9), 769–778.

Zhang, Y., Hirabayashi, Y., & Liu, S. (2012). Catchment-scale reconstruction of glacier mass balance using observations and global climate data: case study of the Hailuogou catchment, south-eastern Tibetan Plateau. *Journal of Hydrology, 444,* 146–160.

Zhang, Y., Liu, S., & Ding, Y. (2007). Glacier meltwater and runoff modelling, Keqicar Baqi glacier, southwestern Tien Shan, China. *Journal of Glaciology, 53*(180), 91–98.

Zhang, Z. et al. (2016). Mass Change of Glaciers in Muztag Ata—Kongur Tagh, Eastern Pamir, China from 1971/76 to 2013/14 as Derived from Remote Sensing Data. PLoS One (pp. 1–19).

Zhao, L., et al. (2010). Thermal state of permafrost and active layer in Central Asia during the international polar year. *Permafrost and Periglacial Processes, 21*(2), 198–207.

Zhao, L., Ding, R., & Moore, J. C. (2014). Glacier volume and area change by 2050 in high mountain Asia. *Global and Planetary Change, 122,* 197–207.

Zhou, G., et al. (2007). Mass balance of the Zhadang glacier in the central Tibetan Plateau. *Journal of Glaciology Geocryol, 29*(3), 360–365.

Zhou, H., Aizen, E., & Aizen, V. (2013). Deriving long term snow cover extent dataset from AVHRR and MODIS data: Central Asia case study. *Remote Sensing of Environment, 136,* 146–162.

Zhou, Y., Li, Z., & Li, J. (2017). Slight glacier mass loss in the Karakoram region during the 1970s to 2000 revealed by KH-9 images and SRTM DEM. *Journal of Glaciology, 63*(238), 331–342.

Water in the Hindu Kush Himalaya

8

Coordinating Lead Authors
Christopher A. Scott, University of Arizona, USA. e-mail: cascott@email.arizona.edu
Fan Zhang, Institute of Tibetan Plateau Research, Chinese Academy of Sciences, China.
e-mail: zhangfan@itpcas.ac.cn
Aditi Mukherji, International Centre for Integrated Mountain Development, Nepal.
e-mail: aditi.mukherji@icimod.org; aditidebroy@gmail.com (corresponding author)

Lead Authors
Walter Immerzeel, Utrecht University, The Netherlands. e-mail: w.w.immerzeel@uu.nl
Daanish Mustafa, King's College London, London, UK. e-mail: daanish.mustafa@kcl.ac.uk
Luna Bharati, International Water Management Institute, Nepal. e-mail: l.bharati@cgiar.org

Contributing Authors
Hongbo Zhang, Institute of Tibetan Plateau Research, Chinese Academy of Sciences, China.
e-mail: zhanghongbo@itpcas.ac.cn
Tamee Albrecht, University of Arizona, USA. e-mail: talbrecht@email.arizona.edu
Arthur Lutz, Utrecht University, The Netherlands. e-mail: a.lutz@futurewater.nl
Santosh Nepal, International Centre for Integrated Mountain Development, Nepal.
e-mail: santosh.nepal@icimod.org
Afreen Siddiqi, Massachusetts Institute of Technology, USA. e-mail: siddiqi@mit.edu
Harris Kuemmerle, Kings College, London, UK. e-mail: harris.kuemmerle@kcl.ac.uk
Manzoor Qadir, United Nations University - Institute for Water, Environment & Health, Canada.
e-mail: Manzoor.Qadir@unu.edu
Sanjeev Bhuchar, International Centre for Integrated Mountain Development, Nepal.
e-mail: sanjeev.bhuchar@icimod.org
Anjal Prakash, International Centre for Integrated Mountain Development, Nepal.
e-mail: anjal.prakash@icimod.org
Rajiv Sinha, Indian Institute of Technology, Kanpur, India. e-mail: rsinha@iitk.ac.in

Review Editors
Jayanta Bandyopadhyay, Observer Research Foundation, India. e-mail: jayanta@iimcal.ac.in
Cecilia Tortajada, National University of Singapore, Singapore. e-mail: cecilia.tortajada@nus.edu.sg

Corresponding Author
Aditi Mukherji, International Centre for Integrated Mountain Development, Nepal.
e-mail: aditi.mukherji@icimod.org; aditidebroy@gmail.com

P. Wester et al. (eds.), *The Hindu Kush Himalaya Assessment*,
https://doi.org/10.1007/978-3-319-92288-1_8

Contents

Chapter Overview

Key Findings

1. **The Hindu Kush Himalaya (HKH) mountains provide two billion people a vital regional lifeline via water** for food (especially irrigation), water for energy (hydropower), and water for ecosystem services (riparian habitats, environmental flows, and rich and diverse cultural values).
2. **Glacier and snow melt are important components of streamflow in the region**; their relative contribution increases with altitude and proximity to glacier and snow reserves. **Groundwater, from springs in the mid-hills of the HKH, is also an important contributor to river baseflow**, but the extent of groundwater contribution to river flow is not known due to limited scientific studies.
3. **Water governance in the HKH is characterised by hybrid formal-informal regimes with a prevalence of informal institutions at the local level and formal state institutions at national and regional levels.** Synergy and support between state and informal water-management institutions is often lacking. Gender inequity is prevalent in both formal and informal institutions and translates into inequity in access to water.

Policy Messages

1. To counter the formidable and immediate threats to water security posed by human drivers including climate change, **equitable, productive, and**

sustainable water use should be promoted through decentralised decision making, effective management of urban pollution, improved infrastructure planning, and enhanced regional cooperation.

2. Ensuring regional and local water security requires **proactive HKH-wide cooperation, specifically in open data sharing among scientists and ministry or agency personnel; conflict management via regional platforms; and investment of public- and private-sector funds for generating and exchanging knowledge, enhancing public awareness, and stimulating action.**
3. **Tradeoffs between upstream and downstream water uses; between rural and urban areas; and among irrigation, energy, industrial and other sectors must be carefully managed in order to enhance water security**, meet the Sustainable Development Goals, and ensure water availability for hydropower that will be essential for HKH countries to achieve (intended) Nationally Determined Contributions for emissions mitigation as established in the 2015 Paris climate accord. This requires **balancing evidence-based policy with political imperatives at the local, national, and HKH regional scales**, while ensuring that mountain communities derive commensurate benefits from HKH water resources in a manner that safeguards downstream water needs.

Commonly described as the "water tower for Asia," the Hindu Kush Himalaya (HKH) plays an important role in ensuring water, food, energy, and environmental security for much of the continent. The HKH is the source of ten major rivers that provide water—while also supporting food and energy production and a range of other ecosystem services—for two billion people across Asia. This chapter takes stock of current scientific knowledge on the availability of water resources in the HKH; the varied components of its water supply; the impact of climate change on future water availability; the components of water demand; and the policy, institutions, and governance challenges for water security in the region.

The monsoon provides the main source of water for the eastern Himalaya; much of this comes as rain between June and September. In winter, the western Himalaya receives at least half of its precipitation from western disturbances (***well-established***).

Knowledge of the amount and distribution of precipitation at higher altitudes (above 5000 m above mean sea level, masl) in the HKH is poor. There are very few meteorological stations at these altitudes and those that exist may not consistently provide data. The lack of reliable data has led to significant anomalies in observed rain and snow data and in observed glacier mass balances. More stations at higher altitudes are urgently needed (***well-established***).

While glacier and snow melt are important components of overall streamflow in the region, their significance varies widely—ranging from very high in western rivers, such as the Indus, to low in eastern rivers, such as the Ganges and the Brahmaputra. In the eastern rivers, rainfall runoff contributes the largest share of streamflow. Still, this share varies substantially within each river basin. The relative contribution of glacier and snow melt, as opposed to rainfall runoff, increases with altitude and proximity to glacier and snow reserves (***well-established***).

Groundwater, from springs in the mid-hills of the HKH, is an important contributor to river baseflow, but the exact extent of this contribution is not known due to limited scientific studies and evidence. The role and contribution of springs to overall water budgets in the region is poorly understood (***well-established***). We urgently need better scientific knowledge of groundwater in the HKH—especially because millions of mountain people depend directly on springs. More is known about groundwater endowments in the plains. Groundwater is overexploited in the western plains, while it remains largely untapped in the eastern plains (***well-established***).

Climate change is expected to drive consistent increases in the total streamflow of the Indus, Ganges, and Brahmaputra rivers. In the Indus, this increase will come for a limited period from increased glacial melt, while in the Ganges and the Brahmaputra, it is expected to come mainly from precipitation (***established but incomplete***). Beyond the mid-century, the Indus Basin may experience decreases in pre-monsoon flow resulting from decreasing glacial melt (***inconclusive***). Changes in future flow volumes will also have a seasonal dimension, with increased peak runoff and decreased low flow in some sub-basins (***established, but incomplete***). Pre-monsoon flows are expected to decline, with implications for irrigation, hydropower, and ecosystem services (***unresolved***).

Disaggregated water-use data are not available for the region defined as the HKH. However, across the entire territory of all eight countries sharing the HKH, about one-fifth of available renewable water resources are being used for human purposes. Countries vary widely in their water-resource endowments and withdrawals (***established, but incomplete***).

For all eight countries—Afghanistan, Bangladesh, Bhutan, China, India, Myanmar, Nepal, and Pakistan—that comprise the region, agriculture (HKH and geographical regions outside HKH) accounts for the largest share of water use—accounting for over 90% of use in Afghanistan and 65% in more industrialised China (***well-established***). India, Bangladesh, Pakistan, and China together account for more than 50% of world's groundwater withdrawals (***well-established***). These withdrawals mostly take place in the plains of river basins that originate in the HKH. Groundwater is used mostly for irrigation and in other sectors like urban water provisioning (***well-established***).

The HKH includes three sub-regions of agricultural water use each with distinct implications in terms of water management: high mountains, mid-hills, and plains. Agriculture in the high mountains and the mid-hills tends to be largely rainfed with supplemental irrigation in the mid-hills. Agriculture in the plains is mostly irrigated (***well-established***). The nature and dynamics of the region's agriculture are shifting in response to climate and demographic changes.

Another use of water—hydropower—is mostly non-consumptive. Yet hydropower can change the timing and location of river flow thereby disrupting natural flow regimes, which can harm other water users, such as local irrigation, capture fisheries and ecosystems (***established***). Such conflicts especially arise in the mid-hills and the mountains—which mark the location of most current and foreseeable hydropower sites. Very often, mountain people do not derive commensurate benefits from these projects (***well-established***). Appropriate benefit-sharing norms are needed to ensure that mountain people also benefit from the region's vast hydropower potential (***established, but incomplete***).

Burgeoning cities and small towns in the HKH confront severe water stress from urbanization, which is often unplanned (***well-established***). This water stress often leads to concerns over water quality, but it also gives rise to practices such as the reuse of partially treated wastewater for agriculture (***established, but incomplete***).

In response to the Millennium Development Goals, the HKH made remarkable strides in achieving access to safe drinking water. The region is also committed to meeting the Sustainable Development Goals (SDGs). Still, much work remains to be done to provide basic sanitation (***well-established***). Rather than managing water for health and sanitation in isolation from water for irrigation, hydropower, municipal supply, and ecosystems, it would be more effective to integrate water management for multiple uses.

The important role of HKH rivers in providing ecosystem services is not well appreciated. Present law and policy frameworks are not adequate to ensure that infrastructure development does not impinge on ecosystem services (***established, but incomplete***).

To ensure water security in the HKH, adequate water availability alone is not enough—what is needed is good water governance. Such governance must be politically and culturally tailored to the local, national, and regional contexts (***well-established***). Water governance in the HKH is characterised by hybrid formal-informal regimes with a prevalence of informal institutions at the local level and formal state institutions at national and regional levels; often with lack of synergy and support between state and informal water-management institutions. Gender inequity is prevalent in both formal and informal institutions. Urban water-supply challenges posed by formal institutional regimes often forced upon pre-existing informal institutions have deleterious consequences for water quality and quantity. Transboundary institutions for water resources are inadequate or non-existent, heightening the risk of conflict while also offering opportunities for HKH-wide cooperation (***well-established***).

Challenges and opportunities vary at different levels: micro (watershed and springshed); meso (river basin); and macro (regional). Among the leading causes of poor water governance in the HKH are constantly changing conditions in the ecologically fragile sloping landscape, dispersed settlements, unequal power dynamics, centralised decision making, inadequate opportunities for local communities to influence their water-security decisions, despite the presence of local institutions (***well-established***). Throughout the HKH, more attention needs to be paid to HKH-specific conditions as well as more general challenges including participatory and cooperative decision making (formal, informal, and hybrid), evidence-based policies, transparent program implementation, accountability at all levels, and transboundary and regional cooperation.

Water Security and the Sustainable Development Goals

SDG Goal 6 is entirely focused on water. While drinking water and sanitation rightly remain central to SDG 6, other considerations have gained importance as well: water quality, wastewater management and reuse, transboundary cooperation, ecosystem services, capacity building, and cooperation.

The SDG-consistent priorities and specific targets for the HKH region (*with our assessment comments in italics*) include the following:

Ensure a year-round secure water supply in the mountains with universal and affordable access to safe drinking water, sanitation, and water for productive purposes.

- Create secure water supply for key development sectors (agriculture, energy) that are viable year-round. *Meeting this target will require that socio-economic and environmental impacts be comprehensively assessed with adequate and timely compensation for mountain communities who are impacted.*
- Build effective and efficient mechanisms to implement and monitor transboundary cooperation

agreements. *Document and assess existing transboundary cooperation agreements in order to apply lessons and expand the scope of future agreements.*

- Achieve universal and equitable access to safe and affordable drinking water for all mountain people by 2030. *As a priority target for mountain communities, this will require that comprehensive programs for spring revival and improvement be taken up, and in urban areas, additional sources of secure and affordable water be made available.*
- Achieve access to adequate and equitable sanitation services and hygiene education for all in mountain regions. *Community-based models with attention to women and marginalized sections of the community must be taken up with support from local and national governments.*
- Reduce the water collecting time and work load of women and children (and of aging males, whenever relevant) by 2030. *Comprehensive programs for spring revival and improvement are urgently required to reduce the burden of water access by all members of mountain communities.*
- Support and strengthen the participation of mountain communities in water management. *Increase decision-making power of local governments and ensure the incorporation of local-knowledge systems and local institutions in water management.*

8.1 Introduction

Water security has emerged as a subset of human security—one that has been raising serious concern throughout the early part of the 21st century. For the purposes of the HIMAP assessment, we use a definition of water security adapted from UN-Water (2013) and Scott et al. (2013) as follows: *Water security is the capacity of HKH populations to safeguard sustainable access to adequate quantities of acceptable quality water for resilient societies and ecosystems, to ensure protection against water-borne pollution and water-related disasters, and to adapt to uncertain global change—in a regional climate of peace and political stability.*

This chapter focuses on current and future water endowments and their spatial distribution (Sect. 8.2), use (Sect. 8.3), and governance (Sect. 8.4). Water quality is recognized to be crucial for human health and ecosystem processes, but the relative lack of observed data and modeled dynamics makes it difficult to systematically address water quality. The chapter focuses primarily on the question of water quantity. Nonetheless, issues related to quality—such as, sediment in large river systems, challenges of wastewater management in HKH urban systems, and major biological and chemical contaminants linked to urbanization—are also discussed.

The major river basins originating in the region are shown in Fig. 8.1. Throughout the chapter, we use specific terms to refer to nested geographical scales: micro (local, springshed, community); meso (river basin, subnational to national); macro (HKH-regional, transboundary); and global (beyond HKH, global).

8.2 Water Availability in the Hindu Kush Himalaya

This section attempts to assess the principal sources of water in the HKH, including precipitation, glacial melt, snowmelt, runoff, river discharge, springs, and groundwater. As already noted, water quantity is the principal focus. Aggregate water availability together with water use data are included in Sect. 8.3. Temporal dynamics are specifically referred to in the section on climate change impacts.

8.2.1 Precipitation

In general, the climate in the eastern part of the Himalayas is characterized by the East Asian and Indian monsoon systems, causing the bulk of precipitation to occur from June to September. The precipitation intensity shows a strong north-south gradient caused by orographic effects (Galewsky 2009). Precipitation patterns in the Pamir, Hindu Kush, and Karakoram ranges in the west are also characterized by westerly and southwesterly flows, causing precipitation to be more evenly distributed throughout the year, as compared with the eastern parts (Bookhagen and Burbank 2010). In the Karakoram, as much as two-thirds of the annual high-altitude precipitation occurs during the winter months (Hewitt 2011). About half of this winter precipitation is brought about by western disturbances, which are eastward propagating cyclones that bring sudden winter precipitation to the northwestern parts of the Indian subcontinent (Barlow et al. 2005).

Meteorological stations are relatively sparse in the HKH (Shea et al. 2015b), in large part due to the poor accessibility of the terrain. Precipitation is especially variable over short horizontal distances due to orographic effects; however, high-altitude precipitation gauge networks are very rare. If there are rain gauges, they are mostly located in the river valleys where precipitation amounts are smaller than at

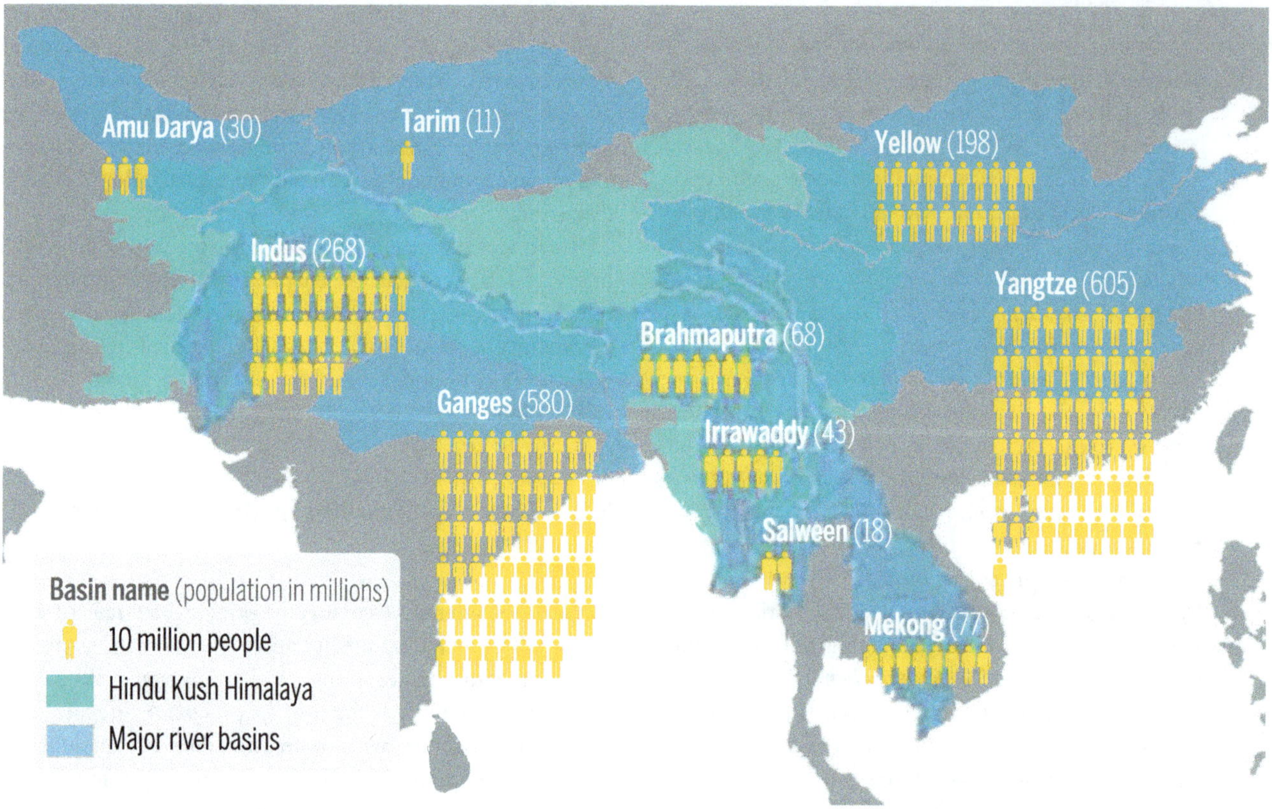

Fig. 8.1 Major river basins originating in the Hindu Kush Himalaya (*Data source* Box 1.1; Table 1.2)

higher altitudes. Furthermore, most gauges have difficulty accurately capturing snowfall. Direct snow-accumulation measurements—using snow pillows, pits, and cores from accumulation zones—are also scarce and usually only account for short periods of time. Therefore, HKH precipitation predictions based on ground observations are not very accurate. In order to obtain more accurate predictions, observed data must be replaced by or supplemented with data gathered through other approaches, including remote sensing and reanalysis techniques to aid in generating gridded climate datasets. Recent research for the upper Indus basin indicates that in order for precipitation data to correspond with observed glacier mass balances and river discharges, the actual amount of precipitation would have to be double the amount estimated from station data (Immerzeel et al. 2015; Dahri et al. 2016).

8.2.2 Cryospheric Contributions to River Flows

At the river-basin scale, in the eastern HKH basins, glaciers play a relatively small role in annual surface runoff. Nevertheless, recent work shows that within each basin there is significant variability (Fig. 8.2); the closer one gets to the glaciers and snow reserves within a basin, the greater the relative importance of glacier and snowmelt runoff (Lutz et al. 2014). Several large-scale benchmark studies have focused on quantifying the importance of glacier and snowmelt runoffs in the overall hydrology of large Asian river basins. Permafrost contributions are addressed in Chap. 5. Glaciers have the potential to provide seasonally delayed meltwater to the rivers. Meltwater can make the greatest contribution to river flow during warm and dry seasons, which is particularly important to the water budget in water-scarce lowlands that are densely populated.

A global study estimating seasonally delayed glacier runoff relative to precipitation input showed that the Indus basin had the greatest human dependence on glacier water within the HKH (Kaser et al. 2010). In another benchmark study, the Normalized Melt Index (NMI) was used to quantify the importance of both glacier and snow meltwater for five major river basins in Asia (the Indus, Ganges, Brahmaputra, Yangtze, and Yellow). The NMI is defined as the volumetric glacier and snowmelt in a basin divided by its downstream natural discharge. This study revealed very large differences among the basins, ranging from 46% snow and 32% glacier contributions in the Indus to 6% snow and 3% glacier contributions in the Ganges, which is largely dependent on the summer monsoon (Immerzeel et al. 2010).

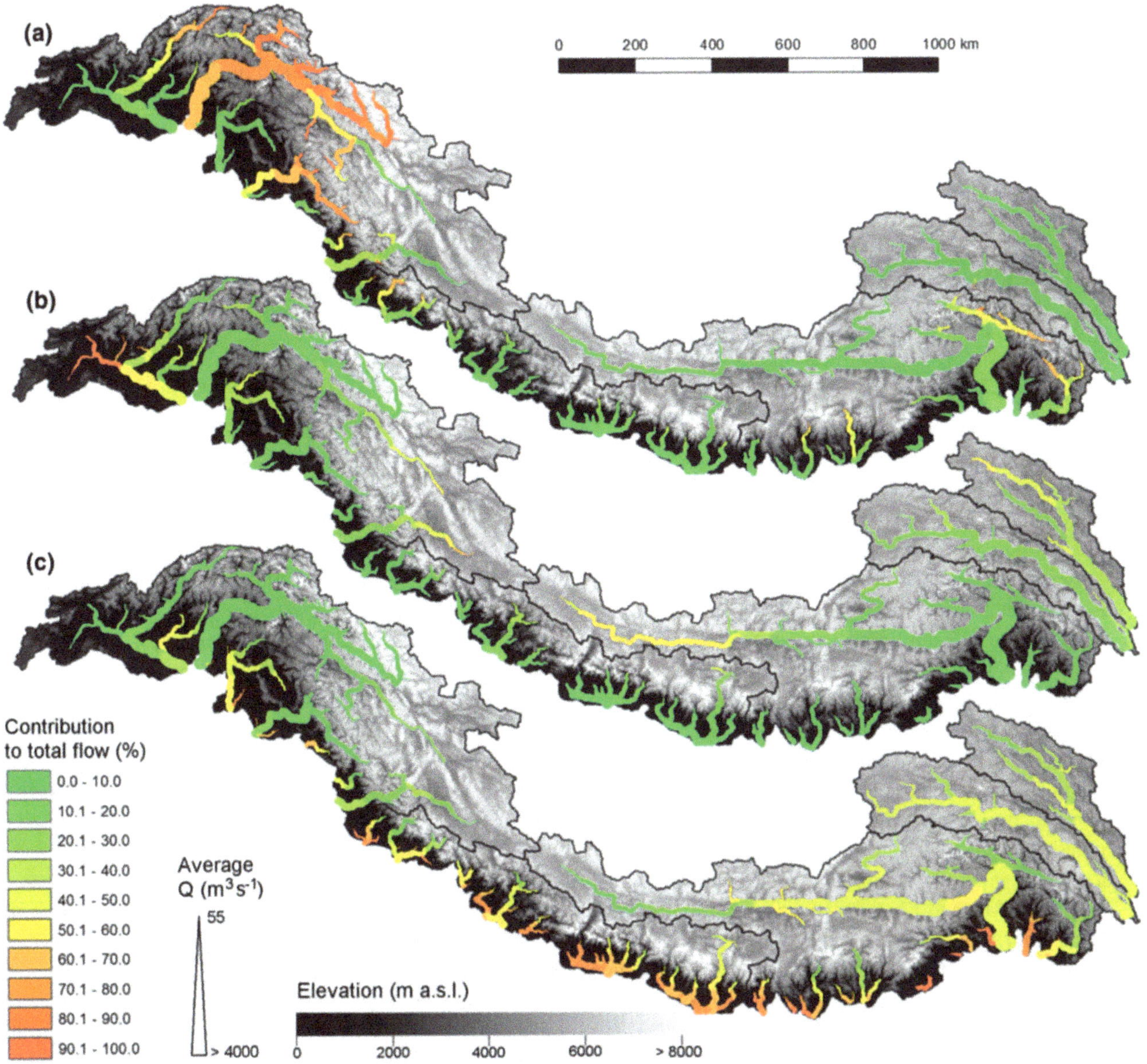

Fig. 8.2 Contribution to total flow by (**a**) glacial melt, (**b**) snowmelt, and (**c**) rainfall-runoff for major streams during the reference period of 1998–2007. Line thickness indicates the average discharge during the reference period (*Source* Reprinted from "Consistent increase in High Asia's runoff due to increasing glacier melt and precipitation" by Lutz, A. F. et al., 2014, Nature Climate Change, 4, p. 590. Copyright © 2014 by Nature Publishing Group)

Another study assessed the upper limit of glacier melt contribution to streamflow to estimate the potential melt contribution by combining energy balance computation—that calculated the amount of energy available for melt—with the Global Land Ice Measurements from Space (GLIMS) database (Schaner et al. 2012). The results once again highlighted the Indus and small basins close to glacier outlets as the most dependent on glacier melt. Yet another study used the Variable Infiltration Capacity (VIC) model to assess the Yellow, Yangtze, Mekong, Salween, Brahmaputra, and Indus rivers. Results showed that these rivers—except for the Indus—were primarily dependent on rainfall runoff. By contrast, the total glacier melt and snowmelt contribution to the Indus streamflow was about 80% (Zhang et al. 2013).

In the headwaters close to the glacier and snow source areas, smaller-scale studies based on either stable isotope analysis (Racoviteanu et al. 2013) or modeling (Immerzeel et al. 2013; Nepal et al. 2014b; Ragettli et al. 2015; Shrestha et al. 2015b; Tahir et al. 2015) showed the significant dependence of river flow on both glacier and snowmelt, even if the larger basins in which the headwaters are located have minimal dependence on meltwater.

8.2.3 Rivers

As indicated in Fig. 8.1, the HKH functions as the water tower (Immerzeel et al. 2010) for much of the southern and eastern Asian continent, serving as the source for ten major river systems. Variations are very pronounced in river discharge, the contributions of different sources, and temporal trends (both seasonal and inter-annual).

A number of studies analyzing observed records have attempted to attribute trends in discharge to meteorological trends:

- A study analyzing streamflow trends from nineteen stations in the upper Indus basin indicated that in highly glaciated catchments the discharge is best correlated to temperature (Archer 2003). According to the analysis, the volume of summer discharge in middle-altitude catchments is predominantly governed by the preceding winter's precipitation, whereas streamflow in catchments further downstream is mainly determined by winter and monsoon rainfall.
- Another study of the upper Indus basin (Khattak et al. 2011) found that increasing trends in streamflow could be related to increases in mean and maximum temperature, particularly in the winter and spring seasons.
- Sharif et al. (2013) concluded that highly glaciated catchments in the upper Indus basin showed decreasing trends in streamflow, whereas streamflow had increased in less glaciated catchments. The study showed flow decreasing in early summer but increasing in the winter.
- Mukhopadhyay and Khan (2014b) showed that runoff in the central Karakoram increased during the melting season from 1985 to 2010. They concluded that increased flow was possible under neutral glacier mass balance conditions as a result of increased temperature and precipitation. This allowed the rate of mass cycling of the glacier to increase even as the mass balance remained neutral.
- Bashir et al. (2017) found an overall decreasing trend in annual accumulated discharge in the Indus river at Tarbela Dam. According to their study, the annual increase in precipitation and decrease in runoff suggested an overall increase in storage of water in the elevated regions of the watershed, mainly in the form of snow and glaciated ice.

A limited number of published studies have estimated the composition of streamflow within different catchments or sub-basins of the Indus, Ganges, and Brahmaputra (Table 8.1). Results are, however, difficult to compare due to the variation in concepts, approaches, and application scales employed.

The findings of each study were largely dependent on the availability of data and application scale (for example, the size of the catchment or basin included in the simulation). The estimates by Immerzeel et al. (2011, 2013), for instance, were made using a distributed model, including a simple ice flow model, whereas the estimates by Soncini et al. (2015) were made using a semi-distributed cryospheric-hydrological model fed and validated with in situ measurements.

River discharge is an essential component of the regional water balance, with important contributions from glacial melt, snowmelt, and spring flow within the HKH. Additionally, river flows play a dominant role in sediment transport and groundwater recharge in the plains. Spatial and temporal trends in river flows are addressed in Sect. 8.2.7.

8.2.4 Sediment Transport

There is a strong relationship between upstream erosion, mass wasting, in-stream transport, and downstream sediment deposition in the HKH. Erosion in the region is strongly determined by young and fragile geological formations, land-management practices, and monsoon precipitation (Nepal et al. 2014a). High sediment flux in the rivers within this region may be largely attributed to the young unconsolidated mountain system with frequent geologic activity that makes the landmass fragile and unstable, in addition to which the intense precipitation of the Indian summer monsoon on the southern side of the main Himalayan ridgeline and the East Asian summer monsoon on the eastern edge of the Tibetan Plateau also contribute a great deal (Bandyopadhyay 2017). Sediment load in rivers can be used as an important proxy for evaluating ecological and environmental conditions as well as the severity of regional erosion (Zeng et al. 2018). The location of villages, type of agricultural land use (including terracing, road access, and other features of the structure), and function of human settlements in the HKH have important mutual effects on erosion, sediment transport, and deposition. An important issue is the interrelation between sediment flux and downstream hazards, such as rapid and frequent channel migration and extensive flooding due to channel instability triggered by channel bed aggradation.

About one-third of the global sediment deposited to the oceans is estimated to be generated from the Tibetan Plateau and its neighbouring regions (Milliman and Meade 1983). The high sedimentation of rivers within this region may be largely attributed to the young, unconsolidated mountain system with more frequent geological activities that make the landmass fragile and unstable as well as the precipitation patterns of the Indian monsoon and the East Asian summer monsoon that fall more intensely on the south side of the main Himalayan ridgeline and the eastern edge of the Tibetan Plateau respectively (Bandyopadhyay 2017). The

Table 8.1 Results of studies estimating streamflow composition at selected locations

Site (river/location)	Country, major river	Reference	Basin area (km^2)	Period	Contribution by component (%)			
					Glacier melt	Snowmelt	Rain runoff	Base-flow
Satluj, Bhakra Dam	India, Indus River	Singh and Jain (2002)	56,874	1986–1996	59		41	–
		Lutz et al. (2014)	–	1998–2007	27.6	20.8	38.6	13
Langtang Khola, Kyangjing	Nepal, Ganges River	Immerzeel et al. (2011)	360	2001–2010	47	6.9	28.8	17.4
		Immerzeel et al. (2013)	–	1961–1990	13	20.4	10	56.6
		Racoviteanu et al. (2013)	352.3	1988–2006	58.3	41.7		
		Ragettli et al. (2015)	350	2012–2013	26	40	34	–
		Lutz et al. (2014)	–	1998–2007	52.5	12.8	25	9.7
Dudh Koshi, Rabuwa Bazar	Nepal, Ganges River	Racoviteanu et al. (2013)	3,711.4	1988–2006	7.4	92.6		
		Lutz et al. (2014)	–	1998–2007	18.8	4.8	64.8	11.6
		Nepal et al. (2014b)	3,712	1985–1997	17	17	46	20
Lhasa Basin	China, Yarlung Tsangpo River	Prasch et al. (2013)	26,339	1971–2000	3	41	56	–
Indus, Besham Qila	Pakistan, Indus River	Mukhopadhyay and Khan (2014a)	164,867	1969–2010	21	49	30	
		Lutz et al. (2014)	–	1998–2007	67.3	17.6	7.1	8
		Mukhopadhyay and Khan (2015)	–	1969–2010	25.8	44.1	–	30.2
Hunza, Dainyor Bridge	Pakistan, Indus River	Mukhopadhyay and Khan (2014a)	13,732	1966–2010	74		26	
		Lutz et al. (2014)	–	1998–2007	80.6	9.6	1.3	8.5
		Mukhopadhyay and Khan (2015)	13,734	1966–2010	42.8	31.3	–	25.9
		Shrestha et al. (2015b)	13,733	2002–2004	33	50	17	–
Baltoro, Baltoro	Pakistan, Indus River	Immerzeel et al. (2013)	–	1961–1990	38.7	21.6	3.5	36.2
Shigar, Shigar	Pakistan, Indus River	Soncini et al. (2015)	~7,000	1985–1997	32.9	39.5	27.6	

"–" indicates not available

Ganges-Brahmaputra River is one of the most sediment-laden rivers in the world, with annual sediment loads of 1235–1670 million tons (Milliman and Meade 1983; Abbas and Subramanian 1984). Approximately, half of this is deposited within the lower basin while the other half is delivered to the ocean (Islam et al. 1999). The annual sediment load of the Ganges River accounts for nearly two-thirds of the total sediment load in the Ganges-Brahmaputra (Abbas and Subramanian 1984). By contrast, the gross sediment load output from the upstream Yarlung Tsangpo River in Tibet is estimated to be just a small fraction (<10%) of the total load in the Ganges-Brahmaputra River (Wasson 2003; Blöthe and Korup 2013), which is due to a large volume of coarse gravel and sand deposited in the upstream river valley (Wang et al. 2016). A recent study on the midstream Yarlung Tsangpo River also indicates that the sediment yield in the catchment is much lower than other major rivers originating from plateaus, such as the upper Yellow River, the upper Yangtze River, the upper Indus River, and the Mekong River (Shi et al. 2018). The sediment load of the Koshi River (the easternmost tributary of the Ganges) is reported to be about 120 million tons per year at Chatra. Because of the high sediment load and the low gradient in the Indo-Gangetic Plain, the river's channel has shifted westward by about 115 km over the past 220 years (Gole and Chitale 1966; Dixit 2009; Chakraborty et al. 2010).

One of the most recent avulsions of the Koshi in 2008 was triggered by a breach in the embankment at Kusaha (12 km upstream of the Koshi barrage), resulting in an approximately 120 km shift of the middle fan region (Sinha 2009). This unprecedented avulsion was primarily attributed to large-scale bed aggradation due to siltation, thereby pushing the river close to the avulsion threshold (Sinha et al. 2014). Recent work has provided an assessment of sediment flux in the Koshi basin in relation to sediment connectivity and the factors controlling the pathways of sediment delivery (Mishra et al. 2016). This assessment has helped characterize sediment dynamics in complex morphological settings and in a mixed environment.

Due to its high-alpine topography, intense meltwater supply, and the summer monsoon, the Indus River transports large volumes of sediment (Nag and Phartiyal 2015), particularly from its upper reaches in northern Pakistan (Meybeck 1976; Ali and Boer 2007; Ashraf et al. 2017). The Yangtze River is ranked globally as the fifth largest river in terms of runoff and the fourth largest in terms of sediment load (Yang et al. 2011). Studies have shown that the upper river basin is the main sediment source for the Yangtze River, while significant deposition occurs in the middle and lower reaches where the slope is gentler (Chen et al. 2001; Wang et al. 2007; Yang et al. 2007). The Yellow River was once the most sediment-laden river in the world, but its sediment concentrations have decreased by approximately 90% since the 1950s due to human activity (e.g., reforestation and engineering measures) and regional climate change (Xu et al. 2004; Wang et al. 2016). Both high sedimentation (i.e., when too much can affect the river's course or result in flooding) and low sedimentation (i.e., when too little may destroy the rich delta ecosystems) present management challenges (Giosan et al. 2014). Thus, the beneficial aspects or risks of sedimentation are largely dependent on the bearing capacity and the degree of change (Wang et al. 2016).

8.2.5 Springs

Mountain springs play an important hydrological role in generating streamflow for non-glaciated catchments and in maintaining winter and dry-season flows across numerous HKH basins. Springs are the primary water source for rural households in the HKH. 80% of rural households in Sikkim rely on springwater (Tambe et al. 2009). Springs also contribute to the baseflow of many rivers in the region. In the Indian Himalaya, 64% of irrigated areas are fed by springs (Rana and Gupta 2009).

Due to factors related to anthropogenic impacts (such as deforestation, grazing, exploitative land use resulting in soil erosion, etc.) and climate change (e.g., highly variable rainfall), springs fed during the monsoon by groundwater or underground aquifers are reported to be drying up and threatening whole ways of life for local communities in most parts of the mid–hills of the HKH (Vashisht and Bam 2013; Mukherji et al. 2016). Springs have been particularly affected by the depletion of shallow water table because of reduced infiltration due to crust formation and by increased intensity of rainfall. They are further impacted by rapid socio-economic growth, demographic changes, and infrastructural developments, such as dams and road building (Mahamuni and Kulkarni 2012; Vashisht and Bam 2013; Mukherji et al. 2016). Due to scarce observation data, the status of most springs in this region is still unknown. According to research, nearly 50% of perennial springs in the Indian Himalaya have dried up or become seasonal (Rana and Gupta 2009). Spring discharges have also significantly declined (Sharda 2005). A case study in the Gaula River Basin in the central HKH showed that, by the late 1980s, spring flow had decreased by at least 25% (Valdiya and Bartarya 1989). In Sikkim, in the eastern HKH, the decrease in spring discharge was found to be over 35% during the 2000s (Tambe et al. 2012). In one of the mid-hills districts of Nepal, as many as 30% of the springs have dried up in the last decade, likely the result of a combination of biophysical, technical, and socio-economic factors (Sharma et al. 2016).

To address the water crisis caused by these dried-up springs, springshed-management strategies and conservation measures should be developed by merging scientific and community knowledge. In doing so, it is important to better establish the relationship between precipitation, recharge, and spring discharge (Negi and Joshi 2004). The few studies published on this matter are based on small, scattered areas (Negi and Joshi 2004; Vashisht and Sharma 2007; Tambe et al. 2012; Tarafdar 2013; Sharma et al. 2016; Kumar and Sen 2017a, b; Paramanik 2017) and reported results show significant variations across the HKH. While spring-discharge variation appears to be consistent with rainfall in Sikkim in the east (Tambe et al. 2012) and Uttarakhand in the central-western Himalaya (Agarwal et al. 2012), it shows an inverse pattern with monthly rainfall in the western Himalayan springs of Kashmir (Negi et al. 2012). These trends suggest that, in addition to precipitation, other causal factors and localized impacts should also be investigated.

Recent studies indicate the importance of developing an improved understanding of the aquifers through which groundwater recharges springs (Jeelani 2008; Mahamuni and Kulkarni 2012). A case study in the western Himalaya shows that spring discharge during the rainy season is very high for Karst springs and much lower for alluvium (fluvio-lacustrine) and Karewa (glacio-fluvio-lacustrine) springs (Jeelani 2008).

The anthropogenic impacts on spring discharge—including those from changes in land use and soil erosion—have been discussed in some studies (Singh and Pande 1989; Valdiya and Bartarya 1989; Tiwari and Joshi 2014). With glacial retreat increasing in this region, the disappearance of small glaciers may be a factor in the drying up of springs (Fort 2015). A workable and realistic management plan for spring watersheds needs both hydrogeological and hydrological characterization of catchments as well as a reliable modeling approach (Kresic and Stevanovic 2009). Thus, additional field investigations of declining springs—along with further research, detailed geohydrology, and modeling studies of well-observed spring catchments—are needed in the HKH.

8.2.6 Groundwater in Lowland Areas of HKH Basins

Hydrogeological characteristics of aquifers remain unknown in most parts of the HKH. Across the region, a number of groundwater studies have been conducted by characterizing aquifer systems in northwestern India (Narula and Gosain 2013; Lapworth et al. 2015) and northeastern India (Michael and Voss 2009; Mahamuni and Kulkarni 2012). In Nepal, studies conducted in Kathmandu Valley provide insight into the geological formation of aquifers (Shrestha et al. 1999), their hydrogeological characteristics (Kc 2003), and their spatial distribution (Pandey and Kazama 2011). These studies constitute a valuable knowledge base for guidance in groundwater management (Pandey et al. 2012). Aquifer mapping and groundwater level information at regular intervals need to be made publicly available.

In South Asia, groundwater constitutes the water source for over 75% of irrigated areas (Shah et al. 2006). Through the use of wells, groundwater also provides drinking water for 85% of rural population in India (Livingston 2009). For the HKH lowlands, too, groundwater is an extremely important source of water. In the plains of South Asia, groundwater is one of the most exploited water resources, of which significant depletions have been observed in the Indus basin of Pakistan (Ashraf and Ahmad 2008), the Kathmandu Valley of Nepal (Dixit and Upadhya 2005; Pandey et al. 2010), as well as northwestern region of India (Rodell et al. 2009).

Groundwater from subsurface recharge and glacier and snow melt can serve as temporary storage for river discharge in the HKH. A model-based study of water budget showed that the contribution of groundwater is about six times higher than that of glacier and snow melt in the central Nepal Himalaya (Andermann et al. 2012). This study also found a significant time lag between rainfall and discharge, indicating the importance of groundwater as temporary subsurface storage for the HKH lowlands. Currently, only a limited number of model-based studies in the HKH (Andermann et al. 2012; Narula and Gosain 2013; Racoviteanu et al. 2013) adequately account for groundwater processes due to data scarcity.

8.2.7 Implications of Climate Change on HKH Water Resources

The implications of climate change on the availability of water resource—spatial distribution, temporal dynamics, and water security in general—are extremely significant. Climate change processes and future projections for the HKH are addressed in detail in Chap. 4 and projections of glacial change are addressed in Chap. 5.

Lutz et al. (2014) showed that, as a result of climate change, a consistent increase in streamflow is expected at large scales for the upstream reaches of the Indus, Ganges, and Brahmaputra rivers until at least 2050. For the upper Indus, this is mainly due to increased glacial melt, whereas for the Ganges and Brahmaputra, the projected increase in streamflow is driven primarily by increased precipitation. These streamflow projections, however, have a large degree of uncertainty, especially for the upper Indus, as projections for precipitation show contradicting patterns.

These studies also show the various responses to climate change among rivers with different streamflow patterns. For example, the Indus River flow is dominated by temperature-driven glacial melt during summer; therefore, the uncertainty in future flow is relatively minor due to the small uncertainty in future temperature changes. On the other hand, the Kabul River has much larger components of rainfall runoff and snowmelt, increasing the uncertainty in future flow due to the large uncertainty in future precipitation. The absolute amounts of glacial melt and snowmelt are not projected to change much in the Brahmaputra and the rivers in the Ganges basin, but their relative contributions are expected to decrease due to increased rainfall runoff. As a result, projections show increased peak discharge in the monsoon season with a large uncertainty in the magnitude of flow increases.

On a smaller scale, projections through the end of the century for the Langtang and the Baltoro catchments (Immerzeel et al. 2013) indicate a consistent increase in total streamflow for both, despite their contrasting climates (RCP4.5 and RCP8.5). These increases range from 172 mm/year (Langtang, 31%) to 278 mm/year (Baltoro, 46%) in 2021–2050 for RCP 4.5 and from 493 mm/year (Langtang, 88%) to 576 mm/year (Baltoro, 96%) in 2071–2100 for RCP 8.5. In the Baltoro catchment, glacial melt is a larger component of total streamflow, and projected increases in melt are expected to be the main cause of the significant increase expected in total streamflow. In the Langtang catchment, projected increases in precipitation account for the increase in total streamflow. Despite the contrasts in climate and hydrological regimes, both catchments are expected to respond similarly to future climate change, especially through the first half of the 21st century.

In the eastern Dudh Koshi catchment in Nepal, Shea et al. (2015a) suggest sustained mass loss from glaciers in the Everest region through the 21st century based on RCP 4.5 and RCP 8.5 climate projections. Similarly, Bajracharya et al. (2014) reported a loss of glacier area of 23% in Bhutan and 25% in Nepal between 1980 and 2010. How and when the loss of glaciers will impact downstream availability of water is an important area for further research.

Hydrological models have been used to assess the likely impact of climate change on water resources in Pakistan (Ashraf 2013) and in the Koshi River Basin in Nepal (Bharati et al. 2014). Bharati et al. (2014) projected increased flow volumes during the monsoon and post-monsoon and decreased flow volumes during the winter and pre-monsoon seasons, with greater impacts likely in certain seasons and sub-basins.

Soncini et al. (2015) have found similar results for the Shigar watershed (which includes the Baltoro watershed), projecting mostly increases in flow until the end of the century and speculating on the potential for slight decreases thereafter, once ice volumes have diminished. In this catchment, changes in precipitation will not compensate for ice loss in the long run. Across the three different RCPs presented in this study, the differences in streamflow change are strikingly small. The authors showed that increases in both temperature and winter precipitation cause streamflow increases to begin earlier, when glacier and snow begin to melt. This is most dramatic for RCP 8.5, in which two of three General Circulation Models (GCMs) show significant flow increases beginning in April instead of June. Other RCPs also show a shift to the earlier onset of increased flow—and this shift gets stronger toward the end of the century. However, one of the GCMs shows a very different pattern, with flows decreasing in spring and increasing slightly in all other months.

Ragettli et al. (2013) have showed that for the Hunza basin simulated decadal mean runoff is relatively constant (with projections until 2050), but strongly contrasting changes occur in some of the sub-basins. Some models showed flow volume decreasing by as much as 50% due to decreases in glacial melt, while others showed flow volume increasing due to increases in snowmelt, precipitation, and temperature. In the basin areas, with projections of decreased flow, the most pronounced reductions occur from June to September. The annual peak runoff is projected to occur in June/July, earlier than the July/August peak of the control period.

In the Lhasa basin, Prasch et al. (2013) have made hydrological projections by forcing a glacio-hydrological model with the IPCC Special Report on Emissions Scenarios (SRES) scenarios. The authors found that the contribution of glacial melt to total runoff will remain almost stable until 2080, although there will be a slight increase during a short period in the spring. By contrast, the contribution of snowmelt to runoff will generally decrease in the Lhasa basin, resulting in changes to water availability. Additionally, the increased evapotranspiration accompanying rising air temperatures will also reduce water availability. However, in contrast, regional vegetation greening observed in Tibetan Plateau (TP) is likely to slow warming rates (Box 8.1).

Box 8.1 Implications of landscape-level vegetation change for evaporation

In recent decades, vegetation changes across the Tibetan Plateau (TP) have shown significant spatial variation. Decreasing trends in Normalized Difference Vegetation Index (NDVI) during the summer growing season have been noted in the southwest, whereas obvious greening was observed in the northeast based on existing global NDVI datasets (Fig. 8.3). Due to warming trends and the grazing-to-grassland project implemented by the Chinese government, regional greening is confirmed by observed NDVI. Despite the warming effects of reduced albedo resulting from increased NDVI, the cooling effects of enhanced evapotranspiration (ET) are dominant in the TP, where ET is believed to be relatively high even at low temperatures (Shen et al. 2015). Greening with increasing NDVI as the proxy is believed to have cooling effects on surface temperatures due to enhanced ET. This is supported by the significant negative correlation between NDVI and daily max temperatures at 55 meteorological stations across the TP. By means of the Weather Research and Forecasting (WRF) model, an increase of NDVI by 0.1 is estimated to result in an increase in ET by ~ 0.5 mm d^{-1} and a decrease in albedo by 0.01. Thus, regional vegetation greening is not only beneficial to ecosystem processes, but also to slowing warming rates Shen et al. (2015).

Based on their review of the impacts of climate change on the Indus, Ganges, and Brahmaputra River Basins, Nepal and Shrestha (2015) noted an increase in glacier melt and snowmelt from 2000s to approximately mid-century, which is then followed by a decrease. Although, increases in meltwater are likely for the next few decades, meltwater volume is likely to decrease abruptly once glacial storage is reduced. Nevertheless, further studies are required to understand intra-annual changes and the impact of extreme events on meltwater volumes. Changes in extreme hydrological events in the Indus, Ganges, and Brahmaputra basins are insufficiently studied. As an increase in precipitation is generally projected, it is highly likely that precipitation extremes—and associated extreme discharges—may increase as well. Soncini et al. (2015) used downscaled GCM data to force a semi-distributed model to conduct a

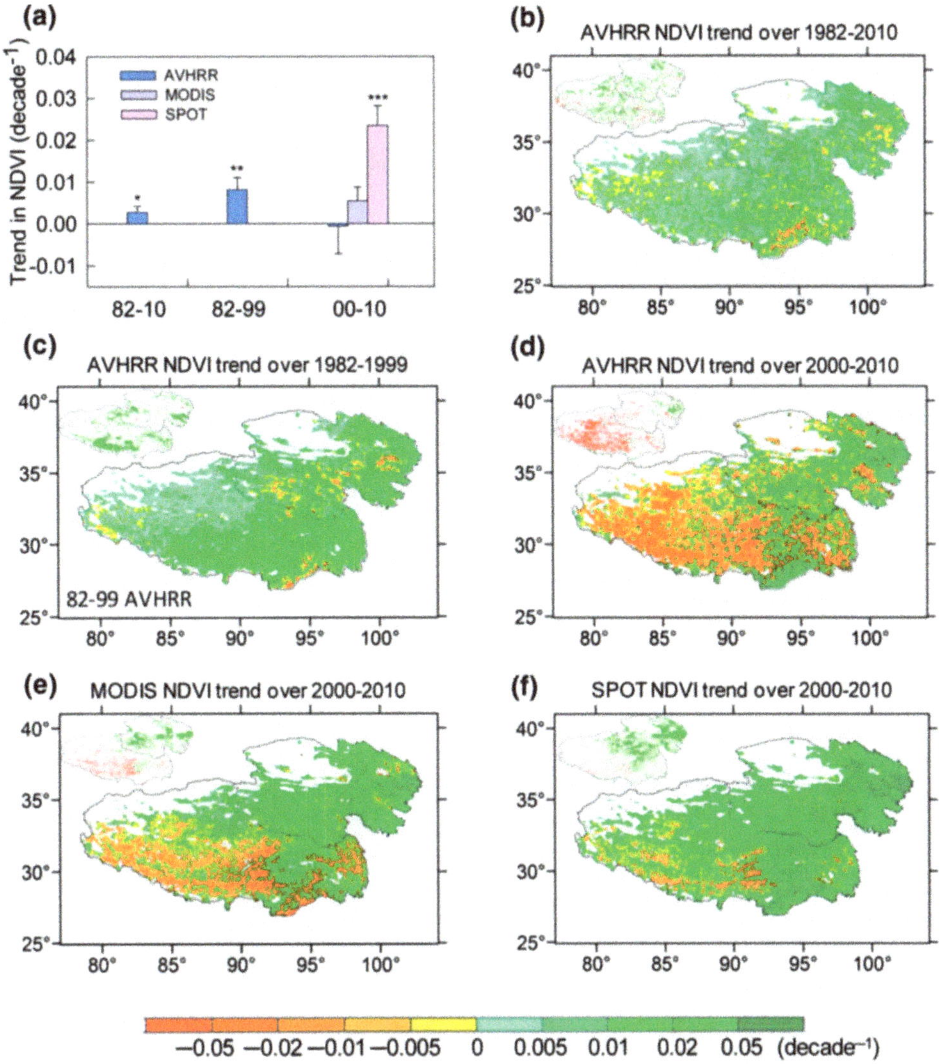

Fig. 8.3 NDVI changes in the growing season (May–September) across the Tibetan Plateau over the past three decades. (**a**) Trend in the growing season NDVI at a regional scale during the periods of 1982–2010, 1982–1999, and 2000–2010. ***$P < 0.01$; **$P < 0.05$; *$P < 0.10$. Trends with no asterisk are not significant ($P > 0.10$). (**b–f**) Spatial distribution of NDVI trends from different datasets examining different periods (*Source* Reprinted from "Evaporative cooling over the Tibetan Plateau induced by vegetation growth" by Shen, M. et al., 2015, Proceedings of the National Academy of Sciences, 112(30), p. 9300. Copyright © 2015 by the National Academy of Sciences of the United States of America)

basic analysis of changes in extreme discharges in the Shigar catchment. Most models indicated increased discharge for the flow-return periods analyzed, indicating the potential for heavier floods during the flood season from June to October.

8.3 Water Use in the Hindu Kush Himalaya

There are wide variations in water endowments in HKH countries in terms of per capita availability, contribution from surface and groundwater sources, as well as with regards to whether or not the water originated within the geographical boundaries of the country or, for that matter, within the HKH. For instance, upstream countries like Bhutan and China generate all their water within their own geographical boundaries, while the downstream country of Bangladesh gets over 90% of its water from beyond its geographical boundaries. The very nature of upstream-downstream linkages and water distribution across the countries makes it imperative that upper and lower riparian communities cooperate in sharing water equitably.

As seen in Table 8.2, total renewable water availability in the eight countries that constitute the HKH is 7745.5 km^3 (AQUASTAT, FAO 2016a). Figures 8.4, 8.5 and 8.6 show the distribution and per capita availability and use across the eight countries of the HKH region. Of the total water resources, 1597.8 km^3 (20.62%) is used annually for various purposes. Some of this usage is consumptive in nature (for agriculture, drinking, and domestic use), as opposed to the largely non-consumptive use in sectors like hydropower.

Table 8.3 shows the sectoral water use in various countries in the region, but for reporting reasons, these do not precisely correspond to the HKH. As with water-resource endowments, there are also wide variations in terms of total volume and per capita water withdrawal, contribution of surface and groundwater to total water withdrawals, and percentage of water withdrawals from the total renewable freshwater available. For instance, per capita water

Table 8.2 Water-resource availability in HKH countries

Country	Long-term average annual precipitation in depth (mm/year)	Long-term average annual precipitation in volume (km^3/year)	Total internal renewable water resources (IRWR) (km^3/year)	Total renewable water resources (km^3/year)	Total renewable surface water (km^3/year)	Total renewable groundwater (km^3/year)	Dependency ratio (%)	Total renewable water resources per capita (m^3/inhab/year)
Afghanistan	327	213.3	47.15	65.370	55.68	10.65	29	2,019
Bangladesh	2,320	334	105	1,226.6	1,206	21.12	91.4	8,343
Bhutan	2,200	84.5	78	78	78	8.1	0	109,244
China	645	6,189	2,812.4	2,839.7	2,739.0	828.8	1	2,079
India	1,170	3,846	1,446	1,911	1,869	432	31	1,582
Myanmar	2,341	1,415	1,002.8	1,167.8	1,157	453.7	14.1	24,537
Nepal	1,500	220.77	198.2	210.2	210.2	20.0	5.71	7,142
Pakistan	494	393.3	55.0	246.8	239.2	55	78	1,474

Source FAO (2016a)

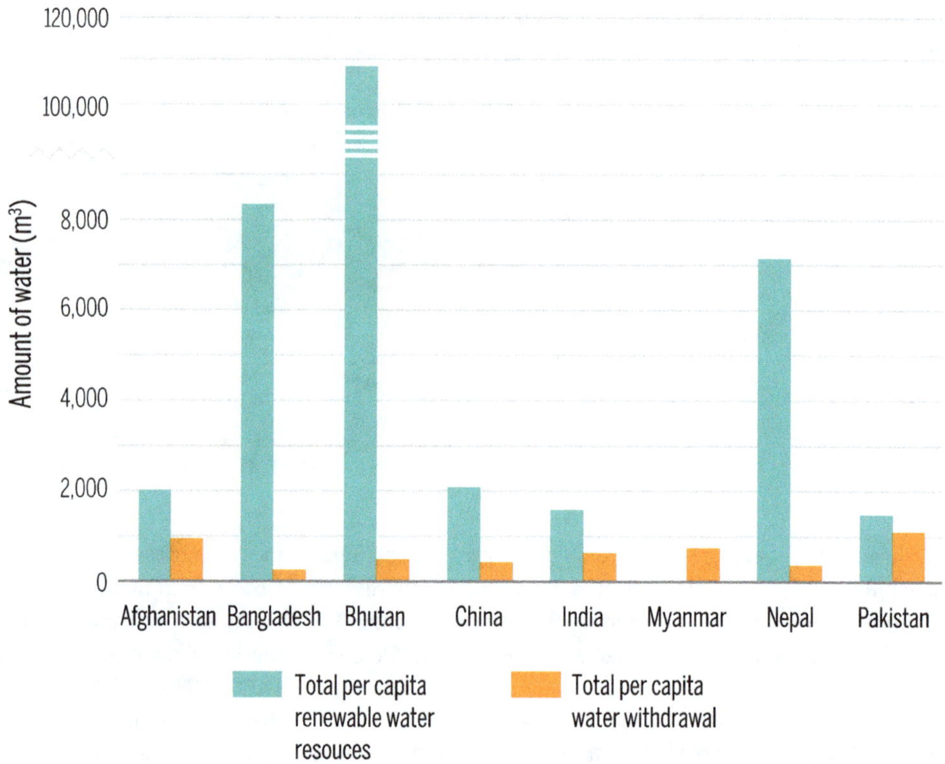

Fig. 8.4 Total per capita renewable water resources and withdrawals by country (*Source* FAO 2016a, AQUASTAT http://www.fao.org/nr/water/aquastat/water_res/index.stm)

withdrawals vary from 1096 m^3/year in Pakistan to a low of 247 m^3/year in Bangladesh. At the same time, Pakistan withdraws 74% of its renewable freshwater resources, while Bhutan withdraws less than 0.5% annually. Groundwater accounts for 79.4% of water withdrawal in Bangladesh and for about 33% of water withdrawal in India and Pakistan.

Figure 8.7 shows the total water withdrawals and withdrawals by sector in the HKH countries. In spite of their varied water withdrawal, what remains constant across all of these countries is that the largest proportion of withdrawals is used for agriculture. Agriculture accounts for close to 90% of water withdrawal in all HKH countries with the exception of China, where 65% of withdrawal is applied to agriculture. By contrast, 25% of China's water withdrawal is used for industrial purposes, while industry accounts for less than 10% of water withdrawal in other countries, reflecting China as the most industrialized country in the HKH. Overall, given the twin effects of anthropogenic and climate induced changes, it is believed that water use in the HKH is at a critical crossroad, and decisions regarding water management and governance taken now will have long-term implications for the future (Mukherji et al. 2015).

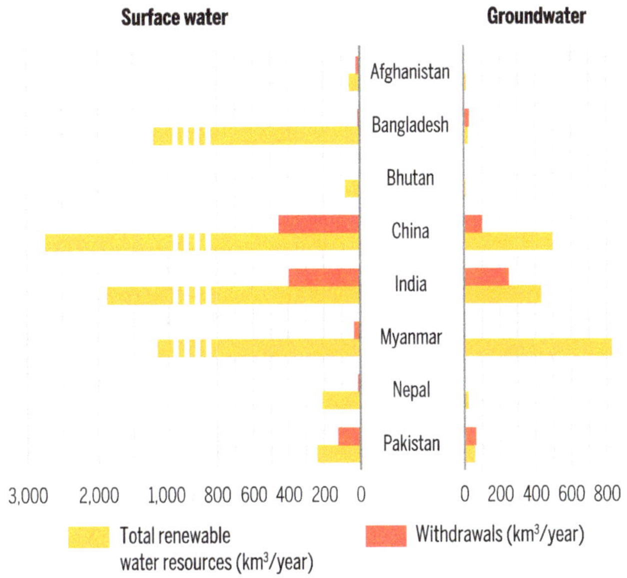

Fig. 8.5 Total renewable water resources and withdrawals by surface water versus groundwater by country (*Source* FAO 2016a, AQUASTAT http://www.fao.org/nr/water/aquastat/water_res/index.stm)

8.3.1 Agricultural Water Use in the Mountains, Hills, and Plains of HKH River Basins

As is the case in most other regions of the world, agriculture accounts for the highest proportion of water withdrawal in the HKH (Table 8.3). Agriculture in the HKH varies according to altitude. Mountains, mid-hills, and plains (including foothills of the Himalayas) offer three distinct agricultural systems. In the high mountains, agriculture is dominated by livestock rearing and orchard cultivation, while in the mid-hills and the plains, cereal crops take precedence. In general, agriculture in the mountains and mid-hills tends to be rainfed, while that of the plains is mostly irrigated (Table 8.4, Fig. 8.8).

8.3.1.1 Hill and Mountain Agricultural Water Use

Most HKH countries maintain the centuries-old tradition of farmer-managed hill and mountain agriculture. While mostly rainfed, these farms are also irrigated seasonally through local streams, springs, and glacier and snowmelt. In the northern mountains of Pakistan (comprised of Gilgit, Baltistan, Chitral, and Upper Dir), there are broadly two types of mountain irrigation systems—those sourced by snowmelt and those by streamflow or springwater (Vincent 1995). There is a large number of detailed and structured case studies of mountain agriculture from different parts of northern Pakistan, e.g. Hunza (Kreutzmann 2011; Parveen et al. 2015), Nanga Parbat (Nüsser and Schmidt 2017) and Chitral (Nüsser 2001), that document continuity and change in mountain agriculture and irrigation. All these studies underpin the critical role of mountain communities in managing their irrigation systems through framing and implementation of context-specific rules.

Snowmelt, streamflow, and/or springwater are diverted through channels along the mountain slopes to the valley bottom where fruit, vegetable, and other crops are grown.

Fig. 8.6 Freshwater withdrawals as proportion of total renewable water resources by country (*Source* FAO 2016a, AQUASTAT http://www.fao.org/nr/water/aquastat/water_res/index.stm)

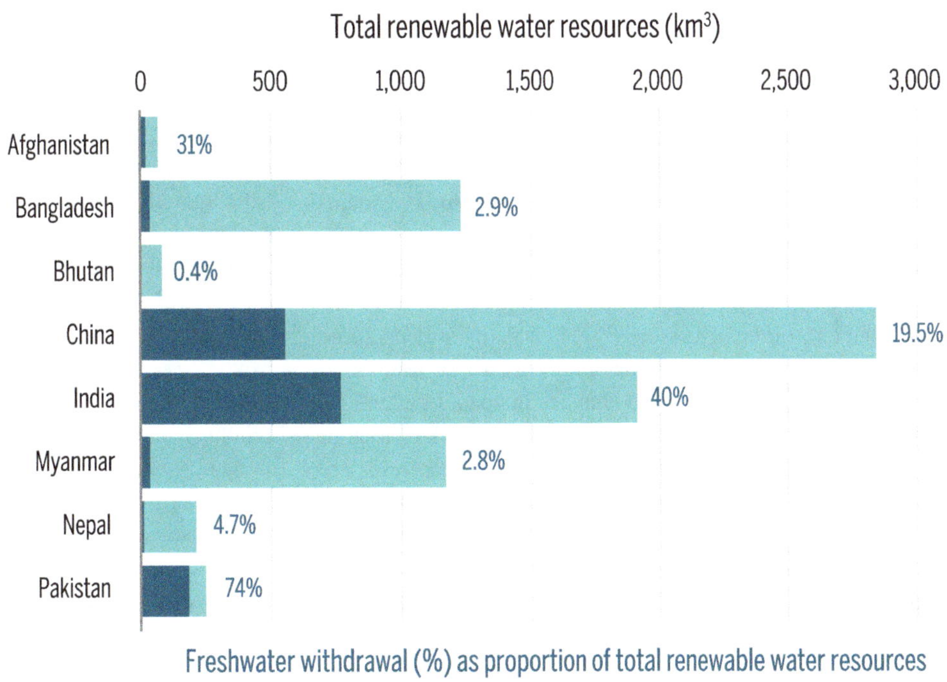

Table 8.3 Sector-wise water withdrawals in HKH countries

Country; Year of data referenced	Total water withdrawal (km^3/year)	Agricultural water withdrawal (km^3/year)	Municipal water withdrawal (km^3/year)	Industrial water withdrawal (km^3/year)	Surface water withdrawal (km^3/year)	Groundwater withdrawal (km^3/year)	Per capita water withdrawal per inhabitant (m^3/year)	Freshwater withdrawal as % of total renewable water resources
Afghanistan; (1998)	20.37	20.00	0.20	0.17	17.24	3.042	937	31
Bangladesh; (2008)	35.87	31.5	3.6	0.77	7.39	28.48	247	2.93
Bhutan; (2008)	0.338	0.318	0.017	0.003	0.338	0	482	0.43
China; (2005)	554.1	358.02	67.53	128.55	452.7	101.4	414	19.5
India; (2010)	647.5	688	56	17	396.5	251	630	40
Myanmar; (2000)	33.23	29.575	3.323	0.332	30.240	2.991	739	2.8
Nepal; (2006)	9.497	9.32	0.147	0.0295	8.444[b]	1.053[a]	359	4.7
Pakistan; (2008)	183.421	172.371	9.650	1.4000	121.9	61.6	1,096	74

Source FAO (2016b)

[a]Government of Nepal 2016, Groundwater Resources Development Board, Ministry of Irrigation, (12 June 2016) retrieved from http://www.gwrdb.gov.np/hydrogeological_studies.php

[b]Derived by subtracting groundwater withdrawals from total water withdrawals

These irrigation systems are unique in that each water-source channel has its own command area, water users frame their own rules, and that these irrigation systems are maintained by active participation of water users with almost no government involvement (Kreutzmann 2011).

While irrigation is reliable during the spring and summer seasons, it is less so in winter due to reduced snowmelt. Cropping patterns are, therefore, adjusted accordingly. From sowing to harvesting, women are actively involved in various aspects of the agricultural practices in these areas (Ishaq and Farooq 2016). Although water distribution may or may not be equitable, customary rules of distribution and maintenance are clearly laid out and followed by water users (Ostrom and Gardner 1993). In general, water utilization in these systems has remained well within the limits of available water resources (Nüsser 2001). There have been some modifications in social organization of water users and associated rules in response to changes in spatial patterns of irrigated mountain farming (Kreutzmann 2011; Thapa et al. 2016). In Afghanistan and highland Balochistan, *karezes* are traditional irrigation systems wherein shallow tunnels tap underground aquifers and convey water to fields downstream (Box 8.2). Another system of irrigation in the hill regions of western HKH is referred to as spate irrigation, in which flood water is harvested and managed for irrigation.

Box 8.2 Glacier-fed irrigation systems in Hunza and Ladakh in Upper Indus; *karezes* and spate irrigation in Afghanistan and Pakistan

In the Upper Hunza region of Pakistan (Kreutzmann 2011; Parveen et al. 2015) and in the trans-Himalayan part of Ladakh in India, glacier melt and snowmelt is the only source of irrigation. Glacio-fluvial dynamics affect these irrigation practices, and local communities adapt to these changes in different ways. In Hopar village in the Karakorum Range in Pakistan, it is not the quantity or timing of meltwater discharge that affects irrigation decisions but other factors such as water quality, reliable access, and control of turbulent flow. In Ladakh, the irrigated area is shrinking—not necessarily as a direct consequence of changing hydrology or glacial melt, but because of changing livelihood strategies involving more off-farm employment (Butz 1989; Nüsser et al. 2012). In recent years, late melting of glaciers due to high climatic snow line has made seasonal water scarcer. In response, communities have come together to create artificial glaciers on southern slopes, both as a strategy to cope with seasonal water scarcity and a long-term adaptation to climate change. Since these glaciers are constructed at lower altitudes and on south-facing slopes, water starts melting from April, thereby affording a longer growing season for the cultivators (Nüsser and Baghel 2016).

In highland Balochistan and Afghanistan, *karez* irrigation has been the backbone of rural water management and agriculture for more than two millennia. A *karez* is an underground aqueduct that passively taps groundwater in the piedmont of the arid and semi-arid highlands. The key physical advantages of the system are that it minimizes evaporative loss and delivers water from mountain aquifers to valley floor communities. *Karezes* require annual maintenance, and *karez* communities have developed strong social capital to enable provision of labor and resources for their upkeep. However, *karezes* have increasingly come under threat across Balochistan because of uncontrolled pumping with high-power electric water pumps. In Afghanistan, more than three decades of war has also taken a toll on the physical and social infrastructure of *karezes*. In the Mastung district in Balochistan, for example, prior to the 1980s there were 365 *karezes*; today there are no more than ten in operation. The drying up of *karezes* has numerous damaging consequences, including the breakdown of the rural social capital anchored in the *karez* infrastructure. Despite a temporary increase in agricultural productivity due to availability of on-demand water, there is a long-term decline of agriculture and rural livelihoods due to groundwater depletion, as in the Quetta valley of Balochistan. This enhances power of the rural elites who own the electric pumps, which mine the groundwater and deprive hundreds of *karez* shareholders of their previously held water rights. These have significantly contributed to rural pauperization and rural-to-urban migration in the region. Furthermore, long-term questions about the sustainability of human life in the arid highlands remain to be addressed, as groundwater depletes from one valley to the next due to over-pumping.

Spate irrigation is common in the drier parts of western HKH, especially in Balochistan province of Pakistan and in Afghanistan. Spate irrigation uses flood water generated from an upstream hill slope that is then stored as soil moisture or collected from ephemeral streams in adjacent low-lying valley bottom. This soil moisture is then channeled through rudimentary and locally constructed infrastructure in order to cultivate crops during dry months (Mehari et al. 2007). These systems are called "*sailaba*" in Balochistan (van Steenbergen 1997). It is estimated that roughly 1.45 million hectares of land is under such seasonal flood irrigation (Khan 1987); yet these systems that cater to the poorest of the poor in the hills and mountains do not receive adequate attention. These systems, much like the other systems of hill and mountain irrigation described above, are managed by farming communities who have adjusted to inherent uncertainty of spate irrigation through rules and regulations that define access and norms for water sharing. Typically, local varieties like drought-resistant sorghum, millets, and wheat are cultivated, and yields are low. This, coupled with inherent uncertainty of occurrences and magnitudes of floods, means that even this minimum yield is not assured every year. Cyclic outmigration of labor is, thus, quite common (van Steenbergen 1997). There are ongoing initiatives to "modernize" these systems, but such modernization efforts will not succeed unless traditional norms and practices are understood and incorporated in new designs.

Nepal is also known for its centuries-old, farmer-managed irrigation systems (Box 8.3). About 70% of irrigation systems in Nepal are operated through farmer-managed irrigation systems (FMIS) (Pradhan 2000). Communities build water channels and weirs to divert water from spring-fed streams for growing paddy in monsoon season and, occasionally, one additional crop during the dry season. Intricate rules govern issues like water distribution, maintenance of infrastructure, and conflict resolution—and evidence shows that these systems have endured for centuries and adapted to changing circumstances (Thapa et al. 2016). Similar spring- and stream-fed irrigation systems are also found in India's western and central Himalayas (Baker 2005). According to Mollinga (2009), the share of irrigated land served by FMISs in India has declined from 18.5% in 1961 to 6.8% in 1991. Other studies have reported that irrigation efficiency is higher in FMISs than in state-managed irrigation systems in the central Indian Himalaya (Kumar et al. 2006).

Box 8.3 Farmer-managed irrigation systems in Nepal

Nepal has a long history of FMISs, in which farmers take sole responsibility for operating and maintaining their irrigation systems. In the absence of strong government intervention in the past, FMISs slowly developed through the collective efforts of farmers looking to irrigate their land. These FMISs provide irrigation services to about two-thirds of the country's total irrigated area—a little more than 1.2 million hectares (Pradhan 2000). FMISs are characterized by the use of low-cost technology appropriate for heterogeneous local conditions, autonomous decision-making suited to local contexts, and collective action by farmers for the operation and maintenance of the irrigation systems (Yoder 1986; Ostrom and Benjamin 1993).

While many FMISs have survived decades of changes to hydro-climatic, social, institutional, and policy conditions (such as, government support for operation, maintenance, and infrastructure, as well as registration in FMIS inventories), their performance is increasingly under stress (Janssen and Anderies 2013). Water availability for irrigation is affected by variability in the intensity and timing of precipitation. Impacts include more flooding and erosion damage to irrigation intake points and canals and, during the dry season, less water available for irrigation and increased competition for it due to prolonged drought (Bastakoti et al. 2015). These challenges are further compounded by socio-economic and institutional changes.

In FMIS, men have traditionally played a dominant role in the maintenance and operation of irrigation systems. But since men are migrating out of the countryside in large numbers and educated youth seem to have less interest in water management, an increasing number of women play a larger role in agriculture and water management, despite being unaccustomed to such tasks and often having limited experience. A recent study by Pokhrel (2014) considered why some FMISs have survived and others have declined or disappeared. The results showed the importance of adaptability in institutions concerned with the use and management of shared resources. This adaptability was characterized by a perceived fairness and bounded flexibility of the institutions—and the survival of an FMIS was dependent on this capacity to adapt to both climate change and to changes in gender relations.

In the northeastern Indian Himalaya and in the highlands of Bangladesh and Myanmar, farming systems are distinct from elsewhere in the HKH and shifting cultivation remains the preferred practice for the numerous ethnic groups in the region (Box 8.4).

Box 8.4 Changing contours of shifting cultivation in North East India, Chittagong Hill tracts, and Myanmar

There are not many studies on water availability and use in the eastern Himalaya, a region known to be abundant in water resources, feeding four major river systems in the HKH—the Brahmaputra, Ganges, Irrawaddy, and Salween. However, we do know that shifting cultivators in this region have for centuries used water resources on a sustainable basis, employing indigenous traditional knowledge and practices—such as the *zabo* farming system in Nagaland, the water management that sustains the rice and fish culture of Apatani tribes in Arunachal Pradesh, the bamboo drip-irrigation system of Meghalaya (Singh and Gupta 2002), or the *Jhiri* system in Chittagong Hill Tracts (CHT) of Bangladesh. Risk-aversion attributes are inherent in shifting cultivation practices. Maintaining high crop diversity as practiced by shifting cultivators helps withstand weather stresses and, hence, reduce the risk of crop failure (Aryal and Choudhury 2015).

In terms of agriculture, many shifting cultivators in the eastern Himalayas are converting to either settled agriculture or to growing more cash root crops on sloping lands. Cultivation on sloping lands without soil and water conservation measures has led to soil erosion and the degradation of ecosystem services. Rasul (2009) reported approximately 89–109 tons/ha/year of soil loss from the cultivation of annual crops (mainly ginger, colocasia, and turmeric) on sloping lands when conventional hoeing tillage methods were applied. With mulching, soil erosion was reduced to about 35 tons/ha/year. In northeast India, the fallow-management practices of shifting cultivators through retention of selective multipurpose species, for example *Alnus nepalensis*, has long-term implications for the provisioning and regulating of ecosystem services like water security that accrue from them (Singh and Choudhury 2015). There are many good sustainable land and water management practices in the region—both traditional and new—but they have yet to be evaluated, documented, and shared.

While the exact contours of mountain and hill irrigation systems may differ in terms of water sources, distribution, and management, there are certain aspects of agricultural water management that are consistent across the entire HKH. For example, indigenous systems of water management have developed effective methods for cooperating, sharing, and resolving disputes—and these local institutions have withstood change and adapted accordingly.

In the past two or three decades, there has been a contraction in hill and mountain agriculture due to a number of factors—both climate and non-climate induced. For instance, in the upper reaches of the Indus, canal infrastructure built for the intake of glacial melt has become

Fig. 8.7 Total water withdrawals and withdrawals by sector by country country (*Source* FAO 2016b, AQUASTAT http://www.fao.org/nr/water/aquastat/water_use/index.stm)

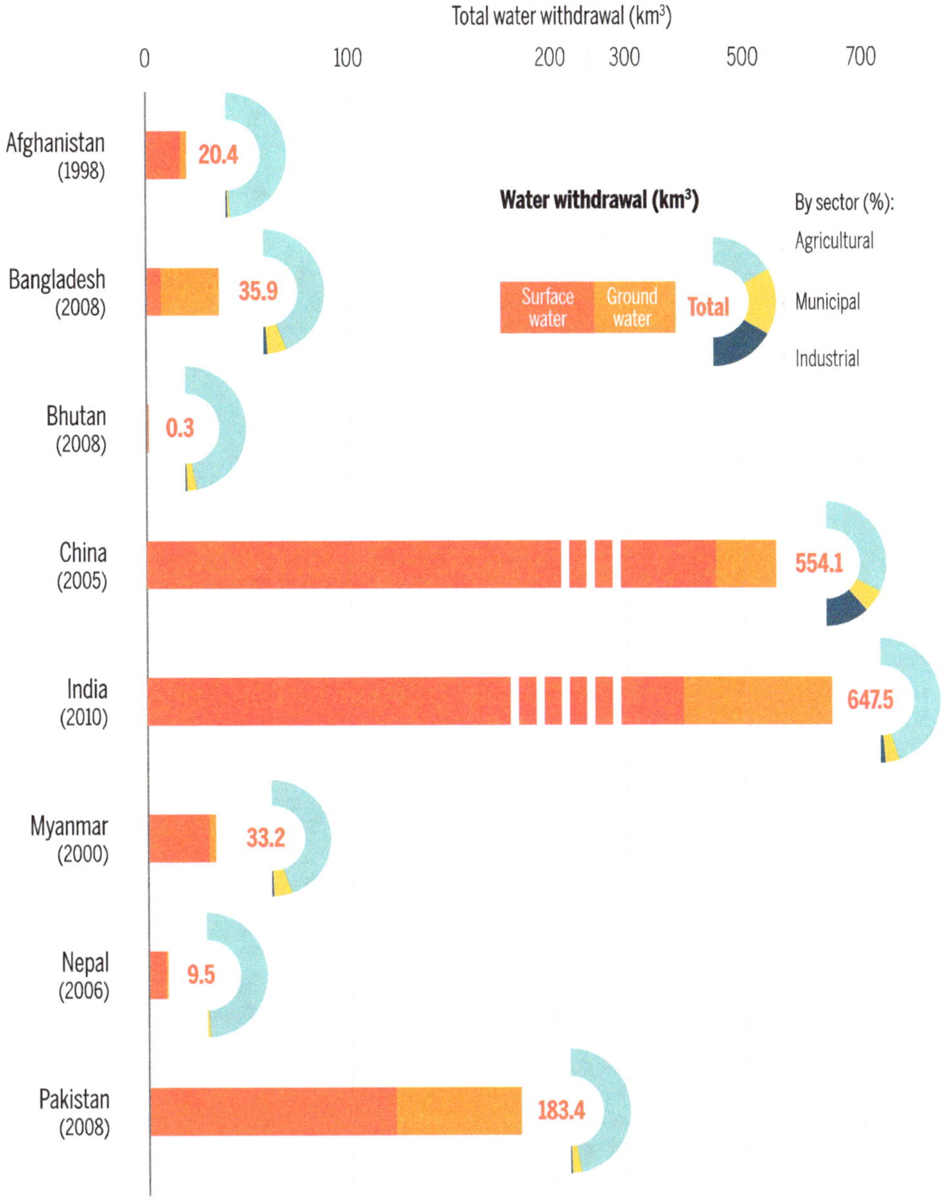

dysfunctional due to glacial retreat in some regions and glacial surge in others. This necessitates the rebuilding of the entire infrastructure, which is both costly and labor-intensive, beyond the reach of many communities, and thus contributing to outmigration. Extreme weather events, such as flash floods, have created additional risks to irrigation infrastructure in these regions. Many FMIS in Nepal and India have also shrunk in size due to urbanization and predominantly male out-migration.

However, irrigation systems are being adapted to changes in various ways. New technologies are being used (including groundwater or surface-water pumps and greenhouses for vegetable-growing); new niche and high-value crops are being introduced (including vegetables, coffee, and nuts). Due to male out-migration, women are increasingly managing these systems, but are yet to receive de-jure land and water rights.

Increased out-migration also offers new opportunities. For example, in some instances, remittances are being used for improving agricultural water management—through investment in vegetable greenhouses, drips, and sprinklers. However, in other instances, entire farms are being abandoned and native vegetation is reclaiming previously cultivated terraces. Hill and mountain irrigation is in transition, and how this transition is handled will be crucial to future water management and to the options for long-term livelihood.

Table 8.4 Rainfed and irrigated areas in the hills, mountains, and plains of the HKH

Country	Area under rainfed and irrigated agriculture in mountains and hills (ha)		Area under rainfed and irrigated agriculture in plains (ha)		Source (year of data) Year of data (Source)
	Rainfed area (% of total cultivated area)	Irrigated area (% of total cultivated area)	Rainfed area (% of total cultivated area)	Irrigated area (% of total cultivated area)	
Afghanistan[a]	3,051,001 (66.7%)	1,522,585 (33.3%)	437,169 (22.3%)	1,526,216 (77.7%)	2001–02 (Maleta and Favre 2003)
Bangladesh[b]	102,790 (72.5%)	38,850 (27.5%)	7,806,393 (52.4%)	7,086,052 (47.6%)	2012–13 (Bangladesh Bureau of Statistics 2015)
Bhutan[c]	50,000 (66.7%)	25,000 (33.3%)	NA	NA	2011 (ADB 2014)
China[d]	14,680,986 (60%)	9,842,614 (40%)	NA	NA	2012–13 (China Statistical Year Book 2015) for rainfed area and total cultivated area; 2005 (FAO AQUASTAT 2016c; Global Map of Irrigated Area) for irrigated area
India[e]	3,216,186 (73.7%)	1,148,459 (26.3%)	12,450,814 (31.2%)	27,396,541 (68.8%)	2011–12 (Land Use Statistics, Directorate of Economics and Statistics, GOI, 2016)
Myanmar[f]	NA	116,075 (NA)	NA	1,988,040 (NA)	2005, (FAO, AQUASTAT, 2016d, Global Map of Irrigated Area, FAO)
Nepal[g]	870,800 (72.6%)	328,700 (27.4%)	338,200 (25.6%)	984,800 (74.4%)	2011–12 (Government of Nepal 2012–13)
Pakistan[h]	753,171 (41.8%)	1,034,994 (57.4%)	3,402,833 (20%)	13,801,888 (80.0%)	2010 (Government of Pakistan, Pakistan Agricultural Census 2010)

[a]In Afghanistan, mountain area includes Badakshan, Central, Eastern, Southern, and Northern mountains; plains include Turkistan, Herat-Farah, and Helmand river valley

[b]In Bangladesh, districts of Bandarban, Khagrachari, and Rangamati are classified as hills; the rest of Bangladesh is classified as plains. In table, gross cropped area has been subtracted from total irrigated area (sum total of irrigated area in different seasons) in order to derive rainfed or non-irrigated area

[c]Bhutan is considered to be comprised of hills and mountains only

[d]In China, Gansu, Qinghai, Sichuan, Tibet, Yunnan, and Xinjiang provinces are considered parts of the HKH

[e]In India, the Indo-Gangetic states (Punjab, Haryana, Uttar Pradesh, Bihar, and West Bengal except the Darjeeling district) and Assam (excluding Karbi Anglong and North Cachar hills) are classified as plains. The states of Jammu and Kashmir, Himachal Pradesh, Uttarakhand, Meghalaya, Tripura, Manipur, Mizoram, Nagaland, Arunachal Pradesh, Darjeeling district of West Bengal, Karbi Anglong, and North Cachar hills of Assam are considered hills and mountains

[f]In Myanmar, Chin, Kachin, and Shan provinces are classified as hills; the rest of the provinces are considered plains. Data on cultivated area in Myanmar's hills and plains provinces are not available

[g]In Nepal, all Terai districts are classified as plains; the rest of the country is classified as hills and mountains

[h]In Pakistan, Khyber Pakhtunkhwa Province is classified as comprised of hills and mountains only; the other provinces (Punjab, Sindh, and Baluchistan) are classified as plains

NA = Not available

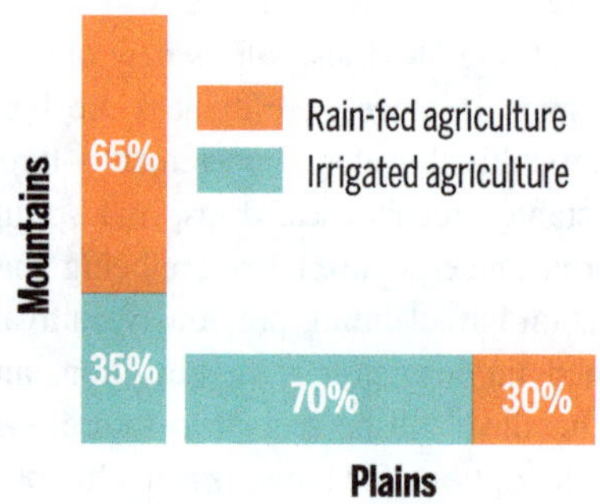

Fig. 8.8 Rainfed versus irrigated agricultural area in the mountains versus the plains. (*Sources* See Table 8.4)

8.3.1.2 Agricultural Water Use in the Plains of HKH River Basins

The extent and sources of irrigation vary; some areas like the Indo-Gangetic Plains in Pakistan, India, and Bangladesh are intensely irrigated, while those in the Nepal Terai are not. While most of these plains have canal infrastructure, groundwater has emerged as the main source of water for irrigation. The associated challenges of irrigation in the plains are:

- over-extraction of groundwater compounded with inefficient use of surface water in areas where water is already scarce (like the Indus and western Ganges);

- under-development of irrigation potential in areas of abundance (like the eastern Ganges, Terai in Nepal, and parts of eastern India); and
- increased frequency and intensity of flood drought cycle.

India is the largest user of agricultural groundwater in the world. It is estimated that there are over 20 million groundwater wells (GOI 2011), of which more than 95% are privately owned by smallholder farmers. These provide a range of livelihoods and productivity benefits to millions of smallholder farmers in India. However, within the overall groundwater story of the Gangetic plains in India, there are two distinct subplots.

The first, and rather well known, is the story of groundwater overexploitation and its consequences. This is broadly the situation in states like Punjab, Haryana, and western Uttar Pradesh. These states have a number of things in common. They receive low to medium rainfall, averaging from 200 to 1000 mm per year. Even though they have alluvial aquifers, recharge is limited by the total amount of rainfall and is, therefore, inherently low. The majority (over 70–80%) of all water extraction mechanisms are operated by electricity. Farmers get electricity either free of cost (Punjab) or at highly subsidized rates (Haryana, Uttar Pradesh). In all of these states, rural poverty is comparatively low and below the all-India average. Groundwater and electricity are major political issues in all of these states, and both remain at the center of vote bank politics. The discourse on overexploitation is fairly well known and documented (Janakarajan and Moench 2006; Moench 2007; Sarkar 2011).

But there is a second, lesser-known subplot to this story —one in which groundwater development falls far short of potential groundwater reserves, even though rainfall and natural recharge is very high. In these areas, abundant groundwater resources coexist with high costs of groundwater extraction, restrictive access policies, and low agricultural growth rates. Here, most pumps run on expensive diesel or whenever farmers get electricity, for which they pay full cost (Shah 2007; Mukherji 2007). These scenarios can be observed in places where rural poverty rates are much higher than the national average and crop productivity is low —more or less all across the eastern Indo-Gangetic belt in India, namely, West Bengal, Bihar, Orissa, and Assam. Much of the eastern Nepal Terai is also part of this story of underdeveloped groundwater resources. However, in some of these parts of eastern HKH, there is a high occurrence of geogenic arsenic and irrigating with arsenic rich water for a sustained period poses the risk of entry of arsenic in the food chain (Box 8.7).

Therefore, agricultural water management in the plains of the HKH requires different policies for regions where water resources are under stress and those where water resources are abundant. In the former, demand-management measures are required; in the latter, larger investments are necessary to tap untapped water resources for future agricultural growth.

8.3.2 Water for Energy

Unlike water used for agriculture, domestic needs, and in urban sectors, use of water for energy production in not consumptive in nature. However, hydropower projects, including the run of the river projects, which are thought to be the most benign in terms of environmental impacts, can lead to extensive changes in river flow regimes, including timing and seasonality of flow. They can also lead to changes in biodiversity (Grumbine and Pandit 2013) and create conflicts with pre-existing systems of water use, including irrigation (Erlewein 2013). The HKH has a total of 500 gigawatts (GW) of hydropower potential, of which only a small fraction is actually developed (Table 8.5). Figure 8.9 shows these data by country.

The hydropower sector in the HKH suffers from the twin challenges of societal pressure and climate change. The sector faces major challenges due glacial melt induced by climate change. Glaciers across the region, except in the Karakoram (Bolch et al. 2017), are retreating, leading to changes in future hydrological regimes. At the same time, risk of glacial lake outburst floods (GLOFs) and landslides are increasing, putting both existing and planned hydropower plants at risk. Nearly as important as climate-related risks are the societal risks of alienating local people in areas where hydropower projects are constructed. These projects are mostly developed in mountain areas, and mountain people fear, and often rightly so, that even as they bear the environmental and social costs of hydropower, the benefits will go to the people in the plains who get electricity (Bandyopadhyay 2002). As a consequence, most hydropower projects have seen widespread protests from the local mountain communities, especially in India where mechanisms of sharing benefits have not been implemented adequately (Diduck et al. 2013). For managing this risk, governments and hydropower companies need to provide direct and tangible benefits to the local mountain communities.

After a hiatus of more than two decades, hydropower is back on the investment agenda of international financial institutions (Baghel and Nüsser 2010). Hydropower investments are also being financed through Clean Development Mechanism (CDM) in parts of India and China, even though it is not entirely clear that large hydropower dams do indeed meet the goals of CDM (Erlewein and Nüsser 2011). New norms for environmental sustainability and benefit sharing with local communities are being developed with the hope that hydropower projects will be better built than in the past

Table 8.5 Hydropower potential in eight HKH countries

Country (1)	Hydropower potential (Mega Watt, MW) (2)	Actual hydropower developed (MW) (3)	Actual generation in GWh (4)	Year (Source of data) (5)
Afghanistan	25,000	1,000	868.6	2006 (Government of Afghanistan, Ministry of Energy and Water 2006)
Bangladesh	1,897	230	590.1	2014 (Bangladesh Power Development Board 2014)
Bhutan	24,000	1,615	7,748	2015 (International Hydropower Association website, 2015)
China	370,000	319,370	1,128,027	2015 (International Hydropower Association website, 2015)
India (all)	148,701	42,848	121,894	Central Electricity Authority (CEA) as of 30th June 2016
Indian Himalayan States				
Assam	680	430	1,011	CEA (2016)
Arunachal Pradesh	50,328	98	366	CEA (2016)
Himachal Pradesh	18,820	1,495	9,451.1	CEA (2016)
Jammu and Kashmir	14,146	2,274.4	4,798.7	CEA (2016)
Meghalaya	2,394	356.6	257	CEA (2016)
Mizoram	2,196	34	NA	CEA (2016)
Manipur	1,784	82	30	CEA(2016)
Nagaland	1,574	53	10	CEA (2016)
Sikkim	NA	270	910	CEA (2016)
Uttarakhand	25,000	3,756	NA	CEA (2016)
Tripura	NA	62	1,025	CEA (2016)
West Bengal	NA	1,328	1,199	CEA (2016)
Nepal	43,000	753	3,496	2015 (International Hydropower Association website 2015)
Myanmar	100,000	3,151 (2015)	9,502	2015 (International Hydropower Association website 2015)
Pakistan	50,000	6,902	33,946.5	WAPDA Annual Report 2013–14

Sources For Col (2) and (3), as mentioned in the last Col (5), for Col (4), United Nations Statistics Division for 2015, available online at https://unstats.un.org/unsd/energy/yearbook/2015/t32.pdf

(Box 8.5). For instance, Nepal has developed multiple mechanisms for sharing benefits with local communities. These include, among others implementations, a formal mechanism of royalty sharing and a unique Nepali home-grown solution of giving hydropower equity shares to local affected local populations (Shrestha et al. 2016). In India, on the other hand, several policies have institutionalized mechanisms of benefit sharing, but lack of implementation means that local communities often protest against these projects. Overall, hydropower can be a win-win development for the concerned region, provided that its negative externalities are managed. One particular area of concern is the irrigation-hydropower tradeoff. Not much is known about the extent to which farmer-managed irrigation systems are affected by hydropower projects, but there is some evidence that with proper planning and local participation, hydropower projects can offset some of the tradeoffs and provide additional irrigation benefits to local people.

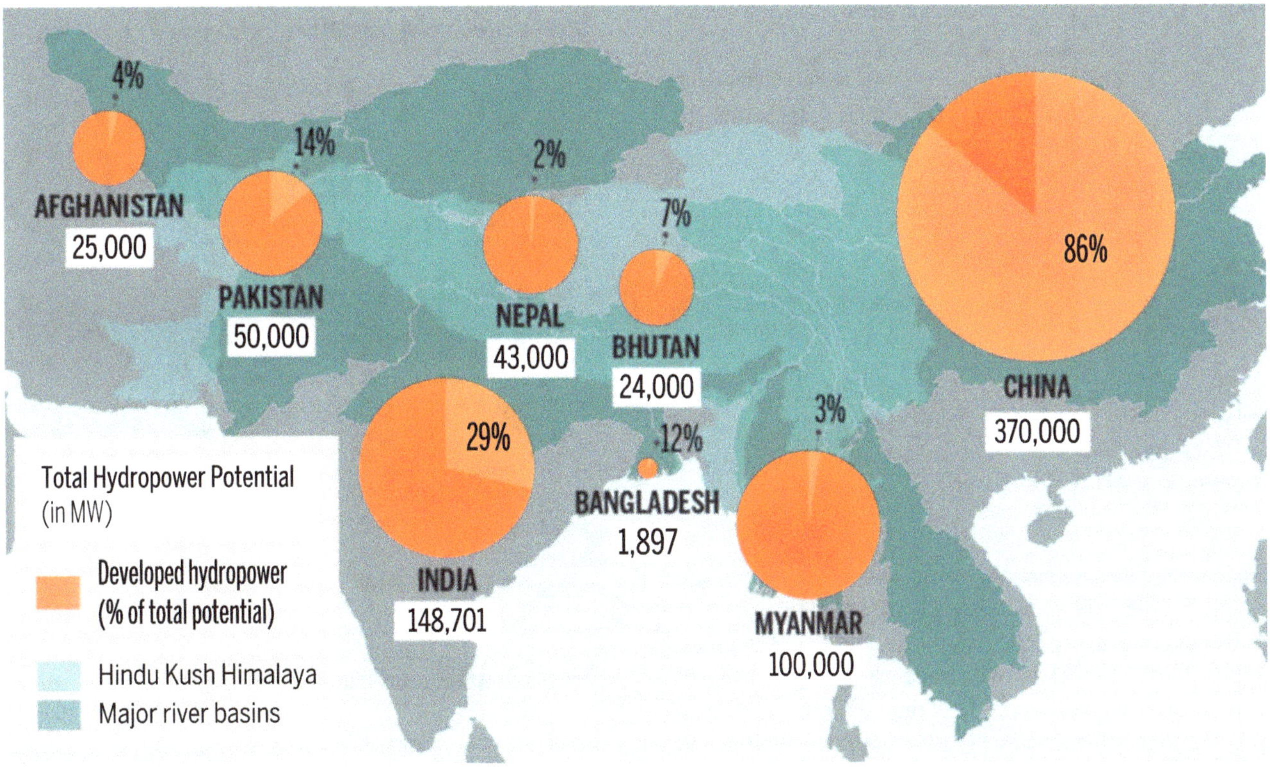

Fig. 8.9 Total hydropower potential and developed hydropower as a percentage of total by country (*Sources* See Table 8.5)

Box 8.5 Water-related benefit sharing in hydropower projects: examples from Nepal

Hydropower development leads to short-term and long-term changes in the hydrology of project-affected areas and often impinges on formal and informal water rights of local populations. Hydropower projects are required to mitigate losses to the local people or compensate for losses related to the reduction of flow in project-affected areas, which are either partially or completely dewatered.

In order to create benefits, hydropower developers invest in improving irrigation systems or fisheries and providing access to drinking water. Several hydropower projects support local farmers in the construction of check dams and irrigation canals and, in some cases, also support the regular maintenance of these facilities. Among these cases, Ridi and Aadhi Khola hydropower projects have been exemplary in showing how small hydropower projects can meet the energy and food-security demands of project-affected communities. The Kali Gandaki-A project recognized local fishing rights of the *Bote* indigenous fishing community, trained the community in new fishing technology, and provided access to government hatchery services. Many hydropower projects have provided drinking water to project-affected citizens as a benefit. Water-supply lines provided by hydropower projects bring clean, reliable water closer to households, reducing the time needed to fetch it from distant sources—a change that has been especially beneficial to women.

8.3.3 Water for Drinking and Sanitation

The Millennium Development Goals (MDGs) set the target of reducing by half the proportion of people without access to safe drinking water and basic sanitation by 2015. As seen in Table 8.6 and Fig. 8.10, most countries of the HKH have performed moderately well in terms of improving access to drinking water but have substantially lagged behind in achieving safe sanitation goals.

In 2015, the global community adopted the SDGs. Unlike the prior Millennium Development Goals (MDGs), which addressed water only in terms of water for sanitation and health, the SDG water-related goals are more comprehensive. Goal 6 focuses on water exclusively. Universal access to drinking water and sanitation correctly remain central, with even more focus needed on quality of service as opposed to just quantitative aspects. At the same time, other considerations are also important—water quality,

Table 8.6 Drinking water and sanitation access in HKH countries

Country	Sanitation access (% of total population)			Drinking water access (% of total population)		
	Total	Urban	Rural	Total	Urban	Rural
Afghanistan	37	60	30	50	78	42
Bangladesh	NA*	57	55	80	85	78
Bhutan	65	87	54	92	99	88
China	NA*	58	52	89	98	82
India	31	54	21	88	96	84
Myanmar	81	86	79	71	75	69
Nepal	62	NA*	NA*	88	93	87
Pakistan	91	NA*	NA*	48	NA*	NA*

NA* not available
Source FAO (2011); WHO/UN-Water (2014)

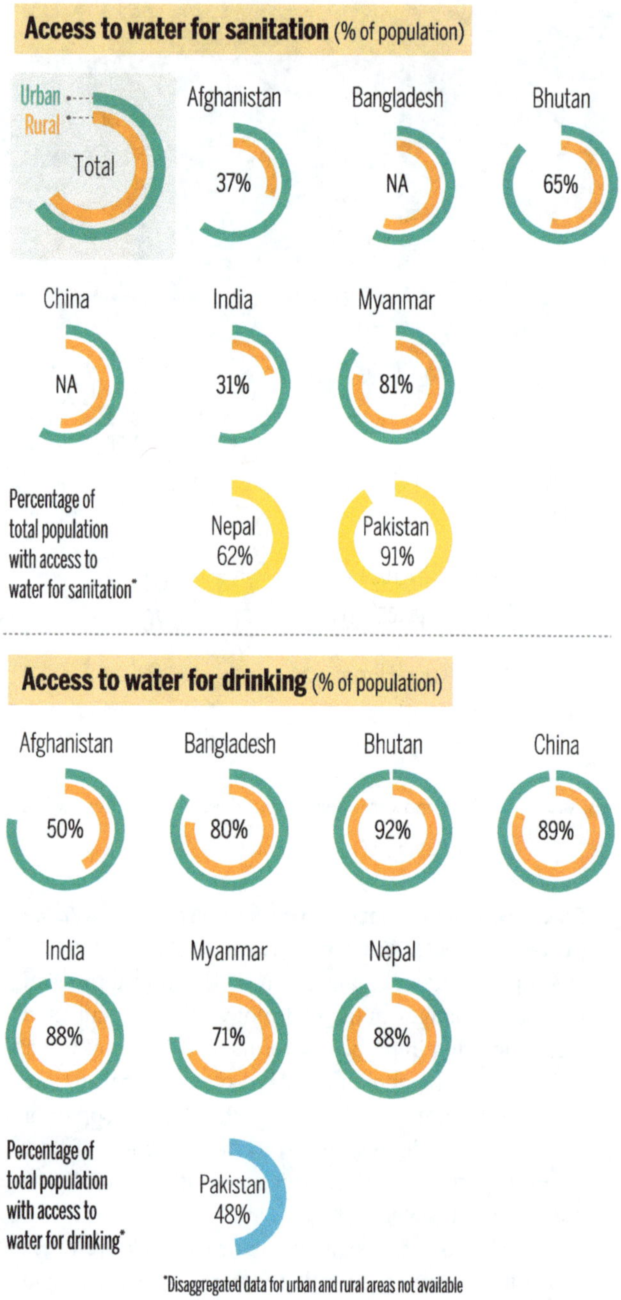

Fig. 8.10 Access to water for sanitation and drinking by country (*Source* FAO 2011; WHO/UN-Water 2014)

wastewater management and reuse, transboundary cooperation, ecosystem services, and capacity building.

Burgeoning urban populations in the HKH will exert further stress on already overstretched urban services. As a result, standard solutions such as providing piped water and building more toilets will add only marginal benefits, unless the realities specific to mountain water resources are taken into account. For instance, tapping mountain springs will become increasingly difficult, given the widespread anecdotal as well as new emerging knowledge that documents drying up of springs (Poudel and Duex 2017; Kumar and Sen 2017a, b). New investments will be necessary for spring revival. ICIMOD and its partners have come up with a comprehensive 8-step methodology for the revival of mountain springs, and it is being tested in few sites in Nepal (Shrestha et al. 2017). This is important because springs are the only source of reliable water supply for large number of mountain communities and they do not get the attention they deserve. Most importantly, communities must be involved and have decision-making authority at all stages of water and sanitation services—from planning and construction to maintenance and management. However, it is important to keep in mind that community itself is a heterogeneous entity, and some members are subject to discrimination due their gender, caste or ethnicity. A study by Coffey et al. (2015) shows that the problem of open defecation in plains of India is deeply related to caste prejudices, whereas in parts of India where caste system is not prevalent (e.g. northeastern states), level of sanitation is much higher than the rest of India (Ghosh and Cairncross 2014).

8.3.4 Urban Water

Following the global trend, all countries in the HKH are urbanizing rapidly. This means, existing urban centers will expand and new urban centers will emerge. Trends of urbanization are somewhat different in each of the HKH countries. In the Pakistan portion of the HKH, the rate of urbanization has been low due to the constraints placed by the terrain and lack of economic opportunities. The urban population in the northern region of Pakistan is less than 20%.

In the Indian Himalayas, the rate of urbanization has been low in the higher altitudes, but it has been more rapid in the foothills (also called the *Siwaliks*). In western Indian

Table 8.7 Gap between municipal water supply and demand in selected cities of the HKH countries of Nepal, India, Bhutan, and Afghanistan

City, country	Average elevation (masl)	Population (year)	Supply (million liters per day MLD)		Demand (MLD)	Demand met (%)		Year of available water supply/demand data
			Wet season	Dry season		Wet season	Dry season	
Kathmandu, Nepal	1350	2,510,000 (2012)	105	86	280	37.5	30.7	2012
Pokhara, Nepal	884	300,000 (2012)	24	21	45	53.3	46.7	2014
Darjeeling, India	2045	132,016 (2011)	8.3	2.3	8.6	96.5	26.7	2002
Mussoorie, India	2005	30,118 (2011)	7.67		14.4	53.3		2014
Shimla, India	2205	171,817 (2011)	54.5		64.7	84.2		2012
Thimphu, Bhutan	2320	79,185 (2005)	NA*		9.9	NA*		2006
Kabul, Afganisthan	1791	3,476,000 (2013)	52.14		NA*	NA*		2013

NA*not available. *Source* City population data from National Population Censuses of respective countries, while municipal water supply and demand statistics have been compiled from various newspaper reports reliable because formal data from municipalities and/or countries/states are not available in the public domain. Darjeeling: TOI (2014); Mussoorie: Pioneer (2013); Shimla: TOI (2013); Thimpu: *The Bhutanese* (2014); Kabul: *The Guardian* (2010); Kathmandu and Pokhara: *Republica* (2014). For Shimla and Mussorie, supply and demand data is aggregated across seasons

Himalaya, Srinagar is the largest urban center, while in the eastern Himalaya, the urban centers of Gangtok, Kalimpong, and Darjeeling have been growing at a very rapid pace. Nepal remains one of the least urbanized countries in South Asia—and also in the world. There are considerable problems in terms of definition in the study of Nepal's urbanization, since the areas designated "urban" have been defined and redefined over the years with evident lack of consistency. In Nepal, Kathmandu is by far the largest urban agglomeration.

Bangladesh occupies a very small section of the Himalayas, represented by the Chittagong Hill Tracts (CHT). In these areas, tourism has flourished and led to the growth of a few small urban centers, namely Rangamati, Bandarban, and Khagrachari. Much of the urbanization in the region has been unplanned and haphazard, leading to serious problems related to water and sanitation.

The fact that mountain towns and cities are also tourist destinations amounts to additional pressure on water resources, and the water needs of the local population often are not met in pursuit of serving the water requirements of tourists. This sometimes leads to social conflicts. Table 8.7 shows that almost no major city in the region is self-sufficient in terms of municipal water supply.

Almost all urban centers suffer from water shortage. Many of these urban centres are hill stations set up by the colonial British government on ridgetops, while water sources are deep down in the valleys. Compounding the problem of water shortage are issues such as neglect of traditional water systems like stone spouts and springs (Molden et al. 2016; Colopy 2012), outdated and poorly constructed water distribution systems that get superimposed on traditional water systems, pipe leakages, and poor governance that puts primacy on piped water supply over other time tested and sustainable sources. Different cities have adopted different coping mechanisms. In Kathmandu and Darjeeling, private water tankers provide water to millions of residents, while in Bhutan, water supply is rationed and people are encouraged to manage their own demand accordingly. In Kathmandu, wastewater generated by city sewage is used to irrigate vegetable crops in peri-urban parts of the valley (Box 8.6).

Water shortages in the urban centers of HKH affect men, women, and marginalized communities differently. In Kathmandu Valley, the poor who live in marginal areas within the city, especially in the peri-urban areas outside the municipal water supply limits, pay a higher price for buying water from informal sources (Raina 2016). The same is true for residents who rent accommodation even in core city centre—they often do not have guaranteed rights to use the municipal connection, which is reserved for the house owner, and end up paying more than double for water (Molden et al. 2016). At the same time, in many Himalayan towns like Kathmandu, urban water security is also closely linked with cultural practice. For example, stone spouts which originate in springs, are seen as a form "of cultural resilience, where people interact with water infrastructure in ways that extend beyond utilitarian concerns" (Molden et al. 2016). This is true for many other urban centres in the Himalayas where traditional water supply systems embedded in local cultural beliefs have been neglected in favor of western patterns of piped water supply. Long term sustainability in urban water in the HKH will require a strategic management of government-sponsored, "modern" piped water systems along with preservation of traditional water systems, particularly in ways that are mindful of gender and class differences in water access, provisioning, and security.

Box 8.6 Wastewater use in Kathmandu Valley

In 2011, Kathmandu had a population of 2.51 million and it has been growing at a rate of 6.6% per year—the fastest urban growth in all of Nepal. It is estimated that a total of 93 million litres per day (MLD) of wastewater is generated from the domestic sector and another 6.5 MLD from the industrial sector. These numbers are growing by the day, but wastewater management facilities have not expanded commensurately. Of the total wastewater generated in the valley, less than 50% is actually collected and treated; the rest is disposed of directly into the rivers.

Wastewater is used extensively for irrigation in the urban and peri-urban parts of the valley. At least one third of the cultivated area in the Valley is irrigated using wastewater, and almost two thirds of this wastewater is used directly in the fields without any kind of treatment (Bastakoti et al. 2014). A majority of the farmers reported using wastewater because there is no source of freshwater for irrigation and also because accessing this wastewater, often used illegally, is free of cost, unlike groundwater, which requires investment in tube wells and diesel pumps. This wastewater also happens to be nutrient-rich, and therefore reduces fertilizer costs. Farmers using this water for irrigation often complain of health issues, such as skin infections, and the indirect health impacts of these pollutants through the consumption of these vegetables and other crops are not trivial.

While it is recognized that negative health and environment impacts of waste use are significant, it is unlikely that policies or infrastructure to deal with the use of wastewater in agriculture will be developed anytime soon. In this context, it is important to also understand the positive contribution of wastewater to the Valley's agricultural economy, while framing adequate policies and institutions to manage the health risks of untreated-wastewater use.

8.3.5 Water Quality: Major Biological and Chemical Contaminants Linked to Urbanization

Due to unsustainable urbanization, industrialization, water abstraction, and agricultural intensification, deterioration of water quality is increasingly becoming a recognized concern in many parts of the HKH (Merz et al. 2003; Mateo-Sagasta and Tare 2016). Water quality degradation has significant impacts on human health and ecosystems and is limiting regional development (Mateo-Sagasta and Tare 2016). Approximately 20% of all deaths among children under five years of age is caused by water-borne diseases (WHO 2006). The water quality challenges are not homogeneous along the river systems and vary between upstream segments in the mountain regions and further downstream segments in the hills and plains (IITC 2010). In the upper segments, the rivers and streams flow on steep and narrow rocky beds, carry cold water, and are subjected to much less anthropogenic pollution. But they also have highly sensitive ecosystems and biodiversity. The lower segments of the river systems, on the other hand, are greatly modified by human interventions in terms of water diversion/abstraction, and they are subjected to a high degree of pollutant loads (IITC 2010; Shah and Shah 2013).

There are no comprehensive studies addressing water quality for the HKH as a whole. However, there have been some studies comparing two or three countries, including the study by Karn and Harada (2001), which looked at surface-water pollution in Kathmandu (Nepal), Delhi (India), and Dhaka (Bangladesh). This study revealed widespread pollution of water resources in all three cities through the presence of organic and pathogenic contaminants, heavy metals, and pesticides (Karn and Harada 2001). For example, in a 13 km stretch in the Bagmati River in the Kathmandu Valley, biochemical oxygen demand (BOD) increased from 3.8 to 30 mg/L moving downstream during 1992–1995. Similarly, BOD in the Yamuna River in Delhi, showed an increase from 1.3 to 17 mg/L in the downstream area (Karn and Harada 2001). Similar situations were found in all of the rivers close to Dhaka (Turag, Buriganga, and Dhaleswori).

Studies have estimated that municipal sewage contributed nearly 85% of all river pollution. This was due to two major factors: first, the unrestricted discharge of raw or partially treated wastewater (of both domestic and industrial origin); and second, the lack of adequate regulatory pollution-control measures and their strict enforcement in real practice (Karn and Harada 2001). The main Ganges stream in India still directly receives at least 2.7 billion m^3 of sewage from medium and big cities every day, of which at least 74% is untreated (Mateo-Sagasta and Tare 2016). Additionally, the number of towns and cities that discharge their untreated wastewater to the tributaries of Ganges, and not only to the main stream, is substantially bigger. The Indian Central Pollution Control Board (CPCB) has identified 138 drains discharging 6 billion liters per day of polluted water into the Ganges (CPCB 2013; Mateo-Sagasta and Tare 2016).

Another issue is the lack of septage management systems. While there is growing interest in septage management, adequate treatment of septage does not currently exist in the region. Dumping of domestic solid wastes—including plastics, glass, and organic waste—also affects the river water quality. The collection capacity of municipal solid waste is limited and the treatment capacity is almost nil (Mateo-Sagasta and Tare 2016).

Industries contributed 14–17% of this river pollution, with pollutants coming from power plants, food-processing, breweries and distilleries, tanneries, as well as the industrial production of fertilizer, insecticides, textiles, carpets, vegetable oil, dairy, pharmaceuticals, and other chemicals (Karn and Harada 2001). The findings of another study suggested that the countries of the Ganges-Brahmaputra-Meghna River Basin are increasing their industrial activities, with approximately 70% of 300–500 million tons of heavy metals, solvents, toxic sludge, and other wastes being discharged untreated into the waterways (Babel and Wahid 2009).

The contribution of non-point source pollution (NPS) from agricultural areas in the HKH region has so far not been documented. Thus, the extent of their contribution to the pollution load is unclear. There are no regionwide comprehensive assessments of non-point source pollution from agriculture. There are some studies which claim that agrochemicals are a key polluter in the rivers (e.g., Gosh, ND), and there are others that believe that it is not a significant source of pollution (e.g., Trivedi 2010). Nevertheless, the observed trends in terms of expansion and intensification of the agricultural sector—including the sharp increase in fertilizer and pesticides use, and the booming development of livestock farms—point to an increasing trend in NPS contribution to degrading water quality (Mateo-Sagasta and Taare 2016).

In terms of groundwater pollution, the urban areas of Kathmandu mostly suffer from infiltration of urban storm water, leakage of wastewaters and septic tanks, and unregulated industrial activities. These wastewaters and septic-system effluent contain high concentrations of dissolved organic carbon, ammonia, pathogens, and organic micro-pollutants, as well as heavy metals and trace elements (Pant 2011). The presence of heavy metals in groundwater in the Swat River was found to vary along different stretches and was attributed to geology, corrosion of plumbing systems, and agricultural and industrial activities (Khan et al. 2013). In this study, the concentration of heavy metals in the groundwater was higher than in surface waters (Khan et al. 2013).

Some studies have also looked at biological pollutants. In groundwater in Kathmandu, maximum coliforms were found in the samples from shallow wells at 267 CFU/100 mL, while the levels in tube and deep-tube wells were 129 and 149 CFU/100 mL respectively. The coliforms detected in shallow wells may be due to poor drainage, the improper construction of septic tanks close to groundwater sources, and the direct discharge of untreated sewage into surface waters (Pant 2011)—all of which further reflect the lack of planning and investment in the region's water infrastructure. Similar results were reported from Rawalpindi, in Pakistan, where Sehar et al. (2011) found municipal water containing fecal coliforms due to leakage of sewage into water supply pipelines. In Srinagar, in India, significant land use changes since the early 1980s have led to pollution of the freshwater Dal Lake, due the discharge of various nutrients and pollutants (Amin et al. 2014).

In some parts of the HKH, such as Afghanistan, infrastructure has been damaged or destroyed by years of war. Only 27% of Afghanistan's population has access to improved water sources, and only 20% have access in rural areas—marking the lowest percentage in the world. While the number of households in urban areas with access to municipal water is growing (35% in the capital city, Kabul), the system for solid-waste collection is limited, with about 70% of the city's solid waste accumulating on roadsides and in drains, rivers, and open spaces—where they pose a significant environmental hazard. In addition, most sewage is disposed of in domestic drainage pits and shallow, open sewage channels that run along the streets, thereby threatening shallow aquifers with pollution from biological and chemical contaminants. A study on heavy metal and microbial loads in sewage-irrigated vegetables in Kabul revealed lead loads and pathogenic contamination higher than the threshold levels (Safi and Buerkert 2011). Considering the high incidences of intestinal diseases and diarrhea, Safi and Buerkert (2011) recommended further detailed surveys and improvements to Kabul's sewage infrastructure to eliminate potential health risks.

Box 8.7 Arsenic in groundwater and its implications for agriculture

Arsenic in groundwater is a significant threat in the lower parts of the Ganges, Brahmaputra, and Mekong basins. In recent years, a growing body of literature has emerged that examines the impact of irrigation with arsenic-rich water on crop production and productivity and looks at the effectiveness of arsenic remediation in agricultural context. A systematic review of 29 high-quality studies (Senanayake and Mukherji 2014) showed no clear relationship between arsenic content in irrigation water—or in soil with arsenic uptake by paddy grains—and crop production. However, there is a near consensus that prolonged cultivation with arsenic-rich water leads to decline in paddy productivity.

The review of literature also shows that there are six broad categories of intervention that can reduce arsenic uptake by crops or prevent its entry into the human food chain. These are: deficit irrigation, soil fertilization, growing alternative field crops (other than paddy), switching to arsenic-tolerant paddy cultivars, reducing arsenic content in rice through cooking methods, and nutritional supplements. Results from these studies show that all of these interventions are successful in preventing excessive arsenic from entering the human food chain, but these practices have yet to be incorporated into mainstream extension activities.

8.3.6 Water Infrastructure

Water storage infrastructure and enhanced management of excess monsoon water would help HKH countries to better meet water requirements during the dry season and to also cope with water scarcity issues. Traditionally, mountain people have found ways to store water by building ponds, terracing fields, harvesting rainwater, and employing small-scale irrigation systems (Molden et al. 2014). Water is also often diverted from mountain springs, which are fed by groundwater and therefore a more reliable source during dry seasons. These methods are still practiced throughout much of the HKH middle hills. However, with increasing demand for agricultural and energy production, the demand for better and larger infrastructure is also increasing.

This has led to construction of various irrigation structures in the region. In Pakistan, two large storage dams situated in the upper Indus basin, Tarbela dam on the Indus, and Mangla dam on the Jhelum now regulate the irrigation system that millions of people downstream rely on. In India alone, there are 4858 completed large dams (and 313 are under construction, of which nearly 100 are located in the mountainous states) (CWC 2014). Most of the rivers in Nepal have little storage compared to the monsoon run-off in the rivers (Bandyopadhyay 2009). Feasibility studies for many large multi-purpose projects with storage have been proposed, but development has been slow, mainly due to lack of common interest and agreement between Nepal and India.

Even in countries like China, India and Pakistan, storage is still quite low, despite the development of extensive irrigation infrastructures. For example, the current storage capacity in the Ganges in large infrastructure is only 10% of total average annual flow. In Nepal, FMIS have, for centuries, been developed and managed by local farmers themselves. Extensive embankment infrastructure has also been built on riverbanks to control floods during the rainy season. In India, about 34,000 km of flood embankments have been constructed, largely in North and Northeast India (Mazumder 2011).

Water-related infrastructure can intensify upstream-downstream linkages, providing benefits and risks to both areas. Structures like dams and reservoirs can store water during flood periods, which can: (a) be available during the dry season through open channels or pipelines for irrigation and other consumptive, recreational, or environmental uses for ecosystem services; (b) produce electricity; and (c) improve navigation. However, these structures come at a high cost to local communities displaced by them as they are forced to relocate and adjust to shifting resources and cultures.

Dams and reservoirs can also be problematic as they can block and store sediment that is transported in river flow. Fine silt and eroded materials are considered beneficial to plains farming; therefore, the blockage of sediment in natural flow can affect agriculture production. Singh (1990) has evaluated the Farakka barrage, determining that it has negatively affected the downstream region of Bangladesh by reducing silt flow, thereby reducing soil fertility, and by increasing the ingress of saltwater up the river. The Brahmaputra leaves behind immense sand deposits in Dhemaji and Lakhimpur districts of Assam, while the Koshi floods in 2008 rendered a large area of fertile lands infertile due to sand deposition. Silt transport in rivers also leads to filling up and reduction of the storage capacity of reservoirs, lakes, and ponds as well as the carrying capacity of canals. The Teesta Barrage in Bangladesh had a provision of a silting tank in the main canal system in order to stop the entry of silt into canals and therefore to the agriculture fields.

In the Indus River Basin, downstream discharge to the sea has decreased significantly due to construction of vast networks of irrigation canals, barrages, and associated structures. Laghari et al. (2015) estimated that these anthropogenic changes have resulted in five times less sediment in downstream areas. Tahmiscioğlu and Anul (2007) highlighted that dam construction and the resulting holding of sediment can lead to changes in the natural water regime, including the composition of soil nutrients downstream.

In Nepal, excessive river sediment has affected most of the power plants in the Himalayas through build-up in reservoirs or by erosion of turbine components, reducing the life of the plants. Al-Faraj and Scholz (2014) highlighted that human-made structures, such as dams and large-scale water systems also decrease water availability in downstream areas of transboundary river basins.

Water storage infrastructure can also include natural structures such as lakes, ponds, groundwater, and soil moisture. Managing natural storage systems are more cost effective and sustainable. Management and enhancement of natural storage systems still need to be explored and if possible, included with the feasibility studies on built infrastructure projects such as large dams.

8.3.7 Ecosystem Processes and Environmental Flows

With the growing degree of human intervention in the rivers of the HKH, concerns have emerged over the future of the aquatic ecosystems in the HKH and continuation of the related ecosystem processes (as has been detailed in Chap. 6). Increasing energy and water demands from the domestic, agricultural, industrial, and commercial sectors are leading to plans for greater exploitation of these rivers.

There are considerable, though poorly understood, implications of climate change on increasing monsoon-season flows and decreasing dry-season flows—particularly coupled with various anthropogenic interventions. The term "environmental flows" (EFs) is now commonly used to refer to a managed flow regime designed to maintain a river in some agreed-upon, non-pristine ecological condition as to make room for the human interventions. Each component of the natural hydrological regime has a certain ecological role to perform. In regulated basins, the magnitude, timing, frequency, and duration of some or all flow components can end up getting modified. A tradeoff has to be arrived at through negotiations among all stakeholders in order to identify the instream flow regime that would support the aquatic ecosystem processes and services in a sub-pristine state. The suite of acceptable flow patterns can ensure a flow regime capable of sustaining target aquatic ecosystem processes and services in a sub-pristine state (Poff et al. 1997; Arthington et al. 2006). EFs can therefore be seen as a way to balance human interventions and the maintenance of river ecology at acceptable levels of degradation.

The knowledge base for operationalizing EF practice is not yet well established in any of the HKH countries, but it is emerging. There has been a particularly significant increase in interest on the matter in China and India (Bandyopadhyay 2017). In the last decade, India has taken a rather mechanical—and not empirically tested—approach for the assessment of EFs, and the government is trying to implement an EF plan under the Ganges rejuvenation program. However, demand for a scientifically tested approach for arriving at an acceptable balance between interventions and EFs is growing. In Nepal, Bhutan, Pakistan, and Bangladesh, EF is beginning to enter discussions on river basin governance. With limited water resource development and with most infrastructure development still in the planning phase (especially in the upper mountain regions), there is still a chance to set up measurement of EF requirements for ecosystem processes and put appropriate policies in place before these rivers are seriously degraded.

Despite the extensive study of environmental water allocations in countries such as the UK, Australia, USA, and South Africa, arriving at managed flow regimes acceptable to all stakeholders in the basin continues to be elusive. In India, the Krishna Water Disputes Tribunal in 2010 had considered the delicate and much awaited issue of EF in a river and implemented it in the Krishna River and its basin area. It awarded separate waters to the river itself in its different reaches, making it incumbent upon the concerned states to make available and maintain the recommended quantities of water in-stream round the year as prescribed by the Tribunal (MoWR 2010). For that purpose, it replaced the EF with Minimum Flows, which incorporates small needs of those who live by the river and whose livelihood depend on that river. It also accounts for the religious need in addition to the EF need. At present, EF are most often justified by ecological concerns—for instance, by the question of preserving ecosystem health for the sake of biodiversity conservation. This approach pays little regard to those whose livelihoods are gained directly from the continuity of ecosystem services of the rivers. For many rural men and women in developing countries, aquatic ecosystems are essential to their wellbeing and livelihood, providing domestic water and also sustaining fisheries, livestock, grazing, and other important resources. Further, as a rule, current EF considerations do not take into account cultural and religious contexts, which are also very important in the HKH. Therefore, there is a need to further develop methodologies for assessing of EFs for the HKH and to explore ways to incorporate consideration of EFs into river-governance practices.

8.4 Water Governance in the Hindu Kush Himalaya

While the previous sections have addressed the physical availability and the uses of water in the HKH, this section will highlight how challenges of governance—as distinct from water scarcity per se—form a crucially important obstacle to achieving water security in the region Biggs et al. (2013a).

As suggested in earlier sections, water scarcity is institutionally mediated across geographical scales within the region. By characterizing the formal, informal, and hybrid water-governance institutions at the local/micro, subnational/national/meso, and regional/international/macro scales, a scalar lens of governance will inform our discussion of HKH water security. The links that connect water security to energy and food security—known as the water-energy-food (WEF) nexus—will also be highlighted. In a critical mode, we question conventional wisdom on the existence of the nexus at the micro scale within the HKH. Scaling up to meso and macro levels, we then critically examine prospective approaches to both basin-scale management within the region and to management of transboundary water relations at the subnational and national scales. We conclude this section with a consideration of pathways toward improved water-related decision-making in the HKH, across micro, meso, and macro scales.

For the purpose of this discussion, water governance is understood to be the mechanism for addressing questions of water access, use, and distribution among social actors, sectors, and across geographical scales. The outcome of good water governance should be social equity and political stability enabled by environmental quality across the HKH. A key premise of our discussion is that water governance is a deeply political enterprise. If water is life, and human life is steeped in politics, then water use and distribution are inevitably steeped in politics as well. This section is organized around geographical scales, not only to illustrate how politics impact water governance at the micro, meso, and macro scales, but also how water politics might be moderated and informed by evidence-based policy.

8.4.1 Characterization of Existing Water Governance Institutions

The HKH is characterized by relatively weak penetration of formal state (national) institutions. This is due to the remoteness of much of the region, owing mainly to topography. Issues over water are no exception to this general lack of strong state presence, although modern state institutions have recently started becoming more influential, especially in terms of infrastructure development at the meso and macro scales. Informal customary water governance at the micro-scale, with its marked gender and other inequalities, has been the predominant institutional norm in the region. The recent rise in state penetration has not replaced existing governance mechanisms, but it has spawned hybrid governance regimes with informal structures heavily mediating state intervention, rather than the reverse. The state has, however, indirectly contributed to profound transformation of informal water governance through the provision of energy and technology for harnessing and managing water supply as well as through investments in infrastructure for irrigation and energy (particularly hydropower).

The region's political geography is dominated by nation-states, which must be the arbiters of any water governance at the country level. At the moment, there are just a few examples of multilateral or regional water-governance structures, e.g., Indus River Commission (and Indus Waters Treaty) and Mekong River Commission for the lower Mekong basin. Two regional multilateral institutions—South Asian Association for Regional Cooperation (SAARC) and the International Centre for Integrated Mountain Development (ICIMOD)—have not been involved in water governance. While comprehensive review of SAARC is beyond the scope of this chapter, suffice it to say that SAARC has maintained a strict neutrality and a studied silence on the subject. On the other hand, ICIMOD has been quite proactive in generating knowledge on water governance and related issues in the region, but it has limited its activities to research and dissemination. This lack of a multilateral or regional governance framework for water is largely a result of the nationalization of water by the HKH countries, as the power disparities among these countries cause them to guard individual sovereignty over water.

Sovereign control over water has not prevented some countries in the region from entering bilateral treaties, which require regulating the exercise of sovereignty over domestic water resources to satisfy treaty obligations. Such treaties and their accompanying institutional structures are discussed in Sect. 8.3.2.

At the subnational scale, Shah (2009) describes the existing water management structures in northern Afghanistan as community-based water-management systems that pivot around the institution of an elected or selected *mir-e-aab* (water master) with minimal or absent state presence in managing most canal and *karez* systems (underground aqueduct, also known as *qanat* in west Asia). Local water-allocation systems were largely disrupted during the long Afghan civil war, and the inability of the post-Taliban regime to restore to original claimants their abandoned or appropriated water rights is a source of considerable resentment among the Afghan populace.

Since 2003, through the Ministry of Rural Rehabilitation and Development, the Afghan government has inserted itself into local-level water management by availing funds for participatory water-infrastructure development through Community Development Councils operating under the National Solidarity Programme. This has led to more of a hybrid institutional regime, with the balance of power resting with the local informal regime more than the state. Due to the contours of local power relations, the results of these hybrid management systems have not been equitable in all cases in terms of gender or class (McCarthy and Mustafa 2014).

In the Pakistani-administered part of the Karakoram range, constituting Gilgit-Baltistan, water-management institutions have also been largely local and community-based—and they have gotten a considerable boost from the investments in community organization and water-infrastructure development undertaken by the Aga Khan Rural Support Programme (AKRSP) and similar NGOs (Box 8.8).

Box 8.8 Aga Khan rural support programme

AKRSP provides an illustrative evidence of a productive and cooperative relationship between state and non-state (or more informal) actors. The AKRSP's primary approach is to promote the participation of local stakeholders in water management through the creation and support of village organizations, which then provide the structure and resources necessary for these villages to effectively manage their own water supplies. This model has enjoyed some success in promoting water access, particularly in more rural and difficult-to-access areas. One of the biggest successes of the program, however, has been its relatively extensive engagement with state authorities, providing logistical, financial, and technical resources in support of these activities (De Spoelberch 1987; Ehsan-ul-Haq 2007). This example illustrates the potential for greater state cooperation with a variety of actors in promoting effective and sustainable water-management activities.

Gilgit-Baltistan might be an exception in the HKH, as it demonstrates synergy and cooperation between state and non-state actors, including the AKRSP, Aga Khan Foundation (AKF), and other related institutions. By contrast, in most of the HKH, typical relations between state and non-state water managers are indifferent, if not downright hostile. In Pakistan, as in India, water is primarily a provincial/state subject, with the central governments only intervening in the financing of large-scale infrastructure projects, such as hydropower, deemed to be of national interest. However, within the HKH areas of both countries, water management, for all practical purposes, remains local and community-based.

Many of these local-scale water-management systems limit access to safe water by gender, causing serious consequences for women and the health and wellbeing of girls. In Nepal, Udas and Zwarteveen (2010) documented that the central irrigation bureaucracy is unable to systematically address issues of gendered access to water because of the country's entrenched patriarchal ethos, confirming the earlier review by Chandra and Fawcett (1999), who documented how lack of participation by women in water-supply projects ultimately increased their workloads and diminished their prospects of benefiting from improved water infrastructure.

In the Indian Himalayas, the central government's Ministry of Water Resources, River Development & Ganga Rejuvenation, and Central Water Commission play strong, engineering-focused, state-led roles in the development of water resources, investment in infrastructure, and data collection and monitoring, with primary influence at the macro scale. Without adequate attention to meso and macro scales, national involvement may tend to overlook local water management, extraction, and allocation practices—especially in the case of FMIS, which have a centuries-long tradition and form the backbone of livelihoods and food-security in rural mountain communities. Additionally, globalization, market integration, the penetration of contract farming, and seasonal to permanent out-migration (especially of working-age males) are having profound impacts on irrigated agriculture.

Within the region, Himachal Pradesh and Sikkim present more dynamic and transformative instances of active state involvement in water management—ranging from irrigation and potable-water supply to hydropower. By contrast, the states of Jammu and Kashmir and Uttarakhand have resorted to a more conventional approach, with central government making a strong imprint on infrastructure and water management. This is due, in part, to territorial and strategic concerns. In J&K, concerns ostensibly include the integration of local communities into mainstream Indian polity; in Uttarakhand, there are governmental concerns over nationalist sentiments for Mother Ganga and for the historical marginalization of hill districts, which formed part of Uttar Pradesh state before breaking off as the new state of Uttarakhand (previously Uttaranchal).

The states of Northeast India present an entirely different picture, with local practices and traditions holding sway, marking a governance system that is less of a hybrid than one in which state and central government institutions are largely absent. With the advent of hydropower in the Northeast—and the perception of a large gap between generation potential and installed capacity on high-volume tributaries to the Brahmaputra—the region is witnessing greater involvement of central government, including investments in infrastructure through public and private capital, as well by multilateral institutions like the Asian Development Bank.

In terms of domestic water supply, the expanding urban areas of the HKH are largely serviced by either centrally or provincially controlled agencies (such as the centrally controlled Kathmandu Valley Water and Sanitation Board). In HKH rural areas, the main government agency responsible for domestic water supply is Public Health Engineering (PHE). Most of the time, however, the domestic water supply is actually serviced through community-based initiatives. PHE has an infrastructure bent and often assumes responsibility for supplying domestic water to larger and medium sized cities in the region, such as Gilgit, Muzaffarabad, and Srinagar.

In all of the urban water-supply situations, the emphasis is on networked, piped water systems, replicating the infrastructure and institutional models of the western and plains cities of South Asia—but without regard to topography, cultural particularities, or the institutional history of water supply in HKH cities. The consequences go beyond an uneven water supply, inequitably rendered in terms of class and location across the urban areas of the HKH; serious health hazards also arise from water-supply contamination.

In sum, the key features of the institutional water landscape in the HKH are as follows:

- Water management is characterized by a hybrid formal-informal regime, with the balance of power in favor of informal institutions, particularly at the local level.
- At the macro scale—and certainly for meso- and macro-scale infrastructural development—the balance of power is in favor of formal state institutions.
- There is a disconnect between the macro/national, meso/regional, and micro/local water-governance institutions, which is largely a function of the political marginality of national terrain within the HKH. This is especially true of larger nation-states. Nepal and Bhutan are the exceptions, with their national boundaries falling predominantly within the HKH.
- There is a need for greater synergy between state and informal water-management institutions without the strict institutional boundaries that exist at present. Local water management and its informal institutions could benefit from state support instead of the antagonism that is present today.
- The gender inequities often witnessed in informal and formal institutions are a matter of serious concern and should be a priority area for reform.
- The urban water-supply systems in emerging cities in the HKH need to be more attuned to the particularities of the topography and the organic growth of the cities where formal institutional regimes uncomfortably preside over the informal institutional landscape, with deleterious consequences for water quality and quantity.

These characteristics of local and national water institutions exist in juxtaposition to governance at the level of transboundary river basins; thus, we turn to conflict and cooperation across geographical scales in the HKH.

8.4.2 River-Basin Approaches and Transboundary Conflict and Cooperation

Countries throughout the HKH face similar challenges of increasing water demand due to economic growth. Availability of and access to water resources vary dramatically throughout the region due to seasonal precipitation patterns, the geographic distribution of glaciers, and, importantly, a lack of adequate governance. In addition, rising uncertainty in water availability and increases in extreme weather are both likely due to climate change (Molden et al. 2014). Management of water resources at a river-basin scale may help in maximizing benefits of infrastructure projects, negotiating competing water and energy uses, and minimizing risk of water-related hazards. However, a river-basin approach is challenging at macro (or meso) scales that are either international or interstate (subnational).

At both transboundary and subnational levels, coordination throughout shared river basins requires increased institutional capacity, particularly across scales, and may require a decoupling of national political aims from shared resource-management objectives. The river-basin approach is particularly relevant in the context of upstream-downstream benefit sharing between HKH areas and downstream populations in the plains. Within the HKH, however, much of the demand for water, especially for drinking water, is met by groundwater from springs and handpumps—and tubewells in urban areas—where the most relevant geographical unit for effective management is not the river basin but the springshed, which do not follow the river valley contours. Springs draw upon mountain aquifers, which may be shared among multiple valleys; therefore, holistic water management must integrate mechanisms across river basins and springsheds in order to better coordinate surface and groundwater resource management.

In the HKH, a river-basin approach would harness the full potential of water resources while managing competing uses in the face of rising demand and increasing uncertainty (Shrestha et al. 2015b). Both infrastructure and institutional water-management approaches benefit from a river-basin perspective. Building institutional capacity at a river-basin scale can improve coordination between upstream and downstream areas. It can also improve cross-sectoral policies for water and energy. However, the river basins originating in the HKH often cross state or national borders, making coordinated basin-wide water management a question of riparian states' ability, institutional arrangements, and above all, political commitment to cooperate or at least creatively address conflicts.

Building transboundary institutional capacity is challenging due to the different needs and priorities of riparian states. River-basin management can even be difficult at the subnational level due to a lack of interstate institutional mechanisms, a predominance of local and community-based water-management schemes, and a lack of alignment among hydrologic boundaries and administrative management units.

At the local scale, much of the agricultural and domestic water supply is dependent upon mountain springs, whose aquifers do not necessarily follow the basin's surface boundaries. The basin approach may be useful at the meso and macro scales, but at the micro/local scale, formal institutions can prove useful by helping to link micro water-management institutions across valleys in order to address common issues of spring recharge, zone protection, and water quality. However, at the international transboundary level, the rivers originating within the HKH continue to be strongly contested, as outlined in Sect. 8.4.2.1.

8.4.2.1 Transboundary Waters

Transboundary resource sharing in South Asia has historically been fraught with contentious relationships, characterized by a focus on national interests, a lack of trust, and hegemonic power play. National interests and international power relations have played a significant role in hydro-politics in the region (Asthana and Shukla 2014). Despite being connected by hydrologic flows (Amu Darya, Indus, Ganges, Brahmaputra, Irrawaddy, Salween, and Mekong basins across national borders), states have often taken unilateral action on water-management decisions, leading to fragmented management of transboundary resources, narrow (albeit understandable) focus on national interests, and negative consequences for neighbouring riparian countries and communities (Asthana and Shukla 2014; Rasul 2014; Giordano et al. 2016).

Where international cooperation on water management exists, agreements are typically made between only two countries, and water disputes are often entangled with other political issues (Shah and Giordano 2013; Giordano et al. 2016). Several governments in the region have multiple challenges of achieving political consent for international water negotiations, whereas others are seen as regional hegemons. Bilateral water treaties often involve nation-states with disparate levels of political power. For example, in the Ganges-Brahmaputra-Meghna basins, India holds separate bilateral treaties with Nepal and Bangladesh, despite the fact that these three countries are all part of a larger shared basin. These water treaties are often inflexible and lack adequate mechanisms for negotiation of inter-party conflicts.

Bilateral water treaties in the region have resulted in varied outcomes for downstream states. Sometimes, transboundary water treaties have improved shared-resource management, but in other cases, the lack of an adequate—or any—agreement has contributed to contentious state relations. The Indus Water Treaty, signed by India and Pakistan in 1960 and currently in effect, secured a significant apportionment of 80% of Indus River Basin flows for Pakistan, the lower riparian state (Shah and Giordano 2013). India and Bhutan were able reach a mutually beneficial agreement on hydropower development in shared river basins; Bhutan earns over 60% of its national GDP from hydropower sales to India (Shah and Giordano 2013).

By contrast, treaties developed for joint hydropower projects on the Gandak, Koshi, and Mahakali Rivers (in 1952, 1954, and 1996, respectively) have tended to exacerbate tensions between India and Nepal. The Koshi agreement provided compensation to Nepal for land inundation, irrigation flows, and benefit sharing from a hydropower and flood-control project constructed by India within Nepalese territory. However, Nepal does not feel that the agreement has been upheld fairly (Shrestha et al. 2012; Giordano et al. 2016). Further, lack of bilateral agreement on required modifications of the Koshi project led to failure of the embankment of 2008, causing major flooding with severe damage and loss of life—and compounding the existing mistrust between these nations.

Water projects within the Brahmaputra and Ganges basins have also led to increased tensions between India and Bangladesh. India constructed the Farakka Barrage on the Ganges to divert dry-season flow for drinking-water and irrigation and to prevent sedimentation in Kolkata port. However, Bangladesh perceives the project as negatively impacting downstream water flows and delta productivity in Bangladesh.

Other projects have been developed unilaterally. India constructed a series of run-of-the-river hydropower projects and a diversion barrage on the Teesta River. Both projects negatively impact downstream Bangladesh. Although agreements on minimum flows and dispute resolution were reached between India and Bangladesh on these projects in 1996 and 1998, in both cases, Bangladesh remains dissatisfied with India's fulfillment of the terms of the agreements. To further complicate international agreement on water sharing, domestic protests within India have weakened the central government's ability to achieve an equitable arrangement with Bangladesh over the Farakka Barrage, an example of how international water cooperation is subverted by domestic political aims.

To redirect water management from conflict among international riparian states toward productive cooperation, joint water projects and research efforts are two ways to engage multiple players and build trust—while also serving to increase the knowledge base on resource issues, improve evidence-based decision making, identify mutually beneficial goals, and leverage cooperation within the scientific community to promote cooperation at other levels (Asthana and Shukla 2014).

Cooperation will also lead to better planning of infrastructure projects, reducing impacts on resources, livelihoods, and ecosystems. Finally, regional information-sharing systems need to be established to facilitate open data exchange within river basins. Data sharing will help facilitate disaster management, increase capacity for information dissemination, improve regional resilience to climate change, and improve early-warning systems for floods—especially for glacial lake outbursts. However, questions about conflict and cooperation at multiple scales must also take into account interdependent resource systems, illustrated through the WEF security nexus and its unique manifestations within the HKH.

8.4.3 The Water-Energy-Food Security Nexus

There is a growing recognition of the important links between water, energy, and food. This triad is such that the security of one is impacted by or influences the others. This so-called WEF security nexus has emerged as an important conceptual paradigm for sustainable resource management. The nexus is considered to be a set of synergies and tradeoffs resulting from the inter-relations among resources, institutions, and security and the linkages between resource use and development, whereby interconnections can allow multiple needs to be addressed simultaneously. The tradeoffs and synergies are multi-dimensional, spanning physical and social spheres across multiple scales (Rasul 2014), both rural and urban (Scott et al. 2016).

The critical links among groundwater, energy, and irrigation have been highlighted in a number of studies. The proliferation of electric pumps for extracting groundwater in India (Mukherji 2007; Shah 2009; Bassi 2017) and of diesel pumps in Pakistan (Siddiqi and Wescoat 2013) has led to an extensive increase in energy consumption for agricultural production in the plains. Policy instruments, such as power tariff reforms, have been identified as interventions that can simultaneously reduce power demand (thereby improving energy supply for non-farm power needs), improve agricultural productivity, promote equity, and allow for more sustainable use of groundwater in agriculture (Kumar et al. 2013).

Within the mountainous regions of the HKH, the key WEF nexus linkages are principally through hydropower-electricity generation and irrigation, with important urban and rural implications that raise questions of political power and access (Allouche et al. 2015). In rural HKH regions especially, fuelwood for heating and cooking as well as rainfed agriculture for food production have WEF nexus implications. In the case of hydropower, off-grid, small-scale systems serving local communities have played an important development role in some areas (Pervaz and Rahman 2012). On the other hand, large, grid-connected hydropower systems serve regional energy demands with a distinct advantage for downstream agriculture and urban demands. In 2013, hydropower constituted a significant portion of total electricity production ($\sim$77% in Nepal; 32% in Pakistan; and 12% in India (IEA 2016)).

Overall, the WEF nexus at the micro/local scale in the HKH is not constituted by critical tradeoffs; rather it exists due to the essential need for water in both the energy and agricultural sectors. However, the prevailing system of access to and use of resources is vulnerable to disturbances in climate—and WEF security in rural areas will be significantly impacted (with the poor being exposed to higher risks). Small-scale hydropower systems will likely be the first affected by changes in streamflow due to climate change (as compared to large systems with significant storage capacity). Furthermore, landslides and floods that disrupt road connectivity in remote mountains will impact food imports and distribution to local markets.

The HKH has extensive hydropower-generation potential (estimated at 500 GW), and several large-scale systems are in operation or in stages of planning and development (Vaidya 2012). Power-generation revenue accrues at the provincial level (where power plants are situated); however, the electricity generated is largely used in the densely populated plains—as is stored water (for irrigated agriculture). For instance, Khyber Pakhtunkhwa (KPK) and Gilgit-Baltistan collectively host 76% (46 GW) of the 60 GW total estimated hydropower potential in Pakistan.

The irrigation benefits of the Indus waters are largely derived in the plains of Punjab and Sindh, whereas energy-generation revenues are accrued in KPK and Gilgit-Baltistan (Siddiqi and Wescoat 2013). An improved evaluation of the distribution of costs and benefits can allow for creating inclusive and equitable arrangements. For instance, a fraction of hydropower revenue could be provisioned for rural development of the province, and some of the new development projects could be directed toward adaptations for climate change impacts, such as deployment of photovoltaic (rather than micro-hydro) systems.

System-level modeling and analysis have largely focused on meso/basin- or macro/national-scale profits and benefits (Yang et al. 2014). As large projects are planned and funded by national agencies, the relevant scale of analysis has been at the provincial (meso) or national (macro) level. These approaches can be improved by incorporating methods and metrics that compute national-level net benefits in conjunction with local costs borne at smaller levels.

8.4.4 Decision-Making Improvements

Water-related decision making across micro, meso, and macro scales is mediated by the relative social and political power of stakeholders and interest groups—and by complex and often involuted institutional histories and designs. At the micro level, the main conduits for decision making are predominantly informal local institutions. These institutions are embedded in the local-level geographies of power organized around class, ethnicity, and gender. To facilitate more inclusive decision-making structures, there is a need for greater synergy between the formal state and informal local institutions. The key political challenge is to make the informal local-level institutions accountable and equitable, without undermining their efficacy or legitimacy. Decision making by informal institutions at the local level is, indeed, based on local knowledge; therefore, there is a need for formal state institutions to learn from the local decision-making bodies.

At the meso/subnational scale, the balance of power is in favor of formal state institutions. Here, the key challenge is to steer what is generally politicized decision-making toward evidence-based decision-making, appropriately informed by science and local knowledge. At this scale, the importance of research and knowledge-generating actors cannot be overemphasized. Scientifically rigorous, socially informed, and locally relevant research could be made available to local level decision-makers, who, in turn, could also be made more accountable. As science points the way toward ecological and economic sustainability, accountability will ensure that attention is paid to social justice and sustainability.

At the macro/international scale, there is an obvious need for greater trust between the nation-states of the HKH. Water conflicts rarely play out in isolation from the range of other issues between nation-states—but water can be a conduit for trust-building. Again, development that goes beyond design and construction could bring dimension to the largely engineering-focused perspective of the national water bureaucracies in the countries of the HKH.

Educating citizens and the press on water issues within the HKH could also provide a counterpoint to the focus on infrastructure and engineering that is currently predominant in addressing water problems in the region. Communicating to all stakeholders the importance of the socio-economic, cultural, spiritual, and ecological dimensions of water resources development is essential to maintain healthy and productive river basins. National water policies informed by multidisciplinary perspectives could help both national and regional initiatives find innovative solutions to seemingly intractable water conflicts, serving such programs as: ICIMOD's Himalayan Adaptation, Water, and Resilience; Ganges Basin program of Water, Land, and Ecosystems under the Consultative Group for International Agricultural Research (CGIAR); and the South Asia Water Initiative. Finally, the link between subnational and national water politics cannot be overlooked. The aforementioned regional-level interventions could also serve to make national water policies more regionally and ecologically sensitive and less narrowly nationalistic.

8.5 Challenges and Ways Forward

Water availability, use, and governance in the HKH are in a constant state of flux. In terms of availability, annual river flows across the HKH during this century in general will not undergo great change because increased precipitation and runoff will tend to counteract reduced flow from glacial melt, except in the Indus and other western basins where contributions from monsoon precipitation are low. However, pre-monsoon flows are expected to decline, impacting irrigation, hydropower, and ecosystem services. Data uncertainties are high, and they cannot easily account for spatial and temporal heterogeneity. Projections indicate that intra-annual variability in surface water will increase. There is anecdotal evidence that springs in the mid-hills of the Himalaya are drying up. These contribute to lean season baseflow in local streams, which are the primary source of drinking water in the HKH mid-hills before joining the rivers fed by glacial melt. Despite high proportions of water use in agriculture throughout the region, increased urban demand for water will continue, based on population growth and the concentration of economic and political power in cities and towns. Meeting SDG targets focused on urban water supply, wastewater, and sanitation will place unprecedented pressure on water resources in HKH cities. In the future, it is likely that larger cities in and around the HKH will resort to long-distance water transfers from HKH highlands to satisfy increasing demand, possibly designing suitable institutions for payment to upstream communities for ecosystem services. It is also likely that some of the bigger cities will invest in workable wastewater-treatment infrastructure. However, the real crisis will occur in smaller emerging towns, which have inadequate funds for infrastructure upgrades, particularly for water supply and wastewater treatment, and lack suitable governance institutions.

Reduced lean-season river flow coupled with increasing urban and upstream demand will result in reduced availability of surface water to downstream farmers for irrigation. Agriculture and irrigation will become increasingly feminized, and formal and informal institutions must respond to this reality, or mountain agriculture and irrigation systems will stagnate or shrink. For example, most FMIS still consider members to be landowners, who are men, many of whom have migrated. If institutions become inclusive, and if remittance money is invested in agriculture, then it is possible that the shift to remunerative crops like coffee, orchards, and mountain niche crops (instead of rainfed cereals) will become even more pronounced.

Hydropower is rapidly emerging as the main source of energy and revenue for Himalayan states, but changing river flow regimes will require hydropower projects to be constructed and managed in order to account for pre-existing water use for irrigation. Without mechanisms to negotiate water sharing between existing agricultural and emerging hydropower uses, the water available for agriculture will decline in localized areas. There is a need for improved benefit-sharing norms that enable the preservation of water flows for agricultural use while allowing substantial, but mediated, hydropower development.

Ecosystem flows in Himalayan rivers and streams are subject to flow regimes that are heavily impacted by human water uses. The greatest impacts on fish, macroinvertebrates, and other riverine flora and fauna result from increasing intra-annual variability of river flows, even though inter-annual variability in flood and drought cycles certainly affects riparian ecosystems. Increasing intra-annual variability will reduce lean-season flows and diminish the high-monsoon flows characteristic of HKH regimes.

Acknowledgements The participation of Christopher Scott and Tamee Albrecht was made possible through the support of the International Water Security Network, a project funded by Lloyd's Register Foundation, a charitable foundation helping protect life and property by supporting engineering-related education, public engagement and the application of research.

References

Abbas, N., & Subramanian, V. (1984). Erosion and sediment transport in the Ganges river basin (India). *Journal of Hydrology, 69*(1–4), 173–182.

ADB. (2014). *Sector Assessment (summary): Agriculture, natural resources, and rural development. Country partnership strategy: Bhutan, 2014–2018*. Manila: The Asian Development Bank.

Agarwal, A., Bhatnagar, N., Nema, R., & Agrawal, N. K. (2012). Rainfall dependence of springs in the Midwestern Himalayan hills of Uttarakhand. *Mountain Research and Development, 32*(4), 446–455.

Al-Faraj, F. A. M., & Scholz, M. (2014). Impact of upstream anthropogenic river regulation on downstream water availability in transboundary river watersheds. *International Journal of Water Resources Development, 31,* 28–49.

Ali, K. F., & De Boer, D. H. (2007). Spatial patterns and variation of suspended sediment yield in the upper Indus River basin, northern Pakistan. *Journal of Hydrology, 334*(3), 368–387.

Allouche, J., Middleton, C., & Gyawali, D. (2015). Technical veil, hidden politics: Interrogating the power linkages behind the nexus. *Water Alternatives, 8,* 610–626.

Amin, A., Fazal, S., Mujtaba, A., & Singh, S. K. (2014). Effects of land transformation on water quality of Dal Lake, Srinagar, India. *Journal of the Indian Society of Remote Sensing, 42,* 119–128.

Andermann, C., Longuevergne, L., Bonnet, S., Crave, A., Davy, P., & Gloaguen, R. (2012). Impact of transient groundwater storage on the discharge of Himalayan rivers. *Nature Geoscience, 5*(2), 127–132.

Archer, D. (2003). Contrasting hydrological regimes in the upper Indus Basin. *Journal of Hydrology, 274*(1–4), 198–210. http://linkinghub.elsevier.com/retrieve/pii/S0022169402004146. Accessed 11 Oct. 2013.

Arthington, A. H., Bunn, S. E., Poff, N. L., & Naiman, R. J. (2006). The challenge of providing environmental flow rules to sustain river ecosystems. *Ecological Applications, 16,* 1311–1318.

Aryal, K., & Choudhury, D. (2015). Climate change: Adaptation, mitigation and transformations of swidden landscapes: Are we throwing the baby out with the bathwater? In M. F. Cairns (Ed.), *Shifting Cultivation and environmental change: Indigenous people, agriculture and forest conservation* (pp. 281–288). New York: Routledge.

Ashraf, A., & Ahmad, Z. (2008). Regional groundwater flow modeling of Upper Chaj Doab of Indus Basin, Pakistan using finite element model (Feflow) and geoinformatics. *Geophysical Journal International, 173,* 17–24.

Ashraf, A. (2013). Changing hydrology of the Himalayan watershed, current perspectives, in contaminant hydrology and water resources sustainability, Dr. P. Bradley (Ed.). ISBN: 978-953-51-1046-0, InTech, https://doi.org/10.5772/54492.

Ashraf, A., Abuzar, M. K., Ahmad, B., Ahmad, M. M., & Hussain, Q. (2017). Modeling risk of soil erosion in high and medium rainfall zones of Pothwar region, Pakistan. *Proceedings of the Pakistan Academy of Sciences: Pakistan Academy of Sciences: B. Life and Environmental Sciences*, *54*(2), 67–77.

Asthana, V., & Shukla, A. C. (2014). *Water security in India: hope, despair, and the challenges of human development*. Bloomsbury Publishing USA.

Babel, M. S., & Wahid, S. M. (2009). Freshwater under threat South East Asia: Vulnerability assessment of freshwater resources to environmental change: Ganges-Brahmaputra-Meghna River Basin, Helmand River Basin, and Indus River Basin. United Nations Environment Programme.

Baghel, R., & Nüsser, M. (2010). Discussing large dams in Asia after the World Commission on Dams: Is a political ecology approach the way forward? *Water Alternatives, 3*(2), 231–248.

Bajracharya, S. R., Maharjan, S. B., Shrestha, F., Bajracharya, O. R., & Baidya, S. (2014). *Glacier status in Nepal and decadal change from 1980 to 2010 based on landsat data*. Kathmandu: ICIMOD.

Baker, J. M. (2005). *The Kuhls of Kangra: Community managed irrigation in the Western Himalaya*. New Delhi: Permanent Black.

Bandyopadhyay, J. (2017). Restoration of ecological status of Himalayan rivers in China and India: The case of the two mother rivers—The Yellow and the Ganges. In S. Dong, J. Bandyopadhyay, & S. Chaturvedi (Eds.), *Environmental Sustainability from the*

Himalayas and the oceans: Struggles and innovations in China and India (pp. 69–98), Chap. 4. Switzerland: Springer.

Bandyopadhyay, J. (2002). A critical look at the report of the World Commission on Dams in the context of the debate on the large dams on the Himalayan rivers. *Water Resource Development, 18*(1), 127–145.

Bandyopadhyay, J. (2009). Climate change and Hindu Kush-Himalayan waters–knowledge gaps and priorities in adaptation. *Sustainable Mountain Development, 56*(56), 17–20.

Bangladesh Bureau of Statistics. (2015). Yearbook of agricultural statistics-2013, 25th Series, Bangladesh Bureau of Statistics (BBS), Statistics and Informatics Division (SID), Ministry of Planning, Government of the People's Republic of Bangladesh.

Bangladesh Power Development Board. (2014). Annual Report, downloaded from http://www.bpdb.gov.bd/download/annual_report/Annual%20Report%202014-15.pdf 12th June 2016.

Barlow, M., et al. (2005). Modulation of daily precipitation over East Africa by the Madden—Julian oscillation. *Monthly Weather Review, 133,* 3579–3594.

Bashir, F., Zeng, X., Gupta, H., & Hazenberg, P. (2017). A hydrometeorological perspective on the Karakoram Anomaly using unique valley-based synoptic weather observations. *Geophysical Research Letters, 44,* 10470–10478.

Bassi, N. (2017). Solarizing groundwater irrigation in India: A growing debate. *International Journal of Water Resources Development*. https://doi.org/10.1080/07900627.2017.1329137.

Bastakoti, R. C., Ale, M., & Shivakoti, G. (2015). Robustness-vulnerability characteristics of irrigation systems in Nepal. In P. Pradhan, U. Gautam, & N. M. Joshi (Eds.), *Small scale irrigation systems: Challenges to sustainable livelihood* (p. 374). Kathmandu: Farmer Managed Irrigation Systems Promotion Trust.

Bastakoti, R. C., Maskey, N., Drechsel, P., & Prasad, S. (2014). Wastewater irrigation in urban agriculture.

Bharati, L., Gurung, P., Jayakody, P., Smakhtin, V., & Bhattarai, U. (2014). The projected impact of climate change on water availability and development in the Koshi Basin, Nepal. *Mountain Research and Development, 34*(2), 118–130.

Biggs, E. M., Duncan, J. M., Atkinson, P. M., & Dash, J. (2013a). Plenty of water, not enough strategy: How inadequate accessibility, poor governance and a volatile government can tip the balance against ensuring water security: The case of Nepal. *Environmental Science & Policy, 33,* 388–394.

Biggs, E. M., Tompkins, E. L., Allen, J., Moon, C., & Allen, R. (2013b). Agricultural adaptation to climate change: Observations from the Mid-Hills of Nepal. *Climate and Development, 5*(2), 165–173. https://doi.org/10.1080/17565529.2013.789791.

Blöthe, J. H., & Korup, O. (2013). Millennial lag times in the Himalayan sediment routing system. *Earth and Planetary Science Letters, 382,* 38–46.

Bolch, T., Pieczonka, T., Mukherjee, K., & Shea, J. (2017). Brief communication: Glaciers in the Hunza catchment (Karakoram) have been nearly in balance since the 1970s. *The Cryosphere, 11,* 531–539. https://doi.org/10.5194/tc-11-531-2017, 2017.

Bookhagen, B., & Burbank, D. W. (2010). Toward a complete Himalayan hydrological budget: Spatiotemporal distribution of snowmelt and rainfall and their impact on river discharge. *Journal of Geophysical Research: Earth Surface, 115*(F3).

Butz, D. (1989). The agricultural use of melt water in Hopar settlement, Pakistan. *Annals of Glaciology, 13,* 35–39.

Central Electricity Authority. (2016). http://www.cea.nic.in/index.html data downloaded on 12th June, 2016. Government of India.

Central Water Comission. (2014). *National register of large dams. Dams of national importance (completed and under construction).* Central Water Commission. Ministry of Water Resource. Government of India, 207 pp.

Chakraborty, T., Kar, R., Ghosh, P., & Basu, S. (2010). Kosimegafan: Historical records, geomorphology and the recent avulsion of the Kosi River. *Quaternary International, 227,* 143–160.

Chandra Regmi, S., & Fawcett, B. (1999). Integrating gender needs into drinking water projects in Nepal. Gender and Development (Vol 7.3, pp. 62–72). Oxfam, Oxford.

Chen, Z., Li, J., Shen, H., et al. (2001). Yangtze river of China: Historical analysis of discharge variability and sediment flux. *Geomorphology, 41*(2), 77–91.

China Statistical Year Book. (2015). http://www.stats.gov.cn/tjsj/ndsj/2015/indexeh.htm on 15th June 2016, published by China Statistical Press, Government of People's Republic of China, Beijing.

Coffey, D., Gupta, A., Hathi, P., Spears, D., Srivastav, N., & Vyas, S. (2015). *Culture and the health transition: Understanding sanitation behaviour in rural north India.* IGC Working Paper, April 2015. https://pdfs.semanticscholar.org/f63e/c230b37b7275185c8a35b8bc551af0808754.pdf.

Colopy, C. (2012). *Dirty, sacred rivers: Confronting South Asia's water crisis.* New York, NY: Oxford University Press.

CPCB. (2013). *Pollution assessment: River Ganga.* Delhi: Central Pollution Control Board.

Dahri, Z. H., et al. (2016). An appraisal of precipitation distribution in the high-altitude catchments of the Indus basin. *Science of the Total Environment*, 548–549, pp. 289–306. http://www.sciencedirect.com/science/article/pii/S0048969716300018.

De Spoelberch, G. (1987). A model: The Aga Khan rural support program. *Challenge, 29*(6), 26–31.

Diduck, A. P., Pratap, D., Sinclair, A. J., & Deane, S. (2013). Perceptions of impacts, public participation, and learning in the planning, assessment and mitigation of two hydroelectric projects in Uttarakhand, India. *Land Use Policy, 33*, pp.170–182.

Dixit, A. (2009). Kosi embankment breach in Nepal: Need for a paradigm shift in responding to floods. *Economic and Political Weekly, 44*(6), 70–78.

Dixit, A., & Upadhya, M. (2005). *Augmenting groundwater in Kathmandu Valley: Challenges and possibilities.* Nepal water conservation foundation, Kathmandu, Nepal.

Ehsan-ul-Haq. (2007). Community response to climatic hazards in Northern Pakistan. *Mountain Research and Development, 27*(4), 308–312.

Erlewein, A. (2013). Disappearing rivers—The limits of environmental assessment for hydropower in India. *Environmental Impact Assessment Review, 43*, 135–143.

Erlewein, A., & Nüsser, M. (2011). Offsetting greenhouse gas emissions in the Himalaya? Clean development dams in Himachal Pradesh, India. *Mountain Research and Development, 31*(4), 293–302.

FAO. (2016a). AQUASTAT. http://www.fao.org/nr/water/aquastat/water_res/index.stm downloaded on 11th June, 2016, Food and Agricultural Organization, Rome.

FAO. (2016b). AQUASTAT, FAO. (http://www.fao.org/nr/water/aquastat/water_use/index.stm) downloaded on 12th of June 2016. Food and Agricultural Organization, Rome.

FAO. (2016c). AQUASTAT. http://www.fao.org/nr/water/aquastat/irrigationmap/chn/index.stm downloaded on 15th of June, 2016. Global Map of Irrigated Area Statistics for China. Food and Agricultural Organization, Rome.

FAO. (2016d). AQUASTAT.http://www.fao.org/nr/water/aquastat/irrigationmap/MMR/index.stm downloaded on 15th of June, 2016. Global Map of Irrigated Area Statistics for China. Food and Agricultural Organization, Rome.

Fort, M. (2015). Natural hazards versus climate change and their potential impacts in the dry, northern Himalayas: Focus on the upper Kali Gandaki (Mustang District, Nepal). *Environmental Earth Sciences, 73*(2), 801–814.

Galewsky, J. (2009). Orographic precipitation isotopic ratios in stratified atmospheric flows: Implications for paleoelevation studies. *Geology, 37*(9), 791–794.

Ghosh, A., & Cairncross, S. (2014). The uneven progress of sanitation in India. *Journal of Water, Sanitation and Hygiene Development, 4* (1), 15–22.

Giordano, M., Gyawali, D., Nishat, A., & Sinha, U. (2016). Can there be progress on trans-boundary water cooperation in the Ganga? In L. Bharati, B. R. Sharma, & V. Smakhtin (Eds.), *The Ganges river Basin: Status and challenges in water, environment and livelihoods*. London: Routledge.

Giosan, L., Syvitski, J., Constantinescu, S., et al. (2014). Climate change: Protect the world's deltas. *Nature, 516*(7529), 31–33.

Gole, C. V., & Chitale, S. V. (1966). Inland delta building activity of Kosi river. *Journal of the Hydraulics Division ASCE, 91,* 111–126.

Gosh (ND) Pollution from non-point sources in Ganga Basin: A Community-based potential for managing the unmarked crisis.

Government of Afghanistan, Ministry of Energy and Water, 2006.

Government of India, GOI. (2011). *Report of the 4th Minor Irrigation Census*. Ministry of Water Resources Development, New Delhi, Government of India.

Government of Nepal. (2016). Groundwater resources development board, Ministry of Irrigation, Government of Nepal, data downloaded on 12th June, 2016. http://www.gwrdb.gov.np/hydrogeological_studies.php.

Government of Pakistan. (2010). Agricultural census 2010—Pakistan Report. http://www.pbs.gov.pk/content/agricultural-census-2010-pakistan-report. 12 June 2016.

Grumbine, R. E., & Pandit, M. K. (2013). Threats from India's Himalaya Dams. *Science, 339,* 36–37.

Hewitt, K. (2011). Glacier change, concentration, and elevation effects in the Karakoram Himalaya, Upper Indus Basin. *Mountain Research and Development, 31*(3), 188–200.

ICIMOD. (2014). International conference on mountain people adapting to change: Solutions beyond boundaries bridging science, policy, and practice. Kathmandu, Nepal, 9–12 November 2014.

ICIMOD. (2011a). *Glacial lakes and glacial lake outburst floods in Nepal*. Kathmandu: ICIMOD.

ICIMOD. (2011b). Mountain, green economy for sustainable development: A concept paper for Rio + 20 and beyond. In *International Conference on Green Economy and Sustainable Mountain Development Opportunities and Challenges in View of Rio* + 20 (p. 31).

IEA. (2016). https://www.iea.org/statistics/. International Energy Association.

IITC. (2010). River Ganga at a glance: Identification of issues and priority actions for restoration; 001_GBP_IIT_GEN_DAT_01_Ver 1_Dec 2010.

Immerzeel, W. W., Van Beek, L. P. H., & Bierkens, M. F. P. (2010). Climate change will affect the Asian water towers. *Science, 328* (5984), 1382–1385. https://doi.org/10.1126/science.1183188.

Immerzeel, W. W., et al. (2011). Hydrological response to climate change in a glacierized catchment in the Himalayas. *Climatic Change, 110,* 721–736. http://www.springerlink.com/index/10.1007/s10584-011-0143-4. Accessed 2 Aug. 2011.

Immerzeel, W. W., et al. (2015). Reconciling high altitude precipitation with glacier mass balances and runoff. *Hydrology and Earth System Sciences, 12,* 4755–4784.

Immerzeel, W. W., Pellicciotti, F., & Bierkens, M. F. P. (2013). Rising river flows throughout the twenty-first century in two Himalayan glacierized watersheds. *Nature Geoscience, 6,* 742–745. http://www.nature.com/doifinder/10.1038/ngeo1896. Accessed 5 Aug. 2013.

International Hydropower Association. (2016). https://www.hydropower.org/country-profiles downloaded on 12th June, 2016.

Ishaq, M., & Farooq, A. (2016). Farming in the northern mountains of Pakistan: Role of women. Retrieved from http://lib.icimod.org/record/12328/files/1006.pdf on 12 June 2016.

Islam, M. R., Begum, S. F., Yamaguchi, Y., & Ogawa, K. (1999). The Ganges and Brahmaputra rivers in Bangladesh: Basin denudation and sedimentation. *Hydrological Processes, 13*(17), 2907–2923.

Janakarajan, S., & Moench, M. (2006). Are wells a potential threat to farmers' well-being? Case of deteriorating groundwater irrigation in Tamil Nadu. *Economic and Political Weekly, 41*(37), 3977–3987.

Janssen, M. A., & Anderies, J. M. (2013). A multi-method approach to study robustness of social–ecological systems: The case of small-scale irrigation systems. *Journal of Institutional Economics, 9*(04), 427–447. http://dx.doi.org/10.1017/S1744137413000180.

Jeelani, G. (2008). Aquifer response to regional climate variability in a part of Kashmir Himalaya in India. *Hydrogeology Journal, 16*(8), 1625–1633.

Karn, S. K., & Harada, H. (2001). Surface water pollution in three urban territories of Nepal, India, and Bangladesh. *Environmental Management, 28*(4), 483–496.

Kaser, G., Großhauser, M., & Marzeion, B. (2010). Contribution potential of glaciers to water availability in different climate regimes. *Proceedings of the National Academy of Sciences, 107* (47), 20223–20227.

Kc, K. (2003). Optimizing water use in Kathmandu valley (ADB TA-3700), Final Draft Report on Groundwater/Hydrogeology in Kathmandu Valley.

Khan, A. N. (1987). Spate irrigation in Pakistan. In *FAO/UNDP, Spate irrigation: Proceedings of the Subregional Expert Consultation on Wadi Development for Agriculture in Natural Yemen*, 6–10 December 1987 (pp. 167–170). Rome: FAO.

Khan, K., Lu, Y., Khan, H., Zakir, S., Khan, S., Khan, A. A., et al. (2013). Health risks associated with heavy metals in the drinking water of Swat, northern Pakistan. *Journal of Environmental Sciences, 25*(10), 2003–2013.

Khattak, M. S., Babel, M. S., & Sharif, M. (2011). Hydro-meteorological trends in the upper Indus River basin in Pakistan. *Climate Research, 46*(2), 103–119. http://www.int-res.com/abstracts/cr/v46/n2/p103-119/. Accessed 29 Nov. 2012.

Kresic, N., & Stevanovic, Z. (2009). *Groundwater hydrology of springs*. Butterworth-Heinemann.

Kreutzmann, H. (2011). Scarcity within Opulence: Water management in the Karakoram mountains revisited. *Journal of Mountain Science, 8,* 525–534.

Kumar, D., Scott, C. A., & Singh, O. P. (2013). Can India raise agricultural productivity while reducing groundwater and energy use? *International Journal of Water Resources Development, 29*(4), 557–573.

Kumar, K., Satyal, G. S., & Kandpal, K. D. (2006). Farmer and state managed hill irrigation systems in Kumaun Himalayas. *Indian Journal of Traditional Knowledge, 5*(1), 132–138.

Kumar, V., Sen, S. (2017a). *Assessment of spring potential for sustainable agriculture: A case study in lesser Himalayas.* Agricultural Water Management. (under review).

Kumar, V., Sen, S. (2017b). Evaluation of spring discharge dynamics using recession curve analysis: a case study in data-scarce region, Lesser Himalayas, India. *Sustainable Water Resources Management*, 1–19. (in press).

Laghari, A. N., Abbasi, H. U., Aziz, A., & Kanasaro, N. A. (2015). Impact analyses of upstream water infrastructure development schemes on downstream flow and sediment discharge and subsequent effect on Deltaic region. *Sindh University Research Journal, 47,* 805–808.

Lapworth, D., MacDonald, A., Krishan, G., Rao, M., Gooddy, D., & Darling, W. (2015). Groundwater recharge and age-depth profiles of intensively exploited groundwater resources in northwest India. *Geophysical Reseach Letters, 42*(18), 7554–7562.

Livingston, M. (2009). *Deep wells and prudence: Towards pragmatic action for addressing groundwater overexploitation in India, Report*. World Bank.

Lutz, A. F., et al. (2014). Consistent increase in High Asia's runoff due to increasing glacier melt and precipitation. *Nature Climate Change, 4,* 587–592.

Mahamuni, K., & Kulkarni, H. (2012). *Groundwater resources and spring hydrogeology in South Sikkim, with special reference to climate change, climate change in Sikkim patterns, impacts and initiatives*. Information and Public Relations Department, Government of Sikkim, Gangtok.

Maleta, H., & Favre, R. (2003). Agriculture and food production in post-war Afghanistan: A report on the winter agricultural survey, 2002–03, p. 29; Joint report of Ministry of Agriculture and Animal Husbandry, FAAHM, Government of Afghanistan and Food, Agriculture and Animal Husbandry Information Management and Policy Unit of the Food and Agriculture Organization of the United Nations.

Mateo-Sagasta, J., Tare, V. (2016). Ganga water quality: Dirty past, promising future? In L. Bharati, B. R. Sharma, V. Smakhtin (Eds.). *The Ganges River Basin: Status and challenges in water, environment and livelihoods* (pp. 222–237). Oxon, UK: Routledge—Earthscan. (Earthscan Series on Major River Basins of the World).

Mazumder, S. K. (2011). Protection of flood embankments by spurs with reference to Kosi River. In *Proceedings of HYDRO-2011*, SVNIT. Surat.

Mehari, A., van Steenbergen, F., & Schultz, B. (2007). Water rights and rules, and management in spate irrigation systems in Eritrea, Yemen and Pakistan. In B. van Koppen, M. Giordano, & J. Butterworth (Eds.), *Community-based Water Law and Water Resource Management Reform in Developing Countries (Comprehensive assessment of water management in agriculture; 5)* (pp. 114–129). Oxfordshire: CABI Publishers.

McCarthy, J., & Mustafa, D. (2014). Despite the best intentions? Experiences with water resource management in northern Afghanistan. In E. Weinthal, J. Troell, & M. Nakayama (Eds.), *Water and post-conflict peacebuilding*. London: Earthscan.

Merz, J., Nakarmi, G., Shrestha, S. K., Dahal, B. M., Dangol, P. M., Dhakal, M. P., et al. (2003). Water: A scarce resource in rural watersheds of Nepal's Middle Mountains. *Mountain Research and Development, 23*(1), 41–49.

Meybeck, M. (1976). Total annual dissolved transport by world major rivers. *Bulletin Hydrological Sciences, 21,* 265–289.

Michael, H. A., & Voss, C. I. (2009). Controls on groundwater flow in the Bengal Basin of India and Bangladesh: Regional modeling analysis. *Hydrogeology Journal, 17*(7), 1561–1577.

Milliman, J. D., & Meade, R. H. (1983). World-wide delivery of river sediment to the oceans. *The Journal of Geology, 91*(1), 1–21.

Molden, O., Griffin, N., & Meehan, K. (2016). The cultural dimensions of household water security: The case of Kathmandu's stone spout systems. *Water International*. https://doi.org/10.1080/02508060.2016.1251677.

Ministry of Water Resources (MoWR). (2010). The report of the Krishna Water Disputes Tribunal. MoWR, Government of India. http://wrmin.nic.in/writereaddata/Inter-StateWaterDisputes/KWDTReport9718468760.pdf. 1 Sept. 2017.

Mishra, K., Sinha, R., Manohar, K. V. S., & Jain, V. (2016). *Sediment dynamics and sediment connectivity in the Koshi basin: implications for river hazards*. Kathmandu: Unpublished Report, ICIMOD.

Moench, M. (2007). When the wells run dry but livelihood continues: Adaptive responses to groundwater depletion and strategies for mitigating the associated impacts. In M. Giordano & K. G. Villholth (Eds.), The agricultural groundwater revolution: Opportunities and threats to development. CABI Publishers, UK (pp. 173–192) (Comprehensive Assessment of Water Management in Agriculture Series No. 3).

Molden, D. J., Vaidya, R. A., Shrestha, A. B., & Shrestha, M. S. (2014). Water infrastructure for the Hindu Kush Himalayas. *International Journal of Water Resources Development, 30*(1), 60–77.

Mollinga, P. P. (2009). Water rights in Farmer Managed Irrigation Systems in India: Equity, rule making, hydraulic property and the ecology. *SAWAS South Asia Water Studies, 1*(1), 1–18.

Mukherji, A. (2007). The energy-irrigation nexus and its impact on groundwater markets in eastern Indo-Gangetic basin: Evidence from West Bengal, India. *Energy Policy, 35,* 6413–6430.

Mukherji, A., Pradhan, N., Shrestha, R., Bhuchar, S., Dhakal, M., Gurung, K. et al. (2016). *Springs and springsheds: ICIMOD position paper*. Unpublished internal document available on request.

Mukherji, A., Molden, D., Nepal, S., Rasul, G., & Wagnon, P. (2015). Himalayan waters at the crossroads: Issues and challenges: Editorial. *International Journal of Water Resources Development, 31*(2), 151–160.

Mukhopadhyay, B., & Khan, A. (2014a). A quantitative assessment of the genetic sources of the hydrologic flow regimes in Upper Indus Basin and its significance in a changing climate. *Journal of Hydrology, 509*, 549–572. http://linkinghub.elsevier.com/retrieve/pii/S0022169413008834. Accessed 6 Feb. 2014.

Mukhopadhyay, B., & Khan, A. (2014b). Rising river flows and glacial mass balance in central Karakoram. *Journal of Hydrology, 513,* 192–203. http://linkinghub.elsevier.com/retrieve/pii/S0022169414002273. Accessed 17 July 2014.

Mukhopadhyay, B., & Khan, A. (2015). A reevaluation of the snowmelt and glacial melt in river flows within Upper Indus Basin and its significance in a changing climate. *Journal of Hydrology, 527*, 119–132. http://dx.doi.org/10.1016/j.jhydrol.2015.04.045.

Nag, D., & Phartiyal, B. (2015). Climatic variations and geomorphology of the Indus River valley, between Nimo and Batalik, Ladakh (NW Trans Himalayas) during Late Quaternary. *Quaternary International, 371,* 87–101.

Narula, K. K., & Gosain, A. (2013). Modeling hydrology, groundwater recharge and non-point nitrate loadings in the Himalayan Upper Yamuna basin. *Science of the Total Environment, 468*, S102–S116.

Negi, G. C., & Joshi, V. (2004). Rainfall and spring discharge patterns in two small drainage catchments in the Western Himalayan Mountains. *India, Environmentalist, 24*(1), 19–28.

Negi, G., Samal, P., Kuniyal, J., Kothyari, B., Sharma, R., & Dhyani, P. (2012). Impact of climate change on the western Himalayan mountain ecosystems: An overview. *Tropical Ecology, 53*(3), 345–356.

Nepal, S., Flügel, W.-A., & Shrestha, A. B. (2014a). Upstream-downstream linkages of hydrological processes in the Himalayan region Ecol. *Process, 3,* 1–16.

Nepal, S., Krause, P., Flügel, W.-A., Fink, M., & Fischer, C. (2014b). Understanding the hydrological system dynamics of a glaciated alpine catchment in the using the J2000 hydrological model. *Hydrological Processes, 28,* 1329–1344.

Nepal, S., Shrestha, A. B. (2015). Impact of climate change on the hydrological regime of the Indus, Ganges and Brahmaputra river basins: A review of the literature. *International Journal of Water Resources Development*, 1–18.

Nüsser, M., Schmidt, S., & Dame, J. (2012). Irrigation and development in the Upper Indus Basin: Characteristics and recent changes of a socio-hydrological system in Central Ladakh, India. *Mountain Research and Development, 32*(1), 51–61.

Nüsser, M. (2001). Understanding cultural landscape transformation: A re-photographic survey in Chitral, Eastern Hindukush, Pakistan. *Landscape and Urban Planning, 57*(241–255).

Nüsser, M., & Baghel, R. (2016). Local knowledge and global concerns: Artificial glaciers as a focus of environmental knowledge and development interventions. In Meusburger et al. (Eds.), *Ethnic and cultural dimensions of knowledge* (pp. 191–209). Dordrecht: Springer.

Nüsser, M., & Schmidt, S. (2017). Nanga Parbat revisited: Evolution and dynamics of sociohydrological interactions in the Northwestern Himalaya. *Annals of the American Association of Geographers, 107* (2), 403–415.

Ostrom, E., & Benjamin, P. (1993). Design principles and the performance of farmer managed irrigation systems in Nepal. In S. Manor & J. Chambouleyron (Eds.). *Performance measurement in farmer managed irrigation systems,* (IIMI, 1993) (pp. 53–62).

Ostrom, E., & Gardner, R. Coping with asymmetries in the commons: Self-Governing Irrigation Systems Can Work. *Journal of Economic Perspectives, 7,* 93–112. Number 4-Fall 1993.

Pandey, V. P., & Kazama, F. (2011). Hydrogeologic characteristics of groundwater aquifers in Kathmandu Valley, Nepal. *Environmental Earth Sciences, 62*(8), 1723–1732.

Pandey, V. P., Chapagain, S. K., & Kazama, F. (2010). Evaluation of groundwater environment of Kathmandu Valley. *Environmental Earth Sciences, 60*(6), 1329–1342.

Pandey, V., Shrestha, S., & Kazama, F. (2012). Groundwater in the Kathmandu Valley: Development dynamics, consequences and prospects for sustainable management. *European Water, 37,* 3–14.

Pant, B. R. (2011). Ground water quality in the Kathmandu valley of Nepal. *Environmental Monitoring and Assessment, 178*(1–4), 477–485.

Paramanik, S. K. (2017). *Analysis of discharge variation and estimation of recession coefficients for different spring systems in Himalayan terrain*. Indian Institute of Technology Roorkee: Master Degree.

Parveen, S., Winiger, M., Schmidt, S., & Nüsser, M. (2015). Irrigation in Upper Hunza: Evolution of socio-hydrological interactions in the Karakoram, Northern Pakistan, *Erdkunde*, Bd. 69, H. 1 (January–March 2015, pp. 69–85).

Pervaz, M., & Rahman, M. L. (2012). Review and evaluation of successful and unsuccessful renewable energy projects in South Asia. In *International Conference on Life Science and Engineering*, Singapore. https://doi.org/10.7763/ipcbee. V45. 2.

Poff, N. L., Allan, J. D., Bain, M. B., Karr, J. R., Prestegaard, K. L. et al. (1997). The natural flow regime: a paradigm for river conservation and restoration. *BioScience*, *47*, 769–784.

Pokhrel, A. (2014). A theory of sustained cooperation with evidence from irrigation institutions in Nepal. Ph.D. dissertation, Massachusetts Institute of Technology (MIT)—Urban Studies and Planning, International Development, Political Economy. USA.

Poudel, D. D., & Duex, T. W. (2017). Vanishing springs in Nepalese mountains: Assessment of water sources, farmers's perceptions, and climate change adaptation. *Mountain Research and Development, 37*(1), 35–46.

Pradhan, P. (2000). Farmer managed irrigation system in Nepal at the Crossroads. In *Paper presented at the 8th Biennial Conference of the International Association for the Study of Common Property (IASCP)*, Bloomington, Indiana, May 30 to 4 July, 2000. http://dlc.dlib.indiana.edu/dlc/bitstream/handle/10535/331/pradhanp041500.pdf?sequence=. Accessed 13 August 2016.

Prasch, M., Mauser, W., & Weber, M. (2013). Quantifying present and future glacier melt-water contribution to runoff in a central Himalayan river basin. *The Cryosphere*, *7*(3), 889–904. http://www.the-cryosphere.net/7/889/2013/. Accessed 24 Sept. 2013.

Raina, A. (2016). *Equity in urban water service delivery and the role of informal water vendors: The case of Kathmandu valley, Nepal (Unpublished Doctoral Dissertation)*. Singapore: National University of Singapore.

Racoviteanu, A. E., Armstrong, R., & Williams, M. W. (2013). Evaluation of an ice ablation model to estimate the contribution of melting glacier ice to annual discharge in the Nepal Himalaya. *Water Resources Research*, *49*(9), 5117–5133. http://doi.wiley.com/10.1002/wrcr.20370. Accessed 23 Dec. 2013.

Ragettli, S., et al. (2013). Sources of uncertainty in modeling the glacio-hydrological response of a Karakoram watershed to climate change. *Water Resources Research*, *49*, 1–19. http://doi.wiley.com/10.1002/wrcr.20450. Accessed 24 Sept. 2013.

Ragettli, S., et al. (2015). Unraveling the hydrology of a Himalayan watershed through integration of high resolution in-situ data and remote sensing with an advanced simulation model. *Advances in Water Resources*, *78*, 94–111. http://dx.doi.org/10.1016/j.advwatres.2015.01.013.

Rana, S., & Gupta, V. (2009). Watershed management in the Indian Himalayan region: Issues and challenges. In *Paper presented at World Environmental and Water Resources Congress 2009@ sGreat Rivers*, ASCE.

Rasul, G. (2009). Ecosystem services and agricultural land-use practices: A case study of the Chittagong Hill Tracts of Bangladesh. *Sustainability: Science, Practice and Policy*, *5*(2).

Rasul, G. (2014). Food, water, and energy security in South Asia: A nexus perspective from the Hindu Kush Himalayan region. *Environmental Science & Policy, 39,* 35–48.

Rodell, M., Velicogna, I., & Famiglietti, J. S. (2009). Satellite-based estimates of groundwater depletion in India. *Nature, 460*(7258), 999–1002.

Safi, Z., & Buerkert, A. (2011). Heavy metal and microbial loads in sewage irrigated vegetables of Kabul, Afghanistan. *Journal of Agriculture and Rural Development in the Tropics and Subtropics, 112,* 29–36.

Sarkar, A. (2011). Socio-economic implications of depleting groundwater resource in Punjab: A comparative analysis of different irrigation systems. *Economic and Political Weekly, 46*(7), 59–66.

Schaner, N., et al. (2012). The contribution of glacier melt to streamflow. *Environmental Research Letters, 7,* 034029. http://stacks.iop.org/1748-9326/7/i=3/a=034029?key=crossref.1d60a5b89febe92ea66fea367de538be. 22 Accessed Sept. 2013.

Scott, C. A., Meza, F. J., Varady, R. G., Tiessen, H., McEvoy, J., Garfin, G. M., et al. (2013). Water security and adaptive management in the arid Americas. *Annals of the Association of American Geographers, 103*(2), 280–289.

Scott, C. A., Crootof, A., Thapa, B., & Shrestha, R. K. (2016). Water-food-energy nexus: Challenges and opportunities. In L. Bharati, B. R. Sharma, & V. Smakhtin (Eds.), *The Ganges River Basin: Status and challenges in water, environment and livelihoods*. London: Routledge.

Sehar, S., Naz, I., Ali, M. I., & Ahmed, S. (2011). Monitoring of physico-chemical and microbiological analysis of underground water samples of district Kallar Syedan, Rawalpindi, Pakistan. *Research Journal of Chemical Sciences, 1,* 24–30.

Senanayake, N., & Mukherji, A. (2014). Irrigating with arsenic contaminated groundwater in West Bengal and Bangladesh: A review of interventions for mitigating adverse health and crop outcomes. *Agricultural Water Management, 135*(2014), 90–99.

Shah, R. D. T., & Shah, D. N. (2013). Evaluation of benthic macroinvertebrate assemblage for disturbance zonation in urban rivers using multivariate analysis: Implications for river management. *Journal of Earth System Science, 122*(4), 1125–1139.

Shah, T. (2007). Crop per drop of diesel: Energy squeeze on India's small holder agriculture. *Economic and Political Weekly, 42*(39), 4002–4009.

Shah, T. (2009). Climate change and groundwater: India's opportunities for mitigation and adaptation. *Environmental Research Letters, 4*.

Shah, T., & Giordano, M. (2013). Himalayan water security: A South Asian perspective. *Asia Policy, 16*(1), 26–31.

Shah, T., Singh, O. P., & Mukherji, A. (2006). Some aspects of South Asia's groundwater irrigation economy: analyses from a survey in India, Pakistan, Nepal Terai and Bangladesh. *Hydrogeology Journal, 14*(3), 286–309.

Sharda, V. (2005). Integrated watershed management: managing valleys and hills in the Himalayas. *Watershed Management Challenges, 61*.

Sharif, M., et al. (2013). Trends in timing and magnitude of flow in the Upper Indus Basin. *Hydrology and Earth System Sciences, 17*(4), 1503–1516. http://www.hydrol-earth-syst-sci.net/17/1503/2013/. Accessed 22 Jan. 2014.

Sharma, B., Nepal, S., Gyawali, D., Pokharel, G. S., Wahid, S. M., Mukherji, A., et al. (2016). *Springs, storage towers, and water conservation in the midhills of Nepal*. Nepal Water Conservation Foundation and International Center for Mountain Development. ICIMOD Working Paper 2016/3. Kathmandu: Nepal.

Shea, J. M., Immerzeel, W. W., Wagnon, P., Vincent, C., & Bajracharya, S. (2015a). Modelling glacier change in the Everest region, Nepal Himalaya. *The Cryosphere, 9,* 1105–1128.

Shea, J. M., Wagnon, P., Immerzeel, W. W., Biron, R., Brun, F., & Pellicciotti, F. (2015b). A comparative high-altitude meteorological analysis from three catchments in the Nepalese Himalaya. *International Journal of Water Resources Development, 31*(2), 174–200.

Shen, M., Piao, S., Jeong, S.-J., Zhou, L., Zeng, Z., Ciais, P., et al. (2015). Evaporative cooling over the Tibetan Plateau induced by vegetation growth. *Proceedings of the National Academy of Sciences, 112*(30), 9299–9304.

Singh, L. J., & Choudhury, D. (2015). Fallow-management practices among the Tangkhuls of Manipur: Safeguarding provision and regulatory services from shifting-cultivation fallows. In F. C. Malcolm (Ed.) Shifting cultivation and environmental change: Indigenous people, agriculture and forest conservation. Routledge (pp. 449–467), New York.

Shi, X., Zhang, F., Lu, X., Wang, Z., Gong, T., Wang, G., et al. (2018). Spatiotemporal variations of suspended sediment transport in the upstream and midstream of the Yarlung Tsangpo River (the upper Brahmaputra), China. Earth Surface Processes and Landforms, *43* (2), 432–443. https://doi.org/10.1002/esp.4258.

Shrestha, R., Desai, J., Mukherji, A., Dhakal, M., Kulkarni, H. & Acharya, S. (2017). Application of eight-step methodology for reviving springs and improving springshed management in the Mid-hills of Nepal, Research Highlight, ICIMOD. http://lib.icimod.org/record/32564/files/icimodWLE017.pdf.

Shrestha, A. B., Agrawal, N. K., Alfthan, B., Bajracharya, S. R., Maréchal, J., & van Oort, B. (Eds.). (2015a). The Himalayan climate and water Atlas: Impact of climate change on water resources in five of Asia's major river basins. GRID-Arendal and CICERO: ICIMOD.

Shrestha, M., Koike, T., Hirabayashi, Y., Xue, Y., Wang, L., Rasul, G., et al. (2015b). Integrated simulation of snow and glacier melt in water and energy balance-based, distributed hydrological modeling framework at Hunza River Basin of Pakistan Karakoram region. *Journal of Geophysical Research: Atmospheres, 120*(10), 4889–4919.

Shrestha, O. M., Koirala, A., Hanisch, J., Busch, K., Kerntke, M., & Jäger, S. (1999). A geo-environmental map for the sustainable development of the Kathmandu Valley, Nepal. *GeoJournal, 49*(2), 165–172.

Shrestha, P., Lord, A., Mukherji, A., Shrestha, R. K., Yadav, L. & Rai, N. (2016). Benefit sharing and sustainable hydropower: Lessons from Nepal, ICIMOD Research Report, 2016/2. International Centre for Integrated Mountain Development, Kathmandu, Nepal.

Shrestha, R. K., Ahlers, R., Bakker, M., & Gupta, J. (2012). Institutional dysfunction and challenges in flood control: A case study of the Kosi Flood 2008. *Economic and Political Weekly, 2,* 45–53.

Siddiqi, A., & Wescoat, J. L. (2013). Energy use in large-scale irrigated agriculture in the Punjab province of Pakistan. *Water International, 38*(5), 571–586.

Singh, A. K., & Pande, R. K. (1989). Changes in spring activity: Experiences of Kumaun Himalaya, India. *Environmentalist, 9*(1), 25–29.

Singh, P., & Jain, S. K. (2002). Snow and glacier melt in the Satluj River at Bhakra Dam in the western Himalayan region. *Hydrological Sciences Journal*, *47*(1), 93–106. http://www.tandfonline.com/doi/abs/10.1080/02626660209492910. Accessed 2 Oct. 2013.

Singh, R. A., & Gupta, R. C. (2002). Traditional land and water management systems of North-East Hill regions. *Indian Journal of Traditional Knowledge, 1*(1), 32–39.

Singh, S. K. (1990). Evaluating Large Dams in India. *Economic and Political Weekly, 25,* 561–574.

Sinha, R. (2009). The Great avulsion of Kosi on 18 August 2008. *Current Science, 97*(3), 429–433.

Sinha, R., Priyanka, S., Jain, V., & Mukul, M. (2014). Avulsion threshold and planform dynamics of the Kosi river in north Bihar

(India) and Nepal: A GIS framework. *Geomorphology, 216,* 157–170.

Soncini, A., et al. (2015). Future hydrological regimes in the Upper Indus Basin: A case study from a high-altitude glacierized catchment. *Journal of Hydrometeorology, 16*(1), 306–326. http://journals.ametsoc.org/doi/abs/10.1175/JHM-D-14-0043.1.

Tahir, A. A., Chevallier, P., Arnaud, Y., et al. (2015). Snow cover trend and hydrological characteristics of the Astore River basin (Western Himalayas) and its comparison to the Hunza basin (Karakoram region). *Science of the Total Environment, 505,* 748–761.

Tahmiscioğlu, M. S., & Anul, N. (2007). *Positive and negative impacts of dams on the environment.* Ankara: International Congress on River Basin Management.

Tambe, S., Arrawatia, M., Kumar, R., Bharti, H., & Shrestha, P. (2009). Conceptualizing strategies to enhance rural water security in Sikkim, Eastern Himalaya, India. *Integrated Water Resource Management, 1.*

Tambe, S., Kharel, G., Arrawatia, M. L., et al. (2012). Reviving dying springs: Climate change adaptation experiments from the Sikkim Himalaya. *Mountain Research and Development, 32*(1), 62–72.

Tarafdar, S. (2013). Understanding the dynamics of high and low spring flow: A key to managing the water resources in a small urbanized hillslope of Lesser Himalaya, India. *Environmental earth sciences, 70*(5), 2107–2114.

Thapa, B., Scott, C. A., Wester, P., & Varady, R. (2016). Towards characterizing the adaptive capacity of farmer-managed irrigation systems: Learnings from Nepal. *Current Opinion in Environmental Sustainability, 21,* 37–44. https://doi.org/10.1016/j.cosust.2016.10.005.

Tiwari, P. C., & Joshi, B. (2014). Environmental changes and their impact on rural water, food, livelihood, and health security in Kumaon Himalaya. *International Journal of Urban and Regional Studies on Contemporary India, 1,* 1–12.

Trivedi, R. C., (2010). *Water quality of the Ganga rive–An overview.* Aquatic Ecosystem Health & Management.

Udas, P. B., & Zwarteveen, Margreet. (2010). Can water professionals meet gender goals? A case study of the Department of Irrigation in Nepal. *Gender and Development, 18*(1), 87–97.

UN-Water. (2013). *Water security and the global water agenda: a UN-water analytical brief.* Hamilton, Ontario: United Nations University.

van Steenbergen, F. (1997). Understanding the sociology of spate irrigation: Cases from Balochistan. *Journal of Arid Environments, 35,* 349–365.

Vaidya, R. (2012). Water and hydropower in the green economy and sustainable development of the Hindu Kush-Himalayan Region. *Hydro Nepal: Journal of Water, Energy and Environment,* 10.

Valdiya, K., & Bartarya, S. (1989). Diminishing discharges of mountain springs in a part of Kumaun Himalaya. *Current Science. Bangalore, 58*(8), 417–426.

Vashisht, A., & Bam, B. (2013). Formulating the spring discharge-function for the recession period by analyzing its recession curve: A case study of the Ranichauri spring (India). *Journal of Earth System Science, 122*(5), 1313–1323.

Vashisht, A., & Sharma, H. (2007). Study on hydrological behaviour of a natural spring. *Current Science, 93*(6), 837–839.

Vincent, L. (1995). *Hill irrigation. Water and development in mountain agriculture.* London: Intermediate Technology Publications.

Wang, S., Fu, B., Piao, S., et al. (2016). Reduced sediment transport in the Yellow River due to anthropogenic changes. *Nature Geoscience, 9*(1), 38.

Wang, Z. Y., Li, Y., He, Y. (2007). Sediment budget of the Yangtze River. *Water Resources Research, 43*(4).

Wasson, R. J. A. (2003). sediment budget for the Ganga-Brahmaputra catchment. *Current Science, 84*(8), 1041–1047.

WHO. (2006). *Guidelines for the safe use of wastewater in agriculture.* Geneva, Switzerland: World Health Organisation.

Xu, X., Lin, H., & Fu, Z. (2004). Probe into the method of regional ecological risk assessment—a case study of wetland in the Yellow River Delta in China. *Journal of Environmental Management, 70* (3), 253–262.

Yang, S. L., Zhang, J., Xu, X. J. (2007). Influence of the Three Gorges Dam on downstream delivery of sediment and its environmental implications. Yangtze River. *Geophysical Research Letters, 34*(10).

Yang, S. L., Milliman, J. D., Li, P., et al. (2011). 50,000 dams later: Erosion of the Yangtze River and its delta. *Global and Planetary Change, 75*(1), 14–20.

Yang, Y. E., Brown, C., Winston, Y., Wescoat, J., & Ringler, C. (2014). Water governance and adaptation to climate change in the Indus River Basin. *Journal of Hydrology, 519,* 2527–2537.

Yoder, R. (1986). *The performance of farmer-managed irrigation systems in the hills of Nepal.* Unpublished Ph.D. dissertation, Cornell University, Ithaca, NY.

Zeng, C., Zhang, F., Lu, X., Wang, G. X., & Gong, T. L. (2018). Improving sediment load estimations: The case of the Yarlung Zangbo River (the upper Brahmaputra, Tibet Plateau). *CATENA, 160,* 201–211.

Zhang, L., Su, F., Yang, D., et al. (2013). Discharge regime and simulation for the upstream of major rivers over Tibetan Plateau. *Journal of Geophysical Research: Atmospheres, 118*(15), 8500–8518.

Food and Nutrition Security in the Hindu Kush Himalaya: Unique Challenges and Niche Opportunities

Coordinating Lead Authors
Golam Rasul, International Centre for Integrated Mountain Development (ICIMOD), Kathmandu, Nepal, e-mail: Golam.Rasul@icimod.org (corresponding author)
Abdul Saboor, PMAS Arid Agriculture University, Rawalpindi, Pakistan, e-mail: drabdul.saboor@uaar.edu.pk.

Lead Authors
Prakash C. Tiwari, Kumaon University, Nainital, Uttarakhand, India, e-mail: pctiwari@yahoo.com
Abid Hussain, International Centre for Integrated Mountain Development (ICIMOD), Kathmandu, Nepal, e-mail: Abid.Hussain@icimod.org
Nilabja Ghosh, Institute of Economic Growth, Delhi, India, e-mail: nila@iegindia.org
Ganesh B. Chettri, Department of Agriculture, Ministry of Agriculture and Forests, Bhutan, e-mail: gbchettri@gmail.com

Contributing Authors
Tiina Kurvits, GRID-Arendal, Norway, e-mail: Tiina.Kurvits@grida.no
Ayesha Mazhar, PMAS Arid Agriculture University, Rawalpindi, Pakistan, e-mail: aisha_elitian@hotmail.com
Touqeer Ahmad, PMAS Arid Agriculture University, Rawalpindi, Pakistan, e-mail: taqi_pooh@hotmail.com
Manbar S. Khadka, Asian Development Bank, Nepal Resident Mission, Kathmandu, e-mail: skmanbar@gmail.com
Narendra Dangol, Sunrise Nepal Food and Beverage Pvt. Ltd., Nepal, e-mail: narendangol@yahoo.com
Nilhari Neupane, International Centre for Integrated Mountain Development (ICIMOD), Kathmandu, Nepal, e-mail: Nilhari.Neupane@icimod.org.
Sakhie Pant, International Centre for Integrated Mountain Development (ICIMOD), Kathmandu, Nepal, e-mail: Sakhie.Pant@icimod.org
Omaid Najmuddin, University of Chinese Academy of Sciences, Beijing, China, e-mail: omaid_najmuddin@yahoo.com

Review Editor
Krishna Prasad Pant, Food and Agriculture Organization (FAO) of the United Nations, Kathmandu, Nepal, e-mail: kppant@yahoo.com

Corresponding Author
Golam Rasul, International Centre for Integrated Mountain Development (ICIMOD), Kathmandu, Nepal, e-mail: Golam.Rasul@icimod.org

P. Wester et al. (eds.), *The Hindu Kush Himalaya Assessment*,
https://doi.org/10.1007/978-3-319-92288-1_9

Contents

Chapter Overview

Key Findings

1. **Food and nutrition insecurity remains a serious challenge in the Hindu Kush Himalaya (HKH) region; more than 30% of the population suffers from food insecurity and around 50% face some form of malnutrition, with women and children suffering the most.** The insecurity is more severe in remote mountain areas. Challenges to food security in the mountain areas differ from those in the plains due to inaccessibility, fragility, seasonality, limited economic opportunities, poor market access, and harsh biophysical conditions.
2. **The causes of food and nutrition insecurity in the HKH are multifaceted and complex, and influenced by a range of factors including high** poverty, natural resource degradation, climate change, low level of market development, uncertain food support, and inadequate policy and institutional support.
3. **Traditional mountain food systems are currently under threat from rapid socioeconomic and environmental changes including** changing dietary habits, changes towards mono-cropping and commodity crops, loss of water sources, soil degradation, and decline in market value. Mountain agriculture is becoming relatively less competitive and the youth are increasingly abandoning agricultural livelihoods, leading to decreased food production and adversely affecting local food systems.

Policy Messages

1. **To address the unique challenges faced by mountain communities regarding food security, national governments in the region need to pay special attention to integrating a mountain perspective into national policies related to food and nutrition security.** The mountain perspective should take into account exploiting agro-ecological potential and mountain-specific niche opportunities while protecting the environment, delivering institutional services, and improving market access.
2. **Governments in the region can be more effective in addressing issues related to food and nutrition security by adopting a holistic approach that includes** revitalizing local food systems, strengthening social safety nets, enhancing knowledge and awareness about nutrition, and reducing physical and socioeconomic vulnerability. Efforts are also needed to diversify livelihood options and develop non-farm sectors such as tourism and handicrafts to enhance household food purchasing power. Attention also needs to be given to increasing the productivity of traditional crops and local breeds of livestock, and to the development of non-timber forest products (NTFPs), medicinal and aromatic plants, mountain niche cash crops, and organic agriculture.
3. **Increased investment in the management of natural resources, including soil, water, and energy, is critical to increase agricultural production, diversify local food systems, and improve nutrition.** Major investments are needed in soil and water management to revitalize springs, ponds, and other water bodies and to develop irrigation facilities and improve the domestic water supply in an environmentally responsible manner in hill and mountain areas.

The mountain people of the Hindu Kush Himalaya (HKH) face large challenges in food and nutrition security. Although progress has been made in calorie intake, malnutrition remains a serious challenge (***well-established***). About 50% of the population suffers from malnutrition, and women and children suffer more. Ending hunger and achieving food and nutrition security—as articulated in the Sustainable Development Goals—is an urgent need for the governments of the region. Agriculture is one of the region's main livelihood options, but many people depend on limited natural resources and poor soils with agroecosystems of low carrying capacity. The region's population is increasing quite fast at close to 1.4% annually, and the per capita availability of land and other natural resources is declining. Traditional agricultural systems are coming under pressure and failing to provide adequate food and income. As a result, nearly one-third of the population is suffering from food insecurity, and between one-fifth and one-half of children (<5 years of age), depending on country, suffer from stunting, with a high incidence of wasting and underweight. A significant share of women also suffer from various forms of anaemia (***well-established***). Some mountain areas exhibit high nutrition insecurity compared to the national average for the whole country, for example eastern Afghanistan, Meghalaya state in India, Chin and Rakhine states in Myanmar, the high mountains of Nepal, and Balochistan province in Pakistan (***well-established***).

Addressing the challenge of food and nutrition security in the HKH has become increasingly complicated by the rapid socioeconomic, demographic, and environmental changes, including migration and climate change. The prevalence of high poverty, youth out-migration, poor infrastructure and market access, depletion of natural resources, and the increased brunt of climate change are key factors, among others, affecting food and nutrition security.

In the remote rural areas where a majority of the HKH population lives—depending heavily on agriculture for food and nutrition—farming and food production are highly susceptible to climate change, due in part to poor irrigation facilities and a high dependence on precipitation. The drying up of springs and water bodies, erratic rainfall, increased floods, increased dry spells, land degradation, and a rising incidence of pests and disease in crops and livestock all pose additional challenges to food and nutrition security in the region.

Despite these numerous challenges, the HKH region also enjoys comparative advantages in certain products and has a good potential and opportunities for revitalizing local food systems, developing mountain niche products and services, and promoting non-farm livelihood options (***established but incomplete, see Sect.*** 9.4). Local food systems—including neglected and underutilized species (NUS) and local breeds of livestock—have a huge potential to diversify the supply of food and micronutrients in the mountains, while enhancing farmers' incomes and enabling them to access nutritious food (see Sects. 9.4.1 and 9.4.2). The nutritional value of many NUS crops such as millets, sorghum, and buckwheat are high (***well-established, see Sect.*** 9.4.1). These crops and local breeds of livestock are also highly adapted to mountain conditions and resilient to climate-induced stresses like drought and frost. In view of the nutritional, environmental, and economic benefits, NUS are recently relabelled as 'Future Smart Foods'. A growing literature suggests that if provided with the required inputs and market links, mountain areas also offer a good agro-ecological potential for growing various cash crops including tea, coffee, nuts, fruit, and vegetables, which can contribute to food and nutrition security (***well-established, see Sect.*** 9.4.3).

The natural resources of the mountains, such as forests, rangelands, and water resources, also provide many opportunities to increase household income and food security if managed and harnessed sustainably (***well-established, see Sect.*** 9.4.4). Timber and non-timber forest products (NTFPs), medicinal and aromatic plants, other varieties of plants and herbs and honeybees can provide additional sources of income. The potential of rangelands in high mountain areas—the basis for livestock production and food security in pastoral communities—could be developed further by enhancing livestock productivity. Mountains are rich in water resources, but these are not harnessed properly for the benefit of local people (see Chap. 8). Properly managing and harnessing the water in springs, streams, snow, and glaciers could greatly increase agricultural production and help to diversify local food systems. Other opportunities exist in tourism, handicrafts, food processing, and medicinal plants as potential sources of income, which would, in turn, improve food and nutrition security in the mountains (***established but incomplete, see Sect.*** 9.4.5).

Achieving sustainable food and nutrition security in the mountains requires a balanced approach between food self-sufficiency and market dependency and an integrated strategy that entails production enhancement and increasing household income, along with improving rural infrastructure for market access and food transportation. The strategy should focus on addressing existing challenges while helping the region to seize opportunities and realize the potential. Due to different ecological and environmental conditions and different levels of access to institutional services result in different challenges and opportunities across the region, the strategies for food and nutrition security need to consider local agro-ecological features and socioeconomic conditions, including access to markets and other basic services. In this chapter, we identify four types of HKH mountain area based on agro-ecological potential and access to markets, information, and institutional services, and suggests area-specific strategies to enhance food and nutrition security for each of these types:

- High agro-ecological potential, good access to markets and institutional services
- High agro-ecological potential, poor access to markets and institutional services
- Low agro-ecological potential, good access to markets and institutional services
- Low agro-ecological potential, poor access to markets and institutional services.

In areas with high agro-ecological potential and good market access, strategies could focus on tapping the potential through land use intensification and growing cash crops in line with market demand; while in areas with high agro-ecological potential and poor market access, strategies could focus on improving market access, developing local food systems, and promoting high value but low volume and non-perishable products. In areas with low agro-ecological potential but good access to markets and institutional services, strategies could focus on promotion of non-farm and off-farm activities to enhance purchasing power; while in areas low in agro-ecological potential and with poor market access, strategies could focus on subsistence use of resources, developing incentive mechanisms for conservation of resources, and support for out-migration to increase purchasing power and thus access to nutritious food. Targeted food safety net programmes such as subsidized food may also be required in low potential areas to ensure food security.

In addition, special attention needs to be paid to the following:

- Establishing community food banks to store food at village level so that households in need can borrow from the banks and return the loaned amount after the harvest,
- Managing water resources, especially, springs in midhills and mountains and promoting efficient and equitable hill irrigation infrastructure, where feasible,
- Establishing mechanisms for cross-border food trade and internal movement of food products, particularly cross-border localized trade/food exchange between border communities,
- Strengthening knowledge on nutrition, child care, and food preparation, including traditional Himalayan fermented foods which are nutritionally rich and healthy, and
- Empowering women by improving knowledge and their control of resources to enable them to take decisions on matters related to family health, education, and feeding.

Food and Nutrition Security in the HKH and the Sustainable Development Goals

Six Sustainable Development Goals (SDGs) are closely related to food and nutrition security:

- Goal 1—*End poverty in all its forms everywhere.*
- Goal 2—*End hunger, achieve food security and improved nutrition, and promote sustainable agriculture.*
- Goal 3—*Ensure healthy lives and promote well-being for all at all ages.*
- Goal 6—*Ensure availability and sustainable management of water and sanitation for all.*
- Goal 12—*Ensure sustainable consumption and production patterns.*
- Goal 13—*Take urgent action to combat climate change and its impacts.*

In addition, achieving food and nutrition security has direct and indirect implications for achieving a number of other SDG goals and targets, such as SDG 1 (eradicating poverty) and SDG 5 (achieving gender equality), which depend on eliminating food insecurity and malnutrition.

Though diverse, the SDGs are also mutually related. For example, ending hunger and achieving food security will depend on achieving water and energy security for food production. Similarly, ensuring healthy lives will depend on achieving food and nutrition security. A failure to ensure food and nutrition security in the HKH would thus imperil the region's pursuit of other SDGs such as ending poverty, ensuring healthy lives, achieving gender equality, and reducing inequalities as well as adaptation and mitigation of climate change.

9.1 Introduction

Food and nutrition security is fundamental for living a healthy and productive life and is essential for socioeconomic development. The system of food production and consumption has shaped human society and the environment for millennia (Desor 2017). Food and nutrition security is thus critical for any society and a top priority in national and global development and environmental agendas. It is also a building block for achieving Sustainable Development Goal (SDG) 2 as well as instrumental for achieving other SDGs such as 1 (no poverty), 3 (good health and wellbeing), 6 (clean water and sanitation), 12 (responsible consumption and production), and 13 (climate action). Hunger and malnutrition are widespread in the Hindu Kush Himalayan (HKH) countries. Of the 795 million people undernourished globally, 52% (415 million) are from HKH countries (FAO-IFAD-WFP 2015). In the HKH region, more than 30% of people face food insecurity (Table 9.1) and one-third to one-half of children (below five years of age) suffer from stunting, with a variation across countries (Rasul et al. 2018). Children and women suffer more, and children's growth and development are affected.

9.1.1 Mountain Specificities and Food Security

The nature and causes of food security in mountain areas differ from those in the plains as a result of differences in the physical environment, transportation and communication facilities, remoteness, and seasonality. Mountain environments in the HKH are characterized by limited accessibility, a high degree of fragility, and marginality (Jodha 2000). The people in the HKH face severe challenges in terms of food and nutrition security due to the harsh biophysical environment and relatively low socioeconomic development. The

Table 9.1 Food (in)security in the Hindu Kush Himalaya region

HKH country	Incidence of food insecurity/food poverty in mountain areas (%)	Approximate number of people faced with food insecurity[a] (millions)	Data source
Afghanistan	33	National level statistic in ALCS (2014) used as a proxy for mountains; 29 of Afghanistan's 34 provinces are mountainous or hilly	7.76
Bangladesh	53	Majumder et al. (2012) estimated % food insecure households. Using household size in Chittagong Hill Tracts to convert households to population won't change the incidence of food insecurity	0.95
Bhutan[b]	5.9	GoB (2012)	0.05
China	11	National level prevalence of undernourishment in FAO 2013 taken as proxy for mountains	3.66
India	18	Statistics taken from Rasul et al. (2018) which uses data from two mountain states as a proxy for the mountain area	15.53
Myanmar	09	Mountain statistics aggregated from the state level statistics in IHLCA (2011)	1.08
Nepal[b]	51	NDHS (2011)	14.61
Pakistan	57	Mountain aggregated statistics approximated based on the statistics from FSA (2009) and population projections for mountain areas	29.45
Total	**31**[c]	–	**73.09**[c]

[a]Population data in Box 1.1; Table 1.1 used to approximate food poor population
[b]Entire territories of Bhutan and Nepal included in the HKH region despite Nepal having some plains areas
[c]Approximated number and percentage of food insecure populations in the HKH region are indicative and based on the individual country level statistics on mountain food (in)security
Note Data presented in the table are not recommended for use in cross-country comparisons because all countries follow different methods for food (in)security assessment

Himalaya is one of the youngest mountain ranges in the world with a prehistoric marine origin. The soils are among the most fertile in the world, but fragile and degradable. Steep slopes and unsustainable terrain present severe challenges in the region and make it unsuitable for the conventional green revolution agriculture practised in the plains areas of the HKH countries. This has made mountain communities highly dependent on the plains for food. Thus approaches based on production alone cannot address the mountain food security issues; access to food through means is also important. While theoretically, in a well-functioning competitive market, access to food through cash income is straightforward, mountain communities often face physical difficulties and uncertainties in accessing food due to their physical remoteness, dispersed settlements, high transport costs, imperfect market conditions, high variability in the food price, and frequent natural disasters such as floods, landslides, avalanches, and earthquakes (Rasul and Hussain 2015). In mountain areas, natural disasters often disturb the fragile communication system and hinder food transportation and access to food. Food prices are also influenced by oil prices and transport bottlenecks caused by natural disasters or political turmoil (Hussain and Routray 2012). The challenges faced by mountain communities are often not adequately understood, and the perspectives of mountain communities are not fully recognized in national agricultural development policies in the HKH countries (Jodha 2000, 2009). As a result, despite good intentions, the policies and strategies pursued by the countries to increase food production using the green revolution approach has failed to yield the desired outcome in the HKH region (Rasul 2010).

Agriculture is one of the primary livelihood options available in the HKH region, and key to achieving food and nutrition security. But a considerable proportion of the population depends on limited natural resources with low carrying capacity. Agriculture in the HKH region has remained largely traditional. Research on rainfed agriculture and mountain niche products has been limited and thus agricultural productivity has remained low, while rapid population growth (close to 1.4% annually) is placing additional pressure on food and nutrition security. Limited industrialization and slow growth of the non-farm sector mean that the increasing labour force is also adding to the pressure on the limited and fragile land resources. The per capita availability of natural resources, particularly land, is decreasing steadily, while traditional agricultural systems are gradually degrading (Adhikari et al. 2017). This has implications for food and nutrition security in the region. Further, the rapid climatic changes currently taking place in the HKH ecosystem may compound the effects on the traditional food and agricultural systems in the region (Aase et al. 2009).

The mighty glaciers and mountain ranges of the HKH are the source of some of the biggest rivers in South Asia (for more, see Chap. 7 Status and Change of the HKH Cryosphere). But the water resources of the region are not fully harnessed, and the irrigation systems are poorly developed (Bandyopadhyay and Perveen 2008). As a result, agriculture in the HKH has remained largely rainfed, making it prone to the vagaries of the weather and highly vulnerable to climate variability and climate change. Natural water bodies like streams and ponds, on which mountain agriculture and livestock depend heavily, are increasingly drying up. The water regime of the HKH is likely to change rapidly with respect to discharge rates, volume, and availability, which will have adverse impacts on the subsistence agricultural economy (Viviroli et al. 2003) (see Chap. 8). This has serious implications for agricultural productivity and food security, and particularly for food availability, across the entire region (Hussain et al. 2016).

In the HKH agro-ecological system, forests are pivotal to the maintenance of crop production levels. The people of the HKH are traditionally forest-dependent communities, but numerous factors are making such areas prone to risk and uncertainty. High population growth (Table 1.1) and resultant changes in land use are decreasing the forested area and leading to unsustainability of HKH agriculture. The rapidly changing patterns of land use, the resultant decrease in forest area, and decreasing annual rainfall have also disrupted the hydrological regime, with water resources diminishing rapidly, mainly due to reduced groundwater recharge (Tiwari and Joshi 2012a). Such conditions have an impact on most aspects of food and nutrition security.

Transportation and other infrastructure in the HKH is relatively weak and poorly developed, thus there is a frictional market and marketing system. The lack of modern facilities (such as cold storage for short-term storage of fruit, processing, and export quality packaging) combined with neglect of traditional storage and processing methods are having a negative impact on food production. Quality control for grading and facilities for washing and disinfection also need to be developed.

9.1.2 Concept of Food and Nutrition Security

The concept of food and nutrition security is still evolving (Box 9.1). Food security is the fundamental precept of the right to adequate food, which was adopted by the Food and Agriculture Organization (FAO) Council in its 127th session held in November 2004. In that session, governments agreed on some voluntary guidelines to support the realization of economic, social, and cultural rights to food and recommended some workable actions to be pursued during the course of the coming decade for the recognition of the right to food. More than a decade has passed, but the need for practical guidance on developing effective institutional and legal frameworks for the right to food, establishing independent monitoring mechanisms, and implementing these frameworks has not been addressed in mountain regions.

Box 9.1 What is food and nutrition security?

"Food security exists when all people, at all times, have physical and economic access to sufficient, safe and nutritious food that meets their dietary needs and food preferences for an active and healthy life" (World Food Summit 1996). Food security has four dimensions: availability, accessibility, utilization, and stability (FAO 2008). Food availability refers to the physical availability of adequate levels of food in a particular area. Cultural acceptability of food is also an important aspect as different cultures have different food preferences (Jones et al. 2013; Hussain and Routray 2012). Food accessibility refers to the physical and economic access to food. Food utilization refers to food quality, safety, and absorption, supported by an adequate health status. Food stability is ensured when food availability, accessibility, and utilization remain secure throughout the year and over a long period of time (see Fig. 9.1). The concept of food security also needs to be understood from both the national and household perspectives. Although national food security is important as providing a foundation, it may not guarantee household food security. It is therefore important that each and every household—and within it every member—has access to safe, nutritionally adequate, and culturally acceptable food (Gillespie and Mason 1991). Hunger and food insecurity can exist at the household level even when at the national level there is enough food. Food security is a multi-dimensional concept that includes (a) physical availability, which involves food production, stocks, and reserves across multiple scales; (b) physical and economic access, which depends on purchasing power, incomes, food prices, transport, and market infrastructure; (c) food utilization, or the capacity to absorb nutrition according to health, dietary diversity, and intra-household distribution; and (d) stability of food supply and access over time in cases of weather variability, price fluctuations, and other transitory shocks or periodic stresses (FAO 2008).

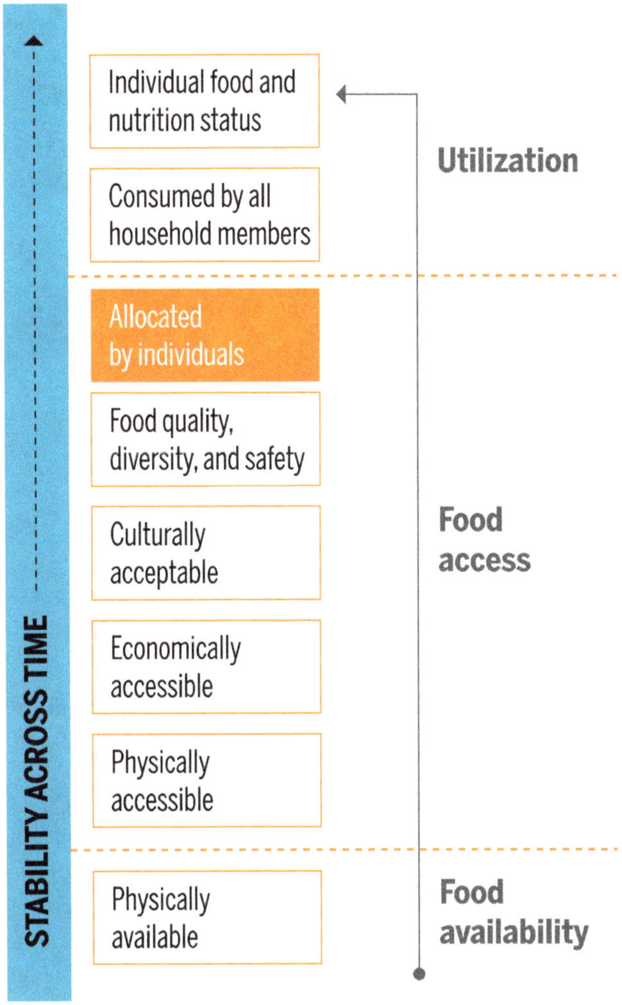

Fig. 9.1 Food security framework (*Source* Adapted from Jones et al. 2013)

The term nutrition security is sometimes used interchangeably with food security, even though nutrition security is much broader as it includes healthcare and hygiene practices (Fig. 9.1). Food security is necessary but not sufficient for nutrition security (Jones et al. 2013). In October 2012, FAO's 'Committee on World Food Security' attempted to define food and nutrition security as *the state that exists when all people at all times have physical, social, and economic access to food, which is safe and consumed in sufficient quantity and quality to meet their dietary needs and food preferences, and is supported by an environment of adequate sanitation, health services, and care, allowing for a healthy and active life*," (FAO 2012). Undernourishment is a state of food insecurity, where caloric intake is below the minimum dietary energy requirement (Jones et al. 2013).

9.1.3 Objective and Scope of this Chapter

As outlined above, the nature of and factors contributing to food and nutrition security in mountain areas are more complex than those in the plains, in particular, due to the harsh biophysical environment, poor accessibility, and weak market infrastructure (Rasul and Hussain 2015; FAO 2015), and a variety of research questions relating to food and nutrition security in mountain regions still need to be addressed. The core questions are as follows:

- How far do the dimensions and underlying factors affecting food and nutrition security vary across the mountain and plains regions of the HKH?
- What is the gap between food supply and demand in the HKH, and how is this gap being filled? What are the prospects and opportunities for traditional and non-traditional food production systems, and how can the strength of indigenous resources be capitalized on through unique policy options?
- What are the key challenges in the arena of food accessibility, distribution, and utilization, and what ways and means will be available to cope with these challenges through public policies and development programmes?
- What best practices exist at regional and international levels that can be tested and replicated for specific communities in the HKH?
- How can the national and international goals of food and nutrition security be achieved for the mountain regions of the HKH?
- Why should the attention of the international community be drawn to the marginalized communities in the HKH, who are often among the most vulnerable people in the world due to cross-border conflicts and political disturbances?

In order to address these questions using an integrated and holistic approach, this chapter aims to understand the causes of food and nutrition insecurity in its four dimensions and to suggest some actionable policy measures to improve the nutrition status of the people living in the HKH. Both qualitative and quantitative data and information have been collected, assessed, and synthesized, and policy issues and their implications for the mountain regions are briefly examined. The first section of the chapter addresses the status of food and nutrition security in the HKH, the second section discusses key issues and challenges in food and nutrition security, the third section describes emerging opportunities and potential in the region, and the final section presents a strategic approach to achieve sustainable food and nutrition security in the mountain areas. The synthesis aims to provide state-of-the-art knowledge on food and nutrition security in

Table 9.2 Food and nutrition (in)security in Afghanistan

Indicators	National	Region						
		Northeast	Northwest	East	Central	West	Southeast	Southwest
% of households not meeting their caloric needs	33.0	n.d.	n.d.	n.d.	n.d.	n.d.	n.d.	n.d.
Stunting[a] (%)	40.9	44.6	45.7	52.6	37.4	31.0	35.0	39.9
Wasting[b] (%)	9.5	8.5	5.9	18.0	8.4	5.6	7.9	10.1
Underweight[c] (%)	25.0	27.8	22.6	41.3	24.1	16.7	27.6	23.6
Underweight women[d] (%)	9.2	13.7	14.8	6.2	7.9	10.9	4.4	9.1
% of women (age 12–49) who received vitamin A postpartum	18.6	16.8	16.1	31.5	6.4	5.6	9.4	22.4
% of women (age 12–49) who had no postnatal check-up	71.5	81.2	72.6	70.3	78.0	77.1	69.8	57.2

[a]Height-for-age: children under 5 years < −2 SD from the international reference median value
[b]Weight-for-height: children under 5 years < −2 SD from the international reference median value
[c]Weight-for-age: children under 5 years < −2 SD from the international reference median value
[d]Weight-for-age: women 15–49 years < −2 SD from the international reference median value
Data source ALCS (2014); AMS (2010)
n.d. = no data available

the HKH region which can be used to inform policy decisions on improving food and nutrition security.

9.2 Status of Food and Nutrition Security in the Mountains of the HKH

9.2.1 Prevalence of Food and Nutrition Insecurity

The prevalence of food insecurity and malnutrition in the HKH countries is very high. In view of the differences in the nature and underlying factors of food and nutrition security in the mountains compared to the plains, it is important to examine the disaggregated status of food and nutrition security in these areas. The statistics show that overall mountain areas are more vulnerable to food and nutrition insecurity, although in certain indicators, some mountain areas can be relatively better than plains areas. The mountain areas are also not homogeneous, and variation across the different mountain regions within a particular country should also be considered. The available information about food and nutrition insecurity in the individual countries within the region is summarized in the following.

Almost all of Afghanistan is hilly or mountainous and no comparison can be made between plains and mountain/hill areas. The prevalence of stunting, wasting, and underweight in children is high in almost all regions of the country (Table 9.2), with stunting particularly high in the north and east and underweight and wasting in the east. The percentage of underweight women aged 15–49 years is significantly higher in the north than elsewhere, while close to 80% of women aged 12–49 years in the northeast, central, and west regions did not receive any postnatal check-up or treatment (Table 9.2). This indicates women's limited access to health services.

Bhutan is also almost entirely hilly or mountainous. Only 6% of the population lives below the food poverty line (Nu. 689 per person per month) (GoB 2012), while 96% of households have access to improved sources of drinking water and 63% to improved sanitation facilities. Nevertheless, the prevalence of stunting and underweight is still high, indicating a lack of diversity in diets and deficiencies of micronutrients (Table 9.3).

In China, the average intake of calories is below the national average in the mountain provinces, except in Yunnan (Table 9.4). The prevalence of stunting in children is relatively high in Sichuan and Yunnan, and the prevalence of anaemia in children is relatively high in Sichuan and Gansu (in both cases the only two mountain provinces for which data is available (Table 9.4).

Table 9.3 Food and nutrition (in)security in Bhutan

Indicators	National
% of population below food poverty line[d]	5.9
Stunting[a] (%)[e]	21.2
Wasting[b] (%)[e]	4.3
Underweight[c] (%)[e]	9.0

[a]Height-for-age: children under 5 years < −2 SD from the international reference median value
[b]Weight-for-height: children under 5 years < −2 SD from the international reference median value
[c]Weight-for-age: children under 5 years < −2 SD from the international reference median value
Data source [d]GoB (2012); [e]NNS (2015)

Table 9.4 Food and nutrition (in)security in China

Indicators	National	Mountain provinces				
		Sichuan	Yunnan	Gansu	Qinghai	Xinjiang
Dietary energy intake (kcal/capita/day)	2172	1966	2231	2011	1986	2021
Protein intake (g/capita/day)	65.0	61.4	67.9	60.1	67.2	63.6
Stunting[a] (%)	3.2	8.0	11.5	n.d.	n.d.	n.d.
Wasting[b] (%)	3.2	3.3	n.d.	n.d.	n.d.	n.d.
Underweight[c] (%)	9.0	15.4	8.4	n.d.	n.d.	n.d.
Prevalence of anaemia (Hb < 12 g/dl) in pregnant women above 18 years of age (%)	17.2	10.0	10.0	17.0	15.2	20.1
Prevalence of anaemia (Hb < 12 g/dl) in children (below 5 years of age) (%)	5.0	13.7	n.d.	24.4	n.d.	n.d.

[a]Height-for-age: children under 5 years < −2 SD from the international reference median value
[b]Weight-for-height: children under 5 years < −2 SD from the international reference median value
[c]Weight-for-age: children under 5 years < −2 SD from the international reference median value
Data source NHPC (2015); Ministry of Health (2013); Xiaoping and Minru (2014); Gansu Province Government (2015)

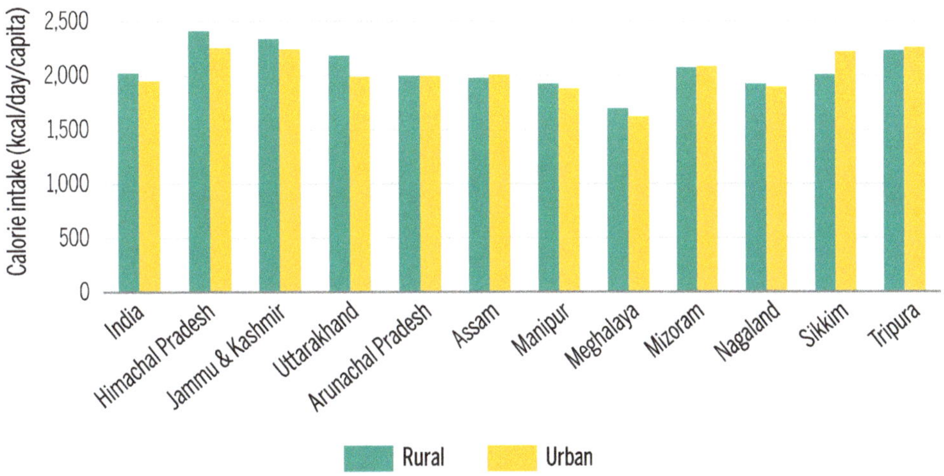

Fig. 9.2 Daily calorie intake in India (*Source* GoI 2010; Schedule Type 1 data from NSS 66th Round, 2010)

In India, the average calorie intake in urban and rural areas of mountain states tends to be slightly lower than the national average, particularly in Manipur, Meghalaya, and Nagaland (Fig. 9.2; Table 9.5). However, the prevalence of stunting, wasting, and underweight in children is lower in the mountains than the national average, possibly due to the presence of extremely poor plains states such as Bihar, Jharkhand, Chhattisgarh, and Orissa in national-level assessments. These states are poor in terms of people's financial capacity (GoI 2014a) as well as health and nutrition status (NFHS 2006). Meghalaya is the only mountain state where the prevalence of stunting, wasting, and underweight children is significantly higher than the national average (Fig. 9.3). In Meghalaya, the area under cultivation of key agricultural crops had declined by nearly 14% in 2011/12 compared to 2002/03 (Roy et al. 2015) which may indicate one factor. Moreover, the shift in Meghalaya from subsistence to commercial farming, multi-cropping to mono-cropping, traditional to modern crops, and food to non-food crops has had an impact on patterns of production and consumption and led to a decline in food diversity that affects nutritional security (Behera et al. 2016). The prevalence of anaemia (by Hb level) in women aged 15–49 years is notably higher in the states of Assam, Sikkim, and Tripura (Table 9.5). Although the overall situation of nutrition security seems slightly better in the mountain states than the national average, the prevalence of malnutrition is still very high.

In Myanmar, the incidence of food poverty in the mountain states of Chin, Rakhine, and Shan is very high compared to the national average (Fig. 9.4). The prevalence of stunting, wasting, and underweight are also substantially higher in Chin, Rakhine, and the northern parts of Shan (Fig. 9.4).

In Nepal, the proportion of food insecure households in the mountains is significantly higher than in the plains (Terai) and compared to the national level (Table 9.6). The

Table 9.5 Food and nutrition (in)security in India

Indicators		National	Mountain states										
			Himachal Pradesh	Jammu and Kashmir	Uttarakhand	Arunachal Pradesh	Assam	Manipur	Meghalaya	Mizoram	Nagaland	Sikkim	Tripura
Calorie intake (kcal/capita/day)	Rural	2020	2407	2334	2179	1995	1974	1919	1692	2067	1918	2003	2223
	Urban	1946	2253	2247	1984	1992	2003	1876	1619	2080	1891	2214	2254
Protein intake (g/capita/day)	Rural	55.0	68.6	61.8	58.6	52.9	48.7	46.4	41.3	49.4	54.2	51.2	55.9
	Urban	53.5	63.1	60.4	55.5	53.2	52.6	45.0	40.7	52.6	55.5	57.8	59.6
Fat intake (g/capita/day)	Rural	38.3	58.9	52.5	48.9	18.1	25.3	14.1	21.2	24.5	13.7	41.5	26.3
	Urban	47.9	61.9	54.7	48.4	26.0	33.5	15.9	22.9	35.1	17.5	44.1	36.1
Stunting[a] (%)		48.0	38.6	35	44.4	43.3	46.5	35.6	55.1	39.8	38.8	38.3	35.7
Wasting[b] (%)		19.8	19.3	14.8	18.8	15.3	13.7	9	30.7	9	13.3	9.7	24.6
Underweight[c] (%)		42.5	36.5	25.6	38	32.5	36.4	22.1	48.8	19.9	25.2	19.7	39.6
Prevalence of anaemia (by Hb level) in children aged 6–59 months		69.5	54.7	58.6	61.4	56.9	69.6	41.1	64.4	44.2	n.d.	59.2	62.9
Prevalence of anaemia (by Hb level) in women aged 15–49 years		55.3	43.3	52.1	55.2	50.6	69.5	35.7	47.2	38.6	n.d.	60.0	65.1

[a]Height-for-age: children under 5 years < −2 SD from the international reference median value
[b]Weight-for-height: children under 5 years < −2 SD from the international reference median value
[c]Weight-for-age: children under 5 years < −2 SD from the international reference median value
Source GoI (2010) (Schedule type 1 data of NSS 66th Round, 2010); NFHS (2006)

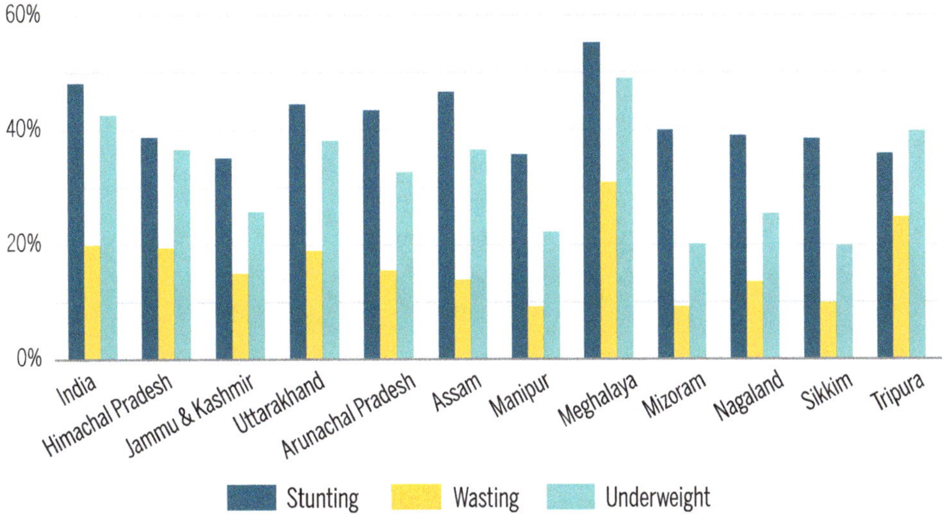

Fig. 9.3 Prevalence of stunting, wasting, and underweight in children under five years of age in India (*Source* NFHS 2006)

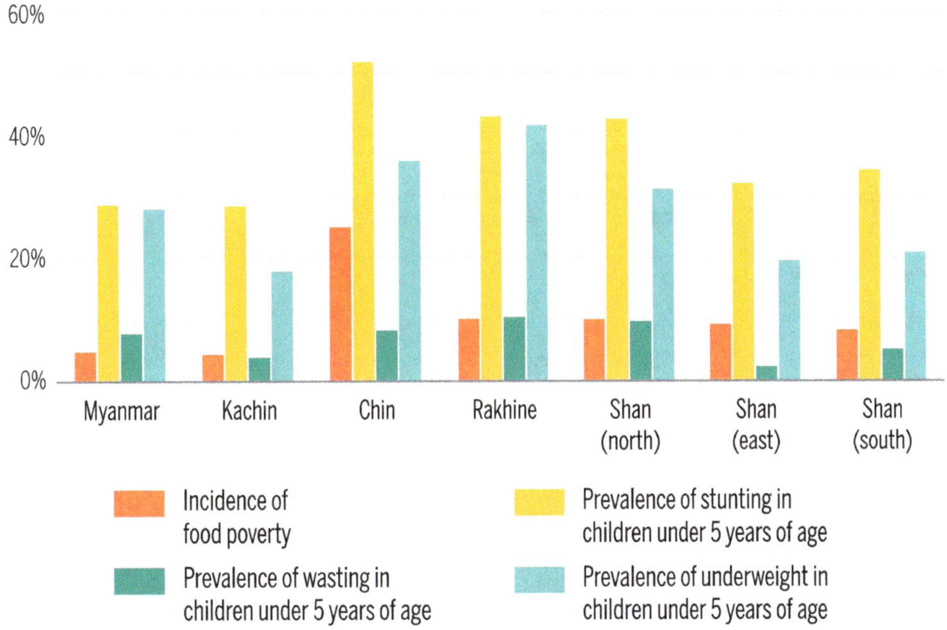

Fig. 9.4 Food poverty and undernutrition in children under five years of age in Myanmar (*Sources* IHLCA 2011; MoNPED and MoH 2011)

Table 9.6 Food and nutrition (in)security in Nepal

Indicator	National	Mountains	Hills	Terai
% of food insecure households	50.8	59.5	52.8	47.9
Prevalence of underweight women (aged 15–49 years)	18.2	16.5	12.4	22.7
Stunting[a] (%)	40.5	52.9	42.1	37.4
Wasting[b] (%)	10.9	10.9	10.6	11.2
Underweight[c] (%)	28.8	35.9	26.6	29.5
Prevalence of anaemia (by Hb level) in children aged 6–59 months	46.2	47.7	41.0	50.2
Prevalence of anaemia (by Hb level) in women aged 15–49 years	34.8	26.7	26.5	41.9

[a]Height-for-age: children under 5 years < −2 SD from the international reference median value
[b]Weight-for-height: children under 5 years < −2 SD from the international reference median value
[c]Weight-for-age: children under 5 years < −2 SD from the international reference median value
Source NDHS (2011)

Table 9.7 Food and nutrition (in)security in Pakistan

Indicators		National	Mountain province/administrative unit[b,c]				
			AJK	Balochistan	FATA	GB	KPK
Prevalence of food insecurity (%)[d]		49	47	61	68	52	56
% of women (aged 15–49 years) facing micronutrient deficiency	Iron deficiency anaemia (NPW)[a]	20	19	16	16	10	5
	Iron deficiency anaemia (PW)	26	28	31	n.d.[c]	30	15
	Vitamin A deficient (NPW)	43	13	50	83	39	72
	Vitamin A deficient (PW)	49	31	62	n.d.	45	85
	Calcium deficient (NPW)	51	6	60	81	45	71
	Calcium deficient (PW)	58	4	63	n.d.	71	61
	Vitamin D deficient (NPW)	85	95	83	85	96	80
	Vitamin D deficient (PW)	86	85	78	n.d.	96	77
% of children (0–59 months) facing growth problems	Severe stunting	22	12	32	36	22	25
	Wasting	17	18	18	10	9	18
	Underweight	31	n.d.	42	13	n.d.	24
% of children (0–59 months, except iodine) facing micronutrient deficiency	Vitamin A deficient	56	37	81	100	82	78
	Zinc deficient	37	49	34	34	33	34
	Vitamin D deficient	41	33	40	26	32	30
	Iodine deficient (6–12 years)	37	65	35	10	70	26
	Iron deficiency anaemia	33	28	23	26	21	13

[a]NPW = non-pregnant women; PW = pregnant women
[b]AJK = Azad Jammu and Kashmir; FATA = Federally Administered Tribal Areas; GB = Gilgit-Baltistan; KPK = Khyber Pakhtunkhwa
[c]n.d. = no data
Source Bhutta et al. (2011); FSA (2009)[d]

prevalence of stunting and underweight in children (<5 years) is also higher in the mountains compared to the plains and national statistics. On the other hand, the prevalence of anaemia and underweight in women (15–49 years) is slightly lower in the mountains.

In Pakistan, the prevalence of food insecurity is significantly higher in the mountain areas when compared to national statistics, with the exception of Azad Jammu and Kashmir (AJK). Almost two-thirds of the population in Balochistan and the Federally Administered Tribal Areas (FATA) is food insecure (Table 9.7), mainly due to lack of financial resources to purchase adequate food (FSA 2009; Hussain and Routray 2012). In line with the higher food insecurity, the majority of mountain people face a higher deficiency in most micronutrients. For example, in Balochistan, FATA, and Khyber Pakhtunkhwa (KPK), vitamin A deficiency among non-pregnant women (aged 15–49) stands at 50%, 83%, and 72%, respectively, while calcium deficiency stands at 60%, 81%, and 71%, respectively. The majority of pregnant women also face deficiencies of vitamin A and calcium in Balochistan and KPK. In Gilgit-Baltistan (GB) and KPK, almost all non-pregnant women have a vitamin D deficiency. The percentage of pregnant women with calcium and vitamin D deficiency in GB is also extremely high (Table 9.2). Almost one-third of children (<5 years) faced severe stunting in Balochistan and FATA. The prevalence of underweight children in Balochistan is also very high in comparison to the national average. All the children in FATA and the majority in Balochistan, GB, and KPK face vitamin A deficiency; and a high proportion of children in AJK face deficiency in zinc and in AJK and GB in iodine.

9.2.2 Seasonality and Vulnerability of Food and Nutrition Security

Mountain communities in the HKH region are highly vulnerable to food and nutrition insecurity as a result of a range of climatic, physical, and socioeconomic factors (Dame and Nüsser 2011; Hussain et al. 2016). A considerable proportion of the population live in remote, high altitude areas and dispersed settlements with limited physical infrastructure and market opportunities (Romeo et al. 2015). People living in such areas often face difficulties in accessing food due to poor infrastructure, weak communications, and limited transportation systems. There is also a seasonal variation in food security; in the winter and rainy seasons, road networks are often disconnected from the plains and food transportation becomes very difficult and highly expensive, particularly if food needs to be airlifted. High transportation costs also increase food prices to levels that may go beyond the purchasing power of poor people and compromise food quality and quantity, affecting nutritional outcome (Ghosh and Sharma 2016). As a result, mountain regions often experience seasonal shortfalls of food with poor people particularly vulnerable to food and nutrition insecurity (Dame and Nüsser 2011; Hussain et al. 2016). Food security tends to be most compromised during the monsoon and snowy winter months, as well as during periods following natural disasters (Box 9.2).

Box 9.2 Seasonality of food and nutrition security in high mountains

High mountain areas in the HKH often experience heavy snowfall during the winter months (December to March), which result in blocked roads and physical isolation of local communities from other areas (e.g., in Upper Rasuwa in Nepal, Upper Mansehra in Pakistan, Northern Afghanistan, and the northwestern Himalayas in India). As an adaptation measure, people often reduce their food intake in terms of both quantity and frequency, which has serious implications for food and nutrition security in winter. Similar situations may occur after disasters such as landslides and floods. The main reason for food stress at these times is the lack of local storage and food processing facilities. To cope with such situations, governments could establish food storage centres to store both local produce and items brought in from other areas when the roads are functional (before winter and monsoon).

9.3 Issues and Challenges to Food and Nutrition Security

Mountain communities face multiple challenges in achieving nutrition security. In most mountain areas, the amount of available land per capita is too low to support sustainable livelihoods (Tiwari and Joshi 2012b; Hussain et al. 2016). Both the total area available per capita and the net sown proportion will become even smaller as the population increases. But the nature and causes of food and nutrition security are complex and multifaceted and mountain communities face several other biophysical, environmental, and socioeconomic constraints and challenges that affect food and nutrition security (Table 9.8). The key challenges are discussed in the following sections. They include deterioration of local food systems, changing dietary habits, climate change impacts, high rates of poverty, increased rates of outmigration, abandonment of cultivable land, rapid urbanization, inadequate infrastructure and market access, depletion of natural resources, barriers to food movement and trade, inadequate access to improved drinking water and sanitation, and gaps in existing policies and programmes. Climate change-induced hazards and biophysical constraints are further adding to the overall problems of inaccessibility and fragility in mountain areas. The factors constraining and facilitating food and nutrition security in the HKH region can be better understood using the conceptual framework shown in Fig. 9.5. They are discussed in more detail in the following.

9.3.1 Deteriorating Local Food Systems

Local biophysical and socioeconomic conditions play an important role in mountain agricultural systems. The huge shift from traditional to new crop varieties and from traditional to modern farming practices has led to a decline in local food systems (Adhikari et al. 2017; Gautam and Andersen 2017). A wide range of traditional crops used to be part of the food basket in mountain areas, including millets (*Eleusine* spp., sorghum (*Sorghum bicolor*), buckwheat (*Fagopyrum* spp.), amaranth (*Amaranthus* spp.), sea-buckthorn (*Hippophae* spp.), barley (*Hordeum vulgare*), naked barley (*Hordeum himalayens*), legumes (*Vigna* spp.), yam (*Dioscorea* spp.), sesame (*Sesamum indicum*), niger (*Guizotia abyssinica*), kaphal (*Myrica esculenta*), chiuri (*Diploknema butyracea*), amala (*Phyllanthus emblica*), pomelo (*Citrus maxima*), and jamun (*Syzygium cumini*) (Padulosi et al. 2012). These traditional crops are also known as neglected and underutilized species (NUS) due to the lack of recognition and underutilization of their potential in terms of their biophysical suitability in mountain ecosystems and their high nutritional value. People often consider NUS as 'foods of the poor' (Adhikari et al. 2017; Mal et al. 2010) and are not aware of their potential for production, income, and nutrition security. Previously widespread crops such as millet, barley, oats, and beans are being replaced with rice, wheat, maize, and high-yielding cash crops such as potato, vegetables, and even non-food crops) (Adhikari et al. 2017; Apetrei 2012).

Some of the most important factors contributing to the decline in NUS production are increasing food demand, declining market value of NUS, increasing market value and demand for dominant crops, increasing standardization of agricultural practices and trends towards mono-cropping, inadequate policy support, changing food preferences in response to urbanization, changes in income levels, lack of nutritional knowledge among consumers, and gradual loss of traditional NUS-based recipes (Adhikari et al. 2017). Disappearance or limited cultivation of NUS is leading to a decline in agricultural diversity in agriculture ecosystems

Table 9.8 Challenges to food and nutrition security

Challenge	Consequences	Dimensions of food and nutrition security likely to be affected negatively
Deterioration of local food systems	• Reduced food production • Reduced production diversity	• Food availability • Food utilization
Changing diets	• Reduced dietary diversity	• Food utilization
Climate change	• Risks to agricultural production • Reduced production diversity • Reduced farm income • Constrained food supply from plains due to climate hazards	• Food availability and stability • Food utilization • Food accessibility • Food stability
Lingering poverty	• Reduced food intake • Reduced dietary diversity	• Food accessibility • Food utilization
Increased outmigration	• Labour shortages in agriculture leading to reduced production	• Food availability
Abandonment of cultivable land	• Low returns, labour shortages, and others leading to land being abandoned and loss of production	• Food availability
Rapid urbanization	• Encroachment of agricultural land leading to reduced agricultural production	• Food availability
Inadequate infrastructure and market centres	• Inadequate food distribution • Higher post-harvest losses • Higher prices of external food items	• Food accessibility • Food availability
Depletion of natural resources	• Loss to water resources • Reduced supply of biomass manure from forests • Reduced supply of edible wild plants and fruit from forest, leading to reduced dietary diversity • Degradation of rangelands and pastures resulting in reduced livestock production and income	• Food availability • Food availability • Food availability and utilization • Food availability and accessibility
Constraints to internal food movement and cross-border trade	• Reduced food supply to mountains • Higher prices of available food	• Food availability • Food accessibility
Inadequate access to improved drinking water, sanitation, and hygiene	• Higher prevalence of diseases	• Health status

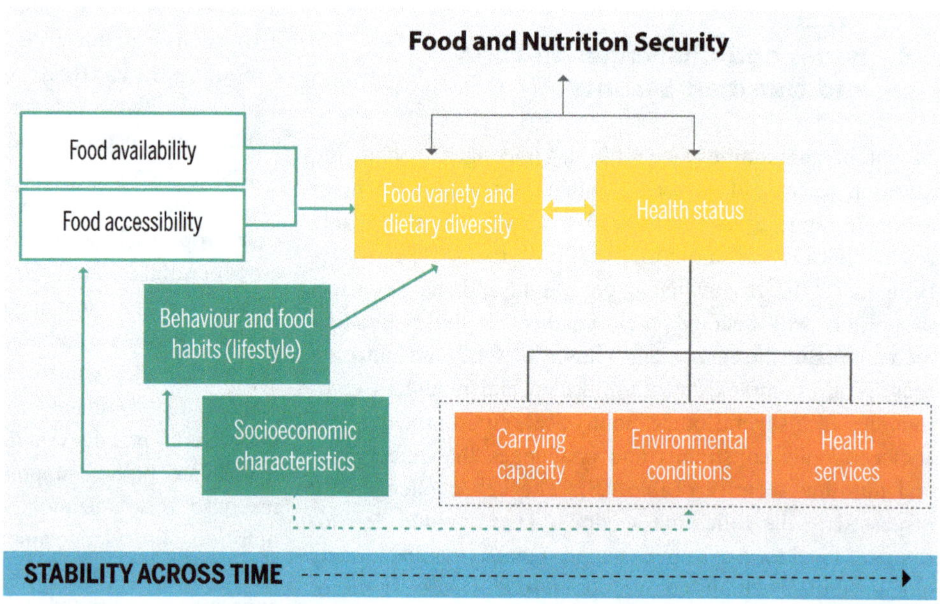

Fig. 9.5 Food and nutrition security: a conceptual framework (*Source* Modified from Gross et al. 2000)

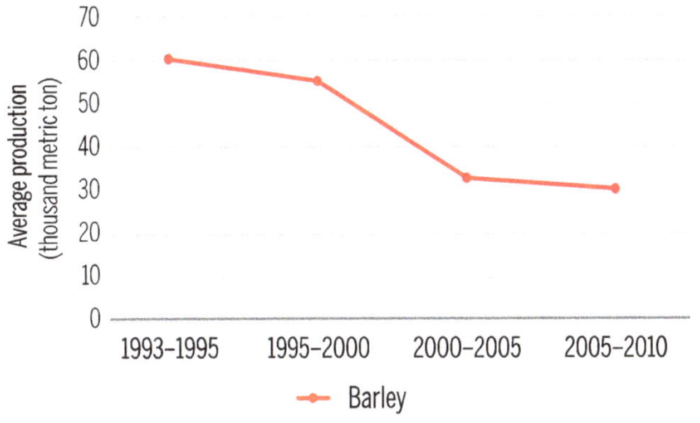

Fig. 9.6 The trend in barley production in KPK, Pakistan (*Source* ASP 2012)

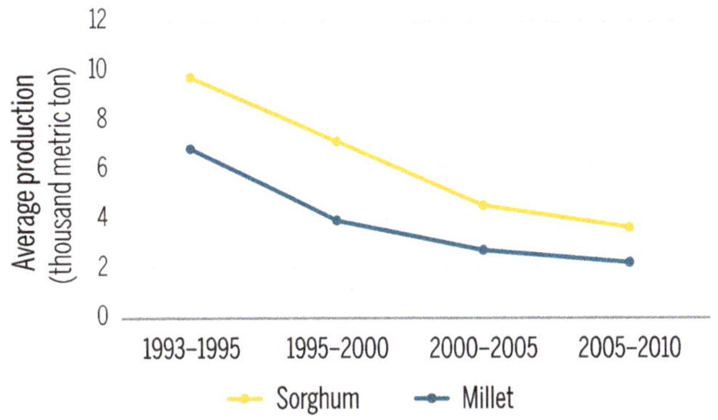

Fig. 9.7 The trend in sorghum and millet production in KPK, Pakistan (*Source* ASP 2012)

and dietary patterns (Mayes et al. 2012), particularly in the HKH (Adhikari et al. 2017; Rasul et al. 2018), with a few commodity crops now dominating food systems at all levels. The loss of agrobiodiversity has become a serious concern and may undermine long-term agricultural sustainability and food security in the region.

The decrease in NUS production has been observed in many areas (Adhikari et al. 2017). For example, the average production of barley in KPK, Pakistan, declined to only 30 thousand metric ton in the late 2010s from more than 60 thousand metric tons in the mid-1990s; while the average production of sorghum declined by nearly 62% and of millet 67% over the same period (Figs. 9.6 and 9.7). In Nepal, the change has already taken place in many areas; the per capita production of millet in 2013/14 was only 10.7 kg, compared to 66, 177, and 80 kg for wheat, rice, and maize, respectively (estimates based on data from MoAD 2015). The change is continuing, although the (low) production of NUS showed a moderate increase between 2007 and 2014 in absolute terms (MoAD 2015), there was a decline or negligible change in terms of per capita production.

At present, protection and promotion of traditional crops is not a top priority in many HKH countries (Hagen 2004) and institutional mechanisms to help local communities either to realize and use the benefits of local agro-biodiversity or to provide market incentives for producers of NUS are inadequate. Policies on food pricing and farm subsidies do not take traditional crops into account, and trade and market policies rarely reflect their nutritional and ecological value (Williams and Haq 2002; Padulosi and Hoeschle-Zeledon 2004). If farmers, do not have supporting policies, appropriate transportation facilities, good market value and chains, or demand, they will prefer to grow dominant crops rather than cultivating traditional crops, particularly when landholdings are very small. This has been the situation over the last two to three decades in the HKH (Adhikari et al. 2017; Sharma et al. 2016). At the same time, local food systems in mountain areas are being further affected indirectly by subsidies provided for new production technologies and inputs, the relatively higher returns of dominant crops such as rice and wheat resulting from the minimum support price policy, and frequent food aid and food supply as remuneration for work (Pingali et al. 2017; Rasul et al. 2018; Maikhuri et al. 2001).

Local food systems face another challenge in human-wildlife conflict. A considerable number of people in high mountain areas live in the proximity of national parks and other protected areas. In these areas, increasing depredation of crops and livestock by wild animals has become a serious issue (Ogra 2008; Manral et al. 2016), again with important implications for people's livelihoods and food and nutrition security.

The deterioration in local food systems is resulting in increased dependence on external food crops (such as fine rice, wheat, and non-organic vegetables) and processed snacks and drinks, in place of traditional crops. This has made mountain people highly vulnerable to food and nutrition insecurity. Price shocks in food, and natural hazards such as floods and landslides, can further restrict food supplies and result in price hikes, leading to short-term food insecurity with long-term impacts. Price hikes and restricted food supplies can lead to a shift from a diverse food culture to a monoculture, reduced overall consumption, high dependence on assistance and aid, and reliance on credit leading to indebtedness (Hussain et al. 2016). Debts may further affect community resilience and sustainable food security in the long-term (Milbert 2009). Further, most food items coming from the plains are non-organic and produced with much use of pesticides and chemical fertilizers. In the recent past, use of chemicals has also been introduced in some mountain areas (e.g., Meghalaya in India) with the shift from traditional to commercial farming (Behera et al. 2016). These non-organically produced food items pose a threat to food safety in the mountains, where people historically cultivated and consumed mainly safe organic food items (Rasul and Thapa 2003; Atreya et al. 2012; Giri et al. 2014).

9.3.2 Changing Dietary Habits

Food habits and diets in the HKH have been undergoing changes in recent years due to socioeconomic developments linked to increased access to roads, schools, and markets, as well as radio, television, and other media (Dame and Nüsser 2011; Finnis 2007). Changes are more prominent in middle and lower elevation villages, where road connections are better and market connections have been established. The consumption of traditional coarse grains, which contain an abundance of micronutrients and fibre, is often considered to be 'backward' in the sociocultural value system (Maikhuri et al. 2001) and refined rice and wheat have become the main food items. The process has been further reinforced by low prices and inadequate incentives for traditional crops and a decrease in production as well as availability of high-yielding varieties of seeds at subsidized prices, and distribution of free or subsidized food (often rice and wheat) by the government (Maikhuri et al. 2001). The replacement of NUS such as millet, barley, beans, and coarse rice, with fine grains and low nutrition processed foods and drinks has important implications for nutrition status in the HKH. While total calorie intake has increased over the years, nutritional status has deteriorated due to micronutrient deficiencies (Rasul et al. 2018). Reduced consumption of diverse foods and decreasing activity levels lay the foundation for obesity and related chronic diseases (Kumanyika et al. 2002; Hill et al. 2003). Foods from NUS, wild vegetable, and fruit species, as well as animal sources, are important sources of micronutrients, yet their consumption has been reduced due to the increasing uniformity of food patterns and changes in dietary habits (Rasul et al. 2016). Globally, changes in lifestyle influencing changes in diet composition and food demand are expected to increase significantly, even with no further population growth (Pradhan et al. 2013). This might have serious implications for the HKH region, as mountain communities depend considerably on national and regional markets for access to food.

9.3.3 Changing Climate

The average ambient temperature in the mountains of the HKH is rising at a rate of 0.06 °C per year, higher than the global average, which has resulted in loss of snowfall and snow cover and rapid melting and shrinking of the majority of glaciers in the region (Shrestha and Aryal 2011) (see Chap. 3 Climate Change in the HKH and Chap. 7 Status and Change of the HKH Cryosphere). Loss of the cryosphere is changing the amount and timing of melt, impacting water availability, and leading to a reduction in food production and food insecurity in the mountains as well as downstream (Rasul 2014). In addition, increasing variability in precipitation patterns and the increased incidence of hazards such as floods and droughts are seriously affecting agricultural productivity and income across the region (Hussain et al. 2016, 2018). Studies conducted in the mountain states of India such as Himachal Pradesh, Jammu and Kashmir, and Uttarakhand (Vishvakarma et al. 2003; Kumar et al. 2006; Negi et al. 2012) have shown that precipitation and temperature patterns have changed in a way that is likely to affect the discharge, volume, and availability of water (Bandyopadhyay and Perveen 2008; Viviroli et al. 2007). For example, in Ladakh, India, farmers are already experiencing a shortage of water for irrigation due to reduced snow and ice (Clouse et al. 2017). Similar experiences are also reported from other high mountain areas in the HKH region (Tiwari and Joshi 2012b). Loss of the cryosphere may have a serious impact on water availability for agriculture (Aase et al. 2009). The changes in key climatic phenomena are also likely to result in more frequent droughts, increased incidence of high intensity rainfall, and more frequent floods (Ailikun 2015; UCCRN 2015), all of which will lead to increased vulnerability and uncertainty in food and nutrition security in the region.

A large survey-based study (over 8,000 households) conducted in the four river basins of the HKH—Upper Indus (Pakistan), Koshi (Nepal), Eastern Brahmaputra (India), and Salween and Mekong (China)—found that the majority of households had perceived an increase over time in the incidence of climate-induced extreme events, including floods, erratic rainfall patterns, high variations in temperature, landslides/erosion, dry spells and droughts, and livestock disease and crop pests (Fig. 9.8). These extreme events affected their agricultural production and income (Hussain et al. 2016).

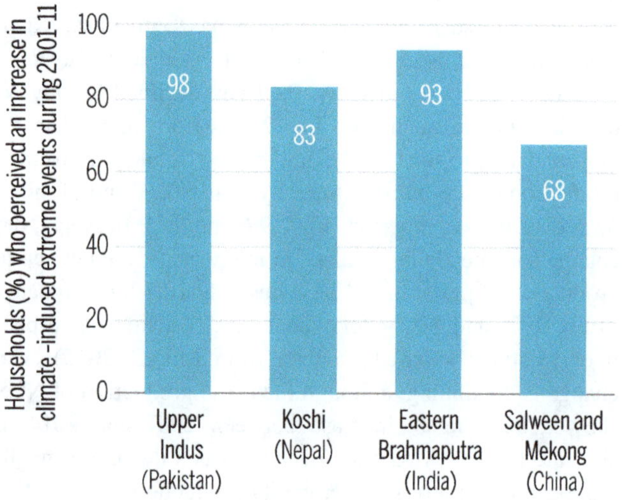

Fig. 9.8 Incidence of climate-induced extreme events during 2001–11 in the HKH (*Source* Hussain et al. 2016)

Some extreme events, such as droughts and floods, have long-term impacts on the food and nutrition security of local communities. Drought is directly and indirectly affecting livelihoods in the mountains and increasing food insecurity through reduced water availability for both agriculture and rangeland production (Xu et al. 2009; Ebi et al. 2007; Joshi et al. 2013; Rasul 2014). Most of the mountain rangelands, particularly in the western parts of the HKH (e.g., Balochistan Province, Pakistan), are facing degradation due to the increased incidence of drought compounded by overgrazing. Degradation of the vegetation cover and deterioration of soil with erosion and loss of nutrients is resulting in a reduction of livestock productivity that further impoverishes pastoral communities (Afzal et al. 2008). For example, a drought from 1998 to 2002 in Balochistan Province of Pakistan resulted in a serious water shortage for agriculture, affecting nearly two million acres of arable land and 9.3 million livestock. Drastic changes in the composition of plant species in the degraded rangeland ecosystem and shortages of fodder and water led to the death of 1.8 million livestock and the destruction of nearly 80% of apple orchards. Food prices rose, and the food security and livelihoods of nearly two million mountain people were affected, resulting in reduced food consumption and the migration of people from drought-affected areas to areas downstream (Shafiq and Kakar 2007). The continued water stress in Balochistan has led to excessive pumping of groundwater, resulting in a further lowering of the water table. In certain parts of Balochistan, this has had a significant impact on local food systems and food security (Rasul et al. 2014).

In contrast, in 2010, frequent fluctuations in the intensity of precipitation in some mountain areas of Pakistan led to an increase in flooding, especially flash flooding, affecting the resource base and infrastructure in both upstream and downstream areas. The devastating floods left around eight million people vulnerable to long-term food insecurity, the majority of them in mountain areas (Rasul et al. 2014). Similar kinds of floods have been observed frequently in Uttarakhand (India) in recent years (2001–2013). Flood events are also increasing in the drier mountain areas, such as Ladakh in India (Ziegler et al. 2016) and Gilgit-Baltistan in Northern Pakistan (Hashmi et al. 2012).

9.3.4 Lingering Poverty

In the HKH, high poverty, both unidimensional and multidimensional, is one of the major causes of food and nutrition insecurity (see Chap. 12 Poverty and Vulnerability in Mountain Livelihoods). Financial resources are needed to afford an adequate diet and ensure food and nutrition security. The choice and consumption of food is largely determined by household income and price of products (Lo et al. 2012). Generally, low income groups prefer low-price high-energy food regardless of its nutritional value (Monsivais and Drewnowski 2007; Kettings et al. 2009). In many mountain areas of the HKH, the incidence of poverty is higher than in the adjacent plains (Khan et al. 2015; Saboor et al. 2015), and local people have a very limited choice of food items due to their low income levels, leading to food and nutrition insecurity. A study conducted in the Mugu district of Nepal revealed that only a few of 198 children studied were getting a nutritious diet including such items as meat, egg, milk, vegetables, and fruit. The main reason for the inadequate dietary intake was the lack of economic affordability of diverse food items (Sharma 2012). Poverty affects all aspects of nutrition—food security, access to safe drinking water, hygiene, housing, health services, and education—further trapping families in poverty from one generation to the next (CPRC 2004).

9.3.5 Increasing Rate of Out-Migration

The inflow of remittances from out-migrants is undoubtedly a potential source for improving local food security and livelihoods through enhanced income, growth of local small businesses, transfer of new technologies, and creation of job opportunities for local skilled and unskilled labour. But out-migration for work has also added to the challenges in mountain areas. Increased out-migration and decreased interest of the younger generation in farming also contribute to low productivity in agriculture as well as abandoning of farmland in some parts of the region, as discussed below. The HKH region faces frequent labour shortages during the critical periods of agricultural activities due to out-migration of active household members, particularly youth, with 70%, 26%, 37%, and 38% of farm households affected in the Upper Indus (Pakistan), Koshi (Nepal), Eastern Brahmaputra (India), and Salween and Mekong (China) river basins (Hussain et al. 2016). The absence of young people is impeding efforts to develop mountain agriculture as a business enterprise. The involvement of youth is very important because they are generally more progressive and more willing to adopt new technologies related to production, harvesting, post-harvest handling, and marketing than older farmers. However, a growing number of reports suggest that despite the problems, migration has become an important source of livelihoods and contributes positively to food security (Crush and Caesar 2017; Gautam 2017). More empirical evidence is required to better understand the relationship between migration and food security.

9.3.6 Abandonment of Cultivable Land

A considerable proportion of cultivable land in the HKH has been abandoned and remains fallow as a result of the low returns, shortage of labour caused by rural out-migration for work, highly unequal distribution of land, limited scope for mechanization, and lack of land ownership by tenants (ICIMOD 2008). In some areas of the Nepalese Himalaya, more than 30% of total cultivated land has been abandoned in the mid-hills districts (Jackson et al. 1998; Thapa 2001; Khanal 2002); close to 30% of agricultural land has been abandoned in the Sikles area of the Gandaki basin in Nepal (Khanal and Watanabe 2006); and cultivated terraced lands decreased by 36% between 1978 and 2014 in the western middle hills of Nepal (Jaquet et al. 2015). Hussain et al. (2016) found that more than a quarter of agricultural land in the Upper Indus was left to serve as pasture and grassland for livestock, because households find it easier and more profitable compared to cultivating crops in the face of labour shortages, while across the Upper Indus, Koshi, Eastern Brahmaputra, and Salween and Mekong river basins, 6% of cultivable land was left fallow without any productive use at all.

9.3.7 Rapid Urbanization

The problems with agricultural land have been further compounded by the fast rural-urban migration and unplanned urbanization. In recent years, several areas in the HKH have experienced rapid, unregulated, and unplanned urban development due to population growth and inadequate land use policies. Expansion of urban areas on agricultural land, especially in the more accessible and fertile areas, placing further pressure on the already shrinking landholdings (Nüsser et al. 2015). Although urbanization is likely to improve market access in adjacent mountain areas, farmers are not receiving any additional benefits due to the lack of production technology and continuation of old practices under the same ecological conditions (Stone 2001). Moreover, urbanization also contributes to changing food habits and dietary patterns with increased consumption of refined cereals, rice and wheat.

9.3.8 Inadequate Infrastructure and Market Access

Adequate and appropriate infrastructure facilities, such as all-weather road connectivity, an easily accessible network of food distribution centres, food storage facilities, and market/collection centres, particularly during the monsoon and snowy winter months (see Box 9.2), are of critical importance for ensuring access to food in the HKH (Partap 1998). Unfortunately, across the region there is still a lack of the infrastructure required for the efficient distribution of food and improving food access, as well as a lack of adequate food storage facilities to maintain food reserves during the rainy and winter months. During the 2013 floods in Uttarakhand in India, food supplies were secured by helicopters and stored in schools due to the lack of storage facilities (Awasthi 2016).

Marketing and processing facilities for agricultural, dairy, and horticultural products are also inadequate in the mountain areas. This not only results in huge post-harvest losses of food crops, fruit, and vegetables, it also makes mountain farming and horticulture economically unviable, particularly for poor and marginalized households. Gilgit-Baltistan (GB) in Pakistan, for example, is an important producer of fruit, but the lack of storage, packing, and transportation facilities, as well as improper handling, result in post-harvest losses of up to 59% of fruit crops, with a particularly high proportion of wastage for perishable commodities like mulberry, peach, and pomegranate (Fig. 9.9). Post-harvest losses of vegetables are also high at up to 40% (Rasul and

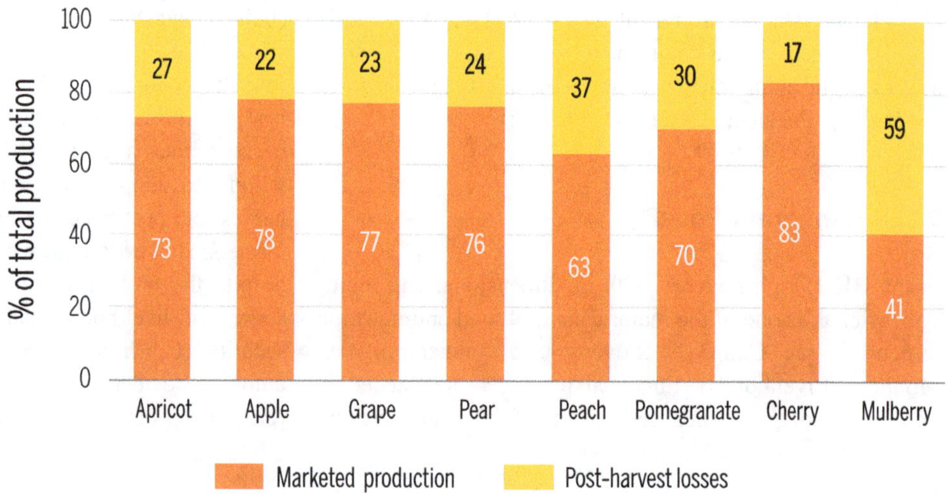

Fig. 9.9 Post-harvest losses in Gilgit-Baltistan, Pakistan (*Source* Rasul et al. 2014)

Hussain 2015). In Balochistan Province, people feed a substantial portion of harvested fruit and vegetables to their livestock due to the limited access to markets (Rasul et al. 2014).

In Nepal, the private sector is heavily engaged in the food commodity trade in the plains, but its role in hill and mountain areas is limited due to the poor transportation infrastructure. Many mountain districts are not yet well-connected to central and regional markets such as Kathmandu and Biratnagar, and even district headquarters are often not well-linked to remote areas in the district. The poor connectivity means that the food trade in hill and mountain areas has been minimal, while the high cost of transportation and relative isolation from functioning markets, means that the prices of major cereals, such as rice and pulses, are generally higher in mountain areas than in the Terai (plains). Due to the lack of functioning markets, low food availability, and low purchasing power, household participation in food markets in hill and mountain areas is limited (WFP and FAO 2007), which makes them vulnerable to food and nutrition security.

A similar situation exists in Bhutan, where much of the agricultural production has very poor access to market outlets. Access to markets is mainly hindered by the remoteness of producers, small production, and difficulty in reaching major market centres. This poor market integration particularly affects the horticultural sector with production of potatoes, beans, chillies, vegetables, and mushrooms, and orchard products such as apples, mandarins, and walnuts.

In the mountain states of India, the viability of food markets also remains challenging (Ghosh and Sharma 2016). First, most mountain areas are poorly connected to the plains. Second, mountain people have low purchasing power, leading to poor access to food. Third, farmers from the mountain states lack market information related to the prices of agricultural products. And finally, natural hazards such as landslides and excessive rain frequently block access to the market systems, affecting food accessibility for many poor and vulnerable communities (Ghosh and Sharma 2016).

The constraints of infrastructure and market services reduce the productivity of agricultural resources, resulting both in seasonal food shortages and restrictions in the development of value-added products. This ultimately restricts both economic growth and improvement of livelihoods in mountain areas (Partap 2011).

9.3.9 Depleting Natural Resources

A large number of people living in rural areas in the HKH region depend heavily on natural resources for their livelihoods and food security (Rasul 2010; Rasul et al. 2014), but rapid urbanization, changes in land use patterns, and increasing rainfall variability have affected the natural resources across the region. Water resources, especially springs, are being depleted rapidly across the HKH region (Tiwari and Joshi 2012a). Although the causes are not yet clear, a range of factors are thought to be responsible, including reduced infiltration and increased runoff, changes in land use (e.g., mixed forest to plantation, forest to agriculture, agriculture to infrastructure, wetlands to drained land), changes in agricultural practices (leading to soil degradation), and changes in precipitation patterns (e.g., from snow to rain, increase in cloudbursts and extreme events) (Tiwari and Joshi 2012a, b; Gautam and Andersen 2017). The amount of surface runoff is also much higher from built-up land than from other land categories, particularly forest and areas under cultivation. The land use changes and resultant hydrological disruptions have had a direct adverse impact on water availability and irrigation potential, which has been considerably reduced over the last three decades, with reduced groundwater recharge leading to drying of springs and streams across the HKH (Rawat 2009). For example, in the Hilkot catchment in Pakistan, 52 of 152 springs were found to have dried up over a five-year period (Merz et al. 2004), and Tambe et al. (2012) observed that spring yields in Sikkim in India had declined by 50%. The loss of basic ecosystem services, and water in particular, has had a negative impact on agricultural productivity across the region.

Forests are critical for maintaining soil fertility and water recharge, but they are depleting all over the HKH. The depletion of forest resources has also had an adverse effect on local agricultural systems through reduced supply of manure (from livestock fed on forest fodder). A study carried out in Uttarakhand, India found that the average supply of manure to agricultural land had declined from 15 tonnes/ha/year in 1980 to 9 tonnes/ha/year in 2010. As a result, the productivity of agriculture had declined by nearly 125 kg/ha (25%). Deforestation has accelerated the process of soil erosion due to increased runoff, and the deforested slopes have become highly vulnerable to slope failures and landslides contributing to the degradation of arable land and declining food production (Tiwari and Joshi 2012b). Depletion of forests has also resulted in reduced availability and consumption of edible wild plant species and fruit in mountain areas.

The increase in prolonged droughts is also contributing to the degradation of rangelands and pastures by reducing vegetation cover and exposing soil, resulting in reduced livestock productivity and further impoverishment of pastoral communities (Afzal et al. 2008).

9.3.10 Hindrances to Internal Food Movement and Cross-Border Trade

Since mountain regions are highly dependent on the plains for food, unrestricted flow of food is critical. However, there are some constraints on internal movement of food and cross-border trade in the HKH that are hindering efforts to achieving food and nutrition security. The constraints include high transportation costs and hazards such as landslides and floods which restrict movement from the plains to the mountains. The high transportation costs lead to higher prices for food items in the mountains, often three to four times as high as for the same items in the plains. The prices of essential food commodities such as pulses, wheat flour, and rice, increased tremendously between 2005 and 2015, reducing the access to adequate food. The food purchasing power of rural people in Uttarakhand decreased by 30–35% over this period, exacerbating food insecurity, and poor and socially marginalized communities either stopped or drastically reduced their purchase of nutritious food products (Tiwari and Joshi 2012b). In Pakistan, the interprovincial movement of food is sometimes halted due to the internal food demand of food-producing provinces (Box 9.3). While food trade and movement between the HKH countries is sometimes affected by economic and geopolitical events, again with serious implications for food and nutrition security as the disruptions induce food shortages and price hikes in the mountain areas (Box 9.4).

Box 9.3 Interprovincial movement of food in Pakistan

In Pakistan, provincial governments sometimes restrict the movement of food to other provinces if provincial production is only sufficient for local consumption. In 2014, the Food Department of Sindh Province banned the interprovincial movement of wheat flour to Balochistan and Punjab in order to fulfil the local requirements in Sindh (The Nation 2014a). The Punjab government also imposed a ban on the interprovincial movement of wheat in 2014 to achieve its wheat procurement target (a repeat of the ban in 2003 and 2004 [Dawn 2004]); the pace of procurement was far below the wheat procurement target of 4 million tonnes set by the provincial government (The Nation 2014b). These bans on food movement resulted in food shortages and price hikes in the wheat-demanding provinces, particularly in the mountains. Gilgit-Baltistan and adjacent mountain regions have suffered severely in terms of food availability and accessibility amidst such restrictions and legal constraints.

Box 9.4 Constraints to cross-border food trade

Food trade between the HKH countries is sometimes constrained by changes in bilateral diplomatic relationships. For example, the movement of food and non-food items from India to Nepal was affected during the last quarter of 2015 due to political unrest and road blockade in the Terai belt, resulting in food and nutrition vulnerability in remote mountain areas as basic food items, medicines, and cooking fuel became unavailable (WSJ 2015). There is some evidence that food movement from Pakistan to Afghanistan is also occasionally affected due to changes in bilateral relationships. In May 2016, the Torkham border between these two countries was shut, resulting in stoppage of trade of food and non-food items (Aljazeera 2016). A regional mechanism is needed to ensure that bilateral diplomatic relationships between countries do not affect the cross-border movement of food items in order to avoid food crises.

9.3.11 Inadequate Access to Improved Drinking Water, Sanitation, and Hygiene

Nutritional status is strongly influenced by access to adequate quantity and quality of water, as well as to sanitation. Water and sanitation are critical for healthy lives, but despite many efforts, access to safe drinking water and sanitation is still limited in the HKH (Table 9.9). Except in Bhutan, a significant portion of the population in the HKH countries depends on unimproved and unsafe water sources. The situation is quite serious in Afghanistan, where 45% of the population still use unsafe water, and in Myanmar, where 19% of the population use unsafe water.

A large part of the population in the HKH is also deprived of improved sanitation facilities (Table 9.9), especially in Afghanistan where more than half of the population have no or unimproved sanitation facilities. Open defecation also remains common in many countries; and is the common practice for 30% of the population in the mountains of India and Nepal and 23% in Pakistan.

Table 9.9 Access to safe drinking water and sanitation

Country/region	Source of drinking water (%)				Sanitation facilities (%)			
	Improved		Unimproved		Improved		Not improved	
	Piped to premises	Other	Surface water	Other	Individual	Shared	Unimproved	Open defecation
Afghanistan	12	43	6	39	32	12	43	13
Bhutan	58	42	0	0	50	28	20	2
Myanmar[a]	8	73	5	14	80	12	4	4
Nepal	24	68	2	6	46	18	4	32
Mountain states India	–	87	–	13	43	2	23	32
Mountain provinces/administrative units Pakistan	37	48	10	5	61	8	7	23

[a]No mountain specific data for Myanmar available; national level data used as a proxy
Source Rasul et al. (2016); MoNPED and MoH (2011); WHO-UNICEF (2015); UNICEF and PBS (2011)

Water and sanitation are linked to human health and nutrition. Inadequate access to safe drinking water and poor hygiene may lead to water-borne and infectious diseases, which can seriously affect food absorption (Pinstrup-Andersen 2009). It is the poor who do not have capacity to invest in water and sanitation, and they suffer the most in relation to water-borne diseases and malnutrition.

9.3.12 Inadequate Recognition of Mountain Specificities in Existing Policies and Programmes

All the HKH countries have been pursuing policies and programmes aimed at achieving food and nutrition security. Although these policies have contributed to increased food production in the plains areas, poor understanding and inadequate recognition of mountain specificities means that they have been less effective in achieving food and nutrition security in the mountains. There is limited investment in research and development on mountain crops particularly on traditional nutritional crops. As a result, their productivity has remained low compared to that of rice and wheat. Research on traditional crop varieties has largely been neglected by both national and international research agencies and traditional crops have been deprived of the development of improved varieties (Padulosi et al. 2012). Moreover, the limited understanding of the different constraints and opportunities faced by mountain communities due to the specific biophysical environment and socio-economic conditions has resulted in a tendency to use a one-size-fit-all approach to plains and mountain areas (Sati 2015; Ghosh and Sharma 2016). Detailed analysis of existing policies and programmes is beyond the scope of this assessment; however, the following provides a brief country-wise overview of some of the most relevant policies and programmes with implications for food and nutrition security in the mountain areas, starting with India which has one of the more extensive systems.

India has the world's largest social security system and has been implementing a number of food safety programmes including providing subsidized food to poor and vulnerable people through a public distribution system and fair price shops. Although the public distribution system has played a great role in improving household food security in the plains, these programmes have had little impact on improving food and nutrition security in mountain areas due, among others, to insufficient infrastructure, high transportation costs, and poor targeting (Dame and Nüsser 2011; Sati 2015; Ghosh and Sharma 2016). The cost of transporting food to remote mountain areas can be prohibitively high, and weak transportation facilities and poor information exchange between mountain communities and administrative centres in the lowlands often hinder the delivery of subsidized food to remote mountain areas (Sati 2015). A recent study in North East India (Mizoram) found that a large number of households were dissatisfied with the way the fair price shops functioned; of 16 fair price shops, only five were functioning regularly (Sati 2015).

Moreover, the focus of the public distribution system is on meeting calorie targets with less attention paid to nutritional diversity. The aim of the National Food Security Act (NFSA) of India is to provide rice and wheat to low income groups (i.e., people below the poverty line). It does not address the issue of nutrition insecurity by providing diverse food items such as pulses and other traditional food grains, or fruit and vegetables (Burchi et al. 2011; Beddington et al. 2012). The NFSA seems to target eradication of hunger but not malnutrition. Providing only wheat and rice at subsidized prices does not address micronutrient deficiency or hidden hunger and can be counterproductive, as farming households often give up cultivation of nutritious NUS such as millet, barley, and sorghum due to the availability of subsidized wheat and rice, as reported by Dame and Nüsser (2011) in Ladakh. Further, the national food grain sector will have to grow by 3.8% annually to meet the quantity of food grain required under the NFSA, either through increased production or increased imports (Sengupta and Mukhopadhyay 2016). If India chooses to increase domestic production of rice and wheat, land use patterns are likely to change and a significant proportion of the area under cultivation of crops such as pulses, vegetables, fruit, and traditional crops could be allocated to rice and wheat. It is likely that traditional food crops would be most affected. Another important aspect of food and nutrition insecurity in the mountain areas in India is the inefficiency of food grain procurement, transportation, and distribution via the Central Pool by the Food Corporation of India (Likhi 2014). In the past, food grain under the public distribution system failed to reach the targeted beneficiaries owing to the lack of proper identification of recipients and mismanagement of subsidized food grain at various layers of distribution.

At the same time, there are a number of promising institutional innovations taking place in India that aim to stabilize food supplies and ensure food availability during seasonal shortfalls. For example, Uttarakhand is establishing local grain banks to create buffer stocks against seasonal shortfalls; the banks store food at village level and are managed by trained women from the community. Households in need can borrow from the banks during the lean season, paying the cost of interest on the loan, and return the loaned amount after the next harvest. So far, 55 grain banks have been set up in food-insecure villages in the state (Ghosh and Sharma 2016).

India is a large country with great variation in ecology and topography, and the drivers and level of poverty across the plains, mountains, and deserts vary significantly. Hence, a single solution with uniform policy instruments (like the NFSA) may not be suitable for attaining sustainable food and nutrition security across all regions (Pingali et al. 2017; Landy 2017).

In Afghanistan, the Public Nutrition Policy and Strategy (PNPS) 2009–13 set objectives to increase awareness about nutrition, reduce micronutrient deficiencies, improve health care, and ensure that responses to treat or prevent moderate, acute, and/or chronic malnutrition are timely and appropriate (GoA 2009). The PNPS focused on the supplementation of key micronutrients rather than a sustainable solution through balanced diets. The policy draft does not plan to diversify agriculture systems, despite the huge potential for this.

In Bangladesh, the government supports communities in the Chittagong Hill Tracts (CHT) through the Food Ration Programme, providing 30 kg of rice per family per month at a cost of 10 Bangladeshi taka per kg. The National Food Policy (2006) and the National Food Policy Plan of Action (2008–15) do not, however, suggest specific strategies to promote sustainable agriculture or non-agricultural income opportunities in the region. The CHT is endowed with immense biological, cultural, and environmental resources, but the area remains one of the most disadvantaged and vulnerable in the country and lags behind in almost all key development indicators. The food and nutrition security scenario is weaker in the CHT than in the rest of the country; overall agricultural production is low and requires special attention in future policies and programmes (Rasul 2015).

Food and nutritional security is a national priority for the Bhutanese government. Food security in Bhutan has improved over the years due to land and food acts implemented under the government's Vision 2020 and the National and Nutrition Security Policy of Bhutan 2012. Although Bhutan has very limited arable land (8% of total land), subsistence agriculture and crop diversification are very common resulting in greater food security at the household level (Tobgay 2005). Food and nutritional security at national level mostly depends on imports. Future challenges for food security could come from conversion of the limited productive agricultural land into residential land, loss of productivity, increased reliance on imported food, and loss of food diversity in the local food system (Ura and Kinga 2004). Studies have highlighted Bhutan's vulnerability to these in the context of future food security and policy directions (De Janvry and Sadoulet 2008).

Food production, distribution, and marketing systems are relatively better in China. Recently, China announced its National Programme for Food and Nutrition (2014–20) to further improve the food supply, facilitate balanced nutrition, and coordinate food production and consumption. The aim is to improve the overall health of all Chinese people and lay a solid foundation for building China into a moderately prosperous society (GoPRC 2014). However, the policies and programmes do not suggest separate policy steps for mountain areas, nor do they address standards for food and nutrition intake for different groups of people (based on age, occupation, and gender).

In Myanmar, a National Plan of Action for Food and Nutrition was formulated in 1994, but the prevalence of food and nutrition insecurity remains very high, particularly in mountain areas such as northern Rakhine, Chin, Kachin, and Shan. Food deprivation is inherently linked to issues related to land entitlement, poor agricultural roads and other infrastructure, technology transfer, markets, and even storage facilities (Pedersen 2014).

Nepal has a Multi-sector Nutrition Plan (MSNP) 2013–2017, and a Food and Nutrition Security Plan of Action 2016, as well as various nutrition-related programmes, but a substantial proportion of the population remains food and nutrition insecure. The twenty or so agriculture-related policies and programmes have also failed to address the loss of 10–20% of food grain due to poor handling and lack of proper storage facilities (Bhandari et al. 2015), and virtually no policies or programmes have region-specific (plains, hills, and mountains) measures. Many hill and mountain areas are suitable for the production of nutritionally rich cereals like maize, millet, barley, and sorghum, as well as other high nutrition crops, but as a result of inadequate supporting mechanisms, farmers do not choose to grow them. The mid- and far-western mountain regions are more food insecure than others owing to their remoteness with poor road networks and poorly developed markets. There is no specific provision such as ration cards for the food insecure population (FAO/Nepal 2010). The government provides food grain in the form of food-for-work programmes with support from the World Food Programme (WFP) and other development partners, but this type of assistance has only increased the dependence of people on food aid, which in the long run is detrimental to food security. Food safety nets such as the Nepal Food Cooperation (NFC) supply cereals and pulses to food insecure areas in Nepal, and there are programmes to provide subsidies on fertilizers, seeds, and transportation of agricultural inputs. The private sector is also involved in distribution of agricultural inputs and services. For example, the Nepal Seed Company supplies improved varieties of crop seeds, the Agriculture Inputs Company supplies subsidized chemical fertilizers, and recently, financial institutions have started providing livestock and crop insurance services. Despite many efforts and an overall increase in calorie intake, stunting and underweight have remained high in mountain areas relative to the national average (Rasul et al. 2018). Development interventions have been less effective in many remote districts, and these regional disparities, coupled with social discrimination against women and other marginalized groups, have remained a major challenge to improving food and nutrition security in many hill and mountain areas.

Pakistan has made significant progress in wheat, rice, and livestock production over the years (ASP 2012). The national Food Security Policy has not yet been implemented, but the country has a number of food and nutrition security programmes implemented by various ministries. Programmes like school food, safe motherhood, and child nutrition implemented by the provincial education and health departments with the assistance of WFP, WHO, UNICEF, and UNESCO have been providing assistance to approximately two million households. Pakistan has placed greater emphasis on addressing the supply side of food security, focusing specifically on maintaining wheat self-sufficiency. The government procures wheat from farmers through the Pakistan Agricultural Storage and Services Corporation Ltd. and bears the cost of storage, handling, and other incidentals. The wheat is then resold through government-owned 'utility stores' at a subsidized rate, together with other subsidized food items such as sugar and oil. Highly subsidized wheat and wheat flour are supplied to some mountain areas such as Gilgit-Baltistan. There are two big challenges to food and nutrition security in Pakistan. First, all programmes have focused on selected food items such as wheat, rice, poultry, and sugarcane, in terms of input supply, support prices, and market mechanisms. Second, the production costs of agricultural commodities are very high, which has led to high consumer prices, even with subsidies. The high levels of poverty in vulnerable areas like the mountains mean that people do not have sufficient purchasing power even for subsidized food items, which again leads to higher food and nutrition insecurity (Hussain and Routray 2012).

There is a need in all the HKH countries to address the specific circumstances of mountain regions in the national agenda and establish specific policy instruments for food security in mountain regions that reflect the vulnerability resulting from the specific climatic, environmental, and socioeconomic conditions. Some countries such as Bangladesh, Myanmar, Nepal, and Pakistan are also in the process of implementing national zero hunger initiatives. These initiatives also need a mountain-specific and nutrition sensitive perspective.

9.4 Potentials and Emerging Opportunities in the Mountain Areas

Although there are several challenges to overcome, there is a wide range of opportunities and potentials that could be tapped to achieve sustainable food and nutrition security in the mountains of the HKH. The main possibilities are summarized in Table 9.10 and discussed in more detail in the following sections.

9.4.1 Tapping the Benefits of Local Food Systems

Integrating NUS such as barley, sorghum, millets, buckwheat, pulses, and beans into local food systems will reduce the climatic and economic risks associated with dominant cereals such as wheat, fine rice, and maize, and cash crops. NUS are richer in micronutrients (Table 9.11) and more resilient to climate stresses (Padulosi and Hoeschle-Zeledon 2004). Dominant cereals and cash crops (food and non-food) are often more input intensive and susceptible to crop failure, seasonality, price shocks, and market forces, and can constitute an unadvisable level of risk for poor farmers (Tulachan 2001; Jenny and Egal 2002; ICIMOD 2008). In addition to diversification of food production, the range of food products prepared from NUS could also be increased to make them more inviting and to improve market demand (Adhikari et al. 2017).

Table 9.10 Opportunities and potential for food and nutrition security

Area of opportunity	Potential improvements	Dimensions of food and nutrition security likely to improve
Potential of local food systems	• Improved food production • Improved food production diversity and dietary diversity • Improved farm income	• Food availability • Food utilization • Food accessibility
Potential of local breeds of livestock	• Improved livestock production, conducive to improved food supply and income	• Food availability and accessibility
Potential for vegetables, fruit, nuts, and tea	• Improved production of vegetables, fruit, and nuts • Improved production diversity • Improved farm income	• Food availability • Food utilization • Food accessibility
Potential for efficient use of natural resources	• Improved watershed and springshed management and improved use of water from springs, snowmelt, and glaciers • Improved rangeland management • Improved production of non-timber forest products (NTFPs), conducive to improved income	• Food availability • Food availability • Food accessibility
Non-farm income opportunities: tourism, handicrafts, and others	• Improved income	• Food accessibility
Productive use of remittances	• Improved income and purchasing power for diverse food items • Improved livelihoods • Improved adaptive capacity to climate change	• Food accessibility and utilization • Food availability and health status • Food availability
Prospects for regional connectivity	• Improved food trade	• Food availability

Table 9.11 Nutritional value of neglected and underutilized food crops

Crop	Botanical name	Nutritive value per 100 g							
		kcal	Protein (g)	Dietary fibre (g)	Thiamine (mg)	Riboflavin (mg)	Calcium (mg)	Iron (mg)	Zinc (mg)
Amaranthus (seed, black)	*Amaranthus cruentus*	356	14.6	7.0	0.0	0.0	181.0	9.3	2.66
Pearl millet	*Pennisetum typhoideum*	348	11.0	11.5	0.3	0.2	27.4	6.4	2.76
Barley	*Hordeum vulgare*	316	10.9	15.6	0.4	0.2	28.6	1.6	1.50
Sorghum	*Sorghum vulgare*	334	10.0	10.2	0.4	0.1	27.6	4.0	1.96
Quinoa	*Chenopodium quinoa*	328	13.1	14.7	0.8	0.2	198.0	7.5	3.31
Little millet	*Panicum miliare*	346	10.1	7.7	0.3	0.1	16.1	1.3	1.82
Foxtail millet	*Setaria italica*	332	8.9	6.4	0.3	0.2	15.3	2.3	1.65
Finger millet	*Eleusine coracana*	321	7.2	11.2	0.4	0.2	364.0	4.6	2.53
Maize (dry)	*Zea mays*	334	8.8	12.2	0.3	0.1	8.9	2.5	2.27
Wheat (whole)	*Triticum aestivum*	322	10.6	11.2	0.5	0.2	39.4	4.0	2.85
Rice (raw, milled)	*Oryza sativa*	356	7.9	2.81	0.05	0.05	7.49	0.65	1.21

(*Source* Longvah et al. 2017)

NUS offer a good option in terms of bringing a balance to local food systems and improving farmers' income provided proper value chains are developed for these crops. In recent years, increased dependence on external food crops and processed snacks and drinks in place of NUS has made mountain people more vulnerable to food and nutrition insecurity. Price shocks in food-producing areas and natural disasters (such as floods and landslides) may result in restricted food supplies and price hikes. Strengthening local food systems by promoting NUS can help improve the stability of local food supplies and reduce dependence on external food items. In some cases, they can also generate income from increased demand in local markets (Fig. 9.10).

Fig. 9.10 Cultivation of millets and local beans in Gatlang VDC, Rasuwa District, Nepal. Households in Gatlang still depend on NUS for both food and income; they sell NUS to local resorts and small hotels in Guljung and Saybrubesi to prepare speciality local food for tourists (*Photo* Bhuwan Thapa)

Marketing of NUS products has improved in recent times. For example, in Nepal, products like oatmeal, buckwheat flour, and millet cakes can now be purchased in major supermarkets. The demand for such products is increasing steadily with increasing consumer awareness. NUS are recently relabelled as 'Future Smart Foods (FSF)' due to their potential for nutrition enhancement, climate change resilience, and diversification of cropping systems. The FAO Regional Office for Asia and the Pacific and its partners have agreed to recognize, identify, nand promote the complementarities of NUS with existing staple crops in local food systems (FAO 2017).

The more than 250 different types of fermented foods linked to the diverse ethnic cultures in the Himalayan region (Tamang 2009) also offer a potentially valuable source for improved nutrition. These naturally fermented foods support nutrition in a variety of ways including bio-preservation of perishable food, bio-enrichment of food to increase nutritional value, and health benefits such as protective and therapeutic properties (Tamang et al. 2012).

9.4.2 Local Breed Livestock

Historically, local breeds of poultry and livestock were, and often still are, a very important source of food for mountain people. For example, in mountain areas such as upper Chitral and Gilgit-Baltistan in Pakistan and upper Rasuwa District in Nepal, yak and chauri (a cross breed of yak and cow) are the main sources of milk and cheese. Similarly, local breeds of goats and sheep are still important sources of food security and livelihoods for many mountain people (Rasul and Hussain 2015). Overall, the population of local breeds of livestock has decreased over time in high elevation areas due to changing priorities for types of livestock as well as climate-induced degradation of rangelands and grasslands. In particular, the population of yak is in decline in India, Nepal, and particularly Bhutan (Wu et al. 2016). Policies have favoured the introduction of improved breeds in both mountains and plains, but in mountain areas these breeds can prove more vulnerable to extreme weather conditions, disease, and seasons with poor production of grass and fodder. As with crops, local breeds are adapted to the harsh conditions in mountains and are better integrated into the complex mixed farming system that makes the most of mountain resources while increasing resilience against disaster. In high altitude mountains, agricultural practices can be diversified by including traditional food crops and local breed livestock, which can provide multiple livelihoods to smallholder farmers and secure sustainable and diversified food.

9.4.3 High Value Mountain Crops

Mountain areas have a huge potential for growing vegetables, fruit, nuts, and tea. In Pakistan and Nepal, for example, mountain areas contribute substantially to the national production of several fruits and nuts (Fig. 9.11), and in some mountain areas of China, India, Nepal, and Pakistan, the production of vegetables, fruit, and nuts has increased despite the climatic and socioeconomic changes (Hussain et al. 2016). A substantial proportion of households observed an increase in the production of summer vegetables and fruit, especially apples, cherries, apricots, and walnuts in the Upper Indus basin in Pakistan; of summer potato, onion, and vegetables in the Koshi basin in Nepal; of tea in the eastern Brahmaputra basin in India; and of walnuts, tea, garlic, tobacco, and sugarcane in the Salween and Mekong basin in China. The local people attributed this increase to the changing climate, suggesting that climate change could also have some positive impacts on mountain agriculture.

The climatic conditions in mountain areas are conducive for growing high-quality seed potatoes, vegetable seeds, off-season vegetables, and medicinal plants, which is reflected in the relatively high production. There is also a huge potential for beans, cucumber, ladyfinger, onions, peas, spinach, and tomatoes (Rasul and Hussain 2015). Many mountain areas make a significant contribution to the national production of fruit and vegetables, for example mountain and hill regions in Nepal contribute 42% of total national vegetable production (MoAD 2015), but there is still a huge potential for enhancing fruit and vegetable production and increasing local income across the HKH. Overall, the mountain areas of China (e.g., Yunnan Province) perform much better than mountain areas in the other HKH countries (Hussain et al. 2016). But Balochistan, Khyber Pakhtunkhwa, and Gilgit-Baltistan in Pakistan already contribute substantially to the national production of apples, apricots, cherries, figs, grapes, loquats, peaches, pears, persimmons, pomegranates, plums, walnuts, and almonds, even though the potential is not yet fully realized. Pakistan exports several types of fruit and nuts produced in the mountains, including apricots, cherries, figs, plums, sloes, peaches, pine nuts, and walnuts, and this could increase if production increases (Rasul and Hussain 2015). Particularly in India, the mountain states are still underutilized and their contribution to national fruit and vegetable production is limited (GoI 2014b). There is, for example, a huge scope for the production of apple in Kinnaur district of Himachal Pradesh.

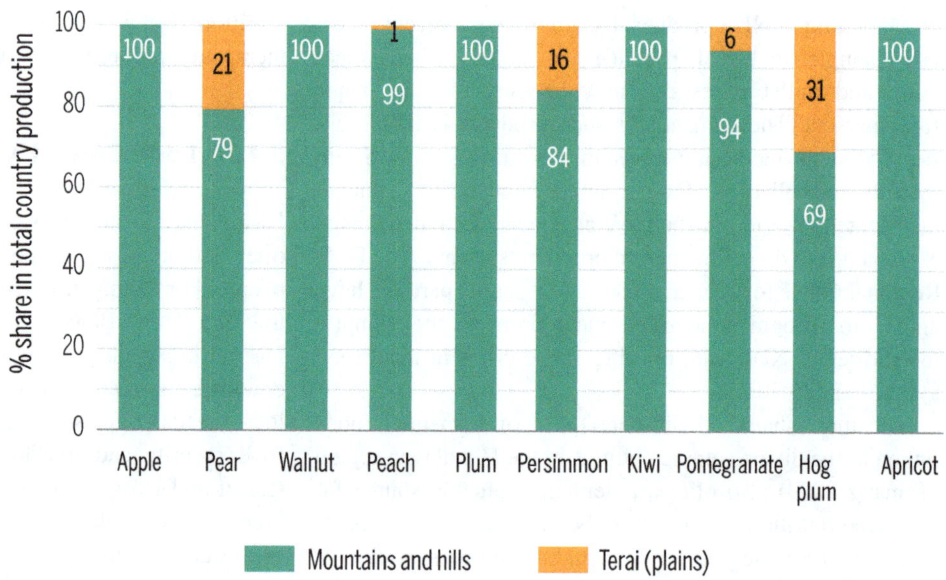

Fig. 9.11 Main fruits grown in the hills and mountains of Nepal (*Source* MoAD 2015)

9.4.4 Sustainable Use of Natural Resources

Mountain areas are rich in natural resources, such as water, forests, rangelands and pasture, biodiversity, and valuable minerals. Sustainable mountain food security cannot be achieved unless the productive resources and the natural environment are conserved and access and optimal utilization ensured.

Rangelands are a very important natural resource for livestock production and food security in high elevation mountain areas, particularly for pastoral communities (Hussain et al. 2016). The substantial rangelands in the mountains of Pakistan, for example, are vital to livestock grazing, one of the main sources of food and income for mountain communities (Table 9.12). Rangelands are the main free-grazing areas for livestock in general and small ruminants (goats and sheep) in particular. In addition, they also provide environmental services such as carbon sequestration, watershed management, biodiversity, and ecotourism (Mirza et al. 2006). However, climate-induced changes in precipitation, combined with overgrazing, are leading to degradation of vegetation cover, deterioration of soil, and lowered productivity for the rangelands themselves and the livestock that feed on them. Corrective actions to conserve and enhance the productivity, sustainability, and ecological health of the rangeland ecosystem would significantly improve food security in the high mountains (Afzal et al. 2008).

Forests are a direct source of food for people in some areas, especially where access to market centres is limited. For example, the tribal communities in Meghalaya and Manipur states in India eat several edible plants and fruits from the forest (Sawian et al. 2007; Gangte et al. 2013). Non-timber forest products (NTFPs) can also play a vital role in food security and livelihoods. The mountains of the HKH are endowed with NTFPs, such as medicinal and aromatic plants and honeybee products. Some mountain plants (e.g., seabuckthorn, wild thyme, black cumin, chamomile, and stevia) and minerals (e.g., salajeet) are widely used for medical purposes across the HKH (Rasul et al. 2014). In the high elevation Skardu valley in Pakistan, 70% of the flora and animal products used for medicinal purposes were from wild species, and 70–80% of local people depended at least to some extent on traditional medicines, using plants to cure common ailments (Bano et al. 2014). Women, especially, can play a pivotal role in preserving unique species of medicinal and other plants if properly trained.

In the HKH, the collection, processing, and marketing of NTFPs suffers from a variety of problems and the problems of poor harvesting practices, poor storage, poor access to market, and over exploitation will need to be addressed in order to tap the full potential of NTFPs and enable mountain communities to reap adequate benefits (Rasul et al. 2008). According to the World Health Organization, the global market for herbal products is over USD 60 billion (Nirali and Shankar 2015). Although a significant portion of the supply of raw materials for these herbal products originates in the HKH, much of the benefit accrues to people and places far away. There is a great potential to generate more income locally by supporting mountain people to develop new livelihood options and derive greater benefits from high value products.

Mountains are also rich in water resources, such as springs, streams, snow cover, and glaciers. These resources are generally underutilized, and there is a vast scope to manage them better to support diversification of agricultural production. For example, construction of micro-irrigation systems for glacial meltwater could provide considerable support for agricultural diversification in Hunza and Nagar (Pakistan). However, meltwater irrigation systems have

Table 9.12 Non-crop options for mountain food and nutrition security in Pakistan

			Mountains					Plains
			Balochistan	FATA	KPK	AJK	GB	Punjab and Sindh
Livestock	Livestock heads per capita		3.7	1.8	1.0	0.9	1.1	0.8
	Livestock units (LUs) per capita[a]		0.6	0.4	0.3	0.2	0.2	0.3
	Sheep and goats	Population ('000)	24,589	4,784	12,962	1,577	1,565	42,723
		Animals per capita	3.2	1.3	0.6	0.5	0.9	0.4
Rangelands	Proportion of rangelands in total geographical area (%)		78.9	48.3		45.1	53.9	46
Forest	Proportion of forest in total geographical area (%)		4.1	13.1		11.6	9.2	4.4
	% share in total revenue generated from forest products		1.9	34.6		13.8	0.1	49.6

(*Source* Rasul and Hussain 2015)
[a]Method for estimating LUs adapted from FAO (2005)

unique characteristics that distinguish them from other types of irrigation system and must be managed with care. Parveen et al. (2015) described a very heterogeneous and unstable situation in Hunza where communities must constantly adapt to drying of channels due to down-wasting, shifting glacier tongues, and sudden changes in flow patterns. Meltwater can also be used for trout fish farming in high mountain areas. Springs and streams are a potential source of irrigation water in various mountain districts of Nepal and India. The most appropriate type of small-scale irrigation system varies across mountain ranges. Piped irrigation systems suit mountain ranges that are steeper and more vulnerable to landslides, while open channel systems suit less steep areas with less chance of landslides. In dry mountain ranges such as Gilgit in Pakistan, Mustang in Nepal, and the western part of Himachal Pradesh in India, solar-powered irrigation pumps may also be used to pump water from streams and springs to upland agricultural fields.

9.4.5 Non-farm Income Opportunities: Tourism, Handicrafts, and Others

It is difficult for mountain regions to become food self-sufficient due to the biophysical constraints (Rasul et al. 2014). To achieve food and nutrition security, it is critical to increase the income of farmers so that adequate food can be purchased in accessible areas. Mountain areas have a huge potential for non-agricultural activities. In areas with low agro-ecological potential, in particular, non-farm income opportunities can be promoted to help secure income and improve purchasing power. Tourism is one of the most viable industries in the HKH, although its full potential has not yet been explored due to lack of know-how, as reflected in the national policy agendas of governments in the region. Other industries may grow together with the development of tourism, especially hotel and restaurant businesses. Tourism can contribute significantly to the local economy if hotels and restaurants promote and offer menus prepared from local NUS, vegetables, fruit, and nuts (Adhikari et al. 2017). Without proper planning, tourism can negatively affect the local economy and food systems, due to the excessive inflow of instant and highly processed food items. But it also offers many opportunities, with a huge potential for handicraft and souvenir cottage industries (embroidery, wood carvings, shawls, blankets, carpets, baskets, gemstones, and many others) if integrated in tourism planning (Rasul and Hussain 2015).

9.4.6 Productive Use of Remittances

Increased out-migration is adding some challenges to mountain agriculture by creating labour shortages and increasing the workload on women (Hussain et al. 2016). However, it also brings new opportunities in the form of higher remittances. There is a huge potential to capitalize on the remittances through effective investment in small and medium local non-farm enterprises, agricultural development (for example, irrigation systems, improved seed and inputs, plantation of fruit trees, fish farming, and medicinal and aromatic plants), and food processing. Such efforts will also generate income opportunities for local youth, which in turn may lower the rate of future out-migration and strengthen food security. A study conducted in the high mountain region of Gojal, Gilgit-Baltistan of Pakistan showed that out-migration and remittances had a positive impact on modernization of farming, improving livelihoods, and achieving sustainable development in the mountains (Benz 2016).

9.4.7 Improving Food Transportation Through Regional Connectivity

Regional connectivity plays an important role in cross-border trade and food transportation and food security. Cross-border trade and transportation facilities can influence food transportation costs and food prices (Pyakuryal et al. 2010). In Nepal, the internal transportation cost of food in the mountain region is seven to eight times higher than that in the plains areas (Pyakuryal et al. 2010). There can also be marked price differences in food across borders in mountain areas. For example, a recent study by Agarwal et al. (2017) found that the price of various vegetables in Agartala in Tripura, a hill state in India, was more than double that across the border in Brahmanbaria, Bangladesh, as a result of inadequate cross-border trade facilities (Agarwal et al. 2017). There are a number of opportunities for localized cross-border trade between mountain communities (Box 9.5).

Box 9.5 Localized cross-border trade

In general, trade between countries is driven by competitiveness, comparative advantage and differences in technology, economies of scale or preferences, natural resources, climatic conditions, and in some circumstances strategic trade policies (Nanda 2012). In mountain areas, the physical isolation of communities due to remoteness and limited accessibility is also a key determinant of localized cross-border trade. Mountain areas, although divided by political borders, are often logistically better connected to each other than to downstream cities in their own countries. Accessibility to main cities is further reduced in winter and following damage caused to infrastructure by landslides, floods, avalanches, and others. Localized cross-border trade could contribute substantially to food and nutrition security, particularly in stressful times, but the exchange of agriculture and livestock goods needs to be facilitated, and if necessary legalized, through formal mechanisms.

People living in the high elevation areas of Rasuwa District in Nepal, for example, may have easier access to Tibet Autonomous Region (TAR) in China for exchange of goods and services than to market centres in Rasuwa such as Syabrubesi and Dhunche. Farmers in the high areas of Rasuwa grow traditional food crops such as barley, millet, and beans, and raise livestock such as goats, sheep, yak and chauri and could either sell or exchange these products for other food products with people in Tibet if cross-border arrangements were in place. This localized trade could reduce the vulnerability of local communities on both sides of the border and help improve food and nutrition security. Similar areas can be identified along the borders of Nepal and India, Pakistan and China, and other HKH countries.

There is also a potential for more distant cross-border trade. A number of fruits, vegetables, handicrafts, and other products from mountain regions of the HKH are quite popular across the world. If incentives for trade liberalization and market competitiveness can maintained through appropriate regulation, and proper value chains established that ensure that an appropriate part of the benefits of products accrue to the producers, then the livelihood patterns of mountain people can be improved. For example, cherries grown in Gilgit (Pakistan) are being exported to the Middle East through the establishment of a well-connected and regulated supply chain. Similarly, there is high potential for creating supply chains for other perishable and non-perishable commodities to a whole range of different markets in Europe, America, the Middle East, and others.

9.5 Sustainable Food and Nutrition Security in the Mountains: Towards a Strategic Approach

Food and nutrition security remains a major challenge in the HKH region. The problem is more severe in remote mountain areas, where agro-ecological potential is low, accessibility is poor, and market infrastructure is weak and fragile. The majority of people live in remote rural areas and depend heavily on agriculture for food and nutrition, but mountain agriculture is highly vulnerable to climate change as a result of erratic rainfall with increased floods and increased dry spells, changing temperatures and seasons, and higher incidence of pests and disease. Rapid socioeconomic and environmental changes, including increased out-migration, shortage of labour, limited mechanization, changing food habits, decline of traditional nutritious crops, and depletion of natural resources, are intensifying the challenge of achieving food and nutrition security.

Challenges to food security in the mountains of the HKH differ from those in the plains as a result of the constraints imposed by harsh biophysical conditions, inaccessibility, fragility, seasonal incomes, low economic opportunities, and poor access to markets and other institutional services. The HKH region needs special attention and mountain-specific approaches to address the issues of food and nutrition insecurity. Achieving food security and addressing malnutrition are fundamental to meeting the SDG-2 goals as well as to achieving the goals of other SDGs, including ending poverty and improving human well-being.

Sustainable food and nutrition security cannot be achieved in the mountains without implementing a holistic approach that works to cope with the challenges specific to the region, while simultaneously realizing its potentials and opportunities. The HKH region needs a balanced strategy between food self-sufficiency and market dependence that not only encompasses a focus on enhancing production and increasing household incomes, but also targets the improvement of rural infrastructure for market access and food transportation. Such a balanced strategy is laid out below.

9.5.1 Location-Specific Approach

The type and level of challenges and opportunities vary across the mountain areas in the HKH due to differences in the ecological and environmental conditions and in access to institutional services. Given the heterogeneity and diversity, formulating strategies for food and nutrition security requires developing a location-specific approach, that takes the ecological and environmental conditions into account and considers access to markets, information, and other institutional services. The mountain areas of the HKH can be divided hypothetically into four classes in terms of agro-ecological potential and access to markets, information, and institutional services; area-specific strategies for these classes are presented in Table 9.13.

In **areas with high agro-ecological potential and good access to markets and services,** the focus should be on exploiting existing potential as much as possible through land-use intensification, efficient water use, improved adaptations to climate change, crop diversification, development of commercial dairies, and growing of cash crops that offer higher incomes. Private investment in production and post-harvest facilities should be encouraged. In **areas with**

Table 9.13 Area-specific approaches based on agro-ecological potential and access to markets, information, and institutional services

Agro-ecological potential and suitability	Access to markets, information, and institutional services	
	Good	Poor
High	**Areas with high potential and good access to markets and services** • Promote intensive food production (where possible multiple cropping), horticulture, and commercial dairy and poultry farming • Enhance support for high value cash crops, e.g., fruit, nuts, tea, and vegetables • Integrate traditional food crops, e.g., millets, barley, beans, local maize, in cropping systems, particularly in areas above 2000 masl • Promote resilient crop varieties and adaptation measures in areas with higher climate change impacts • Establish fruit processing and storage facilities • Encourage private investment in irrigation, fish farming, renewable energy, land management, storage, and the agro-processing industry through institutional support • Improve supply of inputs, credit, insurance facilities, and environmentally-friendly machinery (e.g., solar powered pumps) for agriculture • Encourage spring water management in hills and low mountains, and snowmelt/glacial melt water management in high mountain ranges • Provide incentives for overseas workers to invest remittances in small and medium agribusiness in mountain areas • Encourage women as entrepreneurs/managers in agriculture business	**Areas with high potential but poor access to markets and services** • Improve marketing, storage and transport facilities, information systems, and extension and credit services for fresh fruit, nuts, dried fruit, vegetables, and livestock products • Strengthen local food systems with a focus on NUS • Extend institutional support to promote traditional food crops and support value chain development • Improve transportation facilities, ICT access, and others • Promote high-value non-perishable agricultural products such as pulses, medicinal plants, and honey • Develop infrastructure to enable utilization of the mountains' high agro-ecological potential • Promote livestock and livestock products and by-products • Improve credit, extension, and insurance facilities for livestock
Low	**Areas with low potential but good access to markets and services** • Promote local products such as crafts (e.g., woodcarving, shawls, carpets, caps) and services for markets • Promote conservation technologies that enhance agricultural potential and utilize local niches and provide incentives for conservation • Encourage agroforestry, tree farming for timber and NTFPs, and medicinal plants • Develop local off-farm employment opportunities to reduce outmigration • Encourage local breeds of livestock such as yak, goats, and sheep (mainly pastoralism) in high mountain ranges (>2500 masl)	**Areas with low potential and poor access to markets and services** • Provide incentives for conservation and sustainable use of resources and develop mechanisms for payments for ecosystem services • Establish institutional mechanisms to supply subsidized food items • Encourage non-farm activities, e.g., tourist guides, resorts, hotels, and handicrafts • Promote subsistence agriculture with zero-tillage, mixed cropping, and livestock production • Promote ecotourism and recreation • Develop and harness environmental services

Adapted from Rasul et al. (2014)

high agro-ecological potential but poor access to markets and services, the focus should be on removing marketing constraints and developing infrastructure and institutional support to help exploit existing potential both optimally and sustainably. **In areas with low agro-ecological potential but good access to markets and services,** strategies should focus on better use of existing facilities to promote local breed livestock and non-farm activities, and provision of economic incentives and appropriate regulations to support sustainable use and management of resources. In **areas with low agro-ecological potential and poor access to markets and services,** subsistence use of resources and facilitation of out-migration should be targeted to reduce dependence on local resources and ensure food security. Agricultural extension services and incentive mechanisms need to be structured accordingly in the different areas.

9.5.2 Developing Local Food Systems

A few general recommendations can be made in addition to the area-specific approaches suggested above which may apply to all four classes of mountain area. The first is to strengthen local food systems, which is very important for improving the diversity and quantity of local food production and reducing the dependence on food from outside the region. It may not be possible to achieve complete food self-sufficiency in mountain areas due to land and environmental constraints, but revitalizing the local food systems will reduce the dependence on external food supplies. Emphasis should be placed on nutrition-rich traditional crops (NUS) and hygienic processing of value-added products. Steps should be taken to enhance nutritional knowledge about NUS among both farmers and consumers, to build farmers' capacity for NUS production and value addition, and to establish community seed banks. Formal guidelines could also be given to local hotels and resorts in tourist areas to help them include food items prepared from traditional food crops in their menus. This will enhance the demand for traditional food products and augment income for NUS farmers. Efforts are also needed to promote organic agriculture and establish a participatory guarantee system to support marketing of organic products, and to strengthen support for livestock production. Technical assistance and institutional support need to be strengthened to increase livestock productivity and improve rangeland management in order to improve the food security of livestock-dependent communities. Bio-fortification of selected cereal crops could be considered to meet micronutrient needs, where appropriate.

9.5.3 Strengthening the Agricultural Marketing System and Infrastructure

Local infrastructure and the agricultural marketing system need to be strengthened to reduce crop loss and vulnerability to food insecurity. Approaches include the use of ICTs such as mobile telephones, local FM radio, e-information systems, and other mechanisms to improve farmers' access to market information, and improving processing, storage, and distribution systems to reduce post-harvest food losses (also linked to SDG 12). Local food storage facilities such as community food banks could be established to avoid seasonal food shortfall in high elevation areas where snowfall or hazards can result in physical isolation. Export competitiveness needs to be enhanced through strengthening of technical support and making financial assistance available to improve the production of organic products, and processing, packaging, and marketing. Special incentives and support might be given to transporting high value perishable products that are in high demand in export markets. Attention needs to be paid to establishing a mechanism for cross-border food trade and internal movement of food products to facilitate food trade across borders. It is also necessary to formalize cross-border localized trade/food exchange between border communities with limited access to market centres in their respective countries, particularly during the winter and monsoon seasons and after natural disasters.

9.5.4 Managing Water Resources and Other Ecosystem Services

Forests, rangelands, and water resources are an integral part of the livelihood and food security of mountain people. They provide food, wood, fodder, fuel, medicine, water, and many more goods and services. Mountain watersheds need to be properly managed in order to cope with climate-induced water stresses and ensure the continued flow of water and other ecosystem services to downstream areas. Wherever possible, hill irrigation systems could be developed in the mountains to capitalize on the potential of water from springs, snowmelt, and glaciers. In addition, appropriate incentive mechanisms, such as payments for ecosystem services and access to benefit sharing of genetic resources, need to be developed to encourage mountain communities to use and manage the genetic resources and watersheds sustainably. As we have argued elsewhere (Rasul 2010, 2014), mountain communities also deserve special attention and some compensation for the ecosystem services they provide to downstream through water and ecological services, and their effects on food production in the plains. Efforts need to be made to engage communities in natural resource management, including water, irrigation, forests, and rangelands.

9.5.5 Promoting Rural Non-farm Economic Opportunities

Rural non-farm economic opportunities, such as wage-earning activities, self-employment in commerce, manufacturing, and small businesses involving handicrafts, tourism, and services, are an important source of income for farmers and rural mountain households. Rural non-farm income can be an important part of household income and helps to secure nutrition status as it allows greater access to food. This income may also help to slow rapid or excessive urbanization as well as degradation of natural resources through overexploitation. All possible non-farm income opportunities should be encouraged and supported through institutional mechanisms, including capacity building and vocational training. Efforts could be made to involve local government institutions and the private sector to improve the productive use of remittances as investment in local areas, which can generate further income opportunities and contribute to food security.

9.5.6 Strengthening Knowledge on Nutrition, Child Care, and Food Preparation

Improved knowledge of general health, child care, maternal health, improved drinking water, sanitation and hygiene, food safety, and food preparation can help to improve the nutritional status of mountain people. It would be beneficial to establish a large-scale training programme for women in the preparation of nutritious diets and to promote ethnic Himalayan fermented food. Food and water can carry infectious agents; thus, increasing knowledge about them and improving practices for the safe handling, storage, and cooking of food is also important to reduce risk of food and waterborne diseases. Nutrition education should be included in the curriculum in primary and secondary schools, and special efforts should to be made to empower women, as their knowledge and position in decision-making will improve the nutrition status of children and families.

9.5.7 Strengthening Social Safety Nets for Remote Mountain Areas and Poor and Vulnerable Groups

At present, social safety net program includes mostly rice and wheat commodities. Pulses, millets, and other NUS could be introduced into public (food) distribution systems to improve the nutritional quality of the food provided. Introduction of a minimum support price for nutrition rich mountain crops need to be considered.

References

Aase, T. H., Chaudhary, R. P., & Vetaas, O. R. (2009). Farming flexibility and food security under climatic uncertainty: Manang, Nepal Himalaya, *AREA*. Royal Geographical Society (with The Institute of British Geographers). 1–11. https://doi.org/10.1111/j.1475-4762.2009.00911.x.

Adhikari, L., Hussain, A., & Rasul, G. (2017). Tapping the potential of neglected and underutilized food crops for sustainable nutrition security in the mountains of Pakistan and Nepal. *Sustainability, 9*(2), 291. https://doi.org/10.3390/su9020291.

Afzal, J., Ahmed, M., & Begum, I. (2008). Vision for development of rangelands in Pakistan: A policy perspective. *Quarterly Science Vision 14*(1): 53–58.

Agarwal, T., Biswas, S., Chattopadhyay, S., & Nath, P. (2017) Consumer gains from trade a case study on promoting trade between Bangladesh and Tripura, India in specific agricultural commodities. Discussion Paper. Jaipur, India: CUTS International.

Ailikun. (2015). Climate projection for Nainital (India) and Panchkhal (Nepal) by RMIP products. In: *Public Symposium: Addressing the Fate of Himalayan Sourced Rivers in the Context of Climate Change: Perspectives from China, Nepal, India, Bangladesh and Bhutan, Australian National University, Canberra, Australia.*

ALCS. (2014). *Chapter 7: Food Security. Afghanistan living conditions survey 2013*–14. Central Statistics Organization, Islamic Republic of Afghanistan. Retrieved from http://cso.gov.af/Content/files/ALCS/FOREWORD%20and%20acknowledgments.pdf.

Aljazeera. (2016, May 12). *Border fencing escalates Pakistan-Afghanistan tension*. Retrieved from http://www.aljazeera.com/news/2016/05/border-fencing-escalates-pakistan-afghanistan-tension-160512135324053.html.

AMS. (2010). *Afghanistan mortality survey*. Central Statistics Organization (CSO) Afghanistan, A joint work of USAID, Ministry of Public Health (Afghanistan), WHO, UNICEF, JNFPA and Indian Institute of Health Management Research Jaipur (India). Retrieved from https://dhsprogram.com/pubs/pdf/fr248/fr248.pdf.

Apetrei, C. (2012). *Food security and millet cultivation in Kumaon region of Uttarakhand. Research report for gene campaign*, New Delhi, India. Retrieved from http://genecampaign.org/wp-content/uploads/2014/07/FOOD_SECURITY_AND_MILLET_CULTIVATION.pdf.

ASP. (2012). *Agricultural statistics of Pakistan 2011–12*. Ministry of National Food Security and Research (Economics Wing), Government of Pakistan, Islamabad. Retrieved from http://www.irispunjab.gov.pk/StatisticalReport/Land%20Utilization%20and%20Agricultural%20Statistics/AGRICULTURAL%20STATISTICS%20OF%202011-12.pdf.

Atreya, K., Johnsen, F. H., & Sitaula, B. K. (2012). Health and environmental costs of pesticide use in vegetable farming in Nepal. *Environment, Development and Sustainability, 14*(4), 477–493. https://doi.org/10.1007/s10668-011-9334-4.

Awasthi, I. C. (2016). *Disaster management in mountain economy—A case of Uttarakhand State of India*. Paper prepared for Tenth Annual Himalayan Policy Research Conference 2015. *Himalayan Research Papers Archive, 9* (1). Retrieved from http://digitalrepository.unm.edu/hprc/2015/papers/9/.

Bandyopadhyay, J., & Perveen, S. (2008). The interlinking of Indian rivers: questions on the scientific, economic and environmental dimension of the project. In M. M. Q. Mirza, A. U. Ahmed, & Q. K. Ahmad (Eds.), *Interlinking rivers in India: Issues and concern* (pp. 53–76). Abingdon, UK: Taylor and Francis.

Bano, A., Ahmad, M., Hadda, T. B., Saboor, A., Sultana, S., Zafar, M., et al. (2014). Quantitative ethnomedicinal study of plants used in the Skardu valley at high altitude of Karakoram-Himalayan range, Pakistan. *Journal of Ethnobiology and Ethnomedicine, 10*(43), 1–17.

Beddington, J. R., Asaduzzaman, M., Fernandez, A., Clark, M. E., Guillou, M., Jahn, M. M., et al. (2012). *Achieving food security in the face of climate change*. Final report from the Commission on Sustainable Agriculture and Climate Change. Retrieved from https://cgspace.cgiar.org/bitstream/handle/10568/35589/climate_food_commission-final-mar2012.pdf.

Behera, R. N., Nayak, D. K., Andersen, P., & Måren, I. E. (2016). From jhum to broom: Agricultural land-use change and food security implications on the Meghalaya Plateau, India. *Ambio, 45*(1), 63–77. https://doi.org/10.1007/s13280-015-0691-3.

Benz, A. (2016). Framing modernization interventions: Reassessing the role of migration and translocality in sustainable mountain development in Gilgit-Baltistan, Pakistan. *Mountain Research and Development, 36*(2), 141–152. https://doi.org/10.1659/mrd-journal-d-15-00055.1.

Bhandari, G., Achhami, B. B., Karki, T. B., Bhandari, B., & Bhandari, G. (2015). Survey on maize post-harvest losses and its management practices in the western hills of Nepal. *Journal of Maize Research and Development, 1*(1), 98–105.

Burchi, F., Fanzo, J., & Frison, E. (2011). The role of food and nutrition system approaches in tackling hidden hunger. *International Journal of Environmental Research and Public Health, 8*(2), 358–373. https://doi.org/10.3390/ijerph8020358.

CBS. (2014). *Statistical pocket book of Nepal 2014*. Kathmandu, Nepal: Central Bureau of Statistics, National Planning Commission Secretariat, Government of Nepal. Retrieved from http://cbs.gov.np/image/data/Publication/Statistical%20Pocket%20Book%202014.pdf.

Clouse, C., Anderson, N., & Shippling, T. (2017). Ladakh's artificial glaciers: Climate-adaptive design for water scarcity. *Climate and Development, 9*(5), 428–438. https://doi.org/10.1080/17565529.2016.1167664.

CPRC. (2004). *The chronic poverty report 2004–05*. Manchester, UK: Chronic Poverty Research Centre. Retrieved from http://www.chronicpoverty.org/uploads/publication_files/CPR1_ReportFull.pdf.

Crush, J., & Caesar, M. (2017). Introduction: Cultivating the migration-food security Nexus. *International Migration, 55*(4), 10–17.

Dame, J., & Nüsser, M. (2011). Food security in high mountain regions: Agricultural production and the impact of food subsidies in Ladakh, Northern India. *Food Security, 3*(2), 179–194. https://doi.org/10.1007/s12571-011-0127-2.

Dawn. (2004, July 21). *Punjab yet to notify lifting of ban: Wheat movement*. Retrieved from http://www.dawn.com/news/365079/punjab-yet-to-notify-lifting-of-ban-wheat-movement.

De Janvry, A., & Sadoulet, E. (2008). The global food crisis: Identification of the vulnerable and policy responses. *Agricultural and Research Economics Update, Special Issue: Causes and Consequences of the Food Price Crisis, 12*(2), 18–21.

Desor, S. (2017). Ideas and initiatives towards an alternative food system in India. Kalpavriksh: Deccan Gymkhana, Pune 411004, India.

Ebi, K. L., Woodruff, R., von Hildebrand, A., & Corvalan, C. (2007). Climate change-related health impacts in the Hindu Kush–Himalayas. *EcoHealth, 4*(3), 264–270. https://doi.org/10.1007/s10393-007-0119-z.

FAO. (2017). *Future smart food: Unlocking hidden treasures in Asia and the Pacific. RI-Zero Hunger*—Policy Brief—Agricultural Diversification for a Healthy Diet. Bangkok, Thailand: FAO Regional Office for Asia and the Pacific.

FAO. (2008). An introduction to the basic concepts of food security. *Food Security Information for Action: Practical Guide*. Retrieved from http://www.fao.org/docrep/013/al936e/al936e00.pdf.

FAO. (2012). *Committee on world food security*. A Report of the Meeting held on 15–20 October in Rome Italy. Retrieved from http://www.fao.org/docrep/meeting/026/MD776E.pdf.

FAO. (2005). Livestock sector brief: India, livestock information, sector analysis and policy branch. AGAL. Rome, Italy: Food and Agriculture Organization of the United Nations. Retrieved from http://www.fao.org/ag/againfo/resources/en/publications/sector_briefs/lsb_IND.pdf.

FAO/Nepal. (2010). *Assessment of food security and nutrition situation in Nepal*. Kathmandu, Nepal: FAO. Retrieved from https://webcache.googleusercontent.com/search?q=cache:7GNhFwFNrNAJ:https://www.medbox.org/assessment-of-food-security-and-nutrition-situation-in-nepal/download.pdf+andcd=1andhl=enandct=clnkandgl=np.

FAO. (2015). *Mapping the vulnerability of mountain peoples to food insecurity*. In: R. Romeo, A. Vita, R. Testolin, & Hofer, T. (Eds.), Rome, Italy: Food and Agriculture Organization. Retrieved from http://www.fao.org/3/a-i5175e.pdf.

FAO-IFAD-WFP. (2015). *The state of food insecurity in the world. Meeting the 2015 international hunger targets: Taking stock of uneven progress*. A Joint publication of Food and Agriculture Organization (FAO) of the United Nations, International Fund for Agriculture Development (IFAD) and World Food Programme (WFP). Rome, Italy: FAO. Retrieved from http://www.fao.org/3/a-i4646e.pdf.

Finnis, E. (2007). The political ecology of dietary transitions: Changing production and consumption patterns in the Kolli Hills, India. *Agriculture and Human Values, 24*(3), 343. https://doi.org/10.1007/s10460-007-9070-4.

FSA. (2009). *Food insecurity in Pakistan. Food security analysis*. Joint work of Sustainable Development Policy Institute, Swiss Agency for Development and Cooperation and World Food Program, Pakistan. Retrieved from http://documents.wfp.org/stellent/groups/public/documents/ena/wfp225636.pdf?_ga=2.141433777.2074906472.1523941307-1510118971.1523941307.

Gangte, H. E., Thoudam, N. S., & Zomi, G. T. (2013). Wild edible plants used by the Zou tribe in Manipur, India. *International Journal of Scientific and Research Publications, 3*(5), 1–8.

Gansu Province Government. (2015). *Gansu province food and nutrition development plan (2014–2020)*. Lanzhou, People's Republic of China: Government of Gansu Province.

Gautam, Y. (2017). Seasonal migration and livelihood resilience in the face of climate change in Nepal. *Mountain Research and Development, 37*(4), 436–444.

Gautam, Y., & Andersen, P. (2017). Multiple stressors, food system vulnerability and food insecurity in Humla, Nepal. *Regional Environmental Change, 17*(5), 1493–1504.

Ghosh, N., & Sharma, S. (2016). *Land use, Agriculture and Food security of the Himalayan regions of India*. Delhi, India: Institute of Economic Growth, University of Delhi.

Giri, Y. P., Thapa, R. B., Shrestha, S. M., Pradhan, S. B., Maharjan, R., Sporleder, M., et al. (2014). Pesticide use pattern and awareness of pesticides users with special reference to potato growers in Nepal. *International Journal of Development Research., 4*(11), 2297–2302.

Gillespie, S., & Mason, J. (1991). *Nutrition relevant actions—Nutrition*. Policy Discussion Paper No. 10. World Health Organization. Retrieved from https://www.unscn.org/web/archives_resources/files/Policy_paper_No_10.pdf.

GoA. (2009). *National public nutrition policy and strategy 1388–1392 (2009–2013)*. Ministry of Public Health, Government of Afghanistan. Retrieved from https://extranet.who.int/nutrition/gina/sites/default/files/AFG%202009%20National%20Public%20Nutrition%20Policy%20and%20Strategy.pdf.

GoB. (2012). *Food and nutrition security policy of the kingdom of Bhutan*. Thimphu, Bhutan: Ministry of Agriculture and Forest, Kingdom of Bhutan. Retrieved from http://www.gnhc.gov.bt/en/wp-content/uploads/2017/05/FNS_Policy_Bhutan_Changed.pdf.

GoI. (2010). *Nutritional intake in India*. NSS 66th Round. New Delhi, India: Ministry of Statistics and Programme Implementation, Government of India. Retrieved from http://www.indiaenvironmentportal.org.in/files/file/nutrition%20intake%20in%20india.pdf.

GoI. (2014a). *Report of the expert group to review the methodology for measurement of poverty*. New Delhi, India: Planning Commission, Government of India. Retrieved from http://planningcommission.nic.in/reports/genrep/pov_rep0707.pdf.

GoI. (2014b). *Fruits and vegetables statistics 2013–14*. New Delhi, India: Department of Agriculture and Cooperation Ministry of Agriculture and Farmers Welfare,.

GoPRC (2014). *National programme for food and nutrition (2014–2020)*. Beijing, China: General Office of the State Council of the People's Republic of China.

Gross, R., Schoeneberger, H., Pfeifer, H., & Preuss, H. J. (2000). The four dimensions of food and nutrition security: Definitions and concepts. *SCN News, 20*, 20–25.

Hagen, T. (2004). *Traditional framing practices and farmers' rights in the HKH region*. Policy Brief. No. 6. Lalitpur, Nepal: SAWTEE.

Hashmi, H. N., Siddiqui, Q. T. M., Ghumman, A. R., & Kamal, M. A. (2012). A critical analysis of 2010 floods in Pakistan. *African Journal of Agricultural Research, 7*(7), 1054–1067.

Hill, J. O., Wyatt, H. R., Reed, G. W., & Peters, J. C. (2003). Obesity and the environment: where do we go from here? *Science, 299* (5608), 853–855.

Hussain, A., & Routray, J. K. (2012). Status and factors of food security in Pakistan. *International Journal of Development, 11*(2), 164–185. https://doi.org/10.1108/14468951211241146.

Hussain, A., Rasul, G., Mahapatra, B., & Tuladhar, S. (2016). Household food security in the face of climate change in the Hindu-Kush Himalayan region. *Food Security, 8*(5), 921–937. https://doi.org/10.1007/s12571-016-0607-5.

Hussain, A., Rasul, G., Mahapatra, B., Wahid, S., & Tuladhar, S. (2018). Climate change-induced hazards and local adaptations in agriculture: A study from Koshi River Basin, Nepal. *Natural Hazards, 91*(3), 1365–1383.

ICIMOD. (2008). *Food security in the Hindu Kush Himalayan region*. Position Paper. Kathmandu, Nepal: ICIMOD.

IHLCA. (2011). *Integrated household living conditions survey in Myanmar (2009–10)*. IHLCA Project Technical Unit, Yangon, The Republic of the Union of Myanmar. Supported by Ministry of National Planning and Economic Development, UNDP, UNICEF and SIDA. Retrieved from http://www.mm.undp.org/content/dam/myanmar/docs/Publications/PovRedu/MMR_FA1_IA2_Technical%20Report-Eng.pdf.

Jackson, W. J., Tamrakar, R. M., Hunt, S., & Shepherd, R. K. (1998). Land-use changes in two middle hills districts of Nepal. *Mountain Research and Development, 18*(3), 193–212.

Jaquet, S., Schwilch, G., Hartung-Hofmann, F., Adhikari, A., Sudmeier-Rieux, K., Shrestha, G., et al. (2015). Does outmigration lead to land degradation? Labour shortage and land management in a western Nepal watershed. *Applied Geography, 62,* 157–170. https://doi.org/10.1016/j.apgeog.2015.04.013.

Jenny, A. L., & Egal, F., (2002). *Household food security and nutrition in mountain areas: An often forgotten story.* Rome, Italy: Nutrition Programmes Service, Food and Agriculture Organization of the United Nations (FAO-ESNP). Retrieved from http://peaksmaker.com/files/2010/01/FAO_Nutrition-in-Mountains.pdf.

Jodha, N. S. (2000). Globalization and fragile mountain environments: Policy challenges and choices. *Mountain Research and Development, 20*(4), 296–299.

Jodha, N. S. (2009). Mountain agriculture: Development policies and perspectives. *Indian Journal of Agricultural Economics, 64*(1), 1–14.

Joshi, L., Shrestha, R. M., Jasra, A. W., Joshi, S., Gilani, H., & Ismail, M. (2013). Rangeland ecosystem services in the Hindu Kush Himalayan region. In: W. Ning, G. S. Rawat, S. Joshi, M. Ismail, E. Sharma (Eds.), *High-altitude rangelands and their interfaces in the Hindu Kush Himalayas* (pp. 157–175). Kathmandu, Nepal: ICIMOD.

Jones, A. D., Ngure, F. M., Pelto, G., & Young, S. L. (2013). What are we assessing when we measure food security? A compendium and review of current metrics. *Advances in Nutrition: An International Review Journal, 4*(5), 481–505. https://doi.org/10.3945/an.113.004119.

Kettings, C., Sinclair, A. J., Voevodin, M. (2009). A healthy diet consistent with Australian health recommendations is too expensive for welfare dependent families. *Australian and New Zealand Journal of Public Health 33*(6), 566–572.

Khan, A. U., Saboor, A., Hussain, A., Karim, S., & Hussain, S. (2015). Spatial and temporal investigation of multidimensional poverty in rural Pakistan. *Poverty and Public Policy, 7*(2), 158–175.

Khanal, N. R. (2002). Land use and land cover dynamics in the Himalaya: A case study of the Madi Watershed, Western Development Region, Nepal (PhD dissertation). Kathmandu, Nepal: Tribhuvan University.

Khanal, N. R., & Watanabe, T. (2006). Abandonment of agricultural land and its consequences: A case study in the Sikles area, Gandaki Basin, Nepal Himalaya. *Mountain Research and Development, 26* (1), 32.

Kumanyika, S., Jeffery, R. W., Morabia, A., Ritenbaugh, C., & Antipatis, V. J. (2002). Obesity prevention: The case for action. *International Journal of Obesity, 26*(3), 425–436. https://doi.org/10.1038/sj.ijo.0801938.

Kumar, R., Shai., A. K., Krishna Kumar, K., Patwardhan, S. K., Mishra, P. K., Rewadhar, J. V., et al. (2006). High-resolution climate change scenario for India for the 21st century. *Current Science 90*, 334–345.

Landy, F. (2017). Rescaling the public distribution system in India: Mapping the uneven transition from spatialization to territorialisation. *Environment and Planning C: Politics and Space, 25*(S1), 113–129.

Likhi, A. (2014). *Delivery challenges for India's national food security act 2013*. The World Bank Blog. Retrieved from https://blogs.worldbank.org/publicsphere/delivery-challenges-india-s-national-food-security-act-2013.

Lo, Y. T., Chang, Y. H., Lee, M. S., & Wahlqvist, M. L. (2012). Dietary diversity and food expenditures as indicators of food security in older Taiwanese. *Appetite, 58,* 180–187. https://doi.org/10.1016/j.appet.2011.09.023. Epub 2011 Oct 5.

Longvah, T., Ananthan, R., Bhaskarachary, K., & Venkaiah, K. (2017). *Indian Food Composition Tables*. Hyderabad, India: National Institute of Nutrition.

Maikhuri, R. K., Rao, K. S., & Semwal, R. L. (2001). Changing scenario of Himalayan agro-ecosystems: Loss of agrobiodiversity, an indicator of environmental change in Central Himalaya. *India. Environmentalist, 21*(1), 23–39. https://doi.org/10.1023/A:1010638104135.

Majumder, S., Bala, B. K., & Hossain, M. A. (2012). Food security of the hill tracts of Chittagong in Bangladesh. *Journal of Natural Resources Policy Research, 4*(1), 43–60.

Mal, B., Padulosi, S., & Ravi, S. B. (2010). Minor millets in South Asia: Learnings from IFAD-NUS project in India and Nepal. *Biodiversity International*, Rome, Italy: IFAD and Chennai, India: M. S. Swaminathan Research Foundation.

Manral, U., Sengupta, S., Hussain, S. A., Rana, S., & Badola, R. (2016). Human wildlife conflict in India: A review of economic implication of loss and preventive measures. *Indian Forester, 142* (10), 928–940.

Mayes, S., Massawe, F. J., Alderson, P. G., Roberts, J. A., Azam-Ali, S. N., & Hermann, M. (2012). The potential for underutilized crops to improve security of food production. *Journal of Experimental Botany, 63*(3), 1075–1079. https://doi.org/10.1093/jxb/err396.

Merz, J., Nakarmi, G., Shrestha, S., Dahal, B. M., Dongol, B. S., Schaffner, M., et al. (2004). Public water sources in rural watersheds of Nepal's middle mountains: Issues and constraints. *Environmental Management, 34*(1), 26–37.

Milbert, I. (2009). Policy dimensions of human security and vulnerability challenges: The case of urban India. In H. G. Brauch (Ed.), *Facing global environmental change—Environmental, human, energy, food, health and water security concepts* (pp. 233–242). Berlin, Heidelberg: Berghof Foundation, Springer.

Ministry of Health. (2013). *Year book of health in the people's republic of China 2013*. Statistics Information Centre of Ministry of Health. Beijing, People's Republic of China: People's Medical Publishing House Co. Ltd.

Mirza, S. N., Ahmad, S., Islam, M. (2006). The vagaries of drought in Balochistan and strategies to reduce economic losses. *Journal of Agricultural Research 3*(1), 39–42.

MoNPED and MoH. (2011). *Multiple indicator cluster survey 2009–2010*. Nay Pyi Taw, Myanmar: Ministry of National Planning and Economic Development and Ministry of Health. Retrieved from https://www.unicef.org/myanmar/MICS_Myanmar_Report_2009-10.pdf.

MoAD. (2015). *Statistical information on Nepalese agriculture 2014–15*. Kathmandu, Nepal: Agri-Business Promotion and Statistics Division, Agri. Statistics Section, Ministry of Agricultural Development, Government of Nepal.

Monsivais, P., & Drewnowski, A. (2007). The rising cost of low-energy-density foods. *Journal of the American Dietetic Association, 107*(12), 2071–2076.

Nanda, N. (2012). *Agricultural trade in South Asia: Barriers and prospects*. Working Paper No. 03/12. Kathmandu, Nepal: South Asia Watch on Trade, Economics and Environment (SAWTEE).

NDHS. (2011). *Nepal demographic and health survey*. A joint work of the Ministry of Health and Population (Nepal), New ERA (Nepal), and USAID. Retrieved from https://dhsprogram.com/pubs/pdf/fr257/fr257%5B13april2012%5D.pdf.

Negi, G. C. S., Samal, P. K., Kuniyal, J. C., Kothyari, B. P., Sharma, R. K., & Dhyani, P. P. (2012). Impact of climate change on the western Himalayan mountain ecosystems: An overview. *Tropical Ecology, 53*(3), 345–356.

NFHS. (2006). *National family health survey 2005–2006 (NFHS-3)*. The Demographic and Health Survey Program. Washington, USA: USAID.

NHPC. (2015). *Report on Chinese residents' chronic diseases and nutrition 2015*. Prepared by the Bureau of Disease Prevention and Control, National Health and Family Planning Commission. Beijing, People's Republic of China: People's Medical Publishing House Co. Ltd.

Nirali, B. J., & Shankar, M. B. (2015). Global market analysis of herbal drug formulations. *International Journal of Ayurveda and Pharmaceutical Chemistry, 4*(1), 59–65.

NNS. (2015). *National nutrition survey*. Thimphu, Bhutan: Nutrition Programme, Ministry of Health, Government of Bhutan.

Bhutta, Z. A., Soofi, S., Zaidi, S., Habib, A., & Hussain, I. (2011). *National nutrition survey*. Joint work of Agha Khan University, Karachi, Pakistan Medical Research council, Planning Commission, Planning and Development Division, Government of Pakistan, and Nutrition Wing, Cabinet Division, Government of Pakistan. Supported by UNICEF. Retrieved from https://ecommons.aku.edu/cgi/viewcontent.cgi?referer=https://www.google.com.np/andhttpsredir=1andarticle=1262andcontext=pakistan_fhs_mc_women_childhealth_paediatr.

Nüsser, M., Dame, J., & Schmidt, S. (2015). Urbane Entwicklung im indischen Himalaya. *Geographische Rundschau, 67*(7/8), 32–39.

Ogra, V. M. (2008). Human-wildlife conflict and gender in protected area borderlands: A case study of costs, perceptions, and vulnerabilities from Uttarakhand (Uttaranchal), India. *ScienceDirect, Geoforum, 39*, 1408–1422. https://doi.org/10.1016/j.geoforum.2007.12.004.

Padulosi, S., & Hoeschle-Zeledon, I. (2004). Underutilized plant species: What are they? *LEISA-Magazine*, 5–6. Retrieved from http://lib.icimod.org/record/11453/files/3800.pdf.

Padulosi, S., Bergamini., N., & Lawrence, T. (Eds). (2012). On-farm conservation of neglected and underutilized species: Status, trends and novel approaches to cope with climate change. In: *Proceedings of the International Conference*, Friedrichsdorf, Frankfurt 14–16 June, 2011. Rome, Italy: Biodiversity International.

Partap, T. (1998). Crop productive and sustainability: Shaping the future. In: V. L. Chopra, R. B. Singh, A. Verma (Eds.), *Proceedings of the 2nd International Crop Science Congress*. New Delhi, India: National Academy of Agricultural Sciences and the Indian Council of Agricultural Research.

Partap, T. (2011). Hill agriculture: challenges and opportunities, *Indian Journal of Agricultural Economics, 66*(1), 33–52.

Parveen, S., Winiger, M., Schmidt, S., & Nüsser, M. (2015). Irrigation in Upper Hunza: Evolution of socio-hydrological interactions in the Karakoram, northern Pakistan. *Erdkunde 69*(1), 69–85.

Pedersen, R. (2014. January 24). *Food Security in Myanmar: Opportunities and Challenges*. Retrieved from http://www.bioforsk.no/ikbViewer/page/forside/nyhet?p_document_id=108055.

Pingali, P., Mittra, B., & Rahman, A. (2017). The bumpy road from food to nutrition security—Slow evolution of India's food policy. *Global Food Security, 15*, 77–84.

Pinstrup-Andersen, P. (2009). Food security: definition and measurement. *Food Security, 1*(1), 5–7. https://doi.org/10.1007/s12571-008-0002-y.

Pradhan, P., Reusser, D. E., & Kropp, J. P. (2013). Embodied greenhouse gas emissions in diets. *PLoS ONE, 8*(5), e62228. https://doi.org/10.1371/journal.pone.0062228.

Pyakuryal, B., Roy, D., & Thapa, Y. B. (2010). Trade liberalization and food security in Nepal. *Food Policy, 35*(1), 20–31.

Rasul, G. (2014). Food, water, and energy security in South Asia: A nexus perspective from the Hindu Kush Himalayan region. *Environmental Science & Policy, 39*, 35–48. https://doi.org/10.1016/j.envsci.2014.01.010.

Rasul, G. (2015). *A strategic framework for sustainable development in the Chittagong Hill Tracts of Bangladesh*. ICIMOD Working Paper 2015/3. Kathmandu: ICIMOD.

Rasul, G., & Hussain, A. (2015). Sustainable food security in the mountains of Pakistan: Towards a policy framework. *Ecology of food and nutrition., 54*(6), 625–643. https://doi.org/10.1080/03670244.2015.1052426.

Rasul, G. (2010). The role of the Himalayan mountain systems in food security and agricultural sustainability in South Asia. *International Journal of Rural Management, 6*(1), 95–116.

Rasul, G., & Thapa, G. B. (2003). Shifting cultivation in the mountains of South and Southeast Asia: Regional patterns and factors influencing the change. *Land Degradation and Development, 14*(5), 495–508. https://doi.org/10.1002/ldr.570.

Rasul, G., Hussain, A., Mahapatra, B., & Dangol, N. (2018). Food and nutrition security in the Hindu Kush-Himalayan region. *Journal of the Science of Food and Agriculture, 98*(2), 429–438. https://doi.org/10.1002/jsfa.8530.

Rasul, G., Karki, M., & Sah, R. P. (2008). The role of non-timber forest products in poverty reduction in India: Prospects and problems. *Development in Practice, 18*(6), 779–788. https://doi.org/10.1080/09614520802386876.

Rasul, G., Hussain, A., Khan, M. A., Ahmad, F., & Jasra, A. W. (2014). *Towards a framework for achieving food security in the mountains of Pakistan*. ICIMOD Working Paper 2014/5. Kathmandu: ICIMOD.

Rasul, G., Hussain, A., Sutter, A., Dangol, N., & Sharma, E. (2016). *Towards an integrated approach to nutrition security in the Hindu Kush Himalayan region*. ICIMOD Working Paper 2016/7 Kathmandu: ICIMOD.

Rawat, J. S. (2009). Saving Himalayan Rivers: developing spring sanctuaries in headwater regions. In B. L. Shah (Ed.), *Natural resource conservation in Uttarakhand* (pp. 41–69). Haldwani: Ankit Prakshan.

Roy, A., Singh, N. U., Dkhar, D. S., Mohanty, A. K., Singh, S. B., & Tripathi, A. K. (2015). Food security in north-east region of India—A state-wise analysis. *Agricultural Economics Research Review, 28* (Conference number), 259–266. https://doi.org/10.5958/0974-0279.2015.00041.5.

Romeo, R., Vita, A., Testolin, R., & Hofer, T. (Eds.). (2015). *Mapping the vulnerability of mountain peoples to food insecurity*. Rome, Italy: Food and Agriculture Organisation.

Sati, V. P. (2015). Issues and options of food security and poverty: An empirical study of Mizoram, the Eastern extension of the Himalaya. *Journal of Food Security, 3*(4), 107–114. https://doi.org/10.12691/jfs-3-4-3.

Saboor, A., Khan, A. U., Hussain, A., Ali, I., & Mahmood, K. (2015). Multidimensional deprivations in Pakistan: Regional variations and temporal shifts. *The Quarterly Review of Economics and Finance, 56*, 57–67.

Sawian, J. T., Jeeva, S., Lyndem, F. G., Mishra, B. P., & Laloo, R. C. (2007). Wild edible plants of Meghalaya, North-east India. *Natural Product Radiance, 6*(5), 410–426.

Sengupta, P., & Mukhopadhyay, K. (2016). Economic and environmental impact of national food security act of India. *Agricultural and Food Economics, 4*, 5. https://doi.org/10.1186/s40100-016-0048-7.

Shafiq, M., & Kakar, M. A. (2007). Effects of drought on livestock sector in Balochistan Province of Pakistan. *International Journal of Agriculture and Biology (Pakistan), 9*(4), 1–9.

Sharma, K. R. (2012). Malnutrition in children aged 6–59 months in Mugu district. *Journal of Nepal Health Research Council, 10*(21), 156–159.

Sharma, G., Partap, U., Sharma, E., Rasul, G., & Awasthe, R. K. (2016). *Agrobiodiversity in the Sikkim Himalaya: Sociocultural significance, status, practices, and challenges*. ICIMOD Working Paper 2016/5 Kathmandu: ICIMOD.

Shrestha, A. B., & Aryal, R. (2011). Climate change in Nepal and its impact on Himalayan glaciers. *Regional Environmental Change, 11* (1), 65–77. https://doi.org/10.1007/s10113-010-0174-9.

Stone, G. D. (2001). Theory of the square chicken: Advances in agricultural intensification theory. *Asia Pacific Viewpoint, 42*(2–3), 163–180.

Tambe, S., Kharel, G., Arrawatia, M. L., Kulkarni, H., Mahamuni, K., & Ganeriwala, A. K. (2012). Reviving dying springs: Climate change adaptation experiments from the Sikkim Himalaya. *Mountain Research and Development, 32*(1), 62–72. https://doi.org/10.1659/MRD-JOURNAL-D-11-00079.1.

Tamang, J. P. (2009). *Himalayan fermented foods: Microbiology, nutrition, and ethnic values*. Boca Raton, FL, USA: CRC Press.

Tamang, J. P., Tamang, N., Thapa, S., Dewan, S., Tamang, B., Yonzan, H., et al. (2012) Microorganisms and nutritional value of ethnic fermented foods and alcoholic beverages of North East India. *Indian Journal of Traditional Knowledge, 11*(1), 7–25.

Thapa, P. B. (2001). Land-use/Land cover change with focus on land abandonment in middle hills of Nepal: A case study of Thumki VDC, Kaski District (MA dissertation). Kirtipur, Nepal: Tribhuvan University.

The Nation. (2014a, April 25). *Sindh bans inter-provincial movement of flour*. Retrieved from http://nation.com.pk/business/25-Apr-2014/sindh-bans-inter-provincial-movement-of-flour.

The Nation. (2014b, May 14). *Punjab bans inter-provincial movement of wheat*. Retrieved from http://nation.com.pk/national/14-May-2014/punjab-bans-inter-provincial-movement-of-wheat.

Tiwari, P. C., & Joshi, B. (2012a). Environmental changes and sustainable development of water resources in the Himalayan headwaters of India. *International Journal of Water Resource Management, 26*(4), 883–907. https://doi.org/10.1007/s11269-011-9825-y.

Tiwari, P. C., & Joshi, B. (2012b). Natural and socio-economic drivers of food security in Himalaya. *International Journal of Food Security, 4* (2), 195–207. https://doi.org/10.1007/s12571-012-0178-z.

Tobgay, S. (2005). *Agriculture diversification in Bhutan*. Ministry of Agriculture, Bhutan. Retrieved from http://citeseerx.ist.psu.edu/viewdoc/download?doi=10.1.1.486.7796andrep=rep1andtype=pdf.

Tulachan, P. M. (2001). Mountain agriculture in the Hindu-Kush Himalaya: A regional comparative analysis. *Mountain Research and Development, 21*(3), 260–267.

UCCRN. (2015). *Climate projections for Shimla, India*, Unpublished Document. New York, NY, USA: Urban Climate Change Research Network (UCCRN), Columbia University.

UNICEF and PBS. (2011). *Multiple indicator cluster survey 2010 (State wise report)*. Islamabad, Pakistan: UNICEF and Pakistan Bureau of Statistics.

Ura, K., & Kinga, S. (2004). *Bhutan–sustainable development through good governance*. A case study from reducing poverty, sustaining growth—What works, what doesn't, and why a global exchange for scaling up success. Washington DC, USA: World Bank. Retrieved from http://documents.worldbank.org/curated/en/880451468743944567/pdf/308210BHU0Governance01see0also0307591.pdf.

Vishvakarma, S. C. R., Kuniyal, J. C., & Rao, K. S. (2003). Climate change and its impact on apple cropping in Kullu Valley, North-West Himalaya, India. In: *7th International Symposium on Temperate Zone Fruits in the Tropics and Subtropics, 14–18 October, Nauni-Solan (H.P.), India. ISHS Acta Horticulturae 696*. Leuven, Belgium: International Society for Horticultural Science.

Viviroli, D., Weingartner, R., & Messerli, B. (2003). Assessing the hydrological significance of the world's mountains. *Mountain Research and Development, 23*, 32–40.

Viviroli, D, Dürr, H. H., Messerli, B., Meybeck, M., & Weingartner, R. (2007). Mountains of the world, water towers for humanity: Typology, mapping, and global significance. *Water Resources Research, 43*. W07447. Retrieved from http://www.zora.uzh.ch/id/eprint/109944/.

WFP and FAO. (2007). *Food and agricultural markets in Nepal*. Kathmandu, Nepal: United Nations World Food Programme, Food and Agricultural Organization of the United Nations.

WHO–UNICEF. (2015). In: *Joint Monitoring Programme (JMP) for Water Supply and Sanitation*. Geneva: WHO and New York: UNICEF.

Williams, J. T., & Haq, N. (2002). *Global research on underutilized crops. An assessment of current activities and proposals for enhanced cooperation*. Southampton, UK: ICUC. Retrieved from http://www.fao.org/docs/eims/upload/216780/uoc_assessment_current_activities.pdf.

World Food Summit. (1996). *Rome declaration on world food security*. Rome, Italy: FAO. Retrieved from http://www.fao.org/docrep/003/w3613e/w3613e00.HTM.

WSJ. (2015). The two-month blockade of Nepal explained. *The Wall Street Journal*. Retrieved November 26, 2015, from http://blogs.wsj.com/indiarealtime/2015/11/26/the-two-month-blockade-of-nepal-explained/.

Wu, N., Yi, S., Joshi, S., & Bisht, N. (Eds.). (2016). *Yak on the move: Transboundary challenges and opportunities for yak raising in a changing Hindu Kush Himalayan region*. Kathmandu: ICIMOD.

Xiaoping, L., & Minru, Z. (2014). Survey on the nutrition and health status of Qinghai residents in 2010. *Modern Preventive Medicine, 30*(21), 3879–3886.

Xu, J., Grumbine, R. E., Shrestha, A., Eriksson, M., Yang, X., Wang, Y. U. N., et al. (2009). The melting Himalayas: Cascading effects of climate change on water, biodiversity, and livelihoods. *Conservation Biology, 23*(3), 520–530. https://doi.org/10.1111/j.1523-1739.2009.01237.x.

Ziegler, A. D., Cantarero, S. I., Wasson, R. J., Srivastava, P., Spalzin, S., Chow, W. T., et al. (2016). A clear and present danger: Ladakh's increasing vulnerability to flash floods and debris flows. *Hydrological Processes, 30*(22), 4214–4223. https://doi.org/10.1002/hyp.10919.

Air Pollution in the Hindu Kush Himalaya

10

Coordinating Lead Authors
Eri Saikawa, Emory University, Atlanta, GA, USA, e-mail: eri.saikawa@emory.edu
Arnico Panday, ICIMOD, Kathmandu, Nepal, e-mail: arnico.panday@icimod.org (corresponding author)
Shichang Kang, State Key Laboratory of Cryosphere Science, Northwest Institute of Eco-Environment and Resources, Chinese Academy of Science, Lanzhou, China, e-mail: Shichang.kang@lzb.ac.cn

Lead Authors
Ritesh Gautam, Indian Institute of Technology Bombay, Mumbai, India, e-mail: rgautam.Iitb@gmail.com
Eric Zusman, Institute for Global Environmental Strategies, Hayama, Japan, e-mail: zusman@iges.or.jp
Zhiyuan Cong, Institute of Tibetan Plateau Research, Chinese Academy of Science, Beijing, China, e-mail: zhiyuancong@Itpcas.ac.cn
E. Somanathan, Economics and Planning Unit, Indian Statistical Institute, Delhi, India, e-mail: som@isid.acin
Bhupesh Adhikary, ICIMOD, Kathmandu, Nepal, email: bhupesh.adhikary@icimod.org

Contributing Authors
Robert E. Yokelson, University of Montana, Missoula, MT, USA, e-mail: bob.yokelson@mso.umt.edu
James H. Crawford, NASA Langley Research Center, Hamton, VA, USA, e-mail: james.h.crawford@nasa.gov
Maheswar Rupakheti, Institute for Advanced Sustainability Studies, Potsdam, Germany, e-mail: maheswar.rupakheti@iass-potsdam.de
Wenlu Ye, Emory University, Atlanta, GA, USA, e-mail: wenlu.ye@emory.edu
Md. Golam Saroar, Scientific Officer (Modelling), Clean Air and Sustainable Environment (CASE) Project, Department of Environment Ministry of Environment and Forest, Bangladesh, e-mail: saroar82@gmail.com

Review Editor
Nguyen Thi Kim Oanh, Asian Institute of Technology, Bangkok, Thailand, e-mail: kimoanh@ait.asia

Corresponding Author
Arnico Panday, ICIMOD, Kathmandu, Nepal, e-mail: arnico.panday@icimod.org

P. Wester et al. (eds.), *The Hindu Kush Himalaya Assessment*,
https://doi.org/10.1007/978-3-319-92288-1_10

Contents

Chapter Overview

Key Findings

1. **Air pollution in the Hindu Kush Himalaya (HKH) is on the rise and regional air quality has worsened in the past two decades, with the adjacent Indo-Gangetic Plains (IGP) having become one of the most polluted regions in the world.** The causes include rapid urbanization and population growth, with emissions from diverse pollutant sources—cookstoves, brick kilns, other industries, power plants, and transport. However, major gaps in our understanding remain due to the scarcity of air quality monitoring stations.
2. **Persistent winter fog and haze have increased across the Indo-Gangetic Plains (IGP), leading to reduced visibility and elevated air pollution just south of the HKH and affecting air quality in the HKH as well as in the IGP**. The winter fog reduces crop yields and affects tourism, impacting the lives of millions of people.
3. **The HKH is sensitive to climate change—air pollutants originating within and near the HKH**

amplify the effects of greenhouse gases and accelerate the melting of the cryosphere through the deposition of black carbon and dust, the circulation of the monsoon, and the distribution of rainfall over Asia.

Policy Messages

1. **To mitigate air pollution and its severe socio-economic impacts, investment in clean technologies and infrastructure is essential.** The HKH should focus on leapfrogging directly to environmentally-friendly technologies and energy options in households, agriculture, industry, transport and
2. **Dedicated national institutions are required to address air pollution across multiple sectors and scales and implement air pollution mitigation policies.** These institutions also need to be mandated to cooperate and collaborate regionally to address trans-boundary air pollution.
3. **Education is essential—the HKH needs more mechanisms to enhance knowledge sharing, to increase responsiveness to scientific evidence, and to promote awareness and behavioural change.**

Air pollution has large impacts on the Hindu Kush Himalaya (HKH), affecting not just the health of people and ecosystems, but also climate, the cryosphere, monsoon patterns, water availability, agriculture, and incomes (***established but incomplete***). Although the available data are not comprehensive, they clearly show that the HKH receives significant amounts of air pollution from within and outside of the region, including the Indo-Gangetic Plain (IGP), a region where many rural areas are severely polluted. In addition, the HKH receives trans-boundary pollution from other parts of Asia. This chapter surveys the evidence on regional air pollution and considers options for reducing it, while underlining the need for regional collaboration in mitigation efforts. As described in Chap. 1, the HKH region is fragile and rapidly changing; while the outcome of the interplay of complex drivers is difficult to predict, it will have major consequences. That holds true for air pollution as well.

The past decade has seen a rapid rise not only in air pollution affecting the HKH (***well-established***), but also in our understanding of it. An increasing number of atmospheric monitoring stations have been installed throughout the HKH and the nearby plains, while several important field campaigns have contributed significant knowledge about atmospheric processes. Many collect time series data on particulate matter with a diameter of 10 μm or less (PM_{10}) and with a diameter of 2.5 μm or less ($PM_{2.5}$). Data from these monitoring stations show that air quality has strong seasonal and diurnal cycles in most parts of the HKH, with both meteorology and emission patterns playing important roles in determining concentrations.

Air pollution varies by season: Ground-level particulate matter pollutant concentrations in the HKH are highest during the winter months from December to February, and lowest during the summer monsoon months from June to August (***well-established***). Pollution also varies diurnally, though in different ways for different pollutants. In many plain and valley locations, meteorology and emissions patterns cause morning and evening peaks in PM_{10} and $PM_{2.5}$—while ozone (O_3) peaks during the daytime and sinks to low levels at night, as a result of dry deposition and titration by nitric oxide (NO). Mountaintop and high-altitude locations show an afternoon peak in PM caused by the up-slope arrival of polluted air masses from lower elevations; these higher locations do not exhibit the night-time drop in O_3 levels found in the plains and valleys.

Despite recent improvements in data collection, major gaps in HKH air quality networks persist, with many large cities and even more rural areas within and surrounding the HKH (i.e., IGP) still lack monitoring. These data gaps reflect several challenges. First, the topographical heterogeneity and thus fine-scale atmospheric variations in the HKH means that improved air quality monitoring would require a denser network of stations than is required in flat plains. Second, the current use of different instruments or protocols at different sites means that data from these sites urgently needs to be compared and validated to produce harmonized data bases. Third, in addition to the ground-based observation of air pollution, a full picture would also require the measurement of pollutants' vertical variation. In addition, more advanced instruments with higher sensitivity are essential for advancing our understanding of air pollution.

One of the worsening pollution problems now facing the southern edges of the HKH and the nearby IGP is regional haze during the dry season, augmented by increasing instances of persistent winter fog in the past two decades (***well-established***). In winter months, temperatures over the IGP are cold enough for frequent temperature inversion episodes—a layer of cool air is trapped near the ground under a layer of warm air. This condition suppresses the normal tendency of pollutants to rise and disperse over a wide area, trapping them instead in a relatively shallow boundary layer and causing winter haze to be optically thick (as often seen in satellite imagery). Winter air pollution is also exacerbated by increased biofuel burning for heating, combined with increased open biomass burning: Higher winter-time concentrations have been found at several locations in the IGP for carbonaceous aerosols (black carbon, BC and organic carbon, OC), for sulphate, as well as for nitrate. Finally, dense persistent haze and fog can be self-reinforcing: by reducing the sun's ability to warm the land surface, they can further lower surface temperatures and perpetuate the inversion effect. For all these reasons, poor visibility days can rise as high as 90% during winter throughout the IGP.

The HKH is sensitive to global climate change through its impacts on atmospheric dynamics and thermal forcing (***established but incomplete***). Research also indicates that snow and glacier melt are accelerated by absorbing aerosols by up to 20%. In addition, several climate modeling studies have suggested the importance of aerosol solar absorption in modulating summer monsoon circulation and rainfall distribution over Asia. Water from seasonal snow and glacial melt provides significant resources for regional livelihoods, and monsoon rainfall—particularly over southern Asia—is a crucial freshwater resource for the region (constituting over 70% of annual rainfall). Thus, as aerosol radiative effects and regional warming perturb monsoon circulation and the Himalayan cryosphere, the resulting shifts in rainfall dynamics have critical socio-economic as well as environmental implications.

Air pollutant mitigation is urgently needed in the HKH, given the severe impacts of deteriorating air quality and increasing haze and winter fog across the region—on health, climate, the cryosphere, water resources, agriculture, ecosystems, and livelihoods (***established but incomplete***). Such mitigation will require three elements:

- Investment in clean technologies and green infrastructure.
- Dedicated institutions and policies, at the local and national levels, but also for regional and trans-boundary collaboration and cooperation.
- Behavioural change, supported by knowledge sharing and responsiveness to scientific evidence.

Promising clean technologies exist for mitigation in the household, industrial, transport, and energy sectors. At the household level, steps include chimney installation and the use of cleaner cook stoves—whether with liquefied petroleum gas (LPG), biogas, or electricity. Households may be more likely to adopt these cleaner fuels and technologies with help from government subsidies and credit, as well as through effective awareness building and marketing. At the industry level, brick producers can reduce fuel consumption and mitigate CO_2 and air pollutant emissions by shifting from intermittent or inefficient kilns (such as clamp kilns or bull's trench kilns) to more efficient types (such as zigzag or Vertical Shaft). Finally, green infrastructure policies could significantly reduce short-lived climate pollutants (SLCPs) and greenhouse gases through structural changes to energy and transport systems: For example, the HKH countries could adopt tighter vehicle emissions and fuel quality standards or design cities that promote public and non-motorized transport.

Dedicated institutions and policies are needed, both within countries and across national borders. Because air pollution in the HKH is regional, its mitigation is—critically—a regional responsibility. The region urgently requires institutional arrangements that will enable inter-agency coordination on air pollution, actively engaging multiple stakeholders. To begin with, two constraints need to be removed: the lack of a clear division of labour among government institutions, and the lack of coordinating mechanisms to break down agency silos.

Public awareness and behavioural change are also essential for emissions reduction. For example, people who lack adequate solid waste collection services commonly burn their trash, generating emissions that can be linked to health problems. Raising awareness of these problems can help to build public support and pressure for improving services in urban areas—and for adopting alternative waste disposal methods, such as composting.

Air Pollution and the Sustainable Development Goals

Air pollution mitigation must be part of any effort to meet the Sustainable Development Goals (SDGs), especially three: those on health (SDG 3), sustainable energy (SDG 7), and climate action (SDG 13). To make the connection explicit, we propose three calls to action on air pollutants in the HKH:

- For SDG 3 (*Ensure healthy lives and promote well-being for all at all ages*): Take urgent measures to reduce household air pollution from cook stoves in the HKH.
- For SDG 7 (*Ensure access to affordable, reliable, sustainable, and modern energy for all*): Ensure access to affordable, sustainable, and improved clean fuel for all people in the HKH.
- For SDG 13 (*Take urgent action to combat climate change and its impacts*): Reduce emissions of short-lived climate pollutants (SLCPs)—especially black carbon (BC)—by reducing emissions from brick kilns, cook stoves, open biomass burning, and diesel engines in the HKH.

10.1 Observations of Air Pollution in the HKH

Photographs of snowy peaks against blue skies can give the impression that the Hindu Kush Himalaya (HKH) has pristine air quality. Unfortunately, days when such photographs can be taken are becoming increasingly rare. Recent decades have seen much of the HKH suffering from major air pollution problems. Three urban cities in the HKH, including Peshawar (Pakistan), Mazar-e-Sharif (Afghanistan), and Kabul (Afghanistan), are on the list of the 20 most polluted cities in the world (World Health Organization 2016). During a big fraction of the dry season (see Box 10.1 for definition of seasons), the heavily populated Indo-Gangetic Plain (IGP) just south of the HKH is covered by a thick aerosol haze that reduces visibility and obscures sunlight. The haze often penetrates deep into Himalayan valleys and at times even crosses the Himalaya to reach the Tibetan Plateau.

Across the region, particulate matter (PM)—both primary and secondary aerosols—as well as tropospheric ozone (O_3)—a secondary pollutant—have increased. As illustrated in Fig. 10.1, many cities in the HKH have an annual average $PM_{2.5}$ concentration higher than the World Health Organization (WHO) guideline of 10 $\mu g/m^3$ (WHO 2005). Furthermore, in 12 cities (Allahabad, India; Patna, India; Dehradun, India; Delhi, India; Luchnow, India; Ludhiana, India; Peshawar, Pakistan; Amritsar, India; Rawalpindi, Pakistan; Narayangonj, Bangladesh; Agra, India; Jaipur, India), the annual average concentrations are more than 10 times higher than the guideline value. Meanwhile the IGP has seen an increase in persistent winter fog during the past two decades that is at least partly driven by increased air pollution, in addition to changes in moisture availability. The historical trends of black carbon (BC) reconstructed from lake sediment and ice cores at high altitude sites also show a dramatic increase in recent decades, starting in the 1960s. This section presents our understanding of air pollution in the region based on available observations. Throughout the section, the focus will be on PM and gaseous air pollutants, including carbon monoxide (CO), nitrogen oxides (NO_x), sulphur oxides (SO_x), and O_3.

Box 10.1 Seasons in the HKH
When discussing air pollution in the HKH, we generally talk about two seasons: dry period (October–June) and wet period (June–September), which also corresponds to the monsoon season. The dry period can be further divided into three separate seasons: winter (December–February), pre-monsoon (March–mid-June), and post-monsoon (October–November). These definitions (illustrated in Fig. 10.2) are used throughout this chapter.

In recent years, stations to monitor air pollution have been established at a number of key places within and near the HKH, both in urban and rural areas. In addition, short-duration field campaigns have collected valuable first glimpses into pollution levels and temporal patterns of pollutants that are otherwise not monitored in those locations. While many time-series are short and most datasets do not go back more than a few years, they provide some important observations:

- Air quality in most places has strong seasonal and diurnal cycles, with an important role played by meteorology and emission patterns.
- Air pollution levels are generally high outside of the rainy season.
- Air pollution is very much a regional issue, with widespread haze extending across much of the urban and rural parts of the IGP and southern HKH foothills during the dry season.

10.1.1 Air Pollution Time-Series and Seasonality in the HKH

Many places in the HKH and surrounding regions suffer from unhealthy levels of air pollution. In Bangladesh, we find PM levels more than twice the national annual average ambient air quality standards for PM_{10} (50 $\mu g/m^3$ annual average) and more than 4.5 times higher for $PM_{2.5}$ (15 $\mu g/m^3$ annual average) (Clean Air Asia 2016). Figure 10.3 shows monthly average levels of PM_{10} and $PM_{2.5}$ collected by the only continuous monitoring station in Dhaka, Bangladesh, between 2002 and 2010. We see in the figure a distinct seasonal pattern, with highest concentrations during the winter months from December to February, and lowest concentrations during the summer monsoon months of June to August. A similar seasonal pattern was also found in PM_{10} concentrations in Kathmandu, Nepal (Aryal et al. 2008), while in Varanasi, India, PM_{10} and $PM_{2.5}$ were found to decrease from January to March (Kumar et al. 2015).

Although detailed seasonal data are lacking in many places, the levels of PM are very high in many cities in the region for which data exists, including Islamabad, Pakistan (Parekh et al. 2001) and Delhi and Kolkata, India (Gurjar et al. 2016). A large population in the region is exposed to air pollution that is higher than the WHO annual standard of 20 $\mu g/m^3$ for PM_{10} and 10 $\mu g/m^3$ for $PM_{2.5}$ (WHO 2005). Putero et al. (2015) reported annual PM_{10} concentration of 169 ± 113 $\mu g/m^3$ for Kathmandu during the year 2013. Kathmandu, Nepal has annual average $PM_{2.5}$ concentrations of 49 $\mu g/m^3$ in 2013, exceeding Nepal's own 24-h ambient standard of 40 $\mu g/m^3$ (http://www.who.int/phe/health_topics/outdoorair/databases/cities/en/).

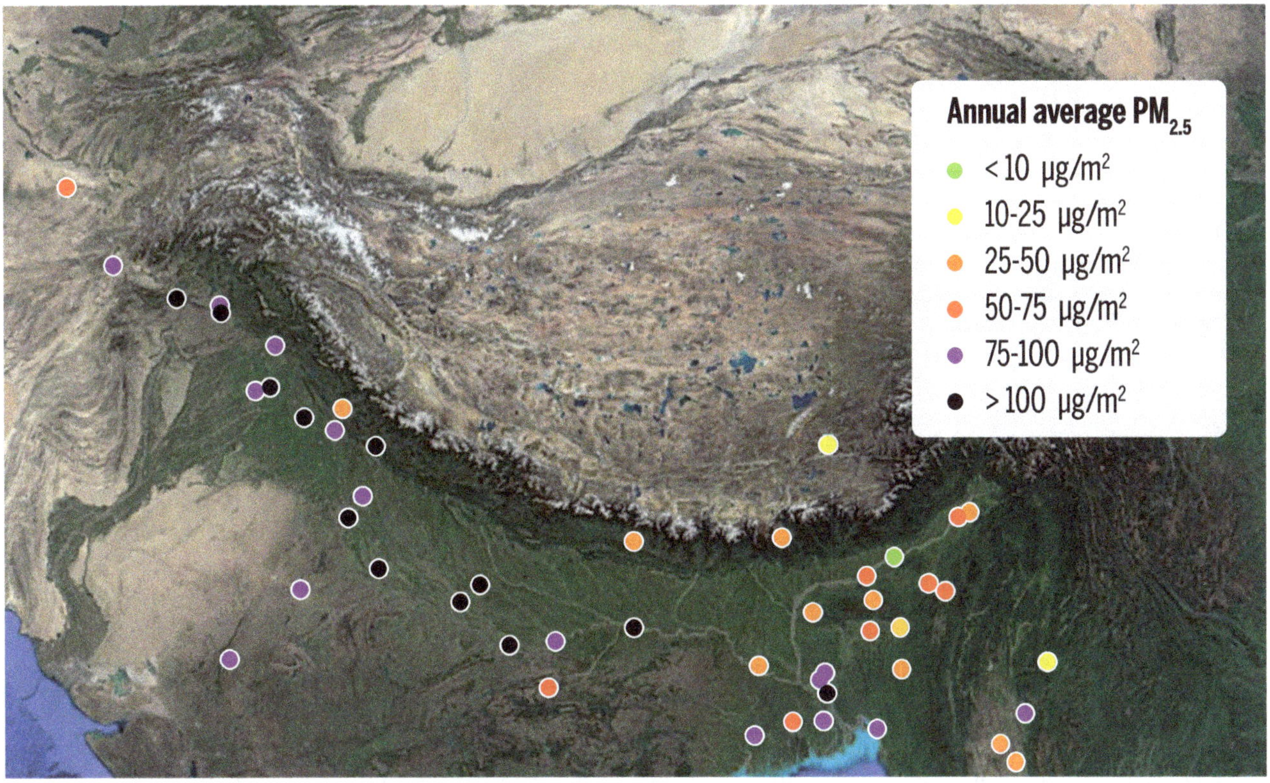

Kabul, Afghanistan	86	Sylhet, Bangladesh	61	Delhi, India	122	Luchnow, India	122	Tezpur, India	6	Namthu, Myanmar	17
Mazar-e Sharif, Afghanistan	68	Thimphu, Bhutan	43	Dibrugarh, India	30	Ludhiana, India	122	Tinsukia, India	30	Kathmandu, Nepal	49
Barisal, Bangladesh	85	Lhasa, China	24	Dimapur, India	48	Meerut, India	69	Tura, India	28	Islamabad, Pakistan	66
Chittagong, Bangladesh	95	Agra, India	105	Guwahati, India	49	Patiala, India	33	Udaipur, India	83	Lahore, Pakistan	68
Dhaka, Bangladesh	89	Aizawl, India	29	Jaipur, India	100	Patna, India	149	Varanasi, India	74	Peshawar, Pakistan	111
Gazipur, Bangladesh	88	Allahabad, India	170	Jammu, India	64	Rishikesh, India	58	Kyauk Pa Taung, Myanmar	50	Rawalpindi, Pakistan	107
Khulna, Bangladesh	61	Amritsar, India	108	Kanpur, India	115	Shillong, India	35	Pyin Oo Lwin, Myanmar	78		
Narayangonj, Bangladesh	106	Chandigarh, India	59	Kohima, India	44	Shimla, India	31	Meikhtila, Myanmar	33		
Rajshahi, Bangladesh	37	Dehradun, India	122	Kolkata, India	61	Silchar, India	48	Taungtha, Myanmar	44		

Fig. 10.1 Annual average $PM_{2.5}$ concentrations in the HKH (*Data source* WHO 2016, http://www.who.int/phe/health_topics/outdoorair/databases/cities/en/)

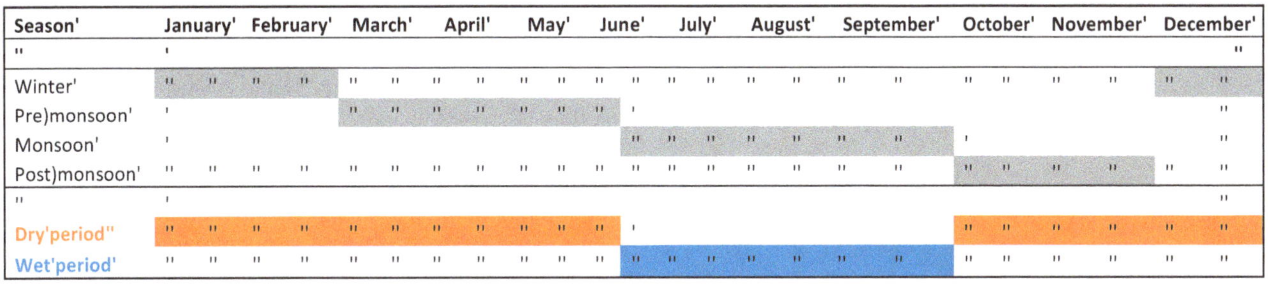

Fig. 10.2 Illustration of seasons for the HKH (*Sources* Bollasina et al. 2002; Hindman et al. 2002; Bonasoni et al. 2012)

$PM_{2.5}$ is usually dominated by smoke particles and secondary aerosols, while PM_{10} is more affected by natural soil dust (Wilson and Suh 1997). Data from both Dhaka and Delhi show a dominated by fine particles—$PM_{2.5}$ is a large fraction of PM_{10} (Gadi et al. 2000). In contrast, Karachi, Pakistan, in the arid south-western corner of the IGP, had $PM_{2.5}$ concentrations of 75 μg/m^3 but PM_{10} concentrations of 437 μg/m^3 in March–April 2009 (Shahid et al. 2016). In addition, it is not only the cities in the region that have high levels of air pollution; rural places in the IGP are also

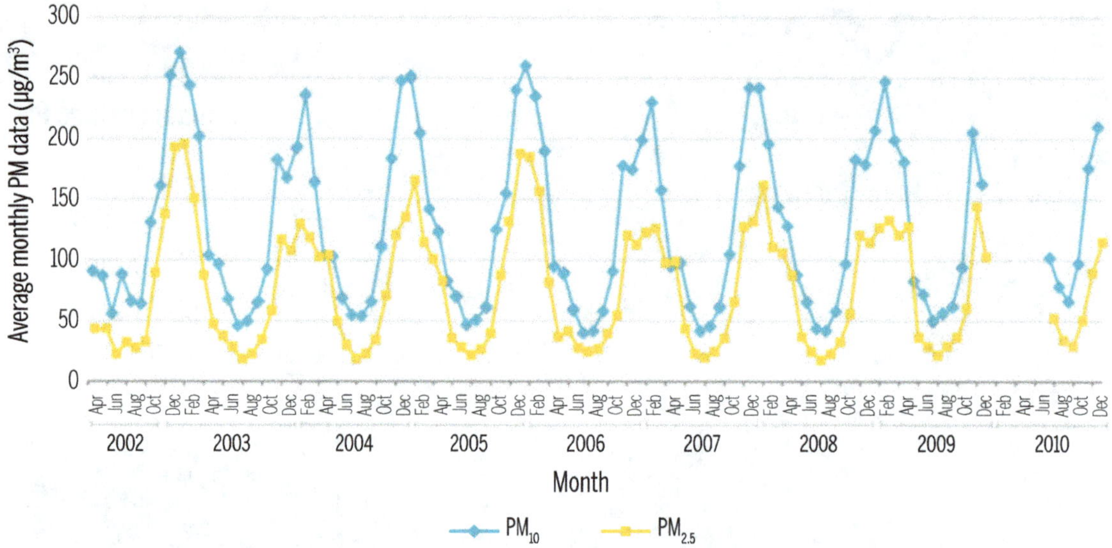

Fig. 10.3 Average monthly PM data in Dhaka, Bangladesh, 2002–2010 (*Data source* Clean Air and Sustainable Environment (CASE), case.doe.gov.bd)

affected, such as Lumbini, Nepal (Rupakheti et al. 2017). In places with a lot of dirty fuel combustion, a significant fraction of $PM_{2.5}$ is composed of BC, the light-absorbing soot that is emitted during incomplete combustion (Shindell et al. 2013).

Although few cities have long enough records to allow us to see trends, the satellite data in Fig. 10.4 shows the increasing aerosol optical depth in the region over time. In Delhi, while levels of PM and NO_x have slightly increased over the past decade, levels of sulphur dioxide (SO_2) and CO have come down as a result of converting public vehicles to compressed natural gas (CNG) (Firdaus and Ahmad 2011; Gurjar et al. 2016). A study on the health effects of the measures found improvements in the respiratory health of people who spent the most time outdoors (Foster and Kumar 2011). Lhasa, Tibet, meanwhile, has seen CO levels remain flat from 1998 to 2012, while NO_2 and SO_2 have increased substantially as a result of a growing vehicle fleet but decreasing biomass combustion (Ran et al. 2014).

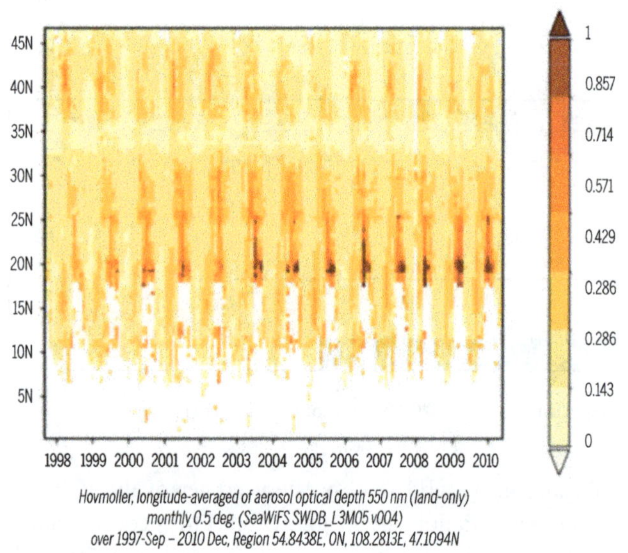

Fig. 10.4 Longitude-averaged aerosol optical depth between 54.8 and 108 E (*Data source* SeaWiFS using NASA's online Giovanni tool (https://giovanni.gsfc.nasa.gov/giovanni/))

10.1.2 Air Pollution Diurnal Cycle in the HKH

In the Kathmandu Valley, Nepal, PM_{10} and $PM_{2.5}$ have been found to exhibit morning and evening peaks that are influenced by a combination of emissions patterns and meteorology (Aryal et al. 2009; Panday and Prinn 2009; Panday et al. 2009). The pattern of morning and evening peaks is also found in CO concentrations in Kathmandu (Panday and Prinn 2009), as shown in Fig. 10.5. CO is another product of incomplete combustion. The fact that the same pattern appeared on weekends and festival days illustrates that emission patterns (such as rush hour) alone cannot explain the observed pattern, which depended on the formation and dissipation of a cold-air pool over the bowl-shaped valley (Panday et al. 2009). Similar diurnal patterns have also been found in CO and NO_x concentrations in Lhasa, Tibet (Ran et al. 2014) and in BC in Kanpur, India (Tripathi et al. 2005) and in Kharagpur, India (Beegum et al. 2009).

Figure 10.6 shows the diurnal cycle of BC during four different seasons in Pantnagar, a semi-urban site at the northern edge of the IGP in Uttarakhand, India, as well as in the city centre of Kathmandu, Nepal (Putero et al. 2015; Joshi et al. 2016). Both places exhibit clear morning and

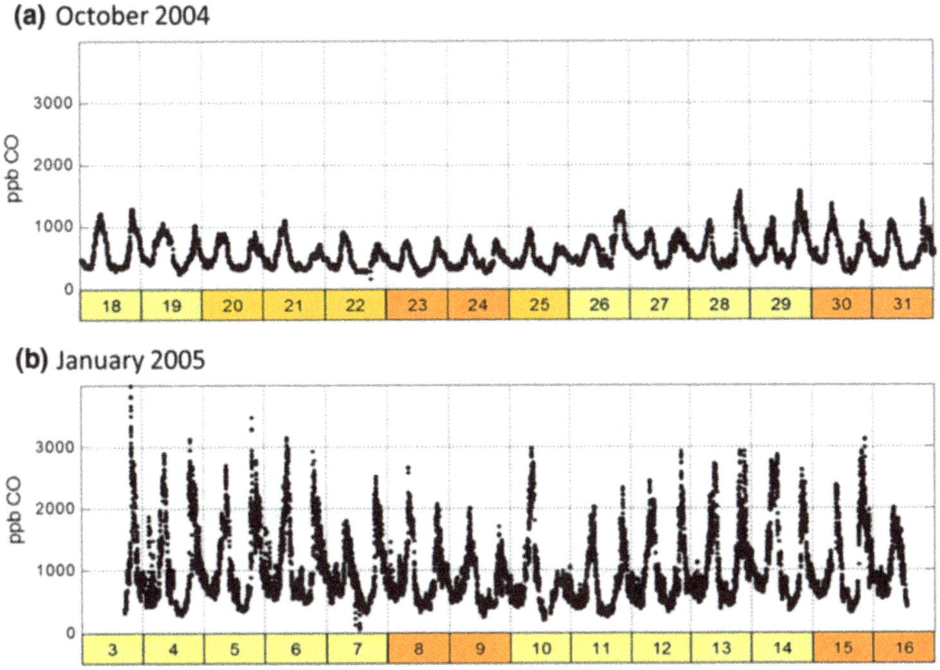

Fig. 10.5 CO in the Kathmandu Valley in October (**a**) and January (**b**) 2005. Weekends are marked in orange and festival days in dark yellow shading (Adapted from Panday and Prinn 2009)

evening peaks in BC. Rupakheti et al. (2017) reported that the evening peak pollution (BC, CO) concentration values In the Kathmandu Valley during the pre-monsoon season in the northern IGP was stronger than morning peak values. Similar patterns were also found in Kullu Valley in India (Sharma et al. 2013). The impact of pollution transport due to wind patterns in mountains and valleys was also observed in the diurnal cycle of the measured aerosol number concentration in the pre-monsoon season, with minimum in the afternoon in central Nepal (Shrestha et al. 2010).

The pollutant gas O_3, in contrast, shows a very different pattern in many cities in and near the HKH (Bonasoni et al. 2012). Tropospheric O_3 is produced by catalytic reactions of NO_x and volatile organic compounds (VOCs) under sunlight. In a number of urban areas, O_3 was found to peak during the day and to sink to low levels at night as a result of dry deposition and titration by NO (Panday and Prinn 2009; Putero et al. 2014; Ran et al. 2014). The daytime peak was a result of replenishment from higher altitudes as well as photochemical production (Sillman 1999; Panday and Prinn 2009). Such patterns have been studied in Kathmandu (Panday 2006; Pudasainee et al. 2010; Putero et al. 2014), as well as in many other cities around the world (Imhoff et al. 1995). In Chandigarh, O_3 was found to exceed India's national ambient air quality standard on all except one day in May 2012 (Sarkar et al. 2016). O_3 diurnal patterns are different on mountain tops and at high elevations, as seen in Fig. 10.7. In Nagarkot, Nepal (Panday and Prinn 2009), as well as in Lhasa, Tibet (Ran et al. 2014) and Nainital, India (Kumar et al. 2010; Sarangi et al. 2013), however, O_3 values stay high at night. That is because they have fewer local NO sources and receive O_3 from the air far off the ground. High-altitude locations also receive O_3 from stratospheric intrusions (Kumar et al. 2010; Cristofanelli et al. 2010).

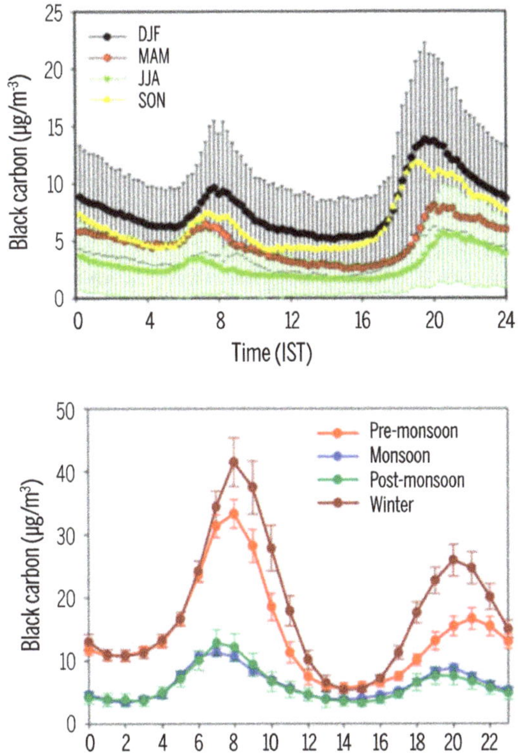

Fig. 10.6 Diurnal concentrations of BC during each season at Pantnagar, Uttarakhand, India (top; *source* Joshi et al. 2016) and at Paknajol, Kathmandu, Nepal (bottom; *source* Putero et al. 2015)

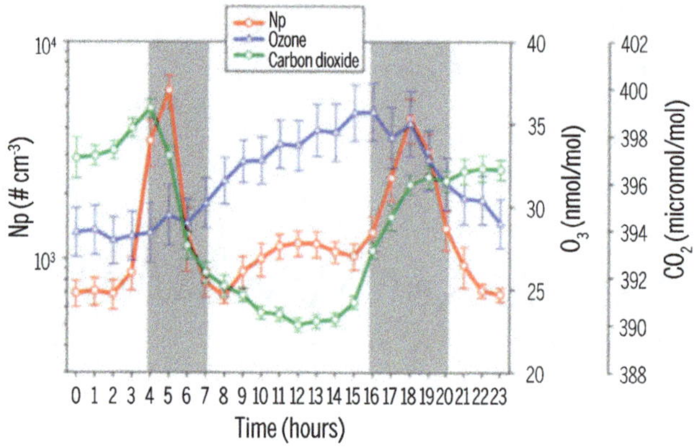

Fig. 10.7 Diurnal variations for aerosol particle number concentration (Np), surface O_3, and CO_2 at Askole in northern Pakistan (3,015 m asl) (*Source* Putero et al. 2014)

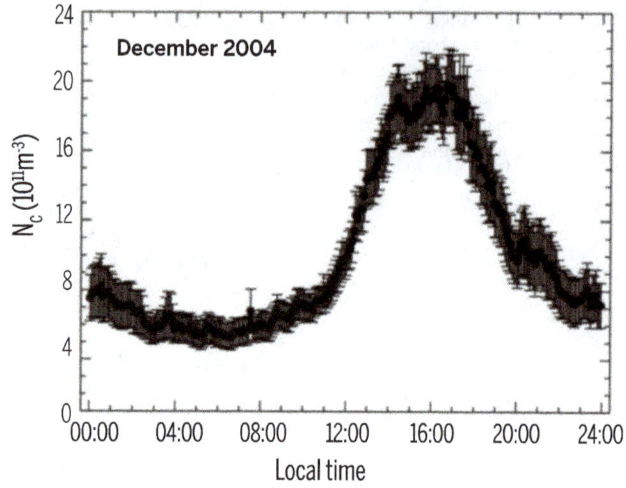

Fig. 10.8 Diurnal cycle of aerosol number concentrations at Manora Peak, Nainital, India (*Source* Pant et al. 2006a)

10.1.3 Air Pollution Observations in High Mountain Areas

While high-altitude sites in the HKH don't exhibit the same alarming levels of air pollution as lowland cities, they are by no means unaffected by the regional pollution. Despite the lack of significant local pollution sources of their own, high mountain areas of HKH receive substantial air pollution from the IGP, as seen in data from stations established along the Himalaya. On the south side of the Himalaya, comprehensive air pollution observations have been carried out at several high-altitude sites, such as Nepal Climate Observatory—Pyramid (NCO-P, 5,079 m asl) in Khumbu, Nepal (Bonasoni et al. 2010), Manora Peak (1,950 m asl) in North India (Ram et al. 2010), and Jomsom (2,900 m asl) in Nepal (Dhungel et al. 2016). To the north of the Himalaya, long-term observations of air pollution include the Qomolangma Station (QOMS) (South Tibetan Plateau, 4,276 m asl) (Cong et al. 2015), Lulang Station (Southeast Tibetan Plateau, 3,326 m asl) (Liu et al. 2013), and Nam Co (inland Tibetan Plateau, 4,730 m asl) (e.g., Wang et al. 2015; Kang et al. 2016; Yin et al. 2017). Parameters monitored at these sites include BC and PM mass concentrations, aerosol optical depth (AOD), CO and O_3 mixing ratios, as well as aerosol and precipitation compositions (OC/elemental carbon [EC], ions, elements, and organic tracers).

Figure 10.8 shows the diurnal cycle of the aerosol number concentrations at Manora Peak (1,954 m asl) near Nainital, India, overlooking the IGP, and depicts an afternoon peak in pollution as aerosols from the IGP travel upslope to reach the site over the course of the day (Pant et al. 2006a). BC in Nainital follows the same diurnal pattern (Beegum et al. 2009). This is another common pattern at mountaintop sites except during the monsoon season (Dumka et al. 2015).

Seasonal patterns are different over the Tibetan Plateau: Illustrating the unusual elevation of coarse particle loading in high mountain areas, high aerosol optical depth (AOD)-an indication of significant presence of aerosols in an atmospheric column-was even observed during monsoon season at NCO-P (median 500 nm AOD was 0.08 compared to 0.04 over winter). Such enhancement is in stark contrast to AOD over Gandhinagar, an IGP site, where AOD during monsoon was at its minimum (Gobbi et al. 2010). As illustrated in Fig. 10.9, during the pre-monsoon season aerosols over the IGP pile up against the slopes of the Himalaya at elevated altitudes. Both in the Kathmandu Valley in the middle hills, and at NCO-P near Mount Everest, O_3 concentrations peak during the pre-monsoon season and are the lowest during monsoon (Fig. 10.10).

Glacial ice cores and lake sediments from alpine areas of the HKH serve as archives for the long-term changes of air pollution (Xu et al. 2009). Kaspari et al. (2011) analysed a Mount Everest ice core spanning 1860–2000. They found that BC concentrations had increased approximately threefold by 1975–2000 compared to 1860–1975, indicating that BC from anthropogenic sources is being transported to high elevation regions of the Himalaya. The timing of the increase in BC is consistent with BC emission inventory data from South Asia and the Middle East. Furthermore, Cong et al. (2013) have reported the atmospheric BC history over the past 150 years using the lake sediments from Nam Co Lake (Tibetan Plateau). The results show that BC was relatively constant until 1900, then began to gradually increase, with a sharp rise beginning around 1960 with a 2.5-fold increase compared to the background level. Clearly, the historical records in both ice core and lake sediment in the HKH reflect an evidently increasing impact of anthropogenic emissions in recent decades, especially from 1960s. Impacts of BC on melting of HKH snow and ice is discussed in Sect. 10.4.3.

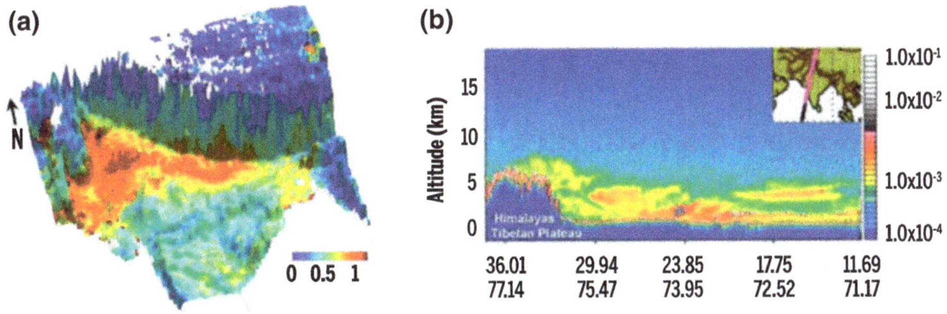

Fig. 10.9 Mean MODIS AOD during pre-monsoon season between 2003 and 2006, projected as a function of surface topography (left) and CALIPSO back-scatter profile from southern India to the Himalaya (right) (*Source* Gautam et al. 2009a)

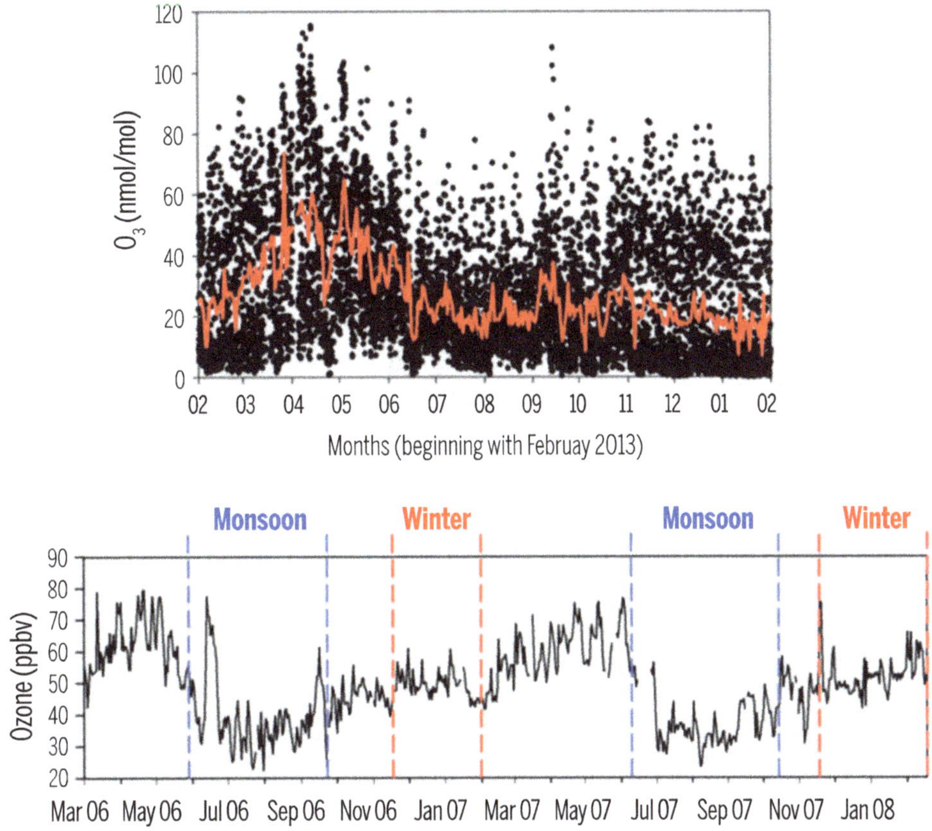

Fig. 10.10 Mean cycle of ozone mixing ratios in Kathmandu (top) and at NCO-P (bottom) (*Sources* Putero et al. 2015; Bonasoni et al. 2010)

10.1.4 Assessment of Existing Observational Networks and Gaps in the HKH for Air Quality

Observations to date have provided valuable information on air quality to assess the anthropogenic influences and evaluate the effects of radiative forcing on the cryosphere. Modellers have also used the observational data to validate model performance. Because most of the high-altitude sites were set up relatively recently (mostly less than 10 years ago), it is still difficult to determine clear long-term temporal trends. However, the archives of ice core and lake sediments provide information going back further in time.

Compared to even just a decade ago, there are many more atmospheric monitoring stations in the HKH and on the nearby plains, while several important field campaigns have increased knowledge about atmospheric processes. There are, however, still major gaps in the air quality networks: Many large cities in the IGP lack air quality monitoring stations, as do smaller urban centres and rural areas in the HKH, as well as many of the fast-growing urban centres in confined valleys within the HKH. There is also an urgent need to compare and validate the monitoring data among different sites that currently have different instruments or protocols to generate a harmonized regional dataset. In addition to ground-based observation of air pollution, measurement of the vertical variation of such pollutants is also necessary to depict the whole picture.

Box 10.2 Chinese Academy of Sciences (CAS)-coordinated monitoring network of atmospheric pollution and cryospheric changes (APCC) over the HKH and surroundings

Discrete and fragmented monitoring based on individual sites has been the source for reporting the status of atmospheric pollution and glacier changes over the HKH region, while a continuous and coordinated monitoring network remains deficient. The resulting gaps limit our understanding of the distribution and origin of atmospheric pollutants and their potential impacts on cryospheric changes over the HKH region.

In early 2013, a coordinated monitoring network was initiated to obtain continuous observational data and systematic samples from cross-sectional regions over the HKH-TP (see Fig. 10.11). The aim was to study the distribution, variation, and transport of atmospheric pollutants and to assess their impacts on cryospheric changes over the region. The network has been completed gradually, covering the large spatial scale of the HKH-TP.

To monitor atmospheric pollutants at each site, the Chinese Academy of Sciences (CAS) operates a total suspended particle (TSP < 100 μm) auto sampler, loaded with a quartz fibre filter (Φ90 mm), running at a flow rate of 100 L/min (LPM) for 24–48 h every six days. During pre-monsoon (from March to May), when the IGP haze reaches higher altitudes, sampling frequency is increased to every two to three days. After transport to the laboratory, filters are punched into small pieces for analysis of multiple parameters including elemental carbon (EC)/OC, soluble ions, elements, and other organic matters. Real time observation (e.g., BC, PM, O_3) has been operated at some superstations (e.g., QOMS, Nam Co, Southeast Tibet). Note that the terms EC and BC both refer to the most refractory and light-absorbing component of carbonaceous particles emitted from combustion. The measurement of elemental carbon (EC) is usually through thermal evolution and BC is determined by optical light-absorption.

Benchmark glaciers have been chosen for pointing observation and periodic sampling to investigate preserved atmospheric pollutants, with special focus on light-absorbing impurities (e.g., BC, OC, and mineral dust) and other relevant snow chemistry. Glacier mass balance and meteorological data are recorded for the assessment of the impacts of light-absorbing impurities on glacier melt. The sites are shown in Fig. 10.11.

Box 10.3 Observatories initiated by ICIMOD in Bhutan and Nepal

Working closely with governments and academic institutions in Bhutan and Nepal, ICIMOD has contributed to establishing a network of observatories across data-sparse regions of the two countries. At the time of writing, this network consisted of air quality stations in Thimphu, Pasakha, and Phuentsholing in Bhutan, and in Lumbini, Chitwan, Ratnapark (central Kathmandu), Pulchowk, and Dhulikhel in Nepal. In addition, a BC observatory was established on Yala Glacier in Nepal In 2016, and its twin was installed at Chele La in Bhutan in 2018. Two comprehensive ridgetop climate and environmental observatories were in preparation in Ichhyakamana, Nepal, and Gedu, Bhutan that, together with Manora Peak in India, form a network of three ridgetop observatories overlooking the IGP.

10.2 Worsening Haze and Winter Fog in and Near the HKH

Before the 1950s, Europe and North America were the largest sources of air pollutant emissions, and they had the haziest skies with significant adverse health impacts (Bell and Davis 2001; Helfland et al. 2001). Since then, the tropics and eastern Asia have dominated (Ramanathan and Carmichael 2008). Two regions in close proximity to the HKH experience particularly heavy haze: eastern China and the IGP (Ramanathan and Carmichael 2008). The IGP stretches out south and west of the Himalaya and is a big source of pollution reaching the high mountains. It is an extensive stretch of highly fertile agricultural plains covering an area of 700,000 km^2. It is also among the world's most densely populated regions, inhabited by over 900 million people, which is ~1/7th of the world's population. Over the past several decades, the IGP has experienced increasing aerosol haze and winter fog. Data going back to 1982 show haze increasing across much of northern India (Sarkar et al. 2006). This section discusses the worsening haze and winter fog in the region.

In recent years, many studies have analysed regional haze in the HKH, using remote sensing data as described below. More attention is also being paid to the recent winter fog occurrences in the region. While a lot of the necessary

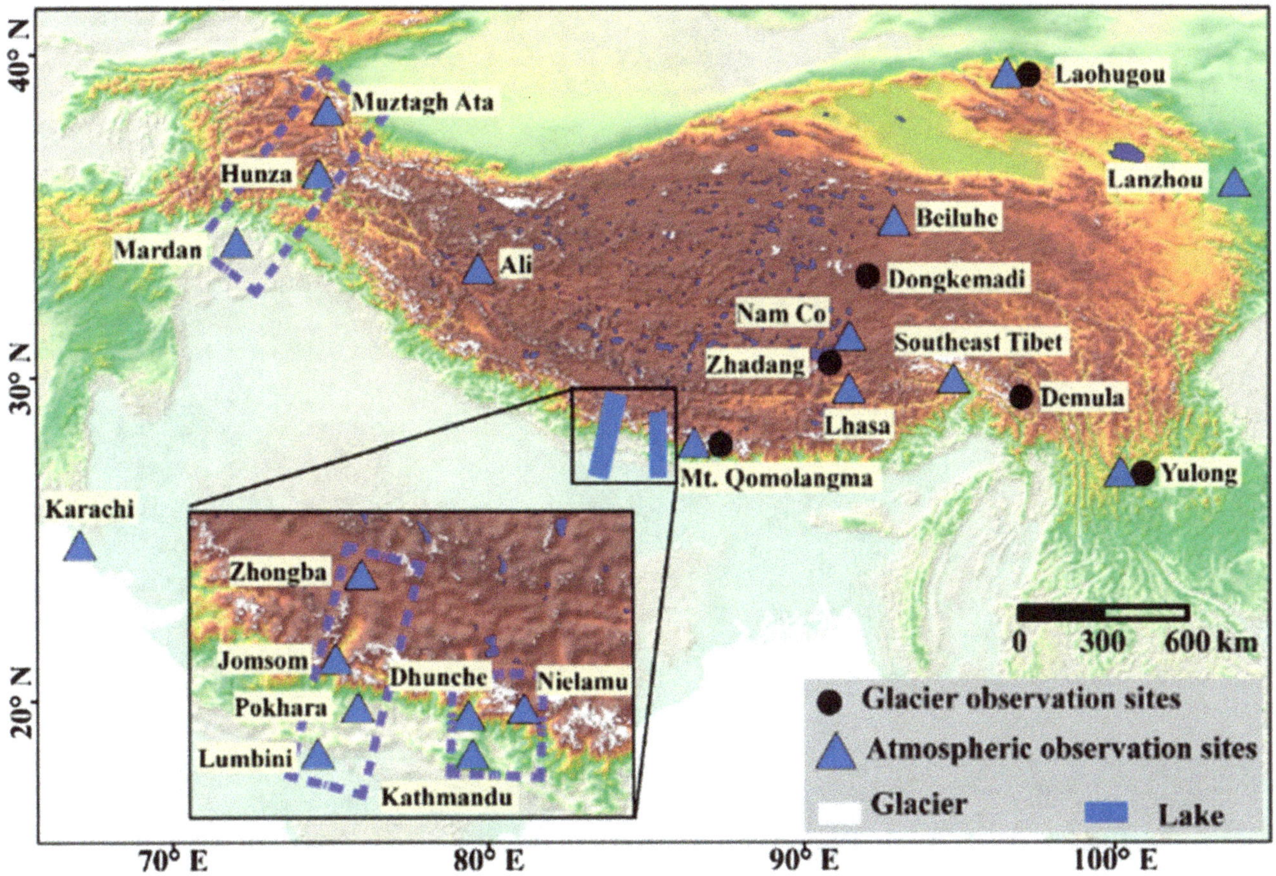

Fig. 10.11 The spatial distribution of observation sites in the APCC network

measurement data are still lacking, the existing data provide some important observations:

- As air pollution worsens in many cities in the HKH and across the IGP (WHO 2016), increased regional haze and winter fog episodes, as well as increased O_3, have been observed in the HKH.
- Many adverse impacts (e.g., visibility, temperature, and health) are associated with regional haze and winter fog.

10.2.1 Regional Haze Pollution in the HKH

During the past decade and a half, the aerosol haze that covers much of the IGP during the dry season has been studied extensively. Measurements of column-integrated aerosol loading retrieved from satellite measurements show the annual build-up of haze over the IGP (Di Girolamo et al. 2004; Jethva et al. 2005; Gautam et al. 2007). Figure 10.12 shows satellite imagery from Terra MODIS, taken on 1 December 2010, with the pollution haze extending across parts of Pakistan, India, Nepal, Bhutan, and Bangladesh.

The build-up of the haze, which is sometimes referred to as atmospheric brown cloud (ABC), starts in the post-monsoon season when, after the rice harvest, paddy residue is burned across large parts of the region. It then intensifies in December and January (Dey and Di Girolamo 2010). Winter temperatures are sufficiently cold within and surrounding the HKH to allow frequent temperature inversion episodes—atmospheric conditions in which a layer of warm air traps a layer of cool air below it close to the ground. This condition suppresses the buoyant vertical transport of pollutants that would otherwise have dispersed over a wider area. Temperature inversion episodes result in the trapping of pollutants within the shallow boundary layer with low wind speeds (Tare et al. 2006). As a result, an optically thick haze layer builds up (as often seen in satellite imagery), with pollution dominated by anthropogenic emissions from vehicles, industries, and open biomass burning, as well as biofuel burning for residential heating and cooking. Elevated concentrations of carbonaceous aerosols (BC and OC), sulphate, and nitrate have been reported over several locations in the IGP (Ganguly et al. 2006). The suppressed vertical mixing in winter leads to plumes of haze getting blown far out over the Bay of

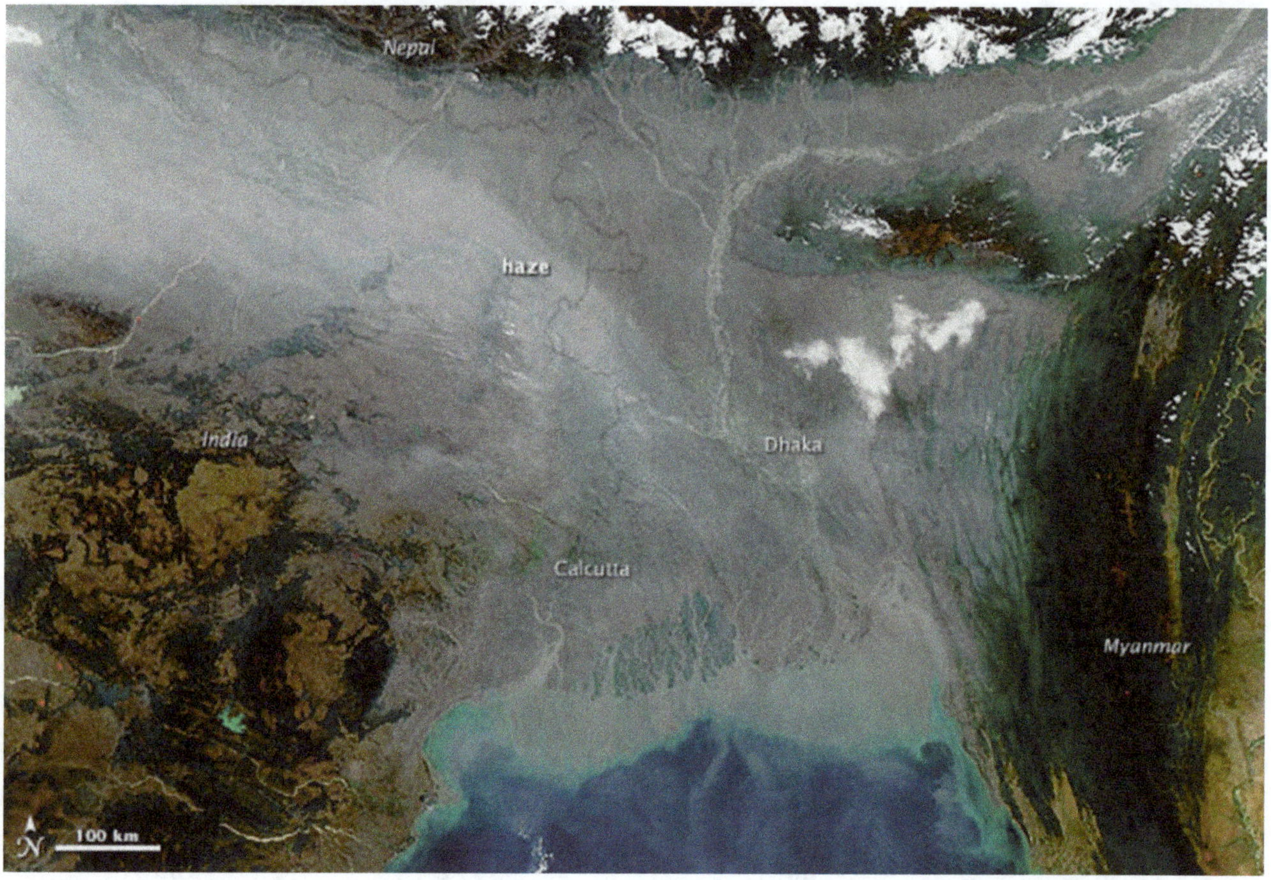

Fig. 10.12 Winter haze blanketing the entire IGP, including regions of Pakistan, India, Nepal, and Bangladesh, south of the Himalaya (right) (*Source* NASA MODIS imagery)

Bengal by low-level westerly winds (Dey and Di Girolamo 2010).

The haze becomes denser in the pre-monsoon months, when there are additional contributions from wind-blown desert dust (Dey and Di Girolamo 2010), most of which originates in the Thar Desert (Gautam et al. 2009b), as well as from extensive agricultural and forest fires. In Delhi, both the total aerosol loading and the proportion due to dust have been found to increase over the course of the pre-monsoon from March to June (Pandithurai et al. 2008). Throughout the dry season, biomass-based cooking fires and increasing amounts of emissions from transport and industries also contribute to the haze. Vertical profile measurements during the pre-monsoon have found high levels of BC up to 4.5 km altitude; layers of enhanced BC have been observed as high as 8 km above sea level (Babu et al. 2011a). Even during the rainy season, significant aerosol build-ups happen during monsoon breaks (Dey and Di Girolamo 2010); however, decreases of 55–70% in season-average aerosol concentrations have been found in India during the monsoon compared to the pre-monsoon (Hyvärinen et al. 2011a). While at a mountain location the combination of cloud and fog droplet activation and rain scavenging was found to remove particles across all sizes; at an urban site near Delhi, where only rain scavenging took place, the finest accumulation mode particles were not removed (Hyvärinen et al. 2011b).

10.2.2 State of Current Knowledge About Haze in the HKH

The Indian Ocean Experiment (INDOEX) field campaign in 1999 discovered the broad extent of regional haze over the Indian Ocean, and in subsequent years there was a realization that the haze was also very widespread over the IGP (UNEP 2002). This realization was further cemented with the advent of NASA's satellite sensors, MODIS and MISR (Mehta 2015), which provided daily views from space (Di Girolamo et al. 2004). In the almost two decades since then, hundreds of papers have been published on various aspects of the haze over India, and a smaller number on the haze over Bangladesh, Nepal and Pakistan. As a result, we have a fairly good understanding of its temporal variations and horizontal extent, chemical composition, and physical and

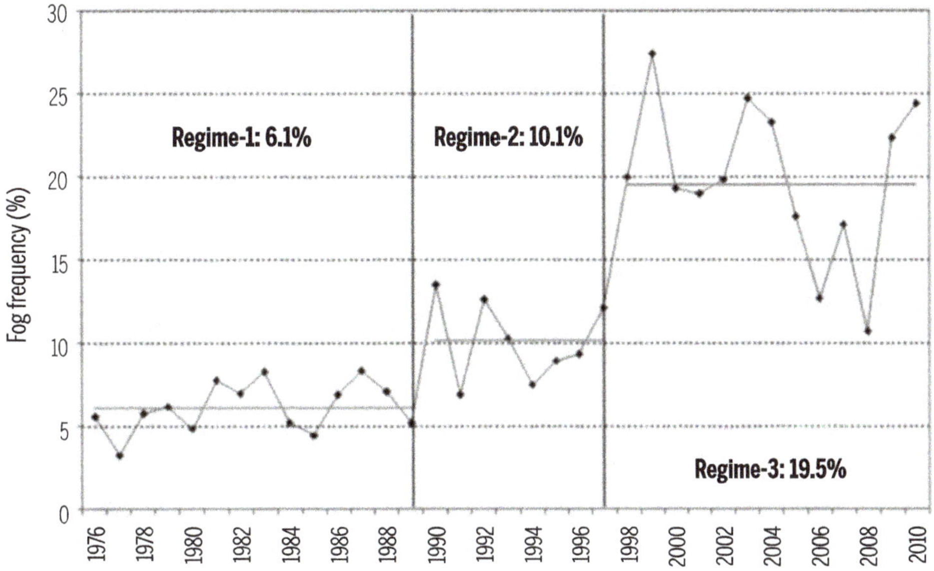

Fig. 10.13 Fog frequency from 82 stations over India and Pakistan from 1976 to 2010 (Syed et al. 2012). The regimes are defined using a mean fog fraction during the time period. Regime-1: 6.1%, Regime-2: 10.1%, Regime-3: 19.5%

optical properties, as well as its radiative impacts. We have some hints, but not yet a good understanding of its impacts on gas-phase photochemistry and its role in changing monsoon precipitation patterns. There is still very little understanding among the general population and policy makers about the causes and costs of the haze.

While the increase in aerosol haze across the IGP has received a lot of attention, there have been far fewer studies about the increase in O_3, which already exceeds air quality standards during the spring in many places in the IGP (Lal et al. 2000; Mittal et al. 2007; Ojha et al. 2012; Sinha et al. 2014). There are only a few papers from a handful of higher-altitude locations including Kullu (1,154 m) (Sharma et al. 2013), Nainital (1,954 m) (Kumar et al. 2010), and NCO-P (5,079 m) (Cristofanelli et al. 2010), as well as several places along the northern side of Mount Everest (Semple and Moore 2008) and Nam Co, Tibetan Plateau (Yin et al. 2017). They all found relatively high levels of O_3 in the spring, with higher altitudes also affected occasionally by stratospheric intrusions.

10.2.3 Increasing Winter Fog over the IGP

Often coinciding with the winter haze is the widespread and persistent occurrence of severe winter fog episodes along the IGP. Note that while haze is commonly defined as an atmospheric condition with reduced visibility, but with visibility remaining above 1 km, fog is defined to have visibility less than 1 km. The majority of the fog episodes last from mid-December into January and result in poor visibility conditions that cause major disruptions to the air and rail transportation sectors, as well as a significant number of deaths in vehicular accidents (Hameed et al. 2000; Ghude et al. 2017). The fraction of days with fog has increased threefold in a 35-year period across the IGP (Ghude et al. 2017; Jenamani 2007; Syed et al. 2012), and tenfold in five decades in Delhi (Mohan and Payra 2014). Figure 10.13 shows that increase has not been linear, but with distinct "regime changes" (Syed et al. 2012). Within each regime, and especially within the third regime, there have been large interannual variations, with a low in 2007–08 of just above 10% and highs exceeding 27%.

Fog conditions usually also vary within each day: At Delhi airport, fog is denser at night than during the day (Jenamani and Tyagi 2011); it appears to be the case across the IGP that visibility improves over the course of the afternoon. MODIS images also show distinct day-to-day variations, with growing, shrinking, and lateral shifting of the IGP fog deck from one day to the next, as shown in Fig. 10.14.

To date there is still a lack of scientific consensus to explain why persistent winter fog has increased across the IGP. The key ingredients needed for fog to form anywhere in the world are well known: Sufficient moisture, airborne surfaces that moisture can condense onto (cloud condensation nuclei, or CCN), and a drop in temperature that allows water vapour to condense (Gultepe et al. 2007). Most of the fog over northern India, and presumably the rest of the IGP, is radiation fog (Syed et al. 2012). This occurs when moist near-surface air cools under clear skies in stagnant, stable air. The available literature points to the role played by synoptic-scale processes in creating conditions that allow the formation of large-scale radiation fog across the IGP within a short time period. The fog not only forms simultaneously over large regions, but inter-annual variations in fog are also coupled across wide areas (Syed et al. 2012).

For a regional-scale fog episode to occur, a high dew point temperature is needed on a regional scale (Sawaisarje

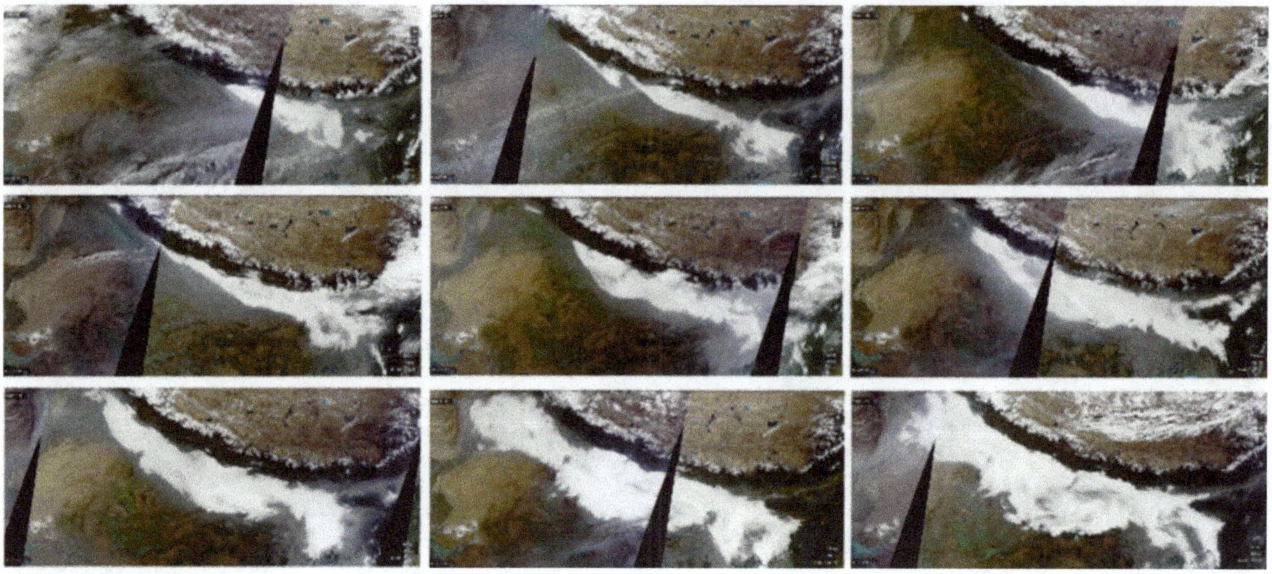

Fig. 10.14 Terra MODIS images of the IGP on 17, 18, 19 (top row), 20, 21, 22 (middle row), and 23, 24, 25 (bottom row) December 2012

et al. 2014)—in other words, a large amount of water vapour needs to be present over a broad region. During the winter season, much of the IGP is subjected to a regional characteristic weather phenomenon known as western disturbances, a series of alternate low- and high-pressure systems. First, a low-pressure system results in enhanced moisture content in the boundary layer; this is subsequently replaced by a high-pressure system with low winds, radiative cooling of the ground, and temperature inversions (Pasricha et al. 2003). The resulting anomalous cold conditions are generally referred to as *sheet lahar* (cold waves) in many parts of the IGP. In addition to western disturbances, it is also possible that the moisture needed for fog in some parts of the IGP could originate in the agricultural fields of Uttar Pradesh, Haryana, and Punjab, where large areas are irrigated in winter (Jenamani 2007).

Moisture and cold temperatures alone are insufficient to form fog if there are no particle surfaces for the fog droplets to condense onto. Los Angeles has seen a decrease in fog as it cleaned up its air pollution problem and as its ambient aerosol concentrations dropped (LaDochy 2013). Cities in many other parts of the world have seen the reverse: They have denser and more persistent fog due to increases in air pollution and thus in the availability of condensation nuclei, resulting in larger numbers of smaller drops with higher optical depth (Syed et al. 2012). Central and eastern China have seen a doubling of fog frequency over the past three decades, as aerosol concentrations have increased (Niu et al. 2010).

Across the IGP the increase in fog appears to be associated with the increasing trends of anthropogenic air pollution (Habib et al. 2006; Sarkar et al. 2006). In Delhi, the onset of fog has been found to match with times of high aerosol number concentration (Mohan and Payra 2014), as well as with high concentrations of CO (Mohan and Payra 2009), a tracer of incomplete combustion. A field campaign at Delhi airport in winter 2015–16 found that fog occurred in a highly polluted environment dominated by combustion and vehicular emissions (Ghude et al. 2017). During winter, aerosols over the IGP are dominated by emissions from biomass burning and agricultural waste burning (Gustafsson et al. 2009; Ram et al. 2012). In Kanpur, a large fraction of measured organic aerosols were found to be from biomass burning. The composition of organic aerosols was different during foggy events, with higher oxygen-to-carbon ratios during foggy events (Chakraborty et al. 2015). Burning of post-harvest paddy residue produces carcinogenic benzenoids that can partition into aerosol phase and oxidize into hydrophilic phenolics and cresols that have increased water vapour uptake and contribute to fog formation (Sarkar et al. 2013).

Analysis of filter samples collected in Hissar and Allahabad, India, found that secondary inorganic aerosol constituents (ammonia, nitrate, sulphate) increased by a factor of two to three during foggy or hazy days compared to clear days (Ram et al. 2012); unfortunately, the study did not distinguish between foggy and hazy days. Acidic gases and ammonia are taken up by fog, as found in a study in Lahore in winter 2005–06 (Biswas et al. 2008). Fog water in Delhi has been found to have a higher concentration of ions than rainwater (Ali et al. 2004).

10.2.4 Causes and Impacts of Haze and Winter Fog

The persistent and dense winter haze and fog reduce the solar insolation available to warm the land surface, which can act to further lower surface temperatures, in turn providing a positive feedback to the persistence of foggy and cold conditions (Gautam 2014). Trends in poor visibility days due to haze and fog during the winter season have been significantly increasing over the IGP, reaching 90% of winter days (De and Dandekar 2001), with serious impacts on air, rail, road, and water-based transportation systems, as well as on agriculture and tourism.

The IGP haze has been found to have increased in the winter and post-monsoon season over the time period from 2001 to 2013 (Mehta 2015), and along with it, global short-wave radiation has been found to have decreased in many cities in India since 1990 (Porch et al. 2007). In China, the increase in haze has been associated with the simultaneous decrease in both cloud cover and solar radiation at the surface (Qian et al. 2006, 2007).

The aerosol haze over the IGP has significant impacts on local and regional climate. It reduces the solar radiation reaching the surface and warms the atmosphere due to absorption of sunlight by BC (Ramanathan and Ramana 2005; Tripathi et al. 2005; Ramanathan and Carmichael 2008). Fifty to 70% of the warming by absorbing aerosols over the IGP takes place above clouds (Satheesh et al. 2008). Based on the observational data, the solar-absorption efficiency has positive correlation with the ratio of absorbing BC to scattering sulphate aerosols (Ramana et al. 2010). While BC from biomass-burning cookstoves was found to directly contribute to regional climate forcing over the IGP (Praveen et al. 2012), fossil fuel-dominated BC plumes have been found to be twice as efficient in their effect on net warming of the atmosphere compared to biomass-burning-dominated plumes (Ramana et al. 2010). A study using multiple sun photometers around Kanpur, India, found that emissions from Kanpur city contributed only 10–15% of the aerosol optical depth observed in Kanpur (Giles et al. 2011), implying a regional origin of the observed haze. Although current state-of-the-art chemical transport models are able to capture seasonality of the pollution haze over low-lying plains, they are still limited in their ability to capture observed seasonality over valleys and mountains (Adhikary et al. 2007). Current mesoscale numerical weather models are not able to adequately predict persistent winter fog over IGP and the foothills area. Thus, more in situ observations are needed in the region to understand the phenomenon and to constrain numerical weather and chemistry models.

Although there seems to be a consensus in the scientific community that winter haze and fog are a strongly coupled system, there is a lack of in-depth and quantitative understanding about the impact of air pollution on fog variability and long-term trends. At a fundamental level, aerosol particles (especially hydrophilic aerosols) act as efficient cloud condensation nuclei, favouring the formation of fog. However, given the complex aerosol composition over the IGP from a wide range of emission sources, the microphysical, chemical, and radiative interactions between aerosols and fog or low-level stratus clouds requires more detailed studies. In this respect, recent ongoing field studies have focused on the winter fog characterization in the IGP led by the Indian Institute of Tropical Meteorology with detailed characterization of fog in Delhi, the Winter Fog Experiment (Ghude et al. 2017), as well as multi-institutional efforts led by ICIMOD at several locations across the IGP. In addition, there are other factors which may be linked to the long-term variability/trends in fog occurrence, such as changes in land use/land cover patterns, increase in the density of irrigation canals, and regional meteorological changes over the IGP.

10.3 Emission Sources and Transport Processes in the HKH

Major emission sources of air pollutants within and upwind of the HKH include cooking and heating fires, open burning of agricultural residue and garbage, forest fires, motor vehicles, and thermal power generation (including with diesel generator sets), as well as brick kilns and other industries. While most of the region's inhabitants still live in homes severely polluted by emissions from traditional biomass cookstoves, outdoor air quality has deteriorated in many urban and rural areas, on mountains, in valleys, and in the IGP. Source regions affecting the HKH include urban and rural places within the HKH and the nearby IGP; smaller amounts of pollution arrive from further away.

To address air pollution in the HKH, it is essential to understand not only which harmful pollutants exist at what concentrations, but also where they come from and how they reach the region. An increasing number of national and regional emissions inventories have been set up that help us assess the magnitude and the source of air pollutant emissions in the HKH countries. In addition, studies using transport and receptor models have provided estimates of where these emissions come from. While more research is essential, they provide us with the basis for two general observations:

- Certain sources—such as cooking and heating, agricultural residue burning, forest fires, brick kilns and diesel pumps—are particularly important for the HKH and the IGP.
- We find a large seasonality in emissions for some of these sources. Agricultural residue burning and brick kilns are especially important seasonal emissions sources, with both taking place in the dry season.

10.3.1 The Major Air Pollutant Emissions Source Sectors in the HKH

Different pollutants may be emitted by the same source, while many pollutants are chemically transformed in the atmosphere, forming secondary pollutants, such as secondary aerosols (both organic and inorganic, such as nitrates and sulphates) and tropospheric O_3 (Fig. 10.15). Emission sources also vary in different countries, as well as the magnitude, depending on factors such as the source of power, industrial activities, and domestic heating/cooking needs. Part of the reason for this is the increase in population density in the HKH, particularly in the lower altitude regions. We illustrate the magnitude and the contribution of the five emission sources (power, industry, transport, domestic, and other) for the five primary pollutants (CO, NO_x, SO_2, BC, and $PM_{2.5}$) estimated in Regional Emissions inventory in ASia (REAS) v2 for each of the HKH countries in Fig. 10.16.

10.3.1.1 Cooking and Heating

Domestic cooking and heating, mainly using biofuels (e.g., wood and dung) is the largest source of CO and PM in the HKH (Venkataraman et al. 2005). Cookstove emissions are also the largest source of BC in South Asia (Gustafsson et al. 2009). This is due to low access to clean fuels and high usage of solid biofuel such as dung, charcoal, and fuelwood in the region, as illustrated in Table 6.1. There are few studies measuring emissions in the field (e.g., Roden et al. 2009; Stockwell et al. 2016) and some studies have attempted to estimate the contributions of indoor air pollution on outdoor air pollution in India (Rehman et al. 2011) and China (Liu et al. 2016). However, more research is essential to not only understand the magnitude of various emissions coming from the residential sector, but also assess the fraction remaining indoors versus the amount leaving the homes to contribute to ambient air pollution.

10.3.1.2 Agricultural Residue Burning

Agricultural residue burning takes place in two distinct forms, as illustrated in Fig. 10.17. The first is the burning of small piles of collected agricultural residue. This may take place on the field, or near the homes, and the purpose might be to get rid of waste material or to generate smoke to chase away mosquitoes and flies from farm animals. Such fires are lit millions of times a day across the IGP and the hilly and mountainous parts of the HKH, in both rural and peri-urban areas. The second is the open-field burning of rice and wheat stubbles in the IGP after mechanical harvest using combine harvesters. The purpose is to clear the fields quickly for the planting of the next crop.

Agricultural residue burning emits OC and BC aerosols, as well as a variety of gaseous air pollutants, including CO,

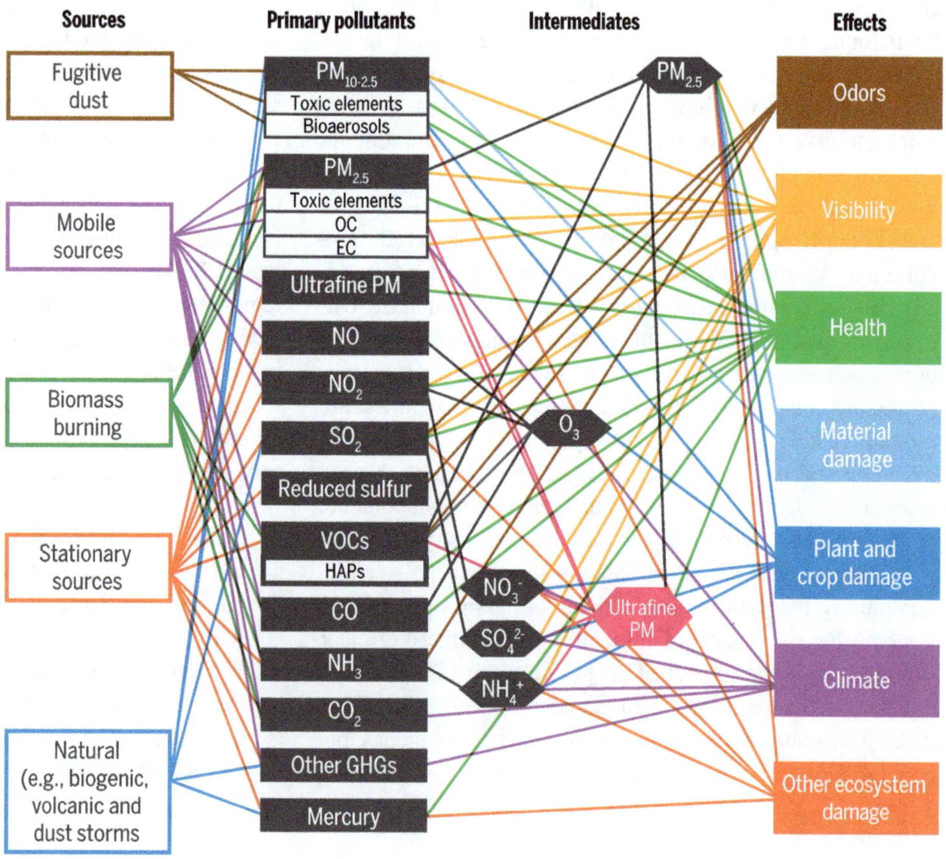

Fig. 10.15 Multiple pollutants and their multiple adverse effects (Cao et al. 2013 adapted by Clean Air Asia 2016)

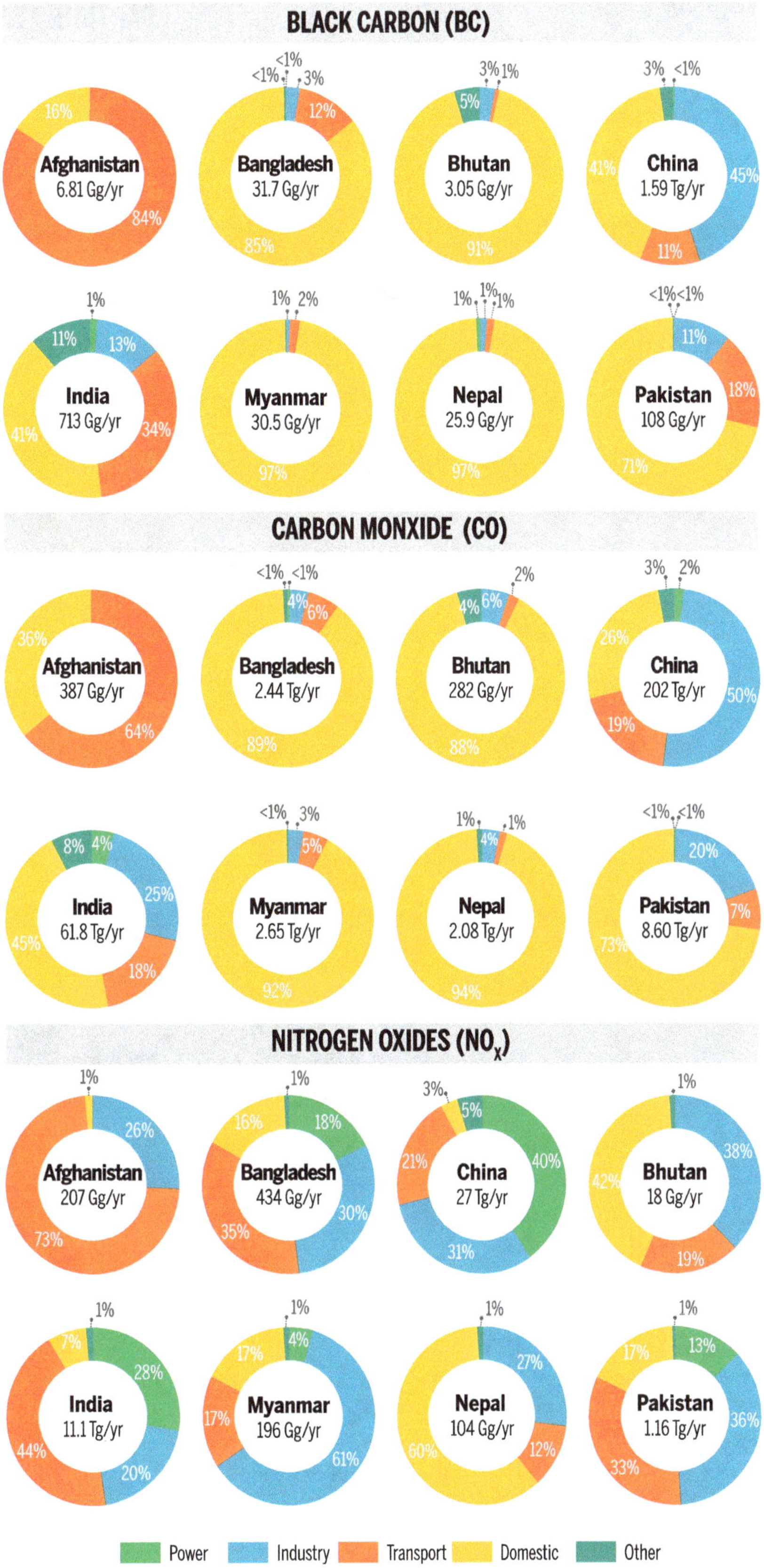

Fig. 10.16 Magnitudes and contribution of five sources (power, industry, transport, domestic, and other) for five pollutants (CO, NO_x, SO_2, $PM_{2.5}$, and BC) for each of the HKH countries estimated in the REAS v2 emissions inventory (Created based on data from Kurokawa et al. 2013)

Fig. 10.16 (continued)

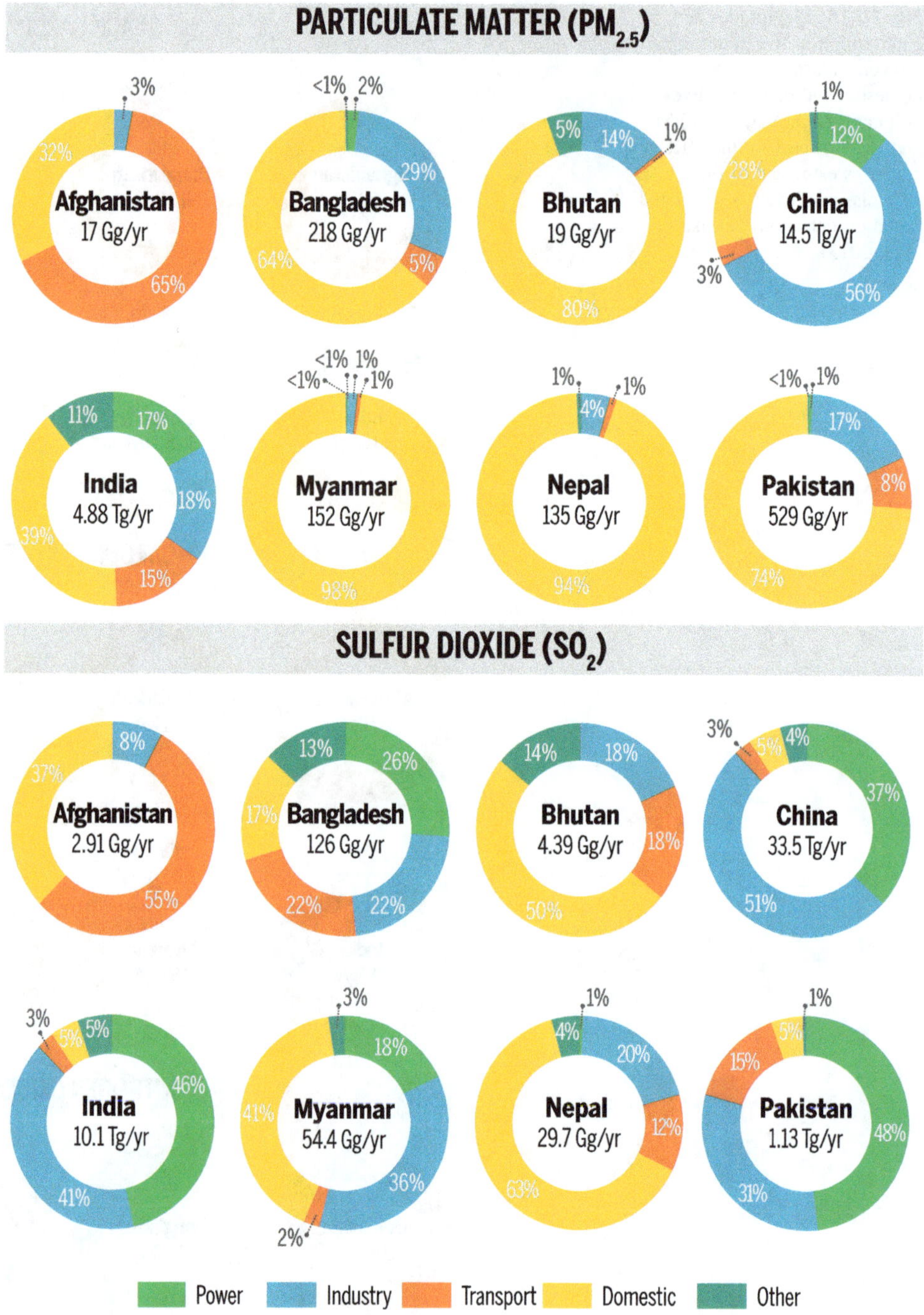

CO_2, CH_4, NO_x, and non-methane VOCs (NMVOCs) (Andreae and Merlet 2001; Guyon et al. 2005; Sharma et al. 2010). Jain et al. (2014) estimated that 21.32 Mt/year of residue is burned in Punjab alone, leading to 1961 Gg/year of CO, 53 Gg/year of NO_x, 335 Gg/year of NMVOCs, 83 Gg/year of $PM_{2.5}$, and 15 Gg/year of BC in the year 2008–09. The Nepal Ambient Monitoring and Source Testing Experiment (NAMaSTE) campaign (see Box 10.4) also found that agricultural residue burning is a large contributor to brown carbon (BrC) and SO_2 (Stockwell et al. 2016). These emissions contribute to the haze (ABC) and to the worsening of winter fog in the HKH. The large agricultural fires detected by MODIS satellite sensors illustrate that burning is most prevalent in Bangladesh, Bhutan, China, India, Nepal, and Pakistan. MODIS is unable to capture the small and short-lived fires outside the passing times, although they are also important sources of air pollution.

Two distinct periods of post-harvest agricultural residue burning are prevalent in the IGP—the burning of paddies between October and November and the burning of wheat residue between April and May. $PM_{2.5}$ mass concentrations of 60–390 $\mu g/m^3$ have been observed during paddy residue

Fig. 10.17 Burning of small piles of agricultural residue (left and centre) versus open-field burning (right). Photos by Arnico Panday

burning, with high OC contribution of more than 30%, in Patiala, Punjab, India. Approximately 250 Gg/year of OC and 60 Gg/year of BC emissions are estimated to be due to agricultural residue burning in the IGP (Rajput et al. 2014). Based on the ratio of the tracers levoglucosan and OC, it was found that in the post-monsoon season, the OC derived from biomass burning contributed a large fraction of the total aerosol OC ($\sim$40%) in Lumbini, Nepal, reconfirming the regional influence of agricultural residue burning (Wan et al. 2017).

10.3.1.3 Garbage Burning

The open burning of waste contributes substantially to local and regional air pollution, and it is a common way to dispose of garbage in South Asia (Asian Productivity Organization 2007). Waste management is considered one of the most important issues in urban areas in the HKH. Indeed, public opinion shows that 59% of urban residents consider solid waste to be the most important problem in Nepal. In the same survey, only 7% listed air pollution as the most important problem; garbage burning is also one of the most poorly characterized emissions sources even in developed countries (USEPA 2006; Stockwell et al. 2016). Garbage burning in the HKH is ubiquitous on a range of scales, and mixed garbage items, including paper products, plastics, and rubber tyres, are often burned together.

Garbage burning is a large contributor to BC, OC, PM, and CO emissions. Wiedinmyer et al. (2014) estimated emissions of various trace gases and PM from the open burning of waste globally and they show a large hotspot of CO emissions within and south of the HKH. Uncertainty is still large in these emissions estimates and more effort to validate and improve these emissions is essential. For example, the NAMaSTE campaign results estimated benzene emissions due to Nepal trash burning to be 1.68 Gg/year (Stockwell et al. 2016), compared to 0.580 Gg/year estimated by Wiedinmyer et al. (2014) using a different methodology.

10.3.1.4 Forest and Scrub Fires

During the dry season, especially from March to May, many forest and scrub fires occur in the southern HKH foothills of India, Nepal, and Bhutan (Rupakheti et al. 2017). Some of these can take place at high altitudes, close to the cryosphere. According to the MODIS fire spot product, an average of 3,908 fire counts per year (2005–10) were recorded in this region. The average burnt areas were estimated to be 1,129 km^2, with BC emissions of 431 t/year (Vadrevu et al. 2012). A vertical profile of aerosols revealed by CALIPSO, as well as the example shown in Fig. 10.18a, suggests smoke plumes can reach altitudes of 4,000–5,000 m, which is much beyond the typical planetary boundary layer height in the region; this gives them a higher chance to undergo long-range transport for dispersion, but also brings them into contact with the lower reaches of the Himalaya cryosphere.

The MODIS fire spot product may still underestimate fires in the HKH due to several reasons. First, it cannot see fires taking place under clouds or heavy haze. Second, MODIS only takes instantaneous snapshots four times a day. Often, the fires take place in the early evenings when there is no MODIS overpass. Fires on steep slopes travel rapidly upwards until they reach the ridgeline or some other barrier, and extinguish themselves when they run out of fuel in their path. The fire photographed in the early evening in Fig. 10.18b did not appear in MODIS, nor did 22 large fires observed by an author at 19:00 on 14 April 2016 from a mountaintop in Gorkha, Nepal.

10.3.1.5 Brick Kilns and Industries

The fired clay brick is one of the most widely used construction materials in the HKH and the IGP, and most of the brick kilns operate only during the dry season. While data on emissions from brick kilns is still lacking, there are estimates of almost 100,000 brick kilns across the IGP. In the Indian states of Uttar Pradesh and Bihar, almost 23,000 brick kilns are known to be in operation. There are no brick kilns in Bhutan and just over 100 and 1,000 brick kilns within the Kathmandu Valley and in Nepal, respectively. Indeed, 75% of global brick production is considered to be concentrated in four countries—China (54%), India (11%), Pakistan (8%), and Bangladesh (4%)—all part of the HKH.

The recent NAMaSTE campaign indicated that brick kilns are a significant source of SO_2, NO_x, BC, OC, and PM, and that the type of kilns, as well as the primary fuels

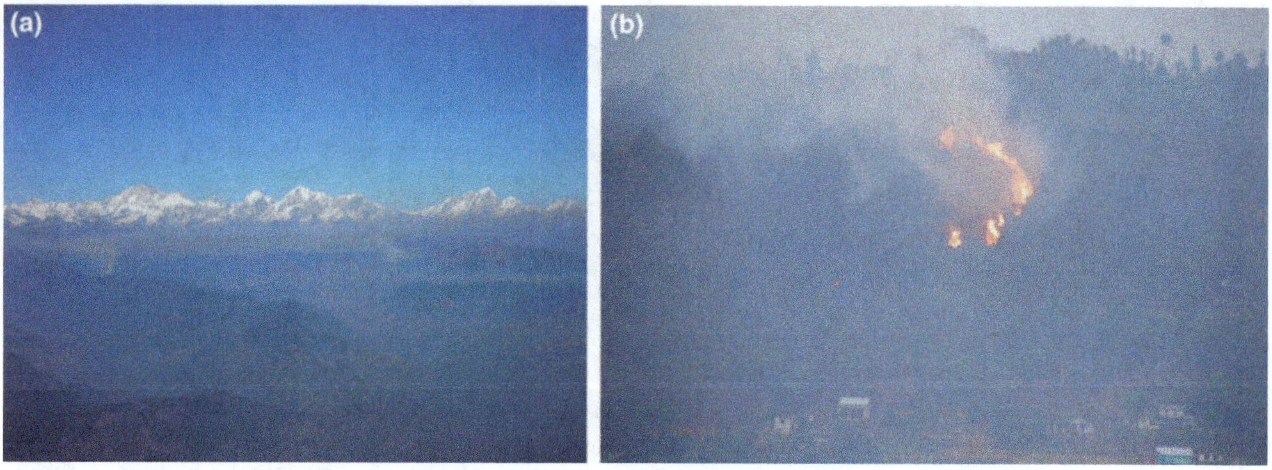

Fig. 10.18 Forest fires in the HKH at high altitude (**a**) and on the outskirts of Kathmandu (**b**). Photos by Arnico Panday

burned, matters in the resulting emissions (Stockwell et al. 2016). Brick production levels are not well understood and more data need to be collected. What seems to be the case is that the number of brick kilns is growing rapidly in many places within the HKH with increasing urban populations (Bisht and Neupane 2015), other than Bhutan, which has no domestic production but imports its bricks.

10.3.1.6 Diesel Generators and Irrigation Pumps

Fossil fuel combustion in the power sector contributes a large portion of SO_2 emissions in the HKH, excluding brick kilns. Diesel generators are an important source of emissions that are often neglected; they are used to meet power demand during frequent power shortages in the region (Laghari 2013). Over ten million diesel pumps are estimated to be in operation in the IGP consumed by these pumps. Distributed generator sets are a significant pollution source and contribute to local and regional air pollution, especially during the winter and pre-monsoon seasons due to the prolonged power shortages (World Bank 2014). BC emissions estimated from diesel generators in the Kathmandu Valley are 75% of diesel emissions estimated in the REAS V2 (Kurokawa et al. 2013), which excludes these emissions. In other words, when power cuts were common in the Kathmandu Valley between 2009 and 2015, emissions from diesel generators were responsible for more than half of the total BC emissions. Although the Kathmandu Valley in recent years reduced power cuts significantly, frequent power cuts are still prevalent in the other parts of HKH and illustrate the importance of these emissions from diesel generator sets in the region.

10.3.1.7 Transport (Road, Rail)

The transport sector contributes to emissions of all pollutants in all of the HKH countries, as shown in Fig. 10.16. It is an especially important source of NO_x emissions. For each of the eight HKH countries, the transport sector contributes from at least 12% up to 73% of total NO_x emissions in a country. For countries with a substantial number of gasoline vehicles, CO emissions from the transport sector are also important. SO_2 emissions from the transport sector illustrate the low fuel quality with high sulphur content used in the engines. Afghanistan, China, India, and Pakistan have substantial contributions by the transport sector to BC emissions, illustrating the existence of polluting diesel vehicles in these countries.

10.3.2 Source Regions and Transport Pathways for Pollution Reaching and Crossing the High Mountains

The HKH receives pollution from sources within the region, both rural and urban, as well as from a number of regions outside. In spring, dust from the Thar Desert reaches the western and central Himalaya (Hegde et al. 2007; Ram et al. 2010). South-eastern Tibet receives biomass-burning emissions from Myanmar, Bangladesh, and eastern India (Zhao et al. 2013), while the northern parts of the Tibetan Plateau receive dust from the Taklimakan Desert (Huang et al. 2007; Cao et al. 2009; Xiao et al. 2010; Chen et al. 2013; Zhao et al. 2013; Wang et al. 2015). Pollution episodes at Mount Waliguang at the north-eastern corner of the HKH, as well as at Mount Muztagh Ata, in the Pamir range and in the Pakistani Karakoram, have been traced to Central Asia and the Middle East (Wang et al. 2006; Che et al. 2011; Gul et al. 2018). Manora Peak in India receives pollution from the IGP, but also from the Middle East (Dumka et al. 2010, 2015).

Pollution episodes at the Nam Co observatory in south central Tibet have been traced to emissions in the IGP (Cong et al. 2007; Xiao et al. 2010; Xia et al. 2011; Wang et al.

2015). The high-altitude areas of HKH were once thought to be relatively pristine, and it was assumed that the Himalaya acted as a barrier to block the transport of pollutants from South Asia to the Tibetan Plateau. However, the in situ observations have pointed out that the high Himalaya are particularly vulnerable to the haze accumulated in the foothills (Li et al. 2016). In addition, high altitude areas have a second vulnerability: incomplete fuel combustion due to decreased oxygen levels, which contributes to pollution in the city of Lhasa (Ran et al. 2014).

The observational data from high altitude sites reveal regional scale pollution processes. For example, the high altitude sites from both sides of the Himalaya (i.e., NCO-P and QOMS) exhibit similar OC and EC abundances. Particularly, OC, EC, and other ionic species (NH_4^+, K^+, NO_3^-, and SO_4^{2-}) at QOMS exhibited a pronounced peak in the pre-monsoon period and a minimum in the monsoon season, similar to the seasonal trends of aerosol composition reported previously from Langtang and NCO-P (Cong et al. 2015). Pollutants from regional haze episodes over the IGP have been found to be transported across the Himalaya to trans-Himalayan valleys (Babu et al. 2011b) and to the Tibetan Plateau (Kusaka et al. 2001; Lüthi et al. 2015).

Li et al. (2016) reported, based on observation at multiple sites on both the north and south sides of the Himalaya and the Tibetan Plateau (TP), equal contributions from fossil fuel (46 ± 11%) and biomass (54 ± 11%) combustion to BC in the Himalaya, consistent with BC source fingerprints from the IGP, whereas BC in the remote northern TP predominantly derives from fossil fuel combustion (66 ± 16%), consistent with Chinese sources. This study and three other studies by Dhungel et al. (2016), Rupakheti et al. (2017), and Lüthi et al. (2015), which reported transport of air pollutants from the IGP to Nepal and then ultimately to the TP crossing the Himalaya under favourable meteorological conditions, indicate the importance of long-range and trans-border transport of air pollution in the region.

A case study combining ground-based and satellite remote sensing data identified a severe aerosol pollution episode over the Tibetan Plateau during 13–19 March 2009 (pre-monsoon) (Lüthi et al. 2015). Trajectory calculations based on the high-resolution numerical weather prediction model, Consortium for Small-scale Modeling (COSMO), were used to locate the source regions and investigate the mechanisms of pollution transport in the complex topography of the Himalaya and Tibet. Lifting and advection of polluted air masses over the great mountain range is enabled by a combination of synoptic-scale and local meteorological processes. During the days prior to the event, winds over the IGP were generally weak at lower levels, allowing for accumulation of pollutants and thus the formation of ABCs. The subsequent passing of synoptic-scale troughs led to south-westerly flow in the middle troposphere over northern and central India, which carried the polluted air mass across the Himalaya. Based on the results of the field observation of aerosols at the north slope of the Himalaya (QOMS), strong positive correlations were observed for dicarboxylic acids with biomass-burning tracers, levoglucosan, and K^+, demonstrating that this area was clearly affected by biomass burning (Cong et al. 2015). The seasonal pattern of dicarboxylic acids is consistent with OC and EC, characterized by a pronounced maximum in the pre-monsoon season. It has been proposed that the local valley breezes and regional atmospheric flow process could facilitate the penetration of the carbonaceous aerosols from South Asia across the high Himalaya.

The mountain/valley wind systems in the southern Himalaya are characterized by up-valley winds in the daytime with a maximum in the afternoon, which delivers substantial pollutants (e.g., BC and O_3) from lowland to higher altitude. This has been studied in situ both in the Kali Gandaki Valley which crosses the high Himalaya at relatively low altitude (Dhungel et al. 2016), as well as at NCO-P, near Everest Base Camp (Bonasoni et al. 2008, 2010; Gobbi et al. 2010; Marcq et al. 2010; Marinoni et al. 2010; Sellegri et al. 2010). Meanwhile, satellite-based tools were used to map the northward slanting haze layers that penetrate into Himalayan valleys from the IGP (Brun et al. 2011). The Karakoram also receives pollution from lower valleys (Putero et al. 2014), while Manora Peak observatory in Nainital, overlooking the IGP, receives substantial afternoon arrival of pollutants from the IGP (Pant et al. 2006b; Ram et al. 2010; Ojha et al. 2012; Dumka et al. 2015).

Chemical transport models, despite their inherent uncertainties in simulating atmospheric processes, are a useful tool to link emission source regions with in situ observations in areas with no local pollution sources. One study used the global chemical transport model GEOS-Chem to trace the origin of BC reaching four different high altitude sites in the HKH, finding sources as far away as Siberia, tropical Africa, and Europe (Kopacz et al. 2011). One of the techniques used by the atmospheric modeling community is to release region-tagged chemical tracers such as CO into the atmosphere and assess their concentration in the area of interest. Another widely used technique is to simply calculate wind back trajectories from the observation site and estimate the contributing region(s) (Jaffee et al. 1999). Numerical simulations have limitations and especially over the extreme topography of the HKH face a trade-off between doing computationally very expensive runs at high spatial and temporal resolution, versus going to lower resolutions that are able to capture key features elsewhere in the world but would miss key features in the HKH. Many atmospheric processes in the HKH are tied closely to the shape of the topography, and if a model does not accurately capture the height of peaks, the depth of valleys, or the cross-section of passes then it might not accurately capture the direction and speed of up- or down-valley flows, the location of cloud formation, or the extent of trans-Himalayan flows.

Satellite-based remote sensing products also provide a good spatial and semi-continuous temporal coverage. However, these products also have severe limitations during cloud cover, over bright reflecting surfaces especially over complex topography, and with regards to the number of tropospheric air pollutants they can observe directly.

The Sulfur Transport and dEposition Model (STEM), an offline regional chemical transport model (Adhikary et al. 2007; Kulkarni et al. 2015) driven by the Weather Research and Forecasting (WRF) model (Skamarock et al. 2008), was used to assess the relative strength of emissions from different geographical regions affecting the high mountain glaciers of the HKH. Anthropogenic CO emissions inventory provided by the Task Force on Hemispheric Transport of Air Pollutants was used to simulate the region-tagged tracers for 2013. For this assessment, region-tagged CO tracers were run on inert mode, thus providing an estimate of the maximum concentration reaching the observation site from the source region. In addition, it is important to note that the emissions are tagged at the time of release and not the actual path that the pollutant follows. Figure 10.19 shows the contribution as a percentage of total for CO pollutant reaching Yala Glacier in the central Himalaya. Although CO is used for this particular analysis, the technique allows for interpolating to other relatively inert species, which have similar spatial emission patterns. Thus, we interpret the results as air pollutants reaching high mountain glacier as opposed to strictly CO tracer.

Analyses of the simulation results show that Yala Glacier is affected mostly by pollutants from Nepal and India (nearby source regions), while occasional contributions come from regions as far away as Africa and the Middle East. Observations of pollutants thought to have originated from the Middle East and Africa over the Langtang region

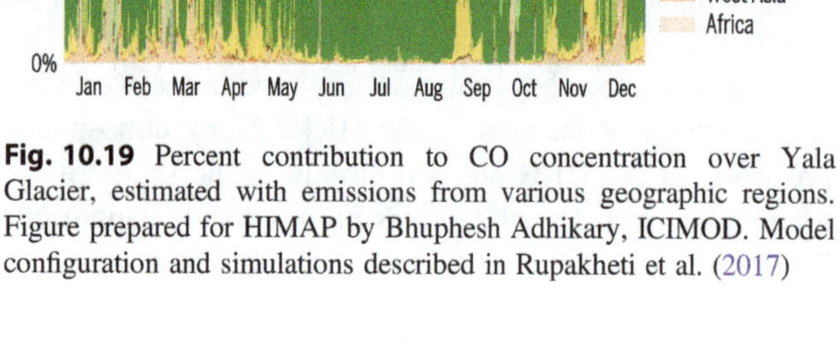

Fig. 10.19 Percent contribution to CO concentration over Yala Glacier, estimated with emissions from various geographic regions. Figure prepared for HIMAP by Bhuphesh Adhikary, ICIMOD. Model configuration and simulations described in Rupakheti et al. (2017)

have been published earlier (Carrico et al. 2003). Since the model is simulated at 25 × 25 km horizontal grid resolution, it is too coarse to capture up-valley winds and other thermally driven microclimates. However, the model does give an overall synoptic picture, consistent with the expected behaviour of pollution transport reaching the high glaciers of the HKH. Several research papers using wind back trajectories analysis report that the source of the observed air pollutants reaching the HKH are primarily from the IGP and as far away as Africa (Carrico et al. 2003; Bonasoni et al. 2012; Ming et al. 2009; Babu et al. 2011b; Lu et al. 2012; Gul et al. 2018).

10.3.3 Assessment of Our Knowledge About Emissions and Processes in the HKH

While some progress has been made in increasing our understanding of the contributions from some sources, much work remains before we can reliably link specific sources to their exact impacts. One of the challenges in the overall quantification of emission transport and climate effects over the IGP is in the general underestimation of aerosol loading in climate and chemical transport model simulations (both global and regional scale models), due to various emission- and meteorology-related factors (Ganguly et al. 2009; Henriksson et al. 2011; Nair et al. 2012; Cherian et al. 2013; Sanap et al. 2014; Pan et al. 2015).

10.3.3.1 Based on Source Apportionment Studies

Source apportionment studies by receptor modeling can assess how much various source categories contribute to ambient pollution levels at a specific site. Figure 10.20 shows the contribution of different sources, including household garbage and municipal solid waste burning to PM_{10} concentrations in the Kathmandu Valley, using a multivariate receptor model (Kim et al. 2015). It is clear that both garbage burning and brick kilns impact the Kathmandu air quality significantly. Kim et al. (2015) conducted measurements for two periods: the first between 21 December 2012 and 3 January 2013 and the second between 13 and 21 February 2013. They analysed BC, OC, ions, and trace metals in the PM_{10} data to assess the quantitative source contribution. They found that brick kilns and biomass/garbage burning contribute to 40% and 22% of BC in the Kathmandu Valley in winter, respectively, while biomass/garbage burning contributes to 32% of OC. Unfortunately, this work was carried out only at a single site, in a single season. Wiedinmyer et al. (2014) estimate that 644 Gg/year of municipal solid waste is burned in Nepal, leading to 6.3 Gg/year (7.7 Gg/year) of $PM_{2.5}$ (PM_{10}). As for BC (OC), they estimate 0.42 Gg/year (3.4 Gg/year) due to garbage burning in Nepal. Recalculating the estimated

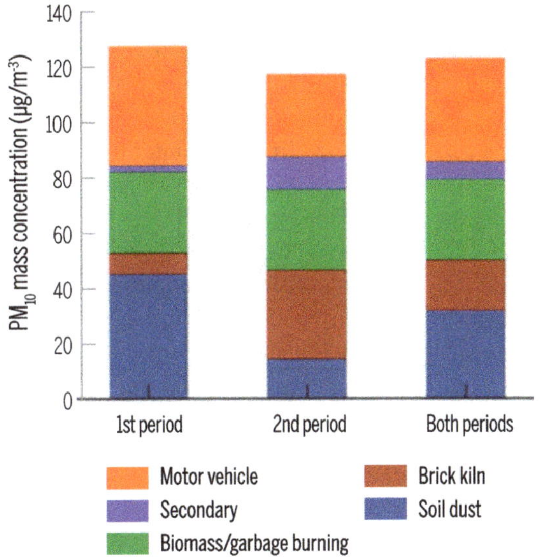

Fig. 10.20 Contribution of motor vehicles, secondary, biomass/garbage burning, brick kilns, and soil dust to PM_{10} concentrations in Nepal (*Source* Kim et al. 2015)

emissions using the emission factors measured in the NAMaSTE campaign, but keeping the same solid waste estimate, garbage burning emits seven, five, and 15 times more $PM_{2.5}$, BC, and OC, respectively, (Stockwell et al. 2016; Jayarathne et al. in press). This translates into revised $PM_{2.5}$, BC, and OC emissions estimates from garbage burning in Nepal being 7.3, 5.1, and 15 Gg/year.

10.3.3.2 Based on Chemical Transport Modeling

The HKH is influenced by both monsoons and westerlies. During the winter and pre-monsoon months of October–April, large-scale circulation patterns (mainly westerlies from central Asia) are the main source of air masses to the region. During the monsoon season of June–September, low pressure systems over the region bring air masses from the Indian Ocean. These result in pollution transport from South Asia (Meitiv 2010). Based on adjoint model results, Kopacz et al. (2011) estimated emissions from northern India and central China to be the major contributor of BC emissions to the Himalaya. For the Tibetan Plateau, they found western and central China, as well as India, Nepal, the Middle East, and Pakistan, to be source origins.

10.3.3.3 Assessment of Existing Emission Inventories

Understanding the spatial and temporal distribution of air pollutant emissions is vital to the implementation of appropriate climate and air quality mitigation measures. Increases in anthropogenic aerosol emissions and loading in South Asia in recent decades have also been well documented (Ohara et al. 2007; Hsu et al. 2012; Babu et al. 2013; Kaskoutis et al. 2013), in contrast to the decreasing emission trends over Europe and North America (Granier et al. 2011). However, large uncertainties exist in emissions inventories, and quantification of these uncertainties is essential for better understanding of the linkages among emissions and air quality, climate, and health. For India, for example, Saikawa et al. (2017a) find large differences among the inventory estimates in the domestic sector for CO_2 (82%), transport for NO_x (57%) and PM (108%), and industry for CO (63%) and SO_2 (44%).

In Nepal, residential sector emissions dominate for many species, and there are significant differences in the inventory estimates for all pollutants. The two emissions inventories, Emissions Database for Global Atmospheric Research (EDGAR) and REAS estimate following percentages due to the residential sector in Nepal: 73% and 90% for PM_{10}, 27% and 83% for NO_x, 67% and 79% for SO_2; and 67% and 98% for CO, respectively. Similarly, estimates for road transport sector emissions vary significantly, depending on the source. Large uncertainties exist in emissions estimates in China and India and more efforts are needed to constrain emissions better in the IGP (Saikawa et al. 2017a, b), as well as in each country of the HKH. There is a need for more ground-based measurements and modeling studies, such as source apportionment, in this region.

Box 10.4 Nepal Ambient Monitoring and Source Testing Experiment (NAMaSTE) campaign

A major input needed by emissions inventories is the "emission factor" for each pollution source—the amount of each pollutant emitted when a certain amount of fuel is burned. For many important pollution sources in the HKH, emission factors either do not exist in global databases or are very different in real life. The Nepal Ambient Monitoring and Source Testing Experiment (NAMaSTE) campaign was implemented in and near Kathmandu and the plains of southern Nepal by ICIMOD and U.S. universities during April 2015. NAMaSTE produced the first, or rare, measurements of aerosol optical properties, aerosol mass, and detailed trace gas speciation for the emissions. The sources of these emissions included many important, under-sampled combustion sources that are widespread in the HKH. Emissions from generators, irrigation pumps, brick kilns, cooking with various stoves and solid fuels, open burning of garbage and crop residue, and motorcycles were measured.

Light absorption by both brown carbon (BrC) and BC was important for many sources. Dung-fuel cooking fires produced higher BrC/BC than fuel-wood cooking fires. BTEX compounds (benzene, toluene, ethylbenzene, xylenes) are air toxins and carcinogens and were major emissions from both types of cooking fires. Biogas was the cleanest cooking fuel of ~12 stove/fuel combinations measured. Crop-residue burning produced high emissions of BrC and SO_2. A clamp kiln (with more biomass fuel) produced large emissions quantities of organic gases, organic aerosol precursors, and BrC, while a zig-zag brick kiln burning mostly coal still emitted large amounts of BC, hydrogen fluoride (HF), hydrogen chloride (HCl), and NO_x, with the halogenated emissions coming from the clay. Both kilns emitted very high SO_2. Garbage burning produced high BC and BTEX emissions. Diesel burned more efficiently than gasoline, but produced larger NO_x and aerosol emission factors. The most polluting common source sampled was gasoline-fueled motorcycles during startup and idling, for which the CO emission factor was ~10 times that of a typical biomass fire (Stockwell et al. 2016). We find that underestimation of the transport sector emissions is possible in the HKH due to the high start-up and idling, combined with increasing number of motorbikes and bad traffic in the region. On the other hand, cooking with biomass is gradually decreasing due to the increasing availability of LPG, particularly in India. Total CO emissions from the rapidly growing transport sector may soon be larger than those from biomass burning.

Box 10.5 Atmospheric Composition and the Asian Monsoon (ACAM)

Understanding the broader impacts of air pollutants in the Himalaya and their interaction with the monsoon system is of great importance. An international group of scientists have been meeting as a part of Atmospheric Composition and the Asian Monsoon (ACAM, https://www2.acom.ucar.edu/acam). Jointly sponsored by the International Global Atmospheric Chemistry project and SPARC (Stratosphere-troposphere Processes and their Role in Climate), ACAM seeks to understand how this rapidly changing part of the world is influencing the atmosphere on regional-to-global scales.

The Himalaya are a data-sparse region in terms of atmospheric composition measurements, and interactions between the monsoon circulation and human emissions have impacts from the surface to the upper atmosphere that are relevant to both air quality and climate. ACAM focuses on four themes: (1) emissions and air quality in the Asian monsoon region; (2) aerosols, clouds, and their interactions with the Asian monsoon; (3) impact of monsoon convection on chemistry; and (4) Upper Troposphere and Lower Stratosphere Response to the Asian Monsoon.

Since 2013, biennial workshops have attracted atmospheric scientists from the international community to share research related to these themes. Participation from Himalayan countries has included scientists from Nepal, China, India, and Pakistan. ACAM offers a venue for community building that emphasizes research cooperation through data sharing, coordinated field observations, and training schools. ACAM representatives also interact with the international chemistry and climate modeling community to understand what observations are most needed to enable better representation of the Himalaya in global models.

10.4 Adverse Impacts of Air Pollution in the HKH

Due to the growing emissions described in the preceding sections, along with rapid urbanisation, industrialization, and motorization, the level of PM (e.g., soot and dust) is very high in many HKH cities. A large population is thus exposed to pollution levels much higher than those recommended by the WHO. But air pollution not only harms the health of people and ecosystems, it also affects the climate, the cryosphere, monsoon patterns, water availability, agriculture, and incomes.

10.4.1 Temperature

Air pollution is responsible for changing the temperature. Absorbing aerosols such as BC and dark dust warm the layers of air where they are located (Ramanathan and Carmichael 2008). In addition, they contribute to cryospheric melting by darkening surfaces onto which they are deposited. Kaspari et al. (2014) estimated (using in situ-observed dust and BC properties and concentrations) the radiative forcing due to BC deposition to be within 75–120 Wm^{-2}, whereas significantly larger forcing (as high as ~500 Wm^{-2}) resulted due to dust deposition. Ménégoz et al. (2014) applied a climate-chemistry global model to evaluate the impact of BC deposition on the snow cover. They estimated that in HKH ranges, BC in snow causes an increase of the net shortwave radiation with an annual mean of 1–3 Wm^{-2}, reducing the snow cover annual duration by one to eight days.

Box 10.6 SusKat-ABC international air pollution measurement campaign 2012–13

Air pollution is a major environmental and societal problem in Nepal that can exert substantial impacts on public health, socioeconomic systems, and national development goals. As the first step towards comprehensive understanding of various aspects of air pollution in Nepal, the SusKat-ABC (Sustainable Atmosphere for the Kathmandu Valley—Atmospheric Brown Cloud) measurement campaign was conducted from December 2012 to June 2013 in Nepal, with a two-month-long intensive measurement phase from December 2012 to February 2013.

SusKat-ABC was jointly coordinated by the Institute for Advanced Sustainability Studies, Germany, and ICIMOD. More than 40 senior scientists and students from 18 research groups in nine countries in Asia, Europe, and the United States deployed state-of-the-art scientific instruments at a network of 23 measurement sites located inside the Kathmandu Valley (with a supersite at Bode), on the valley rim, and in other regions in Nepal, spanning from Lumbini on the northern edge of the IGP, across the Himalaya foothills, to the base of Mount Everest to the north.

The SusKat-ABC campaign was the second-largest international air pollution measurement campaign ever undertaken in South Asia, after the Indian Ocean Experiment (INDOEX) in 1999 (Ramanathan et al. 2001). High-quality ground-based data on air quality and meteorology were collected with unprecedented detail. Measurements of some important species were extended beyond the campaign to provide an opportunity to investigate, for the first time, their seasonal and inter-annual variations at multiple sites. The main objectives of the campaign were to document quantitatively the magnitudes of various air pollutants and to estimate the percentage contributions of major sources and sectors to air quality degradation in the Kathmandu Valley and other regions in Nepal.

In the Kathmandu Valley, wintertime PM_{10} is contributed mainly by local primary sources (motor vehicles: 31%, soil dust including road dust: 26%, biomass/garbage burning: 23%, brick factories: 15%) and about 5% by secondary sources (Kim et al. 2015). PM also contained high levels of toxic compounds such as polycyclic aromatic hydrocarbons (PAHs) (Chen et al. 2016) and mercury (Guo et al. 2017) at levels comparable to those observed in megacities like Delhi and Beijing. The campaign also revealed that gaseous air pollutants such as O_3, SO_2, NO_x, and a variety of volatile organic compounds (VOCs), some toxic to humans, are an important aspect of air pollution problems in the Kathmandu Valley and surrounding regions (Putero et al. 2015; Kiros et al. 2016; Sarkar et al. 2016). Speciated measurements of VOCs with a proton-transfer-reaction-time-of-flight-mass-spectrometry (PTR-ToF-MS), the first PTR-ToF-MS deployment in South Asia, made it possible to identify their sources and estimate their O_3 production potential and contributions to secondary organic aerosol formation (Sarkar et al. 2016).

As a follow-up to the SusKat-ABC campaign, Shakya et al. (2016) conducted a study in spring and summer of 2014 involving traffic police officers in the Kathmandu Valley. They found that lung functions degraded significantly with exposure to increasing levels of $PM_{2.5}$ and BC. Outside the Kathmandu Valley, they found that Lumbini, Pokhara, and Jomsom also suffer from high levels of PM, including BC and PAHs, CO, and O_3 (Chen et al. 2016; Dhungel et al. 2016; Tripathee et al. 2016; Rupakheti et al. 2017).

In summary, the SusKat-ABC results suggest that local action is necessary, but that it alone will not be sufficient to achieve clean air in the region. These key measurements and results provide the strong scientific basis needed for air quality management programmes or pathways to improve air quality and reduce adverse impacts in the region.

Regarding the Tibetan Plateau, Qian et al. (2011) suggested BC in snow increased the surface air temperature by around 1 °C and reduced the spring snowpack, resulting in increased runoff during the late winter and early spring, while the runoff decreased during late spring and early summer. The long-term regional mean radiative forcing via dust deposition on snow showed a rising trend during 1990–2009, which suggested the contribution of aerosols surface radiative effects induced by snow darkening has increased since 1990 (Ji et al. 2016). Furthermore, based on a model simulation, carbonaceous aerosols (BC and OC) increased surface air temperatures by 0.1–0.5 °C over the Tibetan Plateau and decreased temperatures in South Asia during the monsoon season. In the non-monsoon period, temperatures decreased by 0.1–0.5 °C over the southern Tibetan Plateau (Ji et al. 2015).

10.4.2 Precipitation and Monsoon (Cloud Microphysics to Regional Scale)

Air pollution also alters precipitation patterns. Major sources of absorbing aerosols (e.g., dust and BC) are located within Asia, which include both anthropogenic emissions (fossil fuel and biofuel sources) as well as naturally emitted mineral dust aerosols from desert/arid regions. Several climate modeling studies have suggested the importance of aerosol solar absorption in modulating the monsoon circulation and rainfall distribution over Asia (Menon et al. 2002; Ramanathan and Ramana 2005; Chung and Ramanathan 2006; Lau et al. 2006; Meehl et al. 2008; Randles and Ramaswamy 2008; Collier and Zhang 2009; Sud et al. 2009; Wang et al. 2009; Bollasina et al. 2011; Ganguly et al. 2012; Vinoj et al. 2014). The summer monsoon rainfall, particularly over southern Asia, constitutes over 70% of the annual rainfall and is a major freshwater resource to the region. Along with the summer monsoon, the seasonal snow melt from the Himalaya glaciers and snowpack also contributes to the regional socioeconomic livelihood. Therefore, perturbations to the monsoon circulation and Himalayan cryosphere via aerosol radiative effects in terms of regional warming, and subsequent impact on rainfall dynamics, are of critical scientific and societal significance.

One of the physical mechanisms of aerosol-induced radiative impact on the monsoon circulation and long-term effects on monsoon rainfall variability and trends is the "surface dimming effect" proposed by Ramanathan and Ramana (2005). This mechanism focuses on the northern Indian Ocean region, where thick haze, consisting of dust, BC, sulphate, and fly ash aerosols, is transported from southern Asia towards the Indian Ocean. The aerosol transport occurs annually during the dry season, primarily during the winter and early spring months. It has earlier been shown by Satheesh et al. (1999) that the widespread anthropogenic haze results in significant perturbations to the radiation budget, with large reductions in the solar insolation at the ocean surface. The reduction of sunlight reduces the evaporation rates, which further suppresses convection from the Indian Ocean, leading to reduced moisture transport towards the South Asian landmass during the peak monsoon season. This mechanism suggests the weakening of monsoon circulation and reduction of monsoon rainfall.

On an intra-seasonal scale, other mechanisms have been suggested which rest on the heavy and vertically extended aerosol loading over northern parts of south Asia, primarily over the IGP and the foothills of the Himalaya. The so-called elevated heat pump (EHP) hypothesis, proposed by Lau and Kim (2006), focuses on the absorbing aerosol build-up over the IGP prior to the onset of the summer monsoon. The enhanced dust loading, mixed with carbonaceous soot aerosols, is vertically advected to elevated altitudes against the southern slopes of the Himalaya and causes significant warming in the mid-upper troposphere. In their climate model simulations, Lau et al. (2006) demonstrated that the resulting warming creates a temperature anomaly that can amplify the overturning of the meridional circulation, thus causing enhanced moisture influx from the Indian Ocean. This mechanism has been hypothesized in the advancement and intensification of the early summer monsoon. From an observational perspective, few studies have also investigated the various feedback mechanisms of aerosol absorption effects on monsoon circulation and rainfall variability and trends (Lau et al. 2006; Gautam et al. 2009b; Manoj et al. 2010; Bollasina et al. 2011). In addition to the aforementioned direct radiative impacts of aerosols on the monsoon, aerosol particles can also affect cloud microphysics via indirect effects of aerosols by modifying cloud properties; however, the impact of aerosol indirect effects on the system-wide monsoon rainfall are still largely unknown (Rosenfeld 2000).

10.4.3 Cryosphere and Hydrosphere

Glaciers on the HKH hold the largest ice mass outside the polar region, resulting in the name the Third Pole. Ten major Asian rivers originate in this region, including the Yangtze, Yellow, Yarlung Tsangpo (Brahmaputra), Ganges, and Indus. Like a water storage tower, glaciers and snow cover in this region provide fresh water to billions of people downstream. The proximity of the HKH to some of the largest sources of BC makes this area particularly vulnerable. The increasing air pollution in the HKH could impact the cryosphere through a complicated mechanism. First, the light-absorbing particles (e.g., BC, BrC, and iron oxides in dust) could warm the atmosphere through absorbing solar radiation. Ramanathan and Carmichael (2008) suggested that the BC in the high Himalaya may contribute as much as carbon dioxide to the radiative forcing. Such atmospheric warming could subsequently lead to further melting of the snow through thermal transfer. More importantly, the impurities absorb the incoming solar radiation, thereby accelerating snow aging and melting processes and triggering snow-albedo feedback (Flanner et al. 2009; Lau et al. 2010; Yasunari et al. 2010; Qian et al. 2015).

Using MODIS snow albedo data, Ming et al. (2015) investigated the changes of snow and glacier albedo in HKH for the period 2000–11. A general darkening trend of glacier surface was revealed, with the most rapid albedo decrease in the glacial area above 6,000 m. The mass-loss equivalent (MLE) of the HKH glacial area caused by surface shortwave radiation absorption was estimated to be 10.4 Gt/yr, which may contribute to 1.2% of the global sea level rise on annual average (2003–09) (Ming et al. 2015).

Major sources of light-absorbing atmospheric aerosols are located around the HKH snowpack and glaciers, which include mineral dust and carbonaceous aerosols source regions. Mineral dust transport originates over the Thar Desert in India and several southwest Asian desert regions as far away as the Arabian Peninsula regions. The mineral dust aerosol loading peaks during the pre-monsoon and early summer monsoon months (April to July) on an annual basis

(Gautam et al. 2010). Major sources of carbonaceous aerosols are located widely over the IGP. Although the absorption strength of BC aerosols is several orders of magnitude larger than that of dust, the absolute concentrations of dust found in the HKH cryosphere are substantially larger than BC deposition (Ginot et al. 2014; Kaspari et al. 2014). Therefore, in addition to BC, dust aerosol deposition is considered a major agent causing snow albedo reduction in the HKH.

For example, a recently conducted ice core study in the Mera Peak Glacier shows that annual mass fluxes of dust are a few orders of magnitude higher than of BC (Ginot et al. 2014). Longer-term ice core records (on the scale of centuries) have also indicated peak annual dust concentrations in the Tibetan Plateau prior to the summer monsoon season, with enhanced dust deposition during periods of dry/drought conditions over southern Asia (Kang et al. 2000; Thompson et al. 2000). Satellite observations have also found widespread areas with desert dust deposited onto snow cover, and a reduction in snow reflectance in the western Himalaya and Hindu Kush cryosphere (Gautam et al. 2013). The concentrations of both BC and dust are typically found to peak during the pre-monsoon season (Kaspari et al. 2011), although in situ-based measurements of snow impurities are still sparse and limited. The BC deposition over the HKH primarily comes from residential, industrial, and transport sectors, with the largest deposition occurring in summer months (Jenkins et al. 2013).

Modeling results suggested that mineral dust caused a decrease of 5–25 mm in the snow water equivalent over the western Tibetan Plateau, Himalaya, and Pamir Mountains in winter and spring (Ji et al. 2016). Also in the Himalayan region, a 2.0–5.2% albedo reduction from BC deposition could result in an 11.6–33.9% increase in annual discharge if the reduced albedo snow layer remained at the glacier surface (Yasunari et al. 2010; Gertler et al. 2016). It should be noted that, currently, the modeling works for the transport, deposition, and radiative forcing of snow impurities indicate remarkable uncertainties (Qian et al. 2015). Biases may be partly due to the mismatch in the timescale between the modeling and observation. The lack of accurate data on optical and physical properties of impurities (e.g., BC, dust, and organic matters) in the atmosphere and snow is also an important reason. More field observations with higher time resolution (e.g., daily) at representative locations are urgently needed to validate the modeling results.

10.4.4 Ecosystems and Agriculture

Air pollution also has significant impacts on the ecosystem. Acid rain causes damage to forests and acidifies soils, and highly acidic precipitation has been observed in the eastern Himalaya, with pH value as low as 4.2 in Darjeeling, India (Roy et al. 2016). Acid deposition also affects yields for many crops and accelerates the erosion of buildings, statues, and sculptures (Singh and Agrawal 2008). Similarly, leaching of nutrients such as nitrates and phosphates in the rivers has resulted in harmful algal blooms, turning the coastal areas into "dead zones" (Rabalais et al. 2010). Increased nitrogen transport in rivers can also lead to enhanced microbial production of nitrogenous trace gases, including the greenhouse gas (GHG) nitrous oxide (N_2O), both in streams and in receiving coastal areas (Beaulieu et al. 2011; Naqvi et al. 2000).

Exposure to increased ground-level O_3 harms forest, plants, and agricultural crops by penetrating leaves through stomata and by oxidizing plant tissues. Adverse impacts result in impaired photosynthesis, protein and chlorophyll degradation, and reduction in metabolic activity (Booker et al. 2009; Fuhrer 2009). As a result, surface O_3 exposure leads to crop yield losses, increased susceptibility to diseases, and increased senescence (Mauzerall and Wang 2001). Losses of USD 1.2–3.8 billion per year are estimated due to wheat, maize, cotton, and soybean crop damage in India (Avnery et al. 2011; Ghude et al. 2014).

10.4.5 Health

Ambient and indoor air pollution are both large contributors to poor health, especially in developing countries. A recent report by the WHO (2014) estimated the global premature mortality rate linked to these two forms of air pollution at 2.7 million and 4.3 million, respectively. The most recent Global Burden of Disease Study estimates approximately 5.8 million premature deaths due to PM pollution, half due to ambient pollution and half to household air pollution, as well as 217,000 deaths due to ambient ozone (O_3) pollution (Forouzanfar et al. 2016). The largest number of premature mortalities from both ambient and indoor air pollution is considered to be from Asia. For example, in South Asia, household air pollution (HAP) from solid fuel burning is the dominant cause of death (Lim et al. 2012). In East Asia, ambient PM pollution is the fourth largest cause of death, while HAP is ranked fifth (Lim et al. 2012). The HKH is therefore no exception, and there is a large estimated number of premature deaths due to air pollution in each of the countries within the HKH as shown in Table 10.1. The estimates vary, however, depending on the emissions used in the study, as well as the choice of the chemical transport model and the chosen concentration-response functions. For example, the premature mortality from outdoor air pollution in China can range from 300,000 to over a million, depending on the studies (Saikawa et al. 2009; Health Effects Institute 2017). What is important, however, is that these numbers are not negligible and more studies are essential for a better

Table 10.1 Air pollution and impacts of $PM_{2.5}$ exposure in 1990 and 2013 for each of the HKH countries

	Mean annual ambient $PM_{2.5}$ μg/m^3		Total deaths from air pollution		Total welfare losses Million 2011 USD, PPP[1]-adjusted (% GDP equivalent)		Total forgone labour output Million 2011 USD, PPP[1]-adjusted (% GDP equivalent)	
	1990	2013	1990	2013	1990	2013	1990	2013
Afghanistan	N/A	N/A	N/A	N/A	N/A	N/A	N/A	N/A
Bangladesh	29.92	48.36	92880	154,898	6379 (4.66%)	27,452 (6.14%)	1195 (0.87%)	2579 (0.58%)
Bhutan	N/A	N/A	N/A	N/A	N/A	N/A	N/A	N/A
China	39.3	54.36	1,518,942	1,625,164	126,592 (7.35%)	1,589,767 (9.92%)	12,558 (0.73%)	44,567 (0.28%)
India	30.25	46.68	1,043,182	1,403,136	104,906 (6.8%)	505,103 (7.69%)	28,742 (1.86%)	55,390 (0.84%)
Myanmar	N/A	N/A	N/A	N/A	N/A	N/A	N/A	N/A
Nepal	29.68	46.09	16,436	22,038	1033 (4.60%)	2833 (4.68%)	195 (0.87%)	287 (0.47%)
Pakistan	36.55	46.18	103,111	156,191	19,935 (6.06%)	47,713 (5.88%)	4713 (1.43%)	6582 (0.81%)

Source World Bank and Institute for Health Metrics and Evaluation (2016)
Note 1. PPP stands for purchasing power parity

assessment. Available studies on health impacts have focused on exposure to PM, tropospheric O_3, NO_x, and SO_2, as well as polycyclic aromatic hydrocarbons (PAHs) and heavy metals.

10.4.5.1 Particulate Matter and Health

Studies have found that the surface concentration of $PM_{2.5}$ is linearly associated with increased risk of various adverse health impacts, including premature mortality and morbidity (Hogg et al. 2004; Pope et al. 2002, 2004; Brook et al. 2010; Schwartz et al. 2008). Fine particles ($PM_{2.5}$) penetrate deep into the lung and damage cells in the airways, as well as causing lung and cardiovascular diseases. Ambient air pollution and $PM_{2.5}$, in particular, have been classified as carcinogenic to humans (Loomis et al. 2013). There have been 21 studies identified since 1984, which have assessed the health burden of outdoor air pollution in the Kathmandu Valley and Nepal. Despite the limited data and resources, all of these studies have suggested adverse outcomes linked to air pollution exposure.

Key findings from these research projects are threefold. First, the hospital records between 2007 and 2013 show an increase in cases and proportion of respiratory diseases, as well as their contribution to mortality among children under five years old over this period, which corresponded to the deterioration in air quality. Second, survey and observation analyses among people with different occupations, active locations, and age groups show that people who work or live near the main roads in urban areas are exposed to much higher levels of PM concentrations and therefore suffer more from health problems such as respiratory diseases, headache, and chest pain than those who do not. Third, estimated health effects from the dose-response functions show a heavy health burden due to the excessive PM concentrations. Reducing such health burdens can also relieve the country from the economic burden of mitigating those negative health effects.

10.4.5.2 Other Air Pollutants and Health

Tropospheric O_3 also causes adverse health impacts, and studies have linked the exposure to it to higher rates of respiratory and cardiovascular diseases, as well as chronic obstructive pulmonary disease (COPD) (Balakrishnan et al. 2011; Chhabra et al. 2001; Gupta et al. 2007; Jerrett et al. 2009; Pande et al. 2002; Schwartz et al. 1994; Siddique et al. 2010). Asthma mortality and morbidity is also linked to O_3 exposure, and even short-term exposure is known to aggravate existing lung diseases. NO_x is known to have harmful effects on human and ecosystem health and is also linked to the formation of other harmful atmospheric pollutants, such as O_3 and nitrate aerosols (Costa et al. 2014). SO_2 is also harmful to humans (Rall 1974) and ecosystem health and welfare, and it is a precursor to acid rain and sulphate particles (Horton 2012; Loomis et al. 2013; Murray et al. 2012; Pope et al. 2007; Schwartz et al. 1994). Exposure to ambient CO is also harmful to human health (Allred et al. 1989; Haldane 1912; Morris et al. 1995; Stern et al. 1988), and CO emissions are also important precursors to the formation of O_3. Polycyclic aromatic hydrocarbons (PAHs) have also been investigated for possible links to an increased risk of cancers and cardiopulmonary diseases (Armstrong and Gibbs 2009; Burstyn et al. 2003; Mastrangelo et al. 1996). PAHs are produced by incomplete combustion of fuel containing carbon, and there are several hundred compounds in this group. There have so far been only limited studies that have measured indoor PAH concentrations in the HKH (Li et al. 2012; Rantalainen et al. 1999). For the ambient air, during a two-week study period in autumn 2009, Kabul and Mazar-e Sharif in Afghanistan both showed high levels of PAHs from traffic, as well as coal and biomass combustion (Wingfors et al. 2011).

10.4.5.3 Household Air Pollution

HAP contributes to poor health and disproportionately affects those living in developing countries. HAP contains all the pollutants mentioned above and is considered as an additional exposure to ambient air pollution due to the burning within the household. Households that use biomass fuels usually have much higher household exposure than those using cleaner fuels (Shrestha and Shrestha 2005; Xiao et al. 2015). Several acute and chronic health conditions are associated with HAP, including acute respiratory infections (ARI), pneumonia, tuberculosis, chronic lung disease, cardiovascular disease, cataracts, and cancers (Smith et al. 2000; Balakrishnan et al. 2004; Bruce et al. 2007; Dherani et al. 2008; Pope et al. 2010; Epstein et al. 2013). Prolonged exposure to high levels of HAP has been found to increase the frequency of ARI (Ezzati and Kammen 2001). Upadhyay et al. (2015) found that there was a higher probability of life-threatening respiratory illnesses (OR = 1.71–1.78) with the use of solid fuels. It is also important to note that the disease burden is borne predominantly by women and young children, who spend more time indoors (Boy et al. 2002; Smith and Mehta 2003; Bruce et al. 2006).

Approximately 400,000 deaths from acute lower respiratory infection in children under five and 34,000 deaths from COPD in women are attributed annually to HAP in India (Smith et al. 2000; Balakrishnan et al. 2011). India contributes to 24% of the global annual child mortality due to ARI, 50% of which is due to HAP (Upadhyay et al. 2015). In addition, the use of solid fuels is considered to have been responsible for 20% of the total deaths among children between one and four years old (IIPS & ORC Macro 2007).

This high number of premature deaths from HAP in the HKH is due to the high rate of solid fuel use, as shown in Table 6.1. For example, in India, 71% of total households are assumed to use solid fuel, and the rate is 90% in rural areas (Agrawal 2012). Similarly, 50–60% and 98–99% of the population still cooks with biomass fuels in urban and rural Bangladesh, respectively (Bangladesh DHS 2011). In urban and rural Pakistan, respectively, 58% and 94% of the population uses biomass fuel (Qasim et al. 2014). One exception is the urban population in Afghanistan, where 50% use LPG and 10% use electricity. In the rural areas, however, 62% rely on wood and 31% on dung, comprising more than 95% nationally dependent on wood.

Although most research has been focused on health outcomes in adulthood, some recent studies also demonstrate HAP to be associated with low birth weight and infant mortality (Farmer et al. 2014; Pope et al. 2010). There are also some studies that support the hypothesis that exposure to HAP increases the risk of tuberculosis infection and disease, although the data is still too limited to draw firm conclusions (Lin et al. 2007). There is conflicting evidence between HAP and asthma (Bruce et al. 2000), but there is a need for further research, as there has been a recent surge in the childhood asthma prevalence rate in many developing countries, including in the HKH.

Through the $PM_{2.5}$ and BC field campaign conducted in households in Nam Co, Tibet, in 2013, the highest and the third highest 6-h average BC (24.5 $\mu g/m^3$) and $PM_{2.5}$ (873 $\mu g/m^3$) concentrations were observed from a stone house that had a chimney stove, due most likely to limited ventilation in the household (Xiao et al. 2015). Despite the fact that 75% of the residents using a stove with a chimney indicated that they were not worried about indoor air quality, as opposed to only 29% of those using stoves without chimneys, they find that chimney installation did not by itself ensure adequate indoor air quality. Not simply the implementation but also chimney infrastructure is essential for mitigating HAP.

10.4.6 Cost of Air Pollution in the HKH

In addition to adverse health issues, air pollution also results in the loss of labour forces and reduces human welfare. Table 10.1 summarizes the mean annual $PM_{2.5}$, total deaths from air pollution, total welfare losses, and total forgone labour output for years 1990 and 2013 (World Bank and Institute for Health Metrics and Evaluation 2016).

According to the annual report of Nepal's Department of Health Services (DoHS), ARI ranked No. 1 among the top 10 diseases accounting for morbidity in Nepal. The historical data on ARI incidence in Nepal show a significant increase since fiscal year 2007–08. The incidence of ARI per 1,000 children under five years old among new patient visits is as high as 951 during fiscal year 2013–14 (DoHS 2014).

Nepal's annual health cost attributed to urban air pollution is estimated at USD 21 million, equivalent to 0.29% of the nation's GDP (World Bank 2008). The impact of the high air pollution is witnessed and experienced by millions in their daily lives, in terms of worsening air quality (degrading visibility), respiratory health issues, and regionally changing weather/climate patterns.

10.5 Mitigation Efforts for Improving Air Quality in the HKH

For many of the pollution sources detailed earlier, technologies exist that can reduce emissions at the source or that allow carrying out the same activity with lower emissions. The challenge lies with barriers to their uptake: affordability and financing, behavioural inertia, design and durability of technological solutions, and institutional mechanisms to promote them. Infrastructure and energy planning at regional, national, and urban levels has a large impact on reducing emissions in the long term. Many of the countries lack coherent and dedicated institutions that could form and adopt evidence-based policies. Where such institutions exist, implementation and enforcement are weak. The transboundary nature of the problem calls for regional coordination, but this remains an even bigger challenge. This section discusses current and potential mitigating efforts in HKH.

10.5.1 Mitigation Options in the HKH to Reduce Emissions

What technologies can reduce emissions? And what motivates behavioural changes that lead to reduced emissions? This section looks at mitigation efforts and potential future approaches. There are existing technological solutions in various source sectors and we list the main sectors below, including cleaner cookstoves, cleaner brick kilns, diesel particulate filters, the Happy Seeder, solar powered irrigation pumps (SPIP) and others that will most likely be effective in mitigating air pollution in the HKH.

10.5.1.1 Cookstoves

In order to mitigate the prevalent HAP problem in developing countries, a large research effort has centred on intervention studies and the development of so-called "improved cookstoves". These stoves are optimized for fuel efficiency or to minimize HAP (Rehfuess et al. 2014). Installing a stovepipe to vent pollution out of the household is one example. Such stoves can improve health (Ezzati et al. 2004), reduce climate change impact, and be cost effective (Baillis et al. 2009). However, many of these intervention programmes, including those using the Kyoto Protocol's Clean Development Mechanism, have failed (Clark and Peel 2014; Rehfuess et al. 2014). The local residents often do not adopt these improved cookstoves and even if they do, the traditional cookstoves are in many cases used simultaneously, leading to no health benefits or any significant reduction of HAP (Edwards et al. 2007). On the other hand, in some parts of Tibet, there are households that have installed improved cookstoves for comfort or convenience, although awareness of HAP is non-existent and no intervention programmes have taken place. Here again, however, we find that the installation of an improved cookstove by itself does not necessarily lead to adequate indoor air quality. In short, the installation of improved cookstoves has not contributed to HAP reduction as intended. While improved solid-fuel stoves can reduce air pollution, the existing generation of stoves does not do so by enough to provide significant health benefits (Edwards et al. 2007).

There are numerous potential reasons as to why cookstove interventions have failed, including stove design, cultural obstacles, and inappropriateness for local cooking needs. Another major reason is that the prioritization of other basic needs by the local residents (Martin et al. 2011; Mobarak et al. 2012) and the complex interactions between technology, human behaviour, economics, and infrastructure (Jin et al. 2006) have not been sufficiently considered.

Switching to LPG, biogas, or electricity is more promising in some countries where it will be more readily available. Subsidies, credit, and marketing can help in moving households to clean fuels. Subsidies to biogas can be expanded by using a voluntary climate change mitigation approach developed by parties to the United Nations Framework Convention on Climate Change called REDD+ schemes (Somanathan and Bluffstone 2015). The expansion of rural electricity access in Bhutan resulted in the percentage of rural households having electric rice cookers increasing from about 20% in 2003 to about 80% in 2012 (Bhutan Living Standards Survey 2013), presumably substituting some solid fuel with electricity and thus reducing emissions.

10.5.1.2 Brick Kilns

Seven main brick kiln types exist in and surrounding the HKH: clamp kiln, moveable chimney Bull's trench kiln (MCBTK), fixed chimney Bull's trench kiln (FCBTK), zig-zag kiln, vertical shaft brick kiln (VSBK), Hoffman kiln, and tunnel kiln. MCBTK used to be most popular technology in the region but since it was banned across Nepal in 2012, following the ban in the Kathmandu Valley and in India in 2003, FCBTK is now the most widely used technology in the HKH. Bangladesh, however, has not banned MCBTK, which is therefore still ubiquitous in some parts of the country. MCBTK and FCBTK are both continuous, moving fire kiln types, where the fire always stays on and different parts of the kiln are used to warm, fire, and cool bricks simultaneously. Seventy-nine percent of bricks (260 billion) are considered to be produced by this method in India, Pakistan, Bangladesh, and Nepal (Greentech 2014).

This is the most polluting of all the brick kiln types and commonly used fuels include coal, biomass, agricultural residue, and industrial waste.

A zig-zag kiln is very similar to FCBTK with a continuous, moving fire, with the main difference being that the air flow within the kiln follows a zig-zag pattern around the bricks being baked, rather than the oval shape of the kiln. It also uses coal, biomass, and industrial waste as fuel. In VSBK, the ware moves, rather than the fire, creating counter-current heat exchange between air moving upward and bricks moving downward, and it uses coal as major fuel. The Hoffman kiln is a continuous, moving fire kiln that is semi-mechanized. It uses wood and coal as fuel. Like VSBK, the tunnel kiln is a continuous moving ware kiln. Clay products are passed on cars through a horizontal tunnel to be fired in a production process. This is a completely mechanized system and is considered to be the most advanced technology. It uses coal and petcoke as fuel. There are only a limited number of these in the HKH.

Emissions can be reduced significantly by upgrading kiln technology (Schmidt 2013). For example, shifting from the intermittent kilns such as clamp kilns or continuous kilns, including Hoffman or Bull's trench, to a more efficient type, such as vertical shaft (VSBK), would lead to reduced fuel consumption and lower emissions of air pollutants as well as CO_2. Use of alternative building materials, like hollow and perforated bricks or compressed unbaked bricks, should also be encouraged, since they require less raw material and energy to fire. Implementation of the new technologies and building materials, however, is not easy because of the cost associated with such changes. Increased availability of financial incentives for new technologies and stricter policies on emissions from brick kilns is essential for effective mitigation.

10.5.1.3 Mitigating Transport Emissions

There are several approaches to reducing transport emissions in the HKH. One approach is to implement emission standards along with fuel quality standards. For this, catalytic converters in gasoline and diesel vehicles will be particularly useful in mitigating air pollutant emissions. Studies have found that complying with the Euro 3 standards in the HKH would lead to significantly reduced transport emissions (Saikawa et al. 2011; Shrestha et al. 2013).

Another way to mitigate PM emissions effectively from the transport sector is using a diesel particulate filter (DPF). A DPF is a device mounted in the diesel vehicle's exhaust system as an after-treatment device to remove PM from the exhaust. It is usually installed with a catalyst that is programmed to burn off the accumulated PM, when the filter becomes full. A DPF can reduce the amount of particulate emissions from diesel to levels comparable with compressed natural gas (CNG) vehicles. However, DPF is only effective with diesel fuel that has a sulphur content less than 50 ppm. For many countries in the HKH, this fuel quality is currently not available.

To reduce vehicle emissions, a move to electric vehicles, as well as to CNG or LPG vehicles, is a potential option for some countries in the HKH. A decade and a half ago the Kathmandu Valley initiated the installation of 600+ electric three-wheeler vehicles called Safa Tempos to reduce emissions; in recent years, several hundred electric private vehicles have joined. For countries such as Bhutan and Nepal, where its electricity is mostly from renewable sources, this option is extremely beneficial and effective in mitigating air pollution from the transport sector. Pakistan has been leading the installation of CNG vehicles in the world until the 2000s, and several countries in the region have followed suit to convert taxis and some diesel buses in various cities. For countries such as Bangladesh, retrofitting conventional vehicles with CNG will be a better option.

For mitigating emissions from the transport sector, three ingredients are essential: stringent policies to reduce emissions, governmental mechanisms to incentivize the change through tax reduction on cleaner vehicles or subsidies for cleaner fuel, and inspection and maintenance.

10.5.1.4 Other Sources—Agricultural Burning, Burning of Garbage and Emissions for Diesel Powered Irrigation Pumps

Emissions from crop residue burning in the IGP could be reduced via subsidization and expansion of the Happy Seeder, a machine that allows wheat to be planted without removing the residue from the previous rice crop (Gupta 2014). Other alternatives are distinctly more costly for farmers (Pant 2015).

People burn garbage because of the absence or inadequacy of garbage collection services in most less-developed countries (Hoornweg and Bhada-Tata 2012). Public awareness of the health hazards of emissions from garbage burning could lead to alternative disposal via composting, especially in less dense areas. In most urban areas, however, the solution lies in better municipal solid waste collection. Increasing awareness of the problem could help build public support and pressure for financing such services.

Emissions from diesel powered irrigation pumps can be reduced by replacing these with solar powered irrigation pumps.

10.5.2 Incentives and Behaviour Change for Improving Air Quality in the HKH

Emissions from industry and transport require regulations to reduce emissions. Nevertheless, public awareness can also play a vital role. It seems very likely that the deluge of media reports that Delhi had the worst air quality in the world influenced the January 2016 advancement of the time schedule for introducing Euro 6-equivalent standards for fuel and new vehicles in India from 2024 to 2020 (Dallman and Bandivadekar 2016), despite opposition from the automobile industry (Mukherjee 2015). While obvious events like winter smog evoke media coverage and regulation (Wang 2015), without publicly available air quality monitoring data, news reports on Delhi's air quality would not have been possible. Monitoring can, therefore, be important for public and media awareness, which in turn influences regulatory action.

The same is potentially true for HAP reduction. For example, China has pursued improved cookstove programs in the past and the National Improved Stove Program (NISP) of the 1980s and 1990s was successful in distributing 180 million stoves across the country. However, research has indicated that NISP was not successful in adequately mitigating HAP (Edwards et al. 2007). In 2007, the Chinese government launched the One Solar Cooker and One Biomass Stove Program specifically in Tibetan regions and distributed 79,833 biomass stoves and 244,474 solar cookers over four years. In 2012, another China Clean Stove Initiative was launched (World Bank 2013). Yet the literature on interventions that seek to reduce HAP by raising knowledge and awareness is scant (Barnes 2014). It appears essential that cookstove interventions become more community based and community driven so that there is increased awareness and induced behaviour change in order to effectively mitigate HAP in the HKH. Financial incentives and policies that exploit potential co-benefits are also appropriate to nudge individuals to switch to cleaner fuel options.

10.5.3 Planning, Policies, and Institutions for Air Quality Management in the HKH

What infrastructure investments would reduce emissions? What are the existing institutions and policies in HKH countries? What broader frameworks exist today? Many of the technical options surveyed in the previous section focused on abating air pollution at the source level. But there are also upstream structural changes to the energy and transport systems that could lead to significant reductions in air pollution, some of which are SLCPs, as well as GHGs.

10.5.3.1 Clean Energy

Among the most significant changes in recent years has been the sharp increase in renewable energy. As mentioned earlier, clean energy is abundant in Bhutan and Nepal, with most electricity originating from hydropower. Both India and China have been at the forefront of efforts to scale up solar, wind, and hydropower. For instance, China's 11th and 12th Five-Year Plans contain percentage targets for renewable energy (Lin et al. 2007; Guo et al. 2014). These targets are often accompanied by policies that support investments in renewables. These include, for instance, feed-in tariffs (FIT) that pay back consumers for transferring renewable energy into the grid during off-peak times. Other sets of reforms place the onus on local governments to identify context-appropriate solutions. Cities such as Rizhao (Shandong Province)—which has become a pioneer in the use and manufacture of solar technologies—offer good examples of how local governments have innovated even before the central government's energy-saving targets (Starke 2007). There have also been similar efforts to improve energy efficiency in China's five-year plans. Among the more notable is the Top-1,000 Energy-Consuming Enterprises that pushed China's most energy-intensive firms to embrace a series of reforms to firm-level planning and management systems (Price 2008). In India, tariff from solar powered plants is at its historical low and even lower than tariff from thermal plants—thereby making solar energy quite lucrative in the long run.

10.5.3.2 Urban and Transport Planning

Rapid urbanisation and motorization in the HKH have contributed greatly to deteriorating air quality. The transport community divides solutions to these kinds of problems into options that (1) avoid unnecessary travel through land use planning, (2) shift travelers to more efficient modes (often through providing and upgrading public transport), and (3) improve vehicle technologies or fuels (Leather 2009). Many parts of the HKH are developing programs that draw upon the avoid-shift-improve scheme. For instance, India began implementing the Jawaharlal Nehru Urban Renewal Mission, which offered fast-growing cities fiscal transfers to improve the quality of the public transport system and surrounding roads and infrastructure (Agarwal and Zimmerman 2008). The infrastructure improvements are important, as significant levels of pollution can come from the resuspended dust on unpaved or poorly paved roads (Guttikunda and Koppaka 2012).

Some of the transport solutions will need to focus on emissions from diesel vehicles. A well-designed diesel control strategy will require not only tightening emissions standards that enable the installation of control technologies (particulate filters), but also improving fuel quality to ensure that sulphur does not degrade the operation of those technologies (Minjares and Rutherford 2010). In recent years, some countries in the HKH have adopted tighter emissions and fuel quality standards—often with large cities such as Beijing and Delhi leading the way. These reforms will help mitigate emissions from new vehicles; a more challenging issue is the inspection and maintenance (I&M) of existing vehicles. A small percentage of so-called super-emitting vehicles can be responsible for more than 50% of emissions. Unfortunately, I&M programs for diesel have proven notoriously difficult to design and operate effectively in developing country contexts (Hausker 2004).

In Nepal, the ban on the import of new three-wheelers and two-stroke engine vehicles in 1991 and 1992, respectively, marked the initial government efforts to combat air pollution from the road transport sector. The first in-use vehicle emission standard was introduced in 1995 and the vehicle emission testing and control system was implemented in the same year. In 1999, Nepal started importing unleaded fuel, removed highly polluting diesel three-wheelers, and enforced the Green Sticker system to enhance vehicle emission testing and control. In 2004, two-stroke three-wheelers and 20-year-old taxis were removed from Kathmandu, and the ban on the plying of 20-year-old public transport vehicles inside Kathmandu Valley was implemented as well. For emission standards, Government of Nepal (GoN) introduced the Euro 1 equivalent norms for new vehicles in 2000, and started supplying Euro 3 standard fuel in 2010. In addition, since 2007, GoN started collecting a pollution tax of NPR 0.5 from each litre of petrol and diesel sold in Kathmandu Valley. While the necessary policies are in place, compliance remains a challenge.

As growing motorized vehicle fleets have made many streets more dangerous, the fraction of people walking or using bicycles has decreased. Urban transport planning that not only focuses on motorized vehicles, but also provides sufficient safe space and cross-city pathways for non-motorized transport is important both to keep air pollution down and to promote the health benefits of physical exercise.

10.5.4 National Ambient Air Quality Standards and Air Quality Management in the HKH

Table 10.2 illustrates the national ambient air quality standards implemented in each of the countries in the HKH, as well as the three WHO interim targets and the WHO air quality guideline.

10.5.4.1 Air Quality Management Institutions

Institutional arrangements that enable interagency coordination and actively engage multiple stakeholders are critical to managing air pollution problems. One of the main constraints in the HKH (and elsewhere) has been the lack of clear division of roles and responsibilities and the absence of a coordinating mechanism that could break down agency silos (ICIMOD 2007). Many of the HKH countries lack a coherent air pollution institutional framework and clear responsibilities for air quality management of each stakeholder. The institutional framework for air quality management in Nepal (Box 10.7) illustrates why better coordination is needed.

Box 10.7 Nepal's air quality management institutional framework

The institutional framework for air quality management in Nepal involves four ministries and seven departments and centres, as well as four local agencies and institutions such as municipalities and traffic police. The framework is divided into two parts corresponding with respective managing functions—policy making and implementation. Policymaking responsibilities are handled chiefly by the four ministries: the Ministry of Industry (MoI), Ministry of Forest and Environment (MoFE), Ministry of Physical Infrastructure & Transport (MoPIT), and Ministry of Urban Development (MoUD). MoFE is mainly responsible for the environmental protection-related policy making and supervision. MoI and MoPIT are charged with the policy making and supervision related to the industry and transportation sectors, respectively. Since road transport and industries such as brick kilns and cement factories are significant air pollution sources in Nepal, industrial and transportation policies have considerable impacts on air quality. MoUD plays a critical role in the framework, managing urban planning and developing road standards.

Within the ministries, there are several departments and institutions, such as the Department of Environment, which is responsible for the development of vehicle emission standards, and the Department of Transport Management, which is responsible for vehicle inspection and registration. Further, Nepal's Bureau of Standards and Metrology is in charge of fuel quality. Besides ministries and departments, local governments and institutions also contribute to the Kathmandu Valley's air quality control. For example, on-street vehicle emission inspections are often undertaken by the traffic police. Furthermore, although the Environment Protection Council, formed under the chairmanship of the Prime Minister, was intended to be a high-level body that could help with interagency coordination, it did not meet from the early 2000s until 2017.

Table 10.2 National ambient air quality standards in each of the HKH countries

	$PM_{2.5}$		PM_{10}		TSP		SO_2			NO_2			O_3		CO ('000)	
	24 h	Annual	24 h	Annual	24 h	Annual	1 h	24 h	Annual	1 h	24 h	Annual	1 h	8 h	1 h	8 h
Afghanistan	N/A	N/A	N/A	N/A	N/A	N/A	N/A	N/A	N/A	N/A	N/A	N/A	N/A	N/A	N/A	N/A
Bangladesh	65	15	150	50	N/A	N/A	N/A	365	80	N/A	N/A	100	235	157	40	10
Bhutan	N/A	N/A	100	60	200	140		80	60	80	80	60	N/A	N/A	4	2
PR China Grade I	35	15	50	40	120	80	150	50	20	200	80	40	160	100	10	N/A
PR China Grade II	75	35	150	70	300	200	500	150	60	200	80	40	200	160	10	N/A
India	60	40	100	60	N/A	N/A	N/A	80	50		80	40	180	100	4	2
India	60	40	100	60	N/A	N/A	N/A	80	20		80	30	180	100	4	2
Myanmar	N/A	N/A	N/A	N/A	N/A	N/A	N/A	N/A	N/A	N/A	N/A	N/A	N/A	N/A	N/A	N/A
Nepal	40	N/A	120	N/A	230	N/A	N/A	70	N/A	N/A	80	40	N/A	157	N/A	10
Pakistan	35	15	150	120	500	360	N/A	120	N/A	N/A	80	40	130	N/A	10	5
WHO Interim Target 1	75	35	150	70	N/A	N/A	N/A	125	N/A	N/A	N/A	N/A	N/A	160	N/A	N/A
WHO Interim Target 2	50	25	100	50	N/A	N/A	N/A	50	N/A	N/A	N/A	N/A	N/A	N/A	N/A	N/A
WHO Interim Target 3	37.5	15	75	30	N/A	N/A	N/A	N/A	N/A	N/A	N/A	N/A	N/A	N/A	N/A	N/A
WHO Air Quality Guideline	25	10	50	20	N/A	N/A	N/A	20	N/A	N/A	200	40	N/A	100	N/A	N/A

Source Clean Air Asia (2016)

There is some evidence of reforms in the region that have helped overcome some of these institutional challenges. In China, the performance of local officials on many of the air pollution targets is being used to determine whether and where officials will be promoted (Guo et al. 2014; Young et al. 2015). Meanwhile, in Delhi, India, the judicial system played an instrumental role in public transport introducing less polluting CNG (Goyal 2003).

10.5.4.2 Air Quality Management Programs and Initiatives in the HKH

China has instituted a series of increasingly stringent regulations to curb emissions of $PM_{2.5}$. These followed a set of serious smog episodes that beset Beijing and other cities in 2013. In response to the growing public concerns over the smog, the Ministry of Environmental Protection (MEP) published Guidelines for Preparing Emergency Response Plans for Heavy Air Pollution in Cities in May 2013 (General Office of the Ministry of Environmental Protection 2013). Air pollution was also deemed important by the top leaders of the Chinese government, and in September 2013, the State Council issued the Action Plan for Air Pollution Prevention and Control (State Council 2013). One week later, the MEP, together with five other ministries, jointly issued the Detailed Implementation Guidelines for Carrying Out the Plan for Air Pollution Prevention and Control in Beijing-Tianjin-Hebei and Surrounding Areas (Ministry of Environmental Protection 2013).

In addition to the aforementioned initiatives by China, there are also other intergovernmental agreements and frameworks in the HKH to reduce air pollution. We list some of the major ones below to illustrate the current efforts and emphasize the need for more collaboration within the region.

Malé Declaration on Transboundary Air Pollution (1998–2016)

The Malé Declaration on Control and Prevention of Air Pollution and Its Likely Transboundary Effects for South Asia was created in 1998 to bring together Bangladesh, Bhutan, India, Iran, Maldives, Nepal, Pakistan, and Sri Lanka in addressing transboundary air pollution and consequential impacts in South Asia. The agreement was signed by the Governing Council of South Asia Cooperative Environment Program (SACEP) and implemented by the United Nations Environment Programme (UNEP) and SACEP with funding from the Swedish International Development Cooperation Agency. The agreement has made progress on science and gained financial support from the Indian government that could increase ownership and influence. Bangladesh, Bhutan, Nepal, and Sri Lanka have submitted their national emissions inventory as a part of the agreement to prevent transboundary air pollution. There were five phases in total until 2016 and the main activities also included air pollution monitoring and modeling; health, crop, ecosystem, and corrosion impact assessment; integrated assessment modeling; development of policy responses; and awareness raising activities. More information can be found at http://www.sacep.org/?page_id=2725.

The Acid Deposition Monitoring Network in East Asia (EANET) (1993–Current)

EANET covers 13 countries and receives technical support from the Asia Centre for Air Pollution Research as the network centre in Niigata, Japan. The UNEP Regional Resource Centre for Asia and the Pacific serves as the EANET Secretariat. EANET has concentrated on compiling, evaluating, exchanging, and disseminating data. The network has made progress in standardizing and strengthening monitoring and building capacity, but most of this progress has been on science, not policy. Some have concluded that EANET is at the same point as Europe's Convention on Long-Range Transboundary Air Pollution (CLRTAP) in the late 1970s (see discussion of other networks) and that strengthening the implementation of national actions may be more productive than aspiring to emission targets and compliance mechanisms characteristic of a hard-law agreement (Schreurs 2011). More information can be found at http://www.eanet.asia.

Asia Pacific Clean Air Partnership (2014–Current)

The Asia Pacific Clean Air Partnership (APCAP) was launched in 2014 as a mechanism that has as one of its main goals bringing together many regional air pollution agreements in Asia. To help achieve that goal, a Joint Forum was first convened that same year. The Joint Forum aims enable countries and officials associated with different air pollution agreements to discuss pragmatic approaches to improving air quality. A second goal of the APCAP is to equip the region's policy makers with the most recent atmospheric science. To help achieve this goal, an APCAP science panel was created with eminent scientists in the region. Many of the panel members are contributing to a high-profile assessment report on air pollution in the Asia Pacific region. The assessment report will be the shared output of APCAP and the Climate and Clean Air Coalition.

Climate and Clean Air Coalition (2012–Current)

The Climate and Clean Air Coalition (CCAC) was launched in 2012 with seven initial partners (the governments of Bangladesh, Canada, Ghana, Mexico, Sweden, and the United States, along with the UNEP) to improve air quality and climate by reducing SLCPs, especially focusing on BC, methane, and hydrofluorocarbons (HFCs). It has now expanded to more than 50 country partners and even more non-state partners, although to date Bangladesh and Pakistan are the only HKH countries. The CCAC does not intend to promote actions on SLCPs (BC, HFCs, methane and tropospheric O_3) as a substitute to mitigating GHG; rather it recognizes the importance of complementing and supplementing GHG emissions under the United Nations Framework Convention on Climate Change. The CCAC aims to achieve its objectives through a set of seven sector-specific (diesel, waste management, brick kilns) and four cross-cutting (national action planning, urban health) initiatives. Each of these initiatives functions by raising awareness of impacts and mitigation strategies, improving scientific understanding of their impacts, and enhancing and developing national as well as regional actions. More information can be found at http://www.ccacoalition.org/en.

Other Sector-Specific Mechanisms

One of the mechanisms aiming to reduce emissions from residential cookstoves is the Global Alliance for Clean Cookstoves. The Global Alliance was established in 2010 as a private-public partnership with the explicit goal of increasing the number of clean cookstoves to 100 million by 2020. The Global Alliance works with national governments, non-governmental organizations, and the private sector to raise awareness, strengthen scientific evidence, develop standards and labels, and explore innovative financing for clean cookstoves programmes. The alliance shares a close relationship with the Partnership for Clean Indoor Air, which has focused on capacity building, technical assistance, project implementation, and knowledge management on indoor air pollution, household energy, and health. For several years, UNEP has worked with partners on Project Surya on a people-centred approach to support clean cookstoves in India. More information can be found at http://cleancookstoves.org.

There are also several mechanisms with the potential to provide support on clean diesel. The Partnership for Clean Fuels and Vehicles (PCFV) was launched at the World Summit on Sustainable Development in 2002 to assist developing countries in reducing mobile source air pollution. Having achieved success with the removal of lead, the PFCV is now working on lowering sulphur levels in diesel to below 50 ppm. Lowering sulphur levels would be a necessary precondition for installing diesel after-treatment devices. The PCFV has also gathered valuable insights into the aforementioned I&M programmes.

Relevant Experiences from Elsewhere

CLRTAP is one of the best known regional frameworks for reducing transboundary air pollution. It includes the first treaty in 1979 and eight associated protocols between 1984 and 1999 to reduce emissions of selected air pollutants in Europe and North America. CLRTAP offers an example of a science-based, legally binding convention that has been successfully implemented over the years. Each contracting party is responsible for developing the best policies and strategies, and CLRTAP emphasizes the control measures that are both economically feasible and the best available technology. It is also important to recognize the exchange of information that was essential among the countries.

Within Asia itself, another initiative is the ASEAN Haze Agreement. The agreement recognizes the problem of forest fires on air pollution, as well as the economic loss associated with it. Prevention, operational mechanisms, and enforcement are all included in the agreement.

10.6 Key Challenges in Reducing Air Pollution in the HKH

It is clear that rapid urbanisation and other changes in the HKH and surrounding areas have brought increased emissions from industry and transport that require regulation to reduce emissions. There are several key challenges to effectively mitigating the air pollution issues in the HKH, and in this section we list several major challenges in the region.

10.6.1 Recognition of Air Pollution as a Problem and Co-benefits of Mitigation

It is essential for the HKH country governments to recognize air pollution as a problem and to implement effective measures to mitigate it. Health benefits from mitigating air pollution are undeniably the most important and strategies that aim to achieve such benefits from reducing PM emissions from the transport sector and solid fuel use in cookstoves have significant potential in the region (UNEP 2011). Another important source of PM emissions, as illustrated earlier, is brick kilns. Cost-effective measures need to be scaled up to reduce these emissions.

In some cases, measures targeting SLCPs, especially BC, tropospheric O_3, and methane, will lead to co-benefits, including improved public health, reduced crop-yield losses, and reduced climate change impacts (UNEP 2011). There is sufficient evidence of BC concentrations increasing adverse health impacts, including all-cause and cardiovascular mortality, and cohort studies also provide similar evidence of long-term average BC exposure with these mortalities (Janssen et al. 2012). From such measures targeting SLCPs, the greatest benefit for regional climate is expected in the HKH (UNEP/WMO 2011).

10.6.2 Transboundary Pollution Flows, Collaborations, and Cooperation

The regional air pollution problems, including winter haze and fog, have strong implications for the region. As detailed in the previous sections, air pollution is a transboundary issue and has impacts on the region-wide weather, climate, and agriculture, as well as socioeconomic sectors. A coordinated South Asian effort is thus necessary to better understand the processes governing the severity of the regional air pollution in the HKH. Regional coordination and collaboration is essential to enhance the mitigation at the local and national levels (UNEP 2011).

Legally binding regional agreements, similar to the CLRTAP and the ASEAN Haze Agreement, could be used to provide a platform for policy action on controlling air pollution. Intergovernmental initiatives like the Malé Declaration could be an important basis for developing necessary scientific knowledge, in addition to raising awareness and enhancing capacity building for improving air quality in the region. In the HKH, where the impact of BC on climate change is clearly visible, coordinated regional action as seen in the Arctic is particularly important (UNEP 2011).

10.6.3 Implementation Gaps and Challenges

As highlighted in the previous section, there is no shortage of solutions to help manage air pollution in the HKH. Furthermore, in many cases the recommended policy options have already been promulgated in the region. There are, however, several persistent implementation gaps that separate what is written on paper from what is achieved on the ground. This section discusses some of the barriers behind these implementation gaps.

The vast majority of technical solutions require supportive technologies infrastructure to be implemented effectively. This is particularly true for transportation. Improvements in the quality and coverage of public transport, for instance, will fail to generate significant reductions unless there are well-paved and thoughtfully designed roads and hubs. Similarly, the penetration and diffusion of large-scale renewable installations depends heavily on the quality and quantity of the electricity grid. Supportive technologies also play a role in enabling the introduction and diffusion of smaller scale technologies. To illustrate, switching from biomass-based cookstoves to those powered by cleaner fuels requires infrastructure that can deliver a dependable supply of those fuels. In many cases, it is the combined absence of multiple supportive technologies that leads to a condition known as "lock-in," referring to the frequently observed resistance to change that permeates energy systems (Unruh 2000, 2002).

Another set of policy and institutional barriers can also lock in the current technologies and place a drag on reform. Arguably, the most prominent set of policy barriers involve energy subsidies. In many parts of the world, governments provide resources to bring down the costs of fossil fuels. However, in many cases these price supports have had the unintended consequence of placing cleaner fuels and energy sources at a price disadvantage. For example, excessively high subsidies (up to 90% some cases) on solar powered irrigation pumps in some states in India has reduced incentives for technical innovations on the one hand and limited the adoption to only the richest and politically connected farmers (Kishore et al. 2014). More troubling still is that efforts to rein in subsidies are often met with political backlash and reprisals that can force governments to reverse decisions to curb price supports. Further, while subsidies are intended to benefit chiefly the poor, they often bring disproportionate benefits to companies that extract and refine fuels and wealthier segments of the population (Coady et al. 2015).

A related set of barriers involves the institutions that are chiefly responsible for designing and implementing air pollution regulations. As mentioned in the discussion of air quality institutions, interagency coordination can be a major constraint on needed policy reforms. The majority of air pollution problems require more than one agency working together; absent institutional arrangements to facilitate that coordination, policy formulations are likely to be flawed from their inception. More commonly, a lack of institutional coordination surfaces as a barrier during policy implementation. This is particularly true because government agencies in some cases have missions that are closely aligned with those of the polluting entities. The end result can be that while one branch of the government works to enforce regulations, another works at cross-purposes to flout those regulations (Rock 2002; Bhatt et al. 2008).

A third set of institutional barriers relates to administrative capacities. There are multiple ways that a lack of capacity can impinge on regulatory enforcement. Arguably, the most obvious is that an insufficient workforce makes it difficult to enforce regulations. There is, however, also the need not just for sufficient numbers of people, but for people who possess needed skills and knowledge. Well-trained staff are particularly important for a policy area such as air pollution, as the topic is inherently complex and thus requires a clear understanding of the links between source, emissions, and impacts. Finally, it is worth noting that capacity is not simply necessary for effective enforcement; it can also prove useful in conceiving of innovative incentives for behavioural and lifestyle changes that may prevent emissions in the first place (Li and Zusman 2006; World Bank 2007).

A final set of challenges involves some of the regional and global air pollution initiatives discussed previously.

Many scholars of international relations suggest that these institutions would not struggle to regulate the behaviour of governments. The fact that these institutions focus on different countries and pollutants and tend to focus more on information sharing and raising awareness suggests there may be some merit to these claims. At the same time, there is also a body of literature that argues that information sharing and raising awareness play an important though less obvious role in shaping the interests of governments when it comes to transboundary environmental problems (Mitchell 2003, 2006). It is clear that for reducing air pollution in the HKH, policies for mitigation need to be implemented together with enhanced education and awareness-raising campaigns within and surrounding the HKH.

References

Adhikary, B., Carmichael, G. R., Tang, Y., Leung, L. R., Qian, Y., Schauer, J. J., et al. (2007). Characterization of the seasonal cycle of south Asian aerosols: A regional-scale modeling analysis. *Journal of Geophysical Research, 112*(D22), D22S22. http://doi.org/10.1029/2006JD008143.

Agarwal, O. P., & Zimmerman, S. L. (2008). Toward sustainable mobility in urban India. *Transportation Research Record: Journal of the Transportation Research Board, 2048*, 1–7.

Agrawal, S. (2012). Effect of indoor air pollution from biomass and solid fuel combustion on prevalence of self-reported asthma among adult men and women in India: Findings from a nationwide large-scale cross-sectional survey. *Journal of Asthma, 494*(4), 355–365. https://doi.org/10.3109/20770903.2012.663030.

Ali, K., Momin, G. A., Tiwari, S., Safai, P. D., Chate, D. M., & Rao, P. S. P. (2004). Fog and precipitation chemistry at Delhi, North India. *Atmospheric Environment, 38*(25), 4215–4222. https://doi.org/10.1016/j.atmosenv.2004.02.055.

Allred, E. N., Bleecker, E. R., Chaitman, B. R., Dahms, T. E., Gottlieb, S. O., Hackney, J. D., et al. (1989). Short-term effects of carbon monoxide exposure on the exercise performance of subjects with coronary artery disease. *The New England Journal of Medicine, 321*(21), 1426–1432. http://doi.org/10.1056/NEJM198911233212102.

Andreae, M. O., & Merlet, P. (2001). Emission of trace gases and aerosols from biomass burning. *Global Biogeochemical Cycles, 15*(4), 955–966.

Armstrong, B. G., & Gibbs, G. (2009). Exposure–response relationship between lung cancer and polycyclic aromatic hydrocarbons (PAHs). *Occupational and Environmental Medicine, 66*(11), 740–746. https://doi.org/10.1136/oem.2008.043711.

Aryal, R. K., Lee, B. K., Karki, R., Gurung, A., Kandasamy, J., Pathak, B. K., et al. (2008). Seasonal PM10 dynamics in Kathmandu Valley. *Atmospheric Environment, 42*(37), 8623–8633. http://doi.org/10.1016/j.atmosenv.2008.08.016.

Aryal, R. K., Lee, B. K., Karki, R., Gurung, A., Baral, B., & Byeon, S. H. (2009). Dynamics of PM2.5 concentrations in Kathmandu Valley, Nepal. *Journal of Hazardous Materials, 168*(2–3), 732–738. http://doi.org/10.1016/j.jhazmat.2009.02.086.

Asian Productivity Organization. (2007). Solid-waste management issues and challenges in Asia. Report of the APO Survey on Solid-Waste Management 2004–05. Published by the Asian Productivity Organization. ISBN: 92-833-7058-9.

Avnery, S., Mauzerall, D. L., Liu, J., & Horowitz, L. W. (2011). Global crop yield reductions due to surface ozone exposure: 1. Year 2000 crop production losses and economic damage. *Atmospheric Environment, 45*(13), 2284–2296. http://doi.org/10.1016/j.atmosenv.2010.11.045.

Babu, S. S., Moorthy, K. K., Manchanda, R. K., Sinha, P. R., Satheesh, S. K., Vajja, D. P., et al. (2011a). Free tropospheric black carbon aerosol measurements using high altitude balloon: Do BC layers build "their own homes" up in the atmosphere? *Geophysical Research Letters, 38*(8). http://doi.org/10.1029/2011GL046654.

Babu, S. S., Chaubey, J. P., Krishna Moorthy, K., Gogoi, M. M., Kompalli, S. K., Sreekanth, V., et al. (2011b). High altitude (~4520 m amsl) measurements of black carbon aerosols over western trans-Himalayas: Seasonal heterogeneity and source apportionment. *Journal of Geophysical Research: Atmospheres, 116*(D24). http://doi.org/10.1029/2011JD016722.

Babu, S. S., Manoj, M. R., Moorthy, K. K., Gogoi, M. M., Nair, V. S., Kompalli, S. K., et al. (2013). Trends in aerosol optical depth over Indian region: Potential causes and impact indicators. *Journal of Geophysical Research Atmospheres, 118*(20), 11794–11806. http://doi.org/10.1002/2013JD020507.

Balakrishnan, K., Sambandam, S., et al. (2004). Exposure assessment for respirable particulates associated with household fuel use in rural districts of Andhra Pradesh, India. *Journal of Exposure Analysis and Environmental Epidemiology, 1*(14 Suppl), S14–S25.

Balakrishnan, K., Ramaswamy, P., Sambandam, S., Thangavel, G., Ghosh, S., Johnson, P., et al. (2011). Air pollution from household solid fuel combustion in India: An overview of exposure and health related information to inform health research priorities. *Global Health Action, 4*, 1–9. http://doi.org/10.3402/gha.v4i0.5638.

Baillis, et al. (2009). Arresting the killer in the kitchen: The promises and pitfalls of commercializing improved cookstoves. *World Development, 37*(10), 1694–1705.

Bangladesh DHS (Demographic and Health Survey). (2011). Retrieved January, 2013, from http://dhsprogram.com/publications/publication-FR265-DHS-Final-Reports.cfm.

Barnes, B. R. (2014). Behavioural change, indoor air pollution and child respiratory health in developing countries: A review. *International Journal of Environmental Research and Public Health, 11*(5), 4607–4618. https://doi.org/10.3390/ijerph110504607.

Beaulieu, J. J., Tank, J. L., Hamilton, S. K., Wollheim, W. M., Hall, R. O., Mulholland, P. J., et al. (2011). Nitrous oxide emission from denitrification in stream and river networks. *Proceedings of the National Academy of Sciences of the United States of America, 108*(1), 214–219. http://doi.org/10.1073/pnas.1011464108.

Beegum, S. N., Moorthy, K. K., Babu, S. S., Satheesh, S. K., Vinoj, V., Badarinath, K. V. S., et al. (2009). Spatial distribution of aerosol black carbon over India during pre-monsoon season. *Atmospheric Environment, 43*(5), 1071–1078. http://doi.org/10.1016/j.atmosenv.2008.11.042.

Bell, M. L., & Davis, D. L. (2001). Reassessment of the lethal London fog of 1952: Novel indicators of acute and chronic consequences of acute exposure to air pollution. *Environmental Health Perspectives, 109*(3), 389–394.

Bhatt, M. S., Ashraf, S., & Illiyan, A. (2008). *Problems and prospects of environment policy: Indian perspective*. New Delhi: Aakar Books.

Bhutan Living Standards Survey. (2013). Government of Bhutan and Asian Development Bank.

Bisht, G., & Neupane, S. (2015). Impact of Brick Kilns' emission on soil quality of agriculture fields in the vicinity of selected Bhaktapur Area of Nepal. *Applied and Environmental Soil Science*. http://doi.org/10.1155/2015/409401.

Biswas, K. F., Ghauri, B. M., & Husain, L. (2008). Gaseous and aerosol pollutants during fog and clear episodes on South Asian urban atmosphere. *Atmospheric Environment, 42*, 7775–7785.

Bollasina, M., Ramaswamy, V., & Ming, Y. (2011). Anthropogenic aerosols and the weakening of the South Asian summer monsoon. *Science, 334*(6055), 502–505. https://doi.org/10.1126/science.1204994.

Bollasina, M. A., Bertolani, L., & Tartari, G. (2002). Meteorological observations in the Khumbu Valley, Nepal Himalayas, 1994–1999. *Bulletin of Glaciological Research, 19,* 1–11.

Bonasoni, P., Laj, P., Angelini, F., Arduini, J., Bonafè, U., Calzolari, F., et al. (2008). The ABC-Pyramid Atmospheric Research Observatory in Himalaya for aerosol, ozone and halocarbon measurements. *Science of the Total Environment, 391*, 252–261.

Bonasoni, P., Laj, P., Marinoni, A., Sprenger, M., Angelini, F., Arduini, J., et al. (2010). Atmospheric brown clouds in the Himalayas: First two years of continuous observations at the Nepal Climate Observatory-Pyramid (5079 m). *Atmospheric Chemistry and Physics, 10*(15), 7515–7531. https://doi.org/10.5194/acp-10-7515-2010.

Bonasoni, P., Cristofanelli, P., Marinoni, A., Vuillermoz, E., & Adhikary, B. (2012). Atmospheric pollution in the Hindu Kush-Himalaya Region. *Mountain Research and Development, 32* (4), 468–479. https://doi.org/10.1659/MRD-JOURNAL-D-12-00066.1.

Booker, F., Muntifering, R., McGrath, M., Burkey, K., Decoteau, D., Fiscus, E., et al. (2009). The ozone component of global change: Potential effects on agricultural and horticultural plant yield, product quality and interactions with invasive species. *Journal of Integrative Plant Biology, 51*(4), 337–351.

Boy, E., Bruce, N., et al. (2002). Birth weight and exposure to kitchen wood smoke during pregnancy in rural Guatemala. *Environmental Health Perspectives, 110*(1), 109–114.

Brook, R. D., Rajagopalan, S., Pope III, C. A., Brook, J. R., Bhatnagar, A., Diez-Roux, A. V., et al. (2010). Particulate matter air pollution and cardiovascular disease. An update to the Scientific Statement from the American Heart Association. *Circulation, 121*(21), 2331–2378.

Bruce, N., Perez-Padilla, R., & Albalak, R. (2000). Indoor air pollution in developing countries: A major environmental and public health challenge. *Bulletin of the World Health Organization, 78,* 1078–1092.

Bruce, N., Weber, M., et al. (2007). Pneumonia case-finding in the RESPIRE Guatemala indoor air pollution trial: Standardizing methods for resource-poor settings. *Bulletin of the World Health Organization, 85*(7), 535–544.

Bruce, N., Rehfuess, E., et al. (2006). Indoor air pollution. In D. T. Jamison, J. G. Breman, A. R. Measham, et al. (Eds.), *Disease control priorities in developing countries*, Chapter 42 (2nd ed.). World Bank, Washington, DC. Retrieved from: http://www.ncbi.nlm.nih.gov/books/NBK11760/.

Brun, J., Shrestha, P., & Barros, A. P. (2011). Mapping aerosol intrusion in Himalayan valleys using the Moderate Resolution Imaging Spectroradiometer (MODIS) and Cloud-Aerosol Lidar and Infrared Pathfinder Satellite Observation (CALIPSO). *Atmospheric Environment, 45,* 6382–6392.

Burstyn, I., Boffetta, P., Heederik, D., Partanen, T., Kromhout, H., Svane, O., et al. (2003). Mortality from obstructive lung diseases and exposure to polycyclic aromatic hydrocarbons among asphalt workers. *American Journal of Epidemiology, 158*(5), 468–478. http://doi.org/10.1093/aje/kwg180.

Cao, J.-J., Xu, B.-Q., He, J.-Q., Liu, X.-Q., Han, Yo.-M., Wang, G., et al. (2009). Concentrations, seasonal variations, and transport of carbonaceous aerosols at a remote Mountainous region in western China. *Atmospheric Environment, 43*, 4444–4452.

Cao, J., Chow, J. C., Lee, F. S. C., & Watson, J. G. (2013). Evolution of $PM_{2.5}$ measurements and standards in the U.S. and future perspectives for China. *Aerosol and Air Quality Research, 13*(4), 1197–1211. http://doi.org/10.4209/aaqr.2012.11.0302.

Carrico, C. M., Bergin, M. H., Shrestha, A. B., Dibb, J. E., Gomes, L., & Harris, J. M. (2003). The importance of carbon and mineral dust to seasonal aerosol properties in the Nepal Himalaya. *Atmospheric Environment, 37,* 2811–2824.

Chakraborty, A., Bhattu, D., Gupta, T., Tripathi, S. N., & Canagaratna, M. R. (2015). Real-time measurements of ambient aerosols in a polluted Indian city: Sources, characteristics and processing of organic aerosols during foggy and non-foggy periods. *Journal of Geophysical Research, 120,* 9006–9019. https://doi.org/10.1002/2015jd023419.

Che, H., Wang, Y., & Sun, J. (2011). Aerosol optical properties at Mt. Waliguan Observatory. *China Atmospheric Environment, 45,* 6004–6009.

Chen, S., Huang, J., Zhao, C., Qian, Y., Leung, L. R., & Yang, B. (2013). Modeling the transport and radiative forcing of Taklimakan dust over the Tibetan Plateau: A case study in the summer of 2006. *Journal of Geophysical Research: Atmospheres, 118*(2), 797–812. https://doi.org/10.1002/jgrd.50122.

Chen, P. F., Li, C. L., Kang, S. C., Yan, P., Zhang, Q. G., Ji, Z. M., et al. (2016). Source apportionment of particle-bound polycyclic aromatic hydrocarbons in Lumbini, Nepal by using the positive matrix factorization receptor model. *Atmospheric Research, 182,* 46–53.

Cherian, R., Venkataraman, C., Quaas, J., & Ramachandran, S. (2013). GCM simulations of anthropogenic aerosol-induced changes in aerosol extinction, atmospheric heating and precipitation over India. *Journal of Geophysical Research Atmospheres, 118*(7), 2938–2955.

Chhabra, S. K., Chhabra, P., Rajpal, S., & Gupta, R. K. (2001). Ambient air pollution and chronic respiratory morbidity in Delhi. *Archives of Environmental Health, 56*(1), 58–64. https://doi.org/10.1080/00039890109604055.

Chung, C., & Ramanathan, V. (2006). Weakening of the North Indian SST gradients and the monsoon rainfall in India and the Sahel. *Journal of Climate, 19,* 2036–2045.

Clark, M. L., & Peel, J. L. (2014). Perspectives in household air pollution research: Who will benefit from interventions? *Current Environmental Health Reports, 1,* 250–257.

Clean Air Asia. (2016). *Guidance framework for better air quality in Asian cities*. Retrieved from http://cleanairasia.org/ibaq-programme/.

Coady, D., Flamini, V., & Sears, L. (2015). *The unequal benefits of fuel subsidies revisited: Evidence for developing countries*. Washington D.C.

Collier, J. C., & Zhang, G. J. (2009). Aerosol direct forcing of the summer Indian monsoon as simulated by the NCAR CAM3. *Climate Dynamics, 32*(2–3), 313–332. https://doi.org/10.1007/s00382-008-0464-9.

Cong, Z., Kang, S., Liu, X., & Wang, G. (2007). Elemental composition of aerosol in the Nam Co region, Tibetan Plateau, during summer monsoon season. *Atmospheric Environment, 41,* 1180–1187.

Cong, Z., Kang, S., Gao, S., Zhang, Y., Li, Q., & Kawamura, K. (2013). Historical trends of atmospheric black carbon on Tibetan Plateau as reconstructed from a 150-year lake sediment record. *Environmental Science and Technology, 47,* 2579–2586.

Cong, Z., Kawamura, K., Kang, S., & Fu, P. (2015). Penetration of biomass-burning emissions from South Asia through the Himalayas: New insights from atmospheric organic acids. *Scientific Report., 5,* 9580. https://doi.org/10.1038/srep09580.

Costa, S., Ferreira, J., Silveira, C., Costa, C., Lopes, D., Relvas, H., et al. (2014). Integrating health on air quality assessment—Review report on health risks of two major European outdoor air pollutants: PM and NOx. *Journal of Toxicology and Environmental Health.*

Part B, Critical Reviews, *17*(6), 307–340. http://doi.org/10.1080/10937404.2014.946164.

Cristofanelli, P., Bracci, A., Sprenger, M., Marinoni, A., Bonafè, U., Calzolari, F., et al. (2010). Tropospheric ozone variations at the Nepal Climate Observatory-Pyramid (Himalayas, 5079 m a.s.l.) and influence of deep stratospheric intrusion events. *Atmospheric Chemistry and Physics, 10*(14), 6537–6549. https://doi.org/10.5194/acp-10-6537-2010.

Dallmann, T., & Bandivadekar, A. (2016). *India Bharat Stage Vi Emission Standards*. Retrieved from http://www.theicct.org/sites/default/files/publications/IndiaBS%20VI%20Policy%20Update%20vF.pdf.

De, U. S., & Dandekar, M. M. (2001). Natural disasters in urban areas. *Deccan Geogr, 39*(2), 1–12.

Dey, S., & Di Girolamo, L. (2010). A climatology of aerosol optical and microphysical properties over the Indian subcontinent from 9 years (2000–2008) of Multiangle Imaging Spectroradiometer (MSIR) data. *Journal of Geophysical Research, 115*(D15204). http://doi.org/10.1029/2009JD013395.

Dherani, M., Pope, D., et al. (2008). Indoor air pollution from unprocessed solid fuel use and pneumonia risk in children aged under five years: A systematic review and meta-analysis. *Bulletin of the World Health Organization, 86*(5), 390–398.

Dhungel, S., Kathayat, B., Mahata, K., & Panday, A. (2016). Transport of regional pollutants through a remote trans-Himalayan valley in Nepal. *Atmospheric Chemistry and Physics Discussions*, 1–23. https://doi.org/10.5194/acp-2016-824.

Di Girolamo, L., Bond, T. C., Bramer, D., Diner, D. J., Fettinger, F., Kahn, R. A., et al. (2004). Analysis of multi-angle Imaging SpectroRadiometer (MISR) aerosol optical depths over greater India during winter 2001–2004. *Geophysical Research Letters, 31*(23), 1–5. http://doi.org/10.1029/2004GL021273.

DoHS. (2014). Department of Health Services 2070/71 (2013/14) Annual Report. Teku, Kathmandu.

Dumka, U. C., Krishna Moorthy, K., Kumar, R., Hegde, P., Sagar, R., Pant, P., et al. (2010). Characteristics of aerosol black carbon mass concentration over a high altitude location in the Central Himalayas from multi-year measurements. *Atmospheric Research, 96*, 510–521.

Dumka, U. C., Kaskaoutis, D. G., Srivastava, M. K., & Devara, P. C. S. (2015). Scattering and absorption properties of near-surface aerosol over Gangetic-Himalayan region: The role of boundary-layer dynamics and long-range transport. *Atmospheric Chemistry and Physics, 15*(3), 1555–1572. http://doi.org/10.5194/acp-15-1555-2015.

Edwards, R. D., Liu, Y., He, G., Yin, Z., Sinton, J., Peabody, J., et al. (2007). Household CO and PM measured as part of a review of China's National Improved Stove Program. *Indoor Air, 17*(3), 189–203.

Epstein, M. B., Bates, M. N., Arora, N. K., Balakrishnan, K., Jack, D. W., & Smith, K. R. (2013). Household fuels, low birth weight, and neonatal death in India: The separate impacts of biomass, kerosene, and coal. *International Journal of Hygiene and Environmental Health, 216*(5), 523–532. https://doi.org/10.1016/j.ijheh.2012.12.006.

Ezzati, M., & Kammen, D. (2001). Indoor air pollution from biomass contribution and acute respiratory infections in Kenya: An exposure-response study. *Lancet, 358*, 619–624.

Ezzati, et al. (2004). Energy management and global health. *Annual Review of Environmental Resources, 29*.

Farmer, S. A., Nelin, T. D., Falvo, M. J., & Wold, L. E. (2014). Ambient and household air pollution: Complex triggers of disease. *American Journal of Physiology Heart and Circulatory Physiology, 307*(4), H467–H476. https://doi.org/10.1152/ajpheart.00235.2014.

Firdaus, G., & Ahmad, A. (2011). Changing air quality in Delhi, India: Determinants, trends, and policy implications. *Regional Environmental Change, 11*(4), 743–752. https://doi.org/10.1007/s10113-011-0207-z.

Flanner, M. G., Zender, C. S., Hess, P. G., Mahowald, N. M., Painter, T. H., Ramanathan, V., et al. (2009). Springtime warming and reduced snow cover from carbonaceous particles. *Atmospheric Chemistry and Physics, 9*, 2481–2497. https://doi.org/10.5194/acp-9-2481-2009.

Forouzanfar, M. H., & Collaborators, G. 2013 R. F. (2016). Global, regional, and national comparative risk assessment of 79 behavioural, environmental and occupational, and metabolic risks or clusters of risks in 188 countries, 1990–2013: A systematic analysis for the Global Burden of Disease Study 2013. *The Lancet, 386* (10010), 2287–2323. http://doi.org/10.1016/S0140-6736(15)00128-2.Global.

Foster, A., & Kumar, N. (2011). Health effects of air quality regulations in Delhi, India. *Atmospheric Environment, 45*(9), 1675–1683. https://doi.org/10.1016/j.atmosenv.2011.01.005.

Fuhrer, J. (2009). Ozone risk for crops and pastures in present and future climates. *Naturwissenschaften, 96*, 173–194.

Gadi, R., Singh, N., Sarkar, A. K., & Parashar, D. C. (2000). Mass size distribution and chemical composition of aerosols at New Delhi. In K. Siddappa, S. Sadasvian, V. N. Sastri, K. P. Eappen, R. K. Somashekar, & N. Manamoha Rao (Eds.), *Pollution urban environment* (Vol. 488, pp. 30–32). Banglore: Banglore University.

Ganguly, D., Jayaraman, A., & Gadhavi, H. (2006). Physical and optical properties of aerosols over an urban location in western India: Seasonal variabilities. *Journal of Geophysical Research, 111* (D24206). https://doi.org/10.1029/2006jd007392.

Ganguly, D., Ginoux, P., Ramaswamy, V., Winker, D. M., Holben, B. N., & Tripathi, S. N. (2009). Retrieving the composition and concentration of aerosols over the Indo-Gangetic basin using CALIOP and AERONET data. *Geophysical Research Letters, 36* (13), 1–5. https://doi.org/10.1029/2009GL038315.

Ganguly, D., Rasch, P. J., Wang, H., & Yoon, J. H. (2012). Climate response of the South Asian monsoon system to anthropogenic aerosols. *Journal of Geophysical Research Atmospheres, 117*(13), 1–20. https://doi.org/10.1029/2012JD017508.

Gautam, R., Hsu, N. C., Kafatos, M., & Tsay, S. C. (2007). Influences of winter haze on fog/low cloud over the Indo-Gangetic plains. *Journal of Geophysical Research Atmospheres, 112*(5), 1–11. https://doi.org/10.1029/2005JD007036.

Gautam, R., Liu, Z., Singh, R. P., & Hsu, N. C. (2009a). Two contrasting dust dominant periods over India observed from MODIS and CALIPSO data. *Geophysical Research Letters, 36*, L06813. https://doi.org/10.1029/2008gl036967.

Gautam, R., Hsu, N. C., Lau, K.-M., & Kafatos, M. (2009b). Aerosol and rainfall variability over the Indian monsoon region: Distribution, trends, and coupling. *Annales Geophysicae, 27*, 3691–3703.

Gautam, R., Hsu, N. C., & Lau, K.-M. (2010). Premonsoon aerosol characterization and radiative effects over the Indo-Gangetic Plains: Implications for regional climate warming. *Journal of Geophysical Research, 115*(D17), D17208. https://doi.org/10.1029/2010JD013819.

Gautam, R., Hsu, N. C., Lau, W. K. M., & Yasunari, T. J. (2013). Satellite observations of desert dust-induced Himalayan snow darkening. *Geophysical Research Letters, 40*(5), 988–993. https://doi.org/10.1002/grl.50226.

Gautam, R. (2014). Challenges in early warning of the persistent and widespread winter fog over the indo-gangetic plains: A satellite perspective in early warning systems for climate change. In *Reducing disaster: Early warning systems for climate change* (pp. 51–61).

General Office of the Ministry of Environmental Protection. (2013). *Letter on issuing the guidelines for preparing emergency response plan for heavy air pollution in cities* (edited).

Gertler, C. G., Puppala, S. P., Panday, A., Stumm, D., & Shea, J. (2016). Black carbon and the Himalayan cryosphere: A review. *Atmospheric Environment, 125*, 404–417. https://doi.org/10.1016/j.atmosenv.2015.08.078.

Ghude, S. D., Jena, C., Chate, D. M., Beig, G., Pfister, G. G., Kumar, R., et al. (2014). Reductions in India's crop yield due to ozone. *Geophysical Research Letters, 41*(15), 5685–5691. https://doi.org/10.1002/2014GL060930.

Ghude, S. D., Bhat, G. S., Prabhakaran, T., Jenamani, R. K., Chate, D. M., Safai, P. D., et al. (2017). Winter fog experiment over the Indo-Gangetic plains of India. *Current Science, 112*(4). http://doi.org/10.18520/cs/v112/i04/767-784.

Giles, D. M., Holben, B. N., Tripathi, S. N., Eck, T. F., Newcomb, W. W., Slutsker, I., et al. (2011). Aerosol properties over the Indo-Gangetic Plain: A mesoscale perspective from the TIGERZ experiment. *Journal of Geophysical Research, 116*(D18), D18203. http://doi.org/10.1029/2011JD015809.

Ginot, P., Dumont, M., Lim, S., Patris, N., Taupin, J., Wagnon, P., et al. (2014). A 10 year record of black carbon and dust from a Mera Peak ice core (Nepal): Variability and potential impact on melting of Himalayan glaciers. *The Cryosphere*, 1479–1496. https://doi.org/10.5194/tc-8-1479-2014.

Gobbi, G. P., Angelini, F., Bonasoni, P., Verza, G. P., Marinoni, A., & Barnaba, F. (2010). Sunphotometry of the 2006–2007 aerosol optical/radiative properties at the Himalayan Nepal Climate Observatory-Pyramid (5079 m a.s.l.). *Atmospheric Chemistry and Physics, 10*(22), 11209–11221. https://doi.org/10.5194/acp-10-11209-2010.

Goyal, P. (2003). Present scenario of air quality in Delhi: A case study of CNG implementation. *Atmospheric Environment, 37*(38), 5423–5431.

Granier, C., Bessagnet, B., Bond, T., Angiola, A. D., Gon, H. D., Van Der Frost, G. J., et al. (2011). Evolution of anthropogenic and biomass burning emissions of air pollutants at global and regional scales during the 1980–2010 period. *Climatic Change*, 109–163. http://doi.org/10.1007/s10584-011-0154-1.

Greentech. (2014). Factsheets about brick kilns in South Asia and Southeast Asia. Retrieved October 18, 2016, from http://www.ccacoalition.org/sites/default/files/resources/Bricks-SEA.pdf.

Gul, C., Praveen, P. S., Kang, S., Adhikary, B., Zhang, Y., Ali, S., et al. (2018). Concentrations and source regions of light-absorbing particles in snow/ice in northern Pakistan and their impact on snow albedo. *Atmospheric Chemistry and Physics, 18*(4981–5000), 2018.

Gultepe, I., Tardif, R., Michaelides, S. C., Cermak, J., Bott, A., Bendix, J., et al. (2007). Fog research: A review of past achievements and future perspectives. *Pure and Applied Geophysics, 164*, 1127–1159.

Guo, J., Zusman, E., & Moe, E. (2014). Enabling China's low-carbon transition: The 12th Five-Year Plan and the future climate regime. *The political economy of renewable energy and energy security* (pp. 241–257). London: Palgrave Macmillan.

Guo, J., Kang, S., Huang, J., Zhang, Q., Rupakheti, M., Sun, S., et al. (2017). Characterizations of atmospheric particulate-bound mercury in the Kathmandu Valley of Nepal, South Asia. *Science of the Total Environment, 579*. http://doi.org/10.1016/j.scitotenv.2016.11.110.

Gupta, R. (2014). Low-hanging fruit in black carbon mitigation: Crop residue burning in South Asia. *Climate Change Economics, 5*(4), 1450012.

Gupta, S. K., Gupta, S. C., Agarwal, R., Sushma, S., Agrawal, S. S., & Saxena, R. (2007). A multicentric case-control study on the impact of air pollution on eyes in a metropolitan city of India. *Indian Journal of Occupational and Environmental Medicine, 11*(1), 37–40. https://doi.org/10.4103/0019-5278.32463.

Gurjar, B. R., Ravindra, K., & Singh, A. (2016). Air pollution trends over Indian megacities and their local-to-global implications. *Atmospheric Environment, 142*, 475–495.

Gustafsson, Ö., Kruså, M., Zencak, Z., Sheesley, R. J., Granat, L., Engström, E., et al. (2009). Brown clouds over South Asia: Biomass or fossil fuel combustion. *Science, 323*, 495–498.

Guttikunda, S. K., & Koppaka, R. (2012). Co-benefits of an air quality management action plan for transport sector in Hyderabad, India. In *Low carbon transport in Asia: Capturing climate and development co-benefits*.

Guyon, P., Frank, G. P., Welling, M., Chand, D., Artaxo, P., Rizzo, L., et al. (2005). Airborne measurements of trace gas and aerosol particle emissions from biomass burning in Amazonia. *Atmospheric Chemistry and Physics, 5*(11), 2989–3002. http://doi.org/10.5194/acp-5-2989-2005.

Habib, G., Venkataraman, C., Chiapello, I., Ramachandran, S., Boucher, O., & Shekar Reddy, M. (2006). Seasonal and interannual variability in absorbing aerosols over India derived from TOMS: Relationship to regional meteorology and emissions. *Atmospheric Environment, 40*(11), 1909–1921. https://doi.org/10.1016/j.atmosenv.2005.07.077.

Haldane, J. (1912). The dissociation of oxyhaemoglobin in human blood during partial CO poisoning. *Journal of Physiology, 45*, 22–24.

Hameed, S., Ishaq Mirza, M., Ghauri, B. M., Siddiqui, Z. R., Javed, R., Khan, A. R., et al. (2000). On the widespread winter fog in Northeastern Pakistan and India. *Geophysical Research Letters, 27* (13), 1891–1894. http://doi.org/10.1029/1999GL011020.

Hausker, K. (2004). *Vehicle inspection and maintenance programs: International experience and best practices*. Washington D.C.

Health Effects Institute. (2017). *State of global air 2017*. Special Report. Boston, MA: Health Effects Institute.

Hegde, P., Pant, P., Naja, M., Dumka, U. C., & Sagar, R. (2007). South Asian dust episode in June 2006: Aerosol observations in the central Himalayas. *Geophysical Research Letters, 34*, L23802. https://doi.org/10.1029/2007gl030692.

Helfand, W. H., Lazarus, J., & Theerman, P. (2001). Donora, Pennsylvania: An environmental disaster of the 20th century. *American Journal of Public Health, 91*(4), 553.

Henriksson, S. V., Laaksonen, A., Kerminen, V. M., Räisänen, P., Järvinen, H., Sundström, A. M., et al. (2011). Spatial distributions and seasonal cycles of aerosols in India and China seen in global climate-aerosol model. *Atmospheric Chemistry and Physics, 11*(15), 7975–7990. http://doi.org/10.5194/acp-11-7975-2011.

Hogg, J. C., Chu, F., Utokaparch, S., Woods, R., Elliott, W. M., Buzatu, L., et al. (2004). The nature of small-airway obstruction in chronic obstructive pulmonary disease. *The New England Journal of Medicine, 350*(26), 2645–2653.

Hoornweg, D., & Bhada-Tata, P. (2012). *What a waste: A global review of solid waste management*. Washington, D.C.: World Bank.

Horton, R. (2012). GBD 2010: Understanding disease, injury, and risk. *The Lancet*. http://doi.org/10.1016/S0140-6736(12)62133-3.

Hsu, N. C., Gautam, R., Sayer, A. M., Bettenhausen, C., Li, C., Jeong, M. J., et al. (2012). Global and regional trends of aerosol optical depth over land and ocean using SeaWiFS measurements from 1997 to 2010. *Atmospheric Chemistry and Physics, 12*(17), 8037–8053. http://doi.org/10.5194/acp-12-8037-2012.

Huang, J., Minnis, P., Yi, Y., Tang, Q., Wang, X., Hu, Y., et al. (2007). Summer dust aerosols detected from CALIPSO over the Tibetan Plateau. *Geophysical Research Letters, 34*, L18805. https://doi.org/10.1029/2007gl029938.

Hyvärinen, A.-P., Raatikainen, T., Brus, D., Komppula, M., Panwar, T. S., Hooda, R. K., et al. (2011a). Effect of the summer monsoon on aerosols at two measurement stations in Northern India—Part 1: PM and BC concentrations. *Atmospheric Chemistry and Physics, 11* (16), 8271–8282. http://doi.org/10.5194/acp-11-8271-2011.

Hyvärinen, A.-P., Raatikainen, T., Komppula, M., Mielonen, T., Sundström, A.-M., Brus, D., et al. (2011b). Effect of the summer monsoon on aerosols at two measurement stations in Northern India —Part 2: Physical and optical properties. *Atmospheric Chemistry and Physics, 11*(16), 8283–8294. http://doi.org/10.5194/acp-11-8283-2011.

ICIMOD. (2007). *Kathmandu valley environment outlook.* Kathmandu: International Centre for Integrated Mountain Development (ICIMOD).

IIPS & ORC Macro. (2007). *National Family Halth Survey 3 (NFHS3) 2005–06.* Mumbai: International Institute for Population Sciences and Macro International.

Imhoff, R. E., Valente, R., Meagher, J. F., & Luria, M. (1995). The production of O3 in an urban plume: Airborne sampling of the Atlanta urban plume. *Atmospheric Environment, 29*(17), 2349–2358.

Jaffee, D., et al. (1999). Transport of Asian air pollution to North America. *Geophysical Research Letters, 26*(6), 711–714.

Jain, N., Bhatia, A., & Pathak, H. (2014). Emission of air pollutants from crop residue burning in India. *Aerosol and Air Quality Research, 14,* 422–430.

Janssen, N. A., Gerlofs-Nijland, M. E., Lanki, T., Salonen, R. O., Cassee, F., Hoek, G., et al. (2012). *Health effects of black carbon.*

Jayarathne, T., Stockwell, C. E., Christian, T. J., Bhave, P. V., Rathnayake, C., Robiul Islam, Md., et al. (2017). Nepal Ambient Monitoring and Source Testing Experiment (NAMaSTE): Emissions of particulate matter from wood and dung cooking fires, brick kilns, generators, trash and crop residue burning. *Atmospheric Chemistry and Physics* (in press).

Jenamani, R. K. (2007). Alarming rise in fog and pollution causing a fall in maximum temperature over Delhi. *Current Science, 93*(3), 314–322.

Jenamani, R. K., & Tyagi, A. (2011). Monitoring fog at IGI Airport and analysis of its runway-wise spatio-temporal variations using Meso-RVR network. *Current Science, 100*(4), 491–501.

Jenkins, M., Kaspari, S., Kang, S., Grigholm, B., & Mayewski, P. a. (2013). Black carbon concentrations from a Tibetan Plateau ice core spanning 1843–1982: Recent increases due to emissions and glacier melt. *The Cryosphere Discussions, 7*(5), 4855–4880. http://doi.org/10.5194/tcd-7-4855-2013.

Jerrett, M., Burnett, R. T., Pope, C. A., Ito, K., Thurston, G., Krewski, D., et al. (2009). Long-term ozone exposure and mortality. *New England Journal of Medicine, 360*(11), 1085–1095. http://doi.org/10.1056/NEJMoa0803894.

Jethva, H., Satheesh, S. K., & Srinivasan, J. (2005). Seasonal variability of aerosols over the Indo-Gangetic basin. *Journal of Geophysical Research Atmospheres, 110*(21), 1–15. https://doi.org/10.1029/2005JD005938.

Ji, Z., Kang, S., Cong, Z., Zhang, Q., & Yao, T. (2015). Simulation of carbonaceous aerosols over the Third Pole and adjacent regions: Distribution, transportation, deposition, and climatic effects. *Climate Dynamics, 45*(9–10), 2831–2846. https://doi.org/10.1007/s00382-015-2509-1.

Ji, Z., S. Kang, Q. Zhang, Z. Cong, P. Chen, M. Sillanpää. (2016). Investigation of mineral aerosols radiative effects over High Mountain Asia in 1990–2009 using a regional climate model. *Atmospheric Research, 178–179,* 484–496. https://doi.org/10.1016/j.atmosres.2016.05.003.

Jin, et al. (2006). Exposure to indoor air pollution from household energy use in rural China: The interactions of technology, behavior, and knowledge in health risk management. *Social Science and Medicine, 62,* 3161–3176.

Joshi, H., Naja, M., Singh, K. P., Kumar, R., Bhardwaj, P., Babu, S. S., et al. (2016). Investigations of aerosol black carbon from a semi-urban site in the Indo-Gangetic Plain region. *Atmospheric Environment, 125,* 346–359.

Kang, S., Wake, C. P., Qin, D., Mayewski, P. A., & Yao, T. (2000). Monsoon and dust signals recorded in Dasuopu Glacier, Tibetan Plateau. *Journal of Glaciology, 46*(153), 222–226.

Kang, S., Chen, P., Li, C., Liu, B., & Cong, Z. (2016). Atmospheric aerosol elements over the inland Tibetan Plateau: Concentration, seasonality, and transport. *Aerosol and Air Quality Research, 16,* 789–800. https://doi.org/10.4209/aaqr.2015.05.0307.

Kaskoutis, D. G., Sinha, P. R., Vinoj, V., Kosmopoulos, P. G., Tripathi, S. N., Misra, A., et al. (2013). Aerosol properties and radiative forcing over Kanpur during severe aerosol loading conditions. *Atmospheric Environment, 79,* 7–19.

Kaspari, S. D., Schwikowski, M., Gysel, M., Flanner, M. G., Kang, S., Hou, S., et al. (2011). Recent increase in black carbon concentrations from a Mt. Everest ice core spanning 1860–2000 AD. *Geophysical Research Letters, 38*(4), 11–16. http://doi.org/10.1029/2010GL046096.

Kaspari, S., Painter, T. H., Gysel, M., Skiles, S. M., & Schwikowski, M. (2014). Seasonal and elevational variations of black carbon and dust in snow and ice in the Solu-Khumbu, Nepal and estimated radiative forcings. *Atmospheric Chemistry and Physics, 14*(15), 8089–8103. http://doi.org/10.5194/acp-14-8089-2014.

Kim, B. M., Park, J.-S., Kim, S.-W., Kim, H., Jeon, H., Cho, C., et al. (2015). Source apportionment of PM_{10} mass and particulate carbon in the Kathmandu Valley, Nepal. *Atmospheric Environment, 123.* http://doi.org/10.1016/j.atmosenv.2015.10.082.

Kiros, F., Shakya, K. M., Maharjan, R., Byanju, R. M., Regmi, R. P., Naja, M., et al. (2016). Variability of trace gases: Nitrogen oxides, sulfur dioxide, ozone and ammonia in Kathmandu Valley, Nepal. *Aerosol and Air Quality Research, 16,* 3088–3101.

Kopacz, M., Mauzerall, D. L., Wang, J., Leibensperger, E. M., Henze, D. K., & Singh, K. (2011). Origin and radiative forcing of black carbon transported to the Himalayas and Tibetan Plateau. *Atmospheric Chemistry and Physics, 11*(6), 2837–2852. http://doi.org/10.5194/acp-11-2837-2011.

Kulkarni, S., et al. (2015). Source sector and region contributions to BC and $PM_{2.5}$ in Central Asia. *Atmospheric Chemistry and Physics, 15,* 1683–1705.

Kumar, R., Naja, M., Venkataramani, S., & Wild, O. (2010). Variations in surface ozone at Nainital: A high-altitude site in the central Himalayas. *Journal of Geophysical Research Atmospheres, 115* (16), 1–12. https://doi.org/10.1029/2009JD013715.

Kumar, M., Tiwari, S., Murari, V., Singh, A. K., & Banerjee, T. (2015). Wintertime characteristics of aerosols at middle Indo-Gangetic Plain: Impacts of regional meteorology and long range transport. *Atmospheric Environment, 104,* 162–175.

Kurokawa, J., Ohara, T., Morikawa, T., Hanayama, S., Janssens-Maenhout, G., Fukui, T., et al. (2013). Emissions of air pollutants and greenhouse gases over Asian regions during 2000–2008: Regional Emission inventory in ASia (REAS) version 2. *Atmospheric Chemistry and Physics, 13*(21), 11019–11058. http://doi.org/10.5194/acp-13-11019-2013.

Kusaka, H., Kondo, H., Kikegawa, Y., & Kimura, F. (2001). A simple single-layer urban canopy model for atmospheric models: Comparison with multi-layer and slab models. *Boundary-Layer Meteorology, 101,* 329–358.

LaDochy, S. (2013). The disappearance of dense fog in Los Angeles: Another urban impact? *Physical Geography, 26*(3), 177–191.

Laghari, J. R. (2013). Melting glaciers bring energy uncertainty. *Nature, 502,* 617–618. https://doi.org/10.1038/502617a.

Lal, S., Naja, M., & Subbaraya, B. H. (2000). Seasonal variations in surface ozone and its precursors over an urban site in India. *Atmospheric Environment, 34,* 2713–2724.

Lau, K.-M., & Kim, K.-M. (2006). Observational relationships between aerosol and Asian monsoon rainfall, and circulation. *Geophysical Research Letters, 33,* L21810. https://doi.org/10.1029/2006gl027546.

Lau, K. M., Kim, M. K., & Kim, K. M. (2006). Asian summer monsoon anomalies induced by aerosol direct forcing: The role of the Tibetan Plateau. *Climate Dynamics, 26*(7–8), 855–864. https://doi.org/10.1007/s00382-006-0114-z.

Lau, W. K. M., Kim, M.-K., Kim, K.-M., Lee, W.-S., & Lau, W. K. M. (2010). Enhanced surface warming and accelerated snow melt in the Himalayas and Tibetan Plateau induced by absorbing aerosols. *Environmental Research Letters, 5*(5), 25204. https://doi.org/10.1088/1748-9326/5/2/025204.

Leather, J. (2009). *ADB sustainable development working paper series rethinking transport and climate change*. Manila.

Li, C., Chen, P., Kang, S., Yan, F., Hu, Z., Qu, B., et al. (2016a). Concentrations and light absorption characteristics of carbonaceous aerosol in $PM_{2.5}$ and PM_{10} of Lhasa city, the Tibetan Plateau. *Atmospheric Environment, 127,* 340–346.

Li, C. L., Bosch, C., Kang, S. C., Andersson, A., Chen, P. F., Zhang, Q. G., et al. (2016b). Sources of black carbon to the Himalayan-Tibetan Plateau glaciers. *Nature Communications, 7,* 1–7.

Li, C., Kang, S., Chen, P., Zhang, Q., & Fang, G. C. (2012). Characterizations of particle-bound trace metals and polycyclic aromatic hydrocarbons (PAHs) within Tibetan tents of south Tibetan Plateau, China. *Environmental Science and Pollution Research, 19*(5), 1620–1628. https://doi.org/10.1007/s11356-011-0678-y.

Li, W., & Zusman, E. (2006). Translating regulatory promise into environmental progress: institutional capacity and environmental regulation in China. *Environmental Law Reporter, 36*(8), 10616–10623.

Lim, S. S., Vos, T., Flaxman, A. D., Danaei, G., Shibuya, K., Adair-Rohani, H., et al. (2012). A comparative risk assessment of burden of disease and injury attributable to 67 risk factors and risk factor clusters in 21 regions, 1990–2010: A systematic analysis for the Global Burden of Disease Study 2010. *Lancet, 380*(9859), 2224–2260. https://doi.org/10.1016/S0140-6736(12)61766-8.

Lin, H.-H., Ezzati, M., & Murray, M. (2007). Tobacco smoke, indoor air pollution and tuberculosis: A systematic review and meta-analysis. *PLoS Med, 4*(1), e20. https://doi.org/10.1371/journal.pmed.0040020.

Liu, B., Kang, S. C., Sun, J. M., Zhang, Y. L., Xu, R., Wang, Y. J., et al. (2013). Wet precipitation chemistry at a high-altitude site (3,326 m a.s.l.) in the southeastern Tibetan Plateau. *Environmental Science and Pollution Research, 20,* 5013–5027.

Liu, J., Mauzerall, D. L., Chen, Q., Zhang, Q., Song, Y., Peng, W., et al. (2016). Air pollutant emissions from Chinese households: A major and underappreciated ambient pollution source. *Proceedings of the National Academy of Sciences, 113*(28), 7756–7761. http://doi.org/10.1073/pnas.1604537113.

Loomis, D., Grosse, Y., Lauby-Secretan, B., Ghissassi, F. El, Bouvard, V., Benbrahim-Tallaa, L., et al. (2013). The carcinogenicity of outdoor air pollution. *The Lancet Oncology, 14*(13), 1262–1263. http://doi.org/10.1016/S1470-2045(13)70487-X.

Lu, Z., Streets, D. G., Zhang, Q., & Wang, S. (2012). A novel back-trajectory analysis of the origin of black carbon transported to the Himalayas and Tibetan Plateau during 1996–2010. *Geophysical Research Letters, 39,* L01809. https://doi.org/10.1029/2011gl049903.

Lüthi, Z. L., Škerlak, B., Kim, S. W., Lauer, A., Mues, A., Rupakheti, M., et al. (2015). Atmospheric brown clouds reach the Tibetan Plateau by crossing the Himalayas. *Atmospheric Chemistry and Physics, 15*(11), 6007–6021. http://doi.org/10.5194/acp-15-6007-2015.

Marcq, S., Laj, P., Roger, J. C., Villani, P., Sellegri, K., Bonasoni, P., et al. (2010). Aerosol optical properties and radiative forcing in the high Himalaya based on measurements at the Nepal Climate Observatory-Pyramid site (5079 m a.s.l.). *Atmospheric Chemistry and Physics, 10*(13), 5859–5872. https://doi.org/10.5194/acp-10-5859-2010.

Manoj, M. G., Devera, P. C. S., & Safai, P. D. (2010). Absorbing aerosols facilitate transition of Indian monsoon breaks to active spells. *Climate Dynamics, 37,* 2181–2198. https://doi.org/10.1007/s00382-010-0971-3.

Marinoni, A., Cristofanelli, P., Laj, P., Duchi, R., Calzolari, F., Deceseri, S., et al. (2010). Aerosol mass and black carbon concentrations, a two year record at NCO-P (5079 m, Southern Himalayas). *Atmospheric Chemistry and Physics, 10,* 8551–8562.

Martin, et al. (2011). A major environmental cause of death. *Science, 334*(6053), 180–181.

Mastrangelo, G., Fadda, E., & Marzia, V. (1996). Polycyclic aromatic hydrocarbons and cancer in man. *Environmental Health Perspectives, 104*(11), 1166–1170.

Mauzerall, D. L., & Wang, X. (2001). Protecting agricultural crops from the effects of tropospheric ozone exposure: Reconciling Science and Standard Setting in the United States, Europe, and Asia. *Annual Review of Energy and the Environment, 26*(1), 237–268. https://doi.org/10.1146/annurev.energy.26.1.237.

Meehl, G. A., Arblaster, J. M., & Collins, W. D. (2008). Effects of black carbon aerosols on the Indian Monsoon. *Journal of Climate, 21,* 2869–2882.

Menon, S., Hansen, J., Nazarenko, L., Luo, Y., Husar, R. B., Xu, Q., et al. (2002). Climate effects of black carbon aerosols in China and India. *Science (New York, N.Y.), 297*(5590), 2250–2253. https://doi.org/10.1126/science.1075159.

Mehta, M. (2015). A study of aerosol optical depth variations over the Indian region using thirteen years (2001–2013) of MODIS and MISR Level 3 data. *Atmospheric Environment, 109,* 161–170.

Meitiv, D. (2010). SUMMARY: Black carbon and climate change in the Himalayan region prepared for the clean air task force DRAFT: December, 2010. http://www.catf.us/resources/factsheets/files/201012-Black_Carbon_and_Climate_Change_at_the_Third_Pole.pdf.

Ménégoz, M., Krinner, G., Balkanski, Y., Boucher, O., Cozic, A., Lim, S., et al. (2014). Snow cover sensitivity to black carbon deposition in the Himalayas: from atmospheric and ice core measurements to regional climate simulations. *Atmospheric Chemistry and Physics, 14,* 4237–4249. https://doi.org/10.5194/acp-14-4237-2014.

Ming, J., Xiao, C., Cachier, H., Qin, D., Qin, X., Li, Z., et al. (2009). Black carbon (BC) in the snow of glaciers in west China and its potential effects on albedos. *Atmospheric Research, 92*(1), 114–123. https://doi.org/10.1016/j.atmosres.2008.09.007.

Ming, J., Wang, Y., Du, Z., Zhang, T., Guo, W., Xiao, C., et al. (2015). Widespread Albedo decreasing and induced melting of Himalayan snow and ice in the early 21st century. *PLoS ONE, 10*(6), e0126235. https://doi.org/10.1371/journal.pone.0126235.

Ministry of Environmental Protection, National Development and Reform Commission, Ministry of Industry and Information, Ministry of Finance, Ministry of Housing and Urban and Rural Construction, and Energy Bureau. (2013). *Notice on Issuing the Detailed Implementation Guidelines for Carrying out the Plan for Air Pollution Prevention and Control in Beijing-Tianjin-Hebei and Surrounding Areas* (edited).

Minjares, R. J., & Rutherford, D. (2010). Maximising the co-benefits of light-duty dieselisation in Asia. In A. Srinivasan, E. Zusman, & S. Dhakal (Eds.), *Low carbon transport in Asia*. London: Routledge.

Mitchell, R. B. (2003). International environmental agreements: A survey of their features, formation, and effects. *Annual Review of Environment and Resources, 28*(3), 429–461.

Mitchell, R. B. (2006). Problem structure, institutional design, and the relative effectiveness of international environmental agreements. *Global Environmental Politics, 6*(3), https://doi.org/10.1162/glep.2006.6.3.72.

Mittal, M. L., Hess, P. G., Jain, S. L., Arya, B. C., & Sharma, C. (2007). Surface ozone in the Indian region. *Atmospheric Environment, 41,* 6572–6584.

Mobarak, et al. (2012). Low demand for nontraditional cookstove technologies. *Proceedings of the National Academy of Sciences, 109*(27), 10815–10820.

Mohan, M., & Payra, S. (2009). Influence of aerosol spectrum. *Environmental Monitoring and Assessment, 151,* 265–277.

Mohan, M., & Payra, S. (2014). Aerosol number concentration and visibility during dense fog over a subtropical urban site. *Journal of Nanomaterials, 2014,* 495457.

Morris, R., Naumova, E. N., & Munasinghe, R. L. (1995). Ambient air pollution and hospitalization for congestive heart failure among elderly people in seven large U.S. cities. *American Journal of Public Health, 85*(p 1343), 1361–1365.

Mukherjee, S. (2015). *September).* Emission Norms: Vrooming to Bharat VI not an option.

Murray, C. J. L., Ezzati, M., Flaxman, A. D., Lim, S., Lozano, R., Michaud, C., et al. (2012). GBD 2010: Design, definitions, and metrics. *The Lancet.* http://doi.org/10.1016/S0140-6736(12)61899-6.

Nair, V. S., Solmon, F., Giorgi, F., Mariotti, L., Babu, S. S., & Moorthy, K. K. (2012). Simulation of South Asian aerosols for regional climate studies. *Journal Geophysical Research, 117,* D04209. https://doi.org/10.1029/2011jd016711.

Naqvi, S. W. A., Jayakumar, D. A., Narvekar, P. V., Naik, H., Sarma, V. V. S. S., D'Souza, W., et al. (2000). Increased marine production of N_2O due to intensifying anoxia on the Indian continental shelf. *Nature, 408*(6810), 346–349. Retrieved from http://dx.doi.org/10.1038/35042551.

Niu, F., Li, Z., Li, C., Lee, K.-H., & Wang, M. (2010). Increase in wintertime fog in China: Potential impacts of weakening of the Eastern Asian monsoon circulation and increasing aerosol loading. *Journal of Geophysical Research, 115*(D00K20). http://doi.org/10.1029/2009JD013484.

Ohara, T., Akimoto, H., Kurosawa, J., Horii, N., Yamaji, N., Yamaji, K., et al. (2007). An Asian emission inventory of anthropogenic emission sources for the period 1980–2020. *Atmospheric Chemistry and Physics, 7,* 4419–4444.

Ojha, N., Naja, M., Singh, K. P., Sarangi, T., Kumar, R., Lal, S., et al. (2012). Variabilities in ozone at a semi-urban site in the Indo-Gangetic Plain region: Association with the meteorology and regional processes. *Journal of Geophysical Research Atmospheres, 117*(20), 1–19. http://doi.org/10.1029/2012JD017716.

Pan, X., Chin, M., Gautam, R., Bian, H., Kim, D., Colarco, P. R., et al. (2015). A multi-model evaluation of aerosols over South Asia: Common problems and possible causes. *Atmospheric Chemistry and Physics, 15* (10), 5903–5928. http://doi.org/10.5194/acp-15-5903-2015.

Panday, A. K. (2006). *The diurnal cycle of air pollution in the Kathmandu Valley, Nepal.* Cambridge, MA: MIT Center for Global Chance Science.

Panday, A. K., & Prinn, R. G. (2009). Diurnal cycle of air pollution in the Kathmandu Valley, Nepal: Observations. *Journal of Geophysical Research Atmospheres, 114*(9). http://doi.org/10.1029/2008JD009777.

Panday, A. K., Prinn, R. G., & Schär, C. (2009). Diurnal cycle of air pollution in the Kathmandu Valley, Nepal: 2. Modeling results. *Journal of Geophysical Research Atmospheres, 114*(21). http://doi.org/10.1029/2008JD009808.

Pande, J. N., Bhatta, N., Biswas, D., Pandey, R. M., Ahluwalia, G., Siddaramaiah, N. H., et al. (2002). Outdoor air pollution and emergency room visits at a hospital in Delhi. *Indian Journal of Chest Diseases and Allied Sciences, 44*(1), 13–19.

Pandithurai, G., Dipu, S., Dani, K. K., Tiwari, S., Bisht, D. S., Devara, P. C. S., et al. (2008). Aerosol radiative forcing during dust events over New Delhi, India. *Journal of Geophysical Research, 113* (D13), D13209. https://doi.org/10.1029/2008JD009804.

Pant, K. P. (2015). Uniform-price reverse auction for estimating the costs of reducing open-field burning of rice residue in Nepal. *Environmental & Resource Economics, 62*(3), 567–581. https://doi.org/10.1007/s10640-014-9830-8.

Pant, P., Hegde, P., Dumka, U. C., Sagar, R., Satheesh, S. K., Moorthy, K. K., et al. (2006a). Aerosol characteristics at a high-altitude location in central Himalayas: Optical properties and radiative forcing. *Journal of Geophysical Research, 111*(D17), D17206. http://doi.org/10.1029/2005JD006768.

Pant, P., Hegde, P., Dumka, U. C., Saha, A., Srivastava, M. K., & Sagar, R. (2006b). Aerosol characteristics at a high-altitude location during ISRO-GPB Land Campaign—II. *Current Science, 91*(8), 1053–1061.

Parekh, P. P., Khwaja, H. A., Khan, A. R., Naqvi, R. R., Malik, A., Shah, S. A., et al. (2001). Ambient air quality of two metropolitan cities of Pakistan and its health implications. *Atmospheric Environment, 35*(34), 5971–5978. http://doi.org/10.1016/S1352-2310(00)00569-0.

Pasricha, P. K., Gera, B. S., Shastri, S., Maini, H. K., John, T., Ghosh, A. B., et al. (2003). Role of the water vapour greenhouse effect in the forecasting of fog occurrence. *Boundary-Layer Meteorology, 107*(2), 469–482. http://doi.org/10.1023/A:1022128800130.

Pope, C. A., Burnett, R. T., Thun, M. J., Calle, E. E., Krewski, D., Ito, K., et al. (2002). Lung cancer, cardiopulmonary mortality, and long-term exposure to fine particulate air pollution. *JAMA, the Journal of the American Medical Association, 287*(9), 1132–1141. https://doi.org/10.1001/jama.287.9.1132.

Pope, C. A., Burnett, R. T., Thurston, G. D., Thun, M. J., Calle, E. E., Krewski, D., et al. (2004). Cardiovascular mortality and long-term exposure to particulate air pollution: Epidemiological evidence of general pathophysiological pathways of disease. *Circulation, 109* (1), 71–77. https://doi.org/10.1161/01.CIR.0000108927.80044.7F.

Pope, C. A., Rodermund, D. L., & Gee, M. M. (2007). Mortality effects of a copper smelter strike and reduced ambient sulfate particulate matter air pollution. *Environmental Health Perspectives, 115*(5), 679–683. https://doi.org/10.1289/ehp.9762.

Pope, D. P., Mishra, V., Thompson, L., Siddiqui, A. R., Rehfuess, E. A., Weber, M., et al. (2010). Risk of low birth weight and stillbirth associated with indoor air pollution from solid fuel use in developing countries. *Epidemiologic Reviews, 32*(1), 70–81. https://doi.org/10.1093/epirev/mxq005.

Porch, W., Chylek, P., Dubey, M., & Massie, S. (2007). Trends in aerosol optical depth for cities in India. *Atmospheric Environment, 41,* 7524–7532.

Praveen, P. S., Ahmed, T., Kar, A., Rehman, I. H., & Ramanathan, V. (2012). Link between local scale BC emissions in the Indo-Gangetic Plains and large scale atmospheric solar absorption. *Atmospheric Chemistry and Physics, 12*(2), 1173–1187. http://doi.org/10.5194/acp-12-1173-2012.

Price, L. (2008). *China's Top-1000 energy-consuming enterprises program: Reducing energy consumption of the 1000 largest industrial enterprises in China* (June).

Pudasainee, D., Sapkota, B., Bhatnagar, A., Kim, S.-H., & Seo, Y.-C. (2010). Influence of weekdays, weekends and bandhas on surface ozone in Kathmandu valley. *Atmospheric Research, 95,* 150–156.

Putero, D., Cristofanelli, P., Laj, P., Marinoni, A., Villani, P., Broquet, A., et al. (2014). New atmospheric composition observations in the Karakorum region: Influence of local emissions and large-scale circulation during a summer field campaign. *Atmospheric Environment, 97,* 75–82. http://doi.org/10.1016/j.atmosenv.2014.07.063.

Putero, D., Cristofanelli, P., Marinoni, A., Adhikary, B., Duchi, R., Shrestha, S. D., et al. (2015). Seasonal variation of ozone and black carbon observed at Paknajol, an urban site in the Kathmandu Valley, Nepal. *Atmospheric Chemistry and Physics, 15*(24), 13957–13971. http://doi.org/10.5194/acp-15-13957-2015.

Qasim, M., Bashir, A., Anees, M. M., Ghani, M. U., Khalid, M., Malik, J., et al. (2014). Health risk assessment of indoor air quality in developing countries. *Asian Journal Management Sciences and Education, 3*(2).

Qian, Y., Kaiser, D. P., Leung, L. R., & Xu, M. (2006). More frequent cloud-free sky and less surface solar radiation in China from 1955 to 2000. *Geophysical Research Letters, 33,* L01812. https://doi.org/10.1029/2005gl024586.

Qian, Y., Wang, W., Leung, L. R., & Kaiser, D. P. (2007). Variability of solar radiation under cloud-free skies in China: The role of aerosols. *Geophysical Research Letters, 34,* L12804. https://doi.org/10.1029/2006gl028800.

Qian, Y., Flanner, M. G., Leung, L. R., & Wang, W. (2011). Sensitivity studies on the impacts of Tibetan Plateau snowpack pollution on the Asian hydrological cycle and monsoon climate. *Atmospheric Chemistry and Physics, 11,* 1929–1948. https://doi.org/10.5194/acp-11-1929-2011.

Qian, Y., Yasunari, T. J., Doherty, S. J., Flanner, M. G., Lau, W. K. M., Ming, J., et al. (2015). Light-absorbing particles in snow and ice: Measurement and modeling of climatic and hydrological impact. *Advances in Atmospheric Sciences, 32*(1), 64–91. http://doi.org/10.1007/s00376-014-0010-0.

Rabalais, N. N., Díaz, R. J., Levin, L. A., Turner, R. E., Gilbert, D., & Zhang, J. (2010). Dynamics and distribution of natural and human-caused hypoxia. *Biogeosciences, 7,* 585–619. http://doi.org/10.5194/bg-7-585-2010.

Rajput, P., Sarin, M., Sharma, D., & Singh, D. (2014). Characteristics and emission budget of carbonaceous species from post-harvest agricultural-waste burning in source region of the Indo-Gangetic Plain. *Tellus B, 66,* 21026.

Rall, D. P. (1974). Review of the health effects of sulfur oxides. *Environmental Health Perspectives.* http://doi.org/10.1289/ehp.74897.

Ram, K., Sarin, M. M., & Hegde, P. (2010). Long-term record of aerosol optical properties and chemical composition from a high-altitude site (Manora Peak) in Central Himalaya. *Atmospheric Chemistry and Physics, 10*(23), 11791–11803. http://doi.org/10.5194/acp-10-11791-2010.

Ram, K., Sarin, M. M., Sudheer, A. K., & Rengarajan, R. (2012). Carbonaceous and secondary inorganic aerosols during wintertime fog and haze over urban sites in the Indo-Gangetic Plains. *Aerosol and Air Quality Research, 12,* 359–370.

Ramana, M. V., Ramanathan, V., Feng, Y., Yoon, S.-C., Kim, S.-W., Carmichael, G. R., et al. (2010). Warming influenced by the ratio of black carbon to sulphate and the black-carbon source. *Nature Geoscience, 3,* 542–545.

Ramanathan, V., & Carmichael, G. (2008). Global and regional climate changes due to black carbon. *Nature Geoscience, 1,* 221–227. https://doi.org/10.1038/ngeo156.

Ramanathan, V., Crutzen, P. J., Lelieveld, J., Mitra, A. P., Althausen, D., Anderson, J., et al. (2001). Indian Ocean Experiment: An integrated analysis of the climate forcing and effects of the great Indo-Asian haze. *Journal of Geophysical Research, 106*(D22), 28371–28398.

Ramanathan, V., & Ramana, M. V. (2005). Persistent, widespread, and strongly absorbing haze over the Himalaya. *Pure and Applied Geophysics, 162,* 1609–1626.

Ran, L., Lin, W. L., Deji, Y. Z., La, B., Tsering, P. M., Xu, X. B., et al. (2014). Surface gas pollutants in Lhasa, a highland city of Tibet—Current levels and pollution implications. *Atmospheric Chemistry and Physics, 14*(19), 10721–10730. http://doi.org/10.5194/acp-14-10721-2014.

Randles, C. A., & Ramaswamy, V. (2008). Absorbing aerosols over Asia: A Geophysical Fluid Dynamics Laboratory general circulation model sensitivity study of model response to aerosol optical depth and aerosol absorption. *Journal of Geophysical Research, 113,* D21203. https://doi.org/10.1029/2008jd010140.

Rantalainen, A.-L., Hyötyläinen, T., Saramo, M., & Niskanen, I. (1999). Passive sampling of PAHs in indoor air in Nepal. *Toxicological and Environmental Chemistry, 68,* 335–348. https://doi.org/10.1080/02772249909358667.

Rehfuess, et al. (2014). Enablers and barriers to large-scale uptake of improved solid fuel stoves: A systematic review. *Environmental Health Perspectives, 122*(2), 120–130.

Rehman, I. H., Ahmed, T., Praveen, P. S., Kar, A., & Ramanathan, V. (2011). Black carbon emissions from biomass and fossil fuels in rural India. *Atmospheric Chemistry and Physics, 11*(14), 7289–7299. http://doi.org/10.5194/acp-11-7289-2011.

Rock, M. T. (2002). *Pollution control in East Asia: Lessons from newly industrializing economies.* Washington D.C.: Resources for the Future.

Roden, C. A., Bond, T. C., Conway, S., Osorto Pinel, A. B., MacCarty, N., & Still, D. (2009). Laboratory and field investigations of particulate and carbon monoxide emissions from traditional and improved cookstoves. *Atmospheric Environment, 43*(6), 1170–1181. https://doi.org/10.1016/j.atmosenv.2008.05.041.

Rosenfeld, D. (2000). Suppression of rain and snow by urban and industrial air pollution. *Science, 287,* 1793–1796.

Roy, A., Chatterjee, A., Tiwari, S., Sarkar, C., Das, S. K., Ghosh, S. K., et al. (2016). Precipitation chemistry over urban, rural and high altitude Himalayan stations in eastern India. *Atmospheric Research, 181,* 44–53. https://doi.org/10.1016/j.atmosres.2016.06.005.

Rupakheti, D., Adhikary, B., Praveen, P. S., Rupakheti, M., Kang, S., Mahata, K. S., … Lawrence, M. G. (2017). Pre-monsoon air quality over Lumbini, a world heritage site along the Himalayan foothills. *Atmospheric Chemistry and Physics*, 17, 11041–11063. https://doi.org/10.5194/acp-17-11041-2017.

Saikawa, E., et al. (2009). Present and potential future contributions of sulfate, black and organic carbon aerosols from china to global air quality, premature mortality and radiative forcing. *Atmospheric Environment, 4343*(17), 2814–2822.

Saikawa, E., Kurokawa, J., Takigawa, M., Borken-Kleefeld, J., Mauzerall, D. L., Horowitz, L. W., et al. (2011). The impact of China's vehicle emissions on regional air quality in 2000 and 2020: A scenario analysis. *Atmospheric Chemistry and Physics, 11,* 9465–9484.

Saikawa, E., Trail, M., Zhong, M., Wu, Q., Young, C. L., & Janssens-Maenhout, G. (2017a). Uncertainties in emissions estimates of greenhouse gases and air pollutants in India and their impacts on regional air quality. *Environmental Research Letters*, 65002.

Saikawa, E., Kim, H., Zhong, M., Avramov, A., Zhao, Y., & Janssens-Maenhout, G. (2017b). Comparison of emissions inventories of anthropogenic air pollutants and greenhouse gases in China. *Atmospheric Chemistry and Physics*, 6393–6421. https://doi.org/10.5194/acp-17-6393-2017.

Sanap, S. D., Ayantika, D. C., Pandithurai, G., & Niranjan, K. (2014). Assessment of the aerosol distribution over Indian subcontinent in CMIP5 models. *Atmospheric Environment, 87,* 123–137.

Sarangi, T., Naja, M., Ojha, N., Kumar, R., Lal, S., Venkataramani, S., et al. (2013). First simultaneous measurements of ozone, CO, and NOy at a high-altitude regional representative site in the central Himalayas. *Journal of Geophysical Research, 119*, 1592–1611.

Sarkar, S., Chokngamwong, R., Cervone, G., Singh, R. P., & Kafatos, M. (2006). Variability of aerosol optical depth and aerosol forcing over India. *Advances in Space Research, 37*(12), 2153–2159. https://doi.org/10.1016/j.asr.2005.09.043.

Sarkar, C., Kumar, V., & Sinha, V. (2013). Massive emissions of carcinogenic benzoids from paddy residue burning in North India. *Current Science, 104*(12), 1703–1709.

Sarkar, C., Sinha, V., Kumar, V., Rupakheti, M., Panday, A., S Mahata, K., et al. (2016). Overview of VOC emissions and chemistry from PTR-TOF-MS measurements during the SusKat-ABC campaign: High acetaldehyde, isoprene and isocyanic acid in wintertime air of the Kathmandu Valley. *Atmospheric Chemistry and Physics, 16*(6), 3979–4003.

Satheesh, S. K., Krishna Moorthy, K., Babu, S. S., Vinoj, V., & Dutt, C. B. S. (2008). Climate implications of large warming by elevated aerosols over India. *Geophysical Research Letters, 35,* L19809. https://doi.org/10.1029/2008gl034944.

Satheesh, S. K., Ramanathan, V., Xu, L.-J., Lober, J. M., Podgorny, I. A., Prospero, J. M., et al. (1999). A model for the natural and anthropogenic aerosols over the tropical Indian Ocean from Indian Ocean Experiment data. *Journal of Geophysical Research, 104* (D22), 27421–27440.

Sawaisarje, G. K., Khare, P., Shirke, C. Y., Deepakumar, S., & Narkhede, N. M. (2014). Study of winter fog over Indian subcontinent: Climatological perspective. *Mausam, 65*(1), 19–28.

Schmidt, C. (2013). Modernizing artisanal brick kilns: A global need. *Environmental Health Perspectives, 121*(8), a242–a249. https://doi.org/10.1289/ehp.121-a242.

Schreurs, M. (2011). Transboundary Cooperation to Address Acid Rain. In S. Dinar (Ed.), *Beyond resource wars: Scarcity, Environmental degradation, and international cooperation* (pp. 90–116). Boston: MIT Press.

Schwartz, J., Coull, B., Laden, F., & Ryan, L. (2008). The effect of dose and timing of dose on the association between airborne particles and survival. *Environmental Health Perspectives, 116*(1), 64–69. https://doi.org/10.1289/ehp.9955.

Schwartz, J., Dockery, D. W., Neas, L. M., Wypij, D., Ware, J. H., Spengler, J. D., et al. (1994). Acute effects of summer air pollution on respiratory symptom reporting in children. *American Journal of Respiratory and Critical Care Medicine, 150*(5, I), 1234–1242. http://doi.org/10.1164/ajrccm.150.5.7952546.

Sellegri, K., Laj, P., Venzac, H., Boulon, J., Picard, D., Villani, P., et al. (2010). Seasonal variations of aerosol size distributions based on long-term measurements at the high altitude Himalayan site of Nepal Climate Observatory-Pyramid (5079 m), Nepal. *Atmospheric Chemistry and Physics, 10*(21), 10679–10690. https://doi.org/10.5194/acp-10-10679-2010.

Semple, J. L., & Moore, G. W. K. (2008). First observations of surface ozone concentration from the summit region of Mount Everest. *Geophysical Research Letters*, *35*(20), L20818. http://doi.org/Artn. https://doi.org/10.1029/2008gl035295.

Shahid, I., Kistler, M., Mukhtar, A., Ghauri, B. M., Cruz, C. R., Bauer, H., et al. (2016). Chemical characterization and mass closure of PM_{10} and $PM_{2.5}$ at an urban site in Karachi e Pakistan. *Atmospheric Environment, 128,* 114–123.

Sharma, A. R., Kharol, S. K., Badarinath, K. V. S., & Singh, D. (2010). Impact of agriculture crop residue burning on atmospheric aerosol loading—A study over Punjab State, India. *Annales Geophysicae: Atmospheres, Hydrospheres and Space Sciences, 28*(2), 367.

Sharma, P., Chandra, J., Chand, K., Paul, R., Prasad, P., & Chauhan, C. (2013). Surface ozone concentration and its behaviour with aerosols in the northwestern Himalaya. *India, 71,* 44–53.

Shakya, K., Rupakheti, M., Aryal, K., & Peltier, R. (2016). Respiratory effects of high levels of particulate exposure in a cohort of traffic police in Nepal. *Journal of Occupational and Environmental Medicine, 58*(6), 218–225.

Shindell, D., Hicks, K., Ramanathan, V., & Terry, S. (2013). *Integrated assessment of black carbon and tropospheric ozone*. UNEP and WMO.

Shrestha, I. L., & Shrestha, S. L. (2005). Indoor air pollution from biomass fuels and respiratory health of the exposed population in Nepalese households. *International Journal of Occupational and Environmental Health, 11,* 150–160.

Shrestha, P., Barros, A. P., & Khlystov, A. (2010). Chemical composition and aerosol size distribution of the middle mountain range in the Nepal Himalayas during the 2009 pre-monsoon season. *Atmospheric Chemistry and Physics, 10*(23), 11605–11621. http://doi.org/10.5194/acp-10-11605-2010.

Shrestha, S. R., Kim Oanh, N. T., Xu, Q., Rupakheti, M., & Lawrence, M. G. (2013). Analysis of the vehicle fleet in the Kathmandu Valley for estimation of environment and climate co-benefits of technology intrusions. *Atmospheric Environment, 81*, 579–590.

Siddique, S., Banerjee, M., Ray, M. R., & Lahiri, T. (2010). Air pollution and its impact on lung function of children in Delhi, the Capital City of India. *Water, Air, and Soil Pollution, 212*(1), 89–100. https://doi.org/10.1007/s11270-010-0324-1.

Sillman, S. (1999). The relation between ozone, NO_x, and hydrocarbons in urban and polluted rural environments. *Atmospheric Environment, 33,* 1821–1845.

Singh, A., & Agrawal, M. (2008). Acid rain and its ecological consequences. *Journal of Environmental Biology, 29*(1), 15–24.

Sinha, V., Kumar, V., & Sarkar, C. (2014). Chemical composition of pre-monsoon air in the Indo-Gangetic Plain measured using a new air quality facility and PTR-MS: High surface ozone and strong influence of biomass burning. *Atmospheric Chemistry and Physics, 14,* 5914–5941.

Skamarock, W. C., Klemp, J. B., Dudhia, J., Gill, D. O., Barker, D. M., Duda, M. G., et al. (2008). *A description of the advanced research WRF Version 3*. Boulder, CO: NCAR.

Smith, K. R., & Mehta, S. (2003). The burden of disease from indoor air pollution in developing countries: Comparison of estimates. *International Journal of Hygiene and Environmental Health, 206* (4–5), 279–289.

Smith, K. R., Samet, J. M., et al. (2000). Indoor air pollution in developing countries and acute lower respiratory infections in children. *Thorax, 55*(6), 518–532.

Somanathan, E., & Bluffstone, R. (2015). Biogas: clean energy access with low-cost mitigation of climate change. *Environmental & Resource Economics, 62*(2), 265–277. https://doi.org/10.1007/s10640-015-9961-6.

Starke, L. E. (2007). *Cityscape: Rizhao, solar powered city. 2007 State of the world our urban future*. New York: W.W. Norton & Company.

State Council. (2013). *Action plan for air pollution prevention and control* (edited).

Stern, F. B., Halperin, W. E., Hornung, R. W., Ringenburg, V. L., & McCammon, C. S. (1988). Heart disease mortality among bridge and tunnel officers exposed to carbon monoxide. *American Journal of Epidemiology, 128*(6), 1276–1288. Retrieved from http://aje.oxfordjournals.org/content/128/6/1276.abstract.

Stockwell, C. E., Christian, T. J., Goetz, J. D., Jayarathne, T., Bhave, P. V., Praveen, P. S., et al. (2016). Nepal Ambient Monitoring and Source Testing Experiment (NAMaSTE): Emissions of trace gases and light-absorbing carbon from wood and dung cooking fires, garbage and crop residue burning, brick kilns, and other sources. *Atmospheric Chemistry and Physics, 16*(17). https://doi.org/10.5194/acp-16-11043-2016.

Sud, Y. C., Wilcox, E., Lau, W. K.-M., Walker, G. K., Liu, X.-H., Nenes, A. S. B. (2009). Sensitivity of boreal-summer circulation and precipitation to atmospheric aerosols in selected regions—Part 1: Africa and India. *Annals of Geophysics, 27,* 3989–4007.

Syed, F. S., Kornich, H., & Tjernstrom, M. (2012). On the fog variability over south Asia. *Climate Dynamics, 39,* 2993–3005.

Tare, V., Tripathi, S. N., Chinnam, N., Srivastava, A. K., Dey, S., Manar, M., et al. (2006). Measurements of atmospheric parameters during Indian Space Research Organization Geosphere Biosphere Program Land Campaign II at a typical location in the Ganga Basin: 2. Chemical properties. *Journal of Geophysical Research: Atmospheres, 111*(D23). http://doi.org/10.1029/2006JD007279.

Thompson, L. G., Yao, T., Mosley–Thompson, E., Davis, M. E., Henderson, K. A., & Liu, P.-N. (2000). A high-resolution millennial record of the South Asian monsoon from Himalayan ice core. *Science, 289,* 1916–1919.

Tripathee, L., Kang, S. C., Rupakheti, D., Zhang, Q. G., Huang, J., & Sillanpää, M. (2016). Water-soluble ionic composition of aerosols at urban location in the foothills of Himalaya, Pokhara Valley, Nepal. *Atmosphere, 7*(8), 102–114. https://doi.org/10.3390/atmos7080102.

Tripathi, S. N., Dey, S., Tare, V., & Satheesh, S. K. (2005). Aerosol black carbon radiative forcing at an industrial city in northern India. *Geophysical Research Letters, 32,* L08802. https://doi.org/10.1029/2005gl022515.

UNEP. (2002). *The Asian Brown Cloud: Climate and Other Environmental Impacts.* Bangkok: UNEP Resource Center for Asia and the Pacific.

UNEP. (2011). *Near-term climate protection and clean air benefits: Actions for controlling short-lived climate.* Nairobi, Kenya: Forcers United Nations Environment Programme (UNEP).

UNEP/WMO. (2011). *Integrated assessment of black carbon and tropospheric ozone.* United Nations Environment Programme.

Unruh, G. C. (2000). Understanding carbon lock-in. *Energy Policy, 28,* 817–830.

Unruh, G. C. (2002). Escaping carbon lock-in. *Energy Policy, 30,* 317–325.

Upadhyay, A. K., Singh, A., Kumar, K., & Singh, A. (2015). Impact of indoor air pollution from the use of solid fuels on the incidence of life threatening respiratory illnesses in children in India. *BMC Public Health, 15*(300), 2015. https://doi.org/10.1186/s12889-015-1631-7.

USEPA. (2006). *An inventory of sources and environmental releases of dioxin-like compounds in the United States for the years 1987, 1995, and 2000, EPA/600/P-03/002F* (677 pp.). Washington, DC, USA: National Center for Environmental Assessment, Office of Research and Development.

Vadrevu, K. P., Ellicott, E., Giglio, L., Badarinath, K. V. S., Vermote, E., Justice, C., et al. (2012). Vegetation fires in the himalayan region—Aerosol load, black carbon emissions and smoke plume heights. *Atmospheric Environment, 47,* 241–251. https://doi.org/10.1016/j.atmosenv.2011.11.009.

Venkataraman, C., Habib, G., Eiguren-Fernandez, A., Miguel, A. H., & Friedlander, S. K. (2005). Residential biofuels in South Asia: Carbonaceous aerosol emissions and climate impacts. *Science, 307,* 1454–1456.

Vinoj, V., Rasch, P. J., Wang, H., Yoon, J.-H., Ma, P.-L., Landu, K., et al. (2014). Short-term modulation of Indian summer monsoon rainfall by West Asian dust. *Nature Geoscience, 7*(4), 308–313. https://doi.org/10.1038/ngeo2107.

Wan, X., Kang, S., Li, Q., Rupakheti, D., Zhang, Q., Guo, J., et al. (2017). Organic molecular tracers in the atmospheric aerosols from Lumbini, Nepal, in the northern Indo-Gangetic Plain: influence of biomass burning. *Atmospheric Chemistry and Physics, 17,* 8867–8885. https://doi.org/10.5194/acp-17-8867-2017.

Wang, C., Kim, D., Ekman, A. M. L., Barth, M. C., & Rasch, P. J. (2009). Impact of anthropogenic aerosols on Indian summer monsoon. *Geophysical Research Letters, 36,* 1–6. https://doi.org/10.1029/2009GL040114.

Wang, T., Wong, H. L. A., Tang, J., Ding, A., Wu, W. S., & Zhang, X. C. (2006). On the origin of surface ozone and reactive nitrogen observed at a remote mountain site in the northeastern Qinghai-Tibetan Plateau, western China. *Journal of Geophysical Research, 111,* D08303. https://doi.org/10.1029/2005jd006527.

Wang, X., Gong, P., Sheng, J., Joswiak, D. R., & Yao, T. (2015). Long-range atmospheric transport of particulate Polycyclic Aromatic Hydrocarbons and the incursion of aerosols to the southeast Tibetan Plateau. *Atmospheric Environment, 115,* 124–131.

Wang, Y. (2015). Politically connected polluters under smog. *Business and Politics, 17*(1), 97–123. https://doi.org/10.1515/bap-2014-0033.

WHO. (2005). *Air quality guidelines for particulate matter, ozone, nitrogen dioxide and sulfur dioxide, 22.* Retrieved from http://www.euro.who.int/Document/E87950.pdf.

WHO. (2016). *Ambient air pollution: A global assessment of exposure and burden of disease.*

Wiedinmyer, C., Yokelson, R. J., & Gullett, B. K. (2014). Global emissions of trace gases, particulate matter, and hazardous air pollutants from open burning of domestic wastE. *Environmental Science and Technology, 48*(16), 9523–9530. https://doi.org/10.1021/es502250z.

Wilson, W. E., & Suh, H. H. (1997). Fine particles and coarse particles: Concentration relationships relevant to epidemiologic studies. *Journal of the Air and Waste Management Association, 47*(12), 1238–1249.

Wingfors, H., Hägglund, L., & Magnusson, R. (2011). Characterization of the size-distribution of aerosols and particle-bound content of oxygenated PAHs, PAHs, and n-alkanes in urban environments in Afghanistan. *Atmospheric Environment, 45*(26), 4360–4369. https://doi.org/10.1016/j.atmosenv.2011.05.049.

World Bank. (2007). *India: Strengthening institutions for sustainable growth.* Washington D.C.: World Bank.

World Bank. (2008). *Nepal country environmental analysis: Strengthening institutions and management systems for enhanced environmental governance.* Report No. 38984-NP. Environment and Water Resources Management Unit, Sustainable Development Department, South Asia Region. Washington, DC.

World Bank. (2013). *China: Accelerating household access to clean cooking and heating, in East Asia and Pacific clean stove initiative series.* Washington, DC: World Bank.

World Bank. (2014). *Diesel power generation.*

World Bank and Institute for Health Metrics and Evaluation. (2016). *The cost of air pollution: Strengthening the economic case for action.* Washington, DC: World Bank. License: Creative Commons Attribution CC BY 3.0 IGO.

Xia, X., Zong, X., Cong, Z., Chen, H., Kang, S., & Wang, P. (2011). Baseline continental aerosol over the central Tibetan plateau and a case study of aerosol transport from South Asia. *Atmospheric Environment, 45*(39), 7370–7378.

Xiao, H., Kang, S., Zhang, Q., Han, W., Loewen, M., Wong, F., et al. (2010). Transport of semivolatile organic compounds to the Tibetan Plateau: Monthly resolved air concentrations at Nam Co. *Journal of Geophysical Research Atmospheres, 115*(16), 1–9. http://doi.org/10.1029/2010JD013972.

Xiao, Q., Saikawa, E., Yokelson, R. J., Chen, P., Li, C., & Kang, S. (2015). Indoor air pollution from burning yak dung as a household fuel in tibet. *Atmospheric Environment, 102,* 406–412.

Xu, B., Cao, J., Hansen, J., Yao, T., Joswia, D. R., Wang, N., et al. (2009). Black soot and the survival of Tibetan glaciers. *Proceedings of the National Academy of Sciences of the United States of America, 106,* 22114–22118. https://doi.org/10.1073/pnas.0910444106.

Yasunari, T. J., Bonasoni, P., Laj, P., Fujita, K., Vuillermoz, E., Marinoni, A., et al. (2010). Estimated impact of black carbon deposition during pre-monsoon season from Nepal Climate Observatory—Pyramid data and snow albedo changes over Himalayan glaciers. *Atmospheric Chemistry and Physics, 10*(14), 6603–6615. http://doi.org/10.5194/acp-10-6603-2010.

Yin, X., Kang, S., de Foy, B., Cong, Z., Luo, J., Zhang, L., et al. (2017). Surface ozone at Nam Co in the inland Tibetan Plateau: variation, synthesis comparison and regional representativeness. *Atmospheric Chemistry and Physics, 17,* 11293–11311. https://doi.org/10.5194/acp-17-11293-2017.

Young, O. R., Guttman, D., Qi, Y., Bachus, K., Belis, D., Cheng, H., et al. (2015). Institutionalized governance processes: Comparing environmental problem solving in China and the United States. *Global Environmental Change, 31,* 163–173.

Zhao, Z., Cao, J., Shen, Z., Xu, B., Zhu, C., Chen, L. W. A., et al. (2013). Aerosol particles at a high-altitude site on the Southeast Tibetan Plateau, China: Implications for pollution transport from South Asia. *Journal of Geophysical Research Atmospheres, 118* (19), 11360–11375. http://doi.org/10.1002/jgrd.50599.

11 Disaster Risk Reduction and Building Resilience in the Hindu Kush Himalaya

Coordinating Lead Authors
Ramesh Ananda Vaidya, International Centre for Integrated Mountain Development (ICIMOD), e-mail: Ramesh.Vaidya@icimod.org
Mandira Singh Shrestha, International Centre for Integrated Mountain Development (ICIMOD), e-mail: Mandira.Shrestha@icimod.org (corresponding author)
Nusrat Nasab, Aga Khan Agency for Habitat (AKAH), e-mail: nusrat.nasab@akdn.org

Lead Authors
Deo Raj Gurung, Aga Khan Agency for Habitat (AKAH), e-mail: deoraj.gurung@akdn.org
Nagami Kozo, Japan International Cooperation Agency (JICA), e-mail: Nagami.Kozo@jica.go.jp
Neera Shrestha Pradhan, International Centre for Integrated Mountain Development (ICIMOD), e-mail: Neera.Pradhan@icimod.org
Robert James Wasson, National University of Singapore, e-mail: spprjw@nus.edu.sg

Contributing Authors
Arun Bhakta Shrestha, International Centre for Integrated Mountain Development (ICIMOD), e-mail: arun.shrestha@icimod.org
Chanda Goodrich Gurung, International Centre for Integrated Mountain Development (ICIMOD), e-mail: Chanda.Goodrich@icimod.org
Ajay Bajracharya, University of Manitoba, Winnipeg, Canada, e-mail: ajay6985@yahoo.com

Review Editor
Purnamita Dasgupta, Institute of Economic Growth, India, e-mail: pdg@iegindia.org

Corresponding Author
Mandira Singh Shrestha, International Centre for Integrated Mountain Development (ICIMOD), e-mail: Mandira.Shrestha@icimod.org

P. Wester et al. (eds.), *The Hindu Kush Himalaya Assessment*,
https://doi.org/10.1007/978-3-319-92288-1_11

Contents

Chapter Overview

Key Findings

1. **More than one billion people are at risk of exposure to increasing frequency and intensity of natural hazards.** Rising trends appear in the number of disasters reported, the numbers of people killed and affected, and the size of economic losses. This is due to various drivers including climatic change and environmental degradation.
2. **Cascading events resulting from a multi-hazard environment have upstream-downstream linkages, often with transboundary impacts.** The HKH is especially prone to floods, flash floods, avalanches, and landslides, but also to droughts and earthquakes. The number of GLOFs in the region are increasing. There is lack of institutional mechanism and capacity to deal with multi-hazard environment and cascading disasters.
3. **When disasters hit the HKH, they affect more women than men.** Gender is one of the most significant factors affecting vulnerability to disasters in the region. This can be ascribed to women's lack of access to information, power, money, and training, high rates of male outmigration, and associated gender-based norms and barriers.

Policy Messages

1. **Institutions and governments in the HKH urgently need to adopt a standardized, multi-hazard risk assessment approach.** Such an approach should address primary, secondary, and cascading hazards.

2. All stakeholders including governments, individuals, households, and communities need to take urgent action for **enhancing resilience through the four pillars—information, infrastructure, institutions, and insurance**. A balanced use of four key instruments is required: command and control, monetary incentives, persuasion, and nudging. Sensitization to gender differential vulnerabilities within society must be inculcated for reducing mortality and morbidity.
3. **The countries of the HKH need to cooperate more extensively and effectively by sharing data, information, and scientific and indigenous knowledge, and by fostering transboundary disaster risk reduction practices**. Institutional arrangements for collective action should be enhanced and capacity building programmes organized for strengthening regional cooperation.

The Hindu Kush Himalaya (HKH)—covering more than four million square kilometres from Afghanistan to Myanmar—is one of the world's most ecologically diverse mountain biomes, with extreme variations in vegetation. It is also one of the most hazard-prone. Because of its steep terrain, high seismicity, fragile geological formation, and intense and highly variable precipitation, the HKH is especially vulnerable to floods, landslides, avalanches, and earthquakes (***well-established***).

Currently, natural hazards in the HKH are increasing in magnitude and frequency—a trend driven partly by climate change. Environmental degradation generally poses a major threat to lives and livelihoods. However, a community's vulnerability to natural hazards also includes the exposure of people and property to disasters and their impact.

While some of the factors in exposure and vulnerability are physical and environmental, other factors are socioeconomic, such as poverty, human settlement and habitat, lack of preparedness, susceptibility, and adaptive capacity. Poverty leaves many people in the region with few resources when trying to rebuild their homes and livelihoods (see Chap. 12). With the exception of China, the countries of the HKH rank below the global average on the Human Development Index (HDI). Income inequality is also high throughout the region, except in China and Bangladesh. These facts imply high vulnerability to natural hazards (***well-established***).

Gender emerges as one of the most significant socioeconomic factors affecting vulnerability. While both men and women in the HKH have valuable knowledge, skills, experience, and coping capacities, these strengths tend to differ by gender (see Chap. 14). And unlike men's capacities and knowledge, those of women are often ignored in policies and formal arrangements related to development, mitigation and recovery. As we stress in the chapter, policy makers and planners must ensure that women actively participate in capacity building and gain access to information that would prepare them better for disasters. Indicators related to both gender equality and education are even more dramatically low for the HKH compared with world averages, than are income and income equality (***well-established***).

Disaster risks are a function of interplay between three key elements: hazard, exposure, and vulnerability. Natural hazards are increasing in magnitude and occurrence due to various drivers of change including climate change. The susceptibility of a community to the impact of a natural hazard increases due to both exposure of people and property to the hazard, and vulnerability. The IPCC AR5 also elaborates on the interactions among these key elements in addressing the need for risk management and adaptation to the risks of climate change. Based on this concept of disaster risk, hazard, exposure, and vulnerability, this chapter envisions a policy framework to reduce risk and enhance resilience. When seeking ways to increase resilience to disasters in the HKH, policy makers need to consider five key issues:

1. the multi-hazard environment,
2. close links between upstream and downstream hazards,
3. the effects of climate change and variability,
4. the challenge of connectivity and physical access, and
5. governance.

The *multi-hazard environment* is common to many countries of the HKH (***well-established***). In Nepal, for example, the local term for floods is *badhi-pahiro*, 'floods and landslides'—probably because of the way in which flooding mountain streams can erode river banks by undercutting, leading to landslides which in turn can form natural dams that later breach, causing flash floods. Similarly, earthquakes can trigger both landslides and flash floods. Efforts to build resilience thus need to consider not just the primary event, but also secondary hazards involving cascading threats and disasters (Shrestha et al. 2016).

Also important are the *upstream and downstream linkages of hazard events*. Exposure to hazards can extend, though with a time lag, to an area much larger than the site of the primary event, often across international borders. Even within the same country, physical infrastructure that is supposed to increase resilience, such as early warning systems for flash floods, may not be as widely available as necessary. Institutional arrangements for coordination and cooperation at various levels could also be an issue when managing linkages between upstream and downstream hazard events, especially if the communities at each end lack mechanisms for reciprocal cooperation, and this is compounded when the affected areas lie across international boundaries. For example, the outburst of a landslide-dammed lake in the Tibetan Autonomous Region of China could seriously damage a Nepalese hydropower plant. Similarly, events in Nepal could endanger India's densely populated northern states. Communication channels between local authorities in upstream and downstream nations are often poorly developed and central government efforts to establish communication may intervene too late to save human lives or infrastructure (***established but incomplete***).

Among the *impacts of climate change and variability* are the growing number and size of glacial lakes: Himalayan glaciers have retreated rapidly in recent decades, causing many such lakes to form and expand (see Chap. 7) (***well-established***). Climate change is expected to lead to further increases (***established but incomplete***). The instability of the moraine materials holding back these lakes poses a risk of glacial lake outburst floods (GLOF). In addition, shifting monsoon patterns may result in episodes of intense precipitation, leading to further increases in floods, landslides, and soil erosion (***established but incomplete***).

The *challenge of connectivity and physical access* can involve road and air travel and information and communications technology in sparsely-settled and often remote mountain areas. Finally, this very remoteness raises *governance* as a key issue. Compared to better connected areas, local governments and communities need the capacity to make decisions about hazards that commonly affect isolated and remote locations. In addition, national plans and institutional options for strengthening adaptive capacity may not fully reflect local realities and could be more thoroughly informed by local adaptation concerns (***inconclusive***).

To address these five key issues for upstream and downstream communities in the HKH, we present a new disaster risk reduction (DRR) framework that can help in assessing hazard and risks while discussing adaptation and resilience measures. While developed specifically for the region, it draws on two existing frameworks: the Hyogo Framework for Action 2005–15 and the Sendai Framework for Disaster Risk Reduction 2015–2030. It has four elements:

- *Information:* sharing hazard information between upstream and downstream communities, ensuring communication about cascading hazards
- *Infrastructure:* adapting to climatic and seismic risks, investing to enhance connectivity
- *Institutions:* addressing gender and governance dimensions and developing mechanisms to connect national institutions, policies, and actions with local ones
- *Insurance:* insuring, or transferring risk, to build resilience to residual disaster risks (those that may not be eliminated)

We also present a matrix showing how these four disaster risk reduction elements can interact with four components of resilience-building programmes:

1. *command-and-control mechanisms* such as zoning regulations, land use guidelines, and building codes;
2. *monetary incentives* such as subsidies on insurance premiums;
3. *persuasive information* such as risk maps; and
4. *'nudges'* such as early warning systems.

Alongside the new disaster risk reduction framework, standard multi-hazard risk assessment protocols are needed to study the HKH as a multi-hazard environment. Cascading hazards, especially, require a multi-hazard methodology that integrates complex "hazard interactions and interaction networks" and a multi-hazard early warning system. Successful management critically depends on such assessments—but tools for the HKH have not yet been developed. The process of creating these standard protocols must be as participatory as possible, comprised of diverse stakeholder perspectives.

Because natural hazards know no borders, disaster risk reduction in the HKH would benefit greatly from stronger regional cooperation [established but incomplete]. Events such as the 2005 Kashmir earthquake and the 2010 Pakistan floods have prompted calls for increased efforts. In the Asia regional plan for implementation of the Sendai framework for disaster risk reduction 2015–2030 a two-year action plan for 2017 to 2018 seeks to strengthen existing regional mechanisms to reduce the risk of, and enhance early warning and preparedness for, transboundary disasters. A regional approach, with efforts in timely data sharing and modelling, should improve flood management and help mitigate adverse impacts in transboundary basins.

Disaster Risk Reduction (DRR), the HKH, and the Sustainable Development Goals (SDGs)

Building disaster resilience in the mountains requires decision making that is informed by the best available studies of disaster risk reduction (DRR) and climate change adaptation. Because of mountain communities' high vulnerability to extreme weather events and natural hazards such as floods, landslides, earthquakes, and avalanches, disaster risk reduction assumes a high profile in the 2030 sustainable development agenda. Especially relevant to the HKH are Sustainable Development Goals (SDGs) 1 (End poverty), 5 (Gender equality and empowerment of women and girls), 9 (Build resilient infrastructure), 11(Make cities and human settlements inclusive, safe, resilient and sustainable), 13 (Combat climate change), and 15

(Manage forests and combat desertification).To complement the relevant formal SDGs, we propose the following vision for mountain disaster risk reduction: *By 2030, build resilience to reduce disaster risks and losses in lives, livelihoods and assets from natural disasters in mountain regions substantially through informed decision making and enhanced preparedness*. Supporting this vision are four targets:

1. Reduce human deaths, economic loss, and the number of people affected due to disasters and extreme climate events, especially for women and children [SDGs 5, 13].
2. Make human settlements and habitats safe, inclusive, resilient, and sustainable through capacity building, legislation, education, livelihoods, better zoning and building regulations, and a multi-hazard risk reduction approach [SDGs 9, 11, 13].
3. Ensure protection from exposure to extreme events such as floods and droughts [SDGs 1, 13].
4. Provide access to disaster risk reduction and mitigation measures, including finance and technology, with the knowledge and capacity building needed to use them [SDG 13].

11.1 The Hindu Kush Himalaya: An Uncertain, Multi-hazard Environment

Disaster risk depends on how likely different kinds and intensities of natural hazards are to occur, the elements exposed, and their vulnerability (World Bank 2005). Mountain development and disaster risk are inherently linked, as many mountain settlements are located on unstable mountain slopes that are prone to landslide and erosion, or on river terraces and alluvial fans that are susceptible to debris flows and floods. The Hindu Kush Himalaya (HKH) is one of the most fragile mountain regions in the world. Because of its unique geology, steep terrain, intense seasonal precipitation, and high seismicity, the region is highly vulnerable to floods (especially flash floods), landslides, droughts, and earthquakes —a truly multi-hazard environment.

11.1.1 Hazards and Disasters in the HKH

Across the world, the number of extreme geophysical and hydrometeorological events appears to be increasing, resulting in a growing loss of lives and increasing damage to livelihood support systems. The HKH accounted for 21% (4,115 of 18,956) of the major disaster events recorded between 1980 and 2015 in the Em-DAT global database, and 36% of the major events in Asia. Floods and landslides are the most frequently occurring natural hazards, particularly during the monsoon season (Shrestha 2008a, b; Gaire et al. 2015); they accounted for nearly half of the events recorded in the countries of the HKH region (Fig. 11.1).

The number of disaster events from climate, hydrometeorological and geophysical hazards recorded in the countries of the HKH region between 1980 and 2015 is shown in Fig. 11.2. A total of 739 disaster events were recorded in China, 438 in India, and 229 in Bangladesh. Floods are prominent across all countries, landslides in inland countries (Afghanistan, Nepal, and Pakistan), and storms in coastal countries (China, India, and Myanmar). Wildfires are also on the rise across the region.

In a mortality ranking conducted by ADB (2013b), seven of the eight countries in the HKH were included in the 20 Asian countries (out of 44) rated as mortality hotspots (Table 11.1).

Floods and flash floods: Floods, both riverine and flash floods, are the most common hazards in the HKH (Shrestha et al. 2015) and account for 17% of people killed and 51% of the damage (Fig. 11.1). Unlike riverine floods, flash floods occur rapidly with a very short lead time for warning. They can arise following intense rainfall events, or as a result of breaching of natural dams formed by landslides or from glacial lakes formed behind end moraine dams (glacial lake outburst flood or GLOF) (Shrestha 2008a, b). In recent years, increasingly erratic and unpredictable monsoon rainfall patterns and increased climate variability have led to severe and frequent flood disasters in the region. This has adversely impacted lives and livelihoods, agricultural productivity, and hydropower production, among others. For example, the 2010 floods in Pakistan killed more than 2,000 people, with an estimated loss of USD 10 billion (FFC 2010). In India, the 2013 flood in Uttarakhand killed more than 5,000 people (Awasthi et al. 2014; Guha-Sapir et al. 2014; Champati Ray et al. 2016), and possibly as many as 30,000 (Ziegler et al. 2014). Bangladesh is one of the countries in the region most vulnerable to floods, as it is situated on the delta of three major HKH river systems: the Ganges, Brahmaputra, and Meghna (Islam et al. 2010). In Pakistan, flooding is the most frequently occurring hazard; it affects thousands of people and causes millions of dollars of damage annually (Tariq and Giesen 2011). Table 11.2 shows the large flood events reported in the HKH from 1980 to 2015. Figure 11.3 shows the spatial extent and impact of flood disasters occurring between 2010 and 2014.

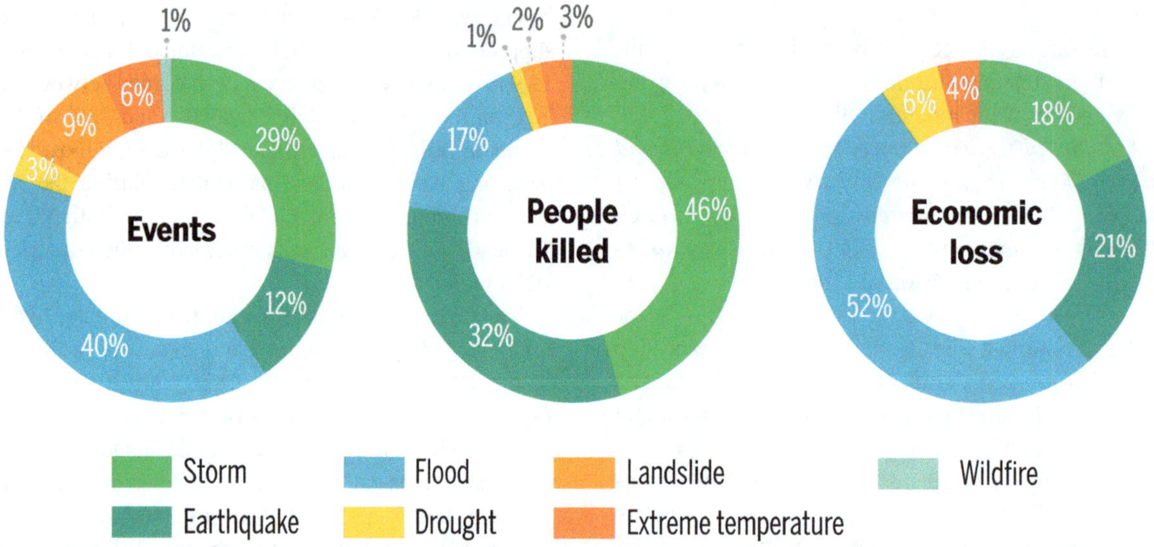

Fig. 11.1 Proportional impact of different types of disaster in HKH countries (whole country, including HKH area) between 1980 and 2015—number of events, persons killed, and economic loss (*Source* EM-DAT: The Emergency Events Database—Université catholique de Louvain (UCL)—CRED, D. Guha-Sapir—www.emdat.be, Brussels, Belgium)

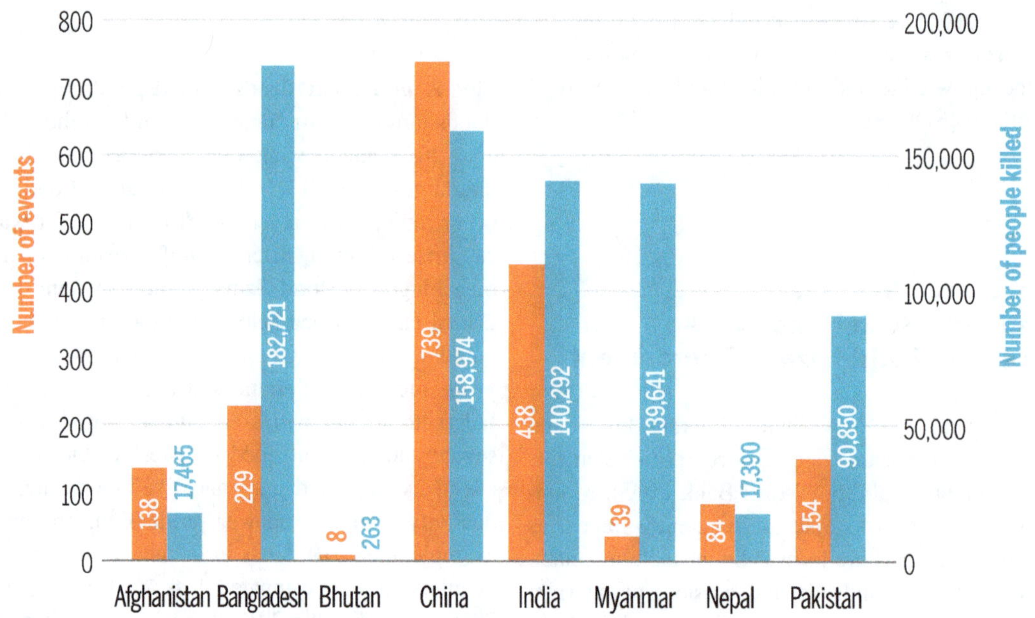

Fig. 11.2 Disaster number (**a**) and people killed (**b**) by climate, hydrometeorological, and geophysical disaster events in HKH countries (whole country, including HKH area) between 1980 and 2015 (*Source* EM-DAT: The Emergency Events Database—Université catholique de Louvain (UCL)—CRED, D. Guha-Sapir—www.emdat.be, Brussels, Belgium)

As of 2000, the HKH has witnessed more than 33 identifiable GLOFs (Richardson and Reynolds 2000). Accelerated glacial thinning, degraded permafrost (Haeberli et al. 2016), and additional retreat in response to rising global temperatures are expected to increase GLOF events in the future (Ives et al. 2010). The most recent recorded GLOFs occurred at Lemthang Tsho in western Bhutan in June 2015 (Gurung et al. 2017) and in multiple locations in Chitral, Pakistan in July 2016. A GLOF upstream of Uttarakhand in 2013 (Allen et al. 2015; Champati Ray et al. 2016) damaged high value infrastructure like hydropower dams (Schwanghart et al. 2016) and impacted the lives of more than 100,000 people. Ice avalanches into expanding lakes can also lead to GLOFs and cause large floods downstream (Haeberli et al. 2016). The number of GLOFs in the HKH is increasing (see Box 11.1).

Table 11.1 High mortality risks from multiple hazards in the HKH

Mortality ranking	Ranked HKH countries	Percent of population in areas at risk	Estimated number of people at risk (millions)	Percent of total area at risk
1	Bangladesh	97.7	139.6	97.1
2	Nepal	97.4	25.9	80.2
5	Bhutan	60.8	0.4	31.2
7	Pakistan	49.6	87.84	22.8
8	Afghanistan	46.0	12.2	7.2
11	China	33.4	450.0	10.6
14	India	27.2	337.8	21.9
16	Myanmar	16.8	10.1	4.5

Source ADB (2013b)

Table 11.2 Severe floods between 1980 and 2015 in the countries of the HKH region

Country	Year	Month	People killed	People affected
India	2013	June	6,453	3,419,473
Pakistan	2010	July	2,113	20,363,496
Bangladesh	2007	September	1,230	13,851,440
India	1998	July	2,131	29,652,200
Bangladesh	1998	September	1,050	15,000,050
India	1997	September	2,357	30,259,020
Nepal	1993	July	1,048	553,268
Pakistan	1992	September	1,446	12,839,868
Afghanistan	1991	May	1,193	139,400

Source EM-DAT: The Emergency Events Database—Université catholique de Louvain (UCL)—CRED, D. Guha-Sapir—www.emdat.be, Brussels, Belgium

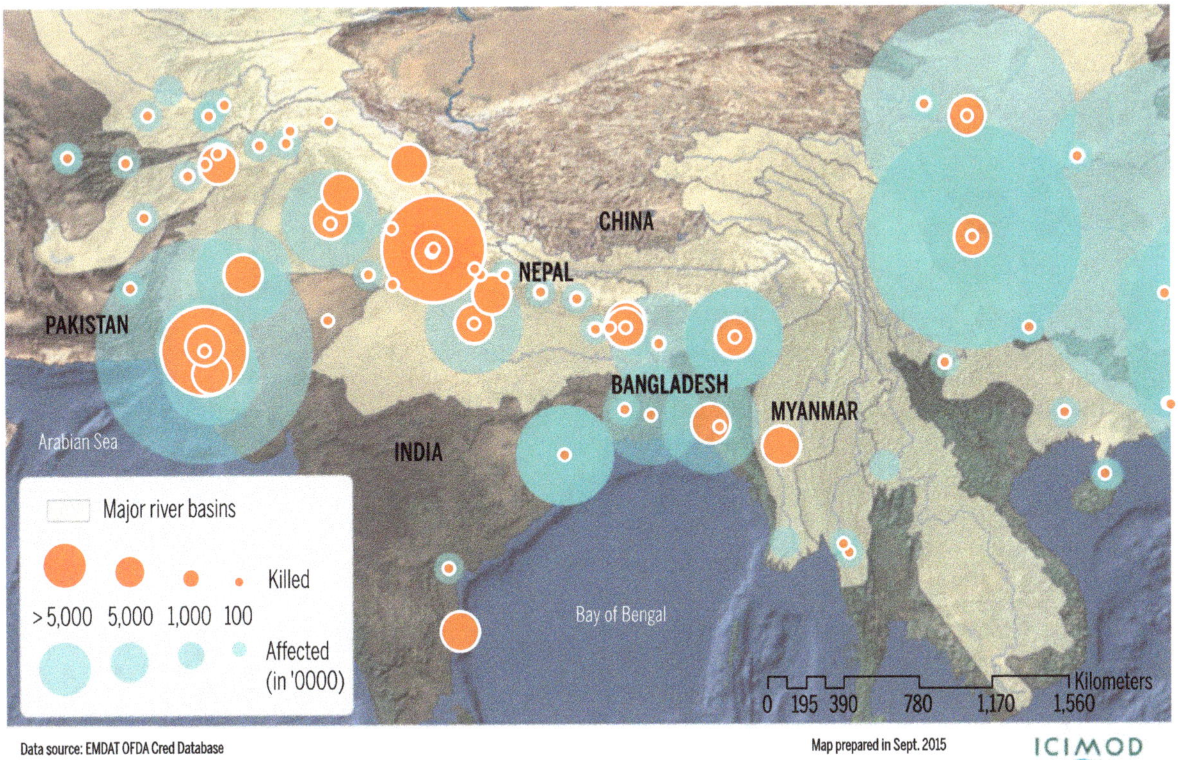

Fig. 11.3 Spatial extent and impact of flood disasters in the major river basins originating in HKH from 2010 to 2014 (*Source* EM-DAT: The Emergency Events Database—Université catholique de Louvain (UCL)—CRED, D. Guha-Sapir—www.emdat.be, Brussels, Belgium)

Box 11.1 Upper Indus glacial lake outburst floods (GLOFs), 1826–2000 CE

Glacial lake outburst floods (GLOFs) can cause considerable damage to life and property. Reviewing 174 years of data from the Karakoram region, Hewitt and Liu (2010) noted that GLOF events have become more frequent in the upper Indus catchment (see Fig. 11.4). Comparison of tree-ring-based precipitation (Singh et al. 2006) and temperature records (Yadav and Singh 2002) in the high western Himalaya, suggests that during period A (1826–1893) temperatures were low and precipitation was decreasing, during period B (1893–1934) temperatures were even lower and precipitation still decreasing, whereas during period C (1934–2000) temperatures were rising and precipitation increasing. The highest frequency of GLOFs occurred when temperatures were at their lowest, particularly in the first part of period B. The climatic controls on the glaciers and these floods are not clear, but there is no evidence of a recent increase of GLOFs that may be attributed to global warming.

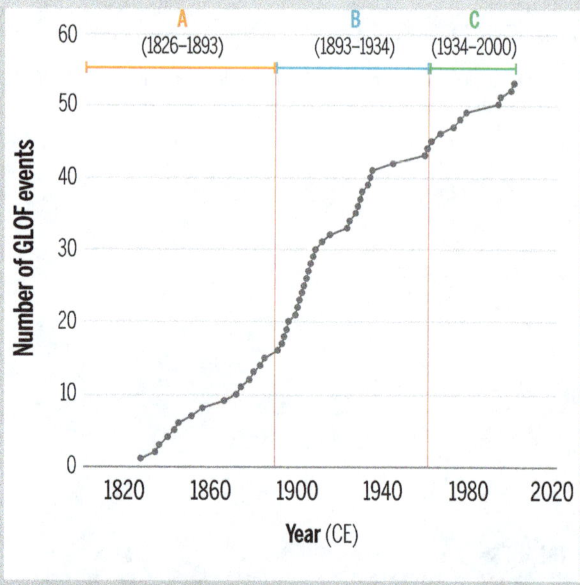

Fig. 11.4 Increasing frequencies of GLOF events in the Upper Indus (*Data source* Hewitt and Liu 2010)

Global flood projections based on the Multiple Coupled Model Intercomparison Project, Version 5-Global Climate Model (CMIP5-GCM) simulations coupled with global hydrology and land surface models show flood hazards increasing over approximately half of the globe, but with great variability at the catchment scale (Dankers et al. 2014). The projected increases in temperature and intense precipitation will induce regional-scale changes in flood frequency and intensity (IPCC 2012), resulting in changes in extreme weather patterns (Elalem and Pal 2015). The Arctic Oscillation and its interaction with the monsoon will also play a role in climate change in the HKH (Joseph et al. 2013); as the Arctic grows warmer, outbursts of cold and dry air are likely to increase, producing more frequent and intense rainfall and triggering increased flooding. Climate change and the accompanying increase in rainfall intensity and alteration of the hydrological cycle have reportedly increased the likelihood of landslides and flooding in HKH countries such as Bhutan and Nepal (Khanal et al. 2007).

Landslides: The HKH is characterized by steep topography: more than 40% of the land area has a slope of 15° or more (Fig. 11.5). With fragile geological formations, a seismically active mountain system, and intense precipitation, the region is a global hot spot for landslides. Hydroclimatic and seismic sensitivity in the area increases the hazard level (Fig. 11.6). Anthropogenic influences like unsustainable development and excessive resource extraction—particularly deforestation and road building—have adversely influenced slope stability and aggravated the possibility of landslide.

Landslides are also common, with the HKH countries accounting for 52% of the landslide events and 61% of deaths recorded in Asia in the Em-DAT global database over the period 1980 to 2015. Global disaster databases tend to grossly underestimate landslide fatalities as these are normally recorded under the primary trigger and not the hazard itself (Nadim et al. 2006). Nevertheless, landslides still

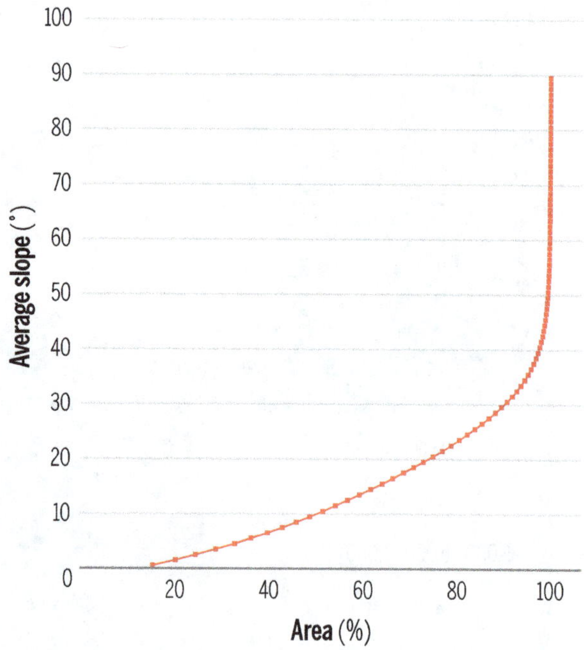

Fig. 11.5 Area characterization of the HKH (*Source* Developed for this chapter by Deo Raj Gurung, using 90 m Shuttle Radar Topographic Mission (SRTM) DEM, available at http://srtm.csi.cgiar.org/)

Fig. 11.6 Landslide hazard **a** induced by precipitation, **b** induced by earthquakes (*Source* Nadim et al. 2013)

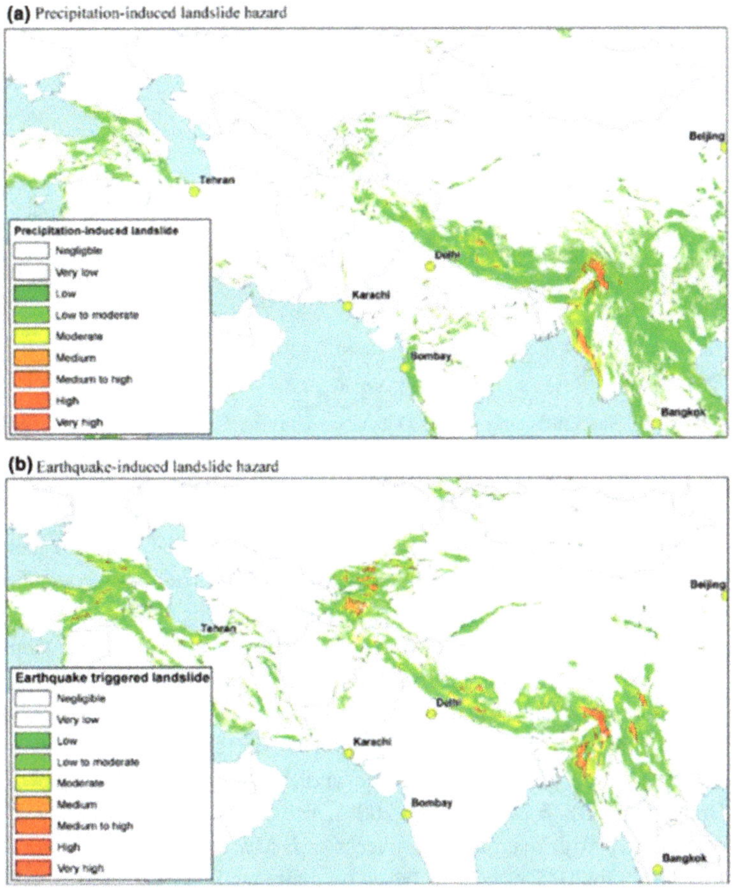

register as among the most catastrophic disasters in the HKH region (Upreti and Dhital 1996; Sarwar 2008; MoSWRR 2009; SAARC 2010; Khan and Khan 2015; Lotay 2015) (Fig. 11.6). Projections give even more cause for concern: a simple analysis using global population density data and a digital elevation model to estimate the number of people in the HKH living on land with a slope of 10–50° shows approximately 5.2 million people at risk of exposure to landslides.

Avalanches: Avalanches are one of the biggest hazards in mountainous terrain where geographic and meteorological conditions give rise to heavy precipitation and accumulations of snow and ice that can hurtle down into inhabited valleys. Avalanches are common in the mountainous areas of the HKH region in winter. Loss of life and property is observed annually in the higher snow-covered areas of Afghanistan, Bhutan, India, Nepal, and Pakistan (SDMC 2008) (see Box 11.2). The 2015 Nepal earthquake triggered several avalanches which killed more than twenty people. In 2014, the Hudhud cyclone initiated in the Bay of Bengal brought heavy precipitation to Nepal resulting in avalanches that killed more than 43 people (Wang et al. 2015).

Box 11.2 Avalanches in Afghanistan

Two million people in Afghanistan are exposed to avalanches, and between 2000 and 2015, more than 153,000 people were affected (World Bank 2017). Heavy snowfall in Panjshir Province in February 2015 triggered 40 avalanches that killed at least 124 people, and in February 2017 avalanches near the Pakistan border killed at least 137 people.

Earthquakes: The HKH is one of the world's youngest mountain belts (GFDRR 2012) and is tectonically active. The major cause of earthquakes in the Himalaya is the subduction of the Indian plate underneath the Eurasian plate, which causes contraction and stress concentration. Seismicity is considered high in this region based on the frequency and intensity of past earthquakes (Rai 2004). Plate

Table 11.3 List of recent large earthquakes in the HKH (1993–2015)

Year	Month	Country	Name	Magnitude (Richter)[a]	Death[a]	Affected[a] (thousand)	Economic loss (thousand USD)[a]	GDP (current million USD)[b]	Loss/GDP (%)
1993	Sep.	India	1993 Latur	6.4	9,748	30	280,000	278,359	0.10
1998	May	Afghanistan	1998 Afghanistan	6.9	4,700	117	10,000	2,912	0.34
2005	Oct.	Pakistan	2005 Kashmir	7.6	73,338	5,128	5,200,000	117,708	4.42
2008	May	China	2008 Sichuan	7.9	87,476	45,976	85,000,000	4,604,285	1.85
2010	Apr.	China	2010 Yushu	6.9	2,968	112	500,000	6,066,351	0.0082
2015	Apr.	Nepal	2015 Nepal	7.8	8,831	5,639	5,174,000	20,801	24.28

[a]EM-DAT: The Emergency Events Database—Université catholique de Louvain (UCL)—CRED, D. Guha-Sapir—www.emdat.be; [b]UN National Accounts Main Aggregates Database; https://unstats.un.org/unsd/snaama/selbasicFast.asp

motion models and GPS measurements indicate that the India-Eurasia convergence continues today at a rate of about 40–50 mm per year (de Mets et al. 1994), while the rate of contraction across the Himalaya is estimated to be only 17.52 ± 2 mm per year (Bilham et al. 1997). The difference in these rates is absorbed by a combination of thrusting, crustal extension, and strike-slip motion within the Eurasian plate (Armijo et al. 1989; Avouac and Tapponnier 1993; Bhatia et al. 2000; Bilham et al. 2001). Table 11.3 lists large earthquakes with magnitude in excess of 6 Mw in the HKH from 1993 to 2015. Scenarios based on mathematical models and inferences based on field investigations indicate that the HKH is a high earthquake risk region (Wesnousky et al. 2017; Bollinger et al. 2004).

Drought: The arid and semi-arid regions of western and northwestern HKH (i.e., the Tibetan Plateau, Afghanistan, northern Pakistan, northwest India, and northwest Nepal) are located in drought-prone areas (Ahmad et al. 2004; Wang et al. 2013). The humid and semi-humid regions also face severe water shortages during the dry months of the year. Drought accounts for only 4% of all disasters reported globally in the Em-DAT database from 1980 to 2015, yet it accounts for 25% of all people affected by climate-related disasters (CRED and UNISDR 2016).

Extreme temperatures: Climatological hazards, including extreme temperatures (heat wave, cold wave, and extreme winter conditions) interacting with exposed and vulnerable human and natural systems, can also lead to disasters (IPCC 2012). Extreme heat is a prevalent public health concern throughout the temperate regions of the world. Extreme heat events have been experienced recently in the HKH (see Box 11.3), and it is likely that the length, frequency, and/or intensity of warm spells, including heat waves, will continue to increase. The factors that contribute to physiological and social vulnerability to heat-related illness and death are age, gender, body mass index, and pre-existing health conditions. A common public health approach, early warning systems, and hazard education can play a significant role in reducing exposure and mortality due to extreme temperatures.

Box 11.3 Heat wave in southern Pakistan

A severe heat wave with temperatures as high as 49 C (120 F) struck southern Pakistan in June 2015. It caused the deaths of about 2,000 people from dehydration and heatstroke, mostly in Sindh province and its capital city, Karachi. The event followed a separate heat wave in India in May 2015 that killed 2,500 people, including 1,735 in the south Indian state of Andhra Pradesh and 585 people in neighbouring Telangana, the most affected areas.

11.1.2 Disaster Trends in the HKH

The Em-DAT database indicates an increasing trend in the HKH in the number of disaster events reported, number of people killed and affected, and economic losses (Fig. 11.7). Between 2000 and 2010, 749 events were reported in the HKH with 399,609 people killed and a huge economic loss. This represents a 143% increase from 1990 to 2000, in agreement with the report by ADB (2013b) of a rising frequency in natural disasters in Asia and the Pacific.

There is an increasing trend in extreme rain events over India (Goswami et al. 2006). An increase in extreme floods is also evident from the historical records for the Alaknanda River in Uttarakhand, India (see Box 11.4).

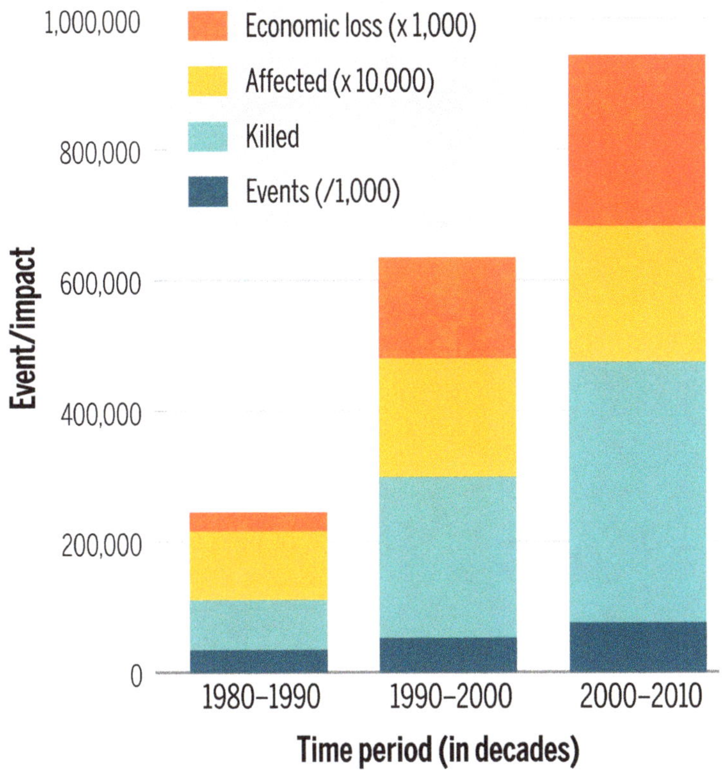

Fig. 11.7 Decadal impact of disasters in the HKH (*Source* EM-DAT: The Emergency Events Database—Université catholique de Louvain (UCL)—CRED, D. Guha-Sapir—www.emdat.be, Brussels, Belgium)

Box 11.4 A history of extreme floods in the Alaknanda River, Uttarakhand, India

Extreme floods occurred in 1894, 1970, and 2013. The most recent flood is fresh in local memory, while the older generation recalls the 1970 flood as one of the forces that kept the Chipko Andolan alive. No one recalls the 1894 deluge. The 2013 flood is thought by some to have been unique and a similar event unlikely to recur. But the history of extreme floods over the past 2000 years, reconstructed from historical accounts and sediments deposited from floodwaters near Srinagar, tells a different story. It is important to note that the oral accounts and sedimentary records are in accordance for the last two extreme floods, so there is confidence that the entire sedimentary record relates to extreme floods. (Wasson et al. 2013; Ziegler et al. 2014).

As can be seen in Fig. 11.8, from roughly 100 to 1700 CE, the frequency of floods was relatively constant at approximately one every 200 years. But in the late 1700 s there was a cluster of five floods, with one every 10 years on average. The most recent three floods in the Alaknanda occurred on average once every 40 years. Most of the floods are likely to have resulted from the confluence of warm and moist monsoon air from the south with cold dry air from the Arctic. This confluence is likely to become more common as the Arctic warms, and thus flood events are also expected to increase.

Large earthquakes are low-probability but high-impact (Fig. 11.9). However, despite the rare occurrence of earthquakes, there has been a gradual temporal increase in the number of fatalities and level of damage from earthquakes, most of which can be attributed to an increasing concentration of economic development and urbanized habitation in the region. For example, loss of life was similar ($\sim$70,000) in the 2005 Kashmir and 2008 Sichuan earthquakes, but the Sichuan disaster, which was closer to urban areas, resulted in an economic loss 16 times greater than in the Kashmir earthquake.

11.1.3 Linking Primary and Secondary Hazards —The Mountain Perspective

In the realm of hazard, the term 'cascading' is used to describe the interconnected nature of natural processes in which a primary event triggers a chain of subsequent (secondary and tertiary) hazard event(s). The cascading nature of hazard, and therefore cascading nature of disaster, was appreciated by the global community after the 2011 Tohoku earthquake (Pescaroli and Alexander 2015), in which an

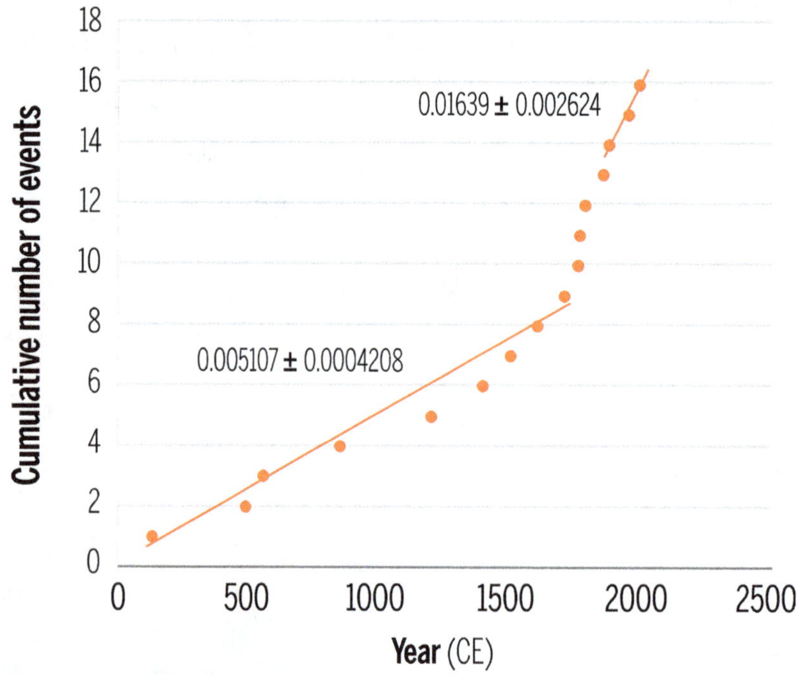

Fig. 11.8 History of extreme floods in the Alaknanda River, Uttarakhand, India (*Data source* Wasson et al. 2013)

earthquake lead to a tsunami which in turn resulted in a nuclear crisis. Earthquakes can generate landslides that block rivers, causing lakes to form, which in turn can generate a landslide lake outburst flood (LLOF) when the dam fails abruptly as a result of overtopping or piping (Wasson and Newell 2015). Earthquakes can also create landslides and dislodge large sections of glaciers or ice walls, which can then cause a GLOF if they fall into a glacial lake.

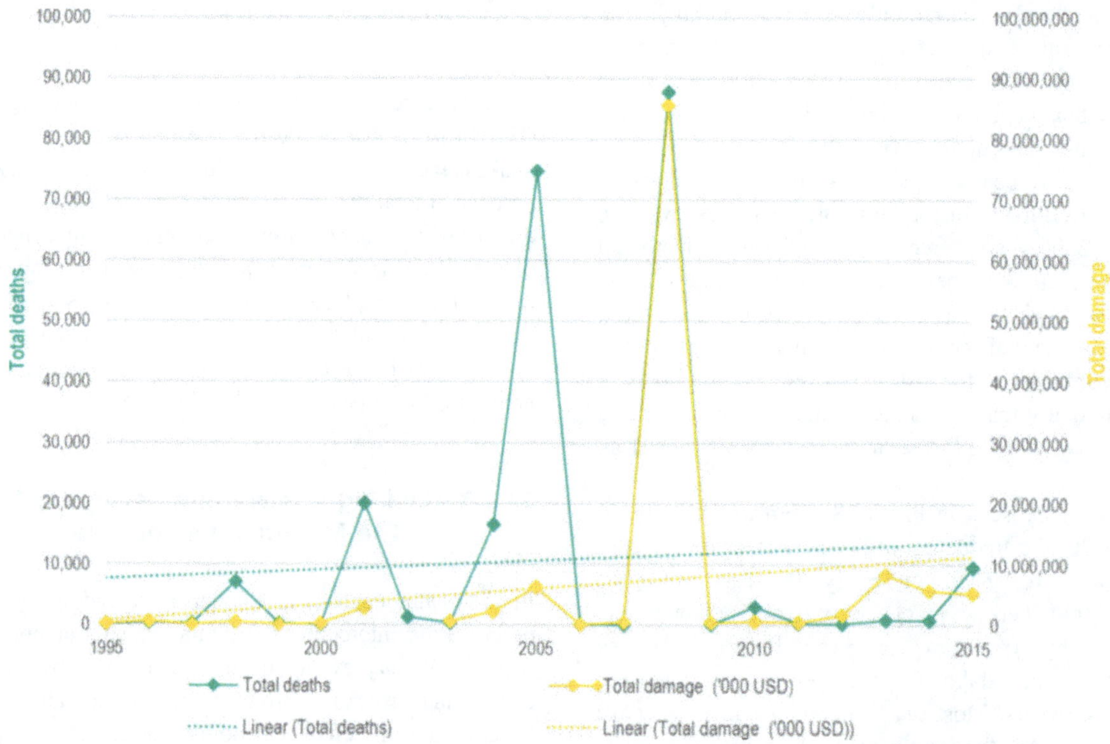

Fig. 11.9 Impact of earthquakes on the number of people killed and total economic losses in the HKH (*Source* EM-DAT: The Emergency Events Database—Université catholique de Louvain (UCL)—CRED, D. Guha-Sapir—www.emdat.be, Brussels, Belgium)

GLOFs can erode the toes of hill slopes downstream, adding sediment to the river and raising the riverbed, which in turn can lead to blockages in the river downstream and the formation of small lakes, which themselves can burst out leading to further toe slope erosion downstream. These types of cascading process have been seen during the 1970 flood in the Alaknanda river in Uttarakhand (Wasson and Newell 2015), the 2015 Gorkha earthquake in Nepal (Kargel et al. 2016), following glacial lake outburst floods in Nepal and Bhutan (Higaki and Sato 2012), and following flash floods (Gupta et al. 2016).

Rainfall can also trigger landslides that dam rivers and lead to LLOFs (Gupta and Chalisgoankar 1999), the flows from which undercut the toes of the hill slopes downstream, adding to the landslide and debris flow material and increasing sedimentation along the river bed, thereby raising it. One result is that settlements that were previously high above the riverbed and therefore safe from floods, become at greater risk from future floods. An example of this can be seen at Sonprayag downstream of the Kedarnath hazard zone in Uttarakhand, where the riverbed is now about 30 m higher than before the 2013 flood. It will be many decades before the riverbed is lowered by sediment evacuation and the flood hazard to local settlements is reduced (Sundriyal et al. 2015; Rautela 2013). Other examples of LLOFs include the Tsatichu landslide in Bhutan (Dunning et al. 2006; Shrestha and Chhophel 2010) and the Jure landslide in the SunKoshi basin, which formed a lake 3 km long that eventually breached.

The context of cascading hazard and thus cascading disaster is particularly relevant in a mountainous setting like the HKH region, where primary and secondary hazards are closely interrelated. Primary hazards may be geophysical or hydrometerological (e.g., landslides), and trigger secondary hazards, such as landslide dams and subsequent outburst floods (Gill and Malamud 2016). For example, the 2015 Nepal earthquake resulted in more than 4,000 landslides (Kargel et al. 2016). Table 11.4 shows some examples of primary and secondary hazards in the HKH region.

The examples of cascading hazards indicate clearly that disaster risk reduction (DRR) plans and policies in the HKH region will be ineffective unless they take a holistic approach which clearly appreciates the interconnectedness of different hazard events. It is important to recognize that a cascading chain of events can unfold immediately, but may also take place after a substantial lapse of time. To address these issues, it is essential to have a multi-hazard early warning system.

11.2 Vulnerability: Physical, Social, Economic, and Environmental Dimensions

Disaster risks are a function of interplay among three key elements: hazards, exposure, and vulnerability (IPCC 2012). The term vulnerability is a state of susceptibility to harm and assumes different connotations depending on the context (Ciurean et al. 2013). A plurality of views and meanings of the term 'vulnerability' are explicit in the different ways the natural science and social systems frame the term and construct measurement frameworks. Similarly, climate change, environmental change, and disaster risk reduction all possess different visions of vulnerability. The 'vulnerability' perspective in disasters is defined as "the characteristics of a person or group and their situation that influences their capacity to anticipate, cope with, resist, and recover from the impact of a natural hazard" (Donner and Rodríguez 2011). The life safety and livelihoods of mountain communities in the HKH region are constantly threatened by multiple geological and hydrological hazards. Climate change, poor land use practices, and forest and land degradation are further exacerbating these risks, especially the risk of hydrological hazards (Shaw and Nibanupudi 2015). There is a clear indication that not only is the frequency of such hazards increasing with time, but also their intensity and impact on the lives and livelihood of people is increasing in severity. The frequency and intensity of disasters is pushing the resilience and recovery capacity of communities,

Table 11.4 Primary and secondary hazards in the mountains of the HKH

Type of hazard		Occurrence
Primary	Secondary	
Earthquake	Landslides	2005 Kashmir earthquake in Pakistan and India
		2008 Wenchuan earthquake in China
		2015 Gorkha earthquake in Nepal
Landslide	Landslide dam and subsequent outburst flood	2014 Nepal: Jure landslide that dammed the Sunkoshi river
		2010 Pakistan: Hunnza Attabad landslide
		2008 China: Landslide-dammed lake at Tangjiashan, Sichuan province
Flood	Erosion and deposition (aggradation and degradation), sand casting (deposition)	2008 Koshi floods in Nepal and India

governments, and institutions to the limit. Climate change is now expected to exacerbate disasters and lead to greater destruction in the future in the HKH region, with potentially profound implications (see Chap. 3: Climate Change in the Hindu Kush Himalaya).

This section focuses on vulnerability in the context of disaster risk management, which is framed as the potential for loss caused by natural hazards, and is a function of exposure, susceptibility, and coping capacity. Birkmann et al. (2013) describe three core thematic dimensions of vulnerability.

Physical dimension: potential for damage to physical assets, including built-up areas, infrastructure, and open spaces.
Social dimension: propensity for human wellbeing to suffer as a result of disruption to individuals (mental and physical health) and the collective social systems (health, education services), and the characteristics of these systems (e.g., gender, marginalization of social groups).
Economic dimension: propensity for loss of economic value from damage to physical assets and/or disruption of productive capacity.

According to UNISDR, vulnerability is "determined by physical, social, economic, and environmental factors or processes, which increase the susceptibility of a[n] [individual] or community [assets or systems] to the impact[s] of hazards" (United Nations 2016). The World Bank and IPCC include governance as the fifth factor influencing vulnerability, and suggest it is particularly important in regions like South Asia, where governance is generally weak. The definition implies that vulnerability is a condition that depends on multiple factors. Mountain systems are inherently challenged by what Jodha (1992) calls 'mountain specificities', which aggravate vulnerability to disaster and include, among others, constraining features such as accessibility, marginality, and fragility. The vulnerability of these countries to disasters is characterized by complex interactions between the natural and socioeconomic conditions (Elalem and Pal 2015).

11.2.1 Physical Factors

Physical vulnerability refers to the vulnerability that stems from the limitations posed by the physical characteristics of the exposed elements, for example, population density, remoteness, limited access to critical amenities, legal challenges, proximity to hazard zones, and design and quality of infrastructure. Table 11.5 shows the percentage of area and population exposed to hazards in the HKH countries. The least vulnerable communities are those with lower levels of exposure that have good access to emergency response services and comparatively high-quality infrastructure.

Table 11.5 Percentage of area and population exposed to hazards by HKH country

Country	Percent of total area exposed	Percent of population exposed	Maximum number of hazard types[a]
Afghanistan	11.1	29.5	3
Bangladesh	35.6	32.9	4
Bhutan	20.1	29.2	4
China	8.4	15.7	3
India	10.5	10.9	4
Myanmar	10.7	10.4	4
Nepal	60.5	51.6	3
Pakistan	5.6	18.2	2

[a]Cyclones, drought, earthquakes, floods, landslides (*Source* World Bank 2005)

Table 11.6 Quality of infrastructure services in the HKH countries (World Economic Forum 2014)

Country	Overall infrastructure	Roads	Air transport	Electricity supply
World	4.23	4.02	4.36	4.50
Bangladesh	2.82	2.88	3.02	2.55
Bhutan	4.63	4.31	3.51	5.85
China	4.36	4.61	4.72	5.22
India	3.75	3.79	4.27	3.43
Nepal	2.93	2.90	2.92	1.83
Pakistan	3.32	3.81	3.92	2.07

Note Scores on a scale of 1 (low) to 7 (high)

The overall quality of infrastructure in Bangladesh, India, Nepal, and Pakistan is lower or much lower than the global average, whereas in Bhutan and China it is slightly better (Table 11.6; World Economic Forum 2014). This is true for roads, air transport, and electricity services. The variation in quality of infrastructure among the HKH countries is partly due to inadequate levels of investment (World Bank 2013). In 1973–2009, Bangladesh, India, and Pakistan spent only six percent of their Gross Domestic Product (GDP) on average on infrastructure, and Nepal only five percent. The World Bank suggested that all these countries will need to invest a higher share of GDP in infrastructure in 2011–2020.

11.2.2 Social Factors

Social vulnerability refers to the human vulnerability resulting from the characteristics inherent in social interactions, institutions, and systems of cultural values which determine the capacity of groups and individuals to deal with disasters and hazards, and is based on the position and situation of people within the physical and social worlds (Dow

1992). Over the years, the term 'social vulnerability' has taken on a broader and increasingly interdisciplinary meaning to incorporate the idea that vulnerability is not just an inherent characteristic of certain groups, but rather it is produced, underlied, and driven by a wide variety of conditions. Therefore, vulnerability is not just defined with respect to exposure to hazards, but also by numerous socioeconomic factors. Some common factors determining social vulnerability include social and economic inequality, marginalization, social exclusion, lack of preparedness and adaptive capacity, and discrimination by gender, social status, disability, and age (Bergstrand et al. 2015). Affluent communities with equity in all spheres of social practice are generally less vulnerable than poor communities where inequality is prevalent. Efforts to reduce vulnerability must not, therefore, be confined to reducing hazard exposure only, but should also include the social systems within which vulnerability is produced (Blaikie et al. 1994).

The UN's Human Development Index (HDI) is one of the most widely-used indicators for measuring quality of life, and provides an interesting starting point for evaluating the HKH countries (Table 11.7). Generally, the HKH countries (except China) have lower HDI values than the world average, which indicates that these populations have higher than average social vulnerability. Education and gender inequality are particularly pronounced in the HKH countries, whereas income inequality is high, except in China and Bangladesh.

Focusing on local and indigenous knowledge is also important for mountain communities, which usually have a rich experience and knowledge linked to their lifestyles and livelihoods. Indigenous knowledge forms the basis of community coping practices, builds up resilience to disasters, and plays an important role in disaster risk reduction. Both in saving lives during a disaster, and helping others recover post disaster, rules built on the basis of indigenous knowledge can help a community to cope more easily (Shaw and Nibanupudi 2015). Combining indigenous and local knowledge with external expertise is vital for resilience.

11.2.3 Economic Factors

The premise that disaster affects rich and poor people differently is based on the idea that economically stronger communities have options to invest in resilient infrastructure, and are economically empowered to invest in better access to emergency services. Thus the level of vulnerability is highly dependent upon the economic status of an individual, the community, and the nation. While disasters cause more economic damage and greater loss to infrastructure in developed nations, they generally take a larger number of human lives in developing countries (Pusch 2004). Economic vulnerability is particularly important in building resilience to disaster and reducing exposure to hazards, and is thus especially important in the HKH, where five of the eight countries (Afghanistan, Bangladesh, Bhutan, Myanmar, and Nepal) are classified as Least Developed Countries (United Nations 2017).

Table 11.8 shows the values for the Multidimensional Poverty Index (MPI) in the HKH countries in 2015. The MPI is measure of acute poverty covering more than

Table 11.7 Human development index and inequalities (UNDP 2016)

Human Development Index and Inequalities									
Country	Human Development Index (HDI)	Inequality in education	Inequality in income	Income inequality			Gender Development Index	Gender Inequality Index	
	Value	(%)	(%)	Quintile ratio	Plama ratio	Gini coefficient	Value	Value	Rank
	2014	2014	2014	2005–2013	2005–2013	2005–2013	20144	2014	2014
China	0.727		29.5	10.1	2.1	37	0.943	0.191	40
India	0.609	42.1	16.1	5	1.4	33.6	0.795	0.563	130
Bhutan	0.605	44.8	19.6	6.8	1.8	38.7	0.897	0.457	97
Bangladesh	0.57	38.6	28.3	4.7	1.3	32.1	0.917	0.503	111
Nepal	0.548	41.4	15.1	5	1.3	32.8	0.908	0.489	108
Pakistan	0.538	44.4	11.6	4.1	1.1	29.6	0.726	0.536	121
Myanmar	0.536	19.4						0.413	85
Afghanistan	0.465	44.8	10.8	4	1	27.8	0.6	0.693	152
Developing countries	0.66	32.3	24.5	–	–	–	0.899	0.478	–
World	0.711	26.8	24	–	–	–	0.924	0.449	–

Source UNDP Human Development Index http://hdr.undp.org/en/content/human-development-index-hdi

Table 11.8 Value of the Multidimensional Poverty Index in the HKH countries, 2015 (UNDP 2016; Alkire and Robles 2017)

Country	MPI
Afghanistan	0.295
Bangladesh	0.196
Bhutan	0.119
China	0.017
India	0.191
Myanmar	0.134
Nepal	0.126
Pakistan	0.230

100 developing countries. It assesses poverty at the individual level and complements traditional income-based poverty measures by capturing the severe deprivations that each person faces with respect to education, health and living standards (Alkire and Robles 2017). With the exception of China, the HKH countries rank high in MPI, which suggests high economic vulnerability.

11.2.4 Environmental Factors

The environmental conditions also play a role in determining a community's vulnerability to disaster. Badly managed environments create unsafe situations and thereby increase vulnerability to disaster. Some of the determinants of environmental vulnerability are poor environmental management, overconsumption of natural resources, degraded ecosystems, decline in risk-regulating ecosystem services, and climate change (ISDR and UNEP 2007). Depletion of natural resources (for example, wetlands) exposes people and infrastructure to natural hazards like floods and storm surges.

The HKH is both a climate change hotspot and a densely populated region, a factor contributing to the depletion and degradation of natural resources, and a pathway to increased vulnerability.

11.2.5 Gender Dimensions

Disasters and climate extremes have differential effects on women, men, and third gender people in all social categories. The pre-existing social structures and norms create greater stress on women and marginalized groups further exacerbating their vulnerability. Records of natural disasters in the Himalayan region over the last few decades show that women are at greater risk of dying than men (Mehta 2007). More women than men die when disasters strike as a result of women's lack of access to information, mobility, and decision-making power and inequitable access to resources and training; as well as gender-based sociocultural norms and barriers, conventional gender responsibilities, and high rates of male outmigration (Mehta 2007; Ariabandhu 2009; Nellemann et al. 2011). In mountain communities, women play a crucial role in protecting, nurturing and sustaining natural resources. At the same time, they are often disadvantaged in terms of benefit sharing, accessing productive resources, and participation in organizational structures and decision making, and are exposed to increased risks associated with climate change during disasters and loss of income from climate shocks (Nibanupudi and Khadka 2015). For example, during the 1991 cyclone in Bangladesh, the mortality rate for women was three times higher than for men (UNEP 1997; Twigg 2009). Gender inequities can be evident in a lack of, or inadequate, early warning information and evacuation procedures and arrangements targeting women. Knowledge of early warnings and the decision to evacuate may be the exclusive domain of men. In some cases, women may be ill-informed about natural hazards and not allowed to make the decision to evacuate (Stark et al. 2013). A UNEP report (UNEP 1997) concluded that the early warning signals had not reached many women downstream.

Vulnerability is particularly high when poverty intersects with discrimination, whether because of gender, caste, ethnicity, or other reasons. This is especially true for women and low caste people (Adger and Kelly 1999; Brooks and Adger 2005; Aguilar et al. 2015). During the 2015 Nepal earthquake, more women than men died in all the affected districts except Kathmandu (Rasul et al. 2015). Fewer opportunities exist for education, political involvement, and access to information, markets, and a myriad of other resources (Ariyabandu and Wickramasinghe 2003). Considering vulnerability factors such as social roles and access to resources and information, women are more vulnerable to climate change and disasters than men. Women also know less than men about their communities' disaster prevention and mitigation projects. Furthermore, natural disasters and climate change often exacerbate existing inequalities and discrimination in such a way that women and girls become more vulnerable and are at higher risk of gender-based violence, sexual harassment, exploitation, abuse, trafficking, and rape during displacement caused by major disasters such as flood, drought, or earthquake.

Men and women possess valuable, but different, knowledge, skills, experience, and coping capacities. However, the strengths and capabilities of women are often ignored in policy decisions and in formal arrangements related to mitigation and recovery. Policy makers and planners generally give little attention to the social barriers and constraints that hinder women's participation in capacity building and their access to information that could help achieve better preparedness. Gender differences are manifested in the disproportionately poorer health and nutritional status, lower levels of access to formal literacy and education, higher levels of economic poverty, higher morbidity/mortality rates, and high workloads of women compared to men, as well as extremely low rates of property ownership, decision making, and representation in governance institutions (Leduc 2011). Adopting a gender-sensitive early warning system approach with appropriate policies in place will help in reducing the disaster mortality of women and contribute to reducing the adverse impact of flood disasters (Shrestha et al. 2014). To have disaster-resilient communities, the participation of both men and women at various levels is essential. Inequalities that exist in society are often strengthened during disaster, and this must be kept in mind when collecting data and analysing and formulating disaster resilience plans and activities (Shaw and Nibanupudi 2015). A gender-sensitive approach that not only recognises the vulnerabilities of women, but also works towards enhancing their resilience and strengthening their ability through awareness raising and capacity building initiatives, is needed to respond effectively to disasters.

11.3 Risk Assessment

Risk is the likelihood of harmful consequences of natural hazards arising from the interaction among hazards, vulnerable elements, and the environment. As discussed earlier, disaster risk depends on the probability that different kinds and intensities of hazards will occur, whether the elements are exposed to these hazards, and the level of vulnerability of the elements exposed to the hazards (World Bank 2005). Risk information forms the cornerstone of any risk reduction agenda; thus, awareness of existing and anticipated risk is essential to guide disaster risk reduction interventions, strategies, and policies. The cost effectiveness of protection measures can be evaluated based on the calculated risk. Risk assessment involves the identification, quantification, and characterization of threats to human health and the environment. But risk analysis is as much a political enterprise as a scientific one, and public perception of risk also plays a role in risk analysis, bringing the issues of values, process, power, and trust into the overall picture (Slovic 1999).

11.3.1 Understanding Risk

Risk assessment to understand the risk situation is the first step to augment risk-informed decision making and development. While many excellent risk assessments exist, there are relatively few risk assessments in practice. Because the HKH is characterized by natural hazard hotspots and with the exception of China, a low HDI, it is considered a high disaster risk region. According to the indicators for hydro meteorological hazards and disaster risk developed in the Asian Water Development Outlook (AWDO) 2013, India in the HKH country is most prone to hydro meteorological hazards, followed by China, Bangladesh, Pakistan, and Nepal—no indicators available for Afghanistan, Bhutan, and Myanmar (ADB 2013a). The AWDO report considers vulnerability as a function of exposure, basic vulnerability, soft coping capacity, and hard coping capacity. Bangladesh faces the highest level of exposure to hazards, followed by Pakistan, Nepal, India, and China. Basic vulnerability, measured by proxies such as poverty levels, among others, is also high in Bangladesh, Pakistan, and Nepal compared to China and India. Soft coping capacity, measured by proxies such as literacy rate, among others, is lowest in Nepal and highest in China, with Bangladesh, Pakistan, and India falling in between. Finally, hard coping capacity, measured by proxies such as infrastructure facilities, is lowest in Nepal, with other countries having similar indicators.

Many global risk assessment exercises mark the HKH countries as highly vulnerable (Garschagen et al. 2016) (Table 11.9). The World Risk Index ranks 171 countries according to their risk of becoming a victim of a disaster as a result of five natural hazards (earthquakes, cyclones, floods, droughts, and sea-level rise). It uses 28 individual indicators, related to exposure and 23 related to elements representing vulnerability (susceptibility, lack of coping capacity, lack of adaptive capacity). The World Risk Index is calculated by multiplying exposure and vulnerability. The higher the risk index value, the greater the risk, and vice versa. Details of the risk calculation approach and indicators used can be found at www.WorldRiskReport.org.

The HKH region already faces a high natural hazard risk, but the impacts of climate change will further aggravate the situation, as a result of the loss and fragmentation of habitats, a reduction in forest biodiversity, the degradation of wetland and riverine island ecosystems, a decline in forage and fodder resources, a reduction in agrobiodiversity, an increase in forest fires, soil fertility degradation, changes in land use patterns, and an increased variability in agricultural productivity (Tse-ring et al. 2010). As in other mountain regions, the Hindu Kush Himalaya have experienced above-average warming (see Chap. 4: Exploring Futures of the Hindu Kush Himalaya Scenarios and Pathways; Nogués-Bravo et al. 2007), which has adversely impacted freshwater, primarily snow, glacier, and permafrost (Yao et al. 2012). Climate change impact modelling projects a scenario of dwindling water availability (Immerzeel et al. 2010) that could undermine the socioeconomic fabric of the downstream societies.

The HKH has experienced rapid environmental changes and it is widely believed that the region will be one of the planet's hot spots for future climate change impacts (Maplecroft 2011). Mountain communities and their livelihoods are sensitive to such changes, which will have a

Table 11.9 Risk index in the HKH countries (2016) (Garschagen et al. 2016)

Country	World Risk Index (%)	Exposure (%)	Vulnerability (%)	Susceptibility (%)	Lack of coping capacity (%)	Lack of adaptive capacity (%)	Rank
Afghanistan	9.50	13.17	72.12	56.05	92.85	67.48	41
Bangladesh	19.17	31.70	60.48	38.23	86.36	56.85	5
Bhutan	7.51	14.81	50.70	29.43	73.77	48.90	60
China	6.39	14.43	44.29	22.81	69.86	40.18	85
India	6.64	11.94	55.60	35.79	80.22	50.78	77
Myanmar	8.90	14.87	59.86	35.63	87.00	56.93	42
Nepal	5.12	9.16	55.91	38.05	81.05	48.64	108
Pakistan	6.96	11.36	61.26	35.04	86.26	62.48	72

variety of impacts on human wellbeing. Primary sector livelihoods such as agricultural livelihoods have become increasingly uncertain and risky and, because of inadequate resources, poor households have especially limited adaptation options and are simply coping (Gentle and Maraseni 2012). Mountain areas are challenging living spaces, and mountain communities have a long history of adapting to extreme conditions. Nevertheless, traditional adaptation mechanisms are often insufficient to cope with recent socioeconomic and environmental changes (Jodha 1997), which have considerably increased the challenges for mountain people in securing their livelihoods (O'Brien and Leichenko 2000).

In developing countries, economic development in mountain regions already lags behind that in the lowlands, foothills, and urban areas (Tanner 2003). Climate change is expected to exacerbate the existing challenges faced by mountain people and their environments, intensify some existing hazards, and result in the emergence of new hazards (O'Brien and Leichenko 2000; Sonesson and Messerli 2002; Macchi et al. 2011). These processes will intensify the exposure component of vulnerability. The sensitivity component will include environmental aspects embedded in the biophysical features of a region and social elements that are closely linked to the nature and range of available livelihood options (Jodha 1997), as well as access to resources (Adger and Kelly 1999; Brooks and Adger 2005; Aguilar et al. 2015).

11.3.2 Risk Informed Decision Making

Risk assessment remains few in practice, which poses a challenge to risk-informed decision making. The Government of Nepal as a party to the United Nations Framework Convention on Climate Change (UNFCCC), has initiated a National Adaptation Plan (NAP) formulation plan, which is an excellent example of moving towards risk-informed decision making (MOPE 2017). The NAP has adopted a vulnerability and risk assessment (VRA) framework based on the Fifth Assessment Report (AR5) of the Intergovernmental Panel on Climate Change (IPCC).

Other challenges in translating risk knowledge to practice are inadequate granularity of information and what has been termed 'spatial scale challenges' by Carr et al. (2015). The Consultative Workshop on Landslide Inventory, Risk Assessment, and Mitigation in Nepal (Gurung et al. 2017) identified spatial scale challenges and inconsistency in methodology as two of the main setbacks to implementing risk assessment results. Strategic decision making at the national and sub-national levels has different information needs compared to local level decision making, which is more operational in nature. There is an obvious gap in national and sub-national risk assessment, as many assessments are done at a micro scale for specific sites/areas. The novel methodology developed and tested in northeast Brazil to reveal regional vulnerability using global level information, using a "combination of both clustering and qualitative dynamics" (Sietz 2014; Sietz et al. 2017), should be adopted. These assessments enable cross-scale comparison of risks and vulnerability and are well-suited to inform decision making at multiple scales.

Another challenge in the HKH is that assessment is skewed more towards hazards than vulnerability and risk. Risk assessment as a process is still far from being mainstreamed into government systems, and is mostly done through project support. This has resulted in risk assessment being conducted in pockets and based on different methods favoured by different project proponents.

There is a need to develop and promote systematic assessment methods and uniform risk assessment protocols (Gaire et al. 2015; ICIMOD 2016; SDMC 2011). The SAARC Comprehensive Disaster Management Framework approved by the Fourteenth SAARC Summit in New Delhi held 3–4 April 2007 identified "developing standards and methodology for hazard and vulnerability assessment…" as one of the pathways necessary to develop and implement risk reduction strategies (retrieved from http://saarc-sdmc.nic.in/framework.asp).

11.4 A Framework for Policies to Reduce Risk and Enhance Resilience

In this section we propose a disaster risk reduction framework to reduce risk and enhance resilience in the HKH, and address the key issues for mountain and downstream communities. The framework we propose is based on the principles of the 1994 Yokohama Strategy and Plan of Action, the 2005 Hyogo Framework of Action, UNDP's 2007 Human Development Report (UNDP 2007), and the Sendai Framework for Disaster Risk Reduction 2015–2030.

The focus in this section is on building resilience to disasters, including those that are climate-induced, and not on climate change adaptation (which is discussed in Chap. 13: Adaptation to Climate Change). The concepts, goals, and processes of climate change adaptation, however, have much in common with disaster risk management (Lavell et al. 2012; Schipper 2009), especially on matters related to managing climate-induced disasters. It may even be possible to conceptualize climate change adaptation starting from disaster risk management, with a clear understanding of the differences between the two (Vaidya et al. 2014). But here we have limited our scope to building resilience to disasters.

11.4.1 A Framework for Reducing Risk and Increasing Resilience to Disasters

Our disaster risk reduction framework for the HKH has four principal elements: Information, Infrastructure, Institutions, and Insurance.

Information: Since the HKH is a hotspot for both hydrometeorological and geophysical hazards, developing a strong knowledge base on extreme weather events and seismic activities in the region is vital to understanding how to increase resilience. In addition, hazard maps for communities and real-time information systems can substantially reduce vulnerability to potential hazards through early warning systems and prudent land use planning, especially in situations where financial protection measures, such as insurance, are not in place. The value a society places on such information may depend on their perception of risk—and their perception may, in turn, depend on the information they have available to sense the likelihood of the hazards.

Often, the government shows a willingness to invest in information systems soon after a hazard event occurs. For example, after a massive earthquake devastated two of its districts in 1993, Maharashtra became the first state in India to implement a comprehensive plan, complete with a state-of-the-art satellite-linked computer network connecting various civic bodies, collectorates, and blocks in the state (Vatsa 2002). It would be much better, however, if such initiatives could be proactive rather than reactive.

Data and information are a prerequisite for informed decision making for disaster risk reduction. Every forecast has some uncertainty and it is important that this is communicated and explained to the decision makers. A broad range of environmental and social data and information may be shared to promote transboundary cooperation for better river basin planning and management and to address climate change (Chenoweth and Feitelson 2001; Grossman 2006; Gerlak et al. 2010). Sharing data and information builds trust and confidence amongst countries and provides a common understanding of the issues, which may result in agreement, joint implementation, and improved transboundary cooperation (Shrestha et al. 2015; Blumstein et al. 2016). In the HKH, there has been some progress with the HYCOS system under the World Meteorological Organization's WHYCOS framework in which countries share real-time hydrometeorological data for flood risk reduction (Shrestha et al. 2015) and work towards an end-to-end early warning system. Working in partnership with several regional and international partners, ICIMOD offers a regional platform for utilizing the latest advances in space technology and GIS (geographic information systems) applications to address disaster challenges and to support risk identification and early warning systems.

It is important to note that after disasters occur, funds for recovery do become available from various sources, both internal and external. But the same quantum of money could be more efficiently and effectively used by mainstreaming it, in part, to development activities that help communities with hazard maps, real-time information systems, and communication channels that reach the last mile before a disaster happens—so that they can be better prepared and thus save lives and livelihoods. The Asia regional plan for implementation of the Sendai framework for disaster risk reduction 2015–2030 includes a two-year action plan for 2017 to 2018 which seeks to strengthen existing regional mechanisms to reduce risk and enhance early warning and preparedness for transboundary disasters. The Sendai Framework may also help to attract funds for generating information on risk. This is a significant change in priority from the 2005 Hyogo Framework, where risk assessment was identified as the second priority (see Box 11.5).

Infrastructure: Investments may also be necessary to create hazard-resilient critical infrastructure such as hospitals for healthcare services and school buildings for use as community shelters after hazardous events. Similar investments in road networks for access to settlements, and communications systems for information flow may also be necessary to ensure connectivity immediately after disasters. Furthermore, critical infrastructure such as water supply systems and electric power plants should be made climate-resilient, and standards for rebuilding structures after earthquakes should be improved.

It is important to note that there was an implicit shift towards a balance between structural and non-structural measures after the announcement of the Hyogo Framework. But the emphasis on the importance of balanced investments in structural measures has reappeared with the promotion of 'Build Back Better' as Priority 4 of the Sendai Framework, and for investment in disaster-resilient critical facilities as Priority 3 on disaster risk reduction. The same priority also highlights investments in ecosystem-based natural resource management approaches.

Institutions: Resources need to be invested in capacity building through training programmes for formal and informal institutions as well as pre-positioning of stockpiles at the local level. Appropriate policies and mechanisms also need to be developed for supportive interfaces between these institutions at both the national and local levels. Institutional arrangements supported by communications technologies and clear message contents must be developed for end-to-end communications up to the last mile. For example, the ability to send alerts for flood hazards or deliver relief measures after earthquakes is crucial.

The Hyogo Framework's first priority was the establishment of institutions for disaster risk reduction: "Ensure that disaster risk reduction is a national and a local priority with a strong institutional basis for implementation." In view of the weaknesses of the institutional basis for converting national policy into local action (Oxley 2013), the Sendai Framework, in its Priority 2 on strengthening disaster risk governance, has emphasized the need "to carry out an assessment of the technical, financial, and administrative disaster risk management capacity to deal with the identified risks—at the local and national levels."

In addition, the Sendai Framework also explicitly mentions the need to develop institutions "to promote transboundary cooperation to enable policy and planning for the implementation of ecosystem-based approaches," which has high relevance in the HKH. Transboundary cooperation can be enhanced at the national and local levels between two or more countries. The Sendai Framework clearly highlights, in Priority 3, the need to promote mechanisms for disaster risk transfer, risk sharing, and retention, and financial protection to reduce the financial impact of disasters. In the Hyogo Framework, Priority 4 discusses the need to develop financial risk-sharing mechanisms (see Box 11.5).

Box 11.5 Priorities of the global agenda: the Hyogo and the Sendai frameworks

Priorities of the Global Agenda	Hyogo Framework of Action 2005–2015: Building the Resilience of Nations and Communities to Disasters	Sendai Framework for Disaster Risk Reduction 2015–2030
Priority 1	Ensure that disaster risk reduction is a national and a local priority with a strong institutional basis for implementation.	Understanding disaster risk
Priority 2	Identify, assess, and monitor disaster risks and enhance early warning.	Strengthening disaster risk governance to manage disaster risk
Priority 3	Use knowledge, innovation, and education to build a culture of safety and resilience at all levels.	Investing in disaster risk reduction for resilience
Priority 4	Reduce underlying risk factors	Enhancing disaster preparedness for effective response and to "Build Back Better" in recovery, rehabilitation, and reconstruction

Source: United Nations (2015); UNISDR (2005)

Insurance: Mechanisms need to be developed before a disaster strikes for raising precautionary funds or for sharing risks in order to provide relief, rehabilitation, and reconstruction efforts. The question is how much residual risk a government can manage itself—and how much residual risk it would need to share or transfer. Governments would need to maintain a pool of reserve funds to address small disasters, and would also need to subsidize insurance premiums, where necessary, for promoting private insurance products, such as index-based weather insurance for drought. Furthermore, beyond a certain level of risk, a government may have to share the indemnities with a private insurer, or the insurer may need to find a reinsurance company for risk-pooling through international markets. Further on, when a risk involves a major catastrophe, a government may have to transfer risk to capital markets through financial instruments such as catastrophe bonds. In such bonds, the issuer is liable to pay interest and principal if the event does not occur during the maturity period, and is not liable to pay back the principal if the event does occur.

It is also critical that special measures and mechanisms of insurance be designed for women, the poor, and marginalized groups. Until such mechanisms are developed, informal institutions like social networks and social capital, where extended families and communities help each other, may be the only forms of insurance available to communities within a reasonable amount of time after an event occurs. Furthermore, even after developing reasonable private insurance products, people may need to be 'nudged' to buy them because of the time inconsistency problem. For example, for drought protection, a farmer would need to decide now to purchase insurance and pay the premiums, but the payout, if any, would take place in the future, which tends to discourage farmers from enrolling in insurance programmes (Banerjee and Duflo 2011).

11.4.2 Relating the Disaster Risk Reduction Elements to Programme Components

In practice, decision makers and governments will ultimately determine if and how the separate elements of disaster risk reduction (information, infrastructure, institutions, and insurance) will be applied to help increase the strength and modalities of resilience to hazards. And that motivation, in turn, depends upon perceptions of risk—by individuals, by communities, by experts, and by society at large (Slovic 1987).

To account for these motivations and perceptions, resilience-building programmes should consider four strategies for changing human behaviour: (1) restrictions on choice through command-and-control mechanisms (e.g., zoning regulations, land use guidelines, and building codes); (2) monetary incentives (e.g., subsidies on insurance premiums); (3) persuasion by providing information (e.g., risk maps); and (4) 'nudging' (e.g., early warning systems). Often, a combination of these methods may be appropriate, and they will, of course, depend on the type of hazard event under consideration for resilience building. Table 11.10 provides some examples of these strategies.

11.4.3 Resilience Building Programmes: Four Examples

When we consider increasing resilience to disasters by reducing the vulnerability of communities through pursuing various measures, we should ask three questions, in this order:

- Whose resilience would we like to enhance? Individuals? Communities? Cities? Or larger units?
- What can be done to increase resilience pursuing one or more of the four disaster risk reduction elements: information, infrastructure, institutions, and insurance?
- How can individuals, communities, or city governments be motivated to adopt the measures that fall into one of these four disaster risk reduction elements?

Below we look at four examples of how these questions are answered using the 4 × 4 matrix in Table 11.10 of the four disaster risk reduction elements (information, infrastructure, institutions, and insurance) and four behavioural strategies (command-and–control, economic incentives, persuasion through information, and nudging).

1. *Index-based weather insurance*

Index-based weather insurance can improve drought resilience. Actions on five major fronts have been identified for such insurance:

a. invest in hydromet networks;
b. engage civil society organizations (CSOs) as social mobilizers to raise awareness;
c. invest in science to understand better the correlation of the index with actual crop yields;
d. invest in risk assessment; and
e. develop reinsurance markets.

In this case, the answer to the first two questions involves talking about increasing the resilience of the farmers using the insurance approach. On the third question, engaging CSOs as social mobilizers to raise awareness would be the persuasion strategy; subsidizing the insurance premium the farmer pays would be the incentives strategy; and offering help with maps to the location where insurance can be purchased would be a nudging strategy. Improving the uptake of crop insurance by farmers is often a real challenge and a combination of a number of strategies may be needed.

2. *Reviving drying springs*

Research suggests five different approaches for reviving dying springs:

a. identify recharge areas accurately;
b. prepare hydrogeological layout maps of the spring aquifer and recharge area;
c. build simple artificial recharge structures (e.g., trenches);
d. incentivize rainwater harvesting in farmers' fields; and
e. build local institutional arrangements to regulate demand.

Table 11.10 Disaster risk reduction elements and behavioural change strategies

	Information	Infrastructure	Institutions	Insurance
Command-and-control mechanisms	Zoning and building code enforcements	• Infrastructure development projects • Technical design standards • Building codes • Land use plan/zoning	Institutionalization of formal and informal institutions	
Incentives		• Rural housing reconstruction program (RHRP): financial support for seismic-resistant housing • Budget for infrastructure development		Subsidizing insurance premium a farmer has to pay for index-based weather insurance for crops
Persuasion	Providing hazard maps	Technical guidelines and dissemination training by engineers regarding infrastructure development	Support from formal and informal institutions	Engaging NGOs as social mobilizers to raise awareness of market insurance for crops
Nudging	Community-based flood early warning systems (CBFEWS)	Promoting retrofitting with nudges to consider traditional and cultural preferences	Institutional arrangement for Community-based flood early warning systems (CBFEWS)Reviving drying springs	Encouraging self-insurance through personal savings motivated by a clearly visible purpose such as loss of crops due to floods

In the case of resilience, we are trying to bolster the capacity of farmers through information, infrastructure, and institutions. On the third question, building local institutional arrangements to regulate demand would be the command-and-control mechanism; grants or subsidized loans for building simple recharge structures such as trenches and ponds would be an incentive method; making the maps available would be persuasion by information; and incentivizing rainwater harvesting in farmers' fields would also be an incentive method, but may also require some nudging.

3. *Building resilience to flash floods*

A number of actions have been identified for building resilience to flash floods: hazard mapping, zoning policies, modern hydromet stations, information and communication technologies, local community involvement, and the national-local supportive interface. In this case, we are trying to build the resilience of the community through information, infrastructure, and institutions. On the third question, zoning policies would be the command-and-control mechanism; such policies would include road alignment and hydropower station location policies. Providing hazard maps would be persuasion and community-based flood early warning systems would be a nudge.

4. *Building resilience to earthquakes*

A number of actions identified for evidence-based analysis of water-related disasters may also be useful for identifying geophysical hazards, especially in the context of earthquakes that lead to landslides, dammed rivers, and flash floods subsequent to the breaching of landslide dams. The actions suggested for evidence-based disaster risk assessment are:

a. conduct risk assessment to identify the nature and magnitude of risk;
b. assess the effectiveness of preventive investment, land use planning, and emergency actions;
c. collect and archive hazard and damage data to develop indicators that make risk assessment evidence-based; and
d. apply the latest science and technology to promote practical risk assessment.

Developing and enforcing land use guidelines with the aim of limiting exposure to geohazards and paying more attention to areas where major infrastructure development projects such as roads and hydropower are proposed (Shrestha et al. 2016) would be a command-and-control mechanism. Similarly, developing applicable project design standards/building codes and communicating them to households and builders to enhance local government management for construction quality control in rural and urban areas (Molden et al. 2016; Shrestha et al. 2016) would be a combination of persuasion and command-and-control mechanisms. Both measures—zoning and building code enforcements—would also require nudges to motivate households and builders to follow the land use guidelines and building codes.

For pursuing these measures, evidence-based analysis could help to build a strong knowledge base. First, land use guidelines based on potential hazard maps could be updated with evidence-based hazard maps. Second, building codes based on potential hazards could be updated with the evidence of the damage to buildings during the earthquake. Evidence-based analysis may also help as a nudge for households occupying existing buildings, because they can see for themselves what could happen if the building codes are not followed.

In a typical case, especially because of the low-probability, high-impact nature of earthquakes, there may be a tendency for households to procrastinate and postpone retrofitting measures for a number of reasons. First, there may be some ambiguity about what constitutes optimal mitigation, because households cannot see what damage could occur if the retrofitting measures are not taken or the building codes not respected. Second, households may have budget constraints for investing in protective measures. In addition, they may see it as affordable or unaffordable based on how they frame it—an improvement similar to installing a leaky roof, which might ultimately lead a house to collapse, or an improvement similar to installing a leaky faucet, which might lead to high water bills. Third, they may also shy away from mitigation efforts because there is uncertainty as to when the next earthquake is likely to occur. It has been found that when making choices for the distant future, we may see the benefits clearly and decide on them, but when the time comes to pay, we tend to focus on costs—leading to procrastination. Therefore, nudges may be necessary to motivate households to invest in retrofitting measures and to sincerely respect building codes (Kunreuther and Michel-Kerjan 2008).

11.4.4 Information Flows Are Crucial for Early Warning Systems

Flood early warning systems are one of the most effective non-structural ways to minimize loss of life and property (Shrestha et al. 2008). Early warnings are transmitted from upstream to downstream communities to minimize the impacts of disasters. Accurate rainfall estimations and sharing of data and information are critical for reliable and timely flood forecasting and warnings. In many regions, operational flood forecasting has traditionally relied on a dense network of rain gauges or ground-based rainfall measuring radar equipment that report in real time. Rapid advances in communication technology are making access to data cheaper. At the same time, hydrological and meteorological monitoring and modelling technologies continue to improve significantly. These technological advances can be exploited to promote regional cooperation for flood risk reduction in the HKH by providing an end-to-end flood information system. The system functions as a decision support tool for decision makers to alert vulnerable communities in a timely and accurate manner.

In the HKH, ICIMOD in partnership with the World Meteorological Organization and the regional member countries of Bangladesh, Bhutan, China, India, Nepal, and Pakistan, has developed a regional flood information system (HKH HYCOS) that allows the visualization and extrapolation of real-time data from gauging stations to any geographical location providing information on the river water levels and the amount of rainfall (Shrestha et al. 2015). Using this real-time data, a flood outlook has been developed for the Ganges Brahmaputra basin. In August 2014, this flood outlook was used by Nepal's Department of Hydrology and Meteorology to issue a flood warning for the rivers of Nepal. It did so by means of a flood bulletin which was widely disseminated through its website (Shrestha and Pradhan 2015).

At the local level, the Hyogo Protocol and the SREX 2012 has identified a gap in getting flood early warnings directly to the communities that are most vulnerable. A community-based flood early warning system (CBFEWS) is an integrated system of tools and plans in which upstream communities, upon detecting flood risk, disseminate the information to vulnerable local communities downstream for preparedness and response to save lives and livelihoods (United Nations 2006). This is done using low-cost technology like wireless and solar-powered transmitters and receiver stations and mobile phone text messaging. Box 11.6 describes an example of this type of system in practice.

Box 11.6 Reaching the most vulnerable across the border

On 12 August 2017, local communities either side of the border crossed by the Ratu river—Bardibas in Mahottari district, Nepal and Bhittamore in Sitamarhi district, India—shared real-time information about an upcoming flood, which helped save lives and livelihoods in the vulnerable downstream communities in Bhittamore. The population in the Indo-Nepal border districts received information from the CBFEWS almost eight hours prior to the event. The upstream-downstream cross-border information flow provided an opportunity for the caretakers in the two countries, local communities, and partner organizations to know about the upcoming flood, prepare themselves and react immediately to save people and property (http://www.icimod.org/?q=28515).

CBFEWS was initiated by ICIMOD in early 2010 under a flash flood project. A human face was given to the technology in 2012 under the HICAP initiative, and CBFEWS was piloted in Assam, India. The impact of CBFEWS was acknowledged by UNFCCC's Momentum for Change 2014 Lighthouse Activity Award as a shining example innovative use of ICT. From 2015 onwards, CBFEWS was out scaled in India (Bihar) and Nepal under the Koshi Basin Initiative, in Pakistan under the Indus Basin Initiative, and in Afghanistan under a special project.

Table 11.11 Comparison of the Pakistan and Nepal earthquakes

	2005 Pakistan earthquake[a]	2015 Nepal earthquake[b]
Total damage and loss	USD 2,851 million (PKR169,333 million)	USD 7,065 million (NPR 706,461 million)
Housing damage and loss	USD 1,152 million (40.41%)	USD 3,505 million (49.62%)
Deaths	73,000	8,702
Injured	70,000	22,303
Houses destroyed	203,579	498,852
Houses damaged	196,574	256,697
Total recovery needs (USD million)	USD 3,503 million	USD 6,695 million
Housing recovery needs (US$ million)	USD 1,552 million (44.30%)	USD 3,278 million (48.96%)

Sources [a]ADB and World Bank (2005); [b]NPC (2015)

11.4.5 Building Critical Infrastructure Which Is Resilient to Disasters

Critical infrastructure is highly vulnerable to, and a major casualty of, natural disasters. Repairing or replacing infrastructure assets after a disaster is often difficult and costly, which can exacerbate the suffering of affected communities. The need to address climate risks in infrastructure projects is becoming increasingly urgent for economic development in emerging markets. The World Bank Group and other international financial institutions are well-placed to address the intersection of climate risks and infrastructure. They are screening investments for climate risks, providing analytical tools to measure risks, and designing measures to respond to risks, including innovative insurance approaches. The private sector can also contribute to disaster risk reduction through corporate social responsibility (CSR) activities. Sudmeier et al. (2013) developed an operational framework to measure resilience and vulnerability to disasters in the mid-hill regions of Nepal by defining resilience indicators based on a literature review, field observations, and a participatory approach with stakeholders. The framework can be used as a tool for guidance, providing direct interventions to reduce the risk of landslides and floods in the vulnerable mountainous regions of Nepal, including building critical infrastructure.

The HKH region is also physically vulnerable to earthquakes. Two major recent earthquakes in the region exemplify the urgent need to enhance physical resilience. On 8 October 2005, Pakistan's northern areas were struck by a 7.6 Mw earthquake. The impact of the 2005 Kashmir earthquake was devastating. More than 73,000 people were killed, 130,000 people were injured, and more than 200,000 houses were destroyed, rendering 3.5 million people homeless. In response, the Government of Pakistan collaborated with international partners to launch a Rural Housing Reconstruction Program (RHRP) at a cost of more than USD 1.5 billion (GFDRR 2013). RHRP relied on an owner-driven mechanism providing multi-tranche financial support to beneficiary households, based on inspection and certification at various stages of construction to ensure compliance with seismic-resistant standards (GFDRR 2013). On 25 April 2015, Nepal was struck by a 7.8 Mw earthquake which affected more than 8 million people. Table 11.11 summarizes the effects of the two earthquakes.

A comparison of the actions taken in the wake of the two earthquakes suggests that building resilience to earthquakes requires taking the following into account:

- Developing seismic-resistant structural designs should be the first important step in developing resilience, but this step needs to reflect on the common vulnerabilities in local practices, and identify the damage patterns and construction materials using damage assessments.
- Evidence-based persuasion and nudging are the keys to communicating the technical requirements to communities and inducing them to apply the designs on the ground.
- Another key factor in developing resilience is to have a transparent mechanism for cash disbursement and technical inspection. Dedicated authority to implement and enforce such standards will help provide a consistent and reliable agent for change in the community's behaviour.

11.4.6 The Role of Institutions Is Critical in Resilience Building Measures

Building resilience to climate change, and its effectiveness, depend on how institutions (formal and informal) at the local and national level structure and internalize incentives for individual and collective action. The role of the institution is important if it is to support vulnerable social groups at the local level and can be considered as a specific component for enhancing capacity and delivering external resources to facilitate resilience and adaptation (Agrawal 2010; Christoplos et al. 2010; Dovers and Hezri 2010).

Pradhan et al. (2012) presented learning from four case studies in the HKH that analyse the role of policy and institutions in local adaptation planning to enhance community resilience. The building of effective resilience is determined by the interface between civic (civil society), public (state/government), and private (market/service organizations) institutions in their formal and informal roles operating at different scales. Agrawal (2010) emphasized that public sector institutions are more likely to facilitate adaptation strategies related to communal pooling, diversification, and storage owing to their command over authoritative action and their ability to channel technical and financial inputs to rural areas. Private sector organizations are more likely to have greater expertise in promoting market exchange and diversification, because of their access to financial resources. Non-profit service organizations may also be able to advance communal pooling. Civic sector institutions can strengthen different responses because of their greater flexibility in redefining goals and adopting new procedures. Depending on the extent to which there is a match or mismatch between the aims and comparative advantages of different institutions, the interface between institutions can be supportive or unsupportive.

A supportive interface is clearly desirable, but rarely found, where formal public institutions are supporting formal and informal institutions at all stages of adaptation planning. An example from China shows how a supportive interface can work. In 2005, after an extreme drought, the Ministry of Water Resources, National Reform and Development Commission, and the Ministry of Civil Affairs jointly issued a "Suggestion on strengthening the establishment of water users' associations" (Policy Decision 10), which recommended the establishment of water user associations to manage rural water infrastructure. The Baoshan Municipality Water Bureau established 520 water user associations between 2006 and February 2009, covering 142,449 households in 306 villages across 65 townships, and managing a total of 13,281 ha of irrigated land. All the counties in Baoshan issued their own implementation guidelines to establish the water user associations, with their own constitutions and regulations governing the operation of the associations. According to the government's report, a supportive interface was achieved between the policy implementation and water user associations, who owned and managed their water infrastructure, promoted water-saving practices, and reduced conflicts in the collection of fees in order to deal with drought. However, some water user associations experienced an unsupportive interface due to lack of funds for their operation, inefficient leadership, and lack of legal clarity regarding their status. The empirical evidence showed that the ability of the communities to maintain a supportive interface largely depended on the relationship between village leaders and local officials. This is an informal mechanism for obtaining a supportive interface from the public sector, which is a barrier to some communities that are not well positioned to procure the support they require from the local government.

In another example from Baoshan after the severe drought of March 2009, the Longyang District Government sent a "Notification on Strengthening Work against the Current Drought" to all government units mentioned in the Plan. The district agriculture bureau submitted a needs assessment and recovery report prepared in consultation with the communities in 18 townships. Based on the report, the provincial committee disbursed funds to those townships which had requested relief such as water pumping machines. The supportive interface between the provincial government and the communities was liaised by the district agriculture bureau to implement the government plan, which resulted in enhanced adaptive capacity and resilience of the communities to address drought.

These examples suggest that resilience building cannot occur in a social vacuum: It needs to be supported by institutions and policies. Planning for resilience building should give greater attention to the development of effective institutional arrangements, which requires supportive interfaces between institutions for building adaptive capacity and enhancing the resilience of communities (Pradhan et al. 2012).

11.4.7 Nudging Could Help Motivate People for Self-insurance

In industrialized countries, market insurance is the primary means of risk management. In the HKH, governments and the private sector are currently trying to promote market insurance for various uses, for example crop insurance. Until market insurance becomes more widely adopted in the region, self-insurance products could be used to help increase resilience to natural disasters. Self-insurance in this context is defined as having adequate personal resources to cope with the consequences of a disaster. Reports suggest that it is easier to motivate people to save when the purpose of saving is clearly visible (Soman and Cheema 2011). This implies that it would be easier to motivate individuals to purchase self-insurance if they live in areas prone to floods and landslides, which occur more frequently (Tversky and Kahneman 1974), than if they live in areas prone to earthquakes and droughts, which occur less frequently. Financial products and institutional mechanisms for saving also need to be simple and practical, including those for small savings by the marginalized and the poor (Dupas and Robinson 2013).

Two field experiments conducted to motivate individuals to save, provide some useful lessons for savings for natural hazards. These experiments are noteworthy because they demonstrate how nudging by developing the appropriate choice architecture could help motivate people to save for self-insurance.

Soman and Cheema (2011) in India demonstrated the importance of clear objectives for which individuals are saving—'earmarking' money in the sense of reserving or setting it aside for a particular purpose. The researchers tested whether households of infrastructure construction workers (146 daily-wage labourers) whose earmarked money envelope was labelled with their children's pictures would save more than participants whose earmarked envelope was not labelled in this way. Regardless of whether the target savings were set at high (INR 80) or low (INR 40), the household savings over 14-weeks were higher for those with a money envelope earmarked with their children's pictures.

Dupas and Robinson (2013) in Kenya demonstrated the importance of a storage mechanism, earmarking, and social commitment in the process of saving for preventative health activities and health emergencies. The study involved participants in 113 local rotating savings and credit associations (ROSCA), in which participants meet periodically and contribute equal amounts to a pot that is taken by one of them. The participants were encouraged to save for health and divided into five groups: two for preventative health with nudging, one for health emergencies only with nudging, one for both preventative and emergency health with nudging, and one without nudging (the savings device). They found that for preventative health, the average impact of earmarking was KES 57.54 and that of social commitment KES 273.46. For health emergencies, they estimated the percentage of participants who could not afford medical treatment for an illness in the past three months but could afford it after participating in the earmarking process. The average impact of the earmarking process was 8% when not monitored and 12% when monitored.

These experiments suggest that a simple savings device, such as an envelope or a storage box, may help to nudge people to save. The savings can be increased by mentally clarifying the purpose of the savings, and further increased by normative pressure through social commitment. In the disaster risk reduction literature, Kunreuther and Michel-Kerjan (2008: p. 60) argue that "recent disasters have provided empirical evidence that a large number of people do nothing in advance of a disaster because they use budgeting heuristics, misperceive the risk, underweigh the future and/or are myopic, fail to learn from past experience, and are influenced by social norms and interdependencies". Two of these issues are addressed in the experiments: (1) earmarking, monitored and unmonitored, to take care of budgeting heuristics by clarifying the purpose of saving; and (2) normative pressures on savers through social commitment, to take care of social norms and interdependencies. Although the experiments were not directly related to saving to cope with natural hazards, they provide important lessons for encouraging self-insurance through savings for vulnerable populations.

11.5 Summary and Way Forward

Mountain communities are threatened by numerous risks from natural hazards and a changing risk pattern. Disaster risk reduction is particularly important in mountain areas for many reasons, including the multi-hazard environment, land use pressure, and the effects of climate change. Flash floods and landslides are the most frequently occurring natural hazards in middle hill terrain in the HKH, particularly during the monsoon season, and flooding in the plains. There is an increasing trend in the number of events reported, people killed, and economic loss due to natural disasters in the region. Records of natural disasters and related studies indicate that more women than men die when disasters strike. This is the result of women's lack of information, mobility, decision-making power, and access to resources and training, as well as gender-based sociocultural norms and barriers, conventional gender responsibilities, and high rates of male outmigration.

Assessing risk without considering the effects of climate change is no longer an option in the mountainous areas, which are particularly sensitive to climate change. Risk-informed planning will help to create safer land use practices and hazard-proof infrastructure and housing. In addition, cross-border cooperation to share information and best practices is necessary for early warning systems and other precautionary measures. Access is important in effective response. Mountain communities are more vulnerable as a result of their remoteness, poor accessibility, and lack of emergency communication. Thus, sustainable mountain development requires a systematic and integrated risk management approach to avoid or reduce future losses.

Disaster risk is expressed as the probability of loss of life, injury, or destroyed or damaged assets which could occur to a system, society, or community in a specific period of time.

Such probability can be estimated by assessing hazards, exposure, and vulnerability. While hazards and exposure can be estimated empirically and quantitatively using historical events, vulnerability assessment has multiple disciplinary theories. Although it is not easy to assess the physical vulnerability in the HKH quantitatively based on data, estimation is possible using national data, such as data on quality of infrastructure services from the World Economic Forum, as proxies. Socioeconomic vulnerability assessments should take into account multiple dimensions, such as income inequality, gender inequality, governance, and national progress for disaster risk reduction in the light of the Sendai Disaster Risk Reduction Framework.

Enhancing community resilience to hazards by reducing vulnerability and pursuing resilience-building measures needs a clear understanding of disaster risks, which can help policy makers to prioritize strategies that increase their population's resilience to these events. A framework is needed for assessing risks due to hazard events and suggesting measures to increase resilience of the communities in the HKH. The framework proposed draws upon the principles of the 1994 Yokohama Strategy and Plan of Action, the 2005 Hyogo Framework of Action, the Sendai Framework for Disaster Risk Reduction 2015–2030, and UNDP's 2007 Human Development Report. It envisions a 4×4 matrix emphasizing the four elements of disaster risk reduction—information, infrastructure, institutions, and insurance—against the four elements for successful planning and execution—command-and-control mechanisms (e.g., zoning regulations; land use guidelines and building codes); monetary incentives (e.g., subsidies on insurance premiums); persuasion by providing information (e.g., risk maps); and nudging (e.g., early warning systems). The framework also helps to address three key questions for pursuing resilience-building measures: Whose resilience would we like to enhance? What can be done to increase resilience? How can the individuals, communities, or city governments be motivated to adopt the measures that fall into one of these four categories? Ultimately, the individual or the group of beneficiaries whose resilience we are trying to enhance must select one or more of these methods to increase resilience. Often, a combination of methods may be appropriate; this will depend on the type of hazard event under consideration for resilience building.

References

ADB. (2013a). *Asian water development outlook 2013: Managing water security in Asia and the Pacific*. Mandaluyong City, Philippines: Asian Development Bank.

ADB. (2013b). *The rise of natural disasters in the Asia and the Pacific*. Mandaluyong City, Philippines: Asian Development Bank. https://www.adb.org/sites/default/files/evaluation-document/36114/files/rise-natural-disasters-asia-pacific.pdf.

ADB and World Bank. (2005). Pakistan 2005 Earthquake Preliminary Damage and Needs Assessment. http://www.recoveryplatform.org/assets/publication/PDNA/CountryPDNAs/Pakistan_Earthquake_2005_Preliminary%20Damage%20and%20Needs%20Assessment.pdf.

Adger, W. N., & Kelly, P. M. (1999). Social vulnerability to climate change and the architecture of entitlements. *Mitigation and Adaptation Strategies for Global Change, 4*(3–4), 253–266.

Agrawal, A. (2010). Local institutions and adaptation to climate change. In R. Mearns & A. Norton (Eds.), *social dimensions of climate change: Equity and vulnerability in a warming world* (pp. 173–198). Washington DC, USA: The World Bank.

Aguilar, L., Granat, M., & Owren, C. (2015). *Roots for the future: The Landscape and way forward on gender and climate change*. Washington, DC: IUCN & GGCA.

Ahmad, S., Hussain, Z., Qureshi, A. S., Majeed, R., & Saleem, M. (2004). Drought Mitigation in Pakistan: Current Status and Options for Future Strategies. Working paper 85. Colombo: Sri Lanka. International Water Management Institute.

Alkire, S., & Robles, G. (2017). Multidimensional Poverty Index Summer 2017: Brief methodological note and results. OPHI Methodological Note 44, University of Oxford.

Allen, S. K., Rastner, I. P., Arora, I. M., Huggel, I. C., & Stoffel, I. M. (2015). Lake outburst and debris flow disaster at Kedarnath, June 2013: Hydrometeorological triggering and topographic predisposition Landslides https://doi.org/10.1007/s10346-015-0584-3.

Ariabandhu, M. M. (2009). Sex, gender and gender relations in disasters. In E. Enarson & G. P. D. Chakrabarti (Eds.), *Women, gender and disaster: Global issues and initiatives*.

Ariyabandu, M., & Wickramasinghe, M. (2003). *Gender dimensions in disaster management: A guide for South Asia*. Practical Action.

Armijo, R., Tapponnier, P., & Han, T. (1989). Late Cenozoic right-lateral strike-slip faulting in southern Tibet. *Journal of Geophysical Research: Solid Earth, 94*(B3), 2787–2838.

Avouac, J., & Tapponnier, P. (1993). Kinematic model of active deformation in central Asia. *Geophysical Research Letters, 20*(10), 895–898.

Awasthi, I. C., Mehta, G. S., & Mamgain, R. P. (2014). Uttarakhand disaster: Lessons and way forward, Occasional-6, Gri Institute of Development Studies. Lucknow, India.

Banerjee, A., & Duflo, E. (2011). *Poor economics: Rethinking poverty and the ways to end it*. Noida, U.P., India: Random House India.

Bergstrand, K., Mayer, B., Brumback, B., & Zhang, Y. (2015, June). Assessing the Relationship between social vulnerability and community resilience to hazards. *Social Indicators Research, 122*(2), 391–409. https://doi.org/10.1007/s11205-014-0698-3.

Bhatia, et al. (2000). A probabilistic hazard map of India and adjoining region, Global Seismic Hazard Program. http://www.seismo.ethz.ch/GSHAP/.

Bilham, R., Larson, K., & Freymueller, J. (1997). GPS measurements of present day convergence across the Nepal Himalaya. *Nature, 386*, 61–64. https://doi.org/10.1038/386061a0.

Bilham, R., Gaur, V. K., & Molnar, P. (2001). Himalayan seismic hazard. *Science, 293*, 1442–1444. https://doi.org/10.1126/science.1062584.

Birkmann, J., Cardona, O. D., Carreño, M. L., Barbat, A. H., Pelling, M., Schneiderbauer, S., et al. (2013). Framing vulnerability, risk and societal responses: The MOVE framework. *Natural Hazards, 67*(2), 193–211. https://doi.org/10.1007/s11069-013-0558-5.

Blaikie, P., Cannon, T., Davis, I., & Wisner, B. (1994). *At risk: Natural hazards, people's vulnerability and disasters*. Routledge.

Blumstein, S., Pohl, B., & Tänzler, D. (2016). *Water and climate diplomacy: Integrative approaches for adaptive action in transboundary river basins*. Germany: Adelphi, Berlin.

Bollinger, L., Avouac, J. P., Cattin, R., & Pandey, M. R. (2004). Stress build up in the Himalaya. *Journal Geophysical Research, 109*, B11405.

Brooks, N., & Adger, W. N. (2005). The determinants of vulnerability and adaptive capacity at the national level and the implications for adaptation. In W. N. Adger, N. Arnell, & E. L. Tompkins (Eds.), *Global environmental change part A* (vol. 15, pp. 151–162).

Carr, E. R., Abrahams, D., De la Poterie, A. T., Suarez, P., & Koelle, B. (2015). Vulnerability assessments, identity and spatial scale challenges in disaster-risk reduction. *Jàmbá: Journal of Disaster Risk Studies, 7*, 1–17. https://doi.org/10.4102/jamba.v7i1.201.

Champati Ray, P. K., Chattoraj, S. L., Bisht, M. P. S., et al. (2016). *Natural Hazards, 81*, 227. https://doi.org/10.1007/s11069-015-2076-0.

Chenoweth, J. L., & Feitelson, E. (2001). Analysis of factors influencing data and information exchange in international river basins. *Water International, 26*(4), 499–512. https://doi.org/10.1080/02508060108686951.

Christoplos, I., Rodriguez, T., Schipper, E. L. F., Narvaez, E. A., Mejia, K. M. B., Buitrago, R., et al. (2010). Learning from recovery after hurricane Mitch. *Disasters, 34*(2), S202–S219.

Ciurean, R. L., Schroter, D., & Glade, T. (2013). Conceptual frameworks of vulnerability assessments for natural disasters reduction. In *Approaches to disaster management—Examining the implications of hazards, emergencies and disasters* (pp. 3–32). https://doi.org/10.5772/55538.

CRED and UNISDR. (2016). The Human Cost of Weather-Related Disasters 1995–2015.

Donner, W., & Rodríguez, H. (2011). Disaster Risk and Vulnerability: The Role and Impact of Population and SocietyEm-DAT (2016). Em-DAT: The international disaster database. Centre for Research on the Epidemiology of Disasters. Universite Catholique de Louvain. Brussels. Retrieved July, 2016, from www.emdat.be.

Dankers, R., Arnell, N. W., Clark, D. B., Falloon, P. D., Fekete, B. M., Gosling, S. N., et al. (2014). First look at changes in flood hazard in the Inter-Sectoral Impact Model Intercomparison Project ensemble. *PNAS, 111*(9), 3257–3261.

de Mets, C., Gordon, R. G., Argus, D. F., & Stein, S. (1994). Effect of recent revisions to the geomagnetic reversal time scale on estimates of current plate motions. *Geophysical Research Letters*.

Dovers, S. R., & Hezri, A. A. (2010). Institutions and policy processes: The means to the ends of adaptation.

Dow, K. (1992). Exploring differences in our common future(s): The meaning of vulnerability to global environmental change. *Geoforum, 23*, 417–436.

Dunning, S., Rosser, N., Petley, D., & Massey, C. (2006). Formation and failure of the Tsatichhu landslide dam, Bhutan. *Landslides, 3* (2), 107–113.

Dupas, P., & Robinson, J. (2013). Why don't the poor save more? Evidence from health savings experiments. *The American Economic Review, 103*(4), 1138–1171.

Elalem, S., & Pal, I. (2015). Mapping the vulnerability hotspots over Hindu-Kush Himalaya region to flooding disasters. *Weather and Climate Extremes, 8*, 46e58. http://dx.doi.org/10.1016/j.wace.2014.12.001.

FFC (Federal Flood Commission). (2010). *Annual flood report*. Islamabad: Ministry of Water and Power, Government of Pakistan.

Gaire, S., Castro Delgado, R., & Arcos González, P. (2015). Disaster risk profile and existing legal framework of Nepal: Floods and landslides. *Risk Management and Healthcare Policy, 8*, 139–149.

Garschagen, M., Hagenlocher, M., Comes, M., Dubbert, M., Sabelfeld, R., Lee, Y. J., et al. (2016). World Risk Report 2016. World Risk Report. Bündnis Entwicklung Hilft and UNU-EHS. http://weltrisikobericht.de/wp-content/uploads/2016/08/WorldRiskReport2016.pdf.

Gentle, P., & Maraseni, T. N. (2012). Climate change, poverty and livelihoods: Adaptation practices by rural mountain communities in Nepal. *Environmental Science & Policy, 21*, 24–34. https://doi.org/10.1016/j.envsci.2012.03.007.

Gerlak, A., Lautze, J., & Giordana, M. (2010). Water resources data and information exchange in transboundary water treaties. *International Environmental Agreements, 11*(2), 179–199.

GFDRR. (2012). Disaster risk management in South Asia: A regional overview. In *Global facility for disaster reduction and recovery*. The World Bank: Washington DC.

GFDRR. (2013). *Learning from the Pakistan experience of post-2005 Earthquake rural housing reconstruction*. Project Highlights. Issue No 7. The World Bank: Washington DC.

Gill, J. C., & Malamud, B. D. (2016). Hazard interactions and interaction networks (cascades) within multi-hazard methodologies. *Earth System Dynamics, 7*(3), 659–679.

Goswami, B. N., Venugopal, V., Sengupta, D., Madhusoodanan, M. S., & Xavier, P. K. (2006). Increasing trend of extreme rain events over India in a warming environment. *Science, 314*, 1442–1445.

Grossman, M. (2006). Cooperation on Africa's international waterbodies: Information needs and the role of information-sharing. In W. Scheumann & S. Neubert (Eds.), *Transboundary water management in Africa: Challenges for development cooperation* (pp. 173–235). Bonn, Germany: German Development Institute.

Guha-Sapir, D., Hoyois, Ph, & Below, R. (2014). *Annual disaster statistical review 2013: The numbers and trends*. Brussels: CRED.

Gupta, V., Nautiyal, H., Kumar, V., Jamir, I., & Tandon, R. S. (2016). Landslide hazards around Uttarkashi township, Garhwal Himalaya, after the tragic flash flood in June 2013. *Natural Hazards, 80*(3), 1689–1707. https://doi.org/10.1007/s11069-015-2048-4.

Gupta, S. K., & Chalisgoankar, R. (1999). Blockade of River Mandakini in Uttar Pradesh-hazard assessment by mathematical approach. *Journal of Indian Water Resources Society, 19*(5)(1), 14–18.

Gurung, D. R., Khanal, N. R., Bajracharya, S. R., Tsering, K., Joshi, S., Tshering, P., et al. (2017). Lemthang Tsho Glacial Lake Outburst Flood (GLOF) in Bhutan: Cause and impact. Geoenvironmental Disaster (accepted).

Haeberli, W., Buetler, M., Huggel, C., Lehmann Friedli, Th, Schaub, Y., & Schleiss, A. J. (2016). New lakes in deglaciating high-mountain regions—Opportunities and risks. *Climatic Change, 139*(2), 201–214. https://doi.org/10.1007/s10584-016-1771-5.

Hewitt, K., & Liu, J. (2010). Ice-dammed lakes and outburst floods, Karakoram Himalaya: Historical perspectives on emerging threats. *Physical Geography, 31*(6), 528–551; Singh, J., Park, W-K., & Yadav, R. (2006). Tree-ring based hydrological records for western Himalaya, India, since A.D. 1560. *Climate Dynamics, 26*, 295–303; Yadav, R., & Singh, J. (2002). Tree-ring-based spring temperature patterns over the past four centuries in Western Himalaya. *Quaternary Research, 57*, 299–305.

Higaki, D., & Sato, G. (2012). Erosion and sedimentation caused by glacial lake outburst floods in the Nepal and Bhutan Himalayas. *Global Environmental Research, 16*, 71–76 (English Edition).

ICIMOD. (2016). Consultation workshop on landslide inventory, risk assessment, and mitigation in Nepal. In *ICIMOD. Proceedings 2016/2*. Kathmandu: ICIMOD.

Immerzeel, W. W., Beek, L. P. H. V., & Bierkens, M. F. P. (2010). Climate change will affect the Asian water towers. *Science, 328* (5984), 4. https://doi.org/10.1126/science.1183188.

IPCC. (2012). Summary for policymakers. In C. B. Field, V. Barros, T. F. Stocker, D. Qin, D. J. Dokken, K. L. Ebi, M. D. Mastrandrea, K. J. Mach, G.-K. Plattner, S. K. Allen, M. Tignor, & P. M. Midgley (Eds.), *Managing the risks of extreme events and disasters to advance climate change adaptation* (pp. 3–21). A Special Report of Working Groups I and II of the Intergovernmental Panel on Climate Change. Cambridge University Press, Cambridge, UK, and New York, NY, USA.

Islam, A. S., Haque, A., & Bala, S. K. (2010). Hydrologic characteristics of floods in Ganges–Brahmaputra–Meghna (GBM) delta. *Natural Hazards, 54*, 797. https://doi.org/10.1007/s11069-010-9504-y.

ISDR and UNEP. (2007). Environment and Vulnerability: Emerging Perspectives.

Ives, J. D., Shrestha, R. B., Mool, P. K., et al. (2010). *Formation of glacial lakes in the Hindu Kush-Himalayas and GLOF risk assessment*. Kathmandu, Nepal: ICIMOD.

Jodha, N. S. (1992). Mountain perspective and sustainability: A framework for development strategies. In N. S. Jodha, M. Banskota, & T. Partap (Eds.), *Sustainable mountain agriculture* (Vol. 1, pp. 41–82), Perspectives and issues. New Delhi: Oxford and IBH Publishers.

Jodha, N. S. (1997). Mountain agriculture. In B. Messerli & J. D. Ives (Eds.), *Mountains of the world: A global priority*. New York: The Parthenon Publishing Group.

Joseph, S., Sahai, A. K., Sharmila, S., Abhilash, S., Borah, N., Chattopadhyay, R., et al. (2013). North Indian heavy rainfall event during June 2013: Diagnostics and extended range prediction. *Climate Dynamics*. https://doi.org/10.1007/s00382-014-2291-5.

Kargel, J. S., Leonard, G. J., Shugar, D. H., Haritashya, U. K., Bevington, A., Fielding, E. J., et al. (2016). Geomorphic and geologic controls of geohazards induced by Nepal's 2015 Gorkha earthquake. *Science, 351*(6269). https://doi.org/10.1126/science.aac8353. http://science.sciencemag.org/content/351/6269/aac8353.

Khan, A. N., & Khan, S. N. (2015). Landslide risk and reduction approaches in Pakistan. In A.-U. Rahman, N. A. Khan, & R. Shaw (Eds.), *Disaster risk reduction approaches in Pakistan* (pp. 145–160). Springer Japan: Tokyo.

Khanal, N. R., Shrestha, M., & Ghimire, M. L. (2007). *Preparing for flood disaster: Mapping and assessing in the Ratu watershed*. Kathmandu: ICIMOD.

Kunreuther, H. C., & Michel-Kerjan, E. O. (2008). A framework for reducing vulnerability to natural disasters: Ex ante and ex post Considerations. Commissioned by the Joint World Bank- UN project on the Economics of Disaster Risk Reduction. Wharton Risk Management and Decision Processes Center. The Wharton School, University of Pennsylvania, Philadelphia, Pennsylvania, USA.

Lotay, Y. (2015). Country Report: BhutanRep., 20 pp.

Lavell, A., Oppenheimer, M., Diop, C., Hess, J., Lempert, R., Li, J., et al. (2012). Climate change: New dimensions in disaster risk, exposure, vulnerability, and resilience. In C. B. Field, V. Barros, T. F. Stocker, D. Qin, D. J. Dokken, K. L. Ebi, M. D. Mastrandrea, K. J. Mach, G.-K. Plattner, S. K. Allen, M. Tignor, & P. M. Midgley (Eds.), *Managing the risks of extreme events and disasters to advance climate change adaptation* (pp. 25–64). A Special Report of Working Groups I and II of the Intergovernmental Panel on Climate Change (IPCC). Cambridge University Press, Cambridge, UK, and New York, NY, USA.

Leduc, B. (2011). *Mainstreaming gender in mountain development: From policy to practice, lessons learned from a gender assessment of four projects implemented in the Hindu Kush-Himalayas*. Kathmandu, Nepal: ICIMOD.

Macchi, M., Gurung, A. M., Hoermann, B., & Choudhury, D. (2011). *Climate variability and change in the Himalayas: Community perceptions and responses*. Kathmandu, Nepal: International Centre for Integrated Mountain Development (ICIMOD).

Maplecroft. (2011). Climate change vulnerability index 2011. https://maplecroft.com/about/news/ccvi.html.

Mehta, M. (2007). *Gender matters*. Kathmandu: ICIMOD.

Molden, D., Sharma, E., & Acharya, G. (2016). Lessons from Nepal's Gorkha earthquake 2015. In S. P. Singh, S. C. Khanal, & M. Joshi (Eds.), *Lessons from Nepal's earthquake for the Indian Himalayas and the Gangetic plains* (pp 1–14). Nainital, Uttarakhand, India: Central Himalayan Environment Association.

MoSWRR. (2009). Myanmar Action Plan on Disaster Risk Reduction (MAPDRR) Rep.

MoPE. (2017). Vulnerability and Risk Assessment Framework and Indicators for National Adaptation Plan (NAP) Formulation Process in Nepal. Ministry of Population and Environment (MoPE), Kathmandu.

Nadim, F., Kjekstad, O., Peduzzi, P., Herold, C., & Jaedicke, C. (2006). Global landslide and avalanche hotspots. *Landslides: Journal of the International Consortium on Landslides, 3*(2), 159–173.

Nadim, F., Jaedicke, C., Smebye, H., & Kalsnes, B. (2013). Assessment of global landslide hazard hotspots. In K. Sassa, B. Rouhban, S. Briceño, M. McSaveney, & B. He (Eds.), *Landslides: global risk preparedness*. Heidelberg: Springer, Berlin.

Nellemann, C., Verma, R., & Hislop, L. (2011). *Women at the frontline of climate change—Gender risks and hopes: A rapid response assessment*. Arendal, Norway: United Nations Environment Programme (UNEP), GRID.

NPC (Nepal Planning Commission). (2015). Nepal Earthquake 2015: Post Disaster Needs Assessment l Vol. B: Sector reports. Government of Nepal.

Nibanupudi, H. K., & Khadka, M. (2015) Gender and disaster resilience in the Hindu Kush Himalayan region. In R. Shaw & H. K. Nibanupudi (Eds.), *Mountain hazards and disaster risk reduction* (pp. 233–250). Springer Japan.

Nogués-Bravoa, D., Araújoc, M. B., Erread, M. P., & Martínez-Rica, J. P. (2007). Exposure of global mountain systems to climate warming during the 21st Century. *Global Environmental Change, 17*, 420–428.

O'Brien, K. L., & Leichenko, R. L. (2000). Double exposure: Assessing the impacts of climate change within the context of economic globalization. *Global Environmental Change, 10*, 221–232.

Oxley, M. C. (2013). Editorial: A "people-centered principles-based" post-Hyogo framework to strengthen the resilience of nations and communities. *International Journal of Disaster Risk Reduction, 4*, 1–9.

Pescaroli, G., & Alexander, D. (2015). A definition of cascading disasters and cascading effects: Going beyond the "toppling dominos" metaphor. *Planet@ Risk, 3*(1).

Pradhan, N. S., Khadgi, V. R., Schipper, L., Kaur, N., & Geoghegan, T. (2012). *Role of policy and institutions in local adaptation to climate change*. Kathmandu, Nepal: ICIMOD.

Pusch, C. (2004). Preventable losses: Saving lives and property through hazard and risk management—A comprehensive risk management framework for Europe and Central Asia. Disaster Risk Management Working Paper 9. World Bank. Washington DC.

Rai, B. (2004). Himalayan seismicity and probability of future earthquake. *IAGA WG, 1*, 20.

Rasul, G., Sharma, B., Mishra, B., Neupane, N., Dorji, T., Khadka, M., et al. (2015). *Strategic framework for resilient livelihoods in earthquake-affected areas of Nepal*. ICIMOD Working Paper 2015/6. Kathmandu: ICIMOD.

Richardson, S. D., & Reynolds, J. M. (2000). An overview of glacial hazards in the Himalayas. *Quaternary International, 65–66*, 31–47.

Rautela, P. (2013). Lessons learnt from the deluge of Kedarnath, Uttarakhand, India. *Asian Journal of Environment and Disaster Management, 5*(2), 43–51.

SAARC. (2010). SAARC Workshop on Landslide Risk Management in South Asia, edited (p. 126).

Sarwar, G. M. (2008, November 2008). Landslide Tragedy of Bangladesh, The First World Landslide Forum, 18–21 Nov 2008, United Nations University (UNU), Tokyo, Japan.

Schipper, E. (2009). Meeting at the crossroads?: Exploring the linkages between climate change adaptation and disaster risk reduction. *Climate and Development, 1*(1), 16–30.

Schwanghart, W., Worni, R., Huggel, C., Stoffel, M., & Korup, O. (2016). Uncertainty in the Himalayan energy–water nexus: Estimating regional exposure to glacial lake outburst floods. *Environmental Research Letters, 11,* 074005. https://doi.org/10.1088/1748-9326/11/7/074005.

SDMC. (2008). *South Asian Disaster Report 2007*. New Delhi, India: SAARC Disaster Management Centre.

SDMC. (2011). SAARC Expert Group Meeting on Landslide Terminology, Classifications, Documentation and Hazard Mapping. SAARC Disaster Management Center, New Delhi, India.

Shaw, R., & Nibanupudi, H. K. (2015). Overview of Mountain Hazards Issues. In *Mountain hazards and disaster risk reduction* (pp. 1–11). Springer Japan.

Shrestha, A. B., Bajracharya, S. R., Kargel, J. S., & Khanal, N. R. (2016). The impact of Nepal's 2015 Gorkha earthquake-induced geohazards. ICIMOD Research Report 2016/1. Kathmandu, ICIMOD.

Shrestha, A. B. (2008a). *Resource manual on flash flood risk management: Module 2 Non-structural Measures*. Kathmandu: ICIMOD.

Shrestha, M. S. (2008b) Impacts of floods in South Asia. *Journal of South Asia Disaster Studies, 1*(1), 85–106 (SAARC Disaster Management Centre, New Delhi, India).

Shrestha, A. B., & Pradhan, N. S. (2015). Strengthening flash flood risk management in the Hindu Kush Himalayas. In *Case studies presented during the INR 1.3 AP session on Climate change adaptation and mitigation in Africa, Americas, Asia-Pacific, Europe and the Mediterranean region/Building resilience to water-related disasters in the Asia-Pacific region of the 7th World Water Forum*. Kathmandu, Nepal: ICIMOD.

Shrestha, A. B., Shah, S. H., & Karim, R. (2008). *Resource manual on flash flood risk management: Community-based management*. Kathmandu, Nepal: ICIMOD.

Shrestha, M. S., & Chhophel, K. (2010). Disasters in the Himalayan Region: A case of Tsatichu Lake in Bhutan. In Behiem, et al. (Eds.), *Integrated watershed management: Perspective and problems, Einar* (pp. 211–222). New York: Springer.

Shrestha, M. S, Kafle, S., Gurung, M., Nibanupudi, H. K., Khadgi, V. R., & Rajkarnikar, G. (2014). *Flood early warning systems in Nepal: A gendered perspective*. ICIMOD Working Paper 2014/4. Kathmandu: ICIMOD.

Shrestha, M. S., Grabs, W. E., & Khadgi, V. R. (2015). Establishment of a regional flood information system in the Hindu Kush Himalayas: Challenges and opportunities. *International Journal of Water Resources Development, 31*(2), 238–252. https://doi.org/10.1080/07900627.2015.1023891.

Sietz, D. (2014). Regionalisation of global insights into dryland vulnerability: Better reflecting smallholders' vulnerability in Northeast Brazil. *Global Environmental Change, 25*(1), 173–185. https://doi.org/10.1016/j.gloenvcha.2014.01.010.

Sietz, D., Ordonez, J. C., Kok, M. T. J., Janssen, P., Hilderink, H. B. M., Tittonell, P., et al. (2017). Nested archetypes of vulnerability in African drylands: Where lies potential for sustainable agricultural intensification? *Environmental Research Letters, 12*(9), 95006. Retrieved from http://stacks.iop.org/1748-9326/12/i=9/a=095006.

Singh, J., Park, W.-K., & Yadav, R. (2006). Tree-ring based hydrological records for western Himalaya, India, since A.D. 1560. *Climate Dynamics, 26,* 295–303.

Slovic, P. (1987). Perception of risk. *Science, 236,* 280–285.

Slovic, P. (1999, August). Trust, emotion, sex, politics, and science: Surveying the risk-assessment battlefield. *Risk Analysis, 19*(4), 689–701.

Soman, D., & Cheema, A. (2011). Earmarking and partitioning: Increasing saving by low-income households. *Journal of Marketing Research, XLVIII*(Special Issue), S14–S22).

Sonesson, M., & Messerli, B. (Eds.). (2002). *The Abisko agenda: Research for mountain area development, Ambio special report 11*. Stockholm: Royal Swedish Academy of Sciences.

Stark, C., Daniel Petz, D., & Ferris, E. (2013). Disaster risk management: A Gender-sensitive approach is a smart approach (Chapter 4). In *The year of recurring disasters: A review of natural disasters in 2012*.

Sudmeier, K. I., Jaboyedoff, M., & Jaquet, S. (2013). Operationalizing "resilience" for disaster risk reduction in mountainous Nepal. *Disaster Prevention and Management, 22*(4), 366–377.

Sundriyal, Y. P., Shukla, A. D., Rana, N, Jayangondaperumal, R., Srivastava, P., Chamyal, L. S., et al. (2015). Terrain response to the extreme rainfall event of June 2013: Evidence from the Alaknanda and Mandakini River Valleys, Garhwal Himalaya, India. *Episodes, 38*(3), 11 pp.

Tanner, T. (2003). Peopling mountain environments: Changing Andean livelihoods in northwest Argentina. *Geographical Journal, 169,* 205–214.

Tariq, M. A. U. R., & Giesen, N. V. D. (2011). Floods and Flood Management in Pakistan. *Physics and Chemistry of the Earth*. https://doi.org/10.1016/j.pce.2011.08.014.

Tse-ring, K., Sharma, E., Chettri, N., & Shrestha, A. (Eds.). (2010). *Climate change vulnerability of mountain ecosystems in the eastern Himalayas—Synthesis report*. Kathmandu, Nepal: ICIMOD.

Tversky, A., & Kahneman, D. (1974). Judgment under uncertainty: Heuristics and biases. Biases in judgments reveal some heuristics of thinking under uncertainty. *Science, 185,* 1124–1131. https://doi.org/Cited By (since 1996) 3914\nExport Date 30 November 2011.

Twigg, J. (2009). *Characteristics of a disaster-resilient community: A guidance note*. London, UK: Department for International Development (DFID).

UNDP (United Nations Development Programme). (2007). *Human development report 2007/2008, fighting climate change: Human solidarity in a divided world*. New York, NY: Palgrave Macmillan.

UNDP. (2016). United Nations Development Programme, Human Development Index. http://hdr.undp.org/en/content/human-development-index-hdi.

UNEP. (1997). Asian disaster management news: Gender and disaster. *A Newsletter of the Disaster Management Community in Asia and the Pacific, 3*(3). Nairobi, Kenya: UNEP.

UNISDR. (2005, January 18–22). Hyogo Framework of Action 2005–2015. Extract from the final report of the World Conference on Disaster Reduction, Kobe, Hyogo, Japan.

United Nations. (2006). *Global survey of early warning systems: An assessment of capacities, gaps and opportunities toward building a comprehensive global early warning system for all natural hazards*. Geneva, Switzerland: UNISDR.

United Nations. (2015). *Sendai framework for disaster risk reduction 2015–2030*. Geneva, Switzerland: UNISDR.

United Nations. (2016). Report of the open-ended intergovernmental expert working group on indicators and terminology relating to disaster risk reduction. http://www.preventionweb.net/files/50683_oiewgreportenglish.pdf.

United Nations. (2017). Development Policy and Analysis Division—Department of Economic and Social Affairs List of Least Developed Countries. https://www.un.org/development/desa/dpad/least-developed-country-category/ldcs-at-a-glance.html.

Upreti, B. N., & Dhital, M. R. (1996). Landslide Studies and Management in NepalRep., 87 pp., ICIMOD.

Vaidya, R. A., Sharma, E., Karky, B. S., Kotru, R., Mool, P., Mukherji, A., et al. (2014). Synthesis: Research insights on climate change and water resources management in the Hindu Kush Himalayas. In R. A. Vaidya

& E. Sharma (Eds.), *Research insights on climate and water in the Hindu Kush Himalayas* (pp. 1–38). Kathmandu, Nepal: ICIMOD.

Vatsa, K. S. (2002). Technological challenges of the disaster management plan for the state of Maharashtra. In S. Bhatnagar & R. Schware (Eds.), *Information and communication technology in development: Cases from India* (pp. 50–62). New Delhi, India: Sage Publications.

Wang, S. Y., Yoon, J. H., Gillies, R. R., & Cho, C. (2013). What caused the winter drought in Western Nepal during recent years? *Journal of Climate, 26*, 8241–8256.

Wang, S. Y., Fosu, B., Gillies, R. R., & Singh, P. M. (2015). The deadly himalayan snowstorm of October 2014: Synoptic conditions and associated trends. *Bulletin of the American Meteorological Society, 96,* S89–S94.

Wasson, R. J., & Newell, E. B. (2015). Links between floods and other water issues in the Himalayan and Tibetan Plateau Region. *Pacific Affairs, 88*(3), 653–676.

Wasson, R. J., Sundriyal, Y. P., Chaudhary, S., Jaiswal, M. K., Morthekai, P., Sati, S. P., et al. (2013). A 1000-year history of large floods in the Upper Ganga catchment, central Himalaya, India. *Quaternary Science Reviews, 77*(2013), 156–166.

Wesnousky, S. G., Kumahara, Y., Nakata, T., Chamlagain, D., & Neupane, P. (2017) New observations disagree with previous interpretations of surface rupture Along the Himalayan frontal thrust during the great 1934 Bihar–Nepal earthquake. *Geophysical Research Letters, 45*(6), 2652–2658.

World Bank. (2005). *Natural disaster hotspots: A global risk analysis*. Disaster Risk Management Series No. 5. Washington, DC: The World Bank.

World Bank. (2013). *Reducing poverty by closing South Asia's infrastructure gap*. The World Bank and Australian Aid.

World Bank. (2017). *Disaster risk profile: Afghanistan*. Washington, DC: The World Bank.

World Economic Forum. (2014). Competitiveness Dataset. http://www3.weforum.org/docs/GCR2014-15/GCI_Dataset_2006-07-2014-15.xlsx.

Yadav, R., & Singh, J. (2002). Tree-ring-based spring temperature patterns over the past four centuries in Western Himalaya. *Quaternary Research, 57,* 299–305.

Yao, T., Thompson, L., Yang, W., Yu, W., Gao, Y., Guo, X., et al. (2012). Different glacier status with atmospheric circulations in Tibetan Plateau and surroundings. *Nature Climate Change, 2*(663–667), 2012.

Ziegler, A. D., Wasson, R. J., Bhardwaj, A., Sundriyal, Y. P., Sati, S. P., Juyal, N., et al. (2014). Pilgrims, progress, and the political economy of disaster preparedness—The example of the 2013 Uttarakhand flood and Kedarnath disaster. *Hydrological Processes*. https://doi.org/10.1002/hyp.10349.

12 Understanding and Tackling Poverty and Vulnerability in Mountain Livelihoods in the Hindu Kush Himalaya

Coordinating Lead Authors

Giovanna Gioli, University of Edinburgh, UK and formerly International Centre for Integrated Mountain Development, ICIMOD, Nepal, e-mail: fraugioli@gmail.com (corresponding author)
Ganesh Thapa, Independent researcher, former IFAD, Italy, e-mail: ganeshbthapa@gmail.com
Fawad Khan, ISET, Pakistan, e-mail: fmkhan@gmail.com

Lead Authors

Purnamita Dasgupta, Institute of Economic Growth, India, e-mail: purnamita.dasgupta@gmail.com
Dev Nathan, Institute for Human Development, India, e-mail: nathandev@hotmail.com
Netra Chhetri, Arizona State University, USA, e-mail: Netra.Chhetri@asu.edu
Lipy Adhikari, International Centre for Integrated Mountain Development, ICIMOD, Nepal, e-mail: lipy.adhikari@icimod.org
Sanjay Kumar Mohanty, International Institute for Population Studies, India, e-mail: Sanjay@iips.net
Elisabetta Aurino, Imperial College London, UK, e-mail: e.aurino@imperial.ac.uk
Laura Mapstone Scott, King's College London, UK, e-mail: lmapstone@gmail.com

Contributing Authors

Jean-Yves Gerlitz, University of Bremen, Germany, e-mail: jgerlitz@bigsss-bremen.de
Bidhubhusan Mohapatra, Population Council, India, e-mail: bbmahapatra@gmail.com
Naresh Saxena, Independent researcher, India, e-mail: naresh.saxena@gmail.com
Marjorie van Strien, Independent researcher, former ICIMOD, The Netherlands, e-mail: marjorie.vanstrien@gmail.com
Amiera Sawas, Grantham Institute, Imperial College London, UK, e-mail: amierasawas@googlemail.com

Review Editors

Enamul Haque, East West University, Dhaka and Asian Center for Development, Dhaka, Bangladesh, e-mail: akehaque@ewubd.edu; akehaque@acdonline.org
Linxiu Zhang, Center for Chinese Agricultural Policy, Chinese Academy of Sciences, China, e-mail: lxzhang.ccap@igsnrr.ac.cn

Corresponding Author

Giovanna Gioli, University of Edinburgh, UK and formerly International Centre for Integrated Mountain Development, ICIMOD, Nepal, e-mail: fraugioli@gmail.com

P. Wester et al. (eds.), *The Hindu Kush Himalaya Assessment*,
https://doi.org/10.1007/978-3-319-92288-1_12

Contents

Chapter Overview

Key Findings

1. **Overall, in the mountains and hills of the HKH region, the poverty incidence is one-third compared to one-fourth for the national average.** Country-level estimates mask significant inequities between mountainous and non-mountainous regions, as well as within mountainous areas. The acute shortage of mountain-specific poverty data makes knowledge-building a high priority.
2. **Poverty reduction approaches/programmes designed at national level are likely to miss crucial subnational and local manifestations of poverty.** Countrywide strategies usually have limited impact on poverty reduction in the mountains and can make mountain livelihoods more vulnerable.
3. **Determinants of vulnerability and of poverty in the HKH overlap substantially.** Cross-countries assessments show similarities in the determinants and patterns of poverty and vulnerability in the HKH. Apart from remoteness, poor accessibility, and high dependence on natural resources, major determinants of poverty and vulnerability in the region are socioeconomic inequities, conflicts, gender inequities and caste/ethnicity-based discrimination. The HKH is home to millions of indigenous peoples who are economically, socially, and politically marginalized.

Policy Messages

1. **Closing the income gap between the plains and mountain regions is not possible without more mountain-specific poverty-reduction programmes.** Global, regional, and national institutions should allocate resources for the development of mountain specific poverty reduction policies to address three key aspects:
   - Social and economic infrastructure for poverty reduction programmes has to take note of the fragile mountain ecosystem.
   - Targeted approaches are required for indigenous and marginalized communities inhabiting mountains.

- New threats from climate change require new forms of social protection.

2. **The most appropriate poverty measures for mountain areas are multidimensional, addressing multiple deprivations in education, health, and living standards**. Emphasis must be laid on regular/systematic collection, collation, and sharing of data among countries in the region on the determinants of poverty, duration of poverty spells, and the causes of transient poverty. The data has to be location-specific and lend themselves to disaggregation by mountain specificities, such as fragility, marginality, and remoteness. Availability of such data can help in designing poverty reduction policies and programmes that are appropriate for mountain areas. Greater investment in generating mountain-specific data is therefore called for.
3. **The international benchmark for poverty in mountain areas needs to be higher than the globally accepted standard of USD 1.90 per capita per day**. Mountain people experience higher costs of living due to factors such as heating costs, food prices, and access to public services. Countries must undertake cost of living surveys targeting mountain areas.

This chapter critically reviews the existing knowledge on livelihoods, poverty, and vulnerability in the Hindu Kush Himalaya (HKH). Development in mountain areas and the practices of people in these areas are uniquely conditioned by distinct characteristics that we term "mountain specificities". Some of these specificities—such as inaccessibility, fragility, and marginality—constrain development. Others—such as abundant biological diversity, ecological niches, and adaptation mechanisms—present development opportunities for mountain people.

Because of these specificities, it is important to ensure that variables and indicators used in estimating poverty capture mountain specificities. For the overall HKH areas, the poverty incidence is one-third compared to one-fourth for the national averages (***established but incomplete***).

This chapter explores the changing contexts of HKH mountain economies and livelihoods, detailing their specific conditions and challenges. How do these specificities affect policies to measure poverty, to reduce it, and to address mountain peoples' vulnerability to climate change? Our analysis leads to policy recommendations for the region.

Estimating mountain poverty using available data

Precise measurement of mountain poverty and vulnerability is a challenge, given the large gaps in mountain-specific data for most HKH countries. National estimates can mask significant inequalities between mountain and non-mountain areas—and even between different mountain areas. The United Nations Development Programme's Multidimensional Poverty Index (MPI), as currently measured, does not take account of mountain specificities. Despite the lack of mountain-specific data on income poverty and multidimensional poverty, we have subjected the available data to as detailed an analysis as possible, resulting in the following findings:

- For income poverty, the national average of incidences vary widely among HKH countries—yet income poverty has declined in all countries over time, most rapidly in Pakistan, India, and Bhutan (***well-established***). National scale poverty estimates may, however, mask significant inequalities between mountain and non-mountain regions, as well as inequalities among mountain areas [*established but incomplete*]. For example, in Nepal, the poverty incidence in 2010/2011 was 42% in the mountains, compared with 23% in the plains and 25% nationally (***well-established***).
- For multidimensional poverty, national average incidences—reflected in the MPI—vary widely, from 5% in China to 59% in Afghanistan. However, if China and Bhutan are excluded, the multidimensional poverty rate is well above 40% across the region. As measured by the MPI, Afghanistan's multidimensional poverty ranks highest (0.35), followed by India's (0.28). China's ranks lowest (0.02), followed by Bhutan's (0.13) (***well-established***).
- The average intensity of deprivation is fairly consistent across HKH countries, varying from 43% in Bhutan and China to 52% in Pakistan (***well-established***). While citations of national poverty rates differ between the MPI and the Human Development Index (HDI), most HKH countries score poorly on both.

The HKH region is rich in cultural heritage, sacred natural sites, and pilgrimage routes that traverse the landscape and constitute the identity of its people. The extension of the argument in the specific case of mountain people, implies an emphasis on preserving the cultural identities of these communities, their cultural bonding and community resilience, that is considered to be beyond the purview of indicators captured in standard multidimensional poverty indices, and yet, these are an essential component of the wellbeing of mountain people (***established but incomplete***). In fact, Bhutan has been estimating Gross National Happiness as an alternative measure to GDP and it has four pillars: good governance, sustainable socioeconomic development, preservation and promotion of culture, and environmental conservation. The question is whether addressing poverty and vulnerability implies the need for public policy that integrates these aspects into response options for poverty and vulnerability reduction in the mountains.

Determinants of poverty and vulnerability in the HKH

Mountain poverty is associated with social markers and inequality at the intersection of class, caste, ethnicity, gender, education, occupation, and employment status (***well-established***). In addition, human wellbeing in mountain areas is inextricably linked to the aforementioned mountain specificities. The determinants of poverty and its persistence in the HKH fall into five categories:

- Remoteness and low access to markets and basic facilities
- Access to natural resources—and high dependence on them
- Demographic factors
- Social and cultural factors
- Marginalization (political and socioeconomic).

All five types of determinants may hinder the conversion of resources, such as income, into desirable outcomes for wellbeing (increasing food and nutrition security; raising educational achievement; improving health) (***established but incomplete***).

Vulnerability—a pervasive aspect of livelihoods in the HKH—is intrinsic to mountain specificities. Mountain systems are fragile. People depend on ecosystem services. The region is highly exposed to weather variability and climate change. Similarly, poverty and vulnerability overlap to a large extent: both are multidimensional, with common causes that manifest in similar risks and outcomes (***established but incomplete***).

Livelihood diversification in the HKH: Causes and context

Mountain livelihoods in the HKH are evolving. The past three decades have seen a significant shift from the agro-pastoral to a combined subsistence-labour system: mountain households no longer rely entirely on their land, though they cannot make do entirely without it. Mountain households increasingly rely on livelihoods that combine farm work with non-farm activities, such as wage labour, circular labour migration, and tourism services.

The determinants of livelihood diversification in the HKH are varied. They include population growth, land fragmentation, and fast-paced urbanization, among other demographic changes. Environmental and climatic change, both global and local, is also driving diversification—as is economic globalization based on the increased connectivity that expands access to markets (***well-established***).

Mountain people experience these changes in a distinctive context. In the mountains of the HKH, land is scarce, for the most part, and formal property rights either do not exist or are ill-defined. Employment is largely informal; access to social and economic services—including financial instruments—is limited and social protection is limited.

Under such conditions, livelihood diversification is expected. Yet diversification alone does not adequately buffer mountain people and enable them to cope with crisis and manage risk (***well-established***). To provide such a buffer, targeted interventions are required. For example, as agricultural patterns shift from traditional multi-cropping to the commercialized mono-cropping of high-value commodities, the preservation of neglected and underutilized food crops (such as millet and buckwheat) can maintain genetic variety and increase food and nutrition security by offering higher resilience to climate change (***established but incomplete***). Such efforts are vital, given increased climate risks and persistent malnutrition. Herein, social groups and cooperatives can play a critical role in managing and using resources.

Two calls for action: Design interventions to target poverty and vulnerability now—and gather data to improve policies in the future

Mountain poverty and vulnerability often respond to the same interventions. For example, mountain-specific policies designed for the water and energy sectors can reduce poverty and livelihood vulnerability. Similarly, mountain-specific social-protection programmes can respond to new challenges driven by global climate change (***inconclusive***). In some cases, the joint provision of health services, decentralized energy, financial services (credit and insurance), and weather information has reduced poverty and climate risks. Furthermore, interventions that tackle challenges facing marginalized groups—including ethnic minorities—can boost economic returns from cultural tourism, alleviating poverty among the poorest, while preserving the region's ecological and cultural diversity (***established but incomplete***).

The above-mentioned examples show that even with existing data, successful mountain-specific interventions can be developed. Accordingly, national, regional, and global institutions should allocate resources to develop mountain-specific policies.

This chapter highlights the urgent need for gathering longitudinal data on poverty and vulnerability determinants, the duration of poverty spells, and the causes of transition in and out of poverty—disaggregated for mountain areas, while harmonized with national surveys and databases that enable the disaggregation of data for mountain areas. If future data gathering is to include information on key potential drivers of poverty in the HKH—such as the impacts of climate and other global drivers of change on mountain-specific livelihoods—then new investments and research on the ground will be needed.

To enable mountain-specific policies and mountain-development pathways that are pro-poor and gender-inclusive, we urge the priority allocation of increased resources to mountain-specific data gathering, with a focus on poverty and vulnerability.

Mountain Poverty and the Sustainable Development Goals

While Sustainable Development Goal (SDG) 1 commits nations to eradicating poverty by 2030, the mountain specificities detailed in this chapter suggest that SDG 1 must be applied in distinct ways to mountain areas.

- First, because mountain people experience higher heating costs, higher calorie requirements, and higher food commodity prices, the international income threshold of extreme poverty for HKH mountain areas should exceed the globally accepted standard of USD 1.90 per capita per day.
- Second, because poverty is generally higher in the mountains of the HKH than in the plains, eradicating national poverty by 2030 will require a higher rate of poverty reduction in mountain areas.
- Third, in meeting SDG Target 1.2—"By 2030, reduce at least by half the proportion of men, women, and children of all ages living in poverty in all its dimensions according to national definitions"—the governments of the HKH should promote the use of multidimensional poverty measures. Such measures are more appropriate generally and, particularly, for the mountains.
- Fourth, poverty reduction programmes must focus more on indigenous peoples, women, and other marginalized social groups, among which the incidence of poverty is most severe. Social protection programmes are needed.

12.1 Anticipating Change: Mountain Livelihoods in a Changing Context

Mountain farmers have continuously adapted their farming systems to a risky and changing environment. Today, multiple transitions are affecting the mountain farming systems in the HKH, presenting both challenges and opportunities. This section provides the context of traditional mountain livelihoods, highlighting the shifts from agro-pastoral subsistence to the multi-local livelihood diversification strategies that integrate on-farm and off-farm activities that are now prevalent in the HKH.

12.1.1 Mountain Livelihoods: Trends, Challenges, and Strengths

For a majority of people in the HKH, crop-livestock agriculture (CLA) has long been a source of livelihood (see Box 12.1). With more than 200 million smallholder farmers, agriculture constitutes 40% of the region's GDP. It also generates the bulk of livestock products—75% of the milk and 60% of the meat—and CLA employs millions of people on farms. In CLA, the livelihood of smallholder farmers strongly depends on natural resources to support the population of humans and their livestock. Overgrazed pastures, soil erosion, forest degradation, reduced recharge of aquifers, and population growth have led to a steady decline in resource endowments (see Chap. 5). As a risk multiplier, global environmental change exacerbates these challenges, but it may also create new opportunities (see Chap. 13).

Box 12.1 Mixed Crop-Livestock Agriculture

Crop-livestock agriculture (CLA) (see Fig. 12.1) is part of a dynamic system of interactions among bio-physical, social, and ecological processes that produces about half of the world's cereals and is a crucial component of food, fibre, and fuel production for developing countries (Herrero et al. 2010). CLA has long offered a sustainable way to make efficient use of available natural resources in the environmentally challenging HKH; CLA farmers rear animals mostly on grass from common-pool resources (rangeland and forest) and non food biomass from crops, while the animals supply manure and traction to agriculture. In addition to the more obvious coupling of crops and livestock, there is also the coupling of common-pool resources and household-level enterprises.

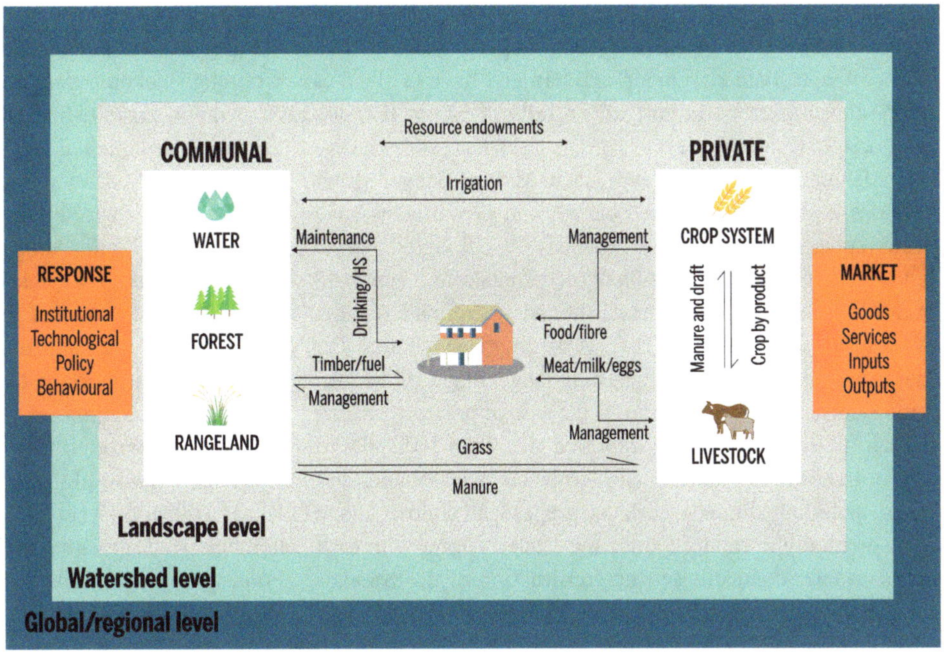

Fig. 12.1 The dynamic system of crop-livestock agriculture in the HKH (*Source* developed by Netra Chhetri for this chapter)

In general, livelihood conditions are strongly linked to the capacity of natural resources to support human populations and their livestock. Sustainable management of water, animal fodder, biomass, and fuelwood derived from common-pool resources is critical for ensuring the continued flow of goods and services for sustaining the livelihoods of smallholder farming communities across the HKH. Overgrazed pastures, soil erosion, forest degradation, and reduced recharge of aquifers have led to a steady decline in resource productivity. As a risk multiplier, climate change exacerbates these challenges. A well-designed research agenda focused on innovation and solutions will help sustain this unique socio-ecological system. Success will require a system-level understanding of the mixed CLA and its vulnerability to socio-ecological stimuli and shocks. To help improve current farming efficiency and the reliability of CLA, science-based and community-driven agricultural innovation and development are needed. Innovation that includes integration of scientifically based best practices with the experience and participation of a broad set of stakeholders is needed to ensure the future resiliency of the system.

By placing the household at the focus of analysis, this assessment highlights both the exogenous (risk factors) and endogenous (coping mechanisms) drivers that make the socio-ecological system of the HKH more vulnerable. Viewing the household as a primary focus of concern is not intended to undermine the importance of intra-household level data in understanding crucial gender and power dynamics, which are covered elsewhere in this assessment (see Chap. 14). Our aim is to demonstrate how the vulnerability of households is enmeshed in a complex of interlinked social, environmental, and market forces. The following sections present a review of the key emerging livelihood trends across the HKH.

12.1.1.1 From Subsistence Farming to High-Value Agriculture

The HKH has undergone a significant transformation in recent decades with respect to land use change, cropping systems, and access to markets. Farmers across the region have been gradually shifting from subsistence to high-value agriculture (Singh et al. 2011). A number of factors have been contributing to this shift, including a growing recognition of niches for high-value crops, such as fruits, vegetables, and spices like ginger, turmeric, and cardamom (Tulachan 2001; Chand et al. 2008; Adhikari 2014). The shift to high-value crops has also been facilitated by other factors, such as improved road networks providing market access for previously isolated communities, growth of remittance inflows, expansion of cooperatives, increased presence of NGOs, and targeted government activities (Kreutzmann 2006; Yi et al. 2007, 2008; Wangchuk and Siebert 2013; Khattri 2012).

Recently, high-value crops have been introduced into crop rotations, especially in the lower-elevation areas with access to markets, technology, and irrigation. In China, high-value cash crops provide farmers four times as much income as rice (Chen 2011). In Nepal, export earnings from the top three high-value crops (lentil, tea, and cardamom) exceed the value of cereal and dairy imports (CBS 2011). Tea, coffee, cardamom, ginger, and turmeric have evolved as important cash crops in part of the hills of Nepal (Pandey et al. 2009; Khattri 2012).

In Bhutan, traditional crops such as buckwheat, sweet potatoes, millet, wheat, naked barley, and green beans have been replaced with potatoes. Intensively cultivated potato fields are heavily fertilized with agrochemicals, and handheld traditional farming tools are substituted with small tractors (Wangchuk and Siebert 2013). A study revealed that the majority of households in the Bumthang village of Bhutan derive most of their income from the sale of potatoes, which is then used for purchasing imported rice from India (Dorji 2011). In northern Pakistan, cash crops such as potatoes and fruits (including almonds, apricots, grapes, and cherries) have been gradually introduced since the 1980s. Along with cash crops, greater connectivity (the construction of the Karakoram Highway, for example) has also greatly improved trade and income in the area (Kreutzmann 2006; Gioli et al. 2014b). A similar phenomenon has been reported in Tibet, where farmers with access to roads grow significantly more cash crops and fewer subsistence food crops, as compared to communities that are not connected to roads (Salick et al. 2005). Despite repeated crop failure, a majority of households in Ladakh, in the Indian Himalaya, continue intensive cultivation of potatoes with the hope of generating and diversifying household income, as on-farm income generation is considered vitally important (Dame and Nusser 2011).

There are several success stories associated with high-value crops. National and provincial governments in the HKH are exploring options for scaling up the successes. Two outstanding stories include vegetable and apple farming in Himachal Pradesh, India and Ningnan County, China.[1] After switching from subsistence crops to high-value crops, their income and quality of life have significantly increased. The sustainability of these emerging agricultural activities—and their competitiveness in the context of a market economy—depends upon effective management of natural resources and technological and social innovation customized to site-specific needs of smallholder farmers. The substitution of subsistence farming by high-value crops has a potential negative impact on livelihoods, as this may lead to the destruction of pasture land in the mountain areas.

While the development of high-value agriculture is crucial in addressing poverty and enhancing livelihoods options, farmers and their supporting institutions should not be complacent, as specialized agriculture is historically known to create vulnerabilities. Livelihood scholars have long viewed diversity as a central determinant of livelihood security (Chambers and Conway 1992). Traditional multiple cropping technologies, tailor-fit to a locality, could be refined and exploited to adapt against the potential threats of global environmental change and to ensure food security (Altieri 1999; Katwal n.d.). Replacing a diverse set of site-specific agricultural practices with intensive, market-driven mono-cropping (such as potatoes) tends to reduce household food self-sufficiency and increase social and ecological risk, and may reduce the wellbeing of rural households in the long run (Yi et al. 2008; Zimmerer 2010).

It is also important to note that farmers' decisions to improve their livelihoods can have negative outcomes—not only locally, through transformations of ecology and social relations, but also globally, through market channels (Adger et al. 2003). Markets in the future will be increasingly homogenized towards global requirements and demands. Food production, distribution, and marketing chains are changing along with improvements in infrastructure, communications, and vertical business structures, integration into the world market, and the rapid rise of supermarkets. New distribution channels, dominated by larger firms including supermarket retailers, will impose high performance demands on their producers and people involved in value chains. Rising energy costs will drive up the costs of fertilizers, irrigation, mechanization, and, thus, food. Systems of agriculture and their resource requirements in the HKH need to be efficient, not heavily reliant on external sources of fertilizer and water. As taught by the tragic 2008 melamine-in-milk episode in Asia, there is a need for strong regulation and governance to avoid inadvertent inclusion of unmonitored agents in agriculture.

12.1.1.2 Shifting Livestock Patterns

Livelihood practices of the people of the HKH fall into three categories (Wu et al. 2016):

- Livelihoods based exclusively on livestock herding
- Livestock herding combined with intensive agriculture
- The practice of mixed crop-livestock systems.

Livestock-based livelihoods (such as the rearing of yak, horse, sheep, and goat) are common in high-altitude areas of the HKH (Berhanu et al. 2007). Access to common-pool resources, such as pasture and water, is fundamental to sustaining this form of livelihood. In the midhills of the HKH, the dominant form of livelihood is the combined practice of rain-fed agriculture (typically maize, wheat, barley, and potatoes) for household consumption and of extensive livestock rearing. Migratory mountain pastoralism, which has been the prevalent form of livestock-keeping in the mountain areas around the Tibetan Plateau, has been

[1]See: http://lib.icimod.org/record/22650/files/c_attachment_184_2113.pdf.

undergoing drastic changes in terms of spatial and temporal migration patterns, livestock management objectives, and the relationship between crop cultivation and livestock (Yi et al. 2007, 2008).

The practice of sedentary mixed crop-livestock (see Box 12.1) is a dominant form of livelihood in the lower part of the HKH, where traditional agricultural practices like multiple cropping, crop diversification, and conservation-tillage farming are prevalent. Multiple cropping is effective in maintaining agricultural production even under unfavourable climatic conditions. Different crops respond to climatic stressors differently, owing to minimum fluctuation in production and production systems. Conservation farming strives to achieve high, sustained agricultural production, while conserving soil and preserving the local environment. The recognition of the multiple functionalities (including recreation and conservation) of rangelands and other ecosystems in the HKH has also created opportunities for livelihood diversification of pastoral communities. In China, pastoralists in the Tibetan Plateau have been receiving cash stipends for using rangeland resources more sustainably (Yi and Ismail 2011).

Livestock is one of the fastest growing agricultural subsectors in the HKH. Although large ruminants such as cows, buffalo, and yaks make up the majority of the livestock population in the mountains, yak numbers are decreasing in India, Nepal, and Bhutan, while increasing in China (Wu et al. 2016). While elite and higher caste farmers raise large ruminants, poorer and disadvantaged ethnic groups tend to raise smaller animals such as goats, pigs, and poultry (Gurung et al. 2005). For poorer households, these smaller animals are considered important because they require a low initial investment and minimal or no input, and offer a quick return on investment on a continuous basis (Brithal and Taneja 2006; Ali 2007; Gerber et al. 2010; Khan and Ashfaq 2010).

Most goats are raised on marginal land and are a crucial source of livelihood for landless (and nearly landless) farmers and for households headed by women. Goats are a valuable commodity and are considered a source of foreign exchange (APHCA and ILRI 2006). In India, over 70% of rural households own livestock, and the majority of livestock-owning households keep sheep, goats, pigs, and poultry due to land scarcity, but also for commercial prospects (Ali 2007). According to the Livestock Census (GoP 2006), among all the domestic livestock in Pakistan, goat had the highest growth rate of 3.98% per year. Pakistan ranks third in Asia in terms of small ruminant population, with an overall annual growth rate of 4%, the highest in Asia (Khan and Ashfaq 2010). In India, between 1992 and 2003, the poultry population increased by an annual growth rate of 5.9%, whereas the pig and ruminant populations showed only marginal increases (Gerber et al. 2010). The growth in livestock population was driven by rapidly increasing demand for livestock products, population growth, urbanisation, and increasing incomes in developing countries (Delgado 2005; Thornton 2010).

12.1.2 Livelihood Diversification in the Mountains

Livelihood diversification is a daily reality for the people of the HKH, and it proves key to buffering against shocks and stresses caused by conflict, market failure, food insecurity, environmental hazards, and climate variability—particularly in a context where informal work and precarious, patriarchal land rights are widespread along with a lack of access to financial instruments and products and other forms of social protection. Population growth has led to land fragmentation and to a significant reduction of grazing area per capita. Over the past three decades, households in the HKH have become somewhat less reliant on land and are increasingly composed of a combination of subsistence farmers and labourers (Herbers 1998). These households no longer rely entirely on their land, yet they cannot subsist without it (Fig. 12.2).

At the macro-economic level, a major shift from agriculture to services and, to a lesser extent, industry is taking place in mountainous countries of the HKH such as Afghanistan, Bhutan, and Nepal and the HKH areas of China and India (Table 12.1). Despite the fact that this trend is coupled with increased urbanisation and human mobility (Hoermann et al. 2010), the percentage of people dependent on agriculture for livelihoods is still very high.

At the household level, the susceptibility or sensitivity of livelihood to climate-dependent natural resources is a

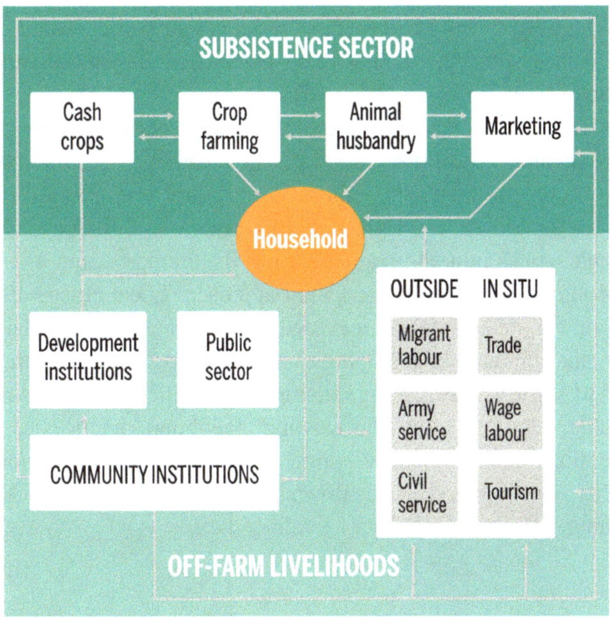

Fig. 12.2 The combined subsistence-labour household

Table 12.1 Structural evolution of HKH mountain economies (in areas where data is available)

Country	Agriculture		Industry		Services	
	2002	2014	2002	2014	2002	2014
Afghanistan	38.5	23.5	23.7	22.3	37.8	54.2
Bhutan	27.4	17.7	38.6	42.9	35.1	39.4
Nepal	40.8	33.7	18.1	15.6	43.3	50.7
HP (India)	23.1	15.9	36.5	41.0	39.5	42.9
Tibet	24.5	10.0	20.2	36.6	55.3	53.5
Yunnan	20.1	15.5	40.4	41.2	39.5	43.3
Sichuan	22.2	12.4	36.7	48.9	41.1	38.7

Source World Development Indicators (World Bank 2017) HP (Himachal Pradesh), India, data from respective Statistical Handbooks. HP first year is 1999–2000. For China: National Statistical Yearbooks, various issues, National Statistics Bureau

concern for a higher number of households than suggested by macro-economic data on GDP. For instance, a significant part of the service industry in the mountain regions is driven by tourism, which is indirectly dependent on the wellbeing of the natural environment (Isaac 2012).

Ongoing social changes, such as fast-paced urbanisation, (predominantly male) outmigration, and population growth have also been key factors inducing change in traditional crop rotations and pastoral practices (Yi et al. 2007, 2008). The growth of small- and medium-sized towns in the HKH is also helping to spur livelihood options, including opportunities for education, waged employment, and trade.

Most HKH communities tend to rely on mixed rural-urban livelihoods rather than pure rural or pure urban characteristics (The Desakota Team 2008). This has also been inspiring young people to migrate (seasonally) to adjoining towns to seek wage employment while maintaining ties with their homes or to undertake international circular migration (see Chap. 15).

Livelihood diversification can vary according to environmental conditions, access to markets, community resource endowments, cultural norms, and resource governance regimes. The HKH's rich diversity—both biophysical and sociocultural—requires that livelihoods cannot be designed with a one-size-fits-all approach. Location-specific policies and strategies are needed to facilitate the engagement of mountain communities and to promote activities that add value while making communities resilient and sustainable. Understanding the existing livelihood practices and options is central to designing future livelihood strategies. No new approach to diversity should undermine the long-term sustainability of livelihoods.

In the next sections of this chapter, we look into livelihood diversification, turning first to on-farm activities and, subsequently, to off-farm work, focusing on labour migration and tourism.

12.1.2.1 On-Farm Activities: The Growing Importance of Medicinal Plants and Future Smart Food

Appropriately designed livelihood diversification can add value to local economies and increase economic opportunities. For example, milk and meat processing, niche-based high-value crops, or sustainable harvest of herbal and medicinal plants can significantly complement livelihood options for the people of the HKH. In high-altitude areas of Bhutan, China, and Nepal the collection of rangeland products, mostly medicinal plants, has been an important source of household income in recent years. From 1999 to 2007, the average annual contribution of income from yarsagumba (*cordyceps sinensis*) to agricultural-sector revenue in the Tibet Autonomous Region (TAR) of China was 10.84% (Wang et al. 2012).

The genetic resources of traditional crops—often referred to as neglected and underutilized food crops (NUFCs) and recently labeled as future smart food (FAO 2017)—are vital for sustainable agriculture (Eyzaguirre et al. 1999; Bhagmal 2007; Padulosi et al. 2011). NUFCs not only play a fundamental role in income generation (Mwangi and Kimathi 2006; Adhikari et al. 2017) and nutrition and food security (Frison et al. 2006; Padulosi et al. 2011; Apeteri 2012; Adhikari et al. 2017), but also hold significant potential for climate-change adaptation (Jarvis et al. 2014).

Some NUFCs are also of great medical importance. For instance, in remote mountain areas, jamun (*syzygium cumini*) is used to treat diabetes. Likewise, in Gilgit-Baltistan in Pakistan, local people have realized the importance of sea buckthorn for nutritional and medicinal purposes and have expanded the cultivation of this crop (Adhikari et al. 2017).

Among many other traditional crops, millets have been cultivated successfully for millennia, indicating resilience to a variety of conditions and some intrinsic potential for continuous production (Apetrei 2012). Finger millet is a particularly rich source of calcium and iron (Singh and Raghuvanshi 2012). By introducing finger millet into the daily diet of mountain people, common problems in the HKH can be addressed, such as the disorders of bone and teeth caused by calcium deficiency and anaemia caused by iron deficiency. This highlights the importance of NUFCs—not only for food security but also for nutrition (see Chap. 9). With respect to seasonality, buckwheat is preferred in the mountains for its fast rate of growth and its weed resistance (Sustainable Agriculture Research and Education n.d.; ICIMOD and GRID Arendal 2014).

NUFCs can make a significant contribution to sustainable nutrition security in the mountains if they are mainstreamed into agriculture, food, and nutrition-security policies and programmes and integrated into local food systems (Adhikari et al. 2017). While not yet a priority for governments in the HKH, the protection and promotion of traditional farming practices and NUFCs (in situ and in vitro) present great opportunities to take advantage of mountain specificities, rather than unsuccessfully seeking to replicate the agricultural intensification approaches designed for the plains.

12.1.2.2 Off-Farm Activities: Labour Migration and Agrarian Transition

Most HKH countries tend to present mixed rural-urban livelihoods rather than those displaying pure rural or pure urban characteristics. A household may well be physically located in a rural area, but it's likely that it relies on an assortment of livelihoods with urban, rural, and, often, transnational components. This fusion, known as desakota, is widespread in the HKH (The Desakota Team 2008). Young people migrate seasonally to adjoining towns seeking wage employment while maintaining ties with their homes. Labour migration is integral to the portfolio of livelihoods that households rely on to deal with everyday struggles and to shield against various economic, environmental, and social risks (Hoermann et al. 2010; Karki 2012; Gioli et al. 2014a; Banerjee et al. 2016).

Circular labour migration at national, regional, and international scales is particularly widely practiced, mostly by young males who migrate for temporary periods (from months to decades) and send back remittances to their countries of origin (see Chap. 15). Given the lack of access to formal risk-mitigation mechanisms in the region, remittances can be described as a "household sponsored insurance system" (Yang and Choi 2007) and as a substitute for social security (Schrieder and Knerr 2000). Moreover, remittance inflows are untouched by perturbations at the local level and thereby constitute a vital channel of income during and after environmental and political crises (Monsutti 2008; Le De et al. 2013).

Labour migration stimulates flow and exchange of not only financial resources but also ideas and capabilities (social remittances), which influence and often challenge traditional structures at home (IOM 2005). Migration generates financial and human capital, which, if leveraged for development, is a proven driver of poverty reduction. Social remittances can play a particularly important role in development (Hoermann et al. 2010). Three countries in the HKH receive more than 10% of their GDP from remittances: Bangladesh (11.6%), Afghanistan (16.3%), and Nepal (28.5%) (IFAD 2013). However, remittance economies and overreliance on labour migration are not a sustainable strategy for development. Labour migration is also an extremely costly strategy, which comes at high financial costs (including permits and travel) and high human costs, especially when people migrate to areas where the rights of workers are not guaranteed, such as the Gulf States.

With the exception of China, labour migration is a highly gendered phenomenon in the HKH, predominantly undertaken by young males. However, women's labour migration is on the rise, with women constituting 13% of the migrant stock in Nepal (CBS 2014; Gioli et al. 2017). While men are absent, women and children take on household responsibilities, including both the productive and domestic roles in tending to livestock and agriculture fields, in addition to taking on work outside the home as labourers (Synnott 2012). The general increase in the workloads of women (resulting from migration and other factors) impinges upon their caregiving roles (Gioli et al. 2014a), exposing women and children to greater threat from water-borne disease (Halvorson 2002). In recounting gendered experiences of change in Baltistan, Azhar-Hewitt (2011) similarly observes the transformation of women's workloads, noting that access to resources and liberties afforded to women in traditional societies are sometimes being denied based upon notions of religiosity and appropriate gender roles, imported in the course of labour migration (see Chap. 14).

Stories of changing gender roles echo across the HKH. In Afghanistan, irrigation water management too becomes a domain with increasing female participation and the intersection of poverty and gender render some women's livelihoods exceptionally vulnerable on account of inability to pay for the petrol for the water pump or community sanction disallowing a woman her share of water (McCarthy and Mustafa 2014).

In the hills of Nepal, most farming households were found willing to neglect farming altogether if alternative sources of income were available to them (Maharjan et al. 2013). Additionally, farmers preferred livestock to crops as a supplement to household income. The share of women in the agriculture sector in Nepal has been 12.6% higher than that of men, and the participation of women in the agricultural labour force increased from 36% in 1981 to 45% in 1991 to 48% in 2001 (FAO 2010).

The loss of agricultural workforce has also increased the phenomenon of land abandonment in the HKH. Farmers in southeastern Nepal are moving toward the non-farm sector due to low returns on investment (Adhikari and Hobley 2012). This has further intensified migration leading to de-intensification of agricultural lands, a trend that is beginning to emerge in other parts of the HKH as well. In Uttarakhand, India, 47% of respondents identified decreasing agricultural productivity as a major reason for migration (Hoermann et al. 2010; Tiwari and Joshi 2016; Mamgain 2004). With men outmigrating in significant numbers, the bulk of the workload and responsibility falls upon women. Much discussion has taken place on the transformational potential of these shifting roles (see Chap. 14). For instance, a study in northern Pakistan showed that positive gender transformative processes are more likely to be intergenerational and driven by increased access to education for girls (Gioli et al. 2014a).

12.1.2.3 Tourism

Within the wider context of searching for livelihood diversification options, some mountain communities have found opportunities in the recreation and tourism industries. The appeal of recreation in the mountains has long been recognized institutionally, for instance, through the designation of national parks. Tourism has played a significant role in transforming mountain communities around the world, by diversifying local economies, generating employment opportunities, and bringing development to these societies. As the sector builds on natural and cultural heritage values, it can bring the additional benefit of positively impacting conservation.

Mountain specificities that are generally considered constraints to development—including poor accessibility, fragility, and marginality—can be transformed into economic opportunities for tourism (Jodha 1992; Sharma 2000; Nepal and Chipeniuk 2005; Kruk 2010). Tourism intersects with and stimulates a wide range of other sectors in the supply chain, especially agriculture, infrastructure, communications, construction, and handicrafts. Tourism also stimulates a new market for local produce, especially high-value crops. As a complementary livelihood option, the development of tourism will not only generate socioeconomic benefits for the region, but may also address wider social and sociocultural concerns (Kruk 2010). Thus, the tourism industry is being strongly pursued and supported by the HKH governments through their National Development Strategies. For example, in 2016 the Government of Nepal launched the National Tourism Strategy 2016–2025, which envisages a fivefold increase in tourist arrivals by 2025 and includes conservation of cultural heritage and a zero-carbon target as important development goals.

Tourism is not a new phenomenon in the HKH (Karki et al. 2012). Several HKH countries have experienced and benefitted from tourism since the 1970s. Though some countries (such as Bhutan and China) have restricted access to tourists for various political reasons, and others (including Afghanistan, Pakistan, Bangladesh, and Nepal) have seen tourism diminish for periods of time due to political unrest and conflict, the region has generally experienced a strong increase in tourism over the past years. In the HKH countries,[2] tourism has become the largest service sector, generating much-needed foreign exchange earnings—currently contributing 2–4% of GDP—and generating 2.8–9.6% of total (direct and indirect) employment. The tourism industry also involves 0.9–8.3% of total investment in these countries. Globally, tourism is predicted to continue increasing steadily over the next 10 years, with major new markets on the rise, particularly in Asia (including locations in India and China). While international tourism is sizable and steadily increasing with economic growth in Asia, intraregional tourism markets represent the greatest opportunities for growth. Due to the growing middle class resulting from increasing wealth and better access to credit, regional markets are showing exponential growth. Political liberalization and border-crossing agreements between neighbouring countries (as with Bhutan and India) are also making regional travel easier.

The revival and strengthening of cultural heritage is considered beneficial to tourism (Ganesh and Madhavi 2007; Isaac 2012). In the HKH, old hospitality traditions have been the basis for lodging and other tourist services (such as the concept of teahouses in Nepal), inspiring a special Himalayan mountain-oriented accommodation sector (Odell and Lama 1998).

Tapping the growth of the tourism market within a framework of sustainable development could be an effective mechanism for both enhancing local livelihoods and conserving heritage. However, many challenges remain in realizing such a vision, given the complex cultural, topographical, and political HKH landscape. Therefore, tourism development in the HKH should be approached in a holistic and comprehensive manner to ensure long-term sustainable growth and the resilience of mountain people. Ecotourism, in particular, has the potential to drive sustainable tourism development while providing opportunities for the development of disadvantaged, marginalized, and rural areas, thereby alleviating poverty and stimulating economic development and social wellbeing (Isaac 2012).

12.1.3 Social Protection Programmes

Communal property remains an important safety net for mountain communities—particularly for the poorest among them. As even rural communities have become integrated with global economies, the rural/urban distinction has become functionally tenuous at best. As communal property

[2]Data based on available national statistics from 2010–12, excluding Afghanistan as tourism in this country is currently minimal (particularly domestic tourism) and limited information is available.

becomes less significant to the livelihoods of more fortunate community members, the strict rules governing use of these assets are eroding (Mustafa and Qazi 2007). The result is that communal land, forest, and rangeland resources are being encroached upon by private interests—to the exclusion of women, children, and poorer members of the community who, lacking food security and social protection, are rendered further vulnerable to various environmental hazards.

The breakdown of traditional risk-sharing mechanisms, along with population growth, urbanisation, and new risks like market volatility and climate change have triggered several formal social-protection initiatives, with social insurance, social assistance, and labour programmes becoming increasingly visible in the HKH's public-policy domain (Wagle 2012). Although these programmes have provided immediate relief to families in need, they are hardly effective in developing the assets and market access necessary for long-term viability of smallholder farmers. There is a strong need to identify innovative, market-based, financial interventions that develop and improve household assets, agricultural entrepreneurial skills, and the resilience of vulnerable communities facing climatic and non-climatic stressors across the HKH.

Social capital is developed by community building achieved through relationships of trust, reciprocity, and exchange, creating links among community members while increasing their access to institutions and government (Ferguson 2012). There are successful examples of development programmes that have leveraged community participation for sustainable use of resources and income-generation activities. For instance, focusing on targeted social assistance, Heifer International Nepal[3] has developed the concept of collective enterprise—values-based holistic community development—wherein group members are jointly engaged in sharing responsibility, effort, risk, and profit.

It is important to recognize the significance of targeted social assistance to the poorest section of society. Nepal's community forest user groups (CFUGs) have been mobilized for poverty-alleviation activities and have pioneered the recognition of community forestry as a pro-poor and pro-women programme (Nightingale 2011). There is increasing global recognition of cooperatives as self-help organizations capable of improving livelihoods and wellbeing (Ferguson 2012). Government, cooperatives, and the private sector have been identified as the three major stakeholders in national economic building in Nepal, where the majority of cooperative members are from poor and middle-class households (Bharadwaj 2012).

12.1.4 From Livelihood Diversification to Poverty and Vulnerability Reduction

Poverty, small holding size, and food insecurity are critical challenges in the HKH, and addressing them calls for a holistic approach. Diversification of farms with high-value crops—while adequately maintaining soil, forests, and other natural resources—is a crucial step in improving the livelihoods of mountain people. In recent years, cash crops such as fruits and vegetables have become prominent in mountain development as a means of improving livelihood options. Livestock, especially small ruminants, have been emerging as a source of cash income for a large number of HKH farmers. At the same time, urbanisation and demographic changes have introduced new geographic exposure to risk, while conversely providing new opportunities to mitigate risk through enhanced capacity building and skill trainings (Hoermann et al. 2010).

The rigid structure of global markets has the potential to create circumstances of nested vulnerability (Eakin and Bojorquez-Tapia 2008), whereby shocks at the global scale cascade down to affect household and community livelihoods at the local level. There is a need for greater effort in identifying the interconnectedness of global and local processes in order to ensure that strategies for enhancing livelihood options in the HKH also increase the resilience of livelihood systems.

12.2 Poverty in the HKH: How to Measure and Tackle It

There are two distinct approaches to measuring human poverty. The first is economic poverty, which is understood as the inability to participate in society due to a lack of resources (Townsend 1979) and is usually measured by household income or consumption. Realizing the limits of such a narrow approach, more holistic ways to measure poverty have emerged over the years, such as multidimensional poverty measures, which is understood to be the failure of basic capabilities to reach certain minimally acceptable levels (Sen 1992) or as a denial of choices and opportunities for living a tolerable life (UNDP 1997).

This chapter considers the multidimensional approach to be more relevant to policy, particularly in the mountain context, as it can more comprehensively capture the complex nature of poverty, thereby better indicating areas for intervention. However, due to data limitations, unidimensional poverty measures based on consumption are used in the first part of this discussion to compare poverty levels across countries and to analyse poverty trends over time. This is followed by a discussion of multidimensional poverty.

[3]http://heifernepal.org/sites/default/files/AR13_0.pdf.

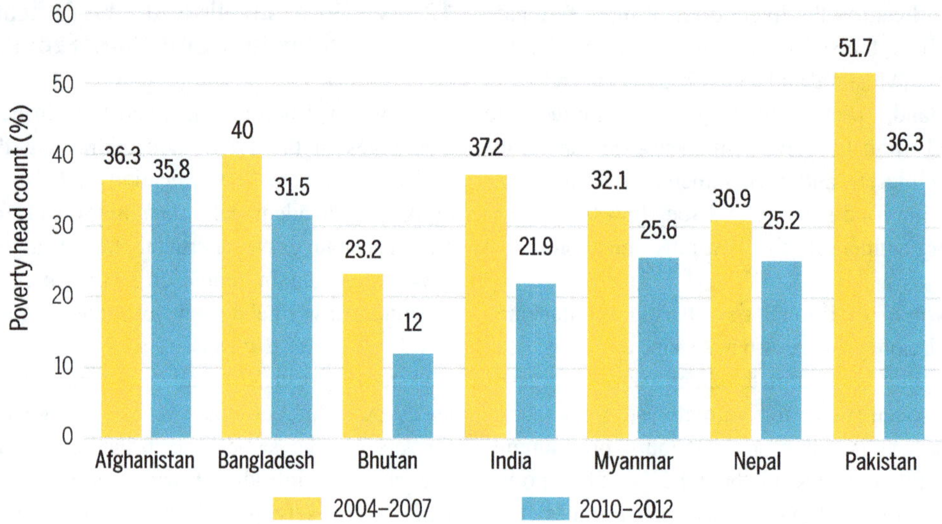

Fig. 12.3 Trend in poverty reduction during the first decade of the 21st century, national standards (World Bank 2017; http://povertydata.worldbank.org/). *Notes* 1. The specific years considered for each country are as follows: Afghanistan: 2007, 2011; Bangladesh: 2005, 2010; Bhutan: 2007, 2012; India: 2004, 2011; Myanmar: 2005, 2010; Nepal: 2010; Pakistan: 2004, 2011. 2. The figures for Myanmar have been taken from a UNDP study conducted for Myanmar using IHLCA (Integrated Household Living Condition Assessment) Survey. (Ministry of National Planning and Economic Development, UNICEF, UNDP, and SIDA 2011: Poverty Profile 2009–2010)

12.2.1 Mountain Poverty: Specificities and Challenges

The characteristics of mountain areas that significantly shape development and human life are referred to as mountain specificities. Among these specificities, inaccessibility, fragility, and marginality[4] are recognized as constraints on development while diversity, biological niches, and adaptation mechanisms are viewed as development opportunities (Jodha 1997). These specificities, combined with the isolated nature of mountain economies, may lead to manifestations of poverty in the mountains that differ from those in the plains. Conventional economic measures of poverty, usually based on income or consumption, fail to capture the complexity of mountain poverty and its noneconomic correlates in these areas. Mountain specificities may hinder the conversion of resources such as income into actual wellbeing outcomes, including adequate nutrition and food security, education, and health (Sen 1999). Thereby, a distinct frame of analysis is needed to understand mountain poverty.

The lack of recognition and understanding of the implications of mountain specificities often leads to misconceptions about the socioeconomic conditions in mountain areas and to misdiagnosis of the sources of poverty (Papola 2002). As a result, the strategies and interventions designed for development in mountain areas tend to be unsuitable and, thereby, ineffective.

In the following discussion, we compare economic poverty levels and trends in HKH countries, contrast poverty levels in the mountain areas and the lowlands, and examine poverty incidence among indigenous peoples and the majority populations in China and India.

12.2.2 Economic Poverty: Levels and Trends in the HKH

Typically, countries have defined national standards for measuring poverty, most commonly a head-count ratio indicating the proportion of the population below a certain threshold of income or consumption. Such measures are commonly defined as a national poverty line. In addition, international agencies estimate international poverty lines, such as USD 1.25 or 1.90 per capita per day in purchasing power parity terms, leading to comparability of head count ratios across countries.

Using the international poverty line of USD 1.25 per capita per day,[5] Fig. 12.3 shows income poverty levels in the HKH countries for two time periods—the mid-2000s and 2010–12. Two observations can be made based on these data. First, the incidence of economic poverty at the national

[4]The mountain areas are marginal and share the attributes of marginal entities due to factors like remoteness and physical isolation, low-productivity resources, or man-made handicaps that prevent mountain areas' participation in 'mainstream' patterns of activities (Jodha 1990a, b).

[5]The international income poverty line has now been revised to USD 1.90 per capita per day. However, due to data unavailability, we are using the old poverty line (USD 1.25 per capita per day).

Table 12.2 Poverty profile of HKH countries (Hunzai et al. 2011)

Country and year of data	Total population (millions)		Population below poverty line (millions)		Population below poverty line (%)	
	Countrywide[a]	HKH areas[b]	Countrywide[c]	HKH areas[c]	Countrywide	HKH areas
Afghanistan 2010	24.5	15.1	8.0	6.3	33.0	42.0
Bangladesh[d] 2009	162.0	1.33	59.9	0.6	37.0	46.0
Bhutan[e] 2009	0.69	0.69	0.19	0.19	23.0	23.0
China 2009	1,331	29.4	220	NA	16.6	NA
India 2009	1,155	72.3	415	24.0	36.0	34.0
Myanmar 2009	49.8	11.0	15.9	NA	32.0	NA
Nepal 2009	29.3	11.8	9.0	4.7	31.0	40.0
Pakistan 2009	169.7	39.3	42.4	12.5	25.0	32.0
HKH total/average	2,921	181	771	61[f]	26.0	31.0[g]

NA = not available
[a]Total population from World Bank 2009 except for Afghanistan, which is from the Central Statistics Organization of Afghanistan
[b]For updated HKH population see Chap. 1, Box 1.1. This Table has to use the 2009 population figures, as updated information on population below the poverty line in the HKH is not available after 2009
[c]Figures for population below poverty line from ICIMOD analysis based on NLSS 2003/04, BLSS 2007, NSS 2003, PSLM 2005/06, HIES 2005/06, and NRVA 2007/08 except for China and Myanmar, which are based on secondary sources
[d]The population of Bangladesh (162 million in 2007) is overestimated, as according to the Bangladesh Bureau of Statistics, the population reached 149.77 million in 2011
[e]Bhutan lies entirely within the HKH, thus the countrywide and the HKH area figures are the same
[f]Accurate figures were not available for China and Myanmar; figures were estimated using the same value for the proportion of population below the poverty line as for each country overall, this is likely to be underestimated
[g]Average of those known and excluding China and Myanmar

level varies widely among the HKH countries. Second, poverty levels are declining over time in all countries, with Pakistan, India, and Bhutan experiencing more rapid decline than the other countries.

12.2.3 Mountains Versus Plains: Implications for Poverty

Table 12.2 summarizes the poverty profile of HKH countries and their mountain areas, with figures showing population and the percentage of population identified as living below the national poverty line (Hunzai et al. 2011). The data presented in the table are based on broad assumptions, as the methodologies used to estimate poverty vary by country, and data on population and the percentage of population living below the poverty line were extracted from various sources referring to different years and, thus, not strictly comparable. Nevertheless, they are sufficiently similar to allow for a broad overview.

In 2009, approximately 2.9 billion people were living in the eight countries of the HKH, of which an estimated 771 million were living below the national poverty line; about 200 million people were living within HKH regions of these countries, of which some 61 million were classified as poor. On average, 31% of the HKH population (excluding China and Myanmar) was below the poverty line, compared with 26% of the total population of these countries.

Among the HKH countries, Bangladesh had the highest incidence of poverty in the mountain areas (46%), followed by Afghanistan (42%), Nepal (40%), India (34%), Pakistan (32%), and Bhutan (23%). In all countries except India, poverty rates were higher in the HKH areas than in the country as a whole. In absolute numbers, the Indian Himalayan region had the highest population of poor people in the mountain areas (24 million), followed by Pakistan (12.5 million), Afghanistan (11.3 million), Nepal (4.7 million), Bangladesh (0.6 million), and Bhutan (0.19 million).

In general, there is a dearth of data on poverty levels and trends in the mountain areas of HKH countries. Gender-disaggregated data are particularly lacking. In the following paragraphs, we compare the incidence of poverty in mountain and non-mountain areas in China, Bangladesh, India, and Nepal with figures based on secondary data.

About 19% of China's population lives in mountain areas. The incidence of income poverty (measured by the World Bank poverty line of USD 1 per day per capita) in these areas is about twice as high as in non-mountain areas (World Bank 2009) (Table 12.3). Similarly, consumption poverty is three times as high in mountain areas as in non-mountain areas. In Bangladesh, more than 61% of households are very poor (those without sufficient monthly income to meet food requirements) in the hilly region of Chittagong Hill Tracts compared with 47% of the households in the valley (ADB 2012) (Table 12.4).

Table 12.3 Poverty incidence in mountainous and non-mountainous regions of China in 2003 (World Bank 2009)

Region	Share of national population (%)	World Bank poverty line			
		Income		Consumption	
		% who are poor	Share of national population of poor (%)	% who are poor	Share of national population of poor (%)
Mountainous	18.5	18.8	50.2	27.9	39.4
Non-mountainous	54.0	6.3	49.0	14.5	59.9

Table 12.4 Poverty incidence by location in Chittagong Hill Tracts, Bangladesh (ADB 2012)

Economic status	Hill		Valley		Total	
	Households	%	Households	%	Households	%
Very poor	1,619	61.3	1,595	47.0	3,214	53.2
Poor	777	29.4	1,476	43.4	2,253	37.3
Better-off	245	9.3	328	9.6	573	9.5
Total	2,641	100	3,399	100	6,040	100

Table 12.5 Poverty incidence in hill states of India for several years between 1973 and 2012 (Saxena 2016)

State	Percentage of population below poverty line			
	1973–74	1993–94	2004–05	2011–12
Arunachal Pradesh	51.93	39.35	31.10	34.67
Assam	51.21	40.86	34.40	31.98
Himachal Pradesh	26.38	28.44	22.90	8.06
Jammu and Kashmir	40.83	25.17	13.20	10.35
Manipur	49.96	33.78	38.00	36.89
Meghalaya	50.2	37.92	16.10	11.87
Mizoram	50.3	25.66	15.30	20.40
Nagaland	50.8	37.92	9.00	18.88
Sikkim	50.8	41.43	31.10	8.19
Tripura	51.0	39.01	40.60	14.05
Uttarakhand	NA	NA	32.70	11.26
India	54.88	35.97	37.20	21.8

NA = not available

Table 12.6 Poverty incidence in Nepal by region 2010–11 (CBS 2011)

Region	Poverty rate (%)	Poverty gap (%)
Mountains	42.27	10.14
Hills	24.32	5.69
Terai (plains)	23.44	4.52
Nepal	25.16	5.13

In India, the picture is different. In 2011–12, the percentage of mountain people living below the poverty line was higher than the national average in only three of the country's 11 hill states (Table 12.5). Although several hill states in India have poverty rates lower than the national average, hill districts generally face significant development deficits compared with the districts in the plains.

A recent study (Dasgupta et al. 2014) attempted to measure the disparity among these 11 hill states and the six other states primarily classified as plains. Using data for the period 2010–11, this study measured five indicator categories: education, health, economics, infrastructure, and basic amenities. Results indicate that, overall, the states with a greater proportion of hilly terrain (more than 75% of the state's total area) fare worse than states with a greater proportion lying within the plains. The highest adverse rankings were observed among the hill states, with the exception of Himachal Pradesh, which scored well.

In India, the hill states also suffer from several challenges that are specific to these states. For example, Nagaland, Arunachal Pradesh, Mizoram, Meghalaya, hill areas of Manipur, and some tribal tracts of Assam have no system of written land records or of land revenue payments. The absence of land records has increased tenure insecurity for the poor due to the growing concentration of land ownership in the hands of a few, resulting in rising rates of tenancy and landlessness, as well as declining output from shifting cultivation. The open-access structural conditions of land cultivation and the fact that the elite are able to corner most government funds have intensified poverty and inequality in these states. The absence of clear property rights has been recognized as a significant cause of degradation of natural resources in hill states.

In Nepal, poverty incidence in mountain areas (42%) is significantly higher than the national average (25%), the midhills (24%), and the terai (plains) (23%) (see Table 12.6). The poverty gap index is also higher in

Table 12.7 Poverty incidence in China by ethnic minority categories in 2003 (World Bank 2009)

Region	Share of population (%)	World Bank poverty line			
		Income		Consumption	
		% who are poor	Share of poor (%)	% who are poor	Share of poor (%)
Ethnic minority	7.7	24.1	26.9	36.6	21.6
Non-ethnic minority	64.8	7.7	72.3	15.7	77.7

mountain areas than in the plains and midhills. The poverty gap index measures the severity of poverty by considering how far, on average, the poor are from poverty line. This figure can be interpreted as the average income shortfall from the poverty line.

12.2.4 Poverty Rates Among Indigenous Peoples in the HKH

The HKH is home to millions of indigenous peoples who are among the region's poorest and are politically and socially marginalised. They are variously known in different countries as ethnic minorities, minority populations, and tribal groups. The terms refer to social groups with a cultural identity distinct from the dominant groups, which makes them vulnerable to disadvantage in the development process (IFAD 2002). Table 12.7 presents poverty rates among indigenous peoples and the overall national populations in China.

In China, ethnic minorities are overwhelmingly concentrated in mountainous areas and are significantly poorer than the Han majority—their consumption poverty level is more than twice as high and their income poverty rate is three times as high as that of Han communities (World Bank 2009). In rural areas, ethnic minorities have less access to wage employment and earn less when they engage in wage employment (Hannum and Wang 2012). Enrolment rates among school-aged children are lower among minority populations than among Han populations. Also, minority areas have less developed healthcare infrastructure and less access to safety nets such as unemployment and pension insurance. Therefore, it is important that poverty reduction efforts complement programmes to improve physical and social infrastructure in the remote areas inhabited by ethnic minorities.

12.2.5 Existing Multidimensional Poverty Measures

Stemming from the pioneering work of Amartya Sen, in the 1980s an increasing number of scholars disputed the idea that command over monetary resources could provide an adequate informational basis to evaluate human wellbeing, arguing that a broader lens was needed in assessing poverty. The UNDP's human development index (HDI) (UNDP 1990) and human poverty index (HPI) (Anand and Sen 1998; UNDP 1998) were the first attempts to formalize the inclusion of non-income components in respectively defining global development and poverty through the inclusion of education and health in the assessment. Subsequently, the multidimensional poverty index (MPI) (Alkire and Santos 2010; UNDP 2010) was developed to assess multiple deprivations in education, health, and the standard of living.

The MPI aggregates household-level data on 10 indicators: years of schooling, child enrolment, child mortality, nutrition, access to and type of electricity, drinking water, sanitation, type of flooring, cooking fuel, and assets. The MPI is derived as the product of the number of people living in multidimensional poverty and the average intensity of deprivation as measured by indicators (Alkire and Santos 2010).

Hence, the MPI presents the percentage of population living in multidimensional poverty adjusted by the intensity of the deprivation. The detailed methodology and global estimates are available elsewhere (Alkire and Foster 2011; Alkire et al. 2015). The use of data from household surveys allows for disaggregation at the subnational level to highlight poverty patterns in terms of geography or household characteristics. Furthermore, the MPI can be deconstructed by dimension, allowing for identification of the main aspect of deprivation in a given population.

Table 12.8 presents the multidimensional poverty head count ratio, the average intensity of poverty, and the multidimensional poverty index in eight HKH countries. Data were drawn from the 2015 Human Development Report (UNDP 2015), except for Myanmar, for which figures have been estimated using the Poverty and Vulnerability Assessment Survey 2013 (Gerlitz et al. 2014). The estimates are compared with the human development index and rank for 2013.

Table 12.8 illustrates three key contributors to multidimensional poverty. As per the case of income poverty, the estimates are not strictly comparable due to dissimilarity of indicators used, varying data sources, geographical coverage within countries, and varying time period. However, all the estimates are based on the Alkire and Foster methodology and provide a broader prospective of multidimensional poverty.

Table 12.8 Multidimensional poverty indices in HIMAP countries

Country	Year and data source	Multidimensional poverty (%)	Intensity of deprivation	Multidimensional poverty index	Contribution of deprivation to multidimensional poverty			2014 HDI value (rank)
					Education (%)	Health (%)	Living standard (%)	
Afghanistan	2010–11, MICS	58.8	49.9	0.353	45.6	19.2	35.2	0.465 (171)
Bangladesh	2011, DHS	49.5	47.8	0.237	28.4	26.6	44.9	0.570 (142)
Bhutan	2010, MICS	29.4	43.5	0.128	33.1	24.8	42.1	0.605 (132)
China	2012, NS	5.2	43.3	0.023	30.0	36.6	33.4	0.727 (90)
India	2005–06, DHS	55.3	51.1	0.282	22.7	32.5	44.8	0.609 (130)
Nepal	2011, DHS	41.4	47.4	0.197	27.3	28.2	44.5	0.549 (145)
Pakistan	2012, DHS	45.6	52.0	0.237	36.2	32.3	31.6	0.538 (147)
Myanmar (Shan and Chin states only)	2013, PVA	53.5	39.3	0.21	29.0	19.5	42.8	0.536[a] (148)

Source Human Development Report 2015; PVA: Poverty and Vulnerability Assessment Survey 2013
Notes MICS: multiple indicator cluster survey; DHS: Demographic and Health Survey; NS: national surveys (only nine provinces)
[a]For the entire country

The extent of multidimensional poverty varied from 5% in China to 59% in Afghanistan. In all countries except China and Bhutan, the multidimensional poverty rate was well above 40%. Though the extent of multidimensional poverty varied widely across countries, the average intensity of poverty was fairly consistent, ranging from 43% in both Bhutan and China to 52% in Pakistan. The multidimensional poverty index was lowest in China (0.02) followed by Bhutan (0.13), and it was highest in Afghanistan (0.35) followed by India (0.28). The comparison of MPI with HDI revealed that the ranking of HDI and MPI are not necessarily similar. Though India ranked second in HDI among eight countries, it ranked seventh in MPI.

Given its global focus, the MPI inevitably loses context-specificity. This issue is particularly salient in the context of mountain areas due to the unique nature of mountain poverty and its variation across countries in the HKH. Country-level estimates may mask significant inequalities between mountainous and non-mountainous regions—as well as inequalities among mountainous areas. While the availability of different, country-specific estimates of multidimensional poverty are increasingly available for developing countries, including some within the HKH (Alkire and Seth 2013 for India; Roche and Santos 2013 for Bangladesh; Santos 2013 for Bhutan; Trani and Bakshhi 2013 for Afghanistan; Mitra 2014 for Nepal; OPHI 2016 for Pakistan), thus far only one study (of Nepal) has addressed the complexity of capturing multidimensional poverty in mountainous regions (Gerlitz et al. 2015a). Box 12.2 presents an example of disparities within a given country, in the context of two provinces in Myanmar. A further gap in the evidence base relates to the decomposition of deprivations within the household based on, for instance, gender and age.

Box 12.2 Multidimensional poverty in Shan and Chin States of Myanmar

In Myanmar, the states of Shan and Chin are of particular interest, as they are largely mountainous and poorly accessible. Shan accounts for 11% of Myanmar's population, while Chin accounts for about 1%. In 2011, the national poverty rate in Myanmar was 26% while the poverty rate was 33% in Shan and 73% in Chin (UNDP 2011).

Using ICIMOD's Poverty and Vulnerability Assessment survey data, Mohanty et al. (2018), estimated the extent of multidimensional poverty in Shan and Chin, measuring education, health, standard of living, energy, water, sanitation, and access to services through 12 indicators. The aggregation of these measurements was based on MPI methodology. Figure 12.4 presents the comparison of consumption poverty and multidimensional poverty in Shan and Chin. The multidimensional head-count ratio and index value were, respectively, 49% and 0.21 in Shan and 75% and 0.32 in Chin. The composition of poverty differed in the two provinces, with education and health contributing one-third to multidimensional poverty in Shan and Chin, respectively. The estimated populations of multidimensional poor and consumption poor were consistent in the case of Chin, but in Shan the multidimensional estimate indicated a much larger poor population than the consumption-based measure, with 28% of the state population considered poor by multidimensional assessment though not considered poor in terms of consumption alone.

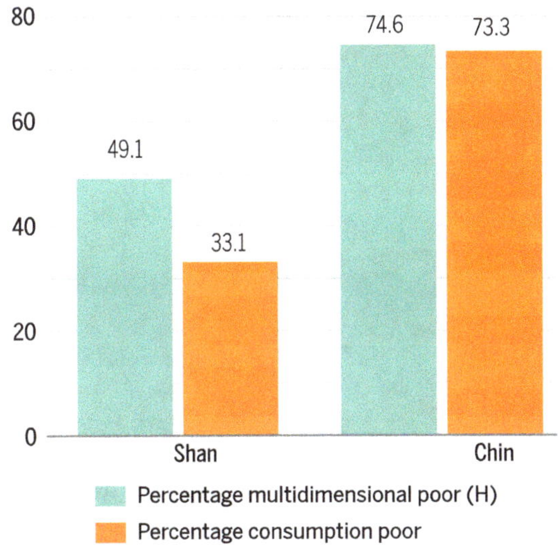

Fig. 12.4 Percentage of multidimensional poor and consumption poor in Shan and Chin in 2013 (*Source* based on data presented in Table 3 in Mohanty et al. 2018)

12.2.6 Determinants of Poverty in the HKH

While poverty is generally associated with social status (class, caste, ethnicity), education, employment status, and occupation, human wellbeing in mountain areas is also inextricably linked to the so-called mountain specificities, conditions including inaccessibility, fragility, marginality, diversity, biological niches, and human adaptation mechanisms (Jodha 1992; Hunzai et al. 2011). Factors that predict poverty and its persistence can be broadly summarized by the following categories (Hunzai et al. 2011; Gerlitz et al. 2012; Gerlitz et al. 2014; See Box 12.3):

- Remoteness and poor accessibility to basic facilities and markets
- Access to and dependence on natural resources
- Demographic factors
- Cultural and social factors
- Lack of empowerment.

In the HKH, mountains provide ecosystem services on which many communities are directly or indirectly dependent. Notably, the most obvious dependence is seen among the rural poor who are often dependent on natural resources to meet their requirements for food, fodder, shelter, and energy. The rural mountain poor include pastoralists, herders, small farmers, and forest dwellers. The cultural and recreational services provided by the surrounding environment also constitute a fundamental aspect of mountain life and culture. Indirect regulating services ensuring hydrological services, soil fertility, protection from erosion, microclimatic stabilization, pollination, and the conservation of biodiversity are also important—not only for those in the immediate area but also for those who are located in relatively distant or downstream areas.

Box 12.3 The multidimensional poverty measure for the Hindu Kush Himalaya: the case of Nepal

The Multidimensional Poverty Measure for the Hindu Kush Himalaya (MPM-HKH) was specifically designed to identify and describe multidimensional poverty in a region that is predominantly rural and mountainous—and covers several of the world's least developed countries (Gerlitz et al. 2015a). It incorporates 16 indicators that capture deprivations in seven dimensions identified through literature review (Gerlitz et al. 2014), data analysis (Hunzai et al. 2011; Gerlitz et al. 2012), and consultation with experts. These dimensions are: education, health, material wellbeing, energy, water and sanitation, social capital, and access to services. They are aggregated using MPI methodology.

The MPM was applied to 23 districts of Nepal, which showed wide variation in the magnitude and composition of multidimensional poverty. The poverty status in the mountains and hills was highly heterogeneous, showing some of the poorest as well as the best-off districts (see Fig. 12.5), which might be explained by the fact that some mountain and hill areas are very remote, while others are well connected or hotspots of tourism. The findings also revealed common patterns in the profile of mountain poverty, such as the frequency with which lack of access to services is the dominant dimension of poverty in mountainous areas (see Fig. 12.6).

The study illustrates the importance of location-specific data in the development of effective poverty reduction strategies. Blanket, country-level approaches are likely to miss crucial local manifestations of poverty and, thus, are likely to be less effective.

According to poverty reports from the HKH countries, households that mainly depend on agriculture face higher risks of falling below the poverty line than households with additional income sources. Due to insufficient agricultural land, changes in agricultural productivity, small and fragmented landholdings, lack of irrigation, lack of mechanization, barriers to market participation, and falling commodity prices, the mountain areas constitute a challenging environment for agriculture (Tulachan 2001; Goodall 2004; Ediger and Huafang 2006). Several studies have mapped the nature and extent of dependence on natural resources among the region's most deprived populations (Shah 2009).

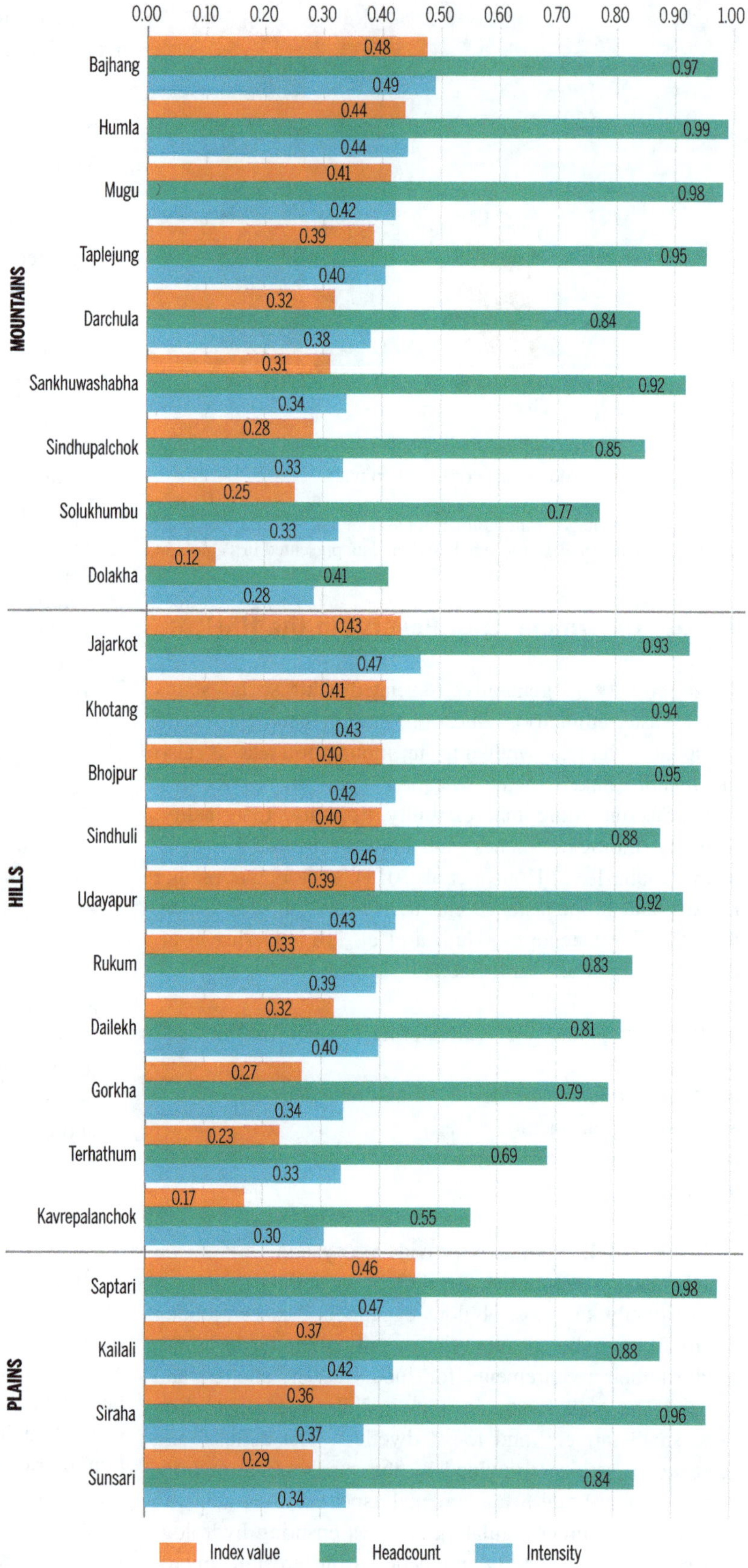

Fig. 12.5 MPM-HKH index value, headcount, and intensity by district (Gerlitz et al. 2015a, p. 283)

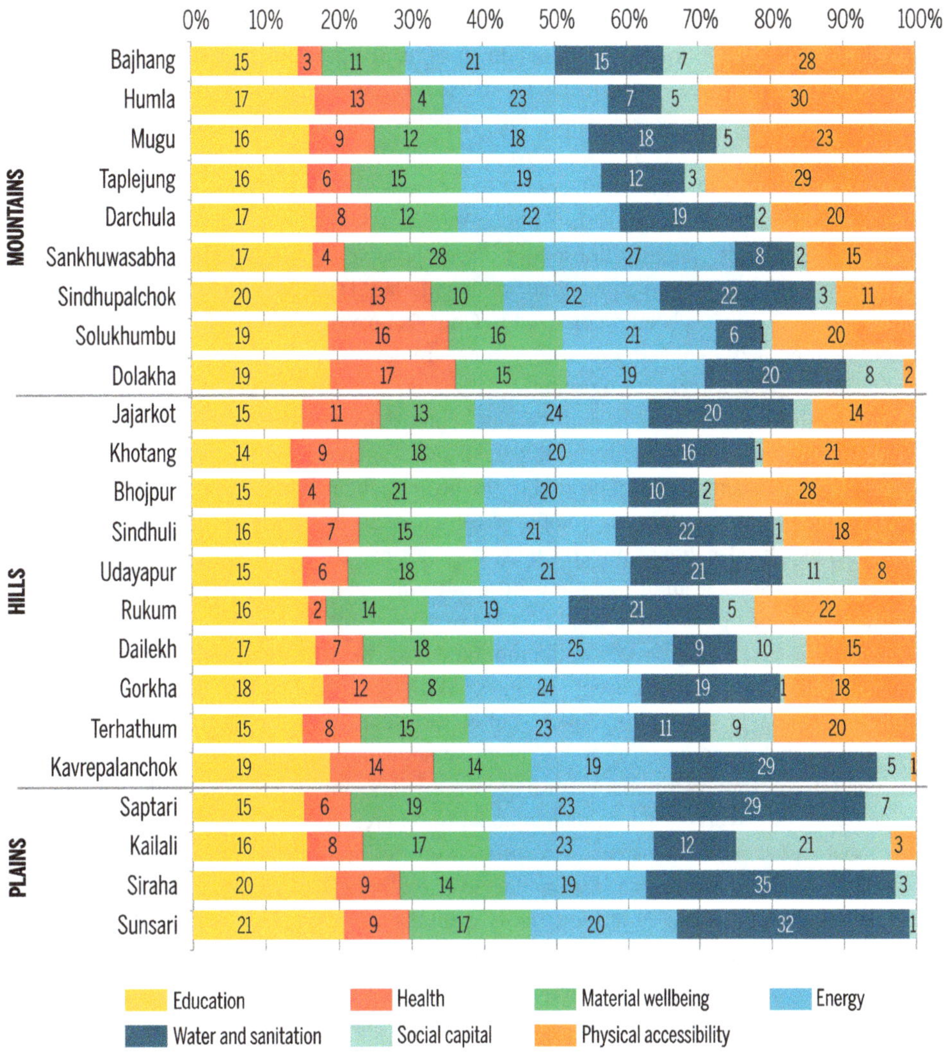

Fig. 12.6 MPM-HKH relative contribution of poverty dimensions by district (Gerlitz et al. 2015a, p. 284)

However, evidence from the region does not conclusively establish how such dependence might be disrupted by human interventions or changes in natural circumstances. Equally debated is the issue of whether the poor's sustained dependence on natural resources works against conservation of natural resources in South Asia, particularly in light of climate change or natural hazards.

Household demographic characteristics, such as adverse dependency ratios (particularly in the presence of migration) (Kaspar 2005; Hoermann et al. 2010; Banerjee et al. 2011; Gerlitz et al. 2012), and cultural norms that limit access by women, minority groups, and lower castes to resources and basic services have been identified as key predictors of poverty and deprivations (de Haan 1999; Bird et al. 2002). A potential, although understudied, determining factor of poverty is the lack of empowerment within remote mountain communities and the district government's lack of accountability to them.

Infrastructure has significant effects on economic well-being (Ali and Pernia 2003). In regions such as the HKH, the availability of public infrastructure is limited by geographical conditions (Escobal and Torero 2005), which also limit the spread of economic activity through the region. For instance, examining the role of geography in regional inequality, welfare, and development, Kanbur and Venables (2005a) found a strong correlation between geography and development. Huge welfare disparities and a high concentration of very poor people exist across the most geographically adverse regions. In summarizing findings from studies in 26 countries, Kanbur and Venables (2005b) found public infrastructure to be a key explanatory factor underlying the level and trend of spatial inequality in a country.

Inaccessibility permeates all of the HKH countries and contributes to the poverty levels observed across economies (Hunzai et al. 2011; Gerlitz et al. 2012). Several studies have noted the impact of remoteness, which often results in poor connectivity in terms of transport and roads, limiting access to markets and to locations with alternate means of employment, better healthcare, and education facilities. Due to elevation alone, mountain areas can face development costs that are two to three times higher than in the plains (Dasgupta et al. 2014). Poverty in mountain areas results not only from poor resource endowments of individual households but also from the severe constraints of unfavourable geographical situations. As a result, poverty usually affects the entire population of a mountain area while it only affects some households in a lowland area. Although there is some variation in income among households and groups in mountain areas, it is less distinct than in the plains. This unique characteristic has been recognized in China's poverty reduction approach, which targets poor areas rather than poor people in development programmes.

Due to the unique terrain and climate conditions in mountain areas, people require a higher caloric intake, warmer clothing, and permanent shelter to protect themselves from extreme weather, as compared to people in the plains. Standardized national poverty lines may not capture these specific needs, leading to failure in identifying many mountain people as poor and in need of additional resources to obtain the same degree of wellbeing (such as being well-nourished or having decent shelter). With greater requirements for food, clothing, and shelter in mountain areas (and the higher cost of goods in these areas), the deficiency in meeting requirements for basic wellbeing—and the incidence of poverty—would understandably be greater.

In most cases, mountain areas also suffer from political, social, and economic marginalization due to their remote location, low population, and indigenous cultures. As a result, mountain peoples are seldom involved in national political and policy-making processes. This results in a lack of representation in the national agenda, leading to a sense of exclusion and lack of empowerment among these mountain populations, which adds a psychological dimension to their poverty (Blaikie and Sadeque 2000). People in mountain areas (particularly women) face severe strain in securing basic necessities, such as water, fuelwood, and fodder for livestock. Furthermore, many agricultural operations must be carried out manually due to the difficult terrain. The resulting hazards, physical strain, and drudgery, which are specific to mountain poverty, are not reflected in the commonly used indicators.

In mountain areas, livelihoods are highly vulnerable due to the limited resource base, fragility of resources and environment, and lack of transport due to difficult terrain. The high incidence of natural hazards often damages the means of livelihoods such as agricultural lands, irrigation channels, and crops, as well as houses, transport, and communication facilities. As a result, the maintenance of livelihoods is highly precarious and the risk of people falling into poverty is much higher than in the lowlands.

There are cultural norms among ethnic minorities in mountain areas that may disfavour particular groups (such as women and indigenous peoples). However, among some ethnic minorities that are matrilineal, women tend to have greater access to resources than men. For example, among the matrilineal Khasi tribe in Meghalaya, India, women have greater access to land and property than men.

Table 12.9 presents the findings from Hunzai et al. (2011) and Gerlitz et al. (2012), who studied selected potential drivers of income poverty, using national-level data disaggregated by mountain and non-mountain areas in Afghanistan, Bangladesh, Bhutan, India, Nepal, and Pakistan. The table highlights remoteness, lack of access to services, and lack of education as key drivers of poverty in mountain regions, but it also shows that mountain regions are not always at a disadvantage in comparison to the plains. These findings also show a lot of variation in the selected indicators—across and within countries (such as India)—which calls for context-specific analysis and policies. This table also shows that surveys and databases are far from harmonized; information on key potential drivers of poverty (such as dependence on natural resources, social status, or empowerment) is lacking, or can be strengthened. Finally, it shows there is a need for longitudinal data and constant monitoring to shed light on poverty in the HKH, allowing for better understanding of such dynamics as the duration of poverty spells and the causes of transitions in and out of poverty.

12.3 Social Vulnerability in a Changing Climate: Why It Matters in the HKH

Mountain livelihoods are highly vulnerable to the impacts of climate and environmental change. Vulnerability—while explained in different ways—generally refers to the extent to which socio-ecological systems are susceptible to and able to cope with the pressures and shocks of climate change (Füssel and Klein 2006). From a hazards perspective (see Chap. 11), vulnerability is the susceptibility of an individual or a group to suffer damage from environmental extremes and from a relative inability to recover from that damage (Mustafa et al. 2011). While the biophysical impacts of climate change have received considerable policy attention, the political, economic, and social impacts are relatively underexplored, yet very important (Adger 2006).

Vulnerability is embedded in everyday power dynamics; thus, its exposure and intensity varies according to political economy, social capital (Pelling and High 2005; Turner 2016), gender (Sultana 2014; Morchain et al. 2015), and ethnicity (Bolin 2007), among other factors. Adger and Kelly (1999) link collective social vulnerability with a lack of institutional and market structures, such as infrastructure, insurance, and social security. An individual's social vulnerability is linked to social status, access to resources, and a diversity of livelihood strategies (Adger and Kelly 1999).

Table 12.9 Factors associated with poverty in mountains and plains in different HKH countries (Hunzai et al. 2011)

	Afghanistan		Bangladesh		Bhutan		India			Nepal		Pakistan	
	Mountain	Plains	Mountain	Plains	Mountain	Plains	Rural Uttakahand	Rural Himalaya West Bengal	Rest of Rural Indian Himalaya	Mountains and hills	Plains	Rural mountains	Rural plains
Population below poverty line (%)	42	23	46	37	34	19	49	59	17	40	28	34	30
Access to basic facilities													
Population with improved source of drinking water (%)	15	16	15	7	46	72	NA	NA	NA	69	90	56	42
Population with electricity (%)	67	87	29	45	59	70	53	22	77	25	35	20	12
Population with improved toilet facilities (%)	37	32	20	31	30	52	NA	NA	NA	40	28	60	56
Accessibility													
Distance to nearest paved road, mean (km)[a]	4	4	NA	NA	4	2	25	1	11	19	1	NA	NA
Time to get to nearest market >1 h (%)	44	39	NA	NA	3	2	56	2	36	7	1	NA	NA
Assets and liabilities													
Land owned per capita, mean (ha)	0	0	0	0	0	0	0	0	0	1	1	1	1
Livestock per capita[c], mean	2	2	1	2	1	2	NA	NA	NA	2	1	2,184	6,832
Loans obtained (%)	62	55	NA	NA	36	40	NA	NA	NA	75	74	NA	NA
Household composition													
HHs with female head (%)	1	1	8	11	27	32	25	10	13	18	11	11	9
Dependency rate, mean	1	1	1	1	1	1	1	1	1	1	1	1	1
HH members in non-agricultural professions, mean (%)	26	27	39	49	15	27	15	11	20	NA	NA	NA	NA
Education													
Uneducated head of HH (%)	NA	NA	NA	NA	85	66	50	61	52	62	59	61	56
Literate HH members >5 years old, mean (%)	NA	NA	NA	NA	49	58	66	66	71	48	44	42	41

Notes
[a]In Bhutan, this indicates distance to next paved road (mean); in India, this indicates percentage of households 10 km or more from next paved road
[b]In Bhutan, this indicates hours to next market centre
[c]In Pakistan, this indicates value of livestock per head in Pakistani Rupees (PKR) (mean)
Comparable data not available for China and Myanmar
NA = not available

This chapter utilizes Cutter's (1996) integrative concept of 'vulnerability of place', which looks at the aggregate of biophysical risk and socioeconomic factors within a geographical area to determine vulnerability. Thus, vulnerability is a function of external factors (such as exposure to environmental shocks and stresses) as well as internal factors (such as sensitivity to change and adaptive capacity). Importantly, this understanding posits vulnerability to climate change as dynamic, varying according to economic, social, geographic, demographic, cultural, institutional, governmental, and environmental factors (IPCC 2012).

12.3.1 How to Measure Vulnerability

To understand the nature of vulnerability to climate change as it impacts local livelihoods, a number of efforts have been made to measure vulnerability. While early vulnerability measures were primarily scientific biophysical assessments of climate change impacts, these measures evolved into more integrated, policy-driven vulnerability assessments that take political, economic, and social drivers into account. This discussion will first review qualitative vulnerability assessments before moving on to quantitative assessments, such the Climate Vulnerability Index (CVI) and the multidimensional livelihood vulnerability index (MLVI), emphasizing work done in mountain areas.

There are several methods of vulnerability and capacity analysis that have been used by international organizations in post-disaster and climate change planning. The Climate Vulnerability and Capacity Analysis (CVCA) is useful as it prioritizes local knowledge on climate change and adaption strategies in the data gathering and analysis process, integrating community knowledge and scientific data to better understand local impacts of climate change. The CVCA tools facilitate a participatory process for multi-stakeholder analysis and collaborative learning to qualitatively address the underlying causes of vulnerability (Dazé et al. 2009).

The Participatory Climate Risk Vulnerability and Capacity Assessment (PCR-VCA) is another useful methodology, as it incorporates an assessment of hazards within the community when evaluating its overall risk context, its livelihood assets base, and the enabling environment (Regmi et al. 2010). ICIMOD's Community-Based Climate Vulnerability and Capacity Assessments in Mountain Areas provides the most relevant analytical framework for understanding vulnerability in the HKH, as it takes mountain specificities into account.

The framework documented in Fig. 12.7 addresses the impacts of climate change variability and non-climatic factors (environmental, economic, social, demographic, technological, and political) and the extent to which these may have adverse or beneficial impacts on a community's exposure, sensitivity, and adaptive capacity (Macchi 2011). This model takes into account an individual's, household's, or community's inherent capacity to adapt to climate change impacts (Füssel and Klein 2006). While adaptive capacity refers to the potential to adjust (see Chap. 13), adaptation is the action of adjustment, which in the context of mountain areas is often a survival practice or a coping strategy (Macchi and Gurung 2015). The Vulnerability and Capacity Assessment (VCA) approach documented above combines a conceptual assessment of vulnerability with the Sustainable Livelihoods Approach (SLA), as shown in Fig. 12.8. The SLA is helpful in understanding vulnerability, as

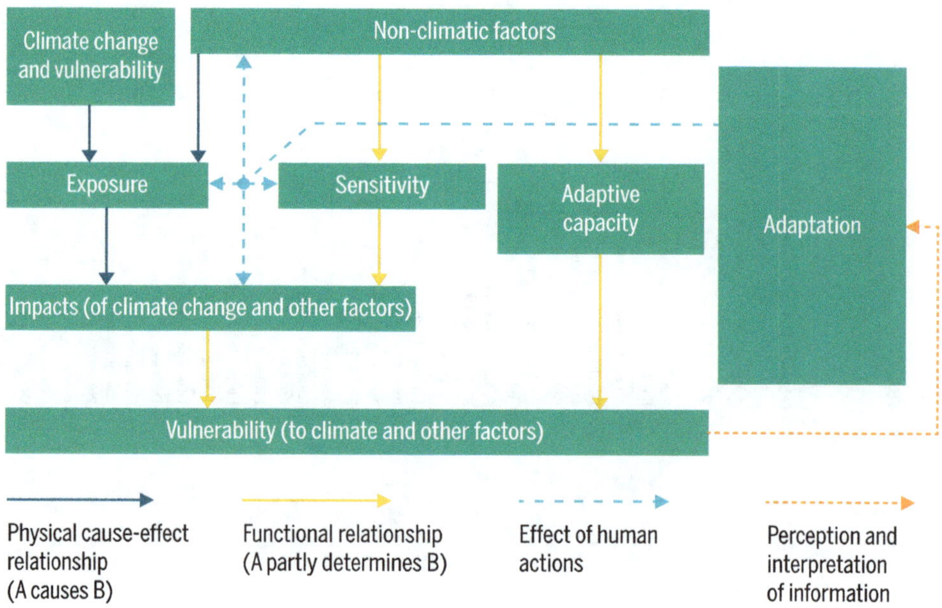

Fig. 12.7 Conceptual framework for vulnerability (Macchi 2011, adapted from Füssel and Klein 2006)

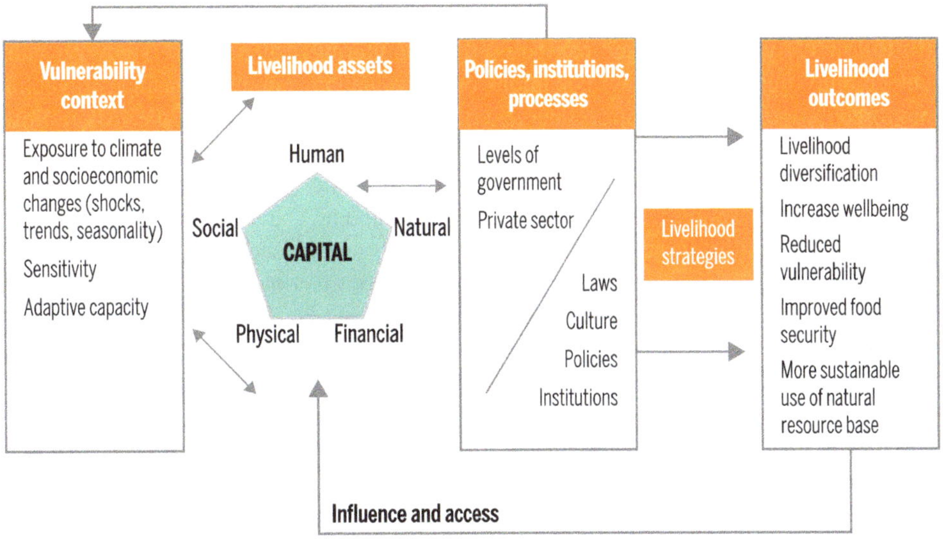

Fig. 12.8 Sustainable livelihoods approach (Macchi 2011, adapted from DFID 1999)

livelihood assets and capabilities are key determinants of a community's capacity to adapt to climate and socioeconomic change (Macchi 2011).

By understanding communities' natural livelihood assets and capacities, it is possible to develop livelihood policies, strategies, and institutions to improve community resilience while fostering livelihood diversification, increased wellbeing, reduced vulnerability, improved food security, and more sustainable use of the natural resource base (Macchi 2011). In Macchi's (2011) VCA, data was collected at community and household levels through participatory rural appraisal, household interviews, and focus group discussions. Attention to gender and marginalized social groups (such as minorities and indigenous peoples) is central to the process, as these groups generally have a weaker livelihood assets base.

Mountain communities have long histories of adapting to extreme environmental conditions and usually base their adaptive strategies on their livelihood assets, particularly their human, social, and natural capital. However, these communities often lack access to financial and physical capital due to their marginalization, isolation, and the fragility of the ecosystems they inhabit, all of which hamper their capacity to adapt (Macchi 2011).

In Uttarakhand, India, the VCA showed that communities already perceived a decrease in rainfall, unpredictable onset of the monsoon, longer dry spells with drought-like conditions, higher temperatures linked with decreased water availability, and warmer winters with less snowfall. These factors have impacted livelihood systems by causing a decline in agricultural productivity, drier streams, less productive lands, and increased incidence of pests and disease. While communities are actively adapting by replacing crops and shifting to smaller livestock, there is future risk of increased insecurity in food and livelihood (Macchi 2011). In a later study conducted in 20 villages in northwest India and across Nepal, Macchi and Gurung (2015) found that many of these coping strategies and adaption mechanisms would not be sustainable in the face of future climate change and were restricted to social groups with appropriate assets. Social markers at the intersection of class, caste, gender, and ethnicity were found to be key factors in determining vulnerability.

12.3.2 Linking Poverty and Vulnerability: Quantitative Vulnerability Assessments

Much of the literature on vulnerability is case-based and discussed in narrative form, as there are many challenges in developing standardized measures of vulnerability across diverse landscapes and social settings. Yet, as in the case of poverty assessment, there have been attempts to measure vulnerability quantitatively in order to achieve a more objective and comparable analysis across spatial and temporal scales.

A number of efforts have been made to measure vulnerability through indices, which use theoretical models to analyse the impacts of composite variables. Many of these studies follow the Alkire-Foster method, discussed earlier, in analysing multidimensional poverty (Alkire and Foster 2011).

Pandey and Jha (2011) proposed a Climate Vulnerability Index (CVI) that incorporates at the household level the three dimensions of vulnerability identified by the IPCC—exposure, sensitivity, and adaptive capability—to assess a community's overall vulnerability to climate change. The index defines and measures exposure as natural hazards and climate variability measurements; sensitivity as health, food, and water variables; and adaptive capability as metrics of socio-demographics, livelihood strategies, and social networks (Pandey and Jha 2011). The CVI ranges from high (0) to low (1) vulnerability and can be used in monitoring vulnerability fluxes under stress conditions or in evaluating proposed programmes or policy interventions by altering variables and comparing the output to the baseline (Pandey and Jha 2011). The study's authors applied the CV to the Indian Himalaya (Srinigar, Uttarkhand) and a major finding was that vulnerability (sensitivity and adaptive capacity) was higher in areas farther from the district headquarters, as compared with closer areas, while exposure was comparable despite distance.

The vulnerability and capacities index (VCI) proposed by Mustafa et al. (2011) may be quite useful in the HKH. Following Woodrow and Anderson (1989) this VCI breaks vulnerability into three broad categories of material, institutional, and attitudinal vulnerability. The VCI focuses on livelihood diversification, infrastructure, and social capital, among other indicators, as factors that mitigate vulnerability and, to a lesser extent, on physical exposure as a driver of vulnerability.

This index may be particularly appropriate for HKH because it is based on insights distilled from vulnerability research in South Asia, specifically Nepal, India, and Pakistan; thus, it was designed with the regional context in mind. The simple additive framework of VCI makes it easy to use, as demonstrated by its use in Sindh, Pakistan (Ghaus et al. 2015). Additionally, the VCI's structure is quite flexible and can be adapted to specific contexts by simply substituting appropriate variables for those outlined in the original formulations. Finally, as the index was tested, it provided substantial insight into how livelihoods, assets, access to services, and a sense of empowerment all intersect in the overall picture of vulnerability.

Another approach to measuring vulnerability recognized the shortcomings of blanket approaches and proposed the inclusion of livelihood types in the analysis, allowing similar livelihoods to be compared in terms of vulnerability through consideration of specific attributes (Kok et al. 2016). This study's approach recognized that different livelihoods, such as pastoralism and sedentary agriculture, are prone to different vulnerabilities and, thereby, multidimensional indicators should use specific variables for each livelihood type. The authors use this method to develop indicators for smallholder farmers, reasoning that specific socio-ecological dimensions would allow for better transfer of the framework to similar livelihoods. There are two key advantages to this pattern approach. Firstly, it can be applied to assess vulnerability at any scale, including household (Sietz et al. 2012) and regional scales (Sietz 2014). Secondly, the similarities depicted by the vulnerability patterns facilitate the transfer of vulnerability reduction strategies based on the assumption that people living in similar socio-ecological conditions would benefit from similar measures to reduce vulnerability.

Gerlitz et al. (2017) conducted the most comprehensive quantitative study on vulnerability in the HKH, creating a multidimensional livelihood vulnerability index (MLVI) based on the Alkire-Foster method for multidimensional poverty. The MLVI measures can identify not only vulnerable people but also areas of intervention—and if conducted over regular intervals, this index can be used to monitor the success of adaptive policies. The assessment covered 6,100 households in almost 280 settlements in the Upper Indus Basin in Pakistan, the Eastern Brahmaputra Basin in India, and the Koshi Basin in Nepal. Twenty-five indicators were used for each of the three domains of vulnerability, where exposure represents the nature of biophysical threat from climate change, sensitivity constitutes the socioeconomic determinants of differentiated impact, and capacity measures the ability to recover from or adapt to such changes.

Figure 12.9 presents the absolute and relative contribution of vulnerability dimensions by district and sub-basin, indicating that the Upper Indus is the least vulnerable of the three basins. Gerlitz et al. (2017) point out that the theory-based index assigns scores based on normative decisions; thus, the final results are influenced heavily by these decisions. They also acknowledge that applying the same values or weights across large areas may obscure crucial local factors affecting livelihood vulnerability. However, this method can be adapted to local realities by adding or eliminating indicators, making this one of the most comprehensive vulnerability assessments in the region.

It is apparent in Fig. 12.9 that Khotang District of the Koshi Basin in Nepal is the most vulnerable; 96% of its population is multi-dimensionally vulnerable in terms of 52% of the 25 vulnerability indicators. Lakhimpur is the most vulnerable in the Eastern Brahmaputra Sub-basin, with 92% of its population vulnerable; and Chitral District in Pakistan marks the highest vulnerability in the Upper Indus Sub-basin, with 65% of its population vulnerable (Gerlitz et al. 2017). Figure 12.10 breaks down the composition of the MLVI, showing the absolute and relative contributions of each vulnerability dimension.

By identifying which indicators are the dominant determinants of vulnerability, this analysis is useful in suggesting targeted policy responses. While a lack of adaptive capacity was the greatest absolute contributor to livelihood vulnerability in Lohit and Udayapur (0.17 and 0.16, respectively), Chitral is highest in relative terms, making up 50% of the MLVI (Gerlitz et al. 2017). To reduce livelihood vulnerability in Chitral, for example, priority may be given to programmes that enhance adaptive capacity or to efforts to improve resources and energy, targeting the specific issues that contribute most to its vulnerability.

The highest absolute contributor to exposure was in Khotang (0.15), whereas exposure marked the highest relative contribution to vulnerability in Hunza-Nagar (38%)

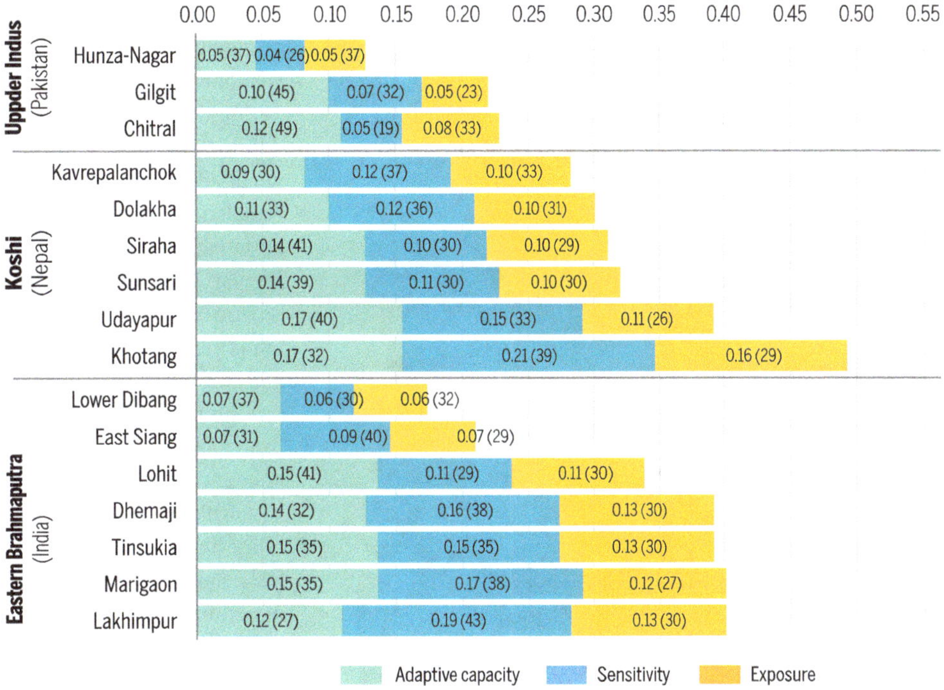

Fig. 12.9 Vulnerability in three sub-basins of the HKH: absolute and relative contribution of vulnerability dimensions by district (Gerlitz et al. 2017)

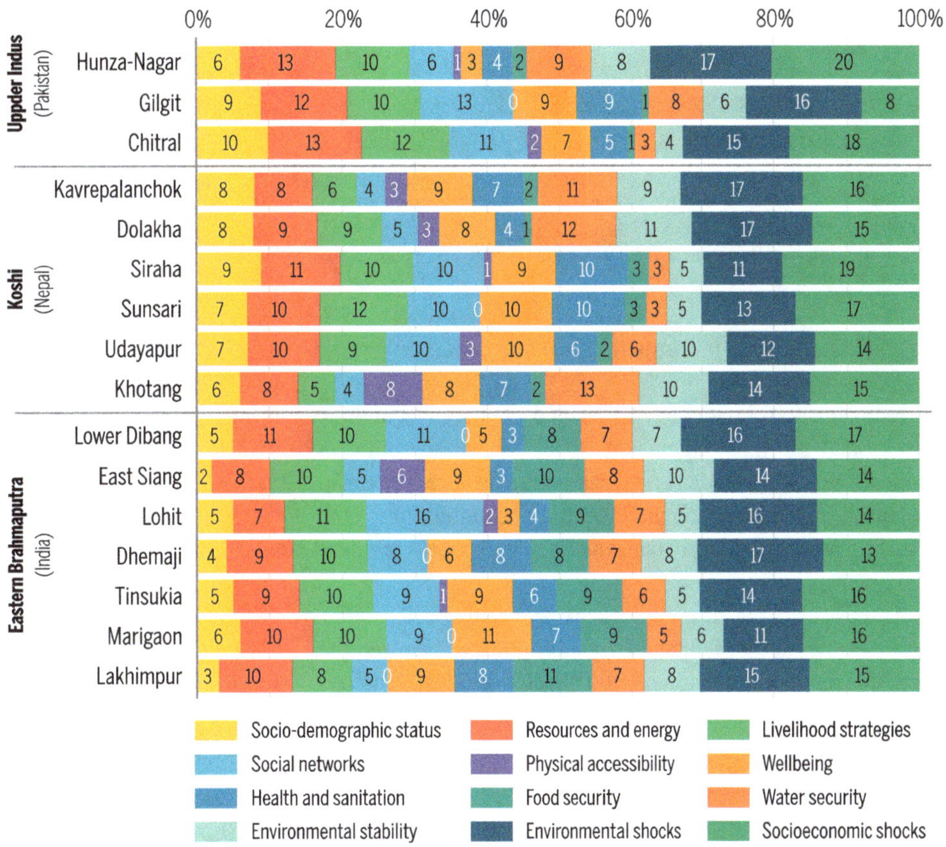

Fig. 12.10 Vulnerability in three sub-basins of the HKH: relative contribution of vulnerability components by district in percentage (Gerlitz et al. 2017)

(Gerlitz et al. 2017). Biophysical risk is the biggest contributor to Hunza-Nagar's vulnerability profile, yet this factor alone doesn't amount to high vulnerability overall. In Khotang, on the other hand, biophysical vulnerability is relatively low, yet the combined risks in terms of infrastructure and livelihoods make it the most vulnerable district in the study. By comparing the drivers of vulnerability between Hunza-Nagar and Khotang, it becomes apparent

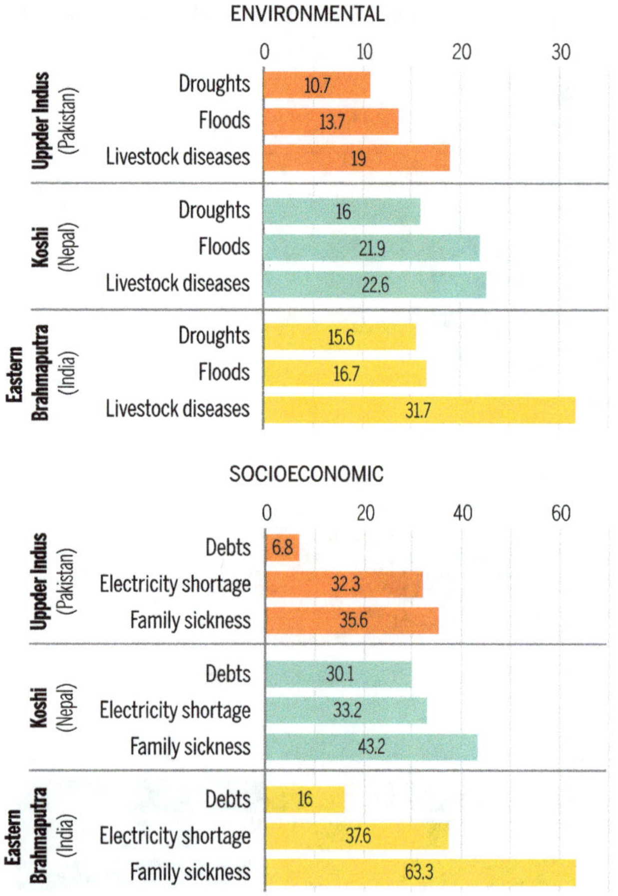

Fig. 12.11 Top three environmental and socioeconomic shocks (Gerlitz et al. 2015b). *Notes* N = 6,096 HH (India = 2,647 HH, Pakistan = 1,139 HH, Nepal = 2,310), weighted analysis, 100%, data: VACA 2011/12

that livelihood and infrastructure are the main drivers of absolute vulnerability.

Exposure to climatic, environmental, and socioeconomic shocks had the most significant impact on vulnerability in each district. Figure 12.11 indicates the top three environmental and socioeconomic shocks to communities, and Fig. 12.12 presents each community's responses to these shocks.

As indicated by these figures, livestock diseases, floods, droughts, erratic rainfall, and irrigation problems caused the most consequential environmental shocks, and communities responded by changing farming practices. In each basin, family illness and electricity shortages caused the most severe socioeconomic shocks, and people within the community often responded by borrowing money.

Such information is extremely useful in suggesting mountain-specific policy options to reduce vulnerability in farming communities, addressing issues within both social and environmental domains to decrease both poverty and vulnerability. Practical measures to support mountain livelihoods include, for example, financial support, decentralized energy production, access to information and research on appropriate farming practices, and timely weather-related data.

Multidimensional vulnerability measures (as described by Alkire and Foster 2011) avoid overly simplistic and overly generalized findings, successfully identifying vulnerable populations and significant areas of intervention, while also allowing the success of adaptive policies to be monitored at regular intervals. The multidimensional vulnerability (and poverty) indices can serve as good baselines for measuring temporal and spatial trends in growth and development—and for attempting to isolate the climate change factor driving this trajectory. In addition, actual empirical evidence from the field (as opposed to predictive theorizing) would help calibrate VCI models to local realities.

On the other hand, participatory vulnerability and capability analyses are more direct in capturing the problems at hand and thereby constitute a better tool for eliciting policy direction. However, such analyses are place-based—determined by local priorities, landscapes, and livelihoods—and thereby not suitable for comparison, failing any promise of universal application.

12.3.3 Evidence of the Intersection of Poverty and Vulnerability in the HKH

This assessment shows that the mountain regions of the HKH have a higher incidence of economic and multidimensional poverty than the plains—and income poverty levels in mountain areas are also higher than in the plains, with the exception of India. Furthermore, there has been less poverty reduction in mountain areas, which has led to increased income inequality between the two regions. Even in India, where several hill states have poverty rates that are lower than the national average, the hill states fare worse than the rest of the country in terms of access to education, healthcare, financial support, infrastructure, and basic amenities. In terms of policy, a significant implication is the need to close the gap in income inequality through more effective poverty-reduction programmes in the mountain areas, including greater investments in education and in development of economic and social infrastructure.

Multidimensional assessment proves to be a better measure of poverty than any single indicator. Livelihoods are intricately—and variously—tied to natural resources and reliant on knowledge of seasonal and climatic change (Elalem and Pal 2015). Therefore, the impacts of climate change on local livelihoods are becoming increasingly severe—and certain livelihood coping strategies are, in turn, impacting climate change. Overall, it is clear that across the HKH inhabitants are experiencing increased vulnerability to climate change, particularly in terms of livelihood and wellbeing.

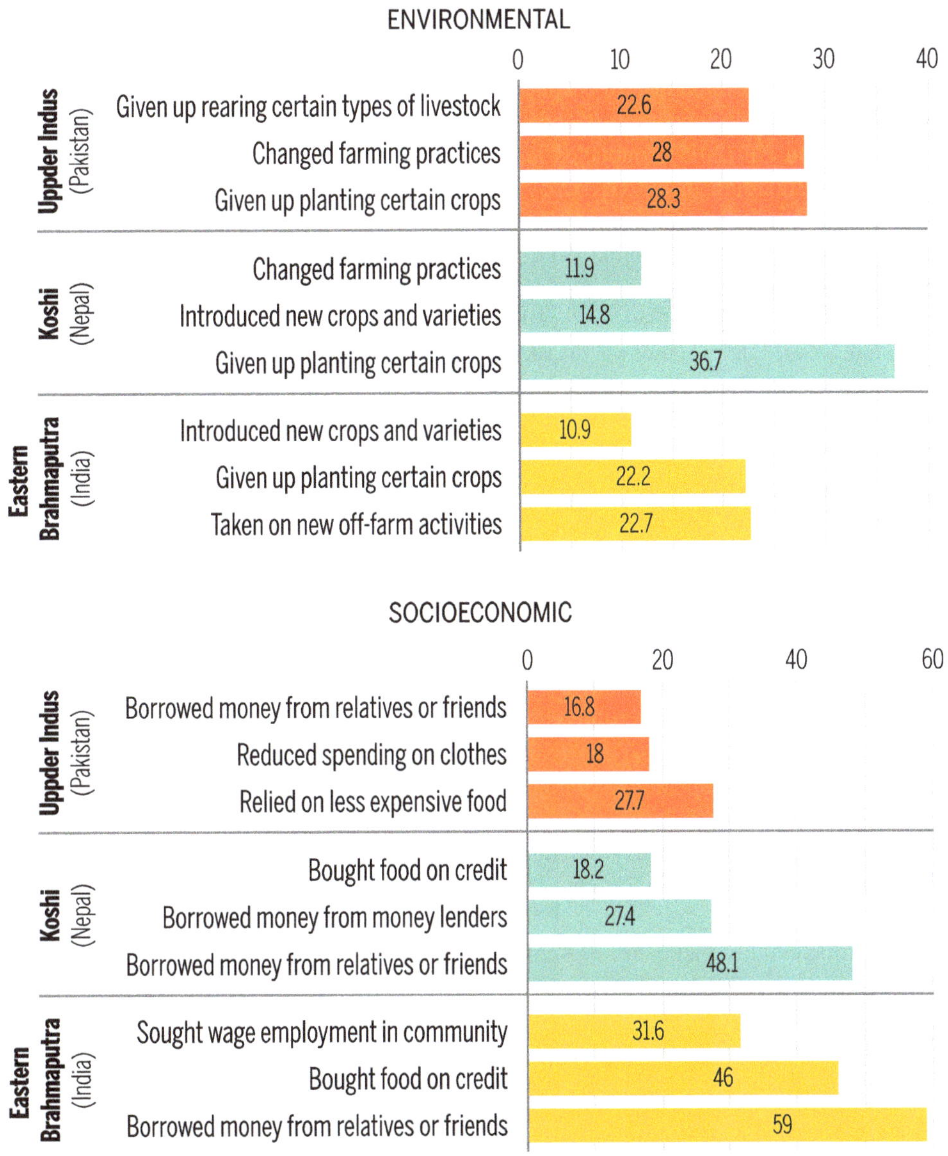

Fig. 12.12 Top three responses to social and environmental shocks (Gerlitz et al. 2015b). *Notes* Top panel; N = 5,630 HH (India = 2,490 HH, Pakistan = 1,124 HH, Nepal = 2,016), weighted analysis, 100%, data: VACA 2011/12. Bottom panel: N = 5,855 HH (India = 2,571 HH, Pakistan = 985 HH, Nepal = 2,299), weighted analysis, 100%, data: VACA 2011/12

Water, land, and forests are essential resources for HKH residents. Quantitative and qualitative studies indicate that these resources are being affected by climate change and the coping strategies used in response to it. Regionally, studies highlight that while the annual average precipitation has not shifted significantly, there have been more frequent extremes, such as droughts and floods, and rainfall patterns have become less predictable overall (EPASSA 2008; Duncan et al. 2013; Wiltshire 2014). Cloudburst incidents are being reported more frequently in the mountainous regions (Shah 2009), and the amount of precipitation falling as snow has decreased while ablation has increased (Wiltshire 2014).

These changes have made it harder for local people to predict and prepare for the onset of monsoon seasons, making the population more vulnerable to flooding disasters, including glacial lake outburst floods (GLOF) (Gentle et al. 2014). With livelihoods dependent on understanding and predicting these weather patterns, HKH residents are also more vulnerable to disasters caused by drought and landslides.

Research shows that in monsoon-affected regions, just one extreme weather event can account for as much as 10% of a catchment's yearly water intake, and 50% of yearly rainfall can occur within a 10-day period (Dahal and Hasegawa 2008; Bookhagen and Burbank 2010). Sediment qualities are also being affected, impacting ecosystems by disrupting irrigation networks, hydropower efforts, and the potable water supply. These extreme weather events not only impact the population's access to ecosystem services but also put residents at risk of physical harm, loss and damage to infrastructure, and food insecurity (see Chaps. 9 and 11).

With global warming continuing at current rates, precipitation projections for the region become increasingly alarming, necessitating the development of adaptation strategies for the area's huge resident population (Gentle et al. 2014).

Elalem and Pal (2015) conducted an assessment of vulnerability to flooding in the HKH. By looking at the historical record of floods spanning 1951–2013, coupled with demographic and socioeconomic data, they assessed flood disaster locations and the resulting economic and human impacts. The authors produced vulnerability maps marking the what, where, and when of the disaster impacts across the HKH. In terms of time, the results indicated a clear trend across the whole region: There were more frequent and more intense disasters throughout 2001–13.

Elalem and Pal (2015) also highlighted economic damages associated with flood disasters. In light of Pakistan's 2010 mega floods, it is not surprising that the authors found that Pakistan experienced the greatest economic damages per disaster, followed by China and India. However, it is notable that the HKH countries with lower GDP (Nepal and Afghanistan, along with Pakistan) were more vulnerable to economic loss throughout the full period of assessment than their wealthier counterparts. This means that poorer countries are less able to respond to extreme events and disasters, indicating that poverty and vulnerability are interlinked. This also suggests a regional strategy of reducing vulnerability by integrating disaster resilience into broader poverty-reduction strategies.

Desertification, in particular, is a concern for HKH populations residing in India and Pakistan. (Ning et al. 2013) concurs that desertification, degradation, and soil erosion are being exacerbated by climate change in the HKH's high-altitude rangelands. Desertification is affecting landscape productivity in many sites, including the Manasarovar catchment in China, various sites in Nepal, and in Sindh, Gilgit-Baltistan, and Balochistan in Pakistan.

In Nepal, the National Action Programme on Land Degradation and Desertification estimates that around 3.3 million hectares (28.2% of the country's total land) are experiencing the process of desertification (Tiwari and Joshi 2012; GoN 2016). In terms of degradation, 70% of the total degraded land is in the forests and the rangelands. Tiwari and Joshi (2012) contend that the drivers of this desertification and degradation are the combination of climate change, infrastructure development, and the coping strategies of populations in poverty, including land-use patterns, settlement, infrastructure development for fuelwood and timber, tree cutting and forest clearing for agriculture, mass wasting, flooding and erosion, and rockslides. Despite these significant figures in Nepal, EPASSA (2008) estimates that the impacts of desertification will be even more severe for Pakistan and India.

Authors highlight several broad linkages between poverty and vulnerability across the HKH, which are exacerbated by climate change impacts. Some of the most consistently reported include:

- Lack of economic diversification resulting in sustained dependence on a degrading and increasingly fragile ecosystem for consumption and livelihood (Suich et al. 2015).
- High dependency of downstream communities on upstream ecosystem services for dry-season water for irrigation, hydropower, drinking water, and soil fertility and nutrients—especially in South Asian HKH countries (Rasul 2014). Moreover, poorer people who depend on mountain ecosystems for survival are not incentivized to conserve resources (Rasul 2014).
- Continued overuse of groundwater-irrigated agriculture, which provides food security for approximately 60–80% of the population, but degrades the groundwater resources needed for other livelihood uses (Shah 2009).
- Growing demand for timber and fuelwood to support industry and livelihoods that has led to heavy degradation of forest resources, upon which poor communities depend (Haigh 1990; Rasul 2014).

12.3.4 Knowledge Gaps and Recommendations

There are some key knowledge gaps in the literature on vulnerability in the HKH that should be explored and illuminated with new empirical research. The most striking gap is in the limited geographical focus of the evidence base. The majority of research focuses on Nepal and India, and there is a need for greater investigation across Pakistan, Afghanistan, Bhutan, Myanmar, Bangladesh, and China. Climate and environment data also needs to be improved and expanded to encompass a comprehensive understanding of the HKH. Data collection is often sparse in the rugged, high-elevation areas of the HKH, which poses a challenge for accurate climate modeling and prediction of climate change effects for such areas. Furthermore, there are also significant gaps in scientific knowledge of climate change, including the effects of black carbon and melting permafrost and the impact of transient groundwater storage on water regimes.

Another key gap in existing literature is in understanding the differentiated impacts of climate change on poverty and vulnerability across different geographical zones within the HKH. Generally, the literature assumes that the nexus between climate change, poverty, and vulnerability can be understood through regional research—but this is debatable, as the studies reviewed highlight highly differentiated impacts depending on local geography, socioeconomic structure, and experienced impacts. Most poverty reduction programmes do not take mountain specificities into consideration; thereby these efforts may actually increase climate risk and fragility in the HKH—or, at best, prove to be less effective in the mountains than in the lowlands. It may be useful to categorize different geographical zones across the HKH according to different factors, and then understand how such differentiated spaces and populations are impacted by climate change.

Participatory research is also needed to understand the intersectional impacts of climate change on poverty and vulnerability. The literature highlights the significance of intersectional categories—especially gender—in accessing livelihood adaptation strategies (Shields 2008). As discussed above, the intersection of gender and economic marginality leads to increased vulnerability and reduced access to sustainable adaptation strategies. Targeted investments informed by this understanding would increase the effectiveness of development efforts and help reduce vulnerability.

Hewitt and Mehta (2012) argued that marginalization in mountain areas is due to subordination of the specific regional socioeconomic planning by larger strategic planning at the state level. In simple terms, mountain-specific socioeconomic planning does not work in service of other and larger geographies—especially not the plains. Therefore, fewer socioeconomic development gains are made in mountain areas. Such trends in mountain-area marginalization have been observed in poverty-reduction strategies as well.

Evidence in this assessment suggests that reduction of both poverty and vulnerability can be achieved through coordinated interventions that are contextualized, intersectional, and mountain specific. Multidimensional poverty measures—which assess multiple deprivations in education, health, and standard of livingvare appropriate for poverty assessment in mountain areas. Since there is an acute shortage of mountain-specific poverty data, an important policy implication is that governments and development partners should allocate more resources to build a comprehensive database using longitudinal surveys at the regional and national levels. Such a database would strengthen, complement, and substantiate the macro-level findings, which are based on the nationally representative datasets. Further assessment could be used to identify and document pockets of poverty and vulnerable communities throughout all HKH countries using a long-term monitoring system.

Where the government cannot provide adequate social protection, investment in social mobilization can strengthen traditional social networks to effectively reduce vulnerability, as has been demonstrated in some areas of Pakistan (Khan 2014). Additionally, income inequality in mountain regions can be addressed through mountain-specific poverty-reduction programmes, including larger per capita investment in education and the development of economic and social infrastructure.

Local promotion of eco-tourism (such as the home-stay model) has significant potential for promoting mountain economies, while supporting women and sustainable, resilient development. By contrast, large direct investment in eco-tourism may increase climate vulnerability by exerting pressures on the environment, though more research needs to be done on different models of tourism and their risks and opportunities across the HKH.

In agriculture, traditional irrigation practices can also be further investigated to identify both risks and opportunities. The application of modern technologies and techniques might allow traditional practice to become sustainable and resilient. Reforesting lands that have become suboptimal for agriculture can encourage groundwater recharge and provide alternate livelihoods based on ecosystem services, such as fibre production and tourism (Chaudhary and Aryal 2009).

Regional and national strategies have had differential impacts on mountain poverty and may even increase vulnerability of mountain livelihoods. Promotion of high-value crops, for example, may be useful by increasing income, but over-promotion will reduce the diversity of income sources, making the endeavour highly susceptible to a covariant risk of large-scale failure due to an unpredicted pest or a shift in rain and temperature patterns.

References

Adhikari, J. (2014). *Agriculture adaptation practices in South Asia: Case of Nepal*. Working Paper No. 01(iii)/14. South Asia Watch on Trade, Economics and Environment (SAWTEE).

Adhikari, L., Hussain, A., & Rasul, G. (2017). Tapping the potential of neglected and underutilized food crops for sustainable nutrition security in the mountains of Pakistan and Nepal. *Sustainability, 9*, 291.

Adhikari, J., & Hobley, M. (2012). *Everyone is leaving—Who will sow our fields?: Impact of migration to Gulf and Malasiya in Khotang district, Nepal*. Kathmandu: NIDS and SDC.

Asian Development Bank, ADB. (2012). *Bangladesh: Chittagong hill tracts rural development project*. Manila: ADB. https://www.adb.org/sites/default/files/evaluation-document/36065/files/pvr-222.pdf.

Anand, S., & Sen, A. (1998). *The income component of the human development index. (mimeo)*. New York: Human Development Report Office, UNDP.

Alkire, S., & Santos, M. E. (2010). *Acute multidimensional poverty: A new index for developing countries*. OPHI Working Papers 38, University of Oxford.

Alkire, S., Foster, J. E., Seth, S., Santos, M. E., Roche, J., & Ballón, P. (2015). *Multidimensional poverty measurement and analysis*. Oxford University Press.

Alkire, S., & Seth, S. (2013). *Multidimensional poverty reduction in India 1999–2006: Slower progress for the poorest groups*. Oxford Poverty & Human Development Initiative.

Alkire, S., & Foster, J. (2011). Understandings and misunderstandings of multidimensional poverty measurement, *9*(2), 289–314.

Adger, W. N., Huq, S., Brown, K., Conway, D., & Hulme, M. (2003). Adaptation to climate change in the developing world. *Progress in Development Studies, 3*(3), 179–195.

Adger, N. (2006). Vulnerability. *Global Environment Change, 16*(3), 268–281.

Adger, N., & Kelly, M. (1999). Social vulnerability to climate change and the architecture of entitlements. *Mitigation and Adaptation Strategies for Global Change, 4,* 253.

Altieri, M. A. (1999). Applying agroecology to enhance the productivity of peasant framing systems in Latin America. *Environment, Development and Sustainability* 1, 197–217. Kluwer Academic Publisher: The Netherlands.

Ali, J. (2007). Livestock sector development and implications for rural poverty alleviation in India. *Livestock Research for Rural Development 19*(2).

Apetrei, C. (2012). Food security and millet cultivation in the Kumaon region of Uttarakhand. Research Report for Gene Campaign.

APHCA and ILRI. (2006). Goats—Undervalued assets in ASIA. In *Proceedings of the APHCHA-ILRI Regional Workshops on Goat Production System and Markets.*

Azhar-Hewitt, F. (2011). *The other side of silence: The lives of women in the Karakoram mountains.* Bloomington: iUniverse.

Ali, I.., & Pernia, E. M. (2003). ERD policy brief series: Economics and research department infrastructure and poverty reduction—What is the connection (no. 13).

Bhagmal. (2007). Neglected and underutilized genetic resources for sustainable agriculture. *Indian Journal of Plant Genetic Resources, 22*(1):1–16.

Bharadwaj, B. (2012). Roles of cooperatives in poverty reduction: A case of Nepal. *Administration and Management Review,* 24.

Banerjee, S., Gerlitz, J. Y., & Hoermann, B. (2011). *Labour migration as a response strategy to water hazards in the Hindu Kush Himalayas.* Kathmandu: ICIMOD.

Banerjee, S., Sijapati, B., Poudel, M., Bisht, S., & Kniveton, D. (2016). Role of remittances in building farm assets in the flood affected households in Koshi sub-basin in Nepal. Springer International Publishing Switzerland 2016. In A. Milan et al. (Eds.), *Migration, risk management and climate change: Evidence and policy responses, global migration issues* (vol. 6). https://doi.org/10.1007/978-3-319-42922-9_2.

Berhanu, W., Colman, D., & Fayissa, B. (2007). Diversification and livelihood sustainability in a semi-arid environment: A case study from Southern Ethiopia. *The Journal of Development Studies, 43,* 871–889. https://doi.org/10.1080/00220380701384554.

Bird, R. B., Bird, D. W., Smith, E. A., & Kushnick, G. C. (2002). Risk and reciprocity in Meriam food sharing. *Evolution and Human Behavior*, 23, 297–321.

Blaikie, P. M., & Sadeque, Z. (2000). *Policy in high places.* Kathmandu: International Centre for Integrated Mountain Development.

Bolin, B. (2007). Race, class, ethnicity, and disaster vulnerability. In *Handbook of Disaster Research* (pp. 113–129).

Bookhagen, B., & Burbank, D. W. (2010). Toward a complete Himalayan hydrological budget: Spatiotemporal distribution of snowmelt and rainfall and their impact on river discharge. *Journal of Geophysical Research.* https://doi.org/10.1029/2009jf001426.

Brithal, P. S., & Taneja; V. K. (2006). Livestock sector in India: Opportunities and challenges for the small holders. In *Workshop on small holder livestock production in India: Opportunities and challenges,* New Delhi.

Central Bureau of Statistics (CBS). (2011). Poverty in Nepal 2010–11. http://cbs.gov.np/image/data/Surveys/poverty%20in%20Nepal%202010-11/CBS%20view%20on%20poverty%20in%20Nepal.pdf.

Central Bureau of Statistics (CBS). (2014). *Population monograph of Nepal: Volume I (population dynamics).* National Planning Commission Secretariat, Government of Nepal.

Chambers, R.., & Conway, G. (1992). *Sustainable rural livelihoods: Practical concepts for the 21st Century.* Discussion Paper 296. Sussex: IDS.

Chand, R., Raju, S. S., & Pandey, L. M. (2008). *Progress and potential of horticulture in India.* Paper prepared for the 68th annual conference of the Indian society of agricultural economics, 28–30 November 2008, Andhra University, Visakhapatnam, India.

Chaudhary, P., & Aryal, K. P. (2009). Global warming in Nepal: Challenges and policy imperatives. *Journal of Forest and Livelihood, 8*(9).

Chen, X. (2011). Agricultural situation and rural policy in today's China. http://www.snzg.cn/article/2011/1117/article_26276.html.

Cutter, S. (1996). Vulnerability to environmental hazards. *Progress in Human Geography, 20,* 529–539.

Dahal, R. K., & Hasegawa, S. (2008). Representative rainfall thresholds for landslides in the Nepal Himalaya. *Geomorphology, 100*(3–4), 429–443. https://doi.org/10.1016/j.geomorph.2008.01.014.

Dame, J., & Nusser, M. (2011). Food security in high mountain regions: Agricultural production and the impact of food subsidies in Ladakh, Northern India (Springer), *3*(2), 179–194.

Dasgupta, P., Morton, J. M., et al. (2014). Rural areas. In C. B. Field, & V. R. Barrros et al. (Eds.), *Climate change 2014: Impacts, adaptation and vulnerability.* Report of the working group II, Intergorvernmental Panel on Climate Change. Cambridge University Press: Cambridge.

Dazé, A., Ambrose, K., & Ehrhart, C. (2009). *Climate vulnerability and capacity analysis (Handbook).* Londong: CARE International.

Delgado, C. (2005). Rising demand for milk and meat in developing countries: Implications for grasslands-based livestock production. In *Grassland: A global resource* (pp. 29–39). The Netherlands: Wageningen Academic Publishers.

Desakota Team. (2008). *Re-imagining the rural-urban continuum: Understanding the role ecosystem services play in the livelihoods of the poor in Desakota regions undergoing rapid change.* Institute for Social and Environmental Transition-Nepal (ISET-Nepal).

DFID. (1999). *Sustainable livelihoods guidance sheets.* London, UK: Department for International Development.

de Haan, A. (1999). Livelihoods and poverty: The role of migration—A critical review of the migration literature. *The Journal of Development Studies, 36*(2).

Dorji, W. (2011). *Opportunities and constraints to community forests for local income generation and livelihoods: A case study of four community forest in Bumthang District, Bhutan.* MS thesis, University of Montana, Missoula.

Duncan, J. M. A., Biggs, E. M., Dash, J., & Atkinson, P. M. (2013). Spatio-temporal trends in precipitation and their implications for water resources management in climate-sensitive Nepal. *Applied Geography, 43,* 138–146.

Eakin, H., & Bojorquez-Tapia, L. A. (2008). Insights into the composition of household vulnerability from multi-criteria decision analysis. *Global Environmental Change, 18,* 112–127.

Ediger, L., & Huafang, C. (2006). Upland China in transition: The impacts of afforestation on landscape patterns and livelihoods. *Mountain Research and Development, 26* (3).

Elalem, S., & Pal, I. (2015). Mapping the vulnerability hotspots over Hindu-Kush Himalaya region to flooding disasters. *Weather and Climate Extremes, 8,* 46–58.

Escobal, J., & Torero, M. (2005). Adverse geography and differences in welfare in Peru. In R. Kanbur & A. J. Venables (Eds.), *Spatial inequality and development.* Oxford University Press. January.

EPASSA. (2008). *Ecosystem services and poverty alleviation study in South Asia. A situation analysis for India and the Hindu Kush Himalayan region (Annexure 3).*

Eyzaguirre, P., Padulosi, S., & Hodgkin, T. (1999). IPGRI's strategy for neglected and underutilized species and the human dimension of agrobiodiversity. In S. Padulosi (Ed.), *Priority setting for underutilized and neglected plant species of the mediterranean region. Report of the IPGRI Conference*, 9–11 February 1998. Aleppo, Syria: ICARDA; Rome, Italy: IPGRI.

FAO. (2010). *Integration of gender in agriculture*. Nepal.

FAO. (2017). *Future smart food. Unlocking hidden treasures in Asia and the Pacific*. Retrieved October 2017, from http://www.fao.org/3/a-i7717e.pdf.

Ferguson, J. A. (2012). Generating sustainable livelihoods; the role of co-operatives. Cooperatives as transformative agents in building sustainable livelihoods and reducing poverty. In *Harnessing the co-operative advantage to build a better world*.

Füssel, H. M., & Klein, R. J. T. (2006). Climate change vulnerability assessments: An evolution of conceptual thinking. *Climatic Change, 75,* 301. https://doi.org/10.1007/s10584-006-0329-3.

Frison, E. A., Smith, I. F., Johns, T., Cherfas, J., & Eyzaguirre, P. B. (2006). Agricultural biodiversity, nutrition, and health: making a difference to hunger and nutrition in the developing world. *Food and Nutrition Bulletin, 27*(2), 167–179.

Ganesh, A., & Madhavi, C. (2007). Impact of tourism on Indian economy—A snapshot. *Journal of Contemporary Research in Management, 1*(1), 2.

Gentle, P., Thwaites, R., Race, D., & Alexander, K. (2014). Differential impacts of climate change on communities in the middle hills region of Nepal. *Natural Hazards, 74*(2), 815–836.

Gerber, P., Mooney, H. A., Dijkman, J., Tarawali, S., & Haan, C. (2010). *Livestock in changing landscape: Experiences and regional perspectives* (vol. 2). Island Press.

Gerlitz, J. Y., Macchi, M., Brooks, N., Pandey, R., Banerjee, S., & Jha, S. K. (2017). The multidimensional livelihood vulnerability index—An instrument to measure livelihood vulnerability to change in the Hindu Kush Himalayas. *Climate and Development, 9*(2), 124–140.

Gerlitz, J. Y., Apablaza, M., Hoermann, B., Hunzai, K., & Bennett, L. (2015a). A multidimensional poverty measure for the Hindu Kush-Himalayas, applied to selected districts in Nepal. *Mountain Research and Development, 35*(3), 278–288.

Gerlitz, J. Y., Banerjee, S., Brooks, N., Hunzai, K., & Macchi, M. (2015b). An approach to measure vulnerability and adaptation to climate change in the Hindu Kush Himalayas. In *Handbook of Climate Change Adaptation* (pp. 151–176).

Gerlitz, J. Y., Banerjee, S., Hoermann, B., Hunzai, K., Macchi, M., & Tuladhar, S. (2014). *Poverty and vulnerability assessment: A survey instrument for the Hindu Kush Himalayas*. Kathmandu: International Centre for Integrated Mountain Development.

Gerlitz, J. Y., Hunzai, K., & Hoermann, B. (2012). Mountain poverty in the Hindu-Kush Himalayas. *Canadian Journal of Development Studies, 33,* 250–265.

Ghaus, K., Manzoor, H., Memon, M. A. I., Nadeem, A., Naveed, A., & Tabinda, A. (2015). *Gender and social vulnerability to climate change: A study of disaster prone areas in Sindh, Pakistan*. Social Policy and Development Centre (SPDC): Karachi, Pakistan. Retrieved 19 March, 2017, from http://www.spdc.org.pk/Publication_detail.aspx?sysID=786.

Gioli, G., Khan, T., Bisht, S., & Scheffran, J. (2014a). Migration as an adaptation strategy and its gendered implications: A case study from the Upper Indus Basin. Special issue "Gender and sustainable development in mountains: Innovative transformations, tenacious resistances". *Mountain Research and Development, 34*(3), 255–265.

Gioli, G., Khan, T., & Scheffran, J. (2014b). Climatic and environmental change in the Karakoram: Making sense of community perceptions and adaptation strategies. *Regional Environmental Change, 14,* 1151–1162.

Gioli, G, Maharjan, A., & Gurung, M. (2017) Neither heroines nor victims: Women migrant workers and changing family and community relations in Nepal. New York: *UN WOMEN Working Paper Series*.

Goodall, S. K. (2004). Rural-to-urban migration and urbanization in Leh, Ladakh: A case study of three nomadic pastoral communities. *Mountain Research and Development, 24,* 220–227.

GoN. (2016). *Nepal: Zero hunger challenge national action plan (2016–2025)*. Kathmandu. http://www.npc.gov.np/images/category/ZHC_NAP_(2016_-_2025).pdf.

GoP. (2006). *Agricultural census organization, statistics division*. Government of Pakistan, Gulberg, Lahore.

Gurung, K., Tulachan, P. M., & Gauchan, D. (2005). *Gender and social dynamica in livestock management: A case study from three ecological zones in Nepal*. Gender Component, LPP Project. CEMORD and NARC.

Halvorson, S. J. (2002). Environmental health risks and gender in the Karakoram himalaya, Northern Pakistan. *Geographical Review*. https://doi.org/10.1111/j.1931-0846.2002.tb00007.x.

Haigh, P. M. (1990). Effect of herbage water-soluble carbohydrate content and weather conditions at ensilage on the fermentation of grass silages made on commercial farms. *Grass & Forage Science, 45,* 263–271.

Hewitt, K., & Mehta, M. (2012). Rethinking risk and disasters in mountain areas. *Journal of Alpine Research* (Online). https://doi.org/10.4000/rga.1653.

Hannum, E., & Wang, M. (2012). China: A case study in rapid poverty reduction. In H. Gillette & P. Harry (Eds.), *In indigenous people, poverty, and development*. Cambridge: Cambridge University Press.

Herbers, H. (1998). Work and nutrition in High Asia. Theoretical framework, methods and main results of a comprehensive study on the production-reproduction-system in northern Pakistan. *EBHR, 14,* 18–30.

Herrero, M., Thornton, P. K., Notenbaert, A. M., Wood, S., Msangi, S., Freeman, H. A., et al. (2010). Smart investments in sustainable food production: Revisiting mixed crop-livestock systems. *Science, 327,* 822–825.

Hoermann, B., Banerjee, S., & Kollmair, M. (2010). *Labour migration for development in the Western Hindu Kush-Himalayas—Understanding a livelihood strategy in the context of socioeconomic and environmental change*. Kathmandu: International Centre for Integrated Mountain Development.

Hunzai, K., Gerlitz, J. Y., & Hoermann, B. (2011). *Understanding mountain poverty in the Hindu Kush-Himalayas: regional report for Afghanistan, Bangladesh, Bhutan, China, India, Myanmar, Nepal, and Pakistan*. Kathmandu: ICIMOD.

ICIMOD and GRID Arendal. (2014). *The last straw: Food security in the Hindu Kush Himalayas and the additional burden of climate change*. Kathmandu: ICIMOD.

IFAD. (2013). *Sending money home to Asia*. ISBN 978-92-9072401-8.

IFAD. (2002). *Regional strategy paper—Asia and the Pacific*. IFAD, March 2002. https://www.ifad.org/documents/10180/3b116bbd-70d6-4219-e4c8b6011866.

IOM. (2005). *World migration 2005: Costs and benefits of international migration*. Geneva: International Organization for Migration.

IPCC. (2012). In C. B. Field, V. Barros, & T. F. Stocker, et al. (Eds.), *Managing the risks of extreme events and disasters to advance climate change adaptation*. Special report of the Intergovernmental Panel on Climate Change. Cambridge and New York: Cambridge University Press.

Isaac, M. (2012). Community-based ecotourism and livelihood enhancement in Sirigu, Ghana. *International Journal of Humanities and Social Science, 2*(18).

Jarvis, D., Sthapit, B., Raj Bhatta, M., & Sthapit, S. (2014). Integrating traditional crop genetic diversity for mountain food security. *Biodiversity International*, 2p.

Jodha, N. S. (1990a). *A framework for integrated mountain development*. Mountain Farming Systems Discussion Paper No. 1. Kathmandu: International Centre for Integrated Mountain Development (ICIMOD).

Jodha, N. S. (1990b). Mountain agriculture: Search for sustainability. *Journal of Farming System Research Extension, 1*(1).

Jodha, N. S. (1992). Mountain perspective and sustainability: A framework for development strategies. In: N. S. Jodha, M. Banskota, & T. Partap (Eds.), *Sustainable mountain agriculture, volume 1: Perspectives and issues*. New Delhi: Oxford and IBH.

Jodha, N. S. (1997). Mountain agriculture. In B. Messerli & J. D. Ives (Eds.), *Mountains of the world: A global priority*. New York/Carnforth Lancashire: The Parthenon Publishing Group.

Kanbur, R., & Venables, A. J. (2005a). *Spatial inequality and development*. Oxford University Press.

Kanbur, R., & Venables, A. J. (2005b). *Rising spatial disparities and development*. UNU-WIDER Policy Brief, United Nations University, November.

Karki, M., Sharma, S., Mahat, T. J., Tuladhar, A., & Aksha, S. (2012). *Sustainable mountain development 1992, 2012 and beyond—Rio +20 assessment report for the Hindu Kush Himalayan Region*. International Centre for Integrated Mountain Development.

Karki, M. (2012). The effects of climate change in the Himalayan regions (Jalabayu paribartanle himali kshetrama pareko prabhav). *Hamro Sampada, 12*(2), 97–98.

Kaspar, H. (2005). *I am the household head now! Gender aspects of out-migration for labour in Nepal*. Kathmandu: NIDS.

Katwal, T. B. (n.d.) *Multiple cropping in Bhutanese agriculture-present status and opportunities*. Paper presented during the "Regional Consultative Meeting on Popularizing Multiple Cropping Innovations as a Means to raise Productivity and Farm Income in SAARC Countries" 31st October–1st November, Peradeniya, Kandy, Srilanka.

Khan, U. F. M., & Ashfaq, F. (2010). Meat production potential of small ruminants under the arid and semi-arid conditions of Pakistan. *Agricultural and Marine Sciences, 15,* 33–39.

Khan, F. (2014). Adaptation vs. development: Basic services for building resilience. *Development in Practice, 24*(4), 559–578. https://doi.org/10.1080/09614524.2014.908823.

Khattri, M. B. (2012). Climate change, millet and ritual relationship with the magars of Argal, Baglung, Nepal. *Dhaulagiri Journal of Sociology and Anthropology, 6*.

Kok, M., Lüdeke, M., & Lucas, P. (2016). A new method for analysing socio-ecological patterns of vulnerability. *Regional Environmental Change, 16,* 229.

Kreutzmann, H. (2006). Introduction. In H. Kreutzmann (Ed.), *Karakoram in transition, culture, development, and ecology in the Hunza Valley* (pp. 1–4). Oxford/Karachi: Oxford University Press.

Kruk, E. (2010). *Two decades of mountain tourism in ICIMOD 1989–2010*. ICIMOD.

Le De, L., Gaillard, G. C., & Friesen, W. (2013). Remittances and disaster: A review. *International Journal of Disaster Risk Reduction, 4,* 34–43.

Macchi, M. (2011). *Framework for community-based climate vulnerability and capacity assessment in mountain areas*. Kathmandu, Nepal: International Centre for Integrated Mountain Development.

Macchi, M., & Gurung. (2015). Community perceptions and responses to climate variability and change in the Himalayas. *Climate and Development. 7,* 5.

Maharjan, A., Bauer, S., & Knerr, B. (2013). *Migration for labor and its impacts on farm production in Nepal*. Centre for the Study of Labor and Mobility, Working Paper IV.

Mamgain, R. P. (2004). *Employment, migration and livelihoods in the hill economy of Uttarakhand*. Ph. D. Thesis, Jawaharlal Nehru University, New Delhi.

McCarthy, J., & D, Mustafa. (2014). Despite the best intentions? Experiences with water resource management in northern Afghanistan. In E. Weinthal, J. Troell, & M. Nakayama (Eds.), *Water and post-conflict peacebuilding*. London: Earthscan.

Ministry of National Planning and Economic Development, UNICEF, UNDP & SIDA. (2011). (2009–2010): Poverty Profile.

Mitra, S. (2014). Synergies among monetary, multidimensional and subjective poverty: Evidence from Nepal. *Social Indicators Research*. 12 December 2014. Retrieved 27 May, 2015, from http://dx.doi.org/10.1007/s11205-014-0831-3.

Mohanty, S. K., Rasul, G., Mahapatra, B., Choudhury, D., Tuladhar, S., & Holmgren, E. V. (2018). Multidimensional poverty in mountainous regions: Shan and Chin in Myanmar. *Social Indicators Research, 138,* 23–44.

Monsutti, A. (2008). Afghan Migratory Strategies and the Three Solutions to the Refugee Problem. *Refugee Survey Quarterly, 27*(1), 58–73.

Morchain, D., Prati, G., Kelsey, F., & Ravon, L. (2015). What if gender became an essential, standard element of vulnerability assessments? *Gender & Development, 23*(3).

Mustafa, D., & Qazi, U. (2007). Transition from Karez to Tubewell irrigation: Development, modernization and social capital in Balochistan Pakistan. *World Development, 35*(10), 1796–1813.

Mustafa, D., Ahmed, S., Saroch, E., & Bell, H. (2011). Pinning down vulnerability: From narratives to numbers. *Disasters, 35,* 62–86.

Mwangi, S., & Kimathi, M. (2006). African leafy vegetables evolves from underutilized species to commercial cash crops. In *Research workshop on collective action and market access for smallholders*, 2–5 October 2006, Cali, Colombia (pp. 1–19).

Nightingale, A. J. (2011). Bounding difference: The embodied production of gender, caste and space. *Geoforum, 42*(2), 15–162.

Ning, W., Rawat, G. S., Joshi, S., Ismail, M., & Sharma, E. (2013). *High-altitude rangelands and their interfaces in the Hindu Kush Himalayas*. Kathmandu: International Centre for Integrated Mountain Development.

Nepal, S. K., & Chipeniuk, R. (2005). Mountain tourism—Toward a conceptual framework. *Tourism Geographies, 7,* 313–333.

Odell, M. J., & Lama, W. B. (1998). Tea house trekking in Nepal: The case for environmentally friendly indigenous tourism. In P. East, K. Luger, & K. Inmann (Eds.), *Sustain ability in mountain tourism: Perspectives for the Himalayan countries* (pp. 191–206). Innsbruck-Vienna: Oeko Himal Publication.

Oxford Poverty and Human Development Initiative (OPHI). (2016). *Multidimensional poverty index in Pakistan*. Oxford: OPHI. http://www.ophi.org.uk/ophi_stories/pakistan-launches-mpi/.

Padulosi, S., Bergamini, N., & Lawrence, T. (Eds.). (2011). On farm conservation of neglected and underutilized species: Status, trends and novel approaches to cope with climate change. In *Proceedings of an international conference, Frankfurt*, 14–16 June 2011. Rome: Bioversity International.

Pandey, R., & Jha, S. K. (2011). Climate vulnerability index—Measure of climate change vulnerability to communities: A case of rural Lower Himalaya, India. *Mitigation and Adaptation Strategies for Global Change, 17*(5).

Pandey, P. R., Pandey, H., & Nakagawa, M. (2009). Assessment of rice and maize based cropping systems for rural livelihood improvement in Nepal. *Journal of Agriculture and Environment, 10,* 57–64.

Papola, T. S. (2002). *Poverty in mountain areas of the Hindu Kush Himalayas: Some basic issues in measurement, diagnosis and alleviation, talking points 2/02*. Kathmandu: ICIMOD.

Pelling, M., & High, C. (2005). Understanding adaptation: What can social capital offer assessments of adaptive capacity? *Global Environmental Change, 15*(4), 308–319.

Rasul, G. (2014). Food, water, and energy security in South Asia: A nexus perspective from the Hindu Kush Himalayas. *Environmental Science & Policy, 26*(2), 174–180.

Regmi, B. R., Morcrette, A., Paudyal, A., Bastakoti, R., & Pradhan, S. (2010). *Participatory tools and techniques for assessing climate change impacts and exploring adaptation options. A community based tool kit for practitioners livelihoods and forestry programme*. Kathmandu, Nepal: UKAID, Livelihoods and Forestry Programme (LFP).

Roche, J. M., Santos, M. E. (2013). In search of a multidimensional poverty index for Latin America. In *Presented at 10th Arnoldshain Seminar*, 29 June–2 July, Gottingen, Alemania; *2013ECINEQ Meeting*, 2–24 July 2013, Bari, Italy.

Salick, J., Yongpin, Y., & Amend, A. (2005). Tibetan land use and change near Khawa Karpo, Eastern Himalayas. *Economic Botany, 59,* 312–325.

Santos, M. E. (2013). Tracking poverty reduction in Bhutan: Income deprivation alongside deprivation in other sources of happiness. *Social Indicators Research, 112*(2), 259–290.

Saxena, N. (2016). *Socio-economic development in the north-eastern states of India*. New Delhi (unpublished): UNDP.

Schrieder, G., & Knerr, B. (2000). Labour migration as a social security mechanism for smallholder households in Sub-Saharan Africa: The case of Cameroon. *Oxford Development Studies, 28,* 223–236. https://doi.org/10.1080/713688309.

Sen, A. (1992). *Inequality reexamined*. Oxford: Clarendon Press.

Sen, A. (1999). *Development as freedom*. Oxford: Oxford University Press.

Shah, T. (2009). Climate change and groundwater: India's opportunities for mitigation and adaptation. *Environmental Research Letters, 4,* 035005.

Sharma, P. (Ed.). (2000). *Tourism as development: Case studies from the Himalaya*. Innsbruck, Wien, Munchen: Himal Books and STUDIENVerlag.

Shields, S. A. (2008). Gender: An intersectionality perspective. *Sex Roles, 59,* 301–311.

Sietz, D., Mamani Choque, S. E., & Lüdeke, M. K. B. (2012). Typical patterns of smallholder vulnerability to weather extremes with regard to food security in the Peruvian Altiplano. *Regional Environmental Change, 12*(3), 489–505.

Sietz, D. (2014). Regionalisation of global insights into dryland vulnerability: Better reflecting smallholders' vulnerability in Northeast Brazil. *Global Environmental Change, 25*(1), 173–185.

Singh, S. P., Khadka, I. B., Karky, B. S., & Sharma, E. (2011). *Climate change in the Hindu Kush-Himalayas the state of current knowledge*. Kathmandu, Nepal: International Centre for Integrated Mountain Development.

Singh, P., & Raghuvanshi, R. S. (2012). Finger millet for food and nutritional security. *African Journal of Food Science, 6*(4), 77–84, 29.

Sultana, F. (2014). Gendering climate change: Geographical insights. *The Professional Geographer, 66*(3), 372–381.

Suich, H., Howe, C., & Mace, G. (2015). Ecosystem services and poverty alleviation: A review of the empirical links. *Ecosystem Services, 12,* 137–147.

Sustainable Agriculture Research and Education. (n.d.). *Nonlegume cover crops*. USDA. http://www.sare.org/Learning-Cent.

Synnott, P. (2012). Climate change, agriculture and food security in Nepal —*Developing Adaptation Strategies and Cultivating Resilience*.

Thornton, P. K. (2010). Livestock production: Recent trends, future prospects. *Philosophical Transactions of the Royal Society B, 365,* 2853–2867. https://doi.org/10.1098/rstb.2010.0134.

Tiwari, P. C., & Joshi, B. (2016). Gender processes in rural out-migration and socio-economic development in the Himalaya. *Migration and Development, 5*(2), 330–350. https://doi.org/10.1080/21632324.2015.1022970.

Tiwari, P. C., & Joshi, B. (2012). Natural and socio-economic factors affecting food security in the Himalayas. *Food Security, 4,* 195–207. https://doi.org/10.1007/s12571-012-0178-z.

Townsend, P. (1979). *Poverty in the United Kingdom: A survey of household resources and standards of living, 1967–1969*. Harmondsworth: Penguin Books.

Trani, J. F., & Bakshhi, P. (2013). Vulnerability and mental health in Afghanistan: Looking beyond war exposure. *Transcultural Psychiatry, 50*(1), 108–139.

Tulachan, P. M. (2001). Mountain agriculture in the Hindu Kush-Himalaya: A regional comparative analysis. *Mountain Research and Development, 21*(3), 260–267.

Turner, M. (2016). Climate vulnerability as a relational concept. *Geoforum, 68,* 29–38.

United Nations Development Programme (UNDP). (1990). *Human development report 1990*. New York: Oxford University Press.

UNDP. (1997). *Human development report 1997. Human development to eradicate poverty*. New York, NY: Oxford University Press.

UNDP. (1998). *Human development report 1998*. New York: Oxford University Press.

UNDP. (2010). *The real wealth of nations: Pathways to human development*. Human Development Report 2010. NewYork: UNDP.

UNDP. (2011). *Integrated household living conditions assessment (IHLCA) survey in Myanmar*. Yangon, Myanmar: United Nations Development Programme.

UNDP. (2015). *Human development report*. New York: UNDP. http://hdr.undp.org/sites/default/files/2015_human_development_report.pdf.

Wang, J., Liu, Y. M., Cao, W., Yao, K. W., Liu, Z. Q., & Guo, J. U. (2012). Anti-inflammation and antioxidant effect of Cordymin, a peptide purified from the medicinal mushroom Cordyceps sinensis in middle cerebral artery occlusion induced focal cerebral ischemia in rats. *Metabolic Brain Disease, 27,* 159–165.

Wagle, U. R. (2012). The food stamps program and economic security among low-income families part I. *Poverty and Public Policy, 4*(4), 223–238.

Wiltshire. (2014). *Wiltshire uncovered report 2014*. Wiltshire Community Foundation, Sandcliff House, 21 Northgate Street, Devizes, Wiltshire SN10 1JT.

Wangchuk, S., & Siebert, S. F. (2013). Agricultural change in Bumthang, Bhutan: Market opportunities, government policies, and climate change. *Society and Natural Resources: An International Journal*. https://doi.org/10.1080/08941920.2013.789575.

World Bank. (2017). *Poverty and equity data, world development indicators*. Retrieved May 2017, from http://povertydata.worldbank.org/.

World Bank. (2009). *India receives world's largest remittance flows*. Washington, DC: Author. Retrieved 14 November, 2012, from http://worldbank.org/E6Y6IE0CQ0.

Woodrow, J., & Anderson, M. (1989) Rising from the ashes: Development strategies in times of disaster. *Community Development*.

Wu, N., Yi, S., Joshi, S., & Bisht, N. (eds) (2016). *Yak on the move: Transboundary challenges and opportunities for Yak raising in a Changing Hindu Kush Himalayan Region*. ICIMOD.

Yang, D., & Choi, H. (2007). Are remittances insurance? Evidence from rainfall shocks in the Philippines. *World Bank Economic Review, 21*(2), 219–248.

Yi, S. L., Ning, W., Peng, L., Qian, W., Fusun, S., Geng, S., et al. (2007). Changes in livestock migration patterns in a Tibetan-style agropastoral system. *Mountain Research and Development, 27*(2), 138–145.

Yi, S. L., Ning, W., Peng, L., Qian, W., Fusun, S., Qiaoying, Z., & Jianzhong M. (2008). Agricultural heritage in disintegration: Trends of agropastoral transhumance on the southeast Tibetan Plateau. *International Journal of Sustainable Development & World Ecology, 15*(3).

Yi, S. L., & Ismail, M. (2011). From pastoral economy to rangeland economy: capturing the multifunctionalities of rangeland resources. In K. Kreutzmann, Y. Yang, & J. Richter (Eds.), *Pastoralism and rangeland management on the Tibet Plateau in the context of climate and global change*. GIZ: Bonn.

Zimmerer, K. (2010). Biological diversity in agriculture and global change. *Annual Review of Environment and Resources, 35*, 137–166.

13 Adaptation to Climate Change in the Hindu Kush Himalaya: Stronger Action Urgently Needed

Coordinating Lead Authors
Arabinda Mishra, International Centre for Integrated Mountain Development (ICIMOD), Nepal, e-mail: Arabinda.Mishra@icimod.org (corresponding author)
Arivudai Nambi Appadurai, World Resources Institute (WRI-India), India, e-mail: Nappadurai@wri.org
Dhrupad Choudhury, International Centre for Integrated Mountain Development (ICIMOD), Nepal, e-mail: Dhrupad.Choudhury@icimod.org

Lead Authors
Bimal Raj Regmi, Oxford Policy Management Limited, Nepal, e-mail: bimal.regmi@opml.co.uk
Ulka Kelkar, World Resources Institute (WRI-India), India, e-mail: ulka.kelkar@gmail.com
Mozaharul Alam, UN Environment, Regional Office for the Asia and Pacific, Thailand, e-mail: alam31@un.org
Pashupati Chaudhary, Agriculture and Forestry University (AFU), Rampur, Nepal, e-mail: pashupatic@hotmail.com; pashupati.nature@gmail.com
Seinn Seinn Mu, Department of Agriculture, Ministry of Agriculture, Livestock and Irrigation, Myanmar, e-mail: seinnmu@gmail.com
Ahsan Uddin Ahmed, Centre for Global Change (CGC), Bangladesh, e-mail: ahsan.ua@gmail.com
Hina Lotia, Leadership for Environment & Development (LEAD) Pakistan, e-mail: hlotia@lead.org.pk
Chao Fu, UN Environment International Ecosystem Management Partnership (UNEP-IEMP), China, e-mail: chao.fu@unep-iemp.org
Thinley Namgyel, National Environment Commission (NEC), Bhutan, e-mail: tn@nec.govt.bt
Upasna Sharma, Department of Humanities and Social Sciences, Indian Institute of Technology Delhi, India, e-mail: upasna@iitd.ac.in; upasna.sharma@gmail.com

Contributing Authors
Luni Piya, Graduate School for International Development and Cooperation (IDEC), Hiroshima University, Japan, e-mail: loonypiya@hotmail.com
Sreeja Nair, School of Social Sciences, Nanyang Technological University, Singapore, e-mail: sreeja.nair@u.nus.edu
Navarun Varma, Residential College 4, National University of Singapore, e-mail: navarun@nus.edu.sg
Chandra Sekhar Bahinipati, Department of Humanities and Social Sciences, Indian Institute of Technology Tirupati, India, e-mail: csbahinipati@iittp.ac.in; csbahinipati@gmail.com
Jyoti Nair, Ashoka Trust for Research in Ecology and the Environment (ATREE), India, e-mail: jyoti.nair@atree.org
Kamal Aryal, International Centre for Integrated Mountain Development (ICIMOD), Nepal, e-mail: Kamal.Aryal@icimod.org
Basharat Ahmed Saeed, Leadership for Environment & Development (LEAD) Pakistan, e-mail: Basharat.a.saeed@gmail.com
Umama Binte Azhar, Leadership for Environment & Development (LEAD) Pakistan, e-mail: umamabinteazhar@gmail.com

Review Editor
Asuncion Lera St.Clair, DNV GL, Norway, and Senior Advisor, Earth Systems Services, Barcelona Super Computer Center, Spain, e-mails: Asun.Lera.St.Clair@dnvgl.com; asun.lerastclair@bsc.es

Corresponding Author
Arabinda Mishra, International Centre for Integrated Mountain Development (ICIMOD), Nepal, e-mail: Arabinda.Mishra@icimod.org

P. Wester et al. (eds.), *The Hindu Kush Himalaya Assessment*,
https://doi.org/10.1007/978-3-319-92288-1_13

Contents

Chapter Overview

Key Findings

1. **Adaptation to climate change is increasingly urgent for the HKH—yet for policy makers it is a complex challenge. Adaptation responses by governments in the HKH are mostly incremental and not yet well integrated with development plans and programmes.** Especially at the subnational level, action suffers from poor understanding of existing autonomous responses, as well as insufficient resources, climate information, and clear adaptation options.
2. **In spite of these challenges, opportunities do exist for a scaled up, inclusive, and more comprehensive climate change adaptation responses in the HKH.** Such opportunities may include improved regional cooperation among HKH countries, incentives to promote policy experimentation, initiatives to develop climate literacy, and enhanced private sector engagement.
3. **Bolstering climate change adaptation in the HKH will require very substantial increases in funding than currently available.** As climate change funding filters down from international and national sources to local government and community levels, the funding loses its "identity" as adaptation finance. With appropriate incentive mechanisms, private financing also might support adaptation.

Policy Messages

1. **Climate change adaptation policies and practices must intensify in the HKH—and must become transformative.** Lessons learned from successful policy instruments, such as the Local Adaptation Plans for Action (LAPAs), should widely inform efforts elsewhere. Governments will benefit from mainstreaming these instruments in their planning and budgeting processes. Institutional capacity on adaptation needs to increase until it fits to purpose at each level of governance.
2. **Local-level autonomous responses to climate variability and extreme events must be systematically studied, documented, and validated.** Such responses need to generate critical, practice-based feedback for adaptation planning at higher governance levels.
3. **HKH countries and institutions must work together to build mechanisms and fora to debate and negotiate key challenges, such as data sharing, and to incentivize regional cooperation and cross-learning at a regional scale.** This cooperation is needed to harness relevant opportunities, expertise and experiences.

Climate change impacts in the mountains of the HKH are already substantive. Increased climate variability is already affecting water availability, ecosystem services, and agricultural production, and extreme weather is causing flash floods, landslides, and debris flow. Climate change is likely to have serious effects in the next decades in the mountains of the HKH ***(well established)***. By 2050, mountain temperatures across the region are projected to increase beyond 2 °C on average, and more at higher elevations. Mountain communities—especially remote ones—are more vulnerable to climate change impacts than non-mountain areas ***(established but incomplete)***. The high mountains are poorly served by life-saving and livelihood-supporting infrastructure. Access to climate information and support services is limited, as is the presence of government extension agencies. Weak institutional links hinder farmers from adopting technology that can contribute to adaptive capacity. For poor and marginalized groups, deep and pervasive structural inequalities make climate change adaptation even more difficult. Although the IPCC Fifth Assessment Report Working Group 2 had four chapters on adaptation, literature on mountain specifics was not prioritized. However, although still insufficient, scientific literature in the region is rapidly emerging, and there is a wealth of information emerging from ongoing adaptation action driven by HKH countries.

This chapter focuses primarily on analysing ongoing adaptation activities and identifying common patterns of adaptation response across the eight countries with territory in the HKH—Afghanistan, Bangladesh, Bhutan, China, India, Myanmar, Nepal, and Pakistan. The chapter distinguishes between adaptation responses that are planned by governments or by non-state actors, and those that are local and autonomous, unplanned. Further, it asks whether the planned responses are sufficiently knowledgeable of autonomous adaptation practices, and it considers the extent of policy support for these practices as well as the need to critically evaluate practices and results. Autonomous adaptation is key in the HKH because it draws on local knowledge systems, recognized as repositories of traditions and management practices that can usefully inform adaptation responses. Yet the systematic documentation of local, autonomous responses to climate change in the HKH is limited, and few attempts have been made to validate these responses scientifically.

Policy makers in the HKH countries are aware of the urgency to act on adaptation but face substantive challenges ***(well established)***. Some of those challenges are a lack of adequate data (both in terms of quantity and quality, and especially at a localized scale) about climate change impacts, weak institutional capacity at various governance levels, social and economic barriers to intervention uptake, and poor infrastructure for development and adaptation purposes.

The chapter finds that government-led planned adaptation responses in the HKH are strongly influenced by the evolving global regime under the United Nations Framework Convention on Climate Change (UNFCCC). Notably, the National Adaptation Plan (NAP) process—established in 2010—emphasizes that to reduce vulnerability and build resilience, countries should integrate climate change adaptation into development planning. This assessment shows that all HKH countries have initiated efforts towards such integration, and established high-level coordinating bodies under political leadership. The government-led responses are at both national and subnational levels through multiple plans, programmes, and projects. In all the countries there is clear identification of priority sectors for adaptation interventions; however, attention to mountain specificities is not common ***(well established)***.

Planned responses to climate change by HKH governments and non-state actors are hindered by large constraints on institutional capacity ***(well established)*** leading to major gaps between policy goals and actual implementation of adaptation programmes. This chapter seeks to identify the most urgent of these constraints—the ones that require immediate action. It also describes many adaptation initiatives undertaken by non-state actors. How can one learn from the most successful of these initiatives, be replicated and scaled up, to inform and create synergies with other coordinated actions? The chapter identifies solutions that could work to better connect adaptation science, policy, and practice.

Given climate change impacts are likely to be substantial in the near future—even though they may be nonlinear and subject to high uncertainty—HKH countries will benefit from moving beyond incremental strategies ***(established but incomplete)*** and scale up transformative adaptation, integrating these with development planning and disaster risk reduction. Assessment of the scientific literature, adaptation policy and action by non-state actors shows there is a large potential for enhanced opportunities through investments for generating science-based climate information and knowledge services, as well as incentives to promote policy experimentation through adaptation pilots. Institutional capacity building on adaptive governance and the creation of knowledge networks are also important. Greater regional cooperation among HKH countries in information and knowledge sharing, particularly in areas such as disaster risk reduction, and food and water security is of critical importance.

There is need for stronger integration of adaptation with national development plans and programmes, including with the ongoing implementation of the Sustainable Development Goals (SDGs). Gaps in a country's adaptive capacity cannot be addressed until political leadership pushes for an intensified adaptation response within the larger development regime. It is also critical to enhance regional collaboration, and substantively increase adaptation finance, including mobilization of funds for greater social protection and risk insurance. Engagement of the private sector in adaptation to climate change and resilience is a critical factor for successful action.

The chapter concludes by putting forward a measurable set of HKH mountain targets consistent with the SDGs, and proposes an SDG-consistent priority for the HKH region on adaptation: *Ensure integration between adaptation to climate change, disaster risk reduction, and sustainable development for the mountains through evidence-based decision making and means of implementation.*

Climate Change Adaptation and the SDGs

HIMAP puts forward a vision for meeting the SDGs in the HKH region. Adaptation to climate change cuts across multiple SDGs (and most HIMAP priorities) and, given climate change impacts in the HKH region, is a necessary condition for socio-ecological resilience. It also deserves specific attention as the result of integrated action with development planning and DRR. First, enhancing resilience in HKH ecosystems, communities, human settlements, and urban areas is intrinsically related to ending poverty in all its forms, ensuring nutrition, access to energy and water, enhanced livelihoods, and freedom from all forms of inequality (in particular gender inequalities). This requires a combination of autonomous and transformative adaptation options, strategically embedded in all development trajectories and in close collaboration with all forms of DRR. Autonomous adaptation builds on ongoing adaptive practices that, without explicit planning, increase adaptive capacity. At the same time, there is a need for transformational adaptation, changing systems and behaviour to generate inclusive change, rather than coping only with climate impacts. There is an urgent need for policy approaches that are pro-poor, socially inclusive, and well integrated with DRR and other resilience-building interventions, while also being forward looking, seeking co-benefits and attentive to trade-offs with the wider sustainable development agenda.

Adaptation to climate change needs to contribute to promoting sustainable production systems to assure food security, nutrition security, and income for mountain people, with particular attention to women's changing roles in agriculture. This encompasses SDGs 2, 5, 6, 7, 9, and 11. Clearly, SDG 13—*Take urgent action to combat climate change and its impacts*—is a central SDG for adaptation to climate change. For the HKH, target 13.2: *Integrate climate change measures into national policies, strategies and planning* is of utmost relevance, and the NAP process can be very effective in this regard.

Further, this assessment experience leads us to pay close attention to SDG 16—*Promote peaceful and inclusive societies for sustainable development, provide access to justice for all, and build effective, accountable, and inclusive institutions at all levels.* Target 16.7: *Ensure responsive, inclusive, participatory and representative decision-making at all levels* is of particular importance. Inclusive institutions are essential in the HKH for adaptation measures to be consistent with social justice.

Finally, SDG 17—*Strengthen the means of implementation and revitalize the global partnership for sustainable development*—includes target 17.6, stressing the need for enhanced regional and international cooperation as well as partnerships with non-state actors, such as business and industry. Adaptation to climate change in the HKH will depend on meeting this target, for practical reasons related to financing, technology transfer, and capacity building.

We propose an SDG-consistent priority for adaptation to climate change in the HKH:

- *Ensure integration between adaptation to climate change, and DRR and sustainable development for the mountains, through evidence-based decision making and means of implementation.*

To materialize this goal, and drawing from the lessons learned in this assessment, we propose 10 indicators of climate change adaptation for the HKH, consistent with SDG priorities and targets, to guide policy makers in implementing the SDG framework in the region:

1. Number of deaths, missing persons, and persons affected by climate hazards per 100,000 people (disaggregated by sex)
2. Economic loss (as a percentage of national GDP) that is averted by climate-proofing critical infrastructure and basic services
3. Percentage of population with access to improved climate information and services
4. Percentage of population with improved access to successful adaptation technologies
5. Proportion of local governments that formulate and implement local adaptation plans aimed at DRR and resilience building for vulnerable population groups
6. Number of cities or urban settlements with access to safe, climate-resilient infrastructure and service delivery systems
7. Percentage of rural population drawing a major part of their household income from climate-resilient livelihood systems
8. Amount of climate financing flowing locally for climate change adaptation (for example, the percentage of the national budget allocated to mountain districts)
9. Access to international funding (for example, from the Green Climate Fund)
10. Number of knowledge institutions actively engaged in adaptation knowledge generation, communication, pilots, and scale-up relevant to the mountain context.

13.1 Adaptation to Climate Change: Stronger Action Urgently Needed

Temperature across the mountains of the HKH is projected to increase by about 1–2 °C by 2050, and by even more at higher elevations (Shrestha et al. 2015). Winters are expected to grow warmer at a faster rate than summers in most places. Precipitation projections are less certain and the spatial variability is high (see Chap. 3). The likely increase in frequency and magnitude of extreme weather events such as high intense rainfall leading to flash floods, landslides, and debris flow (Shrestha et al. 2015) can have serious consequences. These events would most likely affect various productive sectors in the HKH countries, particularly the natural resources systems that provide the livelihoods for poor and marginal communities, and impede the development process (Ahmed and Suphachalasai 2014; Moors and Stoffel 2013; Su et al. 2013). There is also the recognition that climate-related population dislocation will be significant in Bangladesh, China, India, and Pakistan (ADB 2012).

The current level of understanding of adaptation needs and interventions specific to mountain situations continues to be highly limited because of inadequate knowledge of climate change impacts on mountain people and ecosystems. At the same time, adaptation is becoming increasingly urgent for the HKH. A number of research studies (as cited in earlier chapters of this volume) show that increased climate variability is already affecting water availability, ecosystem services, and agricultural production in the region. The huge impact of frequent extreme weather events on life and property in the region underscores the urgency of the situation and the necessity to ramp up adaptation action there.

Five of the eight HKH countries—Afghanistan, Bhutan, Myanmar, Nepal, and Pakistan—are predominantly mountain countries and, at the same time, four of these (excluding Pakistan) are classified as Least Developed Countries (LDCs). Even in case of the remaining two South Asian countries—Bangladesh and India—the mountain states/regions compare poorly with most of their non-mountain counterparts in terms of GDP and human development index (HDI) indicators. In China, most people living in poverty reside in mountain regions that occupy nearly two-thirds of the land.[1] Thus, for mountain people in HKH countries, climate change impacts carry a significant risk of undermining the achievement of fundamental human rights such as those to food, health, adequate housing, and access to safe drinking water and sanitation (Cameron et al. 2013).

While climate change is acting as a natural driver of change, there are several other driving forces such as population change, economic growth, urbanization, and globalization (see Chap. 2) causing rapid socioeconomic transformation in the HKH. This process of change is often accompanied by conditions of high uncertainty due to complex interactions among the driving forces. For adaptation planning and actions under circumstances of large-scale changes (some of which may be nonlinear) and high uncertainty, the HKH countries are required to go beyond incremental strategies and initiate transformative development and adaptation. An example of transformational change urgently required in the HKH is enabling mountain women, left behind because of male out-migration, to emerge from being 'frontline victims' of climate change impacts to become 'risk and resource managers' with control over productive assets (Mishra et al. 2017a). There is definitely a capacity need for policy makers in the HKH to plan and deliver such change, but addressing this need will be possible only when the political leadership pushes for an intensification of the adaptation responses within a larger transformative development regime.

Box 13.1 Approach to this assessment

This assessment seeks to identify patterns in adaptation response common to the eight HKH countries—Afghanistan, Bangladesh, Bhutan, China, India, Myanmar, Nepal, and Pakistan. We look at both autonomous and planned adaptation responses to climate change impacts. In the "planned" category we seek to distinguish between government responses and initiatives taken by non-government actors. We try to understand to what extent adaptation policy formulation is learning from autonomous practices; and, vice versa, we ask to what extent autonomous adaptation is receiving policy support—especially in the case of mountain people's farm-based activities that are of key importance to rural livelihoods in the HKH. In the case of planned adaptation, institutional capacity constraints are many and we seek to identify those that require urgent attention. Adaptation practice in the HKH is dominated by many disconnected activities and initiatives, often mostly proposing incremental change. The question is how to move beyond such fragmentation, and beyond incremental action, to replicate, scale up, and create synergies from mutual learning and coordinated action. This requires transformative approaches that truly generate integrated results with development planning and DRR activities.

13.1.1 Adaptation to Climate Change in the HKH: A Complex Challenge Compounded by Social Differentiation and Poor Development

The mountain communities in HKH countries, particularly those located in remote areas, are most vulnerable to climate change impacts (especially disasters caused by extreme weather events). The adaptation needs of several subsets of highly vulnerable groups in the HKH—such as indigenous peoples, women, migrants and migrant-sending households, urban slum dwellers, and minorities—deserve special understanding and targeted action. Deep and pervasive structural inequalities in HKH societies make adaptation even more difficult for poor and marginalized people. Entitlements to elements of adaptive capacity (for example, ownership of productive assets and access to services of local government agencies) are typically socially differentiated, and this affects the uptake of coping strategies. For instance, in studies conducted in the central and western mountains of Nepal, caste hierarchy and patriarchy-led gender restrictions have been found to act as barriers for socially marginalized groups within a locality to access certain institutions and adopt the adaptation options that are easily accessible to the so-called higher castes (Jones and Boyd 2011; Onta and Resurreccion 2011). For HKH policy

[1]http://www.fao.org/docrep/ARTICLE/WFC/XII/0510-A3.HTM.

makers, such social embeddedness makes adaptation a complex challenge.

The other key constraints to adaptive capacity building of the HKH mountain communities arise because of low availability of relevant scientific information and limited access to climate information and support services. Connectivity through internet and mobile phone services, although improving over time, is still inadequate and not universal in the hill and mountain areas of the region. Because of the difficult terrain, government extension agencies have sparse presence and weak outreach in most parts of the HKH geographical area. Poor information flows and weak support from the extension system results in poor adoption among farmers of new and innovative adaptation measures.

Similar to the thin presence of agricultural extension services in mountain regions, the level of penetration of life- and livelihood-supporting infrastructure is considerably lower in high mountainous terrains compared to other areas within HKH countries. Adaptation measures for infrastructure assets in vital sectors such as transportation, water storage, irrigation, energy, and urban utilities are lacking. Lack of investment in this regard has long remained a challenge in the HKH. The investment bottleneck must be addressed by the leadership of these countries, in order to dent the prevailing development injustice, so that people in the HKH can use such infrastructure and accrue adaptation benefits.

13.1.2 Response of HKH National Governments to Adaptation Challenges, and Influence of Evolving Global Regime

The emergence of adaptation as a response option within the UNFCCC has been marked by several milestones. The first significant action towards adaptation started in 2001 with the establishment of the LDC Expert Group to provide technical support to the LDCs for preparing their National Adaptation Programmes of Action (NAPA); and with the establishment of the LDC Fund and the Special Climate Change Fund (SCCF) to support LDCs and developing countries, respectively, with adaptation action. This was followed by a knowledge and networking platform to support adaptation under the Nairobi Work Plan in 2005. The Adaptation Fund was launched in 2007 under the Kyoto Protocol to fund adaptation action in developing countries under a novel "direct-financing" modality. The Cancun Adaptation Framework in 2010 further increased the focus on adaptation under the UNFCCC process with the establishment of the Adaptation Committee and the launch of the process to formulate and implement National Adaptation Plans (NAPs) for LDCs and other developing countries. Most recently the Paris Agreement in 2015 cemented adaptation under Article 7 which defines an adaptation goal, the approach towards adaptation at the national level, and also provides for "adaptation communications" by parties to submit their priorities and needs for adaptation.

The NAP process established in 2010 emphasizes that reducing vulnerability and building climate resilience requires the integration of adaptation planning with overall development planning. This is further carried into the Paris Agreement with the adaptation goal of "enhancing adaptive capacity, strengthening resilience and reducing vulnerability to climate change, with a view to contributing to sustainable development".[2] All the HKH countries have included adaptation actions in their Nationally Determined Contributions (NDCs),[3] and the majority have explicitly identified the NAP process as the approach for adaptation (Table 13.1).

In their NDCs, some HKH countries emphasize building synergy between climate action and achieving the SDGs. For instance, Myanmar's climate change strategy and action plan identifies adaptation indicators that are linked with the SDG targets in sectors such as health, climate change, biodiversity, and food security. Similarly, Nepal's NDC states that it is "imperative for Nepal to tackle the impact of poverty and climate change simultaneously to achieve Sustainable Development Goals" (p. 9). The NDCs of India and Pakistan also make emphatic reference to the SDG commitment. In its NDC, China has committed to embarking on a sustainable development path that encompasses economic development, social progress, and action on climate change.

Adaptation priorities, based on a review of UNFCCC documents submitted by HKH countries (NAPA, national communications, NDCs), commonly identify agriculture, water, health, forests and biodiversity, and disaster management as key sectors for intervention. Mountain-specific adaptation issues are observed to have received varying degrees of emphasis in country priorities. Thus, in the case of Bhutan (a wholly mountainous country) all its NAPA priorities relate to mountain-specific climate risks and vulnerabilities. Strengthening early-warning systems for floods and glacial lake outburst floods (GLOF) risk reduction are identified as requiring a significantly higher density of stations owing to micro-variation in the topography and climates in a mountain environment. Nepal's NAPA and Local Adaptation Plans of Action (LAPA) specifically recognized the key risks and vulnerabilities of fragile mountain ecosystems such as the rain shadow districts in the mid- and far-western region as the most vulnerable and prioritized the geographic areas for adaptation interventions. Another example of a mountain-specific policy response is India's National Mission for Sustaining the Himalayan Ecosystem, which is one of the eight missions under the country's

[2]https://sustainabledevelopment.un.org/frameworks/parisagreement.

[3]All the HKH country NDCs are available at the public registry (interim) maintained by the UNFCCC Secretariat: http://www4.unfccc.int/ndcregistry/Pages/Home.aspx.

Table 13.1 Summary of adaptation components given in Nationally Determined Contributions, by country

Country	Adaptation vision/goal	Status with respect to NAPA & NAP (or other (sub)national policy frameworks)	Adaptation priorities (in terms of sectors or actions)	Cost of adaptation measures	Target years	Date of INDC submission to UNFCCC
Afghanistan	"to protect the country and its population by enhancing adaptive capacity and resilience, effectively respond to the vulnerabilities of critical sectors, and efficiently mainstream climate change considerations into national development policies, strategies, and plans" (p. 4)	NAPA and NCSA completed in 2009. NAP process under way	• Development of the CCSAP and vulnerability monitoring and assessment system • Mainstreaming of adaptation technologies • Regional and international cooperation for adaptation technology transfer • Meteorological and hydrological monitoring networks and services • Water resources infrastructure and irrigation systems • Community-based natural resources management • Selected species and habitat conservation • Alternative and renewable energy • Regeneration of degraded forests and rangeland areas	USD10.785 billion (of a total financial need of USD17.405 billion)	2020–30	13 October 2015
Bangladesh	"to protect the population, enhance their adaptive capacity and livelihood options, and to protect the overall development of the country in its stride for economic progress and wellbeing of the people" (p. 10)	NAPA submitted in 2005 (revised in 2009); NAP roadmap prepared	• Improved early warning systems • Disaster preparedness and protection measures • Climate-resilient housing, infrastructure, and communication • Urban drainage • River training and dredging • Stress-tolerant variety improvement and cultivation (including livestock and fisheries) • Research and knowledge management • Health • Biodiversity and ecosystem conservation • Institutional capacity building	USD42 billion (of a total financial need of USD69 billion)	2015–30	25 September 2015
Bhutan	"to remain carbon neutral" (p. 1) "Adaptation to adverse impacts of climate change is a priority" and is to be pursued through NAP towards "reducing vulnerability by both integrating climate change adaptation into national development planning and	NAPA prepared in 2006 and updated in 2012; NAP process to begin on receipt of support	• Water security through IWRM • Climate-resilient agriculture and livestock farming • Sustainable forest management and conservation of biodiversity	Costs not indicated	2018–23 (actions integrated in 12th 5-year Plan)	30 September 2015

(continued)

Table 13.1 (continued)

Country	Adaptation vision/goal	Status with respect to NAPA & NAP (or other (sub)national policy frameworks)	Adaptation priorities (in terms of sectors or actions)	Cost of adaptation measures	Target years	Date of INDC submission to UNFCCC
	also implementing priority adaptation actions on the ground" (p. 5)		• Resilience to climate change-induced hazards • Minimize climate-related health risks • Climate-proof transport infrastructure • Climate information services for vulnerability assessment and adaptation planning • Renewable and climate-resilient energy generation			
China	"to proactively adapt to climate change by enhancing mechanisms and capacities to effectively defend against climate change risks in key areas such as agriculture, forestry and water resources, as well as in cities, coastal and ecologically vulnerable areas and to progressively strengthen early warning and emergency response systems and disaster prevention and reduction mechanisms" (p. 5)	National Strategy for Climate Adaptation under implementation	**Measures to enhance overall climate resilience**: • Safe operation of infrastructure of water conservancy, transport, and energy against climate change • Optimal water resources management • Water conservation facilities for farmlands, to vigorously develop water-saving agricultural irrigation and to cultivate heat- and drought-resistant crops • Resilience of coastal areas against climatic disasters • Track, monitor, and assess the impact of climate change on biodiversity • Strengthen forestry infrastructure • Effectively safeguard urban infrastructure • Contingency planning and capacity building for public health services • Improve early warning and communication system • Strengthen disaster risk reduction and emergency response systems	Costs not indicated	By 2030	30 June 2015
India	"To better adapt to climate change by enhancing investments in development programmes in sectors vulnerable to	NAPCC and SAPCC under implementation	The NDC estimate of USD206 billion is for implementing adaptation actions in agriculture, forestry, fisheries	USD206 billion (at 2014–15 prices; mitigation cost estimated around USD834 billion	2015–30	1 October 2015

(continued)

Table 13.1 (continued)

Country	Adaptation vision/goal	Status with respect to NAPA & NAP (or other (sub)national policy frameworks)	Adaptation priorities (in terms of sectors or actions)	Cost of adaptation measures	Target years	Date of INDC submission to UNFCCC
	climate change, particularly agriculture, water resources, Himalayan region, coastal regions, health and disaster management" (p. 29)		infrastructure, water resources and ecosystems. Additional investments would be needed for strengthening resilience and disaster management (p. 31)	until 2030 at 2011 prices)		
Myanmar	"a vision for achieving climate resilient, low-carbon, resource efficient and inclusive development as a contribution to sustainable development" (p. 1)	NAPA under implementation since 2012; NAP to be developed	**NAPA priority sectors**: 1. Resilience in the agriculture sector, developing early-warning systems, and forest preservation measures 2. Public health protection and water resource management 3. Coastal zone protection 4. Energy and industry sectors, and biodiversity preservation	Costs not indicated	Not indicated	28 September 2015
Nepal	"Nepal's Climate Change Policy (2011) envisions a country spared from the adverse impacts of climate change, by considering climate justice, through the pursuit of environmental conservation, human development, and sustainable development … The Policy has objectives of … enhancing the climate adaptation and resilience capacity of local communities for optimum utilization of natural resources and their efficient management" (p. 3)	NAPA prepared in 2010; LAPAs being implemented in 90 village development committees and 7 municipalities, and nearly 2,200 CAPA for community forests developed; NAP preparation ongoing since 2015	• NAP formulation and implementation; implementation of NAPA and LAPA • Strengthening implementation of EFLG framework in village development committees and municipalities to complement climate change adaptation • Study of impacts of climate change in mountains, hills, and lowland ecosystems and landscapes • Research into loss and damage associated with climate change impacts • Sustainable management of forests • Agricultural sector enhancement by adopting climate-friendly technologies and reducing climate change impacts • Climate-induced disasters in earthquake-affected areas • Institutional-level capacity building	Costs not indicated	Varies from sector to sector	11 February 2016
Pakistan	"To build a climate resilient society and economy by ensuring that climate change is	Process of developing a NAP under way	**Medium- to long-term actions (to 2030)**: • Improving irrigation systems	USD7–14 billion per annum	Medium to long-term actions (to 2030);	6 November 2016

(continued)

Table 13.1 (continued)

Country	Adaptation vision/goal	Status with respect to NAPA & NAP (or other (sub)national policy frameworks)	Adaptation priorities (in terms of sectors or actions)	Cost of adaptation measures	Target years	Date of INDC submission to UNFCCC
	mainstreamed in the economically and socially vulnerable sectors of the economy" (p. 15)		• Enhancing water resource management • Strengthening risk management system for the agriculture sector • Implementing a comprehensive climate-smart agriculture programme • Building climate-resilient infrastructure (focus on water) • Improving the emergency response mechanism for managing extreme climate events and strengthening the development of disaster reduction and relief management systems **Near-term actions (2020–25)**: • Developing NAP • Strengthening subnational adaptation planning capacity • Implementation of actions under "National Disaster Management Plan"		near-term actions (2020–25)	

CAPA, Community Adaptation Plans of Action; CCSAP, Climate Change Strategy and Action Plan; EFLG, Environment-Friendly Local Governance; INDC, Intended Nationally Determined Contributions; IWRM, Integrated Water Resource Management; LAPA, Local Adaptation Plans for Action; NAP, National Adaptation Plan; NAPA, National Adaptation Programmes of Action; NAPCC, National Action Plan on Climate Change; NCSA, National Capacity Needs Self-assessment for Global Environmental Management; SAPCC, State Action Plans on Climate Change; UNFCCC, United Nations Framework Convention on Climate Change

National Action Plan on Climate Change (NAPCC). It is the only mission with a geographical focus and its primary objective is to conserve biodiversity, forest cover, and other ecological values in the Himalaya by scientifically assessing the region's vulnerability to climate change.

13.2 Local Autonomous Responses to Climate Variability and Extreme Events in HKH

Autonomous adaptation by mountain communities to change in general has evolved over the ages (Jodha 1992) and can be applied to climate change. This capacity to adjust is based on possibly thousands of reactive (ad hoc, retrofitting, or retrospective) and proactive (anticipating, precautionary, incremental, or prospective) strategies to deal with climate variability and change, but very little is systematically researched and documented, and is specific to the HKH.

13.2.1 Many Autonomous Adaptation Practices Remain Very Relevant Locally and Need Policy Support

In response to climate risks, farmers in the HKH employ a number of reactive and proactive adaptation practices, such as those used by farmers in Nepal (Table 13.2).

Autonomous adaptation practices are rooted in a community's culture, tacit knowledge base, leadership, innovations, collective action, and experiential learning that are mediated by individuals and local institutions (Agrawal

Table 13.2 Examples of reactive and proactive adaptation practices in the agricultural sector in Nepal

Climate risk	Proactive/prospective adaptation	Reactive/retrospective adaptation
Temperature rise	• Mulching of crops to retain moisture • Terrace wall farming to avert risk of crop failure and to check evapotranspiration • Farmer-managed irrigation systems, community-managed drinking water systems, and traditional water mills • Legume integration with maize	• Changing cropping patterns in response to crop failure following climate risks • Introduction of new crops that did not grow well in the past owing to excessive cold (e.g., apple and maize are being introduced to new locations in Mustang)
Erratic rainfall with frequent drought	• Dry seedbeds for sowing rice • Introduction of drought-tolerant crops such as millet, soybean, black gram • Mixing drought-tolerant local varieties with improved varieties of maize for risk distribution • Varying the sowing dates in different plots for spatial risk distribution in case of seasonal droughts	• Direct seeding when drought damages crops • Collection of wild edible foods • Building adaptive local institutional structures to improve distribution, allocation, and efficient use of water resources
Excessive rainfall and flooding	• Raised seedbeds for nursery preparation • Fencing land, orchards, houses • Growing flood-tolerant crops • Hedgerow plantations of deep-rooted grasses along the contours of sloping agricultural lands to prevent loss of topsoil and erosion	• Altering planting time and methods (e.g., double-transplanting of rice) when flood damages crops • Migration to safer places
Hailstones	• Growing hailstone-tolerant crops such as turmeric, ginger, garlic, onion, yam, cardamom, etc.	• Replacing walnuts with almonds to reduce damage caused by hailstones
Biophysical damage	• Shifting cultivation to cultivate less disrupted and deteriorated lands on a rotational basis	• Bamboo fencing to protect land, orchards, and houses • Change of occupation, migration

2008; Smit and Pilifosova 2001). Examples of autonomous adaptation occurring in different sectors are discussed below to illustrate why and how some practices continue to be relevant and need policy support. Some sectors have more documented examples compared to others.

Agriculture

There is sufficient evidence from the HKH countries to conclude that farmers are already adopting a wide variety of autonomous strategies to adapt to a changing climate. To take a few examples:

- In Pakistan, because of increased uncertainty in rainfall patterns, farmers have begun to change their cropping calendar and in some cases have even altered crop varieties. Some field-level studies have documented this practice in some districts in Southern Punjab and Sindh and have written policy recommendations.[4,5]
- In China, drought is compelling farmers to choose a crop that can adapt to the stress, is multifunctional, and at the same time gives better economic returns under such conditions. In addition, farmers are willing to increase investment in irrigation infrastructure and adopt water-saving technologies in response to climate change and the risk of increasing water scarcity (Wang et al. 2010).
- In hill and mountain districts of Nepal, mixing different varieties of beans and maize seeds is a very common risk-aversion strategy, because at least some of the genotypes survive extreme events and environmental stresses. Farmers also alter the sowing dates by a few days in different plots to distribute the risk of short-duration seasonal droughts during peak growing phases (Piya et al. 2012, 2013). People have developed different agroforestry practices to address frequent drought, landslides, and surface soil erosion, and have constructed cost-effective bamboo fencing to protect themselves from recurrent floods (ICIMOD 2007).
- In the hills of Uttarakhand, India, farmers respond to changes in rainfall by shifting to less water-intensive crops and diversifying their sources of livelihood (Kelkar et al. 2008). Farmers in Sikkim have introduced crops (such as maize, cabbage, pumpkin, and carrot) that were previously unable to grow at high altitudes (Ingty and Bawa 2012).
- In Myanmar's mountain areas traditional rotating fallow system or shifting cultivation is the common adaptation of farmers. Terrace farming and irrigated farming from water flow of natural springs are also autonomous adaptation of hill tribes in Northern Shan.

[4]http://www.lead.org.pk/lead/Publications/Thematic%20Agriculture%20LAPA.pdf.

[5]http://www.lead.org.pk/lead/Publications/Thematic%20Water%20LAPA.pdf.

Livestock/Pastoral

- Pastoral communities in Nepal have been using various reactive and proactive adaptation techniques, such as changing grazing areas and transhumant routes[6] and reducing the length of stay at points en route when biophysical conditions deteriorate; stall-feeding animals, growing fodder and forage, reducing herd size, and feeding crop stubbles, by-products, and hay to livestock during dry periods to cope with shortage of feed; preserving native breeds that bear ecological stresses; digging ponds to store water for feeding animals; and moving animal sheds and houses or entire villages to new locations when people face or anticipate some ecological risks (Aryal et al. 2014; Banjade and Paudel 2008; Moktan et al. 2008).
- Pastoralists in Sikkim, India, have responded to changes in snowfall and rainfall by replacing sheep with yak, and by collectively banning the slaughter and sale of sheep for a few years (Ingty and Bawa 2012).
- In the Chinese context Li and Huntsinger (2011) observed how increasing land privatization and the institutionalization of rigid land tenure in the Inner Mongolia region have weakened traditional practices of pasture and herd management, and in the process reduced the resilience of pastoralists to cope with environmental crises like drought. The traditional common property regimes not only allowed for greater mobility over larger areas, which favoured the ecology of natural grasslands, but also offered the benefits of mutual aid and resource pooling (Cao et al. 2013).

Water Resources

- Water is emerging as the most important and pressing climate change impact in Bhutan, with drying sources being reported in many parts of the country (NEC 2011). Norbu and Kusters (2012) reported that small-scale farmers in Punakha Valley were coping with insufficient water for irrigation from traditional sources by pumping rivers.
- In the hills of Nepal, villagers have constructed community-managed water tanks sourced from free-flowing natural springs for water storage and use during recurring dry seasons, which are reported to have intensified in recent years (Piya et al. 2012, 2013). However, with springs drying up, such a practice is at risk of losing relevance and possible alternatives may not be immediately clear to communities.
- In areas of northern Pakistan, glaciers provide water through slope-side channels. To overcome water deficiency local people practise artificial glacier grafting to create new glaciers.[7]

Forests

- A community forestry programme in Nepal has mobilized communities to manage forest resources in a sustainable way.[8] Communities have adopted numerous practices to sustainably manage and benefit from the forest ecosystems, especially in times of stress such as crop failures and droughts.

Disasters

- Over the last few years, long and recurrent droughts in Nepal are making a vast tract of land untended, ultimately leading to substantive yield reduction in several districts across the country. People have responded by migrating to nearby urban centres. Labour migration is a prominent way of diversifying livelihoods and securing a source of income that is not affected by difficult local conditions and shocks.
- In Sikkim, India, indigenous communities are coping with erratic rainfall and unseasonal floods and landslides by using local ecological knowledge and traditional techniques like riverbank retaining walls, terracing, and stabilizing slopes with native plants and rocks (Ingty and Bawa 2012).

Urban Settlements

- Recent research indicates that the urban poor (for example, slum dwellers) are aware of the adversities of climate change and are willing to undertake measures to mitigate the impacts. An example comes from outside the HKH region. In El Salvador, people living in 15 disaster-prone slum areas invested at considerable cost to themselves in multiple strategies aimed at risk reduction, self-insurance and recovery (Wamsler 2007). Similarly, and closer to HKH, the wealthier households in the cities of India, Pakistan, and Vietnam have been observed to

[6]A transhumance system is a strong institution developed to govern animal movements (redirection), prairie management, and rationalization of resources use in response to or anticipated biophysical and ecological changes.

[7]http://lib.icimod.org/record/12783/files/1081.pdf.

[8]More than 25,000 community-based forest management groups across the country are directly engaged in managing about 30% of the country's total forest area. These community-based organizations are not only contributing to sequestering carbon dioxide by sustainable management of forest resources, but also playing effective roles in designing and implementing CAPA based on forest and non-forest benefits (Nepal's INDC, p. 4).

invest in housing improvements to adapt to floods, heat, and typhoons, respectively (Moench et al. 2015).

13.2.2 Autonomous Adaptation May Be Unable to Manage New Risks and 'Surprises' from Climate Change

Although individual and institutional efforts in adapting to climate change are under way, new challenges continuously emerge as "surprises" and many of them are difficult to tackle. Crop failures and life threats caused by new pests and diseases, long droughts, flash floods, GLOF, forest fires, thunderstorms, hail, and other disasters are becoming more unpredictable, irregular, and fierce. Very little is known about the ecological, social, economic, and political conditions required to adapt to such surprises and changes (Varma and Mishra 2017). According to the communities in the eastern Himalaya in India, traditional coping strategies enabled them to withstand environmental stresses, but it is not sufficient to fully insure them from the various threats of climate change likely to occur in the near future (Barua et al. 2014).

Kusters and Wangdi (2013) argued that, as opposed to extreme events such as floods and GLOF, gradual changes such as changes in water supply can be overlooked. They reported that individual efforts by farmers in the Punakha region of Bhutan to adapt to changes in water supply were insufficient and required external support. They also argued that such gradual changes not only lead to loss of income but also impact social cohesion from increasing conflicts where water-sharing rights are being stressed.

Autonomous adaptation within and across communities varies greatly. Systematic documentation of such responses is limited, and attempts at their scientific validation are rare. Moreover, autonomous adaptation is largely a response to current climate variability rather than to future climate change impacts. Even for current climate variability, there is often a time lag in autonomous response which may be explained by structural rigidities. There is also the role of "aspirations" as a driving force that can influence people's decisions on whether to adopt autonomous adaptation measures or, say, to migrate to cities when faced with adversity.

Some autonomous adaptation practices in response to current climate variability can lead to ecosystem imbalance. Robledo et al. (2012) found that small-scale farmers in Tanzania rely heavily on forest resources for subsistence and income-generating activities. Consequently, their short-term coping practices of fuelwood cutting and selling, charcoal and craft making, and forest products hunting are not sustainable and can hasten deforestation. Such an observation is applicable to the HKH as well.

13.2.3 Local Adaptation Knowledge Merits Greater Attention from Science and Policy

In the HKH there are many instances of traditional institutions playing instrumental roles in managing common resources and sharing benefits accruing from it (Bawa et al. 2008; Chaudhary 2011; Ostrom 1990). For instance, the indigenous system of managing local community land and natural resources including forests, water, and biodiversity is the *Dzumsa* system practised by Lachungpas of northern Sikkim, *Na Zong Nyo* by Lepchas of Sikkim, *Kipat* by Rai and Limbus, and *Sok shing* in Bhutan (Chettri et al. 2007). *Kharka pratha* (managing pastures), *mahavir* (bee cliffs), and customary laws for governing fishing are examples of indigenous practices in parts of the Kanchenjunga Landscape (Khattri 2008). The winter grass cutting ceremony practised in the Kanchenjunga region to manage fodder supply is also a local institution evolved to manage forest resources (Brown 1994; Müller-Boker and Kollmair 2000). Pastoralists perform *ubhauli* (upward movement of herds) and *udhauli* (downward movement of herds) celebrations at certain times of the year before herds are moved from one location to another. *Guthi* is practised to collect and save grain in communal storage, to be used when crops fail following epidemics and endemics.

Current research on adaptation in the HKH largely ignores the local-level knowledge systems that are often repositories of traditional knowledge and resource management practices (Kreutzmann 2011; Manandhar et al. 2012). Traditional knowledge-based practices are clearly useful for adapting to the climate variability historically experienced by the communities. There is, however, a need for generating evidence to establish that such knowledge systems are equally relevant when it comes to dealing with the uncertainty and "surprises" likely to arise from future climate change and its impacts.

Blending tacit and scientific knowledge can play an instrumental role in generating location-specific and efficient adaptation innovations (Chhetri et al. 2012). Xu and Grumbine (2014) presented examples from the Asian highlands of "hybrid" forms of adaptation that combine traditional knowledge and bottom-up practices with top-down government-supported strategies, as an alternative to depending on traditional practices alone. In China's Yunnan Province, farmers have been adapting to climate-driven water stress through a combination of local knowledge (changing planting patterns), market dynamics (switching to commercial crop varieties), and government support (state-funded water storage). From Bhutan, Nepal, and Pakistan there are examples of rural communities combining expert opinion on climate change impacts with local knowledge, and engaging with non-governmental organizations (NGOs) for DRR and management. Xu and Grumbine (2014), however, also argued that the choice of such hybrid forms does not seem to follow detailed foresight and adaptive planning, and that they are usually taken up randomly when conditions of climatic stress prevail.

13.3 State-Led Planned Adaptation in the HKH

The HKH countries currently have various adaptation responses at national and subnational levels through policies, programmes, and projects (Table 13.3). The majority of the national adaptation projects and programmes relate to watershed management, climate-resilient agriculture, improved access to information for decision making, and DRR.

13.3.1 Commonalities Among Policy Responses of National Governments

In terms of the policy goal or vision there is a common emphasis across HKH countries on building resilience at the community level and in core economic sectors. Resilience in the agriculture sector is the first priority of Myanmar's NAPA (2012). The 2011 Climate Change Policy of the Government of Nepal aims to build resilience of local communities by enhancing their capacities for efficient natural resource management and use of climate-friendly technologies (MoE 2011). The 2012 National Climate Change Policy of the Government of Pakistan gives particular attention to the needs of economically and socially vulnerable sectors of the economy for the success of climate-resilient development in the country (MoCC 2012).

China's 2013 National Strategy for Climate Change Adaptation aims to significantly enhance the country's capacity to respond to extreme climatic events and thereby build resilience in key sectors ranging from human health to infrastructure (ADB 2015). The whole country is divided into three types of adaptation regions—urbanized, agricultural, and ecological—to undertake specific adaptation tasks. For example, in the Qinghai-Tibet Plateau, one of the five major ecological regions, the tasks include assessment of the plateau's grassland-carrying capacity, grassland enclosure and recovery, glacier monitoring, wetland management, and development of highland valley agriculture.

The immediate to short-term adaptation interventions identified in the national climate policies seem to revolve around strategic knowledge generation accompanied by public awareness and institutional capacity building. For instance, Bhutan's adaptation interventions through the NAPA 1 and NAPA 2 projects focus on reducing the physical risk of climate-induced disasters such as GLOF, landslides, flash floods, and forest fires, and strengthening capacity at national and local levels in early-warning systems and disaster risk management. Livelihoods and agriculture at the district and community levels have also been the focus of recent interventions. The national strategy and adaptation priorities for Bangladesh focus on reducing the risk of climate-induced hazards and extreme events through improved early-warning systems for tropical cyclones, floods, flash floods, and droughts. The priorities also include sector-specific climate-resilient interventions and capacity building at individual and institutional levels to plan and implement adaptation programmes and projects in the country.

Governments in the HKH have started to integrate and mainstream climate change in their development planning and budgeting systems (Ahmed et al. 2018). The establishment of a trust fund in Bangladesh and climate change budget code in Nepal are some examples of these mainstreaming efforts. Afghanistan, Pakistan, Bhutan, and China have also devised policy measures to integrate climate change within ongoing sectoral policies and strategies. Bangladesh's adaptation priorities are well integrated into development plans, and development priorities are discussed in the context of climate change adaptation (Saito 2012). Bhutan has been integrating environment and climate change into development planning, with climate change identified as one of the national key result areas of the 11th Five-Year Development Plan (2013–18) and in planning at sectoral and decentralized levels through the Poverty Environment Initiative. As indicated in the country's NDC, adaptation will be a focus in the next development plan, and the integration of adaptation planning into the development process will be pursued through the NAP process.

There are indications of policy convergence in HKH countries although implementation continues to be sectoral. The majority of the national commitments in the NDCs of Bhutan, India, Pakistan, Nepal, Bangladesh, and Myanmar have included sectoral policy targets, and many of the sectoral policies in these countries—such as agriculture, forestry, water resources, and energy—are strongly interlinked with climate change policies and programmes. Thus, for instance, the National Integrated Water Resources Management Plan 2016 for Bhutan integrates climate change measures into actions and strategies. The main integration component of the Plan is the Bhutan Water Security Index (BWSI). The BWSI will be the basis for measuring water security and for coordinating planning and performance monitoring of water resource management across various agencies. Disaster and Climate Resilience is one of the five key dimensions used to calculate the BWSI.

Table 13.3 Status of government response to adaptation needs in HKH countries

Country	Key adaptation response at policy level and implementation highlights
Afghanistan	**Policy response**: National CCSAP (being finalized); NAPA for Climate Change (NAPA 2009) and National Capacity Needs Self-Assessment for Global Environmental Management in 2009. Integration of climate change in policies and plans (e.g., National Environmental Action Plan) **Implementation highlights**: Support provided by international community and multilateral agencies; GEF has provided support through enabling activities, mid-size projects, and full-size climate change adaptation projects funded by the LDC Fund
Bangladesh	**Policy response**: Bangladesh CCSAP in 2009, BCCTF and Bangladesh Climate Change Resilience Fund; NAPA in 2005 (revised in 2009); adopted a policy decision by an interministerial body to mainstream climate change in project design of any project under ADP, to be guided and coordinated by the Ministry of Planning **Implementation highlights**: Adaptation projects targeting different sectors and vulnerable geographic areas (e.g., coastal afforestation, pilot programme on climate resilience). Over the last three decades, the GOB has invested over USD10 billion (at constant 2007 prices) to make the country more climate resilient and less vulnerable to natural disasters. In recent times GOB has been investing about 23% of its ADP in projects related to climate variability and change. By June 2015 BCCTF had funded over 236 projects of which 41 had been implemented
Bhutan	**Policy response**: NAPA (2006) and updated project profiles (2012), assessments of capacity needs for climate change, and technology needs assessment for adaptation (2013), initiation of national Climate Change Policy in 2016. Framework to mainstreaming environment, climate change and poverty concerns into Five Year Plan (2013–18), the Integrated Water Resources Management Plan 2016, the National Action Program to Combat Land Degradation 2014, and the National Biodiversity Strategy and Action Plan 2014 **Implementation highlights**: Urgent adaptation priority projects such as GLOF risk reduction; reducing risk of landslides and flash floods in key economic and industrial zones; strengthening national capacity for early warning for GLOF, flash floods, and weather forecasting; pilot projects in agriculture, forestry, and capacity building of local and community disaster management groups
China	**Policy response**: National Program on Climate Change, 2014–15; Action Plan for Energy Conservation, Emission Reduction and Low-Carbon Development; National Plan on Climate Change (2014–20); National Strategy for Climate Adaptation, Science and Technology Actions on Climate Change **Implementation highlights**: Climate adaptation plans and activities in key sectors such as agriculture, water resources, terrestrial ecosystems, coastal zone and regions, and human health
India	**Policy response**: NAPCC 2008 under which there are 10 National Missions; seven missions focus on adaptation in sectors like agriculture, water, Himalayan ecosystems, forestry, health, coastal areas, and knowledge management. State Action Plans on Climate Change prepared to mainstream climate change concerns at subnational level, National Adaptation Fund for Climate Change (NAFCC) set up in 2015, Indian Network for Climate Change Assessment (INCCA) set up in 2010 **Implementation highlights**: India's expenditure on programmes with critical adaptation components increased from 1.45% of GDP in 2000–01 to 2.82% in 2009–10. Expenditure on human capabilities and livelihoods (poverty alleviation, health improvement and disease control, and risk management) constitutes more than 80% of total expenditure on adaptation in India
Myanmar	**Policy response**: Draft Myanmar CCSAP 2016, Myanmar Action Plan on Disaster Risk Reduction (MAPDRR 2012), NAPA submitted to UNFCCC in 2012 **Implementation highlights**: Myanmar Climate Change Alliance established in 2013, ongoing major projects funded by the Adaptation Fund and GEF (e.g., "Building Resilience and Adaptation to Climate Extremes and Disasters" (BRACED) Myanmar Programme launched in March 2015; project on "Addressing Climate Change Risks on Water Resources and Food Security in the Dry Zone of Myanmar"; and project on rehabilitation and restoration of degraded land and reserved forest through community participation)
Nepal	**Policy response**: Launch of NAP formulation process in 2015, Climate Change Budget Code 2012, Climate Change Policy 2011, National Framework on LAPA 2011, NAPA 2010 **Implementation highlights**: Nepal Climate Change Support Programme, Community-based Flood Risk and GLOF risk reduction programme, Ecosystem-based Adaptation Programme, Hariyo Ban Project (climate adaptation component), and Multi-stakeholder Forestry Programme (adaptation co-benefits) under various stages of implementation. The Pilot Program for Climate Resilience (PPCR) is ongoing, comprising: (i) Building Climate Resilience of Watersheds in Mountain Eco Regions; (ii) Building Resilience to Climate Related Hazards; (iii) Mainstreaming Climate Change Risk Management in Development; and (iv) Building Climate Resilient Communities through Private Sector Participation. Climate Change Knowledge Management Centre established
Pakistan	**Policy response**: Pakistan Climate Change Act (2017), The 2013 Framework for Implementation of the Climate Change Policy for 2014–30, National Climate Change Policy and National Disaster Risk Reduction Policy (2012), 18th Amendment to the Constitution on devolution of responsibilities for climate action to subnational level **Implementation highlights**: State implementing projects on disaster risk management and climate resilience practices in agriculture. Also efforts by Government of Pakistan to win two Green Climate Fund projects focusing on mountain communities and GLOF. Specific budgetary allocations at national and subnational levels for execution of the 2013 Framework illustrate efforts towards mainstreaming of climate actions. Efforts by the Lahore High Court to oversee the implementation by government departments on priority areas identified under various sectors in the framework for the implementation of climate change. The Climate Change Act mandates the establishment of three important institutions: the Pakistan Climate Change Council, the Pakistan Climate Change Authority, and the Pakistan Climate Change Fund

ADP, Annual Development Plan; BCCTF, Bangladesh Climate Change Trust Fund; CCSAP, Climate Change Strategy and Action Plan; GEF, Global Environmental Facility; GLOF, glacial lakes outburst floods; GOB, Government of Bangladesh; LAPA, Local Adaptation Plans for Action; LDC, Least Developed Countries; NAPA, National Adaptation Programme of Action; NAPCC, National Action Plan on Climate Change; UNFCCC, United Nations Framework Convention on Climate Change

The majority of the adaptation interventions in Nepal, Bhutan, and Myanmar are being guided by their respective NAPA, which by design are meant to focus on current and immediate threats, and to address these rapidly without waiting for lengthy assessment of long-term potential impacts. An anticipatory response to adaptation is reflected in the emphasis on installation of early-warning systems and structural interventions for DRR through some of the NAPA priorities/projects. Thus, for instance, in Bhutan and Nepal, the government is improving infrastructural adaptation by lowering glacier lake levels and deepening river channels, and installing early-warning systems (Adger et al. 2007; Hijioka et al. 2014; Mimura et al. 2014). Early-warning systems have been developed for extreme weather events such as flooding and wildfires (Molden et al. 2016; Shrestha et al. 2015) at many places in the HKH. In Bhutan, early-warning systems are being tested in infectious disease prevention and vector control programmes, as for malaria (Wangdi et al. 2010), and the network of hydrometeorological stations is being strengthened to increase national capacity to provide early warning for various climate-induced disasters. Sustainable land and watershed management interventions are also being emphasized for the anticipatory approach to providing adaptation benefits.[9] In China, the government has attempted to reduce climate vulnerability and impact by adopting adaptation options of early planting, fixing variety growing duration, and late planting (Tao and Zhang 2010). India's Integrated Agro Meteorological Advisory Service programme has provided climate information services to farmers (Tall et al. 2014). Generally, there is a need to strengthen institutional capacity in the HKH countries for more and better anticipatory planning on adaptation.

Building climate resilience in urban areas is emerging as an extremely critical agenda, given the key roles of these areas for larger socioeconomic and political progress, and overall development of HKH countries. Cities are hotspots of population and infrastructure, and thus extremely vulnerable and at risk (Birkmann et al. 2010; Revi et al. 2014). The IPCC AR5 concludes with "very high confidence" that urban climate change-related risks are increasing, with widespread negative impacts on people, economies, and ecosystems. The report warns about amplified risks for people living in informal settlements, hazardous areas, and in areas lacking essential infrastructure and services (Revi et al. 2014). Such a warning is of particular relevance to the growing urban settlements in the HKH, especially because local governments in the region lack the institutional capacity to deal with the twin problems of climate change and providing for rapidly increasing migration into cities (and thus growth of the urban poor). There are some encouraging examples, though, and the city of Chittagong in Bangladesh provides a good example for planned adaptation by the government. Situated on the Bay of Bengal, the city is frequently prone to cyclones and high tides, and institutional mechanisms have been building the city's responsive capacity, despite being constrained by financial limitations. During periods of non-disaster the City Corporation Disaster Management Committee formulates preparedness activities such as establishing mechanisms for forecast dissemination, drills, and shelter preparation. During times of disaster they are able to recruit more members and respond quickly, thus increasing efficiency (Tanner et al. 2008).

A systematic review of national-level climate policy documents in the HKH reveals that they all aim to build the adaptive capacity of people by means of providing livelihood security in the face of climate change risks; in practice, two-thirds of a total of 21 adaptation projects reviewed in the study were found to be focused on disaster risk management and capacity building (Sud et al. 2015). While this may be an indication of policy–practice disconnect, Ford et al. (2014) are right in cautioning that there are many undocumented adaptation actions and many others possibly built into existing mainstreamed programmes to address development priorities. Some adaptation practices may provide unexpected livelihood benefits, such as the introduction of traditional flood mitigation measures in China leading to reductions in both the physical and the economic vulnerabilities of local communities (Yu et al. 2009).

Research into field-level evidence shows that government-run adaptation efforts in the South Asia region are mostly extensions of business-as-usual activities, which might be inadequate in the long run to meet adaptation needs under the post-NDC global emission regime (Ahmed et al. 2018). Exceptions, like GLOF risk mitigation projects, are rare. An enhanced and intensified adaptation response is needed in these countries, and particularly in the mountain regions, to reduce gaps between growing adaptation needs and actual adaptation delivery.

13.3.2 HKH Countries Have High-Level Polities to Coordinate Decentralized Government Action on Climate Change

The overall governance for climate change in Pakistan is under the Prime Minister's Committee on Climate Change, an overarching body formed in 2004. Pakistan was also one of the first countries to set up a Ministry of Climate Change, in 2012. The Pakistan Climate Change Act (17 March 2017) envisages an overarching Pakistan Climate Change Council

[9]Ministry of Agriculture and Forests 2014, in National Action Program to Combat land Degradation 2014 and the National Biodiversity Strategy and Action Plan 2014.

headed by the Prime Minister of Pakistan, with representation of subnational governments at the Chief Ministerial level. The new law also envisages establishment of a high-powered Pakistan Climate Change Authority and Pakistan Climate Change Fund.

India's Prime Minister's Council on Climate Change and Nepal's Climate Change Council, also headed by the Prime Minister, were established in 2007 and 2009, respectively. In Bhutan the high-level National Environment Commission, which is the highest decision-making body in the government on matters related to environment and is chaired by the Prime Minister, functions as the National Climate Change Committee.

In 2007 the Chinese government set up the National Leading Group (NLG) to Address Climate Change, headed by the Chinese premier, to draw up important strategies, policies, and measures related to climate change. The National Development and Reform Commission (NDRC) under the State Council was vested with the authority to undertake the general work in respect of climate change. In 2008 a department was established in the NDRC, responsible for organizing and coordinating action on climate change all over the country. The Department of Climate Change was transferred to the Ministry of Ecology and Environment (MEE) in 2018.

Given the federal structure of government in most HKH countries, there is a trend towards decentralization of government action on climate change in the region. Soon after the approval of Pakistan's 2012 climate policy, a constitutional amendment (18th Amendment) delegated most of the action areas to the provincial governments. This constitutional amendment meant that the provinces had to begin to develop their own projects, allocate their own resources, and strengthen their own institutions. Responding to the challenge, Khyber Pakhtunkhwa, Punjab, and Sindh have started developing their own climate change policies and action plans.

In China, the governance system and working mechanism to address climate change features the unified leadership of the NLG, administration by the MEE and NDRC, division of work among relevant departments under the State Council, and wide participation of various localities and industries. All provinces, autonomous regions, and municipalities directly under the State Council, as well as some sub-provincial or prefectural cities, have established their own institutional mechanisms to address climate change.

In India, subnational policy making has resulted in State Action Plans on Climate Change that identify adaptation interventions relevant to the local context. A number of adaptation projects in Indian Himalayan states are being funded by India's National Adaptation Fund.

The trend towards decentralized policy making on adaptation is justified on the grounds that the response to climate change impacts has to be at the local level. However, there is little understanding and research happening regarding the ideal distribution of adaptation responsibilities among various levels of government. An important aspect of decentralization on adaptation is the extent and nature of decentralized access to adaptation finance by provincial and local governments. As Table 13.4 shows, two HKH countries (India and Bangladesh) have established statutory funding mechanisms for the flow of adaptation finance to local levels of governance; in some others (Bhutan, Bangladesh, Nepal) there is access to international sources.

13.3.3 Public Consultation and Stakeholder Engagement in Adaptation Planning and Implementation Is Very Uneven Among the HKH Countries

Effective adaptation interventions require harnessing synergies among various government schemes along with stakeholder involvement in monitoring and evaluation of policy implementation (Sud et al. 2015). Stakeholder engagement strategies are a crucial aspect of adaptation interventions (Sherman and Ford 2014). However, public consultation and stakeholder engagement in adaptation planning and implementation is highly uneven and, more often than not, ad hoc within HKH countries. For instance, it has been pointed out that in Nepal the NAPA consultations may have missed out on opportunities to promote inclusive climate change responses, particularly to accommodate the concerns of diverse community groups in the country (Ojha et al. 2015).

There are examples showing that in the absence of collaboration among local stakeholders and national actors, many of the adaptation measures failed to deliver and instead led to maladaptive practices (Regmi and Star 2015). Lack of participation can also create mistrust and lack of ownership, leading to new forms of exclusion or reinforcing existing exclusions. Analysis of relevant literature suggests that adaptation planning that relies heavily on select institutions and actors, and powerful social structures at the local level, is likely to create new inequalities. Regmi et al. (2016) in their study of community adaptation planning in Nepal found that in the presence of elite domination of local institutional mechanisms there is no guarantee that poor and vulnerable households will benefit from adaptation interventions. There is instead a new form of dependency of the marginalized groups on the local elites for receiving benefits from the projects or programmes. Nagoda and Nightingale (2017) argue that, in Nepal, there is the risk of climate change adaptation projects reinforcing power relations that perpetuate the vulnerability context. The case studies of adaptation planning in other LDCs and developing countries also illustrate how even participatory methods can fail to genuinely empower or involve marginal communities in adaptation

Table 13.4 Local funding mechanisms for adaptation

Funding mechanism	Managed by	Nature of support	Purpose	Mechanism
National Adaptation Fund for Climate Change (NAFCC), India	Ministry of Environment, Forests and Climate Change (MoEF&CC), Government of India	Projects	To promote subnational level adaptation activities and good practices across the country	Grants are disbursed to state agencies and civil society organizations based on the needs and priorities identified under the National and State action plans
The Local Climate Adaptive Living Facility (LoCAL), International (HKH countries having LoCAL—Bangladesh, Bhutan, Nepal)	United Nations Capital Development Fund (UNCDF)	Projects	To integrate climate change adaptation into local governments' planning and budgeting systems, increase awareness and response to climate change at local level, and increase the amount of finance available to local governments for climate change adaptation	LoCAL combines performance-based climate resilience grants, which ensure programming and verification of climate change expenditures at the local level, with technical and capacity-building support
Bangladesh Climate Change Trust Fund (BCCTF)	An independent trustee board chaired by ministry of environment and forests, government of Bangladesh	Projects	To support the implementation of Bangladesh's Climate Change Strategy and Action Plan (CCSAP)	Grants disbursed through line ministries
Bangladesh Climate Change Resilience Fund (BCCRF)	Government of Bangladesh with technical assistance from the World Bank	Projects and responding to emergency needs	To support the implementation of Bangladesh's Climate Change Strategy and Action Plan (CCSAP)	Grants disbursed through line ministries
Bangladesh Local Disaster Risk Reduction Fund (BLDRRF)	Technical committee	Small and medium-size projects	To support the most vulnerable communities through NGOs and Union Disaster Management Committees	Grants disbursed based on risk reduction action plans

interventions, in both top-down and bottom-up approaches (Dodman and Mitlin 2013; Sherman and Ford 2014).

Regmi et al. (2016) suggested that local institutions could be made more accountable and responsive to vulnerable households by devolving decision-making power to these households and ensuring inclusive provisions in membership, representation, and resource allocation. Wong (2010) suggested that success in dealing with elite capture lies in the flexible use of "counter-elite" and "co-opt-elite" approaches, together with addressing the need to secure alternative livelihoods and to achieve empowerment for the poor.

The multi-stakeholder-led LAPA process in Nepal is considered to be a successful model for mobilizing local institutions and community groups in adaptation planning and recognizing their role in adaptation. This LAPA framework was developed with the participation and support of government, development partners, and civil society organizations. Adaptation investments were costed by the Government of Nepal and integrated into annual budget plans and resource mobilization strategies (Maharjan and Maharjan 2017). At present, Nepal is implementing LAPA in 90 village development committees and seven municipalities (the lowest administrative units in the country). Similarly, about 375 local adaptation plans and nearly 2,200 Community Adaptation Plans of Action for community forests have been developed (Nepal's NDC:6). The experience of making LAPA in Nepal can be considered a replicable model of a bottom-up approach to adaptation planning for mountain communities.

Pakistan, on the other hand, has implemented LAPA at the state and local levels through the support of NGOs. An analysis of LAPA in Nepal and Pakistan concluded that Nepal's focus on official formalization of the process has come at the cost of delayed implementation, while Pakistan's devolved implementation-centric approach lacks official buy-in to nationally scale up the LAPA (Chaudhury et al. 2014).

13.3.4 Limited Tailored Response to Gender and Social Inclusion

Some HKH countries have national-level climate policies that explicitly identify the most vulnerable groups (indigenous peoples, migrants, and women, for instance), but very few that seek to address their specific adaptation needs through tailored responses (see Box 13.2).

Box 13.2 Country policies and treatment of marginalized people

- Pakistan's National Climate Change Policy 2012 explicitly provides for ensuring the rights of indigenous people in the management of rangeland and pastures. It also has detailed provisions for mainstreaming gender perspectives into climate change adaptation, ensuring women's participation in decision making on climate initiatives, and using the indigenous knowledge of women in climate adaptation (MoCC 2012)
- Nepal's Climate Change Policy 2011 provides for the participation of indigenous communities and women in the implementation of climate change-related programmes, and identifies the importance of traditional knowledge in climate adaptation (MoE 2011). Gender is explicitly included as a cross-cutting theme in Nepal's NAPA, and gender-differentiated information was collected through participatory approaches during its formulation (Mainlay and Tan 2012). In the ongoing NAP process, gender and social inclusion is a cross-cutting theme
- Although India's NAPCC does not explicitly mention indigenous peoples, the National Mission for Sustaining the Himalayan Ecosystem under NAPCC addresses the importance of traditional knowledge and community-based natural resource management in the Indian Himalaya (NAPCC 2008). India's NAPCC recognizes the differential impacts of climate change with respect to gender, but incorporates only a few gender-specific measures in any of its national missions (Parikh et al. 2012)
- Bangladesh's Climate Change Strategy and Action Plan (CCSAP) identifies women as a special category of vulnerable group and has special programmes for incorporating gender consideration into climate change management. Within the HKH countries, it is the only climate change policy document that addresses the issue of climate-led forced displacement. The document has provisions to build the capacity for education and training of environmentally displaced people to ease and facilitate their migration and integration into new societies (MoEF 2009)
- Myanmar's CCSAP (2016) specifies implementation plans for poor and landless households in vulnerable areas of dry zone, delta, mountain, and coastal areas in terms of initiating eco-friendly crops, bioenergy schemes, and livelihood diversification activities (MoNERC 2016)

Studies from Nepal and India have reported local cultural and institutional factors that act as barriers to adaptation, particularly for women (Ahmed and Fajber 2009; Jones and Boyd 2011). Because of women's limited ownership of land, they are still largely excluded from training, extension services, irrigation management, and development schemes intended for farmers (Gioli et al. 2014). Despite the policy provisions, realizing adaptation goals for women and socially marginalized groups in the HKH is likely to be very challenging unless there is urgent and more targeted action on the ground to strengthen and improve their access to and control over productive assets, access to formal and non-formal education, mobility, and opportunities to generate income.

13.3.5 Implementing Adaptation Programmes Is Challenged by Institutional Inertia and Inadequate Institutional Capacity

At both national and subnational levels, HKH countries have weak institutional capacity to deal in an effective and timely manner with climate change impacts. Major gaps exist between policy targets and actual implementation of adaptation programmes.

Most institutional capacity needs are related to access to information, knowledge, and resources. National governments have recognized the seriousness of climate change and have submitted NAPA, but significant scientific knowledge and data gaps remain (Davis and Li 2013). Li et al.'s (2012) assessment of adaptation knowledge needs for China concluded that there is need for a knowledge support system to generate, communicate, and manage climate change knowledge and information in support of policy- and decision-making processes at all levels. In the case of India there are financial, technological, and knowledge gaps in adaptation, as well as capacity building and institutional needs (Garg et al. 2015).

The most pervasive of barriers relate to poor coordination within and between organizations responsible for planning and implementing adaptation actions, and a lack of (or irrelevant knowledge/information on) climate change, as well as ineffectual communication between stakeholders involved in adaptation actions (Spires et al. 2014). In Bangladesh, the limited access of local governments to resources has been cited as a barrier to local adaptation (Christensen et al. 2012). Local planners can find it difficult to negotiate the complexity of adaptation without adequate access to guiding information on vulnerabilities and potential impacts. Local institutions are affected by lack of coordination with state and national policies and priorities, as well as by absent cross-cutting institutional coordination that is weakening environmental governance processes in the region (Tiwari and Joshi 2015).

A context-independent "one-size-fits-all" approach to adaptation interventions, and poor implementation, has the potential to result in failure and even maladaptation. Classifying maladaptation is complex, as the effects of adaptation interventions may be spread over time and geography, and across sectors and generations. A few examples that may illustrate potentially maladaptive impacts of adaptation strategies in the HKH are:

a. In the Humla region of Nepal, adaptation strategies were focused on planting drought-resistant crop varieties, which did contribute to reduced climate risk at household and community levels. However, this also led to the reinforcement of gendered roles in agricultural work and resulted in increased pressure for girls to be removed from schooling. A more gender-sensitive response to climate change could have contributed to overall welfare (Onta and Resurreccion 2011).
b. In India and Pakistan, electricity subsidies for pumping groundwater are potentially maladaptive as it can over time lead to aquifer depletion in the absence of strong institutions governing access and water abstraction (Jones et al. 2015).
c. Excessive use of pesticides is prevalent in China and, as a consequence, food production has had a negative impact on health, costing around USD1.4 billion per year. It has also adversely affected farm and off-farm biodiversity (Norse et al. 2001 cited in Rasul and Sharma 2016).

13.4 Adaptation Initiatives by Non-state Actors Shaping Practice in the HKH

Non-governmental actors working on climate change adaptation comprise a wide range of stakeholders including international and national NGOs, civil society organizations, community-based organizations, private business, and media. The initiatives by non-governmental actors are of paramount importance, because they are expected to complement, build synergy, and leverage resources with government programmes. Presently there is great diversity among such initiatives in the HKH.

13.4.1 Multiple Actors and Approaches, but Little Synergy

In many parts of the HKH it is common that, although a number of NGOs may be working on similar adaptation-related issues, more often than not there is very little functional collaboration, coordination, and communication among the actors. This results in isolation, redundancy, and duplication of interventions, and inefficient resource utilization. It has been observed that NGOs follow frameworks and approaches to adaptation that are often similar with respect to goals, tools used, envisioned actors and stakeholders, and design elements. For instance, climate-smart agriculture (CSA) is being promoted by more than a dozen international and national NGOs in Nepal to adapt to climate-led challenges in agriculture, but each tends to adopt or modify their home-grown practices, and little effort is made to cross-fertilize the ideas (Bhatta et al. 2016).

The frequently used frameworks and approaches in the HKH include sustainable livelihoods framework,[10] community-based adaptation[11] (CBA), and ecosystem-based adaptation[12] (EBA). CSA,[13] climate-smart villages, and climate-compatible development adaptation are also concepts that have come to be used in recent years. Adaptation interventions based on these approaches are generally motivated by the argument for "no or low regrets" actions, which are good options regardless of the need to adapt to climate change. Thus, for example, CSA makes sense with

[10] This framework considers financial, social, physical, human, and natural capital (Brooks and Adger 2005).

[11] CARE uses climate-resilient livelihoods, disaster risk reduction, capacity building, and addressing underlying causes of vulnerability as the core themes of its community-based adaptation framework (http://careclimatechange.org/wp-content/uploads/2015/04/CBA_Framework.pdf).

[12] Payment for ecosystem services is widely piloted in the HKH as an EBA approach. Park and Alam (2015) give the example of EBA focused on agro-forest-based adaptation practices to build ecosystem and community resilience in Nepal.

[13] The term CSA was coined by the Food and Agriculture Organization (FAO) at the Hague Conference on Agriculture, Food Security, and Climate Change in 2010. It is defined as agriculture that "sustainably increases productivity, enhances resilience, reduces/removes greenhouse gas emissions, and enhances achievement of national food security and development goals" (FAO 2010). Various international and national NGOs have been implementing the CSA concept in the HKH (ICIMOD 2015; LI-BIRD 2015).

or without climate change: building more productive and resilient agriculture systems will also increase resilience to climate change.

Bangladesh is possibly the global leader in pioneering people-centric, small-scale adaptation initiatives, which are generally planned and executed by non-state actors. The pioneering role of an early initiative called Reducing Vulnerability to Climate Change (RVCC), implemented by CARE Bangladesh with numerous partner organizations including local government and non-governmental partners between 2002 and 2005, laid the foundation towards designing and delivering concepts such as CBA and EBA (Ahmed 2010). Although the pilot-scale project was implemented in the southwestern region of Bangladesh, the concepts have been further examined, extended, and tested repeatedly by both non-state and government actors across Bangladesh, and gradually across more than 70 countries.[14]

As a result of the early successes of CBA methodology and efforts, the Government of Bangladesh allocated about 10% of its climate financing to engage non-state actors towards further refinement of CBA activities. Such activities encompass micro-level agricultural adjustments to ensure continuation of crop production in climate-affected areas by adopting hazard-tolerant crops, adjusting crop calendars based on early signs of climate trends, replicating advanced agronomic practices, and enhancing farmers' skills to switch to livelihood practices that are not directly affected by climate variability or change. Moreover, CBA activities embrace small-scale water management practices, DRR practices, and creation of protection measures either to reduce vulnerability by reducing exposure to known climate-induced hazard(s) or to reduce sensitivity to such known hazards (Ahmed 2010). Although such measures are gaining rapid popularity among users (because of their relative simplicity to replicate at low cost), as well as among non-state actors, it is argued that they might not be adequate in ensuring resilience of affected communities to such a considerable extent that the remaining vulnerability appears only negligible (Ahmed et al. 2018).

In China, numerous innovative and indigenous ecosystem-based approaches to adaptation have been implemented all over the country. In this regard, the Chinese Ecosystem Research Network (CERN) established by the Chinese Academy of Sciences (CAS) in 1988 provides substantial best practices that are collected and documented during its long-term investigation and management of typical ecosystems in China. Demonstration efforts (Table 13.5) have been made to improve low-yield croplands on the North China Plain, soil and water conservation on the Loess Plateau, revegetation of hilly regions in southern China, rehabilitation of eutrophic lakes, protection and recovery of natural vegetation in ecologically fragile areas such as desertified areas, karst systems, agro-pasture ecotones, and permafrost.

Table 13.5 Major technologies and demonstration models for ecosystem-based adaptation by the Chinese Ecosystem Research Network

Demonstration model	Key issues of the ecoregion
Qianyanzhou Model (forest ecosystem)	Comprehensive, controlled and agriculturally sustainable development in hilly areas, south China
Technology for saline–alkali soil treatment (cropland ecosystem)	Comprehensive treatment of middle- and low-yield fields on the Huanghuaihai Plain, north China
Zhifanggou Model (cropland ecosystem)	Water and soil loss and ecosystem restoration on the Loess Plateau
"1/10 Functional Replacement" Model for Degraded Grassland Restoration (grassland ecosystem)	Adaptive management and sustainable development of the Xilingol grassland, north China
Shapotou Model for Desertification Combating (desert ecosystem)	Construction of sand-fixing vegetation protection system in arid sand area, northwest China
"Three Ring" Ecology-Production Paradigm of Ordos Plateau (desert ecosystem)	Sustainable management of desertified land on the Ordos Plateau, north China
Technology for Water Remediation (lake ecosystem)	Treatment of eutrophication and algal blooms of freshwater lakes, Yangtze River Delta
Model of fenced grassland and artificial grassland (grassland ecosystem)	Wise use of natural grassland and seasonal optimization of livestock structures on the Qinghai-Tibetan Plateau

Source Chinese Ecosystem Research Network Annual Report 2012 (in Chinese)

13.4.2 Scaling up and Sustainability: Challenges Common to Projects Implemented by NGOs

Projects implemented by NGOs are too spread out, small scale, short term, and often embarked upon as piloting, testing, and experimentation, with little chance of being scaled up after project funding ends. The support is not adequate to reach out to a large number of households and make a real lasting impact on a wide scale. The technologies and practices to be promoted are poorly scaled up, partly because the projects are rarely integrated into government programmes. On the other hand, scalability via traditional

[14]Since 2007 the London-based International Institute for Development Studies and the Dhaka-based Bangladesh Centre for Advanced Studies have been organizing annual CBA conferences and helping global extension of good lessons learnt.

government or civil society implementation is limited because of institutional inertia and capacity constraints. A recent assessment of adaptation projects in the rainfed agriculture sector in India highlighted four fundamental enabling factors for scalability: resources, partnerships, knowledge management, and understanding of the local context (Appadurai et al. 2015).

There are some success stories of NGO activities being taken up by government and made part of the adaptation policy framework. For instance, the Government of Nepal is scaling up a home garden approach in 20 districts to promote on-farm diversity to contribute to food and nutrition and improve resilience (diversity for adversity) of smallholder farmers with the technical support of an NGO. Similarly, the riverbed farming practice pioneered in Nepal by a local NGO has recently been mainstreamed into national policy, with the Ministry of Local Development producing riverbed farming guidelines in collaboration with the Riverbed Farming Alliance.[15] Gurung et al. (2014)[16] explained how leasehold riverbed vegetable farming is helping landless and land-poor households to form groups and gain economically by using the dry, sandy beds left after the flooded rivers recede in the *terai* plains of Nepal. Among other factors, extreme weather events such as flash floods are linked to the increasing area of sandy riverbeds, and there is potentially an adaptation opportunity in training the landless poor and marginal farmers to take advantage of this underexploited resource (seasonally dry riverbeds) for income diversification purposes.

In Nepal a novel multilevel institutional partnership that includes collaboration with farmers and NGOs has been instrumental in the innovation of location-specific technologies, thereby facilitating the adoption of technologies in a more efficient manner. This alliance has not only improved knowledge networking among institutions, scientists, and farmers, but also led to more efficient adoption of technologies by farmers. This is evident from NGO work on CSA and participatory plant breeding work (Bhatta et al. 2016; Gyawali et al. 2006).

NGOs are good at securing community participation, but they sometimes develop programmes and suggest policies without adequate public consultation and validation. In many cases proposals seem to be developed in accordance with donor requirements and lack on-the-ground reality. There is also research showing that NGO-promoted participation can be limited to those who are elite, educated, male, and higher caste, often leading to marginalization of the poor, women, and vulnerable households (Regmi et al. 2016).

[15]http://www.riverbedfarmingalliance.org.np/news-and-events.html

[16]http://www.satnetasia.org/sites/default/files/SATNET-FS04-Riverbed%20Farming.pdf

The media has often been playing a supporting role, especially in promoting good CBA practices and highlighting the adaptation needs of certain vulnerable groups and locations in the HKH. It has also played a significant role in highlighting the need for participatory governance in climate financing in a few HKH countries such as Bangladesh and Nepal (Ahmed et al. 2018). The inspiring reports of electronic and printed media in Bangladesh have encouraged many non-state actors to emulate CBA activities in different ecosystems, while the media in Nepal has been involved in LAPA processes.

13.5 Financing Adaptation in the HKH

Finance is the biggest challenge to achieving an adequate adaptation response to climate change impacts in the HKH. To date, the quantum of funds accessed for adaptation purposes from international sources by HKH countries has been grossly inadequate when compared to estimates of what is required for the region.

13.5.1 HKH-Specific Estimates of Funding Need and Availability for Climate Change Adaptation Are Absent

Estimating the future costs of adaptation responses to climate change impacts can be highly speculative and subject to many unrealistic assumptions, especially when many of the losses that future climate change may bring are likely to be non-quantifiable. However, it is important to put some numbers to potential adaptation costs as a way to raise urgency in addressing financial needs.

A comprehensive assessment of climate change adaptation finance need and availability is lacking for the HKH—for both the region as a single assessment unit and for any specific geographical area or sector in each country that falls within the HKH.[17] Therefore we have no knowledge of how

[17]The closest financial estimate is found in the 2014 ADB report, “Assessing the Costs of Climate Change and Adaptation in South Asia” (https://www.adb.org/publications/assessing-costs-climate-change-and-adaptation-south-asia). According to the report: “To avoid the damage and economic losses from climate change under the BAU scenario, the region needs to provide an average adaptation expenditure of 0.48% of GDP per annum (USD 40 billion) by 2050 and 0.86% of GDP per annum (USD 73 billion) by 2100. Obviously, regional adaptation costs under the C–C scenario are much lower than the BAU scenario as it only requires an average of 0.36% of GDP per annum (USD 31 billion) by 2050 and 0.48% of GDP per annum (USD41 billion) by 2100. The study took into account investment in building adaptive capacity in anticipation of future climate change as well as climate proofing measures in key sectors toward climate-resilient development” (Ahmed and Suphachalasai 2014).

much finance is available, how much of the available funds has been allocated to mountain regions or spent on addressing mountain-specific issues, the basis of allocation of funds, the scale of mobilization of private sector finance, and so on.

Our assessment, presented below, has applied several assumptions while translating global and country-level assessments for the HKH.

The Fifth Assessment Report by the IPCC (Chambwera et al. 2014) reports global estimates of the costs of adaptation in developing countries ranging between USD70 billion and USD100 billion per year in the period between 2010 and 2050.[18] The aggregated GDP of all developing countries (low income, lower middle income, and upper middle income) comes to USD26.53 trillion (WB 2016). Considering the aggregated GDP along with the estimated cost of adaptation for developing countries, one may argue that the cost of adaptation against each million of GDP would require about USD80,000–110,000 per year by 2030 and would increase to about USD130,000–190,000 by 2050. Considering the number of people living in the HKH and their per capita income by country, and applying a similar argument, one may argue that the region would require USD3.2 billion to USD4.6 billion per year by 2030, which would increase to USD5.5 billion to USD7.8 billion per year by 2050.

It is highly unlikely that international finance will be ever sufficient to address the adaptation needs of HKH countries. Mobilizing funds from domestic sources is imperative, and some countries in the region are taking steps in this regard. For example, Bangladesh has taken the lead in the region by establishing a climate change trust fund, and has already mobilized USD600 million, of which about USD400 million is from internal government resources and about USD200 million is from development partners. The Bhutan Trust Fund for Environmental Conservation also supports small-scale activities (up to USD300,000 each) in Bhutan to address impacts of climate change on ecosystems and species, as well as to support communities in accessing climate change funds from other sources.

The Government of India (GOI) has estimated that it would require about USD230 million over 5 years (2012–17) to support the NMSHE, and key achievements to date include establishing six new centres relevant to climate change in existing institutions in Himalayan states, creating an observational network to monitor the health of the Himalayan ecosystem, and instituting several capacity-building and training programmes. Under the NAFCC established in 2015, more than one-third of the total projects sanctioned (amounting to about USD68 million) have gone to the Indian Himalayan region[19] for projects such as climate resilient sustainable agriculture in rain-fed farming areas in Jammu and Kashmir, climate smart agricultural solutions in Himachal Pradesh, ecosystem management in Assam, sustainable agricultural development in Mizoram, and rejuvenation and climate proofing of springshed in Meghalaya (Gupta et al. 2018).

In addition to mobilizing domestic sources of funding, HKH countries had accessed about USD600 million from different international sources of climate finance (Climate Fund Update,[20] September 2016). Countries in the HKH, except for Bangladesh, are yet to access climate change adaptation funding from the Green Climate Fund, which has allocated 50% of the fund for adaptation. The Government of China has established the USD3.1 billion South–South Cooperation Fund as an independent contribution to global climate finance, particularly for building the capacity of developing countries in adapting to climate change.

Reliable estimates of public expenditure on adaptation are essential for monitoring and evaluation of interventions by governments. Attempts to mainstream climate change financing within the national development process are still evolving. Three of the HKH countries—Bangladesh, Nepal and Pakistan—have conducted Climate Public Expenditure and Institutional Reviews (CPEIR) that provide useful estimates of the country's climate-related expenditure.[21] Pakistan has gone a step ahead to replace CPEIR with an official Climate Expenditure Coding and Tracking System that will help in tracking climate change expenditures on an ongoing basis.

To reemphasize, many of the losses posed by climate change to the HKH countries are not quantifiable; even then it is important to better understand losses to culture, territory, or religious sites due to climate change impacts and to find proxy quantifiers for an eventual valuation. Climate change risks to intangible issues such as values, faith, or tradition should not be underestimated because of the difficulty of attributing a monetary value.

[18]According to UNEP (UNEP 2016) the costs would be much higher—possibly reaching USD300 billion per year by 2030 and USD500 billion per year by 2050.

[19]This region includes 10 hill States i.e. Jammu & Kashmir, Himachal Pradesh, Uttarakhand, Sikkim, Arunachal Pradesh, Nagaland, Manipur, Mizoram, Tripura, Meghalaya, and two partial hill States namely, Assam and West Bengal.

[20]https://climatefundsupdate.org/.

[21]http://www.pk.undp.org/content/dam/pakistan/docs/Environment%20&%20Climate%20Change/UNDP%20Climate%20Report%20V10.pdf.

13.5.2 Business Potential in Adaptation Is High, but Investment Is Limited

Most businesses, while increasingly becoming engaged in mitigation of greenhouse gases, are yet to take a proactive approach to adaptation in the HKH. Few have even assessed the likely effects of climate change on their value chains and operations, although it is understood that adaptation to climate change can add significantly to a company's budget or project cost.[22] The Task Force for Climate Related Financial Disclosures, however, is leading in raising awareness of the risks posed by climate change to investors and, by default, to many large companies with supply chains in the HKH (TCFD 2017).

That business can be a key stakeholder in adaptation policy and practice in the HKH region is demonstrated by the active participation of the Association of Bhutanese Industries and the Bhutan Chamber of Commerce and Industry in the formulation of the country's NAPA. As a result of their inputs, key industrial areas and investments in the Pasakha area in the foothills of the Bhutan Himalaya were identified and prioritized as being at risk from increasing intensity of rainfall, flash floods, and landslides (ADB 2014). The Punatsangchu Hydro Projects I and II in Bhutan are key stakeholders in the GLOF risk reduction project in the Punatsangchu Valley and have made direct contributions to the installation of early-warning systems (UNDP 2012).

Often the focus falls on the role of the private financial sector in providing risk management options, including insurance and financing for large projects (Khattri et al. 2010). However, there are major sectors of climate financing in India—such as renewable energy, energy efficiency, transport sector improvement, clean-tech, and waste management—which are either from international financing companies or from social–environmental safeguard budgets (GIZ 2015). In Bangladesh the private sector is more engaged in agriculture and climate change adaptation (IFC 2010). Case studies show that the role of the private sector can be instrumental during the period of disaster risk and rehabilitation work. It was effective during the case of post-tsunami rehabilitation work in India and other parts of Asia (Chatterjee and Shaw 2015).

An assessment[23] carried out in Pakistan, on public finance support to stimulate private sector investment in building disaster resilience and climate change adaptation, identified the public finance interventions required to support two sectors: agriculture and construction. The study established the absence of private sector involvement in disaster and climate risk management planning frameworks, as well as in specific initiatives within key economic sectors, or at provincial or local levels.

Mobilizing investment that results in win–win outcomes and building public goods seems to be the way forward for getting business engaged in adaptation. Some other areas, such as agribusiness value chains, tourism, insurance, and climate services—in which we see the potential of business-led adaptation in the HKH—are discussed below, with examples:

- As part of the Pilot Programme for Climate Resilience in Nepal executed by the International Financial Corporation, "agribusiness companies (such as processors of rice, maize, and sugarcane) have committed to provide training to farmers to reduce crop losses and increase productivity, with the long-term aim of sustained private sector involvement and transformational change" (ADB 2014).
- Private companies are experimenting with a range of business models for ICT-based climate and market services, either to open new markets or to safeguard their supply chains. One such mobile-based initiative[24] presently serves thousands of farmers in Uttarakhand State in India. Scaling up of such services in the HKH is technically feasible, as much of the forecasting can be based on (private or public European) weather satellites. Of course, to make such products financially sustainable in the HKH, private sector providers will need to address issues of mobile connectivity, low literacy, and handholding support (customization, interactivity, responsiveness, call centres, field extension agents) for farmers.
- For insurance companies, weather index-based crop insurance is potentially a big market opportunity. Recently the Indian state of Himachal Pradesh opened weather-indexed insurance for ginger, potato, tomato, and pea crops, while a private insurer has implemented a pilot scheme in Uttarakhand and Assam States. There is a pilot-scale effort in Bangladesh towards insuring livelihoods against floods.[25] In Nepal, it is reported that the Insurance Board is making preliminary regulatory changes to introduce rainfall and hailstorm insurance for apple

[22]For example, the International Finance Corporation (IFC) did a risk assessment for the Khimti I Hydropower Project in Dolakha District in Nepal using scenarios of future climate change impacts including variable and uncertain streamflow in the dry season and the risks of extreme events like flash floods, landslides, and GLOF. Incorporating the increased likelihood of floods and GLOF into dam design leads to increased project cost (Communication from PCD, DHPS, MoEA cited in ADB 2014).

[23]https://www.gov.uk/government/uploads/system/uploads/attachment_data/file/305412/stimulating-private-sector-engagement-climate-disaster-resilience.pdf.

[24]http://www.icimod.org/?q=23463.

[25]http://oxfamblogs.org/bangladesh/wp-content/uploads/2013/08/Product-Brochure1.pdf.

farming on a pilot basis.[26] Scaling up such pilot schemes remains a challenge and requires investment in a dense network of weather stations and regulatory changes. As climate risks increase under a climate change scenario, such schemes will further need to be backed by a combination of global reinsurance and public climate finance.

- Pakistan's National Disaster Management Authority is working on a disaster risk insurance fund targeting vulnerable populations for protection against climate hazards.[27] In addition to public sector insurance bodies such as National Insurance Corporation Limited and Pakistan Reinsurance Corporation Limited, private sector insurance organizations have also been engaged through the Insurance Association of Pakistan. One of the key fund design recommendations coming out of this project is for a public–private partnership model. Engagement of the private sector in such an initiative is important to fulfil the capital requirements for setting up the fund and ensuring long-term financial sustainability.

13.5.3 Opportunity to Build Adaptive Capacity from Labour Migration and Migrant Remittances

Human mobility for livelihood purpose is a significant socioeconomic phenomenon in the HKH. The mountainous region generally has higher labour out-migration than in-migration. Whether internal or international, migrant remittances are increasingly becoming an important source of income for households in the HKH (see Chap. 15). Apart from financial remittances, social remittances (acquired skill, knowledge, and confidence) is also extremely important, and considerably under-researched.

Research shows that remittance-recipient households may not be using the financial resources in the best possible manner for improving adaptive capacity (Banerjee et al. 2017) although there is emerging evidence of the potential of certain types of remittances to contribute to adaptive capacity building in certain socio-economic contexts (see Chap. 15). Governments have a critical role in creating enabling conditions in origin communities through the promotion of opportunities for skill-building, entrepreneurship and livelihood diversification, and complementing such interventions with investment for improving the required economic infrastructure.

[26]http://sakchyam.com.np/first-weather-index-insurance-product-launched-in-nepal-for-apple-farmers-of-jumla/.

[27]https://cdkn.org/project/disaster-risk-insurance-for-vulnerable-communities-in-pakistan/?loclang=en_gb.

13.6 The Way Forward—Key Areas for Priority Action by HKH Country Governments

Building on the lessons of this assessment exercise, this section puts forward an agenda for action for HKH countries. Opportunities do exist for a scaled-up, inclusive, and more comprehensive adaptation response in the HKH. This assessment suggests that the way forward should, among other things, include: (a) greater regional cooperation among HKH countries in areas such as DRR and food security; (b) stronger integration of adaptation in national development plans and programmes; (c) convergence of adaptation, DRR, SDGs, and resilience-building priorities; (d) investment for generating science-based climate information and knowledge services, creating knowledge networks, and boosting climate literacy; (e) promoting policy experimentation through adaptation pilots; (f) building institutional capacity on adaptive governance; and (g) engaging with private business.

13.6.1 Regional Cooperation on Adaptation Is Essential

Climate change impacts do not recognize national boundaries, and for the HKH—which shares so many resource systems—there can be significant cross-border implications with the potential to induce political instability and conflict at the regional level. Adaptation to the impacts of climate change can be viewed as a "regional public good" (Sandler 1998) and therefore justifies the call for greater cooperation among HKH countries. Even though upstream–downstream linkages between mountains and plains provide a strong basis for regional cooperation, presently there are very few examples of cross-country adaptation projects or programmes from the HKH (Sud et al. 2015). Among the South Asian riparian countries, enhanced regional collaboration for undertaking integrated scientific research, policy making, and implementation of cross-country adaptation measures has been frequently cited as essential (Viviroli et al. 2011; Xu et al. 2009).

There is scope to structure cooperation among HKH countries around the complete food–water–energy nexus (Rasul 2014). The Climate Summit for a Living Himalayas held in Bhutan in 2011 produced a regional Framework of Cooperation in which the agreed areas of cooperation included food security, natural freshwater systems of the Himalaya, biodiversity, and energy security. However, there has been no follow-up action or movement since the summit, which is indicative of the challenges remaining in implementing regional cooperative action on adaptation.

Initiatives for regional cooperation can take advantage of some global processes as well. The UNFCCC's Lima Adaptation Knowledge Initiative is being implemented through its Nairobi Work Programme in collaboration with UN Environment and has taken steps to foster regional cooperation for identifying "priority knowledge gaps" for vulnerable subregions, and soliciting partnerships and cooperation in bridging knowledge gaps. This can be taken forward to stimulate stronger and more effective knowledge-based partnerships for adaptation among HKH countries.

A transformative stimulus from the global regime would be to allow global adaptation funding mechanisms to put in place funding instruments that can accommodate multi-country, transboundary, and regional adaptation action proposals. At the regional level, a potential mechanism could be the SAARC (South Asian Association for Regional Cooperation) Development Fund, and efforts should be made to explore the potential of SAARC in initiating a funding instrument that sets an example in bridging this critical funding gap.

13.6.2 Stronger Integration of Adaptation in Development Plans and Programmes

Given the diverse socioeconomic and biophysical variability across the HKH, scaling up adaptation to achieve transformative results is a huge challenge. It is therefore important for policy makers to embed adaptation to climate change in ongoing large-scale development-based activities. The NAP process and associated funding from the Green Climate Fund is likely to be helpful in this regard.

While aiming for an integrated approach, it is still necessary (though difficult and politically contentious) to retain the distinction between adaptation and development, to prevent diversion of adaptation investments to conventional development activities, as this potentially could lead to maladaptation (Ayres and Huq 2009; Ericksen et al. 2011). Bundling adaptation with traditional development funds would not be desirable from an operational point of view, since the mandate for the latter is too general, and tracking adaptation spending is likely to become a challenge. This approach of distinguishing additional climate change risks or additional environmental benefits has long been practised in the financing of projects by the GEF, where baseline activities need to be demonstrated and additional financing is provided only for the incremental cost of climate change or other environmental action.

13.6.3 Convergence of Adaptation, DRR, SDGs, and Resilience-Building Priorities

Climate change adaptation is identified as the cross-cutting priority with clear linkages to DRR, SDGs, and resilience building in most of the national policies and plans in the HKH that highlight the importance of greater integration. Those developing countries in the process of preparing NAPs and the countries that already have NAPA (like Nepal) have clearly indicated the need to link climate change adaptation with DRR and SDG priorities. In addition to the policies and plans, several adaptation initiatives in the region have piloted projects to facilitate the integration of climate change with DRR and SDGs.

Overarching national polices and frameworks, which can translate the idea of integration of climate change adaptation, DRR, and SDG into action, need to be developed. This will allow the national HKH governments to strategically think of policy, legal, and financial instruments that can facilitate convergence of the three.

13.6.4 Investing to Enhance Synergies Between CBA and EBA

While there is considerable focus on CBA, many EBA projects also exist around the world, including several in the HKH. A recent example is the GEF–SCCF-funded project Enhancing Capacity, Knowledge and Technology Support to Build Climate Resilience of Vulnerable Developing Countries, which has been recognized as the "first mover" in catalysing global collaborations on EBA in the context of South–South cooperation. It is a joint initiative between UNEP and the NDRC of China, consisting of concrete, on-the-ground EBA interventions in Seychelles, Nepal, and Mauritania, representing coastal, mountain, and arid/semi-arid ecosystems, respectively. Another example is the German-funded mountain EBA, which pilots in three mountainous countries: Nepal, Peru, and Uganda (UNDP 2015). Revi et al. (2014:539) provide examples in their global assessment, concluding that EBA is a "key contributor to urban resilience".

These initiatives have demonstrated the significant potential of an EBA approach for addressing climate change in the HKH. Since trade-offs between competing demands for ecosystem services are becoming increasingly significant, strong political leadership and optimal investment are required to enhance the synergies between EBA and other approaches.

13.6.5 Generating New Knowledge and Boosting Climate Literacy

The adaptation policies and plans of all HKH countries emphasize generating science-based climate information, knowledge, and services to enhance the resilience of climate-sensitive sectors and vulnerable households. For example, in Myanmar and Bhutan, their respective NAPAs have identified a lack of locally usable knowledge and information on weather and seasonal forecasting to assist farm production operations. The NAPA of Afghanistan also identifies lack of empirical data on the extent and impact of desertification, thus hindering effective climate change adaptation. The problem of sparse data is fundamental in mountains, a part of the planet that also has limited infrastructure to generate climate data in the first place. Evidence-based decision making on adaptation warrants greater efforts towards strategic knowledge generation and communication through knowledge networks (for example, the Indian Network for Climate Change Assessment, Himalayan University Consortium) and knowledge management platforms.

In general, across the HKH, people's access to climate change information, knowledge, and services continues to be severely limited. The opportunity, however, lies in the fact that there are strong community networks to exchange information, knowledge, and skills relevant to climate change risk identification and adaptation (MoSTE 2015). For example, in Pakistan, a national-level alliance for climate adaptation comprising almost 200 local-level community organizations has been in existence since 2010. This has contributed to the awareness of local organizations and communities of issues related to climate change.[28] Similar networks can be encouraged and supported in other HKH countries. This has to be accompanied by sustained efforts to promote climate literacy among the general public, which would require massive awareness campaigns, incorporation of climate change issues into the educational curriculum, and active engagement of the media.

13.6.6 A Policy Environment for Social Learning

Responding to novel situations requires the capacity to learn. Social learning plays a critical role in building adaptiveness in natural resource management (Lebel et al. 2010) and enabling transitions (Tschakert and Dietrich 2010). Policy experimentation, including piloting, is important in enabling social learning to overcome system lock-in and facilitate restructuring of existing social–technical systems for changes in norms, values, goals, processes, and actors (O'Brien et al. 2012). In discussing the example of specific water resource management pilots such as improved watermill and spring recharging in the Garhwal Himalaya, Uttarakhand, Agarwal (2013) argues that adaptation policies in the Himalayan states should encourage the development and adoption of replicable and scalable models of conservation of natural resources and livelihood support, based on decentralized planning.

13.6.7 Building Institutions and Capabilities

Effective adaptation requires enhancing adaptive capacity: this includes, among other things, change in institutions and institutional processes for better governance. The transformation of local institutions requires targeted resources for human capacity building as well as for setting up technology-enabled decision support systems. Stakeholder partnerships involving communities and local line agencies of the government will play a crucial role in effective adaptation outcomes. Since in mountainous terrain the reach of line agencies is limited when it comes to providing extension services, one approach would be to leverage existing community networks for the purpose. Social capital is one of the factors that is generally associated with mountain people's ability to recover early from disasters (Mishra et al. 2017b) and this can be considered as an asset when planning for adaptation interventions. Local (and national) agencies of the government are receptive to scaling up interventions by NGOs if tangible results can be demonstrated, and more so when communities advocate the benefit of such interventions and seek clear, focused support.

Local governance organizations, which have accommodated and legitimized customary institutions to formulate strategies for local natural resources, can take the role of bridging organizations in South Asia. One such example is the Dzumsa in Sikkim (India), which is an administrative unit continuing with the traditions of chiefdom. It regulates the migratory pattern for agro-pastoralists in the area, as well as governing other aspects like benefit sharing from different livelihood activities, social conduct, and cultural practices. It reserves certain pastures from grazing during seasons of normal snowfall and levies fines as an instrument of regulation. However, during seasons of erratic snowfall when local predictions cannot be successfully applied, the reserved pastures are made accessible for convenience of the agro-pastoralists. Such local governance institutions can fit between resource-using communities, civil society, local universities and technical institutes, government departments, and elected representatives.

Planned adaptation in the HKH will essentially require a mix of policy responses that consider a range of options from incremental to adaptive and potentially to

[28]http://www.lead.org.pk/lead/Pages/naca.aspx.

transformative strategies. Given that climate uncertainty cannot be reduced completely, it is critical to deploy a decision-making framework to identify a suite of policy options instead of an optimal or best solution (Hallegatte et al. 2012; Smith et al. 2010). Conditions of "surprise" might offer little or no scope for decision makers to respond from history or experience (Lempert et al. 2003; Walker et al. 2010). The concept of adaptive policy making (policies adapt over time as conditions change and learning takes place) has received much attention in the past decade as a useful approach to policy formulation under uncertainty (Swanson and Bhadwal 2009). Capabilities required for adaptive governance need to be created in institutions at all levels of governance in HKH countries.

13.6.8 Engaging with Private Business

Business can become a partner in scaling up adaptation efficiently. There is considerable scope for attracting private finance to adaptation, provided the risks and transaction costs of such investments can be managed. Business looks for profits and scalability, and thus tends to be restricted to regions with a large business presence, market potential, and investment climate sustainability (Chin 2014). Since adaptation is about helping the vulnerable, with a focus on poor and marginal communities, novel models of public–private partnerships are required. Apart from protecting current operations, new revenue streams can be identified, such as ICT and insurance. At the same time local enterprises will need to be supported, not just big business.

Corporate social responsibility can play a very powerful role in building infrastructure and human capital. For example, Google has provided funding for disease outbreak monitoring in Myanmar, and the Rockefeller Foundation is working to build urban resilience in a number of HKH cities (Shimla, Shillong, Guwahati, and Leh) through the Asian Climate Cities Resilience Network (WRI 2009).

13.6.9 Engaging with the Emerging Policy Regime on Loss and Damage

Under the UNFCCC a new policy regime to prepare for inevitable losses and damage is covered under the Warsaw International Mechanism on Loss and Damage (L&D) from Climate Change established in November 2013, and subsequently under the Paris Agreement. Policy makers within the UNFCCC are potentially planning to introduce a climate insurance scheme for highly vulnerable countries (Surminski and Oramas-Dorta 2014). The L&D mechanism is premised on the existence of a residual policy gap between climate change adaptation, DRR, and current social protection and risk transfer mechanisms,[29] wherein climate change may generate conditions that cannot be mitigated, adapted to, or insured against (Dow et al. 2013). Thus, for instance, studies have categorically found that existing adaptation measures are insufficient to avoid potential L&D from weather variability in Bhutan (Kusters and Wangdi 2013) and flooding in Nepal (Bauer 2013).

Although efforts are being made by government agencies to estimate post-disaster L&D, there is still controversy over the definition and estimation methodology (Birkmann and Welle 2015). Even though the direction is unclear at the moment, this assessment feels that it is potentially beneficial for HKH countries to stay engaged with the L&D process.

References

ADB. (2015). *Addressing climate change risks, disasters, and adaptation in the People's Republic of China*. Mandaluyong City, Philippines: Asian Development Bank.

ADB. (2014). *Climate risks in the SAARC region: Ways to address the social, economic & environmental challenges. Draft report for SAARC*. Mandaluyong City, Philippines: Asian Development Bank.

ADB. (2013). *Regional economics of climate change in South Asia: adaptation and impact assessment*. Manila: Asian Development Bank.

ADB. (2012). *Addressing Climate Change and Migration in Asia and the Pacific*. Mandaluyong City, Philippines: Asian Development Bank.

Adger, W. N., Agrawala, S., Mirza, M. M. Q., Conde, C., O'Brien, K., Pulhin, J., et al. (2007). Assessment of adaptation practices, options, constraints and capacity. In M. L. Parry, O. F. Canziani, J. P. Palutikof, P. J. van der Linden & C. E. Hanson, (Eds.), *Climate change 2007: Impacts, adaptation and vulnerability. Contribution of working group II to the fourth assessment report of the intergovernmental panel on climate change*. Cambridge University Press, Cambridge, UK, 717–743.

Agarwal, S. K. (2013). Emerging technological intervention models with scalable solutions for adaptation to climate change and livelihood Gains in Indian Himalayan Region: Case studies on action research at the grassroots level. In S. Nautiyal et al. (Eds.), *Knowledge Systems of Societies for Adaptation and Mitigation of Impacts of Climate Change, Environmental Science and Engineering*. Berlin, Heidelberg: Springer. https://doi.org/10.1007/978-3-642-36143-2_33.

Agrawal, A. (2008). The role of local institutions in adaptation to climate change. *Paper prepared for the social dimensions of climate change, social development department, The World Bank*, Washington, DC, March 5–6, 2008. http://siteresources.worldbank.org/EXTSOCIALDEVELOPMENT/Resources/updated_SDCCWorkingPaper_LocalInstitutions.pdf.

Ahmed, A. U. (Ed.). (2010). *Reducing vulnerability to climate change: The pioneering example of community based adaptation in Bangladesh, centre for global change (CGC) and CARE Bangladesh, Dhaka* (p. 156).

[29]As things stand now, the Social Protection Index (SPI) for South Asia, 0.061, is the lowest of any region. Countries such as Bangladesh, Bhutan, and India spend less than 2% of GDP on social protection and have relatively low SPIs of 0.051 or lower. Nepal does moderately better, despite being a low-income country, with an SPI of 0.068 and spending 2.1% of GDP on social protection (ADB 2013).

Ahmed, A. U., Appadurai, N., & Neelormi, S. (2018). Status of climate change adaptation in South Asia region. In M. Alam, J. Lee, & P. Sawhney (Eds.), *Status of climate change adaptation in Asia and the Pacific*. Switzerland: Springer. (in press).

Ahmed, M., & Suphachalasai, S. (2014). *Assessing the costs of climate change and adaptation in South Asia*. Mandaluyong City, Philippines: Asian Development Bank.

Ahmed, S., & Fajber, E. (2009). Engendering adaptation to climate variability Gujarat, India. *Gender & Development, 17*(1), 33–50.

Appadurai, A. N., Chaudhury, C., Dinshaw, A., Ginoya, N., McGray, H., Rangwala, L., et al. (2015). *Scaling success: Lessons from adaptation pilots in the rainfed regions of India* (p. 112). Washington DC: World Resources Institute.

Aryal, S., Maraseni, T. N., & Cockfield, G. (2014). Sustainability of transhumance grazing systems under socio-economic threats in Langtang, Nepal. *Journal of Mountain Science, 11*(4), 1023–1034.

Ayres, J., & Huq, S. (2009). Supporting adaptation to climate change: what role for ODA? *Development Policy Review, 27*(6), 675–692.

Banerjee, S., Kniveton, D., Black, R., Bisht, S., Das, P., & Tuladhar, S. (2017). Does financial remittance build household level adaptive capacity? A case study of the flood affected households of Upper Assam in India. *KNOMAD Working Paper 18*.

Banjade, M. R., & Paudel, N. S. (2008). Mobile Pastoralism in Crisis: Challenges, conflicts and status of Pasture Tenure in Nepal Mountains. *Journal of Forest and Livelihood, 7*(1), 49–57.

Barua, A., Katyaini, S., Mili, B., & Gooch, P. (2014). Climate change and poverty: Building resilience of rural mountain communities in south Sikkim, eastern Himalaya, India. *Regional Environmental Change, 14*, 267–280.

Bauer, K. (2013). Are preventive and coping measures enough to avoid loss and damage from flooding in Udayapur district, Nepal? *International Journal of Global Warming, 5*(4), 433–451.

Bawa, K. S., Balachander, G., & Raven, P. (2008). A case of new institutions. *Science, 319*, 136.

Bhatta, K. P., Chaudhary, P., Khattri-Chhetri, A., Thapa, K., Rana, R. B., Gurung, D., et al. (2016). *Scaling-up climate smart agriculture in Nepal—Technology identification and piloting methodology report. Local initiatives for biodiversity, research, and development (LI-BIRD) and the consultative group for international agricultural research's (CGIAR) Research program on climate change, agriculture, and food security (CCAFS)*, Kaski, Nepal. Birkmann, J., Garschagen, M., Kraas, F., & Quang, N. (2010). Adaptive urban governance: new challenges for the second generation of urban adaptation strategies to climate change. *Sustainability Science, 5* (2), 185–206. http://dx.doi.org/10.1007/s11625-010-0111-3.

Birkmann, J., & Welle, T. (2015). Assessing the risk of loss and damage: Exposure, vulnerability and risk to climate related hazards for different country classifications. *International Journal of Global Warming, 8*(2), 191–212.

Brooks, N., & Adger, W. N. (2005). Assessing and enhancing adaptive capacity. In B. Lim, E. SpangerSiegfried, I. Burton, E. L. Malone, & S. Huq (Eds.), *Adaptation policy frameworks for climate change* (pp. 165–182). New York, NY: Cambridge University Press.

Brown, T. (1994). Livelihood, land and local institutions. a report on the findings of the WWF feasibility study for the conservation of the Kanchen-Junga Area. *WWF Nepal program report series 4*. Kathmandu: World Wildlife Fund (WWF).

Cameron, E., Shine, T., & Bevins, W. (2013). Climate justice: Equity and justice informing a new climate agreement. *Working paper*. World Resources Institute, Washington DC and Mary Robinson Foundation—Climate Justice, Dublin. Retrieved October 7, 2016 from, http://www.wri.org/sites/default/files/climate_justice_equity_and_justice_informing_a_new_climate_agreement.pdf.

Cao, J., Yeh, E. T., Holden, N. M., Qin, Y., & Ren, Z. (2013). The roles of overgrazing, climate change and policy as drivers of degradation of China's grasslands. *Nomadic Peoples, 17*(2), 82–101. https://doi.org/10.3167/np.2013.170207.

Chambwera, M., Heal, G., Dubeux, C., Hallegatte, S., Leclerc, L., Markandya, B. A., et al. (2014) Economics of adaptation. In C. B. Field, V. R. Barros, D. J. Dokken, K. J. Mach, M. D. Mastrandrea, T. E. Bilir, M. Chatterjee, K. L. Ebi, Y. O. Estrada, R. C. Genova, B. Girma, E. S. Kissel, A. N. Levy, S. MacCracken, P. R. Mastrandrea, & L. L. White (eds.), *Climate change 2014: Impacts, adaptation, and vulnerability. Part A: Global and sectoral aspects. Contribution of working group II to the fifth assessment report of the intergovernmental panel on climate change* (pp. 945–977). Cambridge, UK and New York, NY, USA: Cambridge University Press.

Chatterjee, R., & Shaw, R. (2015). Role of regional organizations for enhancing private sector involvement in disaster risk reduction in developing Asia. In *Disaster management and privates*.

Chaudhury, A. S., Sova, C. A., Rasheed, T., Thornton, T. F., Baral, P., & Zeb, A. (2014). Deconstructing local adaptation plans for action (LAPAs)—Analysis of Nepal and Pakistan LAPA initiatives. *Working paper no. 67*. CGIAR Research Program on Climate Change, Agriculture and Food Security (CCAFS), Copenhagen, Denmark. www.ccafs.cgiar.org.

Chaudhary, P. (2011). Transboundary conservation in Kangchenjunga landscape in the Eastern Himalayas. *Graduate Doctoral Dissertations*. Paper 61. http://scholarworks.umb.edu/doctoral_dissertations/61.

Chhetri, N., Chaudhary, P., Tiwari, P. R., & Yadaw, R. B. (2012). Institutional and technological innovation: Understanding agricultural adaptation to climate change in Nepal. *Applied Geography, 33*, 142–150. Chetri D. K., Nepali Karki, D. B., Sah, R. & Devkota N. R. (2011). Transhumance effect on husbandry practices and physiological attributes of Chauri (Yak-Cattle) in Rasuwa District. *Our Nature, 9*, 128–137.

Chettri, N., Thapa, R., & Shakya, B. (2007). Participatory conservation planning in Kangchenjunga transboundary biodiversity conservation landscape. *Tropical Ecology, 48*, 163–176.

Chin, C. (2014). Time for business partnerships in the Himalayas. https://www.thethirdpole.net/2014/01/27/time-for-business-partnerships-in-the-himalayas/.

Christensen, K., Raihan, S., Ahsan, R., Uddin, A. M. N., Ahmed, C. S., & Wright, H. (2012). *Ensuring access for the climate vulnerable in Bangladesh: Financing local adaptation. ActionAid Bangladesh (AAB), Action research for community adaptation in Bangladesh (ARCAB), Bangladesh centre for advanced studies (BCAS), and international centre for climate change and development (ICCCAD), AAB*, Dhaka, Bangladesh (126 pp).

Davis, M., & Li, L. (2013). *Understanding the policy contexts for mainstreaming climate change in Bhutan and Nepal: A synthesis. Regional climate change adaptation knowledge platform for Asia, Partner Report Series No. 10*. Bangkok: Stockholm Environment Institute. www.asiapacificadapt.net or www.weADAPT.org.

Dodman, D., & Mitlin, D. (2013). Challenges for community-based adaptation: Discovering the potential for transformation. *Journal of International Development, 25*(5), 640–659.

Dow, K., Berkhout, F., Preston, B. L., Klein, R. J. T., Midgely, G., & Shaw, M. R. (2013). Limits to adaptation. *Nature Climate Change, 3*, 305–307. https://doi.org/10.1038/nclimate1847.

Eriksen, S., Aldunce, P., Bahinipati, C. S., Martins, R., Sygna, L., Ulsrud, K., et al. (2011). When not every response to climate change is a good one: Identifying principles for sustainable adaptation. *Climate and Development, 3*(1), 7–20.

FAO. (2010). *Climate smart agriculture: Policies, practices and financing for food security, adaptation, and mitigation. Food and agriculture organization (FAO) of the United Nations*, Rome, Italy.

Fenton, A., Gallagher, D., Wright, H., Huq, S., & Nyandiga, C. (2014). Up-scaling finance for community-based adaptation. *Climate and Development, 6*(4), 388–397.

Ford, J.D., Berrang-Ford, L., Bunce, A., McKay, C., Irwin, M., & Pearce, T. (2014). The current status of climate change adaptation policy and practice in Africa and Asia. *Regional Environmental Change*. https://doi.org/10.1007/s10113-014-0648-2.

Garg, A., Mishra, V., & Dholakia, H. (2015). Climate change and India: Adaptation gap (2015)—A preliminary assessment. *Working paper of Indian institute of management Ahmedabad (IIMA)* W. P. No. 2015-11-01.

GIZ. (2015). *The role of the private sector to scale-up climate finance in India: Final report; international climate initiative/German federal ministry of the environment*.

Gioli, G., Khan, T., Bisht, S., & Scheffran, J. (2014). Migration as an adaptation strategy and its gendered implications: A case study from the Upper Indus Basin. *Mountain Research and Development, 34* (3), 255–265.

Gupta, N., Mishra, A., Agrawal, N. K. & Satapathy, S. (2018). Inter-State cooperation to address climate change adaptation in the Indian Himalayan region—need of the hour. *Economic and Political Weekly, LIII*(12), 35–40, March 24.

Gurung, G. B., Pande, D. P., & Khanal, N. P. (2014). Riverbed vegetable farming for enhancing livelihoods of people: a case study in the Tarai region of Nepal. *Communities and livelihood strategies in developing countries* (pp. 97–108). Japan: Springer.

Gyawali, S., Sthapit, B. R., Bhandari, B., Shrestha, P., Joshi, B. K., Mudwori, A., et al. (2006). Participatory plant breeding: A strategy of on-farm conservation and improvement of landraces. In: B. R. Sthapit & Gauchan D. (Eds.), *Proceedings of a national symposium*, 18–19 July 2006, Kathmandu, Nepal (pp 164–173). On-farm Management of Agricultural Biodiversity in Nepal: Lessons Learned. Nepal Agricultural Research Council, LIBIRD, Bioversity and IDRC.

Hallegatte, S., Shah, A., Lempert, R., Brown, C., & Gill, S. (2012). Investment decision making under deep uncertainty—application to climate change. *Policy research working paper series 6193*, The World Bank.

Hijioka, Y., Lin, E., Pereira, J. J., Corlett, R. T., Cui, X., Insarov, G. E., et al. (2014). Asia. In V. R. Barros, C. B. Field, D. J. Dokken, M. D. Mastrandrea, K. J. Mach, T. E. Bilir, M. Chatterjee, K. L. Ebi, Y. O. Estrada, R. C. Genova, B. Girma, E. S. Kissel, A. N. Levy, S. MacCracken, P. R. Mastrandrea & L. L. White (eds.) *Climate change 2014: Impacts, adaptation, and vulnerability. Part B: Regional aspects. Contribution of working group ii to the fifth assessment report of the intergovernmental panel on climate change* (pp. 1327–1370). Cambridge, UK and New York, NY, USA: Cambridge University Press.

ICIMOD. (2015). http://lib.icimod.org/record/30770/files/CSV15.pdf.

ICIMOD. (2007). http://lib.icimod.org/record/22140/files/attachment_143.pdf.

IFC. (2010). A strategy to engage the private sector in climate change adaptation in Bangladesh; Study carried out by Asian Tiger Capital Partners. Ingty, T., & Bawa, K. S. (2012) Climate change and indigenous people. In: M. L. Arrawatia & S. Tambe (Eds.), *Climate change in Sikkim: Patterns, impacts and initiatives* (pp. 275–290). Gangtok, India: Published by: Information and Public Relations Department, Government of Sikkim.

Jodha, N. S. (1992). Mountain perspective and sustainability: a framework for development strategies. In N. S. Jodha, M. Banskota, & T. Partap (Eds.), *Sustainable mountain agriculture*. New Delhi: Oxford and IBH Publishing.

Jones, L., & Boyd, E. (2011). Exploring social barriers to adaptation: Insights from Western Nepal. *Global Environmental Change, 21*, 1262–1274.

Jones, L., Carabine, E., & Schipper, L. (2015). (Re) Conceptualising maladaptation in policy and practice: Towards an evaluative framework. *PRISE working paper*. London: Overseas Development Institute. http://prise.odi.org/wp-content/uploads/2015/06/Reconceptualising-maladaptation.pdf.

Kelkar, U., Narula, K. K., Sharma, V. P., & Chandna, U. (2008). Vulnerability and adaptation to climate variability and water stress in Uttarakhand State, India. *Global Environmental Change, 18*, 564–574.

Khattri, A., Parameshwar, D., & Pellech, S. (2010). *Opportunities for private sector engagement in Urban climate change resilience building* (p. 94). Bangkok, Thailand: Rockefeller Foundation.

Khatri, N. K. (2008). Traditional practices and customary Laws of the Kirat people of Eastern Nepal and comparison with Nepal's statutory Laws. In N. Chettri, B. Shakya, & E. Sharma (Eds.), *Biodiversity conservation in the Kanchenjunga landscape*. Kathmandu, Nepal: International Center for Integrated Mountain Development (ICIMOD).

Kreutzmann, H. (2011). Scarcity within opulence: Water management in the Karakoram Mountains revisited. *Journal of Mountain Science, 8*(4), 525–534.

Kusters, K., & Wangdi, N. (2013). The costs of adaptation: Changes in water availability and farmers' responses in Punakha district, Bhutan. *The International Journal of Global Warming, 5*(4), 387–399.

Lempert, R. J., Popper, S. W., & Bankes, S. C. (2003). *Shaping the next one hundred years: New methods for quantitative, long-term policy analysis*. Santa Monica, CA: RAND.

Li, L., Fei, X., Xu, J., & Slater, H. (2012), *Scoping assessment of knowledge needs in climate change adaptation in China, adaptation knowledge platform, partner report series no. 2*. Bangkok: Stockholm Environment Institute. http://www.sei-inter.

Li, W., & Huntsinger, L. (2011). China's grassland contract policy and its impacts on herder ability to benefit in Inner Mongolia: tragic feedbacks. *Ecology and Society, 16*(2), 1. www.ecologyandsociety.org/vol16/iss2/art1/.

LI-BIRD. (2015). Scaling up climate smart agriculture in Nepal. *Project brochure*. Pokhara, Nepal: LI-BIRD.

Maharjan, S. K., & Maharjan, K. L. (2017). Review of climate policies and roles of institutions in the policy formulation and implementation of adaptation plans and strategies in Nepal. *Journal of International Development and Cooperation, 23*(1 & 2), 41–54.

Mainlay, J., & Tan, S. F. (2012). Mainstreaming gender and climate change in Nepal. *IIED climate change working paper no. 2*. London: International Institute for Environment and Development (IIED).

Manandhar, S., Pandey, V. P., Ishidaira, H., & Kazama, F. (2012). Perturbation study of climate change impacts in a snow-fed river basin. *Hydrological Processes, 27*(24), 3461–3474.

Mimura, N., Pulwarty, R. S, Duc, D. M., Elshinnawy, I., Redsteer, M. H., Huang, H. Q., et al. (2014). Adaptation planning and implementation. In C. B. Field, V. R. Barros, D. J. Dokken, K. J. Mach, M. D. Mastrandrea, T. E. Bilir, M. Chatterjee, K. L. Ebi, Y. O. Estrada, R. C. Genova, B. Girma, E. S. Kissel, A. N. Levy, S. MacCracken, P. R. Mastrandrea & L. L. White (eds.) *Climate change 2014: Impacts, adaptation, and vulnerability. Part A: Global and sectoral aspects. Contribution of working group II to the fifth assessment report of the intergovernmental panel on climate change* (pp. 869–898). Cambridge, UK and New York, NY, USA: Cambridge University Pres.

Mishra, A., Agrawal, N. K., Thorup, G., Bisht, S., Lelkanger, I. C. P., & Gupta, N. (2017a). *Strengthening women's roles as risk and resource managers at the frontline of climate change, Adaptation Solution Brief 1*. Kathmandu: ICIMOD. http://lib.icimod.org/record/32584/files/icimodASB.pdf.

Mishra, A., Ghate, R., Maharjan, A., Gurung, J., Pathak, G., & Upraity, A. (2017b). Building ex ante resilience of disaster-exposed mountain communities: Drawing insights from the Nepal earthquake recovery, *International Journal of Disaster Risk Reduction, 22*, 167–178. Elsevier. http://dx.doi.org/10.1016/j.ijdrr.2017.03.008.

Moors, E. J., Stoffel, M. (2013). Changing monsoon patterns, snow and glacial melt, its impacts and adaptation options in northern India: Synthesis. Science of the Total Environment. http://dx.doi.org/10.1016/j.scitotenv.2013.11.058.

MoCC. (2012). *National climate change policy. Government of Pakistan, ministry of climate change*, Islamabad, Pakistan. Retrieved October 20, 2016, from http://wp.cedha.net/wp-content/uploads/2014/05/Pakistan-Naitonal-Climate-Change-Policy.pdf.

MoE. (2011). *Climate change policy, 2011. Ministry of environment, government of Nepal*, Kathmandu, Nepal. Retrieved October 20, 2016, from http://www.climatenepal.org.np/main/?p=research&sp=onlinelibrary&opt=detail&id=419.

MoEF. (2009). *Bangladesh climate change strategy and action plan 2009. Ministry of environment and forests, government of the people's republic of Bangladesh*, Dhaka, Bangladesh (xviii + 76 pp.). Retrieved October 20, 2016, from http://www.climatechangecell.org.bd/Documents/climate_change_strategy2009.pdf.

Moench, M., Khan, F., MacClune, K., Amman, C., Tran, P., & Hawley, K. (2015). *The sheltering from a gathering storm research team: Transforming vulnerability: Shelter, adaptation, and climate thresholds, climate and development*. https://doi.org/10.1080/17565529.2015.1067592.

MoNERC. (2016). *Myanmar climate change strategy and action plan*. Naypyitaw, Myanmar: Final Draft.

MoSTE. (2015). *Indigenous and local knowledge and practices for climate resilience in Nepal, mainstreaming climate change risk management in development, ministry of science, technology and environment (MoSTE)*, Kathmandu, Nepal.

Moktan, M. R., Norbu, L., Nirola, H., Dukpa, K., Rai, T. B., & Dorji, R. (2008). Ecological and social aspects of transhumant herding in Bhutan. *Mountain Research and Development, 28*(1), 41–48.

Molden, D. J., Shrestha, A. B., Nepal, S., & Immerzeel, W. W. (2016). Downstream implications of climate change in the Himalayas. *Water security, climate change and sustainable development* (pp. 65–82). Singapore: Springer.

Müller-Böker, U., & Kollmair, M. (2000). Livelihood strategies and local perceptions of a new nature conservation project in Nepal: The Kanchenjunga conservation area project. *Mountain Research and Development, 20*, 324–331.

Nagoda, S., & Nightingale, A. J. (2017). Participation and power in climate change adaptation policies: Vulnerability in food security programs in Nepal. *World Development, 100*, 85–93.

NAPCC. (2008). *National action plan on climate change. Government of India, Prime Minister's Council on climate change.* Retrieved October 20, 2016, from http://www.moef.nic.in/downloads/home/Pg01-52.pdf.

NEC. (2011). *Second national communication to UNFCCC. National environment commission, royal government of Bhutan*, Thimphu.

Norbu, W., & Kusters, K. (2012). *The costs of adaptation in Punakha, Bhutan: Loss and damage associated with changing monsoon patterns, climate development knowledge network (CDKN).* http://loss-and-damage.net/download/6902.pdf.

Norse, D., Li, J., Jin, L., & Zhang, Z. (2001). *Environmental costs of rice production in China, lessons from Hunan and Hubei*. Bethesda, MD: Aileen International Press.

O'Brien, K., Pelling, M., Patwardhan, A., Hallegatte, S., Maskrey, A., Oki, T., et al. (2012). Toward a sustainable and resilient future. In C. B. Field, V. Barros, T. F. Stocker, D. Qin, D. J. Dokken, K. L. Ebi, M. D. Mastrandrea, K. J. Mach, G.-K. Plattner, S. K. Allen, M. Tignor, & P. M. Midgley (Eds.), *Managing the risks of extreme events and disasters to advance climate change adaptation. A special report of working groups I and II of the intergovernmental panel on climate change (IPCC)* (pp. 437–486). Cambridge, UK, and New York, NY, USA: Cambridge University Press.

Ojha, H. R., Ghimire, S., Pain, A., Nightingale, A., Khatri, D. B., & Dhungana, H. (2015). *Policy without politics: technocratic control of climate change adaptation policy making in Nepal Clim. Policy* (pp. 415–433).

Onta, N., & Resurreccion, B. P. (2011). The role of gender and caste in climate adaptation strategies in Nepal. *Mountain Research and Development, 31*(4), 351–356.

Ostrom, E. (1990). *Governing the commons: the evolution of institutions for collective action* (P. 280). Cambridge University Press.

Parikh, J., Upadhyay, D. K., & Singh, T. (2012). Gender perspectives on climate change and human security in India: An analysis of national missions on climate change. *Cadmus, 1*(4), 180–186.

Park, J., & Alam, M. (2015). Ecosystem-based Adaptation Planning in the Panchase Mountain Ecological Region. *Hydro Nepal: Journal of Water, Energy and Environment., 17*, 34–41.

Piya, L., Maharjan, K. L., & Joshi, N. P. (2012). Comparison of adaptive capacity and adaptation practices in response to climate change and extremes among the Chepang households in rural Mid-Hills of Nepal. *Journal of International Development and Cooperation, 18*(4), 55–75.

Piya, L., Maharjan, K. L., & Joshi, N. P. (2013). Determinants of adaptation practices to climate change by Chepang households in the rural Mid-Hills of Nepal. *Regional Environmental Change, 2013*(13), 437–447. https://doi.org/10.1007/s10113-012-0359-5.

Rasul, G. (2014). Food, water, and energy security in South Asia: A nexus perspective from the Hindu Kush Himalayan region. *Environmental Science & Policy, 39*, 35–48.

Rasul, G., & Sharma, B. (2016). The nexus approach to water–energy–food security: An option for adaptation to climate change. *Climate Policy, 16*(6), 682–702.

Regmi, B. R., & Star, C. (2015). Exploring the policy environment for mainstreaming community based adaptation (CBA) in Nepal. *International Journal of Climate Change Strategies and Management, 7*(4), 423–441.

Regmi, B. R., Star, C., & Filho, W. L. (2016). An overview of the opportunities and challenges of promoting climate change adaptation at the local level: A case study from a community adaptation planning in Nepal. *Climatic Change, 138*, 537–550. https://doi.org/10.1007/s10584-016-1765-3.

Revi, A., Satterthwaite, D. E., Aragón-Durand, F., Corfee-Morlot, J., Kiunsi, R. B .R., Pelling, M., Roberts, D. C., Solecki, W. (2014). Urban areas. In C. B. Field, V. R. Barros, D. J. Dokken, K. J. Mach, M. D. Mastrandrea, T. E. Bilir, M. Chatterjee, K. L. Ebi, Y. O. Estrada, R. C. Genova, B. Girma, E. S. Kissel, A. N. Levy, S. MacCracken, P. R. Mastrandrea, & L. L. White (Eds.), *Climate*

Change 2014: Impacts, Adaptation, and Vulnerability. Part A: Global and Sectoral Aspects. Contribution of Working Group II to the Fifth Assessment Report of the Intergovernmental Panel on Climate Change (pp. 535–612). Cambridge, UK and New York, NY, USA: Cambridge University Press.

Robledo, C., Clot, N., Hammill, A., & Riché, B. (2012). The role of forest ecosystems in community-based coping strategies to climate hazards: Three examples from rural areas in Africa. *Forest Policy and Economics, 24,* 20–28.

Saito, N. (2012). Mainstreaming climate change adaptation in least developed countries in South and Southeast Asia. *Mitigation Adaptation Strategies for Global Change, 18,* 825–849.

Sandler, T. (1998). Global and regional public goods: a prognosis for collective action. *Fiscal Studies, 19*(3), 221–247.

Sherman, M. H., & Ford, J. (2014). Stakeholder engagement in adaptation interventions: an evaluation of projects in developing nations. *Climate Policy, 14*(3), 417–441.

Shrestha, A. B., Agrawal, N. K., Alfthan, B., Bajracharya, S. R., Marechal, J., & van Oort, B. (eds.). (2015). *The Himalayan climate and water atlas: Impact of climate change on water resources in five of Asia's major river basins*. ICIMOD, Grid-Arendal, and CICERO.

Shrestha, M. S., Grabs, W. E., & Khadgi, V. R. (2015b). Establishment of a regional flood information system in the Hindu Kush Himalayas: Challenges and opportunities. *International Journal of Water Resources Development, 31*(2), 238–252.

Smit, B., & Pilifosova, O. (2001). Adaptation to climate change in the context of sustainable development and equity. In J. J. Mcarthy, O. F. Canziani, N. A. Learty, D. J. Dokken, & K. S. White (Eds.), *Climate change 2001: Impacts, adaptation and vulnerability, international panel on climate change* (pp. 877–912). Cambridge, UK and New York, USA: Cambridge University Press.

Smith, M. S., Horrocks, L., Harvey, A., & Hamilton, C. (2010). Rethinking adaptation for a 4 °C world. *Philosophical Transactions of the Royal Society, 2011.*

Spires, M., Shackleton, S., & Cundill, G. (2014). Barriers to implementing planned community based adaptation in developing countries: A systematic literature review. *Climate and Development, 6*(3), 277–287.

Su, Y., Lu, J., Manandhar, S., Ahmad A., & Xu J. (2013). Policy and institutions in adaptation to climate change: Case study on tree crop diversity in China, Nepal, and Pakistan. *ICIMOD working paper 2013/3.*

Sud, R., Mishra, A., Varma, N., & Bhadwal, S. (2015). Adaptation policy and practice in densely populated glacier-fed river basins of South Asia: a systematic review. *Regional Environmental Change, 15,* 825–836. https://doi.org/10.1007/s10113-014-0711-z.

Surminski, S., & Oramas-Dorta, D. (2014). Flood insurance schemes and climate adaptation in developing countries. *International Journal of Disaster Risk Reduction, 7,* 154–164.

Swanson, D., & Bhadwal, S. (Eds.). (2009). *Creating adaptive policies: A guide for policymaking in an uncertain world.* Ottawa: Sage, New Delhi/IDRC.

Tall, A., Hansen, J., Jay, A., Campbell, B., Kinyangi, J., Aggarwal, P. K., et al. (2014). *Scaling up climate services for farmers: Mission possible. Learning from good practice in Africa and South Asia. CCAFS report no. 13. Copenhagen: CGIAR research program on climate change, agriculture and food security (CCAFS).* www.ccafs.cgiar.org.

Tanner, T. M., Mitchell, T., Polack, E., & Guenther, B. (2008). *Urban governance for adaptation: Assessing climate change resilience in Ten Asian Cities, IDS Working Paper 315.* Brighton: IDS.

Tao, F., & Zhang, Z. (2010). Adaptation of maize production to climate change in North China Plain: quantify the relative contributions of adaptation options. *European Journal of Agronomy, 33,* 103–116.

Task Force on Climate-related Financial Disclosures (TCFD). (2017). *Final report of the task force on climate-related financial disclosures.* Retrieved June 2017, from https://www.fsb-tcfd.org/wp-content/uploads/2017/06/FINAL-TCFD-Report-062817.pdf.

Tiwari, P. C., & Joshi, B. (2015). Local and regional institutions and environmental governance in Hindu Kush Himalaya. *Environmental Science & Policy, 49,* 66–74.

Tschakert, P., & Dietrich, K. A. (2010). Anticipatory learning for climate change adaptation and resilience. *Ecology and Society, 15* (2), 11.

UNDP. (2015). *Making the case for ecosystem-based adaptation: The Global Mountain EbA Programme in Nepal, Peru and Uganda. United Nations Development Programme*, New York. http://www.pnuma.org/cambio_climatico/publicaciones/UNDP_(2015)-Mt_EbA_report_FINAL2_web_vs(041215).pdf.

UNDP. (2012). *Technical and social impact assessment: Reducing climate change-induced risk and vulnerabilities from glacial lake outburst floods in Punakha, Wangdue and Chamkhar Valleys.*

UNEP. (2016). *Adaptation gap report update, the united nations environment programme (UNEP)*, Nairobi, Kenya.

Varma, N., & Mishra, A. (2017). Discourses, narratives and purposeful action- Unraveling the social-ecological complexity within Brahmaputra basin in India. *Environmental Policy and Governance.* https://doi.org/10.1002/eet.1745.

Viviroli, D., Archer, D. R., Buytaert, W., Fowler, H. J., Greenwood, G. B., Hamlet, A. F., et al. (2011). Climate change and mountain water resources: overview and recommendations for research, management and policy. *Hydrology and Earth System Sciences, 15*(2), 471–504.

Walker, W., Marchau, V., & Swanson, D. (2010). Addressing deep uncertainties using adaptive policies. *Technological Forecasting & Social Change, 77*(6), Special Section 2.

Wamsler, C. (2007). Bridging the gaps: stakeholder-based strategies for risk reduction and financing for the urban poor. *Environment & Urbanization, 19*(1), 115–142. https://doi.org/10.1177/0956247807077029.

Wang, J., Huang, J., & Rozelle, S. (2010). Climate change and China's agricultural sector: An overview of impacts, adaptation and mitigation, ICTSD–IPC platform on climate change, agriculture and trade, issue brief no. 5. In *International Centre for Trade and Sustainable Development, Geneva, Switzerland, and International Food & Agricultural Trade Policy Council*, Washington, DC.

Wangdi, K.,. Singhasivanon, P., Silawan, T., Lawpoolsri, S., White, N. J., & Kaewkungwal J. (2010). Development of temporal modelling for forecasting and prediction of malaria infections using time-series and ARIMAX analyses: A case study in endemic districts of Bhutan. *Malaria Journal, 9*(251), pp. 251–259.

WB. (2016). World Development Indicators database, World Bank, October 11, 2016.

Wong, S. (2010). Elite capture or capture elites? Lessons from the 'counter-elite' and 'co-opt-elite' approaches in Bangladesh and Ghana (No. 2010, 82). Working paper//World Institute for Development Economics Research.

WRI. (2009). http://pdf.wri.org/making_climate_your_business.pdf.

Xu, J., & Grumbine, R. E. (2014). Integrating local hybrid knowledge and state support for climate change adaptation in the Asian Highlands. *Climatic Change*, 93–104.

Xu, J., Grumbine, R. E., Shrestha, A., Eriksson, M., Yang, X., Wang, Y., et al. (2009). The Melting Himalayas: Cascading effects of climate change on water, Biodiversity, and Livelihoods. *Conservation Biology, 23*(3), 520–530.

Yu, X., Jiang, L., Wang, J., Wang, L., Lei, G., & Pittock, J. (2009). Freshwater management and climate change adaptation: Experiences from the central Yangtze in China. *Climate and Development, 1,* 241–248.

In the Shadows of the Himalayan Mountains: Persistent Gender and Social Exclusion in Development

14

Coordinating Lead Authors
Bernadette P. Resurrección, Stockholm Environment Institute, Bangkok, Thailand,
e-mail: bernadette.resurreccion@sei.org
Chanda Gurung Goodrich, International Centre for Integrated Mountain Development, Kathmandu, Nepal,
e-mail: Chanda.Goodrich@icimod.org (corresponding author)
Yiching Song, Centre for Chinese Agricultural Policy (CCAP), Chinese Academy for Science, Beijing, China,
e-mail: songyc.ccap@igsnrr.ac.cn

Lead Authors
Aditya Bastola, South Asia Consortium for Interdisciplinary Water Resources Studies (SaciWATERs), Hyderabad, India, and International Centre for Integrated Mountain Development, Kathmandu, Nepal,
e-mail: Aditya.Bastola@icimod.org
Anjal Prakash, International Centre for Integrated Mountain Development, Kathmandu, Nepal,
e-mail: Anjal.Prakash@icimod.org
Deepa Joshi, Centre for Agroecology, Water and Resilience (CAWR), Coventry University,
e-mail: deepa.joshi@coventry.ac.uk
Janwillem Liebrand, Water Resources Management Group, Wageningen University, Wageningen, The Netherlands, e-mail: janwillem.liebrand@wur.nl
Shaheen Ashraf Shah, USAID-funded Pakistan Reading Project (PRP), Islamabad, Pakistan,
e-mail: shani.ashraf9@gmail.com

Review Editor
Sara Ahmed, Centre for Heritage Management, Ahmedabad University, India,
e-mail: sara.ahmed@ahduni.edu.in

Corresponding Author
Chanda Gurung Goodrich, International Centre for Integrated Mountain Development, Kathmandu, Nepal,
e-mail: Chanda.Goodrich@icimod.org

P. Wester et al. (eds.), *The Hindu Kush Himalaya Assessment*,
https://doi.org/10.1007/978-3-319-92288-1_14

Contents

Chapter Overview

Key Findings

1. **Polices and responses in HKH countries overlook women's multiple forms of oppressions and exclusions.** Since women and men in the region are a heterogeneous group, they have overlapping ethnic, class, and caste identities that result in multiple forms of marginalization and exclusion. However, too many simplifications and one-dimensional framings around women's identities—and people—persist, and policies and responses end up overlooking multiple forms of oppression and exclusions.
2. **Existing laws and policies do not support the multiple ways in which women negotiate their roles in households, communities, and the market.** This is because men from the mountain regions are out-migrating in large numbers leaving women to manage not just household work but also other work related to agriculture, natural resource management, community and even other public sphere related work—markets or public institutions—that were traditionally men's role. As an example, land tenure and employment policies undervalue rural women's critical roles in food security, sustainable agriculture, and natural resource management despite women taking on the major role in these sectors.
3. **Women throughout the HKH do not have corresponding decision-making rights or control over resources despite shouldering both productive and reproductive workloads and responsibilities.** This results in undervaluation of their work in society and therefore not taken into account in most policy making processes and outcomes.

Policy Messages

1. **Policies that support adaptation to climate change will not succeed unless they consider gender and how it interacts with other factors such as class/caste, ethnicity, and geography, which will require disaggregated data.** The available national data on women in HKH countries does not reflect the diversity and intersectionality, because they rely on aggregate measurements and aggregated country data that may not be representative of the mountain areas. It is critical that disaggregated mountain specific data be collected and analysed to inform policies that are suitable.
2. **Policies to improve women's participation in decision making and climate governance must go beyond numbers and quotas to create mechanisms that ensure empowerment and promote women's rights and agency.** Given the fragile and dispersed nature of settlements in the HKH, facilitating women's active voice and

agency will equally require working with gender responsive grassroots organizations and nurturing local and inclusive leadership, (i.e. top-down and bottom up approaches).

3. **All levels of government must allocate resources —financial and human—for gender responsive interventions at scale and adopt clear accountability mechanisms, such as gender budgeting, to demonstrate their commitment to gender equality as indicated in the SDGs.** Most countries in the HKH lack gender budgeting and accountability mechanisms on gender—almost all government agencies are male-dominated and gender neutral.

Climate change and extreme weather events in combination with socioeconomic processes and opportunities have an especially severe impact on people living in remote mountain areas of the Hindu Kush Himalaya (HKH). For instance, growing urbanisation of mountain towns and cities and globalisation has led to aspirations and consumerism, which is a significant push factor in migration. At the same time, unreliable rural and agricultural outcomes are seen as increasing outmigration of men seeking better livelihoods for their families. This situation can leave women with heavier work burdens, but might also provide women with a complex, newfound 'independence' (***well-established***).

This shift in women's and men's responsibilities has not been matched by a corresponding shift in policies and attitudes about gender. For example, policies on land tenure and employment disadvantage rural women and men—especially those in poor remote mountainous areas—by undervaluing their critical roles in food security, sustainable agriculture, and natural resource management (***established but incomplete***).

Furthermore, even as gender inequalities develop in an increasingly complex manner with contextual political and economic situations intersecting with class, caste, religion, age, and ethnicity, there are continued assumptions made around a single homogenous class of 'mountain' women.

Although less is known about the gender-differentiated effects of climate change, particularly in the context of limited economic opportunities in mountain areas, studies illuminate the relative quality of women's and men's lives, livelihoods, and access to resources under equally changing socioeconomic and political conditions (***established but incomplete***).

The linear, techno-managerial approaches to climate governance with simplistic one-size-fits-all solutions and 'quick fixes' for gender equality and women's empowerment fail to recognize the complexity of women's and men's realities (***established but incomplete***).

Problematic understanding of gender

Most often, understandings of gender are simplistic: Gender is equated with women, particularly poor rural women. Notions of gender are simplified in policy making, and reduced to the inclusion of some "poor women". This simplistic and apolitical interpretation and way of integrating gender into climate interventions and policies poses large problems, which manifests in the assumption that engaging women on projects is taking care of women's needs and will lead to women's empowerment (***well-established***).

Furthermore, there is the paradoxical positioning of homogenous categories of "mountain women" as being both "vulnerable victims" of climate change as well as "formidable champions" of climate adaptation. This has led to extreme approaches in policies: from a welfare approach in which women are taken as passive beneficiaries to one where they are seen as "fixers" of environmental problems based on assumptions about their "volunteer" time in projects as "natural care-givers" ***(well-established)***.

This dichotomization of women into one of two identifiers creates policymaking foci on numbers and quotas as measures of change and progress, rather than on the structural issues of inequality and discrimination. A major problem is the overlooking of gender relations between women and men, between women, and between men (***well-established***).

Related to the above is the focus on numbers and quotas as measures of change and progress, rather than on the structural issues of inequality and discrimination. A major problem is the overlooking of gender relations between women and men, between women, and between men (***well-established***).

The empirical problem: How data aggregation obscures women's complex experiences in the HKH?

Gender differences intersect with other dimensions of social and geographical differentiation. Class, caste, ethnicity, and age intersect with different geographical and sociocultural settings such as upstream, midstream, and downstream communities in mountain contexts to produce differential access to resources. Women and men are thus marked by multiple, coexisting identities that create overlapping—and often conflicting—relations of inequality and hierarchy, inclusion and exclusion.

However, the available national data on women in HKH countries does not reflect this diversity and intersectionality, because they rely on aggregate measurements. Part of the problem is that aggregated country data may not be representative of the mountain areas, which (except in Bhutan and Nepal) form only a portion of HKH countries' national territories. This lack of mountain specificity could help to explain, for example, why the 2014 Gender Inequality Index[1] varies so widely across the HKH, from China with the lowest inequality to Afghanistan with the highest (***established but incomplete***).

Questions about interventions: How to engage women and men of disadvantaged groups?

Policies and programs have long focused on the functional rather than the structural aspects of gender relations. In most sectors, gender mainstreaming policies have applied the concept of gender narrowly, often as a synonym for "poor [rural] women"—and without further differentiation of these women's needs, interests, emotions, identities, and roles. As a result, these policies produce technocratic quick fixes that place unrealistic burdens on women already in poverty (***well-established***).

To be sure, today's technocratic programs include women in economic development. Yet this inclusion is rationalized only by an appeal to economic gain. While the pursuit of economic efficiency can offer women economic opportunities, it does not fully address their unequal power relations with men, vis-à-vis equal wages for equal work, or the sharing of domestic work responsibilities (***established but incomplete***).

Exploring 'feminisation of responsibilities' and addressing a masculine working culture

As more men migrate in search of livelihood options, rural women assume a disproportionate share of responsibilities—agricultural labour, reproductive work, and other labour that supports community welfare, as well as responsibilities in the public sphere, giving rise to a gendered rhetoric of 'feminisation of responsibilities'. Within this women may be assigned new 'caring' roles as 'climate agents', expected to adapt to climate change and cushion its adverse effects on their households and communities. Such rhetoric has the effect of adding climate change adaptation to the list—already long—of women's caring roles (***established but incomplete***).

As important as it is to analyse the rhetoric of women's roles, what is considered or played out as roles for and by men—and the nature of these gendered divisions of labour and how these vary across socio-political and socio-ecological contexts—requires attention, especially at the science and policy levels. Yet, there is often little attention to these, and assumptions are made, especially of what women [can do] and do in policy and practice. It is important to add here that since men continue to be key actors in environmental science, policy, and intervention across South Asia, the manner in which masculinity mediates environmental science, policy, and governance is important to know and address. Progressive policies in irrigation and water planning, for example, will require that we address the 'masculine' working culture in this sector. Often, engineers and experts adhere to a hegemonic culture of science and technology, upholding 'masculine' professional norms ensures credibility—and yet, paradoxically, these same professionals may be the ones tasked with formulating and implementing policies that promote diversity, gender equality, and social inclusion (***established but incomplete***).

Gender, Inclusive Development and the Sustainable Development Goals

The SDGs are broader than the MDGs as they include environmental, social, and economic sustainability, and are also "potentially more transformative" better reflecting complex challenges and the need for structural reform (Sakiko 2016, p. 47). Sakiko further points out that the gender-specific SDG 5—"Achieve gender equality and empower all women and girls"—has multidimensional implications that integrate targets "related to gender-based violence, harmful practices, unpaid care work, voice, sexual and reproductive health and rights, economic resources, technology, and legislative change". Furthermore, gender issues and targets are not limited to SDG 5 only but are also reflected in other goals, such as those related to poverty, hunger, education and health, water and sanitation, and work.

The underlying transformative agenda and incorporation of the complex and multifaceted realities of women's lives in the SDGs are all the more relevant to the HKH, where social and gender inequalities persist such as groups of women in mountain areas who perform multiple roles: in the household, agriculture, natural resource management, communities, and the markets and other public spaces at higher scales, and particularly with the trend of large-scale male outmigration.

Consequently the nine HKH priorities clearly envision sustainable development of the region and its peoples with a transformative agenda of equity and inclusion. Priority 3—"Achieve gender and social equity through inclusive and transformative change in the mountains"—links directly with SDG 5 and goes a step ahead in its transformative agenda as the targets address the many facets of the complex realities of women's lives—gender-based violence, participation in decision making, gender equality and disparity, and

[1]GII is an index developed by UNDP that measures gender inequality on three counts – health, empowerment and participation in labour markets (UNDP 1995).

practices of social and gender exclusions, policies, and institutions. Priority 3, while being a standalone goal, also links with all the other priorities as each priority has social and gender dimensions as well as targets, which has further implications for gender equality and transformative change. The need of the hour is for development actors and policy makers to integrate gender and social inclusion targets and indicators in all actions and policies.

This new transformative agenda points out the need to do much more than the usual rhetoric on gender and unravel the complex and interrelated dimensions of women's and men's lives.

14.1 Context and Particularities to Gender and Social Inclusion in the HKH

The Hindu Kush Himalaya (HKH) are considered one of the most ecologically sensitive areas in the world. Although data is limited and contested, it is generally agreed that the effects of climate change are predicted to happen here "first" and with the greatest intensity (Singh et al. 2011, p. iv). What is less well known is how the effects of climate change will impact the geographically diverse mountain ecosystems and, in turn, the lives, livelihoods, and resources of diverse communities in the region.

In this chapter we draw attention to the fact that climate change problems and solutions are largely techno-centric in design. Jasanoff (2010) notes that "modern science" has framed climate change as a global phenomenon that "detaches knowledge from meaning". Technical observations of more easily measurable phenomena such as changes in temperature and precipitation undermine attention to the understanding of the embedded local experiences of people as well as "social institutions and ethical commitments at four levels: communal, political, spatial and temporal" (Jasanoff 2010, p. 233). In other words, little is known about how changes in climate will result in complex changes in the quality of lives, livelihoods, and resources of diverse groups of people. We argue that, especially in the HKH, it is important and necessary to creatively link "abstract generalizations, specificity and objectivity" of climate science and climate interventions with contextually relevant "scales of social meanings", experiences, and subjectivities (Jasanoff 2010, p. 235). Similar concerns were echoed by Forsyth (2001), who questions how far the so-called scientific descriptions of environmental change have been accepted as factual without this being reflected by the experiences and values of marginalised people.

In this chapter, we focus on unpacking popular assumptions related to climate change and proposed climate interventions using a social relations approach (SRA) and feminist political ecology (FPE) framework. For this work, we provide case studies that demonstrate the complex workings of gender relations in the context of climate change in the HKH. Specifically, these case studies highlight the unique, embodied experiences of climate change and how gender power relations affect green economy interventions.

Box 14.1 What is gender?

Gender refers to socially constructed roles, responsibilities, and opportunities associated with men and women, as well as hidden power structures that govern the relationships between them. Inequality between the sexes is not due to biological factors, but is determined by the learnt, unequal, and inequitable treatment socially accorded to women (UNDP 2010).

As social differentiations and identities based on class, caste, ethnicity, age, and other factors intersect with gender, addressing gender requires understanding that women [as well as men] are not homogeneous categories.

This chapter aims to critically assess two primary issues:

- The intersections between gender and social equity with climate change in the context of changing realities across the HKH
- How climate-related institutional interventions respond —or do not respond—to the complex and diverse realities of people's lives on the ground.

In this assessment we combine the SRA and FPE approaches (Box 14.2).

Box 14.2 Social relations approach and feminist political ecology frameworks

The Social Relations Approach (SRA) focuses on the nature and construct of inequality as determined through distributions of resources, responsibilities, and power. It emphasizes the analysis of relationships between people, their relationship to resources and activities, and how these are reworked across institutional levels in specific contexts—from the household to formal and informal institutions including the state and the market. SRA also emphasizes the idea that the overall goal of development interventions is and should be human wellbeing and not just economic growth (Kabeer 1994; Kabeer and Subrahmanian

1996, p. 25). Thus the social relations approach departs from narrow, technical interpretations of gender as women and of women and men as isolated categories, thus shifting away from the rather "impersonal, apolitical, and universal imaginary of climate change [impacts and interventions], projected and endorsed by science" (Jasanoff 2010, p. 235).

Feminist political ecology (FPE) recognises the close interlinkages of gender with other social categories and differences in gender-environment relations, and points out that resource-related relationships relate to "women's particular circumstances" (Molyneux 2007, p. 231). These circumstances not only interact with class, caste, race, culture, and ethnicity to shape processes of ecological change; they also differ in different social, political, and economic settings dynamically shaping "gender as a critical variable in shaping resource access and control" (Rocheleau et al. 1996, p. 4). FPE recognizes the importance of conducting 'science from below' or examining people's embodied experiences of resource degradation, disasters, mobility, and displacement as these connect with other scales of power and decision making (Hanson 2015; Harding 2008). In short, feminist political ecology is concerned with an intersectional analysis on gender-environment relations that actively considers gender in combination with social factors such as race, ethnicity, caste, class, age, disability, and others.

There are thus several similarities between the SRA and FPE frameworks in their aim to capture the complexity of gender-power relations, to unpack the gendered nature of institutions at scale, and to map the interactions between policies, practices, and ground realities at different institutional locations (Hillenbrand et al. 2014).

Mountain peoples, characterised by socio-ecological differences, have a crucial role in natural resource management and climate change adaptation. To evolve a more contextualised understanding of the HKH, it is important to examine the drivers of change behind gendered lives and livelihoods, as gender relations vary in complex ways. Overall the livelihoods of mountain communities in the HKH are largely agrarian: agriculture, livestock, and management of natural resources. Coping strategies in response to themyriad drivers of change include migration, wage and casual labour, and labour-intensive household management and income generation through small-scale trade (Leduc 2009). Natural resource management also figures heavily into community-based and individual/household coping strategies, drawing from a rich traditional culture and knowledge about this topic.

The processes of globalisation, regionalisation, and economic liberalisation are connecting markets and reconfiguring economic relations, interactions, and dependencies. While global birth rates are levelling, population continues to grow rapidly in the HKH, placing additional pressure on urban environments and infrastructure in contexts in which government policies alone are unable to curb the trends (Karki et al. 2011; UN-HABITAT 2007). These trends have also meant opening up mountain communities to a wider world of institutional arrangements, relationships, and opportunities, and the emergence of a consumer class that is shaping new aspirations and desires, sculpted by a culture of money (GoN 2014). In tandem, these trends alter land-human relationships, affecting how people use, access, control, and manage natural resources (Jodha 2007). In this process, local knowledge systems are rendered obsolete while giving rise to new bodies of information, creating new livelihood systems, and setting in motion new patterns of consumption and acquisition, as well as "reconfiguring people's relationships to one another, within and across households and communities … within and among state institutions and other macro agencies" (Gurung and Bisht 2014, p. 5). Furthermore, rural to urban migrants—largely young men—seek off-farm employment leading to changing demographic patterns, with growing elderly rural populations and the 'feminisation' of farm and non-farm activities in terms of production, exchange, and distribution (also see Chap. 15).

These factors or drivers are affecting women and men differently and changing gender roles and relations, leading to a widening of the differences between women's and men's income-earning and asset-controlling possibilities (Bastola et al. 2015; Nellemann et al. 2011; Nibanupudi and Khadka 2015; Sogani 2013).

14.2 Climate Change and Gender: Experiences from Below

Few studies focus on understanding gendered impacts due to changing climate in the HKH (Ogra and Badola 2015). In this context, it is critical to recognize that gender does not equate to women: There is no single class of HKH women and men, and no universality of experience in regard to climate change impacts. Understanding the complexities of diverse nature-society interrelations in the context of climate change in the HKH requires a viewpoint from below.

It is widely recognized that, on the one hand, climate change results in higher risks and greater burdens for women, and on the other hand, due to their local knowledge and management of households, women play critical roles in adaptation. Such generalizations are especially prominent in popular discourse on climate and mountain women in the

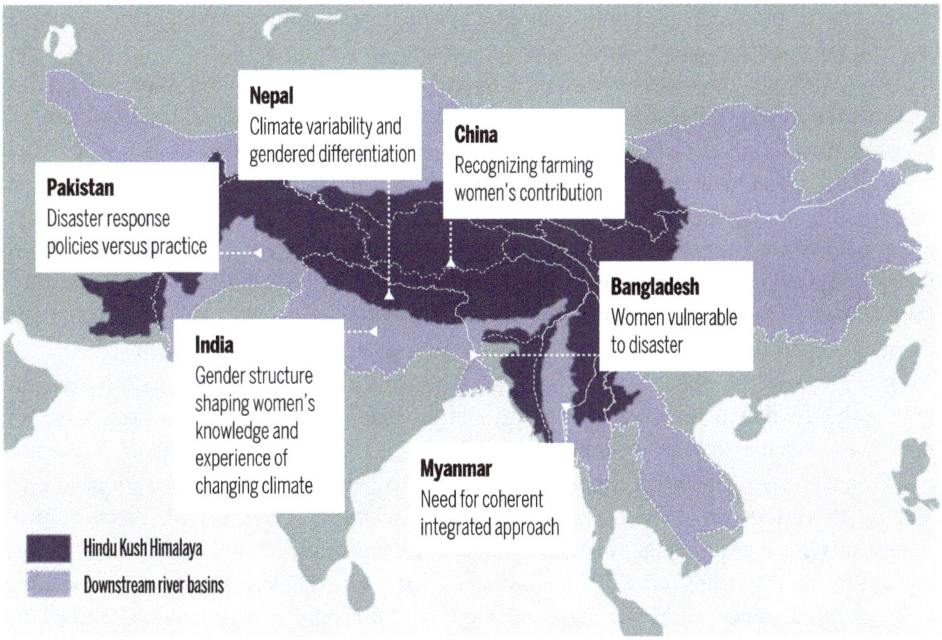

Fig. 14.1 Case studies of climate change and gender—experiences from below

HKH (Joshi 2014, p. 247). There are countless case studies of mountain women who, through their experience, responsibilities, and strength, are reported to play a much stronger role than men in the management of ecosystem services and food security, and therefore in climate change adaptation (Nellemann et al. 2011). Mountain women's knowledge, capability, and commitment to the environment and their families are used to highlight their ability to adapt in extreme situations such as conflict, natural disasters, and displacement (Leduc 2010). Such narratives create persuasive arguments that mountain women are critical actors in mitigating and adapting to climate change. But what is also important to consider is that despite the criticality of their roles, women in the HKH face challenging situations and obstacles as they carry out these roles.

This section of the chapter focuses on how gender and other social relations are negotiated with increasing climate variability. Within similar geographical regions, cultures, ethnicities, castes, and ecological settings, discriminative practices are likely to intensify with increasing environmental change (Bhattarai et al. 2015). In this context, we explore gender-differentiated access and control over resources and its impact on women and men in the HKH (Carvajal-Escobar et al. 2008).

14.2.1 Case Studies

This subsection problematizes the interplay of gender and other social differences that are superimposed on climate change impacts and create multiple vulnerabilities for the poor and marginalised women and men. Due to a lack of large-scale data, and a lack of cases from all HKH countries, we focus on illustrative case studies to show the interlinkages between gender and climate change. Some specific case studies presented from India, Nepal, Pakistan, Bangladesh, Myanmar, and China highlight the manifestation of gender vulnerability and women's roles from the HKH (see Fig. 14.1).

14.2.1.1 Bangladesh—Women Vulnerable to Disaster

The case of Bangladesh demonstrates how water-related hazards interact with gender to create situations where men and women differently experience, suffer, and cope with hazards and disasters (Sultana 2010). Women are feared to be more vulnerable to climate-induced water-related stresses and extreme events like floods (Brody et al. 2008; Dankelman 2008), and there is ample evidence that during natural disasters women and girls are more prone to mortality compared to men and boys. In Bangladesh, the 1991 cyclones and floods claimed more female lives than male: Amongst females 10 and older, girls and women were three times more likely to have perished (Bern et al. 1993; Parikh 2007; Roehr 2007; Twigg 2004;). In contrast, the mortality rate of men in both flood- and salinity-prone areas was only 17% of the total (Golam et al. 2009). This higher female death rate has been attributed to gender norms on what men or women should do in a disaster or the resources they have at their disposal (Chowdhury et al. 1993; Dasgupta et al. 2010; Nelson et al. 2002; Neumayer and Plümper 2007; Sharmin and Islam 2013; WEDO 2008).

For example, early warning signals had not reached large numbers of women because the information had been disseminated primarily in public places to which many women do not have easy access; and even when women received warnings they were constrained by cultural norms that restrict women's freedom of movement in public—that is, women were not allowed to leave their houses without a male relative, and many women waited for their husbands to return home to take the decision to evacuate, thereby losing precious time that might have saved their lives and those of their children (D'Cunha 1997; Fordham 2001; Mehta 2007; Parikh 2007; Sharmin and Islam 2013; UNEP 1997). Furthermore, conditions in cyclone shelters were not suited to women's needs, for example there were no separate toilets for men and women, poor water, and no toiletries like sanitary pads and the shelters were designed in a manner that was insensitive to gender specific needs and cultural values leading to various problems that enhanced the discomfort particularly for menstruating, pregnant, and lactating women (Baden et al. 1994; D'Cunha 1997; Fordham 2001). Similarly, a rapid gender field survey conducted in 2007 on Cyclone Sidr relief efforts found that after the storm, women were vulnerable to harassment, violence, famine, and human trafficking; the study also found that women were less likely to take off clothing, such as their long saris, during floods and these got caught in the floating debris, increasing their chances of drowning (Lauren Khan 2012). Thus, the differences in the gendered divisions of roles and labour, gendered rights, decision making, and women's weak bargaining power within the household result in women facing more suffering—including sexual assault—before, during, and after each disaster event.

A recent study by the Bangladesh Centre for Advanced Studies (Mallick et al. 2016) found that poor women and their livelihoods are badly affected by heavy rain, landslides, and water logging in Rangamati District in the Chittagong Hill Tracts (CHT). Small-scale agriculture led by women, water sources, and livelihoods of the poor women and men are severely affected by climate variability and extremes like flash flood, landslides, and water logging. These are causing economic hardship, deepening poverty, and increased food insecurity in the region.

Climate-related water stresses and gender inequalities aggravate the situation for women and girls in a number of ways. For example, poor families have been forced to marry off very young girls, resulting in many dropping out of school and others experiencing violence (Alston et al. 2014; Atkinson and Bruce 2015). Khan et al. (2016) and Khan and Kumar (2010) show how peri-urban communities in the coastal city of Khulna are vulnerable to water insecurity because of their exposure to different stressors including urbanisation. Additionally, climate change has led to rising sea levels and, consequently, saline contamination. As a result, pre-eclampsia, eclampsia, and hypertension among pregnant women is reported to have increased due to increasing salinity of drinking water (Khan et al. 2008, 2014).

The authors report that vulnerability of the peri-urban communities in Khulna varies across the sites as well as gender and different social groups within the communities. Prakash (2016) concludes that disaggregating vulnerability with an intersection of class, caste, and gender is important for understanding the complex issues related to climate change and its implications on gender. Lack of data means a lack of understanding and evidence on gendered vulnerabilities.

In Bangladesh, even when there is severe disruption of local freshwater sources following floods, cyclones, and saline intrusion, women are responsible, irrespective of their physical condition, to provide drinking water for their families for which they must walk long distances—sometimes up to 10 kilometres every day over difficult terrain—in search of water, consuming an enormous amount of their time and effort (WEDO 2008). After floods, day-to-day tasks such as cooking and cleaning the house become more time-consuming due to rising water levels. Women are often compelled to raise their stoves or go to neighbours' houses to prepare food (World Bank 2010). When the flood waters are particularly high, women go out on shallow-bottomed boats some distance to find the privacy to relieve themselves. Lack of supplies, the impossibility of disposal, and the problems of keeping oneself clean make menstruation particularly challenging. For many, the trauma of past experiences with the spate of water-borne diseases continues to linger long after the event (Mehta 2007).

14.2.1.2 China—Recognizing Farming Women's Contributions

China's rapid industrialization, urbanisation, and marketization since the 1980s has meant farmers cannot survive on farming alone due to small average landholdings (0.6 ha). Therefore, many have adapted their coping strategies. For example, the "one household, two sectors" approach (husband in the city, wife on the farm) is common to many families. In this situation, rural women have come to assume greater responsibility for agricultural production on top of their domestic and childcare duties. This is especially true in the poorer remote mountain areas in southwest China, which has a rich agricultural biodiversity that lends itself to using bio-culture farming systems as an adaptation strategy to support rural livelihoods and strengthen food security for the region.

Research carried out by the Centre for Chinese Agriculture Policy (Song et al. 2016) of the Chinese Academy of Sciences revealed that globalisation, rapid development, and climate change had delivered serious impacts to local food systems: severe droughts, increased temperature, and extreme weather. As a result, local farming species and

Table 14.1 Migrants in total labour force and women migrants in Guangxi and Yunnan

Year	2002	2007	2012
% migrants in total labour force	42.56	55.94	62.09
% women out of total migrants	38.48	39.84	42.06

Data source Survey of 320 rural households in Guangxi and Yunnan Provinces in 2013 (Song et al. 2016)

landraces were disappearing at an alarming rate, and the existing bio-culture landscape and local seed systems were threatened. These developments have precipitated a rise in social challenges such as extreme poverty, food security issues, increasing environmental degradation, and more frequent natural disasters to small farmers (primarily ethnic women) in remote mountainous areas. Women, as the main cultivators, seed savers, and users, are the most affected by climate changes and at the same time are the key custodians for farmer seeds.

Two village case studies from the rural areas in Guangxi and Yunnan Provinces illustrate some of the challenges facing mountain households. Situated in southwest China, Guangxi and Yunnan are home to most of China's rural poor mountain ethnic minority communities and have a rising trend of male migration. Over the past decade, the percentage of migrants in the total labour force in the communities has grown from 42.56% (2002) to 62.09% (2012)—a 20% increase (Table 14.1). Men comprise the majority of migrants, though many young women migrate as well.

Research shows that women are playing a key role in enhancing agricultural production for their households when offered the opportunity to make important decisions about the farm. In order to help farmers in remote mountain villages conserve seeds, improve their preferred landraces, and redirect benefits, the CCAP team initiated community-based conservation, participatory variety selection trials, and seed production in a number of trial villages in Guangxi and Yunnan through woman farmer–to–woman farmer exchanges facilitated by the project team. This initiative led to increased income and food quality for households participating in the programme (Song et al. 2016).

The participatory plant breeding activities continue today in the villages and to date have conserved more than 100 food crop varieties, improved 15 drought-resistant or quality landraces, and generated significant value for the women's group from seed production and other value-adding activities. A women's group in Guzhai village has developed into women-led farmer cooperatives in 10 years. A women's group in a Yunnan village created an idea exchange with the women's cooperative in Guangxi for learning organic farming. These activities have increased women's income three times more than "modern" agriculture promoted in the last 30 years. More significantly, the process has enhanced women's community-based collaboration in sharing productive activities, managing natural resources, and linking to external information, markets and society. This is quite helpful for empowering women and their self-organization economically and socially (Song and Vernooy 2010; Song et al. 2016).

14.2.1.3 India—Gender Structures Shaping Women's Knowledge and Experience of Changing Climate

Drawing on evidence from four case studies across the Indian Himalaya, this case assesses how gendered institutions shape women's knowledge, experience, and adaptation in changing climatic contexts. In the western Himalayan state of Uttarakhand, Moitra and Kumar (2016) explore the role of community radio in addressing the risks and challenges posed by climate change. Women reported an increase in the intensity of both summer heat and winter cold combined with sustained periods of fog; an increase in the frequency of natural disasters, including drought; and, consequently, more hardships for them in terms of water collection and agriculture. Established in 2000 in the township of Chamba, Garhwal, Henvalvani Community Radio has adopted a variety of broadcasting options combined with innovative activities to create spaces for communities, particularly women, to voice their issues. Engaging scientists in programming on low-cost and accessible solutions to agricultural problems (weeds, pests) for example, has helped women farmers in remote mountain areas sustain their crops and fields.

In their work on Uttarakhand's Nanda Devi Biosphere Reserve, Ogra and Badola (2015) explore the implications of climate change amongst indigenous communities, the Bhotiyas and the Garhwali *paharis* (mountain folk) whose traditional livelihood strategies based on seasonal migration have given way to agriculture, supplemented by a variety of secondary income-generating opportunities, including eco-tourism. However, while there are increasing opportunities for men as porters and guides, their absence on long treks, for example, poses an additional burden on women's work with limited recognition (Ogra and Badola 2015, p. 513). Impacts on women differ by age and poverty—younger women (daughters-in-law), smaller households, and households without men typically face greater levels of vulnerability reflected in terms of access to food and labour. While economic incentives provided by eco-tourism tend to benefit better-off households that can offer homestay facilities, wider multiplier effects in villages through established social networks have been noted. Additionally, infrastructure improvements at the village and household levels are having positive impacts for women and children, such as improved access to water and sanitation or waste management. Access to information through social interactions with visitors led many women to report an increase in self-awareness and knowledge on conservation and development (Ibid, p. 516).

Across the Himalaya, in the north-eastern state of Nagaland, Singh and Singh (2015) focus on climate-induced water-stressed communities in four districts: Mon, Mokokchung, Tuensang, and Kohima. The authors explore how climatic factors inducing water stress interact with non-climatic features in this mountainous region, producing new challenges that are gendered in nature. They report that domestic water access has always been a gendered problem, especially during the dry winter months when natural springs and streams dry up or reduce discharge. Women and children are forced to walk long distances downhill in search of water.

Meanwhile, Barua et al. (2014) argue that in Sikkim, vulnerability to climate change is not only about physical or geographical factors, but also structured by socioeconomic factors such as lack of education, health care, and limited livelihood opportunities, which constrain households in overcoming poverty and "weaken their resilience to manage climate risks" (Barua et al. 2014, p. 274). Despite equal opportunities for women in Sikkim in terms of access to employment, education, and health care, deep-rooted gender biases persist. During a crisis, it is the girl child who is withdrawn from the school first and girls are typically discouraged from pursuing higher education if it is at the expense of their domestic responsibilities.

In sum, the case studies show gendered impacts of climate change cut across social differentiated categories that are based on ethnicity and caste. This is particularly true for women from agriculture-dependent households. Furthermore the dimensions that appear crucial for sustainable livelihood development are income, migration status, and household size, as these relate directly to household assets and capacities of the household members. These case studies also reveal that gender division of labour along with gender norms and practices strongly shape women's knowledge and experiences of changing climate. For women in poor families, the hardship is compounded for example they do not have the option of purchasing food, fuelwood, fodder, among others. Thus, the risks and labour costs that women in the study reported as inherent to their work increase in the absence of alternative assets.

14.2.1.4 Myanmar—Need for a Coherent and Integrated Approach

Myanmar is vulnerable to a wide range of hazards including floods, cyclones, earthquakes, landslides, and tsunamis. Over the last decade the country has dealt with the devastating effects of Cyclone Nargis in May 2008, which severely impacted the Ayeyarwady and Yangon Divisions, and Cyclone Giri, which hit Rakhine State in October 2010 (IFRC 2015).

Cyclone Nargis hit Ayeyarwady, killing an estimated 130,000 people, of which 60% were women (CARE Canada 2010). The impact was immense and led to massive displacement and decimation of agriculture and infrastructure across the region. An estimated 2.4 million people lost their homes and livelihoods: the cyclone caused devastating damage to the environments of Ayeyarwady and Yangon, where local livelihoods are heavily reliant on natural resources (Pender 2009).

In this region, women traditionally play an important role in income generation including small-scale trade, shop keeping, fish processing, and crafts. They also play a key role in subsistence agriculture, fishing activities, and maintaining food security for their households. However, despite these responsibilities, women have less control over resources than men.

Women's Protection Technical Working Group (2010) point out that despite having a range of livelihood opportunities, the overwhelming majority of the people living in areas affected by Cyclone Nargis survive from harvest to harvest, sometimes relying on loans from moneylenders to tide them over to the next season. Of the loans taken out in the region, 50% go to households that have rights to paddy land, 20% to households that derive their primary income from fishing, and 30% to landless labourers. In such a situation, women who survived the cyclone were left more vulnerable—without family, incomes, livelihoods, homes, or assets, and with little access to quality sexual and reproductive health care or psychosocial support services (Women's Protection Technical Working Group 2010).

In the wake of the cyclone, the UN set up the Protection of Children and Women (PCW) cluster under which both children's and women's protection would be addressed (among other protection issues) as autonomous issues, in separate sub-clusters. The Women's Protection technical working group transitioned into a sub-cluster (WPSC), with the focus on multi-sectoral (protection, gender-based violence, livelihoods, education, health, and reproductive health) and cross-cutting (health, psychosocial and legal support) issues faced by women in the context of the cyclone-affected areas (Pender 2009). Pender (2009, p. 4) report states that the main reason for this separate cluster was that the PCW addressed women's issues "only in relation to the relationship of women to children". The report further indicates that the assessments conducted by WPSC show that despite women (particularly young widows, women separated from their families, and single female heads of households) being identified as the most vulnerable by community members, there was lack of targeting the specific needs and experiences of women—for instance, there was little to no gender or age disaggregated data, and there were hardly any stand-alone women's protection programmes. This was largely due to the absence of an overall coordinated structure in the PCW. Consequently, despite the program being community based, it overlooked the women's specific needs and experiences. Furthermore, strategies relating to the different sectors such as agriculture and fisheries, were gender blind, and unable to recognising women's contribution in these sectors (such as paddy transplanting and fish processing) (p. 6).

The Pender (2009) report concludes that the formation of the WPSC led to a holistic view, including gender-based violence, whereby the WPSC prioritized the need for holistic support for survivors of gender-based violence, including health, psychosocial, and legal resources. Not only did the program increase the scope of gender-based violence work, but also mainstreamed gender in a practical and tangible way. This contributed to women's empowerment through livelihoods, education, and comprehensive health care services. This is an example of an efficient and coordinated structure that mainstreamed gender across sectors.

A key lesson from this case is that sustainable livelihoods require a coherent and integrated approach across a number of sectors, including water, shelter, livelihoods and food security, education and training, sanitation and hygiene, and disaster risk reduction. Complementing and combining the sectoral approaches with capacity building and institutional strengthening of national and local governments and civil society is critical for the development of laws and policies that support sustainable development (UNEP 2009, 2012).

14.2.1.5 Nepal—Climate Variability and Gendered Differentiation

Bastola et al. (2015) recently conducted a study to understand perceptions of climate change impacts across the mountain, hill, and terai regions in the Central Development Region of Nepal. They posed important questions on how the adverse impacts of a changing environment shaped gender and social relations and added more weight to women's workloads to provide care for family members while also managing the production spheres.

This study illustrates the lack of effort undertaken by intervention agents to understand how climate change differentially impacts women and men, and how little is known about the links between climate change, gender, and other social intersections. This situation is exacerbated by the fact that little gender disaggregated data is available. Moreover, this study supports the findings of other studies (Dhimal 2015; Eriksson et al. 2008; Goh 2012; Leduc et al. 2008; Leduc 2009; Massey et al. 2010) that show how climate variability is more likely to increase women's workloads across sectors. Figure 14.2 shows gender disaggregated agriculture work in the three geographical regions, indicating that women are increasingly involved in agriculture work (such as ploughing, sowing, weeding, irrigating, preparing field channels, harvesting, threshing, and selling of agriculture produce).

The ratio of women to men in agricultural activities is particularly pronounced in the mountain regions, due primarily to male outmigration for work. Gender disaggregated data on work allocation show two major points that have both positive and negative outcomes. On the positive side, women take control on household economic affairs (selling of agriculture produce) that traditionally were controlled by men. On the other side, there is an increasing feminisation of labour where women are forced to bear more responsibility for agriculture activities, particularly in the mountain region where male outmigration is high (Bettini and Gioli 2015). Bhattarai et al. (2015) share similar findings that at least one man from each household in the mountain region has migrated to Gulf Cooperation Council countries or Malaysia.

In cases where both women and men have out-migrated, elderly family members assume the responsibility of farm management, and agricultural land was often left barren (Bastola et al. 2015). In the case of the population which has stayed behind, we see more severe effects on women, as they suffer more often from psychological impacts while adapting to climate change. Since the publication of Census 2011, it is estimated about 8% of the total population have migrated for employment to foreign countries. The total remittances contribute about 29.1% of the national economy (in 2013/2014) (GoN 2014). To build local resilience through income diversification, the Government of Nepal under the Foreign Employment Policy 2012 directs establishment of labour banks. These institutions are yet to be implemented on the ground. Unless policy frameworks with institutional arrangements are in place, outmigration will cause those staying behind, particularly women, children, the elderly, and the physically disabled, to remain vulnerable to climate variability (Bettini and Gioli 2015).

In the male outmigration situation, some women have begun to strengthen their access and control over the economic gains from agriculture (across all central regions as shown in Fig. 14.2), and the driver may not necessarily be a consequence of male outmigration. There are several other factors that are important determinants of gendered vulnerability to climate variability, including geography, religion, class, and ethnic divisions (Bastola et al. 2015; Maraseni 2012). Until now there has been little focus on women's capacity, their critical role in managing production in the absence of the male population (Skinner 2011). Therefore, experiences of climate variability are not always disproportionately negative or only negative for women, but in most situations women are likely to experience disproportionately the negative impacts due to the existing social gender structures that favour men (Bhattarai et al. 2015, p. 122).

14.2.1.6 Pakistan—Disaster Response Policies Versus Practice

Case studies from Pakistan reflect the lack of focus on gender dimensions during disaster response implementation despite gender sensitive policies and guidelines. The disaster management authorities have taken useful steps for making

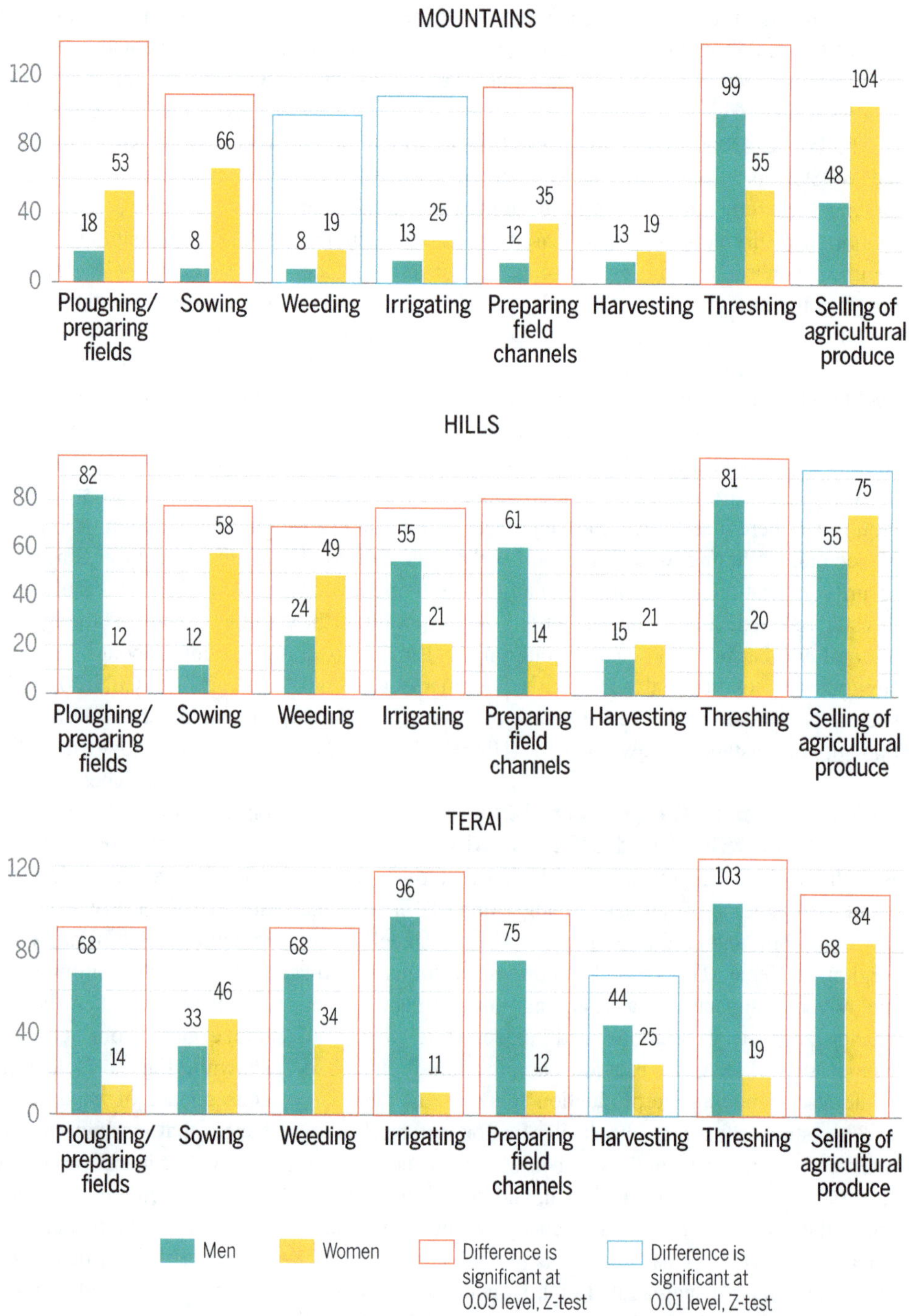

Fig. 14.2 Number of women and men engaged in various agricultural activities by region (*Primary data source* Bastola et al. 2015). *Notes* The number denotes total number of respondents for a given activity

disaster response and recovery gender sensitive. For instance, the National Disaster Management Authority (NDMA) has developed guidelines, 'Disaster Risk Reduction (DRR), Gender and Environment', with an objective to provide disaster managers with the initial tools for taking care of the infrastructure and social vulnerability during the disaster. At the provincial level, the Provincial Disaster Management Authority (PDMA) have established 'Gender and Child Cells' and developed various codes of conduct to support gender integration during a humanitarian response.

However, despite this progress, humanitarian response implementation to date has not sufficiently focused on significant gender dimensions (Hamid and Afzal 2013). Gilgit-Baltistan's Contingency Plan for Floods reveals that vulnerable groups (women, children, elderly, and disabled persons) are neglected in the planning and conduct of relief operations (GDMA 2015, p. 7).

A survey of the literature in Pakistan on this topic stresses the need for gender disaggregated data. Though effective adaptation strategies need to address fundamental gender disparities in the disaster relief process, in the absence of robust data, gender analyses will remain inadequate and establishing gender-sensitive needs, response, recovery, and rehabilitation almost impossible (IFRC 2007). At present, only Khyber Pakhtunkhwa PDMA has begun to collect gender disaggregated data. Among their early findings, they state that more women than men died in the rains and cyclone of 2015 (PDMA/KP 2015).

Studies also show that post-disaster processes of consultation were also poor in terms of collecting gender disaggregated data. Only men were solicited for information. Consultations with children, pregnant women, the disabled, and other vulnerable groups were insignificant. Most explanations for this oversight highlighted by various studies included masculine culture and male domination in most formal government institutions (including N/PMDAs, irrigation, WAPDA, and disaster-related district administrations), lack of capacity in terms of integrating gender, and sociocultural barriers that restrict men's access to vulnerable groups particularly women (Shah 2012; Hamid and Afzal 2013; Shah and Memon 2012).

Responses from the field, especially poor communities affected by floods, also reveal that gender is a largely neglected aspect of community infrastructure planning and provision (Shah 2012). However, women, children, the elderly, and the disabled pay a particularly high price for this lack of sensitive infrastructure development. After the floods of 2010, it was observed that floods came suddenly, and without an early warning system there was hardly any time for communities to make a timely and planned evacuation (UNIFEM 2010). An analysis of early warning systems conducted by LEAD Pakistan (2015) found that 88 out of the 145 districts of Pakistan were at risk for flood, but only 39 districts were covered by early warning systems. It is noteworthy that the majority of women and girls had no independent access to the tools used for communicating flood warnings, such as loudspeakers in mosques, mobile phones, and FM radio. Therefore, merely providing early flood warnings through various media sources may not necessarily ensure that women and other vulnerable groups have information on which to take action (Mustafa et al. 2015).

14.2.2 Social Structure, Gender, and Climate Change: Differential Vulnerabilities

Climate change conditions have also intensified competition over water and agricultural resources. Tetlay and Raza (1998) examined growing outmigration for employment with households increasingly becoming "female managed", especially in the HKH. The livelihoods of the people in the region are based on agriculture, livestock raising, management of natural resources, migration, labour-intensive household management, and income generation through small-scale trade and wage and casual labour (Leduc and Shrestha 2008). The diminishing subsistence prospects and underemployment in rural areas have pushed (mainly) men into seeking alternative livelihoods in off-farm domains resulting in an increase in, and intensification of, women's work and role as primary supporters of homesteads and family farms (Bose 2000; Gurung 1999; Leduc 2009; Mehta 1996; Sidh and Basu 2011). Thus, over the course of the past decades, male outmigration in mountain agriculture has emerged as one of the most pressing issues facing productivity and food security. But the ownership of natural resources like land and forest is confined to more powerful segments of society, primarily men. The agricultural, technical, and institutional support such as extension, credit, and subsidies are offered mainly to men, who are household heads in most cases.

Women and minority groups across the HKH are the most affected by the impacts of climate change. In China, the "one household, two sector" phenomenon is often seen as an outcome of globalisation and a changing environment. Women who stay behind take on the agricultural work in addition to other domestic responsibilities to secure the food basket. Case studies from the HKH clearly indicate there is an increasing proportion of 'feminisation' of labour as men out-migrate in search of work in response to climate variability as well as other socioeconomic factors, which makes agriculture less predictable and reliable. Therefore, cases from China show that women are starting to play key roles in managing natural resources for food security, climate change adaptation, and transition to a green economy.

Because the HKH stands at a heightened risk for natural disasters due to climate change impacts, it has once again raised serious concerns on gender narratives. The question is how such factors intersect with differential vulnerabilities in the event of floods, cyclones, landslides, earthquakes, and other disasters. Drawing on experiences from the cases of Myanmar and Pakistan, the impacts of disasters have a detrimental effect on different groups of women and marginalised groups which adds to an already formidable burden to provide food, water, and health care. Also, these events expose vulnerable groups to a high risk of violence and bring added livelihood insecurity (Brody et al. 2008; IFRC 2007; Neumayer and Plümper 2007; Parikh 2007; Vincent et al. 2010).

The differential impact of climate on gender is apparent in the HKH. For example, a first form of 'gendered' vulnerability to climate change relates to labour (Sugden et al. 2014, p. vii)—women's workload increases as distances travelled by women increase to access natural resources (such as water, fuelwood, fodder, food, pastures, medicinal plants, fuel, and crops) and as production schedules are affected due to changing environments and climate conditions (Bhattarai et al. 2015; Sugden et al. 2014). Thus it is important to mainstream gender-sensitive approaches when addressing environmental issues in the HKH. Several crucial points need to be addressed. First, there is a need to identify vulnerable groups in terms of age, disabilities, and social and religious groups and cater to their needs in disaster response to avoid further marginalisation in the process. Second, strengthening of the collaboration between government authorities and humanitarian organizations is required for the gender responsiveness of the relief programs. I/NGOs can play a role to build the capacity of government, as well as communities, especially in gender-inclusive development.

In spite of these recognised needs, gender-inclusive planning and implementation are not integral to development processes in the developing world (Moser 2012). The HKH is not devoid of this phenomenon. It is also important to recognise the traditional cultures that are women centric and female managed, and if we are to properly address the challenges to mountain communities, this means robust data will be required to confront fundamental gender disparities (Gurung 1998; Tulachan 2001).

What is most critical to consider is context—the context in which the changes are taking place and that is resulting in diverse vulnerabilities and impacts. This means there is a need to think more critically and creatively about the broader implications of different vulnerabilities and impacts and ways to address them.

14.3 Environmental Governance and Gender in the HKH: A Cautious View on Green Growth Visions

Environmental governance in the HKH is increasingly tailored towards an outlook of green economic growth. The impetus towards green economic growth has been accelerated as a mitigating response to climate change. This poses old and new questions regarding the shaky connections between social wellbeing, equity, and efforts to 'green' economies and sustainable development as a whole. Policies and programs have long focused on the technical aspects of climate change, privileging, for instance, an efficient use of natural resources over social aspects, such as issues of equality, discrimination, and empowerment. They have given importance to engaging women to make development work efficient rather than to bring about change in unequal power and gender relations by addressing the power relations between men and women. Therefore, it is important that the nature of interventions shift from "technocratic quick-fix[es] to gender inequalities [—] interventions that make unrealistic expectations on women [already] in poverty —to interventions that understand history and context better, … [and] recognise not only women and their vulnerabilities, but equally men and their masculinities. … [T]his calls for a 'long-haul, deeply political challenge'" (ODI 2007).

From a gender perspective, the two-way relationship between gender relations and environmental change needs to be understood in the context of the green economy. Gender relations have a powerful influence on how environments are used and managed, and hence on patterns of ecological change over time (Leach et al. 1995), and similarly environmental trends and shocks also have an impact on gender relations. Most of the time, there are direct implications such as forest or water degradation and depletion that alter the gendered distribution of resources or encourage gender-based coping strategies.

The term 'green economy' is defined as an economy that results in improved human wellbeing and social equity, while significantly reducing environmental risks and ecological scarcities. It is low carbon, resource efficient, and socially inclusive (Stone 2011, p. 1–2). Advocates state that a move towards a green economy can be profitable where economic growth will be healthier, stronger, and more vigorous with this transformation rather than without it (Brockington 2012). However, critiques of the green economy (Bullard and Müller 2012; Gupta and Agarwal 2013; Kosoy et al. 2012) argue that politico-economic and cultural constraints need to be considered to create strategies that are successful in achieving the goal of ending environmental degradation and reducing poverty.

'Green growth' has evolved into a number of contemporary forms in the region, such as heightened forest conservation efforts to spur carbon trading, bio-energy development, natural park enclosures, increased water regulation to ensure efficient uses largely for productive ends, and payments for ecosystem services. Research increasingly shows there may be difficult trade-offs between green growth, environmental sustainability, and social wellbeing (Fairhead et al. 2012; Harcourt and Nelson 2015). As a result, there is growing ambivalence around some so-called green projects. For instance, hydropower development is being promoted as a source of clean energy, but research has demonstrated it may in fact sidestep the wellbeing needs of dislocated communities. Forestry options such as payment-for-ecosystem services and REDD+ , as well as standards and certifications applied to the carbon trade and offset projects are envisaged to ensure fair practices, but may implicitly tap women as a reserve, but cheap, army of labour with benefits that remain unclear to them.

What seems to be emerging is a new regime of appropriating and managing both nature and society for so-called 'green' ends. This new green regime builds on earlier weaknesses of community-based natural resource management. It fails to correct social/gender exclusionary practices and continues to appropriate women's undervalued labour for green projects. This regime also builds on earlier tokenistic practices to involve women as a social group, often resorting to 'ticking the box' exercises to legitimize them as an accounted-for constituency. Despite earlier intentions to apply bottom-up approaches, the state continues to employ community-based natural resource management as an instrument of control of both nature and society, an effort that was once touted as "bureaucratizing communities" (Gauld 2000), increasingly turning these communities into corporate appendages of the state. In today's natural resource management regimes, more emphasis is also turning towards employing techno-scientific approaches in mitigating climate change and addressing green growth goals (MacGregor 2010).

Current efforts to mitigate climate change and spur green growth emphasize efficient management based on scientific, financial, and market-based goals and principles to drive natural resource management (Paudel and Paudel 2013). Green growth projects and natural resource management today attempt to mitigate and adapt to climate change in ways that are de-politicized, 'masculinized', and male dominated, in an effort to appropriate and 'tame unruly nature' (MacGregor 2010; Taylor 2014; Tschakert 2012). This approach in many ways creates persistent silences around the political economic drivers of climate change, focusing almost entirely on its technical aspects and solutions.

Therefore, the 'eagle-eye' science of popular climate knowledge, science, and environmental governance interventions needs to be complemented by locally contextual 'toad-eye' science and interventions (Gyawali and Thompson 2016). Sustainable development in the HKH demands socially and gender-inclusive climate science and environmental governance policies and strategies.

To assess the status of governance of the environment and green economic growth from a gender and social inclusion perspective, this section highlights first the status of gender mainstreaming at the policy and institutional levels and second how these policies tend to unfold on the ground. Then, the current promotion of hydropower development in the HKH as green development is critically assessed, and a plea is made for a more conscious reflection on the performance of professionals and the linkages between men, 'masculinity', and power in knowledge production in environmental governance itself.

14.3.1 Gender Mainstreaming Policies and Institutions

Most countries in the HKH have called for or claimed to have mainstreamed gender in the various sectors within the domain of environmental governance. Even in sectors in which inequities and social exclusion have long been rendered invisible—such as irrigation and water resources planning—gender and women have now earned a legitimate place in research and policy agendas (Zwarteveen 2006). However, there is little to celebrate when we consider the manner in which gender is interpreted and integrated in climate interventions and policies. The attempts to 'gender mainstream' in climate policies, strategies, and interventions remain plagued by simplistic, apolitical interpretations of gender: 'gender as women', the paradoxical positioning of homogenous categories of 'mountain women' as being both 'vulnerable victims' of climate change as well as 'formidable champions' of climate adaptation, and the idea that engaging women on projects is taking care of women's needs and empowering women.

Thus, the current status of gender mainstreaming in environmental governance is not only a clear measure of the progress that has been made, but also a reason for deep concern. Two cases—one from Pakistan and one from China—show the largest reason for concern is that gender mainstreaming appears to have been achieved in environmental governance by adopting a very narrow and simplified concept of 'gender'. The term is essentially used as a synonym for 'poor rural women'. The habit of equating gender issues with grassroots women's issues in development and the modernist idea that women's empowerment can be implemented top down, seems to do more harm than good (Liebrand 2014). Overall, the trend is that women are treated as victims—not as agents.

The case of Pakistan highlights especially that most progressive gender policies in environmental governance in the HKH are accompanied by a structural lack of financial resources or (human) skills and capacities In other words, gender is often not considered a priority in environmental management and climate change policies. Generally, it can be observed that state agencies in various HKH countries have committed themselves to the promotion of gender equality, and they make a proclaimed effort to make it happen, but simultaneously, there is a persistent impression that gender mainstreaming is (also) promoted as a form of window dressing.

14.3.1.1 Pakistan

Pakistan launched its Climate Change Policy in 2012, with an aim to ensure that climate change is mainstreamed in the economically and socially vulnerable sectors of the economy and to steer Pakistan towards climate resilient development. In spite of numerous challenges, Pakistan has initiated many other policies to address climate change and natural resources management, such as the National Water Policy, National Drinking Water Policy, National Climate Change Policy, the National Sanitation Policy, and others. All these policies present frameworks to address the key challenges of climate change and natural resource management at a national level and serve as guiding principles to the provinces to initiate their own policies to protect natural and environmental resources.[2] A similar trend is visible in other HKH countries.

Recognising women and other vulnerable groups as powerful agents of change, and the differential impact of climate change on gender, most of the aforementioned policies do suggest various measures for gender mainstreaming (Hamid and Afzal 2013; SPDC 2015). For instance, the third objective of the climate change policy focuses on pro-poor gender-sensitive adaptation while also promoting mitigation to the extent possible. The National Drinking Water Policy, the National Sanitation Policy, and others acknowledge, in particular, women's active role in water management. Furthermore, Pakistan is also a signatory to several international norms and standards that lay the foundation for gender equality. The Government of Pakistan has also allocated employment quotas for women in provincial and federal institutions (Rai et al. 2007).

But Pakistan ranks as the world's second-lowest country—144 out of 145 countries, according to the 2015 Global Gender Gap report—in terms of gender equality and the equitable division of resources and opportunities among men and women. Major fields like climate change, disaster, water, irrigation, mitigation, and the environment in Pakistan are still considered a male domain, outside the purview of women (Shah and Memon 2012). The fact is, the majority of formal government institutions and structures dealing in climate change and natural resource management are highly male dominated. The marginalisation of women was evident in the staffing patterns of these organizations (Shah 2012; Hamid and Afzal 2013). There are no formal mechanisms to ensure a gender balance in higher level positions in climate change, water, and irrigation bureaucracies. Therefore, in the absence of strong implementation, gender mainstreaming of governance structures does not, for the most part, translate into practice.

It is also important to understand that most government institutions have limited financial and technical resources/capacity available especially in terms of integrating gender into climate change planning (SPDC 2015). This omission can further result in poor response in terms of meeting the needs of marginalised groups. The literature on gender, climate change, and disaster in Pakistan includes various well-documented experiences that highlight the victimisation of women and other vulnerable groups during emergencies, as well as the undermining of their productive role in community building after disasters (Akçar 2001; Bari 1998; Enarson 1999; Morrow and Phillips 1999; Shah 2012).

[2]After the 18th constitutional amendment, provinces in Pakistan are now empowered to initiate their own provincial policies and institutional arrangements.

14.3.1.2 China

The Chinese government has increasingly acknowledged how poverty intertwines with biological and cultural diversity through the government's ecological civilization strategy and green social transition. Although China has achieved significant poverty reduction in the last decade, poverty levels remain high, at roughly 200 million people, according to international poverty standards. There are 14 state-identified poverty areas, mostly in remote mountain areas, which are at the same time well-known for their rich biological and ethnic diversity, diversified landscapes, and valuable bio-cultural heritage. China is active in climate change adaptation as well and launched the South-South Programme during COP 21 in Paris in 2015 to support other developing countries through the South-South Collaborative Fund for Climate Change Adaptation. China has a large national Climate Change Adaptation Plan using ecosystem-based adaptation as a major methodology and scientific technologies as key tools. Yet, consideration and integration of gender analysis for inclusion in community-based adaptation and women's roles in adaptation is limited. Further study on the links between and integration of ecosystem-based adaptation and community-based adaptation is urgently needed. The previously described case studies and the examples from Guangxi and Yunnan (Sect. 2.1) provide both strong examples for implementing community-based gender-sensitive biodiversity management and important survey data at the policy and action levels.

14.3.2 How Policies for Gender Mainstreaming and Social Inclusion Unfold on the Ground

Policies for gender mainstreaming and social inclusion in environmental governance on the ground are complex and diverse. As illustrated by cases from China (community-based biodiversity management) and Nepal (community forestry), current policies for user participation and community-based management inadequately address gender concerns, although these policies often explicitly seek to address them. The inclusion of some women as representatives of user committees has, by and large, not altered the marginalisation and social exclusion of women from these groups. One reason is that government agencies use policies of decentralization, user participation, and community-based management as a means to exercise control, regulation, and state power.

14.3.2.1 Women's Role in Seed and Food Security in China's Mountains

Farming women play crucial custodial roles in seed and food diversity all over the world. Women's reproductive roles as mothers and family keepers build their interest, expertise, and knowledge in seeds and food biodiversity issues. They are making an essential contribution to the resilience and continuity of the world's ecologic and food systems. Survey data in southwest China confirms this: Amongst small holding farmers, 62% of women play a role in seed selection and storage (Song and Zhang 2015).

In order to help farmers in remote villages conserve seeds and improve their preferred landraces and participatory plant breeding (PPB) varieties, both to save seed cost and to create incentive and redirect benefits to PPB farmers, a team initiated community-based conservation and a participatory variety selection trial and seed production of a PPB variety, Guinuo 2006, in a number of trial villages in Guangxi. This PPB hybrid seed production has been carried out by women farmers' groups in this area since 2005 and has expanded to women's groups in Stone Village in Yunnan through farmer-to-farmer exchanges facilitated by the project team.

The PPB activities in Stone Village have conserved more than 50 food crop varieties and improved 10 drought-resistant or quality landraces, and the women's groups have generated a significant amount of money from seed production. The groups have also started learning ecological and organic farming practices from Guzhai village and plan to register a women's farmer cooperatives next year.

Both village case studies illustrate an important rural development pathways in mountainous areas of China: community-based and women-led cooperative, diversified agriculture combined with strong horizontal integration (Song et al. 2014). Together, these ideas represent "a locally-driven empowerment process in which farmers, led by women, have improved their capacity to deliberate about choices of action, experiment with options, create new practices, and enlarge the network of horizontal relationships, and thus obtain more autonomy in realizing their aspirations according to their own agendas" (Song and Vernooy 2010).

Both cases benefit from strong technical support and capacity building, accompanied by targeted research. The project support for both the community cases in Guangxi and Yunnan is more focused on broad rural development than on commercial motives. The work also benefits from interactions and collaborations with other cooperatives, restaurants, NGOs, research centres and universities, and the government's agricultural extension service. The process of expansion is a capacity-building and empowering process for these women-led cooperatives and self-directed communities.

14.3.2.2 Community Forestry and Hydropower in Nepal

(a) Community Forestry Programme

Nepal's Community Forestry Programme (CFP) has been hailed as an environmentally and socially transformative initiative, but in fact, has in many ways not benefited the poor. Some scholars allege that community forest user groups (CFUGs) have actually resulted in "restricting access to resources by the poor" (Gupta et al. 2011) and as such, CFP has not really ensured equity, inclusion, or gender sensitivity (Neupane 2003; Paudel 2012).

Following the conception of the community participatory approach, the early 1990s saw a wave of popularity in decentralized formal arrangements for forest governance as governments realized the need to transfer responsibilities to local institutions and bestow decision-making powers on communities for better resource management outcomes. Hence, over the past few decades, there have been many devolutionary initiatives in a number of the HKH countries to bring about gender and social inclusion. However, in tracing the trajectory of these initiatives, we find that policies and programmes establishing local forest governance institutions were gender blind and communities were treated as an ungendered entity (Agarwal 2000; Das 2011; Arora-Jonsson 2014). Only in later years, with changes in approaches to decentralized governance and feminist criticism of such policies, were steps taken by the government to integrate gender equality concerns in programmes and schemes at different levels (Tyagi and Das 2017, p. 239).

From amongst such initiatives, Nepal's model of community forestry as a green governance initiative has been hailed as the most successful, transformative, and people-oriented model of local-level forest governance for its social as well as environmental objectives. CFP in Nepal started in the 1970s and strengthened through later policy and legal instruments such as the Master Plan for the Forestry Sector 1988, the Forest Act 1993, and the Forest Regulation of 1995. This legislation provided favourable conditions for the successful handover of national forests to local communities. As such, the CFP has been the largest and longest participatory green initiative, with 40% of Nepal's population belonging to more than 15,000 CFUGs which are involved in managing 25% of the country's forest area (Gupta et al. 2011; Karki et al. 2011). With the shift in property rights from the state to communities, CFUGs have been able to exercise a bundle of property rights regarding access, use, and management of national forests. The CFP goes beyond managing forests for environmental and economic benefits; it has become an important instrument and process for social change—empowering the marginalised (Gupta et al. 2011).

In Nepal, the Poverty Reduction Strategy Paper of 2002 and the Millennium Development Goals regard the CFP as a suitable instrument for achieving the country's poverty reduction goal (Kanel 2007), while international development agencies and governments view it as a tool for poverty reduction and sustainable natural resource management (Gupta et al. 2011). The programme has been hailed as "inclusive and equitable" as well as "able to address socio-political and environmental concerns at the national and regional levels" (Karki et al. 2011, p. 22). CFUGs were first set up as projects but are now recognized as institutions. A number of cases have been celebrated for successfully arresting deforestation, helping to improve forest cover, and fomenting "genuine local participation and support" (Ibid, p. 23). Where there has been a greater presence of women in community forestry institutions, many statistically demonstrable benefits, such as enhancement of women's effective participation in decision making, women's stronger influence in the nature of decisions made, and women's roles in improving forest conservation outcomes, have been noted (Agarwal 2010).

However, the CFP has been critiqued at two levels: its process and its inclusion and participation. Agrawal et al. (1999, p. 2) describe the CFP as a "highly political process since it seeks to redistribute power and resources within the territorial confines of a given nation-state". Although the CFP seems progressive in nature the government still holds the power. Sections 67 and 68 of the Forest Act state that the government has the ownership of all types of community-managed forests and has discretionary power to alter the use of forestland and to withdraw the community forest on certain conditions (HMGN 1993). The management plan of CFUGs is a contractual document giving tenure rights over forest resources, and violation of any provision of the management plan by any member of the CFUG can affect the tenure rights of all members of the group. These administrative powers are held by the District Forest Officer (Gupta et al. 2011). Ultimately, this means that communities have only usufruct rights over the forests they nurture and guard; the government has agreed only to hand over the degraded hill forests for restoration and conservation, while it maintains control over the richer forests of the terai. The Federation of Community Forest Users-Nepal (FECOFUN), a network of CFUGs and forestry-related NGOs, has criticized the government on this point. These conditions have made CFUGs insecure about their tenure in community forestry and many communities have lost interest in participating.

On the inclusion front, numerous studies indicate that the extent of change is actually limited within large numbers of women, the poor, and excluded caste and ethnic groups who participated in community forestry processes and institutions (Agarwal 2009; Bushley 2002; Chhetri 2001; Lama and Buchy 2002; Nightingale 2002; Parajuli et al. 2010; Paudyal 2008; Uprety et al. 2012; Winrock 2002; Yadav et al. 2008). These groups benefited less from community forestry than wealthier and influential households: They could obtain free fuelwood and other non-timber forest products from the same forests before the introduction of the CFP; once the CFP declared these forests as community forests their access was limited and, therefore, many communities and groups were not interested in participating (Maharjan 1998; Malla et al. 2003). Similarly, procedures for electing the committee and decision makers through consensus and voting also resulted in well-off male and upper caste people dominating the CFUGs, which meant that powerful elites of the community shaped the rules of access to forest resources due to the prevailing sociocultural norms and barriers that influence participation in these institutions along social axes of differentiation such as class, caste, ethnicity and age (Agarwal 1997; Nightingale 2002). The Ministry of Forests and Soil Conservation in its 2013 review report cites prevailing cultural norms as the reason women's access to and influence of decision-making processes is muted, despite CFP's efforts to ensure representation of women in key decision-making positions. Furthermore, time constraints on women for domestic work limit their participation (Agarwal 2010). Staddon et al. (2015, p. 268) document a participatory community forestry project in the middle hills of eastern Nepal as a case of a well-intentioned development gone wrong. Although the aim was to invite local communities to participate, what unfolded were "multiple tyrannies" (Ibid, p. 274, 276). They found "uneven participation that provided minimal benefits to the most marginalized (women and those who are illiterate)" and that while many did "participate" as per the terms of engagement defined in a rather top-down fashion, there was—as in many other projects—an "inadvertent reinforc[ing of] existing power relations, diverting control away from communities and towards forestry authorities" (Ibid.).

We can also see that the experiences of 'women's only' CFUGs are not altogether positive. Studies have shown there is often increased marginalisation of these organizations and little to no increase in the empowerment of women (Buchy and Rai 2008; Rai and Buchy 2004; Seeley 1996). Agarwal (2010) found that 'women's only' CFUGs were allocated poorer forests compared to mixed-sex groups.

When analysed from a gender lens, institutions that seem be equitable, efficient and sustainable are not really so (Agarwal 2000) and that efforts towards gender mainstreaming in forest governance and policy have been far from desirable (Tyagi and Das 2017, p. 235). Meanwhile, current policy thrusts on gender mainstreaming in natural resource governance have considered gender as synonymous to women (Arora-Jonsson 2014). This understanding is problematic in that it implies gender as an issue for women only. Finally, gender mainstreaming as applied mostly results in adding more female members to local governance systems (Mukhopadhyay 2004). In this way, gender mainstreaming has been limited by its efficiency and functional approach rather than a structural approach towards empowerment. The entire structural and power relationship between genders remains almost untouched in policy and practice.

Gupta et al. (2011) argue that the since 2001, when discourse on environmentalism became a global issue, the involvement of non-state actors, such as donors and NGOs, increased and that these actors changed their operational strategy during Nepal's decade-long political insurgency. Although these influential non-state actors were aware of power relationships, they did not challenge them but rather used them for two purposes: to establish and advance their organization, and to establish their role as service providers to the forest sector and donors. On the other hand, non-state actors such as Federation of community Forestry Users Nepal (FECOFUN), who have been supported by donors for policy activities, had little ability to raise the agenda of challenging power issues within forestry governance because they depended on donors for financial and intellectual support. In this way, the "dynamics and complexity of actors' interactions, perceptions and power/knowledge in participatory forestry play a role in the exclusion of the poor, dalits,[3] and other disadvantaged social groups" (Gupta et al. 2011).

(b) **Hydropower**

The recent surge in hydropower development as a climate-mitigating strategy makes for an interesting case to analyse the 'depoliticised' framing and positioning of gender in two processes currently emblematic for the HKH. Hydropower development is articulately positioned and presented as being climate mitigating and, as such, hydropower projects [producing renewable and clean energy] qualify for top-up funding through the Clean Development Mechanism. Hydropower has emerged in the region as an economically viable and sustainable energy option and country governments and donor agencies are increasingly in support of hydropower, citing other numerous benefits, apart from energy generation, provided by hydropower dams such as flood control and irrigation, which would also contribute to poverty alleviation and sustainable development (Shrestha et al. 2016).

However, hydropower development in the region has also led to "adverse socio-environmental impacts. … particularly common at the local level", primarily because hydropower projects are more concerned with "national and regional economic priorities" and pay "little attention to the adverse impacts on affected local populations (mostly mountain communities)" (Shrestha et al. 2016, p. 1). Furthermore, hydropower development policies and strategies in the region pay little attention to gender, regardless of the climate merits assigned to clean energy development. While there is significant attention to the risks of hydropower development in these regions for local communities, the analytical scale of the 'local' remains essentially unpacked. While hydropower is indeed renewable, the waterscape is often irreversibly changed by the processes of generating hydropower. What might be the social, economic, and environmental costs of large dam development in a region that is said to be not only geologically and ecologically unique but also politically fragile, with ethnic and cultural tensions and political faults corresponding to international and national boundaries?

The Shrestha et al. (2016) report on the benefit sharing mechanisms[4] in Nepal's hydropower sector show that "while benefit-sharing programmes generally seek to share benefits equally across project-affected populations. … certain kinds of people have less access to the benefits of hydropower development than others" (p. 41). Women are among these groups who are under-represented in the process of hydropower development in terms of stakeholder consultation, local hiring and employment, establishing local development priorities, and local governance. Women from marginalised groups (Janajati,[5] dalit, ultra-poor, and disabled women) experience further disadvantage due to social power hierarchies.

The Shrestha report examines the benefit sharing mechanisms of hydropower development and concludes that

[3]*Dalit* is the name given to the lowest class of people in traditional Indian society, who were formally referred to as "the untouchables", falling altogether outside the Hindu caste categories and subject to extensive social restrictions.

[4]In the hydropower sector, benefit sharing mechanism would mean ways and methods to share the profits, but more important, the advantages derived from the hydropower development by the companies. The early practice of benefit sharing was 'trickle down' to local communities, whereby broader national and regional economic development was expected to bring the dividends of development to local citizens. Later, the practice moved on to compensation and mitigation for minimizing the negative impacts of projects, and mechanisms were designed to ensure that individuals and communities adversely impacted by hydropower development were compensated for any losses sustained. More recently, the emphasis is on sustainable development, therefore, benefit sharing mechanisms now go beyond mitigation and compensation "to maximize development benefits and more equitable outcomes, and working directly with local communities to increase investment effectiveness" (Shrestha et al. 2016:7).

[5]Adivasi/Janajatis: People or communities having their own mother tongue and traditional customs, distinct cultural identity, social structure, and written or oral history. They are interchangeably referred to as "Janajatis," "ethnic groups," and "indigenous nationalities." The government identified 59 groups as indigenous nationalities and these are categorized into five groups based on their economic and sociocultural status. These include "endangered," "highly marginalized," "marginalized," "disadvantaged," and "advanced" groups. Newars and Thakalis are the only two groups classified as "advanced" groups (ADB 2010).

"patterns of social exclusion based on gender, caste, ethnicity, and class were apparent" (41). The authors note that women receive considerably fewer direct benefits in both employment and training. Furthermore, women are provided training in gender-stereotypical skills such as knitting, cooking, and weaving. The report states that "these issues reflect a larger problem with gender inequity: The collective voice of women is routinely subjugated in local processes of decision-making about hydropower projects" (25).

The report (40) outlines the following gender and social inequities that are evident during the process of hydropower development:

- Uneven patterns of awareness and information about stakeholder rights
- Unequal participation in community consultations and decision making about hydropower development, leading to the prioritization of some agenda and the subjugation of others
- Uneven distribution of impacts related to hydropower development, as certain subpopulations and social groups are disproportionately affected or economically dislocated
- Unequal ability to mobilize for individual benefits (i.e., seeking employment, participating in training programmes, or purchasing project shares)
- Informal inequalities in the distribution of benefits notionally shared by communities (e.g., community development programmes, royalties) due to unequal access to public facilities and patterns of social hierarchy.

14.3.3 Professionals, Knowledge, and Masculinities: A Scale Challenge in NRM Governance

In environmental governance, many knowledge and policy domains are treated as fields of engineering and technology, especially in the field of natural resource management (Adhikary 1995). The domain of irrigation and water governance is one example (Liebrand 2014; Zwarteveen 2006). It is an area of expertise that historically is managed as a field of irrigation engineering, particularly in Asia (Ongsakul et al. 2012). In this context, several fields in environmental governance have been qualified by feminist scholars as 'masculine'. For clarification, masculinity is not just about 'men'; it is also about ideas of supremacy and modernist convictions in engineering to which notions of power are attached. It is true that male professionals have traditionally dominated the field of engineering and technology—their presence in the profession is considered 'normal' (Gupta 2007; Kulkarni et al. 2009; Nair 2012; Parikh and Sukhatme 2004). But there also exists a strong epistemic tradition in environmental governance, in fields such as irrigation and water expert thinking, that sees the world as uniform, makeable, and manageable (Zwarteveen 2011). Consequently, it has long rendered thinking and speaking about women (and men) irrelevant. It is also in this 'masculine' professional culture that policies such as Integrated Water Resources Management (IWRM) gain meaning and that knowledge on climate change adaptation acquires legitimacy.

The domination of engineering knowledge and the 'masculinity' of professional cultures in natural resources management (NRM) can be considered a scale challenge[6] in environmental governance (Liebrand 2010). For the HKH, this is reason for great concern. It means that there is a structural mismatch between actual realities in the field and expectations and administrative realities at the policy level. For instance, the policy objective in water governance to support livelihoods and create opportunities for all is undermined by a culture of 'masculinity' among water professionals. This means that complex 'problems' in the field, such as gender inequities in irrigation, are conceptualized as 'technical' or 'engineering' problems, based on ideals of the universal application of science and technology, and because 'hardware' solutions (e.g., canals, roads, pumps, and other infrastructure) are associated with 'real' development and with those professionals who hold this knowledge, i.e., mostly male engineers. The associations between men, 'masculinity', and power/knowledge in water governance partially explains why current measures to bridge the gap, between field and policy levels, continue to fall short of expectations.

In Nepal, for instance, the Department of Irrigation has adopted policies and programmes in the past two decades to improve irrigation and water resource management. Recurring elements of these programmes include decentralization, user participation, women's inclusion and, more recently, public-private partnerships (Gautam 2006; Shukla and Sharma 1997; Singh et al. 2014). Yet Nepal's irrigation and water resources development sector continues to be characterized by infrastructure interventions that are typically proposed and executed by an elite class of male engineers; and because such interventions tend to be designed in such a way that they fit the existing national socioeconomic context (for instance, the construction of an irrigation system without implementing land reforms), persistent and historic injustices and social inequities along divisions of class, caste, and ethnicity and gender (DFID/WB 2006) are being reproduced through such hardware interventions.

More generally, more than three decades of mainstreaming gender in (water) development research and

[6]A scale challenge represents a situation in which the current combination of cross-scale and cross-level interactions threatens to undermine the resilience of a human-environment system (Cash et al. 2006).

policy have failed to come to grips with 'masculinities' in the profession (Laurie 2005; Liebrand 2014). In spite of repeated calls by feminist researchers to address masculinities in NRM, engineers and experts in water planning still tend to be men and modernist convictions are rarely discussed and challenged in the promotion of science and technology (Zwarteveen 2008, 2011). Hence, in spite of critique, a question of '*what* does it deliver', continues to guide most discussions on planned development—not *if* it delivers at all or *what* it means for different people and *how* it actually works in the field. A knowledge tradition that sees the world as uniform, makeable, and manageable remains the dominant model; and most development interpretations in research and policy, for instance, made by irrigation and water experts today continue to emphasise and attach greater value to knowledge and experiences that present the world as rational, universal, and genderless (Liebrand 2014).

As noted, 'masculinity' is not just about men. Women professionals in engineering can be gender-insensitive and male engineers can be gender champions. For example, some male engineers in the Department of Irrigation are found today who discuss gender issues professionally and rationally, working hard to meet social equity and gender goals in development (see Udas and Zwarteveen 2010). It also is visible that gender and women have earned a legitimate place in water research and irrigation policy agendas, but at the same time there is to date little reflection among professionals on development norms and 'masculinity' in environmental governance and how these influence, for instance, water expert thinking and the way experts see irrigation development and water resources management. To address this concern, a way forward is to increase a capacity in experts to reflect on their knowledge and professional practises; they need to internalize a conviction that knowledge and development is never neutral and value-free. One promising way to achieve this is to integrate social sciences knowledge more firmly in engineering and technical education, and to ensure a more diverse representation among professionals themselves, in terms of social background (e.g., more women professionals, more experts of marginalized classes, castes, and ethnicities) and disciplinary education (hiring more sociologists in engineering departments, for instance).

14.4 Conclusion

This chapter has shown that climate change and extreme weather have differential impacts on women and men in the HKH. The case studies confirm that women's experiences in the HKH are multiple, differentiated, and sometimes contradictory and, in some cases, effect new chains of vulnerability.

Women across different socioeconomic categories are disproportionately affected because of structural inequalities in the distribution of rights, assets, resources, and power based on repressive cultural rules and norms. As a result, women are often poorer and less educated than men and excluded from political and household decision-making processes at various scales that affect their lives. Today's technocratic programs include women in economic development; this inclusion is rationalized only by an appeal to economic gain. While the pursuit of economic efficiency can offer women economic opportunities, it does not fully address their unequal power relations with men, vis-à-vis equal wages for equal work, or the sharing of domestic work responsibilities. Accordingly, the focus needs to be more on context. This means not only recognition of women's vulnerabilities but also how gender roles and identities constructed for women as well as for men contribute to gender imbalances, inequalities, power, and privilege. Interventions must shift their attention to the structures that underlie gender inequality. This work calls for long-term political engagement.

A major problem is that empirical data on the impacts of climate change are often presented in aggregate terms, reflecting an unfounded assumption that climate change affects people uniformly. Policy makers need to acknowledge that experiences of climate change, and responses to climate change, are defined by intersecting factors of age, caste, class, gender, and ethnicity. Therefore, policy needs to support mechanisms for collection of disaggregated data. This will require the development of such mechanisms, which can be shared and communicated effectively.

At the same time it must be understood that gender disaggregated data have their uses and limits: They can indicate disparities and differences (uses), but there is a need to explore and understand the multi-scalar drivers of inequality. This requires highly textured studies—nested, multi-scalar, and historical—that ask compelling questions about power, identity, struggles, and exclusions using mixed quantitative and qualitative methods. Such studies are urgent if we want to address the gender gap and track progress according to the SDG targets.

Far too often, when programmes talk about addressing gender, they imply acting at the household or community levels by taking the poor women as 'volunteers'. They typically address women's practical gender needs, by associating women's needs and aspirations as small in size and subsistence-oriented, which, while an important entry point, often do not question the gender division of labour or access to resource rights and ownership. The consequence is that this creates an unsurpassable binary of scale between what happens in households and communities and the broader arena of strategic environmental decisions (Lahiri-Dutt 2014). Furthermore, such an approach ignores women's strategic needs that are critical to women's empowerment and ultimately to transformative change. Interventions must shift their attention to the structures that underlie gender inequality.

While the creation of space for women to participate in natural resources management and governance is critical in terms of exercising voice and agency, policies must not further feminize responsibilities in ways that will add to, rather than reduce, environmental burdens for women. Women's engagement and participation in climate change policy making and on-the-ground interventions should follow a logic that is empowering and promotes women's rights—not one that is dictated solely by efficiency.

Making development and adaptation efforts more gender inclusive and socially inclusive will require addressing contextually determined relations of inequality at the household, community, and other institutional levels. Such efforts are more likely to lead to sustained, transformative outcomes when based on longitudinal studies that explore how climate change affects various groups of women and men, both separately and jointly. This work calls for long-term political engagement and will require consistent attention to addressing these complex relations of inequality to ensure and enable change. Unfortunately, there is no simple, easy way to reduce the complexities of gender inequality. Attention needs to be paid to such challenging dimensions of development practice if preparedness for climate and natural resource management is to address gendered divides and disparities. Without professional and critical self-reflection, we have little reason to assume that new policies will succeed any more than past efforts to increase gender equality, women's empowerment, social inclusion, and climate change adaptation.

Underlying all this is that policies need to allocate resources—financial and human—for gender responsive interventions at scale and adopt clear accountability mechanisms, such as gender budgeting, to demonstrate their commitment to gender equality as indicated in the SDGs.

In conclusion, we set forth a vision of inclusive development for the HKH complementing, and in the spirit of, the Sustainable Development Goals adopted by the global community in 2015: By 2030, environmental governance processes, policies, and strategies at scale (from local to global) are gender inclusive and cognizant of the mosaic of nested, uniquely diverse, dynamic, and mostly gender-inequitable socio-ecological systems in the HKH.

References

ADB. (2010). *Overview of gender equality and social inclusion in Nepal*. Manila: Asian Development Bank.

Adhikary, M. (1995). Women graduates in agriculture and forestry development in Nepal (No. REP-7905. CIMMYT).

Agarwal, B. (1997). Environmental action, gender equity and women's participation. *Development and Change, 28*(1), 1–44.

Agarwal, B. (2000). Conceptualising environmental collective action: Why gender matters. *Cambridge Journal of Economics, 24*(3), 283–310.

Agarwal, B. (2009). Gender and forest conservation: The impact of women's participation in community forest governance. *Ecological Economics, 68*(11), 2785–2799.

Agarwal, B. (2010). The impact of women in Nepal's community forestry management. Gender perspectives in mountain development. In *Sustainable Mountain Development*, No. 57, ICIMOD, Summer 2010 (pp. 26–29).

Agrawal, A., Britt, C., & Kanel, K. (1999). *Decentralization in Nepal: A comparative analysis*. A report on the participatory district development program. ICS Press Institute for Contemporary Studies.

Akçar, S. (2001). Grassroots women's collectives–Roles in post-disaster effort: Potential for sustainable partnership and good governance (lessons Learned from the Marmara Earthquake in Turkey). United Nations Division for the Advancement of Women.

Alston, M., Whittenbury, K., Haynes, A., & Godden, N. (2014, December). Are climate challenges reinforcing child and forced marriage and dowry as adaptation strategies in the context of Bangladesh? In Women's Studies International Forum (Vol. 47, pp. 137–144). Pergamon.

Arora-Jonsson, S. (2014, December). Forty years of gender research and environmental policy: Where do we stand? In Women's Studies International Forum (Vol. 47, pp. 295–308). Pergamon.

Atkinson, H. G., & Bruce, J. (2015). Adolescent girls, human rights and the expanding climate emergency. *Annals of Global Health, 81*(3), 323–330.

Baden, S., Green, C., Goetz, A. M., & Guhathakurta, M. (1994). *Background report on gender issues in Bangladesh* (Vol. 26). University of Sussex. IDS.

Bari, F. (1998). *Gender, disaster, and empowerment: A case study from Pakistan* (pp. 1–8). The gendered terrain of disaster: Through women's eyes.

Barua, A., Katyaini, S., Mili, B., & Gooch, P. (2014). Climate change and poverty: Building resilience of rural mountain communities in South Sikkim, Eastern Himalaya, India. *Regional Environmental Change, 14*(1), 267–280.

Bastola, A., Hynie, M., & Poudel, S. (2015). Gender and Ethnic Practices: Can gender based practices in ethnic groups illustrate the impacts of climate change? In *World Symposium on Climate Change Adaptation*, 2–4 September 2015, Manchester. Grantham Research Institute (GRI) (2015). Myanmar. http://www.lse.ac.uk/GranthamInstitute/legislation/countries/myanmar/. Accessed 26 April 2016.

Bern, C., Sniezek, J., Mathbor, G. M., Siddiqi, M. S., Ronsmans, C., Chowdhury, A. M. et al. (1993). Risk factors for mortality in the Bangladesh cyclone of 1991. *Bulletin of the World Health Organization, 71*(1), 73.

Bettini, G., & Gioli, G. (2015). Waltz with development: insights on the developmentalization of climate-induced migration. *Migration and Development, 5*(2), 171–189.

Bhattarai, B., Beilin, R., & Ford, R. (2015). Gender, agrobiodiversity, and climate change: A study of adaptation practices in the Nepal Himalayas. *World Development, 70,* 122–132.

Bose, A. (2000). Demography of Himalayan villages: Missing men and lonely women. *Economic and Political Weekly*, 2361–2363.

Brockington, D. (2012). A radically conservative vision? The challenge of UNEP's towards a green economy. *Development and Change, 43* (1), 409–422.

Brody, A., Demetriades, J., & Esplen, E. (2008). Gender and climate change: Mapping the linkages, a scoping study on knowledge and gaps. BRIDGE, Institute of Development Studies, University of Sussex.

Buchy, M., & Rai, B. (2008). Do women-only approaches to natural resource management help women? The case of community forestry in Nepal. Gender and natural resource management: Livelihoods, mobility and interventions (pp. 127–147).

Buchy, M., & Subba, S. (2003). Why is community forestry a social-and gender-blind technology? The case of Nepal. *Gender, Technology and Development, 7*(3), 313–332.

Bullard, N., & Müller, T. (2012). Beyond the 'Green Economy': System change, not climate change? *Development, 55*(1), 54–62.

Bushley, B. (2002). Users perceptions of the impacts of community forestry on community development in Nepal's western Terai region. Care Nepal.

Canada, C. A. R. E. (2010). *Cyclone Nargis: Myanmar two years later*. Ottawa: CARE Canada.

Carvajal-Escobar, Y., Quintero-Angel, M., & Garcia-Vargas, M. (2008). Women's role in adapting to climate change and variability. *Advances in Geosciences, 14,* 277–280.

Cash, D., Adger, W. N., Berkes, F., Garden, P., Lebel, L., Olsson, P., et al. (2006). Scale and cross-scale dynamics: Governance and information in a multilevel world. *Ecology and Society, 11*(2).

Chhetri, G. (2001). A social and cultural perspective of women and community forestry in Nepal. *Contributions to Nepalese Studies, 28* (1), 53–72.

Chowdhury, A. M. R., Bhuyia, A. U., Choudhury, A. Y., & Sen, R. (1993). The Bangladesh cyclone of 1991: Why so many people died. *Disasters, 17*(4), 291–304. climdev15.pdf. Accessed 6 June 2017.

D'Cunha, J. (1997). Engendering disaster preparedness and management. *Asian Disaster Management News, 3*(3), 2–5.

Dankelman, I. E. M. (2008). Gender, climate change and human security: Lessons from Bangladesh, Ghana and Senegal.

Das, N. (2011). Can gender-sensitive forestry programmes increase women's income? Lessons from a forest fringe community in an Indian province. *Rural Society, 20*(2), 160–173.

Dasgupta, S., Huq, M., Khan, Z. H., Ahmed, M. M. Z., Mukherjee, N., Khan, M. F., et al. (2010). Vulnerability of Bangladesh to cyclones in a changing climate: Potential damages and adaptation cost.

DFID/WB. (2006). Unequal citizens. Gender, caste and ethnic exclusion in Nepal. Summary. Kathmandu, Nepal: Department for International Development (DFID) and World Bank (WB).

Dhimal, M. L. (2015). Gender differentiated health impacts of environmental and climate change in Nepal. In S. Pradhananga, J. Panthi, & D. Bhattarai (Eds.), *Proceedings of International Conference on Climate Change Innovation and Resilience for Sustainable Livelihood* (pp. 96–100), 12–14 January 2015, Kathmandu, Nepal. Kathmandu: Small Earth Nepal. https://smallearthnepal.files.wordpress.com/2014/06/proceeding.pdf.

Enarson, E. (1999). Women and housing issues in two US disasters: Case studies from Hurricane Andrew and the Red River Valley flood. *International Journal of Mass Emergencies and Disasters, 17* (1), 39–63.

Eriksson, M., Fang, J., & Dekens, J. (2008). How does climate change affect human health in the Hindu Kush-Himalaya region. *Regional Health Forum, 12*(1), 11–15.

Fairhead, J., Leach, M., & Scoones, I. (2012). Green Grabbing: A new appropriation of nature? *Journal of Peasant Studies, 39*(2), 237–261.

Fordham, M. (2001) Challenging boundaries: A gender perspective on early warning in disaster and environmental management. United Nations, Division for the Advancement of Women (DAW), International Strategy for Disaster Reduction (ISDR), Expert group meeting on "Environmental management and the mitigation of natural disasters: A gender perspective", 6–9 November 2001, Ankara, Turkey. https://www.unisdr.org/files/8264_EP52001Oct261.pdf.

Forsyth, T. (2001). Critical realism and political ecology. In A. Stainer & G. Lopez (Eds.), (2001) *After postmodernism: Critical realism?* (pp. 146–154). London: Athlone Press.

Fukuda-Parr, S. (2016). From the millennium development goals to the sustainable development goals: Shifts in purpose, concept, and politics of global goal setting for development. *Gender & Development, 24*(1), 43–52. https://doi.org/10.1080/13552074.2016.1145895.

Gauld, R. (2000). Maintaining centralized control in community-based forestry: Policy construction in the Philippines. *Development and Change, 31*(1), 229–254.

Gautam, S. R. (2006). Incorporating groundwater irrigation: Technology dynamics and conjunctive water management in the Nepal Terai.

GDMA (Gilgit-Baltistan Disaster Management Authority). (2015). Gilgit Baltistan's (GB) Contingency Plan for Floods.

Goh, A. H. (2012). *A literature review of the gender-differentiated impacts of climate change on women's and men's assets and well-being in developing countries*. Washington, DC: International Food Policy Research Institute.

Golam Rabbani, M. D., Atiq Rahman, A., & Mainuddin, K. (2009). Women's vulnerability to water-related hazards: Comparing three areas affected by climate change in Bangladesh. *Waterlines, 28*(3), 235–249.

GoN. (2014). Labour migration for employment: A status report for Nepal: 2013/2014, Ministry of Labour and Employment, Department of Foreign Employment, Government of Nepal.

Gupta, N. (2007). Indian women in doctoral education in science and engineering a study of informal milieu at the reputed Indian Institutes of Technology. *Science, Technology and Human Values, 32*(5), 507–533.

Gupta, S., & Agarwal, B. (2013). *Gender and green governance: The political economy of women's presence within and beyond community forestry*.

Gupta, S. P., Bhattarai, H. K., & Sah, R. (2011). Exclusion in community forestry of Nepal. A term paper on exclusion in community forestry of Nepal, Kathmandu University, Nepal.

Gurung, J. D. (1998). *Mountain women of the Hindu Kush-Himalayas: The hidden perspective*.

Gurung, J. D. (1999). *Searching for women's voices in the Hindu Kush-Himalayas*. International Centre for Integrated Mountain Development (ICIMOD).

Gurung, D. D., & Bisht, S. (2014). *Women's empowerment at the frontline of adaptation: emerging issues, adaptive practices, and priorities in Nepal*. ICIMOD Working Paper, (2014/3).

Gyawali, D., & Thompson, M. (2016). Restoring Development Dharma with Toad's Eye Science? In *IDS Bulletin- 'States, Markets and Society—New Relationships for a New Development Era'* (Vol. 47, No. 2A November 2016, pp. 1979–1989). https://opendocs.ids.ac.uk/opendocs/bitstream/handle/123456789/12691/IDSB472A_10.190881968-2016.192.pdf?sequence=1&isAllowed=y. Accessed 31 Sept. 2017.

Hamid and Afzal. (November, 2013). *Gender, water and climate change: The case of Pakistan*.

Hanson, A. M. S. (2015). Shoes in the seaweed and bottles on the beach: Global garbage and women's oral histories of socioenvironmental change in coastal Yucatan. In S. Buechler & A.-M. Hanson (Eds.), *A political ecology of women, water and global environmental change* (pp. 165–184).

Harcourt, W., & Nelson, I. L. (Eds.). (2015). *Practising feminist political ecologies: Moving beyond the 'green Economy'*. Zed Books.

Harding, S. (2008). *Sciences from below: Feminisms, postcolonialities, and modernities*. Duke University Press.

Hillenbrand, E., Lakzadeh, P., Sokhoin, L., Talukder, Z., Green, T., & McLean, J. (2014). Using the social relations approach to capture complexity in women's empowerment: Using gender analysis in the

Fish on Farms project in Cambodia. *Gender & Development, 22*(2), 351–368.

HMGN. (1993). Forest Act 2049, Ministry of Law and Justice, His Majesty's Government of Nepal. http://www.wedo.org/wp-content/uploads/hsn-studyfinal-may-20-2008.pdf.

IFRC. (2007). World disaster report. Focus on discrimination. http://www.ifrc.org/Global/Publications/disasters/WDR/WDR2007-English.pdf.

IFRC. (2015). Epidemic preparedness in Myanmar. https://www.preparecenter.org/sites/default/files/epdemic_preparedness_in_myanmar.pdf.

Jasanoff, S. (2010). A new climate for society. *Theory, Culture & Society, 27*(2–3), 233–253.

Jodha, N. S. (2007). Mountain commons: Changing space and status at community levels in the Himalayas. *Journal of Mountain Science, 4* (2), 124–135.

Joshi, D. (2014). Feminist solidarity? Women's engagement in politics and the implications for water management in the Darjeeling Himalaya. *Mountain Research and Development, 34*(3), 243–254.

Kabeer, N. (1994). *Reversed realities: Gender hierarchies in development thought*. Verso.

Kabeer, N., & Subrahmanian, R. (1996). *Institutions, relations and outcomes: Framework and tools for gender-aware planning*. Brighton, UK: Institute of Development Studies.

Kanel, K. R. (2007). Economic impacts of forest policy changes: A perspective from Nepal. *The Initiation, 1,* 36–42.

Karki, M., Sharma S., Mahat, T., Aksha, S., & Tuladhar, A. (2011). *From Rio 1992 to 2012 and beyond: Sustainable mountain development Hindu Kush Himalaya (HKH) region*. Draft for Discussion. Regional Assessment Report for Rio+20: Hindu Kush Himalaya and SE Asia Pacific Mountains. Kathmandu: ICIMOD.

Khan, M., & Kumar, U. (2010). Water security in Peri-urban South Asia. Adapting to climate change and urbanization, scoping study report, IWFM, BUET, Khulna. http://saciwaters.org/periurban/Scoping_Study_Report_Khulna.pdf. Accessed 30 Sept. 2017.

Khan, A., Mojumder, S. K., Kovats, S., & Vineis, P. (2008). Saline contamination of drinking water in Bangladesh. *The Lancet, 371* (9610), 385.

Khan, A. E., Scheelbeek, P. F. D., Shilpi, A. B., Chan, Q., Mojumder, S. K., Rahman, A., et al. (2014). Salinity in drinking water and the risk of (pre) eclampsia and gestational hypertension in coastal Bangladesh: a case-control study. *PLoS One, 9*(9), e108715.

Khan, M. S. A., Shahjahan Mondal, M., Rahman, R., Huq, H., Datta, D. K., Kumar, U., et al. (2016). Urban burden on Peri-urban areas: Shared use of a river in a coastal city vulnerable to climate change. In V. Narain & P. Anjal (Eds.), *Water security in Peri-urban South Asia: Adapting to climate change and urbanization* (pp. 286–296). New Delhi, India: Oxford University Press.

Kosoy, N., Brown, P. G., Bosselmann, K., Duraiappah, A., Mackey, B., Martinez-Alier, J., et al. (2012). Pillars for a flourishing Earth: Planetary boundaries, economic growth delusion and green economy. *Current Opinion in Environmental Sustainability, 4*(1), 74–79.

Kulkarni, S., Bhat, S., Majumdar, S., & Goodrich, C. G. (2009). *Situational analysis of women water professionals in South Asia*.

Lahiri-Dutt, K. (2014). *Experiencing and coping with change: Women-headed farming households in the Eastern Gangetic Plains*. Australian Centre for International Agricultural Research.

Lama, A., & Buchy, M. (2002). *Gender, class, caste and participation: The case of community forestry in Nepal. Bulletin (Centre for Women's Development Studies), 9*(1), 27–41.

Lauren Khan, A. (2012). Chapter 9 creative adaptation: Bangladesh's resilience to flooding in a changing climate. In *Climate Change Modeling For Local Adaptation In The Hindu Kush-Himalayan Region* (pp. 159–175). Emerald Group Publishing Limited.

Laurie, N. (2005). Establishing development orthodoxy: Negotiating masculinities in the water sector. *Development and Change, 36*(3), 527–549.

Leach, M., Joekes, S., & Green, C. (1995). A predominant set of perspectives highlights, *26*(1).

LEAD Pakistan. (March, 2015). Early warning system and disaster risk information, leadership for environment & development (LEAD) Pakistan.

Leduc, B. (2009). Gender and climate change in the Himalayas, background paper for e-discussion. In *Climate Change in the Himalayas: The Gender Perspective,* 5–25 October, ICIMOD. Kathmandu: ICIMOD and Asia Pacific Mountain Network (APMN).

Leduc, B. (2010). Adaptation to climate change-why gender makes a difference? *Adaptation to climate change-why gender makes a difference?* (57), 8–11.

Leduc, B., & Shrestha, A. (2008). *Gender and climate change in the Hindu Kush-Himalayas: Nepal case study*. Kathmandu, Nepal: International Centre for Integrated Mountain Development (ICIMOD).

Leduc, B., Shrestha, A., & Bhattarai, B. (2008). *Gender and climate change in the Hindu Kush Himalayas of Nepal*. International Centre for Integrated Mountain Development (ICIMOD) for WEDO, 8.

Liebrand, J. (2010). Masculinities: A scale challenge in irrigation governance in Nepal. In *Dynamics of farmer managed irrigation systems: Socio-institutional, economic and technical contexts* (pp. 53–72).

Liebrand, J. (2014). *Masculinities among irrigation engineers and water professionals in Nepal*. Ph.D. thesis, Wageningen University, Wageningen, NL. ISBN: 978-94-6257-141-9.

MacGregor, S. (2010). 'Gender and climate change': From impacts to discourses. *Journal of the Indian Ocean Region, 6*(2), 223–238.

Maharjan, M. R. (1998). The flow and distribution of costs and benefits in the Chuliban community forest, Dhankuta District, Nepal. *Rural Development Forestry Network, 23,* 1–12.

Malla, Y. B., Neupane, H. R., & Branney, P. J. (2003). Why aren't poor people benefiting more from community forestry. *Journal of forest and livelihood, 3*(1), 78–92.

Mallick, D., Golam Jilani, K., & Akand, E. S. (2016). Vulnerability of women and poor to climate change; gender responsive adaptation in Bangladesh. Bangladesh Centre for Advanced Studies and Christian Aid, Dhala, Bangladesh, July.

Maraseni, T. N. (2012). Climate change, poverty and livelihoods: Adaptation practices by rural mountain communities in Nepal. *Environmental Science & Policy, 21,* 24–34.

Massey, D. S., Axinn, W. G., & Ghimire, D. J. (2010). Environmental change and out-migration: Evidence from Nepal. *Population and Environment, 32*(2–3), 109–136.

Mehta, M. (1996). Our lives are no different from that of our buffaloes: Agricultural change and gendered spaces in a central Himalayan valley. Feminist political ecology: Global issues and local experiences (pp. 180–210).

Mehta, M. (2007). *Gender matters: Lessons for disaster risk reduction in South Asia*. International Centre for Integrated Mountain Development (ICIMOD).

Moitra, A., & Kumar, A. (2016). Hill women's voices and community communication about climate change: The case of Henvalvani Community Radio in India. The praxis of social inequality in India: A global perspective (pp. 137–159).

Molyneux, M. (2007). 18| The chimera of success: Gender ennui and the changed international policy environment. *Feminisms in Development: Contradictions, Contestations and Challenges, 9,* 227.

Morrow, B. H., & Phillips, B. (1999). What's gender "Got to do with it"? *International Journal of Mass Emergencies and Disasters, 17* (1), 5–11.

Moser, C. (2012). *Gender planning and development: Theory, practice and training*. Routledge.

Mukhopadhyay, M. (2004). Mainstreaming gender or "streaming" gender away: Feminists marooned in the development business. *IDS Bulletin, 35*(4), 95–103.

Mustafa, D., Gioli, G., Qazi, S., Waraich, R., Rehman, A., & Zahoor, R. (2015). Gendering flood early warning systems: The case of Pakistan. *Environmental Hazards, 14*(4), 312–328.

Nair, S. (2012). Women in Indian engineering: A preliminary analysis of data from the graduate level engineering education field in Kerala and Rajasthan.

Nellemann, C., Verma, R., Hislop, L. (Eds.), (2011). Women at the frontline of climate change: Gender risks and hopes. A rapid response assessment. United Nations Environment Programme, GRID-Arendal.

Nelson, V., Meadows, K., Cannon, T., Morton, J., & Martin, A. (2002). Uncertain predictions, invisible impacts, and the need to mainstream gender in climate change adaptations. *Gender & Development, 10* (2), 51–59.

Neumayer, E., & Plümper, T. (2007). The gendered nature of natural disasters: The impact of catastrophic events on the gender gap in life expectancy, 1981–2002. *Annals of the Association of American Geographers, 97*(3), 551–566.

Neupane, H. (2003). Contested impact of community forestry on equity: Some evidence from Nepal. *Journal of Forest and Livelihood, 2*(2), 55–61.

Nibanupudi, H. K., & Khadka, M. (2015). Gender and disaster resilience in Hindu Kush Himalayan Region. In H. K. Nibanupudi & R. Shaw (Eds.), Mountain hazards and disaster risk reduction, disaster risk reduction (pp. 233–249). https://doi.org/10.1007/978-4-431-55242-0_13.

Nightingale, A. J. (2002). Participating or just sitting in? The dynamics of gender and caste in community forestry. *Journal of forest and livelihood, 2*, 1.

ODI (Overseas Development Institute). 2007. Gender fatigue: What can we do to overcome it? (7 March 2007).

Ogra, M. V., & Badola, R. (2015). Gender and climate change in the Indian Himalayas: Global threats, local vulnerabilities, and livelihood diversification at the Nanda Devi Biosphere Reserve. *Earth System Dynamics*, *6*(2).

Ongsakul, R., Resurreccion, B., & Sajor, E. (2012). Normalizing masculinities in water bureaucracy in Thailand. *International Journal of Public Administration, 35*(9), 577–586.

Parajuli, R., Pokharel, R. K., & Lamichhane, D. (2010). Social discrimination in community forestry: Socio-economic and gender perspectives. *Banko Janakari, 20*(2), 26–33.

Parikh, J. (2007). *Gender and climate change framework for analysis, policy & action.*

Parikh, P. P., & Sukhatme, S. P. (2004). Women engineers in India. *Economic and Political Weekly*, 193–201.

Paudel, G. (2012). *Forest size, income status and inequality*. Doctoral dissertation, Tribhuvan University.

Paudel, D., & Paudel, G. (2013). Corpo-bureaucratizing community forest: Commercialization and increased financial transaction in community forestry user groups in Nepal. *Journal of Forest and Livelihood, 9*(1), 1–15.

Paudyal, B. R. (2008). *Agrarian structures and distributive outcomes: A study of community forestry in Nepal*. Doctoral dissertation, International Institute of Social Studies of Erasmus University (ISS).

PDMA/KP. (2015). *Report on Torrential Rains and Mini Cyclone in Peshawar Division (April 26)*. Provincial Disaster Management Authority Relief, Rehabilitation & Settlement Department, Government of Khyber Pakhtunkhwa.

Pender, E. (2009). Opportunities for coordination: Protection, women's protection and gender based violence. GenCap.

Prakash, A. (2016). Towards Peri-urban water security- implications for adaptive capacity, public policy, and future research and action. In V. Narain & P. Anjal (Eds.), *Water security in Peri-urban South Asia: Adapting to climate change and urbanization* (pp. 286–296). New Delhi, India: Oxford University Press.

Rai, B., & Buchy, M. (2004, August). Institutional exclusion of women in community forestry: Is women-only strategy a right answer. In *Twenty-five years of community forestry: Proceedings of the Fourth National Workshop on Community Forestry* (pp. 4–6).

Rai, S., Shah, N., & Avaz, A. (2007). *Achieving gender equality in public offices in Pakistan.*

Rocheleau, D., Thomas-Slayter, B., & Wangari, E. (1996). *Feminist political ecology: Global issues and local experience*. Routledge.

Roehr, U. (2007). Gender, climate change and adaptation. Introduction to the gender dimensions. In *Both Ends Briefing Paper Series.*

Seeley, J. (1996). Women, Men and Roles in Forestry. *Banko Janakari, 6*(1), 39.

Shah, S. A. (2012). Gender and building homes in disaster in Sindh, Pakistan. *Gender & Development, 20*(2), 249–264.

Shah, S. A., & Memon, N. (2012). Entering male domain and challenging stereotypes. In A. Prakash, G. C. Goodrich, & S. Singh (Eds.), *Informing Water Policies: Case Studies from South Asia.* Rutledge, New Delhi.

Sharmin, Z., & Islam, M. S. (2013). Consequences of climate change and gender vulnerability: Bangladesh perspective.

Shrestha, P., Lord, A., Mukherji, A., Shrestha, R. K., Yadav, L., & Rai, N. (2016). *Benefit sharing and sustainable hydropower: Lessons from Nepal*. ICIMOD Research Report 2016/2. Kathmandu: Nepal.

Shukla, A., & Sharma, K. R. (1997). *Participatory irrigation management in Nepal: A monograph on evolution, processes and performance.*

Sidh, S. N., & Basu, S. (2011). Women's contribution to household food and economic security: A study in the Garhwal Himalayas, India. *Mountain Research and Development, 31*(2), 102–111.

Singh, N., & Singh, O. P. (2015). Climate change, water and gender: Impact and adaptation in North-Eastern Hills of India. *International Social Work, 58*(3), 375–384.

Singh, S. P., Bassignana-Khadka, I., Karky, B. S., & Sharma, E. (2011). *Climate change in the Hindu Kush-Himalayas: The state of current knowledge*. International Centre for Integrated Mountain Development (ICIMOD).

Singh, M., Liebrand, J., & Joshi, D. (2014). Cultivating "success" and "failure" in policy: Participatory irrigation management in Nepal. *Development in Practice, 24*(2), 155–173.

Skinner, E. (2011). *Gender and climate change: Overview report.* Bridge.

Sogani, R. (2013). Climate change: A Himalayan perspective 'Local Knowledge–The Way Forward'. In *Research, action and policy: Addressing the gendered impacts of climate change* (pp. 265–275). The Netherlands: Springer.

Song, Y., & Vernooy, R. (2010). Seeds of empowerment: action research in the context of the feminization of agriculture in southwest China. *Gender, Technology and Development, 14*(1), 25–44.

Song, Y., & Zhang, Y. (2015). SIFOR China project base-line study report.

Song, Y., Qi, G., Zhang, Y., & Vernooy, R. (2014). Farmer cooperatives in China: Diverse pathways to sustainable rural development. *International Journal of Agricultural Sustainability, 12*(2), 95–108.

Song, Y., Zhang, Y., Song, X., & Vernooy, R. (2016). *Access and benefit sharing in participatory plant breeding in Southwest China. Farming matters*. Special Issue, Access and benefit sharing of genetic resources (pp. 18–23).

SPDC. (2015). *Gender and social vulnerability to climate change; A study of disaster prone areas in Sindh*. Social Policy and Development Centre (SPDC).

Staddon, S. C., Nightingale, A., & Shrestha, S. K. (2015). Exploring participation in ecological monitoring in Nepal's community forests. *Environmental Conservation, 42*(03), 268–277.

Stone, S. (2011). Towards a green economy: Pathways to sustainable development and poverty eradication—A synthesis for policy makers. *Environment*, 1–34. http://doi.org/10.1063/1.3159605.

Sugden, F., Silva, S. D., Clement, F., Maskey-Amatya, N., Ramesh, V., Philip, A., et al. (2014). *A framework to understand gender and structural vulnerability to climate change in the Ganges River Basin: Lessons from Bangladesh, India and Nepal*. IWMI Working Paper, (159).

Sultana, F. (2010). Living in hazardous waterscapes: Gendered vulnerabilities and experiences of floods and disasters. *Environmental Hazards, 9*(1), 43–53.

Taylor, M. (2014). *The political ecology of climate change adaptation: Livelihoods, agrarian change and the conflicts of development*. Routledge.

Tetlay, M., & Raza, A. (1998). From small farmer development to sustainable livelihood: A case study on the Evolution of AKRSP in Northern Pakistan. Aga han Khan Rural support program, Gilgit, Northern Areas. Pakistan.

Tschakert, P. (2012). From impacts to embodied experiences: Tracing political ecology in climate change research. *Geografisk Tidsskrift-Danish Journal of Geography, 112*(2), 144–158.

Tulachan, P. M. (2001). Mountain agriculture in the Hindu Kush–Himalaya: A regional comparative analysis. *Mountain Research and Development, 21*(3), 260–267.

Twigg, J. (2004). Disaster risk reduction: Mitigation and preparedness in development and emergency programming. Humanitarian Practice Network, Overseas Development Institute.

Tyagi, N., & Das, S. (2017). Gender mainstreaming in forest governance: Analysing 25 years of research and policy in South Asia. *International Forestry Review, 19*(2), 34–44.

Udas, P. B., & Zwarteveen, M. Z. (2010). Can water professionals meet gender goals? A case study of the Department of Irrigation in Nepal. *Gender & Development, 18*(1), 87–97.

UNDP. (1995). *Human Development Report 1995*. New York: Oxford University Press.

UNDP. (2010). *Gender, climate change and community-based adaptation*. New York: UNDP.

UNEP. (1997). *Global Environmental Outlook*. Oxford, UK: United Nations Environmental Programme and Oxford University Press.

UNEP. (2009). *Learning from Cyclone Nargis: Investing in the environment for livelihoods and disaster risk reduction: A case study*. Nairobi, Kenya: United Nations Environment Programme (UNEP).

UNEP. (2012). Myanmar's national adaptation programme of action (NAPA) to climate change. http://unfccc.int/resource/docs/napa/mmr01.pdf. Accessed 26 April 2016.

UN-HABITAT. (2007). Global report on human settlement 2007. Enhancing Urban Safety and Security. EarthScan. London. Sterling VA.

UNIFEM. (2010). *Pakistan floods 2010—Preliminary rapid gender assessment of Pakistan's flood crisis* (pp. 1–32). UNIFEM, United Nations Development Fund for Women.

Uprety, D. R., Gurung, A., Bista, R., Karki, R., & Bhandari, K. (2012). Community forestry in Nepal: A scenario of exclusiveness and its implications. *Frontiers in science, 2*(3), 41–46.

Vincent, K., Wnajiru, L., Aubry, A., Mershon, A., Nyandiga, C., Cull, T., et al. (2010). Gender, climate change and community-based adaptation: A guidebook for designing and implementing gender sensitive community-based adaptation programmes and projects.

WEDO. (2008). *Gender, climate change, and human security: Lessons from Bangladesh, Ghana, and Senegal*.

Winrock. (2002). *Emerging issues in community forestry in Nepal*. Kathmandu: Winrock International, Nepal.

Women's Protection Technical Working Group. (2010). Women's protection assessments: post cyclone Nargis, Myanmar. http://reliefweb.int/sites/reliefweb.int/files/resources/F6CAA5ECD57BC6054925776100202353-Full_Report.pdf. Accessed 27 April 2016.

World Bank. (2010). *The social dimensions of adaptation to climate change in Bangladesh*. Discussion paper, No. 12. Washington, DC: World Bank.

Yadav, B. D., Bigsby, H. R., & MacDonald, I. A. (2008). Who are controlling community forestry user groups in Nepal? Scrutiny of elite theory.

Zwarteveen, M. Z. (2006). *Wedlock or deadlock? Feminists' attempts to engage irrigation engineers*. Ponsen & Looijen.

Zwarteveen, M. Z. (2008). Men, masculinities and water powers in irrigation. *Water Alternatives, 1*(1), 111–130.

Zwarteveen, M. (2011). Questioning masculinities in water. *Economic and Political Weekly*, 40–48.

Migration in the Hindu Kush Himalaya: Drivers, Consequences, and Governance

15

Coordinating Lead Authors
Tasneem Siddiqui, University of Dhaka, Dhaka, Bangladesh, e-mail: tsiddiqui59@gmail.com
Ram B. Bhagat, International Institute of Population Sciences, Mumbai, India, e-mail: rbbhagat@iips.net
Soumyadeep Banerjee, International Centre for Integrated Mountain Development, Kathmandu, Nepal, e-mail: soumyadeep.banerjee@icimod.org (corresponding author)

Lead Authors
Chengfang Liu, China Center for Agricultural Policy, School of Advanced Agricultural Sciences, Peking University, Beijing, China, e-mail: cfliu.ccap@pku.edu.cn; liucf@pku.edu.cn
Bandita Sijapati, Centre for the Study of Labour and Mobility, Social Science Baha, Kathmandu, Nepal, e-mail: bsijapati@ceslam.org
Rashid Memon, Lahore University of Management Sciences, Lahore, Pakistan, e-mail: rashid.memon@lums.edu.pk
Pema Thinley, Ministry of Agriculture and Forests, Thimphu, Bhutan, e-mail: pemathinley@moaf.gov.bt
Michiko Ito, International Organization for Migration, Yangon, Myanmar, e-mail: mito@iom.int
Orzala Nemat, Afghanistan Research and Evaluation Unit, Kabul, Afghanistan, e-mail: orzala.an@gmail.com
Ghulam Muhammad Arif, Independent expert, Islamabad, Pakistan, e-mail: gmarif@pide.org.pk

Review Editor
Richard Black, College of Social Sciences, University of Birmingham, Birmingham, United Kingdom, e-mail: r.black@bham.ac.uk

Corresponding Author
Soumyadeep Banerjee, International Centre for Integrated Mountain Development, Kathmandu, Nepal, e-mail: soumyadeep.banerjee@icimod.org

P. Wester et al. (eds.), *The Hindu Kush Himalaya Assessment*,
https://doi.org/10.1007/978-3-319-92288-1_15

Contents

Chapter Overview

Key Findings

1. **Migration drives a broad range of economic, social and political changes throughout the HKH, while migration itself is determined by multiple factors.** The decision to migrate is influenced by demographic, economic, environmental (including climate change), political and social drivers, as well as by individual and household characteristics, and intervening obstacles and facilitators. In most situations in the HKH region, economic drivers will continue to be the

most influential among the drivers. The aspirations of youthful populations, concentration of higher education opportunities in urban centres, ever widening transportation infrastructure and continuous improvement in digital communication will bolster the influence of the economic driver.

2. **Labour migration contributes significantly to poverty reduction in the HKH region, although this depends on who is able to move, and under what conditions**. The effects of migration in sending areas in the HKH region depend on the characteristics of sending household, financial institutions, markets, migration type, public amenities, social networks and social structures. **Migration can be seen as a way to promote resilience to climate change, but investment in agriculture or climate adaptation is rarely the first priority of migrant households in mountainous area**. Generally, remittances are invested in consumer goods, education, food, health care, and housing.
3. **Issues associated with internal migration remain peripheral to the policy discourse of most HKH countries, even though more than three times as many people migrate internally compared to international migration**. Moreover, policymakers and planners tend to perceive rural to urban migration as a challenge to urban and rural development. There is a need to mainstream migration in development policies and programmes.

Policy Messages

1. **HKH countries focusing on development will be more successful if policy makers approach migration not as a challenge, but instead seek ways to mainstream it into adaptation, development and risk reduction processes**. Such mainstreaming can occur within national policy processes and sub-national programmes associated with the Sustainable Development Goals, the Sendai Framework for Disaster Risk Reduction, and the United Nations Framework Convention on Climate Change. Governments could create enabling conditions that help people to choose whether they migrate to another place or stay where they are, rather than explicitly or implicitly attempting to prevent migration. For this, **policy makers, researchers as well as grass roots NGOs in the HKH countries urgently need quality data on seasonal migration, remittances, and reintegration of returnee migrants**. To understand better the effects of migration in mountain areas, the census and national sample surveys need to provide mountain-specific data.
2. **Social protections—such as the right to access public amenities and services—should be made portable across administrative boundaries**. To ensure portability of rights and entitlements, special efforts may be needed in countries with a federal model of governance to address the problems faced by internal migrants in accessing social security programmes in destination communities.
3. **Measures to reduce migration costs and decentralize migration governance are required if low-income households are to benefit fully from migration**. For instance, opening labour offices (that issue work permits or have help-desks) in the district centres could improve accessibility among migrant workers from low-income households and female migrants. This would make the migration process safer and strengthen migration governance. Such measures will also enhance the benefits of migration for the sending families—especially female family members.

For the countries of the Hindu Kush Himalaya (HKH) region, the importance of migration continues to be significant for livelihoods. Migration governance, therefore, is a critical priority (***well-established***). This chapter focuses on labour migration in the eight HKH countries. It explores the countries' overall migration experience and, where possible, highlights findings specific to mountain areas of the HKH.

Migration generally leads to changes in sending households and origin communities (***established but incomplete***). Assets are created; livelihoods are diversified. People in these households and communities gain better access to food, and they are more likely to have a safety net during a crisis. They are better able to access information. They acquire new knowledge and skills. Their social networks expand.

Remittances from migrants can have both positive and negative effects on the remittance-receiving households and origin communities, depending on context-specific factors (***established but incomplete***). Among these factors are the type of migration; the stage in the migration cycle; the asset base of the sending household; and the institutions and generic development conditions present in the origin community.

Migration governance in the HKH comprises various national, regional and international policies and frameworks (***well-established***). Both national polices and regional and international instruments govern migration from and within HKH countries. The range and coverage of these policies vary significantly by country. Where international migration is seen as a significant issue—as in Bangladesh, India, Nepal, and Pakistan—it is governed by comprehensive policies and regulations and by nodal ministries.

Over time, major origin countries have increasingly recognized that migration can promote economic development —through decent wages for migrant labour, and also through the earning of foreign exchange (***established but incomplete***). Accordingly, some countries have acted in various ways to facilitate international migration (***established but incomplete***). These measures include establishment of a migration governance system; reducing migration costs; streamlining the remittance transfer process; assisting the reintegration of returnee migrants; and engaging the diaspora in national development.

Earlier, international migration governance was based on regulation and control (***well-established***). Governments have sought to manage recruitment, introduced restriction on movements of certain categories of workers (such as unskilled women). Some of the HKH countries have criminalised irregular and undocumented migration. Nonetheless, protection mechanism in both origin and the destination countries remained weak. In recent times some of the HKH governments have introduced new migration polices, framed new laws, created a separate ministry, among other actions (***established but incomplete***). For example, the Government of Nepal is trying to set a basic condition for a job contract whereby the employer is responsible for bearing the cost of the visa, ticket, and all other related expenses for the worker. The prospective migrant requires to pay only for services such as pre-departure orientation training, medical check-ups, the government-mandated worker's welfare fund and insurance.

Along with international migration, the HKH also experiences internal and cross-border migration. Urban centres are attractive to the migrants due to access to employment opportunities, urban amenities and services (e.g. education, electricity, healthcare and water), and opportunities for participating in the market. A lack of adequate education facilities, particularly opportunities for higher education, is a common reason for the youth to migrate to urban centres. Mountain communities are increasingly better connected with major market centres due to development in communication and transportation.

On internal migration, most HKH countries have public policies that reflect a strong sedentary bias: migration is perceived as a challenge to urbanization and planning processes. Discussion on internal migration in the context of urban development mostly concentrates on measures to reduce migration from rural to urban areas. This negative attitude towards internal migrants is often supplemented by a "sons of the soil" ideology (***well-established***).

Internal migrants to urban areas of HKH countries—who are relatively less educated, less skilled, and employed in the informal sector—experience exclusion of various nature (***established but incomplete***). Denied their rights, these internal migrants hardly enjoy social security such as public food distribution. They lack access to education and health care. Most importantly, they lack entitlement to housing at their migration destination, because they lack proof of identity and residence. In many cases they and their families end up living in informal settlements, with limited access to public amenities.

These forms of exclusion limit the benefits of rural to urban migration (***well-established***). Moreover, they create new risks for internal migrants and their families (***established but incomplete***). Accordingly, vulnerable internal migrants in HKH countries—who work in marginalised areas such as domestic work, construction, hawkers, and security guards—should be supported with new social protection measures.

Many regional and international instruments are important for the governance of migration. The countries of the HKH have acceded to some of these instruments more than others. Recently, major sending countries in the region—Afghanistan, Bangladesh, China, India, Nepal, Pakistan—have joined regional consultative forums, such as the Colombo Process and Abu Dhabi Dialogue. A few of them have ratified the 1990 UN Convention on the Protection of All Migrant Workers and Members of Their Families.[1] But none of these countries has ratified the ILO Conventions that are considered significant for migrants: the Domestic Workers' Convention (2011) (C189),[2] the Migrant Workers (Supplementary Provisions) Convention (1975)[3] (No. 143), and the Private Employment Agencies Convention (1997)[4] (or C181) (***well-established***).

A significant development for the HKH in recent years has been the expansion of laws and policies against human trafficking (***established but incomplete***). Countries in the region have shifted away from a narrow definition of trafficking—limited to intended commercial sexual exploitation —to a broader interpretation that includes labour trafficking, slavery, bonded labour, organ trade, and drug trafficking. Laws in Bangladesh and Nepal have enabled the creation of special funds to support victims. Although gaps remain in

[1]http://www.ohchr.org/EN/ProfessionalInterest/Pages/CMW.aspx.

[2]http://www.ilo.org/dyn/normlex/en/f?p=NORMLEXPUB:12100:0::NO::P12100_ILO_CODE:C189.

[3]http://www.ilo.org/dyn/normlex/en/f?p=NORMLEXPUB:12100:0::NO::P12100_ILO_CODE:C143.

[4]http://www.ilo.org/dyn/normlex/en/f?p=NORMLEXPUB:12100:0::NO::P12100_INSTRUMENT_ID:312326.

the implementation of these anti-trafficking laws and policies, the changes are welcome.

A challenge to analysing migration's effects across the HKH is that certain countries in the region, such as Nepal and Afghanistan, are mostly mountainous, whereas others are not. For example, only a small part of Bangladesh (the Chittagong Hill Tracts) belongs to the Himalaya mountain system. Whereas Nepal's experience—as a country—of short-term international contractual migration can generally be applied to its mountain areas, Bangladesh does not allow for such generalization: its national data may not reflect any conditions specific to the mountains (***well-established***).

Migration and the Sustainable Development Goals

Two Sustainable Development Goals (SDGs) directly pertain to migration governance. The first of these is SDG 8: *Promote sustained, inclusive and sustainable economic growth, full and productive employment and decent work for all.* Under this goal, Target 8.8 especially highlights the need to protect migrant workers: "Protect labour rights and promote safe and secure working environments for all, including migrants, in particular women migrants, and those in precarious employment." Other targets under SDG 8 with implications for migrants include 8.2, 8.5, and 8.10.

The second SDG with targets focused on migrants is SDG 10: *Reduce inequality within and among countries.* Two targets under this goal call for efforts specifically related to migration:

- Target 10.7—"Facilitate orderly, safe, regular and responsible migration and mobility of people, including through the implementation of planned and well-managed migration policies."
- Target 10.C—"By 2030, reduce to less than 3% the transaction costs of migrant remittances and eliminate remittance corridors with costs higher than 5%."

Beyond SDGs 8 and 10, migration will also be affected by initiatives under the SDGs to end hunger (Targets 2.4 and 2.C); to educate and develop skills equitably, including for women and the marginalized (Targets 4.4, 5.A, 13.B); to promote inclusive cities, with safe, affordable housing and secure living conditions (Targets 11.1. and 11.5); and to meet a range of other development objectives (Targets 1.5, 6.4. 6.5, 6.6, 9.1, 9.C, 13.1).

Households in the HKH adopt migration as a strategy to increase their income, diversify their livelihoods, seek a better life, and manage risks arising from various stresses and shocks. As countries in the region pursue their development priorities, they should emphasize programmes that provide for social inclusion, social protection, and adaptation planning. Countries can unlock the potential of migration by reducing its risks—protecting households against the possibility that migration will erode their assets. Other recommended steps are to make the policy response on human mobility comprehensive; to enable migrants to benefit from government development initiatives, including the response to climate change; and to bring remittances into development planning, exploring how they may enhance capacity in communities of origin.

15.1 Labour Migration in the HKH: A Challenge and Opportunity for Inclusive Growth

Migration has become an integral part of current global development process. As per United Nations Department of Economic and Social Affairs (2013a, p. 25) in 2005, 763 million people have moved within their own countries and another 232 million people have moved outside their countries of origin (United Nations Department of Economic and Social Affairs 2013b). This chapter is on internal and international labour migration from the Hindu Kush Himalayan (HKH) region. Across the HKH countries population movement is widely perceived as a challenge. However, there is a growing understanding that it can also open up new opportunities for development for those who migrate, their left behind families, communities of their origin as well as to the national economies of origin and destination areas and countries. Emerging research shows that a large number of climate change affected households in this region have been using labour migration as one of the many adaptation tools. Remittances sent by the migrants are being used in disaster risk reduction. In the past mainly men migrated for work, and women were left behind, or, accompanied the men as spouse. Since the beginning of the new millennium, an ever increasing number of women are participating in labour migration in some of the HKH countries. This chapter attempts to understand the dynamics of labour migration in this region. One of the major goals of the chapter is to consolidate the current state of knowledge on migration, and where possible, collate mountain-specific information on migration. It concentrates on three areas of migration research: drivers, consequences, and governance of labour migration.

There are several challenges in analysing migration in the HKH region. For one, most parts of Afghanistan, Bhutan, and Nepal are considered to be within the HKH whereas only a small part of Bangladesh belongs to the Himalaya. Therefore, while the experience of short-term international contractual migration of Nepal, for example, can be generalized for the mountain and mid-hill areas of the country, such a generalization is not possible in the Chittagong Hill Tracts of Bangladesh. Unlike other parts of Bangladesh, three districts of Chittagong Hill Tracts (i.e., Bandarban, Khagrachhari and Rangamati) hardly had any experience of short-term international migration. In the case of China, India, and Myanmar, different patterns and characteristics of population movements are observed in different parts of mountain regions of these countries. Hence, it is challenging to meaningfully analyse the implications of labour migration, particularly for the areas belonging to the HKH region. Absence of mountain specific data in nationally representative sample surveys is another challenge. Data that are generated at the country level are not gathered following common standardized definition and method. This limits the scope of cross-country comparisons. Besides, the available migration data from country level census is unlikely to cover circular and temporary migration. Though the data on international remittances have been improving over the past decade, data on domestic remittances remains scarce and scattered in several HKH countries.

The chapter is divided into five sections including this introductory one. Section two presents the country-specific migration patterns. Section three looks into what drives labour migration from the mountain areas of these countries. Section four presents consequences of labour migration for mountain communities. Section five analyses the governance of labour migration in the HKH region. Section six draws some key conclusions.

15.2 Diverse Trends, Consistent Importance of Migration

This section provides a brief country-specific overview of the diverse patterns of migration and remittance flow in the HKH. All HKH countries, except Afghanistan and Bhutan, show an increasing stock of emigrants (see Fig. 15.1). Figure 15.2 shows that the countries of the HKH region received a considerable amount of remittances. There has

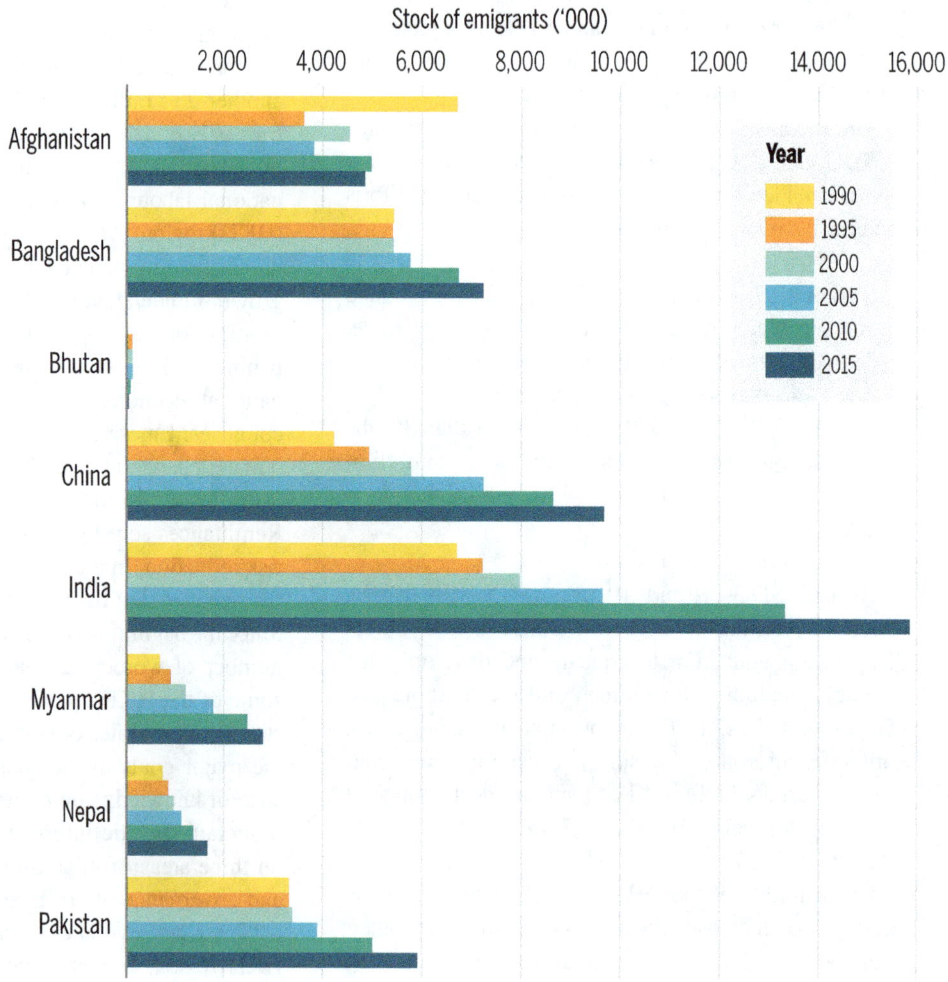

Fig. 15.1 Changes in the emigrant stock of the HKH countries (*Source* United Nations Department of Economic and Social Affairs 2015; http://www.un.org/en/development/desa/population/migration/data/estimates2/estimates15.shtml)

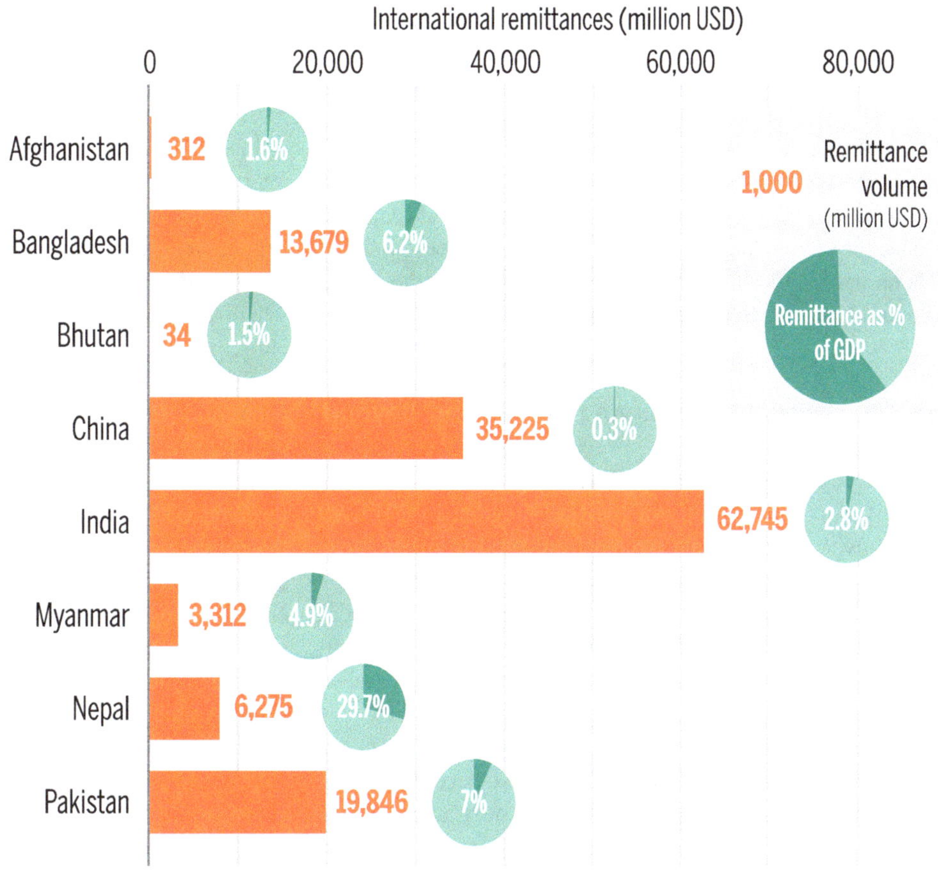

Fig. 15.2 International remittances received by HKH countries in 2016 (in Million US Dollar) (*Source* World Bank 2017; https://data.worldbank.org/indicator/BM.TRF.PWKR.CD.DT)

been a sharp rise in remittances in countries like Bhutan and Myanmar in recent years (see Fig. 15.3).

15.2.1 Afghanistan: Mixed Flows

Afghanistan's migration patterns are perhaps the most unique across the world. The country in its recent history produced one of the highest number of refugees and internally displaced

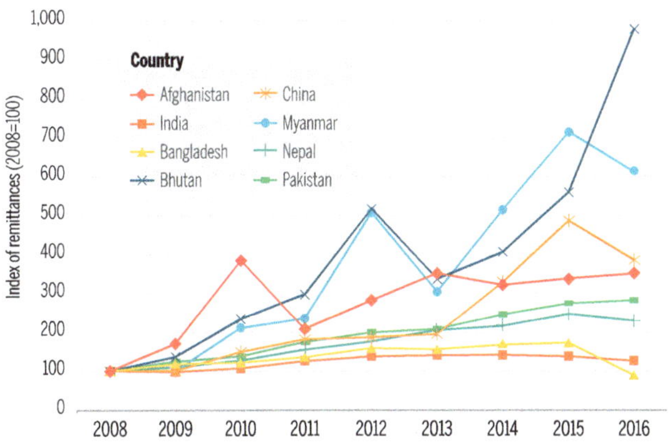

Fig. 15.3 Changes in the index of international remittances received by HKH Countries in 2016 (2008 = 100) (*Source* World Bank 2017; https://data.worldbank.org/indicator/BM.TRF.PWKR.CD.DT)

persons, while it has also become the largest recipient of returnees (Norland 2016). The history of internal, regional and international displacement in Afghanistan is directly linked to its political, economic and social formations, and in particular, it is linked with protracted years of war and conflict. Since 1979 when Afghans left the country due to Soviet invasion, the net negative migration rate was 56.7/1000 persons; between 1990 and 1995 it reversed to positive net migration of 44.4/1000 persons, and under the Taliban regime (1995–2000) this rate sank below parity to −6.5/1000 persons (International Organization for Migration 2014). Over 320,000 Afghans returned in 2016 alone. The country has become the largest recipient of returnees in comparison to all European and South Asian countries put together (Norland 2016). There are 2.5 million Afghan refugees still living in Pakistan; of them 1 million are registered, and 1.5 million are unregistered (or undocumented) and hence cannot access United Nations High Commissioner for Refugees cash grants or get verified as refugees for getting support (Ahmadi and Lakhani 2017). There are 950,000 Afghan refugees in Iran and returns from Iran during 2016 were 436,236 undocumented and 2305 documented (Ahmadi and Lakhani 2017). Many of returnees from both Iran and Pakistan are considered 'voluntary returnees'. However, in reality the repatriations are often coercive with refugees left with little choice to remain

(Ahmadi and Lakhani 2017). Despite this fact, Pakistan still remains the first destination for the Afghan refugees. Over 1.2 million persons are displaced internally. The majority of internally displaced Afghans come from four provinces: Kundoz (northern), Uruzganan and Helmand (southern) and Farah (western). During 2016, more than half of the IDP population (56%) were children (Bjelica 2016). Afghanistan is also the second largest source country for refugees globally. There are 2.7 million Afghan refugees across the world (United Nations High Commissioner for Refugees 2017).

15.2.2 Bangladesh: Moving Within and Beyond

With a population of 160 million, Bangladesh is the most densely populated country of the HKH region. In 2015, 90 out of every 1000 people moved from rural to urban areas (Bangladesh Bureau of Statistics 2016). Bangladesh is rapidly becoming urban. Rural-urban population movement is one of the major reasons behind such urbanization. Seasonal migration is also extremely common in Bangladesh. However, there is little data on this.

Since 1970 Bangladesh has been participating in the short-term labour market of the Gulf and other Arab countries, and Southeast Asia. Traditionally only men participated in this market. Since the lifting of the ban in 2003, the number of female migrants also started increasing. In 2016 around 750,000 persons migrated overseas for employment. Sixteen percent of them were women. Bangladeshi workers mostly get jobs under unskilled and semi-skilled categories. However, as the Bureau of Manpower Employment and Training of the Government of Bangladesh considers domestic workers as skilled, government statistics show quite a high proportion of skilled workers. Bureau of Manpower Employment and Training data indicate that among the total migrant workers who went abroad in 2016, 43.1% were skilled, 40.08% were less skilled, 15.83 were semi-skilled, and 0.61% were professionals (Bureau of Manpower Employment and Training 2016). The education level of those who participate in the short-term international labour market is lower than that of people who migrate from rural to urban areas for work (Siddiqui and Mahmood 2015). Eighty-one percent of the total workers who migrated in 2016 moved to Gulf and other Arab countries, and the remainder migrated to Southeast Asia. Oman, Saudi Arabia, Qatar, Bahrain, and Singapore were the major destinations of Bangladeshi migrant workers in 2016 (Bureau of Manpower Employment and Training 2017). However, international short-term migration from three hill districts of the Chittagong Hill Tracts region is almost non-existent. The flow from Bandarban, Khagrachhari and Rangamati constituted 0.12%, 0.07% and 0.06% respectively of the total flow in 2016. (Refugee and Migratory Movements Research Unit 2017). More importantly, people from these three districts who participate in migration are from the Bengali community; the ethnic hill population does not have representation in international migration. Since the Peace Accord of 1997, internal labour migration from the CHT has been increasing. Indigenous male population from this area mostly migrate to export processing zones of Chittagong Division to work in the manufacturing sector and to the mega city of Dhaka where they have created a niche as security guards. Women participate in the manufacturing sector and domestic work.

Bangladesh is a major remittance receiving country. As per World Bank data from 2014, its position was 10th in the world (The World Bank 2016a). The flow of remittance dropped substantially in 2016. That year Bangladesh received USD 13.6 billion. The figure was almost 11% less than in the previous year (Bangladesh Bank 2017). A similar decline in remittance inflow was experienced by India and Nepal. Drop in oil prices, low economic growth in the Gulf region, and depreciation of Euro and Pound Sterling have been identified as major reasons (The World Bank 2016a). Saudi Arabia is still the most important source of remittance for Bangladesh, followed by the UAE, USA, and Malaysia.

15.2.3 Bhutan: A Closed Mountain State?

Bhutan has an estimated population of 764,667 people of whom 30.6% live in the urban areas and the remainder continue to live in the rural areas (Ministry of Labour and Human Resources 2015). A study by the Ministry of Agriculture and Forest (2013, p. 1) estimated that 21.4% of the total population are migrants. Between 2000 and 2013, the migration rate in Bhutan was estimated to be 10.9/1000 people (Ministry of Agriculture and Forests 2013). One of the key issues facing Bhutan is the migration of people from rural to urban areas. It has been reported that nearly 65.6% of rural households have at least one member who has migrated to an urban centre (Ministry of Agriculture and Forests 2013). Urban centres in Thimphu dzongkhag were destinations of one-third of the migrants. Previous research has shown that lack of employment opportunities, limited access to education, small landholding, limited access to market, and limited access to other services (e.g. health and motor-roads) are the major reasons why people leave their rural homes (Ministry of Agriculture 2006; Ministry of Agriculture and Forests 2013). Out-migration from Bhutan is minimal and consists mostly of Bhutanese students studying abroad. Bhutan also receives immigrants, particularly temporary migrant workers from India, who mostly work in the construction sector (Ministry of Labour and Human Resources 2015).

15.2.4 China: Inter-provincial Migration

China's rapid development and urbanization has induced large numbers of rural residents to migrate from their homes in the countryside to urban areas (Hu et al. 2008; Ministry of Human Resources and Social Security 2013; Wen and Lin 2012). The 2016 Migrant Worker Monitoring Survey Report shows that there were 281.7 million migrant workers, which account for about three quarters of its rural labour force (National Bureau of Statistics of China 2017). Among those migrant workers from rural areas, 45% migrate to places within their own provinces whereas the rest migrate outside of their provinces. However, there are significant variations across regions. The same report reveals that eastern and northeastern regions see 82% and 77% of their migrant workers migrate within provinces. In contrast, these numbers are 48% and 38% for western and central regions, respectively.[5]

Disaggregation by gender shows that women account for 35% of the total migrant workers with the rest 65% being men. Among those who migrate within their provinces, the share of women is 37%, five percentage points higher than the share of women among those who migrate outside their provinces (National Bureau of Statistics of China 2017).

If we look at the migrant workers by birth cohorts, the National Bureau of Statistics of China (2017) report shows that while the share of the 21–30 and 41–50 cohorts of migrant workers are 27% and 29% respectively, the percentages are 19% and 22% for 50 and above and 31–40 cohorts respectively. Only a small share (3%) of migrant workers fall into the 16–20 cohort.

As with the rest of the developing countries in this region, it is difficult to get a good estimate of the total volume of remittances sent home by migrant workers. According to Gong et al. (2008), migrants from rural areas in China sent about USD 30 billion remittances in 2005.

In China, the pattern of migration in its mountainous regions looks somewhat different than the pattern in China as a whole. Taking Sichuan province in the southwest part of China, for example, a recent survey conducted in rural Sichuan by Zhang et al. (2017) shows that in 2015, 79% of rural labour force had found some off-farm employment. Of them, 24% migrated within Sichuan province while the rest migrated to places outside of Sichuan. As for the gender composition of those migrants, 49% are women with the rest 51% being men. When it comes to age, while the share of the 50 years old and above and 41–50 cohorts of migrant workers are 31% and 26% respectively, the percentages are 18% and 17% for the 31–40 and 21–30 cohorts respectively. Almost 8% of migrant workers fall into the 16–20 cohort. When examining remittances, the survey reveals that 66% of migrants sent remittances home. Among those who did send remittances, an average migrant sent 20,463 yuan (about USD 3,200) home in 2015.

15.2.5 India: Long-Distance Internal Migration

Internal migrants account for 37% (453 million) of the country's population as per the 2011 census of India. Compared to internal migration, the stock of emigrants was 15.5 million i.e., about 1.2% of India's population (United Nations 2015). Increasing urbanization and development of growth centres in urban locations are contributing to internal migration in India. Rural to urban migration has been contributing substantially to urbanization and the rural-urban demographic composition of households. The rural households are increasingly dependent on urban resources and non-farm jobs as urbanization has been shaping rural-urban relationship through various flows of goods and services, financial flows, and movement of people. Several parts of the Indian Himalayan region have been urbanizing fast, which influences mobility of labour (Lusome and Bhagat 2013; Mohanty and Bhagat 2013). Increasing urbanization is also accompanied by a change in mobility among women. Although women predominantly migrate due to marriage and family related reasons, a significant proportion of them are now joining the workforce after migration (e.g. domestic work and construction sector). A large number of placement agencies are involved in the recruitment process (Neetha 2003; Srivastava 2012).

Emigration from the Indian Himalayan region is lower compared to that from the plains. The emigration is largely influenced by a higher socio-economic status, a network of emigrants and emigration infrastructure (Bhagat et al. 2013). On the other hand, inter-state migration from the Himalayan region is higher than that from the plains.

India is the world's largest remittance recipient. As per the World Bank, in 2014 India received USD 70 billion, which declined to USD 62 billion in 2016.[6] The survey-based estimates show that household remittances sent by internal migrants in 2007–2008 were twice those sent by

[5]The east region includes 10 provinces/municipalities: Beijing, Tianjin, Hebei, Shanghai, Jiangsu, Zhejiang, Fujian, Shandong, Guangdong and Hainan. The northeast region includes 3 provinces: Liaoning, Jilin and Heilongjiang. The central region includes 6 provinces: Shanxi, Anhui, Jiangxi, Henan, Hubei and Hunan. The west region includes 12 provinces/autonomous regions/municipalities: Inner Mongolia, Guangxi, Chongqing, Sichuan, Guizhou, Yunnan, Tibet, Shaanxi, Gansu, Qinghai, Ningxia and Xinjiang. Statistics show that in 2015, the per capita GDP in the east region ranks the first at 71,013 Yuan, followed by the northeast region (52,814 Yuan), the central region (40,274 Yuan), and the west region (39,056 Yuan) (National Bureau of Statistics of China 2016).

[6]http://data.worldbank.org/indicator/BX.TRF.PWKR.DT.GD.ZS.

international migrants for the same period (National Sample Survey Organization 2010). Further, in the Indian Himalayan region a relatively higher proportion of households (12%) received remittances compared to the non-Himalayan part of the country (9%), according to National Sample Survey Organization 64th conducted in 2007–2008.

15.2.6 Myanmar: Through the 'Open Doors'

Migration has been a prominent feature of Myanmar's socio-economic makeup for a long time (e.g., Andrus 1948). Nonetheless, the scale and dynamics of migration increased in the past couple of decades due to political and economic changes both within the country and in neighbouring countries. The 2014 Myanmar Population and Housing Census (Department of Population 2015, 2016) counted over 9.39 million internal migrants, and based on the number of international migrants counted during the Census, it was estimated that 4.25 million Myanmar nationals live abroad. Together, they represent 25% of Myanmar's total population of 51 million. The Census revealed that recent internal movement revolved around Yangon, Myanmar's former capital and the largest city, and suggests that expanding industrial development is a powerful instrument influencing the direction of migration. Overall, the eastern part of Myanmar with geographical proximity to Thailand and China experienced high levels of recent in-migration, while the central and western part of Myanmar where poverty is more widespread experienced net-population loss from migration.

When it comes to internal migration, female migrants outnumber male migrants with 53% of migrants being female. While females migrated to follow family (49%), for employment (23%) or for marriage (19%), employment was the largest most common reason behind male migration (47%). The vast majority of recent migrants concentrated at ages around 25–30 years, and they were generally better educated than non-migrants, and were less likely to be unemployed (Department of Population 2016).

Of the estimated 4.25 million Myanmar nationals living abroad, over 70% are reported to have migrated to Thailand, followed by Malaysia (15%), and China (5%) (Department of Population 2015). 75% of Myanmar migrants in Thailand come from the eastern part of Myanmar, which creates a vacuum in the labour market in this region, and this in turn attracts in-migration of internal migrants from central and western Myanmar. Origin states and regions of migrants going to other destination countries are more diverse, with Yangon being the largest source region. Contrary to internal migration, there are fewer female international migrants (39%); however, this varies from one destination country to the other (e.g., 97 and 81% of Myanmar migrants to South Korea and Malaysia are male, while 51% to Singapore are female).

It is difficult to have a good sense of the total volume of remittances to Myanmar due to the predominantly informal nature of the transfer. The World Bank estimated the total remittances to Myanmar in 2015 to be USD 3.2 billion, which contributes to 5% of national GDP (The World Bank 2016b). The Ministry of Labour, Employment and National Security (MoLES) was reported to have estimated the annual official and unofficial inflow of remittances at USD 8 billion in 2015 (Kyaw 2015). International Organization for Migration estimated that USD 2.17 billion were sent back from Thailand to Myanmar in 2012–2013, of which 83% were sent through unofficial channels (Ito 2016).

15.2.7 Nepal: A Remittance Economy?

As reported in the Nepal Living Standards Survey (NLSS) of 2010/11, 53% of households in Nepal have at least one absentee living within or outside the country (Government of Nepal 2011, p. 133). As per the 2011 Census, this indicates more than two-fold increase in the number of Nepalis living away from the country from 2001 to 2011 (Government of Nepal 2012). The figures for labour migrants are equally significant with a total of 2,723,587 labour permits issued by the Department of Labour and Employment from 2008/2009 through 2014–2015 (Government of Nepal 2016). While men account for approximately 96% of the labour permits issued between 2008 and 2015, the number of females seeking employment has increased by 2.5 times over the same period (Government of Nepal 2016, pp. 7–8). According to data available at the Department of Foreign Employment (DoFE), 47% of migrants are from the age group 26–35 followed by age group 36–45 (26%). There is no significant variation in the age categories of male and female migrants (Government of Nepal 2016).

In terms of destination, the data available from DoFE does not include migrant workers going to India since labour permits are not required for India, and to the Republic of Korea since workers migrate through the government-to-government agreement (Government of Nepal 2016). However, if one were to include these countries, the 2011 census figures indicate that the percentage of Nepalis going to India is equal to those headed towards the Gulf (approximately 38%) followed by the ASEAN countries (13%) (Government of Nepal 2013). The proportion of Nepali migrants going to India has however considerably decreased, from 77.2% in 2001 to 37.6% in 2011 (Government of Nepal 2014). Primary destinations include Malaysia and the Gulf countries, which account for 85% of the labour permits issued during the same period (Government of Nepal 2016). There is however a slight variation in the destination countries for women and men with a higher percentage of men going to Malaysia, Qatar, and Saudi Arabia, where the demand for construction workers is higher, while the proportion of females is higher for countries like the UAE and Kuwait, where women are mostly employed as domestic workers (Government of Nepal 2016).

Overall, this increasing trend of Nepalis migrating for employment abroad has meant that the percentage of remittance-recipient households has increased from 23% in 1995–1996 to 56% in 2010–2011 (Government of Nepal 2011). Currently, remittances contribute about 30% of the country's GDP—a significant increase from 10.9% in 2003/2004 (Nepal Economic Forum 2014), making Nepal the third-highest recipient of remittances as a share of GDP among all countries in the world (The World Bank 2014). Remittances to Nepal rose dramatically in response to the earthquake, growing by 27.6% in three months from April to June 2015 compared to the same period the year before. However, following this peak, inflow of remittances has slowed down registering a single digit growth by January 2016. Between February and April 2016, remittances have contracted by 5.3% (in USD terms) compared to the same period of the previous year (The World Bank 2016c).

15.2.8 Pakistan: Diverse Flows

Labour migrants moving internally in Pakistan account for approximately 2% of the population—a rate that has been roughly constant over the last twenty years.[7] Given the country's size, however, this suggests a stock of 36 million migrant workers at any given period of time. The stock of international migration is much smaller at 9.6 million (1971–2016).[8] That being said, the flow of international migration has seen a strong uptick since 2005. The number of workers emigrating annually has steadily increased from roughly 150,000 in 2005 to roughly 850,000 in 2016.[9] In terms of areas of origin, nearly 45% of emigrants are from the 10 HKH districts of Abbotabad, Bannu, Lower Dir, Kohat, Mansehra, Mardan, Mirpur, Muzaffarabad, Poonch, and Swat. In terms of gender, only 4% of internal migrants are women while the gender composition of international migrants is unknown (Pakistan Bureau of Statistics 2011).

In 2016, 90% of international flows were towards just two countries, UAE and Saudi Arabia, popular destinations for Pakistani workers since the 1970s. Anecdotal evidence suggests that these flows generally consist of young single men. Moreover, the skill levels of the emigrants also appear to be improving over the years, suggesting an impact on the magnitude of the remittances (Amjad et al. 2015). In fact, remittances have increased from roughly USD 1.5 billion in 1998 to USD 13 billion in 2012, i.e. from 2.0% (1998) of GDP to 6.2% (2012). Perhaps more importantly, data from 2012 suggest that remittances constitute roughly half of the exports of goods and services and thus provide critical balance of payments support (Amjad et al. 2015).

Few studies have looked into the impact of remittances at the household level. Ahmed et al. (2010) show that the mean expenditure of a migrant household is 41% higher than that of a non-migrant household. Remittances have a positive effect on the shares of household expenditures on food, education, clothing and recreation. The highest increase is in the expenditure share on durables at 74%. The budget share of education increases only by 2.9%. Moreover, the impacts of remittances on household welfare were strongest in rural rather than urban Pakistan.

15.3 Diverse Trends in Migration

Migration decision depends upon individual, household, and community characteristics, interplay of intervening obstacles, and influence of demographic, economic, environmental, political and social factors. Due to complex interactions between these drivers, it is rarely possible to identify individuals whose migration decision was solely influenced by one particular driver. Despite the existence of migration drivers, whether migration occurs or not depends on a series of intervening factors and personal household characteristics (Black et al. 2011).

15.3.1 In Search of a Better Life

The economic prospects of the plains and urban areas contrasts with the slower economic development of mountainous areas; with livelihood opportunities in the latter, particularly in the rural areas, generally limited to primary sector (Kollmair and Banerjee 2011). The farm sector in this region is constrained by low productivity and underemployment. It is undermined by environmental shocks and stresses, land degradation, land-use regulations, price fluctuations, lack of access to formal credit and lack of market access (Bohle and Adhikari 1998; Huo et al. 2006; Liang and Ma 2004; Olimova and Olimov 2007). The urban and peri-urban areas provide easier access to non-farm income opportunities, especially the informal sector, to rural households. In Bhutan, internal migration flows primarily to the economic hubs of the country—Paro, Chukha, and Thimphu—from the rural areas (Ministry of Agriculture and Forests 2013; Ura 2013). Similarly, in Afghanistan, Kabul receives the largest number of migrants (mostly men, aged 15–45) followed by Jalalabad et al. (Opel 2005).

[7]Pakistan Bureau of Statistics, Labour Force Surveys various rounds.
[8]http://beoe.gov.pk/reports-and-statistics. This is an overestimate since it is a sum of outflows over the years. No data is available on return migration.
[9]Ibid.

Respondents of two Chittagong Hill Tracts districts of Bangladesh identified a wide range of economic reasons from low wage in the area of origin and better income opportunities at the destination to desire for economic betterment and coming out of poverty (Siddiqui and Billah 2014). Unlike in most other developing countries, in Bhutan people migrating to urban areas are relatively well-off and have a certain level of qualification (Ura 2013). In Afghanistan, the deterioration of the overall security situation as well as lack of job opportunities across the country has also led to an increase in migration from the villages to urban centres (Opel 2005). People who leave rural areas find limited opportunities in the urban centres due to their limited capacities. Despite this, the rural-urban migration in Afghanistan is rising rapidly (Government of the Islamic Republic of Afghanistan 2015).

Migration outcome is also influenced by pre-migration wealth and skill level. Rich households have more choices with respect to migration destination. Those who migrate to developed countries belong to the rich and middle-income households or to the highly skilled category. A majority of those who migrate to the Gulf region, other Arab countries and Southeast Asia to do unskilled work are not from rich or middle-income households but they are mostly from households above the poverty level. Poorer people are likely to opt for rural-to-rural and rural-to-urban, seasonal and temporary internal migration with lower salaries and wages and thus low remittances.

15.3.2 Better Educated, Better Connected

Education has emerged as an important social determinant of migration. The lack of adequate education facilities and limited access to better education was the most commonly cited reason for leaving rural homes in Bhutan (Ministry of Agriculture 2006; Ministry of Agriculture and Forests 2013). The same is true with mountainous areas of upper Myanmar where there is critical shortage of education infrastructure coupled with ongoing instability; as a result, larger proportions of people migrate for education (Department of Population 2015). At the same time, most of the migrants from this region have some form of education. The literacy rate of the migrating population in Bhutan is considerably higher than the national average (Ministry of Agriculture 2006). In Nepal, Massey et al. (2007) found that each additional year of schooling raised the odds of long distance migration by a highly significant 5%.

Migration requires certain resources to meet the costs; therefore, it may not be a feasible strategy for all households. In Uttarakhand state of India, persons belonging to socially dominant castes, which have better access to education, financial resources and social networks, are more likely to migrate (Jain 2010). Further, as the experience of Nepal suggests, the choice of destination is determined by several factors such as the socio-economic status of households (Bhandari 2004; Jain 2010), stage in migration cycle, type of aspired job, labour requirements at home, familiarity with destination, and migrant networks (Bruslé 2008). Social networks based on familial affiliation to a social or cultural group influence people's decision to migrate and choice of destination. These networks support migration by extending loans, assisting with logistics, arranging jobs and accommodation, and providing emotional support to the migrant or the family left behind. Benz (2016) reported that assistance from one generation of migrants from Gilgit-Baltistan allowed subsequent migrants to find employment or to enroll in a higher-level education. In a study in Chittagong Hill Tracts in Bangladesh by Siddiqui and Billah (2014), 37% of the respondent households reported that although they wanted to send family members to seek work elsewhere, they could not since they did not have information on work opportunities outside the village. They also did not know anyone at the destination (Siddiqui and Billah 2014).

Though mountain communities are increasingly better connected to large market areas of the region due to development in communications, electrification, and transportation (see Benz 2016; Kreutzmann 2004; Olimova and Olimov 2007) but the relationship of this progress to migration is still ambiguous. Migration may be a consequence of the structural changes in mountain areas. Urban centres are attractive to potential migrants because of access to employment opportunities, access to urban amenities and services such as education, health, electricity and water, and opportunities for participating in the market. According to the Ministry of Agriculture (2006), rural to urban migration is a significant contributor to the urbanization process in Bhutan as 72% of all urban dwellers could be classified as migrants from rural areas. Increased urbanization has been a significant cause of rural to urban migration in the Indian Himalayas (Lusome and Bhagat 2013; Mohanty and Bhagat 2013). In Afghanistan, though, Ghobadi et al. (2005) found households that resided in communities with more irrigated land and services (e.g. markets, public transportation or health facilities) showed lower probability of migration.

15.3.3 Young Men and Women on the Move

Demographic factors such as age, gender, and household composition influence the migration process. Since the establishment of the current political setting in 2001, patterns of international migration from Afghanistan have changed in terms of gender and age group. For instance, most migrants from Afghanistan to the United Kingdom in 2001 were young men in their 20s and 30s. In 2006 the number of unaccompanied minors increased. Although the majority of Afghan migrants are still young males, the number of female migrants—who are married to Afghan residents in other countries—has increased since 2008 (Hall 2013). Ministry of Agriculture and Forests (2013) reports that the economically and physically active population migrated the most in Bhutan. Ura (2013) indicates that 61.2% of migrants in Bhutan comprised young people in the age group of 10–34 years, majority of whom were men. Although men still dominate migration flows in China, the gap between male and female participation in the migrant labour force is narrowing, especially among cohorts aged between 16 and 20 years old (Li et al. 2013). Male migrants accounted for over 95% of the labour permits issued by Department of Foreign Employment in Nepal between 2008 and 2015 (Government of Nepal 2016). A study in Chitwan district of Nepal by Massey et al. (2007) reports that while the likelihood of local mobility and long-distance migration declined with rising age; this effect was more pronounced for the former rather than the latter. The low participation of women in labour migration in this region (especially Afghanistan, India, Nepal, and Pakistan) can be explained by women's lack of education and exposure to migration opportunities as well as the traditional division of labour by household, which requires women to oversee domestic responsibilities (Hoermann et al. 2010). Marriage is a major reason for female migration in the entire HKH region. Gender and other forms of social inequities are some notable drivers of migration; severe exclusion is one of the factors that compel people to migrate. A few studies on female migration have identified family violence and/or broken homes as a reason for women's migration from Nepal (Bhadra 2013; International Labour Organization 2015). In Bangladesh some women migrated to accumulate resources to bear the cost of their own or their family members' wedding including dowry (Siddiqui 2008).

15.3.4 States Policies Shape Migration

States policies seek to either explicitly or implicitly control migration, or may have an independent effect on whether people move or not. A hukou classifies China's citizens as either rural or urban residents. Without an urban hukou, migrants and their families have limited access to urban public services, including housing, healthcare, social security, and above all, education. The education of migrant children has become one of the major challenges for both migrant families and the Chinese education system. In China, public schools are supposed to provide free education to children. However, the free education is only guaranteed to children whose hukou matches the school's location (Fu and Ren 2010). Since migrant children in cities still retain their rural hukou, they are allowed to enroll in urban public schools only if there is available space. In many cases migrant parents can only enroll their children if they are willing and able to pay steep out-of-district tuition fees. Though in a case study in Uttarakhand state of India, Jain (2010) suggests that Mahatma Gandhi National Rural Employment Guarantee programme had reduced the need for seasonal migration, mainly among unskilled or less educated persons, the impacts of this programme on migration remains ambiguous. The Government of Nepal in August 2012 placed another ban on women less than 30 years of age from migrating abroad to work as domestic help. The ban was lifted in 2016, but on the condition that the minimum wage requirements for domestic workers have been met (Government of Nepal 2015; International Labour Organization 2015).

In turn, other policies seek to facilitate migration. The open border between India and Nepal was recognized by the India-Nepal Treaty of Friendship of 1950. This permitted the citizens of either country a visa-free entry into the other country and stay as long as desired (Adhikari et al. 2008; Subedi 1991). For the poor, acquiring official documents such as a passport could often be an insurmountable challenge (Hoermann and Kollmair 2008). However, the open border between India and Nepal allows them to use other national identification documents. For example, an Indian citizen could use any one of the following documents to enter Nepal: Driving license with photo, election commission card with photo, identity card issued by Embassy of India in Kathmandu, identity card with photo issued by Sub-Divisional Magistrate or any other higher ranking officials, passport, photo identity card issued by a government agency or ration card with photo.[10] A Nepali citizen could establish his/her identity as a Nepali citizen using a citizenship certificate, limited validity photo-identity certificate issued by Nepalese Mission in India passport or voter's identification card.[11]

There is no legal regime for labour migration between India and Bangladesh. However, migration for work does take place between the two countries, but flows are mostly irregular. Those who migrate from Bangladesh are less skilled workers. Men mostly work as agricultural and day labourers, waste collectors, rickshaw pullers, and cleaners; women mostly work as domestic workers. Those who come to work in Bangladesh from India are mostly skilled workers. They work in the garment and other manufacturing sectors as well as health and service sectors (Siddiqui 2006). There is no authentic data available on Indians working in Bangladesh and Bangladeshis working in India. The issue of

[10] https://www.nepal.gov.np/NationalPortal/view-page?id=131.
[11] https://boi.gov.in/content/nepalese-passengers.

migration between these two countries is highly politicised. India does not figure in the remittance inflow data of the central bank of Bangladesh (Bangladesh Bank 2017).

15.3.5 Migration: A Strategy for Climate Change Adaptation

Migration patterns are shaped by interaction between demographic, economic, environmental, political, and social drivers (Intergovernmental Panel on Climate Change 2014). Mere existence of these migration drivers does not ensure migration as an outcome, which is determined by intervening factors as well as household and personal characteristics (Black et al. 2011). Though climatic events and trends are likely to influence these patterns (Intergovernmental Panel on Climate Change 2014), it is extremely complex to establish a causal relationship between environmental change (including climatic stressors) and migration. Moreover, various forms of human mobility (e.g. displacement, migration, and relocation) could be observed in the communities affected by disasters and environmental change. Some case studies (Banerjee et al. 2011; Bohra-Mishra and Massey 2011; Massey et al. 2007, 2010; Mueller et al. 2014; Shrestha and Bhandari 2007; Siddiqui and Billah 2014) specifically focus on the relationship between environmental change and migration in the HKH region. Migrant and non-migrant households of two Chittagong Hill Tracts districts highlighted 28 types of environmental and climatic hazards that have led some households or their members to migrate. These are irregular rainfall, temperature rise, deforestation, river erosion, landslide, drought/lowering of water level and flash floods. These hazards had profound impact on their ecosystem, resulting in the drying of mountain streams, shortage of water for irrigation and drinking, increased sedimentation and reduction in *Jhum* production. Over time many of the people left their villages in search of employment; many households now have one or more members working in other places (Siddiqui and Billah 2014). A case study in the Chitwan valley of Nepal reported that local mobility (within the Chitwan valley) was more likely to be influenced by environmental change rather than long-distance mobility (outside the Chitwan valley) (Massey et al. 2010). Another case study in the same area by Bohra-Mishra and Massey (2011) found a strong and consistent relationship between a local move (within Chitwan district) with neighbourhood density, increase in time required to collect fodder and firewood, and decline in agricultural productivity. This study also reported that environmental deterioration had little influence on migration outside Chitwan district (other districts in Nepal or overseas). However, an increase in time to collect firewood did increase the likelihood of male migration to other districts of Nepal or overseas (Bohra-Mishra and Massey 2011). A study by Mueller et al. (2014) found that heat stress increased long-term migration of men in rural Pakistan between 1991 and 2012. This migration was mainly driven by negative effects of heat stress on farm and non-farm income. A study by Banerjee et al. (2011) in four countries (China, India, Nepal, and Pakistan) across the HKH region reported that it was more likely that a household member would migrate in search of work in rural communities affected by rapid onset water hazards (e.g., riverine and flash floods) than those affected by slow onset water hazards (i.e., drought and water scarcity). A passing reference to the effects of environmental stressors on migration could be found in some studies (e.g., Ghobadi et al. 2005; Ministry of Agriculture and Forests 2013). For example, Ghobadi et al. (2005) reported that drought-affected households in Afghanistan are more likely to have members migrate for work than those households unaffected by drought. A report on migration in Bhutan suggested that factors such as predation by wildlife, drying up of water sources, outbreak of pests and diseases, and other calamities contributed to migration (Ministry of Agriculture and Forests 2013).

15.4 Consequences of Labour Migration for Mountain Communities

In this section, we analyse the ways in which migration outcomes have influenced the drivers of migration or intervening obstacles. In other words, we will be exploring how migration has acted as a driver of change. Lots of evidence on the consequences of labour migration are provided by country level studies. Findings from Afghanistan, Bhutan, and Nepal could be treated as mountain- or hill- specific experience. In the case of other countries, it is a challenge to infer HKH specific findings as only certain areas of these countries belong to the HKH region.

15.4.1 Migration Reduces Mountain Poverty and Creates New Livelihood Opportunities

The economic consequences of migration vary across countries and communities. A majority of the migrant-sending households, irrespective of whether they have an internal, cross-border, or international migrant workers, benefit economically from their investment in migration. Migration increases livelihood and employment opportunities, and more importantly, their financial income

(Government of Nepal 2016; International Organization for Migration and Bangladesh Bank 2009; Mahmood 2011; Park and Wang 2010; Siddiqui and Mahmood 2015; Srivastava 2011; Zhao et al. 2012). About 59.1% of rural population in Bhutan perceived remittances as the main benefit of having some of their household members living and working in urban areas (Ministry of Agriculture and Forests 2013). Using household panel data from China, Du et al. (2005) estimates that a household's income per capita increases by 8.5–13.1% if a member migrates for work, but the overall impact on poverty is modest because most poor people do not migrate. According to the Myanmar Labour Force Survey of 2015, 85.1% of the international migrants remit money or goods with the average of USD 6,669 per annum. Urban households receive USD 33,330 from each international migrant family member. Rural households only receive the average of USD 1,661. This creates significant income gaps even among international migrants (Ministry of Labour, Immigration and Population 2016), let alone the income gaps between internal migrant households and non-migrant households. Griffith's (2016) analysis of the dataset from the Rural Household Survey found that remittances are not consistently used for improvements in the socio-economic conditions—levels of investments in livelihoods are higher among households whose remittance income acts as a supplement, whereas remittances-dependent households had higher proportion of their remittances spent on food, education, and health but less on livelihood and savings. A study that compared results from the Nepal Living Standards Survey found that one-fifth of poverty reduction in Nepal that occurred between 1995 and 2004—from 42% below poverty line to 31% in 2003/2004—can be attributed to increased levels of work-related migration and remittances sent home (Lokshin et al. 2010). The same study also found that while the increase in migration abroad was the leading cause of this poverty reduction, internal migration also played an important role (Lokshin et al. 2010). The difference between the per capita remittances received by an individual in the poorest and richest consumption quintiles in Nepal is large (Government of Nepal 2011). In the Gojal region of Gilgit-Baltistan, migration contributed to the sectoral and spatial livelihood diversification, leading to economic upliftment of the people (Benz 2016). Black et al. (2006) suggests that migration can increase inequality, especially if there are restrictions on who can migrate, and to where, as this can lead to 'capture' of profitable migration routes by wealthier groups, and/or limitation of migration benefits for the poor.

Though remittances are spent on consumption, these may also contribute to investment in agriculture or business. A number of studies suggest that remittances play a crucial role in rural economic development worldwide (Hugo 2002). Remittances can provide flows of capital into small farms in the peripheral rural areas (Ratha 2003). Furthermore, these remittances are more dependable and less volatile than other flows from abroad such as direct investment or official development projects (Ratha et al. 2011). Siddiqui and Mahmood (2015) found that short-term international migrants from Bangladesh contributed more to agricultural development than internal migrants by using improved seeds, adequate fertilizer, regular irrigation and insecticides. One of the Chittagong Hill Tracts districts was part of the study. This group also made investment in irrigation pump, power tiller, tractor, paddy separators and portable rice processing machines.

Cost has direct impact on the outcome of migration. Those who pay high costs are likely to gain less economic benefits. As the cost of migration was low in the 1980s and 1990s, short-term contract migration produced positive economic and social results for the majority of migrant-sending households. Studies have shown that over the years the cost of migration has become exorbitantly high and it is near impossible to reap sustainable economic benefit from migration (Siddiqui 2017). Visa trading in destination countries, the existence of tiers of intermediaries in both the countries of origin and destination, lack of accountability of recruiting agencies, lack of efficient governance system, and lack of information on safe migration among potential migrants have contributed to the soaring cost of migration. Among migrants from the HKH region, Bangladeshi migrants bear the highest migration cost. According to a report by the Migrant Forum Asia (n.d.), Bangladeshi migrant workers pay USD 2500–5000 to go to Gulf countries while their Indian and Nepali counterparts pay USD 1000–3000.

In a study conducted by the Ministry of Agriculture and Forests (2013), about 49.2% of rural households in Bhutan reported farm labour shortage as a result of out-migration. An analysis of the National Labour Force Survey from 2010 to 2012 reveals a different scenario, suggesting that rural employment expanded by 3.6% annually. However, the expansion is due to increased employment of people over 45 years of age; it is also important to note that rural agricultural employment of age group 15–24 declined by 24.7% per annum. It is evident that the elderly and women are now engaged in agriculture. Given the composition of farm labour, they are not able to optimally utilize the resources. This has direct impact on agricultural production and poses a risk to the national goal of food security and food self-sufficiency. However, these studies on labour shortage make an assumption that factors such as age structure, culture, labour market, and social norms remain constant. In the context of Bangladesh, international migration did result in labour shortage in some of the high migration areas, particularly during sowing and harvesting seasons. Interestingly it did not affect the agricultural production because internal migrants from other regions filled the labour shortage. In reality, then, benefits of international migration got distributed among the internal migrants in the form employment opportunities (Siddiqui and Mahmood 2015).

In recent times, improved access to technology has significantly shaped migration outcomes. Siddiqui and Mahmood (2015) found that in Bangladesh, 98% of migrant households own at least one mobile phone. In the past migrants had little control over the remittances they sent, as it was difficult for them to communicate with the recipients on a regular basis. Once the family members received the remittances, they could get away with not using the money according to the migrant's wishes. Now that migrants own mobile phones, they can monitor how the remittance they send is being used on a day-to-day basis. One-fourth of the migrants interviewed in the study by Siddiqui and Mahmood (2015) reported they directly oversee the investment of remittance. For instance, they proactively shape decisions related to the purchase of land and agricultural equipment, the number of day labourers to be employed, and the type of seeds to be bought.

Migrant workers also bring back knowledge and skills from destination to origin communities. There is a lack of literature on social remittances and the contribution of returnees in the HKH region. For the international labour migrants, return is an integral part of their migration cycle. However only a small number of migrants plan their post return activities in advance (Wickramasekara 2011). A study on the contribution of returnees found that a large number of returnee migrants want to re-migrate as they feel that upon return they no longer have the contact with networks that would help them to establish a business or find jobs (Siddiqui and Abrar 2002). More recent studies, however, have found that many of the returnees have contributed in developing enterprises in rural areas (Siddiqui 2017).

15.4.2 Increased School Enrolment, Better Outcomes for Children

Ratha et al. (2011) highlighted that international migration contributes to the formation of human capital. Various studies showed that a disproportionately higher portion of remittances from international migration is spent on education and health than on everyday consumption (Adams and Page 2005; Adams et al. 2008; Nagarajan 2009; Ratha et al. 2011). Evidence from Nepal and Pakistan suggest that short-term international migration is associated with increased school enrolment (Bhadra 2007; Mansuri 2006). In China, some studies have examined the impact of parental migration on the development outcomes of children. For example, from a comprehensive dataset covering 141,000 children in ten provinces (from 27 surveys conducted between 2009 and 2013), Zhou et al. (2015) analysed nine indicators of health, nutrition, and education. They found that for all nine indicators, children left behind by their migrant parents with a caregiver (typically paternal grandparents—in their home communities) performed as well as or better than children living with both parents in their home communities. Similar studies are hardly available for most parts of the HKH region. Migration of a family member provides income to the elderly but also creates a vacuum for care (Hoang 2011). Age selective nature of migration means that elderly and children would be left behind and this may increase the workload. Evidence on these issues in the HKH region is sparse.

15.4.3 Left Behind or Left in Charge

Consequences of migration also depend on personal and household characteristics such as age, ethnicity, marital status, religion, sex, and wealth. Siddiqui and Abrar (2003) found that whether remittance will be used as current income or a portion of it will be invested in enterprises for further income generation depends on household members' age. Families with male members in the age group 25 to 45 years invested in different business enterprises, whereas most families with female members who had to be married-off spent more on dowry. Families who only had elderly members did not invest in business enterprises. Along with day-to-day consumption they invested a portion of the remittances in buying land. Siddiqui and Mahmood (2015) reported that male migrants invested more in agricultural development compared to female migrants. High-income households have more choices with respect to migration destination. Those who migrate to developed western countries belong to the richest economic quartile or to the highly skilled category. A majority of those who migrate to the Gulf region, other Arab countries and Southeast Asia to do unskilled work are not from high or middle-income households but they are mostly from households just above the poverty level. Poorer people are likely to opt for rural-to-rural and rural-to-urban, seasonal and temporary internal migration with lower salaries and wages and thus low remittances.

Beyond economic returns, recent studies also indicate that migration has yielded significant social benefits. For example, in the case of Nepalese Dalits, urban migration has not only provided them social, economic and educational opportunities, but also the possibility of escaping traditional caste-based discrimination (Pariyar and Lovett 2016). Siddiqui (2001) showed that majority of divorced and separated women became both economically and socially empowered through migration. Some found new partners and some others came out of abusive marriage. On the other hand, some of the stable marriages broke down because in the absence of their wives, migrants enter into new relationships. Similarly religious background also influences migration outcome. In Bangladesh, compared to the dominant religious group, minority communities have less access to international short-term contract migration (Siddiqui and Abrar 2003).

15.4.4 Diaspora

The general understanding is that migrants often contribute to the welfare of their communities of origin. The Afghan diaspora in the Netherlands helped in designing education curricula, translation of text books and training and skill development of manpower in Afghanistan (United Nations University 2017). The Nepal Institute of Development Studies and the World Bank (2009) survey reveals that collective remittances from both the Nepali diaspora and short-term migrants have contributed to the establishment of public libraries, trade schools, health posts in schools, water supply in remote areas, tower clocks in village centres, and computers in schools in Nepal. Migrants from India and Bangladesh have established orphanages and faith-based schools, mostly in their own villages (Siddiqui 2004; Singhvi 2001).

15.4.5 Disaster Preparedness Is Not a Top Priority

Recent studies that explore the role of migration in the context of vulnerability to environmental change in the HKH region of China (Banerjee et al. 2018; Yahui and Banerjee 2017), India (Banerjee et al. 2017), Nepal (Chapagain and Gentle 2015; Jaquet et al. 2016), and Pakistan (Gioli et al. 2014) suggest that households receiving remittances prioritise immediate consumption and economic needs initially, and then health and education. Remittances are used to respond to extreme events, providing finance for relief and reconstruction in the face of extreme events (e.g., earthquakes, floods). For example, remittances to Nepal increased in the aftermath of the 2015 earthquake, growing by 27.6% within three months from April to June 2015 compared to the same period the year before (The World Bank 2016c). Remittances are used to procure food and other basic needs during or in the aftermath of a disaster and rebuild livelihoods and lost assets (Banerjee et al. 2011; Suleri and Savage 2006). An increase in remittances in the aftermath of an extreme event is often temporary. In Nepal, inflow of remittances had slowed down, registering a single digit growth by January 2016. In three months from February to April 2016, remittances had contracted by 5.3% (in USD terms) compared to the same period of the previous year (The World Bank 2016c).

The extent to which remittances would have a positive or negative role in remittance-recipient households and origin communities is context specific (Barnett and Webber 2009; De Haas 2012). Banerjee et al. (2018) found that remittance-recipient households in Baoshan County, Yunnan province, China have less adaptive capacity in response to drought than non-recipient households. The authors also reported that the adaptive capacity of those households who had received remittances over longer periods were found to have improved, and their exposure to such events was also lower (Banerjee et al. 2018). A case study in Assam state of India found that the duration for which remittances are received by a household has significant and positive association with the structural changes made in the house to address flood impacts (Banerjee et al. 2017). Chapagain and Gentle (2015) reported that remittances had been used by families in landslide prone communities in Jumla and Syangja districts of Nepal to buy land or house in a safe area, which permitted these families to leave environmentally fragile area and even find off-farm opportunities. Investment in climate change adaptation measures at the household level is often not made until later. Based on a study in three districts of Bangladesh, Siddiqui et al. (2018) found that a higher percentage of migrant households perceived that their economic status had improved over the preceding five years in comparison to non-migrant households. This along with other findings, led the researchers to conclude that those households that incorporate migration of a member with other in-situ practices adapt better to climate change stresses.

Out-migration has non-economic consequences in origin communities located in environmentally fragile areas. A study in Kaski district of western Nepal found that migration from uphill to downhill communities had resulted in land abandonment and an increase in forest cover in the upper part of the Harpan watershed (Jaquet et al. 2016). However, this process had also increased pressure on the land and exposure to flooding in the lower part of this watershed (Jaquet et al. 2016). There has been little research on migration, environment and climate change in mountain regions of Asia from gender perspective. Based on a study in West Karakoram region of Pakistan, Gioli et al. (2014) reports that there was an increase in female school enrollment due to remittances. However, male out-migration did not result in significant long-term changes in the intra-household and community-level decision-making power of women; women were not entitled to manage a household's economic assets (including remittances), and male out-migration limited the mobility of women left-behind.

15.4.6 New Types of Politics

Migration has contributed to increased political demands, including voting rights for migrants from the HKH countries. Since 2008 Bhutan has permitted its overseas citizens to vote and Afghanistan has allowed it since 2004 (The Record Nepal 2017). Changes in electoral laws had been approved by the Government of India that will permit Non-Resident Indians to cast their vote in state assembly and national parliamentary elections from overseas (The Indian Express 2017). The Election Commission of India had launched a web portal to register Non-Resident Indians, who would be able to cast their votes in their registered constituencies through a proxy.[12] However, the inter-state migrants in India would still have to visit their registered constituencies to cast their vote (Bhatnagar 2017). At a public hearing, the National Election Commission of Bangladesh stated that it would explore a mechanism for allowing absentee voting for those citizens who are residing outside the country for work during the elections (The Daily Observer 2017). The Government of Nepal was ordered by the Supreme Court to draft a law that enables Nepali migrants abroad to vote (The Kathmandu Post 2018).

On important national matters, the diaspora sometimes plays an important role in their countries of destination. The Singhvi report (2001) highlights the role of Non-Resident Indians in the aftermath of Pokhran nuclear tests and during the Kargil conflict. This high-profile government report acknowledged that the Indian diaspora extended a valuable service to the Indian government by explaining its position to the policy makers as well as the public of their countries of destination, particularly in the United States of America.

In some HKH countries, irregular cross-border movement of people has become a major source of political tension. Migration between India and Bangladesh is a good example. Different actors operating at federal and state levels in India have securitized migration from Bangladesh to India by portraying migrants as a 'threat to national security', 'infiltrators from the east', 'terrorists', 'weapon traffickers', 'demographic invaders', and 'criminals'. This securitization and criminalization have resulted in border fencing and push back of less skilled Bangladeshi men and women (Siddiqui 2006).

[12]http://eci.nic.in/OverseasVoters/home.html.

15.5 The Governance of Labour Migration in the HKH Region

15.5.1 National Policies, Laws and Acts

In order to govern international labour migration, most of the HKH countries have framed different policies, enacted laws and established institutions. However, except China and to some extent India, none of these countries have such instruments or policies concerning internal migration.

In 2006 Bangladesh framed its first national policy on overseas employment. A major objective of these policies is to expand short-term contract labour migration (Wickramasekara 2011). The title of Pakistan's migration policy explains this point adequately—"*National Emigration Policy: Promoting Regular Emigration and Protecting Emigrants.*" Nepal had enacted the Foreign Employment Act 2007 with the same objective. Along with promotion of labour migration, protection of the workers is also an important part of these policies. Under the Decent Work project of International Labour Organization, in 2016 the second overseas employment policy of Bangladesh had been drafted.

Afghanistan, Bangladesh, India and Nepal also have specific policies on female migration. Most of these countries initially either imposed a restriction or ban on migration of low skilled women. Over the years, due to demand in certain sectors of the labour market, which requires participation of women, most of these countries withdrew the bans or loosened the restrictions on female migration. For example, in 1997, Bangladesh placed a complete ban on all female migration for work. That same year it withdrew the ban on professionals but imposed restrictions on migration of unskilled female workers. But by 2003 it more or less withdrew the restrictions on female migration (Siddiqui 2008). Starting in 1985, the Government of Nepal has introduced various types of bans and restrictions on women's migration, the latest being in August 2012 when a ban was placed on migration of women under 30 years to be employed in domestic work (Sijapati et al. 2015).

Since 2000, some HKH countries have established new ministries or line agencies to govern international labour migration (Wicramasekera 2011). Afghanistan governs international migration through the Ministry of Labour, Social Affairs, Martyrs and Disabled (MoLSAMD). In India, emigration for overseas employment is regulated by the Protectorate of Emigrants (POE) as per the provisions of Emigration Act 1983. The Ministry of External Affairs has been the nodal agency supervising and controlling the emigration regulation and governance. However, in 2004, a new Ministry of Overseas Indian Affairs was created to look after the problems of emigrants and diaspora. It was merged with

Ministry of External Affairs in 2016. The office of the POE, however, continues to function as before. The Department of Foreign Employment (DoFE) of Nepal and the Bureau of Emigration and Overseas Employment (BEOE) in Pakistan regulate labour movements from respective countries. In Bangladesh, the Ministry of Labour (MoL) oversees international labour migration. In 2001, the Ministry of Expatriates' Welfare and Overseas Employment (MoEWOE) was created by the Bangladesh government.

Until the early 1980s emigration from HKH countries that were part of the British Empire was governed by the 1922 Emigration Act of British India. After the oil price hike in the Middle East, when the number of migrants grew dramatically, governments of many countries in the HKH region enacted new laws. The main objectives of these laws were to facilitate the outflow of workers through regulation of recruitment agencies and protection of unskilled workers (Abrar 2005). Since 2000 a new set of laws emerged. For example, Nepal framed the Foreign Employment Act, 2007 replacing the Foreign Employment Act of 1985. The Emigration Ordinance (1979) and the Emigration Rules (1979) constitute the legal framework to safeguard the rights of Pakistani Migrant workers and regulate the activities of overseas employment promoters and recruiting agencies (Pakistan Institute of Legislative Development And Transparency 2008). Afghanistan framed the Labour Code of Afghanistan: Regulation for Sending Afghan Workers Abroad by Directorate of Rights and Labour Law, 2005 (Government of Afghanistan 2005). India framed a new law titled the Emigration Act 1983, replacing the Act of 1922.[13] In 2013, Bangladesh replaced its Emigration Ordinance 1982 (Number 29 of 1982) with the Migration and Overseas Employment Act 2013. The influence of the 1990 UN Convention is visible in the newly framed documents. Rights and welfare of migrants are more pronounced. Moreover, offences are also redefined and new types of crimes and fraudulences have been included. Punishments for committing those crimes have also been increased. The preamble to the 2013 Migration and Overseas Employment Act of Bangladesh clearly states that the new law is framed to make national law consistent with the 1990 Convention and other International Labour Organization Conventions.

The origin countries of the HKH region acknowledge the contribution of the diaspora in home country development. For example, India has undertaken a comprehensive policy to engage the diaspora in its development process. Some of these include: establishment of the then Ministry of Overseas Indian Affairs in 2004 to coordinate activities aimed at reaching out to the Indian diaspora and to build and foster networks with overseas Indians. Establishment of the Overseas Indian Facilitation Centre (OIFC), a not-for-profit public-private initiative of the Ministry of Overseas Indian Affairs, and the Confederation of Indian Industry (CII), is important in this respect. Currently, Overseas Indian Facilitation Centre's activities include responding to queries on various issues raised by the Indian diaspora, management of online business networking portal, and market forums in India and overseas.[14] Since 2003, the Government of India has been hosting an annual diaspora conference, the Pravasi Bharatiya Divas. The event is designed to serve as a platform for interaction between overseas Indians, the Indian government, and interested segments of the Indian society, such as businesspersons and cultural and charity organizations.[15]

Policy provisions vary significantly in the Hindu Kush region with regard to internal migration. For instance, the Constitution of India provides for "freedom of movement" as well as "freedom of employment" anywhere in India (Government of India 1950). Additionally, there is also a specific law, the Inter State Migrant Workmen (Regulation of Employment and Conditions of Service) Act of 1979 (Government of India 1979) and Central Rules of 1980 (Government of India 1980) that is meant to guide internal labour migrants. The Inter-state Migrant Workmen Regulation Act 1979, deals with contractor-led movements of inter-state migrant labour. This law provides protection to internal migrants. But many workers do not move with contractors. They find work independently through the network of family, friends and kin and so do not fall under the purview of the Inter-state Migrant Workmen Regulation Act. Further, many schemes and programmes for workers in the informal sector are also applicable to migrant workers. But most of the time internal migrants cannot avail of those as they require registration and in some cases, identity cards which they often do not possess.

The right to work is a fundamental civil right in China. There are a number of laws and regulations applicable to the migrant workforce, such as the Labour Law of the People's Republic of China, which became effective in 1995, Law of the People's Republic of China on Employment Contracts, which became effective in 2008 (Standing Committee of the National People's Congress 1994, 2007) and Regulations on Work-Related Injury Insurances, which became effective in 2011 (State Council 2010). Moreover, the Interim Regulations on Residence Permits adopted by the State Council in 2015 contain a set of specific regulations that provide

[13]https://indiankanoon.org/doc/99408546/.

[14]'Overseas Indian Facilitation Centre (OIFC)', http://moia.gov.in/services.aspx?ID1=205&id=m2&idp=205&mainid=196 (accessed 3 August 2014).

[15]'Pravasi Bharatiya Divas', http://www.pbd-india.com/about.html (accessed 6 August 2014).

migrant workers access to basic public services and facilities in the destination city. Due to the household registration (hukou) system, however, citizens' freedom of movement within China is somewhat restricted. For example, migrants and their families have limited access to urban public services (e.g. healthcare, social security and education) if they do not have an urban hukou (Lai et al. 2014).

There is no policy on internal migration in Bangladesh. Although there is no restriction on any type of internal migration, urban development strategies, environmental policies such as the National Adaptation Programme of Action 2006 (NAPA), initial drafts of Bangladesh Climate Change Strategy and Action Plan 2009 (BCCSAP) or Disaster Risk Reduction Strategy perceive rural to urban migration as a problem of urbanization. Accordingly, most of these plans aim to reduce rural to urban migration through local level adaptation to climate change. While there are no explicit policies regulating internal migration in Pakistan, Gazdar (2003) makes an interesting case connecting urban planning and regulation with internal migration. Gazdar (2003) shows that the city managers follow 'settle first, regularize later' policy in Pakistan. When workers migrate to urban centres, they require access to housing and public goods, the provisioning of which is the hallmark of good urban planning.

15.5.2 International Instruments

With respect to international migration there are quite a few multilateral as well as bilateral instruments through which migration is governed. Most important of International Labour Organization instruments relating to migrant workers are the Migration of Employment Convention (Revised) 1949 (No: 97); the Migrant Workers' (Supplementary Provisions) Convention, 1975 (No. 143); and the Migrant Workers' Recommendation (No 151). The 1990 UN International Convention on the Protection of the Rights of all Migrant Workers and the Members of their Families (ICRMW) is the most comprehensive instrument relating to migrant workers. Multilateralism is important for protection of the rights of the migrant workers; unfortunately none of the countries of origin of the HKH region, except Bangladesh, have ratified the 1990 Convention, and nor has any destination country. Therefore, due to lack of ratification the rights assigned under the international instrument cannot be exercised.

Of late, major origin countries of the region such as Afghanistan, Bangladesh, China, India, Nepal, and Pakistan have joined regional consultative forums such as the Colombo Process and Abu Dhabi Dialogue. Further, most countries in the region have ratified the Convention on All Forms of Discrimination on Women (CEDAW), which includes specific provisions on women migrant workers, especially as part of General Recommendation No. 26.[16] All the HKH countries espoused Sustainable Development Goals (SDGs) in pursuing their development agenda. It has major ramifications for both internal and international migrants of these countries. Target 8.8 of Goal 8 specifically mentions the migrants when it says, "Protect labour rights and promote safe and secure working environments for all workers, including migrant workers, in particular women migrants, and those in precarious employment."

In the absence of enforcement of multilateral instruments, bilateral agreement or memorandum of understanding are the only avenues under which labour migration is facilitated. Bilateral instruments focus on mutual intent to enhance employment opportunities in the destination countries; measures that destination countries will take for the protection and welfare of workers in the organized sector; regulation of the recruitment process in both countries; and establishment of a joint working group to ensure the implementation of the memorandum of understanding and bilaterally resolve any labour concerns. Barring a few, most of these bilateral instruments again focus on managing the recruitment and employment processes, with few, if any, provisions for protecting the workers or ensuring decent work for the migrants during their sojourn abroad (Siddiqui 2006; Wicramasekera 2015). Unfortunately destination countries are not keen on even signing bilateral agreements with such minimum rights. That is why the number of bilateral agreements is also low.

Labour migration in the region is characterized not only by short-term contract international migration but also by cross-border migration. For historical reasons, cross-border migration remains a sensitive issue. The Nepal-India Peace and Friendship Treaty 1950 and the open border between Nepal and India are perhaps exceptions. There are some regulations on cross-border movements. They mostly focus on regulating irregular movements, including cross-border trafficking. The Bangladesh-India Agreement for Mutual Legal Assistance on Criminal Issues and Transfer of Sentenced Persons of 2010 and the Bangladesh-India Agreement for Combating Terrorism, Transnational Organized Crimes, and Illegal Drug Trafficking of 2010 are examples of this regulation.

15.5.3 Innovation in Governance and Migration

Countries of the HKH region have innovated new methods in providing services to the migrants. Generating funds

[16]CEDAW (2009), /C/2009/WP.1/R General Recommendation no. 26 on Women Migrant Workers.

outside government revenue and development budgets is one of them. As early as 1990, Bangladesh had innovated a method of generating fund to support migrant workers: Wage Earners' Welfare Fund. This fund is generated through subscription by migrant workers, consular fee charged in the embassies and interests charged from the bond money kept from the recruiting agency as license fee. The government has created a huge fund through this initiative. The line ministry uses this fund in providing welfare services to the migrants. The services include pre-departure training and briefing, 21-day training of female migrants, repatriation of deceased bodies, compensation to the injured and families of deceased, and managing shelter homes in destination (RMMRU 2016). Different countries are now replicating this model. Under the Foreign Employment Act of 2007, Nepal established the Foreign Employment Welfare Fund, which is a public fund managed by the Foreign Employment Promotion Board, in 2008 (Article 32 and Article 33, FEA 2007).[17] Each out-bound worker pays a fee of NPR 1000 (approximately USD 9.6) to the welfare fund. This welfare fund also receives license fees from recruitment agencies and other institutions associated with foreign employment and grants provided by local or foreign institutions. To date, the welfare fund has primarily been used for repatriating seriously injured workers, repatriating mortal remains of deceased workers, providing compensation to the injured workers or families of deceased workers (Paoletti et al. 2014). However, despite the provisions in the law, initiatives from the Foreign Employment Promotion Board to provide skill-oriented training to migrant workers, reintegration programmes for returnees are limited alongside the low awareness about the welfare fund among migrant workers (Office of the United Nations High Commissioner for Human Rights 2018; Paoletti et al. 2014).

India has also established a fund to support the migrants. However, it does not charge the workers, rather it takes a security deposit of USD 2500 per worker from the foreign employer who hires an Indian worker.[18] India also has an insurance scheme titled, the Pravasi Bharatiya Bima Yojana, which was launched in 2003 and amended in 2006, 2008 and 2017 (Government of India 2006). It is compulsory for all migrant workers who have obtained a clearance from the POE. Under this scheme, the migrant workers are insured for a minimum coverage of INR 0.3 million for the entire period of the employment contract. Some of the salient features of the scheme include: (i) cost of transporting the dead body, in case of any such eventuality; (ii) transportation costs for workers who are stranded or experience substantive changes in the employment contract; (iii) travel support to migrant workers who fall sick or are declared medically unfit to work; and (iv) medical coverage of a minimum of INR 50,000. India has also innovated a method of generating welfare fund for internal migrants who work in the construction sector. Under the Building and Other Construction Workers Act, 1996 it has created a provision for levying a tax for construction. Substantial funds have been collected by the Construction Welfare Boards in many states, but implementation of the programme is poor due to the lack of registration of workers (Bhagat 2014).

The Nepal government has undertaken a policy not to allow a worker to migrate through purchase of a work visa. The government is trying to set a basic condition for a job contract whereby the employer is responsible for bearing the cost of the visa, ticket, and all other related expenses of the worker. The recruitment agencies (hereafter RAs) are permitted to charge a maximum of NPR 10,000 (approximately USD 96) as a service and promotional fee only when the employing company overseas provides in writing that they will not bear the travel and recruitment costs of the concerned worker(s). Accordingly, the 'free-visa, free-ticket' or the 'zero-cost migration', as the scheme has come to be known, requires the prospective migrant to pay only for services such as pre-departure orientation training, medical check-ups, the government-mandated workers' welfare fund, and insurance. The policy is applicable to seven countries, namely Bahrain, Kuwait, Oman, Qatar, Saudi Arabia, the United Arab Emirates, and Malaysia. So far there are indications that while the policy measure is significant in principle, implementation has been weak, leading, in some instances, to more abuse of the workers (Amnesty International 2017; Sijapati et al. 2017). However, if Nepal does succeed, it will be a great innovation in workers' recruitment.

15.5.4 Challenges of Migration Governance

HKH countries face some major challenges in governing both internal and international migration. In the case of international migration, some common challenges that these countries face include high cost of migration, informality of the recruitment process, visa trading and lack of protection from abusive labour conditions. Some of the challenges are specific to women migrants, particularly for those who go abroad as domestic workers. Over the last ten years the governments of this region have been trying to reduce the cost of migration through bilateral agreements. In recent years, the origin country governments have placed the issue of cost of migration in multilateral fora such as Colombo Process, Abu Dhabi Dialogue and annual meetings of the UN Global Forum on

[17]Government of Nepal. 'Foreign Employment Act, 2007' Available at http://dofe.gov.np/new/uploads/article/foreign-employment-act_20120420110111.pdf.

[18]http://mea.gov.in/rajya-sabha.htm?dtl/27853/QUESTION+NO3226+INDIAN+MIGRANT+DOMESTIC+WORKERS+IN+GULF+COUNTRIES.

Migration and Development. Some of these governments have tried to reduce the cost of migration by fixing the upper limit. However, none of these measures have produced their desired results yet. Although the governments have framed formal laws and created institutions to regulate short-term contract international migration most of the tasks of recruitment are done informally by the sub-agents. Most governments have attempted to abolish the system of sub-agents, but little headway has been made in this regard. Visa trading is another challenge. In some labour receiving countries, work visas change hands at least two to three times before they are made available to the recruiting agencies of origin countries through bidding. This issue is also being discussed in various fora. Governments of some destination countries have begun taking punitive measures against visa traders. Still the practice is pervasive. Restriction over movement, long working hours and in some cases physical and sexual abuse are some of the major challenges to overcome for the governments of origin and destination countries.

The lack of policy on internal migration in most HKH countries keep internal migrants outside the purview of government programmes and entitlements. In countries such as India, which has taken some policy measures to protect internal migrants, a negative attitude and hostility towards internal migrants persists, sustained by the '*sons of the soil*' ideology despite the fact that the Indian Constitution guarantees the right to move as a fundamental principle under Article 19 (Weiner 1978). A group of internal migrants who are illiterate, low skilled, temporary and seasonal migrants are more vulnerable to various kinds of exclusion and denial of their rights in urban areas. These categories of migrants are excluded from social security programmes such as public distribution of food, access to education and healthcare and, most importantly, entitlement to housing at the destination because they lack proof of identity and residency. Social security programmes are place-bound and the implementation of such programmes fall under the purview of the state governments. Inter-state migrants face more hardships as social security programmes are not portable (Bhagat 2012). A major challenge for Afghanistan is to resettle the returnee refugees as well as the internally displaced persons ensuring their work opportunities and social protection.

15.6 Conclusion

This chapter has provided an outline of the patterns and trends of labour migration from the HKH region. It shows that migration trends are diverse. Internal migration is a feature across the region and the flow is towards major urban centres, as well as to areas of commercial agriculture. Some countries have a longer history of short-term international labour migration to the Gulf and other Southeast Asian countries and some others are recent joiners. There have been significant experiences of cross-border movements for work between India and Nepal, and India and Bangladesh, and from India to Bhutan.

Much migration in the region is gender specific. In the case of Afghanistan it is difficult to disentangle labour migration from refugee flows. Two-thirds of rural families of Bhutan have a member in an urban centre. Labour migration has grown mainly since the early 1990s in China and most migration here is intra-provincial. Long distance internal migration is an important characteristic of labour migration in India. International migration is almost non-existent from HKH region of Bangladesh. Myanmar has both internal and international migration and more than half the internal migrants are women.

The main driver of labour migration in the HKH region is 'search for a better life'. Limited opportunities in mountains reflect high dependence on agriculture, lack of market access and limited livelihood security. Economic growth of urban centres within and beyond the HKH region exerts a strong influence on migration. There is a strong demand for male workers from the mountain but also a growing demand for female workers in the domestic care and hospitality industries. Access to education and social networks are strong drivers of migration. Migration is highly age selective. Environmental degradation and climate stressors can weaken agricultural productivity or heighten risks contributing to labour migration, especially over shorter distance.

Country evidence shows that migration reduces mountain poverty. A significant portion of rural households benefit from migrant remittances. Remittances are spent on consumption, but may also contribute to investment in agriculture and other businesses. Despite fears of negative impacts on left behind children, migration helped increase school enrollment. Migration has contributed to increased political demands including voting rights.

An assessment of governance of labour migration in the HKH region shows that countries with significant international labour migration have developed comprehensive policies and regulations to govern international migration, often with dedicated ministries. However, policies on internal migration are yet to be framed except for a few countries. None of these countries, barring Bangladesh, have ratified the 1990 United Nations Convention on the Protection of the Rights of All Migrant Workers and the Members of their Families. Therefore, labour migration mostly takes place through bilateral agreements or memorandums of understanding. The SDGs specifically endorse policies to reduce the cost of migrant remittances and some HKH governments are actively pursuing these although there are many obstacles. Several countries in the region have specific institutions for the diaspora including overseas offices, welfare funds and diaspora focused financial instruments.

HKH countries face many challenges in governing labour migration. Absence of a comprehensive policy on internal migration in some of the HKH countries keeps the internal migrants outside the purview of government programmes and entitlements. In some areas, polices are restrictive. Countries which have polices on internal migration do not always honour portability of rights. Whether countries have a regulated (India, China) or unregulated (Pakistan) approach to internal migration, there is a strong sedentary bias in policies that affect internal migrants, in contrast to international migration. In the case of international labour migration some common challenges that HKH countries face include high cost of migration, informality of the recruitment process, visa trading and lack of protection from abusive and forced labour conditions. Women who migrate as domestic workers face strict restrictions on movement, long working hours, and in some cases, physical and sexual abuse.

15.7 Policy Recommendations and Way Forward

On basis of the conclusions above, this section attempts to provide some recommendations. As all HKH countries have now reshaped their development goals in line with SDGs, the recommendations have been devised to correlate with relevant SDGs as well. Two SDGs apply directly to migration: Decent Work and Economic Growth (Goal: 8) and Reduce Inequality (Goal: 10). Several other SDGs have relevance for migration in terms of policy development.

15.7.1 Regular, Safe and Orderly Migration

SDG 8 talks about promoting inclusive and sustainable economic growth. Target 8.8 specifically underscores the need to protect the rights and promote safe working environments for migrant workers, including women migrants. Targets 8.5 (*achieve full and productive employment and decent work*) and 8.10 (*access to banking, insurance and financial services for all*) also have implications for migrants.

Target 10.7 specifically calls for facilitating orderly, safe, regular and responsible migration. We have attempted to demonstrate in this chapter that internal and international migrants are quite different and require different forms of support and interventions to ensure safety and stability. In respect to internal migrants, policies regarding urban planning, growth sector development, and climate change adaptation operational in rural and urban areas have to be targeted. Some of these countries impose barrier on internal, rural to urban migration. Relevant policies need to be reshaped on the basis of the demand of labour market.

Restrictive policies in destination countries remains the main hurdle for regular, safe and orderly migration. Destination countries need to frame more pragmatic policies that take into consideration the demand for different types of skills and services. We have argued that, along with bilateral negotiation, countries should encourage multilateralism to set common minimum standards for the migrant workers. The UN convention on Protection of Rights of All Migrant Workers and their Family Members (1990) and International Labour Organization instruments should be honoured in both origin and destination countries. The Kafala system in the Middle Eastern countries and the outsourcing system in some of the Southeast Asian countries need substantial reform. Origin countries also bear responsibility for insuring safe and orderly migration for their citizens. HKH countries are currently experimenting with different governance models in this regard and they can learn much from cross-country exchange among HKH neighbours and beyond.

15.7.2 Protection of Migrants' Right and Social Security

Target 8.8 of the SDGs highlights protecting labour rights and promoting a safe and secure working environment. The 1990 UN International Convention on the Protection of the Rights of All Migrant Workers and the Members of their Families (ICMW) is the most comprehensive instrument of its kind, and has only been ratified by one HKH country: Bangladesh.

This important agreement needs to be ratified by all HKH countries without any further delay. The UN needs to launch a global campaign to encourage ratification of the convention by the destination countries. The UN also needs to ensure that none of the rights of the migrants is compromised in the global compact for migration. HKH countries should advocate for more rigorous protection needs in the consultative forums of Colombo Process, Abu Dhabi Dialogue, and Bali Process. Most countries in the region have ratified the Convention on All Forms of Discrimination Against Women (CEDAW), which includes specific provisions for women migrant workers. During bilateral agreements, HKH countries can pursue protection of women migrants under the auspices of CEDAW.

The Rights of Internal Migrants are protected by the national laws of HKH countries. Decent work conditions entail the right to work, rights at work, social protections, and social dialogue. HKH countries need to provide protection from vulnerabilities and contingencies that take individuals out of work at old age, at times of sudden loss of livelihood, sickness, accident or death. In the case of informal work, these protections are almost non-existent. Attaining the SDG 8 would require establishing these rights.

15.7.3 Mainstreaming Migration into Development

The section on the consequences of migration demonstrated the role of migration in reducing poverty, increasing living standards, and enhancing the social and economic development of migrants, their families, and their communities. To enhance the development impacts and to reduce negative consequences associated with migration, measures should be taken to optimize the potential gains. If low-income households are to fully benefit from migration this is an absolute necessity. For example, the effectiveness of Nepali policy on 'free-visa, free-ticket' for migrant workers should be evaluated, and, if it is successful, replicated in other HKH countries.

Almost all HKH countries have taken different steps to integrate international migration into their development policies. However, most of these policies look at international migrant workers and their remittances as of valuable tool for attaining national development goals. They rarely look for avenues that create space for migrants or their left-behind families to benefit from on-going national development interventions. For example, in the area of financial inclusion, migrants are seen as potential savers. Programmes need to be developed that would provide access to credit, training, business advisory services, and others incentives so that migrants or their families have the tools to turn themselves into investors (Goal 8 & 9: Industry, Innovation and Infrastructure). Integrating return migrants in the development process is another strategy to use their accumulated skills and knowledge and, in many cases, their newly-acquired assets and wealth. Only China and India have specific policies in respect to internal migrants. In some cases, we have noted, these polices on internal migration work as barriers to integration rather than facilitators. Right-based and inclusive policies need to be framed targeting the internal migrants.

15.7.4 Climate Change, Migrants and Urban Cities

Each HKH country has developed its own policies, strategies, and action plans for climate change. These policies still regard internal migration as a threat. Policy planners need to transform their optics on labour migration and remittance into one of opportunity and as an added potential for adapting to climate change. This chapter suggests that the majority of migrants from communities affected by the climate change and variability move within their country of origin. Their major flows are from rural to urban centres, particularly to major urban centres. It was also observed that new migrant populations in urban centres face significant challenges associated with violence, social exclusion, discrimination in the labour market, human and physical insecurities, lack of access to health care, and exposure to new types of environmental risk. SDG 11 targets the creation of sustainable cities and community to enhance inclusive and sustainable urbanization. In order to attain SDG 11, HKH countries need to integrate housing, health, hygiene, sanitation, and security for migrants into their city planning efforts.

15.7.5 Collection and Compilation of Migration Data

Target 16.10 of SDG 16 deals with standardizing and public access to information and data. Migration data are highly deficient in HKH countries. In several HKH countries, labour force surveys provide data on emigration. But these are not available on a regular basis. On the other hand, some countries like India use the census to provide data on immigration rather than emigration. There is also a considerable delay in the publication of migration data by censuses compared to other population characteristics.

Sometimes migration data are provided on stock but not on flow and vice versa. There is a need for country-specific migration surveys at regular intervals in HKH countries. This information would help cross-country comparisons when trying to assess mountain economies. Furthermore, policy makers, researchers and NGOs could benefit from this data to fine tune and enhance their programmes and outreach on migration issues. However, in order to understand better the effects of migration in mountain areas, this data needs to have separate consideration for mountain areas in HKH countries. Data from other sources, such emigration at ports of departure, migrant registration prior to departure, insurance plans taken in connection with overseas employment, and remittances sent through banking channels would be useful to explore various dimensions of migration.

References

Abrar, C. R. (2005). *Study on labour migration from SAARC countries: Reality and dynamics, A study commissioned by BATU-SAARC*. Dhaka: Refugee and Migratory Movement Research Unit.

Adams, R. H., & Page, J. (2005). Do international migration and remittances reduce poverty in developing countries? *World Development, 33*(10), 1645–1669.

Adams, R. H., Cuecuecha, A., & Page, J. (2008). *Remittances, consumption and investment in Ghana*. World Bank Policy Research Working Paper, No. 4515. Washington D.C.: The World Bank.

Adhikari, K., Suwal, B., & Sharma, M. (2008). *Foreign labour migration and remittance in Nepal: Policy, institution and services*. Washington D.C.: International Food Policy Research Institute.

Ahmadi, B., & Lakhani, S. (2017) *The Afghan refugee crisis in 2016*, Peace Brief, US Institute of Peace, Feb, 2017. Retrieved May 12, 2017, from https://www.usip.org/publications/2017/02/afghan-refugee-crisis-2016.

Ahmed, V., Sugiyarto, G., & Jha, S. (2010). *Remittances and household welfare: A case study of Pakistan*. ADB Economics Working Papers No. 194. Manila: Asian Development Bank. http://hdl.handle.net/11540/1536.

Amjad, R., Irfan, M., & Arif, G. M. (2015). An analysis of the remittances market in Pakistan. In: R. Amjad, & S. Burki (Eds.), *Pakistan: Moving the economy forward* (pp. 280–310). Cambridge, United Kingdom: Cambridge University Press. https://doi.org/10.1017/CBO9781316271711.012.

Amnesty International. (2017). *Turning people into profits: Abusive recruitment, trafficking and forced labour of Nepali migrant workers*. London: Amnesty International.

Andrus, J. R. (1948). *Burmese economic life*. Stanford, California, CA: Stanford University Press.

Banerjee, S., Black, R., Mishra, A., & Kniveton, D. (2018). Assessing vulnerability of remittance-recipient and non-recipient households in rural communities affected by extreme weather events: Case studies from south-west China and north-east India. *Population, Space and Place*. https://doi.org/10.1002/psp.2157.

Banerjee, S., Kniveton, D., Black, R., Bisht, S., Das, P. J., & Tuladhar, S. (2017). *Does financial remittance build household level adaptive capacity? A case study of the flood affected households of Upper Assam in India*. KNOMAD Working Paper 18. Washington D.C.: The World Bank.

Banerjee, S., Gerlitz, J. Y., & Hoermann, B. (2011). *Labour migration as a response strategy to water hazards in the Hindu Kush-Himalayas*. Kathmandu: International Centre for Integrated Mountain Development.

Bangladesh Bank. (2017). Monthly data on Wage earner's remittance. www.bangladeshbank.org.

Bangladesh Bureau of Statistics. (2016). *Bangladesh sample vital statistics*. Dhaka, Bangladesh: BBS. https://www.google.com/url?sa=t&rct=j&q=&esrc=s&source=web&cd=11&cad=rja&uact=8&ved=0ahUKEwiGgoDwkK7XAhVBqI8KHVV4DAYQFghDMAo&url=http%3A%2F%2F203.112.218.65%3A8008%2FWebTestApplication%2Fuserfiles%2FImage%2FLatestReports%2FSVRS_2015.pdf&usg=AOvVaw2Z_Xp9SMVvFwIEgxVfIsDh.

Barnett, J. R., & Webber, M. (2009). *Accommodating migration to promote adaptation to climate change*. The Commission on Climate Change and Development. Policy Research Working Paper 5270. Washington D.C.: The World Bank.

Benz, A. (2016). Framing modernization interventions: Reassesing the role of migration and translocality in sustainable mountain development in Gilgit-Balitstan, Pakistan. *Mountain Research and Development, 36*(2), 141–152.

Bhadra, C. (2013). *Final report on impact of foreign labour migration to enhance economic security and address VAW among Nepali women migrant workers and responsiveness of local governance to ensure safe migration, submitted to MoWCSW, MoLE, MoFALD of GoN.*

Bhadra, C. (2007). *Women's international labour migration and impact of their remittance on poverty reduction: Case of Nepal*. Seminar presentation on Labour Migration, Employment and Poverty Alleviation in South Asia, Kathmandu, 9–10 August.

Bhagat, R. B. (2014). *Urban migration tends, challenges and opportunities in India*. Background Paper to World Migration Report 2015. Geneva: International Organization for Migration.

Bhagat, R. (2012). A turnaround in India's urbanization. *Asia-Pacific Population Journal, 27*(2), 23–39.

Bhagat, R. B., Keshri, K., & Ali, I. (2013). Emigration and flow of remittances. *Migration and Development, 2*(1), 93–105.

Bhandari, P. (2004). Relative deprivation and migration in an agricultural setting of Nepal. *Population and Environment, 25*(5), 475–499.

Bhatnagar, G. V. (2017, August 3). Modi government's decision to allow proxy voting by NRIs may kick up a hornet's nest. *The Wire*. https://thewire.in/diplomacy/centres-decision-allow-proxy-voting-rights-nri-may-kick-hornets-nest.

Bjelica, J. (2016). Over half a million Afghans flee conflict in 2016: A look at the IDP statistics, rights and freedoms, Afghan Analysts Network, December 2016.

Black, R., Bennett, S. R. G., Thomas, S. M., & Beddington, J. R. (2011). Climate change: migration as adaptation. *Nature, 478*, 447–449. https://doi.org/10.1038/478477a.

Black, R., Natali, C., & Skinner, J. (2006). *Migration and inequality*. Background Paper for World Development Report 2006. Washington DC: The World Bank.

Bohle, H. G., & Adhikari, J. (1998). Rural livelihoods at risk: How Nepalese farmers cope with food insecurity. *Mountain Research and Development, 18*, 321–332.

Bohra-Mishra, P., & Massey, D. S. (2011). Environmental degradation and out-migration: New evidence from Nepal. In E. Piguet, A. Pécoud, & P. de Guchteneire (Eds.), *Migration and climate change* (pp. 74–101). Cambridge, CAMBS: Cambridge University Press.

Bruslé, T. (2008). Choosing a destination and work. *Mountain Research and Development, 28*(3), 240–247.

Bureau of Manpower, Employment and Training. (2017). *Overseas employment in 2017*. Dhaka, Bangladesh: BMET. http://www.bmet.org.bd/BMET/stattisticalDataAction.

Bureau of Manpower, Employment and Training. (2016). *Overseas Employment in 2016*. Dhaka, Bangladesh: BMET. http://www.bmet.org.bd/BMET/stattisticalDataAction.

Chapagain, B., & Gentle, P. (2015). Withdrawing from Agrarian livelihoods: Environmental migration in Nepal. *Journal of Mountain Science, 12*(1), 1–13.

De Haas, H. (2012). The migration and development pendulum: A critical view on research and policy. *International Migration, 50*(3), 8–25.

Department of Population. (2016). *The 2014 Myanmar population and housing census: Thematic report on migration and urbanization*. Nay Pyi Taw, Myanmar: DoP.

Department of Population. (2015). *The 2014 Myanmar population and housing census: The union report*. Nay Pyi Taw, Myanamar: DoP.

Du, Y., Park, A., & Wang, S. (2005). Migration and rural poverty in China. *Journal of Comparative Economics, 33*(4), 688–709.

EC working to ensure voting rights of NRBs. (2017, September 24). *The Daily Observer*. http://www.observerbd.com/details.php?id=96519.

Fu, Q., & Ren, Q. (2010). Educational inequality under China's rural-urban divide: The Hukou system and return to education. *Environment and Planning A, 42*(3), 592–610.

Gazdar, H. (2003). A review of migration issues in Pakistan. Paper prepared and presented at the Regional Conference on Migration, Development and Pro-Poor Policy Choices in Asia. The Conference was jointly organised by the Refugee and Migratory Movements Research Unit, Bangladesh, and the Department for International Development, UK.

Ghobadi, N., Koettl, J., & Vakis, R. (2005). *Moving out of poverty: Migration insights from rural Afghanistan*. Working Paper Series. Kabul: Afghanistan Research and Evaluation Unit (AREU).

Gioli, G., Khan, T., Bisht, S., & Scheffran, J. (2014). Migration as an adaptation strategy and its gendered implications: A case study from the Upper Indus Basin. *Mountain Research and Development, 34*(3), 255–265.

Gong, X., Kong, S. T., Li, S., & Meng, X. (2008). Rural–urban migrants. *China's dilemma* (pp. 110–152).

Government clears proxy vote move for NRIs. (2017, August 3). *The Indian Express*. http://indianexpress.com/article/india/government-clears-proxy-vote-move-for-nris-4779759/.

Government of Afghanistan. (2005). Regulation for sending Afghan workers to abroad (unofficial English translation) in Official Gazette, no. 906, (Kabul, Ministry of Labour, Social Affairs, Martyrs and Disabled (MoLSAMD), Directorate of Rights and Labour Law).

Government of the Islamic Republic of Afghanistan (GIRoA). (2015). The state of Afghan Cities 2015, GIRoA Kabul, 2015.

Government of India. (2006). *Pravasi Bharatiya Bima Yojana*. New Delhi: Ministry of External Affairs. http://www.mea.gov.in/oe-faq.htm.

Government of India. (1980). Inter-state Migrant Workmen (Regulation of Employment and Conditions of Service Conditions) Central Rules. New Delhi: Ministry of Labour.

Government of India. (1979). The Inter-state Migrant Workmen (Regulation of Employment and Conditions of Service) Act of 1979. Office of Chief Labour Commissioner (Central). New Delhi: Ministry of Labour. http://clc.gov.in/clc/clcold/Acts/shtm/ismw.php.

Government of India. (1950). *The Constitution of India*. New Delhi.

Government of Nepal. (2011). Nepal Living Standards Survey 2010/11, vol. 2. Kathmandu: Central Bureau of Statistics.

Government of Nepal. (2013). Nepal thematic report on food security. Kathmandu: National Planning Commission.

Government of Nepal. (2015). Guidelines related to managing domestic workers going for foreign employment, 2015.

Government of Nepal. (2016). Labour migration for employment. A status report for Nepal: 2014/2015. Kathmandu: Ministry of Labour and Employment.

Griffiths, M. (2016). Remittance economy in Rural Myanmar. In M. Griffiths & M. Ito (Eds.), *Migration from Myanmar: Perspectives from current research* (pp. 99–116). Yangon, Myanmar: Social Policy and Poverty Research Group.

Hall, S (2013). Complexities and challenges in Afghan migration: Policy and research event, IS Academy Policy Brief | No. 14. Retrieved May 12, 2017, from http://samuelhall.org/REPORTS/Complexities%20and%20Challenges%20in%20Afghan%20Migration.pdf.

Hoang, L. (2011). *Impact of internal and international labour migration: South-East Asia*. Migrating out of Poverty Research Programme Consortium Working Paper 28. Brighton: University of Sussex.

Hoermann, B., Banerjee, S., & Kollmair, M. (2010). *Labour migration for development in the western Hindu Kush-Himalayas*. Kathmandu: International Centre for Integrated Mountain Development.

Hoermann, B., & Kollmair, M. (2008). *Labour migration and remittances in the Hindu Kush-Himalayan region*. Kathmandu: International Centre for Integrated Mountain Development.

Hu, X., Cook, S., & Salazar, M. A. (2008). Internal migration and health in China. *The Lancet, 372*(9651), 1717–1719.

Hugo, G. (2002). Effects of international migration on the family in Indonesia. *Asian and Pacific Migration Journal, 11*(1), 13–46.

Huo, W., Deng, G., Wang, P., & Yang, X. (2006). The pull of inter-provincial migration of rural labor force and its influence on policy making. *Ecological Economy, 2,* 25–31.

Intergovernmental Panel on Climate Change (IPCC). (2014). *Climate change 2014: Impacts, adaptation and vulnerability*. Contribution of Working Group II to the Fifth Assessment Report of the Intergovernmental Panel on Climate Change. Cambridge university press.

International Labour Organization. (2015). *No easy exit—Migration bans affecting women from Nepal*. Labour Migration Branch (MIGRANT)—Geneva: ILO.

International Organization for Migration. (2014). Afghanistan Migration Profile, International Organisation for Migration, p. 22. Retrieved February 27, 2017, from http://samuelhall.org/wp-content/uploads/2016/07/IOM-Afghanistan-Migration-Profile.pdf.

International Organization for Migration and Bangladesh Bank. (2009). *Nationwide household remittance survey in Bangladesh*. Dhaka: IOM.

Ito, M. (2016). Myanmar migrant workers in Thailand—Their characteristics, migration experiences and aspirations on return. In M. Griffiths & M. Ito (Eds.), *Migration in Myanmar: Perspectives from current research* (pp. 88–98). Yangon, Myanmar: Social Policy and Poverty Research Group.

Jain, A. (2010). *Labour migration and remittances in Uttarakhand*. Kathmandu: International Centre for Integrated Mountain Development.

Jaquet, S., Shrestha, G., Kohler, T., & Schwilch, G. (2016). The effects of migration on livelihoods, land management, and vulnerability to natural disasters in the Harpan watershed in Western Nepal. *Mountain Research and Development, 36*(4), 494–505.

Kollmair, M., & Banerjee, S. (2011). *Drivers of migration in mountainous regions of the developing world: A review*. Report submitted to Foresight Environmental Migration Project. London: UK Government Office of Science (GO-Science).

Kreutzmann, H. (2004). Accessibility for high Asia: Comparative perspectives on Northern Pakistan's traffic infrastructure and linkages with its neighbors in the Hindukush-Karakoram-Himalaya. *Journal of Mountain Science, 1*(3), 193–201.

Kyaw, A. T. (2015, September 2). Informal cash flows threaten kyat policy. *Myanmar Times*. https://www.mmtimes.com/business/16267-informal-cash-flows-threaten-kyat-policy.html.

Lai, F., Liu, C., Luo, R., Zhang, L., Ma, X., Bai, Y., et al. (2014). The education of China's migrant children: The missing link in China's education system. *International Journal of Educational Development, 37,* 68–77.

Li, Q., Huang, J., Luo, R., & Liu, C. (2013). China's labor transition and the future of China's rural wages and employment. *China & World Economy, 21*(3), 4–24.

Liang, Z., & Ma, Z. (2004). China's floating population: New evidence from the 2000 census. *Population and Development Review, 30*(3), 467–488.

Lokshin, M., Bontch-Osmolovski, M., & Glinskaya, E. (2010). Work-related migration and poverty reduction in Nepal. *Review of Development Economics, 14*(2), 323–332.

Lusome, R., & Bhagat, R. B. (2013). Migration and workforce participation in the Himalayan states. In: T. V. Sekher, A. Singh, & S. Parsuraman (Ed.), *Population, health and development* (pp. 435–446). Delhi, India: Academic Foundation.

Mahmood, R. (2011). *Cross labour mobility, remittances and economic development in South Asia*. Washington D.C., United States of America: The World Bank.

Mansuri, G. (2006). *Migration, sex bias, and child growth in rural Pakistan*. Policy Research Working Paper No. 3946. Washington D.C.: The World Bank.

Massey, D. S., Axinn, W. G., & Ghimire, D. J. (2010). Environmental change and out-migration: Evidence from Nepal. *Population and Environment, 32*(2), 109–136.

Massey, D., Axinn, W., & Ghimire, D. (2007). *Environmental change and out-migration: Evidence from Nepal*. Population Studies Center Research Report 07–615. Ann Arbor, MI: Population Studies Center.

Migrant Forum Asia. (n.d). Open Working Group on Labour Migration and Recruitment, Recruitment Fees and Migrants' Rights Violation Policy Brief #1, MFA, Manila.

Millions still waiting on a chance to vote. The Record (2017, May 11). *The Record Nepal* http://www.recordnepal.com/wire/news-analysis/millions-still-waiting-on-a-chance-to-vote/.

Ministry of Agriculture. (2006). *Rural-urban migration in Bhutan*. Bhutan: Thimphu.

Ministry of Agriculture and Forests. (2013). *Migration in Bhutan (Its Extent, Causes and effects)*. Bhutan: Thimphu.

Ministry of Human Resources and Social Security. (2013). *Annual statistical bulletin of human resources and social security development in 2013*. Beijing, China: MHRSS [in Chinese]. Retrieved from http://www.mohrss.gov.cn/SYrlzyhshbzb/zwgk/szrs/ndtjsj/tjgb/201405/t20140529_131147.htm .

Ministry of Labour and Human Resources. (2015). *Labour force survey report*. Bhutan: Thimphu.

Ministry of Labour, Immigration and Population. (2016). *Report on Myanmar labour force survey: January–March 2015*. Myanmar: Nay Pyi Taw.

Mohanty, S., & Bhagat, R. B. (2013). Spatial pattern of urbanisation and urban growth in Western Himalayan region in India. In: T.V. Sekher, A. Singh, S. Parsuraman, (Eds.), *Population, health and development* (pp. 413–434). Delhi, India: Academic Foundation.

Mueller, V., Gray, C., & Kosec, K. (2014). Heat stress increases long-term human migration in rural Pakistan. *Nature Climate Change, 4*, 182–185. https://doi.org/10.1038/nclimate2103.

Nagarajan, S. (2009). *Migration, remittances, and household health: Evidence from South Africa (Unpublished doctoral thesis)*. United States of America: George Washington University.

National Bureaus of Statistics of China. (2017). *The 2016 migrant worker monitoring survey report*. Beijing, China: NBSC. Retrieved November 8, 2017, from http://www.stats.gov.cn/tjsj/zxfb/201704/t20170428_1489334.html.

National Bureau of Statistics of China. (2016). *China statistical yearbook 2016*. Beining, China: NBSC.

National Sample Survey Organization. (2010). *Migration in India 2007–08*. New Delhi, India: NSSO.

Neetha, N. (2003). *Migration, social networking, and employment: a study of domestic workers in Delhi*. NLI Research Studies Service No. 2002/37. Noida, UP: VV Giri National Labour Institute.

Nepal Economic Forum. (2014). *Docking Nepal's economic analysis*. Lalitpur: Nefport.

Nepal Institute of Development Studies and World Bank. (2009). *Nepal migration survey*. Kathmandu: NIDS. (unpublished).

Nordland, R. (2016, November 4). Afghanistan itself is now taking in the most Afghan migrants. *The New York Times*. http://www.nytimes.com/2016/11/05/world/asia/afghanistan-migrants.html?_r=0.

Office of the United Nations High Commissioner for Human Rights. (2018). End-of-mission statement of the UN Special Rapporteur on the human rights of migrants on his visit to Nepal,' 5 February 2018. http://www.ohchr.org/en/NewsEvents/Pages/DisplayNews.aspx?NewsID=22632&LangID=E.

Olimova, S., & Olimov, M. (2007). Labor migration from mountainous areas in the Central Asian region: Good or evil? *Mountain Research and Development, 27*(2), 104–108.

Opel, A. (2005). *Bound for the city: A study of rural to urban labour migration in Afghanistan*. Working Paper Series. Kabul: Afghanistan Research and Evaluation Unit.

Pakistan Bureau of Statistics. (2011). *The labor force survey*. Islamabad, Pakistan: PBS.

Pakistan Institute of Legislative Development and Transparency. (2008). *Overseas Pakistani workers: Significance and issues of migration*. Briefing Paper No. 34. Islamabad: PILDAT.

Paoletti, Sarah, Taylor-Nicholson, Eleanor, Sijapati, Bandita, & Farbenblum, Bassina. (2014). *Migrant workers' access to justice at home: Nepal*. New York: Open Society Foundations.

Park, A., & Wang, D. (2010). Migration and urban poverty and inequality in China. *China Economic Journal, 3*(1), 49–67.

Pariyar, B., & Lovett, J. C. (2016). Dalit identity in urban Pokhara. *Nepal. Geoforum, 76*, 134–147.

Ratha, D. (2003). *Worker's remittances: The important and stable source of external development finance*. Global Development Finance. Washington D.C.: The World Bank.

Ratha, D., Mohapatra, S., & Scheja. E. (2011). *Impact of migration on economic and social development: A review of evidence and emerging issues*. Policy Research Working Paper No. 5558. Washington DC: The World Bank.

Refugee and Migratory Movements Research Unit. (2017). *Labour migration from Bangladesh 2016: Achievements and challenges*. Dhaka: RMMRU.

Refugee and Migratory Movements Research Unit. (2016). *Institutional capacity building for management of the wage earners' welfare fund*. Policy Brief 17, Dhaka: RMMRU. https://www.dropbox.com/s/po9xaq6420tzl4o/Policy%20Brief%2017.pdf?dl=0.

SC orders govt to ensure migrants' right to vote. (2018, March 23). *The Kathmandu post*. http://kathmandupost.ekantipur.com/news/2018-03-23/sc-orders-govt-to-ensure-migrants-right-to-vote.html.

Shrestha, S. S., & Bhandari, P. (2007). Environmental security and labor migration in Nepal. *Population and Environment, 29*(1), 25–38.

Siddiqui, T. (Ed.). (2017). *Untold stories of migrants: Dreams and realities*. Dhaka, Bangladesh: Refugee and Migratory Movement Research Unit.

Siddiqui, T. (2008). Migration and gender in Asia. United Nations Economic and Social Commission for Asia and the Pacific Population Division, Department of Economic and Social Affairs Bangkok, Thailand. http://www.un.org/esa/population/meetings/EGM_Ittmig_Asia/P06_Siddiqui.pdf.

Siddiqui, T. (2006). Securitization of irregular migration: The South Asian Case. In: R. Emmers, M. C. Anthony, & A. Acharya (Ed.), *Studying non traditional security in Asia* (pp. 143–167). Singapore: Marshall Cavendish Academic.

Siddiqui, T. (2004). *Institutional diaspora linkage: The emigrant Bangladeshis in UK and USA*. Dhaka: GOB and IOM.

Siddiqui, T. (2001). *Transcending boundaries: Labour migration of women from Bangladesh*. Dhaka: University Press Limited.

Siddiqui, T., & Abrar, C. R. (2003). *Migrant workers' remittances and micro finance in Bangladesh*. Geneva: International Labour Organization.

Siddiqui, T., & Abrar, C. R. (2002). *Contribution of returnees: An analytical survey of post return experiences*. Dhaka: International Organization for Migration.

Siddiqui, T., & Billah, M. (2014). *The role of labour migration in climate change adaptation in Bangladesh with a special focus on the Chittagong Hill Tract region*. Working Paper Series no. 47. Dhaka: RMMRU.

Siddiqui, T., Bhuiyan, M. R. A., Kniveton, D., Black, R., Islam, M. T., & Martin, M. (2018). Situating migration in planned and autonomous practices in Bangladesh. In S. I. Rajan & R. B. Bhagat (Eds.), *Climate change, vulnerability and migration* (pp. 119–146). Oxon, United Kingdom: Routledge.

Siddiqui, T., & Mahmood, R. (2015). Impact of migration on poverty and local development in Bangladesh, SDC and RMMRU.

Sijapati, B., Ayub, M., & Kharel, H. (2017). In: S. Irudaya Rajan (Ed.), *Making migration free: An analysis of Nepal's 'free-visa, free-ticket' scheme. South asia migration report 2017: Recruitment, remittances and reintegration.* Abingdon and London: Routledge.

Sijapati, B., Bhattarai, A., & Pathak, D. (2015). *Analysis of labour market and migration trends in Nepal*. Kathmandu: Deutsche Gesellschaft für Internationale Zusammenarbeit and International Labour Organization.

Singhvi, L. M. (2001). *The Indian diaspora*. The Report of the High Level Committee on Indian Diaspora (HLCID), 19th December 2001.

Srivastava, R. (2012). Internal migration in India: An overview of its features, trends and policy challenges. In: *UNESCO/UNICEF*

National Workshop on Internal Migration and Human Development in India, 6–7 December 2011, Workshop Compendium (Vol. 2). Workshop Papers. New Delhi, UNESCO/UNICEF.

Srivastava, R. (2011). Internal migration in India. Background Paper Prepared for the Migrating Out of Poverty Research Programme Consortium, University of Sussex.

Standing Committee of the National People's Congress. (2007). Law of the People's Republic of China on Employment Contracts. Retrieved November 8, 2017, from http://www.gov.cn/jrzg/2007-06/29/content_667720.htm. China.

Standing Committee of the National People's Congress. (1994). Labor Law of the People's Republic of China. Retrieved November 8, 2017, from http://www.gov.cn/banshi/2005-05/25/content_905.htm. China.

State Council. (2010). Regulations on work-related injury insurances. China. Retrieved November 8, 2017, from http://www.gov.cn/gongbao/content/2011/content_1778064.htm. China.

Subedi, B. (1991). International migration in Nepal: Towards an analytical framework. *Contribution to Nepalese Studies, 18*(1), 83–102.

Suleri, A., & Savage, K. (2006). *Remittances in crisis: A case study of Pakistan. Humanitarian policy group background paper*. London: Overseas Development Institute.

The World Bank. (2016a). *Migration and remittances: Recent developments and outlooks*. Migration and Development brief No. 26. Washington D.C: World Bank.

The World Bank. (2016b). *A country on the move: Domestic migration in two regions of Myanmar*. Yangon: World Bank.

The World Bank. (2016c). *Remittances at risk*. Nepal Development Update. Kathmandu: World Bank. http://documents.worldbank.org/curated/en/564551468198011442/pdf/106393-WP-PUBLIC-ADD-SERIES-Nepal-Development-Update-2016.pdf.

The World Bank. (2014). Migration and development brief No. 23, 2014, Washington, D.C.

United Nations. (2015). *The world population prospects: The 2015 revisions*. UN: Population Division.

United Nations Department of Economic and Social Affairs. (2013a). Cross-national comparisons of internal migration: An update on global patterns and trends. Population Division Technical Paper 2013/1. New York, United Nations.

United Nations Department of Economic and Social Affairs. (2013b). *International migration report 2013*. New York: United Nations.

United Nations Department of Economic and Social Affairs. (2015). Trends in international migration stock: Migrants by destination and origin (United Nations database POP/PB?MIG?Stock/Rev 2015). http://www.un.org/en/development/desa/population/migration/data/estimates2/estimates15.shtml.

United Nations University. (2017). Highly skilled Afghan diaspora contributes to innovation and change, United Nations University. http://unu.edu/publication.

United Nations High Commissioner for Refugees. (2017). Facts and figures about refugees. Retrieved February 27, 2017, from http://www.unhcr.ie/about-unhcr/facts-and-figures-about-refugees.

Ura, K. (2013). Migration in Bhutan: To the west and towns, The Bhutanese.

Weiner, M. (1978). *Sons of the soil: Migration and ethnic conflict in India*. Princeton, NJ: Princeton University Press.

Wen, M., & Lin, D. (2012). Child development in rural China: Children left behind by their migrant parents and children of nonmigrant families. *Children Development, 83*(1), 120–136.

Wickramasekara, P. (2015). *Bilateral agreements and memoranda of understanding on migration of low skilled workers: A review. Report prepared for the Labour Migration Branch*. Geneva: International Labour Organization.

Wickramasekara, P. (2011). *Labour migration in South Asia: A review of issues, policies and practices*. International Migration Paper No. 108. Geneva: International Labour Organization.

Yahui, Z., & Banerjee, S. (2017). The role of out-migration in villagers' adaptation to drought—A case of Baoshan, Yunnan, China. *Advances in Climate Change Research, 13*(4), 398–404.

Zhang, L., Liu, C., Bai, Y., Dong, Y., & Wang, W. (2017). *Mid-term progress report submitted to the National Natural Science Foundation of China. The Center for Chinese Agricultural Policy*. Beijing: Chinese Academy of Sciences.

Zhao, R., Liu, Y., & Yan, Y. (2012). *Labor migration choice and its impacts on households in rural China*. The Agricultural and Applied Economics Association's 2012 AAEA Annual Meeting. Seattle: The Agricultural and Applied Economics Association.

Zhou, C., Sylvia, S., Zhang, L., Luo, R., Yi, H., Liu, C., et al. (2015). China's left-behind children: Impact of parental migration on health, nutrition, and educational outcomes. *Health Affairs, 34*(11), 1964–1971.

16 Governance: Key for Environmental Sustainability in the Hindu Kush Himalaya

Coordinating Lead Authors
Hemant Raj Ojha, University of New South Wales, Institute for Studies and Development Worldwide (IFSD), Australia, e-mail: ojhahemant1@gmail.com
Rucha Ghate, International Centre for Integrated Mountain Development, Kathmandu, Nepal, e-mail: ruchaghate@gmail.com (corresponding author)
Lam Dorji, Centre for Environment and Development, Thimphu, Bhutan, e-mail: ldorjie@gmail.com

Lead Authors
Ankita Shrestha, Southasia Institute of Advanced Studies, Kathmandu, Nepal, e-mail: anki.shrestha@gmail.com
Dinesh Paudel, Appalachian State University, USA, e-mail: paudeld@appstate.edu
Andrea Nightingale, Swedish University of Agricultural Sciences, Uppsala, and Norwegian University of Life Sciences, Sweden, e-mail: andrea.nightingale@slu.se
Krishna Shrestha, University of New South Wales, Australia, e-mail: krishna.shrestha@unsw.edu.au
Muhammad Arif Watto, University of Agriculture, Faisalabad, Pakistan, e-mail: arif.watto@uaf.edu.pk
Rajan Kotru, International Centre for Integrated Mountain Development, Nepal, e-mail: rajan.kotru@icimod.org

Contributing Authors
Jun He, Yunnan University, China, e-mail: Jun.he@ynu.edu.cn
Aung Kyaw Naing, RECOFTC - The Center for People and Forests, country program, Myanmar, e-mail: aungkyawnaing@recoftc.org
Tashi Wangmo, National Council of Bhutan, Bhutan, e-mail: tashiwangmo2002@yahoo.com
Vishal Narain, Management Development Institute, Gurgaon, India, e-mail: vishalnarain@mdi.ac.in
Anamika Barua, Dept of Humanities and Social Sciences, IIT Guwahati, India, e-mail: anamika.barua@gmail.com
Babar Shahbaz, University of Agriculture, Faisalabad, Pakistan, e-mail: bsuaf@yahoo.com
Bimal Regmi, International Centre for Integrated Mountain Development, Nepal, e-mail: bimalrocks@yahoo.com
Niaz Ahmed Khan, University of Dhaka, Bangladesh, e-mail: khan.niaz.ahmed@gmail.com
Binod Chapagain, RECOFTC - The Center for People and Forests, Bangkok, Thailand, e-mail: binod.chapagain@recoftc.org

Review Editor
Philippus Wester, International Centre for Integrated Mountain Development, Kathmandu, Nepal, e-mail: philippus.wester@icimod.org

Corresponding Author
Rucha Ghate, International Centre for Integrated Mountain Development, Kathmandu, Nepal, e-mail: ruchaghate@gmail.com

P. Wester et al. (eds.), *The Hindu Kush Himalaya Assessment*,
https://doi.org/10.1007/978-3-319-92288-1_16

Contents

Chapter Overview

Key Findings

1. **There are few existing regional policies and processes for environmental governance in the Hindu Kush Himalaya (HKH)—most are national and subnational.** If this imbalance persists, it will undermine sustainable natural resource management in the region. Furthermore, it should be stressed, the implementation of existing policies and legislations in the HKH remains ineffective.
2. **Environmental governance reforms in the HKH emphasize decentralization, often creating positive local outcomes—yet these local initiatives are not adequately supported through subnational and national governance systems.** Successful local initiatives involve local communities, local governments and—increasingly—local and small-scale business groups.
3. **HKH countries lack institutions to link upstream and downstream communities in river basins and mountain landscapes.** Environmental institutions need to address the complex geography

of the region. Collaboration is hampered by limited data and knowledge sharing, by weak local political representation at higher levels, and by insufficient attention to social equity and inclusion.

Policy Messages

1. **Governments and environmental institutions in the HKH need to act now to strengthen the interface among science, policy, and practice.** Urgent actions in this direction are needed to improve policymaking, regional development planning, and adaptive environmental governance in the face of growing climate change impacts and persistent livelihoods challenges.
2. **Transboundary cooperation is crucial for improving environmental governance in the HKH.** However, it is lacking mainly because the focus of intergovernmental initiatives has been on political and economic interests, rather than social and environmental wellbeing at the regional scale. Progress requires the establishment of formal, mutually agreeable frameworks for cooperation that are translated into action. HKH countries should build on ongoing collaboration initiatives to build and expand cooperation on complex transboundary environmental issues such as water basin management and energy security.
3. **Environmental policy implementation in HKH countries will improve only if national governments recognize the multi-sectorial and cross-scalar nature of environmental governance.** Implementation depends on the engagement of various stakeholders, including local communities. There is also a need for facilitating upstream and downstream interactions for improving landscape level governance. For this, **governments need to create regulatory frameworks and local institutional arrangements to enable the expansion of successful initiatives to empower community action and inspire community-government partnerships.** There is an opportunity to learn from the past four decades of decentralization and community based resource management policies and programs and upscale and institutionalize the successful initiatives across the region.

This chapter identifies four governance keys to the sustainable future of the HKH:

1. *Institutional innovation—for landscape level governance, upstream-downstream linkages, and for translating policy goals into action;*
2. *Upscaling and institutionalizing decentralized and community based resource management practices;*
3. *Transboundary cooperation for managing connected landscapes; and*
4. *Science–policy–practice interface for decision making, learning and effective implementation of policies and programs.*

The governance of environmental resources holds the key to the future of sustainable development in the Hindu Kush Himalaya (HKH). Environmental resources in the region are diverse and include forest, water, biodiversity, and agriculture. The governance of these resources involves a complex ensemble of policies, institutions, policy-making practices, and implementation procedures (***well-established***). More broadly, the term *environmental governance* denotes the ways in which both formal and informal institutions act to control and manage the environment in light of various social, cultural, economic, and ecological values. It also entails questions like who benefits and who loses, and finding ways to ensure equitable sharing of benefits, costs, and risks.

While disproportionate in their influence, national environmental policies and institutions are complex and still evolving in response to many challenges of HKH environmental governance. A number of state agencies have emerged in the HKH countries, but they face challenges to achieve coordination and linkages across different levels: local, sub-national, national, regional, and global (***well-established***). At present, environmental governance in the HKH remains distributed unevenly across these various levels, with limited mechanisms and processes in place to build linkages across scales. Much of the planning and decision making power rests with national and subnational authorities, while regional and local authorities have much less. This imbalance in the vertical distribution of governing power is often incompatible with sustainable natural resource management (***well-established***).

Institutional diversity in environmental governance is growing, creating opportunities for innovation. State agencies and local communities, too, have a long history in HKH environmental governance. And non-governmental organizations, private organizations, and knowledge communities have recently become more active. In general, the space for multi-actor engagement is now expanding across the region, but this opportunity still remains underutilised (***established but incomplete***).

Environmental governance faces other challenges besides the dominance of national and subnational authorities. One is that environmental institutions have not yet fully adapted to the complex geography of the Hindu Kush Himalaya. As a result, the region lacks appropriate and context-specific institutions to link upstream and downstream communities in river basins (established but incomplete). Another institutional challenge is the sectoral fragmentation of environmental governance, which impedes coordination (***well-established***).

The HKH region has begun to see reforms in environmental governance, especially through decentralization and devolution. Many of these initiatives address new and recent concerns in the region—climate change, disaster risks, and threats to biodiversity. Other reforms have promoted shifts toward market mechanisms in resource governance, especially for the water, energy, and agriculture sectors (***established but incomplete***).

Although many of the reforms at the national and sub-national level aiming at decentralization and devolution have succeeded locally, they cannot have a more substantial impact without concurrent reforms in national and sub-national governance. These are needed to ensure an adequate institutional set up and linkages at various levels of governance. The reason is that local initiatives tend to prove limited when brought to scale. The limitations arise not only from poorly conceived policy frameworks, but from inadequate attention to inequalities of power and from fundamental deficits in accountability and representation. Unclear lines of authority and accountability often hinder effective devolution, and significant policy reforms are not translated into practice (***established but incomplete***).

Other failures of institutional effectiveness and implementation also persist. Cooperation among HKH country governments is hampered by limited cross-border knowledge sharing, as well as by recurrent geopolitical standoffs in the context of growing global significance of the region. Within the national level, governance systems characterize weak cross-scale political representation and insufficient attention to social equity and inclusion (***established but incomplete***). Another weakness is a continued disregard of scientific evidence which is itself limited in the context of HKH environmental systems. Although rich and abundant, local knowledge also remains underutilized in policy-making processes (***established but incomplete***).

Increasingly, a need is being recognized for a polycentric approach to HKH environmental resource governance: one that would emphasize new partnerships and cross-scale linkages with improved knowledge-sharing platforms for diverse stakeholders (***established but incomplete***). Also urgently needed are improved political articulations in the region that is undergoing rapid socioeconomic change and emerging geopolitical dynamics, and at the same time confronting serious environmental risks (***established but incomplete***).

While taking a broad view of environmental governance in the HKH, this chapter specifically examines three major resource sectors in the region: forest, water, and rangeland. These are the most prominent resources capturing a large part of the environmental governance space in the region-for example, 60% of the HKH region is under rangeland and 12 major river systems originate from the Himalayas. It documents the current state of governance in these sectors and describes emerging trends in environmental policy and practice. It identifies both gaps and opportunities in policy-making, in regulatory arrangements and enforceability, and in the adaptability of environmental governance to rising climate change impacts. Its central claim is that managing the resources of the HKH sustainably will depend largely on the emergence of regional and innovative institutional arrangements—to foster intergovernmental dialogue, to further common policy initiatives, and to enable collaborative trans-border community practices.

Environmental Governance and the Sustainable Development Goals

Promote a mountain-specific agenda for achieving the SDGs through increased regional cooperation among and between mountain regions and nations.

It is increasingly being recognized that international cooperation is essential for global or regional environmental governance. In HKH region such cooperation needs to be based on sharing of scientifically collected information, developing comprehensive policies and accepting innovative practices leading to sustainable and mutual benefits. There is a need to revitalize the regional partnership in decision making institutions and processes by prioritizing the uniqueness of the HKH mountains and its people and ensuring their representation in higher platforms of decision-making. It is also important to recognize the need for allocating sufficient resources for maintaining mountain ecosystems, and at the same time develop incentive mechanisms to encourage private as well as collective efforts towards conservation and sustainable use.

16.1 Exploring Changes in Environmental Governance in the HKH

This chapter focuses on assessing environmental governance and highlighting key trends and practices in policymaking, institutional dynamics, and resource management practices at local, national, sub-national, and regional levels. In Chap. 2

we have seen how environmental challenges cannot be addressed without strong and timely government interventions. It also elaborated on the consequences of poor environmental governance or a lack of it (see Sect. 2.3.3 for more details). The present chapter establishes a link between socioenvironmental change and the underlying structure of governance, which is itself changing over time. This link is important because, as we have seen in Chap. 4, researchers and practitioners consider poor governance as the third most important factor, after disasters and climate change, that poses the greatest threat to the prosperity of the region. The intention is not to make an exhaustive assessment of environmental governance in the HKH, but to identify and present specific forms of evidence that demonstrate the current state of governance and emerging trends in environmental policy and practice. In doing so, we identify gaps and opportunities related to policy making, regulatory arrangements and enforceability, and the overall adaptability of the governance system in the face of growing climate change impact in the region. We recognize that the term "environment" is broad, comprising the totality of the natural system, and we focus on three specific resource sectors—forest, water, and rangeland—to highlight key trends and practices in governance.

We take a broad view of environmental governance as an ensemble of state policies, formal and informal institutions, and practices related to the control and management of the environment for a variety of economic, cultural, ecological, and social values. In essence, environmental governance involves defining and enacting rules related to the use, control, and management of environmental resources. Such rules emerge either inside the formal system of the state (such as regulatory arrangements) or under traditional and indigenous arrangements of resource control. *Regulations* are defined as legally binding rules formed by the state, including constitutions, statutes, common laws, and governmental regulations, which are externally enforced (Bacho 2005).

We also consider policy-making processes and stakeholder engagement as essential components of environmental governance. *Policies* are primarily seen as statements of intent to guide actions in both the short and long term. *Institution* is used to include organizations and entities that are involved in the process of translating or otherwise modifying policy and regulatory arrangements in the practice of governance. Policies and institutions are closely linked to practices. They determine who is eligible to make decisions, which actions are allowed, which rules will be used, what procedures will be followed, what information must be produced, what payoffs will be assigned to individuals, and how outcomes and processes will be monitored (Ostrom 1991). We cover both formal (statutory) and informal (traditional) institutions in environmental governance. Formal institutions are generally linked to official, governmental, or bureaucratic formalities and are usually legally defined. At the local level, institutions also include those based on social norms and rules that are not formally coded (Leach et al. 1999). They include traditional authorities, indigenous groups (chiefs, clan heads, family heads, and others), and organizations, as defined in local societal norms, values, and beliefs.

Overall, governance processes and practices are conceptualized as multi-scalar phenomena; thus, the assessment presented here covers local, sub-national, national, and regional levels (Fig. 16.1). The assessment is based on secondary data and literature review, capturing the latest research while drawing on policy literature, while also drawing on the longstanding research of the lead and contributing authors. The evidence presented is not drawn equally from all countries of the region but is instead based on the availability of published sources. We did not aim for a balanced representation of countries or sectors; rather, we focused on presenting the diversity of situations and trends in environmental governance. We have used case studies as illustrations, not for comparative analysis.

The chapter is organized by scale—national, local, and regional—followed by an analysis of cross-scale interactions and how practices across scales are either linked or remain disjointed. Following this Introduction, the next section focuses on environmental governance at the national and sub-national levels by presenting a broad review of the status of policy and institutions governing natural resources and analyses policy implementation at this level. We indicate how policies, regulations, and institutions are changing over time and highlight important gaps related to the lack of coherence, deficits in implementation, and tendency to ignore evidence in policy-making processes.

Thereafter, we assess local level environmental governance, focusing on community-level institutions and local government. As we demonstrate, decentralized and community-based approaches to environmental management have emerged as a global agenda for advancing sustainable development, and considering this shift in global policy debate, we provide an assessment of this phenomenon in the HKH. We assess the extent and trends of devolution and decentralization of power to communities or local governments, drawing on indicative evidence on the outcomes of this important policy change in terms of livelihood security and ecological sustainability. We show that ambiguous lines of authority, representation, and accountability too often hinder effective decentralization and devolution in practice. We also explore opportunities for the HKH countries to strengthen local governance systems for natural resource management, to empower communities, and to involve local governments and authorities.

This is followed by an assessment of regional-level cooperation in environmental governance in the HKH. We assess the evolving mechanisms and processes through

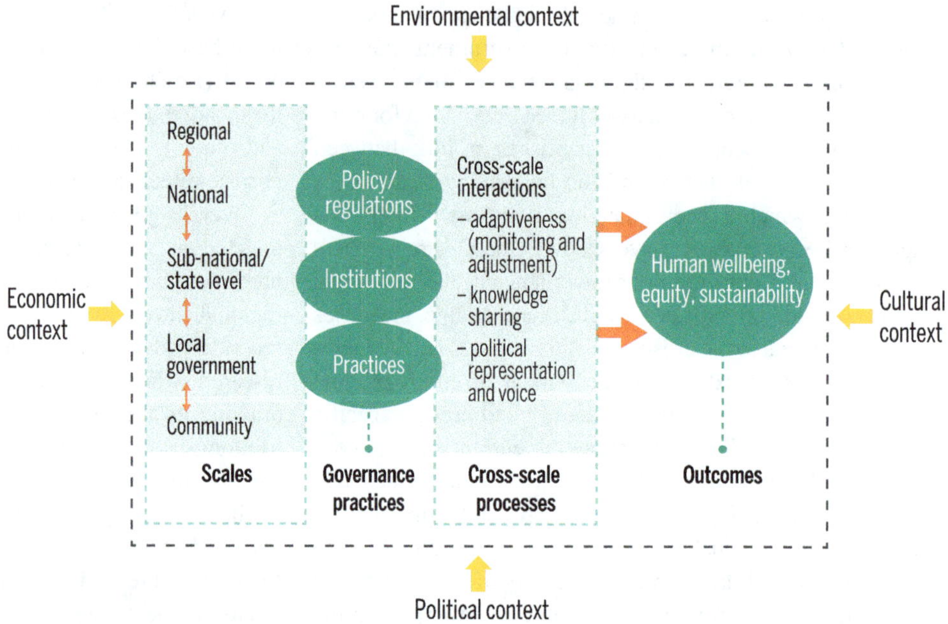

Fig. 16.1 Conceptual framework for multidimensional governance

which HKH countries engage in regional cooperation, and then identify their promise and gaps. Highlighting the context of globalization and increasing interconnectedness among the HKH countries, we demonstrate that the prospect for regional environmental cooperation has increased. Lastly, the chapter presents a cross-scalar analysis of governance focusing on knowledge interfaces, vertical interaction, social inclusion, and adaptive governance systems. Here, we explore gaps and opportunities for linking local, national, and regional processes for better vertical cooperation using the lens of polycentric and adaptive governance. We demonstrate that although environmental governance in the HKH is a cross-scalar practice, deliberate attempts to foster cross-scale learning and adaptation are far from adequate. Finally, we summarize the situation of environmental governance and the opportunities presented for improvement.

Overall, the chapter provides a strategic assessment of environmental governance in the HKH and identifies opportunities for improved cross-scalar governance in the context of the growing need for tackling the twin challenges of ongoing climate change and socioeconomic transformation in the region.

16.2 Predominance of National and Sub-national Policies and Institutions

Although there is increasing acknowledgement of the HKH as a region of interconnected transboundary landscapes, governance and policy processes remain primarily at the level of individual countries. Despite being at different levels of economic development and diverse in their political, administrative, and legal setup (Chettri et al. 2008), all the countries in the region have formulated policies and legal instruments to safeguard the environment and foster sustainable utilization of natural resources (see Table 16.1). Environmental governance, in particular, has gained momentum under the increasing authority and role of central governments in policy making and development planning since the 1950s.[1] From that point, and especially after the 1992 United Nations Conference on Environment and Development, countries started formulating policies concerning the environment and sustainable management of natural resources. The region has seen increasing policy reforms to regulate the environment and common pool resources (CPRs), particularly forests, rangelands, and water. Overall, there has been a gradual paradigm shift from a species- and habitat-focused approach to a participatory livelihood-based landscape approach to conservation policies and practices (Sharma et al. 2010).

16.2.1 Recent History of National and Sub-national HKH Environmental Policies and Institutions

Environmental policies and laws in the HKH countries initially evolved under central governments and were later expanded through donor support and NGO and community

[1]The first national Five-Year-Plans (FYPs) were implemented by India from 1951 to1956, China from 1953 to 1957, Pakistan from 1955 to 1960, Afghanistan and Nepal from 1956 to 1961, Bhutan from 1961 to 1966, Bangladesh from 1973 to 1978, and Myanmar from 2011/12 to 2015/16.

Table 16.1 Primary environmental and natural resource policies and laws in the HKH

Countries	Environment (General)	Forest	Rangeland	Water
Afghanistan	National Environmental Protection Agency (NEPA) 2005, Islamic Republic of Afghanistan, Environmental Law 2007	Policy and Strategy for the Forestry and Range Management Sub-sectors 2003	Rangeland Law 2012	Strategic Policy Framework for the Water Sector 2006
		Forest Law 2006		Water Law of Afghanistan 2009 (revised from 1991)
Bangladesh	Bangladesh Environment Conservation Act 2010; National Environment Policy 1992, and Bangladesh Environmental Conservation Rules 1997	Forest Act 1927 and Amendment	N.A.	Bangladesh Water Act 2013
Bhutan	National Environment Protection Act 2007	Forest and Nature Conservation Act 1995	Forest and Nature Conservation Act 1995	Water Policy 2008
		Biodiversity Act 2003	Land Act of Bhutan 2007	Water Act 2011
				Water Regulations 2014
China	Environmental Protection Law 1989, revised 2014, effective 1 January 2015	Forest Law 1998	Rangeland Law 2003	Water Law 1988
India	Environment Protection Act 1986	Forest (Conservation) Act 1980	N.A.	The Water (Prevention and Pollution Control) Act 1974
	National Environment Policy 2006			
Nepal	Environment Protection Act 1997	Forest Act 1993 Forest Regulations 1995	Rangeland Policy 2012	Water Resources Act 1992
Myanmar	Environmental Conservation Law 2012	Forest Law 1992 Forest Policy 1995 Forest Rules 1995	N.A.	Canal Act 1905
				Water Power Act 1927
				Groundwater Act 1930
				Conservation of Water Resources Law 2006
Pakistan	Environment Protection Act 1997	National Forest Policy 2015	Khyber Pakhtunkhwa Rangeland Policy 2014	Draft National Water Policy 2005
	National Environment Policy 2005			National Sanitation Policy 2006
				National Drinking Water Policy 2009

movements. All HKH countries have a long history of community dependence on and engagement in the management of natural resources, whereby the natural resources were controlled, appropriated, and managed through locally evolved norms, indigenous practices, and institutions (Fisher 1989; Shrestha 2016) (see Sect. 16.3). With the increasing involvement of the state in modern times, natural resources have increasingly been governed through policies and laws made centrally.

In some of the HKH countries, the emergence of national environmental policies and the expansion of institutions have been spurred in part by democratic changes. In Bangladesh, Bhutan, India, Nepal, Pakistan, and, recently, Myanmar, elected parliaments provide policy directions and make laws. In Afghanistan, presidential decrees and laws passed by the national parliament continue to be influenced by tribal or customary laws (Khan 2015). In China, the National People's Congress (NPC) and its Standing Committee are mandated bodies that approve policy directives and enact laws. Today, all HKH countries have a plethora of policies and legal instruments (see Table 16.1). Forest regulations were the first among the environmental laws to be made in the region. In India, the Forest Act 1865 that marked the beginning of state control of forest was formulated under the pre-partition British regime, primarily for the colonial interest in using the forests for industry (Baginski and Blaikie 2007; Muhammed et al. 2008). In Nepal, forests were nationalized in 1957, prior to which many areas were converted to farmland and timber extracted for export (Gautam et al. 2004). In Bhutan, state control of forests came with the enactment and revision of the Forest and Nature Conservation Act of 1969 (later revised in 1995) and the Land Act of 1979, which reduced the status of local communities from "owners" to "proprietors" (Dorji et al. 2006). All countries accord high priority to conservation of forests and biodiversity; 39% of the region's estimated 3,441,719 km^2 is designated as protected area (Chettri et al. 2008).

Box 16.1 Factors affecting the governance of natural resources in the Hindu Kush Himalaya

1. The remoteness of mountain communities and their distance from decision-making centres reduces coordination of formal and informal institutions.
2. Physical fragility has a high impact on natural resources, necessitating a mechanism for quick response, which local institutions may lack.
3. Being highly dependent on nature for subsistence, mountain communities do not respond well to restrictions on resource use, which are difficult to impose and monitor.
4. Ethnic diversity, resulting in the heterogeneity of cultures, traditions, and practices, can hinder consensus in decision making.
5. Lack of clear property rights over high-value niche products in mountains may lead to over-exploitation.
6. The transboundary quality of natural resources places limits on the effective monitoring, patrolling, and use of sanctions across borders.
7. The governance of resources that are transboundary in nature is limited by conflicting policies and institutions among the countries sharing the resources.

Alongside policies and regulations, the national institutional landscape has also grown over the past several decades. The institutional setup for regulation and implementation of the environmental policies and laws is centralized, with coordination authority vested in ministries, commissions, and departments. Water resources are regulated by councils or commissions in Afghanistan, Bhutan, and Nepal, and by ministries in China, India, Pakistan, and Myanmar. Forest resources in the region are regulated primarily by forest departments at the central, state, and provincial levels. There are no institutions solely dedicated to regulating rangelands. Except for China, which has a separate law, rangelands in most HKH countries come under the purview of the forest department.

16.2.2 The Recent Shift Toward Decentralization and Community Participation

Although there was an initial transfer of authority and control of natural resources, especially forests and rangelands, from traditional communities to central agencies, environmental and natural resource policies are now moving toward decentralization and devolution to communities of power and rights over natural resource access and management (Agrawal 2001). The HKH countries are progressively embracing good governance based on the principles of decentralization, gender-sensitive participation, and equitable access to resources. Governments are increasingly finding ways to promote local management of natural resources, especially forest resources, while ensuring central government control and authority.

According to Sharma et al. (2010), along with the integration of many global conventions, there has been a gradual paradigm shift in conservation policies and practices encompassing acceptance of local communities as an integral part of national conservation initiatives. This shift has been triggered in part by the acknowledgement of continued environmental degradation when resources were exclusively under centralized government authority. Forests and rangelands in all HKH countries were initially brought under state control through exploitative forest policies and laws primarily to meet state interests in exploiting forest to generate revenue and support industry (Muhammed et al. 2008). Responding to deterioration in the condition of forests, and acknowledging the increasing cost of managing forest without community involvement, governments have implemented processes, sometimes through donor support, that allow local community access and engagement in forest and rangeland conservation and management. This has come about in the form of social forestry in Bangladesh, community forestry in Nepal and Bhutan, joint forest management in Pakistan, and the Scheduled Tribes and Other Traditional Forest Dwellers (Recognition of Forest Rights) Act, 2006 in India (Baginski and Blaikie 2007). The move towards decentralization of authority under community forestry programmes is reported to have contributed immensely to the economic advancement of isolated rural communities in Nepal (Dahal and Chapagain 2005) and Bhutan (Namgyel 2005) (discussed in more detail in Sect. 16.3). In Bangladesh, for example, six major social forestry projects have together generated a total of 80.55 m man/days of work opportunities and raised some 97,584 ha of plantations between 1980 and 1999. FD office records suggest that Taka 700 million worth of products were harvested from social forestry plantations in about a decade, and Taka 5000 crore worth of assets generated (Chowdhury 2001; Mowla 2001; Khan et al. 2004).

The shift toward decentralized management is also apparent in the water sector. Water use and management in all HKH countries is governed specifically by water policies and acts initiated at different times that provide for state control over all sources of water (Table 16.1). But in recent times, governments have been taking the initiative to move toward decentralized management of water resources through the establishment and empowerment of river basin committees and water users' associations for the purpose of equitable benefit sharing. Since the first half of the 1990s, several programmatic efforts (notably, the Third and Fourth Fisheries Projects, Management of Aquatic Resources through Community Husbandry—MACH) have been made to engage local communities in water co-management in Bangladesh with varying degrees of success. Despite many limitations, these efforts have refocused attention on community engagement and documented the benefits arising from such engagements (USAID 2010; Halder and Thompson 2006).

The situation for rangelands is different. While China has a separate policy for rangelands, Nepal finalized its rangeland policy only in 2015. In the case of Pakistan, rangelands are under the jurisdiction of provincial governments, and the province of Khybar Pakhtunkhwa adopted a rangeland policy in 2014. In the rest of the HKH countries, rangeland laws are integrated into the provisions under forest and conservation laws.

While the trend toward decentralized management is welcome (see Box 16.1 for examples), it is now widely agreed that local institutions in the HKH countries lack coordination with the state and with policies at the national level. This lack of cross-cutting institutional coordination weakens the environmental governance processes within the countries of the region.

Box 16.2 Forests as a source of income

The majority of the countries in the HKH have evolved from being poverty-indifferent to being poverty-responsive, with policies that now take into account the needs of the poor by clearly defining their rights to resources. For example, Nepal's community forestry regulations allow community forest user groups to freely fix prices, market, and make profits, with 25% of the income from forestry invested for forest development (Ojha et al. 2009). From 2007 onward, it was made mandatory to invest 35% of income generated from community forests in pro-poor programmes and to include women in the key posts of the user group executive committees (Pokharel et al. 2012). In Bhutan, the Gross National Happiness Commission adopts a budget allocation formula that allows for resources to be allocated on the basis of population and poverty incidence. In India, policy and legislation provide quotas for employment in the government and other sectors (Venkateswarlu 2013). Further, the recent grant of authority to highland nomadic communities for the harvest of the medicinal *Cordyceps sinensis* has transformed their poverty and hardship to prosperity (Namgyel 2005).

16.2.3 The Emerging Focus on Multi-stakeholder Engagement

In global environmental governance reforms, stakeholder participation is widely acknowledged as an important aspect of policy making, development planning, and implementation of sustainable natural resource management (Larson and Soto 2008). This trend finds resonance in environmental governance in the HKH, and can be described as emerging hybrid modes of governance across the state-market-community. There are three distinct levels with two interfaces (Khan and Hossain 2006):

1. *international and regional level,* where the stakeholders are primarily international governing bodies, donor agencies, international NGOs, and governments;
2. *national level,* where the players are government and national-level NGOs, together with the private sector; and
3. *grass-roots level,* where the main actors are community-based organizations (CBOs), local NGOs, local governments, and development agencies.

Each of these levels has distinct functions and forms of partnership among the three types of social institutions, i.e., the state, private sector, and civil society. Coordination and collaboration between these operational levels occurs across two interfaces: *national–international* and *national–grass-roots* (Khan and Hossain 2006).

In the HKH, the public sector, communities, and private sector are emerging as partners, though not equal, in the governance of resources. For example, in Bhutan, sectoral policies are subjected to vigorous stakeholder consultations and a policy screening tool to ensure that the proposed policy upholds the values and principles of Gross National Happiness (GNH). Gross National Happiness is the fundamental development philosophy providing the framework for sectoral policies and initiatives. Essentially, it calls for a balanced and integrated approach to socioeconomic development, preservation of the environmental and cultural heritage, and promotion of good governance. In India, the media, judiciary, and civil society are playing increasingly vocal roles in advocating for mountain environments, and all development projects that affect natural resources require clearance from local panchayat institutions (these local institutions are for conflict resolution, rural development, and natural resource governance in many South Asian countries, including India, Bangladesh, Nepal, and Pakistan) (Ghosh 2008). Communities and advocacy groups also engage in public consultations over new government policies. In Nepal, stakeholder engagement has been expanding with increasing recognition of the private sector (Timsina and Gotame 2014).

16.2.4 The Role of Governance in Mainstreaming Climate Change in Development Plans

All countries in the HKH have heeded the United Nations call for mainstreaming environment into development planning and decision making in order to achieve the dual purpose of poverty reduction and environmental sustainability. A key aspect of mainstreaming environment is to orient governance, institutions, and political systems aimed at improving the environment in ways that also benefit the poor and reduce vulnerability to the effects of climate change (UNDP-UNEP 2009). Mainstreaming of environment and climate change through projects has begun in the region through the adoption of the Sustainable Development Goals and the ratification of the 2015 Paris Agreement and with support from international donors (UNFCCC 2015). However, the policies often adopt a linear process with discrete steps of policy formulation and implementation, without considering the governance challenges that exist (Friend et al. 2013). As a result, the process is often driven by global imperatives rather than through national-level political forums, as reported in the case of climate policy development in Nepal (Nightingale 2016; Ojha et al. 2016). While adaptation remains a priority, the HKH countries are also pursuing mitigation policies to reduce carbon emissions. Bangladesh, China, India, and Nepal were among the first globally to ratify the Paris Agreement.

Experiences in mainstreaming climate change into development plans reveal implementation challenges due to the disconnect between various tiers of governance. In Nepal, projects—often donor supported—have had a limited effect on the ground due to the highly centralized design and implementation process, engagement of only government stakeholders, limited time frames, and fragmented activities. Regmi and Star (2015) analysed how climate change is mainstreamed in Nepal in two case studies. The findings suggest that one of the governance challenges to mainstreaming of community-based adaptation into the development process is the gap in implementation approaches. Regmi and Star's findings show that bottom-up approaches, for example the Local Adaptation Plan of Action (LAPA), were successful in mobilizing local community groups and increasing their awareness. However, they failed to influence the government institutions, which resulted in a lack of support and continuity from the centre, highlighting the critical disconnect between policy and practice.

16.2.5 Identifying Barriers to Policy Implementation

The policies and laws for the management of natural resources in the HKH are fairly comprehensive including provisions for decentralization and the need for mainstreaming wider environmental concerns in development planning. Table 16.1 summarizes the major policies and legislation formulated for environmental protection in general, and forests, water, and rangeland, in particular. All eight countries have an overarching environmental protection act. The forest related acts are the oldest, and rangeland related acts and policies the most recent (2015 in Nepal). There have been conflicting views on having integrated policy and legislation for all-natural resources versus having separate policies and acts with a focus on specific resources.

The experience suggests that mainstreaming at policy level has been successful, but that gaps between policy formulation and implementation are common in all the HKH countries. A number of studies, as cited below, have noted barriers and challenges in implementing policies and legislation, including the lack of clarity and appropriateness of policies and the inadequacy of the regulatory and institutional arrangements. In China, the Environmental Protection Law of 2014, though perceived as the most progressive and stringent, was considered by some to be inadequate in terms of the overriding power of other laws such as the water law, as well as other factors, including the fragmented and overlapping structure of environmental governance; shielding of the government from lawsuits by citizens, civil groups, and NGOs for breach of environmental standards; and lack of capacity and conflicts of interest that may impede implementation (Zhang and Cao 2015). In Pakistan, policies were described as rhetoric, theoretical, and politics-driven, and were often associated with change of government (Shabaz et al. 2007). In Bangladesh, the policy response to forestry sector development was described as loaded with rhetoric (Khan and Harriss-White 2012). In countries like China and India, where mountain areas constitute only a fraction of the entire country, overall national policies do not adequately address mountain-specific needs (Sharma and Chettri 2005), and this disregard of mountain specificities reduces options for communities, rendering them vulnerable to environmental and economic distresses (Jodha 2005). In Nepal, one of the key factors thought to be hindering successful implementation of pro-poor community forestry was that both projects and government line agencies had historically been compartmentalized and were restricted in the range of services they could offer to user groups (Pokharel et al. 2007).

Another aspect of the policy implementation gap is the donor dependence of the developing economies in the region. Countries like Afghanistan, Bangladesh, Bhutan, Myanmar, Nepal, and Pakistan are still dependent on external assistance for their development needs. Donors are assisting these countries in developing exhaustive policies and legal instruments, but they are not adequately followed through for a number of reasons related to relevance and inadequate institutional, human, and financial capacity.

16.2.6 Gaps in Building Adequate and Responsive Institutions

Administrative agencies form the fundamental institutional setting for policy implementation and thereby influence outcomes (Maynard-Moody and Herbert 1989). Inappropriate institutional settings and lack of strategies and actions for proper implementation can be as much a cause for policy-related failures as inappropriately contextualized policies (Peters 2015; Zittoun 2015). Faulty implementation can also be the result of the inadequacy of the administrative agencies represented by the public institutions, or their ineffectiveness in delivering public services and implementing programmes. Institutional weaknesses have been identified as the direct result of bureaucratic political weaknesses and non-accountability in South Asian states like Afghanistan, Bangladesh, India, Nepal, and Pakistan (Robbins 2000; Wirsing 2007; Barnes and Laerhoven 2014; Fleischman 2014).

Limited institutional capacity has affected the natural resource sector in many HKH countries, and this in turn has compromised the implementation of policies and the provision of timely and useful feedback to policy development. In Nepal, the functions of government agencies, particularly sectoral organizations such as the district forest offices (DFOs) and the district livestock service offices (DLSOs), have often been influenced by the availability of programme funds. While institutions supporting two related programmes —community forestry and leasehold forestry—lacked coordination to enhance overall effectiveness (Bhattarai et al. 2007; Ojha 2014). In Bangladesh, several managerial and technical problems have hindered policy and programme implementation in forest management. The problems include lack of skilled manpower, minimal capital investment, and lack of relevant expertise for translating international forest policies into national forestry plans at the national and regional levels (Muhammed et al. 2008). In Afghanistan, centralized state institutions have co-existed uneasily with fragmented, decentralized traditional society since attempts at state-building began (See Box 16.2).

While rangeland institutions are well-developed in countries like China and Bhutan, they have only recently started to take shape in India and Nepal. In Nepal, the policies and governance necessary for promoting rangeland management have only been developed recently (Ministry of Livestock Development 2017). Previously, public services and technical support were either unavailable or inaccessible to local herders in the mountainous rangelands of northern Nepal, and research, development, and provision of extension services were lacking (Dong et al. 2009).

Similarly, institutions for water management in many parts of the region lack the capacity to design and deliver effective management functions. For example, Pakistan lacks investment in storage capacity, water-use efficiency, and sustainable management of surface water and groundwater resources to avoid problems of soil salinization and waterlogging (Watto and Mugera 2016). In Afghanistan, institutions do not have access to suitable or sufficient information and data for planning, and there is a shortage of efficient institutions, organizational capabilities of staff, and effective rules and regulations regarding water use, in part due to persistent wars. In Bhutan, lack of human resources and capacity remains a major constraint on effective inter-agency coordination for implementing integrated water resource management required by the Water Act. Mahmoodi (2008) found that only 30% of the country's potential water resources were being used, while the rest remained inaccessible. The current status of policy and institutions for upstream-downstream linkages is also limited in the HKH, although the importance of these linkages is increasingly recognized in recent research on transboundary water governance (Pigram 2000; Nepal et al. 2014; Rasul 2014; Shrestha and Ghate 2016).

Box 16.3 Changing institutional structure due to conflicts

For centuries, tribal and religious leaders in Afghanistan have created "microsocieties", which are related to central and other powers on the basis of negotiation and patronage. However, the years of conflict and civil war have changed the nature of politics both at the local level and between the local and national levels.

The co-existence broke down as power became highly decentralized and factional leaders, operating in relatively distinct geographic areas, organized loose alliances to gain control of, or resist, the centre. Hierarchies of commanders, so-called "warlords", came to dominate large areas, linked in some areas to tribal structures. Structures of power are dynamic, however, and evidence from Afghanistan suggests that they continue to change, and the complex ways that power holders interact with the state institutions continue to be modified (Lister 2007).

The war in Afghanistan has also caused disintegration of the traditional arrangements of rangeland institutions and led to the lack of an enabling policy environment for sustainable rangeland use. Further, the lack of coherent legislation on land rights has generated conflicts between nomads and sedentary farmers due to conflicts of interest between winter grazing and crop cultivation. In some cases, conflicts have arisen between government authorities and local communities, since government-owned land was not distinguished from publicly-owned land, and common resources were taken away from communities by those in power (Ali and Shaoliang 2013).

The lack of institutional connectivity in the region has also hindered the ability of nation states to respond to crises. These constraints severely affected the relief and rehabilitation programmes after the Kedarnath disaster in the Uttarakhand Himalaya in June 2013, in which several thousand people died and the livelihoods and assets of a large proportion of the regional population were devastated (Tiwari and Joshi 2015). At present, some HKH countries are renewing their approach toward local government structures; for example, in Afghanistan they are linking village institutions, government ministries, and international donors (Noelle-Karimi 2006).

16.3 Empowering Local Institutions Through Decentralization and Devolution

In practice, environmental governance in the HKH is a profoundly local phenomenon, involving local communities, local governments, and increasingly local and small-scale business groups. Over the centuries, a variety of community-based natural resource management systems evolved in the region, based on traditional norms and indigenous institutions (Denholm 1991; Gilmour 1990; Farooquee et al. 2004; Sharma et al. 2006; Kreutzmann 2012). In Nepal, Farmer Managed Irrigation Systems (FMIS) are known to exist going back centuries, and these continue to irrigate almost 70% of farmland in the country. (For more details, see Chap. 8.) However, since the 1950s, almost all the HKH countries have seen increased state engagement in policy formulation, enactment, and enforcement of legislation and implementation of state-sponsored programmes (Blaikie and Sadeque 2000). The era of direct state control over environmental resources led to a crisis of environmental degradation in the late 1980s (Ives and Messerli 1989), prompting national governments to adopt policies for participatory management, joint management, and community management, particularly for forest resources (Hobley 1996). Alongside the age-old tradition of community-based environmental management systems, all eight countries in the region have undertaken formal decentralization and devolution reforms, especially after 1990, transferring some power of governance to local communities and elected local governments (Baginski and Blaikie 2007; Pasakhala et al. 2017).

Globally, the move toward devolution and decentralization has been driven by multiple forces, including the loss of legitimacy and/or credibility of the centralized state (Bardhan 2002). This has led to demands for deregulation and an increased role for market players (Mohan 1996), escalating concerns about poverty reduction (Crook 2003), increasing awareness about the need for environmental conservation (Agrawal 2001), and the growing consensus that local stakeholder participation is required to achieve sustainability goals, which implies a need to improve governance through community involvement, engagement, and ownership (Fung and Wright 2001; Ribot 2003, 2007).

In the HKH, governments in collaboration with donors have invested heavily in strengthening local and subnational systems and developing new institutions, often reviving the indigenous institutions. Both customary and formal community management institutions, "facilitated" by state and development agencies, are important elements in the local governance of environmental resources, despite the diverse historical trajectories of environmental governance across the region (Mosse 1997; Agrawal and Gibson 1999; Blaikie 2006). For example, Nepal has a strongly decentralized community forestry system which blends traditional institutions and state-sponsored decentralization reform (Gilmour and Fisher 1991; Ojha 2014). On the other hand, published literature indicates that the efforts of establishing a strong centralized state have not been complemented by sufficient attention to local government at the provincial and district level. As a result, governance at the local level is very complex and confused in Afghanistan. The Indian Himalaya have a local government-centric forest management system which engages local people while retaining control within the government (Hobley 1996; Agrawal and Yadama 1997; Sivaramakrishnan 1999), and the Chinese Himalaya have also seen increasing trends towards decentralization, which is akin to de-concentration (Larson and Ribot 2004). In contrast, in Pakistan, the authority of the forest department has barely been decentralized, despite the adoption of a participatory or joint forest management strategy (Shahbaz et al. 2007).

The degree of decentralization also varies across the resource sectors within a country. For example, in contrast to forest and biodiversity management, the governance of water resources remains predominantly a national and sub-national issue in most HKH countries. On the other hand, rangeland (which covers 60% of the entire HKH) is primarily under community management even when formal rights are retained by the government (Sharma et al. 2007). While these examples suggest a common trend in transferring power from higher to lower levels of governance, there is no coherent body of evidence from the region to suggest that decentralization always has a positive outcome for people and the environment (Jütting et al. 2004). Rather, the impact varies in line with the geographic and political diversity in and within the individual countries.

Many of the decentralization and devolution policies are sponsored by aid agencies. International support has been crucial in providing resources for decentralization projects and, in particular, in promoting mechanisms to enhance gender equality and social inclusion. Nevertheless, donor influence in decentralized governance has not always produced sustainable change and innovation (Guthman 1997; Sarin 2001, cited in Sarkar 2008; Rangan 1996; Nautiyal 2011). In Bangladesh, the imposition of a generic co-management "model" has caused local-level problems in negotiating popular acceptance of the programme and long-term institutional sustainability. In Pakistan, decentralization and collaborative governance agendas are driven by international agencies and the national administration, causing tensions with more local-level forest departments in the north of the country (Ali and Nyborg 2010).

Furthermore, the effectiveness of local institutions is shaped by external factors (Ojha et al. 2016a). For example, *van panchayats* in Uttarakhand in India are facing tremendous demographic and economic pressure, which is undermining the effectiveness of these institutions (Sarkar 2008). On the other hand, it is yet to be ascertained if the Biodiversity Management Committees (BMC) in the state have as much acceptability among communities as *van panchayats* or *gram panchayats* that have statutory mandate similar to BMC. The increasing professionalization of resource governance is leading to further democratic deficits. Policies become stronger in terms of technical considerations but lack the flexibility required for successful implementation in real life situations. Consequently, stakeholders often find themselves in a situation where state policies do not address their interests, leading to increased conflicts over the proper governance of local-level resources (Nightingale 2005). Therefore, community-based natural resource management is not only about community-level decision makers; it also requires inputs from government actors and agencies, and sometimes the involvement of market actors where forest products can be exchanged for cash (Agrawal and Verma 2017).

16.3.1 The Status of Local Informal Institutions Under Increasing Decentralization

Growing socio-anthropological research clearly states that communities possess unique local information which distant state agencies generally do not possess, and which may help solve complex environmental problems (Ostrom 1990, 2010). Community-based institutions are vital in the enforcement of shared norms of behaviour and in resolving conflicts (North 1990; Ghate and Chaturvedi 2016). Moreover, the formal involvement of community members is likely to enable a more equitable allocation of benefits from environmental resources because community members have a better understanding of their needs and can pressure state officials to distribute benefits more equitably (Agrawal and Lemos 2007). Chapter 8 indicates how traditional institutions are marked with gender inequality when it comes to access to water.

In Afghanistan, the new Water Sector Policy has adopted principles for implementing integrated water resource management (IWRM) and decentralizing activities gradually to the river-basin and sub-basin levels. According to Mahmoodi (2008), programmes have been developed to achieve the strategic goals outlined in the policy in terms of institutional development, capacity building, poverty alleviation, modernizing irrigation systems, prevention of water loss, water resource development, expanding rural water supply, improving sanitation system, and river bank protection. The government now prioritizes water and power development through IWRM, and water has been heralded as the means to achieve other economic and developmental goals.

Box 16.4 Community Forestry in Nepal: An example of successful community-based environmental management

Nepal's community forestry has a formal history of more than 40 years; it was first triggered by the Himalayan degradation perceived in the 1970s (Ives and Messerli 1989). With more than one-third of the population involved, and over a dozen international and bilateral agencies supporting the process through at least four different regimes, community forestry is now deemed successful in terms of scale of activity, length of experience, diversity of local management regimes, and richness in facilitative experience in devising ways to achieve both poverty reduction and conservation goals (Acharya 2002; Ojha et al. 2009). However, gender mainstreaming in forest governance and policy remains far from desirable (for details, see Chap. 8).

16.3.2 The Effectiveness of Formal Community-Based and Decentralized Environmental Management

Assessing the effectiveness of decentralized programmes in the HKH is difficult, as indicated by the challenges outlined in the previous section. However, some key insights emerge, even if the picture is mixed. The most notable case of community-based resource management is Nepal's community forestry, which started in the late 70s in the hills of Nepal (Gilmour and Fisher 1991). This programme is considered successful in (a) bringing local communities back into the management process by recognizing their rights over forests; (b) fostering institutional innovation based on community or participatory forest management mechanisms; (c) transforming the rules of resource management to favour forest sustainability; (d) developing a self-sustaining system by broadening the range of actors to undertake diverse functions; and (e) balancing discursive and practical aspects of the democratic movement (Pokharel et al. 2007). These categories are useful in terms of assessing community forestry in other parts of the HKH, but the institutional and policy diversity across the region makes direct comparisons difficult.

Empirical evidence suggests Nepal's community forestry has been a success in terms of enhancing the flow of forest products, improving livelihood opportunities for forest-dependent people (Pokharel and Nurse 2004; Subedi 2006; Chapagain and Banjade 2009), transforming institutions and social capital (Nightingale 2005, 2006; Bhattarai et al. 2009), and improving the ecological condition of forests (Gautam et al. 2002). The success of community forestry in Nepal served as a major impetus for Bhutan, Myanmar, and Pakistan to initiate community forestry. However, the effectiveness of community forestry in these new contexts has been mixed, in part due to differences in the decentralized policies.

Efforts at the decentralized management of other resources also suffer from institutional diversity, overlaps with traditional systems, and lack of integration with the government departments that control the same resource under centralized mechanisms. The Indian Himalaya have a wide range of local governance systems across forest, water, and rangelands, with marked differences between the eastern and western sections. They include van panchayats in Uttarakhand and Himachal Pradesh, traditional forest management institutions in North East India, and joint forest management across India. These programmes have been only partially successful, and have been criticized in particular for failing to protect the rights of tribal groups. The Scheduled Tribes and Other Traditional Forest Dwellers (Recognition of Forest Rights) Act 2006 was introduced in an attempt to decentralize forest resources to these groups, but its implementation has been resisted by government departments at various levels (Sarker 2011; Kumar et al. 2015). Since the 1990s, several efforts have been made in the water management sector in India, focusing on the role of local communities (Agarwal and Narain 2002). These efforts involve forming user groups called water users' associations (WUAs) for irrigation systems in an attempt to ensure more effective distribution of limited irrigation water.

Nepal has developed water governance mechanisms at the local level, but at the same time, the governance of watersheds and rivers is not systematic; moreover, it is managed mostly centrally. At the local level, most places have systems of *kulos*, or canals, for irrigation, and in some cases for drinking water, that are collectively dug and mutually maintained. In many places, *kulo* systems stretch for dozens of kilometres. Many of these systems are very old, while others are new innovations spurred on by development projects and government provision of drinking water systems. *Kulos* tend to have "invisible" management institutions in that they lack formal written rules or regular meetings. Rather, they have established mechanisms for getting all user households to contribute labour for maintenance of the canals, and meetings only take place when conflicts arise or the system breaks down.

In Pakistan, a decentralized local government system was introduced in 2001 and most of the provincial departments were handed over to local (district) governments. Forest departments, however, were among the few departments that were not devolved, even though participatory or joint forest management and social forestry had already been initiated in Khyber Pakhtunkhwa (KPK) (Shabaz et al. 2007). A study based in Baltistan and Kargil suggests that interventions by government and non-government agencies alter water-users' institutional arrangements, and that such interventions should build on, rather than erode, existing arrangements (Hill 2014).

In Myanmar, the Forest Department issued the Community Forestry Instructions (CFIs) in 1995 and has been encouraging the development of community forestry in collaboration with international donor organizations. According to the data for 2015 from the Forest Department Statistics Division (available at its website), a total of 2,023 community forest user groups with 54,041 members have established 113,016 ha (279,268 acres) of community forest plantations. Although the Forest Department is trying to revise the existing 1995 CFI in consultation with multi-stakeholders, the degree of devolution to local communities is limited. Decentralization was mandated by the 2008 Constitution of Myanmar (GOM 2008), but has been met by challenges at both the local and central level. Myanmar is a highly centralized country with a history of subnational conflicts and relatively weak local institutions and communities, all of which has hampered the decentralization effort (Nixon et al. 2013).

In Bangladesh, since the early 2000s, there has been an organized attempt toward the consolidation of decentralized "community-focused" governance, especially in the forest and wetlands sectors. These programmes have mainly been dispensed under the broad rubric of "co-management" targeting the country's protected areas (PAs). The adoption of this co-management approach is seen as an attempt to influence the governance process toward a transformation from the conventional custodian system to a more participatory management regime. While an encouraging trend, problems in implementation abound (Jashimuddin and Inoue 2012).

16.3.3 The Insecurity of Local Resource Rights Despite Increasing Decentralization

Resource rights and tenure have been identified as a critical aspect of decentralized natural resource management that is needed to ensure that users feel ownership in the long-term sustainability of their resources (Acharya et al. 2008; Paudel et al. 2009; Larson 2010, 2011; Larson et al. 2013). Yet regularizing property and tenure rights is never straightforward. Decentralization often ignores traditional institutions and crafts new ones, leading to longer-term conflicts both within and between communities. At present, a variety of tenurial arrangements are evolving across the region with varying degrees of effectiveness.

In Pakistan, forest management at the local institutional level is represented by the joint forest management committees (JFMCs); while at the provincial level, the Forest Department of Khyber Pakhtunkhwa gives the power to divisional forest officers to assign state forest to a JFMC. The stated function of the JFMCs is to protect and manage their piece of forest. However, there is a huge gap between the *de jure* and de facto status. The *de jure* status of JFMCs is democratic, but many research studies have argued that most JFMCs have been created exclusively for timber harvesting rather than forest management and/or protection (Shahbaz 2009).

In China, there is a high degree of devolution underway, as power is shifted from the state to village collectives and individual ownership of forests (Xu 2010). However, despite a significant decentralization of forest rights, farmers feel that their property rights are still insecure (Qin et al. 2011). Likewise, the arrangements for payment for ecosystem services in China are notable for governance innovation in terms of public-private partnership, but confusion still persists about the role of the private sector (Zhen and Zhang 2011).

In India, institutional arrangements for environmental services have often been made without an adequate understanding of the local politics of negotiation, leading to a number of conflicts and institutional breakdown. In North East India, very few people within the local communities actually own the resources; thus, any new attempts at tenure reform encounter strong resistance. In western India, local-level water users' associations (WUAs) were hijacked by the local elite, who were able to evade accountability because of their location in village networks (Narain 2003). Within the larger policy discourse on irrigation reform, WUAs came to be seen as some kind of panacea for the ills of the irrigation sector, but there was little effort to integrate these measures with other aspects of irrigation reform, such as pricing, cost recovery, or improvement in water supplies. Thus, WUA formation remained something of a piecemeal strategy carried out at the persuasion of donors and funders, with little reflection on what changes were being sought at the field level.

In Pakistan, the conservation focus in decentralized forest management has led to only limited benefit for livelihoods from participatory forestry, partly because there is still limited devolution of power and a failure to address pre-existing structures such as the traditional systems of *riwaj* and *jirga*, which exclude women and marginalized people (Shahbaz et al. 2011).

16.3.4 Emerging Institutional Ambiguity in the Shift Toward Decentralization

Despite the adoption by the HKH countries of a broad policy vision of decentralized environmental governance, the implementation of this policy is characterized by contradictions and lack of coherence, a situation which at times is akin to "giving with one hand while taking with the other" (Capistrano 2012: 210). The evidence shows that in practice, the decentralization processes across the HKH are not smooth as a result of many challenges, including deep-rooted centralized governance, sub-national conflicts, and the limited capacity of local institutions (Nixon et al. 2013).

In China, confusion over the direction of policy has led to either inappropriate policies or poor implementation of existing policies—as, for example, in the case of rangeland policy (Gongbuzeren et al. 2015). In Bhutan, central agencies continue to dominate control and allocation of natural resources. Provision of forest resources such as timber and sand for rural community needs are regulated by the Department of Forest and Park Services through its divisional offices at the district level. Urban demands for timber and sand are regulated and rationed through the government-owned Natural Resource Development Corporation Limited (NRDCL). In Myanmar, policy has shifted toward decentralization and devolution, but as in most places, government departments still compete in actual implementation, undermining policy goals. For example, Section 17(a) of the National Constitution states that "the executive power of the Union is shared among the Pyidaungsu, Regions, and States; Self-Administrative power shall be shared between Self-Administered Areas as prescribed by this Constitution". In terms of forest resource management, the Community Forestry Institution (1995) allows localities to establish community forestry in areas adjacent to or currently relied on by the local communities. However, the Forest Department does not allow local communities to implement community forestry within protected areas, and it grants permission only in protected public forests and reserve forests. Thus, in terms of power sharing, local communities are not empowered to manage their forest resources, even though they were granted the right to forest tenure 30 years ago.

In the highlands of Pakistan, forest departments were not decentralized during the 2001 decentralization reforms, although the district governments were given the mandate to take care of the interests of local communities who depend on forest resources for their livelihoods. These ambiguous lines of authority led to communication gaps and eventually poor interactions between the stakeholders in forest management interventions (Shabaz et al. 2007; Watto et al. 2010).

In Nepal, there is continued confusion between the roles of communities and local government in the local-level governance of natural resources, as the Forest Act 1993 and Local Self Governance Act 1999 created authority overlaps among forest officers, local governments, and community groups. Six different regimes of decentralized and/or collaborative governance have emerged, partly because of the need to accommodate the diverse and contested interests of communities, state agencies, and other players. Due to unclear laws and overlapping functions, the central government continues in its role of providing public services like health, education, and drinking water through its line agencies. Such functions overlap to a considerable extent between different levels of government and are ambiguous in terms of which level is responsible for regulation, financing, and implementation (Gautam and Pokhrel 2011).

In Bhutan, there is a tendency to restrict legislative rights through bylaws or directives that undermine the original objectives of decentralized policies (Chhetri et al. 2009). For example, the Bhutan Observer reported research results in November 2012 which showed that contributions from community forestry account for only 0.3% of household income because of the restrictions on the use of forest resources.

In Afghanistan, efforts to rehabilitate depleted watersheds are constrained by poor security, uncertain land tenure and use rights, limited local environmental management capacity and focus on agriculture to meet immediate local economic needs (Groninger 2012). Environmental management is embroiled in cultural, transnational, military, and developmental worldviews and institutions that leave little room for the perspectives of local people.

16.3.5 The Potential and Limitations of Local Institutions in Confronting Social, Environmental, and Political Challenges

Evaluating the outcomes of devolution is complicated by the kinds of diversity in policy and practice and the challenges outlined above. Nevertheless, several achievements can be identified which provide examples of good practice.

Nepal's community forestry has been particularly strong in attempting to address the well-known problems of elite capture and exclusion of women and marginalized people. In terms of gender and equity, many gains have been made. Innovative practices include the following (Acharya 2004; Hobley and Jha 2012):

- Use of quotas on management boards and within organizations and user groups to try to ensure the inclusion of socially marginalized people. In Nepal, as more widely in the HKH, this most often means quotas for women, Dalits, and indigenous people. While these provisions may not safeguard the rights of marginalized people, they do provide legal and normative grounds for such people to demand a greater stake in decision-making processes.
- The promotion of marginalized community members into leadership roles within organizations. This is sometimes done by quotas and sometimes promoted by donors as a desirable practice. We see more of this practice within community-level institutions than at other levels.
- The formation of community-level management groups that are restricted to women or Dalit or indigenous members. These groups are intended to help overcome some of the broad discrimination that all members of these groups feel. Groups composed of marginalized people often find it difficult to retain control over their resources, however, and face greater challenges in negotiating with powerful government actors (Nightingale 2006).

Alongside these optimistic trends, growing evidence indicates that the success of Nepal community forestry is mixed in terms of both livelihoods (Malla 2000, 2001;

Acharya 2002, 2004; Thoms 2008; Shrestha 2016) and ecological systems (Acharya 2004). Multiple processes are shaping these issues: agricultural intensification in localities close to urban areas (Raut et al. 2010), increased market demand for timber and forest products without effective evolution of local institutions (Pandit and Thapa 2004), outmigration, and climate change. Indeed, community forestry is not free from issues related to distributional injustice, recentralization, and ineffective use of forest resources under community management (Ojha 2006; Thoms 2008; Shrestha 2012; Nightingale 2016). In particular, local-level institutions face challenges related to elite capture and what we might call "participatory exclusion", whereby marginalized people might be given a place within a group and even a voice, but in practice they are unable to influence group decisions (Nightingale 2005, 2006; Ojha 2006; Shrestha 2012).

In other parts of the HKH, similar attempts have been made to address social inequalities and tenure insecurity for marginalized people, again with mixed results. The achievements of the community forestry programmes in Bangladesh over the past three decades include efforts to include women and disadvantaged groups, reforestation on marginal lands, and improvements to degraded forest and community lands (Jashimuddin and Inoue 2012). Recent research suggests that (1) despite its limitations, co-management as a concept has gradually taken root in Bangladesh; (2) the degree and level of active community participation remains low and limited, and there is a clear weakness in orienting local communities to the key legal and policy issues related to protected area governance in the country; (3) the key drivers of success in co-management are poverty alleviation through livelihood creation, capacity building, equity in benefit sharing, recognition of tenure rights, and shared governance that involves the devolution of power to transparent participatory local institutions; and (4) in many instances, local site-specific needs and demands have not been adequately addressed.

In Myanmar, the establishment of community forests supports a wide range of benefits for local communities in terms of their basic needs (poles/posts, fuelwood, other non-timber forest products) and ecosystem services (recreation, mitigating climate change, conserving natural water springs for drinking water and other uses), as well as developing social capital at the community level (Tint et al. 2011). The Forest Law 1992 (GOM 1992), Forest Policy 1995 (GOM 1995a), and CFI 1995 (GOM 1995b) barely recognize gender and equity issues in managing and utilizing natural resources, and the policy emphasis is still on revenue/economic benefit and managing environmental sustainability. More recently, however, the Land Use Policy (2016) has recognized the rights and benefits of local communities, particularly for women, marginalized people, and ethnic minorities.

Analysis of the forests of Himachal Pradesh suggests that the tension represented by the co-governance management regime contrasted with the indigenous system is associated with a worsening condition of the forests (Agrawal and Chhatre 2006).

Overall, despite its challenges, decentralization has helped to legitimize local-level institutions across the HKH, both nationally and internationally. The formal processes of decentralizing environmental resource management are underpinned by changes in the constitutional, legal, policy, and procedural arrangements in all the countries. However, decentralization faces many challenges in terms of acceptance at different levels of governance within countries and the ability of established institutions to confront social, environmental, and political change.

In recent years, pre-existing ("traditional") institutions have either been replaced or modified through the introduction of formal institutions. For example, in Afghanistan, despite strong community-based arrangements, attempts have been made by national and international organizations to modernize the irrigation system using centralized approaches. While the indigenous technology did not allow the measurement of water consumption, the externally-funded projects for modernization of water management which can do that are seen as potentially violating the local water governance systems and causing conflict (Wegerich 2009). In northern China, participatory water users' associations (WUAs) are replacing traditional irrigation water management systems in order to promote economically and ecologically beneficial water management options (Zhang et al. 2013). In Bhutan, the enactment of forest laws has led to nationalization of forests and loss of traditional rights over forest, resulting in diminished incentives to protect (Dorji et al. 2006) and non-compliant behavior at the local level (Webb and Dorji 2007). The shifting (devolution) of roles from formal to informal institutions is best exemplified in the forestry sector in Nepal. Forest policy and regulations related to community forestry allow the government to hand over the whole or parts of a national forest area to the local community (see Box 16.3).

Studies show that traditional institutions have been crucial in resource conservation and local livelihoods, but problems of inequity and elite domination also abound. In many parts of Khyber Pakhtunkhwa in northwestern Pakistan, decisions regarding access to forest resources and other socioeconomic aspects are rooted in sociocultural mechanisms called *riwaj* or customary law and *jirga* (assembly of tribal elders) (Rome 2005). These institutions have been criticized by civil society groups for failing to safeguard the rights of women and other marginalized community members (Shahbaz and Ali 2003). In Nepal, in situations where the formal mandate for newly decentralized institutions has lapsed, local elites tend to fill the vacuum of control, most often to the detriment of the rights of women and marginalized resource users (Nightingale 2006; Nightingale and Sharma 2014). It is therefore crucial that both the potential and the limitations of local-level institutions are considered when promoting various forms of decentralization.

16.4 Regional Cooperation Is Key for Environmental Governance in the HKH

Regional or transnational cooperation has emerged as a key form of global environmental governance, as opportunities for international cooperation and transboundary environmental management expand rapidly. Although the Himalaya is probably one of the most politically complex and ecologically fragile regions in the world (Ives 2012), it is also an example of the potential for provision of a range of environmental systems and services (Grumbine and Pandit 2013; Blaikie and Muldavin 2004). With the rise of the Chinese and Indian economies—located to the north and south of the mountain chain—the region has also become a geopolitical and political economic centre of global importance (Wirsing 2013; Zhang 2016), simultaneously posing both ecological and political opportunities and challenges (Goldstein et al. 2006). The increasing level of upstream-downstream interrelationships, such as those pertaining to transboundary water systems, also demands greater international cooperation in the region. As a result, achieving sustainable natural resource management in the HKH will require a reasonable degree of cooperation among national governments and communities across borders (Karki and Gurung 2012), as well as large-scale investments for prosperity in the future (Chap. 4). The need for regional collaboration is most vividly demonstrated by the case of water management.

Water "ignores political boundaries, fluctuates in both space and time, and has multiple and conflicting demands on its use", and its international law is poorly developed, contradictory, and unenforceable (Wolf and Hamner 2000: 123). With ten river systems flowing across the boundaries of the HKH, bilateral and multilateral treaties and agreements for the allocation and distribution of water will be indispensable in governing the ecosystems at a regional scale. Several such agreements have been signed over the past five decades (Table 16.2). Some bilateral programmes (for example, the Koshi River Watershed Management Programme between Nepal and India) have been implemented since the 1980s and 1990s with the aim of regulating the water supply and minimizing flooding. Afghanistan and Pakistan are in the process of reaching a similar water-sharing agreement for the Kunar River in the Kabul Basin (Vick 2014). Others, such as the Koshi River Agreement, the Gandak River Water Sharing Agreement, and the Mahakali River Agreement between Nepal and India (Uprety and Salman 2011), the Ganges Water Treaty between India and Bangladesh (Brichieri-Colombi and Bradnock 2003), and the Indus River Water Treaty between India and Pakistan (Sahni 2006) aim primarily at ensuring the generation and sharing of power, and are often considered to lack fair, comprehensive, and equitable arrangements for resource use and allocation (Mustafa 2010; Uprety and Salman 2011; Butler 2016). Formal cooperation between China and India on transboundary water management remains limited (Rahaman and Varis 2009).

16.4.1 The Emerging Emphasis on Transboundary Cooperation for Regional Wellbeing

The practices of cooperation among the HKH countries are being driven increasingly by economic interests or political bargaining. The economic interests of the larger economies are competing for the extraction of natural resources, at times leading to a deterioration in the livelihoods and ecological integrity of smaller countries and communities. For example, upstream communities are being displaced in the process of construction of large-scale river dams to serve downstream populations (Blaikie and Muldavin 2004). The cross-border power trade has been growing and is one of the main contested issues, especially from the perspective of smaller countries (Crow and Singh 2000; Rest 2012). India has established the power trading companies specifically to develop and trade hydropower in the region (Karki and Vaidhya 2009). However, power trading agreements are not promoted through regional cooperation; rather, they are dealt with primarily at the bilateral level. The rationale is that the huge energy market in India can provide a source of income to countries such as Nepal and Bhutan and eventually promote environmental and conservation efforts (Biswas 2011). In most cases of transboundary cooperation, however, the regional economic powers such as China and India are often unable to reach agreements with the smaller countries (which in fact hold the larger section of the HKH) that consider the wider social, cultural, and environmental interests of the smaller countries and the upstream communities (Blaikie and Muldavin 2004; Mustafa 2010).

Subnational-level collaborations along the borders are becoming effective, especially in dealing with issues related to transborder wildlife movements, grazing management, soil and water management, and maintaining ecological hot spots along the transboundary rivers. Even though these practices are ad hoc and irregular, they are a good example of how transboundary natural resource management could be implemented by mobilizing local authorities and subnational landscapes. These practices are common in areas with

Table 16.2 Transboundary rivers in the HKH and related water treaties

Transboundary river (Riparian countries)	Treaties, agreements, and institutions	Nature of water sharing and use as per treaty	Outcomes
Mahakali [1] (Nepal and India)	1996 Treaty of Mahakali Mahakali River Commission	Guided by principles of equality, mutual benefit, and no harm to either party; equal number of representatives from both parties; equal partnership to define obligations and corresponding rights and duties regarding water use, water distribution, hydroelectricity, irrigation, and flood control	Arbitration tribunal dedicated to resolving differences arising under the Treaty; equal representation of national arbitrators with a third neutral arbitrator presiding over the tribunal (as written in Treaty agreement); provision for planning, surveying, development, and operation of any work on the tributaries of the Mahakali river to be carried out independently by either party in its own territory
Koshi [2] (Nepal and India)	The 1954 Agreement on the Koshi Project (revised in 1996) Indo-Nepal Koshi Project Commission acting as the coordination committee for the Koshi Project	Flood control; irrigation; generation of hydroelectricity and prevention of erosion in Nepal areas on the right side of the river, upstream of the barrage; surveys and investigations necessary for proper design, construction, and maintenance of the barrage and connected works; navigation and fishing rights reserved by Nepal	Disputes or differences resolved through arbitration in which arbitrators are nominated from each side within 90 days of delivery of notice of dispute arising from either party, and the arbitrators' decision is binding; if the arbitrators disagree, both parties appoint an umpire whose decision is final and binding
Gandaki [3] (Nepal and India)	Gandak River Treaty of 1959 followed by the Agreement of Gandak Irrigation and Power Project 1975	Irrigation and hydropower development in both countries; investigation and surveys; communication	No assessments yet
Teesta and other water resources [4] (Bangladesh and India)	Indo-Bangladesh Treaty of Friendship, Cooperation, and Peace; 25-year treaty signed in 1972; also known as Indira-Mujib Treaty The Joint Rivers Commission (JRC) of 1972 established by the Government of Bangladesh, Ministry of Water Resources	JRC addresses issues related to sharing water of common rivers; transmission of flood-related data from India to Bangladesh; construction and repair of embankment and bank protection work along common/border rivers	The Indo-Bangladesh Treaty of 1972 was not renewed in 1997, as both riparian countries declined renegotiation; India's role was seen as excessively imposing and Bangladesh's share was deemed unequal; disputes over water resources at Farakka Barrage, and India's perceived delay in withdrawing troops added to the contention; other political externalities
Ganga-Brahmaputra-Meghna [5] (Bhutan and India)	Chukka Project of circa 1980, based solely on generating hydropower from the Wanchu Cascade at Chukka	No assessments yet	Project deemed highly successful; covered its cost by 1993 and increased capacity to 370 MW; provided impetus for Bhutan's industrialization and commercial development; Bhutan moved from country with the lowest per capita income in South Asia during the 1960s to the highest in the Ganga-Brahmaputra-Meghna region (includes Bangladesh, Bhutan, India, Nepal, and Pakistan), with foreseeable stipulated earnings of more than USD100 million per annum with similar hydropower projects Kuri Chu, Chukka II, and Chukka III by 2015

(continued)

Table 16.2 (continued)

Transboundary river (Riparian countries)	Treaties, agreements, and institutions	Nature of water sharing and use as per treaty	Outcomes
Brahmaputra/Yalu Zangbu [6] (China and India)	MOU on hydrological data sharing on the Brahmaputra/Yalu Zangbu River signed in 2001, renewed in 2008 and 2013	Data and information sharing on water level, discharge, and rainfall every year	No assessments yet
Sutlej/Langquin Zangbu [7] (China and India)	No treaty as such; China agreed to provide hydrological information	Information for flood control, data sharing, and transmission	No assessments yet
Indo-China river-basins [8] (China and India)	Expert-level mechanism formed in 2006 in which expert group, made up of representatives from both sides, discusses interaction and cooperation on provision of flood season hydrological data, emergency management, and other issues on an annual basis	Data sharing for flood control, emergency management, and other issues	No assessments yet
Indus [9] (India and Pakistan; Afghanistan and China are upper riparian but not members)	Indus Treaty of 1960	Pakistan claimed historical rights and "equitable apportionment" and India claimed prior use and preservation of status quo	Disputes over how water will be utilized and allocated were later resolved through the World Bank's involvement as a mediator and arbitrator financier of the partition projects, which meant that the final treaty was planned and formulated by the Bank; involvement of only a few riparian countries may have led to a less effective alliance; only treaty of its kind to arise post-partition in the Indian subcontinent; the partition of the Indus happened after attempts at basin-wide development and planning failed; the treaty is still considered a sub-optimal solution to the management of the Indus
Mekong [10] (Thailand, Laos, Viet Nam, Cambodia, China, Myanmar)	The Mekong Committee (MC) established in 1957; in 1995 the Mekong River Commission (MRC) was developed	Hydropower, flood control, fishing, irrigation, navigation, and water supply	Dam building a threat to lower riparian countries; growing conflict between upper and lower riparian; legal and political differences and complexities in implementation; involvement of only a few riparian countries led to a less effective alliance
			Data collection; coordination; training programmes; planning studies; management of water for developmental uses; ministerial-level reach in the Commission for political influence

Source Shrestha, A. and Ghate, R., 2016

[1] Stiftung (2001); [2, 3] GON (1975); [4, 9] Kliot et al. (2001); [5] Biswas (2011); [6–8] Cumming (2011); [10] Guttman (2003)

open and semi-open border controls, such as those between Nepal and India, Nepal and China, Bhutan and India, and Bangladesh and India.

In recent years, donor-funded transboundary projects have become a new category of regional-level collaboration for the management of natural resources and ecological systems in the HKH. For example, the International Centre for Integrated Mountain Development (ICIMOD) is facilitating the Kailash Sacred Landscape Conservation and Development Initiative (KSLCDI) as a collaborative effort between China, India, and Nepal, with financial support from the German Federal Ministry for Economic Cooperation and Development (BMZ) and the United Kingdom Department for International Development (DFID). The Initiative is a collaborative programme for transboundary cooperation and ecosystem management across a culturally and ecologically important landscape within the three countries. ICIMOD has planned or is implementing a number of other transboundary landscape and trans-Himalayan transect programmes in the HKH, including in the Mt. Everest region (Sherpa et al. 2003), the Hkakabo Razi Complex (Guangwei 2002), and the Kanchenjunga Landscape (Chettri et al. 2008), which have shown that regional cooperation for biodiversity conservation and management is both feasible and necessary (Chhetri et al. 2009; Schild and Sharma 2011).

These donor-initiated collaborative programmes have developed useful methodologies for transboundary governance of a particular landscape, promoted operational-level networks, and provided opportunities for regional-level policy discussions. However, there is a long way to go in crafting a systematic and comprehensive framework so that the countries in the HKH can collaborate and implement regional-level activities that collectively promote the Himalayas as a sustainable eco-region.

Trans-community collaborations based on informal understanding, historic cultural ties, traditional trade practices, mutual cattle herding systems, and watershed connectivity are common. They are based on organic transboundary mechanisms for governing natural resources at the local level. For example, communities across the border between Nepal and China, Nepal and India, and Bhutan and India have collaborated historically on pasture land management, promotion of forest diversity, regulation of wells for water, religious protection of ecological landscapes, and mutual development of livelihood opportunities (Ning et al. 2013, 2016). Chapter 5 elaborates why cooperation among regional member countries is essential for dealing with cross-border wildlife trade and human-wildlife conflict. These historical and indigenous trans-community practices can provide a basis for a larger eco-Himalaya-level collaboration and regional policy framework that ensures community participation and addresses both livelihoods and national-level economic interests.

Developing vertical and horizontal linkages and collaboration among intergovernmental, national, and local institutions is essential in fostering transboundary partnerships and cooperation. The existing regional-level institutions have generated some ecological knowledge and project-based partnerships. However, one of the main reasons for the lack of positive environmental outcomes has been that the focus of intergovernmental initiatives and bilateral treaties has been on economic interests and profitability rather than social and environmental wellbeing at a regional scale. To quote from Pasakhala et al. (2017), "in the HKH, classical transboundary cooperation will grow organically, triggered by common management objectives and common livelihoods opportunities, to constructively forge conservation and development across scales, for instance, common value chains and common branding of products, responsible and cultural heritage tourism."

16.4.2 The Growing Need for Formal Frameworks as a Foundation for Regional Cooperation

While transboundary landscape management initiatives are emerging and the age-old practices of trans-community collaboration at the community level are advancing, these collaborations are not directly endorsed or informed by formal treaties between the participating nations, and the small-scale transboundary activities are not strongly linked to the larger processes of cooperation emerging at the regional level. One of the more important regional collaborative platforms with potential for improving collaborative governance of natural resources is the South Asian Association for Regional Cooperation (SAARC). SAARC has identified environmental restoration, disaster risk reduction, and climate change as priority areas for regional cooperation, and the SAARC countries have collectively agreed that sustainable development and environmental management are the most significant issues in the region (Dorji 2007; Wyes and Lewandowski 2012). The SAARC Development Goals include conservation of land, forest, biodiversity, and water resources; reduction of natural disasters; and climate change mitigation and adaptation (Wyes and Lewandowski 2012). Other networks like Bhutan, Bangladesh, India, and Nepal (BBIN) and The Bay of Bengal Initiative for Multi-Sectoral Technical and Economic Cooperation (BIMSTEC), which includes Bangladesh, India, Myanmar, Sri Lanka, Thailand, Bhutan, and Nepal, may not provide viable institutional modalities, as these are mainly driven by regional geopolitical interests and exclude China and Pakistan, for example. Similarly, networks that support on-the-ground institutional mechanisms for natural resource governance, like the South Asia Wildlife Enforcement Network (SAWEN) and South Asia Watch on

Trade, Economics, and Environment (SAWTEE), have a specific but limited agenda. What is needed for a sustainable HKH is a platform shared by all eight countries as an eco-region and at that scale.

The SAARC Comprehensive Framework on Disaster Management and Disaster Prevention is another initiative on regional-level collaborative governance. The SAARC Centre for Disaster Management and Preparedness, SAARC Coastal Zone Management Centre, and SAARC Meteorological Research Centre are some examples that can provide instruction on how regional institutional mechanisms could be established for natural resource management. However, notwithstanding the enormous potential, SAARC has remained a weak and ineffective regional body and has not been able to develop trust and mutual understanding on landscape-level collaboration and governance. The non-binding nature of SAARC-level agreements, lack of trust among the member countries, and growing geopolitical tension in the region have all turned this regional institution into a formality rather than a commitment for regional collaboration (Tiwari and Joshi 2015).

Scholars argue that the lack of horizontal and vertical linkages among diverse institutions and actors is hindering efforts towards collaborative natural resource governance in the HKH (Kohler et al. 2012). There is still a lack of serious recognition by the national-level actors of the need for regional-level cooperation on natural resource governance (Tiwari and Joshi 2015). Efforts from intergovernmental institutions such as SAARC and ICIMOD are non-binding and have been found insufficient to truly generate concrete collaborative commitments and programmes (Wyes and Lewandowski 2012). The potential for horizontal linkages among regional institutions such as ICIMOD and SAARC remains underutilized, as the processes are driven by different actors, use approaches that are either too informal or too rigid, or remain limited to academic debates without engaging the national decision makers.

16.4.3 The Emergence of Knowledge-Sharing Platforms to Facilitate Regional Cooperation

HKH countries have varied and extremely rich experiences in relation to environmental management, often with a potential for sharing and adoption across borders. However, existing practices remain isolated inside individual countries and are not much shared across the region (Tiwari and Joshi 2015). Community-level institutions such as community forestry in Nepal (Ojha 2014; Paudel 2016) and joint forest management in India (Bhattacharya et al. 2010) have been effective as bottom-up institutions in promoting social and ecological wellbeing. They present an opportunity for communities, local governments, small businesses, and policy makers to learn through comparative and cross-sharing of experience and learning. Policy makers could also learn how local practices can inform national policy development processes.

Developing trans-community collaboration among local actors and fostering vertical linkages between the regional HKH-level and the local practices of resource governance across multiple HKH countries could provide an opportunity for multi-scalar governance of the HKH as a comprehensive eco-region (see next section for a more detailed discussion of multi-scalar governance). However, such linkages are currently confined to national-level actors and have varying degrees of effectiveness, while linkages within countries still tend to be top-down (Malla 2000; Ojha 2014). There is a tremendous opportunity for recognition and encouragement of knowledge-sharing platforms at a regional level to pool insights and experience from across the region.

The various processes and mechanisms related to knowledge-sharing at the regional level provide important insights into how such sharing can be strengthened. The work of ICIMOD demonstrates the case for an intergovernmental mechanism to implement transboundary action research projects and generate and share knowledge among both state and non-state actors in the region. However, regional processes led by civil society tend to be confined to the South Asian belt of the Himalaya, mainly Bangladesh, India, and Nepal. The civil society interface between China and South Asia remains limited (Blaikie and Muldavin 2004).

Given the ecological complexity and political sensitivities of regional environmental governance, it is important to envision multiple pathways and approaches for regional knowledge-sharing and policy engagement platforms. Such platforms could be helpful not only in shaping common policy agendas, but also in fostering regional exchange for the development and implementation of common methodologies at the landscape level. Systematic documentation and sharing of local-level good practices—such as those related to watershed management, poverty reduction, and community-based environmental management and climate change adaptation—would greatly enrich the regional knowledge-sharing platform and help in crafting a regional-level collaborative mechanism.

16.4.4 Emerging Mechanisms for Managing Conflicts and Equitably Distributing Benefits

The issue of upstream and downstream relationships for the allocation of resources, especially of water and responsibility for managing the watershed, is challenging, and the future possibility of collaboration, especially along the Himalayan rivers, will depend on how the distributional issue is resolved in the region. Power imbalances among the participating countries and the domination of bigger economies in extracting natural resources have historically been a barrier to identifying just and fair institutional mechanisms. Smaller economies, which are primarily dependent on subsistence relationships with natural resources, are in conflict with the dominant political and economic interests in the HKH. Disputes and contentions over transboundary water management and allocations persist (Uprety and Salman 2011).

Studies at multiple levels have shown that there is an urgent need for a shared framework and understanding on natural resources governance at the regional scale, supported by appropriate institutions (Blaikie and Muldavin 2004; Tiwari and Joshi 2015). These institutions could include (1) special provisions for mountain regions—especially in countries such as China, India, and Pakistan, where mountains are not the focus of policy processes, and (2) mechanisms and processes for collaboration through which the interests and priorities of the smaller HKH countries, which are caught between larger geopolitical interests, could be protected to facilitate the fair distribution of the costs and benefits of regional transboundary resource management.

Institutional innovations, an eco-regional development framework, and ecological connectivity across national borders are paramount for ensuring the sustainable future of the HKH. Indeed, a range of institutions for transboundary collaboration has emerged in recent years (Tiwari and Joshi 2015). Included among existing institutions are the age-old practices of transboundary community cooperation in natural resource management, such as communities involved in cross-border herding along the mountain frontiers of China, India, and Nepal (Sharma and Chettri 2005). However, there is still no robust regional framework for collaborative resource management among the HKH countries (Tiwari and Joshi 2015), and current practices of regional cooperation are not driven by the goal of enhancing ecological integrity or long-term social wellbeing (Bawa et al. 2010). With two of the major economies—China and India—located on the two sides of the Himalayan range, there is an immense possibility for regional cooperation to help "mitigate climate change, environmental damage, and biodiversity loss both regionally and globally" (*ibid.*).

The existing regional governance and collaboration efforts are based mainly on geopolitical and political economic interests rather than conservation and social interests. In developing the Himalaya as an eco-region, beyond the political boundaries, the role and involvement of Chinese institutions and communities will be decisive. Half of the HKH eco-region lies within China, but the role of China in the region remains limited to natural resource trade and a few donor-designed conservation projects. Common resource management in the current situation of geopolitical tension and political and economic competition remains uncertain, and there are no substantial efforts underway to generate political will at the national and intergovernmental levels. The future for regional governance of natural resource management in the region will depend to a large extent on whether innovative institutional arrangements emerge to facilitate intergovernmental dialogue, common policy initiatives, and collaborative transborder community practices beyond the state of conflict, as demonstrated in various forms of transboundary water interactions (Sahni 2006). The pathway to prosperity in the region, as elaborated in Chap. 4, Scenarios, is through large-scale investments and regional cooperation across multiple scales.

16.5 The Need for Strengthening Cross-Scalar Interfaces and Adaptive Governance

With greater recognition of the HKH's complex geographical and political environment, diverse social systems, and upstream–downstream interconnectedness, attention should be paid to strengthening linkages among institutions across scales, building an effective interface among the various knowledge-based systems (Weiss et al. 2012), and creating new bridging institutions for transformative governance (Huitema et al. 2009). Given the wider concerns about environmental resilience and social equity outcomes, there is a growing consensus in the scientific community, which manifests in the environmental governance literature, on two processes identified as crucial. First is the need to embrace an adaptive approach to environmental governance (Cooper and Wheeler 2015), and second is the need to ensure an inclusive participatory process in decision making and benefit sharing (Blaikie and Sadeque 2000; Saravanan 2009; Ojha and Hall 2013). These requirements indicate the desirability of a multi-level governance approach, as this approach underscores four underlying principles of adaptive governance: management on a bioregional scale, polycentric governance, public participation, and experimentation (Huitema et al. 2009). To support these processes, a complex combination is required of "openness of practices, active involvement of key actors, strong but inclusive leadership, and a knowledge-based hybrid multi-level network combining horizontal and vertical network governance" (Naustdalslid 2015: 913).

Scholarly research on multi-scalar governance in the context of the developing world found it "too underdeveloped to make any serious evaluation" (Stephenson 2013: 830). Research in global governance supports the idea that centralized governments must give way to co-management that opens up different levels of control and governance (Pahl-Wostl et al. 2008) and helps tackle issues related to uncertainties and the persistent lack of communication among environmental stakeholders (Weiss et al. 2012). Furthermore, the interactions between policy-making arrangements (Piattoni 2015) and institutions at multiple levels and scales are continuously shaping both development and governance outcomes (Cash et al. 2006). This means that achieving more politically and ecologically sustainable solutions to problems requires addressing scale issues and creating dynamic linkages across levels (Cash et al. 2006; Piattoni 2015). Multi-level governance involves vital linkages connecting mutually dependent governance levels with cross-scalar processes of policy making, planning, and implementation activities (Stephenson 2013), and is also seen as a form of polycentric governance involving co-management (Andersson and Ostrom 2008). Such framing of governance helps to understand how, where, and for whom governance decisions affecting sustainable development are made (Wilbanks 2007) and how they can be improved.

Below, we examine the key aspects of current multilevel and cross-scalar governance, both worldwide and in the HKH, and further explore why multilevel governance is crucial for the region.

16.5.1 Emerging Mechanisms for Managing Conflicts and Equitably Distributing Benefits

While in some environmental sectors, institutional governance is lacking, others are marred by excessive or overly complex institutional arrangements. Our analysis revealed a disconnect in all the HKH countries between policies and institutions for forest, rangeland, and water (see Sects. 16.3 and 16.4). The issue of institutional fragmentation is not new; the persistent failure to ensure coordination and build linkages and synergies among the wide range of environmental institutions is widespread in the HKH. The challenge is further compounded by the fact that, historically, the environmental sector in the region has been organized under separate knowledge disciplines, and institutions have mostly evolved across resource sectors as a result of organization according to discipline and sectoral divisions. Accordingly, a number of resource specific institutions have emerged in various sub-sectors of the environment, including forest, rangeland, and water, across the community, public, and private domains.

There are reported cases of institutions in each sector operating mostly in silos, often in contradiction to one another, as the mechanisms and strategies for integration and coordination remain weak. There is also a lack of effective institutions for resource management. This is exemplified in the HKH by upstream–downstream river basin management (Shrestha et al. 2013). The lack of such institutions is now felt, as there is a vacuum in catering to the various management needs which arise in different parts of these basins. Meeting these needs requires local and national governments to co-manage cross-border water resources (Pigram 2000; Moellenkamp 2007), but the formal institutional settings, policies, and research in the HKH are far too limited for this. Further, cross-border river basins place national and regional governance in juxtaposition with local-level governance, necessitating more integration (Molle 2009; Pahl-Wostl et al. 2008).

16.5.2 The Imbalance Between Vertical and Horizontal Distribution Under Multilevel Governance

There is an asymmetrical distribution of power and authority across different levels (as well as domains) of governance, as the central governments in all eight HKH countries play the primary role of formulating policies and enacting laws without necessarily engaging the stakeholders in lower levels of governance. The currently dominant normative paradigm of governance emphasizes the state as the legitimate site of authority; unless this authority is delegated or enacted through polycentric institutions, there is a tendency to overlook the importance of decision making at lower scales of governance. At present, however, both local and national jurisdictions are difficult to exercise, as local- and national-level environmental challenges become increasingly intertwined (Rosenau et al. 2004).

While the strengthened role of central and subnational-level authorities has made it possible for governments to adapt policies and practices to suit their own context and priorities (Jörgensen et al. 2015), central-level policymaking and implementation have largely occurred in isolation from, or in contradiction to, both supranational and local levels. State-centred governments are known to prioritize large-scale infrastructure, have vested political and financial interests, ignore local processes and hydrological interconnectedness, and neglect environmental degradation (Molle and Mamanpoush 2012). For example, the governance of water resources in Bangladesh is biased toward structural solutions of flood control and irrigation using a centralized approach that ignores the many other uses of water, such as drinking and sanitation, fisheries, navigation, and ecology (Chowdhury and Rasul 2011: 44).

Research has shown that addressing the complex problems of the environment requires a mix of institutions and designs that facilitate experimentation, learning, and change (Dietz et al. 2003). This entails a governance structure based on collaboration, which occurs when different government bodies work together with non-governmental stakeholders and interest groups to manage issues that cross jurisdictional boundaries and fall into different natural resource management policy sectors (Huitema et al. 2009). Multilevel bioregional approaches to water management have been institutionalized in the EU's Water Framework Directive (WFD). Research suggests that the concept of multilevel governance could be useful in cultural contexts outside the EU, if it takes indigenous governance concepts and ideas of scale into account and maps power relations in the making and implementation of policies (Hensengerth 2015).

The HKH countries have yet to capitalize on the opportunities created by the proliferation of stakeholders in environmental governance, and in the worst cases, the centralized and state-centric systems continue to function alongside the novel arrangements. The perils of a strong centralized system are exemplified in the institutional settings in the State of Uttarakhand in India. Studies show that linkages among public institutions like the *jal nigam*, *jal sansthan*, Water Directorate, and Urban Development Directorate are not only poor but also conflicting, as their functions overlap with other governing bodies like the community-level water users' groups and municipal corporations, which also have some authority for water management, especially on the supply side.

Further experience suggests that the desired environmental and social outcomes are not necessarily achieved when national-level institutions have a dominant influence. For example, Bangladesh's overarching policy on natural resource management has traditionally been focused on earning revenue for the government. In pursuing this policy, natural resource management typically depended on public bureaucracy, which essentially followed a policing, exclusionary, and non-participatory approach to operations (Khan 1998; Chowdhury and Khan 2017). This trend has led to a number of outcomes: (1) resultant alienation of the local communities from the management and use of natural resources; (2) creation and patronization of an elitist aristocracy who enjoyed de facto privatized control and authority over funds and acted as intermediaries between the state and local communities; (3) systematic undermining of local institutions, community initiatives, customary rights, and wisdom in natural resource management; and 4) overshadowing of broader sustainability and environmental considerations in the management of funds (Khan and Harriss-White 2012: 103). Often, the access to water resources and the costs and benefits of water resource projects are distributed unequally; while the rich gain more access to water resources, the poor bear the cost (Chowdhury and Rasul 2011: 44).

Evidence on the outcomes of decentralization and devolution in the HKH also underscores both the potential and, more importantly, the limits of local governance in the Himalaya. Research reveals that co-governance, in place of local autonomy, can still lead to inferior outcomes when the central actors wish to retain supervision of the decentralized institutions, which contradicts the concept of organic linkages between indigenous management and knowledge systems and causes tension in the politics of decentralization of natural resources. One analysis suggests that because of this tension, the forests in Himachal are in a worse condition under the co-governance management regime than under the indigenous system (Agrawal and Chhatre 2006). Therefore, it is important to incentivize policies that encourage the development of proper polycentric systems for natural resource governance to maintain local benefits while increasing the capacity to deal with socioecological challenges (Bixler 2014). Lessons from co-management arrangements in the forest and wetlands in Bangladesh indicate that links with local government and formal recognition of CBOs are important in establishing their legitimacy to represent community interests and in overcoming conflicts (Khan and Harriss-White 2012).

The multilevel governance framework provided in the case of Chinese hydropower suggests that even strongly nation-state-centric governance regimes need not automatically be top-down, but can be highly fragmented and subject to complex and multi-scalar decision-making processes. In China, informal networks between the energy bureaucracy and hydropower developers determine the hydropower decision-making process. These informal networks sit starkly at odds with China's state-centric governance system. By putting authoritarian and indigenous governance concepts together, vibrant and reflexive systems of governance with adaptive skills are finding their way into the hitherto state-dominated Chinese hydropower governance, with a dominance of informal networks in the decision-making process compared to the formal bureaucratic approach (Hensengerth 2015).

16.5.3 The Emerging Role of Knowledge Interfaces in the Shift Towards Adaptive Governance

The building of co-management and cross-border management structures depends heavily on knowledge co-production, mediation, translation, and negotiation (Cash et al. 2006).

The first point is that improving communication links between knowledge producers and policy makers in environmental management is important for evidence-based decision making (Weiss et al. 2012). The use of feedback and reflections to improve and prepare a governance system to tackle unprecedented risks and shocks is a key part of adaptive governance. However, policy-monitoring mechanisms vary across the HKH countries, as does their use of research and feedback from practice in policy making and governance decision making. In water resource management, China has adopted the practice of first piloting, learning from the experience, and then applying the revised policies and laws to the entire country. In contrast, in Nepal's forest sector, policy-making processes are not informed by scientific research (Ojha et al. 2016b); despite the growing body of research, policy review and revisions are ad hoc and often driven by external donors or international agencies. In Bhutan, proposed policies are passed through a policy-screening tool developed on the principles of gross national happiness. Midterm and end of Five-Year Plan reviews provide an opportunity to review policies and implementation. The guidelines for development of Five-Year plans to some extent incorporate revisions based on these reviews.

The second point is that knowledge is produced and understood at different scales, and bringing clarity to a mismatch of scales is integral to knowledge use. Scalar issues pertaining to knowledge in natural resource management and environmental assessments deeply influence governance and policy making. Research suggests that the integration of local and scientific knowledge must be open to questioning how different knowledge systems may be fundamentally embedded in different scales, influencing problem definition and solutions. Both local traditional knowledge and scientific knowledge are contextual and applied, diverse and inherently multi-scalar. Therefore, "scale politics and mismatch between scales of knowledge exist within local communities", and scientific knowledge will inevitably interfere politically with local knowledge. Ahlborg and Nightingale (2012: 16) argue that "it is not simply the case that one needs to confer with local actors or elites or those considered local experts; rather, one needs to attend to how scales of knowledge produce a politics of knowing that can have real implications for on-the-ground management".

Insufficient data and lack of monitoring capability are major obstacles to proactively seeking, synthesizing, and using both experiential and scientific knowledge to support decision making in conditions of uncertainty. The ability to do this is a key part of adaptive governance. At times, decisions are made when contradictory claims to knowledge exist, resulting in unclear and confusing policy directions. For example, in Pakistan, many researchers have argued that the actual forest area of the country is much lower than that officially claimed, as a result of definitional and jurisdictional issues (Shahbaz 2009). In Nepal, despite a plethora of studies conducted into forestry and agriculture in the hills, there is still a lack of information about what is happening on the ground at different locations, or analysis of the impact of different community-based resource management regimes. Studies are also divided across three key sectors—land management, forestry, and agriculture—and there is limited knowledge on cross-sectoral interactions.

Critical insights have, however, emerged as to how research can be linked to practice and policy processes. For example, in Nepal's forestry system (Banjade 2013; McDougall et al. 2013; Ojha et al. 2010), studies show that alliance-led resistance and research-informed deliberation can overcome threats of recentralization (Sunam et al. 2013). However, even with relevant and emerging research, the quality and form of presentation is often not strong enough to attract the attention of, or to be compatible with, policy actors and community leaders.

There are a few systematic, longitudinal studies that have used robust methodologies and have intimately engaged with the contexts, but these studies are often shaped by the interests of the sponsors. Studies are framed either as global knowledge questions or donor project objectives, and there have been few engaged research attempts to critically expose or contribute to local decision systems that impact smallholder agriculture and forestry. This is because the agenda of sustainability in development is promoted by international aid agencies. There is a long history of scientific claims that development interventions have failed to achieve their intended goals, a matter that has become a key base to advance theorizing on the problems of and potential solutions for the Himalaya (Thompson and Warburton 1985; Ives et al. 1987; Ives and Messerli 1989; Blaikie et al. 2002; Ives 2004). The Himalayan myth (that the region was facing an environmental disaster due to rapid population growth, causing extensive deforestation) was based on a discredited western scientific construct (Ives et al. 1987; Bruijnzeel and Bremmer 1989; Ives 1989, 2004; Ives and Messerli 1989; Guthman 1997; Forsyth 1998), but it has created confusion and led to assumptions and misinterpretations about the Himalaya that have affected development planning and implementation and led to misguided research efforts (Ives et al. 1987; Ives 2012). The overgeneralization of complex problems and their solutions in the HKH has led to a failure in development-oriented research (Ives 2012).

China's adoption of the practice of piloting laws and revising them on the basis of implementation experience is an interesting move toward adaptive governance. For example, the Environmental Protection Law of 1979 was enacted on a trial basis and, because it proved more solution-oriented than based on proactive prevention, was amended and reenacted in December 1989. Likewise, under the perceived situation of the tragedy of the commons, reforms in cropland regions saw a transformation from communal management of rangelands to autonomous individual household responsibility over farm management (Banks et al. 2003). In Bhutan, natural resources, especially Sokshing (homestead leaf litter forests), Tsamdrog (rangelands), and water have been nationalized through enactment and revision of the Land Act (RGOB 2007), the Forest and Nature Conservation Act (RGOB 1995), and the Water Act (RGOB 2011) for the sole purpose of conservation and sustainable utilization. In recent times, degradation of the natural environment resulting from mining activities, with no tangible economic benefits to the government and communities, has led to the emergence of policies to limit the private sector and initiate state engagement in harnessing the economic potential of natural resources, demonstrating that adaptive policy making can work in a complex political environment, as exemplified earlier by the evolution of community forestry policy in Nepal. On the other hand, there have also been significant developments in the area of water resource management. In its effort to implement the Water Act of 2014, Bhutan adopted the National Integrated Water Resource Management Plan (NIWRMP), which calls for water security as a national goal to be achieved through employment of the Bhutan Water Security Index (BWSI) and associated interagency coordination framework for planning, implementation, monitoring, and reporting progress.

Cote and Nightingale (2012) argue that the conceptualization of social change in environmental research is itself problematic, as it emphasizes a focus on the structures and functionality of an institutional system devoid of political, historical, and cultural meaning. Greater efforts at situating definitions and questioning formulations about resilience within political and cultural heterogeneities can help address both this issue and the underlying normative concerns. We believe that the question of adaptability and resilience of the environmental governance system depends to a large extent on the role of knowledge at the intersection between social and environmental dynamics, and the ways in which the politics of knowledge helps to address normative questions in the context of power asymmetry and competing value systems—all of which are not external to, but rather integral to the development and functioning of an environmental management system. Fundamentally speaking, it is important to consider how various systems of knowledge are mobilized and an effective interface is created. Researchers should not overlook the importance of engagement with policy actors at different stages of research.

16.5.4 The Emerging Emphasis on Multilevel Learning for Inclusive Policies and Transformational Governance

Evidence suggests that the efforts to ensure inclusion in governance are not fully informed by the unfolding social and political dynamics. We argue that inclusive policy processes should not only be restricted to cross-scalar policy and institutions, but also that they be prioritized within scales and at all levels by increasing the level of participation and democratic representation of stakeholders, individuals, and marginal social groups. Further, institutions need to be flexible and to encourage reflection, learning, experimentation, and innovative responses to specific local capabilities and needs. Institutional arrangements which facilitate multilevel learning are critical for exploring uncertainties and finding solutions for climate change adaptation (Huntjens et al. 2012), which is now a global preoccupation.

Sustainable development also requires transformational adaptation to changes in underlying cultural and political systems, not just adaptation of specific management practices (Pelling 2011). This approach requires environmental policy and institutions to embrace social learning while undertaking the governance and management of natural resources (Olsson et al. 2006; Leach et al. 2007; Plummer et al. 2013). Indeed, governance itself is a process of learning in relation not only to management operations but also to fundamental cultural values and political standpoints (Dressler et al. 2010).

While it is clear that inclusive policy processes help to ensure a more just governance process, it is not clear which governance structures are best for doing this. Research suggests that monocentric perspectives are better at ensuring public participation, as it is easier to provide feedback to the public through a central structure. Similarly, management as experimentation may work better in a monocentric system than in a polycentric one. However, in the case of river basins, unitary authorities at the sub-basin or watershed level may be better at connecting the public with the central authorities (Huitema et al. 2009). Local-level institutions can provide space for the empowerment of historically marginalized groups, but this representation can sometimes be tokenistic at best (Cooke and Kothari 2001; Nightingale 2002; Sikor and Lund 2009). Social inclusion can be freed from rhetoric if the monitoring mechanisms prioritize communication and the use of feedback and reflections to improve governance.

Finally, the shift in gender roles in both mountainous and peri-urban contexts has potentially profound implications for inclusion in governance. As an example, urbanization or rural-urban migration can lead to a transformation of gender relations around water in particular, as well as around natural resources more broadly. It is also argued that with the acquisition of village grazing lands to support urban expansion, gender relations around natural resources can be transformed (Vij and Narain 2016). In India, studies have shown that men performed the task of water collection in upper-caste households, but with occupational diversification, this task is increasingly being performed by women (Ranjan and Narain 2012). In other areas, efforts have been made to reduce women's drudgery in water collection in the mountains by bringing water sources closer to homes, but these efforts proved limited in their impact on improving women's quality of life. Along with climate change, both urbanization and rural-urban migration create new demands on women's time in mountain contexts, while the expansion of a water supply may itself create new water collection tasks (Narain 2014). Thus, focusing on changing gender relations around water may yield a better understanding of the impacts of drinking water supply interventions on women's quality of life than estimating impact simply in terms of conventional indicators of women's water burdens.

16.6 Opportunities for Improving Environmental Governance in the HKH

This chapter has provided an assessment of environmental governance in the Hindu-Kush Himalaya, encompassing processes at local, national, and regional scales. It shows that the governance of environmental resources in the region involves a complex ensemble of state policy frameworks, policy making and implementing organizations, knowledge communities, traditional institutions, and the private sector. The field of environmental governance entails pluralistic and highly dispersed authorities at local, subnational, national, regional, and global levels. The distribution of governing power across various levels of governance is unbalanced and often incompatible with the sustainable management of natural resources. Environmental institutions have yet to fully recognize the complex Himalayan geography, which is characterized by the lack of appropriate, context-specific institutions to link upstream and downstream communities in river basins.

National environmental policies and institutions in the region are complex and still evolving. Governance reforms include the adoption of new environmental concerns such as climate change, disaster risks, and biodiversity; decentralization and devolution; and new shifts to adopt market mechanisms in resource governance. While policy visions and frameworks for improved environmental governance have emerged in response to the change in contexts, there is a lack of effective implementation. This is because developing new policy visions is not concurrently linked to a process of institutional reform. The link between policy and practice is poor, and many of the lessons learned from the field—as well as evidence from research—goes unused in national policy processes.

A variety of decentralization initiatives have emerged with significantly positive outcomes in different localities, yet these are not adequately supported or institutionalized in the national and subnational systems of governance. Regional cooperation on environmental issues remains limited, and there is an urgent need for the HKH countries to engage more meaningfully in devising frameworks for fair cooperation, as countries have intimate upstream and downstream linkages, complementary expertise and experiences, and differing levels of national capability and negotiating power. The future of regional governance of natural resource management depends to a large extent on how innovative institutional arrangements emerge in the region to facilitate intergovernmental dialogue, common policy initiatives, and collaborative trans-border community practices, beyond the dichotomy of cooperation and conflict which has existed in varied forms of transboundary management of natural resources.

Given the rich experiences and policy experiments in diverse sub-regions and localities in the HKH, there is an important need to strengthen the cross-scalar knowledge interface and inclusive governance processes—both within and between the nation-states—to ensure fair and adaptive governance in the face of the growing risks related to climate change and disaster in the Himalaya.

The findings show that there are at least four opportunities for improving environmental governance in the region: (1) strengthening the interface between science, policy, and practice; (2) strengthening the institutional capacity to implement new policies; (3) scaling up community-based environmental management systems by creating more enabling regulatory frameworks and appropriate local institutional arrangements; and (4) strengthening transboundary cooperation among the HKH countries.

References

Acharya, K. P. (2002). Twenty-four years of community forestry in Nepal. *International Forestry Review, 4*(2), 149–156.

Acharya, K. P. (2004). Does community forest management support biodiversity conservation: Evidences from two community forests from the middle hills of Nepal. *Journal of Forest and Livelihood, 4*(1).

Acharya, K. P., Adhikari, J., & Khanal, D. R. (2008). Forest tenure regimes and their impact on livelihoods in Nepal. *Journal of Forest and Livelihood, 71,* 6–18.

Agarwal, A., & Narain, S. (2002). *Community and household water management: The key to environmental regeneration and poverty alleviation*. New Delhi: Concept Publishing Company.

Agrawal, A. (2001). The regulatory community: Decentralization and the environment in the van panchayats (forest councils) of Kumaon, India. *Mountain Research and Development, 21*(3), 208–211.

Agrawal, A., & Yadama, G. (1997). How do local institutions mediate market and population pressures on resources? Forest Panchayats in Kumaon, India. *Development and Change, 28,* 435–465.

Agrawal, A., & Gibson, C. C. (1999). Enchantment and disenchantment: The role of community in natural resource conservation. *World Development, 27*(4), 629–649.

Agrawal, A., & Chhatre, A. (2006). Explaining success on the commons: Community forest governance in the Indian Himalaya. *World Development, 34*(1), 149–166.

Agrawal, A., & Lemos, M. C. (2007). A greener revolution in the making? Environmental governance in the 21st century. *Environment: Science and Policy for Sustainable Development, 49*(5), 36–45. http://dx.doi.org/10.3200/ENVT.49.5.36-45.

Agrawal, A., & Verma, R. (2017). *Natural resource governance at multiple scales in the Hindu Kush Himalaya*. International Centre for Integrated Mountain Development (ICIMOD) Working Paper 2017/4.

Ahlborg, H., & Nightingale, A. J. (2012). Mismatch between scales of knowledge in Nepalese forestry: Epistemology, power and policy implications. *Ecology and Society, 17*(4), 16. https://doi.org/10.5751/ES-05171-170416.

Ali, A., & Shaoliang, Yi. (2013). Highland rangelands of Afghanistan: Significance, management, issues and strategies. In W. Ning, G. S. Rawat, S. Joshi, M. Ismail, & E. Sharma (Eds.), *High-altitude rangelands and their interfaces in the Hindu Kush Himalayas* (pp. 15–24). Kathmandu: ICIMOD.

Ali, J., & Nyborg, I. L. P. (2010). Corruption or an alternative system? Re-assessing corruption and the role of the forest services in the northern areas. *Pakistan. International Forestry Review, 12*(3), 209–220.

Andersson, K. P., & Ostrom, E. (2008). Analyzing decentralised resource regimes from a polycentric perspective. *Policy Sciences, 41*(1), 71–93.

Bacho, F. Z. L. (2005). Decentralization in a pluralist state: Ethnic identity, resource conflicts and development in the East Gonja district of Ghana. *Ghana Journal of Development Studies, 2*(1).

Baginski, O. S., & Blaikie, P. (Eds.). (2007). *Forests, people and power: The political ecology of reform in South Asia*. London: Earthscan.

Banjade, M. R. (2013). Learning to improve livelihoods: Applying adaptive collaborative approach to forest governance in Nepal. In H. Ojha, A. Hall, & V. R. Sulaiman (Eds.), *Adaptive collaborative approaches in natural resource governance: Rethinking participation, learning and innovation*. London: Routledge.

Banks, T., Richard, C., Ping, L., & Zhaoli, Y. (2003). Community-based grassland management in western China. *Mountain Research and Development, 23*(2), 132–140.

Bardhan, P. (2002). Decentralisation of governance and development. *The Journal of Economic Perspectives, 16*(4), 185–205.

Barnes, C., & Laerhoven, F. V. (2014). Making it last? Analysing the role of NGO interventions in the development of institutions for durable collective action in Indian community forestry. *Environmental Science & Policy, 53,* 192–205.

Bawa, K. S., Koh, L. P., Lee, T. M., Liu, J., Ramakrishnan, P., Douglas, W. Y., et al. (2010). China, India, and the environment. *Science, 327*(5972), 1457–1459.

Bhattacharya, P., Pradhan, L., & Yadav, G. (2010). Joint forest management in India: Experiences of two decades. *Resources, Conservation and Recycling, 54*(8), 469–480.

Bhattarai, B., Dhungana, S. P., & Ojha, H. (2007). Poor-focused common forest management: Lessons from leasehold forestry in Nepal. *Forest and Livelihood, 6*(2), 20–29.

Bhattarai, S., kumar Jha, P., & Chapagain, N. (2009). *Towards pro-poor institutions: Exclusive rights to the poor groups in community forest management*. Discussion paper. Kathmandu: ForestAction.

Biswas, A. K. (2011). Cooperation or conflict in transboundary water management: Case study of South Asia. *Hydrological Science Journal, 56*(4), 662–670.

Bixler, R. P. (2014). From community forest management to polycentric governance: Assessing evidence from the bottom up. *Society and Natural Resources, 27*(2), 155–169. http://doi.org/10.1080/08941920.2013.840021.

Blaikie, P. (2006). Is small really beautiful? Community-based natural resource management in Malawi and Botswana. *World Development, 34*(11), 1942–1957.

Blaikie, P. M., & Sadeque, S. Z. (2000). *Policy in high places: Environment and development in the Himalayan region*. ICIMOD.

Blaikie, P., Cameron, J., & Seddon, D. (2002). Understanding 20 years of change in West-Central Nepal: Continuity and change in lives and ideas. *World Development, 30*(7), 1255–1270.

Blaikie, P., & Muldavin, J. S. S. (2004). Upstream, downstream, China, India: The politics of environment in the Himalayan region. *Annals of the Association of American Geographers, 94*(3), 520–548.

Brichieri-Colombi, S., & Bradnock, R. W. (2003). Geopolitics, water and development in South Asia: Cooperative development in the Ganges-Brahmaputra delta. *The Geographical Journal, 169*(1), 43–64.

Bruijnzeel, L. A., & Bremmer, C. N. (1989). Highland-lowland interactions in the Ganges Brahmaputra river basin: a review of published literature. ICIMOD Occasional Paper No. 11. Kathmandu: ICIMOD.

Butler, C. J. (2016). *Knowledge, nature, and nationalism: The Upper Karnali dam in Nepal*. Doctoral dissertation, University of California, Santa Cruz.

Capistrano, D. (2012). Decentralization and forest governance in Asia and the Pacific: Trends, lessons and continuing challenges. In C. J. P. Colfer, G. R. Dahal, & D. Capistrano (Eds.), *Lessons from forest decentralization: Money, justice and the quest for good governance in Asia-Pacific* (pp. 211–232). London: Earthscan.

Cash, D. W., Adger, W. N., Berkes, F., Garden, P., Lebel, L., Olsson, P., et al. (2006). Scale and cross-scale dynamics: Governance and information in a multilevel world. *Ecology and Society, 11*(2), 8.

Chapagain, N., & Banjade, M. R. (2009). Community forestry and local development: Experiences from the Koshi hills of Nepal. *Journal of Forest and Livelihood, 8*(2), 78–92.

Chettri, N., Shakya, B., Thapa, R., &Sharma, E. (2008). Status of a protected area system in the Hindu Kush-Himalayas: An analysis of PA coverage. *International Journal of Biodiversity Science and Management, 4*,164–178.

Chhetri, B. B., Schmidt, K., & Gilmour, D. (2009). Community forestry in Bhutan-exploring opportunities and facing challenges. In *Community forestry international workshop*, Pokhara, Nepal.

Chowdhury, J. K. (2001). *Does forestry pay in Bangladesh? Lessons from ADB's forest sector projects*. Paper presented in the Forestry Sector Workshop, organized by the Forest Department and the ADB, December 3-5, 2001, Dhaka.

Chowdhury, A. A. M., & Khan, N. A. (2017). Participatory forestry in Bangladesh: Practices among selected ethnic communities in Chittagong Hill Tracts. Dhaka: OSDER Publications.

Chowdhury, A. K. M. J. U., & Rasul, G. (2011). Equity and social justice in water resource governance: The case of Bangladesh. *South Asia Water Studies, 2*(2), 44–58.

Cooke, B., & Kothari, U. (2001). *Participation: The new tyranny?*. New York: Zed Books.

Cote, M., & Nightingale, A. J. (2012). Resilience thinking meets social theory: Situating social change in socio-ecological systems (SES) research. *Progress in Human Geography, 36*(4), 475–489.

Cooper, S. J., & Wheeler, T. (2015). Adaptive governance: Livelihood innovation for climate resilience in Uganda. *Geoforum, 65,* 96–107.

Crook, R. C. (2003). Decentralisation and poverty reduction in Africa: The politics of local–central relations. *Public Administration and Development, 23*(1), 77–88.

Crow, B., & Singh, N. (2000). Impediments and innovation in international rivers: The waters of South Asia. *World Development, 28*(11), 1907–1925.

Cumming, G. S. (2011). The Resilience of Big River-basins. *Water International, 36*(1), 63–95.

Dahal, G. R., & Chapagain, A. (2005). Community forestry in Nepal: Decentralized forest governance. In C. J. P. Colfer, G. R. Dahal, & D. Capistrano (Eds.), *Lessons from forest decentralization: Money, justice, and the quest for good governance in Asia-Pacific* (pp. 67–82). London: Earthscan.

Denholm, J. (1991). Agroforestry in mountain areas of the Hindu Kush–Himalayan Region. *ICIMOD Occasional paper* (17).

Dietz, T., Ostrom, E., & Stern, P. (2003). The struggle to govern the commons. *Science, 302*(5652), 1907–1912.

Dong, S., Lassoie, J., Shrestha, K. K., Yan, Z., Sharma, E., & Pariya, D. (2009). Institutional development for sustainable rangeland resource and ecosystem management in mountainous areas of Northern Nepal. *Journal of Environmental Management, 90*(2), 994–1003. http://doi.org/10.1016/j.jenvman.2008.03.005.

Dorji, C. (2007). Mountain development in South Asia. *Sustainable Mountain Development, 53.*

Dorji, L., Webb, E. L., & Shivakoti, G. P. (2006). Forest property rights under nationalized forest management in Bhutan. *Environmental Conservation, 33*(2), 141–147.

Dressler, W., Bram, B., Michael, S., Brockington, D., Hayes, T., Kull, C., et al. (2010). From hope to crisis and back again? A critical history of the global CBNRM narrative. *Environmental Conservation, 37*(1), 1–11.

Farooquee, N. A., Majila, B., & Kala, C. (2004). Indigenous knowledge systems and sustainable management of natural resources in a high altitude society in Kumaun Himalaya, India. *Journal of Human Ecology, 16*(1), 33–42.

Fisher, R. J. (1989). *Indigenous systems of common property forest management in Nepal*. Issue 18 of Working paper. Hawaii: East-West Environment and Policy Institute.

Fleischman, F. D. (2014). Why do foresters plant trees? Testing theories of bureaucratic decision-making in Central India. *World Development, 62,* 62–74.

Forsyth, T. (1998). Mountain myths revisited: Integrating natural and social environmental science. *Mountain Research and Development, 18*(2), 107–116.

Friend, R., Jarvie, J., Reed, S. O., Ratri, S., Pakamas, T., & Canh, T. V. (2013). Mainstreaming urban climate resilience into policy and planning: Reflections from Asia. *Urban Climate, 7,* 6–19.

Fung, A., & Wright, E. O. (2001). Deepening democracy: Innovations in empowered participatory governance. *Politics and Society, 29*(1), 5–41.

Gautam, M. S., & Pokhrel, B. (2011). *Foreign aid and public policy process in Nepal: A case of forestry and local governance.* Discussion Paper 3. Kathmandu: Southasia Institute of Advanced Studies.

Gautam, A. P., Webb, E. L., & Eiumnoh, A. (2002). GIS assessment of land use/land cover changes associated with community forestry implementation in the Middle Hills of Nepal. *Mountain Research and Development, 22*(1), 63–69.

Gautam, A. P., Shivakoti, G. P., & Webb, E. L. (2004). A review of forest policies, institutions, and changes in the resource condition in Nepal. *International Forestry Review, 6*(2), 136–148.

Ghate, R., & Chaturvedi, R. (2016). Unpacking the governance conundrum for better natural resource management. *ICIMOD Working Paper* (2016/2).

Ghosh, D. K. (2008). Governance at local level: the case of panchayats in West Bengal. *The Indian Journal of Political Science*, 71–88.

Gilmour, D. A. (1990). Resource availability and indigenous forest management systems in Nepal. *Society & Natural Resources, 3*(2), 145–158.

Gilmour, D. A., & Fisher, R. J. (1991). *Villagers, forests, and foresters: The philosophy, process, and practice of community forestry in Nepal.*

Goldstein, A., Pinaud, N., & Reisen, H. (2006). *The rise of China and India: What's in it for Africa?.* OECD Development Centre Policy Insights, No. 19. Paris: OECD Publishing. http://dx.doi.org/10.1787/246616177271.

GOM. (1992). Forest Law 1992. Forest Department, Ministry of Natural Resources and Environmental Conservation. Government of Myanmar.

GOM. (1995a). Forest Policy 1995. Forest Department, Ministry of Natural Resources and Environmental Conservation. Government of Myanmar.

GOM. (1995b). *Community forestry instructions.* Forest Department, Ministry of Natural Resources and Environmental Conservation, Government of Myanmar.

GOM. (2008). *Constitution of Myanmar 2008.* Government of Myanmar [pdf] Available at: http://asean-law.senate.go.th/files/Myanmar2008.pdf.

GON. (1975). *Agreement between His Majesty's Government of Nepal and the Government of India on the Gandak Irrigation and Power Project.* Kathmandu: Ministry of Water and Power, Government of Nepal.

Gongbuzeren, Yanbo Li, & Li, Wenjun. (2015). China's rangeland management policy debates: What have we learned. *Rangeland Ecology & Management, 68*(4), 305–314.

Groninger, J. W. (2012). Reforestation strategies amid social instability: Lessons from Afghanistan. *Environmental Management, 49*(4), 833–845.

Grumbine, R. E., & Pandit, M. K. (2013). Threats from India's Himalaya dams. *Science, 339*(6115), 36–37.

Guangwei, C. (2002). Biodiversity in the eastern Himalayas: Conservation through dialogue. Summary reports of workshops on biodiversity conservation in the Hindu Kush-Himalayan ecoregion. In *Biodiversity in the eastern Himalayas: Conservation through dialogue. Summary reports of workshops on biodiversity conservation in the Hindu Kush-Himalayan ecoregion.* ICIMOD.

Guthman, J. (1997). Representing crisis: The theory of Himalayan environmental degradation and the project of development in Post-Rana Nepal. *Development and Change, 28*(1), 45–69.

Guttman, H. (2003). The Mekong river-basin: Practical experiences in transboundary water management in *free flow: Reaching water security through cooperation.* In J. Griffiths & R. Lambert (Eds.), *Free flow: Reaching water security through cooperation.* UNESCO.

Halder, S., & Thompson, P. (2006). *Community based co-management: A solution to wetland degradation in Bangladesh, MACH Technical Paper 1.* Management of Aquatic Ecosystem through Community Husbandry Project, Dhaka: Winrock International.

Hensengerth, O. (2015). Multi-level governance of hydropower in China? The problem of transplanting a Western concept into the Chinese governance context. In *Multi-level governance: The missing linkages* (pp. 295–320). Emerald Group Publishing Limited.

Hill, J. (2014). Farmer-managed irrigation systems in Baltistan and Kargil. *Ladhak Studies, 31,* 4–23.

Hobley, M. (1996). *Participatory forestry: the process of change in India and Nepal.* Overseas Development Institute (ODI).

Hobley, M., & Jha, C. (2012). *Persistence and change: Review of 30 years of community forestry in Nepal.* Kathmandu: HURDEC Nepal.

Huitema, D., Mostert, E., Egas, W., Moellenkamp, S., Pahl-Wostl, C., & Yalcin, R. (2009). Adaptive water governance: Assessing the institutional prescriptions of adaptive (co-)management from a governance perspective and defining a research agenda. *Ecology and Society, 14*(1).

Huntjens, P., Lebel, L., Pahl-Wostl, C., Camkin, J., Schulze, R., & Kranz, N. (2012). Institutional design propositions for the governance of adaptation to climate change in the water sector. *Global Environmental Change, 22*(1), 67–81.

Ives, J. D. (1989). Deforestation in the Himalayas: The cause of increased flooding in Bangladesh and northern India? *Land Use Policy, 6*(3), 187–193.

Ives, J. D. (2004). *Himalayan perceptions: Environmental change and the well-being of mountain peoples.* Routledge.

Ives, J. D. (2012). Environmental change and challenge in the Himalaya. A historical perspective. *Pirineos, 167,* 29–68.

Ives, J. D., & Messerli, B. (1989). *The Himalayan dilemma: Reconciling development and conservation*. Psychology Press.

Ives, J. D., Messerli, B., & Thompson, M. (1987). Research strategy for the Himalayan Region conference conclusions and overview. *Mountain Research and Development*, 332–344.

Jashimuddin, M., & Inoue, M. (2012). Community forestry for sustainable forest management: Experiences from Bangladesh and policy recommendations. *Formath, 11,* 133–166.

Jodha, N. S. (2005). Adaptation strategies against growing environmental and social vulnerabilities in mountain areas. *Himalayan Journal of Sciences, 3*(5), 33–42.

Jörgensen, K., Mishra, A., & Sarangi, G. K. (2015). Multi-level climate governance in India: The role of the states in climate action planning and renewable energies. *Journal of Integrative Environmental Sciences, 12*(4), 267–283.

Jütting, J. P., Kauffmann, C., McDonnell, I., Osterrieder, H., Pinaud, N., & Wegner, L. (2004). *Decentralization and poverty in developing countries: Exploring the impact*.

Karki, R., & Gurung, A. (2012). An overview of climate change and its impact on agriculture: A review from least development country. *Nepal. International Journal of Ecosystem, 2*(2), 19–24.

Karki, M. B., & Vaidya, R. (2009). Adaptation to climate change impacts and regional cooperation on water and hazards in the Himalayan Region. *South Asia Water Journal, 1*(2), 117–131.

Khan, N. A. (1998). *A political economy of forest resource use: Case studies of social forestry in Bangladesh*. Aldershot (UK): Ashgate.

Khan, N. A., Choudhury, J. K., Huda, K. S., & Mondal, M. I. (2004). *An overview of social forestry in Bangladesh*. Dhaka: Bangladesh Forest Department in collaboration with University of Chittagong.

Khan, N. A., & Harriss-White, B. (2012). The ghost in the machine: Deconstructing forest policy discourse in Bangladesh. *Economic and Political Weekly, 47*(17), 100–110.

Khan, H. M. (2015). *Islamic law, customary law and the Afghan informal justice'*. USIP.

Khan, M. R., & Hossain, A. (2006). *Regulatory vs participatory governance and environmental sustainability in Asia*. Dhaka: North South University.

Kliot, N., Shmueli, D., & Shamir, U. (2001). Institutions for management of transboundary water resources: Their nature, characteristics and shortcomings. *Water Policy, 3*(3), 229–255. http://www.fao.org/fileadmin/user_upload/mountain_partnership/docs/LOW_Global_Green_Economy_RIO20.pdf

Kohler, T., Pratt, J., Debarbieux, B., Balsiger, J., Rudaz, G., & Maselli, D. (2012). Sustainable mountain development, green economy and institutions. *From Rio*, 1992 to Rio 2012 and beyond. Final Draft for Rio 2012.

Kreutzmann, H. (2012). Pastoralism: A way forward or back?. In *Pastoral practices in High Asia* (pp. 323–336). Springer Netherlands.

Kumar, K., Singh, N. M., & Kerr, J. M. (2015). Decentralisation and democratic forest reforms in India: Moving to a rights-based approach. *Forest Policy and Economics, 51,* 1–8.

Larson, A. M. (2010). *Forests for people: Community rights and forest tenure reform*. London: Earthscan.

Larson, A. M. (2011). Forest tenure reform in the age of climate change: Lessons for REDD+. *Global Environmental Change, 21*(2), 540–549.

Larson, A. M., & Ribot, J. (2004). Democratic decentralisation through a natural resource lens: An introduction. *The European Journal of Development Research, 16*(1), 1–25.

Larson, A. M., & Soto, F. (2008). Decentralization of natural resource governance regimes. *Annual Review of Environment and Resources, 33,* 213–239.

Larson, A. M., Brockhaus, M., Sunderlin, W. D., Duchelle, A., Babon, A., Dokken, T., et al. (2013). Land tenure and REDD+: The good, the bad and the ugly. *Global Environmental Change, 23*(3), 678–689.

Leach, M., Mearns, R., & Scoones, I. (1999). Environmental entitlements: Dynamics and institutions in community-based natural resource management. *World Development, 27*(2), 225–247.

Leach, M., Bloom, G., Ely, A., Nightingale, P., Scoones, I., Shah, E., et al. (2007). *Understanding governance: Pathways to sustainability*. Brighton: STEPS Centre.

Lister, S. (2007). *Understanding state-building and local government in Afghanistan*. Crisis States Research Working Paper 14. London: Crisis States Research Centre.

Mahmoodi, S. M. (2008). Integrated water resources management for rural development and environmental protection in Afghanistan. *Journal of Developments in Sustainable Agriculture, 3*(1), 9–19.

Malla, Y. B. (2000). Impact of community forestry policy on rural livelihoods and food security in Nepal. *Unasylva, 51*(202), 37–45.

Malla, Y. B. (2001). Changing policies and the persistence of patron-client relations in Nepal: Stakeholders' responses to changes in forest policies. *Environmental History*, 287–307.

Maynard-Moody, S., & Herbert, A. W. (1989). Beyond implementation: Developing an institutional theory of administrative policy making. *Public Administration Review, 49*(2), 137–143.

McDougall, C. L., Leeuwis, C., Bhattarai, T., Maharjan, M. R., & Jiggins, J. (2013). Engaging women and the poor: Adaptive collaborative governance of community forests in Nepal. *Agriculture and Human Values, 30*(4), 569–585.

Ministry of Livestock Development. (2017). *Rangeland policy, 2010*. Retrieved January 17, 2017, from http://www.npafc.gov.np/content.php?id=261.

Moellenkamp, S. (2007). The "WFD-effect" on upstream-downstream relations in international river basins—Insights from the Rhine and the Elbe basins. *Hydrology and Earth System Sciences Discussions, 4,* 1407–1428.

Mohan, G. (1996). Adjustment and decentralization in Ghana: A case of diminished sovereignty. *Political Geography, 15*(1), 75–94.

Molle, F. (2009). River-basin planning and management: The social life of a concept. *Geoforum, 40,* 484–494.

Molle, F., & Mamanpoush, A. (2012). Scale, governance and the management of river basins: A case study from Central Iran. *Geoforum, 43*(2), 285–294.

Mosse, D. (1997). The symbolic making of a common property resource: History, ecology and locality in a tank-irrigated landscape in south India. *Development and Change, 28*(3), 467–504.

Mowla, G. (2001). *Evolution of forest management and Today's Forest Department: A perspective analysis* (in Bengali), *Sharanika* (pp. 46–47). Dhaka: Forest Department.

Muhammed, N., Koike, M., & Haque, F. (2008). Forest policy and sustainable forest management in Bangladesh: An analysis from national and international perspectives. *New Forests, 36*(2), 201–216.

Mustafa, D. (2010). *Hydropolitics in Pakistan's Indus Basin*. US Institute of Peace.

Namgyel, P. (2005). *Forest policy and income opportunities from NTFP commercialisation in Bhutan*. Doctoral dissertation, The University of Reading.

Narain, V. (2003). *Institutions, technology and water control: Water users association and irrigation management reform in two large-scale systems in India*. New Delhi: Orient Longman.

Narain, V. (2014). Local environment: The international journal of justice and sustainability. *Local Environment, 19*(9), 974–989.

Naustdalslid, J. (2015). Multi-level water governance—The case of the Morsa River Basin in Norway. *Journal of Environmental Planning and Management, 58*(5), 913–931.

Nautiyal, S. (2011). Can conservation and development interventions in the Indian Central Himalaya ensure environmental sustainability? *A socioecological evaluation. Sustainability Science, 6*(2), 151–167.

Nepal, S., Flügel, W. A., & Shrestha, A. B. (2014). Upstream-downstream linkages of hydrological processes in the Himalayan region. *Ecological Processes, 3*(1), 19.

Nightingale, A. J. (2002). Participating or just sitting in? The dynamics of gender and caste in community forestry. *Journal of Forest and Livelihood, 2,* 1.

Nightingale, A. J. (2005). "The experts taught us all we know": Professionalisation and knowledge in Nepalese community forestry. *Antipode, 37*(3), 581–604.

Nightingale, A. J. (2006). The nature of gender: Work, gender, and environment. *Environment and planning D: Society and space, 24* (2), 165–185.

Nightingale, A. J. (2016). Adaptive scholarship and situated knowledges? Hybrid methodologies and plural epistemologies in climate change adaptation research. *Area, 48*(1), 41–47.

Nightingale, A. J., & Sharma, J. R. (2014). Conflict resilience among community forestry user groups: Experiences in Nepal. *Disasters, 38*(3), 517–539. https://doi.org/10.1111/disa.12056.

Ning, W., Rawat, G. S., Joshi, S., Ismail, M., & Sharma, E. (2013). *High-altitude rangelands and their interfaces in the Hindu Kush Himalayas.* International Centre for Integrated Mountain Development (ICIMOD).

Ning, W., Shaoliang, Y., Joshi, S., & Bisht, N. (2016). Yak on the move: transboundary challenges and opportunities for yak raising in a changing Hindu Kush Himalayan region. In *Yak on the move: Transboundary challenges and opportunities for yak raising in a changing Hindu Kush Himalayan region.*

Nixon, H., Joelene, C., Saw, K. P. C., Lynn, T. A., & Arnold, M. (2013). *State and region governments in Myanmar.* Myanmar Development Resource Institute-Centre for Economic and Social Development and the Asia Foundation.

Noelle-Karimi, C. (2006). Village institutions in the perception of national and international actors in Afghanistan. Amu Darya Project Working Paper No. 1. Zentrum für Entwicklungsforschung Center for Development Research, University of Bonn.

North, D. C. (1990). *Institutions, institutional change and economic performance.* Cambridge: Cambridge University Press.

Ojha, H. R. (2006). Techno-Bureaucratic Doxa and the challenges of deliberative governance—The case of community forestry policy and practice in Nepal. *Policy and Society, 252,* 131–175.

Ojha, H. R. (2014). Beyond the 'Local Community': The evolution of multi-scale politics in Nepal's community forestry regimes. *International Forestry Review, 16*(3), 339–353.

Ojha, H., Persha, L., & Chhatre, A. (2009). *Community forestry in Nepal: A policy innovation for local livelihoods* (Vol. 913). International Food Policy Research Institute.

Ojha, H. R., Paudel, N. S., Banjade, M. R., McDougall, C., & Cameron, J. (2010). The deliberative scientist: Integrating science and politics in forest resource governance in Nepal. In *Beyond the biophysical* (pp. 167–191). Springer Netherlands.

Ojha, H. R., & Hall, A. (2013). Adaptive collaborative approaches: Traditions, foundations and frontiers. In H. R. Ojha, A. Hall, & V. R. Sulaiman (Eds.), *Adaptive collaborative approaches in natural resource governance: Rethinking participation, learning and innovation.* New York: Routledge.

Ojha, H. R., Ford, R., Keenan, R. J., Race, D., Vega, D. C., Baral, H., et al. (2016a). Delocalizing communities: Changing forms of community engagement in natural resources governance. *World Development, 87,* 274–290.

Ojha, H. R., Khatri, D. B., Shrestha, K. K., Bhattarai, B., Baral, J. C., Basnett, B. S., et al. (2016b). Can evidence and voice influence policy? A critical assessment of Nepal's Forestry Sector Strategy, 2014. *Society & Natural Resources, 29*(3), 357–373. https://doi.org/10.1080/08941920.2015.1122851.

Ojha, H. R., Ghimire, S., Pain, A., Nightingale, A., Khatri, D. B., & Dhungana, H. (2016c). Policy without politics: Technocratic control of climate change adaptation policy making in Nepal. *Climate Policy, 16*(4), 415–433.

Olsson, P., Gunderson, L., Carpenter, S., Ryan, P., Lebel, L., Folke, C., et al. (2006). Shooting the rapids: Navigating transitions to adaptive governance of social-ecological systems. *Ecology and Society, 11*(1).

Ostrom, E. (1990). *Governing the commons: The evolution of institutions for collective actions.* Camridge: Cambridge University Press.

Ostrom, E. (1991). Rational choice theory and institutional analysis: Toward complementarity. *American Political Science Review, 85* (01), 237–243.

Ostrom, E. (2010). Beyond markets and states: Polycentric governance of complex economic systems. *Transnational Corporations Review, 2*(2), 1–12.

Pahl-Wostl, C., Gupta, J., & Petry, D. (2008). Governance and the global water system: A theoretical exploration. *Global Governance, 14*(4), 419–435.

Pandit, B. H., & Thapa, G. B. (2004). Poverty and resource degradation under different common forest resource management systems in the mountains of Nepal. *Society and Natural Resources, 17*(1), 1–16.

Pasakhala, B., Ghate, R., & Kotru, R. (2017). *Integrating conservation and development in transboundary landscapes: Looking back to move forward.* ICIMOD Working Paper 2017/18.

Paudel, D. (2016). Re-inventing the commons: Community forestry as accumulation without dispossession in Nepal. *The Journal of Peasant Studies, 43*(5), 989–1009.

Paudel, N. S., Banjade, M. R., Dahal, G. R., & ForestAction, N. (2009). *Community forestry in changing context: Emerging market opportunities and tenure rights.* Kathmandu: ForestAction and CIFOR.

Pelling, M. (2011). *Adaptation to climate change: From resilience to transformation.* New York: Routledge.

Peters, B. G. (2015). State failure, governance failure and policy failure: Exploring the linkages. *Public Policy and Administration, 30*(3–4), 261–276.

Piattoni, S. (2015). Multi-level governance: Underplayed features, overblown expectation and missing linkages. In *Multi-level governance: The missing linkages* (pp. 321–342). Emerald Group Publishing Limited.

Pigram, J. J. (2000). Towards upstream-downstream hydrosolidarity Australia's Murray-Darling River Basin. *Water International, 25*(2), 222–226.

Plummer, R., Armitage, D. R., & de Loë, R. C. (2013). Adaptive co-management and its relationship to environmental governance. *Ecology and Society, 18*(1), 21.

Pokharel, B. K., & Nurse, M. (2004). Forests and people's livelihood: Benefiting the poor from community forestry. *Journal of forest and Livelihood, 4*(1), 19–29.

Pokharel, B. K., Branney, P., Nurse, M., & Malla, Y. B. (2007). Community forestry: Conserving forests, sustaining livelihoods and strengthening democracy. *Journal of Forest and Livelihood, 6*(2), 8–19.

Pokharel, B. K., Rayamajhi, S., & Tiwari, K. R. (2012). Nepal's community forestry: Need of better governance. In *Global perspectives on sustainable forest management* (pp. 43–58). Shanghai, China: InTech.

Qin, P., Carlsson, F., & Xu, J. (2011). Forest tenure reform in China: A choice experiment on farmers' property rights preferences. *Land Economics, 87*(3), 473–487.

Rahaman, M. M., & Varis, O. (2009). Integrated water management of the Brahmaputra basin: Perspectives and hope for regional development. *Natural Resources Forum, 33*(1), 60–75.

Rangan, H. (1996). From Chipko to Uttaranchal: Development, environment, and social protest in the Garhwal Himalayas, India. In R. Peet & M. Watts (Eds.), *Liberation ecologies: Environment, development and social movements* (pp. 205–226). London: Routledge.

Ranjan, P. and Narain, V. (2012). *Urbanization, climate change and water security: a study of vulnerability and adaptation in Sultanpur and Jhanjhrola Khera in peri-urban Gurgaon, India.* (Peri-Urban Water Security Discussion Paper Series, Paper No. 3). Retrieved from SaciWATERs, India website: http://www.saciwaters.org/periurban/discussion-paper-3.pdf.

Rasul, G. (2014). Food, water, and energy security in South Asia: A nexus perspective from the Hindu Kush Himalayan region. *Environmental Science & Policy, 39*, 35–48.

Raut, N., Sitaula, B. K., & Bajracharya, R. M. (2010). Agricultural intensification: Linking with livelihood improvement and environmental degradation in mid-hills of Nepal. *Journal of Agriculture and Environment, 11*, 83–94.

Regmi, B. R., & Star, C. (2015). Identifying operational mechanisms for the mainstreaming of climate change adaptation in Nepal. *Climate and Development, 6*(4), 306–317.

Rest, M. (2012). Generating power: Debates on development around Arun-3 hydropower project. *Contemporary South Asia, 20*, 105–111.

RGOB. (1995). *Forest and Nature Conservation Act of Bhutan*. Royal Government of Bhutan (RGOB).

RGOB. (2007). *The Land Act of Bhutan 2007*. Royal Government of Bhutan (RGOB).

RGOB. (2011). *The Water Act of Bhutan 2011*. Royal Government of Bhutan (RGOB).

Ribot, J. C. (2003). Democratic decentralisation of natural resources: Institutional choice and discretionary power transfers in Sub-Saharan Africa. *Public Administration and Development, 23*, 53–65.

Ribot, J. C. (2007). Representation, citizenship and the public domain in democratic decentralization. *Development, 50*(1), 43–49.

Robbins, P. (2000). The rotten institution: Corruption in natural resource management. *Political Geography, 19*, 423–443.

Rome, S. I. (2005). *Forestry in the princely state of Swat and Kalam (North-West Pakistan): A historical perspective on norms and practices* (No. 6). IP6 Working Paper.

Rosenau, J. N., von Weizsäcker, E., & Petschow, U. (Eds.). (2004). *Governance and sustainability. Exploring the roadmap to sustainability after Johannesburg*. Sheffield: Greenleaf Publishing.

Sahni, H. K. (2006). The politics of water in South Asia: The case of the Indus Waters Treaty. *SAIS Review of International Affairs, 26* (2), 153–165.

Saravanan, V. (2009). Decentralisation and water resources management in the Indian Himalayas: The contribution of new institutional theories. *Conservation and Society, 7*(3), 176–191.

Sarkar, R. (2008). Decentralised forest governance in central Himalayas: A re-evaluation of outcomes. *Economic and Political Weekly, 43*(18), 54–63.

Sarker, D. (2011). The implementation of the forest rights act in India: Critical issues. *Economic Affairs, 31*(2), 25–29.

Schild, A., & Sharma, E. (2011). Sustainable mountain development revisited. *Mountain Research and Development, 31*(3), 237–241.

Shahbaz, B. (2009). Dilemmas in participatory forest management in Northwest Pakistan. *A Livelihood Perspective, 25.*

Shahbaz, B., & Ali, T. (2003). Sustainable livelihoods framework and state of the Pakistan forests. In S.A. Mufti S.A., S.S. Hussain, & A. M. Khan (Eds.), *Mountains of Pakistan: Protection, potential and prospects*. Islamabad, Pakistan: Global Change Impact Studies Centre.

Shabaz, B., Ali, T., & Suleri, A. Q. (2007). A critical analysis of forest policies of Pakistan: Implications for sustainable livelihoods. *Mitigation and Adaptation Strategies for Global Change, 12*(4), 441–453.

Shahbaz, B., Ali, T., & Suleri, A. Q. (2011). Dilemmas and challenges in forest conservation and development interventions: Case of Northwest Pakistan. *Forest Policy and Economics, 13*(6), 473–478.

Sharma, E., & Chettri, N. (2005). ICIMOD's transboundary biodiversity management initiative in the Hindu Kush-Himalayas. *Mountain Research and Development, 25*(3), 278–281.

Sharma, E., Chettri, N., & Gyamtsho, P. (2006). Advances in community based natural resources management in the Hindu Kush-Himalayan Region. In *Capitalisation and sharing of experiences on the interaction between forest policies and land use pattern in Asia* (Vol. 2, pp. 9–23).

Sharma, E., Chettri, N., Gurung, J., & Shakya, B. (2007). *The landscape approach in biodiversity conservation: A regional cooperation framework for implementation of the convention on biological diversity in the Kangchenjunga landscape*. ICIMOD.

Sharma, E., Chettri, N., & Oli, K. P. (2010). Mountain biodiversity conservation and management: A paradigm shift in policies and practices in the Hindu Kush-Himalayas. *Ecological Research, 25* (5), 909–923.

Sherpa, L. N., Peniston, B., Lama, W., & Richard, C. (2003). *Hands around Everest: Transboundary cooperation for conservation and sustainable livelihoods*. International Centre for Integrated Mountain Development (ICIMOD).

Shrestha, K. (2012). Towards environmental equity in Nepalese community forestry. In F. D. Gordon & G. K. Freeland (Eds.), *International environmental justice: Competing claims and perspectives* (pp. 97–111). Hertfordshire: ILM Publications.

Shrestha, K. (2016). *Dilemmas of justice: Collective action and equity in Nepal's community forestry*. New Delhi: Adroit Publisher.

Shrestha, A. B., Wahid, S. M., Vaidya, R. A., Shrestha, M. S., & Molden, D. J. (2013). Regional water cooperation in the Hindu Kush Himalayan region. In J. Griffiths & R. Lambert (Eds.), *Free flow: Reaching water security through cooperation* (pp. 65–69). UNESCO.

Shrestha, A., & Ghate, R. (2016). *Transboundary water governance in HKH: Beyond the dialectics of conflict and cooperation.* International Centre for Integrated Mountain Development (ICIMOD), Hi-AWARE Working paper-7.

Sikor, T., & Lund, C. (2009). Access and property: A question of power and authority. *Development and Change, 40*(1), 1–22.

Sivaramakrishnan, K. (1999). *Modern forests: Statemaking and environmental change in colonial Eastern India*. Stanford, CA: Stanford University Press.

Stephenson, P. (2013). Twenty years of multi-level governance: 'Where does it come from? What is it? Where is it going?'. *Journal of European Public Policy, 20*(6), 817–837.

Subedi, B. P. (2006). *Linking plant-based enterprises and local communities to biodiversity conservation in Nepal Himalaya*. New Delhi: Akhil Book Distributors.

Stiftung, F. E. (2001). *Treaty of Mahakali*. Retrieved from http://www.nepaldemocracy.org/documents/treaties_agreements/indo-nepal_treaty_mahakali.htm.

Sunam, R. K., Paudel, N. S., & Paudel, G. (2013). Community forestry and the threat of recentralization in Nepal: Contesting the bureaucratic hegemony in policy process. *Society & Natural Resources, 26* (12), 1407–1421.

Thompson, M., & Warburton, M. (1985). Knowing where to hit it: a conceptual framework for the sustainable development of the Himalaya. *Mountain Research and Development*, 203–220.

Thoms, C. A. (2008). Community control of resources and the challenge of improving local livelihoods: A critical examination of community forestry in Nepal. *Geoforum, 39*(3), 1452–1465.

Timsina, N. P., & Gotame, B. (2014). *Political economy of forestry sector of Nepal: Analysis of actors' engagement and policy processes*. Lalitpur, Nepal: ForestAction.

Tint, K., Baginski, S. O., & Gyi, M. K. K. (2011). *Community forestry in Myanmar: Progress and potentials*. Environment Conservation and Community Development Initiative, Yangon, Myanmar and School of International Development, University of East Anglia, UK.

Tiwari, P. C., & Joshi, B. (2015). Local and regional institutions and environmental governance in Hindu Kush Himalaya. *Environmental Science & Policy, 49,* 66–74.

UNDP-UNEP. (2009). *Guidance note on mainstreaming environment into national development planning*. Nairobi: UNPEI. Retrieved from https://www.cbd.int/doc/meetings/nbsap/nbsapcbw-seasi-01/other/nbsapcbw-seasi-01-undp-unep-guide-en.pdf.

UNFCCC. (2015). *Report of the conference of the parties on its twenty-first session, held in Paris from 30 November to 13 December 2015*. Retrieved from http://unfccc.int/resource/docs/2015/cop21/eng/10.pdf.

Uprety, K., & Salman, S. M. (2011). Legal aspects of sharing and management of transboundary waters in South Asia: Preventing conflicts and promoting cooperation. *Hydrological Sciences Journal, 56*(4), 641–661.

USAID. (2010). *Bangladesh environment sector assessment and strategic analysis*. Arlington: United States Agency for International Development by ECODIT.

Venkateswarlu, C. (2013). Constitutional provisions and legislations for empowering the weaker sections. *Abhinav-International Monthly Refereed Journal of Research in Management & Technology, 2,* 120–129.

Vick, M. J. (2014). Steps towards an Afghanistan-Pakistan water-sharing agreement. *International Journal of Water Resources Development, 30*(2), 224–229.

Vij, S., & Narain, V. (2016). Land, water & power: The demise of common property resources in periurban Gurgaon, India. *Land Use Policy, 50,* 59–66.

Watto, M. A., Ali, M., Khan, A. J., & Shahbaz, B. (2010). Stakeholder's role and interaction in the context of forestry extension interventions in the Northwest Pakistan. *Pakistan Journal of Agricultural Sciences, 47*(2), 173–177.

Watto, M. A., & Mugera, A. W. (2016). Groundwater depletion in the Indus basin of Pakistan. Imperatives and repercussions. *International Journal of River Basin Management, 14*(4), 447–458.

Webb, E. L., & Dorji, L. (2007). The evolution of forest related institutions in Bhutan: Adaptations of local people to the rising state in decentralization, forests, and rural communities. In E. L. Webb & Shivakoti (Eds.), *Policy outcomes in Southeast Asia*. G. SAGE Publications.

Wegerich, K. (2009). *Water strategy meets local reality*. Afghanistan Research and Evaluation Unit.

Weiss, K., Hamann, M., Kinney, M., & Marsh, H. (2012). Knowledge exchange and policy influence in a marine resource governance network. *Global Environmental Change, 22*(1), 178–188.

Wilbanks, T. J. (2007). Scale and sustainability. *Climate Policy, 7*(4), 278–287.

Wirsing, R. G. (2007). Hydro-politics in South Asia: The domestic roots of interstate river Rivalry. *Asian Affairs: An American Review, 34*(1), 3–22.

Wirsing, R. G. (2013). Melting the geopolitical ice in south Asia. *Asia Policy, 16*(1), 19–25.

Wolf, A. T., & Hamner, J. H. (2000). Trends in transboundary water disputes and dispute resolution. In M. R. Lowi & B. R. Shaw (Eds.), *Environment and security: Discourses and practices* (pp. 123–148). London: Palgrave Macmillan.

Wyes, H. W., & Lewandowski, M. (2012). *Narrowing the gaps through regional cooperation institutions and governance systems*. ADBI Working Paper 359. Tokyo: Asian Development Bank Institute.

Xu, J. T. (2010). Collective forest tenure reform in China: What has been achieved so far. In *World.*

Zhang, B., & Cao, C. (2015). Four gaps in China's new environmental law: Implementation and accountability will remain challenging, especially at the local level, warn Bo Zhang and Cong Cao. *Nature, 517*(7535), 433–435.

Zhang, H. (2016). Sino-Indian water disputes: the coming water wars? *Wiley Interdisciplinary Reviews: Water, 3*(2), 155–166.

Zhang, L., Heerink, N., Dries, L., & Shi, X. (2013). Water users associations and irrigation water productivity in northern China. *Ecological Economics, 95,* 128–136.

Zhen, L., & Zhang, H. (2011). Payment for ecosystem services in China: An overview. *Living Reviews in Landscape Research, 5*(2), 5–21.

Zittoun, P. (2015). Analysing policy failure as an argumentative strategy in the policymaking process: A pragmatist perspective. *Public Policy and Administration, 30*(3–4), 243–260.

Annex 1: Glossary

Glossary Editor: Avash Pandey, International Centre for Integrated Mountain Development (ICIMOD), Kathmandu, Nepal (avash.pandey@icimod.org)

Active layer The layer of ground above permafrost subject to annual thawing and freezing (ACGR 1998; Washburn 1979).

Adaptation The process of adjustment to actual or expected climate and its effects. In human systems, adaptation seeks to moderate harm or exploit beneficial opportunities. In natural systems, human intervention may facilitate adjustment to expected climate and its effects.

Adaptation limit The point at which an actor's objectives (or system needs) cannot be secured from intolerable risks through adaptive actions.

Adaptation needs The circumstances requiring action to ensure safety of populations and security of assets in response to climate impacts.

Adaptive capacity The ability of systems, institutions, humans, and other organisms to adjust to potential damage, to take advantage of opportunities, or to respond to consequences.

Autonomous adaptation Adaptation in response to experienced climate and its effects, without planning explicitly or consciously focused on addressing climate change. Also referred to as spontaneous adaptation.

Incremental adaptation Adaptation actions where the central aim is to maintain the essence and integrity of a system or process at a given scale.

Transformational adaptation Adaptation that changes the fundamental attributes of a system in response to climate and its effects.

Afforestation Planting of new forests on lands that historically have not contained forests. For a discussion of the term forest and related terms such as afforestation, reforestation and deforestation, see the IPCC Special Report on Land Use, Land-Use Change, and Forestry (IPCC 2000a). See also information provided by the United Nations Framework Convention on Climate Change (UNFCCC 2013) and the report on Definitions and Methodological Options to Inventory Emissions from Direct Human-induced Degradation of Forests and Devegetation of Other Vegetation Types (IPCC 2003, 2014a, b).

Air pollution Air pollution occurs when harmful substances including particulates, biological molecules and trace gases are introduced into Earth's atmosphere. Human activity and natural processes can both generate air pollution.

Air pollution episode An air pollution episode is a period of abnormally high levels of air pollutants. Usually, it is formed by the combination of emissions and meteorological conditions.

P. Wester et al. (eds.), *The Hindu Kush Himalaya Assessment*,
https://doi.org/10.1007/978-3-319-92288-1

Air quality Air quality describes the concentration and composition of air pollutants in the air. It is closely related to the visibility, public health, and environment impacts (such as crop yield, ecosystems and biodiversity).

Air quality management Air quality management is a set of actions to regulate and work toward the accomplishment of air quality goals and objectives. It requires actions by government, business, industry, NGO's and the population.

Albedo The fraction of solar radiation reflected by a surface or object, often expressed as a percentage. Snow-covered surfaces have a high albedo, the albedo of soils ranges from high to low and vegetation-covered surfaces and oceans have a low albedo. The Earth's planetary albedo varies mainly through varying cloudiness, snow, ice, leaf area and land cover changes.

Baseline/reference The baseline (or reference) is the state against which change is measured. A baseline period is the period relative to which anomalies are computed. In the context of transformation pathways, the term baseline scenarios refers to scenarios that are based on the assumption that no mitigation policies or measures will be implemented beyond those that are already in force and/or are legislated or planned to be adopted. Baseline scenarios are not intended to be predictions of the future, but rather counterfactual constructions that can serve to highlight the level of emissions that would occur without further policy effort. Typically, baseline scenarios are then compared to mitigation scenarios that are constructed to meet different goals for greenhouse gas (GHG) emissions, atmospheric concentrations or temperature change. The term baseline scenario is used interchangeably with reference scenario and no policy scenario. In much of the literature the term is also synonymous with the term business-as-usual (BAU) scenario, although the term BAU has fallen out of favour because the idea of business as usual in century-long socio-economic projections is hard to fathom. See also Emission scenario, Representative Concentration Pathways (RCPs) and SRES scenarios.

Baseline The reference scenario for measurable quantities from which an alternative outcome can be measured, for example, a non-intervention scenario is used as a reference in the analysis of intervention scenarios. A baseline may be an extrapolation of recent trends, or it may assume frozen technology or costs. See also business as usual, models, scenario (Verbruggen et al. 2011).

Benchmark A measurable variable used as a baseline or reference in evaluating the performance of a technology, a system or an organization. Benchmarks may be drawn from internal experience, from external correspondences or from legal requirements and are often used to gauge changes in performance over time (Verbruggen et al. 2011).

Biodiversity The variability among living organisms from all sources including, inter alia, terrestrial, marine and other aquatic ecosystems and the ecological complexes of which they are part this includes diversity within species, between species and of ecosystems (CBD 1992).

Biofuels Fuels derived directly or indirectly from biomass. Biofuels can be split up into three categories: solid biofuels; liquid biofuels; and biogases. Solid biofuels comprise mainly fuelwood, wood residues, wood pellets, animal waste, vegetal material. Liquid biofuels are mainly biodiesel and bioethanol used as transport fuels. Biogas is composed principally of methane and carbon dioxide produced by anaerobic digestion of biomass, comprising: landfill gas, formed by the digestion of landfilled wastes; sewage sludge gas, produced from the anaerobic fermentation of sewage sludge; and other biogas, such as biogas produced from the anaerobic fermentation of animal slurries and of wastes in abattoirs, breweries and other agro-food industries (OECD/IEA 2005).

Biomass Organic, non-fossil material of biological origin (plants and animals) used as a raw material for production of biofuels. It includes wide range of materials harvested from nature or biological portion of waste. The most typical example is wood (firewood, wood residues, wood waste, tree branches, stump, wood pellets), which is the largest biomass energy source. Other examples of biomass are grass, bamboo, corn, sugarcane, animal waste, sewage sludge and algae. Using biomass as a fuel is deemed carbon neutral as carbon was trapped from the atmosphere during the biomass life cycle (Eurostat).[1] In the World Energy Outlook, the International Energy Agency (2010) defines traditional biomass as biomass consumption in the residential sector in developing countries that refers to the often-unsustainable use of wood, charcoal, agricultural residues and animal dung for cooking and heating. All other biomass use is defined as modern biomass differentiated further by IPCC report into modern bioenergy (which encompasses electricity generation and combined heat and power from biomass and municipal solid waste, biogas, residential space and hot water in buildings and commercial applications from biomass, municipal solid waste, and biogas, and liquid transport fuels) and industrial bioenergy applications (which include heating through steam generation and self-generation of electricity and combined heat and power in the pulp and paper industry, forest products, food and related industries) (Verbruggen et al. 2011).

Biotechnology Biological processes or organisms used for the production of materials and services that are of benefit to humankind, including techniques for the improvement of the characteristics of economically important plants and animals and for the development of micro-organisms to act on the environment (Zaid et al. 1999).

Black carbon Black carbon (BC) is a major component of light-absorbing refractory carbonaceous matter, produced by incomplete combustion of fossil fuels, biofuel, and biomass. The measurement of BC is usually based on the optical light absorption. BC can strongly absorb solar radiation at visible wavelengths causing warming effects in the atmosphere. Once deposited on a glacier and snow cover, it can reduce the albedo.

Burden sharing/effort sharing In the context of mitigation, burden sharing refers to sharing the effort of reducing the sources or enhancing the sinks of greenhouse gases (GHGs) from historical or projected levels, usually allocated by some criteria, as well as sharing the cost burden across countries.

Business-As-Usual (BAU) See Baseline/reference.

Business As Usual (BAU) The future is projected or predicted on the assumption that operating conditions and applied policies remain what they are at present. See also baseline, models, scenario (Verbruggen et al. 2011).

Cancún agreements A set of decisions adopted at the 16th Session of the Conference of the Parties (COP) to the United Nations Framework Convention on Climate Change (UNFCCC), including the following, among others: the newly established Green Climate Fund (GCF), a newly established technology mechanism, a process for advancing discussions on adaptation, a formal process for reporting mitigation commitments, a goal of limiting global mean surface temperature increase to 2 °C and an agreement on MRV—Measurement, Reporting and Verification for those countries that receive international support for their mitigation efforts.

Cancún pledges During 2010, many countries submitted their existing plans for controlling greenhouse gas (GHG) emissions to the Climate Change Secretariat and these proposals have now been formally acknowledged under the United Nations Framework Convention on

[1]http://ec.europa.eu/eurostat/statistics-explained/index.php/Glossary:Biomass.

Climate Change (UNFCCC). Developed countries presented their plans in the shape of economy-wide targets to reduce emissions, mainly up to 2020, while developing countries proposed ways to limit their growth of emissions in the shape of plans of action.

Carbon cycle The term used to describe the flow of carbon (in various forms, e.g., as carbon dioxide (CO_2) through the atmosphere, ocean, terrestrial and marine biosphere and lithosphere. In this report, the reference unit for the global carbon cycle is $GtCO_2$ or GtC (Gigatonne of carbon = 1 GtC = 10^{15} grams of carbon. This corresponds to 3.667 $GtCO_2$).

Carbon dioxide Carbon dioxide (CO_2) is a colorless gas with a density about 60% higher than that of air that is odorless at normally encountered concentrations. It occurs naturally in Earth's atmosphere as a trace gas at a concentration of about 0.04% (400 ppm) by volume. It is mainly formed during respiration, biomass/fossil fuel combustion, and organic decomposition (especially caused by land-use change). Carbon dioxide could regulate the temperature of the earth. Nowadays, CO_2 is a dominant greenhouse gas which causes global warming.

Carbon intensity (or an emission intensity) The emission rate of a given pollutant relative to the intensity of a specific activity, or an industrial production process; for example grams of carbon dioxide released per megajoule of energy produced, or the ratio of greenhouse gas emissions produced to gross domestic product (GDP).

Carbon market The set of organised and bilateral transactions by which countries trade credits received for greenhouse-gas emission reductions. The market is used to comply with emission goals, or to voluntarily offset a country's own emissions. The carbon market was launched by the creation of three mechanisms under the Kyoto Protocol: emissions trading, across developed countries; the Clean Development Mechanism, based on projects in developing countries; and Joint Implementation, based on projects in developed countries (OECD/IEA 2018).

Carbon monoxide Carbon monoxide (CO) is a colorless, odorless, and tasteless gas that is slightly less dense than air. It results from the incomplete combustion of carbon-containing fuels such as natural gas, gasoline, or wood, and is emitted by a wide variety of combustion sources, including motor vehicles, power plants, wildfires, and incinerators. Carbon monoxide is harmful for health because it binds to hemoglobin in the blood, reducing the ability of blood to carry oxygen.

Carbon sequestration The process involved in carbon capture and the long-term storage of atmospheric carbon dioxide (CO_2) and may refer specifically to: (1) the process of removing carbon from the atmosphere and depositing it in a reservoir. When carried out deliberately, this may also be referred to as carbon dioxide removal, which is a form of geoengineering; (2) carbon capture and storage, where carbon dioxide is removed from flue gases (e.g., at power stations) before being stored in underground reservoirs; (3) natural biogeochemical cycling of carbon between the atmosphere and reservoirs, such as by chemical weathering of rocks (Sedjo and Sohngen 2012).

Circular migration It is repeated movement of persons between a place of origin and the place of destination. It may involve single destination or multiple before returning to the place of origin.

Clean Development Mechanism (CDM) A mechanism under the Kyoto Protocol through which developed (Annex B) countries may finance greenhouse gas emission reduction or removal projects in developing (Non-Annex B) countries and receive credits for doing so which they may apply for meeting mandatory limits on their own emissions (Verbruggen et al. 2011).

Climate Climate in a narrow sense is usually defined as the average weather, or more rigorously, as the statistical description in terms of the mean and variability of relevant quantities over a period of time ranging from months to thousands or millions of years. The classical period for averaging these variables is 30 years, as defined by the World Meteorological Organization. The relevant quantities are most often surface variables such as temperature, precipitation and wind. Climate in a wider sense is the state, including a statistical description, of the climate system.

Climate change Climate change refers to a change in the state of the climate that can be identified (e.g., by using statistical tests) by changes in the mean and/or the variability of its properties and that persists for an extended period, typically decades or longer. Climate change may be due to natural internal processes or external forcings such as modulations of the solar cycles, volcanic eruptions and persistent anthropogenic changes in the composition of the atmosphere or in land use. Note that the Framework Convention on Climate Change (UNFCCC), in its Article 1, defines climate change as: 'a change of climate which is attributed directly or indirectly to human activity that alters the composition of the global atmosphere and which is in addition to natural climate variability observed over comparable time periods'. The UNFCCC thus makes a distinction between climate change attributable to human activities altering the atmospheric composition and climate variability attributable to natural causes. See also Detection and Attribution.

Climate extreme (extreme weather or climate event) See Extreme weather event.

Climate projection A projection of the response of the climate system to emission or concentration scenarios of greenhouse gases and aerosols, or radiative forcing scenarios, often based upon simulations by climate models. Climate projections are distinguished from climate predictions in order to emphasize that climate projections depend upon the emission/concentration/ radiative forcing scenario used, which are based on assumptions concerning, for example, future socioeconomic and technological developments that may or may not be realised and are therefore subject to substantial uncertainty.

Climate scenario A plausible and often simplified representation of the future climate, based on an internally consistent set of climatological relationships that has been constructed for explicit use in investigating the potential consequences of anthropogenic climate change, often serving as input to impact models. Climate projections often serve as the raw material for constructing climate scenarios, but climate scenarios usually require additional information such as about the observed current climate. A climate change scenario is the difference between a climate scenario and the current climate.

Climate variability Climate variability refers to variations in the mean state and other statistics (such as standard deviations, the occurrence of extremes, etc.) of the climate on all spatial and temporal scales beyond that of individual weather events. Variability may be due to natural internal processes within the climate system (internal variability), or to variations in natural or anthropogenic external forcing (external variability).

Climate finance There is no agreed definition of climate finance. The term climate finance is applied both to the financial resources devoted to addressing climate change globally and to financial flows to developing countries to assist them in addressing climate change. The literature includes several concepts in these categories, among which the most commonly used include:

Incremental costs The cost of capital of the incremental investment and the change of operating and maintenance costs for a mitigation or adaptation project in comparison to a reference project. It can be calculated as the difference of the net present values of the two projects.

Incremental investment The extra capital required for the initial investment for a mitigation or adaptation project in comparison to a reference project.

Total climate finance All financial flows whose expected effect is to reduce net greenhouse gas (GHG) emissions and/or to enhance resilience to the impacts of climate variability and the projected climate change. This covers private and public funds, domestic and international flows and expenditures for mitigation and adaptation to current climate variability as well as future climate change.

Total climate finance flowing to developing countries The amount of the total climate finance invested in developing countries that comes from developed countries. This covers private and public funds.

Private climate finance flowing to developing countries Finance and investment by private actors in/from developed countries for mitigation and adaptation activities in developing countries.

Public climate finance flowing to developing countries Finance provided by developed countries' governments and bilateral institutions as well as by multilateral institutions for mitigation and adaptation activities in developing countries. Most of the funds provided are concessional loans and grants.

Climate model (spectrum or hierarchy) A numerical representation of the climate system based on the physical, chemical and biological properties of its components, their interactions and feedback processes and accounting for some of its known properties. The climate system can be represented by models of varying complexity; that is, for any one component or combination of components a spectrum or hierarchy of models can be identified, differing in such aspects as the number of spatial dimensions, the extent to which physical, chemical or biological processes are explicitly represented, or the level at which empirical parametrizations are involved. Coupled Atmosphere–Ocean General Circulation Models (AOGCMs) provide a representation of the climate system that is near or at the most comprehensive end of the spectrum currently available. There is an evolution towards more complex models with interactive chemistry and biology. Climate models are applied as a research tool to study and simulate the climate and for operational purposes, including monthly, seasonal and interannual climate predictions.

Climate projection A climate projection is the simulated response of the climate system to a scenario of future emission or concentration of greenhouse gases (GHGs) and aerosols, generally derived using climate models. Climate projections are distinguished from climate predictions by their dependence on the emission/concentration/radiative forcing scenario used, which is in turn based on assumptions concerning, for example, future socio-economic and technological developments that may or may not be realized.

Climate-resilient pathways Iterative processes for managing change within complex systems in order to reduce disruptions and enhance opportunities associated with climate change.

Climate response See Climate sensitivity.

Climate sensitivity In IPCC reports, equilibrium climate sensitivity (units: °C) refers to the equilibrium (steady state) change in the annual global mean surface temperature following a doubling of the atmospheric equivalent carbon dioxide (CO_2) concentration. Owing to computational constraints, the equilibrium climate sensitivity in a climate model is sometimes estimated by running an atmospheric general circulation model coupled to a mixed-layer ocean model, because equilibrium climate sensitivity is largely determined by atmospheric processes.

Efficient models can be run to equilibrium with a dynamic ocean. The climate sensitivity parameter (units: °C $(W\ m^{-2})^{-1}$) refers to the equilibrium change in the annual global mean surface temperature following a unit change in radiative forcing.

The effective climate sensitivity (units: °C) is an estimate of the global mean surface temperature response to doubled CO_2 concentration that is evaluated from model output or observations for evolving non-equilibrium conditions. It is a measure of the strengths of the climate feedbacks at a particular time and may vary with forcing history and climate state and therefore may differ from equilibrium climate sensitivity.

The transient climate response (units: °C) is the change in the global mean surface temperature, averaged over a 20-year period, centered at the time of atmospheric CO_2 doubling, in a climate model simulation in which CO_2 increases at 1%/yr. It is a measure of the strength and rapidity of the surface temperature response to greenhouse gas (GHG) forcing.

Climate system The climate system is the highly complex system consisting of five major components: the atmosphere, the hydrosphere, the cryosphere, the lithosphere and the biosphere and the interactions between them. The climate system evolves in time under the influence of its own internal dynamics and because of external forcings such as volcanic eruptions, solar variations and anthropogenic forcings such as the changing composition of the atmosphere and land-use change.

Climate variability Climate variability refers to variations in the mean state and other statistics (such as standard deviations, the occurrence of extremes, etc.) of the climate on all spatial and temporal scales beyond that of individual weather events. Variability may be due to natural internal processes within the climate system (internal variability), or to variations in natural or anthropogenic external forcing (external variability). See also Climate change.

Co-benefits The positive effects that a policy or measure aimed at one objective might have on other objectives, irrespective of the net effect on overall social welfare. Co-benefits are often subject to uncertainty and depend on local circumstances and implementation practices, among other factors. Co-benefits are also referred to as ancillary benefits. Appropriate consideration of co-benefits of greenhouse gas mitigation can greatly influence policy decisions concerning the timing and level of mitigation action, and there can be significant advantages to the national economy and technical innovation.

Confidence

Well established: comprehensive meta-analysis or other synthesis or multiple independent studies that agree.

Established but incomplete: general agreement although only a limited number of studies exist; no comprehensive synthesis and/or the studies that exist address the question imprecisely.

Unresolved: multiple independent studies exist but conclusions do not agree.

Inconclusive: limited evidence, recognizing major knowledge gaps.

Cost-effectiveness A policy is more cost-effective if it achieves a given policy goal at lower cost. Integrated models approximate cost-effective solutions, unless they are specifically constrained to behave otherwise. Cost-effective mitigation scenarios are those based on a stylized implementation approach in which a single price on carbon dioxide (CO_2) and other greenhouse gases (GHGs) is applied across the globe in every sector of every country and that rises over time in a way that achieves lowest global discounted costs.

Cryosphere The cryosphere is the part of the Earth system that contains ice, for example snow on the ground, glaciers, ice sheets, lake ice, river ice, sea ice, seasonally and perennially frozen ground (GCW 2016).

Debris-covered glacier A glacier that is covered at its tongue with supra-glacial debris across its full width (Kirkbride 2011).

Decentralized energy/distributed energy Energy systems are considered to be distributed if (1) the systems of production are relatively small and dispersed (such as small-scale solar PV on rooftops), rather than relatively large and centralized; or (2) generation and distribution occur independently from a centralized network (Cleanleap 2016).[2]

Deforestation Conversion of forest to non-forest. For a discussion of the term forest and related terms such as afforestation, reforestation and deforestation, see the IPCC Special Report on Land Use, Land-Use Change, and Forestry (IPCC 2000a). See also information provided by the United Nations Framework Convention on Climate Change (UNFCCC 2013) and the report on Definitions and Methodological Options to Inventory Emissions from Direct Human-induced Degradation of Forests and Devegetation of Other Vegetation Types (IPCC 2003, 2014a, b).

Detection and attribution Detection of change is defined as the process of demonstrating that climate or a system affected by climate has changed in some defined statistical sense, without providing a reason for that change. An identified change is detected in observations if its likelihood of occurrence by chance due to internal variability alone is determined to be small, for example, <10%. Attribution is defined as the process of evaluating the relative contributions of multiple causal factors to a change or event with an assignment of statistical confidence (Hegerl et al. 2010).

Diaspora Diaspora is broadly defined as individuals and members of networks, associations and communities who have left their country of origin, and maintain links with their homelands. This concept covers not only emigrants but also erstwhile emigrants with the citizenship of the host country, dual citizens, and second-/third-generation born persons in the host countries.

Disabled Adjusted Life Year (DALYS) DALYs for a disease or health condition are calculated as the sum of the Years of Life Lost (YLL) due to premature mortality in the population and the Years Lost due to Disability (YLD) for people living with the health condition or its consequences. One DALY can be thought of as one lost year of "healthy" life. The sum of these DALYs across the population, or the burden of disease, can be thought of as a measurement of the gap between current health status and an ideal health situation where the entire population lives to an advanced age, free of disease and disability (WHO 2018).[3]

Disaster A serious disruption of the functioning of a community or a society at any scale due to hazardous events interacting with conditions of exposure, vulnerability and capacity, leading to one or more of the following: human, material, economic and environmental losses and impacts (UNISDR 2017).

Disaster risk The potential loss of life, injury, or destroyed or damaged assets which could occur to a system, society or a community in a specific period of time, determined probabilistically as a function of hazard, exposure, vulnerability and capacity (UNISDR 2017).

---

[2]http://cleanleap.com/where-are-we-renewable-energy/05-distributed-renewable-energy-developing-countries.
[3]http://www.who.int/healthinfo/global_burden_disease/metrics_daly/en/.

Disaster risk governance The system of institutions, mechanisms, policy and legal frameworks and other arrangements to guide, coordinate and oversee disaster risk reduction and related areas of policy (UNISDR 2017).

Disaster risk management Disaster risk management is the application of disaster risk reduction policies and strategies to prevent new disaster risk, reduce existing disaster risk and manage residual risk, contributing to the strengthening of resilience and reduction of disaster losses (UNISDR 2017).

Disaster risk reduction Disaster risk reduction is aimed at preventing new and reducing existing disaster risk and managing residual risk, all of which contribute to strengthening resilience and therefore to the achievement of sustainable development (UNISDR 2017).

Dispersion model Atmospheric dispersion models are computer programs that use mathematical algorithms to simulate how pollutants in the ambient atmosphere disperse and, in some cases, how they react in the atmosphere.

Downhill scenario Downhill scenario represents strong climate change, a socially economically and politically unstable region and strong ecosystem degradation. In the downhill scenario. regional conflicts over resource sharing persist—and even multiply as scarcity increases. People and institutions do not benefit from emerging opportunities for efficient resource use. Communities remain isolated from the larger market systems. Mountain livelihoods do not enjoy inclusive growth through new, scientifically proven skills and practices. Ecosystems are degraded, mitigation efforts fail, and fossil fuels remain the dominant energy source. Climate change impacts reflect the Intergovernmental Panel on Climate Change (IPCC)'s worst case scenario—global temperature rising by substantially more than 2.0 °C.

Downscaling Downscaling is a method that derives local- to regional-scale (10–100 km) information from larger-scale models or data analyses. Two main methods are distinguished: dynamical downscaling and empirical/statistical downscaling. The dynamical method uses the output of regional climate models, global models with variable spatial resolution or high-resolution global models. The empirical/statistical methods develop statistical relationships that link the large-scale atmospheric variables with local/regional climate variables. In all cases, the quality of the downscaled product depends on the quality of the driving model.

Drivers In a policy context, drivers provide an impetus and direction for initiating and supporting policy actions. The deployment of renewable energy is, for example, driven by concerns about climate change or energy security. In a more general sense, a driver is the leverage to bring about a reaction, for example, emissions are caused by fossil fuel consumption and/or economic growth. See also opportunities (Verbruggen et al. 2011).

Drought A period of abnormally dry weather long enough to cause a serious hydrological imbalance. Drought is a relative term; therefore any discussion in terms of precipitation deficit must refer to the particular precipitation-related activity that is under discussion. For example, shortage of precipitation during the growing season impinges on crop production or ecosystem function in general (due to soil moisture drought, also termed agricultural drought) and during the runoff and percolation season primarily affects water supplies (hydrological drought). Storage changes in soil moisture and groundwater are also affected by increases in actual evapotranspiration in addition to reductions in precipitation. A period with an abnormal precipitation deficit is defined as a meteorological drought. A megadrought is a very lengthy and pervasive drought, lasting much longer than normal, usually a decade or more.

Drivers of change Natural or human-induced factors that directly or indirectly cause changes in the system. A direct driver of change unequivocally influences ecosystem processes and/or sociocultural and economic characteristics, and can be identified and measured to differing degrees of accuracy; an indirect driver of change operates by altering the level or rate of change of other, more direct drivers (MA 2005).

Early warning system The set of capacities needed to generate and disseminate timely and meaningful warning information to enable individuals, communities and organizations threatened by a hazard to prepare to act promptly and appropriately to reduce the possibility of harm or loss.

Early warning systems (Chap. 11) An integrated system of hazard monitoring, forecasting and prediction, disaster risk assessment, communication and preparedness activities systems and processes that enables individuals, communities, governments, businesses and others to take timely action to reduce disaster risks in advance of hazardous events (UNISDR 2017).

Economic loss Total economic impact that consists of direct economic loss and indirect economic loss.
Direct economic loss: the monetary value of total or partial destruction of physical assets existing in the affected area. Direct economic loss is nearly equivalent to physical damage.
Indirect economic loss: a decline in economic value added as a consequence of direct economic loss and/or human and environmental impacts (UNISDR 2017).

Ecosystem An ecosystem is a functional unit consisting of living organisms, their non-living environment and the interactions within and between them. The components included in a given ecosystem and its spatial boundaries depend on the purpose for which the ecosystem is defined: in some cases they are relatively sharp, while in others they are diffuse. Ecosystem boundaries can change over time. Ecosystems are nested within other ecosystems and their scale can range from very small to the entire biosphere. In the current era, most ecosystems either contain people as key organisms, or are influenced by the effects of human activities in their environment.

Ecosystem approach The ecosystem approach is a strategy for the integrated management of land, water and living resources that promotes conservation and sustainable use in an equitable way (CBD 1992).

Ecosystem services The benefits people obtain from ecosystems. These include provisioning services such as food and water; regulating services such as flood and disease control; cultural services such as spiritual, recreational, and cultural benefits; and supporting services such as nutrient cycling that maintain the conditions for life on Earth. The concept "ecosystem goods and services" is synonymous with ecosystem services (MEA 2005).

Electricity The flow of passing charge through a conductor, driven by a difference in voltage between the ends of the conductor. Electrical power is generated by work from heat in a gas or steam turbine or from wind, oceans or falling water, or produced directly from sunlight using a photovoltaic device or chemically in a fuel cell. Being a current, electricity cannot be stored and requires wires and cables for its transmission (see grid). Because electric current flows immediately, the demand for electricity must be matched by production in real time (Verbruggen et al. 2011).

Electricity generation The total amount of electricity generated by power only or combined heat and power plants including generation required for own use (also referred to as gross generation). Electricity production on the other hand is the total amount of electricity generated by a power plant which includes own-use electricity, as well as transmission and distribution losses (OECD/IEA 2018).

Elemental carbon Elemental carbon (EC), similar to black carbon, also refers to the major component of light-absorbing refractory carbonaceous matter. In most cases, these two terms are interchangeable. The measurement of EC is usually through thermal evolution and chemical determinations.

Emigration The act of leaving from one country with a view to settling in another. The person who emigrates is called an emigrant.

Emissions Direct emissions are released and attributed at points in a specific renewable energy chain, whether a sector, a technology or an activity. For example, methane emissions from decomposing submerged organic materials in hydropower reservoirs, or the release of CO_2 dissolved in hot water from geothermal plants, or CO_2 from biomass combustion. Indirect emissions are due to activities outside the considered renewable energy chain but which are required to realize the renewable energy deployment. For example, emissions from increased production of fertilizers used in the cultivation of biofuel crops or emissions from displaced crop production or deforestation as the result of biofuel crops. Avoided emissions are emission reductions arising from mitigation measures like renewable energy deployment (Verbruggen et al. 2011).

Emission control technology In order to reduce emissions of different pollutants, various technologies have been adopted, which are termed as emission control technologies. For example, catalysis technology was developed to reduce and detoxify the gasoline and diesel vehicle exhaust. In the industrial sectors, emission control technologies were also used to constrain emissions (SO_2 or VOCs) from coal fuel combustion.

Energy The amount of work or heat delivered. Energy is classified in a variety of types and becomes available to human ends when it flows from one place to another or is converted from one type into another. Daily, the sun supplies large flows of radiation energy. Part of that energy is used directly, while part undergoes several conversions creating water evaporation, winds, etc. Some share is stored in biomass or rivers that can be harvested. Some share is directly usable such as daylight, ventilation or ambient heat. Primary energy (also referred to as energy sources) is the energy embodied in natural resources (e.g., coal, crude oil, natural gas, uranium, and renewable sources). Primary energy is transformed into secondary energy by cleaning (natural gas), refining (crude oil to oil products) or by conversion into electricity or heat. When the secondary energy is delivered at the end-use facilities it is called final energy (e.g., electricity at the wall outlet), where it becomes usable energy in supplying services (e.g., light) (Verbruggen et al. 2011).

Energy access People are provided the ability to benefit from affordable, clean and reliable energy services for basic human needs (cooking and heating, lighting, communication, mobility) and productive uses (Verbruggen et al. 2011).

Energy efficiency The ratio of useful energy or other useful physical outputs obtained from a system, conversion process, transmission or storage activity to the input of energy (measured as kWh/kWh, tonnes/kWh or any other physical measure of useful output like tonne-km transported, etc.). Energy efficiency is a component of energy intensity (Verbruggen et al. 2011).

Energy intensity The ratio of energy inputs (in Joules) to the economic output (in dollars) that absorbed the energy input. Energy intensity is the reciprocal of energy productivity. At the national level, energy intensity is the ratio of total domestic primary (or final) energy use to gross domestic product (GDP). Energy intensity is also used as a name for the ratio of energy inputs to output or performance in physical terms (e.g., tonnes of steel output, tonne-km transported, etc.) and in such cases, is the reciprocal of energy efficiency (Verbruggen et al. 2011).

EnergyPlus EnergyPlus refers to projects, programmes and interventions that promote basic, social (community) and productive uses of energy. The EnergyPlus approach promotes the productive use of energy: for generation of equitable employment and additional income; for meeting needs of existing and new enterprises; for community needs such as strengthened security and better access to education and health care, including through electricity for street lighting, clinics and schools; and for lifestyle needs to improve living standards. In turn, the formula of EnergyPlus = Energy Access + Empowerment can be seen as contributing to overall sustainable human development and poverty reduction (UNDP 2015).

Energy poverty It is defined in its broadest terms as the absence of sufficient choice in accessing adequate, affordable, reliable, high quality, safe and environmentally benign energy services to support economic and human development (Reddy 2000). There are several different approaches to define and measure energy poverty ranging from simple access based (lack of access to modern energy services) to more complex quantity based approach (minimum amount of physical energy necessary for basic needs such as cooking and lighting) including expenditure based (households that spend more than a certain percent—generally above 10%—of their expenditure on energy) and consensual based perceived deprivation (households facing difficulties in order to meet basic energy services) with each having strengths and weakness. For the sake of simplicity, the Energy Outlook 2015 defines energy poverty as a lack of access to modern energy \services using two indicators at household level: lack of access to electricity and reliance on the traditional use of biomass for cooking. These services are defined as household access to electricity and clean cooking facilities (e.g. fuels and stoves that do not cause air pollution in houses) (OECD/IEA 2018).[4]

Energy savings Decreasing energy intensity by changing the activities that demand energy inputs. Energy savings can be realized by technical, organizational, institutional and structural actions and by changed behavior (Verbruggen et al. 2011).

Energy security The goal of a given country, or the global community as a whole, to maintain an adequate energy supply. Measures encompass safeguarding access to energy resources; enabling development and deployment of technologies; building sufficient infrastructure to generate, store and transmit energy supplies; ensuring enforceable contracts of delivery; and access to energy at affordable prices for a specific society or groups in society (Verbruggen et al. 2011).

Energy services Energy services are the tasks to be performed using energy. A specific energy service such as lighting may be supplied by a number of different means from daylighting to oil lamps to incandescent, fluorescent or light-emitting diode devices. The amount of energy used to provide a service may vary over a factor of 10 or more, and the corresponding greenhouse gas emissions may vary from zero to a very high value depending on the source of energy and the type of end-use device (Verbruggen et al. 2011).

Energy transition It refers to the shift from current energy production and consumption systems, which rely primarily on non-renewable energy sources such as oil, natural gas and coal, to a more efficient, lower-carbon energy mix. According to the energy trilemma concept, the energy transitional challenges basically involves balancing three interwoven objectives: energy security (the reliability of energy supply must be ensured to meet current and future demand); energy equity (energy must be accessible around the world, particularly in emerging markets, at an affordable cost) and environmental sustainability (global warming calls for improved energy efficiency and the development of renewable and low-greenhouse gas (GHGs) energy sources (WEC 2016).

Exposure The situation of people, infrastructure, housing, production capacities and other tangible human assets located in hazard-prone areas (UNISDR 2017).

El Niño-Southern Oscillation (ENSO) The term El Niño was initially used to describe a warm-water current that periodically flows along the coast of Ecuador and Peru, disrupting the local fishery. It has since become identified with a basin-wide warming of the tropical Pacific Ocean east of the dateline. This oceanic event is associated with a fluctuation of a global-scale

[4]http://www.iea.org/topics/energypoverty/.

tropical and subtropical surface pressure pattern called the Southern Oscillation. This coupled atmosphere–ocean phenomenon, with preferred time scales of two to about seven years, is known as the El Niño-Southern Oscillation (ENSO). It is often measured by the surface pressure anomaly difference between Tahiti and Darwin or the sea surface temperatures in the central and eastern equatorial Pacific. During an ENSO event, the prevailing trade winds weaken, reducing upwelling and altering ocean currents such that the sea surface temperatures warm, further weakening the trade winds. This event has a great impact on the wind, sea surface temperature and precipitation patterns in the tropical Pacific. It has climatic effects throughout the Pacific region and in many other parts of the world, through global teleconnections. The cold phase of ENSO is called La Niña.

Emission scenario A plausible representation of the future development of emissions of substances that are potentially radiatively active (e.g., greenhouse gases (GHGs), aerosols) based on a coherent and internally consistent set of assumptions about driving forces (such as demographic and socio-economic development, technological change, energy and land use) and their key relationships. Concentration scenarios, derived from emission scenarios, are used as input to a climate model to compute climate projections. In IPCC (1992) a set of emission scenarios was presented which were used as a basis for the climate projections in IPCC (1996). These emission scenarios are referred to as the IS92 scenarios. In the IPCC Special Report on Emissions Scenarios (IPCC 2000b) emission scenarios, the so-called SRES scenarios, were published, some of which were used, among others, as a basis for the climate projections presented in Chaps. 9–11 of IPCC WGI TAR (IPCC 2001) and Chaps. 10 and 11 of IPCC WGI AR4 (IPCC 2007) as well as in the IPCC WGI AR5 (IPCC 2013). New emission scenarios for climate change, the four Representative Concentration Pathways, were developed for, but independently of, the present IPCC assessment. See also Baseline/reference, Mitigation scenario and Transformation pathway (IPCC 2014a, b).

Energy access Access to clean, reliable and affordable energy services for cooking and heating, lighting, communications and productive uses (AGECC 2010).

Energy intensity The ratio of energy use to economic or physical output.

Energy security The goal of a given country, or the global community as a whole, to maintain an adequate, stable and predictable energy supply. Measures encompass safeguarding the sufficiency of energy resources to meet national energy demand at competitive and stable prices and the resilience of the energy supply; enabling development and deployment of technologies; building sufficient infrastructure to generate, store and transmit energy supplies and ensuring enforceable contracts of delivery.

Ensemble A collection of model simulations characterizing a climate prediction or projection. Differences in initial conditions and model formulation result in different evolutions of the modeled system and may give information on uncertainty associated with model error and error in initial conditions in the case of climate forecasts and on uncertainty associated with model error and with internally generated climate variability in the case of climate projections.

Equilibrium Line Altitude (ELA) The spatially averaged altitude of the equilibrium line. The equilibrium line is a set of points on the surface of the glacier where the climatic mass balance is zero at a given moment. The equilibrium line separates the accumulation zone (where the glacier is gaining mass) from the ablation zone (where the glacier is losing mass): (modified after Cogley et al. 2011).

Eutrophication Over-enrichment of water by nutrients such as nitrogen and phosphorus. It is one of the leading causes of water quality impairment. The two most acute symptoms of eutrophication are hypoxia (or oxygen depletion) and harmful algal blooms.

Exposure The presence of people, livelihoods, species or ecosystems, environ- mental functions, services, and resources, infrastructure, or economic, social, or cultural assets in places and settings that could be adversely affected.

External forcing External forcing refers to a forcing agent outside the climate system causing a change in the climate system. Volcanic eruptions, solar variations and anthropogenic changes in the composition of the atmosphere and land-use change are external forcings. Orbital forcing is also an external forcing as the insolation changes with orbital parameters eccentricity, tilt and precession of the equinox.

Extreme weather event An extreme weather event is an event that is rare at a particular place and time of year. Definitions of rare vary, but an extreme weather event would normally be as rare as or rarer than the 10th or 90th percentile of a probability density function estimated from observations. By definition, the characteristics of what is called extreme weather may vary from place to place in an absolute sense. When a pattern of extreme weather persists for some time, such as a season, it may be classed as an extreme climate event, especially if it yields an average or total that is itself extreme (e.g., drought or heavy rainfall over a season).

Feedback See Climate feedback.

Flood The overflowing of the normal confines of a stream or other body of water, or the accumulation of water over areas not normally submerged. Floods include river (fluvial) floods, flash floods, urban floods, pluvial floods, sewer floods, coastal floods and glacial lake outburst floods.

Food accessibility It is defined as the access by individuals to adequate resources (entitlements) for acquiring appropriate foods for a nutritious diet. Entitlements are defined as the set of all commodity bundles over which a person can establish command given the legal, political, economic and social arrangements of the community in which they live (including traditional rights such as access to common resources) (FAO 2006).

Food and nutrition security In 2012, Committee on World Food Security defined 'food and nutrition security' as the state that exists when all people, at all times, have physical, social and economic access to food which is consumed in sufficient quantity and quality to meet their dietary needs and food preferences, and is support by an environment of adequate sanitation, health services and care, allowing for a healthy and active life (FAO 2012a, b).

Food availability Food availability addresses the "supply side" of food security and is determined by the level of food production, stock levels and net trade (FAO 2008).

Food security Food security exists when all people, at all times, have physical and economic access to sufficient safe and nutritious food that meets their dietary needs and food preferences for an active and healthy life (FAO 1996).

Food stability To be food secure, a population, household or individual must have access to adequate food at all times. They should not risk losing access to food as a consequence of sudden shocks (e.g. an economic or climatic crisis) or cyclical events (e.g. seasonal food insecurity). The concept of stability can therefore refer to both the availability and access dimensions of food security (FAO 2006).

Food utilization Utilization of food through adequate diet, clean water, sanitation and health care to reach a state of nutritional well-being where all physiological needs are met. This brings out the importance of non-food inputs in food security (FAO 2006).

Forest A vegetation type dominated by trees. Many definitions of the term forest are in use throughout the world, reflecting wide differences in biogeophysical conditions, social structure and economics. For a discussion of the term forest and related terms such as afforestation, reforestation and deforestation, see the IPCC Special Report on Land Use, Land-Use Change, and Forestry (IPCC 2000a). See also information provided by the United Nations Framework Convention on Climate Change (UNFCCC 2013) and the Report on Definitions and Methodological Options to Inventory Emissions from Direct Human-induced Degradation of Forests and Devegetation of Other Vegetation Types (IPCC 2003, 2014a, b).

Fossil fuel It is a generic term for non-renewable energy sources such as coal, coal products, natural gas, derived gas, crude oil, petroleum products and non-renewable wastes. These fuels originate from plants and animals that existed in the geological past (for example, millions of years ago). Fossil fuels can be also made by industrial processes from other fossil fuels (for example in the oil refinery, crude oil is transformed into motor gasoline) (OECD/IEA 2018).[5]

Gender analysis The systematic gathering and examination of information on gender differences and social relations in order to identify, understand and redress inequities based on gender.

Gender mainstreaming An organisational strategy to bring a gender perspective to all aspects of an institution's policy and activities, through building gender capacity and accountability.

General circulation model See *Climate model.*

Glacier A perennial mass of ice, and possibly firn and snow, originating on the land surface by therecrystallization of snow or other forms of solid precipitation and showing evidence of past or present flow (Cogley et al. 2011).

Glacierized Region or terrain, containing glaciers or covered by glacier ice today (Cogley et al. 2011).

Glacier mass balance The change in the mass of a glacier over a stated span of time. The span of time is often a year or a season. The annual mass balance is the sum of accumulation and ablation over the mass-balance year, equivalent to the sum of annual accumulation and annual ablation (modified after Cogley et al. 2011).

Global climate model (also referred to as general circulation model both abbreviated as GCM) See Climate model.

Geothermal energy Accessible thermal energy stored in the Earth's interior, in both rock and trapped steam or liquid water (hydrothermal resources), which may be used to generate electric energy in a thermal power plant, or to supply heat to any process requiring it. The main sources of geothermal energy are the residual energy available from planet formation and the energy continuously generated from radionuclide decay (Verbruggen et al. 2011).

Governance Governance is a comprehensive and inclusive concept of the full range of means for deciding, managing and implementing policies and measures. Whereas government is defined strictly in terms of the nation-state, the more inclusive concept of governance, recognizes the contributions of various levels of government (global, international, regional, local) and the contributing roles of the private sector, of nongovernmental actors and of civil society to addressing the many types of issues facing the global community (Verbruggen et al. 2011).

Greenhouse Gases (GHGs) Greenhouse gases are those gaseous constituents of the atmosphere, both natural and anthropogenic, that absorb and emit radiation at specific wavelengths within the spectrum of thermal infrared radiation emitted by the Earth's surface, the atmosphere and clouds. This property causes the greenhouse effect. Water vapour (H_2O), carbon dioxide (CO_2), nitrous oxide (N_2O), methane (CH_4) and ozone (O_3) are the primary greenhouse gases in the Earth's atmosphere. Moreover, there are a number of entirely human-made

---

[5]http://www.iea.org/about/glossary.

greenhouse gases in the atmosphere, such as the halocarbons and other chlorine- and bromine-containing substances, dealt with under the Montreal Protocol. Besides CO_2, N_2O and CH_4, the Kyoto Protocol deals with the greenhouse gases sulphur hexafluoride (SF_6), hydrofluorocarbons (HFCs) and perfluorocarbons (PFCs) (Verbruggen et al. 2011).

Grid (electric grid electricity grid, power grid) A network consisting of wires, switches and transformers to transmit electricity from power sources to power users. A large network is layered from low-voltage (110–240 V) distribution, over intermediate voltage (1–50 kV) to high-voltage (above 50 kV to MV) transport subsystems. Interconnected grids cover large areas up to continents. The grid is a power exchange platform enhancing supply reliability and economies of scale (Verbruggen et al. 2011).

Gross Domestic Product (GDP) The sum of gross value added, at purchasers' prices, by all resident and non-resident producers in the economy, plus any taxes and minus any subsidies not included in the value of the products in a country or a geographic region for a given period, normally one year. It is calculated without deducting for depreciation of fabricated assets or depletion and degradation of natural resources (Verbruggen et al. 2011).

Hard infrastructure Hard infrastructure is defined as transportation, energy, water and sanitation, and telecommunication as opposed to soft infrastructure (the institutional facilities used to deliver hard infrastructure through market and non-market economic, social, and political interactions) (Khan and Weiss 2006).

Hazard The potential occurrence of a natural or human-induced physical event or trend or physical impact that may cause loss of life, injury, or other health impacts, as well as damage and loss to property, infrastructure, livelihoods, service provision, ecosystems and environmental resources. In this report, the term hazard usually refers to climate-related physical events or trends or their physical impacts.

Heat wave A period of abnormally and uncomfortably hot weather.

Human Development Index (HDI) The HDI allows the assessment of countries' progress regarding social and economic development as a composite index of three indicators: (1) health measured by life expectancy at birth; (2) knowledge as measured by a combination of the adult literacy rate and the combined primary, secondary and tertiary school enrolment ratio; and (3) standard of living as gross domestic product per capita (in purchasing power parity). The HDI only acts as a broad proxy for some of the key issues of human development; for instance, it does not reflect issues such as political participation or gender inequalities (Verbruggen et al. 2011).

Hydrological cycle The cycle in which water evaporates from the oceans and the land surface, is carried over the Earth in atmospheric circulation as water vapour, condenses to form clouds, precipitates over ocean and land as rain or snow, which on land can be intercepted by trees and vegetation, provides runoff on the land surface, infiltrates into soils, recharges groundwater, discharges into streams and ultimately flows out into the oceans, from which it will eventually evaporate again. The various systems involved in the hydrological cycle are usually referred to as hydrological systems.

Hydropower The energy of water moving from higher to lower elevations that is converted into mechanical energy through a turbine or other device that is either used directly for mechanical work or more commonly to operate a generator that produces electricity. The term is also used to describe the kinetic energy of stream flow that may also be converted into mechanical energy of a generator through an in-stream turbine to produce electricity (Verbruggen et al. 2011).

Immigration The act by which residents of a country move into another country for the purpose of settlement. The person who immigrates is called an immigrant.

Impacts (consequences, outcomes) Effects on natural and human systems. In this report, the term impacts is used primarily to refer to the effects on natural and human systems of extreme weather and climate events and of climate change. Impacts generally refer to effects on lives, livelihoods, health, ecosystems, economies, societies, cultures, services and infrastructure due to the interaction of climate changes or hazardous climate events occurring within a specific time period and the vulnerability of an exposed society or system. Impacts are also referred to as consequences and outcomes. The impacts of climate change on geophysical systems, including floods, droughts and sea level rise, are a subset of impacts called physical impacts.

Indigenous people Indigenous people are variously known in different countries as ethnic minorities, minority populations, and tribal groups. The terms refer to social groups with a cultural identity distinct from the dominant groups, which makes them vulnerable to disadvantage in the development process (IFAD 2002).

Indo-Gangetic plains The Indo-Gangetic Plains is a 255 million hectare (630 million acres) plain region in the Indian subcontinent, stretching westward from (and including) the combined delta of the Brahmaputra River valley and the Ganges (Ganga) River to the Indus River valley.

Industrial revolution A period of rapid industrial growth with far-reaching social and economic consequences, beginning in Britain during the second half of the 18th century and spreading to Europe and later to other countries including the United States. The invention of the steam engine was an important trigger of this development. The industrial revolution marks the beginning of a strong increase in the use of fossil fuels and emission of, in particular, fossil carbon dioxide (CO_2). In this report the terms pre-industrial and industrial refer, somewhat arbitrarily, to the periods before and after 1750, respectively.

In-migration Migration of a person who enters into a geographically defined area from another within a country.

Integrated assessment A method of analysis that combines results and models from the physical, biological, economic and social sciences and the interactions among these components in a consistent framework to evaluate the status and the consequences of environmental change and the policy responses to it.

Intensive agriculture Intensive agriculture involves various types of agriculture with higher levels of input and output per unit of agricultural land area. It is characterized by a low fallow ratio, higher use of inputs such as capital and labor, and higher crop yields per unit land area. Source: https://en.wikipedia.org/wiki/Intensive_farming.

Internal migration Migration that occurs within the political boundaries of a country.

Internally displaced persons "Internally displaced persons are persons or groups of persons who have been forced or obliged to flee or to leave their homes or places of habitual residence, particularly as a result, or in order to avoid the effects, of armed conflict, situations of generalized violence, violations of human rights, or natural/human-made disasters, and who have not crossed an internationally recognized State border" (UN 1998).

International migration Migration that occurs across the political countries of a country. It is a movement from one to another country.

In-migrant A migrant person defined in relation to the place of destination.

Internal variability See Climate variability.

Intersectionality This is the term used to define the intersecting of social differentiations and identities based on class, caste, ethnicity, age, and other factors with gender.

Irregular migration International migration that takes place in violation of rules and laws of the origin, transit and destination countries.

Karez Underground aqueduct that passively taps groundwater in the piedmont of the arid and semi-arid highlands.

Kyoto protocol The Kyoto Protocol to the UNFCCC was adopted at the Third Session of the Conference of the Parties in 1997 in Kyoto. It contains legally binding commitments, in addition to those included in the UNFCCC. Annex B countries agreed to reduce their anthropogenic greenhouse gas emissions (CO_2, methane, nitrous oxide, hydrofluorocarbons, perfluorocarbons and sulphur hexafluoride) by at least 5% below 1990 levels in the commitment period 2008 to 2012. The Kyoto Protocol came into force on 16 February 2005 (Verbruggen et al. 2011).

Labour migration Migration of persons from one administrative area to another for the purpose of employment.

Landscape approach Dealing with large-scale processes in an integrated and multidisciplinary manner, combining natural resources management with environmental and livelihood considerations (FAO 2012a, b).

Land use and land-use change Land use refers to the total of arrangements, activities and inputs undertaken in a certain land cover type (a set of human actions). The term land use is also used in the sense of the social and economic purposes for which land is managed (e.g., grazing, timber extraction and conservation). In urban settlements it is related to land uses within cities and their hinterlands. Urban land use has implications on city management, structure and form and thus on energy demand, green-house gas (GHG) emissions and mobility, among other aspects.

Land-Use Change (LUC) Land-use change refers to a change in the use or management of land by humans, which may lead to a change in land cover. Land cover and land-use change may have an impact on the surface albedo, evapotranspiration, sources and sinks of greenhouse gases (GHGs), or other properties of the climate system and may thus give rise to radiative forcing and/or other impacts on climate, locally or globally. See also the IPCC Special Report on Land Use, Land-Use Change, and Forestry (IPCC 2000a).

Last Glacial Maximum (LGM) The Last Glacial Maximum refers to the time of maximum extent of the ice sheets during the last glaciation, approximately 21 ka. This period has been widely studied because the radiative forcings and boundary conditions are relatively well known and because the global cooling during that period is comparable with the projected warming over the 21st century.

Leakage Phenomena whereby the reduction in emissions (relative to a baseline) in a jurisdiction/sector associated with the implementation of mitigation policy is offset to some degree by an increase outside the jurisdiction/sector through induced changes in consumption, production, prices, land use and/or trade across the jurisdictions/sectors. Leakage can occur at a number of levels, be it a project, state, province, nation or world region.

In the context of Carbon Dioxide Capture and Storage (CCS), CO_2 leakage refers to the escape of injected carbon dioxide (CO_2) from the storage location and eventual release to the atmosphere. In the context of other substances, the term is used more generically, such as for methane (CH_4) leakage (e.g., from fossil fuel extraction activities) and hydrofluorocarbon (HFC) leakage (e.g., from refrigeration and air- conditioning systems).

Likelihood The chance of a specific outcome occurring, where this might be estimated probabilistically. Likelihood is expressed in this report using a standard terminology (Mastrandrea et al. 2010). See also Confidence and Uncertainty.

Little Ice Age (LIA) An interval between approximately AD 1400 and 1900 when temperatures in the Northern Hemisphere were generally colder than today's, especially in Europe.

Lock-in Lock-in occurs when a market is stuck with a standard even though participants would be better off with an alternative. In this report, lock-in is used more broadly as path dependence, which is the generic situation where decisions, events or outcomes at one point in time constrain adaptation, mitigation or other actions or options at a later point in time.

Low regrets policy A policy that would generate net social and/or economic benefits under current climate and a range of future climate change scenarios.

Maladaptive actions (or maladaptation) Actions that may lead to increased risk of adverse climate-related outcomes, increased vulnerability to climate change, or diminished welfare, now or in the future.

Methane Methane is a chemical compound with the chemical formula CH_4. It is a group-14 hydride and the simplest alkane, and is the main constituent of natural gas. It could be produced from the decomposition of organic matter. It is one of the major greenhouse gases in the atmosphere, causing the global warming.

Migration Migration is a change in the usual place of residence from one geographically defined territory to another either on permanent or semi-permanent basis. While migration is a move, migrant is a person who undertakes the move.

Mitigation (of climate change) A human intervention to reduce the sources or enhance the sinks of greenhouse gases (GHGs). This report also assesses human interventions to reduce the sources of other substances which may contribute directly or indirectly to limiting climate change, including, for example, the reduction of particulate matter emissions that can directly alter the radiation balance (e.g., black carbon) or measures that control emissions of carbon monoxide, nitrogen oxides, Volatile Organic Compounds and other pollutants that can alter the concentration of tropospheric ozone which has an indirect effect on the climate.

Mitigation scenario A plausible description of the future that describes how the (studied) system responds to the implementation of mitigation policies and measures. See also Baseline/reference, Emission scenario, Representative Concentration Pathways (RCPs), SRES scenarios and Transformation pathway.

Monsoon A monsoon is a tropical and subtropical seasonal reversal in both the surface winds and associated precipitation, caused by differential heating between a continental-scale land mass and the adjacent ocean. Monsoon rains occur mainly over land in summer.

Mountain specificities The concept of mountain specificities is tailored to capture the particular challenges of mountain ecosystems (Jodha 1992). These specificities are classified as either constraining features, such as accessibility, marginality, and fragility; or enabling features such as diversity, niche, and human adaptation capacity. Within the mountain specificity framework, the term accessibility captures elements of distance, mobility, and availability of risk management options. Marginality refers to the relative endowments of a system. In a mountain system, marginality is created by slope/altitude, low resource productivity and reinforced by lack of social and political capital. Fragility can best be understood as the diminished capacity of a social or ecological system to buffer shocks. Diversity, niche, and adaptation capacity capture different coping abilities and strategies that emerge from NRM patterns, livelihood endowments, and cultural practices.

Multidimensional Poverty Index (MPI) The multidimensional poverty index (MPI) is an index of acute multidimensional poverty (Alkire and Santos 2010). It reflects deprivations in education to health outcomes to assets and services. It reveals a different pattern of poverty than income poverty, as it highlights deprivations directly. The MPI has three dimensions: health, education, and standard of living. These are measured using different indicators.

Multi-hazard The selection of multiple major hazards that the country faces, and (2) the specific contexts where hazardous events may occur simultaneously, cascadingly or cumulatively over time, and taking into account the potential interrelated effects (UNISDR 2017).

Neglected and Underutilized Food Crops (NUFCs) These are food crops which are part of a larger diversity portfolio, were once more popular but today are neglected by the people (Adhikari et al. 2017). These crops continue to be grown, managed and collected in marginal locations because of their usefulness for local populations. The NUFCs in the HKH are mainly millets, sorghum, buckwheat, barley, beans, black gram, horse crop, taro, yam, amala and mammon.

Off-grid Off-the-grid is a system and lifestyle designed to help people function without the support of remote infrastructure, such as an electrical grid (Vanini 2014).

Organic carbon Organic carbon is the amount of carbon found in complex mixture of chemical compounds containing carbon-carbon bonds produced from fossil fuel and biofuel burning and natural biogenic emissions.

Out-migrant A migrant person defined in relation to the place of origin.

Out-migration Migration of a person who leaves a geographically defined area to another within a country.

Ozone Ozone is an inorganic molecule with the chemical formula O_3 and is a pale blue gas with a distinctively pungent smell. It is formed from dioxygen by the action of ultraviolet light and also atmospheric electrical discharges. Ozone is a powerful oxidant, making it a potent respiratory hazard and pollutant near ground level. On the other hand, the ozone layer (a portion of the stratosphere with a higher concentration of ozone) could prevent damaging ultraviolet light from reaching the Earth's surface, to be beneficial for both plants and animals.

Particulate matter Particulate matter, also known as atmospheric particulate matter or particulates, or suspended particulate matter, are microscopic solid or liquid matter suspended in Earth's atmosphere.

Permafrost Ground (soil or rock and included ice and organic material) that remains at or below 0 °C for at least two consecutive years.

$PM_{2.5}$ $PM_{2.5}$ are fine particles with an aerodynamic diameter of 2.5 μm or less. It is also known as fine particulate matter. $PM_{2.5}$ particles are small enough to be penetrated deep into the lungs.

PM_{10} PM_{10} are particulate matters with an aerodynamic diameter of 10 μm or less.

Potential pathways Potential pathways constitute a set of actions and combinations of actions that a decision maker (individual, country, business, policy maker) can take. Pathways usually differ in terms of trade-offs, opportunities and challenges, but still lead to the same outcome.

Poverty Poverty is a complex concept with several definitions stemming from different schools of thought. It can refer to material circumstances (such as need, pattern of deprivation or limited resources), economic conditions (such as standard of living, inequality or economic position) and/or social relationships (such as social class, dependency, exclusion, lack of basic security or lack of entitlement).

Practical Gender Needs (PGNs) The immediate needs identified by women to assist their survival in their socially accepted roles, within existing power structures. PGNs do not directly challenge gender inequalities.

Pre-industrial See Industrial Revolution.

Preparedness The knowledge and capacities developed by governments, response and recovery organizations, communities and individuals to effectively anticipate, respond to and recover from the impacts of likely, imminent or current disasters (UNISDR 2017).

Prevention Activities and measures to avoid existing and new disaster risks (UNISDR 2017).

Primary pollutants Primary pollutants are usually produced from a process directly, such as carbon monoxide gas from motor vehicle exhaust, or the sulfur dioxide released directly from factories.

Private costs Private costs are carried by individuals, companies or other private entities that undertake an action, whereas social costs include additionally the external costs on the environment and on society as a whole. Quantitative estimates of both private and social costs may be incomplete, because of difficulties in measuring all relevant effects.

Productive role/work The productive role/work relates to work performed by women and men for pay in cash or kind (market production, informal production, home production, subsistence production).

Projection A projection is a potential future evolution of a quantity or set of quantities, often computed with the aid of a model. Unlike predictions, projections are conditional on assumptions concerning, for example, future socio-economic and technological developments that may or may not be realized. See also Climate projection.

Prosperous scenario Prosperous scenario represents HKH facing weak climate change, a socially, economically and politically stable region, and low ecosystem degradation. In the prosperous scenario, regional cooperation across sectors and across governing institutions enables mountain and downstream people to utilize a full range of ecosystem services people to enjoy sustainable livelihoods and economic growth. The diversity and uniqueness of the region's natural resource assets, political life, and collaborative capacities are embraced. Biodiversity flourishes and the health of ecosystems improves. Climate change mitigation efforts largely succeed, as the regional economy shifts to clean and renewable sources for most of its energy needs. The impact of climate change reflects the IPCC's moderate scenario.

Purchasing Power Parity (PPP) The rate of currency conversion that equalizes the purchasing power of different currencies. It makes allowance for the differences in price levels and spending patterns between different countries (OECD/IEA 2018).

Radiative forcing The strength of drivers is quantified as Radiative Forcing (RF) in units watts per square meter (W/m^2) as in previous IPCC assessments. RF is the change in energy flux caused by a driver and is calculated at the tropopause or at the top of the atmosphere.

Receptor model Atmospheric receptor model is multivariate statistical method to understand the nature of the source/receptor relationship, thereby identifying and apportioning sources of contaminants. The fundamental principle of receptor modeling is that mass conservation (balance) can be assumed. Currently, the Chemical Mass Balance (CMB), Principal Component Analysis (PCA) and Positive Matrix Factorization (PMF) methods are the most widely used receptor models.

Reducing emissions from deforestation and forest degradation (REDD) An effort to create financial value for the carbon stored in forests, offering incentives for developing countries to reduce emissions from forested lands and invest in low-carbon paths to sustainable development (SD). It is therefore a mechanism for mitigation that results from avoiding deforestation. REDD+ goes beyond reforestation and forest degradation and includes the role of conservation, sustainable management of forests and enhancement of forest carbon stocks. The concept was first introduced in 2005 in the 11th Session of the Conference of the Parties (COP) in Montreal and later given greater recognition in the 13th Session of the COP in 2007 at Bali and inclusion in the Bali Action Plan which called for 'policy approaches and positive incentives on issues relating to reducing emissions from deforestation and forest degradation in developing countries (REDD) and the role of conservation, sustainable management of forests and enhancement of forest carbon stock in developing countries'. Since then, support for REDD has increased and has slowly become a framework for action supported by a number of countries.

Reforestation Planting of forests on lands that have previously contained forests but that have been converted to some other use. For a discussion of the term forest and related terms such as afforestation, reforestation and deforestation, see the IPCC Special Report on Land Use, Land-Use Change, and Forestry (IPCC 2000a). See also information provided by the United Nations Framework Convention on Climate Change (UNFCCC 2013). See also the Report on Definitions and Methodological Options to Inventory Emissions from Direct Human-induced Degradation of Forests and Devegetation of Other Vegetation Types (IPCC 2003, 2014a, b).

Renewable energy Any form of energy from solar, geophysical or biological sources that is replenished by natural processes at a rate that equals or exceeds its rate of use. Renewable energy is obtained from the continuing or repetitive flows of energy occurring in the natural environment and includes low-carbon technologies such as solar energy, hydropower, wind, tide and waves and ocean thermal energy, as well as renewable fuels such as biomass. For a more detailed description see specific renewable energy types in this glossary, for example, biomass, solar, hydropower, ocean, geothermal and wind (Verbruggen et al. 2011).

Representative Concentration Pathways (RCPs) Scenarios that include time series of emissions and concentrations of the full suite of greenhouse gases (GHGs) and aerosols and chemically active gases, as well as land use/land cover (Moss et al. 2008). The word representative signifies that each RCP provides only one of many possible scenarios that would lead to the specific radiative forcing characteristics. The term pathway emphasizes that not only the long-term concentration levels are of interest, but also the trajectory taken over time to reach that outcome (Moss et al. 2010).

RCPs usually refer to the portion of the concentration pathway extending up to 2100, for which Integrated Assessment Models produced corresponding emission scenarios. Extended Concentration Pathways (ECPs) describe extensions of the RCPs from 2100 to 2500 that were calculated using simple rules generated by stakeholder consultations and do not represent fully consistent scenarios. Four RCPs produced from Integrated Assessment Models were selected from the published literature and are used in the present IPCC Assessment as a basis for the climate predictions and projections presented in WGI AR5 Chaps. 11–14 (IPCC 2013):

RCP2.6 One pathway where radiative forcing peaks at approximately 3 W/m^2 before 2100 and then declines (the corresponding ECP assuming constant emissions after 2100).

RCP4.5 and RCP6.0 Two intermediate stabilization pathways in which radiative forcing is stabilized at approximately 4.5 W/m^2 and 6.0 W/m^2 after 2100 (the corresponding ECPs assuming constant concentrations after 2150).

RCP8.5 One high pathway for which radiative forcing reaches >8.5 W/m^2 by 2100 and continues to rise for some amount of time (the corresponding ECP assuming constant emissions after 2100 and constant concentrations after 2250) (IPCC 2014a, b).

Reproductive role/work The reproductive role/work includes the care and maintenance of the actual and future workforce of the family (childbearing responsibilities and domestic tasks), which is mostly assigned to women (EIGE 2018).

Remittances The earnings acquired by migrants/emigrants that are transferred back to the households at place of origin. It includes transfers from both domestic as well as abroad.

Refugees A person who, "owing to a well-founded fear of persecution for reasons of race, religion, nationality, membership of a particular social group or political opinions, is outside the country of his nationality and is unable or, owing to such fear, is unwilling to avail himself of the protection of that country or who, not having a nationality and being outside the country of his former habitual residence as a result of such events, is unable or, owing to such fear, is unwilling to return to it." (UNHCR 2010).

Resilience The capacity of social, economic and environmental systems to cope with a hazardous event or trend or disturbance, responding or reorganizing in ways that maintain their essential function, identity and structure, while also maintaining the capacity for adaptation, learning and transformation.

Return migration It denotes the movement of a migrant returning to his/her place of origin. The return may or may not be voluntary.

Risk The potential for consequences where something of value is at stake and where the outcome is uncertain, recognizing the diversity of values. Risk is often represented as probability or likelihood of occurrence of hazardous events or trends multiplied by the impacts if these events or trends occur. In this report, the term risk is often used to refer to the potential, when the outcome is uncertain, for adverse consequences on lives, livelihoods, health, ecosystems and species, economic, social and cultural assets, services (including environmental services) and infrastructure.

Risk management The plans, actions or policies to reduce the likelihood and/or consequences of risks or to respond to consequences.

Scenario A plausible description of how the future may develop based on a coherent and internally consistent set of assumptions about key driving forces (e.g., rate of technological change (TC), prices) and relationships. Note that scenarios are neither predictions nor forecasts, but are useful to provide a view of the implications of developments and actions.

Seasonal migration Seasonal migration is a movement which is undertaken during the part of a year depending upon seasonal requirement of work and conditions in the place of origin.

Secondary pollutants Secondary pollutants are not emitted directly. Rather, they form in the air when primary pollutants react or interact.

Sequestration The uptake (i.e., the addition of a substance of concern to a reservoir) of carbon containing substances, in particular carbon dioxide (CO_2), in terrestrial or marine reservoirs. Biological sequestration includes direct removal of CO_2 from the atmosphere through land-use change (LUC), afforestation, reforestation, revegetation, carbon storage in landfill and practices that enhance soil carbon in agriculture (cropland management, grazing land management). In parts of the literature, but not in this report, (carbon) sequestration is used to refer to Carbon Dioxide Capture and Storage (CCS).

Shared Socio-Economic Pathways (SSPs) Currently, the idea of SSPs is developed as a basis for new emissions and socio-economic scenarios. An SSP is one of a collection of pathways that describe alternative futures of socio-economic development in the absence of climate policy intervention. The combination of SSP-based socio-economic scenarios and Representative Concentration Pathway (RCP)-based climate projections should provide a useful integrative frame for climate impact and policy analysis.

Short-lived climate pollutants SLCPs are defined as gases and particles that contribute to warming and that have a lifetime of a few days to approximately 10 years. These include black carbon (BC), tropospheric ozone (O_3) and its precursors CO, nmVOC and NOx, methane (CH_4), and some hydrofluorocarbons (HFCs). A characteristic of the climate effects of the short-lived climate pollutants, with the exception of the HFCs and to a certain extent CH4, is that it matters where in the world the emissions are released. They also significantly impact food, water and economic security for large populations throughout the world, both directly through their negative effects on public health, agriculture and ecosystems, and indirectly through their impact on the climate.

Snow Water Equivalent (SWE) A measurement of the amount of water contained in a snowpack. It can be considered as the depth of water that would theoretically result if the entire snowpack was melted instantly. SWE is the product of snow depth and snow density.

Social costs See Private costs.

Socio ecological systems Social-ecological systems are complex, integrated systems in which humans are part of nature (Berkes and Folke 1994).

Solar energy Energy from the Sun that is captured either as heat, as light that is converted into chemical energy by natural or artificial photosynthesis, or by photovoltaic panels and converted directly into electricity. Concentrating solar power (CSP) systems use either lenses or mirrors to capture large amounts of solar energy and focus it down to a smaller region of space. The higher temperatures produced can operate a thermal steam turbine or be used in high-temperature industrial processes. Direct solar energy refers to the use of solar energy as it arrives at the Earth's surface before it is stored in water or soils. Solar thermal is the use of direct solar energy for heat end-uses, excluding CSP. Active solar needs equipment like panels, pumps and fans to collect and distribute the energy. Passive solar is based on structural design and construction techniques that enable buildings to utilize solar energy for heating, cooling and lighting by non-mechanical means (Verbruggen et al. 2011).

Solar Home System (SHS) A stand-alone system composed of a relatively low-power photovoltaic module, a battery and sometimes a charge controller that can power small electric devices and provide modest amounts of electricity to homes for lighting and radios, usually in rural or remote regions that are not connected to the electricity grid. A SHS typically includes one or more PV modules consisting of solar cells, a charge controller which distributes power and protects the batteries and appliances from damage and at least one battery to store energy for use when the sun is not shining.

Springshed Land area that contributes flow to a spring via recharge.

SRES scenarios SRES scenarios are emission scenarios developed by IPCC (2000a) and used, among others, as a basis for some of the climate projections shown in Chaps. 9–11 of IPCC WGI TAR (IPCC 2001), Chaps. 10 and 11 of IPCC WGI AR4 (IPCC 2007), as well as in the IPCC WGI AR5 (IPCC 2013, 2014a, b).

Storm surge The temporary increase, at a particular locality, in the height of the sea due to extreme meteorological conditions (low atmospheric pressure and/or strong winds). The storm surge is defined as being the excess above the level expected from the tidal variation alone at that time and place.

Storyline A narrative description of a scenario (or family of scenarios), highlighting the main scenario characteristics, relationships between key driving forces and the dynamics of their evolution.

Strategic Gender Needs (SGNs) Chap. 14 The needs identified by women that require strategies for challenging male dominance and privilege. These needs may relate to inequalities in the gender division of labour, in ownership and control of resources, in participation in decision-making, or to experiences of domestic and other sexual violence.

Structural change Changes, for example, in the relative share of gross domestic product (GDP) produced by the industrial, agricultural, or services sectors of an economy, or more generally, systems transformations whereby some components are either replaced or potentially substituted by other components.

Sustainable energy Sustainable energy is a form of energy that meet our today's demand of energy without putting them in danger of getting expired or depleted and can be used over and over again.[6]

Sustainability Sustainability is generally defined as a system's overall ability to meet present human needs without compromising the ability of future generations to meet their needs. Sustainability has three main pillars: social, environmental, and economic. These three pillars are often informally referred to as people, planet and profits (Beattie 2017).

Sustainable development Development that meets the needs of the present without compromising the ability of future generations to meet their own needs (WCED 1987). Sustainable development is characterized by social/sociocultural equity, environmental protection, and economic viability. Sustainable development is the pathway to sustainability.

Sustainable development goals The Sustainable Development Goals (SDGs), also known as Global Goals, constitute a universal call to action to end poverty, protect the planet, and ensure that all people enjoy peace and prosperity. These 17 Goals build on the successes of the Millennium Development Goals, while also including new areas such as climate change, economic inequality, innovation, sustainable consumption, peace, justice, and other priorities. The SDGs are interconnected—often, key to success for one involves tackling issues more commonly associated with another (UNDP 2018).

Temporary migration Temporary migration is a short-term migration including seasonal migration. While all seasonal migrations are temporary migration, all temporary migrations are not seasonal in nature.

[6]https://sustainabledevelopment.un.org/.

Transformation A change in the fundamental attributes of natural and human systems.

Transformation pathway The trajectory taken over time to meet different goals for greenhouse gas (GHG) emissions, atmospheric concentrations, or global mean surface temperature change that implies a set of economic, technological and behavioural changes. This can encompass changes in the way energy and infrastructure are used and produced, natural resources are managed and institutions are set up and in the pace and direction of technological change (TC). See also Baseline/reference, Emission scenario, Mitigation scenario, Representative Concentration Pathways (RCPs) and SRES scenarios.

Tree rings Concentric rings of secondary wood evident in a cross section of the stem of a woody plant. The difference between the dense, small-celled late wood of one season and the wide-celled early wood of the following spring enables the age of a tree to be estimated, and the ring widths or density can be related to climate parameters such as temperature and precipitation.

Tropospheric ozone Tropospheric (or ground-level) ozone (O_3) is the ozone present in the lowest portion of the atmosphere (up to 10–15 km above the ground) that is harmful to the people and the ecosystem.

Uncertainty A state of incomplete knowledge that can result from a lack of information or from disagreement about what is known or even knowable. It may have many types of sources, from imprecision in the data to ambiguously defined concepts or terminology, or uncertain projections of human behaviour. Uncertainty can therefore be represented by quantitative measures (e.g., a probability density function) or by qualitative statements (e.g., reflecting the judgment of a team of experts) (see Moss and Schneider 2000; Manning et al. 2004; Mastrandrea et al. 2010). See also Confidence and Likelihood (IPCC 2014a, b).

Vulnerability The propensity or predisposition to be adversely affected. Vulnerability encompasses a variety of concepts and elements including sensitivity or susceptibility to harm and lack of capacity to cope and adapt.

Water-energy-food nexus Inter-relation among water, energy, and food as resources, institutions for their management, and security.

Water security The capacity of HKH populations to safeguard sustainable access to adequate quantities of acceptable quality water for resilient societies and ecosystems, to ensure protection against water-borne pollution and water-related disasters, and to adapt to uncertain global change—in a regional climate of peace and political stability.

Wind energy Kinetic energy from air currents arising from uneven heating of the Earth's surface. A wind turbine is a rotating machine including its support structure for converting the kinetic energy to mechanical shaft energy to generate electricity. A windmill has oblique vanes or sails and the mechanical power obtained is mostly used directly, for example, for water pumping. A wind farm, wind project or wind power plant is a group of wind turbines interconnected to a common utility system through a system of transformers, distribution lines, and (usually) one substation (Verbruggen et al. 2011).

References

ACGR. (1998). *Associate committee on geotechnical research: Glossary of permafrost and related ground-ice terms*. National Research Council of Canada, Associate Committee on Geotechnical Research (S.A. Harris, H.M. French, J.A. Heginbottom, G.H. Johnston, B. Ladanyi, D.C. Sego, R.O. van Everdingen) Technical Memorandum 142. https://ipa.arcticportal.org/images/Glossary/Glossary_of_Permafrost_and_Related_Ground-Ice_Terms_1998.pdf.

Adhikari, L., Hussain, A., & Rasul, G. (2017). Tapping the potential of neglected and underutilized food crops for sustainable nutrition security in the mountains of Pakistan and Nepal. *Sustainability, 2017*(9), 271.

AGECC. (2010). *Energy for a sustainable future*. Summary Reports and Recommendations. New York. https://doi.org/10.1016/j.enbuild.2006.04.011.

Alkire, S., & M. E. Santos. (2010). *Multidimensional poverty*. Oxford Poverty and Human Development Initiative, University of Oxford. July 2010.

Beattie, A. (2017). *The 3 pillars of corporate sustainability*. Retrieved 11 June, 2018, from https://www.investopedia.com/articles/investing/100515/three-pillars-corporate-sustainability.asp.

Berkes, F., & Folke, C. (1994). Linking social and ecological systems for resilience and sustainability. *Beijer discussion paper series No. 52*. Stockholm.

Cogley, J.G., Hock, R., Rasmussen, L.A., Arendt, A.A., Bauder, A., Braithwaite, R.J., et al. (2011). *Glossary of glacier mass balance and related terms*.

CBD. (1992).*Convention on biological diversity*. United Nations.

Cleanleap. (2016). *Distributed renewable energy in developing countries*. Retrieved 14 June, 2018, from http://cleanleap.com/where-are-we-renewable-energy/05-distributed-renewable-energy-developing-countries.

Convention on biological Diversity, united Nation (1992). Retrieved 10 September, 2017, from https://www.cbd.int/ecosystem/.

EIGE. (2018). *Gender equality glossary and thesaurus*. Retrieved June 14, 2018, from http://eige.europa.eu/rdc/thesaurus/terms/1352.

FAO. (1996). Report of the world food summit. In *World food summit*. Rome: Food and Agriculture Organisation of the United Nations.

FAO. (2006). *Food security*. Policy brief, Issue 2. Agriculture and Development Economics Division, FAO. Retrieved from http://www.fao.org/forestry/13128-0e6f36f27e0091055bec28ebe830f46b3.pdf.

FAO. (2012). *Committee on world food security*. A report of the meeting held on 15–20 October in Rome Italy. Retrieved from http://www.fao.org/docrep/meeting/026/MD776E.pdf.

FAO. (2008). *An introduction to the basic concepts of food security*. Food Security Information for Action: Practical Guides. Food and Agriculture Organization of the United Nations. Retrieved from http://www.fao.org/docrep/013/al936e/al936e00.pdf.

FAO. (2012). Mainstreaming climate-smart agriculture into a broader landscape approach. In *Background paper for the second global conference on agriculture, food security and climate change*. Rome, Italy: Food and Agricultural Organization of the United Nations.

GCW. (2016). *Global Cryosphere watch implementation plan, version 1.7*. WMO. https://library.wmo.int/opac/doc_num.php?explnum_id=3538.

Hegerl, G. C., Hoegh-Guldberg, O., Casassa, G., Hoerling, M. P., Kovats, R. S., Parmesan, C., et al. (2010). Good practice guidance paper on detection and attribution related to anthropogenic climate change. In *Meeting report of the intergovernmental panel on climate change expert meeting on detection and attribution of anthropogenic climate change*. IPCC Working Group I Technical Support Unit, University of Bern, Bern, Switzerland.

IEA. (2010). *World energy outlook 2010*. France: Paris Cedex. Retrieved from https://www.iea.org/newsroom/news/2010/november/world-energy-outlook-2010.html.

IFAD. (2002). *Regional strategy paper—Asia and the Pacific*. IFAD, March 2002. https://www.ifad.org/documents/10180/3b116bbd-70d6-4219-e4c8b6011866.

IPCC. (1992). *Climate Change 1992: The Supplementary Report to the IPCC Scientific Assessment*. [Houghton, J. T., B. A. Callander and S. K. Varney (eds.)]. Cambridge, United Kingdom and New York, NY, USA: Cambridge University Press.

IPCC. (1996). *Climate Change 1995: The Science of Climate Change. Contribution of Working Group I to the Second Assessment Report of the Intergovernmental Panel on Climate Change*. [Houghton, J. T., L. G. Meira, A. Callander, N. Harris, A. Kattenberg and K. Maskell (eds.)]. Cambridge, United Kingdom and New York, NY, USA: Cambridge University Press.

IPCC. (2000a). *Land Use, Land-Use Change, and Forestry*. Special Report of the Intergovernmental Panel on Climate Change [Watson, R. T., I. R. Noble, B. Bolin, N. H. Ravindranath, D. J. Verardo and D. J. Dokken (eds.)]. Cambridge, United Kingdom and New York, NY, USA: Cambridge University Press.

IPCC. (2000b). *Emissions Scenarios. Special Report of Working Group III of the Intergovernmental Panel on Climate Change*. [Nakićenović, N. and R. Swart (eds.)]. Cambridge, United Kingdom and New York, NY, USA: Cambridge University Press.

IPCC. (2001). *Climate Change 2001: The Scientific Basis. Contribution of Working Group I to the Third Assessment Report of thePanel on Climate Change* [Houghton, J.T., Y. Ding, D.J. Griggs, M. Noguer, P. J. van der Linden, X. Dai, K., and C.A. Johnson (eds.)]. Cambridge, United Kingdom and New York, NY, USA: Cambridge University Press.

IPCC. (2003). *Definitions and Methodological Options to Inventory Emissions from Direct Human-Induced Degradation of Forests and Devegetation of Other Vegetation Types* [Penman, J., M. Gytarsky, T. Hiraishi, T. Krug, D. Kruger, R. Pipatti, L. Buendia, K. Miwa, T. Ngara, K. Tanabe and F. Wagner (eds.)] (p. 32). Tokyo, Japan: The Institute for Global Environmental Strategies (IGES).

IPCC. (2007). *Climate Change 2007: The Physical Science Basis. Contribution of Working Group I to the Fourth Assessment Report of the Intergovernmental Panel on Climate Change*. [Solomon, S., D. Qin, M. Manning, Z. Chen, M. Marquis, K. B. Averyt, M. Tignor and H. L. Miller (eds.)]. Cambridge, United Kingdom and New York, NY, USA: Cambridge University Press.

IPCC. (2013). *Climate Change 2013: The Physical Science Basis. Contribution of Working Group I to the Fifth Assessment Report of the Intergovernmental Panel on Climate Change*. [Stocker, T. F., D. Qin, G.-K. Plattner, M. Tignor, S. K. Allen, J. Boschung, A. Nauels, Y. Xia, V. Bex and P. M. Midgley (eds.)]. Cambridge, United Kingdom and New York, NY, USA: Cambridge University Press.

IPCC. (2014a). Annex II: Glossary [K. J. Mach, S. Planton, & C. von Stechow (Eds.)]. In Core Writing Team, R. K. Pachauri, & L. A. Meyer (Eds.), *Climate change 2014: Synthesis report. Contribution of working groups I, II and III to the fifth assessment report of the intergovernmental panel on climate change* (pp. 117–130). Geneva, Switzerland: IPCC.

IPCC. (2014b). *Climate change 2014: Impacts, adaptation, and vulnerability WGII, III*. http://www.ipcc.ch/pdf/assessment-report/ar5/wg2/WGIIAR5-AnnexII_FINAL.pdf.

Jodha, N. S. (1992). Mountain perspective and sustainability: A framework for development strategy. In Sustainable Mountain Agriculture (Eds.), New Delhi.

Kirkbride, M. P. (2011). Debris-covered glaciers. In V. P. Singh, P. Singh, & U. K. Haritashya (Eds.), *Encyclopedia of snow, ice and glaciers* (pp. 190–191). Dordrecht: Springer Science+Business Media B.V.

MA. (2005). *Ecosystems and human wellbeing: Current state and trends*, Vol. 1. Washington, D.C.: Island Press.

Manning, M. R., M. Petit, D. Easterling, J. Murphy, A. Patwardhan, H.-H. Rogner., et al. (eds.), 2004. IPCC Workshop on Describing Scientific Uncertainties in Climate Change to Support Analysis of Risk of Options. Workshop Report (p. 138). Intergovernmental Panel on Climate Change, Geneva, Switzerland.

Mastrandrea, M. D., Field, C. B., Stocker, T. F., Edenhofer, O., Ebi, K. L., Frame, D. J., et al. (2010). *Guidance note for lead authors of the IPCC fifth assessment report on consistent treatment of uncertainties*. Intergovernmental Panel on Climate Change (IPCC). http://www.ipcc.ch.

MEA. (2005). *Millennium ecosystem assessment. Ecosystems and human well-being: Biodiversity synthesis*. Washington, DC : World Resources Institute.

Millennium Ecosystem Assessment. (2005). *Ecosystems and human well-being: Wetlands and water*. Washington, DC 5: World resources institute. https://millenniumassessment.org/documents/document.300.aspx.pdf.

Moss, R. and S. Schneider, 2000. Uncertainties in the IPCC TAR: Recommendations to Lead Authors for More Consistent Assessment and Reporting. In: IPCC Supporting Material: Guidance Papers on Cross Cutting Issues in the Third Assessment Report of the IPCC [Pachauri, R., T. Taniguchi and K. Tanaka (eds.)]. Intergovernmental Panel on Climate Change, Geneva, Switzerland, pp. 33–51.

Moss, R., M. Babiker, S. Brinkman, E. Calvo, T. Carter, J. Edmonds, I. Elgizouli, S. Emori, L. Erda, K. Hibbard, R. Jones, M. Kainuma, J. Kelleher, J. F. Lamarque, M. Manning, B. Matthews, J. Meehl, L. Meyer, J. Mitchell, N. Nakicenovic, B. O'Neill, R. Pichs, K. Riahi, S. Rose, P. Runci, R. Stouffer, D. van Vuuren, J. Weyant, T. Wilbanks, J. P. van Ypersele and M. Zurek. (2008). *Towards new scenarios for analysis of emissions, climate change, impacts and response strategies*. IPCC Expert Meeting Report, 19-21 September, 2007, Noordwijkerhout, (p. 132). Netherlands, Intergovernmental Panel on Climate Change (IPCC), Geneva, Switzerland.

Moss, R., J. A., Edmonds, K. A. Hibbard, M. R. Manning, S. K. Rose, D. P. van Vuuren, T. R. Carter, S. Emori, M. Kainuma, T. Kram, G. A. Meehl, J. F. B. Mitchell, N. Nakicenovic, K. Riahi, S. J. Smith, R. J. Stouffer, A. M. Thomson, J. P. Weyant and T. J. Wilbanks. (2010). The next generation of scenarios for climate change research and assessment. *Nature*, 463, 747–756.

OECD/IEA. (2005). *Energy statistics manual*, Paris Cedex, France.

OECD/IEA. (2018). *Glossary*. Retrieved June 14, 2018, from http://www.iea.org/about/glossary/c/#tabs-2.

Reddy, A. K. N. (2000). Energy and social issues. In *World energy assessment*. New York: UNDP. ISBN 92-1-126-126-0.

Sedjo, R., & Sohngen, B. (2012). Carbon sequestration in forests and soils. *Annual Review of Resource Economics, 4*(1), 127–144.

UN. (1998). Report of the representative of the secretary-general, Mr. Francis M. Deng, submitted pursuant to commission resolution 1997/39. Addendum: Guiding principles on internal displacement. Retrieved from http://www.un-documents.net/gpid.htm.

UNDP. (2015). *EnergyPlus guidelines: Planning for improved energy access and productive uses of energy*, New York.

UNDP. (2018). Sustainable development goals. Retrieved June 11, 2018, from http://www.undp.org/content/undp/en/home/sustainable-development-goals.html.

UNFCCC. (2013). Reporting and accounting of LULUCF activities under the Kyoto Protocol. United Nations Framework Convention on Climatic Change (UNFCCC), Bonn, Germany. Available at: http://unfccc.int/methods/lulucf/items/4129.php.

UNHCR. (2010). *Convention and protocol relating to the status of refugees*. Geneva, Switzerland: UNHCR, Communications and Public Information Service.

UNISDR. (2017). *Terminology on disaster risk reduction*. Retrieved June 11, 2018, from https://www.unisdr.org/we/inform/terminology.

Vanini, P. (2014). *Off the grid: Re-assembling domestic life*, p. 10.

Washburn, A. L. (1979). *Geocryology*. London: Edward Arnold.

Verbruggen, A., Moomaw, W., Nyboer, J. (2011). Annex I: Glossary, acronyms, chemical symbols and prefixes. In O. Edenhofer, R. PichsMadruga, Y. Sokona, K. Seyboth, P. Matschoss, S. Kadner, T. Zwickel, P. Eickemeier, G. Hansen, S. Schlömer, & C. von Stechow (Eds.), *IPCC special report on renewable energy sources and climate change mitigation*. Cambridge, United Kingdom and New York, NY, USA: Cambridge University Press.

WCED. (1987). *Report of the World Commission on Environment and Development: Our common future*. Retrieved from http://www.un-documents.net/our-common-future.pdf.

WEC. (2016). *World energy issues monitor. A climate of innovation responding to the commodity price storm*. London, United Kingdom: World Energy Council 2016. Retrieved from www.worldenergy.org.

WHO. (2018). *Health statistics and information systems*. Retrieved June 11, 2018, from http://www.who.int/healthinfo/global_burden_disease/metrics_daly/en/.

Zaid, A., Hughes, H. G., Porceddu, E., & Nicholas, F. W. (1999). *Glossary of biotechnology and genetic engineering*. Rome: FAO./Section1>

Annex 2: List of Acronyms

AAB	Action Aid Bangladesh
AAGR	Average Annual Growth Rate
ABC	Atmospheric brown cloud
ACAM	Atmospheric Composition and the Asian Monsoon
ACRE	Atmospheric Circulation Reconstructions over the Earth
ADB	Asian Development Bank
ADP	Annual Development Plan
AEPC	Alternate Energy Promotion Centre
AIRD	Adaptation and Impacts Research Division
AJK	Azad Jammu and Kashmir
AKAH	Aga Khan Agency for Habitat
AKF	Aga Khan Foundation
AKRSP	Aga Khan Rural Support Programme
AMP	Aid Management Policy
AMS	Accelerator Mass Spectrometry
ANBSAP	Afghanistan's National Biodiversity Strategy and Action Plan
AOD	Aerosol optical depth
AOGCM	Atmosphere-Ocean General Circulation Models
APCAP	Asia Pacific Clean Air Partnership
APHCA	Animal Production and Health Commission for Asia and the Pacific
AR5	Assessment Report 5
ARCAB	Action Research for Community Adaptation in Bangladesh
ARI	Acute respiratory infections
ATP	Annual total precipitation
ATREE	Ashoka Trust for Research in Ecology and Environment
AWDO	Asian Water Development Outlook
BAU	Business as Usual
BBIN	Bhutan, Bangladesh, India, and Nepal
BBS	Bangladesh Bureau of Statistics
BC	Black carbon
BCAS	Bangladesh Centre for Advanced Studies
BCCRF	Bangladesh Climate Change Resilience Fund
BCCSAP	Bangladesh Climate Change Strategy and Action Plan
BCCTF	Bangladesh Climate Change Trust Fund
BEOE	Bureau of Emigration and Overseas Employment
BIMSTEC	Bengal Initiative for Multi-Sectoral Technical and Economic Cooperation
BLSS	Bhutan Living Standard Survey
BMC	Biodiversity Management Committee
BMET	Bureau of Manpower Employment and Training

P. Wester et al. (eds.), *The Hindu Kush Himalaya Assessment*,
https://doi.org/10.1007/978-3-319-92288-1

BMZ	Federal Ministry for Economic Cooperation and Development (Germany)
BOD	Biochemical Oxygen Demand
BOM	Bureau of Meteorology
BP	Before Present
BRACED	Building Resilience and Adaptation to Climate Extremes and Disasters
BrC	Brown carbon
BRICS	Association of Brazil, Russia, India, China and South Africa
BTEX	Benzene, toluene, ethylbenzene, xylenes
BWSI	Bhutan Water Security Index
CAM	Climate Anomaly Method
CAPA	Community Adaptation Plans of Action
CARE	Cooperative for Assistance and Relief Everywhere
CAS	Chinese Academy of Sciences
CBA	Community Based Adaptation
CBD	Convention on Biological Diversity
CBFEWS	Community-based flood early warning system
CBOs	Community Based Organizations
CBNRM	Community-based Natural Resources Management
CBS	Central Bureau of Statistics
CCA	Community Conserved Areas
CCAC	Climate and Clean Air Coalition
CCAF	Climate Change, Agriculture and Food Security
CCAP	Centre for Chinese Agriculture Policy
CCCma	Canadian Centre for Climate Modelling and Analysis
CCCR	Centre for Climate Change Research
CCD	Climate Change Division
CCN	Cloud condensation nuclei
CCSAP	Climate Change Strategy and Action Plan
CDKN	Climate Development Knowledge Network
CDM	Clean Development Mechanism
CEA	Central Electricity Authority, Government of India
CEDAW	Convention on All Forms of Discrimination on Women
CEMORD	Center for Mountain Research Development
CERFACS	Centre Européen de Recherche et de Formation Avancée en Calcul Scientifique
CERN	Chinese Ecosystem Research Network
CF	Community Forestry
CFIs	Community Forestry Instructions
CFP	Community Forestry Programme
CFUGs	Community Forest User Groups
CGC	Centre for Global Change
CGIAR	Consultative Group for International Agricultural Research
CHEA	Central Himalayan Environment Association
CHT	Chittagong Hill Tracts
CIA	Central Intelligence Agency
CICERO	Center for International Climate and Environmental Research—Oslo
CIFOR	Center for International Forestry Research
CII	Confederation of Indian Industry
CITES	Convention on International Trade in Endangered Species of Wild Fauna and Flora
CLA	Crop-livestock Agriculture
CLRTAP	Convention on Long-Range Transboundary Air Pollution
CMA	China Meteorological Administration

CMIP	Couple Model Intercomparison Project
CMIP5	Coupled Model Intercomparison Project, Version 5
CNG	Compressed natural gas
CNRM	Centre National de Recherches Météorologiques
CNY	Chinese currency
CO_2	Carbon Dioxide
CO	Carbon monoxide
COP	Conference of the Parties
COP21	UNFCCC Conference of Parties 21
COPD	Chronic obstructive pulmonary disease
CORDEX	Coordinated Regional Downscaling Experiment
COSMO	Consortium for Small-scale Modeling
CP	Current Policies
CPCB	Central Pollution Control Board
CPEIR	Climate Public Expenditure and Institutional Reviews
CPRs	Common Pool Resources
CR	Critically endangered
CRED	Centre for Research on the Epidemiology of Disasters
CREP	Community Rural Electrification Programme
CRISPR	Clustered Regularly Interspaced Short Palindromic Repeats
CRU	Climate Research Unit
CSA	Climate Smart Agriculture
CSIRO	Commonwealth Scientific and Industrial Research Organisation
CSOs	Civil Society Organizations
CSR	Corporate Social Responsibility
CUTS	Consumer Unity & Trust Society
CVCA	Climate Vulnerability and Capacity Analysis
CVI	Climate Vulnerability Index
CWC	Central Water Commission, Government of India
DABS	Da Afghanistan Breshna Sherkat
DALYs	Disability-Adjusted Life Years
DEM	Digital Elevation Model
DFID	United Kingdom Department for International Development
DFOs	District Forest Offices
DHPS	Department of Hydropower and Power Systems
DHS	Demographic and Health Survey
DLSOs	District Livestock Service Offices
DNV GL	Det Norske Veritas Germanischer Lloyd
DoE	Department of Energy
DoE	Department of Environment
DoFE	Department of Foreign Employment
DOI	Digital Object Identifier
DoP	Department of Population
DPF	Diesel particulate filter
DPIA	Daily Precipitation Intensity Anomalies
DRD	Department of Rural Development
DRR	Disaster Risk Reduction
DTR	Diurnal Temperature range
EANET	Acid Deposition Monitoring Network in East Asia
EBA	Ecosystem Based Adaptation
EBHR	European Bulletin of Himalayan Research
EC	Elemental carbon
EC	European Consortium

EC	Ecosystems
ECA	European Climate Assessment
EDGAR	Emissions Database for Global Atmospheric Research
EDW	Elevation Dependent Warming
EF	Environmental Flows
EFLG	Environment-Friendly Local Governance
EHP	Elevated Heat Pump
EM-DAT	The Emergency Events Database
EN	Endangered
ENSO	El Niño-Southern Oscillation
ERD	Economics and Research Department
ES	Ecosystem Services
ESAP	Energy Sector Assistance Program
ESCAP	Economic and Social Commission for Asia and the Pacific
ESCO	Energy Service Company
ESGF	Earth System Grid Federation
ESV	Ecosystem Services Value
ET	Evapotranspiration
ETCCDI	Expert Team on Climate Change Detection and Indices
EU	European Union
FAO	Food and Agriculture Organization
FATA	Federally Administered Tribal Areas
FCBTK	Fixed Chimney Bull's Trench Kiln
FECOFUN	Federation of Community Forest Users-Nepal
FD	Frost Days
FD	Forest Department
FFC	Federal Flood Commission
FIT	Feed-in Tariffs
FMIS	Farmer Managed Irrigation Systems
FPE	Feminist Political Ecology
FS	Food Security
FSF	Future Smart Food
FW	Fortress World
FYPs	Five Year Plans
GAINS	Greenhouse Gas—Air Pollution Interactions and Synergies
GB	Gilgit-Baltistan
GBPNIHESD	GB Pant National Institute of Himalayan Environment and Sustainable Development
GCC	Gulf Coast Countries
GCESS	Global Change and Earth System Science
GCM	General Circulation Models
GCM	Global Climate Model
GCOS	Global Climate Observing System
GDP	Gross Domestic Product
GEA	Global Energy Assessment
GEF	Global Environmental Facility
GEH	Gross Enrolment Ratio in Higher Education
GEnS	Global Environmental Stratification
GEO	Global Environmental Outlooks
GFDL	Geophysical Fluid Dynamics Laboratory
GFDRR	Global Facility for Disaster Reduction and Recovery
GGCA	Global Gender and Climate Alliance
GHCND	Global Historical Climatology Network-Daily

GHCM	Global Historical Climatology Network
GHG	Green House Gas
GII	Gender Inequality Index
GIS	Geographic Information System
GIRoA	Government of the Islamic Republic of Afghanistan
GIZ	Deutsche Gesellschaft für Internationale Zusammenarbeit
GLDP	Global Land Daily Precipitation
GLIMS	Global Land Ice Measurements from Space
GLMP	Global Land Monthly Precipitation
GLOFs	Glacial Lake Outburst Floods
GLSAT	Global Land Surface Air Temperature
GM	Genetically Modified Organisms
GMS	Greater Mekong Sub-Region
GNH	Gross National Happiness
GoB	Government of Bangladesh
GoI	Government of India
GoN	Government of Nepal
GoP	Government of Pakistan
GPS	Global Positioning System
GSOD	Global Surface Summary of the Day
GT	Great Transitions
GTN-G	Global Terrestrial Network for Glaciers
GTN-P	Global Terrestrial Network for Permafrost
GTOS	Global Terrestrial Observing System
GW	Gigawatts
GWh	Gigawatt hours
HAP	Household Air Pollution
Hb	Hemoglobin
HCl	Hydrogen Chloride
HDI	Human Development Index
HF	Hydrogen Fluoride
HFCs	Hydrofluorocarbons
HH	Household
HIES	Household Integrated Economic Survey
HIMAP	Hindu Kush Himalayan Monitoring and Assessment Programme
HKH	Hindu Kush Himalaya
HKH-TP	Hindu Kush-Himalaya-Tibetan Plateau-Pamir
HMA	High Mountain Asia
HSE	High School Education
HVDC	High-Voltage Direct Current
HYCOS	Hydrological Cycle Observing System
I&M	Inspection and Maintenance
IAS	Invasive Alien Species
ICARDA	International Center for Agricultural Research in the Dry Areas
ICCCAD	International Centre for Climate Change and Development
ICDP	Integrated Conservation and Development Programme
ICHEC	Irish Centre for High-End Computing
ICIMOD	International Centre for Integrated Mountain Development
ICRAF	International Centre for Research in Agroforestry
ICRMW	International Convention on the Protection of the Rights of all Migrant Workers and the Members of their Families
ICS	Improved Cook Stoves
ICSU	International Council for Science

ICT	Information and Communication Technology
ICTP	International Centre for Theoretical Physics
IDCOL	Infrastructure Development Company Ltd.
IDEC	International Development and Cooperation
IDP	Internally Displaced Population
IDRC	International Development Research Centre
IEA	International Energy Agency
IEMP	International Ecosystem Management Partnership
IFAD	International Fund for Agricultural Development
IFC	International Finance Corporation
IFSD	Institute for Studies and Development Worldwide
IGP	Indo-Gangetic Plains
IGSNRR	Institute of Geographic Sciences and Natural Resources Research
IHCLA	Integrated Household Living Condition Assessment
IIED	International Institute for Environment and Development
IIMA	Indian Institute of Management Ahmedabad
IITM	Indian Institute of Tropical Meteorology
ILO	International Labour Organization
ILRI	International Livestock Research Institute
IMF	International Monetary Fund
INCCA	Indian Network for Climate Change Assessment
INDC	Intended Nationally Determined Contributions
INDOEX	Indian Ocean Experiment
I/NGOs	International/Non-Government Organizations
INM	Institute of Numerical Mathematics
INR	Indian Rupee
IOD	Indian Ocean Dipole
IOM	International Organization for Migration
IPBES	Intergovernmental Science-Policy Platform on Biodiversity and Ecosystem Services
IPCC	Intergovernmental Panel on Climate Change
IPGRI	International Plant Genetic Resources Institute
IPSL	Institut Pierre-Simon Laplace
IR	Intense Rain
ISDR	International Strategy for Disaster Reduction
ISET	Institute for Social and Environmental Transition
ISMWRA	Inter-state Migrant Workmen Regulation Act
ISO	International Organization for Standardization
IT	Information Technology
IUCN	International Union for Conservation of Nature
IWRM	Integrated Water Resources Management
JETC	Joint Expert Technical Committee
JFM	Joint Forest Management
JFMCs	Joint Forest Management Committees
JICA	Japan International Cooperation Agency
JRC	Joint Rivers Commission
KBA	Key Biodiversity Areas
KES	Kenyan Shilling
KPK	Khyber Pakhtunkhwa
KSLCDI	Kailash Sacred Landscape Conservation and Development Initiative
ktoe	kiloton of oil equivalent
kV	kilovolt
kWh	kilowatt hour

L&D	Loss and Damage
LAPA	Local Adaptation Plan of Action
LC	Least Concern
LDCs	Least Developed Countries
LDRRF	Local Disaster Risk Reduction Fund
LEAD	Leadership for Environment & Development
LED	Light-Emitting Diode
LFP	Livelihoods and Forestry Programme
LGM	Last Glacial Maximum
LH	Livelihoods
LHF	Leasehold Forestry
LIA	Little Ice Age
LI-BIRD	Local Initiatives for Biodiversity, Research and Development
LLOF	Landslide Lake Outburst Flood
LoCAL	Local Climate Adaptive Living Facility
LPG	Liquefied Petroleum Gas
LR	Light rain
LULC	Land Use and Land Cover
LULCC	Land Use and Land Cover Change
m^2	square meter
MA	Millennium Ecosystem Assessment
MACH	Management of Aquatic Resources through Community Husbandry
MAPDRR	Myanmar Action Plan on Disaster Risk Reduction
masl	meters above sea level
MC	Mekong Committee
MCBTK	Moveable Chimney Bull's Trench Kiln
MCMC	Markov Chain Monte Carlo
MDGs	Millennium Development Goals
MEs	Medium scale Enterprise
MEP	Ministry of Environmental Protection
MEW	Ministry of Energy and Water
MF	Market Forces
MICS	Multiple Indicator Cluster Survey
MIROC	Model for Interdisciplinary Research on Climate
MJO	Madden–Julian Oscillation
MLD	Million Litres per Day
MLE	Mass-loss Equivalent
MLVI	Multidimensional Livelihood Vulnerability Index
MNRE	Ministry of New and Renewable Energy
MoA	Ministry of Agriculture
MoAF	Ministry of Agriculture and Forests
MoCC	Ministry of Climate Change
MODIS	Moderate Resolution Remote Sensing Imagery
MoE	Ministry of Environment
MoEA	Ministry of Economic Affairs
MoECF	Ministry of Environmental Conservation and Forestry
MoEF	Ministry of Environment and Forests
MoEF	Ministry of Environment Forests & Climate
MoEPC	Ministry of Environmental Protection of China
MoFSC	Ministry of Forests and Soil Conservation
MOHC	Met Office Hadley Centre
MoEWOE	Ministry of Expatriates' Welfare and Overseas Employment
MoHRSS	Ministry of Human Resources and Social Security

MoI	Ministry of Industry
MOIA	Ministry of Overseas Indians
MoL	Ministry of Labour
MoLE	Ministry of Labour and Employment
MoLES	Ministry of Labour, Employment and National Security
MoLHR	Ministry of Labour & Human Resources
MoLIP	Ministry of Labour, Immigration and Population
MoLSAMD	Ministry of Labour, Social Affairs, Martyrs and Disabled
MoNREC	Ministry of Natural Resources and Environmental Conservation
MoPE	Ministry of Population and Environment
MoPIT	Ministry of Physical Infrastructure & Transport
MoSTE	Ministry of Science, Technology and Environment
MoSWRR	Ministry of Social Welfare Relief and Resettlement
MoU	Memorandum of Understanding
MoUD	Ministry of Urban Development
MPI	Multidimensional Poverty Index
MPIM	Max Planck Institute for Meteorology
MPM-HKH	Multidimensional Poverty Measure for the Hindu Kush Himalaya
MR	Moderate rain
MRC	Mekong River Commission
MRRD	Ministry of Reconstruction and Rural Development
$MtCO_2$	Million ton of carbon dioxide
MW	Megawatt
N_2O	Nitrous oxide
NA	Not Available
NAFCC	National Adaptation Fund for Climate Change
NAMaSTE	Nepal Ambient Monitoring and Source Testing Experiment
NAO	North Atlantic Oscillation
NAP	National Adaptation Plan
NAPA	National Adaptation Programme of Action
NAPCC	National Action Plan on Climate Change
NARC	National Agricultural Research Council
NBCI	National Biomass Cookstove Initiative
NBI	Nile Basin Initiative
NBSC	National Bureau of Statistics of China
NCAR	National Center for Atmospheric Research
NCC	Norwegian Climate Centre
NCD	Nature Conservation Division
NCEP	National Centers for Environmental Prediction
NCO-P	Nepal Climate Observatory—Pyramid
NCSA	National Capacity Needs Self-assessment for Global Environmental Management
NDCs	Nationally Determined Contributions
NDMA	National Disaster Management Authority
NDRC	National Development and Reform Commission
NDVI	Normalized Difference Vegetation Index
NE	Not Evaluated
NEA	Nepal Electricity Authority
NEC	National Environment Commission
NEPA	Nepal Environmental Protection Agency
NER	Net Enrolment Ratio
NFCP	Natural Forest Conservation Program
NFSA	National Food Security Act (India)

NGOs	Non-governmental organizations
NIDS	Nepal Institute of Development Studies
NISP	National Improved Stove Programme
NIWE	National Institute of Wind Energy
NIWRMP	National Integrated Water Resource Management Plan
NLG	National Leading Group
NLSS	Nepal Living Standards Survey
NMHS	National Mission for Himalayan Studies
NMI	Normalized Melt Index
NMSHE	National Mission for Sustaining the Himalayan Ecosystem
NMVOCs	Non-methane Volatile Organic Compounds
NO	Nitric oxide
NO_2	Nitrogen dioxide
NOAA	National Oceanic and Atmospheric Administration
NOx	Nitrogen oxides
NP	New Policies
NPC	National Planning Commission, Nepal
NPC	National People's Congress
NPIC	National Programme for Improved Chulhas
NPS	Non-Point Source
NPR	Nepalese Rupee
NRDCL	Natural Resource Development Corporation Limited
NRM	Natural Resource Management
NRVA	National Risk and Vulnerability Assessment
NS	National Surveys
NSF	National Science Foundation
NSS	National Sample Survey
NSSO	National Sample Survey Organization
NT	Near Threatened
NTFPs	Non-Timber Forest Products
NUFCs	Neglected and Underutilized Food Crops
NUS	Neglected and underutilized species
O&M cost	Organization and management cost
O_3	Ozone
OC	Organic Carbon
ODI	Overseas Development Institute
OECD	Organization for Economic Co-operation and Development
OIFC	Overseas Indian Facilitation Centre
OPHI	Oxford Poverty and Human Development Initiative
PA	Precipitation Anomaly
PA	Protected Area
PAHs	Polycyclic Aromatic Hydrocarbons
PCFV	Partnership for Clean Fuels and Vehicles
PCR-VCA	Participatory Climate Risk Vulnerability and Capacity Assessment
PCW	Protection of Children and Women
PDMA	Provincial Disaster Management Authority
PES	Payment for Environmental Services
PFM	Participatory Forest Management
PHE	Public Health Engineering
PJ	petajoule
PKR	Pakistani Rupee
PM	Particulate Matter
PM_{10}	Particulate Matter with a diameter of 10 μm or less

$PM_{2.5}$	Particulate Matter with a diameter of 2.5 μm or less
PMUY	Pradhan Mantri Ujjwala Yojana
POE	Protectorate of Emigrants
PPA	Precipitation Percent Anomaly
PPA’s	Power Purchase Agreements
PPB	Participatory Plant Breeding
PPCR	Pilot Program for Climate Resilience
PPP	Purchasing Power Parity
PRC	Pew Research Center
PRECIS	Providing Regional Climates for Impact Studies
PSA	Precipitation Standardized Anomalies
PSLM	Pakistan Social and Living Standards Measurement
PTR-ToF-MS	Proton-Transfer-Reaction-Time-of-Flight-Mass-Spectrometry
PV	Photovoltaic
PVA	Poverty and Vulnerability Assessment
QCCCE	Queensland Climate Change Centre of Excellence
QOMS	Qomolangma Station
R&D	Research and Development
RCM	Regional Climate Models
RCP	Representative Concentration Pathway
RE	Renewable Energy
REAS	Regional Emissions inventory in Asia
RECAST	Research Center for Applied Sciences & Technology
REDD+	Reduce Emission from Deforestation and Forest Degradation, and foster conservation, sustainable management of forests, and enhancement of forest carbon stocks
REDP	Rural Energy Development Program
RETs	Renewable Energy Technologies
RGI	Randolph Glacier Inventory
RGOB	Royal Government of Bhutan
RHRP	Rural Housing Reconstruction Program
RMMRU	Refugee and Migratory Movements Research Unit
ROSCA	Rotating Savings and Credit Association
RVCC	Reducing Vulnerability to Climate Change
SAARC	South Asian Association for Regional Cooperation
SACEP	South Asia Cooperative Environment Program
SAFTA	South Asian Free Trade Area
SAPCC	State Action Plans on Climate Change
SAPP	Southern African Power Pool
SASEC	South Asia Subregional Economic Cooperation Program
SAWTEE	South Asia Watch on Tradé, Economics, and Environment
SCA	Snow-Covered Area
SCCF	Special Climate Change Fund
SCF	Snow-Covered Fraction
SCD	snow cover duration
SDC	Swiss Development Cooperation
SDG	Sustainable Development Goal
SDGs	Sustainable Development Goals
SDG 2	Sustainable Development Goal 2
SDG 7	Sustainable Development Goal 7
SDII	Simple Daily Intensity Index
SDMC	SAARC Disaster Management Centre
SE4All	Sustainable Energy for All

SF	Social Forestry
SHS	Solar Home Systems
SIDA	Swedish International Development Cooperation Agency
SLA	Snow Line Altitude
SLA	Sustainable Livelihoods Approach
SLCP	Sloping Land Conversion Program
SLCPs	Short-lived Climate Pollutants
SMC	Scalabrini Migration Centre
SMEs	Small and Medium scale Enterprise
SMHI	Swedish Meteorological and Hydrological Institute
SMS	Short Message Service
SO_2	Sulphur Dioxide
SOx	Sulphur Oxides
SPARC	Stratosphere-troposphere Processes and their Role in Climate
SPDC	Social Policy and Development Centre
SPI	Social Protection Index
SPIP	Solar-powered Irrigation Pumps
SRA	Social Relations Approach
SRES	Special Report on Emissions Scenarios
SREX	Special Report on Managing the Risks of Extreme Events and Disasters to Advance Climate Change Adaptation
SRTM	Shuttle Radar Topographic Mission
SSPs	Shared Socio-Economic Pathways
SSR	Surface Solar Radiation
STEM	Sulfur Transport and Deposition Model
SU	Summer Day
SusKat-ABC	Sustainable Atmosphere for the Kathmandu Valley—Atmospheric Brown Cloud
SWE	Snow Water Equivalent
T&D	Transmission and Distribution
TAR	Tibet Autonomous Region
TCFD	Task Force on Climate-related Financial Disclosures
TEPC	Trade and Export Promotion Centre
TERI	The Energy and Resources Institute
TFEC	Total Final Energy consumption
toe	tonnes of oil equivalent
TP	Tibetan Plateau
TPES	Total Primary Energy Supply
TRMM	Tropical Rainfall Measurement Mission
TSP	Total Suspended Particle
UAE	United Arab Emirates
UAV	Unmanned Aerial Vehicle
UIB	Upper Indus basin
UN	United Nations
UNCDF	United Nations Capital Development Fund
UNCED	United Nations Conference on Environment and Development
UNDESA	United Nations, Department of Economic and Social Affairs
UNDP	United Nations Development Programme
UNEP	United Nations Environment Programme
UNESCO	United Nations Educational, Scientific and Cultural Organization
UNFCCC	United Nations Framework Convention on Climate Change
UNHCR	United Nations High Commissioner for Refugees
UNICEF	United Nations Children's Emergency Fund

UNISDR	United Nations International Strategy for Disaster Reduction
UNU	United Nations University
U.S.	United States
USAID	United States Agency for International Development
USD	United States Dollar
VACA	Vulnerability and Adaptive Capacity Assessment
VAT	Value-Added Tax
VCI	Vulnerability and Capacities Index
VIC	Variable Infiltration Capacity
VOCs	Volatile Organic Compounds
VRA	Vulnerability and Risk Assessment
VSBK	Vertical Shaft Brick Kiln
VU	Vulnerable
W	Watt
WAPDA	Water & Power Development Authority
WB	World Bank
WCD	Wildlife Conservation Division
WCMC	World Conservation Monitoring Centre
WCRP	World Climate Research Programme
WD	Wet Day
WDA	Wet-Day Anomaly
WECS	Water and Energy Commission Secretariat
WEF	Water-Energy-Food
WFD	Water Framework Directive
WGMS	World Glacier Monitoring Service
WHO	World Health Organization
WHYCOS	World Hydrological Cycle Observing System
WIDER	World Institute for Development Economics Research
WPSC	Women's Protection Technical Working group transitioned into a sub-cluster
WR	Water Resources
WRF	Weather Research and Forecasting
WRI-India	World Resources Institute, India
WUAs	Water Users' Associations
WWF	World Wildlife Fund for Nature

Annex 3: Reviewers of the HIMAP Assessment Report

Shankar Adhikari, Ministry of Forests and Soil Conservation, Government of Nepal, Nepal
Carolina Adler, Mountain Research Initiative, Switzerland
Muhammad Adnan, Global Change Impact Studies Centre, Pakistan
Nand Kishor Agrawal, International Centre for Integrated Mountain Development, Nepal
Amanullah, Agriculture University Peshawar, Pakistan
Muhammad Amjad, Global Change Impact Studies Centre, Pakistan
Arshad Ashraf, Pakistan Agricultural Research Council, Pakistan
Anamika Barua, IIT Guwahati, Assam, India
Furrukh Bashir, University of Arizona, Tucson, USA and Pakistan Meteorological Department, Pakistan
Bimbika Sijapati Basnett, Center for International Forestry Research, Indonesia
Lynn Bennett, Independent Researcher, USA
Manan Bhan, Independent Researcher, India
Laxmi Dutt Bhatta, International Centre for Integrated Mountain Development, Nepal
Bodo Bookhagen, University of Potsdam, Germany and University of California, Santa Barbara, USA
Suresh Chandra, Central Water Commission (Retd), India
Swapnil A Chaudhari, International Centre for Integrated Mountain Development, Nepal
Nakul Chettri, International Centre for Integrated Mountain Development, Nepal
Kanchan Chopra, Institute for Economic Growth, India
Olivia Conniff Molden, University of Oregon, USA
Juliane Dame, Heidelberg University, Germany
Someshwor Das, Department of Atmospheric Science, Central University of Rajasthan, India
Purnamita Dasgupta, Institute for Economic Growth, India
Maya Daurio, Independent scholar, USA
Brian Dawson, SDIP Climate Change Adviser, Australia
Thakur Prasad Devkota, NAP Process Theme Lead TNCH, Nepal
Mandira Lamichhane Dhimal, Goethe University, Frankfurt am Main, Germany
Priyanka Dissanayake, University of Moratuwa, Sri Lanka
Natali Dologlou, Metsovion Interdisciplinary Research Center, National Technical University of Athens, Greece
John Dore, Department of Foreign Affairs and Trade, Government of Australia, Australia
Mats Eriksson, International Centre for Integrated Mountain Development, Nepal
Maria Josefina Figueroa, Copenhagen Business School, Department of Business and Politics, Denmark
Mark Flanner, University of Michigan, USA
Monique Fort, University Paris Diderot, Sorbonne Paris Cité, France
Tighe Geoghegan, Green Park Consultants GPC Ltd., UK
Don Gilmour, Consultant, Australia

P. Wester et al. (eds.), *The Hindu Kush Himalaya Assessment*,
https://doi.org/10.1007/978-3-319-92288-1

Arif Goheer, Global Change Impact Studies Centre, Pakistan
R. Edward Grumbine, Grand Canyon Trust, USA
Himangana Gupta, Ministry of Environment, Forest and Climate Change, Government of India, India
Gehendra B. Gurung, Practical Action, Nepal
Ngawang Gyeltshen, UNDP, Bhutan
Barry Haack, George Mason University, USA
Wilfried Haeberli, Geography Department, University of Zurich, Switzerland
N. S. K. Harsh, Forest Research Institute, Indian Council of Forestry Research and Education, India
Zia-ur-Rahman Hashmi, Global Change Impact Studies Centre, Pakistan
Saleemul Huq, International Centre for Climate Change and Development, Bangladesh
Hans Hurni, Centre for Development and Environment, University of Bern, Switzerland
Muhammad Ijaz, Global Change Impact Studies Centre, Pakistan
Richard M. Johnson, Bath Spa University, UK
Niraj Prakash Joshi, Hiroshima University, Nepal
Durga D. Kandel, Murray Darling Basin Authority, Government of Australia, Australia
Sarah B. Kapnick, National Oceanic and Atmospheric Administration, USA
Aftab Ahmad Khan, Global Change Impact Studies Centre, Pakistan
Bahadar Nawab Khattak, COMSATS Institute of Information Technology (CIIT), Pakistan
James David Kirkham, International Centre for Integrated Mountain Development, Nepal and the Scott Polar Research Institute, University of Cambridge, UK
Inka Koch, International Centre for Integrated Mountain Development, Nepal
Rajan Kotru, International Centre for Integrated Mountain Development, Nepal
Xuan Li, FAO Regional Office of Asia and Pacific (FAORAP), Thailand
Amina Maharjan, International Centre for Integrated Mountain Development, Nepal
Dwijen Mallick, Bangladesh Centre for Advanced Studies, Bangladesh
Amjad Masood, Global Change Impact Studies Centre, Pakistan
Shahbaz Mehmood, Global Change Impact Studies Centre, Pakistan
Michel Mesquita, Uni Research and the Bjerknes Centre for Climate Research, Norway
Kaleem Anwar Mir, Global Change Impact Studies Centre, Pakistan
Arabinda Mishra, International Centre for Integrated Mountain Development, Nepal
David Molden, International Centre for Integrated Mountain Development, Nepal
Aditi Mukherji, International Centre for Integrated Mountain Development, Nepal
Vaishali Naik, NOAA Geophysical Fluid Dynamics Laboratory, USA
Abdul Karim Nayani, Former Advisor to Chairman NDMA Pakistan and Former UNISDR Country Representative Pakistan, Pakistan
GCS Negi, G.B. Pant National Institute of Himalayan Environment and Sustainable Development, India
Mani Nepal, International Centre for Integrated Mountain Development, Nepal
Santosh Nepal, International Centre for Integrated Mountain Development, Nepal
Marcus Nüsser, Heidelberg University, Germany
Krishna Prasad Pant, Kathmandu University, Nepal
Cui Peng, Institute of Mountain Hazard and Environment, Chinese Academy of Sciences, China
Prajal Pradhan, Potsdam Institute for Climate Impact Research (PIK), Germany
Irudaya Rajan, Centre for Development Studies, India
Brij Mohan Singh Rathore, International Centre for Integrated Mountain Development, Nepal
Golam Rasul, International Centre for Integrated Mountain Development, Nepal
Gopal S. Rawat, Wildlife Institute of India, India
Nadia Rehman, Global Change Impact Studies Centre, Pakistan
Patricia Romero Lankao, National Center for Atmospheric Research, USA
Dipesh Rupakheti, Institute of Tibetan Plateau Research, Chinese Academy of Sciences, China

Sachidananda Satapathy, Ministry of Environment, Forest and Climate Change, Government of India
Susanne Schmidt, Heidelberg University, Germany
Christopher A. Scott, University of Arizona, USA
Nuzba Shaheen, Global Change Impact Studies Centre, Pakistan
Eklabya Sharma, International Centre for Integrated Mountain Development, Nepal
Ramesh C. Sharma, HNB Garhwal University, India
Rajib Shaw, Keio University, Japan
Binaya Raj Shivakoti, Institute for Global Environmental Strategies, Japan
Sabnam Shivakoti, Government of Nepal, Nepal
Bharat Babu Shrestha, Central Department of Botany, Tribhuvan University, Nepal
Diana Sietz, Wageningen University, Netherlands
Dorothea Stumm, International Centre for Integrated Mountain Development, Nepal
Bhim Prasad Subedi, Tribhuvan University, Nepal
Md. Abu Syed, Bangladesh Centre for Advanced Studies, Bangladesh
Gianni Tartari, Water Research Institute, National Research Council, Italy
Prakash C. Tiwari, Kumaun University, India
Pranita Bhusan Udas, International Centre for Integrated Mountain Development, Nepal
Ramesh Ananda Vaidya, International Centre for Integrated Mountain Development, Nepal
Saurabh Vijay, Technical University of Denmark, Denmark
Patrick Wagnon, Institut de Recherche pour le Développement, France
Mengben Wang, Institute of Loess Plateau, Shanxi University, China
Jim Woodhill, Social Economic and Food Security Advisor for Australian Department of Foreign Affairs SDIP Programme, Australia
Jianchu Xu, Kunming Institute of Botany, Chinese Academy of Sciences, China
Qudsia Zafar, Global Change Impact Studies Centre, Pakistan
Lin Zhen, Institute of Geographic Science and Natural Resources Science, Chinese Academy of Sciences, China
Ping Zhao, Chinese Academy of Meteorological Sciences, China
Zong-Ci Zhao, Center for Earth System Science (CESS), Tsinghua University, China
Dan Zhu, Chengdu Institute of Botany, Chinese Academy of Sciences, China

Annex 4: Permissions to Publish

Fig. 2.1 A framework of three pillars of mountain sustainability and the interactive network of drivers

Permission granted Marc Foggin

Fig. 3.1 RegCM4 elevation (in km) with three regions of interest defined by grid cells in each box above 2,500 m a.s.l.

Permission granted by Elsevier and China Meteorological Administration

Fig. 3.2 Annual mean temperature anomaly series (°C) relative to 1961–90 mean values for Tmean (**a**), Tmax, Tmin, and DTR

Permission granted by Springer and China Meteorological Administration

Fig. 3.3 The grid-averaged trends of annual mean temperature in the Hindu Kush Himalaya (HKH) since 1901

Permission granted by Springer and China Meteorological Administration

Fig 3.4 The regional average annual PSA and PPA over 113 years (1901–2013) in the Hindu Kush Himalaya

Permission granted by Springer and China Meteorological Administration

Fig. 3.6 The change trends in annual precipitation percentage anomaly (PPA, unit: % decade −1) (**a**), wet day anomaly (WDA, unit: mm decade−1) (**b**); and daily precipitation intensity anomaly (DPIA, unit: mm/day decade−1) (**c**) in the Hindu Kush Himalaya over 53 years (1961–2013). Filled symbols represent statistically significant data at 0.05 confidence level (*Data source* CMA GLMP; Zhan et al. 2017)

Permission granted by Springer and China Meteorological Administration

Fig. 3.7 Annual mean anomaly series of extreme temperature indices of the Hindu Kush Himalaya (HKH) for 1961–2014 for **a** cold nights (TN10p); **b** cold days (TX10p); **c** warm nights (TN90p); **d** warm days (TX90p); **e** monthly maximum value of daily maximum temperature (TXx); **f** monthly minimum value of daily minimum temperature (TNn); **g** frost days (FD); **h** summer days (SU); and **i** diurnal temperature range (DTR) (relative to 1961–90 mean values). The trends are calculated only for the grid boxes with sufficient data, as explained in the text (*Data source* CMA GLASAT; Sun et al. 2017a)

Permission granted by Springer and China Meteorological Administration

Fig 3.8 The change trends of extreme temperature indices of the Hindu Kush Himalaya for 1961–2014 for **a** cold nights, **b** cold days, **c** warm nights, **d** warm days, **e** frost day, **f** summer day, **g** monthly maximum value of daily maximum temperature, **h** monthly minimum value of daily minimum temperature, and **i** diurnal temperature range (relative to 1961–90 mean values) (*Data source* CMA GLSAT; Sun et al. 2017a)

Permission granted by Springer and China Meteorological Administration

Fig. 3.9 The regional average anomaly time series of extreme precipitation indices for annual amount (unit: mm) (left); days (unit: day) (central); and intensity (unit: mm/day) (right) of light (above), moderate (middle), and intense (below) precipitation over the last 53 years (1961–

P. Wester et al. (eds.), *The Hindu Kush Himalaya Assessment*,
https://doi.org/10.1007/978-3-319-92288-1

2013) in the Hindu Kush Himalaya. Dashed lines represent linear trends (*Data source* CMA GLDP; Zhan et al. 2017)

Permission granted by Springer and China Meteorological Administration

Fig. 3.10 The change trends of annual precipitation amount percent anomaly (PPA) (1/left) and annual precipitation day anomaly (WDA) (2/right) for light (**a**), moderate (**b**), and intense (**c**) precipitation over the last 53 years (1961–2013) in the Hindu Kush Himalaya. Filled symbols represent statistically significant trends at the 0.05 confidence level (*Data source* CMA GLDP; Zhan et al. 2017)

Permission granted by Springer and China Meteorological Administration

Fig. 3.12 Seasonal ensemble mean climate change in the Hindu Kush Himalaya (HKH) in the near future ([2036–65] with respect to [1976–2005]) for (top panels) surface air temperature (°C) and (bottom panels) total precipitation (%), with scenarios (first and third column) RCP4.5 and (second and fourth column) RCP8.5, during (**a–d**) summer monsoon (JJAS) and (**e–h**) winter (DJF) seasons. Ensemble mean of downscaling CMIP5 GCM with CORDEX South Asia RCM (listed in Table 3.5). Striping in bottom panels indicates where at least 10 of the 13 realizations concur on an increase (vertical) or decrease (horizontal) in RCPs. The HKH boundary is shown with dashed line. The boxes represent the three HKH sub-regions used for detailed analysis (see text)

Permission granted by Springer and China Meteorological Administration

Fig. 3.13 Seasonal ensemble mean climate change in the Hindu Kush Himalaya (HKH) in the far future ([2066–2095]–[1976–2005]) for (top panels) surface air temperature (oC) and (bottom panels) total precipitation (%), with scenarios (first and third column) RCP4.5 and (second and fourth column) RCP8.5, during (**a–d**) summer monsoon (JJAS) and (e–h) winter (DJF) seasons. Ensemble mean of downscaling CMIP5 GCM with CORDEX South Asia RCM (listed in Annex "Permissions to Publish" and Table 3.5). Striping in bottom panels indicates where at least 10 of the 13 realizations concur on an increase (vertical) or decrease (horizontal) in RCPs. The HKH boundary is shown by a dashed line. The boxes represent three HKH sub-regions used for detailed analysis (see text)

Permission granted by Springer and China Meteorological Administration

Fig. 5.2 Linkages between ecosystems, biodiversity, and human wellbeing

Permission granted by Elsevier

Fig. 5.4 Vegetation zones and dominant forest types in the HKH (Chettri et al. 2010)

Permission granted by ICIMOD

Fig. 5.5 Climatic records, radiocarbon dates, and charred cereal grain records from 53 investigated sites on the north east Tibetan Plateau (NETP)

Permission granted by The American Association for the Advancement of Science

Fig. 5.12 Trend in number and area coverage of protected areas in the HKH from 1918 to 2007

Permission granted by Taylor and Francis

Fig. 6.1 Framework for Addressing Sustainable Energy Transition

Permission granted by ICIMOD

Box 7.2 Schematic diagrams of five mechanisms of elevation dependent warming. dT/dt is the change in temperature over time (adapted from Mountain Research Initiative EDW Working Group 2015)

Permission granted by Springer Nature

Fig. 7.3 Snow cover fraction trends (%/yr) over the HKH-TP from 2000 to 2014 from (*Source* Li et al. 2017)

Permission granted by John Wiley and Sons

Fig. 7.9 Mass change projections and prevalence of debris cover for all RGI sub-regions (**a–o**) that are encompassed by the RGI regions 13 (Central Asia), 14 (South Asia West), and 15 (South Asia East)

Permission granted by Springer Nature

Fig. 8.2 Contribution to total flow by glacial melt, snowmelt, and rainfall-runoff for major streams
Permission granted by SpringerNature
Fig. 8.3 NDVI changes in the growing season across TP over past three decades
Permission granted by PNAS and National Academy of Sciences
Fig. 9.1 Food security framework (*Source* adapted from Jones et al. 2013)
Permission granted by Oxford University Press
Fig. 9.9 Post-harvest losses in Gilgit-Baltistan, Pakistan (*Source* Rasul et al. 2014)
Permission granted by ICIMOD
Fig. 10.4 CO in the Kathmandu Valley in October (**a**) and January (**b**) 2005. Weekends are marked in light shading and festival days in medium shading. *Source* Panday and Prinn (2009)
Permission granted by John Wiley and Sons
Fig. 10.5 a Diurnal concentrations of BC during each season at Pantnagar, Uttarakhand, India (left, *Source* Joshi et al. 2016) and at Paknajol, Kathmandu, Nepal (right, *Source* Putero et al. 2015)
Permission granted by Elsevier
Fig. 10.6 Diurnal variations for aerosol particle number concentration (Np), surface O_3, and CO_2 at Askole in northern Pakistan (3,015 masl). *Source* Putero et al. (2014)
Permission granted by Elsevier
Fig. 10.7 Diurnal cycle of aerosol number concentrations at Manora Peak, Nainital, India
Permission granted by John Wiley and Sons
Fig. 10.8 Mean MODIS AOD during pre-monsoon season between 2003 and 2006, projected as a function of surface topography (left) and CALIPSO back-scatter profile from southern India to the Himalaya (right). *Source* Gautam et al. (2009a)
Permission granted by John Wiley and Sons
Fig. 10.11 Fog frequency from 82 stations over India and Pakistan from 1976 to 2010 (Syed et al. 2012)
Permission granted by Springer Nature
Fig. 10.13 Multiple pollutants and their multiple adverse effects. (Cao et al. 2013, adapted by Clean Air Asia 2016)
Permission granted by Dr. Hsi-Hsien Yang, Production Center of Aerosol and Air Quality Research
Fig. 10.18 Contribution of motor vehicles, secondary, biomass/garbage burning, brick kilns, and soil dust to PM10 concentrations in Nepal (*Source* Kim et al. 2015)
Permission granted by Elsevier
Fig. 12.4 Percentage of multidimensional poor and consumption poor in Shan and Chin in 2013 (*Source* Gerlitz et al. 2014)
Permission granted by ICIMOD
Fig. 12.7 Conceptual framework for vulnerability (Macchi 2011, adapted from Füssel and Klein 2006)
Permission granted by ICIMOD
Fig. 12.8 Sustainable livelihoods approach (Macchi 2011, adapted from DFID 1999)
Permission granted by ICIMOD
Fig. 12.9 Vulnerability in three sub-basins of the HKH: absolute and relative contribution of vulnerability dimensions by district (Gerlitz et al. 2017)
Permission granted by Taylor & Francis
Fig. 12.10 Vulnerability in three sub-basins of the HKH: relative contribution of vulnerability components by district in percentage (Gerlitz et al. 2017)
Permission granted by Taylor & Francis

www.ingramcontent.com/pod-product-compliance
Lightning Source LLC
LaVergne TN
LVHW062152200926
848484LV00007B/1905

9781013276156